# Lecture Notes in Production Engineering

**Lecture Notes in Production Engineering (LNPE)** is a new book series that reports the latest research and developments in Production Engineering, comprising:

- Biomanufacturing
- Control and Management of Processes
- Cutting and Forming
- Design
- Life Cycle Engineering
- Machines and Systems
- Optimization
- Precision Engineering and Metrology
- Surfaces

LNPE publishes authored conference proceedings, contributed volumes and authored monographs that present cutting-edge research information as well as new perspectives on classical fields, while maintaining Springer's high standards of excellence. Also considered for publication are lecture notes and other related material of exceptionally high quality and interest. The subject matter should be original and timely, reporting the latest research and developments in all areas of production engineering. The target audience of LNPE consists of advanced level students, researchers, as well as industry professionals working at the forefront of their fields. Much like Springer's other Lecture Notes series, LNPE will be distributed through Springer's print and electronic publishing channels. To submit a proposal or request further information please contact Anthony Doyle, Executive Editor, Springer (anthony.doyle@springer.com).

More information about this series at ▶ https://link.springer.com/bookseries/10642

Christian Brecher · Manfred Weck

# Machine Tools Production Systems 3

Mechatronic Systems, Control and Automation

Springer

**Christian Brecher**
Laboratory for Machine Tools and
Production Engineering (WZL)
RWTH Aachen University
Aachen, Germany

**Manfred Weck**
Laboratory for Machine Tools and
Production Engineering (WZL)
RWTH Aachen University
Aachen, Germany

ISSN 2194-0525 ISSN 2194-0533 (electronic)
Lecture Notes in Production Engineering
ISBN 978-3-658-34621-8 ISBN 978-3-658-34622-5 (eBook)
https://doi.org/10.1007/978-3-658-34622-5



Responsible Editor: Eric Blaschke
This Springer imprint is published by the registered company Springer Fachmedien Wiesbaden GmbH part of Springer Nature.
The registered company address is: Abraham-Lincoln-Str. 46, 65189 Wiesbaden, Germany

# Foreword for the Machine Tools and Manufacturing Systems Compendium

Machine tools are among the most important means of production in the metalworking industry. The current high standard of living in industrialized nations would be inconceivable had this category of machinery not been developed. The Federal Republic of Germany is a world leader in the production of machine tools. In the Federal Republic of Germany, approximately 7.4% of the total mechanical engineering production volume comes from the machine tool industry and 7.0% of those employed in mechanical engineering work in the machine tool industry (German Machine Tool Builder's Association (VDW), as of 2018).

The application areas for machine tools are just as diverse as their structural design and degree of automation. According to the technological processes, this extensive field ranges from machine tools for primary forming and forming, through machine tools for cutting (including chip-removing machine tools) to joining machines. These machines have varying degrees of automation and more or less flexibility depending on the workpieces to be machined and the lot sizes. As a result, single-purpose and special machine tools as well as universal machines with an extensive range of potential applications are available on the market.

Due to the increasing demands on performance and accuracy, the designer of these machines must ensure an optimum design of the individual machine components. To do this, the designer must have a comprehensive knowledge of the relationships between the physical properties of the components and of the machine elements. Today, the designer can provide computer-aided designs using a substantial library of programs. The criteria that determine performance and accuracy such as the geometric, kinematic, static, dynamic, thermoelastic and acoustic properties can be evaluated through metrological analyses and objective assessment methods. The specific improvements necessary can then be put in place.

The continuing tendency toward the automation of machine tools has led to a broad spectrum of control system alternatives. In recent years, developments in electrical engineering, electronics and software technology have had a significant influence on machine control systems. Microprocessors and process computers can now provide control-engineering solutions that were previously unthinkable. The mechanization and automation efforts have also included the areas of material transport and machine feeding. Considerations in this area led to transfer lines in mass production and to flexible manufacturing cells and systems in small and medium series manufacturing.

The three volumes that make up this series on the subject of Machine Tools and Manufacturing Systems are aimed at both university students majoring in production engineering and all practicing professionals who must keep pace with the increasingly complex material covering this branch of mechanical engineering. In addition, these volumes will help the user select suitable machines and the relevant control systems. The machine manufacturer is provided with ways of producing the best layout of the machine components, drive systems and control systems together with options for specific improvements based on metrological analyses and objective assessment methods.

The content of the complete work is closely based on the Machine Tools lectures at the RWTH Aachen University and is structured as follows:

Volume 1: Machine tools—Machine types and application areas,
Volume 2: Machine tools—Design, calculation and metrological assessment,
Volume 3: Machine tools—Mechatronic Systems, Control and Automation.

**Christian Brecher**
Aachen
September 2019

# Foreword for Volume 3

The objective of this Volume 3 is to give the reader an insight into the control and automation technology of machine tools. The focuses are placed on showing solutions that have proven themselves over a long period in practical applications and presenting recent developments.

**The first part** of this volume covers mechatronic components for generating the highly dynamic drive motions of machine tools with the emphasis being placed on their mechanical and control-related design. Diagnosing machine and process states is growing in importance for complex and highly automated machines. Metrological signal acquisition, processing, and interpretation are shown in detail for many practical applications.

**The second part** of this volume discusses the different control types for machine tools in particular. The robots used in the peripheral equipment, and the associated robot control systems, are also covered in this part. An introduction to master control systems and an engineering summary complete the analysis of the overall system.

The contents of this volume are divided as follows:

Following a brief introduction (▶ Chap. 1) introducing the basic terminology of mechatronics and control engineering and outlining the historical development of automation, ▶ Chap. 2 explains the underlying structure of a feed axis. Before the mechanical transmission elements are described in detail, an introduction is given to electronic conversion technology. Also, the demands made of motors as the drive systems for feed axes are explained and different motor types presented. The position measuring systems required to control the feed drives are also explained to the reader in this chapter.

The level of production precision is highly dependent on the dynamic behavior of the feed axis. This is discussed in more detail in ▶ Chap. 3. Following an explanation of the fundamentals of control is a presentation of the different methods for improving the dynamic behavior of a feed axis. The overall position control loop of a feed axis is also covered in more detail: The transfer behavior of the mechanical components is described in addition to the most important parameters.

Control of the feed drives is then discussed in ▶ Chap. 4. The general setup of complex tool paths is described in the first part. Modern-day control systems are implemented using computers that, in various modules, perform interpolation and position control and provide a suitable user interface to the worker. Then, the reader is given an understanding of typical path errors and how they arise before measures for reducing them are explained.

In ▶ Chap. 5, the reader may learn about the design of feed drives. After introducing different drive concepts, the procedure for designing a feed drive is described and the demands placed on individual components are explained in more detail. Following the mechanical design, the specification of a suitable measuring system is addressed briefly. Finally, the steps necessary for putting a feed drive into operation, both manually and automatically, are detailed.

Given the high loads put on important components (such as main spindles, feed drives, and tools), machine and process diagnostics are of major importance. The situation is amplified by the fact that systems are frequently operated with little, or no, supervision by the machine operator. ▶ Chap. 6 thus covers this very comprehensively. As, in almost all situations, actual parameters (such as abrasion and wear) cannot be determined directly by means of measurement, indirect methods are applied. In the process, the effects (forces, vibrations, and temperature) are measured and models are used to determine the parameters of interest. In addition to accurately and reliably measuring values, analysis and correct interpretation of the measurement data play a major role. Many practical examples of important machine components and manufacturing processes are presented in this chapter.

▶ Chap. 7 is an introduction to the second part of this volume. In addition to a definition of the term automation, a brief description follows of the historic development of machine tool automation. Following an introduction to the modes of operation used in control systems, some practical examples of automated functions in a machine tool environment are presented.

Next, in ▶ Chap. 8, is the description of mechanical controllers as the oldest form of control engineering. While having lost considerable importance in the era of electronics and informatics, many examples of mechanical automatons can still be found in workshops.

The fundamentals of information processing are covered afterwards. ▶ Chap. 9 forms the basis for the electrical controls discussed in ▶ Chap. 10. In addition to hard-wired logic controllers, programmable logic controllers (PLCs) are introduced and their principles of operation are illustrated using examples. Particular importance is attached to programmable logic controllers in the machine tool industry. This is why the main method for systematic development of complex PLC programs is presented in detail and not just different programming methods.

Numerical controllers (NC) have taken on key importance for machine tool automation. ▶ Chap. 11 discusses in detail the internal design, functional scope, programming, and the diverse programming methods, operating numerical controllers, and development trends.

Covered afterwards separately (▶ Chap. 12) is the geometric data link in NC controllers. This includes the various methods for interpolation, controlled movement, and tool corrections. Handling the feed override (internal to the NC) and externally influencing the speed are explained separately at the end of this chapter.

Nowadays, numerically controlled handling systems and industrial robots frequently constitute parts of autonomous manufacturing and assembly cells and flexible production systems. They are used in a lot of different applications including those for machine feeding and workpiece conveyance. Furthermore, they are often used for welding and cutting work. ▶ Chap. 13 thus covers robot control systems intensively. Different robot kinematics, coordinate transformations, and programming capabilities are the focuses of this chapter.

The trend towards integrated information and data processing systems through end-to-end integration and networking of all company divisions is taking an increasingly greater foothold. This cross-departmental, computer-aided information processing (known as "CIM", Computer Integrated Manufacturing) in production systems connects all automation equipment and computers intending to make the information processing actions more efficient. A model for the CIM structure of a company is presented in ▶ Chap. 14. The control station systems for controlling interlinked systems represent one focus here.

Advancing automation and the merging of in-house IT systems are putting particular demands on the development processes of a system. Processes must be designed to be reliable and controllable, the integration of new functions is a requirement, and safety issues must be taken into account. Design engineering for automation systems is thus discussed in ▶ Chap. 15. Methodology, architecture modeling, systems engineering, design, implementation, and test are the key tasks in engineering explained to the reader concerning software development and electrical design.

This ninth edition was revised with the cooperation of my research assistants. My sincere thanks go to all those involved for their enormous commitment. I would like to express special thank-yous to Dipl.-Ing. Christian Ecker and M.Sc. Christian Fimmers for coordinating and organizing the basic revision of the ninth edition.

I also wish to thank most sincerely the companies who provided illustrations for this volume.

**Christian Brecher**
Aachen
May 2021

# Contents

**Supplementary Information**

# Abbreviation

| | |
|---|---|
| 2D | Two-dimensional |
| **3D** | Three-dimensional |
| **A/D** | Analog to Digital |
| **AC (Chap. 2)** | alternating current |
| **AC (Chap. 11)** | Axes Control |
| **AC (Chap. 6)** | Adaptive control |
| **ACC** | Adaptive control constraint |
| **ACO** | Adaptive control optimization |
| **AI** | Artificial intelligence |
| **AlNiCo** | Aluminum-Nickel-Cobalt |
| **ALU** | Arithmetic Logic Unit |
| **AP** | Application protocol |
| **API** | Application Programming Interface |
| **APR** | Acoustic Pulse Recognition |
| **APT** | Automatically Programmed Tools |
| **AR** | Augmented Reality |
| **ARMA model** | Autoregressive-moving average model |
| **ASCII** | American Standard Code for Information Interchange |
| **ASIC** | Application-Specific Integrated Circuit |
| **ASK** | Amplitude Shift Keying |
| **BCD** | Binary-Coded Decimal |
| **BSD** | Ball screw drive |
| **CAD** | Computer-Aided Design |
| **CAM** | Computer-Aided Manufacturing |
| **CAN-Bus** | Controller Area Network |
| **CAP** | Computer-Aided Planning |
| **CAQ** | Computer-Aided Quality |
| **CAx** | Computer-Aided x/Computer-Assisted x |
| **CCD** | Charge Coupled Device |
| **CCU** | Cognitive Control Unit |
| **CFC** | Continuous Function Chart |
| **CFT** | Conformal field theory |
| **CIM** | Computer Integrated Manufacturing |
| **CLDATA** | Cutter Location Data |
| **CLP** | Current limiting phase |
| **CM** | Condition monitoring |
| **CMOS** | Complementary Metal Oxide Semiconductor |
| **CNC** | Computerized Numerical Control |
| **CNF** | conjunctive normal form |
| **COM** | Component Object Model |
| **CPS** | Cyber-Physical System |
| **CPU** | Central Processing Unit |
| **CRM** | Customer Relationship Management |
| **CSMA** | Carrier Sense Multiple Access |
| **CSMA/CA** | Carrier Sense Multiple Access/Collision Avoidance |
| **CSMA/CD** | Carrier Sense Multiple Access/Collision Detection |
| **D** | Differential transfer behaviour |
| **D/A** | Digital to Analog |
| **DB** | Data block |
| **DBMS** | Database Management System |
| **DC** | direct current |
| **DCC** | Drive Control Chart |
| **DCOM** | Distributed Component Object Model |
| **DDE** | Dynamic Data Exchange |
| **DH-Matrices** | Denavit-Hartenberg-Matrices |
| **DI** | Diffused Illumination |
| **DIN** | German Institute for Standardization |
| **DIN EN** | DIN EN-Norm |
| **DNC** | Distributed Numerical Control |
| **DSP** | Digital signal processor |
| **DST** | Dispersive Signal Technology |
| **DT** | drive telegram |
| **DTL** | Diode-Transistor Logic |
| **DWG** | Drawing |
| **DXF** | Data Exchange Format |
| **EDI** | Electronic Data Interchange |
| **EDIFACT** | Electronic Data Interchange For Administration, Commerce, and Transport |

| | |
|---|---|
| **EDM (Chap. 10)** | Engineering Data Management |
| **EDM (Chap. 6)** | Electrical discharge machining |
| **EDP** | Electronic Data Processing |
| **EEA** | European Economic Area |
| **EEPROM** | Electrically Erasable Programmable Read-Only Memory |
| **EMC** | Electromagnetic Compatibility |
| **ERP** | Enterprise Resource Planning |
| **EWOTeK** | Research project: Increasing the efficiency of machine tools by optimizing component operation technologies |
| **FB** | Function block |
| **FC** | Function |
| **FFT** | Fast Fourier transform |
| **FMC** | Flexible Manufacturing Cell |
| **FMS** | Flexible Manufacturing System |
| **FPGA** | Field Programmable Gate Array |
| **FTIR** | Frustrated Total Internal Reflection |
| **GPRS** | General Packet Radio Service |
| **GSM** | Global System for Mobile Communications |
| **HAC** | Human Axis Control |
| **HART** | Highway Addressable Remote Transducer |
| **HF** | High frequency |
| **HiL** | Hardware-in-the-Loop |
| **HM** | Hard metal |
| **HMC** | Human Machine Control |
| **HMI** | Human Machine Interface |
| **HR** | Human Resources |
| **HRC** | Human-Robot-Collaboration |
| **HRM** | Human Resource Management |
| **HSC** | High Speed Cutting |
| **HTTP** | Hypertext Transfer Protocol |
| **HWD** | Head Worn Display |
| **I** | Integrating transfer behaviour |
| **IC** | Integrated Circuit |
| **IEC** | International Electrotechnical Commission |
| **IEEE** | Institute of Electrical and Electronics Engineers |
| **IGBT** | Insulated Gate Bipolar Transistor |
| **IGES** | Initial Graphics Exchange Specification |
| **IL** | Instruction List |
| **IP** | Internet Protocol |
| **IPC** | Industrial Personal Computers |
| **IR** | Industrial Robot |
| **IRL** | Industrial Robot Language |
| **ISM** | Industrial, Scientific and Medical Band |
| **ISO** | International Organization for Standardization |
| **IT** | Information Technology |
| **JT** | Jupiter Tesselation |
| **K-Map** | Karnaugh Map |
| **LC** | Logic Control |
| **LCD** | Liquid Crystal Display |
| **LD** | Ladder Diagram |
| **LED** | Light Emitting Diode |
| **LS** | Limit Switch |
| **LSB** | Least Significant Bit |
| **Macros** | Programming automation |
| **Max.** | Maximum |
| **MC** | Motion Control |
| **MDA** | Machine Data Acquisition |
| **MDT** | Master data telegram |
| **MES** | Manufacturing Execution System |
| **MIT** | Massachusetts Institute of Technology |
| **MLP** | Multilayer perceptron |
| **MMC** | Man Machine Control |
| **MMS** | Manufacturing Message Specification |
| **MOS** | Metal Oxide Semiconductor |
| **MOSFET** | Metal Oxide Semiconductor Field-Effect Transistor |

| | |
|---|---|
| **MQTT** | Message Queuing Telemetry Transport |
| **MRP** | Manufacturing Resources Planning |
| **MSE** | Mean Squared Error |
| **MST** | Master synchronous telegram |
| **MTTF** | Mean Time To Failure |
| **N** | Magnetic north pole |
| **NC** | Numerical Control |
| **NCK** | NC kernel |
| **NCU** | Numerical Control Unit |
| **NdFeB** | Neodymium Iron Boron |
| **NFC** | Near Field Communication |
| **NMOS** | N-doped Metal Oxide Semiconductor |
| **NURBS** | Non Uniform Rational Basis Spline |
| **OB** | Organization block |
| **OEE** | Overall equipment effectiveness |
| **OLED** | Organic Light Emitting Diode |
| **OPC** | Open Platform Communications (formerly: OLE for Process Control) |
| **OPC AE** | OPC Alarms and Events |
| **OPC DA** | OPC Data Access |
| **OPC HDA** | OPC Historical Data Access |
| **OPC UA** | OPC Unified Architecture |
| **OSI** | Open Systems Interconnection |
| **P** | Proportional transfer behaviour |
| **PAL** | Programmable Array Logic |
| **PC** | Personal Computer |
| **PCH** | Pallet Changer |
| **PCI** | Peripheral Component Interconnect |
| **PD** | Proportional differential transfer behaviour |
| **PDA** | Production Data Acquisition |
| **PDM** | Product Data Management |
| **PII** | Process Image Input |
| **PIO** | Process Image Output |
| **PI** | Proportional integrating transfer behaviour |
| **PI-Controller** | Proportional-Integral-Controller |
| **PID** | Proportional integrating differential transfer behaviour |
| **PID-Controller** | Proportional-Integral-Differential-Controller |
| **PL** | Performance-Level |
| **PLA** | Programmable Logic Array |
| **PLC** | Programmable Logic Controller |
| **PLD** | Programmable Logic Device |
| **PLM** | Product lifecycle management |
| **PMOS** | P-doped Metal Oxide Semiconductor |
| **PRC** | Pseudo-Random-Code |
| **PROFIBUS** | PROcessFIeld BUS |
| **PROFIBUS DP** | PROcessFIeld BUS Decentralised Peripherals |
| **PSD** | Position Sensitive Diode |
| **PT** | Proportional dead-time transfer behaviour |
| **PT1** | First-order lag element |
| **PTP** | Point to Point |
| **PWM** | Pulse width modulation |
| **QA** | Quality Assurance |
| **RAM** | Random Access Memory |
| **RBF** | Radial basis function network |
| **RC** | Robot Control |
| **RFID** | Radio-Frequency Identification |
| **RGB** | RGB color space |
| **RNN** | Recurrent neural network |
| **ROM** | Read-Only Memory |
| **RRS** | Realistic Robot Simulation |
| **RSD** | Roller screw drive |
| **S** | Magnetic south pole |
| **s chart** | Standard deviation chart |
| **S1** | Continuous duty |
| **S2** | Short time duty |
| **S3** | Intermittent periodic duty |
| **S6** | Continuous operation periodic duty |
| **SAW** | Surface Acoustic Wave |
| **SC** | Spindle Control |
| **SCADA** | Supervisory Control and Data Acquisition |

**SCARA** Selective Compliance Assembly Robot Arm
**SCM** Supply Chain Management
**SERCOS** Serial Realtime Communication System
**SFC** Sequential Function Chart
**SiL** Software-in-the-Loop
**Simodrive 611D** Product (Low voltage inverter) of the company Siemens AG
**Sinumerik 840D** Product (Numerical Control) of the company Siemens AG
**SISO** Single Input, Single Output
**SmCo** Samarium-Cobalt
**SME** Small and Medium-Sized Enterprises
**SOAP** Simple Object Access Protocol
**SPC** Statistical process control
**ST** Structured Text
**STEP** Standard for the Exchange of Product Model Data
**STL** Stereolithography File Format

**T** Tool
**TC** Tool Changer
**TCP (Chap. 6, 11 and 13)** Tool Center Point
**TCP (Chap. 9, 11, 13 and 14)** Transmission Control Protocol

**TDM** Tool Data Management
**TDO** Technology Data Organization
**TFT** Thin Film Transistor
**TPM** Total productive maintenance
**TQM** Total quality management
**TTL** Transistor-Transistor Logic

**UMTS** Universal Mobile Telecommunications System
**UWB** Ultra-Wide-Band
**VDAFS** Verband der Automobilindustrie (Association of the German Automotive Industry)—surface interface
**VDE** Verband der Elektrotechnik Elektronik Informationstechnik (Association of electrical engineering electronics information technology)
**VDI** Verein Deutscher Ingenieure (Association of German Engineers)
**VDW** Verein Deutscher Werkzeugmaschinenfabriken (Association of German machine tool factories)
**VLSI** Very Large Scale Integration

**WISA** Wireless Interface for Sensors and Actors
**WLAN** Wireless Local Area Network
**WMAN** Wireless Metropolitan Area Network
**WOP** Workshop-oriented Programming
**WPAN** Wireless Personal Area Network
**WWM** Width of wear marks
**WZL** Laboratory for machine tools and production engineering

**x chart** Mean value chart
**XML** Extensible Markup Language

**ZVEI** Zentralverband Elektrotechnik- und Elektronikindustrie (Central Association of the Electrical Engineering and Electronics Industry)

# 1 Introduction

I thank my staff members Christian Fimmers M. Sc. and Simon Storms M. Sc. for their contribution in revising this chapter.


C. Brecher and M. Weck, *Machine Tools Production Systems 3*,
Lecture Notes in Production Engineering, https://doi.org/10.1007/978-3-658-34622-5_1

Contents

## 1.1 The Machine Tool as a Mechatronic System

Since the 1970s, the sustained optimization of microelectronics and the microcomputer has deeply changed and continually improved automation technology, in particular the scope of functionality and performance of machine tools. It thus became possible to create flexible, programmable control technology that could perform complex switching and control operations using software systems tailored specifically for the purpose.

Nowadays, therefore, interdisciplinary teams made up of members from mechanical engineering, electrical engineering, and IT are involved in the development of modern machine tools and their components. The overlapping of individual disciplines thus offers considerable synergy potential, for example, to improve the functionality of components or to physically integrate technical systems into compact, *smart* units [ISER99]. The term "mechatronics," derived from the disciplines involved, is generally used for components whose function is based upon this integration. (see ◘ Fig. 1.1).

The mechatronics term used today, however, extends beyond the component-only reference and includes, in particular, the interdisciplinary development methodology, that is, the characteristic procedure for creating mechatronic systems [JANS10] (see *Engineering*, ► Sect. 1.2).

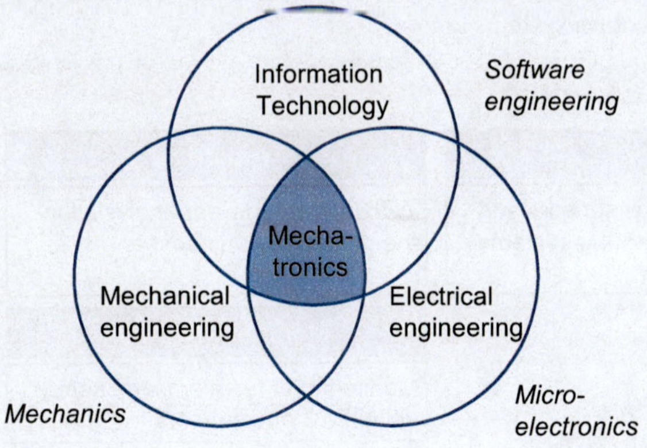


◘ **Fig. 1.1.** Specialist disciplines involved in developing mechatronic components

A mechatronic system is characterized by its structure comprising a base system, various sensors and actuators and one or more components for information processing (see ◘ Fig. 1.2). The base system here represents the interface to system functions visible from the outside by exerting influence on the external material and energy flows. Inside the system, this manipulation is by controlling actuators, i.e., the driving elements of the system, that are powered by a separate energy supply to carry out the functions. Information on the current status of the base system is determined by sensors that record different parameters for this (such as electrical current and voltage). After all, the sensor data are interpreted and the actuator control signals are provided by an integrated computer system connected via an IT link. The computer system comprises one or more processors and performs all sorts of control tasks.

Also, the computer system generally has external interfaces to exchange information with other technical systems. Different protocols are used for communication here. The properties of these protocols (such as communication speed (real-time capability) and reliability (redundancy)) need to be tailored to and selected for the specific application. It is common for information only to be preprocessed in a mechatronic system. More complex calculations and the determination of reference values, on the other hand, are usually the purview of more powerful or more comprehensively networked computer systems.

In addition to industrial machine-to-machine communication using industrial communication protocols, input options for a system operator can also be provided. Input can come directly from the mechatronic component, such as by pressing an OK button, or initially via an operator terminal that formats and sends the input values of the operator for the technical system.

◘ **Fig. 1.2.** General base structure of a mechatronic system

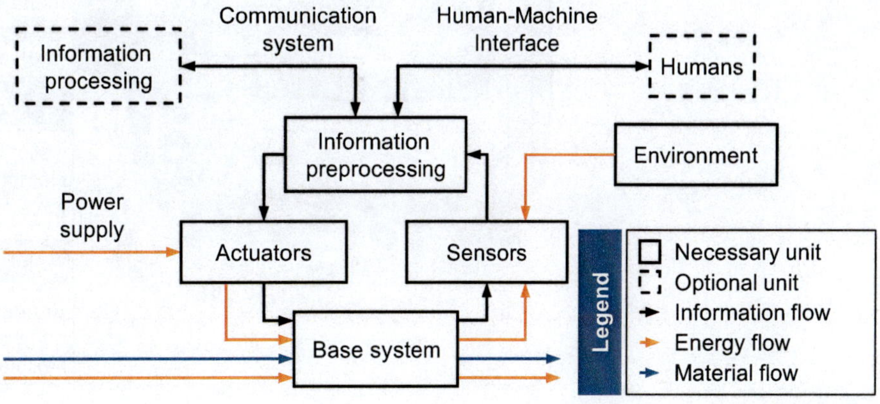

The machine tools of today, and the machines and systems of other sectors, comprise many different mechatronic units whose demands on the processing and provision of information differ greatly. ◘ Figure 1.3 provides an overview of the mechatronic subsystems of machine tools. Today, for example, there are several sensors in main spindle drives, for position and speed monitoring of the drive, for temperature monitoring, or for implementing diagnostics functions, that provide information on the current state of the spindle or protect it from damage. For feed drives that move the clamped tool along defined paths relative to the main spindle, the greatest demands are made on communication and control response - especially when multiple feed axes are involved in the movement and need to be synchronized.

To aid the machining process, automated handling systems are used that, for example, change the tool, store tools in magazines, or chuck and re-clamp workpieces. Here, too, strict demands are placed on the interface communication between the control unit of the handling system (internal to the machine) in each case and the master machine controller, for example, to prevent collisions and to coordinate the sequence of processes.

Furthermore, a machine tool itself is linked into the in-house production topology as a complex, all-in-one mechatronic system. On the one hand, communication for triggering jobs and interlinking takes place here vertically, i.e., with a master control system. On the other, different equipment (external to the machine), such as conveyor and test systems having their own controllers, can be part of the process.

## 1.2 Engineering

Manufacturing companies are faced with the challenge of responding quickly to new trends while keeping manufacturing costs low. This puts great demands on the flexibility and ability to reconfigure production systems, resulting in a high level of system complexity. Given that short development and commissioning times continue to be demands made on system manufacturers, the efficient mastery of engineering processes for manufacturing machine tools and production-related systems is a key core competency to compete internationally.

For most machine manufacturers, the development process for mechatronic systems takes place across multiple departments in line with the interdisciplinary problem in question. The departments typically involved in the development of machine tools and production systems are:

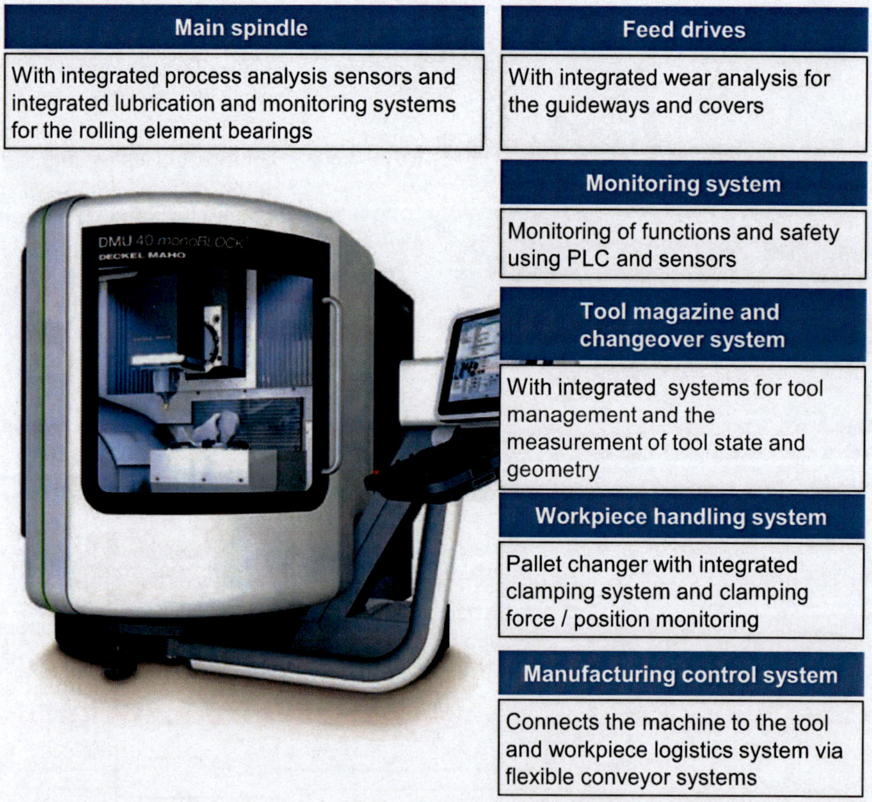


◘ **Fig. 1.3.** Mechatronic components of machine tools. *Source* DMG

- Mechanical engineering
- Electrical engineering
- Fluidics (often linked with electrical engineering)
- Control engineering (linked in part with electrical engineering)
- Software development
- Assembly/commissioning

While the processes of development previously followed a strictly sequential cycle, the transition nowadays is more towards the increased parallelization of different steps [VDI11]. This leads to shorter cycle times overall for order processing, especially for customized machines and systems. It is also clear that parallelism means that, for highly integrated systems, in particular, the need for communication between departments rises sharply when the disciplines are tightly networked. Potential interdependencies here between individual phases represent high risks. If, for example, design decisions for control-related programming necessitate different component sizing from fluid design, this might also entail changes in mechanical design.

To efficiently master the development process while system complexity increases, it is possible, on the organizational side, for example, to use different process models in system design and in working with change iterations, such as the V-model and the waterfall model [BRÖH95, ROYC87]. On the engineering side, the development process can be aided in particular by the systematic and cross-departmental use of suitable software. A particularly high level of integration of different development tools can be attained by using tools from a single source. For the machine manufacturer, however, this brings with it the risk of becoming dependent on the products from the software vendor. This can lead to high risk and poor negotiation position on the company level.

Modeling represents another base technology for mastering interdisciplinary systems. Be it a central documentation medium of a systems engineering process or the behavior model for physical simulation, interdisciplinary models especially represent in a centralized manner the most diverse aspects of a system and permit in principle partially or fully automated, computer-based interpretation and processing.

A widespread example of the expedient use of models is virtual commissioning. This permits the implementation of control programs prior to building the actual system. Depending on the availability of models, simulation of the system is either entirely virtual (software-in-the-loop) or real control hardware is used to actuate virtual components (hardware-in-the-loop).

## 1.3 Current Trends

Much research and development work is currently grouped under the "Industry 4.0" buzzword in industrial production technology. Industry 4.0 is synonymous here with the technical and organizational upheavals brought about by state-of-the-art information and communication technologies incrementally finding a foothold in manufacturing. Along the lines of the three previous deep-rooted diversifications in production technology over the last 300 years, the fourth industrial revolution is already being announced in advance in many discussions (see ◘ Fig. 1.4). It marries production technology operating across the globe with informatics and information technology (IT), with the role of humans receiving special consideration.

As information technologies have been enhanced, many process models for development have arisen, such as the spiral model, V, W, double-W, and Y-models, and software solutions for individual sectors and specific problems [VOGE03, BOEH88]. The increasing number of standalone solutions, however, is causing severe disruption to in-house information flow. In the 1980s, the endeavors in developing an overarching integration concept for information processing in production companies were summarized under the term Computer Integrated Manufacturing (CIM). The primary task was to unite production planning and control, with CAx technologies taking center stage.

The fact that many CIM projects did not succeed as hoped was attributable to various organizational and technical factors:
- Lack of regard for individuals with their creative skills
- Low maturity levels of information and communication technologies
- Lack of digital data for networking
- Lack of openness of software tools
- Lack of availability of communication standards and exchange formats
- Lack of IT expertise outside specialist departments
- Lack of consideration of organizational framework requirements
- High IT infrastructure costs
- Low weighting of complexity reduction as defined for lean approaches

Although these factors meant the CIM vision was deemed a failure at the end of the 1990s, the basic idea of integrated information management was still relevant because the problems of the heterogeneous IT landscape remained unsolved [ABRA05]. From the perspective of engineering, the integration thought process was pursued

1

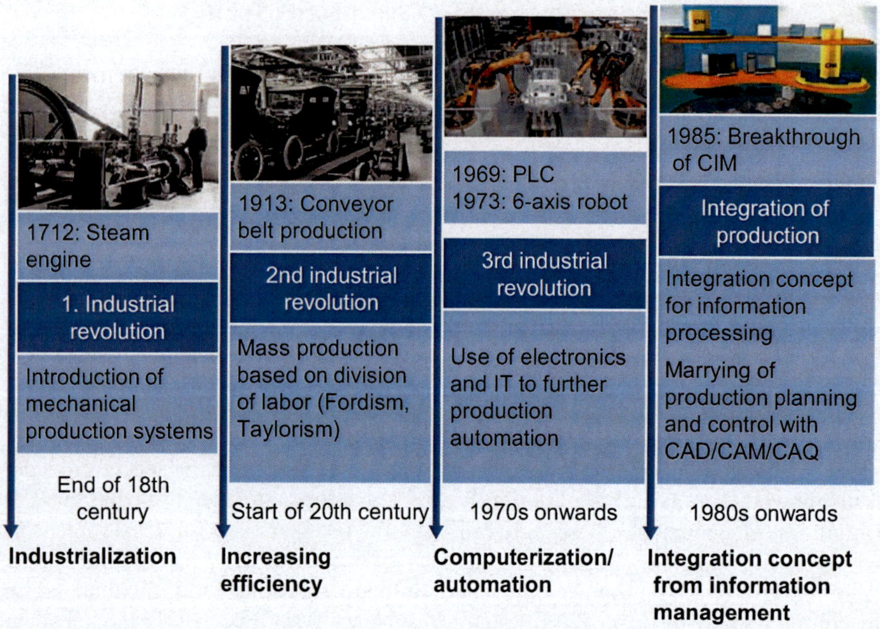


**Fig. 1.4.** Important milestones on the path to modern production systems. *Source* [BREC14]

under the term Product Life-Cycle Management (PLM). The starting point here was product data management, i.e., the administration of CAx data and technical documents using master systems that modeled the product structures, workflows, access permissions, etc. The transition from PDM to PLM is characterized by the fact that integration solutions were created between Design engineering and other in-house departments (Project Management, Production, Service, Controlling, etc.). Another approach with which parts of the CIM vision are being pursued and developed is the Digital Factory. This was introduced as an umbrella term for networked, digital models of the factory and of the associated resources and processes [VDI11]. The Digital Factory thus comprises many application areas - from layout planning, logistics and assembly planning, to virtual commissioning.

## References

[ABRA05] Abramovici, M., Schulte, S.: PLM-Neue Bezeichnung für alte CIM-Ansätze oder Weiterentwicklung von PDM. Konstruktion **1**(1), 2 (2005)

[BOEH88] Boehm, B.W.: A spiral model of software development and enhancement. Computer **21**(5), 61–72 (1988)

[BREC14] Brecher, C., Behnen, D., Brumm, M., Carl, C., Ecker, C., Herfs, W., Klement, R., Königs, M., Komma, T., Lohse, W., Malik, A., Müller, S., Özdemir, D.: Virtualisierung und Vernetzung in Produktionssystemen. In: Brecher, C., Klocke, F., Schmitt, R., Schuh, G. (Hrsg.) Industrie 4.0: Aachener Perspektiven, 1. Aufl. Herzogenrath: Shaker (2014)

[BRÖH95] Bröhl, A.-P., Dröschel, W.: Das V-Modell. Der Standard für die Softwareentwicklung mit Praxisleitfaden, 2. Aufl. Oldenbourg, München (1995)

[ISER99] Isermann, R.: Mechatronische Systeme. Grundlagen. Springer, Berlin (1999)

[JANS10] Janschek, K.: Systementwurf mechatronischer Systeme. Methoden – Modelle – Konzepte. Springer, Berlin (2010)

[ROYC87] Royce, W. W.: Managing the development of large software systems. In: Proceedings of the 9th International Conference on Software Engineering. ICSE '87. Monterey, California, USA, IEEE (Veranst.), S. 328–338 (1987)

[VDI11] Richtlinie VDI 4499 Blatt 2: Digitale Fabrik – Digitaler Fabrikbetrieb (2011)

[VOGE03] Vogel-Heuser, B.: Systems software engineering. Angewandte Methoden des Systementwurfs für Ingenieure. München, Oldenbourg (2003)

# 2 Feed Axes in Machine Tools

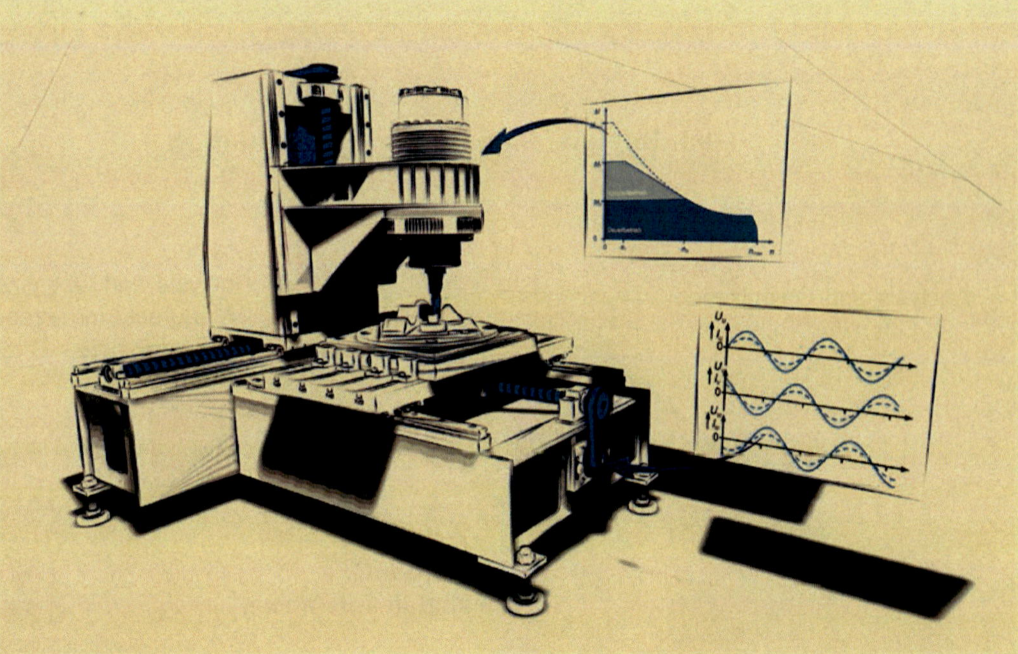

---

I thank my staff members Igor Medeiros Benincá M. Sc., Christian Fimmers M. Sc., Sebastian Kehne M. Sc. M. Sc., Simon Storms M. Sc. and Stephan Wein M. Sc. for their contribution in revising this chapter.


C. Brecher and M. Weck, *Machine Tools Production Systems 3*,
Lecture Notes in Production Engineering, https://doi.org/10.1007/978-3-658-34622-5_2

## Contents

The most important elements of automatic production facilities include feed drives that generate the contour of workpieces according to specified movement instructions in conjunction with tools. On purely mechanical and electrical/hydromechanical automatic production machines, information on the travel and speed required for machining a workpiece is available in the form of mechanical storage, or in cam plates or templates (refer to ► Chap. 8). Numerically controlled machine tools receive the relevant information in the form of NC data (see ► Chap. 11).

The control unit of the machine analyses the corresponding signals and relays them via the interpolator to the drive unit as reference variables. It is the actuator mechanism for relative motions between the tool and workpiece. Its motor then performs a change in rotation angle at the rotational speed specified. This change is converted by mechanical transmission elements to a corresponding position change of the machine component to move (see ► Sect. 2.1) (◘ Fig. 2.1).

To perform the relative motion specified externally between the tool and workpiece, one or more movements of individual axes, controlled in a coordinated way, are superimposed [BREC02]. The workpiece clamp or tool carriage can be driven for this depending on the machine variant [GROS06].

Each of the individual axes form a closed system comprising drive, power conversion and position determination. The basis of its structural composition is a general cause and effect relationship between the different system components that can be shown schematically in the form of a control loop (see ◘ Fig. 2.2). It defines both the information flow as well as the interfaces between internal and external elements of the axis system. While converter, motor and mechanical transmission elements form the control path and are less effective because of the positioning process, appropriate measuring systems are used for feedback of the individual control variables.

The differences between measured actual values and specified reference values give the control deviations that are converted to new values for the relevant actuators by a controller. Control theory methods can be used for the definition and design of the overall dynamic system (see ► Chap. 3).

Feed drives are differentiated into direct and indirect drives based on their principle of operation. While conventional drive systems, including mechanical–electrical drive systems, that generate translational motion indirectly, such as with mechanical conversion of a rotary drive motion, power generation for linear direct drives is directly between the primary and secondary parts, that can be moved relative to each other. Taking the electrical linear motor drive as an evenly wound rotary drive, the primary part corresponds to the stator of an equivalent rotating machine.

To check the suitability of a corresponding drive principle for the system of a single machine tool, its properties must be contrasted with the task and system-specific requirements (such as positioning accuracy, jerk-free motion, and mechatronic compatibility with other system components). Here, different parameters each define one aspect of system behavior that is given from the combination of the individual component properties. For feed drives for machine tools, the attainable speed, maximum acceleration, attainable

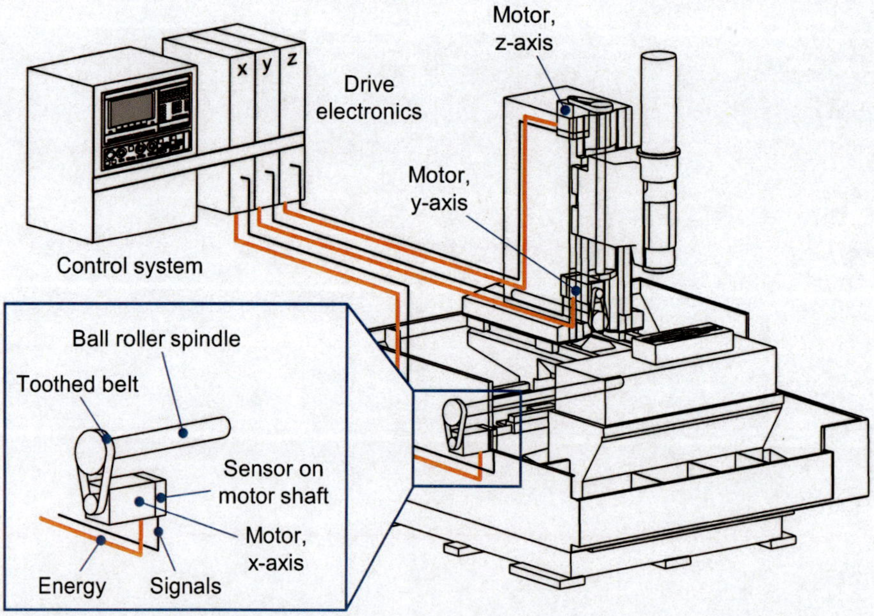


◘ **Fig. 2.1** Feed motion generation in a 3-axis machine tool

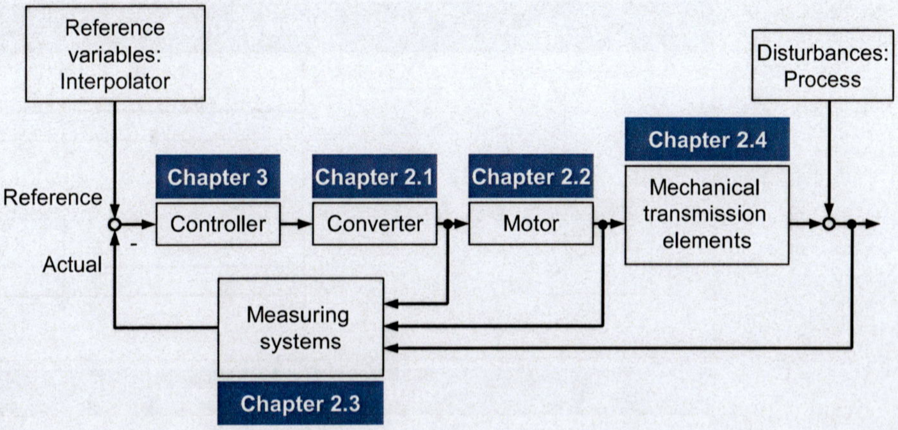


**Fig. 2.2** Schematic layout of a feed axis system and the chapter structure

feed forces, kinematic accuracy, and static and dynamic stiffnesses are typical characteristics for behavior [BREC02]. The nature and assembly of the drive train have a particular bearing on these characteristics.

Below, the individual system subsections are described according to the relationship in Fig. 2.2 and relevant design parameters for implementing the function are explained.

## 2.1 Converters for machine tool feed axes

Servo amplifiers have a control/regulator unit as well as a power unit. To take full advantage of the dynamic properties of feed servo motors, state-of-the-art power electronic components are used.

Numerically controlled machine tools require activation that permits operation of the servo motor in all four quadrants of the speed-torque chart:

1. Clockwise rotation, motoring
2. Clockwise rotation, braking (regeneration)
3. Counterclockwise rotation, motoring
4. Counterclockwise rotation, braking (regeneration)

Transistor and thyristor amplifiers are the best technologies available today. Transistor amplifiers are built for small and medium-sized drive systems rated up to 200 kW. The range of thyristor amplifiers extends up to power ratings of 10 MW.

### 2.1.1 Design of converter systems

If the requirement is to activate d.c. motors, a high power d.c. voltage source that can be regulated, and the polarity of which is reversible, is required. The energy necessary is taken from the a.c. or 3-phase supply and, after transformation to the voltage required as necessary, is converted in a controllable rectifier. For drive systems with d.c. motors having a low power rating, an output voltage proportional to the controller input voltage can be realized with power operational amplifiers (Fig. 2.3, top). For analog d.c. amplifiers, a high dynamic range can be attained. The power loss in this amplifier type can, however, be far greater than the nominal power of the motor.

For pulse width and frequency modulators, the transistors or thyristors act as electronic switches that apply a pulse-like d.c. voltage to the motor. To deliver high powers, the intermediate circuit voltage can be generated directly using rectification from the 400 V supply. The average value of the output current is controlled by the pulse width or frequency (Fig. 2.3, middle). Insulated Gate Bipolar Transistor (IGBT) technology is normally used because, compared to the bipolar transistors used in the past, it features higher possible switching frequencies (above 100 kHz), shorter delay times and simpler activation. These amplifiers have virtually the same dynamic properties as d.c. amplifiers yet the power losses in the transistor are considerably lower. The output current is smoothed sufficiently by the motor inductance rendering additional smoothing chokes superfluous in most cases.

A 3-phase servo motor is usually controlled using sinusoidal voltages of variable frequency and amplitude. A high-resolution position encoder is used to determine the exact rotor position required here. Lower torque ripple and higher dynamic range are attained compared to the conventional supply with rectangle wave currents. Pulse-controlled inverters are used to supply 3-phase servo motors (Fig. 2.3, bottom). The intermediate circuit voltage is usually generated with direct rectification or with conversion from the 400 V supply. Intermediate circuit voltages to 750 V are also

2

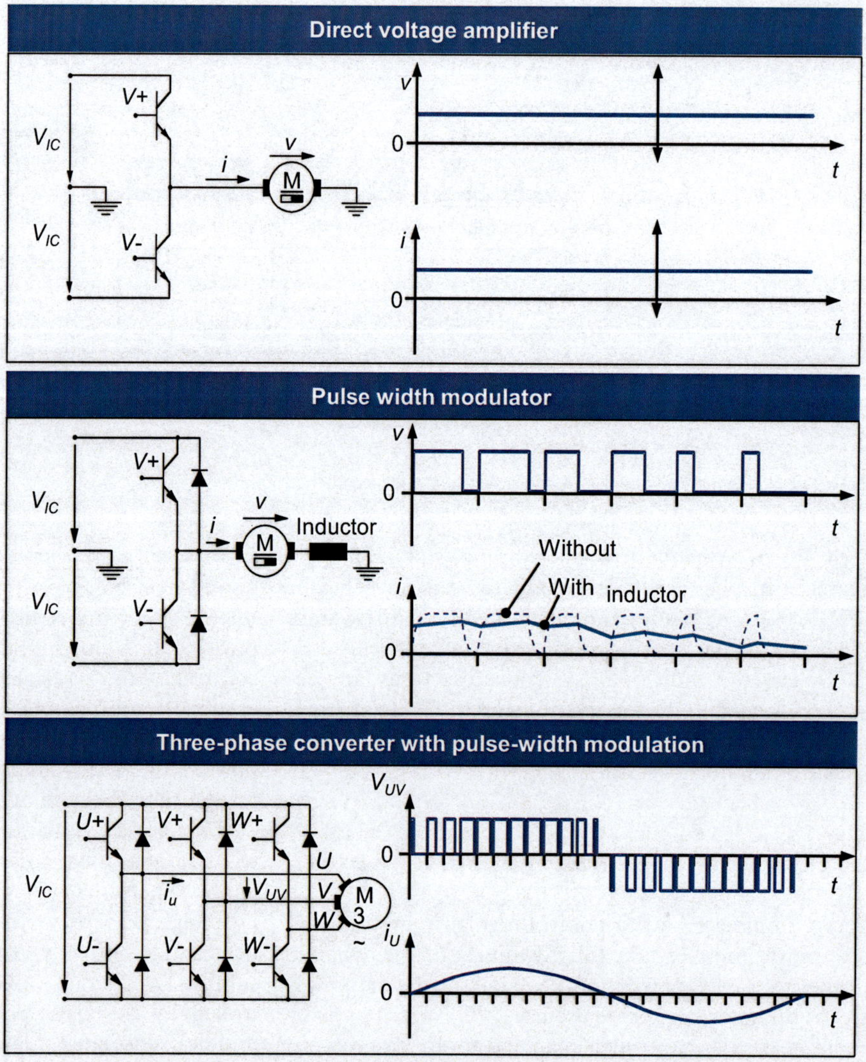


**Fig. 2.3** Basic circuitry of electronic control amplifiers

used. For the downstream three-phase inverter, IGBT technology is best suited for the motor power ratings customary for servo motors because the high switching frequencies resulting in only low torque ripple and noise on the servo motor.

Protecting the power stack from overloading, short circuits, ground faults, electrical surges and voltage failures is state-of-the-art nowadays. Notifications about such fault conditions and other signals, and a connection to Fieldbus systems, are now the standard for advanced drive amplifiers. Fault conditions can also be reported directly by the drive and be retrieved by the controller (over the SERCOS interface, for example)—see ► Sect. 2.1.3.2.

As a rule, asynchronous motors used mainly for main spindle drives can operate the same way as permanent magnet synchronous motors with the same power electronics that are typical for servo axes. By exchanging the software, the same hardware can be used for the other motor type. To generate the switching signals for actuating the IGBT, integrated modules and microcontrollers are available for analyzing the current measurement and measuring the rotor position [SCHM09].

### 2.1.2 Control electronics in converters

Nowadays, frequency converters for electric drive systems in machine tools are normally modular in design. The drive module, for example, usually comprises a power stack with a plug-in control card. This enables the same control electronics to be used for different converter sizes. Conversely, the controller feature set required can be determined individually by using different control cards.

The controller hardware in the converter systems has a different interface to the power electronics. This is dependent on the manufacturer. The PWM (pulse width modulation) interface is usually used here to take effect as deep as possible in the functional chain. This means the power electronics no longer have any kind of intelligence or computing power of their own. Instead, they receive the signals necessary for interconnecting the IGBTs (Insulated Gate Bipolar Transistors) and make available measuring signals (such as current, temperature and intermediate circuit voltage) for analysis on the control card.

The control card is primarily responsible for controlling a motor. This comprises current, rotational speed or linear speed control as well as position control depending on build level and the interface to the controller. The control loops can be either analog (in older systems) or digital.

The basis for control is the measurement of actual values. This too is usually performed in the drive controls. The input to the current control is the actual current value that can be measured in the converter. On older (analog) systems, speed control is based on measuring the voltage of the tachogenerator. For digital drive controls, the time-discrete differentiated signal of the rotary encoder or linear gauge is used here. This also serves as the actual signal for the position control. Drive control is covered in detail in ▶ Chap. 3 and 4.

Drive controls nowadays perform a number of other functions in addition to actually controlling the motor. These are monitoring and safety functions, commissioning aids and measurement operations—presented in more detail in ▶ Sect. 2.1.2.3.

#### 2.1.2.1 Analog control

Analog controls are characterized by the internal parameters (such as speed difference and reference current value) being represented by analog signals, such as voltages (◘ Fig. 2.4). Operational amplifiers are used to process the signals. Controllers with PI behavior are used predominantly. PID controllers can also be found in isolated cases.

On analog controllers, the control factors are normally adjusted with mechanical actuators. This is relatively complex and inflexible. During commissioning, every axis needs to be adjusted individually by a technician.

Nowadays, analog control only plays a minor role in machine tools. Advances in computer technologies have made processors available that are capable of meeting the strict requirements of control in machine tools.

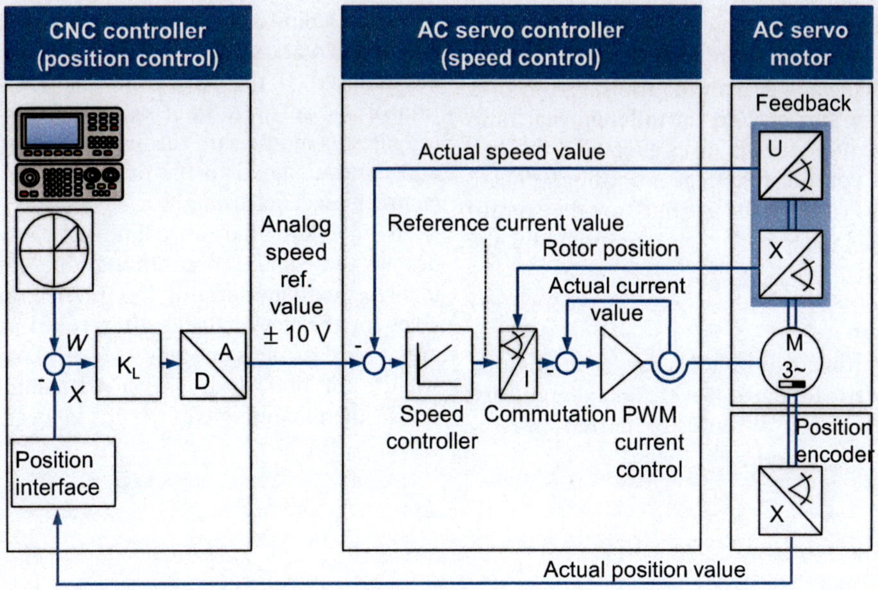


◘ **Fig. 2.4** Diagram of an analog feed control system

#### 2.1.2.2 Digital control

Digital processors are also used in fast speed and current control loops. Digital signal processors (DSPs) are normally used here given the high computing power necessary. These CPUs are specially designed for the calculation of complex operations and high data throughput.

Speed and current control loops standard now are sampled at a frequency of up to 16 kHz (clock rate 62.5 μs) depending on application. For the current control of an asynchronous main spindle for example, field control means more computing time is required than for controlling a synchronous motor.

The measuring signals must also be analyzed in this fast clock rate. The actual current must be available for the current controller and, in every rotational speed and speed control cycle, the actual position must be analyzed and the actual speed calculated. Here, the axis computer must in part also compensate the non-linear behavior of the conversion and amplifier electronics to calculate the exact position value.

#### 2.1.2.3 Additional functions of digital drive controllers

The advent of microprocessors in drive controllers has made it possible for many functions to be moved from the controller to the drives. The goals here include better bundling of drive-specific functionality and minimizing the required data interchange with a simpler structure. Some additional functions in drives are described in brief below.

##### Parameter set change

Drive controllers normally support the management of multiple parameter sets. Changing between these settings can be relatively straightforward during operation, due to a change of gear in the transmission of a main spindle, for example. Here, the different gear ratio means another reduced inertia takes effect and this can necessitate other controller settings accordingly. These controller settings can then be activated by the control system by selecting a parameter set appropriate for the gear currently set.

##### Threshold monitoring

In individual controllers, thresholds can be configured so as not to overload the downstream components. These are, for example, a transistor current threshold specified by the size of the converter, a current or torque threshold and the threshold rotational speed of the motor.

Also, the output parameters of the controllers are monitored to see if they stay at the maximum value over a longer period. In such cases, a parameter configuration error or disruption is assumed and the scheduled machining is stopped.

##### Monitoring of motor and converter temperatures

The drive uses sensors to monitor the temperatures in the motor coils and in the converter to prevent these components from overloading. Two temperature thresholds can normally be set. When the first one is exceeded, a warning is issued to the control system that can then stop the specific drives. If the second threshold is exceeded, the drive system is switched off to prevent damage.

In addition to monitoring the motor temperatures, the power introduced theoretically into the motor is also monitored to prevent a momentary overload. The high momentary overload capacity is the reason here (five times the level for synchronous motors). If the motor is run for too long at the current limit, permanent thermal damage can occur even before the temperature sensors have reached the switch-off threshold.

##### Traverse range monitoring

Many drive controllers already monitor the traverse range of the axis, and realize the software limit switches necessary to protect the mechanics in this way. This enables a 3-level safety concept to be implemented in machine axes (◘ Fig. 2.5).

The control system only issues reference values within the allowed axis traverse range. At the boundaries of the working area, the path velocity is intentionally reduced. If the drive controller ascertains a violation of the software limit switches, it immediately decelerates its motor with the maximum torque permitted and issues a signal to the other drive controllers that can then react accordingly.

If one of the hardware limit switches is still passed despite these safety precautions (due, for example, to incorrect configuration of the traverse ranges), the enables for the power stacks are revoked and the brakes are applied at the same time. Motors coast to a standstill if not braked. If the speed cannot be reduced within the remaining gap to the stopper, an axis might hit it.

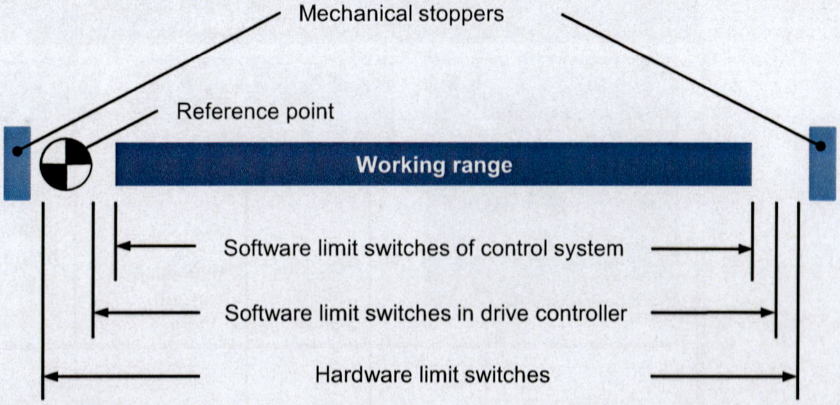


**Fig. 2.5** Traverse range limitation in feed axes

For practical purposes, the reference point for incremental measuring systems (see ▶ Sect. 2.3) is outside the traverse range of an axis. This enables the reference point to be approached from every position in the traverse range and in the same direction.

#### Plausibility checks

The tracking error is also often monitored in addition to the traverse range. If the maximum permitted tracking error is exceeded, the drive assumes overloading of the axis or incorrect configuration and switches itself off.

Seize monitoring is a similar function. If, after sufficient motor torque is built up, the axis does not move within a specified time, the axis is also switched off.

Motors on vertical axes generally have a brake. This is applied after the holding torque in the motor is built up. This function can also be performed by the drive to ease the load on other controller systems. The brake can be switched back on automatically by the drive as soon as the motor torque is no longer being applied (to prevent the axis from falling, for example).

#### Compensation of geometrical errors

Drive controllers also containing the position control loop can compensate spindle pitch errors and separation errors of the direct measuring system. A compensation table is defined for this in the drive controller. The values for compensation can be determined for example using a laser interferometer to measure the axis.

#### Commissioning support

Modern digital axis controllers support commissioning by calculating the start settings for the controllers from the available motor size data. The machine can be moved with these controller settings. Optimization by a technician is generally advisable, however, to improve the dynamic properties of the axis.

### 2.1.3 Interfaces to the control system

The interface between drive and NC is realized on different levels depending on manufacturer and application (◘ Fig. 2.6). Reference values for the control loops realized in the drives are sent over the interface. The control loops are closed in the drive or control system depending on the position of the interface. Realization of a control loop in the control system presupposes that the actual values required for the control are transferred from the drives to the control system (provided they are measured in the drives and not directly in the control system).

In ◘ Fig. 2.6 on the left, the rotational speed and current control loops are closed in the controller system. The position is also measured in the controller. Reference current values are transferred via the interface. One such interface can be found, for example, in the Ω systems from Baumüller Nürnberg GmbH.

The reference rotational reference speed interface is shown in the center of ◘ Fig. 2.6. The position controller is closed in the control system, and the other control loops are closed in the drive. Actual values are measured in the drive, meaning the current values for actual position are sent via the interface to the controller to close the position control loop. The reasoning behind this procedure is that the rotational speed signal for digital drive controllers is usually derived from the position signal of the motor encoder. An optional, direct measuring system is also connected to the drive given the simpler scalability of the number of axes. The rotational speed interface is used in the Siemens Simodrive 611D system, for example.

◘ Figure 2.6 shows on the right the position of an interface for reference position values. Here all control loops are closed in the drive. This greatly reduces interface data traffic, and high controller clock rates

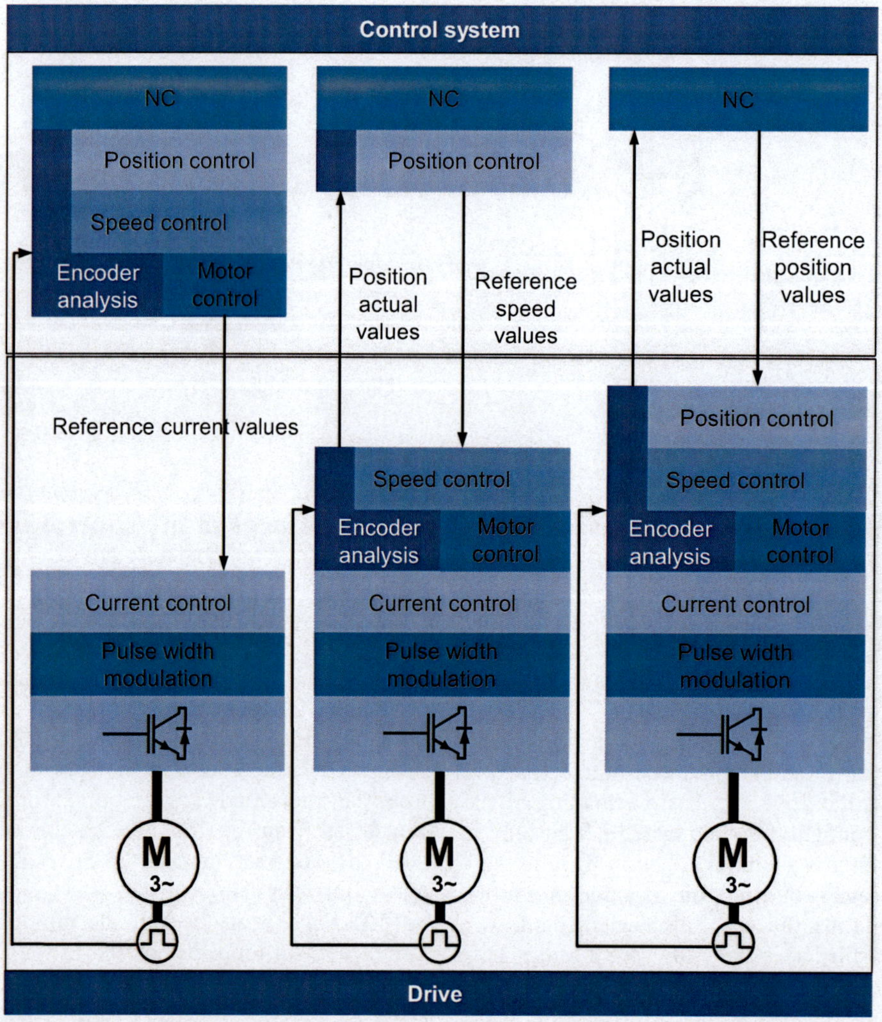


**Fig. 2.6** Different positions of the interface between control system and drive

are possible because quick data flows need only be realized in one device. This also opens up the possibilities of performing fine interpolation in the drive, and to only passing updated reference position values to the drives in the interpolation cycle. This reduces the data rate further. One such interface is usually used in Sercos systems, for example (► Sect. 2.1.3.2).

It is not possible at this time to identify a clear trend towards one particular position of the interface between drive and controller. All of the interfaces have specific benefits and downsides depending on the application. A downstream interface, for example, (e.g., reference current values) enables comprehensive coordination of the control of different drives because the data in the control system can be exchanged quickly between the individual axes. This solution is complicated, however, and can be expensive since a lot of data need to be transported within short times. On the other hand, an upstream interface (e.g., reference position values) enables quick sampling rates of the control loops because no external communication is necessary inside the control loops. Coordinating the different drives with each other is difficult, however, in the fast control loops.

The interfaces for reference current value and reference rotational speed value were realized previously using analog voltage signals. The inadequate resolution and accuracy of the analog signals meant, however, that this method was not suitable for a position interface.

The analog ±10 V and digital SERCOS interfaces are described below.

#### 2.1.3.1 Analog interface

Expanding the servo amplifier with a rotational speed control loop transformed the amplifier into a drive control unit with the internationally standardized, analog ±10 V rotational speed interface for the amplifier

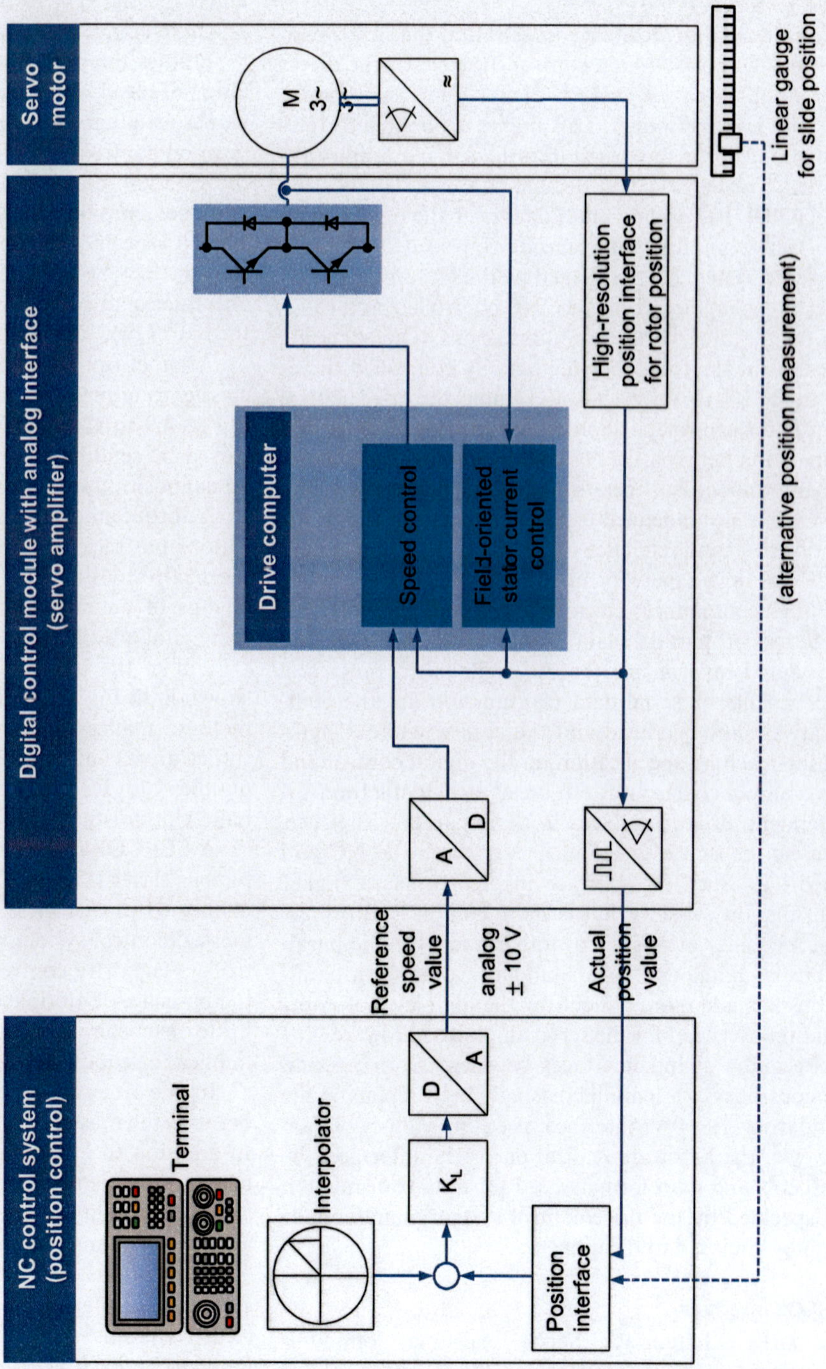


**Fig. 2.7** Structure of a drive controller with analog rotational speed interface according to Bosch Rexroth

input. This interface enabled drives and controls from different manufacturers to interact perfectly.

Since 1990, digital technology has been used increasingly for feed drives to replace the analog input interface, inaccurate and susceptible to interference for the standards of today, with a digital interface with high levels of reproducibility, reliability and flexibility.

The best available technology enables both position and rotational speed control to be digital. In the lowermost (current) control loop, good dynamic characteristics mean analog circuits are still being used frequently because fast, and thus expensive, computers are required for digital control. They is only justified if the requirements call for it. Figure 2.7 shows the structure of a drive controller with conventional analog ± 10 V interface. Both position control in the NC and rotational speed control in the servo amplifier are performed digitally here already.

#### 2.1.3.2 Digital interface

A high degree of flexibility in selecting the best possible open and closed-loop control strategy for the drives can be achieved by using a digital interface between NC and servo amplifier. This makes it possible to integrate control for the drives into the servo amplifiers to a greater or lesser extent depending on the application [SCHÖ09]. By sending more data over the digital drive interface, commissioning, monitoring and logging of the servo system are performed by the NC controller.

The cycle times of modern NC controllers and PLCs are on the order of a few milliseconds. The sampling rates in the control loops of digitally controlled drives are in the 100 μs range. If, for example, the drive control loop is opened, or position control decentralized, communication between the NC and the decentralized components must also operate with the same cycle times. Given that no undefined delays may occur in the transfer of actual and reference values, a transmission system with real-time capability in the sub-millisecond range, and high communication capacity, is required.

Serial or parallel data transmission methods are considered in principle for the digital drive interface. The benefits of serial data transmission are the comparatively low overhead and the ability, without making far-reaching specification on the digital control and servo amplifier hardware, to create a manufacturer-independent drive interface. With a parallel interface, data can be sent over a data bus between the NC and the drives. Such an interface inherently has a higher transmission capacity, but is more complex. Transmission reliability means the overall capacity of the parallel bus is limited ($\leq 10$ m). Standardization for a parallel bus is made more difficult by the high level of coordination overhead for the servo amplifier hardware.

Recently, digital interfaces between the drives and the control system have increasingly been replacing the standardized $\pm 10$ V interface used previously. Often however, the hardware realization of the interface, the protocols and data formats used for data transmission are specified by the drive/control system manufacturer and not disclosed to the user.

**SERCOS interface**

The SERCOS interface (Serial Real-time Communication System) is a standardized digital interface for serial communication between control systems and drives. It was specified by a study group at the German Machine Tool Builder's Association (VDW) and the Electrical Drive Systems Section at the German Electrical and Electronic Manufacturers' Association (ZVEI). The aim of the agreements is a manufacturer-independent interface solution having considerably extended functionality compared to the analog $\pm 10$ V rotational speed interface in general use previously [LARS03]. Particular importance was attached to time equidistance and the synchronization of measuring times, applying reference values and interpolation.

Besides the predefined, standard modes (torque, rotational speed and position specification by the NC), numerous other extended modes are supported, such as position control with pilot control. ◘ Figure 2.8 depicts the structure of a digital drive control where the servo amplifier contains the entire position control system. The NC specifies reference position values that are sent over the SERCOS interface to the drive. Actual position and rotational speed values can be read by a sensor on the motor or with a high-resolution linear scale.

Another option is the digital implementation of the classic rotational reference speed interface (compare ◘ Fig. 2.9 to ◘ Fig. 2.7). The position control loop is closed by sending rotational speed reference values or actual position values over the digital interface.

A torque interface is also possible for special applications but requires a high communication capacity in the relatively short current control cycle due to the exchange of actual and reference values necessary over the digital interface.

Unlike the usage of the serial SERCOS interface shown in ◘ Fig. 2.9, a parallel drive bus is used by Siemens to realize this structure in the 840D/611D controller/drive combination. Short cycle times for a higher number of axes can be achieved owing to the higher bandwidth of the parallel drive bus.

A SERCOS ring comprises a bus master and multiple bus slaves (◘ Fig. 2.10). Several bus masters can be connected to one SERCOS master controller. The numerical control system is the bus master or master controller. The drive controllers for the individual axes are the bus slaves. Multiple drives (slaves) can be connected up for each physical interface by using powerful communication modules.

In one cycle, all data regularly required is exchanged between the master and the drives (cyclic transmission). In addition to cyclic data, the flexible transmission of demand data is possible (acyclic transmission). The real-time capability of the SERCOS architecture is characterized by a maximum jitter time of $\pm 7$ μs.

The stations and slaves are arranged in a ring and connected using fiber optics. When the master broadcasts telegrams, they are relayed by all stations and finally arrive back at the master. The slaves pass on the data received (repeater function). The repeaters mean that a time delay of the data flow occurs at every station for the transmission time of a bit.

The exact maximum number of drives to be services by each optical fiber ring depends on the cycle time required for the operating mode selected, the data volume and the potential gross data rate. Also, the response and processing times internal to the drives and control system can have a major bearing on the communication performance attainable.

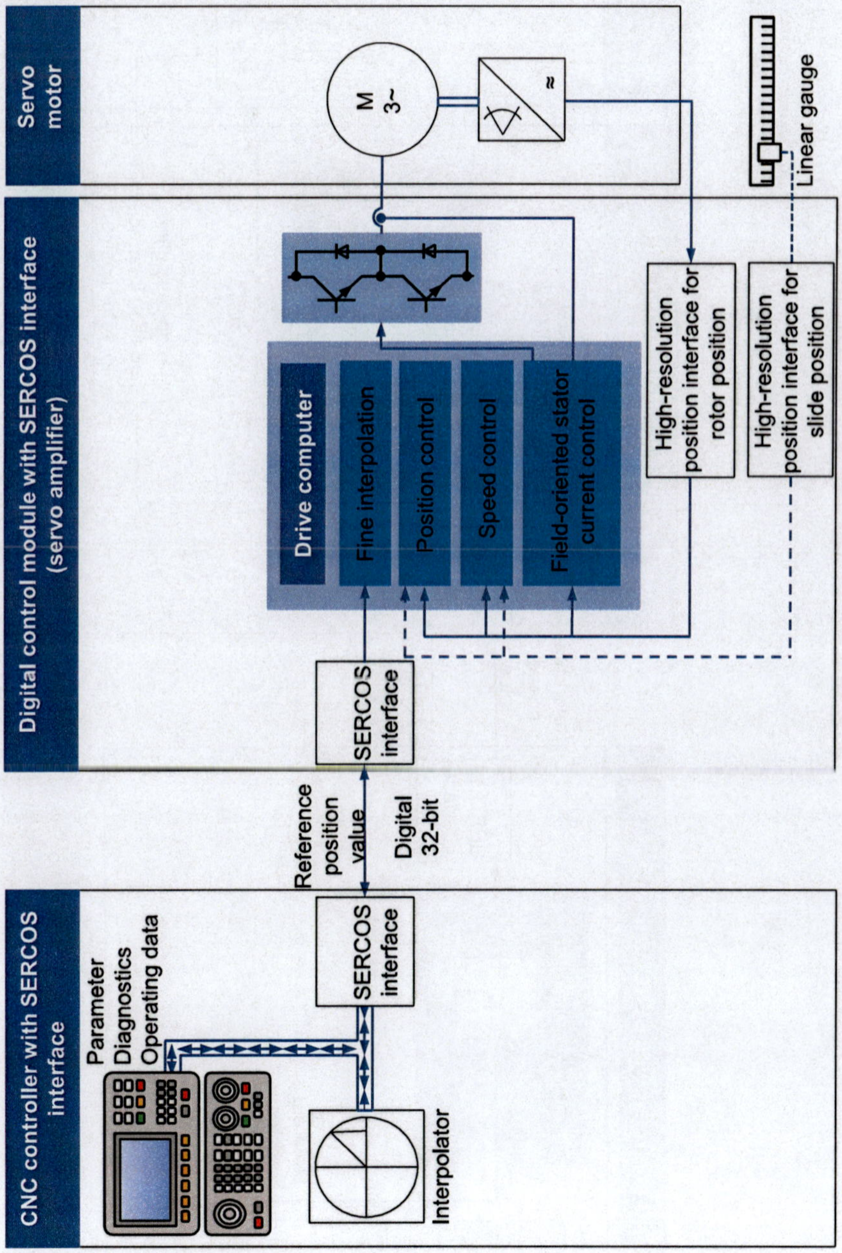


■ **Fig. 2.8** Structure of a drive control with digital position interface according to Bosch Rexroth

### Initialization procedure and communication phases

All communication parameters are specified and exchanged during the 5-phase system initialization procedure. The master specifies the communication cycle time by broadcasting a master synchronous telegram (MST) at equidistant time intervals. It only contains a single data byte, that in turn contains the current communication phase (0 to 4). Cycle times can have values of 62.5 µs, 125 µs, 250 µs, 500 µs, and every integer multiple of 1 ms (up to a maximum of 65 ms). The MST is received by all slaves, and starts the time counters for the transmission times on the slaves.

During Phase 0, the master checks whether the fiber glass ring is complete by broadcasting the MST. In Phases 1 and 2, the master sends, in addition to the MST, a master data telegram (MDT) with only one data record. This telegram does not yet contain any cyclic data. By specifying an individual slave address, the master selects every single slave in succession to identify the drives and transfer communication parameters.

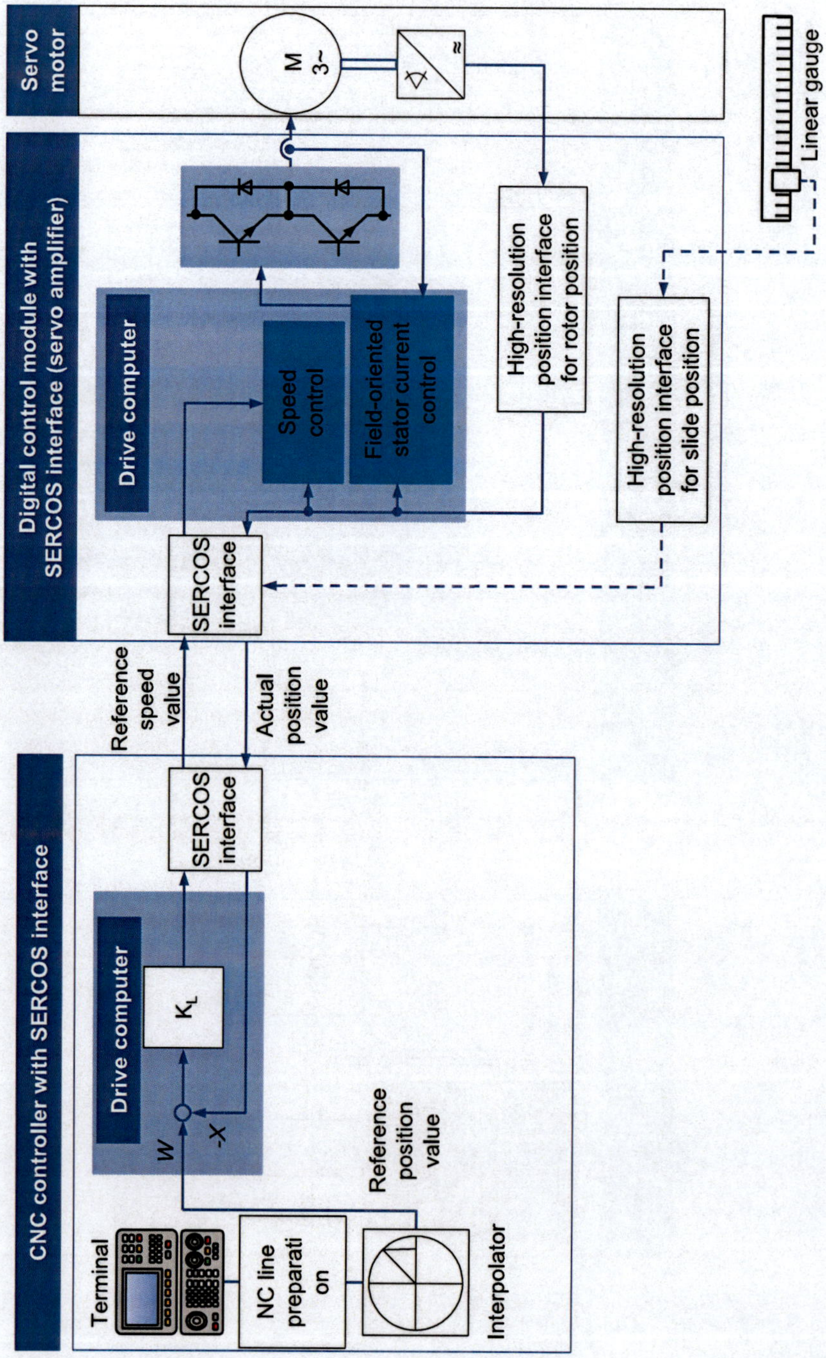


**Fig. 2.9** Structure of a drive control with digital a rotational speed interface according to Bosch Rexroth

Every slave responds with a drive telegram (DT), that also does not yet contain any cyclic data. The system is then ready to switch to cyclic mode to transfer real-time data. The cycle time and send time slots are still subjected to weak requirements in these phases.

In Phase 3, additional parameters are first transferred and final checks performed. The cyclic data still has no validity. Cyclic operating phase 4 follows. Transitions between individual phases are only permitted in ascending order (0–1, 1–2, 2–3, 3–4). Should a fatal communication error occur, however, the system resets to Phase 0 and a new initialization phase is performed.

The sequence of synchronous and data telegrams mentioned repeats in every communication cycle. The

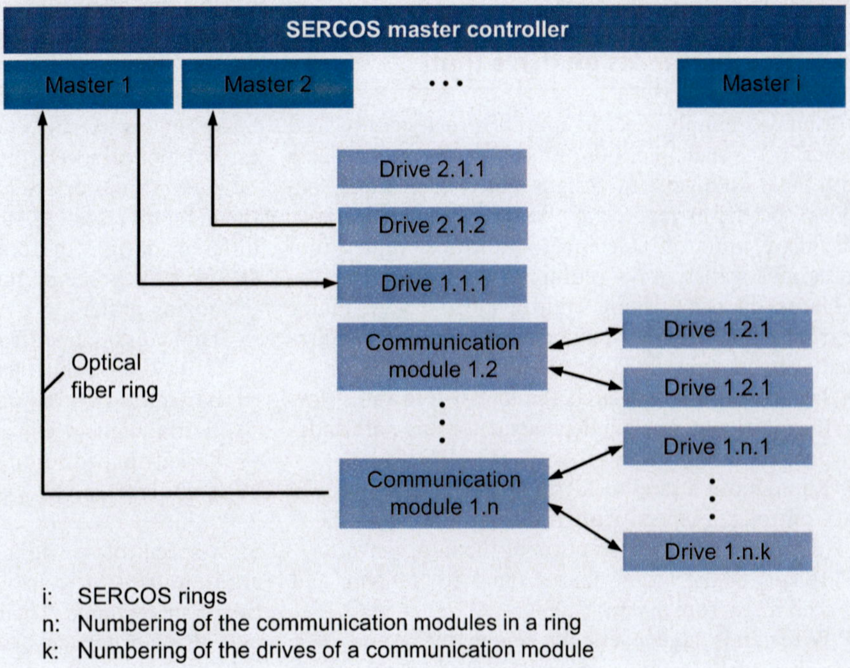


**Fig. 2.10** SERCOS architecture according to Schneider Group

fixed communication protocol means a high cyclic communication performance is attained. It is, however, to the detriment of a slightly restricted flexibility. Direct communication between the slaves is not possible, for example.

### Transmission of cyclic data

Reference value data and the control word are sent in the MDT to the drive as cyclic data. The drive responds with the DT containing the cyclic actual value data and the drive status. The relevant actual and reference values are exchanged depending on the mode of the axis controller as a position, speed or torque controller, and configuration by the user. If, for example, the axis controller contains the entire position control system, the NC need only specify the reference position values ($x_{set}$,$y_{set}$,...). For the pilot control, a speed modulation factor ($n_{x,on}$,$n_{y,on}$,...) can be sent as an additive reference value. Transferring the actual parameters is not required as a rule but is also possible for monitoring or display purposes. The NC uses a control word to specify the axis controller mode. This allows, for example, the drive to switch between position and speed control, also during operation. The drive uses the status word to report status changes (such as switch-off and overload), and the current readiness state to the NC.

### Transmission of demand data

Acyclic data transmission enables the display and entry of virtually any data, parameter and diagnostics information, internal to the drive and less time-critical, via the NC control panel. Demand data (such as parameters and manufacturer-specific information of any kind) is sent over several consecutive cycles and transferred together with the cyclic data. The information word, sent in the data records of the drive telegrams and master data telegrams, is provided for this purpose.

The master is able to read and write slave parameter data when requested by the control system. Requests and acknowledgments for acyclic data are secured using handshake bits in the control or status word. The NC is able to specify thresholds, for example, for every drive, send commands and change the drive properties via the parameters (such as to change the control parameters and adapt the drive to different encoders). The drive can report to the NC when specific axis positions are reached or more detailed information on its status (such as motor temperature). During the initialization phase, the transmission properties are specified via the acyclic data channel. The contents of telegrams are stipulated here for example. Also, the NC can request the supported data and commands for a particular drive.

## 2.2 Motors in feed axes

### 2.2.1 Requirements on drive units

Nowadays, mainly electric motors are used for feed drives in the machine tool industry. They are available with the most diverse of designs, power classes and special properties as regards application in machine tools. The drive units of machine tools with complex tool paths must satisfy strict requirements because they, as actuators in the position control loop, have a major bearing on machining quality. The following requirements are accordingly made on drives:

- Jerk-free running even at the lowest feed rates (low torque ripple, angle-independent torque, path-independent feed force)
- Rapid motion rates to 120 m/min
- Controllable speed with ratio of at least 1:10,000 (depending on the resolution of the D/A converter)
- Torque at the motor shaft from 5 to 100 Nm, and feed force from 500 to 20,000 N
- Withstands high overloads momentarily and high pulse torque for acceleration and deceleration (briefly 4 to 10 times the values)
- Highest possible acceleration behavior over the entire speed range
- High positioning accuracy, executing the shortest displacement increments (< 1 µm)
- High dynamic quality, i.e., low time constants (rise time $t_{up} = 5$ ms to 50 ms, dead time $t_t <$ sampling time of NC controller), to attain startup and delay times that are as short as possible
- High cutoff frequency ($\omega_E \geq 100$ to $600\ s^{-1}$) where the lower bound is required for large amplitudes and the upper bound for small amplitudes so that highly dynamic path movements and exact positioning are possible
- Low size and weight
- Low power loss (temperature rise)
- High reliability

The aforementioned requirements can be realized using electrical and hydraulic motors specially developed for highly dynamic drives in conjunction with suitable open and closed-loop control equipment. Motors optimized in this way are known as servo motors.

For (rotatory) electric motors, a distinction is made between direct current, synchronous, asynchronous and stepper motors. In addition, the linear motor represents a drive that performs a translation motion in line with the same work principles.

The dynamic properties of the drives, such as low mass and low moment of inertia, high acceleration behavior and extremely uniform running, even at the lowest rotational speeds and feed rates, are characteristic of servo motors. Servo motors are manufactured in the most diverse of custom designs. The fundamentally different designs of rotatory motors are contrasted in ◘ Fig. 2.11. Some of the measures to reduce the mass of moving parts:

- Thin, disk-shaped rotor for axial field profile, low rotor weight, short length (disk rotor)
- Bar-shaped rotor, slimline design, low moment of inertia, high-speed rotor (bar-wound rotor)
- Basket-shaped rotor and additional stator core (hollow rotor)

Slow-speed rotors with comparably high moment of inertia are used to realize direct drives with a wide rotational speed range. Their high mass means they have a high thermal capacity. Accordingly, they can usually be subjected to high overloads momentarily and can deliver a high peak torque.

#### Disk rotors

Disk motors (see ◘ Fig. 2.11, top left) are direct current servo motors with permanent field excitation where a disk (rotor) comprising insulation material and with bonded copper wires is pivot-mounted in the air gap. The magnetic field generated by typically eight to 10 pole pairs propagates evenly inside the motor. The iron parts of the motor housing form the magnetic inference. Current is passed directly to the conductor tracks via special carbon brushes. The interaction with the magnetic field causes a force to act upon the rotor disk so that the disk rotates.

The lack of iron in the armature not only reduces the moment of inertia. This also results in low values for the armature induction. This enables high current rise rates to be realized so the motor can be accelerated to the required rotational speed within a few milliseconds.

#### Bar-wound rotors (high-speed rotors)

Bar-wound or inertia motors (see ◘ Fig. 2.11, top right) are direct current servo motors with a bar-shaped armature that exhibit low moments of inertia and high acceleration behavior. A grooveless silicon steel plate

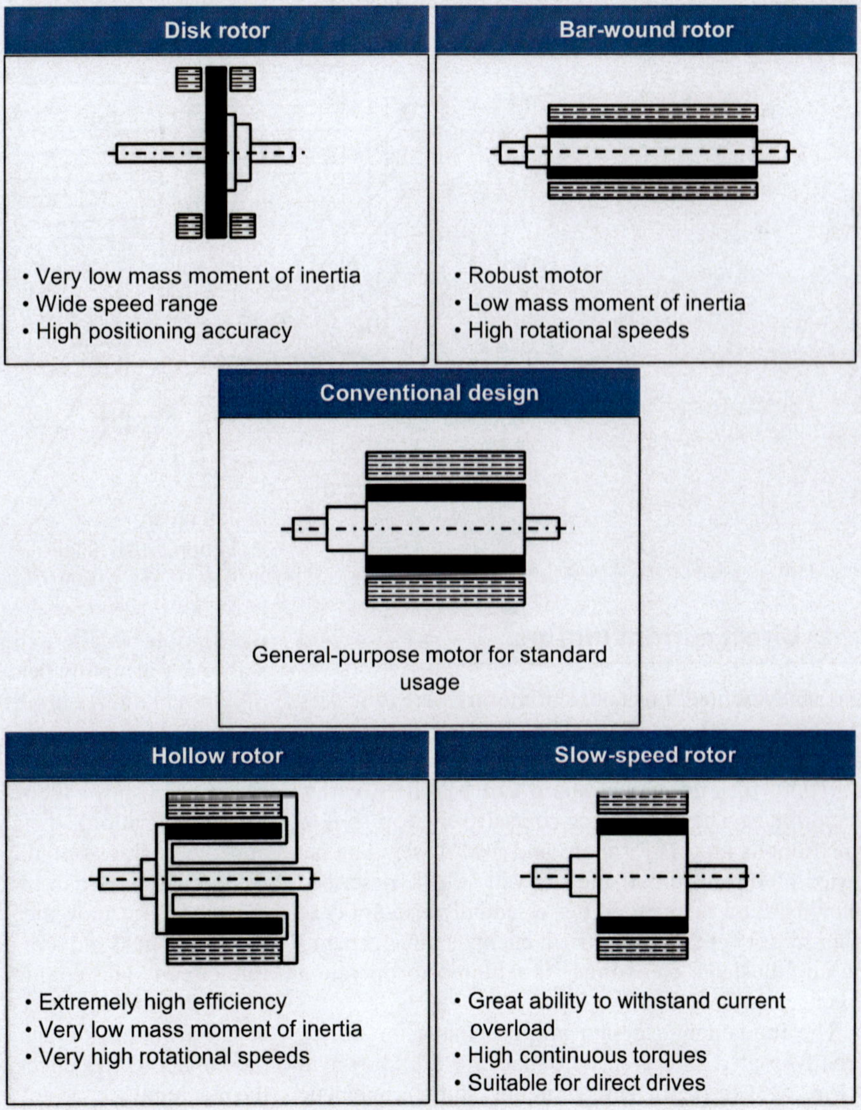


**Fig. 2.11** Designs of servo motors

core is pressed on to the motor shaft as a carrier for the armature winding. The winding is secured by glass fiber bindings.

### Hollow rotors (bell-type armature rotors)

On a hollow rotor motor (see Fig. 2.11, bottom left), the armature windings are secured on a bell-shaped winding basket joined permanently to the motor shaft. In contrast to a conventional direct current motor, the armature iron is not part of the shaft. Instead, it is fixed and forms part of the housing. This gives a low mass moment of inertia for the rotor.

### Slow-speed rotors

These kinds of servo motors are direct current motors excited by permanent magnets and with a high number of poles (refer to Fig. 2.11, bottom right). They are characterized by high torques, even at low rotational speeds. The thermal losses occurring are absorbed by the high heat storage capacity of the large rotor. The motor is suitable for the direct drive of feed spindles. The maximum rotational speed usually extends to 3000 $min^{-1}$. Excitation by permanent magnets means the efficiency level is high.

2

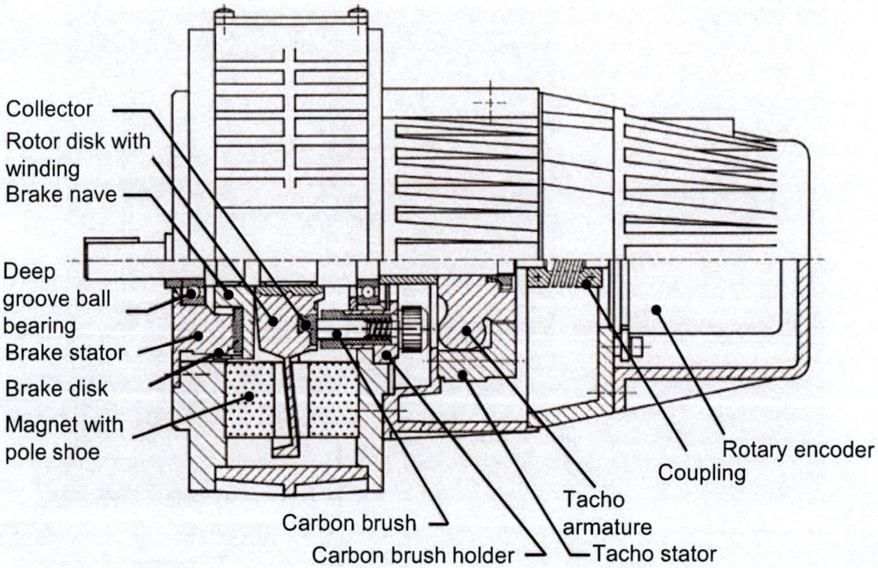


**Fig. 2.12** Design of a disk motor with speed indicator and rotary encoder according to INFRANOR

### 2.2.2 Direct current motors

Externally-excited direct current motors were long used as main drives on machine tools because they exhibit high torsional rigidity under load in the controlled state, and their rotational speed can be varied within broad limits. They have since conceded the top spot to synchronous motors for main and feed drives. The behavior of direct current motors will still be described below because their properties essentially also apply to other motor types. An electrical energy source generating an adjustable d.c. voltage is required to operate a direct current machine.

The fundamental design and schematic for an externally-excited direct current machine are shown in Fig. 2.13. In the machine column (stator), magnetic flux $\Phi$ is generated by the excitation current $I_f$, that permeates the armature (rotor) through the air gap. The armature current flows in the windings of the armature (shown in the figure as a dual conductor loop perpendicular to the drawing plane). In line with the physical principle dictating that a magnetic field exerts a force upon a conductor carrying a current, a torque is generated in the direction indicated. For the direction of the torque to remain constant (rotation), the direction of the armature current in the corresponding part of the coil must be reversed (commutated) after half a revolution by the rotor. The collector fins on the shaft are used for this. These fins conduct the armature current supplied via the carbon brushes through the rotor windings in alternating directions so that a spatially stationary armature field is generated.

On real direct current machines, the stator windings and the rotor windings are spread around the circumference to generate a uniform torque. The bottom of Fig. 2.13 shows the rotor windings (shown unwound for simplification), their connections, the collector fins and the positions of the brushes. Another physical effect is plays a role in the behavior of the direct current machine. The movement of the armature winding in the stator field induces a counter voltage in the armature circuit. This counter voltage increases proportionally with rotational speed and is in the reverse direction of the armature voltage. The following section explains the impact of this effect on the operational behavior of the machine.

#### Basic and system equations

The basic equations of the direct current motor:

$$V_i = k_1 \phi n \text{ Induced counter voltage} \tag{2.1}$$

$$V_A = V_i + R_A I_A$$
$$V_A = k_1 \phi n + R_A I_A \text{ or}$$
$$V_A = k_2 \phi \omega + R_A I_A \text{ where } \omega = 2\pi n \text{ and} \tag{2.2}$$
$$k_2 = k_1 \frac{1}{2\pi}$$

$$T = k_2 \phi I_A = \frac{k_1}{2\pi} \phi I_A \tag{2.3}$$

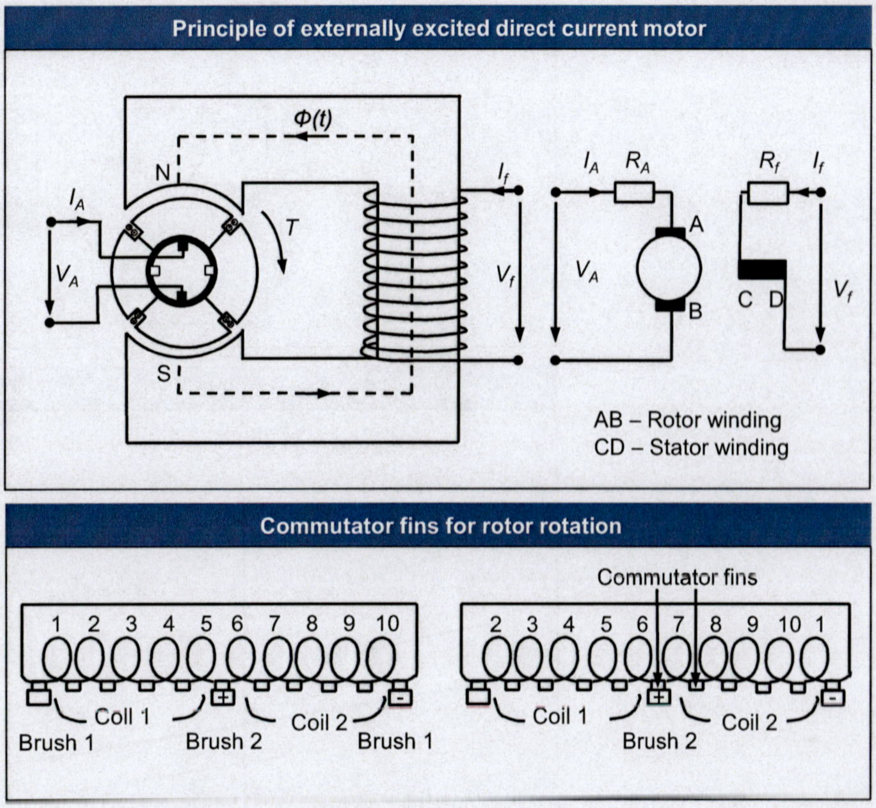


**Fig. 2.13** Principle and schematic for an externally-excited direct current machine

Here, $V_A$ is the armature voltage, $R_A$ the ohmic internal armature resistance, $I_A$ the armature current and $\phi$ the magnetic flux generated by the excitation winding. $V_i = k_1\phi n$ is the induced counter voltage of the motor and $k_1$ and $k_2$ are the constants specific to the motor, derived from the setup, size and winding arrangement for the machine in question. Relation $k_1 = 2\pi k_2$ applies. Applicable on startup ($n = 0$) for the induced counter voltage $V_i$:

$$n = 0 \rightarrow V_i = k_1\phi n = 0 \tag{2.4}$$

The maximum armature current $I_{A\,\max}$ flows. According to Eq. (2.2), this is only limited by the internal resistance of the armature:

$$n = 0 \rightarrow \mathrm{I}_{A\ \max} = \frac{V_A}{R_A} \tag{2.5}$$

The motor torque

$$T = k_2\phi I_{A\,\max} = k_2\phi\frac{V_A}{R_A} \tag{2.6}$$

is correspondingly high on startup for accelerating the rotor and coupled load. The induced countervoltage grows proportionally with increasing rotational speed as in Eq. (2.1) until the value of the armature voltage applied, reduced by the load-dependent armature voltage loss ($I_A R_A$), is reached. Equation (2.3) results for the idle mode of the motor (no frictional losses):

$$T = 0 \rightarrow I_A = 0 \tag{2.7}$$

For the armature current of $I_A = 0$, this applies according to Eq. (2.2) for the idle speed $n_0$:

$$V_A = V_i \rightarrow n_0 = \frac{1}{k_1\phi}V_A \tag{2.8}$$

For operation, Eqs. (2.2) and (2.3) give the load-dependent.

speed equation for the uncontrolled direct current motor:

$$n = \frac{V_A}{k_1\phi} - \frac{R_A}{k_1 k_2 \phi^2}T \tag{2.9}$$

This relation is shown in the form of a speed-torque family of characteristics in Fig. 2.14. These also show the transition to generator mode. As can be seen from Eq. (2.9), there are two ways to change the rotational speed of the direct current motor.

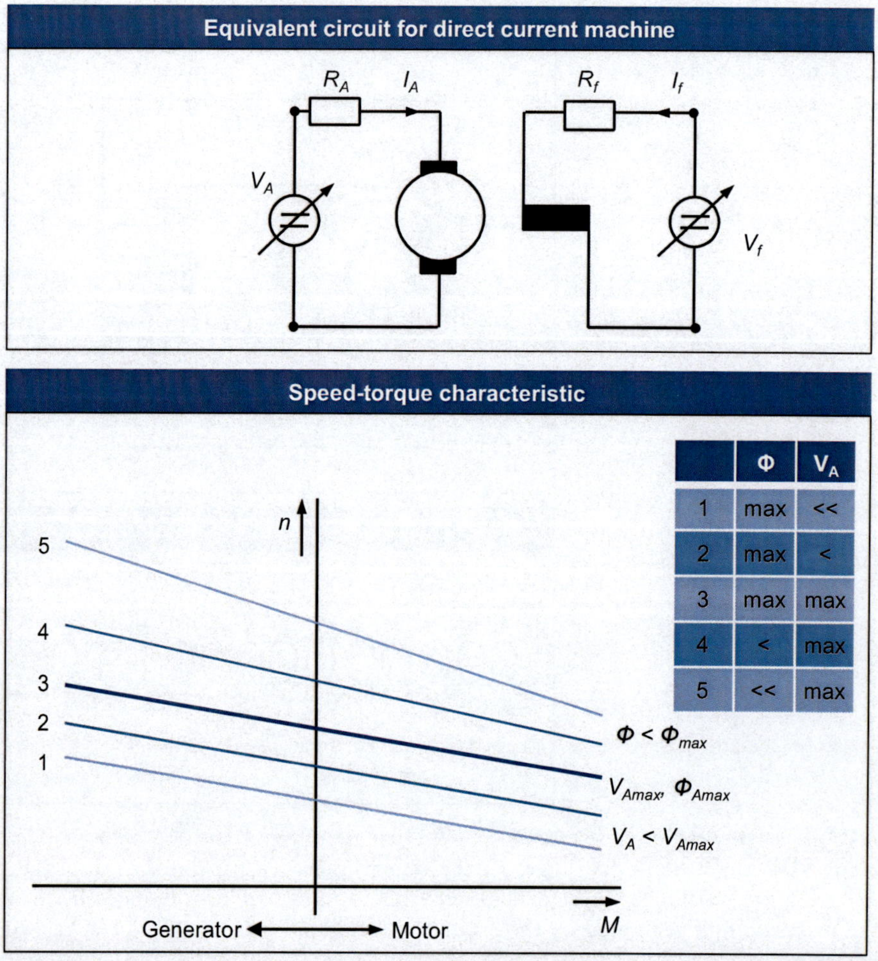


**Fig. 2.14** Speed-torque characteristic according to [HENN99]

First, within the speed range ($n = 0$ to $n_0$), the armature voltage $V_A$ is adjusted for the maximum excitation field $\phi_{max}$. For another speed increase beyond $n_0$, the $\phi$ field is weakened. The load-dependent ($T$) speed drop is determined by the second term in Eq. (2.9): $\frac{R_A}{k_1 k_2 \phi^2} T$. It is caused by the power loss in the armature resistance $R_A$. The negative slopes of straight lines 1 to 3 are parallel because $\phi = \phi_{max}$ is set as a constant. The field strength $\phi$ is reduced in the speed range of the field weakening $n > n_0$. This means the slope of load lines 4 and 5 change with $\frac{1}{\phi_2}$, i.e., the load-dependent speed drop increases.

### Speed adjustment and load limits

The speed of a direct current machine can be adjusted using the armature voltage $V_A$ (armature or primary adjustment) or the magnetic flux $\phi$ (field or secondary adjustment) (**◘** Fig. 2.15). As in Eq. (2.9), increasing the armature voltage increases the rotational speed. In contrast, increasing flux $\varphi$ reduces the rotational speed. Different load limits arise in both adjustment ranges. These are explained by way of **◘** Fig. 2.15. The top half depicts the two setting parameters (armature voltage and magnetic flux) shown versus the rotational speed. The lower part has the associated set of characteristics $T(n)$ and the threshold characteristics for load torque $T_{max}$ and permitted armature current $I_{A\,max}$.

Also entered are the limits for the maximum available power $P_{mech}$ at the motor shaft and the electrical power drawn from the power line $P_{max}$. The maximal permitted continual armature current (operating time (OT) = 100%), limited by the thermal resilience of the winding, and by commutator sparking at higher speeds, is assumed to be constant in both ranges:

$$I_{A\,max} = \text{const} \neq f(n) \qquad (2.10)$$

Momentary overloading is possible depending on the thermal capacity of the rotor and the initial temperature. Commutator sparking denotes spark generation at the edges of the carbon brushes. This phenomenon re-

**Fig. 2.15** Primary and secondary rotational speed adjustment ranges

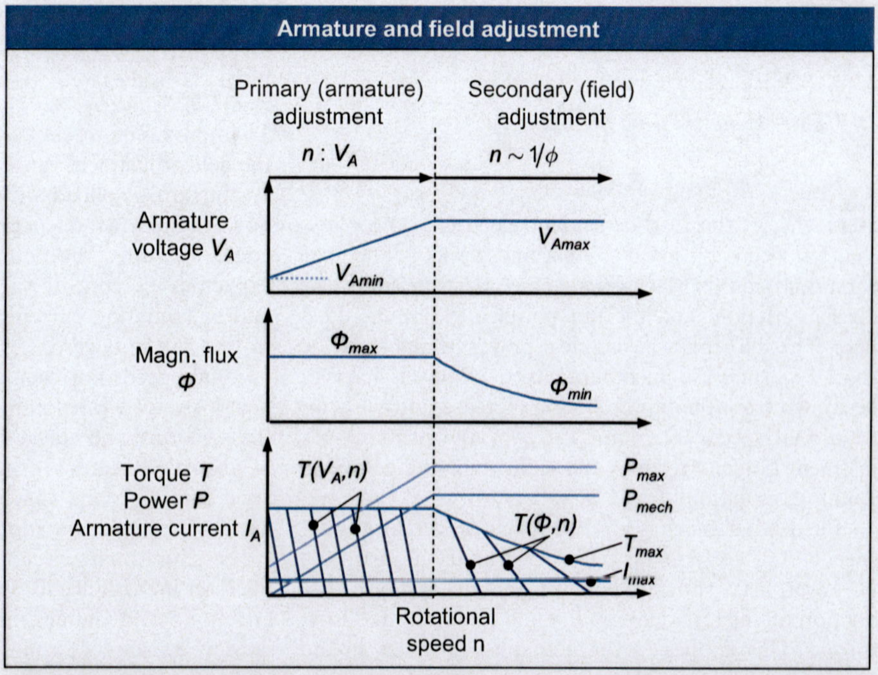


**Usual values**

$$\frac{V_{A\min}}{V_{A\max}} = \frac{1}{50}$$

$$\frac{\phi_{A\min}}{\phi_{A\max}} = \frac{1}{4}$$

**Load characteristics**

$$n = \frac{V_A}{k_1\phi} - \frac{R_A}{k_1 k_2 \phi} T$$

or

$$T(n) = \frac{V_A}{R_A} k_2 \phi - \frac{k_1 k_2 \phi^2}{R_A} n$$

**Thresholds**

$$I_{A\max} = const$$

$$T_{\max} = k_2 \phi I_{A\max}$$

$$P_{\max} = P_{mech,\max} + P_{loss,\max} = 2\pi n T_{\max} + R_A I_{A\max}^2$$

$$P_{loss,\max} = R_A I_{A\max}^2 = \frac{R_A T_{\max}^2}{k_2^2 \phi^2}$$

sults from a self-induced voltage that slows the transition of current from one fin to the next. This generates sparks starting at a particular voltage.

The armature adjustment covers the range from low to medium rotational speeds at high motor torque loads. The set maximum excitation guarantees the highest possible torsional rigidity.

$$V_{A\min} \leq V_A \leq V_{A\max} \tag{2.11}$$

$$\phi = \phi_{\max} = \text{const} \tag{2.12}$$

The Eqs. (2.2) and (2.9) for the dependency of the rotational speed from the armature voltage yield:

$$n = \frac{V_A}{k_1\phi_{\max}} - \frac{R_A I_A}{k_1\phi_{\max}} = \frac{V_A}{k_1\phi_{\max}} - \frac{R_A T}{k_1 k_2 \phi_{\max}^2} \tag{2.13}$$

Since the magnetic flux in the armature adjustment range is constant (Eq. (2.12)), the rotational speed is a linear function of the armature voltage being applied. The rotational speed is reduced by a load-dependent fraction:

$$n\colon V_A - CT \tag{2.14}$$

The continual load torque $M_{max}$ permitted follows directly from Eq. (2.3):

$$T_{\max} = k_2 \phi_{\max} I_{max} \tag{2.15}$$

A constant, maximum load torque is given by the conditions in Eqs. (2.10) and (2.12). To determine the maximum electrical continuous power drawn, Eq. 6.2 is multiplied by $I_{A\max}$:

2

$$P\max = V_A I_{A\max} = k_1\phi_{\max} n I_{max} + RAI^2_{\max} \quad (2.16)$$

This corresponds to a power balance of:

$$P_{\max} = P_{mech} + P_{verl} \quad (2.17)$$

$$P_{\max} = 2\pi n M_{\max} + R_A I^2_{A\max} \quad (2.18)$$

Here, $P_{loss}$ is the load-dependent electrical power loss that has been converted to heat and is independent of rotational speed. Mechanical losses from bearing and air gap friction, and the fan power, are not included here. The maximum mechanical power available at the shaft $P_{mech}$ and the maximum electrical power that can be drawn from the power system increase linearly with rotational speed. The limit $V_{A\max}$ of the primary adjustment range is given by the withstand voltage of the winding insulation.

Field adjustment is used to reach higher rotational speeds than are possible with the primary adjustment. However, as is shown below, this entails increased reduction of the torsional rigidity and the permitted load torque. This applies to the actuating variables in the field adjustment range:

$$V_A = V_{A\max} = const, \quad (2.19)$$

$$\phi_{\min} \le \phi \le \phi_{\max} \quad (2.20)$$

From Eq. (2.2) comes the dependency of the rotational speed of the magnetic flux $\phi$:

$$n = \frac{1}{k_1\phi}(V_{A\max} - R_A I_A) : \frac{1}{\phi} \quad (2.21)$$

Factoring in the load torque means Eq. (2.9) applies accordingly:

$$n = \frac{V_{A\max}}{k_1\phi} - \frac{R_A}{k_1 k_2 \phi^2} T. \quad (2.22)$$

In contrast to the primary adjustment, in which the gradient value of the characteristics $\frac{-R_A}{k_1 k_2 \phi^2_{\max}}$ is constant, there is no parallel set of characteristics M(n) for the secondary adjustment. With decreasing flux, the characteristics $\frac{1}{\phi^2}$ become flatter, i.e. the motor becomes "softer" (■ Figs. 2.14 and 2.15). The maximum permitted continual load torque decreases with increasing rotational speed, in line with the flux reduction:

$$M_{\max} = k_2\phi I_{A\max} \sim \phi \quad (2.23)$$

According to Eq. (2.16), this applies to the power:

$$P_{\max} = U_{A\max} I_{A\max} = k_1\phi I_{A\max} n + RAI^2_{A\max}, \quad (2.24)$$

$$P_{\max} = const, \quad (2.25)$$

$$P_{loss,\max} = const. \quad (2.26)$$

To improve the torsional rigidity of the main drives, in the field adjustment range in particular,

they are provided with speed control in many cases nowadays. A tachogenerator or incremental rotary encoder measures the actual rotational speed. In their capacities as control variables, the armature voltage and/or excitation current keep changing until the control deviation is zero, i.e., the actual and reference rotational speeds correspond. To keep the eddy current heat losses generated in the power converter supply low, the armature and stator winding cores are often made of insulated plates. In addition to the externally excited direct current machine where the armature and exciter windings are supplied separately, other circuit types (shunt, main circuit wiring) are common, but are of lesser importance for the machine tool industry. Relatively good dynamics, high synchronized speed quality, low noise, wide rotational speed adjustment range and straightforward actuation are benefits of the direct current motor. In contrast, the downsides are the maintenance overhead for the brushes, commutator sparking (the reason why this motor cannot be used in potentially explosive environments) and rotational speed limitation by "mechanical commutation". As a result, the dynamics of the motor are inherently limited because only a certain amount of current may flow via the brushes and collectors without causing damage to the contact elements.

It was for these reasons that the direct current motor gradually started to be replaced years ago by the three-phase a. c. motor. For the two motor principles used, synchronous and asynchronous machines, the aforementioned problems do not occur due to use of "electronic commutation".

### Characteristic and current limiting

As, for the direct current motor, current is routed via brushes to the collector, there is, for this motor type, a natural limit for the maximum current that can still be carried without damage to the contact elements. The primary graph for this "commutation limit" is shown in ■ Fig. 2.16. The maximum commutation current is a function of the rotational speed and is inversely proportional to this speed. To prevent damage to the contact elements, a rotational speed-dependent current limiter is generally integrated into the servo amplifier.

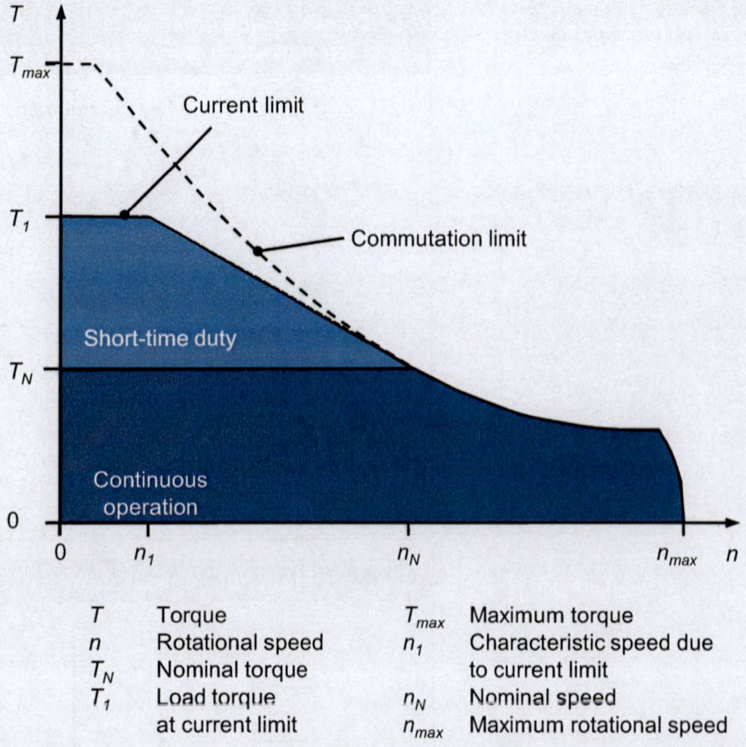


**Fig. 2.16** Motor characteristic and graph of a rotational speed-dependent current limiter for a direct current motor

This results in the ratio of maximum available torque to nominal torque effectively being reduced. The motor characteristic represents nominal operation. All currents above this line are in the overload area of the motor and may only be attained momentarily.

### 2.2.3 Synchronous motors

In contrast to the direct current machine, the magnetic excitation field for the synchronous machine is generated by the rotating component, the rotor. For a rotor rotation with angular velocity ω, a rotary field with the same angular velocity is also generated in the air gap between the stator and rotor. For a stationary torque to be generated, the stator coil must also generate a rotary field. This is achieved by a suitably positioned 3-phase winding and is proportional to the supply voltage frequency applied. Synchronous machines used in the machine tool and industrial robot sectors are only permanent magnet synchronous motors. Figure 2.17 shows the basic setup.

The rotor carries several permanent magnets depending on the number of pole pairs. These provide a constant armature field. The current required for the torque direction and level is applied to the coils in the stator in line with the angle position of the rotor (magnetic flux direction by means of the stator). A force is generated between the coils (stator) and the field (rotor) according to the Lorentz principle. Commutation is achieved by the change in direction of the current in the stator, without brushes and collectors being necessary as on a direct current motor. The rotor turns at the frequency of the supply voltage (synchronous rotational speed $n_S$):

$$n_S = f \tag{2.27}$$

If the rotor has multiple poles (p pole pairs), the synchronous rotational speed is reduced accordingly:

$$n_S = \frac{f}{p} \tag{2.28}$$

2

**Fig. 2.17** Rotation speed controller of a synchronous machine

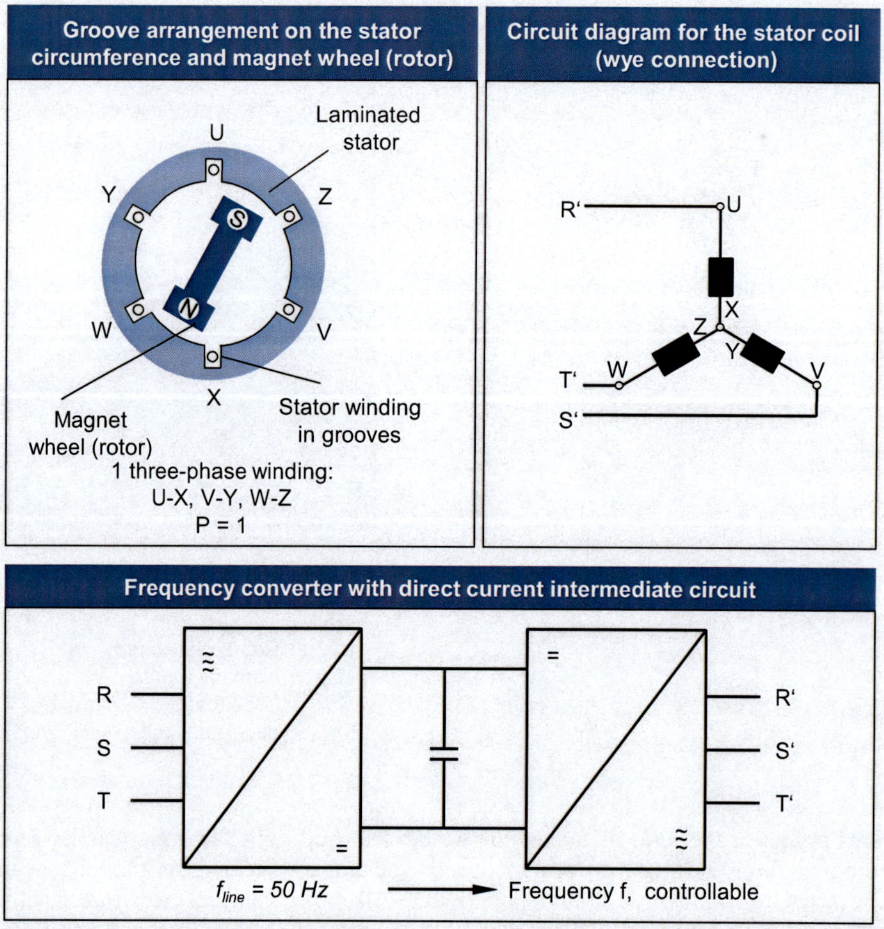


To set different rotational speeds for a synchronous machine, a frequency conversion of the supply system is required. Figure 2.17 shows in schematic form the converter operated by the 3-phase power supply network. The 50 Hz supply voltage with phases R, S and T is first rectified and stored in an energy storage medium (capacitor, inductor) in the intermediate circuit. From the d.c. voltage, the downstream inverter generates a three-phase system R', S', T' whose frequency f can be adjusted. Control of the current supplied to the stator windings must be in relation to the rotor angle position (current direction) and the torque required (amperage) and is provided by "electronic commutation". The three stator windings of a three-phase synchronous machine are positioned in the grooves of the stator, as in the top left of Fig. 2.17. Compared to direct current motors, synchronous motors exhibit higher dynamics because there is no commutation limit. As a result, they can be momentarily subjected to higher currents even at high speeds. In machine tool applications, synchronous motors are used mainly as feed drives.

**Supply types and functional principle**

A fundamental difference is made between supplying with sinusoidal and rectangle wave currents for synchronous motors (Fig. 2.18). The benefits of supplying rectangle wave currents are simpler signal processing and the use of a simpler sensor to determine the rotor position. The sensor, fitted with only three sensor elements, detects the start points of the U, V and W supplies. In special cases, the position encoder can be eliminated entirely if the rotor position is calculated from the induced counter voltage during the phases that are at times not supplied with current. For supplies with sinusoidal currents, more complex, cyclically operating sensors are used for exact detection of the absolute rotor position. Supplying with sinusoidal currents can generally cause damping of the harmonics, and so a high synchronism quality to be attained for the drive. Advances in power electronics and for absolute rotor position encoders mean only sine activation is applied nowadays for new developments.

The two supply types are contrasted in Fig. 2.18. Each of the top three rows shows graphs of air gap

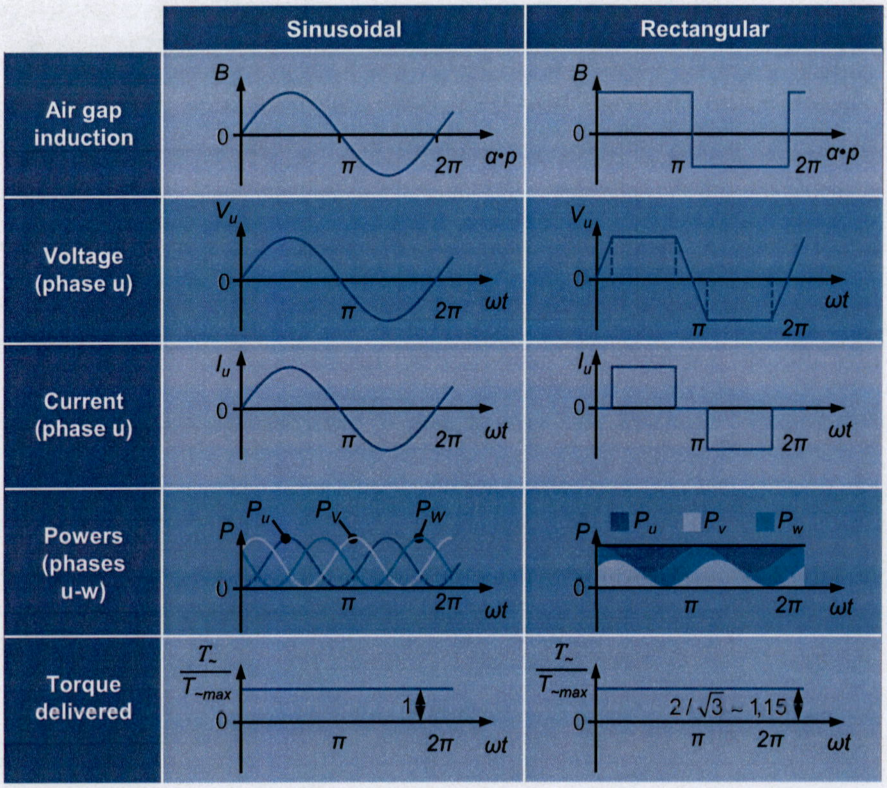


**Fig. 2.18** Supply types for permanent magnet synchronous motors according to Henneberger

induction $B$, and voltage and current in phase $u$ for both supply types. The profiles in the two other phase windings are derived from this profile by a phase shift of 120° (phase $v$) or 240° (phase $w$). The superposition of the powers in the individual phases is shown underneath. Both the sinusoidal and rectangle-wave supplies give overall a constant motor power output. For the rectangle-wave supply, the rectangles have a pulse-interval ratio of 120°: 60° and a phase shift of 120° in relation to each other, meaning exactly two phases are supplied with current at any one time. The motor delivers a constant torque. The trapezoidal profile of the induced voltage comes from slanting the grooves. This is to prevent torque rippling.

Figure 2.19 shows once again the sinusoidal supply of the synchronous motor for all phases individually. A 6-pole motor serves as an example. The term 6-pole ($p=3$, $p$: number of pole pairs for the motor) means that the supply frequency in the phases is three times higher than the motor speed. In rotor position 0°, current flows through phase $v$ in the positive direction and through phase $w$ in the negative direction, while phase $u$ has no current. In position 30°, positive current is applied to phase $u$ and negative current to phases $v$ and $w$. The current directions activated this way over time in the stator windings generate a torque on the magnetized rotor in the same direction that sets the rotor in motion clockwise. Supply to the phases follows the rotor position. For the 6-pole motor, every phase supply is sinusoidal and offset by 60° or 120° in phase. The curves of the induced voltages in the individual phases and at the terminals are also shown in Fig. 2.19.

Figure 2.20 shows a synchronous motor and a rotor shaft with the number of pole pairs $p=5$. Every pole is made up of segments, arranged next to each other, in the axial direction. The segments in turn comprise small plates that follow the curvature of the rotor circumference with single elements. Aluminum nickel cobalt (AlNiCo), samarium cobalt (SmCo), neodymium iron boron (NdFeB) and ferrite are used as magnetic materials. Figure 2.21 shows the typical structure of a synchronous motor.

A particular problem with most synchronous motors is developing a uniform torque that is independent of the rotor position. Harmonics in the field curve and in the distribution of the electric loading can cause alternating torques that have an interfering effect during the positioning process of the drive. The slanting of grooves can help here. This is usually performed in

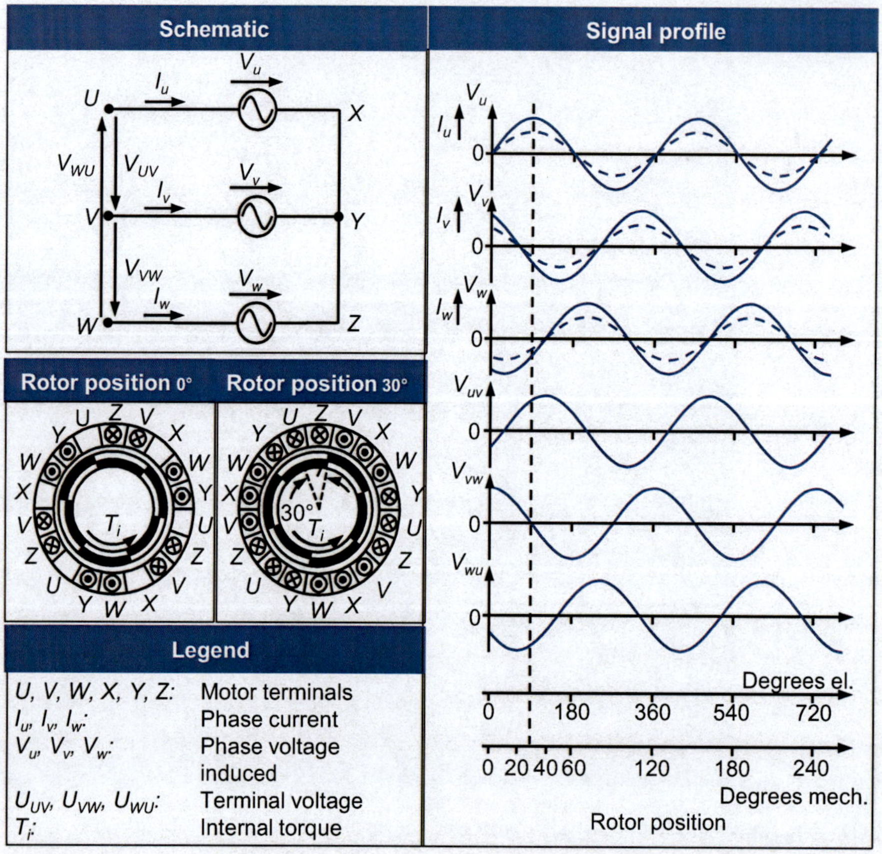


**Fig. 2.19** Functional principle and sinusoidal supply of a permanent magnet synchronous motor according to Bosch Rexroth

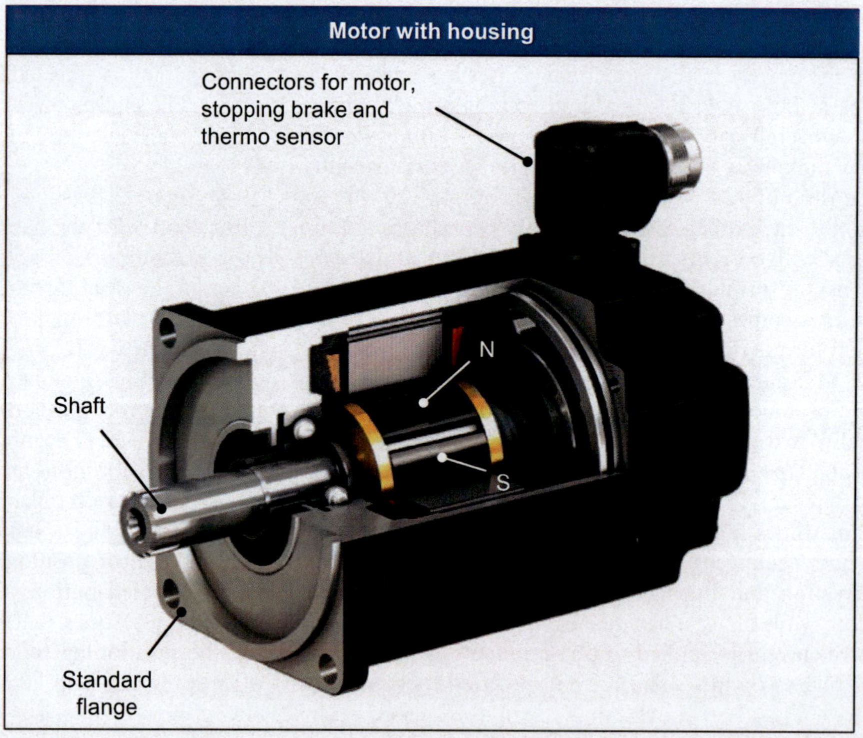


**Fig. 2.20** Synchronous motor and rotor shaft with 5 pole pairs according to Bosch Rexroth

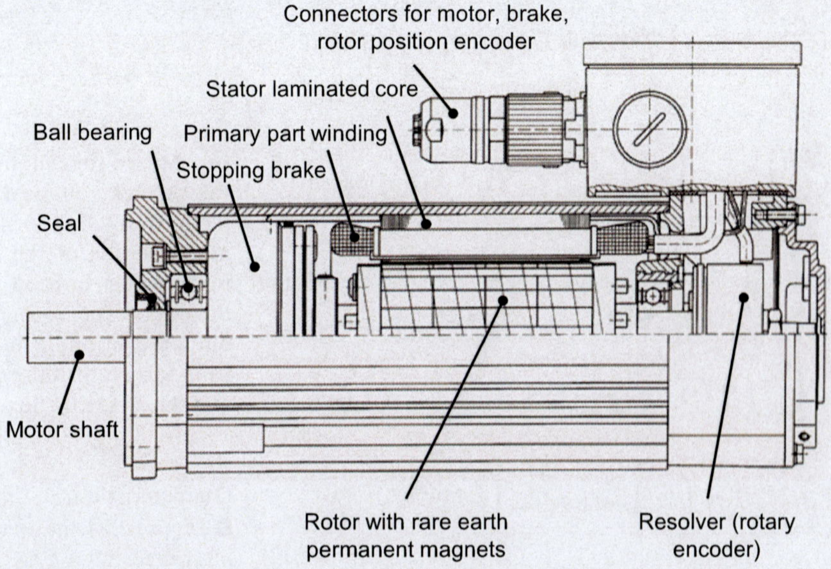


**Fig. 2.21** Structure of a synchronous motor. *Source:* GEORGII KOBOLD

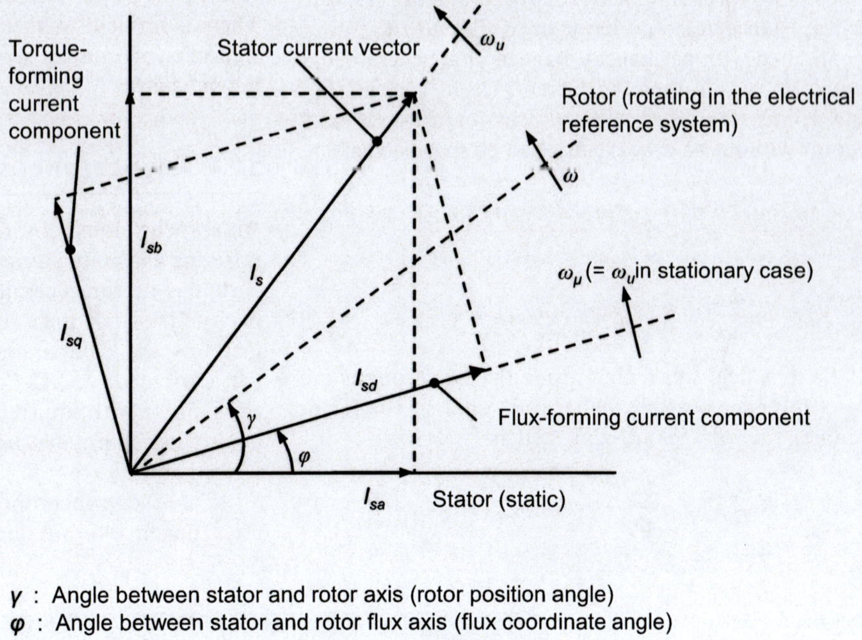


**Fig. 2.22** Vector representation of the transformation of stator currents

the stator. Increasing the number of pole pairs for the winding can also change the harmonics in the electric loading.

### Equivalent circuit and basic equations of the synchronous motor

The 3-phase stator current system was introduced in the previous section (also refer to Fig. 2.19). To facilitate simpler calculation of the currents and the resultant motor torques, the currents are usually transformed to a coordinate system rotating with the stator flux.

Figure 2.22 illustrates this transformation of the currents in the stator of a synchronous motor. A conversion can be performed here from the actual 3-phase stator current system ($i_u$, $i_v$, $i_w$) to an equivalent 2-phase system ($i_{sa}$, $i_{sb}$) in the coordinate system of the stator at rest using a transformation calculation. The three currents ($i_u$, $i_v$ and $i_w$) are linearly dependent, $\boldsymbol{i_w = -i_u - i_v}$, the reason why $T_{32}$ is only a $2 \times 2$ matrix. The angle $\boldsymbol{\varphi(t)}$ between the coordinate systems, changing over time, is an input into the conversion from ($i_{sa}$, $i_{sb}$) to the rotating coordinate system of the stator flux using the currents ($i_{sd}$, $i_{sq}$):

2

$$\begin{pmatrix} i_{sa} \\ i_{sb} \end{pmatrix} = T_{32} \cdot \begin{pmatrix} i_u \\ i_v \end{pmatrix}, \text{ with } T_{32} = \begin{pmatrix} \sqrt{\frac{3}{2}} & 0 \\ \frac{1}{\sqrt{2}} & \sqrt{2} \end{pmatrix} \quad (2.29)$$

$$\begin{pmatrix} i_{sd} \\ i_{sq} \end{pmatrix} = T_{\varphi} \cdot \begin{pmatrix} i_{sa} \\ i_{sb} \end{pmatrix}, \text{ with } T_{\varphi} = \begin{pmatrix} \cos\varphi(t) & \sin\varphi(t) \\ -\sin\varphi(t) & \cos\varphi(t) \end{pmatrix} \quad (2.30)$$

The reverse transformation uses the inverse of the transformation matrix:

$$\begin{pmatrix} i_u \\ i_v \end{pmatrix} = T_{32}^{-1} \cdot \begin{pmatrix} i_{sa} \\ i_{sb} \end{pmatrix}, \text{ with } T_{32}^{-1} = \begin{pmatrix} \sqrt{\frac{2}{3}} & 0 \\ -\frac{1}{\sqrt{6}} & \frac{1}{\sqrt{2}} \end{pmatrix} \quad (2.31)$$

$$\begin{pmatrix} i_{sa} \\ i_{sb} \end{pmatrix} = T_{\varphi}^{-1} \cdot \begin{pmatrix} i_{sd} \\ i_{sq} \end{pmatrix}, \text{ with } T_{\varphi}^{-1} = \begin{pmatrix} \cos\varphi(t) & -\sin\varphi(t) \\ \sin\varphi(t) & \cos\varphi(t) \end{pmatrix} \quad (2.32)$$

◘ Figure 2.23 shows an equivalent circuit of a synchronous motor. For simplicity, the stator comprises electrical resistance $R_{Str}$ and inductance $L_D$ and the rotor is configured with permanent magnets that generate magnetic flux $\psi_M$. Taking the currents $(i_{sd}, i_{sq})$ as the starting points, the associated voltages for a synchronous motor without field weakening can be specified as:

$$V_d = -p\omega_M L_D i_{sq} \quad (2.33)$$

and

$$V_q = L_D \frac{di_{sq}}{dt} + R_{Str} i_{sq} + p\omega_M \psi_{PM} \quad (2.34)$$

If there is field weakening inside the synchronous motor (to render possible higher rotational speeds for example), the voltages are expressed as:

$$V_d = R_{Str} I_d + L_D \frac{di_{sq}}{dt} - p\omega_M L_D i_{sq} \quad (2.35)$$

and

$$V_q = R_{Str} i_{sd} + L_D \frac{di_{sq}}{dt} + p\omega_M(\psi_{PM} + L_D i_{sd}). \quad (2.36)$$

For a synchronous motor, field weakening can only be realized by an increased flux-forming current $i_{sd}$ that lowers voltage $V_q$. This means the magnetic flux induced by flux-forming current $i_{sd}$ counteracts the magnetic flux of the permanent magnets on the rotor, resulting, in turn, in a weaker magnetic field between the stator and rotor. For both cases (with and without a reduction of flux), the motor torque is calculated from torque-forming current $i_{sq}$, number of pole pairs $p$ and the magnetic flux.

$$T_M = 3p\psi_{PM} i_{sq} \quad (2.37)$$

**Characteristic**

◘ Figure 2.24 shows the typical load curve and operating ranges of a synchronous motor. There is no commutation limit here as there is for direct current motors. Rather, the power is limited by the servo amplifier used. There is no field weakening mode for the standard design of synchronous servo motors as permanent magnet machines.

### 2.2.4 Asynchronous motors

An asynchronous motor is superior to all other motors in terms of simplicity and robustness. Also, an asynchronous motor generally exhibits a high overload capacity. The dynamics attainable with special control methods are comparable to those achieved with direct current drives. ◘ Figure 2.25 shows an asynchronous motor with squirrel-cage design. Externally, this is barely distinguishable from the servo motor types described above.

Like the synchronous machine, the asynchronous machine is an induction machine because the

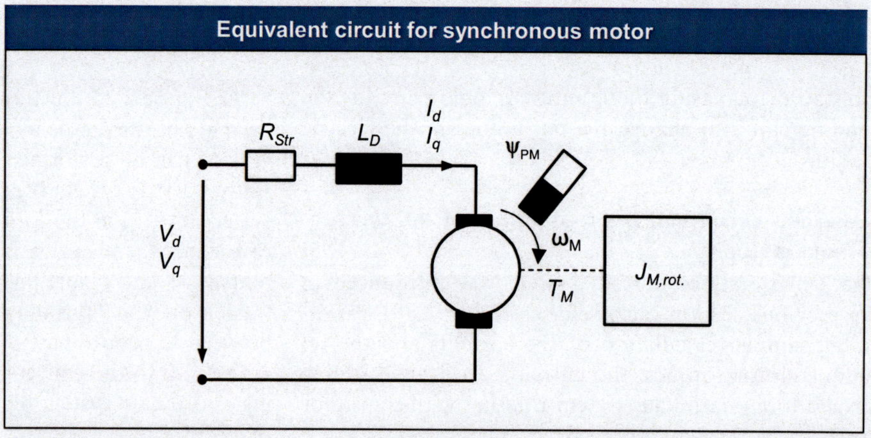


◘ **Fig. 2.23** Equivalent circuit of the synchronous motor according to [GROS06]

**Fig. 2.24** Operating ranges of a synchronous motor

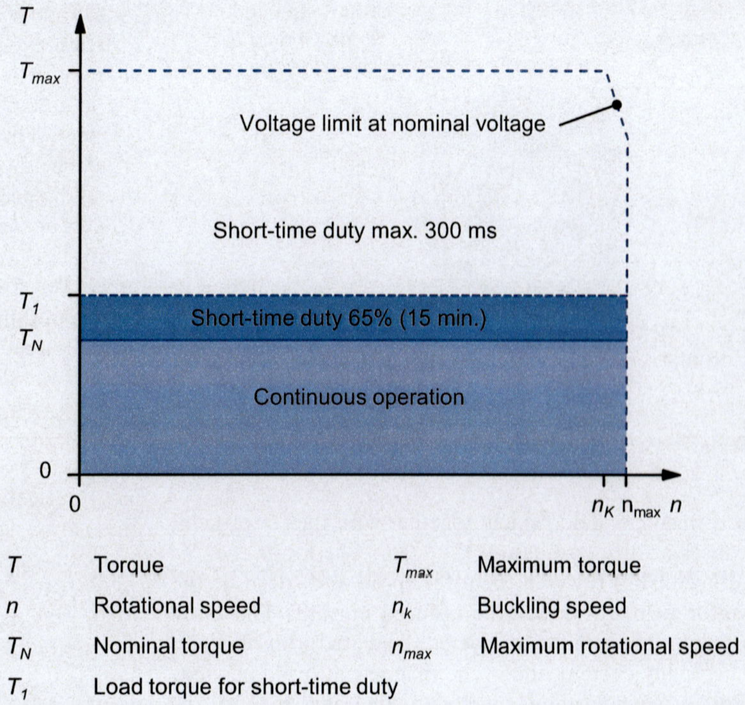


**Fig. 2.25** Structure of a squirrel-cage asynchronous motor according to ABB

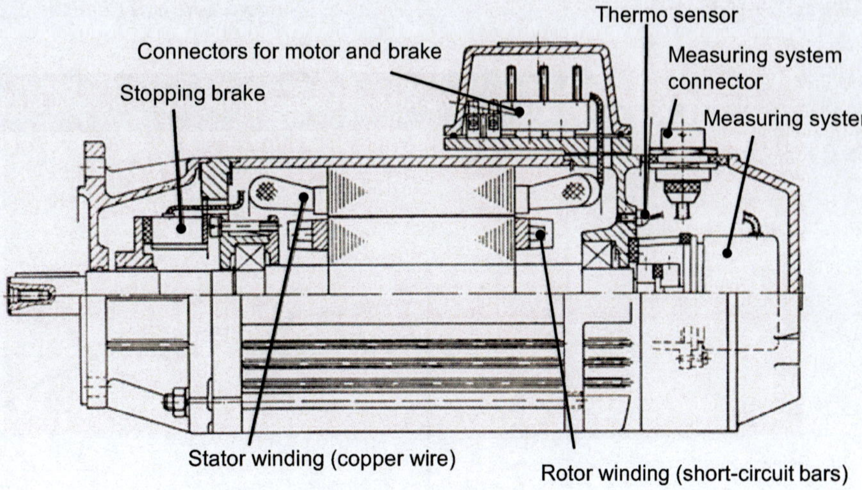


field does not come to a stop like on the d.c. machine. Rather, it rotates. The stator of an asynchronous machine is designed like that of a synchronous machine. A 3-phase winding is spread around the circumference (Fig. 2.26).

Unlike a synchronous motor, the rotor has a 3-phase winding. It is either short-circuited (short circuit or squirrel cage rotor) or is routed to the outside via three slip rings or contact brushes (slip ring rotor). Only the squirrel-cage design is used in the machine tool industry. The rotor current required for the rotor field is induced in the rotor via the stator in line with the transformer principle. For this to function, a rotational speed delta (slip) is necessary between the stator rotary field and the rotor speed. The motor features a simple and sturdy design. As a result, it is very resistant to interference and requires virtually no maintenance. Downstream gear drives are used to adjust the speed of these motors that are connected to the 50 Hz power system.

The speed of this type of motor can be adjusted nowadays. A 3-phase frequency converter is required for this, as for the synchronous motor. The function and design of the motor are explained briefly below. The rotary field in the stator induces a voltage in the rotor conductors. This generates a current, the

2

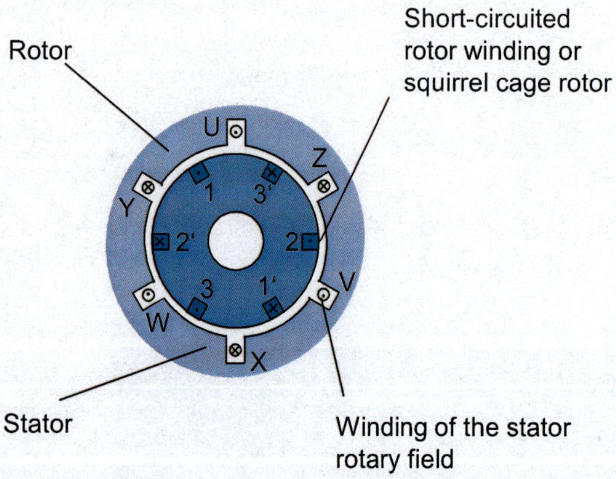


**Fig. 2.26** Structure of an asynchronous machine

self-magnetic field, which, together with the stator field, causes a torque to act on the rotor. The lower the difference (slip) between the rotor speed and speed of the stator field, the lower the induced current. The body of the rotor is made of layered, insulated iron plates to keep eddy current and cyclic magnetization losses low. For the asynchronous machine with a squirrel cage rotor usually used, the rotor winding uses conductor rods having a relatively large cross section.

Figure 2.27 shows the speed and torque characteristics, as well as the current and representative load characteristics of an asynchronous machine, for a motor at a constant nominal speed (power system frequency). The preferred operating range of this motor is just below the synchronous speed. This motor has the high speed rigidity under load required for a main drive. The starting torque is low. The motor generates a torque that increases with increasing speed until the breakover speed is reached. The synchronous speed $n_S$ corresponding to the speed of the rotating stator field is virtually reached only in the no-load state. As load increases, the slip

$$s = \frac{n_S - n}{n_S}, \tag{2.38}$$

i.e., the relative difference between synchronous speed and actual speed, increases. Slip $s_N =$ is 3 to 5% in nominal operation. Equation (2.38) gives the nominal speed of the motor ($f = 50Hz$), at a slip of 3%, of 2910, 1455 or 727.5 $\text{min}^{-1}$, depending on the number of pole pairs for the rotor—Eq. (2.28). Another downside besides the low starting torque is the high starting current in the rotor. This can reach five times the nominal current. The torque characteristic in relation to breakdown torque $T_k$ can be expressed as a function of slip

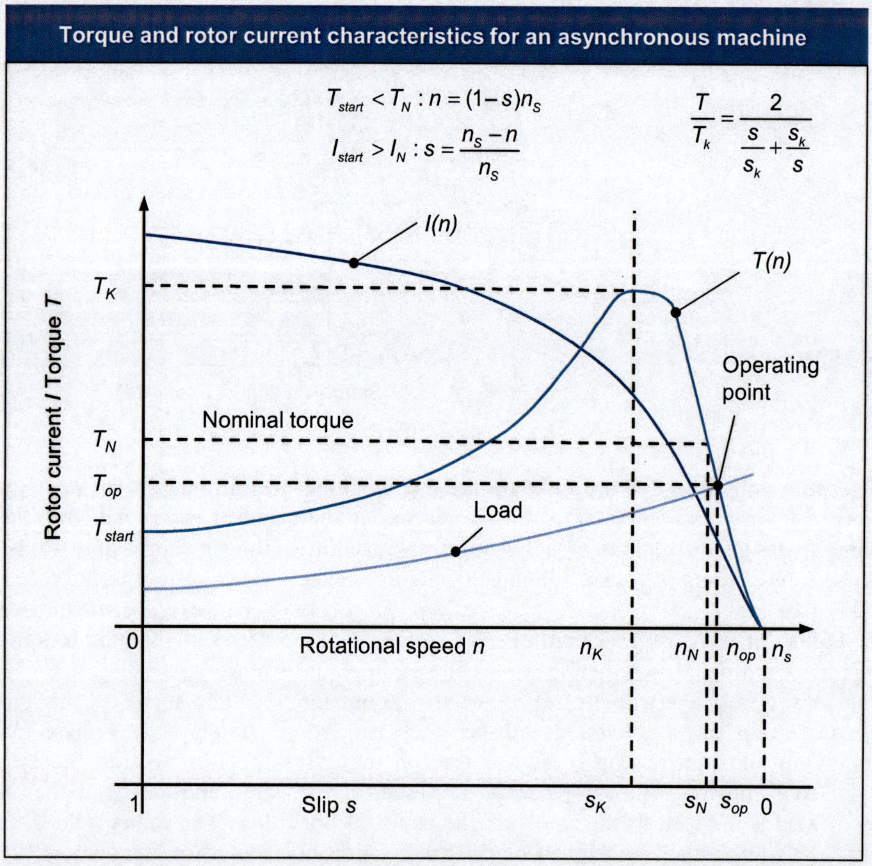


**Fig. 2.27** Torque and rotor characteristics for an asynchronous machine

**Fig. 2.28** Delta-wye switch for the stator winding

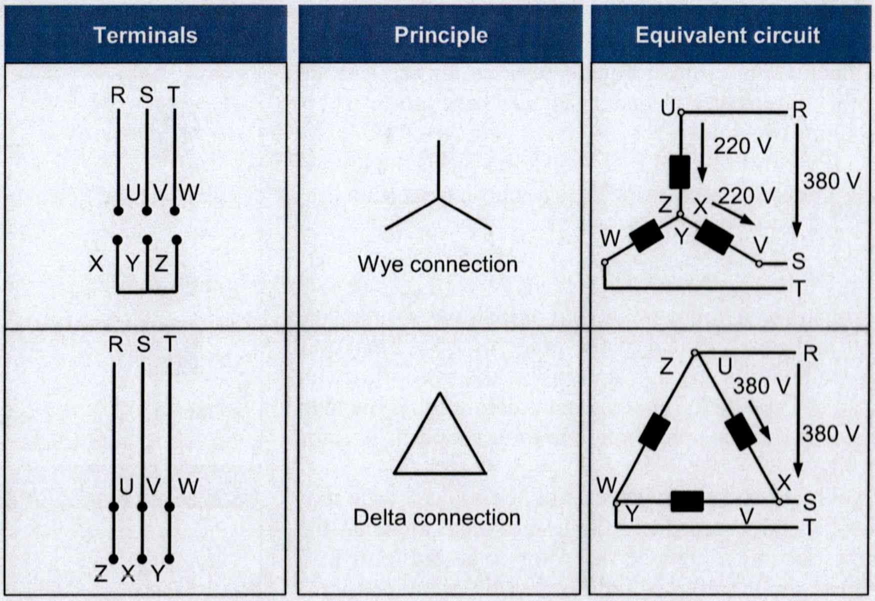


**Fig. 2.29** Operating modes and torque characteristic of the asynchronous machine

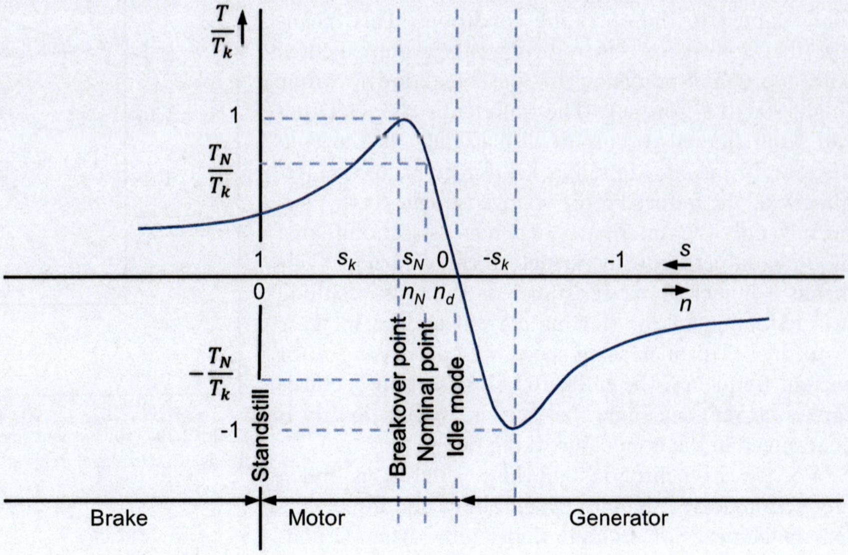


$s$ and the slip at the breakdown torque $s_k$ (Kloss formula):

$$\frac{T}{T_k} = \frac{2}{\frac{s}{s_k} + \frac{s_k}{s}} \tag{2.39}$$

For motors that are not speed-controlled, i.e., those powered from the 50 Hz factory power system, it is possible to reduce the starting current using a delta-wye switch for the stator field (Fig. 2.28).

For wye operation in the starting phase, the stator windings are at 220 V. This means the current only reaches about a third of the value compared to delta operation. However, the starting torque falls by the same amount, meaning this switch is only suitable for driving loads that start up easily without appreciable breakaway or friction torque. After starting up, the motor is switched to delta mode.

An asynchronous machine can work in three modes. A differentiation is made here between Brake, Motor and Generator. Figure 2.29 shows the three modes and the corresponding graph of the torque characteristic. Knowledge of the range over which the asynchronous machine operates stably is important for operating it. The following applies in general:

$$\frac{dT_M}{dn} < \frac{dT_{Load}}{dn} \tag{2.40}$$

This combination satisfies the requirements made on a stable equilibrium. If the motor speed drops due to a fluctuation, motor torque is released for acceleration and, conversely, a torque deficit slows the motor if it is rotating too quickly.

If a constant load is assumed, this yields the following stability condition for the asynchronous machine:

$$\frac{dT_M}{dn} < 0 \Rightarrow -s_k < s < s_k \tag{2.41}$$

From this, it follows that the asynchronous machine can only be operated in the range between breakover speed and idle speed. The term breakdown torque can also be explained from these considerations. If the load exceeds this maximum value in motor mode, the rotor stops.

The torque characteristics of the squirrel cage motor depend, especially in the lower speed range, on the cross-section shapes of the rotor rods. ◘ Figure 2.30 shows commonly used rod cross sections and the associated torque and speed characteristics. The different cross sections make it possible to realize torque profiles adapted to the operating conditions. This means also that motors for heavy starting can be implemented with a certain reduction of the breakdown torque (high starting torque). The different characteristics can be attributed to current displacement in the conductors. In the startup range, i.e., for high slip, the frequency of the induced rotor voltage is high. At this frequency, only the outer parts of the cross-sectional area of the conductors in the armature carry current. This means a higher ohmic resistance of the rods, resulting in a reduction of the starting current and an increase in torque. With increasing speed, i.e., decreasing rotor current frequency, the entire rod cross section becomes carries current, meaning that high torsional rigidity is guaranteed in the nominal working range.

For an asynchronous machine, control is much more complicated than for synchronous and direct current machines. For accurate and highly dynamic control of the torque, i.e., to control two freely changeable stator currents, the size and angular position of the rotor magnetic field must be measured. On the synchronous machine, they are directly related to the rotor position. Conversely, on the asynchronous machine, this information needs to be calculated from the stator currents and machine speed. Apart from these directly measurable parameters, the rotor resistance and inductance motor parameters are also inputs into the calculation. The influence of temperature and magnetic saturation means these parameters are not constant.

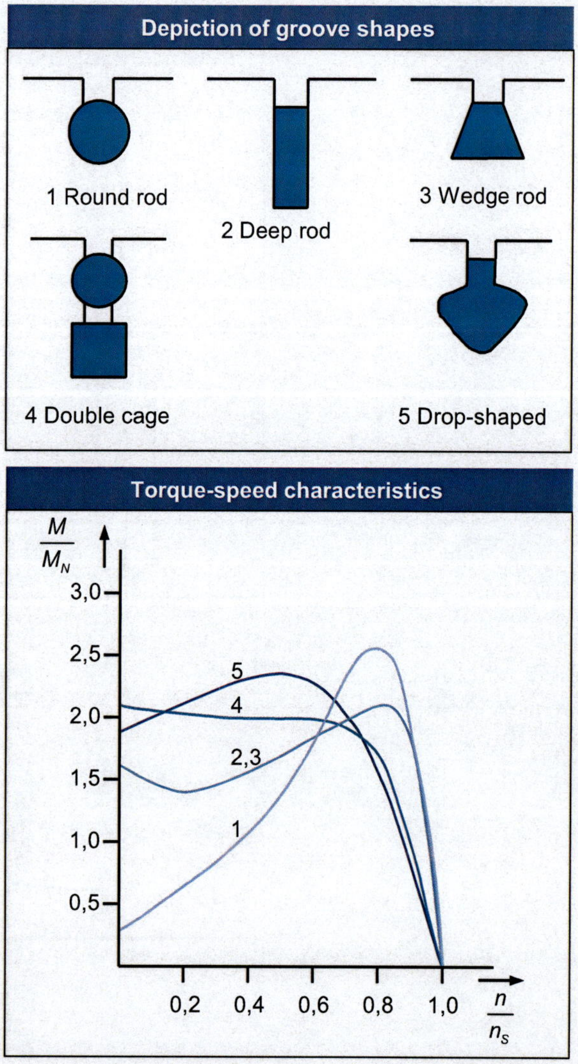


◘ **Fig. 2.30** Torque and speed characteristics for different groove forms of the squirrel cage rotor

### Field-oriented control

"Field-oriented control" forms the basis of speed control for asynchronous motors [SCHÖ09]. This is based upon transformation of the stator currents (already described) to a coordinate system rotating with the magnetic flux. The controller decouples the currents necessary for establishing the field strength and torque to control the field-forming and torque-forming currents independently as in the direct current motor. Control here runs in two different coordinate systems for the currents. The 3-phase stator current system or stator coil coordinate system ($i_u$, $i_v$, $i_w$) is converted to the 2-phase stator coordinate system ($i_{sa}$, $i_{sb}$). This in turn is transformed into a field coordinate system ($i_{sd}$, $i_{sq}$) rotating at the frequency of the rotor flux. ◘ Figure 2.22 illustrates the 2-phase depiction of the stator currents in the stator coordinate system and in the rotating rotor coordinate system. The following control strategy can now be used in the field coordinates: keeping the rotor flux constant. The torque can then be set using only the torque-forming cross-flow current component $i_{sq}$. For power controllers with high switching frequencies, this independent control is possible because there is a guarantee that the rotor flux is only subjected to marginal fluctuation. The functional diagram for field-oriented control is shown in ◘ Fig. 2.31. To the right of the transformation blocks, the currents are the physical phase currents of the motor (stator coil coordinates). The current controllers, based mostly on pulse width modulation (PWM), work with the actual phase currents, meaning they can be controlled and limited. Essentially only two currents need be controlled because the third is derived from the other two in a symmetric three-phase current system. To the left of the transformation blocks is the decoupled control of the field, the torque-forming current and the rotational speed in field coordinates. Rotor flux $\Phi r$ is proportional to longitudinal current component $i_{\mu}$ in the field coordinate system and to magnetizing inductance $L_{1h}$:

$$\phi_r \sim L_{lh}\left(i_{sd} + T_r \frac{di_{sd}}{dt}\right) = L_{lh} i_{\mu} \tag{2.42}$$

Magnetic flux $\Phi r$ is kept at constant with a flux regulator. According to Equation

$$T = \frac{pL_{lh}}{1+\sigma_2} i_{\mu} i_{sq} \qquad \textit{with } \sigma_2 \text{ Dispersion}, \quad p \text{ Poles} \tag{2.43}$$

for the motor torque, the torque can now be set with just the cross-flow current component $i_{sq}$ for constant flux $\Phi_r$. The following relationships apply to the calculation of angle $\phi(t)$ from the stator currents in field coordinates ($i_{sd}$, $i_{sq}$) and for machine speed $\omega$:

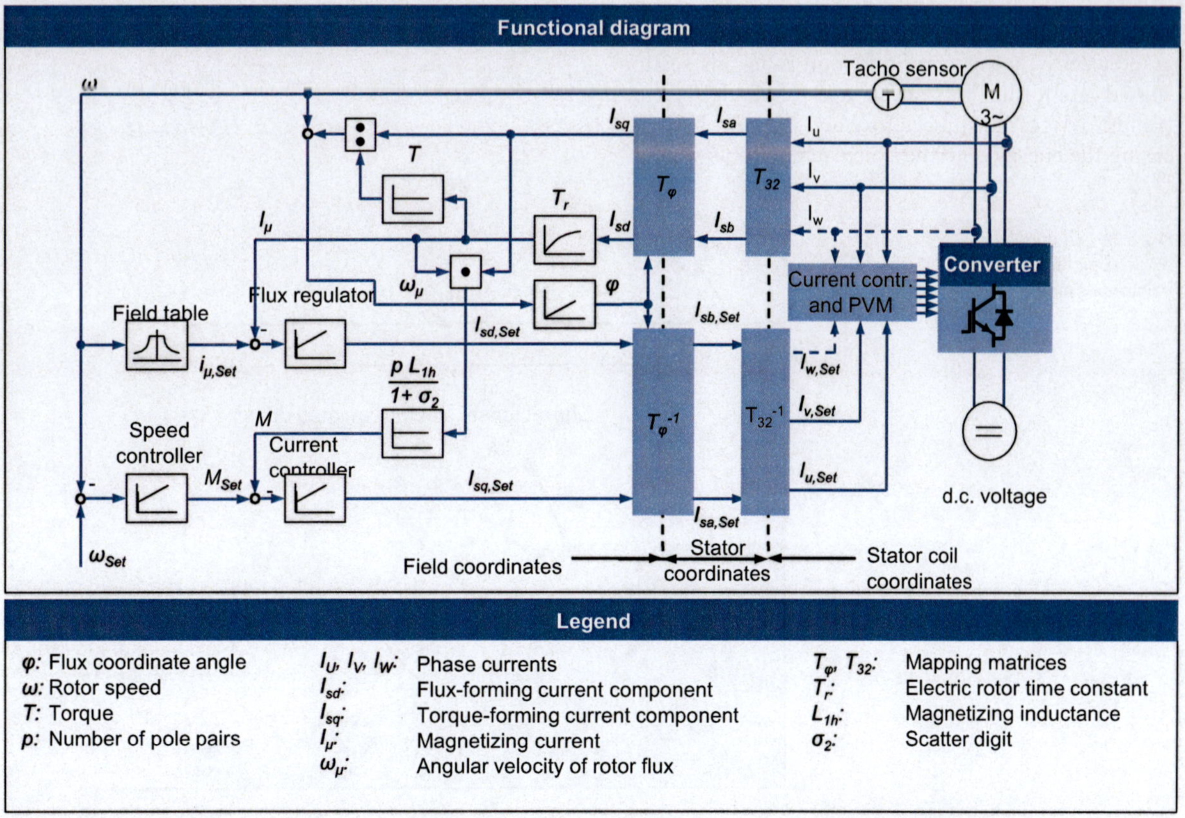


◘ **Fig. 2.31** Control of an asynchronous motor using the principle of field-oriented control according to [HENN99]

$$\varphi = \int_0^t (\omega + \omega_R)dt$$

with $\omega = \frac{d\gamma}{dt}$ Angular Speed of the Rotor
$\omega_R = \frac{i_q}{T_r \cdot i_\mu}$ Circular frequency of the rotor currents
$T_r$ Electrical Rotor Time Constant

(2.44)

### Characteristics

Field-oriented control means the behavior of the asynchronous machine is similar to that of the direct current machine. The speed-torque characteristic in particular becomes a straight line.

One benefit in using an asynchronous motor is the large field weakening range. Here, the rotational speed can be adjusted over a large range with the power output staying constant, Fig. 2.32. Startup of the motor with nominal slip, and so nominal current, continues to be possible by using a variable frequency in the supply. These properties accommodate the requirements for main spindle motors.

### Energy efficiency by reducing the magnetization current

The representation of flux reduction to increase the energy efficiency of asynchronous motors in this section is based upon the illustration and research findings from the EWOTeK project [o.V.12]. Measures for increasing the energy efficiency of machine tools were investigated as part of the project funded by the German Federal Ministry for Education and Research. Asynchronous motors, due to their design, require part of the input current to set up the magnetic rotor field with no torque being yet exerted on the motor shaft. Once rotor magnetization is sufficient, a torque can be built up using the Lorentz force principle. The required strength of this magnetization is specified by the motor design. The basic equation for increasing energy efficiency by reducing the magnetization current forms the relation between the torque of a field-oriented, controlled asynchronous motor and the flux-forming current component $i_d$:

$$T_{el} = \frac{pL_h}{1+\sigma_2} i_\mu i_q, \quad (2.45)$$

where $p$ is the number of pole pairs, $L_h$ the magnetizing inductance, $\sigma_2$ the scatter digit and $i_\mu$ the magnetization current.

The rotor magnetization is responsible for the strength of the magnetic field in the air gap between rotor and stator. The current $i_\mu$ required for the magnetization is derived from the flux-forming current component $i_d$ as

$$T_2 \frac{\partial i_\mu}{\partial t} + i_\mu = i_d. \quad (2.46)$$

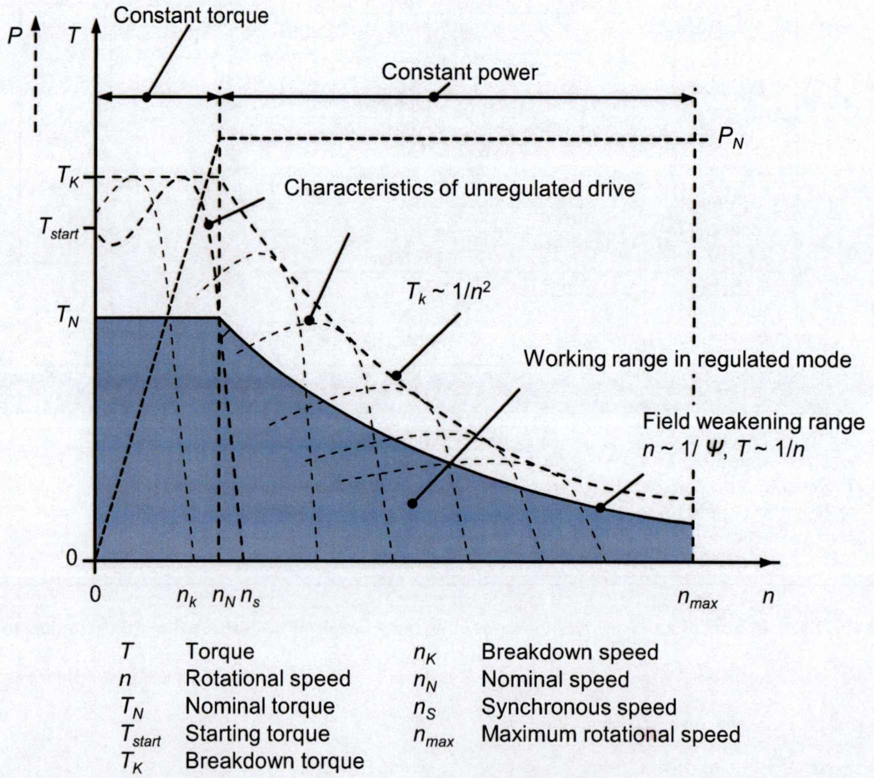


Fig. 2.32 Characteristics and operating ranges for an asynchronous motor

Two key relations are shown with Eqs. (2.45) and (2.46).

First, the torque of an asynchronous machine is directly proportional to the torque-forming current $i_q$ and magnetization current $i_\mu$. It follows that lowering the magnetization current reduces the maximum torque. Torque-forming current $i_q$ can be applied with a short delay time allowing highly dynamic control of the torque. The $PT_1$ behavior of the magnetization current can be seen in Eq. (2.46). It is only dependent on flux-forming current component $i_d$ and rotor time constant $T_2$ of the motor. Rotor time constant values are usually between 150 and 250 ms. A relationship with the rotor inductance and resistance is given by Eq. (2.47).

$$T_2 = \frac{(1+\sigma_2)L_h}{R_R'} = \frac{L_2'}{R_R'} = \frac{L_2}{R_R}, \tag{2.47}$$

$L_2'$ and $R_R'$ specify the rotor inductance and resistance respectively. The two parameters are rotor parameters converted to the number of stator windings.

If used in machine tools however, just a torque below the nominal torque is often required. This provides an energy saving potential in reducing the rotor magnetization. A load-independent decrease of magnetization would, however, prolong the chip-to-chip time because the main spindle is decelerated with maximum torque and then accelerated again after the change. If the maximum torque is not available here due to a low level of magnetization, deceleration and acceleration take longer, generally resulting in longer tool change times. Another application is the start of the material contact by the tool. A significant drop in spindle speed must be avoided in this case in particular. A high torque must be provided to counter the additional mechanical load torque.

Another aspect in intervening in the control structure is that the additional computing power required to determine the load-dependent rotor magnetization does not cause the maximum computing time in the current regulator cycle to be exceeded. One way to adapt the magnetization current to the given load conditions is derived below. The power drawn by an asynchronous machine is the starting point of the derivation with the objective of energy efficiency.

The power drawn by an asynchronous machine is calculated according to [HAME06] as:

$$P_c = P_V + P_{FE} + \frac{\partial W_M}{\partial t} + P_{mech}, \tag{2.48}$$

where $P_{FE}$ specifies the iron losses (comprising eddy current and hysteresis losses) and $\partial W_M/\partial t$ the change of magnetic energy in the motor. Mechanical power $P_{mech}$ is calculated from the motor torque as:

$$P_{mech} = 2\pi \cdot n \cdot T_{el}. \tag{2.49}$$

The copper losses $P_V$ due to ohmic resistances are a decisive factor for increasing energy efficiency by lowering the magnetization current:

$$P_V = R_S\left(i_d^2 + i_q^2\right) + \frac{i_q^2 L_h^2}{T_2^2 R_R'}. \tag{2.50}$$

For the static case, the optimum magnetization current can be calculated as a function of the torque from Eqs. (2.45), (2.47) and (2.46).

$$i_q = \frac{T_{el} T_2 R_R'}{p L_h^2 i_d} \tag{2.51}$$

Inserting Eq. (2.50) into Eq. (2.51) enables the power loss to be determined for a given motor flux.

$$P_V = \frac{R_S\left(i_d^4 p^2 L_h^4 + M_{el}^2 T_2^2 R_2'^2\right) + T_{el}^2 R_R' L_h^2}{p^2 L_h^4 i_d^2} \tag{2.52}$$

An optimum magnetization current is present at a minimum of the power loss in relation to the flux-forming current component.

$$\frac{\partial P_V}{\partial i_d} = 0 \Rightarrow i_\mu = \sqrt{\frac{|T_{el}|}{p}} \cdot \frac{1}{L_h} \cdot \sqrt[4]{\frac{R_R' L_h^2 + L_2^2 R_S}{R_S}} \tag{2.53}$$

So, Eq. (2.53) permits at run-time the calculation of the best stationary magnetization current $i_\mu$ as a function of known motor parameters and the reference motor torque $M_{soll}$.

### 2.2.5 Design and calculation of electric drives

After previously discussing the static characteristics of motors resulting from the setting parameters after the settling time in stationary mode, the speed ramp-up over time for a drive under load from the point of switch-on will be examined now. Note here that the maximum current is limited to protect the motor and power amplifier ($I_{max}$ = is normally 2 to 3 $I_N$). This current limit is effective for large rotational speed changes in particular, such as for ramp-up from zero to the nominal speed. For minor rotational speed changes requiring small current changes, the current limit is not effective. Both cases are shown below using a direct current machine as an example. Determining the ramp-up behavior for non-linear motor and load characteristics without a current limit for a direct current motor drive, and ramp-up for an asynchronous motor, are also described.

#### 2.2.5.1 Ramp-up without a current limit

Deriving closed-form solution equations for ramp-up is only a sensible approach for simple, linear motor and load characteristics. Numerical solutions lend themselves to complex, non-linear motor and load profiles. In the following example, an externally excited direct current motor (◘ Fig. 2.33) drives a load with characteristic $T_L = T_{L0} + c\omega$. In this example, the armature voltage is to take a step-like change from zero to $V_{A1}$. The total moment of inertia applied to the motor shaft is specified as $J_{total}$.

Applicable to the rotation according to the equation of motion:

$$T_a(\omega) = T_M(\omega) - T_L(\omega) \\ = J_{total}\dot{\omega} \text{ where } T_a \text{ is the acceleration torque} \tag{2.54}$$

and

$$T_L(\omega) = T_{L0} + c\omega, \text{ and } \omega = 2\pi n \tag{2.55}$$

Following on from Eqs. 6.2 and 6.3:

$$T_M(\omega) = \frac{V_{A1}}{R_A}k_2\phi - \frac{k_2^2\phi^2}{R_A}\omega \tag{2.56}$$

$T_a$ is the acceleration torque generated by the motor to overcome the inertia of the masses moved. The following function results depending on angular frequency $\omega$:

$$T_a(\omega) = \left(\frac{V_{A1}}{R_A}k_2\phi - \frac{k_2^2\phi^2}{R_A}\omega\right) - (T_{L0} + c\omega) = J_{total}\dot{\omega}$$
$$\Rightarrow J_{total}\dot{\omega} + \left(\frac{k_2^2\phi^2}{R_A} + c\right)\omega = \frac{V_{A1}}{R_A}k_2\phi - T_{L0} \tag{2.57}$$

The differential equation is as follows in its general form:

$$T\dot{\omega} + \omega = \omega_{WP} \tag{2.58}$$

where $\omega_{WP}$ is the rotational speed reached at the working point. The following also applies:

$$\frac{J_{total}}{\frac{k_2^2\phi^2}{R_A} + c}\dot{\omega} + \omega = \frac{\frac{V_{A1}}{R_A}k_2\phi - T_{L0}}{\frac{k_2^2\phi^2}{R_A} + c}$$
$$= \frac{T_{1\max} - T_{L0}}{\frac{T_{1\max}}{\omega_{1\max}} + c} = \omega_{WP} \text{ where } \omega_{1\max} = \frac{V_{A1}}{k_2\phi} \tag{2.59}$$

$T_{1\max}$ and $\omega_{1\max}$ here are the coordinate intersection values of the motor characteristic. The solution to this equation is:

$$\omega(t) = \omega_{WP}\left(1 - e^{-t/T}\right). \tag{2.60}$$

Comparing the coefficients in Eqs. (2.58) and (2.59) gives ramp-up time constant T and the angular frequency of the stationary working point $\omega_{WP}$:

$$T = \frac{J_{total}}{\frac{k_2^2\phi^2}{R_A} + c} = \frac{J_{total}}{\frac{T_{1\max}}{\omega_{1\max}} + c} \tag{2.61}$$

The ramp-up graph and the graphs of the armature voltage ($V_A$), armature current ($I_A$) and motor torque ($T_M$) versus time are at the bottom of ◘ Fig. 2.33. The rotational speed is zero at the point of switch-on. The switch-on current is expressed as:

$$I_{A\max} = \frac{V_{A1}}{R_A} > I_{max,zul}, \tag{2.62}$$

that is, generally much greater than the maximum permitted current.

#### 2.2.5.2 Ramp-up with a current limit

The motor amplifiers have an electronic current limit to protect the motor from overload, in particular when there are significant rotational speed steps, for example, during startup. This means the drive behavior inevitably becomes discontinuous. Using the previous example (◘ Fig. 2.33), the maximum motor current is to be limited to

$$I_{max,zul} = 2I_N \tag{2.63}$$

where nominal current $I_N$ is a motor constant the motor can handle continuously without suffering thermal damage. It makes sense to split the calculation operation for the ramp-up into two sections:

*1st section:* Ramp-up phase with a current limit, i.e. the time of the ramp-up in which the motor would draw a current greater than .

*2nd section:* Enabling a current limit starting at the rotational speed at which the motor inherently draws a current lower than $2I_N$.

The top left part of ◘ Fig. 2.34 uses the static motor torque and load torque characteristic to show the discontinuous curve of the acceleration torque. Up to the switchover point at rotational speed $\omega_{UP}$, the motor torque is limited to

$$T_{M\max} = 2T_N = 2k_2\phi I_N = \text{const.} \tag{2.64}$$

. The torque equilibrium for the current limiting phase (SB) where

$$T_B(\omega) = T_{M\max} - (T_{L0} + c\omega_{CL}) = J_{total}\dot{\omega}_{CL} \tag{2.65}$$

yields the differential equations for the movement:

$$\frac{J_{total}}{c}\dot{\omega}_{CL} + \omega_{CL} = \frac{T_{M\max} - T_{L0}}{c} \tag{2.66}$$

$$T_{CL}\dot{\omega}_{CL} + \omega_{CL} = \omega_{CL0} \tag{2.67}$$

where
$$T_{CL} = \frac{J_{total}}{c}, \tag{2.68}$$

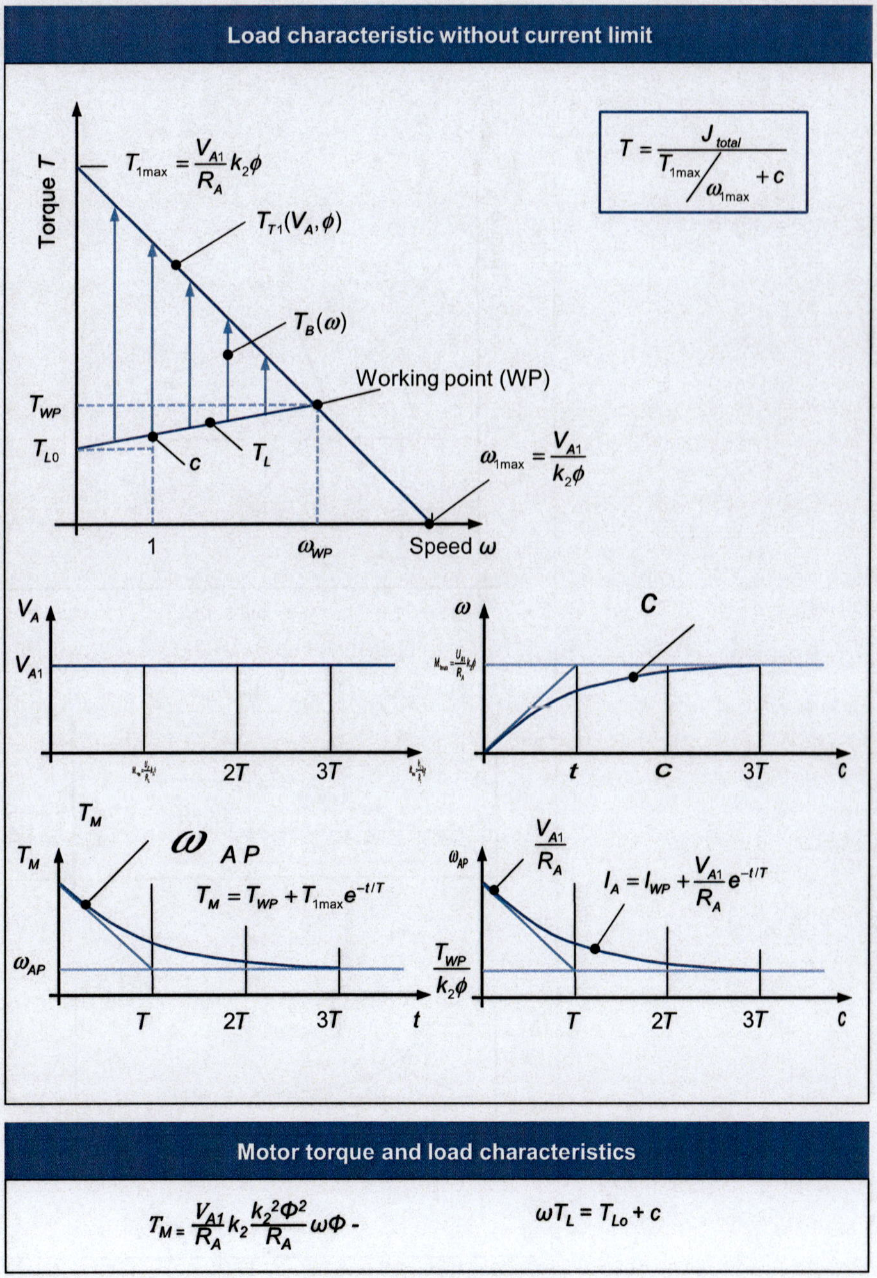


**Fig. 2.33** Ramp-up of a direct current drive with linear load characteristic without a current limit

$$\omega_{CL0} = \frac{T_{M\,max} - T_{L0}}{c} > \omega_{WP} \tag{2.69}$$

with the solution for the ramp-up

$$\omega_{CL}(t) = \omega_{CL0}\left(1 - e^{-t/T_{CL}}\right) \tag{2.70}$$

where $\omega_{CL0}$ represents a theoretical working speed that would be reached if $T_{M\,max} = 2I_N$ were to be effective over an infinite speed range (Fig. 2.34).

Switchover speed $\omega_{UP}$, where the motor current falls below $2I_N$ due to the influence of armature voltage $V_{A1}$, is calculated from the static characteristics as:

$$\omega_{UP} = \omega_{1\,max}\left(1 - \frac{T_{M\,max}}{T_{1\,max}}\right) \tag{2.71}$$

where $T_{M\,max}$ is the maximum torque with current limit and $T_{1\,max}$ the maximum torque without current limit

**Fig. 2.34** Ramp-up of a direct current drive with a linear load characteristic

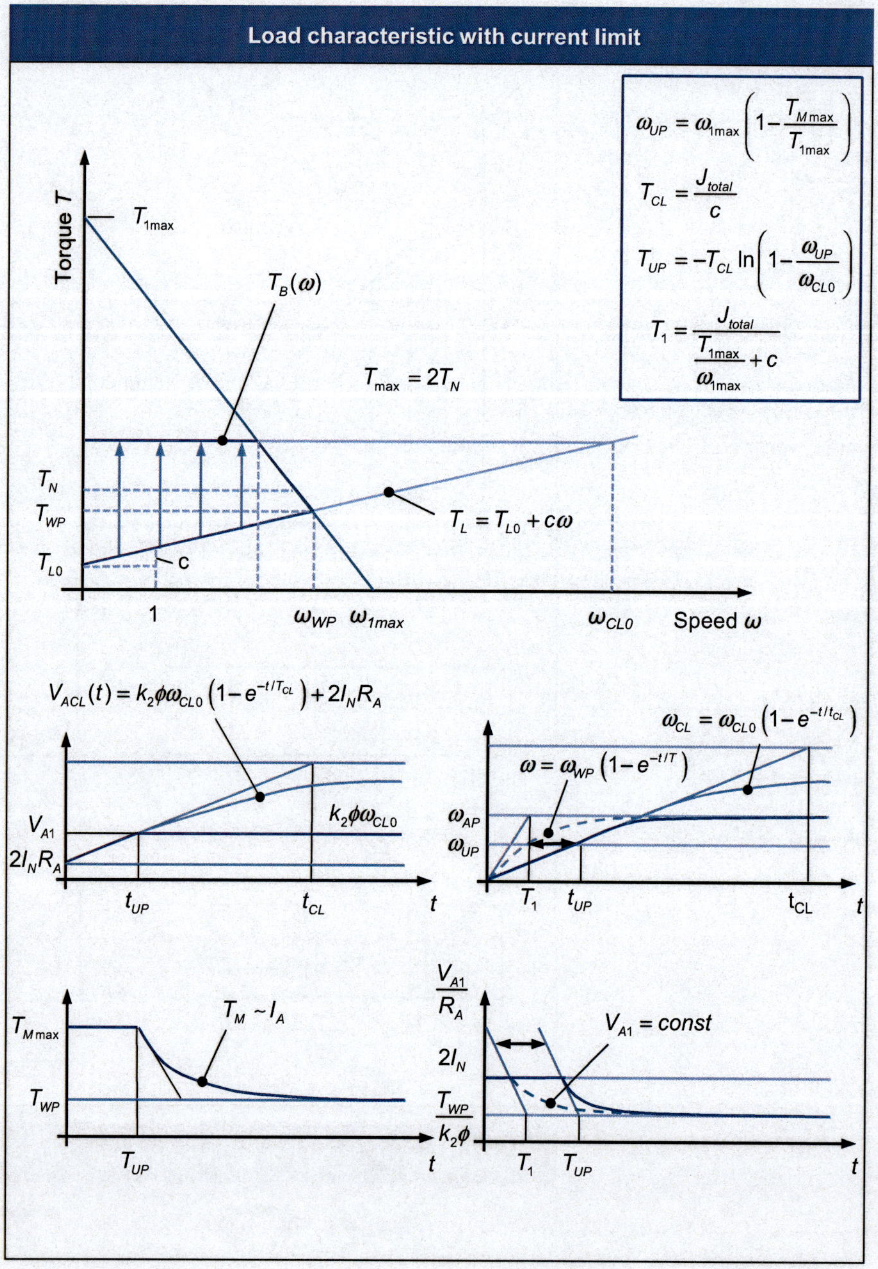


for $n = 0$. The time of the ramp-up to time $t = t_{UP}$ comes from Eq. (2.70):

$$t_{UP} = -T_{CL}\ln\left(1-\frac{\omega_{UP}}{\omega_{CL0}}\right). \tag{2.72}$$

In the current limiting phase, the armature voltage must be reduced. When the motor is stationary, i.e., when ramp-up begins, the voltage is specified by the armature resistance and the current limit $2I_N$:

$$V_A(t=0) = 2I_NR_A \tag{2.73}$$

As motor speed ω increases, armature voltage $V_A(t)$ must also overcome the electromotive countervoltage of the motor. It thus ramps up according to the laws of Eqs. (2.2) and (2.70) where

$$V_A(t) = k_2\phi\omega(t) + 2I_NR_A \tag{2.74}$$

$$V_A(t) = k_2\phi\omega_{CL0}\left(1-e^{-t/T_{CL}}\right) + 2I_NR_A \tag{2.75}$$

until $V_A(t) = V_{A1}$ is reached. This would have been set at the start without a current limit (Fig. 2.33). In the second phase, i.e., for $\omega_{UP} \leq \omega$, the relationships without a current limit derived before apply. For this reason, the relevant characteristics are also shown in ◘ Fig. 2.34 in the same way as ◘ Fig. 2.33. The profiles of the characteristics for the second part of the ramp-up phase are given with a parallel translation of these curves along the time axis through the point at $t = t_{UP}$, as indicated by the arrows in ◘ Fig. 2.34. For the case that the load torque characteristic is $T_L = T_{L0} = const$ and has no rotation speed-dependent term "c", Eq. (2.65) simplifies to:

$$T_a(\omega) = T_{M\max} - T_{L0} = J_{total}\dot{\omega}_{CL} = \text{const} \tag{2.76}$$

This means there is a constant acceleration torque in the current limiting phase. The first part of the ramp-up curve becomes a straight line:

$$\omega_{CL}(t) = \frac{T_{M\max} - T_{L0}}{J_{total}} t \text{ for } \omega_{CL} \leq \omega_{UP}. \tag{2.77}$$

The relationship for $\omega_{UP}$ according to Eq. (2.71) stays the same. The time until the switchover point is given from Eq. (2.77) as

$$t_{UP} = \frac{J_{total}}{T_{M\max} - T_{L0}} \omega_{UP} \tag{2.78}$$

The armature voltage also ramps up linearly until $\omega = \omega_{UP}$ or $t = t_{UP}$:

$$\begin{aligned} V_A(t) &= k2\phi\omega(t) + 2I_N R_A \\ &= k_2\phi\left(\frac{T_{M\max} - T_{L0}}{J_{total}}\right)t + 2I_N R_A \end{aligned} \tag{2.79}$$

What is described above applies beyond $t_{UP}$ or $\omega_{UP}$, i.e., during the phase without a current limit.

### 2.2.5.3 Numeric determination of the ramp-up for non-linear and discontinuous characteristics

If the characteristics of motor and load are not linear, an approximation method can be applied to determine the ramp-up (◘ Fig. 2.35).

Within specified angular velocity ranges $\Delta\omega_i$, the actual function progressions, $T_M(\omega)$ and $T_L(\omega)$, are replaced by their mean values $\overline{T}_M(\Delta\omega_i)$ or $\overline{T}_L(\Delta\omega_i)$ (step curve). For these angular frequency intervals, the equation of motion extends over into the linear relation:

$$T_a(\Delta\omega_i) = T_M(\Delta\omega_i) - T_L(\Delta\omega_i) = J_{total}\frac{\Delta\omega_i}{\Delta t_i} \tag{2.80}$$

It is possible from this to calculate time $\Delta t_i$ that the drive requires to traverse angular velocity range $\Delta\omega_i$:

$$\Delta t_i = \frac{J_{total}\Delta\omega_i}{T_a(\Delta\omega_i)} \tag{2.81}$$

The ramp-up curve can be constructed by stringing together the linear segments for the individual $\Delta\omega_i$ values with the determined time segments $\Delta t_i$ on the time axis. If the time segments $\Delta t_i$ are selected to be short enough, the accuracy of the method is certainly adequate.

### 2.2.5.4 Selection of motors according to static considerations

Main drives are used to generate the working motions in the form of cutting and forming movements relatively between tools and workpieces. To arrive at optimum tool service lives and surface qualities on the workpiece, the speeds, adapted to the different tool and workpiece diameters and their materials, must be appropriately set. The two most important static design criteria to specify a main drive for a particular machine are the power and rotational speed adjustment ranges. The motor power is given here by the product of the cutting force and cutting speed at the machining point. When motors are selected, the extreme values of machining situations occurring are used as the basis in most cases and a safety margin then added. Factoring in efficiency $\eta_{mech}$ of the mechanical transfer elements and motor gives the electric nominal motor power required as

$$P_M = P_{Process}\frac{1}{\eta_{mech}}. \tag{2.82}$$

The load over time is also included in rare cases to prevent over engineering the motors. The nominal power specified for a motor is generally the power the motor is able to deliver continuously without going into thermal overload. The momentary peak powers may be considerably higher. The powers of electrical main drives can thus be selected in line with their operating mode, i.e., their load profile as in DIN VDE 0530 [DIN02]. The motor load over time must, of course, be known here. The most important operating modes for main drives are shown in ◘ Fig. 2.36.

In addition to continuous operation (operating mode S1) where the motor attains thermal equilibrium under a constant load, short-time duty S2 and periodic operating modes S3 and S6 are also relevant. Short-time duty S2 comprises the load profile from a load time and a cool-down phase. Within load time $t_B$, the motor temperature, starting from the ambient temperature, must not reach the state of thermal equilibrium. The following cool-down phase must at least be long enough for the motor temperature to be able to return to the initial temperature.

Load time $t_B$ is also specified as part of the power details for operating mode S2 (example: 40 kW S2 30 min). The load profile in periodic intermittent mode S3 also comprises a load time $t_B$ and a standstill time $t_{St}$. Here however, the sequence of load and standstill is repeated cyclically with the motor temperature not

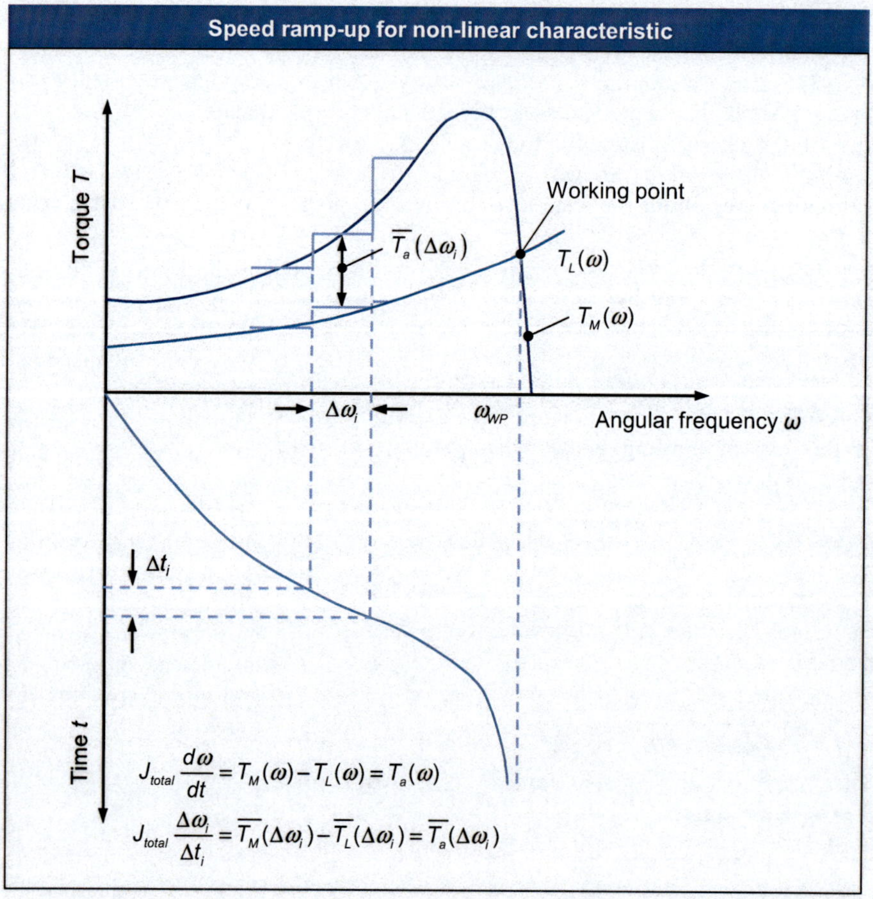


**Fig. 2.35** Determination of the ramp-up for non-linear motor and load characteristics

reaching the state of thermal equilibrium either in the load time or standstill time. The parameters for this operating mode are cycle time $t_S$ (understood to be the total of load time $t_B$ and subsequent standstill time $t_{St}$) and relative duty cycle $t_r$ (the $t_B$ to $t_S$ ratio specified in %).

So, specifying a motor power as "30 kW S3 40% 15 min" means that the motor can be loaded cyclically with 30 kW when the load times are 6 min each and the following standstill times are 9 min long. If the cycle is not in the power specification, it is stipulated as 10 min. For operating mode S6 (uninterrupted periodic duty with intermittent loading), the load profile comprises the periodically alternating load and no-load times. The power specifications are as those for operating mode S3.

### 2.2.6 Designs based on direct current and 3-phase servo drives

Up to a few years ago, direct current motors were almost the only motors being used as servo drives. This was attributable to the straightforward and affordable supply and control ability by controlling only one current. In contrast, three phase currents, depending on the rotor field, must be regulated to set the torque required for 3-phase drives. Measuring the stator currents and rotor position is also required for this more complex control. Since modern power semiconductors and the advancement of microelectronics have enabled the affordable realization of complex controller and converter modules, the direct current servo motor, in the face of the fundamental inferiority of its drive properties, has largely been replaced by 3-phase drives [ISER99].

The all-important downside of the d.c. motor is the commutator. It is subjected to wear and restricts the motor dynamics. Also, the risk of the brushes burning in during a no-load cycle means only a comparatively low torque can be delivered. Sliding contacts between stator and rotor are redundant for three-phase a.c. motors used for servo drives, meaning the motors are maintenance-free.

One key benefit of the synchronous motor compared to the asynchronous motor is far easier controllability. On the other hand, the asynchronous machine can be operated over a wide speed range with field

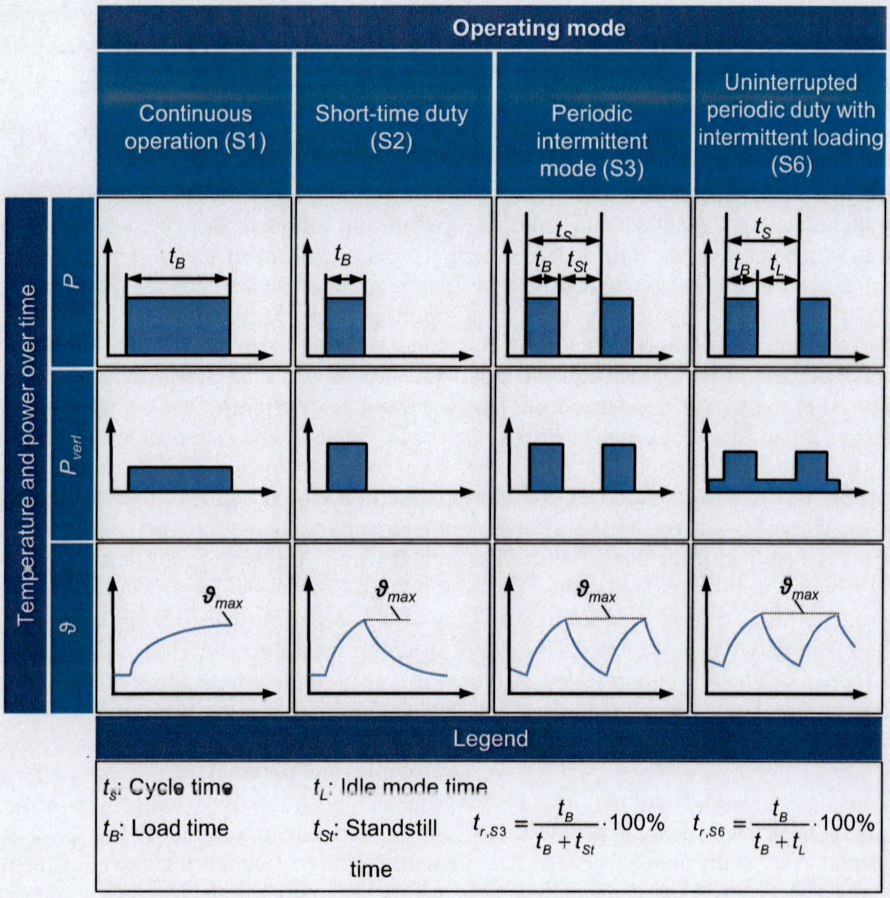


**Fig. 2.36** Operating modes as per DIN VDE 0530

weakening. For the synchronous machine, this mode is not possible in the standard, permanent magnet-excited design. For the synchronous machine, permanent excitation means there is no appreciable heating of the rotor while considerable heating here is likely for direct current and asynchronous motors.

Dynamically, asynchronous and synchronous drives are virtually equivalent, and are superior to direct current drives. Apart from improving control, the objectives of other developments for all motor types are higher integration density and, in conjunction with drive amplifiers, easier-to-use diagnostics, operating capabilities, greater levels of networking and coupling of multiple drives.

### 2.2.6.1 Stepper motors

The characteristic behavior of stepper motors is rotation of the motor shaft in discrete steps. In terms of its principle, the stepper motor is a synchronous motor with stator coils that are supplied by a square-wave current alternating cyclically. This forms a magnetic field whose orientation jumps when the stator currents are switched over. The rotor follows this field into its new position and so performs a discrete angle step.

Stepper motors enable quick, step-accurate positioning without feedback on the rotor position, i.e., no control loop with complex measuring system is required as a rule. If the motor is used as a positioning drive, the step angle should be kept as small as possible. An upstream gear unit can be used to tailor the number of steps per revolution to requirements.

The uncontrollable falling out of sequence when loading is too high and the tendency towards vibrations when operating near the natural frequencies are downsides. Straightforward activation yet high positioning accuracy, as well as robustness and freedom from maintenance, are points favoring this solution.

#### Stepper motor designs

The laminated stator of a stepper motor has pronounced, usually serrated, stator poles and concentrated exciter windings. The serration of the pole ends in the longitudinal direction is used to reduce the step angle. The different designs are distinguished mainly by the diverse rotor variants.

Reluctance stepper motors have an unexcited rotor with tooth structure. The torque is generated by the magnetic conductivity of the rotor different around the circumferential direction [ISER99].

Stepper motors excited with permanent magnets correspond to the conventional cylindrical rotor synchronous machine with permanent excitation. The rotor reaches coincidence with the excited stator winding. When excitation is switched off, these motors develop a self-holding torque, i.e., the non-excited motor can be loaded statically without causing rotation.

To arrive at greater resolutions and higher torques, the benefits of reluctance and permanent-excited motors can be combined in the hybrid stepper motor. The rotor here comprises a permanent magnet magnetized in the axial direction, with two serrated soft iron disks attached to the ends. The magnetic circuit of the motor is closed by the stator, giving rise to a concentration of the field in the teeth. Fitting the two toothed washers offset by half a tooth spacing with respect to each other results in a fine grid of north and south poles around the circumference. Serration of the stator poles means the air gap field is concentrated in the areas where the stator and rotor teeth face each other. The standard number of pole pairs $p$ is in the range of 25 to 125.

Serration of the stator poles also allows there to be a large number of pole pairs in the stator. The number of steps is proportional to the number of pole pairs $p$, as well as to the number of electrically independent phases. It is for this reason that hybrid stepper motors are manufactured using more than two phases (although additional overhead for control electronics is required).

#### Operating modes of stepper motors

Operating modes are divided into full-step, half-step and micro-step modes, depending on the current feed to the stator coils. ◘ Figure 2.37 shows the basic setup for activating a three-phase stepper motor and the stepwise advancement in half-step mode for the permanently excited stepper motor.

In full-step mode, several adjacent windings are fed with current because a higher field vector results than when feeding current to a single winding. In the stepper motor model, two south poles and one north pole form in the stator at time $t_1$. The rotor moves into the aligned position shown. A full-step mode is reached here when only the odd-numbered step sequences are selected in the current supply diagram in ◘ Fig. 2.37. By reversing the polarity of voltage $V_B$ at time $t_3$, the rotor makes the first step with step angle 60°. Further steps are controlled by reversing the polarity again. Rotating left is achieved with the reverse sequence of actuation. In half-step mode, another step is inserted between the full steps. One winding is switched off each time, causing the rotor to settle directly on the poles of the winding supplied with current and so to perform half-steps [SCHÖ13].

Another way is to lower step-by-step the excitation of a phase while simultaneously increasing the excitation in the second phase. Theoretically, a full step here is split into any number of individual steps. This operating mode is called micro-step mode. In this mode, a high angle resolution can be attained even with a low number of pole pairs. In practice, the number of intermediate steps is limited by the angle errors caused by the load, the static friction strip and imprecise control of the phase currents.

In isolated cases, stepper motors are also operated like an electronically commutated direct current machine in a closed control loop. They are fitted with an incremental encoder for this. By controlling the stator currents, the transient that otherwise would occur in an uncontrolled manner can be compensated when an angle step is performed. This makes a just-in-time positioning process possible because the uncontrolled falling out of sequence of the stepper motor is prevented when the step sequence is too quick.

#### Examples and parameters

The technical data of a stepper motor and the torque-speed dependency are in ◘ Fig. 2.38. For operation at higher step frequencies, the maximum torque with which the stepper motor may be loaded is smaller than the holding torque. This stems from the fact that the currents can no longer form in full as the frequency increases. The inductances of the windings are the primary cause of this. The drop in the characteristic can be shifted towards higher frequencies using constant current mode in the power stack of the electronic actuator.

The abrupt frequency changes used to accelerate or decelerate a stepper motor from a dead stop without step errors are called the start and stop frequencies. If the motor must be operated at a higher step frequency, this frequency must be increased to the desired frequency using a time function. The acceleration process can be well matched to the profile of the operating threshold torque using an exponential progression of the ramp-up curve.

### 2.2.6.2 Linear motors

Advancements in the cutting materials used in metal cutting over recent years allow high-speed machining with cutting and feed rates that are orders of magnitude higher than those of conventional processes. To generate translational feed motions, conventional rotatory drives are combined with a mechanical conversion gear unit, such as a ball screw drive, pinion rack or

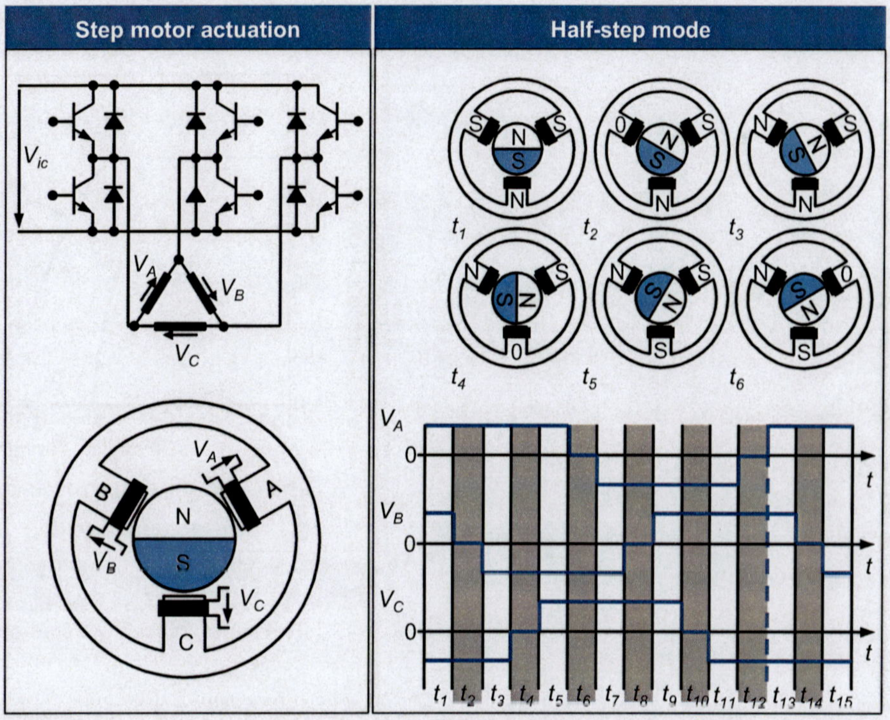


**Fig. 2.37** Functional principle and operating modes for a three-phase stepper motor according to Berger-Lahr

worm rack. Downsides here are the unfavorable properties of the mechanical transmission elements, such as play, elasticity, friction and the additional moment of inertia. They represent the limiting factors in the drive train with regard to speed, load stiffness, travel distance and dynamics.

When linear motors are used, mechanical transmission elements can be eliminated to overcome these disadvantages. The force flow from the motor to the mechanics on the power take-off side takes the shortest route here so there is no reverse backlash or elasticities of the drive train. In conjunction with digital control systems, a linear motor enables a high level of control quality to be attained with a large $K_v$ factor (see ▶ Chap. 3). This factor also enables low tracking error and high positioning accuracy even at high traversing speeds. As a result, the demands placed on modern feed drives, such as higher machining speed with equal or improved raceway accuracy, high static and dynamic load stiffnesses, and essentially any long travel distance, can be met better. The drive train of a linear drive can be designed in this way to be free of wear, play and friction.

### Designs of linear motors

Linear motors are linear variants of rotating machines. Intuitively, they can be imagined as a version of a rotation motor cut open up to the axis. A linear motor can, as a rule, be a stepper, asynchronous, synchronous or direct current motor. Realization as a direct current motor is, however, not feasible because contact elements are necessary to transfer the current. The two subsystems of a linear motor are called the primary and secondary parts. The primary part is derived from the stator of the appropriately rotating machine. ◘ Figure 2.39 contrasts the basic designs of the asynchronous and synchronous linear motors. Essentially the same design of the primary parts, comprising a laminated core with inlaid 3-phase windings, can be seen. In an asynchronous linear motor, the secondary part comprises a squirrel cage with iron core. In a synchronous linear motor, it carries the permanent magnets.

For the usual design of a synchronous linear motor with iron bonded to the primary part, the permanent magnetic excitation causes high attraction forces between the primary and secondary parts that can be up to 10 times the continual feed force. These forces must be absorbed by the linear guide, potentially resulting in friction and wear. The dimensions of the permanent magnets on a synchronous motor are comparable to the air gap. It follows from this that the feed force is not only dependent on the current but also on the feed path. This force ripple (cogging) can be greatly reduced with a slanted arrangement of the winding, a slanting of the iron core grooves and also by attaching the permanent magnets at an angle. For the feed direction, the force ripple is on the order of 5% of the feed force. The force ripple is also acts in the normal direction. Also

**Fig. 2.38** Technical data of a five-phase stepper motor according to Berger-Lahr

| Parameter | Value |
|---|---|
| Step angle [°] | 0,36 |
| Starting torque [Nm] | 6,80 |
| Max. torque [Nm] | 6,00 |
| Holding torque, excited [Nm] | 6,78 |
| Rotor moment of inertia [kgm²] | 3,3 ·10⁻⁴ |
| Steps per revolution | 1000 |
| Power class [W] | 350 |
| Max. step frequency [kHz] | 60 |
| Max. rotational speed [s⁻¹] | 120 |
| Max. start-stop frequency [kHz] | 5,3 |
| Nominal current / supply [A] | 2,3 |
| Resistance / winding [©] | 6,5 |
| Weight [kg] approx. | 4,2 |

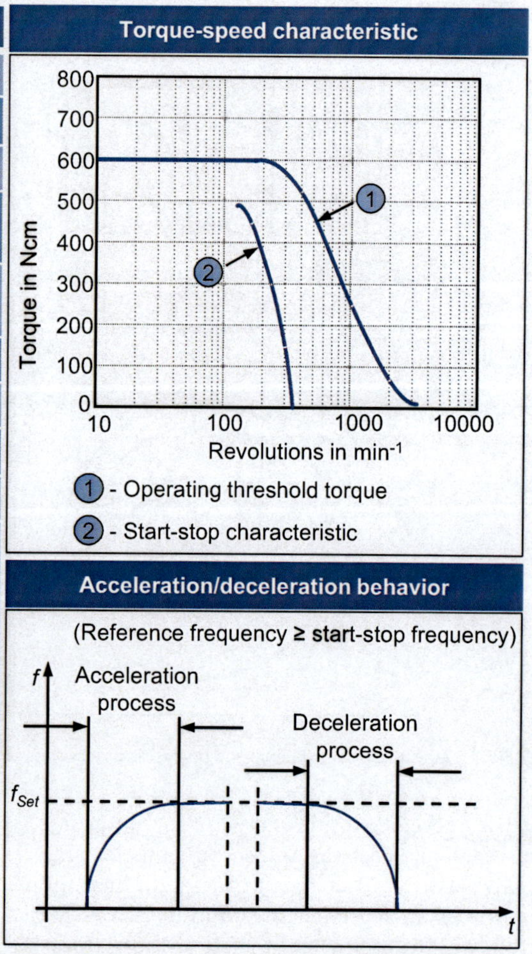


resulting from the slanting mentioned is that a low lateral force (if not eliminated by an appropriate symmetric design) can be assumed in addition to the feed force and attraction force. Detecting the position of the poles of the permanent magnets is required to actuate the synchronous motor. A Hall probe or absolute path measuring system can be used for this.

To perform relative motions, either the primary or secondary part needs to be extended, resulting in a long or short stator design. The long stator motor exhibits increasing power loss that is proportional to the travel distance and so its use is only viable for short travel distances. Highly dynamic linear motors use the short stator design almost exclusively. The secondary part of the motor is extended. Given that, for this type of linear motor, the surface of the secondary part grows in proportion to the travel distance, many expensive permanent magnets are required to realize the synchronous linear motor.

In addition to the single-comb arrangement of the primary and secondary parts of a linear motor, shown in Fig. 2.39, the dual-comb linear motor and solenoid motor are are other versions (Fig. 2.40). The solenoid motor is tube-shaped and is also called a housing linear motor. Production and design-related downsides of the solenoid motor mean it only plays a minor role in practical applications.

The single-comb arrangement shown in Fig. 2.39 is the most cost-effective solution with the most straightforward design and can be applied to travel distances of virtually any length. For the dual-comb arrangement, two individual drives are positioned symmetrsically such that the attraction forces cancel each other for the most part. The attraction forces can be routed by way of the housing as internal forces. Also, the weight of the secondary part can be reduced, allowing considerably greater accelerations. The dual-comb design is thus used for short-stroke movements when high accelerations are required.

Another way to reduce the mass of synchronous linear motors is to use the iron-free design of the primary part. Although this reduces the action of force, higher dynamics of the drive and so controller bandwidths to 100 Hz can be attained. This continues to keep cogging negligibly low.

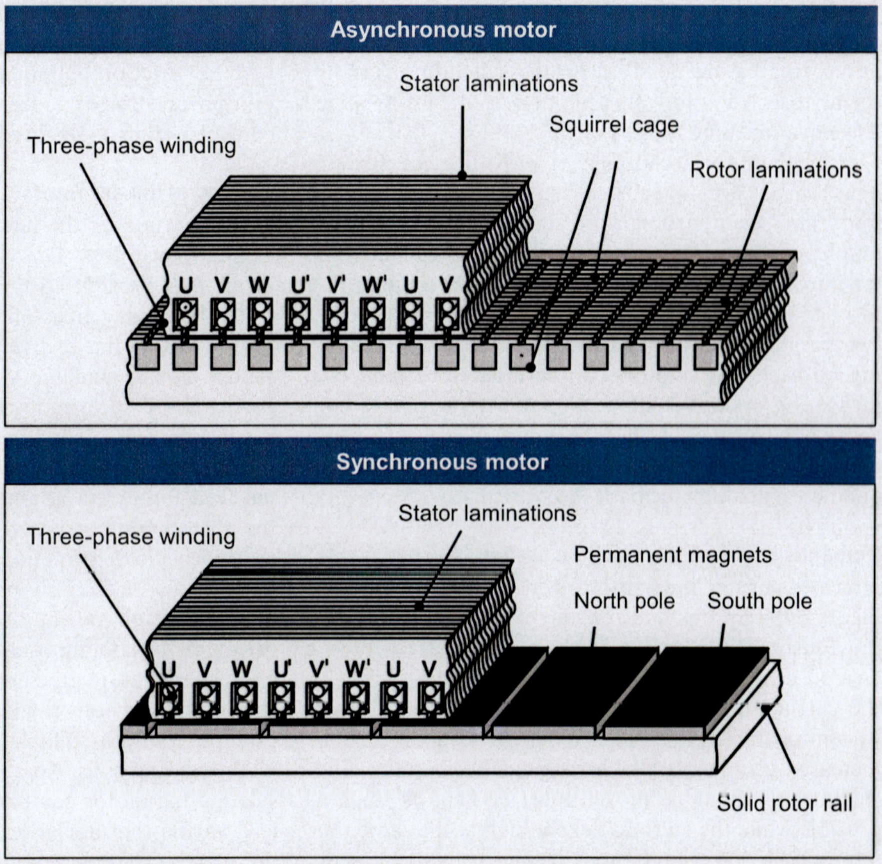


**Fig. 2.39** Cross sections of asynchronous and synchronous linear motors

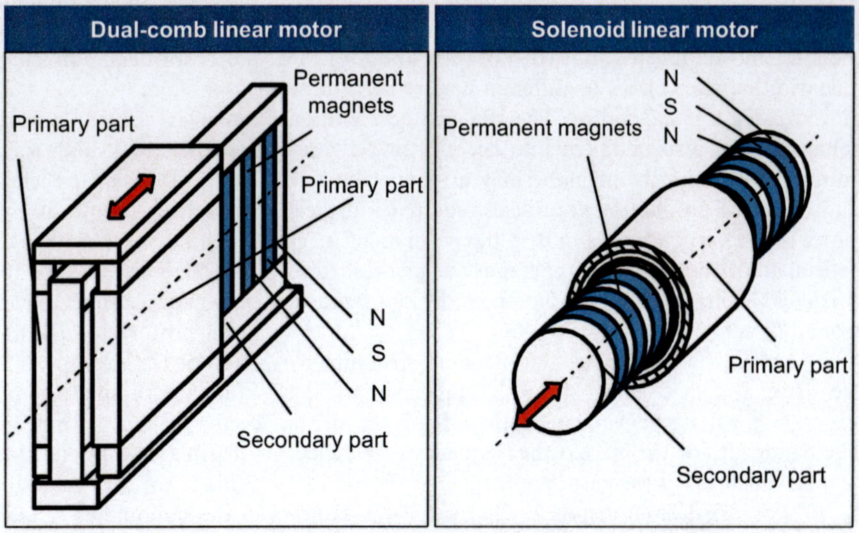


**Fig. 2.40** Dual-comb and solenoid linear motors

A benefit of asynchronous linear motors is that no magnetic forces act when no power is being applied to the machine. A substantial disadvantage to the asynchronous linear motor compared to the synchronous version, however, is the high level of heating in the secondary part rail. Major thermoelastic machine structure deformations may result. Key benefits of the synchronous motor compared to the asynchronous motor are the considerably higher continual feed forces attainable, no electrical losses in the secondary part and its ability to withstand high overloading (up to a factor of 30:1). Also, a speed range from 1 μm/s to 10 m/s is possible with modern synchronous linear motors. Feed forces up to about 20 kN can be attained and accelerations to 10 g are possible. So the synchronous motor seems to be the more suitable design for use as a feed drive in a machine tool.

Linear motor technology is providing an array of new avenues for machine design. Using multiple primary and secondary parts for every drive, axis for example, enables the driving force to be introduced better into the mechanical design and the maximum feed force to be increased. It is possible, for example, to use two parallel secondary parts with a primary part moving on each. The motors can be fitted with their own measuring systems and be powered by separate converters or be operated by one common measuring system on a single converter. This enables gantry designs and new machine kinematics to be realized.

### Demands made on the machine design

However, using linear motors in machine tools also places different demands on the machine concept than previously (◘ Fig. 2.41). As the magnetic feed force is created directly between the primary part attached to the carriage and the stationary secondary part, force transmission is not possible. While the feed force can be increased using a parallel arrangement of multiple linear motors, the mass moved must be kept as small as possible while the stiffness is kept high at the same time to attain high accelerations. This means lightweight design measures make sense. The fact that the available force depends solely on the motor used and the acceleration behavior is directly inversely proportional to the mass moved linearly must be borne in mind for the scenario that workpieces of different masses need to be moved using a linear motor. The changed acceleration behavior must also be taken into account in NC programming because the machine may no longer be able to provide the production accuracies required. Furthermore, for the synchronous motor, the permanent magnetic state of the secondary part must be mentioned in particular. This makes installation and chip removal more difficult.

To prevent excitation of the machine structure by the drive, the mechanical natural frequencies of the structure must lie above the bandwidth of the drive. The bandwidth of the drive is the frequency of a deflection at which the travel response measured should still be 70% of the desired response. Given that the bandwidth of a linear motor is higher than that of a conventional electromechanical drive, the requirements of the structure design become much stricter. For safety reasons, brakes or a counterweight must also be provided for vertical axes. Without compensation, a high constant part of the feed force would be necessary to overcome the weight, resulting in a high temperature rise and direction-dependent acceleration behavior of the linear motor. Linear rolling element guide systems with low friction mainly are used as guides. A high-resolution position encoder, in conjunction with a digital control system, is required to increase control quality.

### Cooling of linear motors

One downside of the linear motor is its low efficiency, or high power loss. This can generate considerable heat. While it is possible to move the heat-producing motor to the boundary areas of the structure for conventional feed drives, in linear direct drives, heat builds up in the center of the machine structure. Effective cooling, in particular of linear motors, is thus required for their use in machine tools. The first purpose of cooling is to cool the motor windings themselves such that higher continual feed forces are attainable. Secondly, heat flow into the surrounding structure must be prevented to avert thermoelastic deformations. The power loss arising can be dissipated effectively using liquid cooling, for example. In addition to heat dissipation, it is important in the design of the cooling concept to ensure that there is only a low temperature gradient across the cooler. Otherwise, structure distortions can result. This can be achieved, for example, using a suitable arrangement and orientation of the cooling coils. Apart from active thermal steps, insulating the motor compartment can have a positive effect on the thermoelastic behavior of a directly-driven linear axis. For example, introducing a laminated fabric layer makes the system thermally more inert reducing the effects of load fluctuations. ◘ Figure 2.42 shows the cross section of a thermally enclosed synchronous linear motor for demanding applications.

**Examples** Linear motors are used in highly dynamic machines with high traversing speeds and limited load forces. Typical application areas are laser machining, jet cutting and automatic pick-and-place machines. Machine tools with synchronous linear motors have also become available recently. In a vertical machining center for example, having three axes with synchronous linear motors, a rapid motion rate of 2 m/s and acceleration of the axes up to 2 g are achieved. In other applications with single-comb linear motors, $K_v$ factors exceeding 300 $s^{-1}$ for a measuring system resolution to 0.1 μm are attained. Linear motors are also used to advantage in ultra-precision machines, where resolutions in the nanometer range are attained for small strokes. One typical application for the dual-comb linear motor is out-of-round turning. For a moved mass of approx. 5 kg, workpiece accuracies of about 2 μm are attained for 6000 stroke movements per minute, with $K_v = 2000 s^{-1}$ and continual accelerations of 110 $m/s^2$.

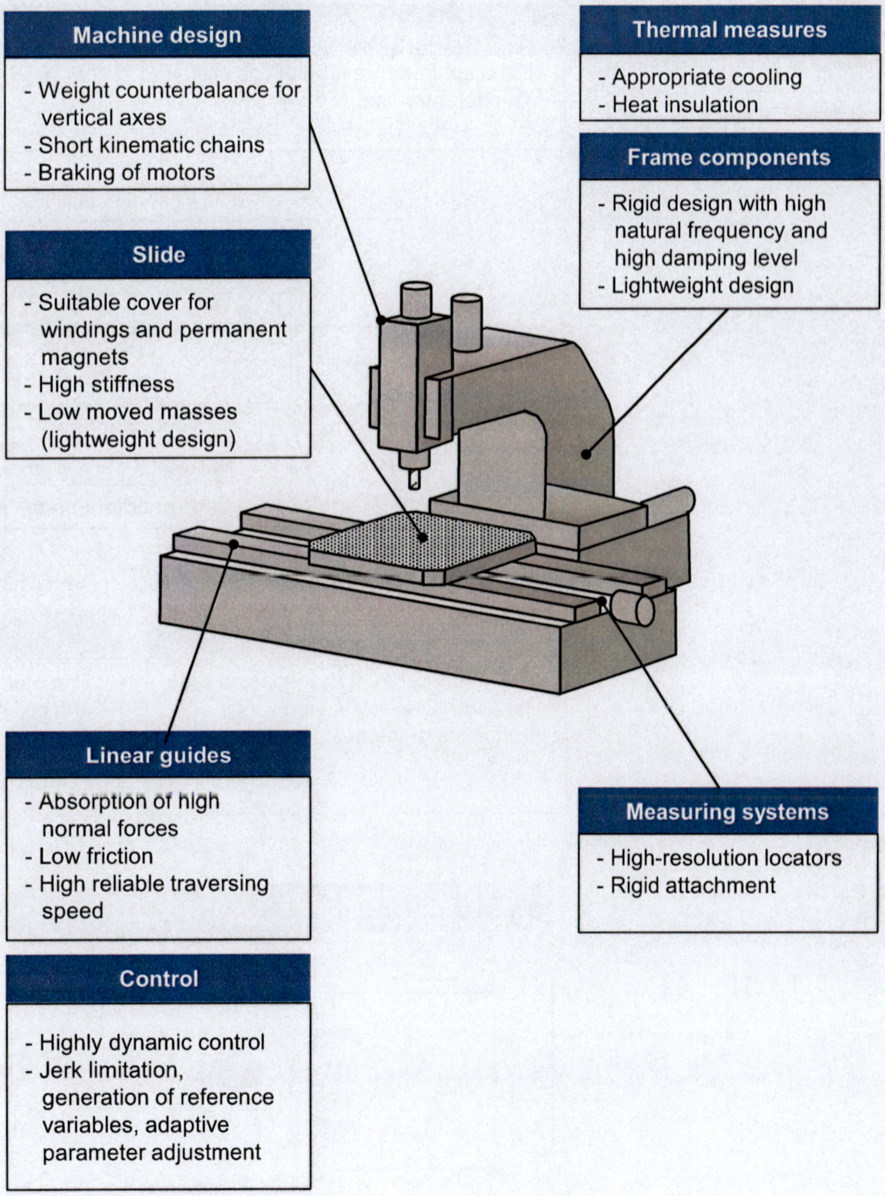


**Fig. 2.41** Demands on the machine concept when linear motors are used

### 2.2.7 Current and power measurements on electric motors

All the power in a machine tool, be it required to move the carriage or for cutting, must ultimately be delivered by a drive. Consequently, changes to the production process must also be reflected in changes to the electrical power drawn by the motor (whereby only the active power is meaningful). For d.c. motors, this his proportional to the product of the armature current and the armature voltage. In three-phase a.c. motors, the phase shift between current and voltage must also be taken into account. If only one axis is connected to the intermediate circuit of the drives, or the requirement is to only monitor the superimposed power of multiple axes, it is also possible to determine the power by measuring the current and voltage in the intermediate circuit of the drive amplifier.

For d.c. motors, the current is proportional to the force, meaning direct conclusions can be drawn about the feed force or torque.

Different principles are applied for measuring the motor current. Previously, a shunt resistor was often inserted into the motor feed. The motor current generates a proportional voltage drop across the resistor (Fig. 2.44). This measurement setup can be used for both d.c. and a.c. motors. The drawback of the shunt is the power loss across the resistor. In addition, the measured voltage is not potential-free and entails ad-

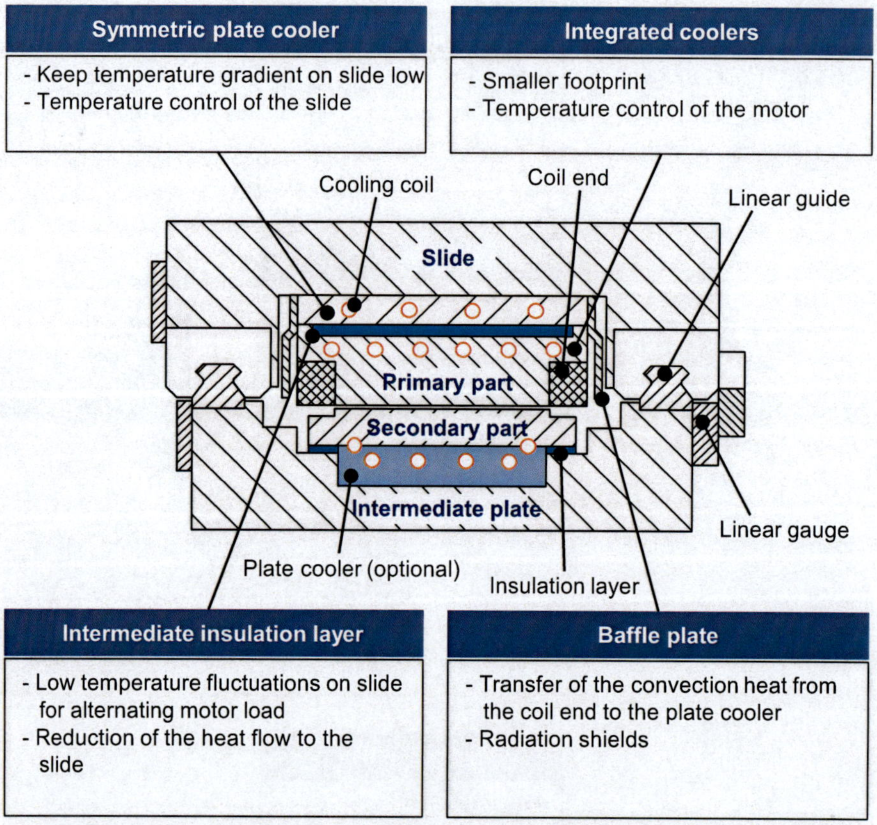


**Fig. 2.42** Measures for minimizing the temperature influence of a linear motor on the mechanical structure according to [Eun99]

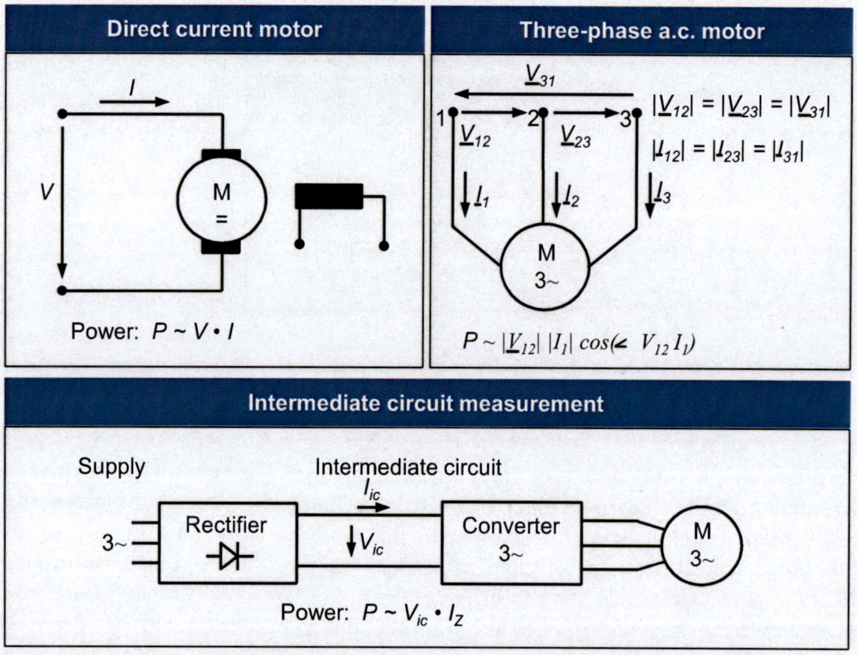


**Fig. 2.43** Calculation of active power for d.c. and a.c motors

ditional work for subsequent processing in the downstream electrical circuit.

The transformer principle can only be used for a.c. measurements. The motor feed is routed through a toroidal core where the alternating magnetic field tangential to the conductor is concentrated (◘ Fig. 2.44). According to the principle of induction, this field generates a voltage in the coil wound around the core. This voltage is proportional to the current through the corresponding conductor. No additional power loss is generated with this measuring procedure. The voltage tapped at the secondary coil is potential-free.

A Hall element can be used to directly measure the magnetic field around a conductor of the motor feed line. A Hall element comprises a semiconductor chip through which a constant current flows in one direction. If a magnetic field permeates the element, the charges flowing are deflected from their path by the Lorentz force. As a result, a voltage is generated perpendicular to the direction of current and magnetic field that is proportional to the strength of the magnetic induction.

To measure the motor current, the magnetic field of the feed line is in turn concentrated into a toroidal core. The core is cut open at one place and the Hall element is inserted into this air gap (◘ Fig. 2.44). As Hall elements can also be used to measure the strength of static magnetic fields, this setup can also measure direct currents. This is now the most commonly used method for measuring current.

The measurement setups described above for current and power measurements are easy to retrofit on machines. As a result, the power or current sensor has become the most widespread sensor in the field of monitoring systems. In the future, information internal to the controller, instead of external sensors, will be used more often for these applications.

However, a problem common to both systems is that dynamic processes in the production process can not be determined or can be but only with strong attenuation in the motor power. The mechanical elements connected between the process location and drive motor (such as gear units, spindles and carriages, as well as the motor itself) exhibit power losses and, due to their inertial masses for dynamic signals, act like low-pass filters with low cutoff frequencies. Due to these influences, minor process-dependent force and torque fluctuations (e.g., when smoothing) are hard to detect from the motor current measurement.

## 2.3 Position measuring systems for NC machines

A key component of the position control loop of numerically controlled machine tools is the measurement equipment used to measure the actual positions of machine components. Examples of this are the path measurement for carriages and determination of the angular position of drive spindles and rotating tables. Analog and digital measuring systems with electronic analysis units facilitate the fast and accurate reading of machine positions and automatic analysis of the measurement

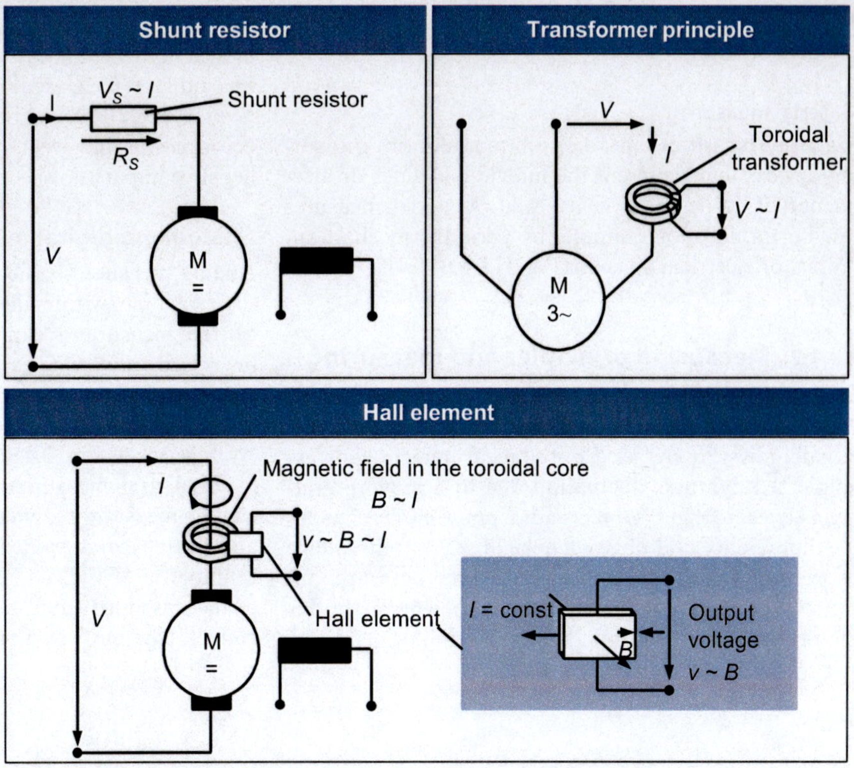


◘ **Fig. 2.44** Method of motor current measurement

results. Using these kinds of measuring systems means high positioning accuracy and associated high manufacturing quality are attained.

## 2.3.1 Basics of path and angle measurement

The terms of measurement technology are defined in VDI/VDE 2600 and DIN 1319 [VDI73, WECK99]. Important basic terms are explained below.

### 2.3.1.1 Basic terms

#### Measurand and measured value

The measurand is the physical parameter whose value is determined through measurement (such as travel or angles). The measured value is the value of that measurand to be determined specifically. It is specified as a combination of numerical value and unit (e.g., 345.125 mm or 10.32°).

#### Sensor

A sensor is a measuring device that records the measurand at its input and outputs a relevant measuring signal at its output. The term "encoder" is also commonly used in place of "sensor" that is recommended in VDI/VDE 2600.

#### Measurement location

The place where the sensor is located is called the measurement location.

#### Material measure

Material measures are the sub-modules in measuring devices that represent the individual values or a sequence of values of a measurand. Material measures can be formed, for example, by periodic, mechanical, optical or electrical divisions [VDI73, DIN71].

### 2.3.1.2 Measuring principles and measuring methods

A measuring principle is understood to be the characteristic, physical law used in the measurement. For example, a brightness fluctuation due to a moving grating between a light source and a photodiode changes the illuminance and thus a change in the output voltage of the photodiode. This can be used for path and angle measurement equipment. The measuring method is the practical application of the measuring principle for reading measured values. Simple examples are used below to explain the different types and features of measured-value acquisition.

#### Direct and indirect measured-value acquisition

In the direct method, the measured value is obtained by directly comparing the measurand to a corresponding reference value. In the indirect method, the measurand of interest is converted to a different kind of physical parameter. The measured value is then determined from this using physical relationships.

Direct and indirect measured-value acquisition are contrasted in ◘ Table 2.1. The linear scale in the measurement setup shown in the table part on the left is used as the material measure. The position of the carriage can be read directly from the gauge by the sensor. This is an indirect measurement if the linear movement of the carriage is detected by the detour of the feed spindle rotation. An angle step transducer on the feed spindle is used to record the measured value. The path of the carriage can be measured indirectly using the angle measurement and the familiar relationship between the rotational motion of the ball screw and the linear carriage movement (◘ Tab. 2.1on right). Transfer errors of the mechanical system are not recorded, however. As a result, they can also not be corrected by the position control loop.

Typical error factors for the various measured-value acquisition methods are summarized in the lower part of ◘ Tab. 2.1 [BRAA97, DIN88, WALC74]. While for direct measured-value acquisition, defective production and setup accuracies of the gauges and their thermoelastic behavior can falsify the measurement result, for indirect measured-value acquisition, pitch errors, reverse backlash, static and thermal-related elastic deformation and wear of the feed spindle also have a negative impact on the measurement accuracy.

#### Analog and digital measured-value acquisition

Analog measured-value acquisition is characterized by the fact that, within the measuring range, every value of the measurand can be assigned a corresponding measured value. One example of this is the slide potentiometer that provides a continual assignment between the measurand (travel $x$) and analog output variable (voltage V), ◘ Fig. 2.45.

In digital measured-value acquisition, the parameter to measure is divided into equal-sized increments and the measured value is shown as their multiples. The example in ◘ Fig. 2.45 shows the division of travel $x$ to measure into equal sections (spacing of lines on the gauge). The measured value is determined by counting

**Table 2.1** Principles and error factors in direct and indirect measured-value acquisition

| Direct measured-value acquisition (drive component and converter separate) | Indirect measured-value acquisition (drive component and converter identical) |
|---|---|
| Measuring system operates translationally and is coupled directly to the longitudinal motion of the machine slide | Detecting of longitudinal motion from the rotational motion of the working spindle |
| 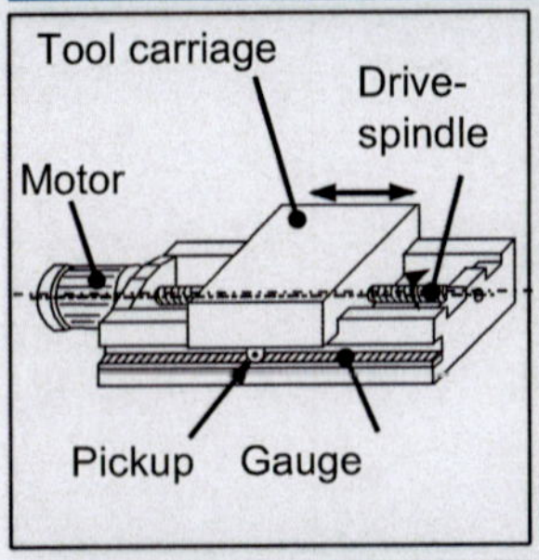 | 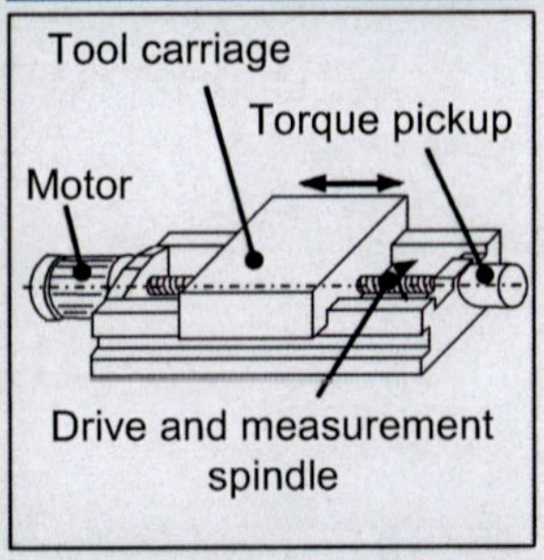 |
| **Error factors** | **Error factors** |
| • Temperature<br>• Separation errors<br>• Distance and angle errors<br>• Faults at the joints of the gauges | • Elastic deformation of the spindle<br>• Pitch error<br>• Play<br>• Spindle wear<br>• Pickup faults<br>• Temperature |

the individual sections or by directly reading the place values, and can be analyzed and displayed using analysis electronics (refer to ▶ Sect. 2.3.2).

## Relative and absolute measured-value acquisition

Another differentiating feature in conjunction with forming measured values is the nature of the gauge used. Digital measurement can be differentiated as follows:

- Digital-relative (incremental) measured-value acquisition
- Digital-absolute (coded) measured-value acquisition

In digital-incremental measured-value acquisition, the periodic measuring signals are counted and then totaled while a distance is being traversed (Fig. 2.47). The design of the lined gauge means the absolute place

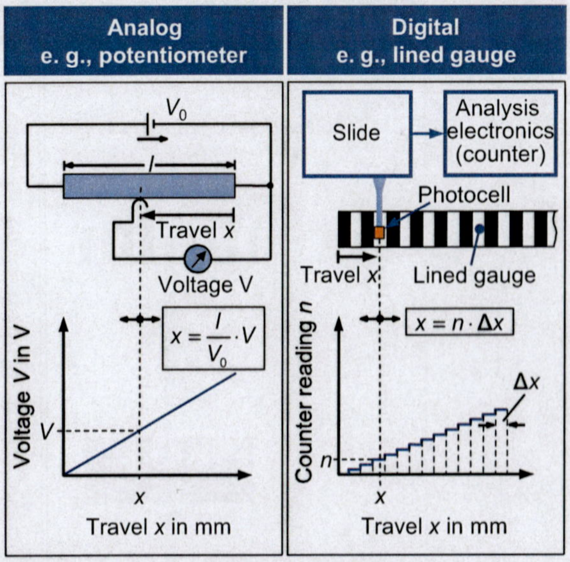


**Fig. 2.45** Principles of analog and digital measured-value acquisition

of the moving machine carriage can not be measured with this method. Rather, only the position relative to the place where the counting process is started. This method is thus called the relative or incremental measuring method. A continuous, fixed assignment between the absolute position of the carriage and the measured value characterizes digital-absolute measured-value acquisition. In the example in Fig. 2.46, every position on the gauge is denoted by the unique combination of coded marks.

The most significant benefits of absolute over relative/incremental measured-value acquisition are:

- Every measured value is relative to a specified zero point and can be read directly. Cumulative (chain) measurement errors do not occur.
- The correct measured value is retained even after a fault (such as a voltage outage) without any particular aids.

But absolute-digital measured-value acquisition also has downsides. If the requirement is for high measurement accuracy over a larger measuring range, the gauge and sampling unit for binary or Gray-encoded measuring devices have a much more complex design due to the high number of code tracks required and they have a larger footprint (see ▶ Sect. 2.3.2.1.2).

Analog systems as in Fig. 2.45 are absolute measuring systems. Their relative resolution (from $10^{-3}$ to $10^{-4}$) means the measurement accuracy of analog-only measured-value acquisition is not adequate for larger measuring ranges. But if the measurement path

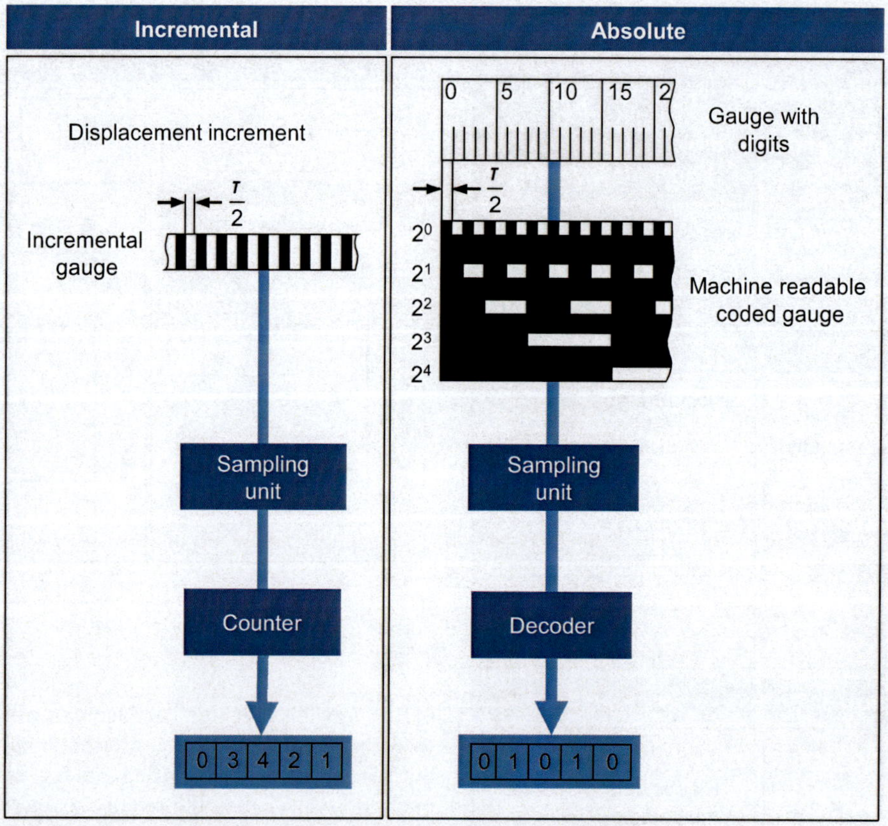


**Fig. 2.46** Examples of incremental and absolute path measurement

is divided up into many short sections, using the analog method is a sensible option to obtain the cyclic-absolute measured values. Cyclic-absolute measured-value determination combines the benefits of analog (high enough resolution) and incremental (straightforward, fail-safe measured-value determination) measurement technologies. The functional principle of cyclic-absolute measurement is shown in Fig. 2.47. Within recurring period lengths $\tau$ (measurement cycles), an absolute measurement of travel $x$ is made by an analog system. The number $N$ of cycles passed through over distance $x$ is also counted. Then, $h(x^*)$ is the analog measurement (such as voltage $V$), $h_0$ is the maximum value of the analog measurement and $\tau$ is the period length of the measurement cycle. Many path measuring systems in use function on the cyclic-absolute method so the requirement for high resolution over a long travel distance can be satisfied. Since the NC control system of the machine tool can only process digital signals internally, digitizing the analog measuring signals is always required.

## 2.3.2 Measuring systems

### 2.3.2.1 Photoelectric measuring methods

The high resolution required has meant that general measuring systems using photoelectric sampling of linear scales and index plates for rotatory systems have established themselves today.

The main components of photoelectric measuring systems are the gauge or index plate, and the sampling unit. The reflected or transmitted light method is used for sampling (Fig. 2.48). In the reflected light method, the gauge has alternate reflecting and non-reflecting zones. In the transmitted light method, the gauge comprises transparent and opaque zones. The sampling unit comprises the light source (almost always an LED), optics, sampling grating and photo receiver. When there is relative motion between the sampling unit and gauge, the photo receiver receives light of modulated brightness. The receiver circuitry converts the light intensity to electrical signals that can be analyzed for path measurement [ERNS98, MILL90].

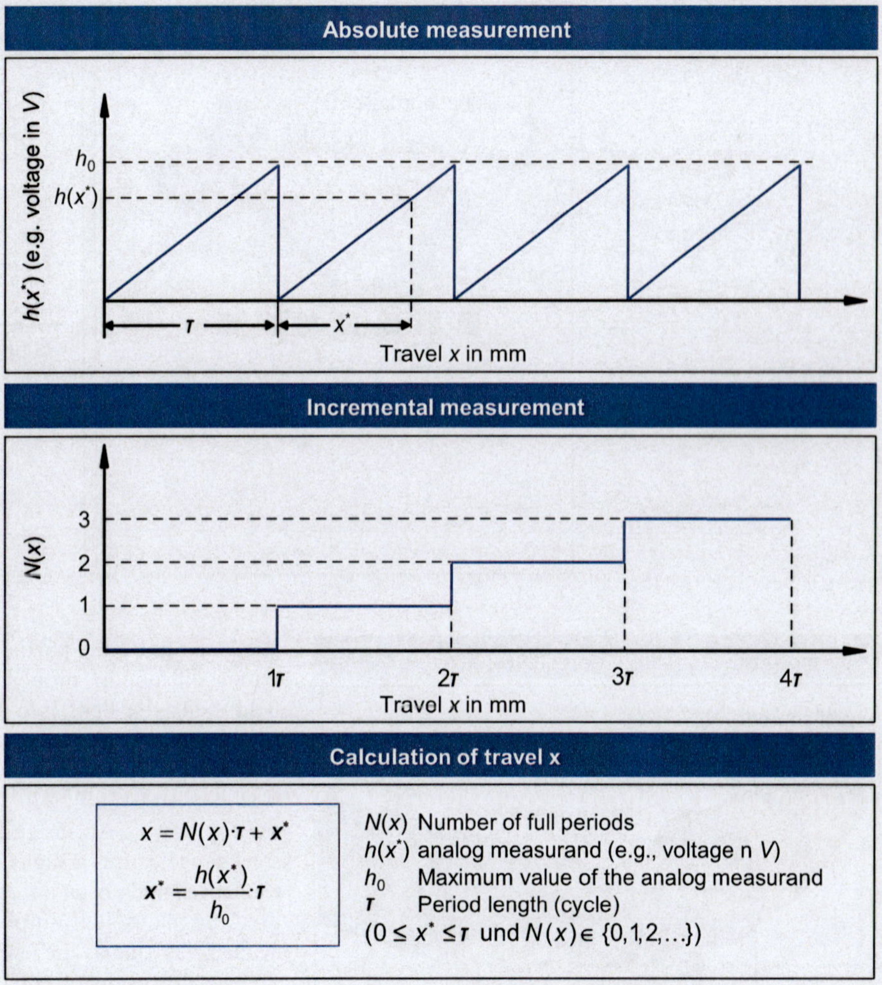


**Fig. 2.47** Principle of cyclic-absolute path measurement

## Digital-incremental measuring systems

Figures 2.49 and 2.50 show the setups of two commonly used incremental measuring systems for length and angle measurements. Figure 2.49 shows a photoelectric linear measuring system in which the transmitted light method is used for gauge sampling. The incremental rotary encoder in Fig. 2.50 also samples the index plate using the transmitted light method. The measurement is taken with photoelectric sampling of the precision glass scale or index plate using a sampling plate and photo elements. The finite expansion of the active surface of the photo element covers several graduation lines on the disk, meaning a sampling grating is positioned between the light source and receiver. When the gauge moves or the sensor shaft rotates relative to the sampling unit, approximate sinusoidal signals are generated at the output of the photodiodes. These signals are converted to a square wave signal sequence in the analysis unit.

At least two photo elements are required to generate direction-inducing signals (refer to Sect. 2.3.3.4). However, there are four photo elements in the linear measuring system of Fig. 2.49 (4-field sampling). The arrangement of four measuring elements offset by 90° allows a special evaluation of the measurement signals (Fig. 2.51). The 0° signal ($I_1$) has the opposite phase to the 180° signal ($I_2$), i.e., the signals are subtracted from each other. The same process applies to the 90° ($I_3$) and 270° ($I_4$) signals. This yields the sine and cosine signals that fluctuate periodically around the zero line. An alternating current without direct current component, where a square pulse is generated for every zero-crossing, is beneficial for subsequent evaluation. Interpolation to increase the resolution is performed using the method described in ► Sect. 2.3.3.

The 4-field sampling in Fig. 2.49 is sensitive to dirt on the gauge. This is because the light intensity fluctuations lead to offset errors in the individual phase positions (0°, 180°, 90° and 270°) and so to erroneous measurements. When the sampling unit moves over dirt on the gauge, the dirt is detected by the photoelements

2

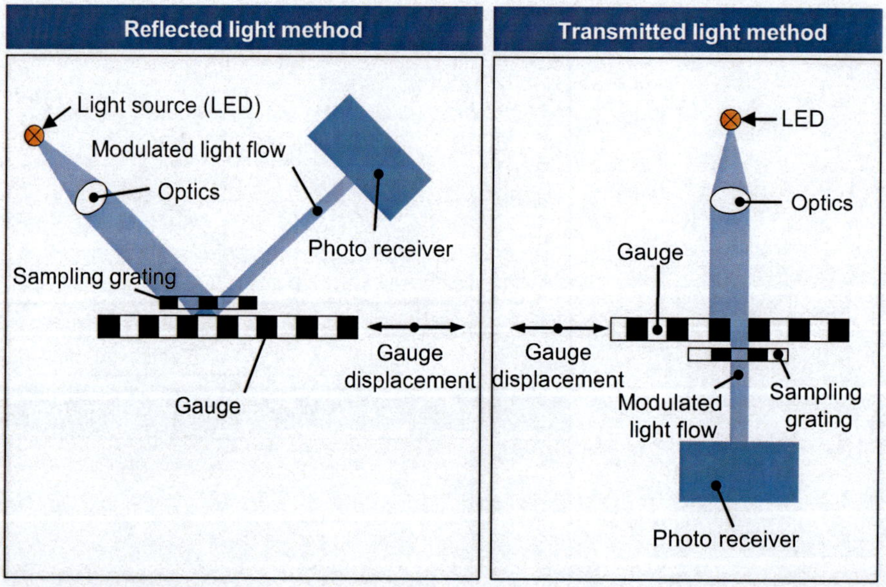


**Fig. 2.48** Principle of photoelectric sampling for incremental methods

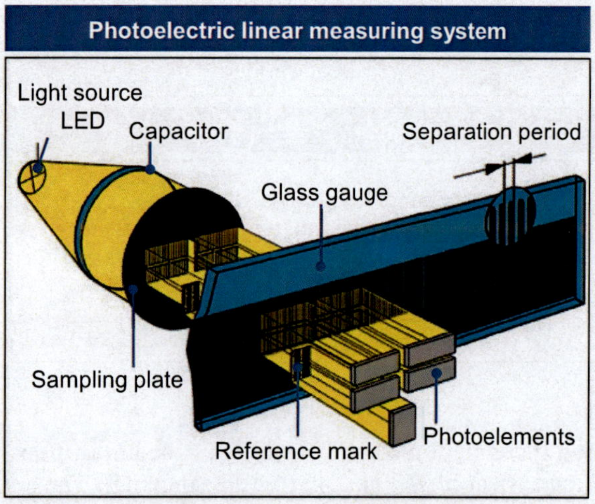


**Fig. 2.49** Photoelectric linear measuring system using the transmitted light principle according to HEIDENHAIN

in succession such that the erroneous measurement occurs at four positions.

Quasi 1-field sampling using the reflected light method is an improvement (Fig. 2.52) [BRAA00]. Reflecting steel tape measures with divisions of 40 µm, for example, are used on the gauge. The sampling plate has a transparent sampling grating with two different subfield types. Both subfield types have a grating that is parallel to the gauge graduation and has the same divisions, the amplitude grating. Parallel to the direction of motion, they have a diffraction grating with two different grating spacings (e.g., 5 µm on the light-colored fields and 7 µm on the dark fields of the sampling plate in Figs. 2.52 and 2.48). The different diffraction grating spacings on the partial sampling fields result in different diffraction properties. Every partial sampling field splits the light into two inversely phased partial ray beams that are collected on the photo elements as in Fig. 2.52. Accordingly, the photo elements provide signals with 0°, 90°, 270° and 180° phase shifts (from the top down). An insignificant amount of dirt on the steel gauge causes a slight intensity change of all photo element signals because only a single sampling field is used. So, sinusoidal output signals of high quality are available as long as the level of dirt is low.

The downside of all incremental methods is that they require known start positions where the counting of increments for left and right movement begins. If there is only one reference mark (see Fig. 2.49), referencing to this start position is required after the measuring system is switched on.

Thus, multiple reference marks having different spacings to each other can be applied to linear scales (Fig. 2.53). After two such spacing-encoded reference marks are traversed, the absolute position on the gauge is determined by the analysis unit counting the increments between the marks.

The divisions on the gauge or disk feature the maximum manufacturing accuracy and exceptional definition of optical edges. They are made by vapor-depositing a very thin, hard, metallic, graduation-forming layer in a vacuum or by scoring glass or metal gauges. For the digital resolution of a distance or angle change, the spacing or number of graduation lines on the measurement element is the decisive factor.

For the transmitted or reflected light method shown in Fig. 2.48 to work according to the laws of geometric optics, the divisions on a gauge must not be very much less than 10 µm. To attain a higher resolution,

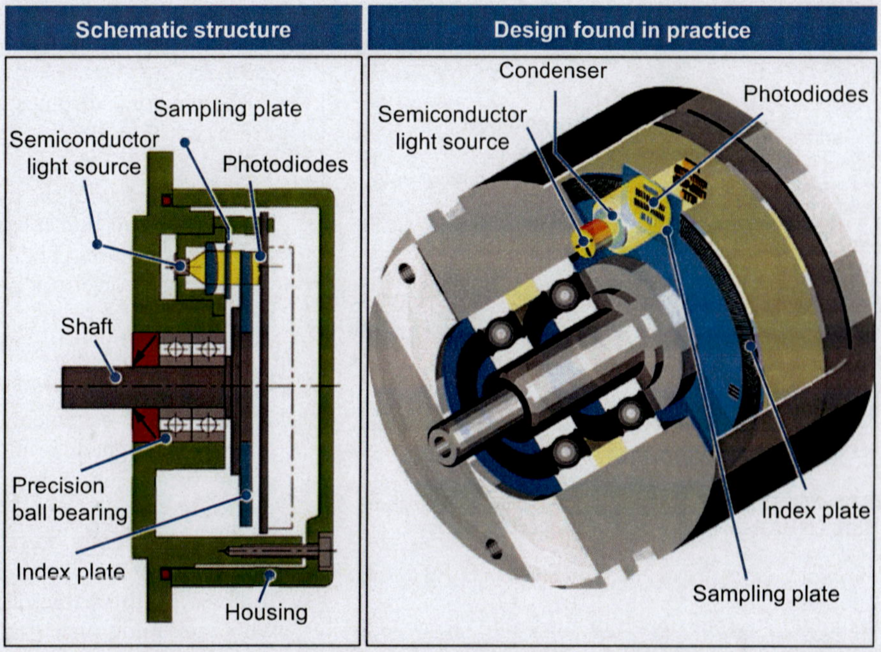


**Fig. 2.50** Inside structure of a rotary encoder according to HEIDENHAIN

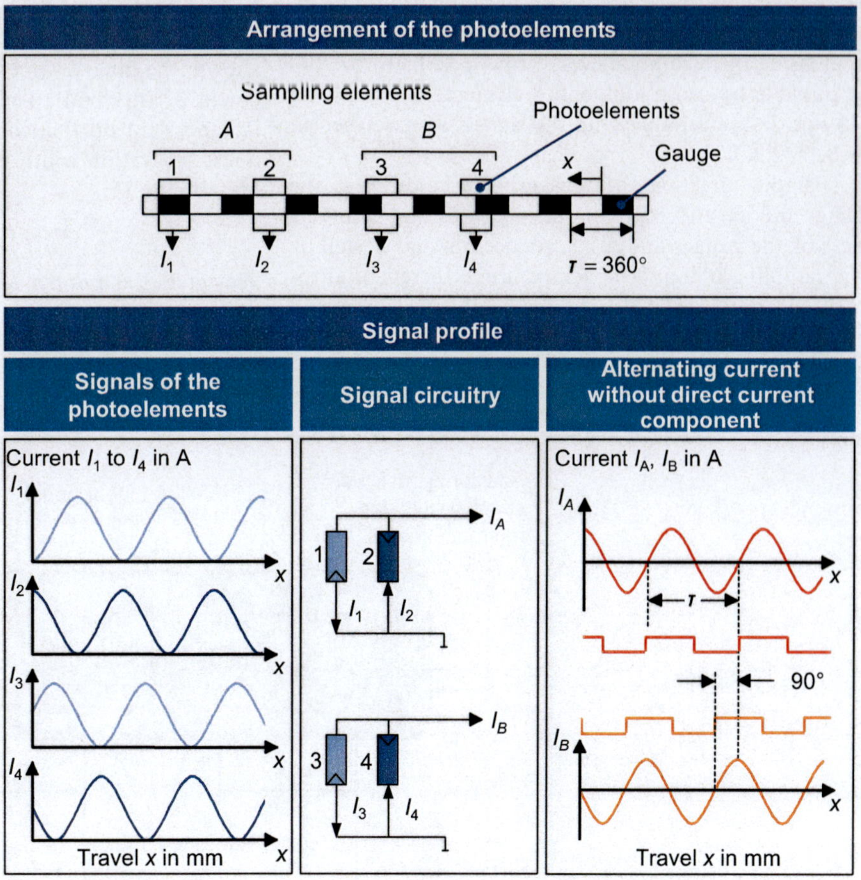


**Fig. 2.51** Signal profiles when sampling an incremental gauge

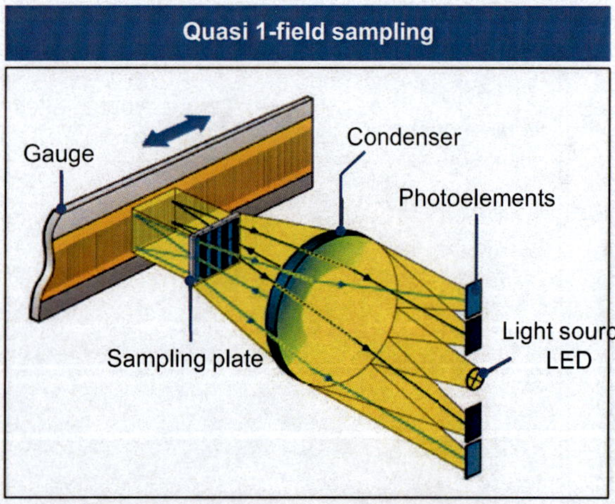


■ **Fig. 2.52** Principle of quasi 1-field sampling according to HEIDENHAIN

the analog output signals of the photo elements are interpolated using the methods described in ▶ Sect. 2.3.3. Because the signals are only approximately sinusoidal, i.e., they have a super positioned harmonic component, interpolation by a factor greater than 25 is not viable. Increasing the resolution even further to a factor of 100 is possible by using four photo elements for sampling (■ Fig. 2.49) and analyzing the signals as described in ▶ Sect. 2.3.3.

Factors affecting the measuring accuracy are the gauge holder and disk mount, evenness and straightness of the gauges and disks, connection and design of the sampling unit and the temperature. Incremental rotary encoders are also used for length measurements in conjunction with mechanical transmission elements (such as rack and pinion, spindle and nut). See ▶ Sect. 2.3.1.2.1.

Incremental distance and angle measuring systems are now the preferred choice in machine tools. Often quoted as disadvantages of these incremental measuring systems are that the measured value is lost when the power fails and that it can be distorted in the event of interference pulses. The latter no longer occurs due to the advanced levels of development for electronic devices.

## Digital-absolute measuring systems

In contrast to incremental methods, every distance element is assigned a unique numerical value in encoded measuring systems. ■ Figure 2.54 shows schematically a binary-encoded gauge with five code tracks. There are different ways to encode a gauge interval. For binary-encoded gauges, the separation width of a track (width of a track field) reduced by a factor of 2 with each additional track. As in ■ Fig. 2.54, assigning every light-colored field a binary value of "1" and every dark field a binary value of "0" allows a decimal number to be read for every simultaneous sampling of all tracks at every point of distance $x$. In the example shown, the binary values read (10,110) give distance element 22 (decimal). The number of required tracks $N$ is dependent on the length of the gauge $x_{total}$ and the finest separation width $\Delta x$ (track $k=0$). The following law applies: $H\_i$

$$x_{ges} = \Delta x \cdot 2^N (N = 1, 2, 3, \ldots) \qquad (2.83)$$

This gives the number of tracks $N$ required for the precision required $\Delta x$:

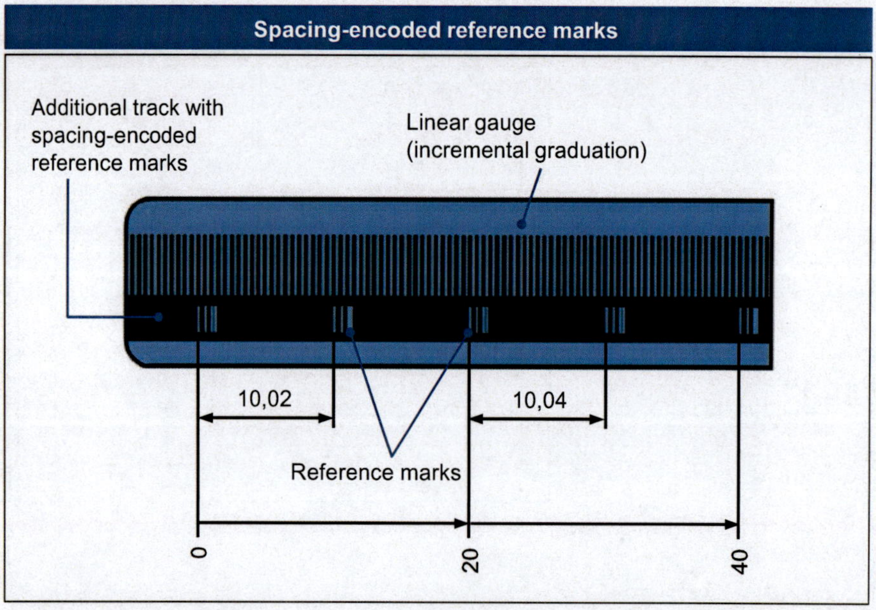


■ **Fig. 2.53** Spacing-encoded reference marks according to HEIDENHAIN

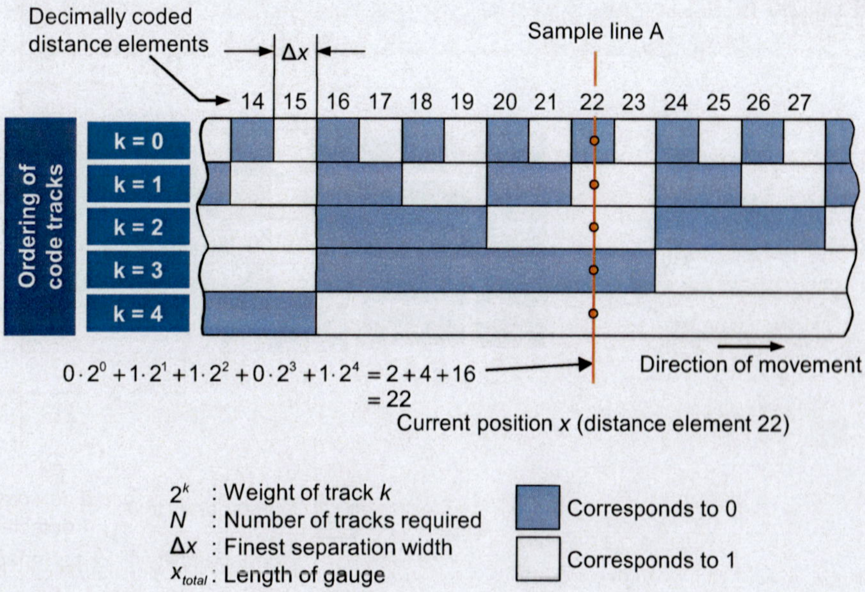


**Fig. 2.54** Binary-encoded gauge with 5 tracks

$$N = \left\lceil \lg\left(\frac{x_{total}}{\Delta x}\right) \cdot \frac{1}{\lg 2} \right\rceil \qquad (2.84)$$

Sampling the gauge using the method described above can also lead to errors, however. The actual positions of the sampling elements can, for example, differ from their reference positions due to precision thresholds in production. During a transition from one distance element to the next, it might be the case that some sampling elements still read the old values and some read the new values. At the transition point between distance elements 23 and 24 for example, a slight skew in the sampling elements of tracks $k=3$ and 4 means that, instead of the correct value being read,

$$10111 \triangleq 23 \text{ oder } 11000 \triangleq 24$$

the incorrect value is read

$$11111 \triangleq 31$$

These measurement errors are particular serious when the higher-value tracks are sampled incorrectly. Such misinterpretations can be prevented like this:

- Sampling every track (except the finest one) of a dual-encoded gauge using sensor heads at gap $2^k \cdot \tau/4$ (Fig. 2.55). The principle is that every time value "0" is read in the lower-value track, the value of the leading sampling element in the next higher-value track is valid. If a value of "1" is read, the lagging sampling element of the higher-value track is valid. In this way, the measured value is clearly read. This process is called "V-sampling" due to the arrangement of the sensor heads [DIN88, MILL90]. The high number of sensor heads required for this solution entails high cost, however.
- Encoding the gauge using the Gray code (refer to ▶ Chap. 9, "Basics of information processing"). During a transition from one value to another, the binary state only changes for this encoding method in one track of the gauge.

Figure 2.56 shows the main setup of a digital-absolute rotary encoder for measuring angles. The drive shaft carries a circular code disk made of break proof plastic that faces, with surfaces parallel, a stationary counter grating (sampling grating) a short distance away. As light and dark-colored fields, the code disk bears the signal key that is read by an emitting light source and receiver and made available for subsequent processing.

To increase the measuring range, as shown in Fig. 2.56, several groups of sampling elements can be accommodated in one housing and connected to each other with a precision gear unit. The backlash in the gear unit between the sampling elements must be eliminated here for the measurement accuracy to be constant over the entire measuring range. Figure 2.57 shows a device implementation.

### Incremental-absolute measuring systems

A combination of the incremental and absolute measuring methods offers the benefits of both methods, Fig. 2.58 [MILL90]. For incremental-absolute sampling, an absolute track with a pseudo random code (PRC) is applied on a transmitted light gauge, e.g., with divisions of 75 μm (low resolution) and an incremental track, for example, with divisions of 20 μm (high res-

2

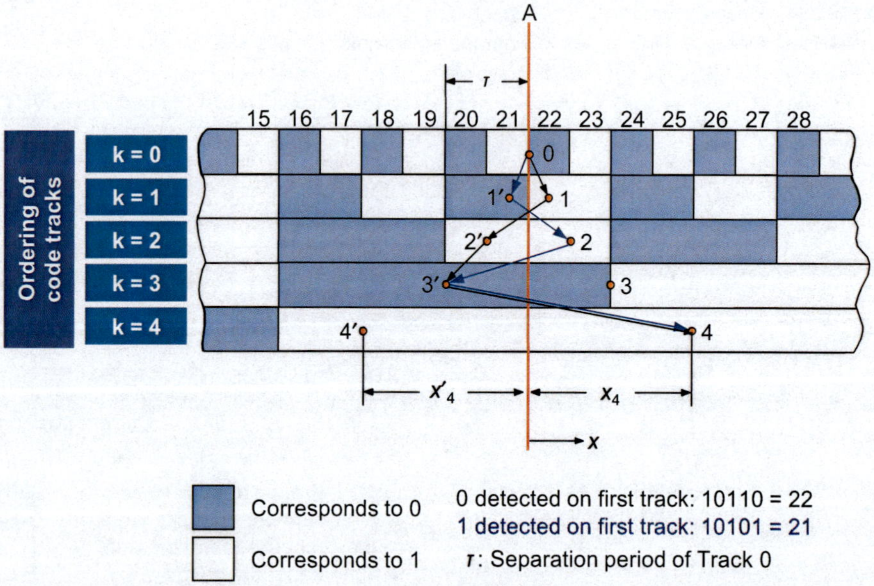


**Fig. 2.55** Dual sampling for a binary-encoded gauge

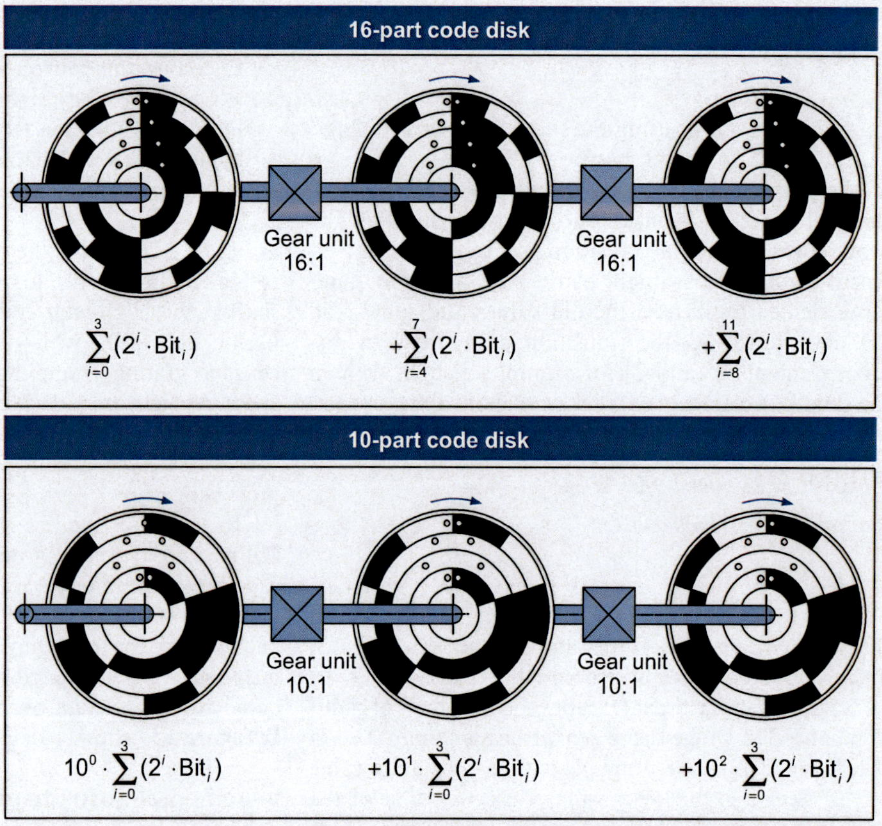


**Fig. 2.56** Principle of a binary-encoded and a binary-decimal encoded angle encoder with a measuring range of more than one revolution according to halstrup-walcher, Kirchzarten

olution). The pseudo random code encodes the absolute positions using a word length of 16 bits. A separate reader is installed for both of the tracks. The tracks are thus read independently of each other.

The absolute track is read with a line sensor comprising a field of pixels (photoelements, e.g., CCD sensors) that is exposed momentarily by the absolute measuring track. The light/dark encoding of the

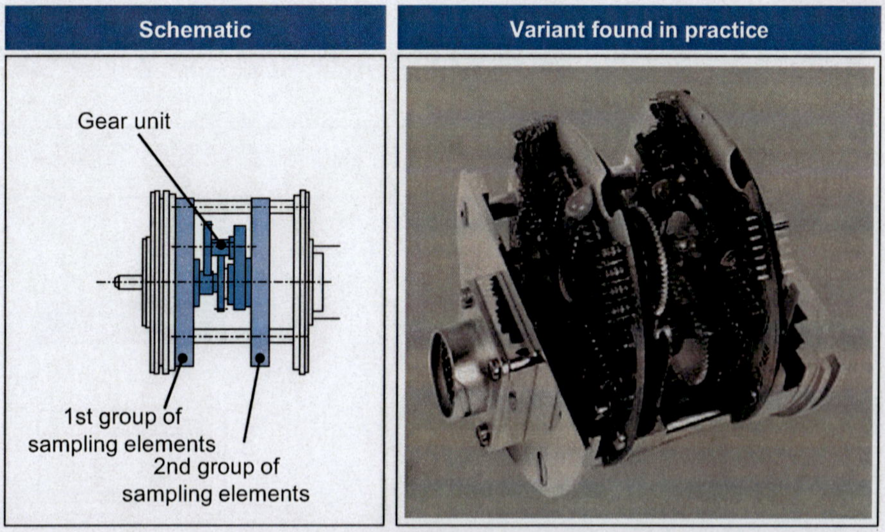


**Fig. 2.57** Rotary encoder with two groups of sampling elements and interconnected gearboxes to increase the resolution for every revolution according to POSITAL FRABA

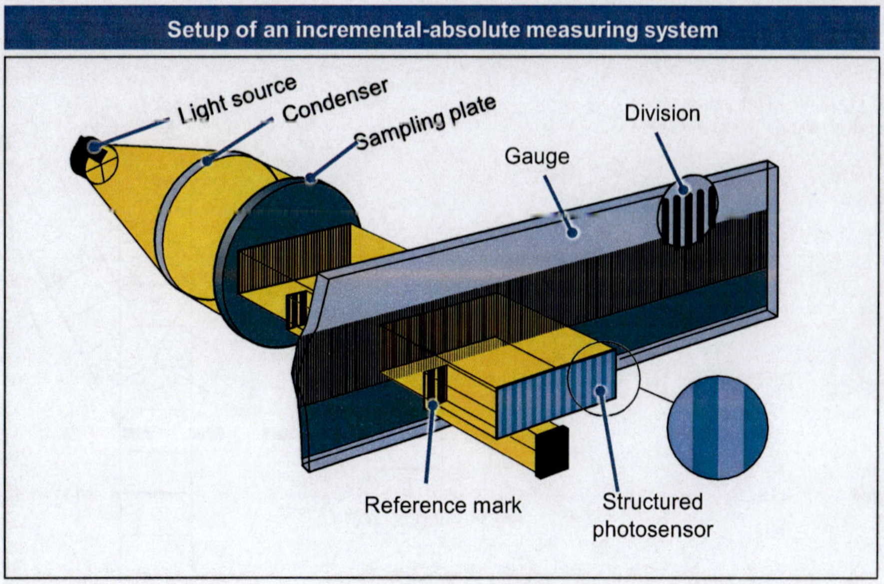


**Fig. 2.58** Setup of an incremental-absolute measuring system according to HEIDENHAIN

pattern stored is saved like on a photo, in the form of charges in the pixels of the line sensor. At the same time, the incremental track is interpreted with 4-field sampling to determine the exact position with high resolution.

A microcontroller then determines the absolute position by decoding the pattern from the pixel image. The positions of the individual bright/dark transitions are analyzed to increase the accuracy. It is possible from the absolute track to determine with sufficient accuracy the position so as to determine the ordinal number of the incremental period. In the last step, the incremental signals are interpolated and added to the absolute position. This makes it possible to calculate the position with a maximum resolution of 0.1 μm. The method in Fig. 2.59 is applied to prevent read errors due to dirt on the absolute track. For this, the PRC is read at four positions having the same spacing. As the four values must have exactly the same spacing, erroneous values can be detected easily and excluded from the evaluation.

### Interferential path measuring systems

For use in ultra-precision machines and in high-accuracy systems for microchip production, there are length measuring systems having resolutions down to a few nanometers. These systems make use of the superposition of light wave trains that are diffracted at optical gratings [ERNS98, HEID00].

2

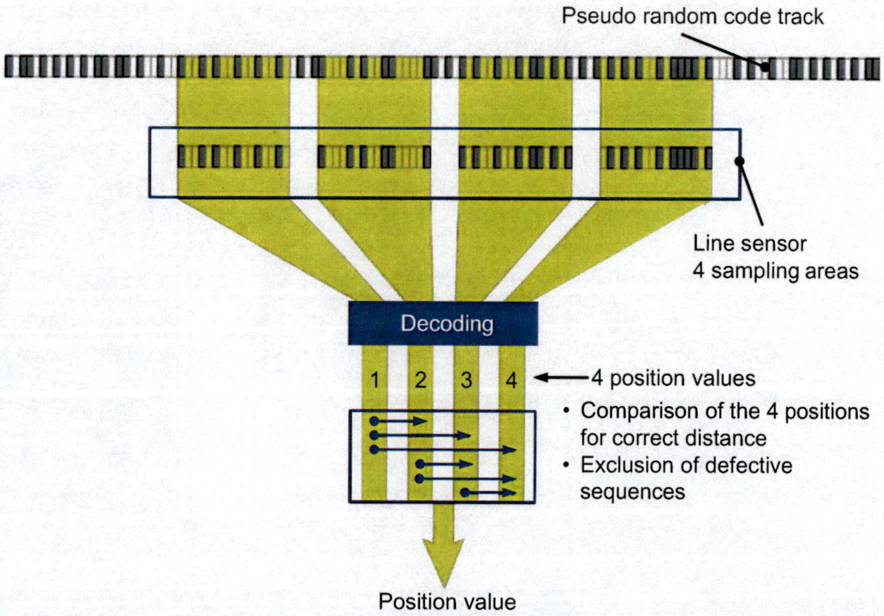


**Fig. 2.59** Increasing the level of error protection by measuring four absolute positions according to HEIDENHAIN

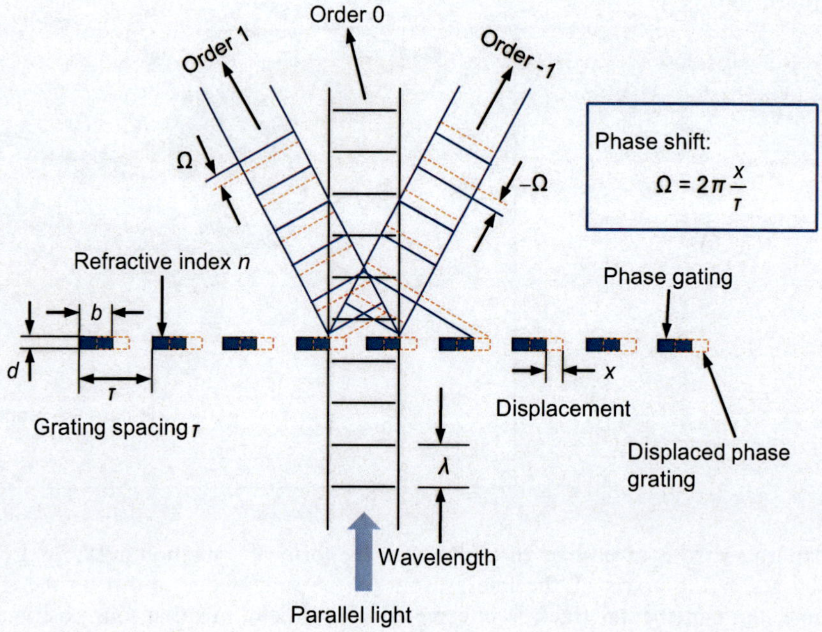


**Fig. 2.60** Diffraction at optical grating (orders −1, 0 and + 1)

When a flat light wave front hits an optical phase grating, the light is diffracted in different directions. Figure 2.60 shows such a diffraction phenomenon. Only the -1, 0 and + 1 diffraction orders are shown here and further on because only these are usually used for the technical realization of length measuring systems. The intensity distribution and phase position of the wavefront behind the phase grating are dependent on grating width $b$, thickness of the webs $d$, refractive index $n$ of the grating rod material and grating spacing $\tau$. An exact intensity distribution can be calculated using the Fourier transform [SESS91].

Neglecting losses and reflections, the intensity of the order 0 diffraction tends to 0 provided the grating webs and openings are equally wide and the waves passing through the translucent webs have a path difference of $\lambda/2$ compared to waves passing through the opening. Moving the grating by distance $x$ changes

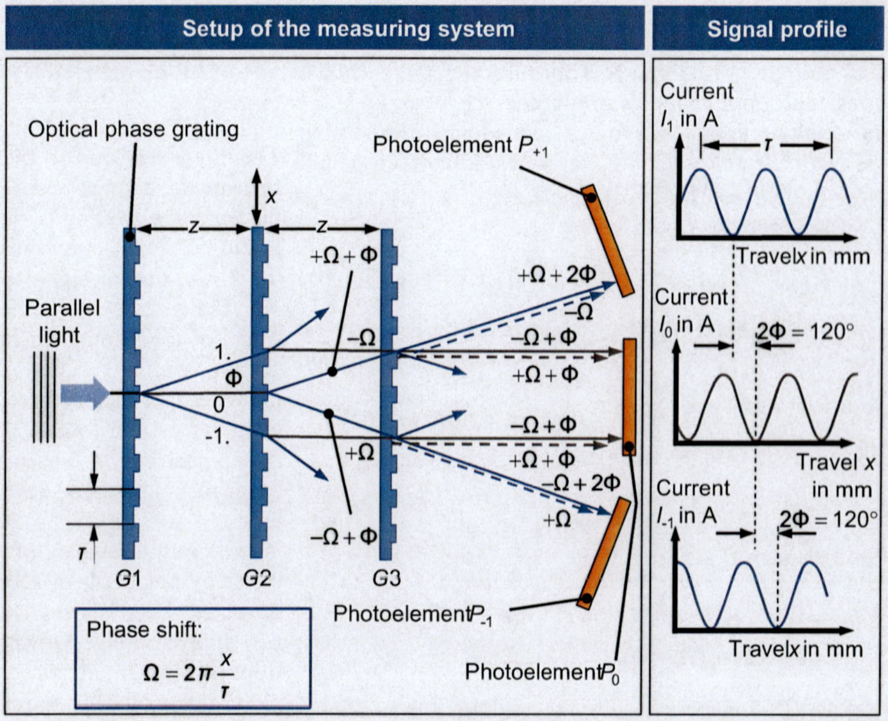


**Fig. 2.61** Operation of interferential path measuring systems according to HEIDENHAIN

the phase position of the diffracted waves of orders -1 and + 1 by angle $-\Omega$ or $+\Omega$. Order 0 waves are not phase-shifted (Fig. 2.60).

Figure 2.61 shows three phase gratings positioned in a row at equal distance $z$. All three gratings have divisions of $\tau$. The gratings $G1$ and $G3$ are identical. They are selected such that diffraction order 0 has constant phase shift $\Phi$ in relation to orders $-1$ and $+1$. Grating $G2$ can be move with respect to $G1$ and $G3$.

Figure 2.61 shows, in simplified form, only the propagation directions of the waves as rays. If a beam of parallel light waves hits grating $G1$, it is diffracted into orders $-1$, 0 and $+1$. The 0 order waves are offset in relation to the 1 order by constant angle $\Phi$.

At grating $G2$, the light beams impacting there are diffracted again to orders -1 and + 1. Grating $G2$ does not generate any 0 order diffraction. When the gauge is moved, the waves of the individual light beams are also phase-shifted by angle $+\Omega$ or $-\Omega$ (Fig. 2.60).

At grating $G3$, having identical properties to grating $G1$, the light is diffracted again. Individual light beams having different phase positions in relation to each other are now superimposed behind this grating. These phase positions are dependent on the displacement of grating $G2$ in the direction of the X-axis. Interferences result behind grating $G3$. These can be measured as intensity fluctuations as a function of travel distance x. The currents ($I_{-1}$ to $I_{+1}$) measured at photoelements $P_{-1}$ to $P_{+1}$ can be used here in place of light intensities $\hat{I}_{-1}$ to $\hat{I}_{+1}$ because the two parameters are proportional to each other. The relationship between travel distance $x$, the currents measured ($I_{-1}$ to $I_{+1}$) and their phase positions is shown in the graphs on the right side of Fig. 2.61.

As an electromagnetic wave, light can be defined by its electrical field strength $E$ and its magnetic field strength $B$. In the following calculations, the light waves are defined in a simplified form (ideal, plain wave) by their electrical field strength $E$ as a cosine function with phase shift $\varphi$, frequency $f$ and angular frequency $\omega = 2\pi f$.

$$E = A\cos(\omega t + \varphi) \tag{2.85}$$

To simplify the calculations with the angle functions, the oscillation is expressed in complex notation, its real part corresponds to the oscillation observed. In the following, complex numbers ($\underline{Z}$) are always expressed with an underline.

$$\begin{aligned} \underline{E} &= A(\cos(\omega t + \varphi) + i\sin(\omega t + \varphi)) \\ &= A e^{i(\omega t + \varphi)} \qquad (\textit{euler equation}) \\ &= A e^{i\varphi} e^{i\omega t} \\ \textit{with } &\underline{E} \in E = \mathrm{Re}(\underline{E}) \end{aligned} \tag{2.86}$$

The component $\underline{V} = Ae^{i\varphi}$ is called the complex amplitude (phasor) of $\underline{E}$.

The amplitudes of the light waves (in ◘ Fig. 2.61) of the individual, non-interfering light beams are identical and are normalized to 1 for all following calculations. The three complex amplitudes ($\underline{V}_{-1}$ to $\underline{V}_{+1}$) of the resulting light waves are calculated each time from the two superimposed light waves by adding their complex amplitudes [MENZ73].

Photoelement $P_{+1}$:

$$\underline{V}_{+1} = 1 \cdot e^{-i\Omega} + 1 \cdot e^{i(2\Phi+\Omega)} = e^{i\Phi} \cdot \left(e^{-i(\Omega+\Phi)} + e^{i(\Omega+\Phi)}\right) = e^{i\Phi} \cdot 2\cos(\Omega + \Phi) \tag{2.87}$$

Photoelement $P_0$:

$$\underline{V}_0 = 1 \cdot e^{i(\Phi-\Omega)} + 1 \cdot e^{i(\Phi+\Omega)} = e^{i\Phi} \cdot \left(e^{-i\Omega} + e^{i\Omega}\right) = e^{i\Phi} \cdot 2\cos\Omega \tag{2.88}$$

Photoelement $P_{-1}$:

$$\underline{V}_{-1} = 1 \cdot e^{i\Omega} + 1 \cdot e^{i(2\Phi-\Omega)} = e^{i\Phi} \cdot \left(e^{-i(\Omega-\Phi)} + e^{i(\Omega-\Phi)}\right) = e^{i\Phi} \cdot 2\cos(\Omega - \Phi) \tag{2.89}$$

$|\underline{V}_{-1}|$ to $|\underline{V}_{+1}|$ correspond to the real light wave amplitudes and $|\underline{V}_{-1}|^2$ to $|\underline{V}_{+1}|^2$ the intensities of the light waves. For the calculation, the following generally applicable formula

$$|\underline{C}|^2 = \underline{C} \cdot \underline{C}^* \tag{2.90}$$

is used. $\underline{C}^*$ denotes the conjugated complex number of $\underline{C}$.

Continuing to use this for the angle

$$\Omega = 2\pi x/\tau, \tag{2.91}$$

gives the light intensities $\hat{I}_{-1}$ to $\hat{I}_{+1}$ at the individual photoelements in relation to travel distance $x$ as:

Photoelement $P_{+1}$:

$$\hat{I}_{+1} = V_{+1} \cdot V_{+1}^* = 4\cos^2(\Omega + \Phi) = 2\,[1 + \cos 2(\Omega + \Phi)] = 2\left\{1 + \cos[2\pi\left(\frac{2x}{\tau}\right) + 2\Phi]\right\} \tag{2.92}$$

Photoelement $P_0$:

$$\hat{I}_0 = V_0 \cdot V_0^* = 4\cos^2\Omega = 2\,[1 + \cos 2\Omega] = 2\left\{1 + \cos[2\pi\left(\frac{2x}{\tau}\right)]\right\} \tag{2.93}$$

Photoelement $P_{-1}$:

$$\hat{I}_{-1} = V_{-1} \cdot V_{-1}^* = 4\cos^2(\Omega - \Phi) = 2\,[1 + \cos 2(\Omega - \Phi)] = 2\left\{1 + \cos[2\pi\left(\frac{2x}{\tau}\right) - 2\Phi]\right\} \tag{2.94}$$

The intensities, and so the output currents of the photoelements, change accordingly depending on the displacement of grating $G2$ following a cosine function. If grating $G2$ is offset by the distance of one grating spacing $\tau$, the output signal passes through two full periods (◘ Fig. 2.61). Thus, the period length of the output signal of the photoelements is shorter by factor 2 than grating spacing $\tau$. The light wave length has no bearing here. This means there are also no specific measurement errors due to changes in the refractive index of the air such as those occurring for laser interferometers (refer to Volume 2, ► Sect. 9.7.6.1, "Laser interferometers").

The output signals of the three photoelements are offset by angle $2\Phi$ in relation to each other. Angular offset $2\Phi = 120°$ is set by arranging gratings $G1$ and $G3$ appropriately. Downstream electronics generate from this two square wave pulse sequences again with a phase-offset of $\pm 90°$ in relation to each other for direction detection.

The light intensity fluctuations are generated by the interference of optical waves. So the intensity profile versus travel distance $x$, and thus also the output signal of the photoelements, are almost perfectly sinusoidal. The electrical signals can thus undergo fine interpolation in line with the methods specified in Sect. 2.3.3. It is possible to interpolate the analog signals of the photoelements by a factor of 100. So, together with the 4 x analysis of the digitalized signals, a resolution can be attained that is finer than the signal period by a factor of 400, or finer than the grating spacing $\tau$ by a factor of 800. This is how measurement increments with divisions under 10 nm can be realized. Resolutions of even 1 nm are attained with technically modified systems [SPIE90].

The function and measurement accuracy are not impacted when spacing $z$ between the gratings changes within a tolerance range. This property is useful, in the practical application of measuring systems in particular, because production and installation tolerances always need to be factored in. It is not until there are relatively large deviations, when the tolerance range is exceeded, that there is no more superposition behind grating $G3$ due to the associated geometric displacement of the beam of rays or the interfering light beam is no longer projected onto the photoelements.

In the technical implementation of interferential length measuring systems, the arrangement shown in ◘ Fig. 2.61 comprising three gratings is folded around the plane by center grating $G2$. This means identical gratings $G1$ and $G3$ coincide physically. For grating $G2$, a grating with reflecting webs and recesses is used. ◘ Figure 2.62 shows such an arrangement.

Since the principle also works with non-ideal monochromatic light, an LED is used as the light source. A condenser lens is used to generate a parallel light beam. This passes through the transmitted light phase grating, is reflected on the gauge grating and passes again through the sampling grating. The corresponding refractions occur at the individual grating. The interfering light beams are mapped onto the photoelements by the condenser lens.

The gauge is made of steel. As a result, the accuracy of the measuring system is only dependent on the temperature expansion of the steel and this is simple to compensate mathematically. The reflecting gauge grating is formed from gold layers applied with high precision. The grating of the sampling plate is etched in glass. The air gap between the phase grating and the gauge is about 1 mm, for a tolerance of approx. ±0.5 mm.

Their specific attributes (straightforward setup, high resolution, high accuracy, easy to compensate measurement errors and broad measurement tolerances in the installation process) make interferential length measuring systems favored alternatives to laser interferometers.

### 2.3.2.2 Interferometers

Interferometers are used for very high precision in path measurement. ◘ Figure 2.63 shows the function of a Michelson interferometer. A monochromatic light beam hits half-silvered mirror $S_1$ and splits. One part of the light beam hits mirror $S_3$, the distance of this is to be measured. The other part hits reference mirror $S_2$ that stays at the same position. The reflections cause both light beams to hit the photodetector. The interference effect causes the light beams to cancel out or intensify each other, depending on the distance.

In practice, homodyne as well as heterodyne laser interferometers are used because they also determine the direction of motion. Both systems use polarized light. Homodyne laser interferometers work with one frequency. Their heterodyne counterparts, on the other hand, use two close frequencies (refer to Volume 2, ► Sect. 9.7.6.1, "Laser interferometers").

### 2.3.2.3 Electromagnetic sensors

Electromagnetic measuring equipment is based on the induction principle (◘ Fig. 2.64). A conductor $a$ carrying an alternating current $i_E(t)$ generates an alternating field, known as magnetic flux $\Phi$. The magnetic field lines induce, in a second conductor $b$, voltage $v_A(t)$ whose value is dependent, among other factors, on the distance between the conductors and their position in relation to each other. When conductor $b$ moves by $\Delta x$, the effective value of the induced voltage changes by absolute value $\Delta v_A$. This amount is a measure of the distance change.

Two alternating voltages with a phase offset of 90° are often induced by a conductor carrying an alternating current in offset conductor loops. When there is movement, the positions of the induction windings mean the amplitudes of the voltages relative to each other change so that their evaluation can provide information on distance. Inductosyn and resolvers are the most well-known measuring systems using the induction principle.

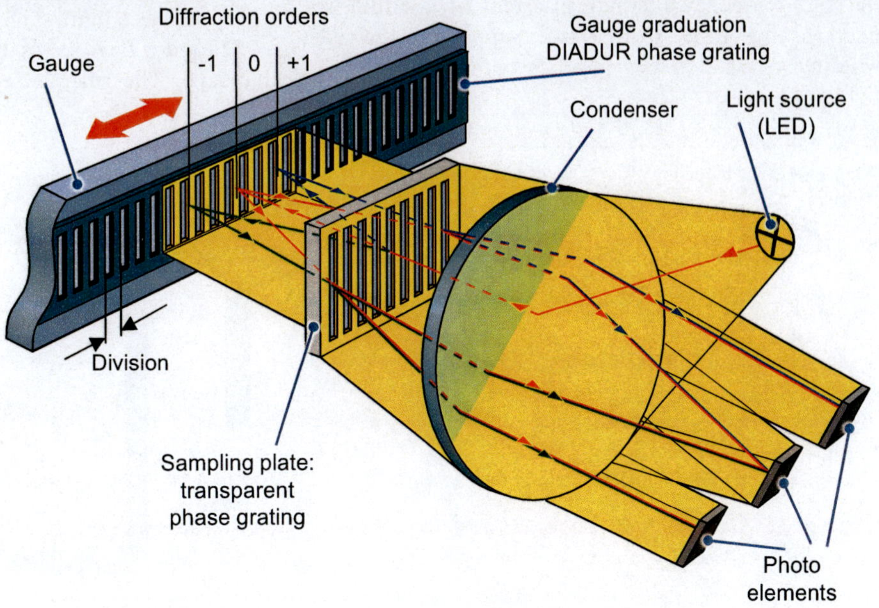


◘ **Fig. 2.62** Structure of an interferential measuring system according to HEIDENHAIN

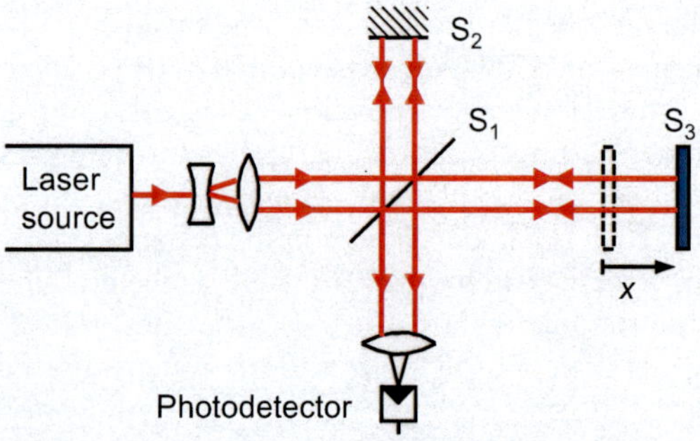


Fig. 2.63 Principle of the Michelson interferometer

## Inductosyn

There are two Inductosyn variants. The linear Inductosyn is used for length measurements and the rotary Inductosyn for angle measurements. Both systems are cyclic-absolute measuring systems working on the same physical principle. They only differ in mechanical design.

The components of a linear Inductosyn, the ruler and rider, contain meander conductor windings that are applied to the base body in the form of a printed circuit (Fig. 2.65). So that the induced magnetic field is subjected to changes, the base body must not be made of ferromagnetic material. Glass or ceramic is normally selected as the base material. Steel with synthetic material insulation is also used, however. The windings on the ruler and rider have the same separation period. Two winding groups, offset by $(N+1/4)\cdot\tau$, $N=0, 1, 2,\ldots$, i.e., spatially by 90° in relation to each other, are on the rider. This arrangement enables, with phase and amplitude measurements, detection of a direction of motion and precise position analysis (refer to Sect. 2.3.3.3).

High-frequency alternating current $i_E(t)$, within frequency range 1 kHz to 10 kHz, is supplied to the ruler winding.

$$i_E(t) = \hat{i}_E \sin \omega t \tag{2.95}$$

This causes an alternating magnetic field that induces alternating voltages $v_{A1}(t)$ and $v_{A2}(t)$ in the moving rider windings. Figure 2.66 shows the spatial arrangement of the ruler and rider, and the curves of the a.c. voltages induced in the rider. The instantaneous amplitude or phase positions of $v_{A1}(t)$ and $v_{A2}(t)$ are dependent on the position of the rider.

The spatial offset of the two rider windings by $(n+1/4)\cdot\tau=\pi/2$ (electric) means amplitudes $A_1$ and $A_2$ of induced a.c. voltages $v_{A1}(t)$ and $v_{A2}(t)$ are proportional to the cosine and sine of electrical rotation angle $\alpha$ (see Fig. 2.65):

$$A_1 = K \cdot \hat{i}_E \cdot \cos\alpha \tag{2.96}$$

$$A_2 = K \cdot \hat{i}_E \cdot \sin\alpha \tag{2.97}$$

$$v_{A1}(t) = K \cdot \hat{i}_E \cdot \sin\omega t \cdot \cos\alpha \tag{2.98}$$

$$v_{A2}(t) = K \cdot \hat{i}_E \cdot \sin\omega t \cdot \sin\alpha \tag{2.99}$$

$$\alpha = \frac{2\pi x}{\tau} \text{ for } 0 \le x \le \tau \tag{2.100}$$

where $K$ is the electromagnetic coupling factor.

For high-frequency voltage profiles, a displacement of the rider on the ruler results in phase jumps of 180° where the envelope curves cross zero. The rider position is uniquely determined by evaluating phase $\alpha$ versus the curve of the enveloping voltage profiles in the rider windings and using the phase relationship (in-phase or phase opposition) of the high-frequency, induced signals. The amplitude analysis method described in Sect. 2.3.3.3 is used to convert rider voltages into digital pulse sequences that are processed in the NC control system as counting increments.

Figure 2.67 shows the structure of a linear Inductosyn. The ruler is secured to the stationary ma-

Fig. 2.64 The induction principle

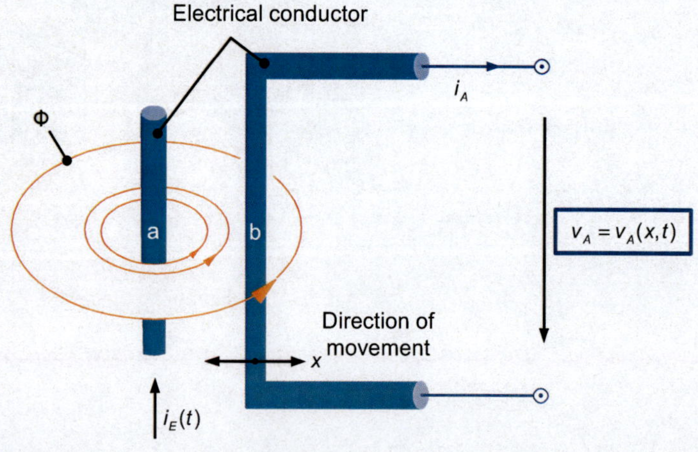

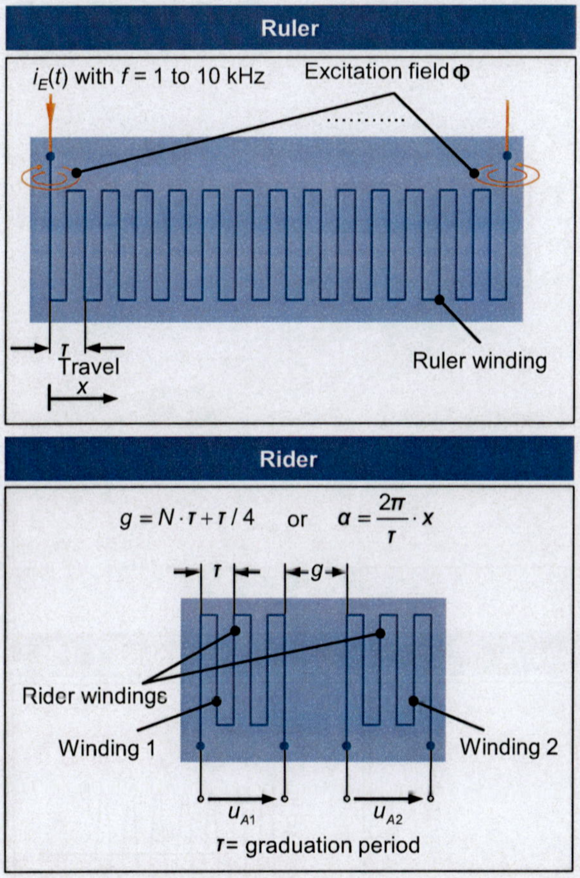


**Fig. 2.65** Winding arrangement of the linear Inductors

chine component (such as the bed). The rider is on the moving machine component (such as the carriage or table). A relatively small air gap (about 0.25 mm) must be maintained between the two components.

Individual Inductosyn rulers generally have divisions of 2 mm or 1/10″ and lengths of 250 to 1000 mm. For larger distances, several ruler elements are lined up together. Careful adjustment is required to prevent errors at the joints.

## Resolvers

A resolver is an electromagnetic angle measuring sensor that can also be used, for example, as a path measuring system in conjunction with a ball screw. In terms of the structure, a resolver resembles a small electric motor with precision windings. The stator carries two winding groups. The planes of these groups are one above the other vertically. This means they are physically rotated 90° with respect to each other. The basic arrangement of the windings is shown in Fig. 2.68.

High-frequency alternating current $i_E(t)$ (about 10 kHz) is supplied to the rotor winding, causing alternating magnetic field Φ. This field, in turn, induces alternating voltages $v_{A1}(t)$ and $v_{A2}(t)$ in the stator windings. These voltages are proportional to the cosine and sine of rotation angle $\alpha$ (as the vector diagram shows). The voltage profiles induced in the stator windings are evaluated in the same way as for the Inductosyn measuring system [HEWL76, WILD59]. Angle $\alpha$ determined during this evaluation corresponds to the rotation angle of the resolver rotor. The periodicity of the signal profile only enables unique measured value as-

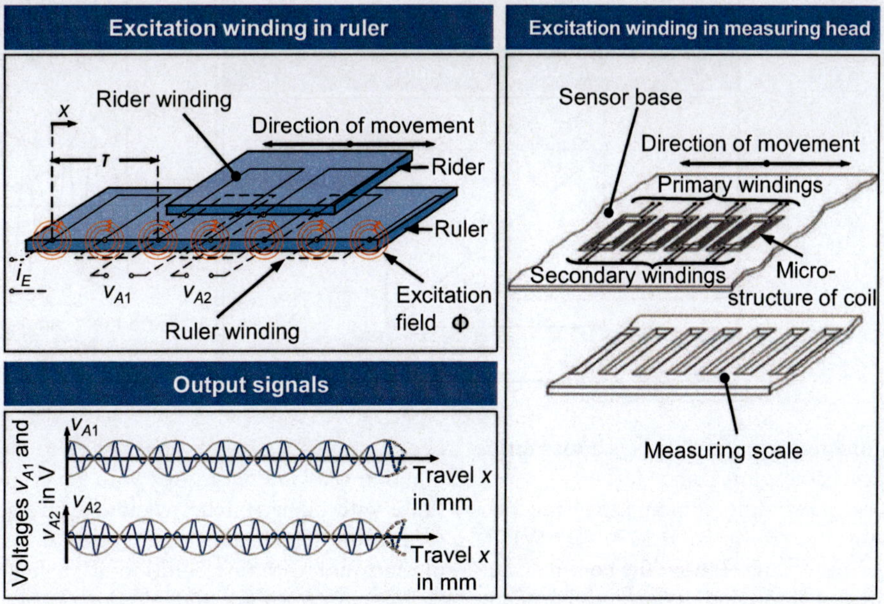


**Fig. 2.66** Basic diagram and function of the linear Inductors according to AMO Messtechnik

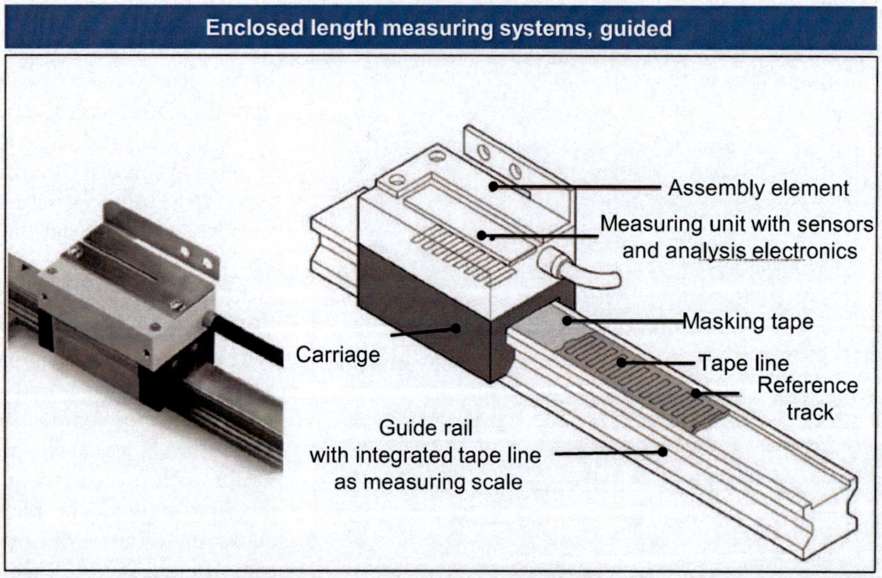


**Fig. 2.67** Structure of the linear Inductosyn according to AMO Messtechnik

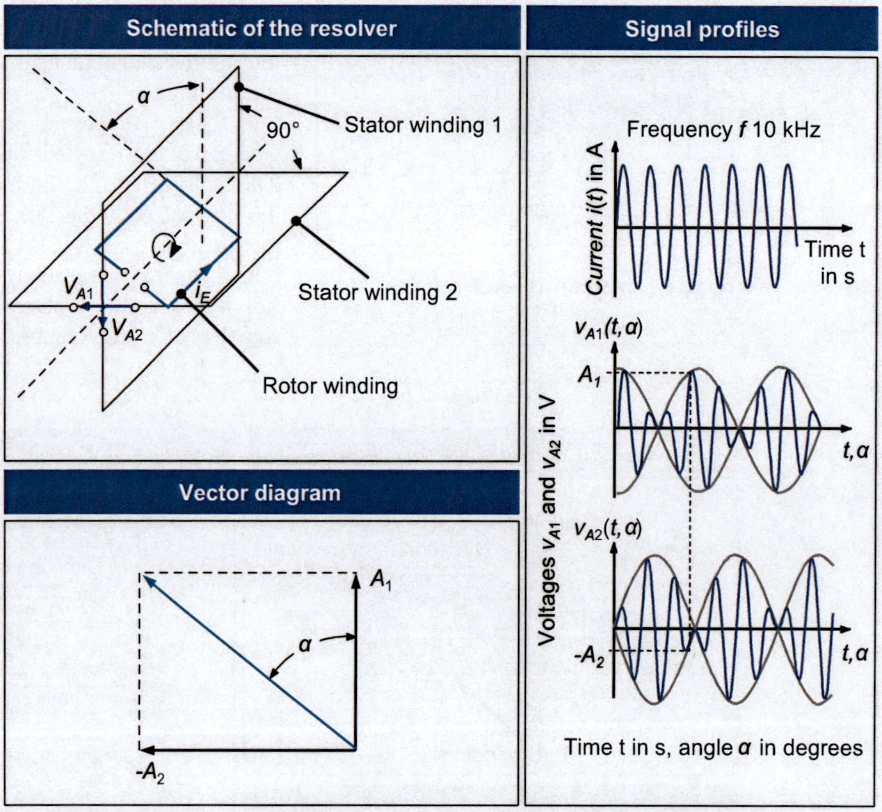


**Fig. 2.68** Functional principle of a resolver

signment within one rotor spacing. This is a measuring system using the cyclic-absolute principle.

Different types of winding combinations are possible in the stator and rotor [AULM70, HEWL76, WILD59]. To increase the resolution, the complete stator winding is divided into winding groups. The number of groups is given by pole pitch $p$. Using the combination of winding groups with pole pitch 1, and of those with higher pitches, yields resolvers that are able to measure the absolute angle position within one revolution and with a resolution of a few arc seconds. Figure 2.69 shows the physical structure of a revolver. As with most resolvers, the rotor winding here is also supplied by a rotary transformer without a commutator ring. Hollow shaft variants are widespread for direct placement onto a shaft.

Compact design, high accuracy and extreme robustness are particular benefits of the resolver measuring system. They are commonly used in military and aerospace applications. In machine tools, resolvers have been broadly displaced by optical rotary encoders. They are, however, being integrated increasingly in speed-controlled synchronous and asynchronous motors. In such applications, they measure the angular position of the rotor to ensure stator field control has the correct phase relationship.

### 2.3.2.4 Magnetic sensors

In magnetic path measuring systems, the material measure is formed by the periodic division of the gauge into north and south poles. The gauge comprises a hard magnetic material with high coercive field strength and with permanent magnetization. The magnetic field is scanned by magneto-resistive sensors. The resistance of such sensors change under the influence of a magnetic field. One scanning unit contains four sensors connected to form a Wheatstone bridge (refer to Volume 2, ► Sect. 9.7.5 "Inductive sensors"). The sensors are offset with respect to each other as in the arrangement shown in ◘ Fig. 2.70. This causes bridge output voltage $v_A(x)$ to fluctuate sinusoid ally around the zero line over travel distance $x$. Another scanning unit, offset with respect to the first by $N\tau+\tau/4$ ($N=1$, 2, 3, …), generates the second output signal shifted by 90°. Both signals serve as input parameters for the interpolation unit.

Magnetic path measuring systems are not susceptible to impurities caused by oils and greases. Their compact volume also means they lend themselves to direct installation into the rail of a rolling element guide system (◘ Fig. 2.71). Any clinging magnetically conductive metal particles are removed by chip scrapers, that constitute part of the guide unit anyway. Together with an interpolation unit, the measuring systems offer a resolution of usually 1 µm, sufficient for machine tools. To determine the absolute measuring position, there is a reference mark on the gauge.

A magnetized synthetic strip can also be used as a gauge. This strip is protected from mechanical influences by a non-magnetic metal strip. Magnetic path measuring systems are affordable because the synthetic strip is low-priced and installation is simple. One drawback of these measuring systems is their low resolution (about 10 µm). Their use is thus preferred in woodworking machines, in guillotine shears for section steel partitioning [SIMO71] and for handling tasks.

Another rotary variant of a magnetic measuring system is described in ◘ Fig. 2.72. A disk made of a ferromagnetic material (such as steel) has fine, precision toothing around its circumference. The scanning elements are arranged at a short distance (0.2 to 0.5 mm) from the crown circle diameter of the sprocket. The scanning elements generate a permanent magnetic field that also permeates the teeth of the index plate. The magnetic flux changes depending on whether there is a tooth or a gap in front of the scanning element. The magnetic flux is detected by magneto-resistive sensors and converted to an analog, approximately sinusoidal voltage signal [HEID89, HEWL92]. A second scanning element, offset by $N\tau+\tau/4$, generates a signal phase-shifted by 90° allowing detection of forwards and backwards. To increase the resolution, the output signals are fed into an interpolation unit (see Sect. 2.3.3).

Standard modules (e.g., $m=0.3$ or $m=0.5$) are used for the teeth spacing. Sprockets with diameters from 77

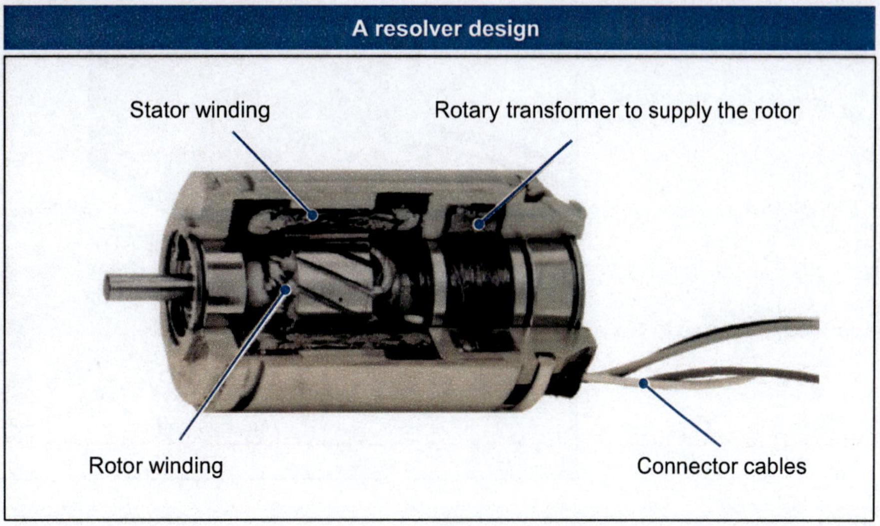


◘ **Fig. 2.69** Structure of a resolver. *Source:* SIEMENS

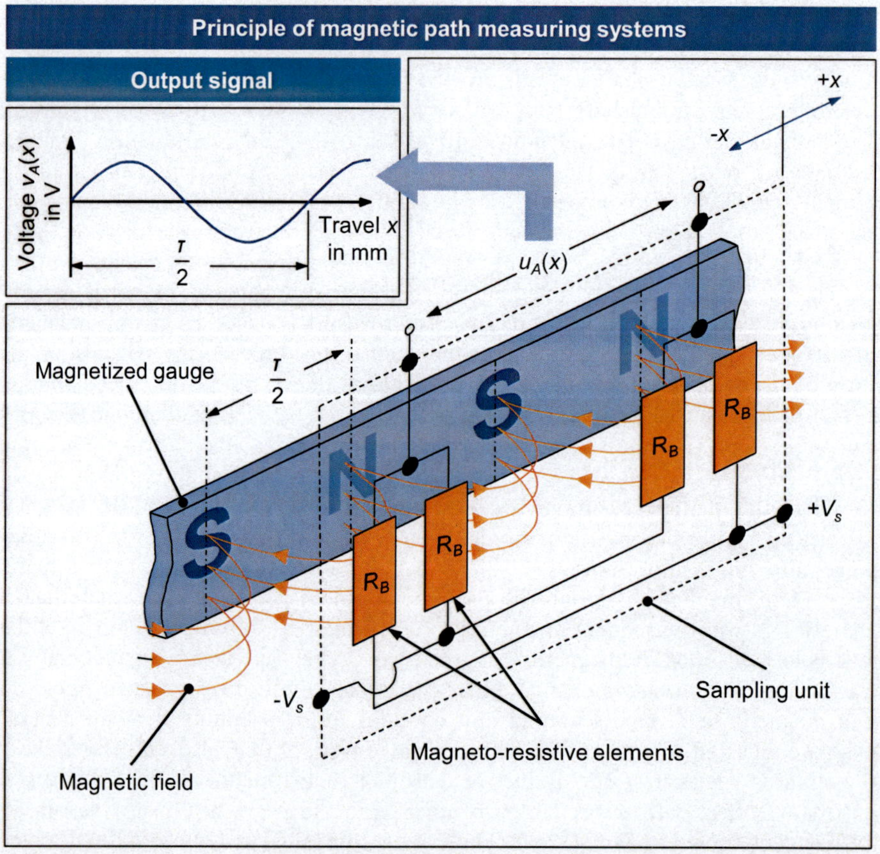


**Fig. 2.70** Principle of magnetic path measuring systems according to HEIDENHAIN

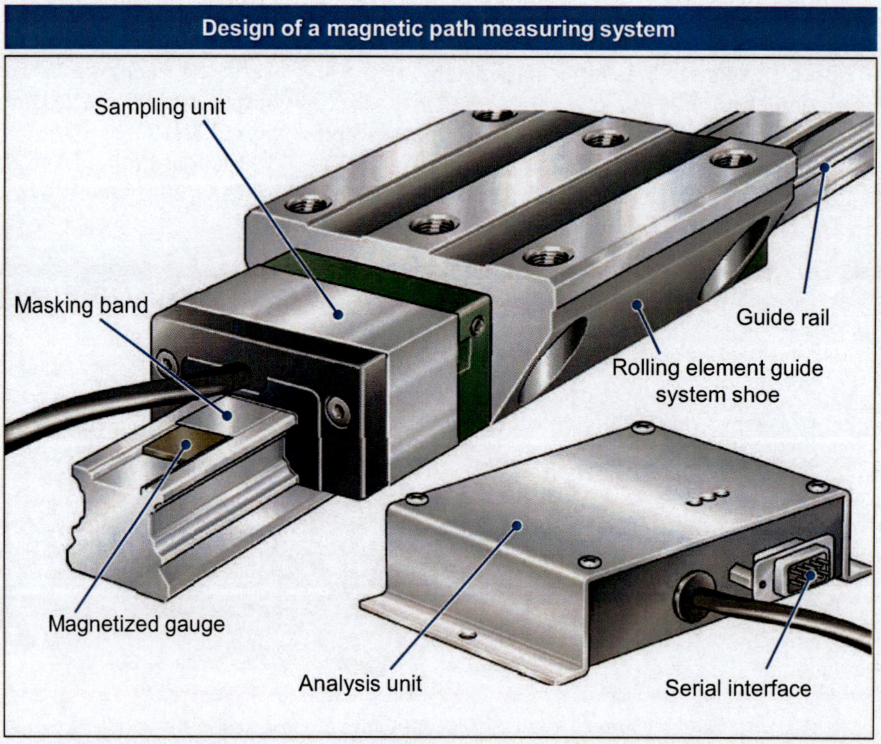


**Fig. 2.71** Structure of a magnetic path measuring system. *Source:* Schaeffler Technologies AG & Co. KG

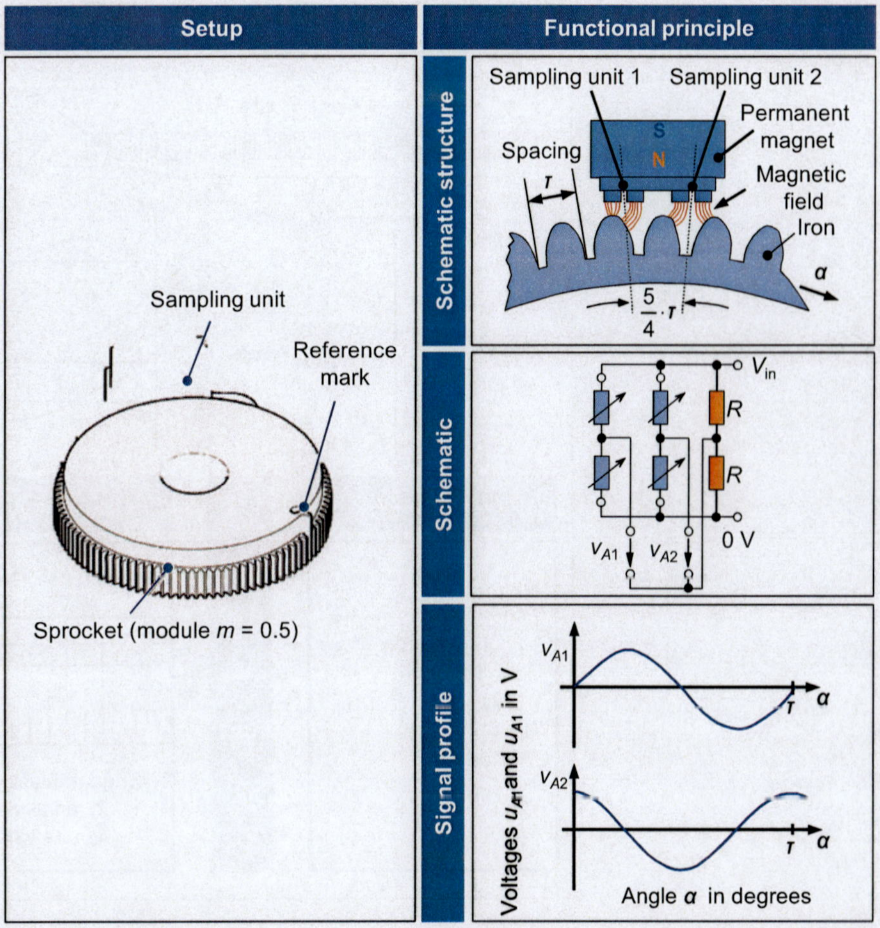


**Fig. 2.72** Scanning teeth spacing using magneto-resistive sensors according to SIEMENS, Lenord + Bauer

to 257 mm, having 256 or 512 teeth, are available. The angle can be measured to a precision of 1/100°. A resolution of up to 18,000 measurement increments per revolution can be attained depending on interpolation.

The magnetic measuring system is very sensitive to mechanical damage to the teeth spacing, but very tolerant of dirt caused by oil and grease. The preference for measuring the angle position of the main spindle for lathes, therefore, is to use an NC-controlled C-axis without any further encapsulation. Previously, optical rotary encoders were used. These were fitted next to the spindle and driven by a toothed belt. The accuracy of the angle measurement using the system shown in Fig. 2.72 is far higher due to the lack of transmission errors of the toothed belt drive. Fitting the sprocket on the spindle is critical, however. Low concentricity and runout tolerances are required for the system to function precisely.

### 2.3.3 Interpolation methods and direction detection

The output values of distance and angle measuring systems are generally available in the form of periodic, analog signals with amplitudes that change sinusoidally with travel distance. The period length is dependent on the graduation of the measuring system. The period length is generally too approximate, meaning just counting does not render the resolution required for machine tools. Travel distances are usually measured in measurement increments of 1 μm. For ultra-precision machines, resolutions of even a few nanometers are required. The requirement, depending on the type of the analog output signals, is to apply interpolation methods (Fig. 2.73). Interpolation divides the period of the measuring signal into integer multiples. The output signal of the interpolation unit comprises two rectangle sequences that are phase-offset by 90° in relation to each other. The leading sign of the phase shift of the signal profiles in relation to each other indicates the direction of motion. The resolution can be increased by counting the edges of a rectangle sequence (2 × analy-

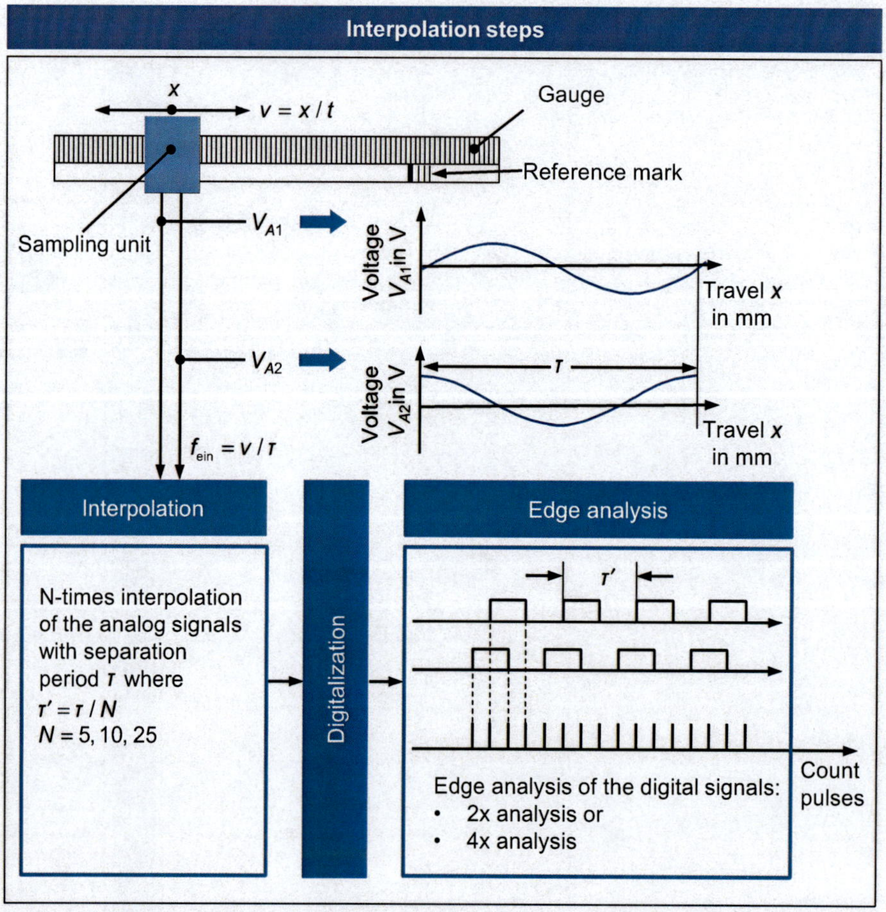


**Fig. 2.73** Interpolation steps and methods for increasing the resolution

sis) or both sequences (4 × analysis) within graduation period τ' of the digital signal.

Input frequency $f_{in}$ of the interpolation unit is calculated as in Fig. 2.73 using:

$$f_{in} = \frac{v}{\tau} \tag{2.101}$$

where

$v$ – is the travel speed
$\tau$ – is the graduation period

The input frequency of the interpolation unit is limited. Increasing the resolution often requires a reduction in the maximum possible travel speed.

#### 2.3.3.1 Interpolation with auxiliary phases

If the output signals of a measuring system are available in the form of two sinusoidal voltages offset by 90°, the interpolation method can be applied with auxiliary phases or frequency multiplication [MILL90]. The two voltages follow the curves

$$V_{A1} = k \cdot \cos\alpha \tag{2.102}$$

$$V_{A2} = k \cdot \sin\alpha \tag{2.103}$$

where

$$\alpha = \frac{2\pi x}{\tau}$$

$x$ = distance or angle traveled within one graduation
$\tau$ = graduation period

If the two voltages are superimposed using the resistor network in Fig. 2.76 as

$$V_n = A_n V_{A1} + B_n V_{A2} = A_n \cdot k \cdot \cos\alpha + B_n \cdot k \cdot \sin\alpha \tag{2.104}$$

, signal $V_n$ results. This is phase-shifted by $\lambda_n$ in relation to $V_{A1}$ depending on how large $A_n$ and $B_n$ are chosen.

$$V_n = \sqrt{A_n^2 + B_n^2} \cdot k \cdot \cos(\alpha - \lambda_n) \tag{2.105}$$

**Fig. 2.74** Adding element in detail

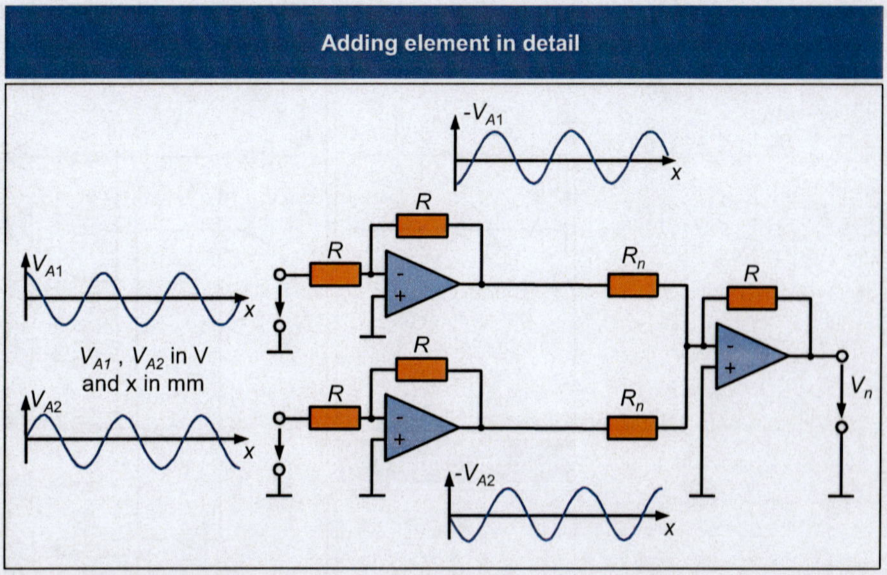


For the calculation of values $A_n$ and $B_n$ for specified phase shift $\lambda_n$, an adding element of the resistor network is shown in detail in Fig. 2.74.

Output voltage $U_n$ in Fig. 2.74 is then calculated as in Eq. (2.106).

$$\begin{aligned} V_n &= -R \cdot \left(\frac{-V_{A1}}{R_n} + \frac{-V_{A2}}{R'_n}\right) \\ &= \frac{R}{R_n} V_{A1} + \frac{R}{R'_n} V_{A2} \\ &= \frac{R}{R_n} k \cos(\alpha) + \frac{R}{R'_n} k \sin(\alpha) \end{aligned} \tag{2.106}$$

As the phase shift is specified by $\lambda_n = n \cdot 18°$ ($n = 1, 2, 3,$, etc.) and the amplitudes of the input and output signals are identical, the following applies:

$$V_n = k \cos(\alpha - \lambda_n),$$

with the addition theorem for trigonometrical functions, this leads to

$$V_n = k \cos(\lambda_n)\ \cos(\alpha) + k \sin(\lambda_n)\ \sin(\alpha)$$

and a coefficient comparison with (2.104) and (2.106) yields

$$A_n = \cos(\lambda_n) = \frac{R}{R_n} \Leftrightarrow \boxed{R_n = \frac{R}{\cos(\lambda_n)}} \tag{2.107}$$

$$B_n = \sin(\lambda_n) = \frac{R}{R'_n} \Leftrightarrow \boxed{R'_n = \frac{R}{\sin(\lambda_n)}}. \tag{2.108}$$

The following then applies to the phase shift

$$\lambda_n = \arctan\frac{B_n}{A_n} = \arctan\frac{R_n}{R'_n} = \arctan\frac{R/\cos\lambda_n}{R/\sin\lambda_n}. \tag{2.109}$$

From the two sinusoidal voltages of the photoelements, the circuit in the middle of Fig. 2.75 generates a number of phase-shifted output signals ($V_n$) that are shown in the bottom left of Fig. 2.75 as a vector diagram. The X and Y components of the voltage of an output correspond to the voltages at its upstream resistors (trigonometrical Pythagoras).

The zero-crossings of the output signals are interpreted by downstream logic circuitry resulting in two pulse sequences, having a phase offset of 90° to each other. Graduation $\tau'$ of these pulse sequences must be an integer fraction of the graduation $\tau$ of the input signal. Increasing the resolution of a gauge or rotary encoder by a maximum factor of 25 is possible using this interpolation method. The number of measurement increments can be increased again by a factor of 4 by counting every edge of the two pulse sequences. This means it is possible, with sufficient accuracy, to increase the resolution 100 times compared to the gauge spacing.

Voltages $V_{A1}$ and $V_{A2}$ returned by the measuring system are not ideally sinusoidal. They contain harmonics as well as offset and phase errors. These errors mean increasing the resolution further with interpolation is not viable. Also note that interpolation only increases the resolution, not the accuracy of the measuring system.

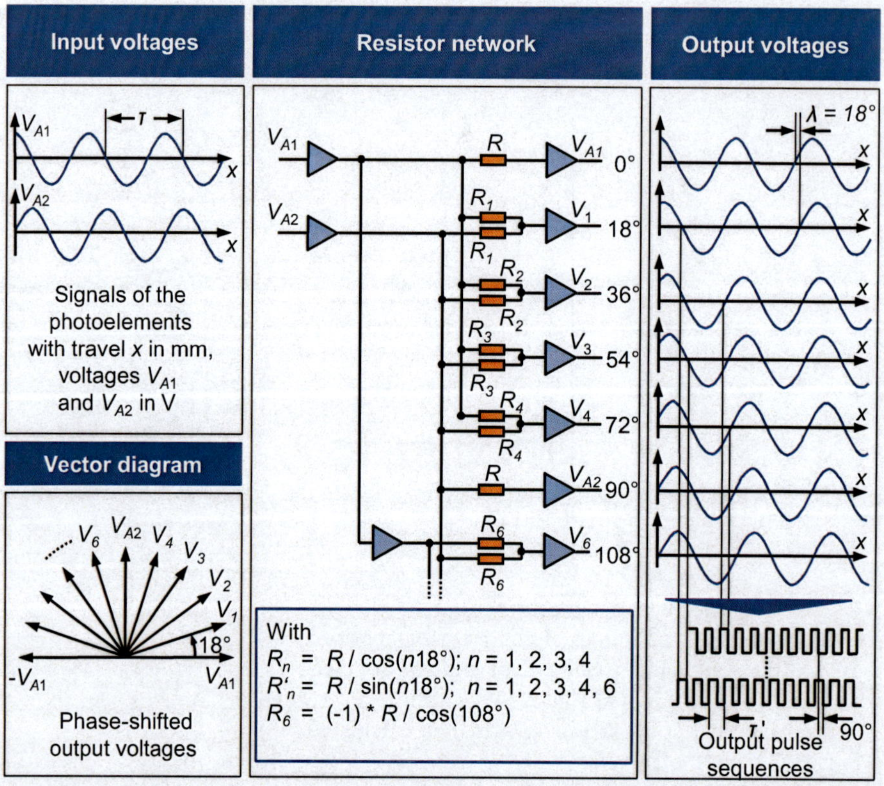


**Fig. 2.75** Principle of interpolation (5x) with auxiliary phases or frequency multiplication according to HEIDENHAIN

### 2.3.3.2 Digital interpolation

Interpolation can also be realized with an arc tangent table in the analysis electronics. If the amplitudes of voltages $V_{A1}$ and $V_{A2}$ are identical, the following apply, where $k$ is a constant (amplitude factor) and $\alpha = 2\pi x/\tau$

$$V_{A1} = k \cdot \cos\alpha$$

$$V_{A2} = k \cdot \sin\alpha$$

$$\frac{V_{A2}}{V_{A1}} = \frac{k \cdot \sin\alpha}{k \cdot \cos\alpha} = \tan\alpha \quad (-\pi/2 < \alpha \le \pi/2) \tag{2.110}$$

Applicable for path traversed $x$:

$$x = \frac{\tau}{2\pi}\arctan\frac{V_{A2}}{V_{A1}} + n\frac{\tau}{2} \; (n = 1, 2, 3 \ldots) \tag{2.111}$$

Figure 2.76 shows the block diagram of a digital interpolation unit that uses an arc tangent table. Analog signals $V_{A1}$ and $V_{A2}$ from the measuring system are first digitized. These digital values are used as addresses for an interpolation table from which the associated arc tangent value is read. To generate the pulse sequences, a lag counter is activated with the table value. This circuit contains a controllable pulse generator, a difference calculator and a forward/backwards counter that form a closed control loop. The difference calculator compares the entered value from the interpolation table with the reading of the forwards/backwards counter.

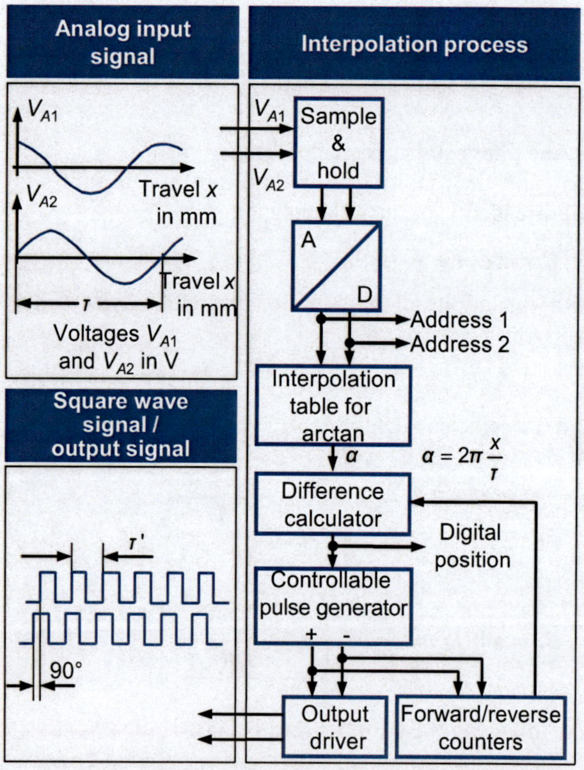


**Fig. 2.76** Block diagram of a digital interpolator

The difference value is used to control the pulse generator. Depending on how big the value is, the pulse generator feeds downward or upward counting pulses at the appropriate frequency to the counter. The generator does not output pulses if the table value and counter reading match. The count pulses are converted by the output driver to two square wave signals that have an electrical phase offset of 90° in relation to each other.

More complex digital interpolators include additional microprocessors and memory areas storing correction tables for the compensation of systematic measuring errors. Subdividing the signal period of the measuring system by a maximum of 4096 is possible using this kind of digital analysis unit [TR14]. This extremely fine subdivision is used mainly to generate a sufficient number of pulses, even when the traversing movement is very slow. The current speed value can then be determined from the displacement signal by differentiation with sufficient accuracy. This eliminates an additional tachogenerator for the velocity control loop that is secondary to the position control loop.

#### 2.3.3.3 Amplitude analysis

Amplitude analysis is applied when the measuring system outputs two signals at constant frequency $f$ (carrier frequency) with an amplitude that varies sinusoidally in the measurand (travel distance, rotation angle). The two a.c. signals have the form

$$v_1 = A \cdot \sin \omega t \cdot \cos \alpha \qquad (2.112)$$

$$v_2 = A \cdot \sin \omega t \cdot \sin \alpha \qquad (2.113)$$

$$\alpha = \frac{2\pi x}{\tau} \qquad (2.114)$$

$$\omega = 2\pi f \qquad (2.115)$$

where
$x$ = path or angle traversed
$\tau$ = graduation
$f$ = carrier frequency

The signals are amplified and aligned with the correct phase relationship. Interpolation is then possible.

#### 2.3.3.4 Direction detection

In an incremental measuring method, the periods of a measuring signal are counted during the process. Two periodic signals, offset by 90° in relation to each other, are required for direction detection. When the direction of motion changes, the phase shift jumps from +90° to −90°. ◘ Figure 2.77 illustrates clearly the principle of direction detection. A gauge bears a light and dark raster with graduation period $\tau$. The raster is scanned by two photoelements $A$ and $B$ having widths that are are very small compared to the gauge graduation.

Depending on the direction of movement of the gauge (to the left or right), different signal sequences $a$ and $b$ appear at the output of the scanning elements $A$ and $B$. In the analysis circuitry (top right of ◘ Fig. 2.77), square wave sequence $b$ and its inverse $\bar{b}$ are differentiated resulting in the signals $b'$ and $\bar{b}'$. Using the AND gate operation of $b'$ and $\bar{b}'$ with $a$ means only the positive parts of signals $b'$ and $\bar{b}'$ are forwarded to the counter as count pulses. Depending on whether the gauge moves to the left or right, the counter connected is decremented or incremented and so corresponds to the discrete distance length or rotation angle.

### 2.3.4 Selecting and installing measuring devices

A number of selection options address the various requirements placed on functionality and capability of different position measuring devices for lengths and angles. For measurement accuracy, i.e., the deviation of the position value determined from the actual position value, manufacturing accuracy is the first key factor. Secondly, the mounting location in the machine and compliance with installation tolerances exert considerable influence over the accuracy of the position value determined.

#### 2.3.4.1 Selecting a measuring device

Machines and systems used to install and transport components usually place less strict requirements on a position measuring device—such as manufacturing machines facilitating a chip removal process for narrow tolerance components.

Spindle / rotary encoder systems and length measuring devices are used to measure lengths and linear travel. Spindle / rotary encoder solutions use the spindle pitch of the drive spindle as a material measure, that is rather broad, i.e., in range 5 to 40 mm. The rotary encoder is used for the mechanical-electronic subdivision of the spindle pitch, meaning measurement increments down to the micrometer can be realized. Heat sources originating in the normally greatly pre-stressed spindle nut, or radiating from the drive and causing an increase in spindle length, generally have a detrimental effect on positioning accuracy. Local heating of the spindle, difficult to detect at points, results in considerable, difficult to predict deviations in the positioning of feed axes. This ultimately leads to scrap in production.

Direct length measuring devices record traversing movements of the axis carriage directly and nowadays are used in the vast majority of cases in machine tools. So the control loop includes the feed mechanics, in con-

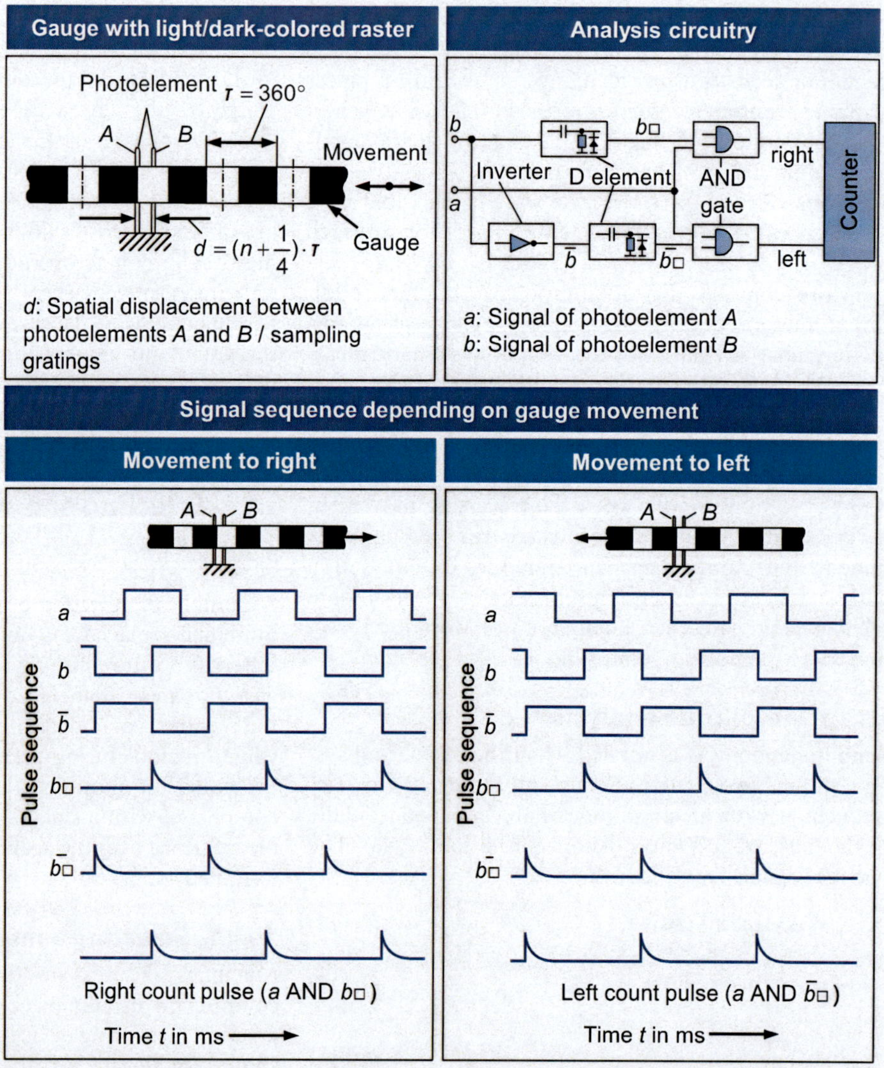


**Fig. 2.77** Direction detection for an incremental measuring system according to Herold, Maßberg, Stute

trast to the spindle / rotary encoder solution. Spindle expansion due to thermal factors has no bearing on the accuracy of the position value determined. For enclosed length measuring devices, a housing protects the gauge and scanning unit from chips, dust and splash water. These variants are particularly suited to usage in machine tools. Enclosed length measuring devices generally have inherent guiding, i.e., the scanning unit is guided by a bearing relative to the gauge. This can limit the accelerations in certain circumstances. For such cases, open length measuring devices are available that work without mechanical contact between the scanning unit and gauge.

Rotational motion and angles, for rotating tables and spindle axes for example, are measured by two device types: rotary encoders and angle measuring devices. These differ in terms of measurement accuracy. Rotary encoders generally have measurement accuracies of $\pm 10$ arc seconds. Measurement accuracies to $\pm 0.2$ arc seconds can be attained with angle measuring devices.

Enclosed rotary encoders and angle measuring devices always have inherent bearings for the rotors. There are differences in shaft variants. Devices with a solid shaft usually require a coupling to connect to the shaft to be measured and this coupling compensates radial and axial shaft movements. The use of leaf spring and tube bellow couplings is widespread here (Fig. 2.50).

Devices with a hollow shaft have an integrated or attached stator coupling. The stator coupling need only accommodate the torque resulting from the bearing friction and compensate the eccentricity movements and wobbling of the shaft end where the rotary encoder is coupled. These devices enable direct, and so rigid, coupling to the shaft to be measured and thus minimize static and dynamic measurement errors. The relatively short bearing diameters available on the market mean the hollow shaft diameters are limited for variants with inherent bearing. If the machine design requires larger internal diameters, devices without an inherent bearing are used. As is the case for open length measuring devices, open angle measuring devices are also contactless, i.e., frictionless, and permit higher accelerations and rotational speeds.

Position measuring devices are normally available with various electronic interfaces standard on the market. They can be connected in a modular manner to the control and regulation architecture used.

### 2.3.4.2 Place of installation in a system or machine

To measure position information with small deviations, the place of installation inside the machine or system is key in addition to the accuracy of the measuring device.

When lengths are measured, the errors in axial guidance for example negatively affect the measurement result. If the measurement axis does not run exactly parallel to the material measure, tilting errors (Abbe errors) occur for example. These can be minimized however with suitable selection of the installation location. So the length measuring device should be placed as close as possible to the test object or tool contact.

When angle measuring devices are fitted or added, mainly eccentricity errors occur, arising between the machine bearing and index plate when the devices are open. Deviations occurring due to eccentricity can be mainly eliminated by a second diametrically affixed sensor head. For devices with inherent bearing resembling the coupling affixed on the shaft side, and on hollow shaft variants, the integrated stator coupling compensates misalignments.

The ambient conditions prevalent at the measurement location (such as chips, dust and splash water) that are detrimental to the correct functioning of the device must be taken into consideration in all cases. Liquid media such as coolant can collect inside enclosed devices. Suitable protective measures can aid here, such as covers and drain holes enabling the media to drain off. Injecting compressed air into the inside of the housing provides additional protection from dust and splash water entering. Cleaning the machine must also be considered when selecting the installation location. All too often, machines are cleaned after use with aggressive cleaning agents or compressed air.

If the measuring devices, operating predominantly with photoelectric sampling, can not be protected adequately, magnetic and inductive measuring devices represent an alternative. Lower accuracy must be accepted as a downside here.

One key influencing factor on positioning accuracy is temperature. Measuring devices should never be fitted in the immediate vicinity of temperature sources.

### 2.3.4.3 Installation notes

Once the place of installation for the position measuring device is set, the mounting tolerances stipulated by the manufacturer of the measuring device in the connection dimensional drawings must be met. These include permitted guidance deviations, squareness errors and parallelisms of mounting surfaces. Failing to comply with the specified tolerances reduces positioning accuracy and negatively impacts the function and service life of the measuring device.

Enclosed length measuring devices can be fitted easily in consideration of the mounting tolerances because the scanning unit is guided on the gauge and so no adjustment of the scanning unit in relation to the gauge is needed (◘ Fig. 2.78). This additional adjustment work is always required for open length measuring devices. Mechanical aids such as spacer plates and angle brackets and electronic aids providing quantitative or qualitative information about the measuring signal, simplify this work.

Enclosed angle measuring devices with inherent bearing, and rotary encoders with a solid shaft, are secured on the stator side using their mounting flange, and in compliance with the mounting tolerances. For connection on the shaft side, these devices require a tube-gaiter type, diaphragm or flat coupling (◘ Fig. 2.50) that compensates alignment errors and skewing of the rotary encoder shaft and motor shaft / ball screw. Additional kinematic errors can occur for this device type.

Another design is the hollow shaft variant with the fitted stator coupling, as shown in ◘ Fig. 2.79. The rigid joint on the shaft side is torsionally stiff. The stator coupling must allow movement of the housing caused by axis errors.

Open angle measuring devices have no inherent bearing. Attention needs to be paid to eccentricity errors here. Strip solutions are also possible for open angle measuring devices. Here, internal or external grooves for the strip must be prepared with relatively tight tolerances. For external grooves, the installation space for the strip lock must also be considered.

### 2.3.4.4 Electrical connection

Automated machines and systems are mechatronic systems in which considerable capacities are installed. Capacitive and inductive couplings can cause disruption to the measuring signals. This crossfeed is mainly from cables and the inputs/outputs of equipment. Typical interference sources include the strong magnetic fields of transformers and electric motors, relays, contactors,

2

**Fig. 2.78** Installation dimensions and tolerances on an enclosed length measuring system according to HEIDENHAIN

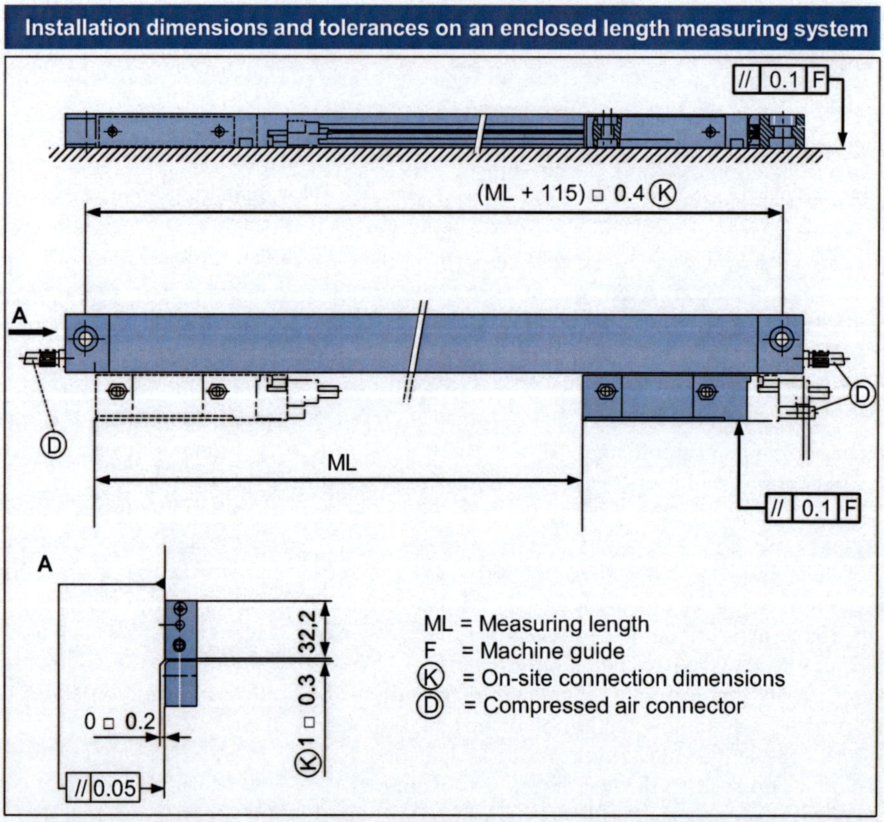


**Fig. 2.79** Installation dimensions of an angle measuring system with fitted stator coupling. *Source:* Heidenhain

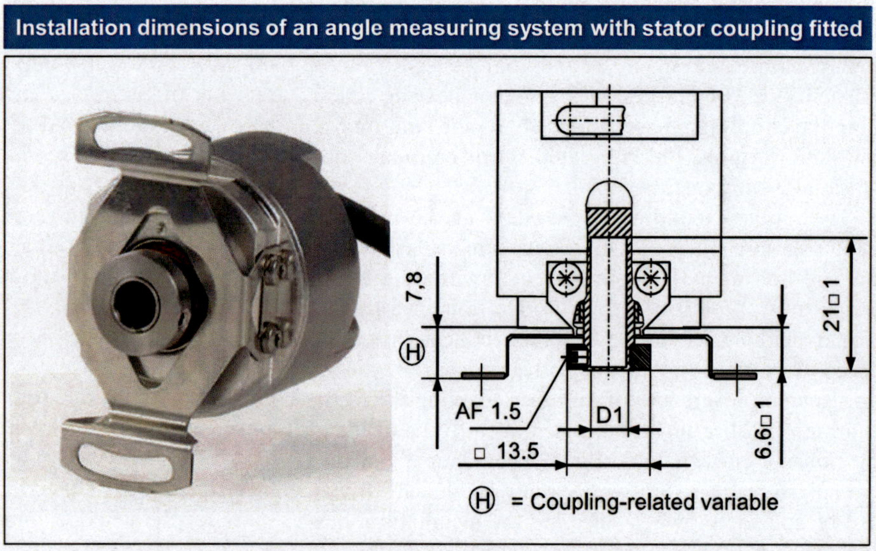


solenoid valves, high frequency equipment, pulsing devices and the stray magnetic fields of switch-mode power supplies, frequency converters and their power and supply cables.

To prevent electrical disturbances, only specially shielded measuring device cables and connectors should be used (EN 50,178). These cables should be as short as possible and not be run near the power cables of other electrical systems.

Attention must be paid to the suitability of connector cables used. Bending radii, drag chain suitability and resistance to media must be taken into account here. To keep within the supply voltage range for measuring devices, the voltage drop over the cable length must be borne in mind.

## 2.4 Mechanical transmission elements

All the components of a feed drive that are in the force flow between motor and tool or workpiece are mechanical components. In addition to the components for converting a rotary motion to a translation motion, such as ball screw drives, these include feed gear units, power transmission components and couplings. Additional components are tool and workpiece holders, linear guide systems, bearing arrangements, guideway covers and cable drag chains (see ► Sect. 15.1). ◘ Figure 2.80 shows the schematic setup of a feed drive with electrical servo motor, toothed belt drive, ball screw, threaded nut and workpiece table.

The feed tasks in machine tools differ greatly in part as regards travel distance/speed/acceleration and available space. Inevitably resulting from this, there are clear differences pertaining to the drive principle to choose and the variants of mechanical components used.

The general development trend for ball screw drives is towards solutions without gear unit. For machines with long travel distances, rack and pinion/worm drives and linear motor drives are being used. The linear motor drive is also the solution of choice now for medium-sized machines with very high acceleration requirements for feed motion.

The feed function can be realized differently in many cases as regards design solution. Then, reliability and cost considerations are paramount in the selection of feed principle.

### 2.4.1 Components for converting rotation motion to translation motion

Key for electromechanical feed drives are the components for converting the rotational motion of the electric drive to a linear feed motion for table or carriage. Ball screw/nut systems are used predominantly for this in small and medium-sized machine tools. For long travel distances, there is a risk that feed spindles will buckle under load due to their length and the spindle rotational frequency will drop into the range of the spindle's natural frequency of bending vibration. Feed drives with ball screw drive and rotating spindle are thus only generally used up to a travel distance length of approx. 4 m. In heavy-duty lathes and long table milling machines, therefore, rack and pinion or worm and rack systems are used.

For feed drives having less strict requirements, especially as regards stiffness, toothed belt drives are also used due to their lower costs.

#### 2.4.1.1 Ball screw/nut drives

Feed spindles today are almost exclusively designed as ball screw spindles driven by an upstream gear unit or directly by a feed motor (◘ Fig. 2.81). Trapezoidal-thread spindles with bronze nuts or hydrostatic nuts, rollers and rolling ring threaded spindles are also in use.

**Ball screw drives**

A ball screw drive (BSD), also called a ball screw, is used, like all nut-bolt systems, to convert a rotational motion to a linear motion, or vice versa. Feed drives in machine tool axes and axes of motion in measuring equipment are the primary application areas.

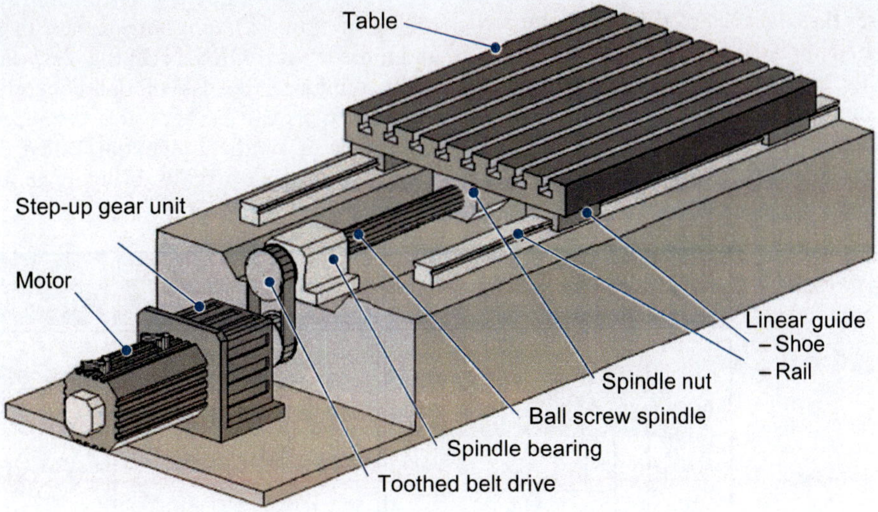


◘ **Fig. 2.80** Schematic setup of a feed drive with electrical servo motor and ball screw / nut system

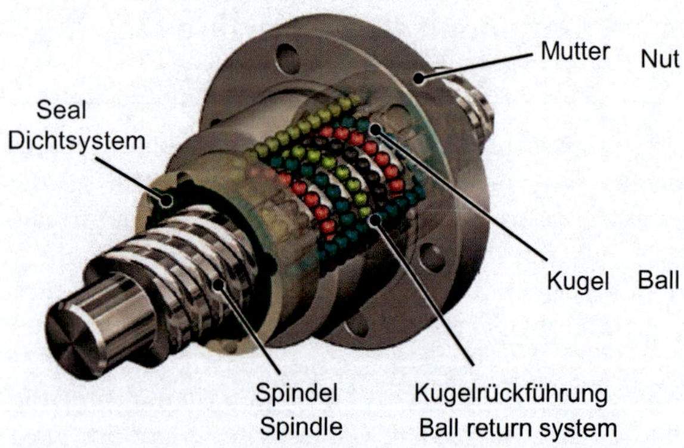


**Fig. 2.81** Setup of a ball screw drive. *Source:* Bosch Rexroth

Compared to the two other widespread systems for feed generation, rack and pinion drive and linear motor, the ball screw drive features very high efficiency on account of the rolling friction, no stick–slip effect, zero clearance and sufficiently high stiffness. The low level of damping, heating of the system at high speeds and varying axial stiffness in relation to the nut position are downsides. Today, ball screw drives can be used with axial speeds up to 300 m/min and an axial acceleration of about 30 m/s$^2$.

A ball screw drive essentially comprises a spindle, a nut and balls (Fig. 2.81). The spindle and nut have a thread profile, with the edges emulating a ball race. Spindles and nuts are often simplex. But it is possible to realize systems having multiple threads, usually up to four, as shown in Fig. 2.81. The multi-thread variant enables load carrying capacity and service life to be extended for certain BSD variants. While in a motion screw thread the thread surfaces slide directly on top of each other. They are separated from each other by balls on the ball screw drive so that the sliding friction is replaced mainly by the rolling friction present. This significantly increases the efficiency of the BSD compared to the motion screw thread. Usually the spindle has a rotary drive and the nut is secured against twisting. The thread pitch means there is relative axial motion between the nut and spindle. Using the nut to drive the ball screw drive is also possible, but is not as common.

As the balls roll between the thread flanks during use, they would find their way out of the thread groove. This is prevented by ball recirculation. There is essentially a difference between through-the-nut ball return where the balls are routed back to the starting position after completing a full cycle through the entire nut length and track-to-track internal ball return where balls are returned after every thread winding. The options for ball recirculation are shown in Fig. 2.82.

The tangential inlet of the balls into the ball return system, the level of quietness even at high dynamics, and a higher number of load-bearing balls, are features of the through-the-nut ball return system. Downsides of this system are the longer ball chain requiring a higher level of manufacturing precision and greater sensitivity of the ball screw drive to radial forces. The through-the-nut ball return system can be realized using a return duct integrated into the nut or a tube. Using a tube has the benefit of optimized shaping of the ball inlet zone, but is to the detriment of the nut diameter, which increases as a result.

Despite the ability to shape the ball inlet zone, impacts by the balls are unavoidable at high speeds. So in addition to the geometric design, importance is attached to the tube material in regard to noise levels, warming and wear of the nut at high speeds. Compared to the through-the-nut external ball return system with tube, the integrated return duct is more complex to manufacture and install but does have the benefits of quieter operation and a smaller nut diameter. The track-to-track internal ball return is less expensive to manufacture compared to the through-the-nut external ball return system, is more tolerant to radial forces and is easier to install. A downside of the track-to-track internal ball return is the lower number of load-carrying balls, something that reduces the load carrying capacity of the ball screw drive. Furthermore, the number of ball inlets and outlets is detrimental to the uniform rolling motion, entailing an increase in friction, wear and noise levels [FORS10] (Fig. 2.82 on right).

To minimize the risk of ball congestion in the return, the balls can be kept at a distance by a sort of cage made of synthetic material, called the ball chain. The lower number of balls, bringing about a reduction

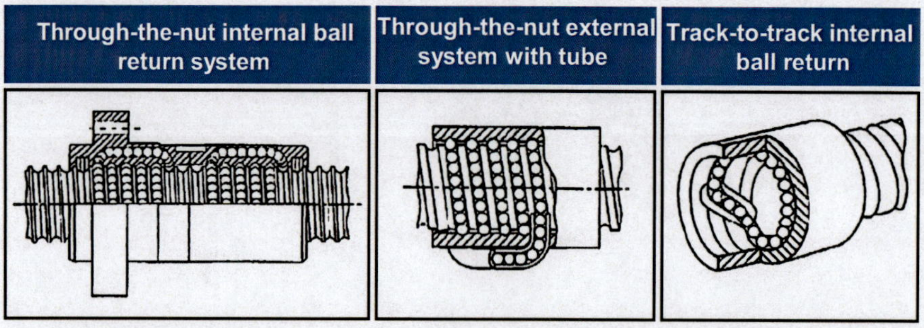


**Fig. 2.82** Options for a ball return system according to MANNESMANN, REMSCHEID

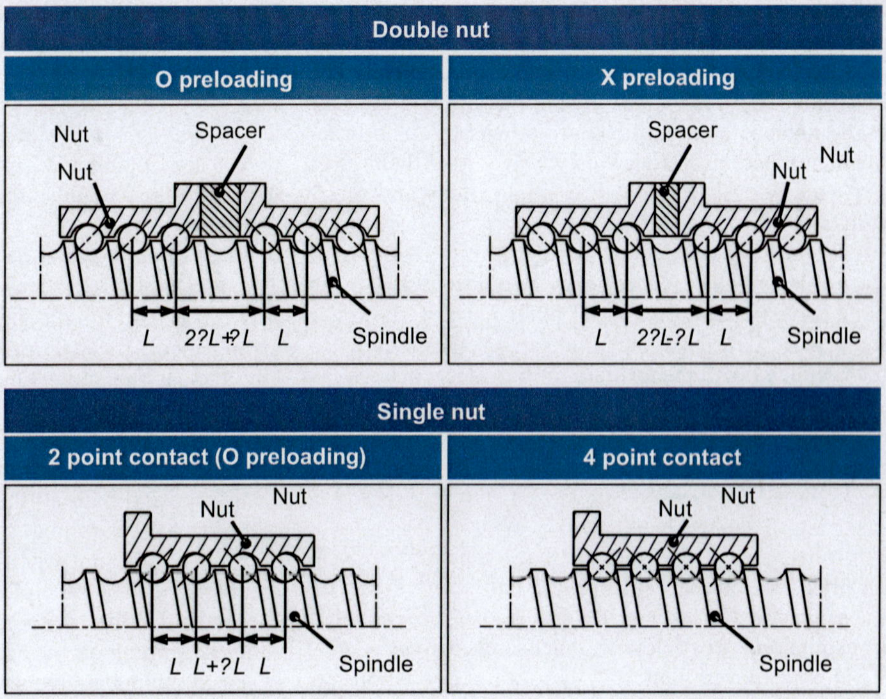


**Fig. 2.83** Options for preloading according to Steinmeyer

of the load carrying capacity, is a downside here however.

To satisfy the strict requirements of positioning accuracy, high stiffness and/or zero clearance (no reversal error) are stipulated for ball screw drives. This is achieved with preloading that can be realized either by splitting the spindle nut into two (double nut) or by having the ball race in the single nut. Figure 2.83 shows the options for preloading the single and double nuts. The top part of the figure shows two double nuts where the spindle is under tensile stress (left) and compressive stress (right). A calibrated spacer ring is inserted between the nuts. This presses the halves of the nut together or apart.

More detailed information on tensioning, designing and calculations, as well as on seals and lubrication, for ball screw drives is in Volume 2— ► Sect. 6.6.4.

### Roller screw drives

Roller screw drives (RSD), like ball screw drives, convert a rotational motion to an axial motion. All roller screw drives essentially comprise a thread-profiled spindle, a thread-profiled nut and thread or groove-profiled planetary rollers that are seated in planetary gear carriers (guiding disks), Fig. 2.84. The planetary rollers here too are used to minimize friction in the contacts. For certain designs and variants of roller screw drives, the planetary geometry also contributes towards the overall pitch of the system. When a roller screw drive is powered by the spindle, the planetary rollers rotate about their own axis due to friction at the contact points, and, at the same time, follow a planetary circulation movement around the spindle being supported by the planetary gear carriers. The interaction of the profiles of spindle, nut and planetary rollers, and the planetary circulation movement of the planetary rollers around the spindle, cause the nut to move along the spindle.

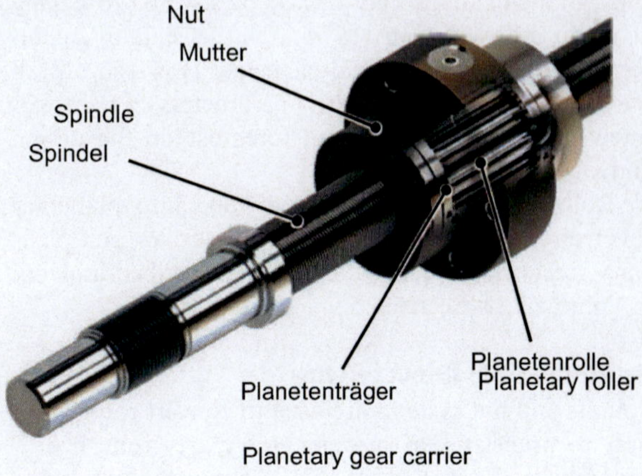


**Fig. 2.84** Structure of a roller screw drive. *Source:* Bosch Rexroth

To ensure the point contacts of the edges at all times and to prevent "edge loading", the planetary thread flanks are spherical. The high number of contact points and the large radius of curvature of the thread flanks have both benefits and downsides compared to ball screw drives. The far higher load carrying capacity, the stiffness and the realizable axial force from a given drive torque are beneficial. The lower efficiency, associated greater heating and the generally higher purchase price, are downsides. The benefits mentioned mean roller screw drives lend themselves particularly well to heavy-duty applications and precision equipment.

Similar to the ball screw drive, the nut traverses relative to the spindle for every drive revolution exactly the path corresponding to overall pitch $P_{total}$ of the roller screw drive. The axial force that can be created can also be calculated with the formula for ball screw drives:

$$v_{ax} = n_{Drive} \cdot P_{total} \tag{2.116}$$

$$F_{ax} = \frac{M_{Drive} \cdot 2\pi \cdot \eta_{PRG}}{P_{total}} \tag{2.117}$$

For roller screw drives, the fact that a much lower overall pitch can be realized than for ball screw drives of the same size applies in general. A higher level of force transmission is possible with roller screw drives.

However, overall pitch $P_{total}$ does not correspond to the spindle pitch for all designs and variants of roller screw drives as is the case for ball screw drives. The spindle pitch can also not always be selected to be any value for roller screw drives because they exhibit a complex interplay of individual elements. How the overall pitch made from the geometric parameters of the roller screw drive depends first and foremost on the design and variant selected.

Roller screw drives can be divided into planetary roller screws, recirculating roller screw drives, and planetary screw drives. These designs are described in detail in Volume 2, ► Sect. 6.6.5.

### Hydrostatic spindle nut systems

Spindle and nut systems are used to convert rotary motion to translational movement. Apart from hydrodynamic lead screw drives for secondary drive units and ball screw drives in feed drives, hydrostatic screw drives are used especially in heavy-duty machine tools, broaching machines and in the fields of precision and high-precision machines. Here they are used to move the carriage, for example. The main advantages of a hydrostatic screw drive over a ball screw drive are its high damping, zero-clearance (a function of its operating principle) and its wear-free operation.

◘ Figure 2.85 illustrates the setup and design of a hydrostatic screw drive. The cross-sectional drawing (lower image) shows the operating principle of a hydrostatic screw drive. The pockets are in the thread flanks in the nut. They are supplied with pressure oil via capillaries which here are also housed inside the nut. The return oil can flow away either via the threads or via holes in the root and head of the flanks. In the functional principle shown, the bearing pockets of both halves of the nut are positioned in the opposite direction for preloading the system. The oil supply system for *common pump and capillaries* is advisable here due to the high number of pockets, e.g., four pockets per flank for four to eight threads.

### Ball screw bearing arrangement

The third important component of the feed spindle drive, after the spindle and nut, is the spindle bearing arrangement. Its function is to guide the spindle radially and, at the same time, to accommodate the feed forces in the axial direction. Spindle deformations and displacements must stay within permitted limits here. So the requirements pertaining to high axial load carrying capacity, high stiffness, zero clearance, low level of bearing friction, high speed and high running accuracy are paramount when selecting a bearing arrangement for ball screw spindles.

The individual criteria take on particular significance depending on the application. For the feed drives of large milling machines having high cutting forces for example, the stiffness of the bearing as part of the overall system plays a major role. For the benefit of a high stiffness value, the friction of strongly preloaded bearings for example is neglected compared to other friction losses. In contrast, this factor dominates for feed drives of grinding machines where very high positioning accuracies are required for low moving masses. A bearing arrangement with low friction is required here.

For ball screw drives to attain the strict positioning accuracies, the bearing arrangements must exhibit in particular high axial stiffness over the entire travel distance. For conventional bearing arrangements with a fixed bearing and a floating bearing, the stiffness of the arrangement decreases hyperbolically on removing the carriage from the fixed bearing. Also implementing the second bearing point as a fixed bearing enables a mirror image of the stiffness curve to be superposed so as to give a symmetric curve.

In ◘ Fig. 2.86, the top bearing arrangement has one fixed bearing and one floating bearing, and the lower bearing arrangement has two fixed bearings. It can be seen that the overall stiffness for two axial bearings is increased greatly, and is approximately constant over a greater range in the center of the spindle.

Due to the friction between the spindle and spindle nut, moving the feed slide causes heating of the spindle, resulting in spindle expansion. For a dual-sided clamped spindle, additional pressure forces result that cause compression of the spindle. A spindle subjected to high temperature rises due to great frictional loading in continuous operation is thus preloaded when cold.

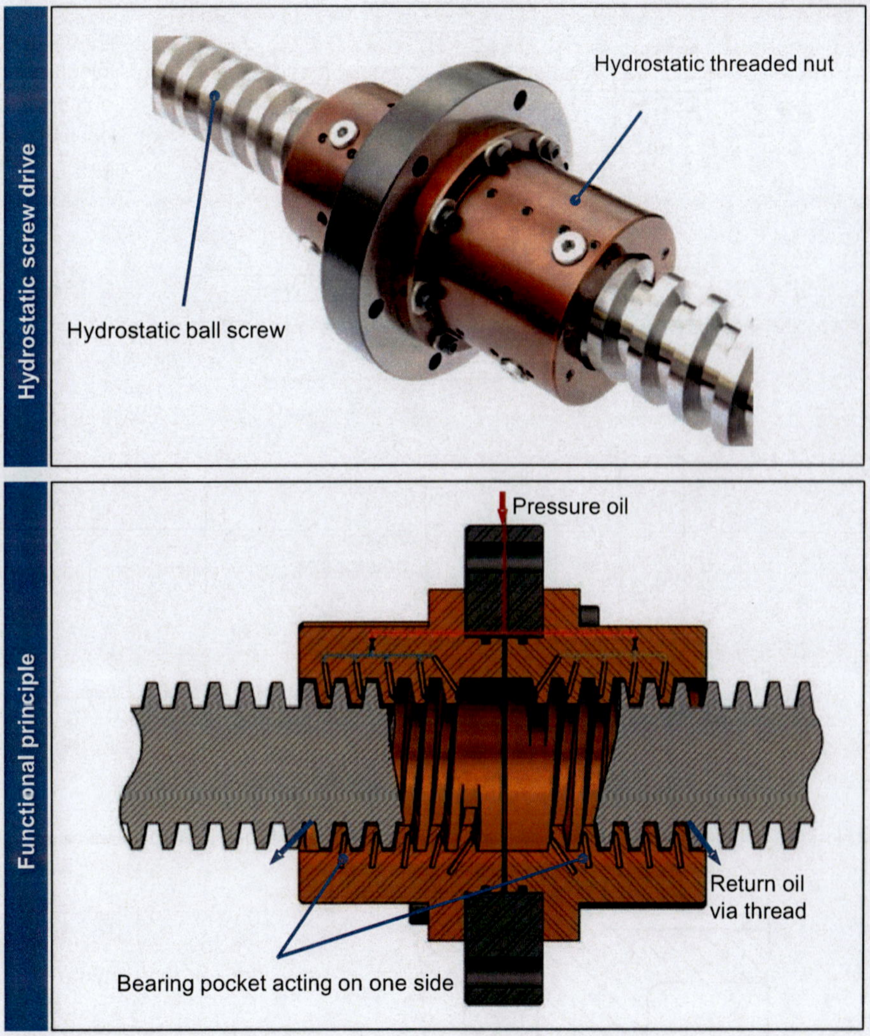


**Fig. 2.85** Operating principle of a hydrostatic screw drive according to ZOLLERN

When it expands under heat, the preload of the spindle is reduced initially. Axial pressure forces can only occur when the spindle is heated further. These are tolerable however provided the correct preload is selected.

The following design rules must be observed for rigid ball screw bearing arrangements:

- Needle and roller bearings are preferred to ball bearings due to their line contact, and so their greater stiffness.
- Bearing rings and spacer rings should be avoided if possible because the contact surfaces reduce stiffness.
- Fitting and distance surfaces should be ground. A high roughness level requires a low contact area portion and reduces stiffness. Parallel contact surfaces prevent the bearing from canting.
- Preloading the axial bearings is absolutely necessary.
- Rigid joints between separable surfaces should be provided if possible. Attention must be paid in particular here to a design of rigid bolted connections.
- In general, bearing modification parts are to be designed very carefully in regard to engineering and production.

Figure 2.87 shows an installation example of a combined radial-axial roller bearing. Preloading the axial bearing can be achieved with a suitable selection of the spacer bush length.

Installation examples of ball screw bearing arrangements with axial angular contact ball bearings are shown in Fig. 2.88. A fixed non-locating bearing as shown at the top of Fig. 2.88 is standard for medium stiffness requirements. Installation as in Fig. 2.89 (at the bottom) is recommended for feed drives with strict stiffness requirements. The spindle is preloaded under

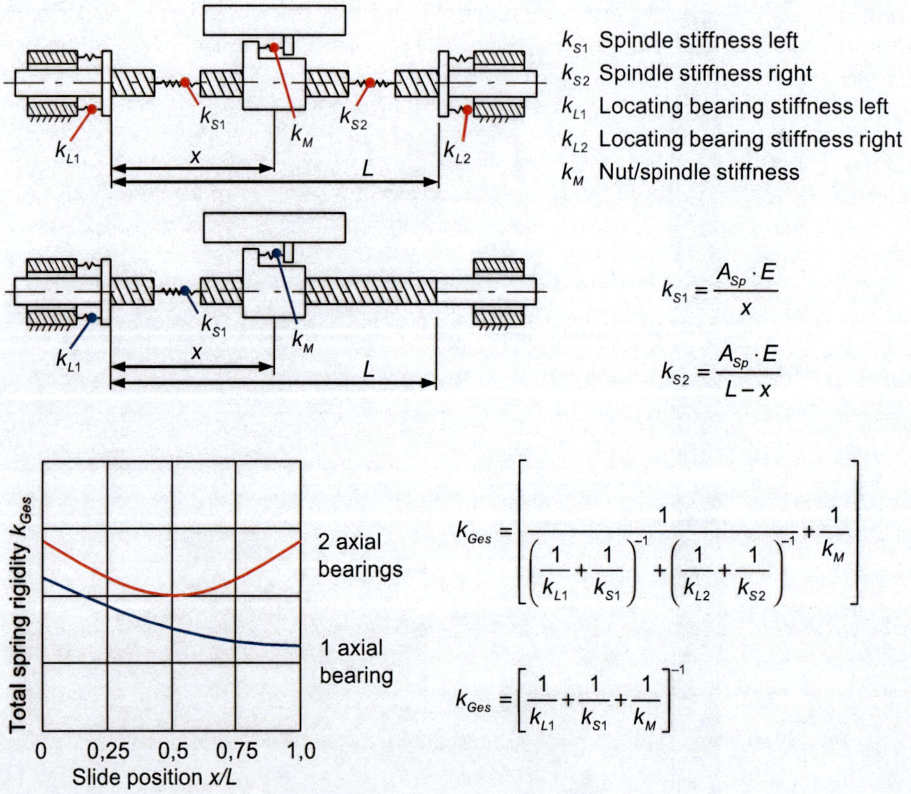


**Fig. 2.86** Stiffness behavior of a spindle drive with a single-sided and dual-sided axial fixed bearing

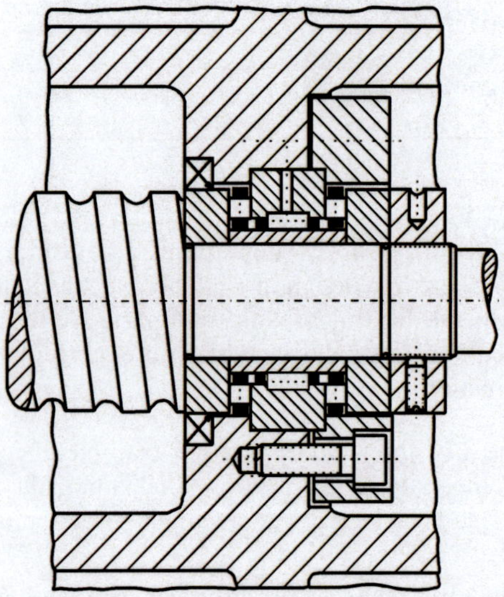

**Fig. 2.87** Installation example of a combined radial-axial roller bearing for ball screws

tension at both ends by two angular contact ball bearings in a dual O arrangement. One angular contact ball bearing seated opposite transfers the force into the frame when tensile preloading changes over to compressive stress as a result of the spindle heating and the expansion that generates. In addition to the bearing forms already mentioned, ball screws mounted on one side with free ends are also usual for short spindle lengths and when the stiffness requirements are less strict.

Spindle nut drives with stationary ball screw are used so that ball screw drives can also be used for long travel distances allowing high stiffnesses (Fig. 2.89). While, in conventional systems, the ball screw is driven, in this example it is the spindle nut that rotates, driven by an on-board motor via a toothed belt.

In this way, spindles with large diameters can be used. Also, preloading the spindles using disk springs and/or clamping elements is very easy. Spindle supports can also be seen in Fig. 2.89. These prevent vibration of the spindle at the natural bending frequency, even for long spindle lengths. Here, the spindle supports that move on guide rails are pushed by the spindle nut. As a result, the spindle supports are not moved in the traverse range of the spindle nut from the end position on one side until the opposite spindle support is reached. As with the motor spindles for main drives, solutions that integrate the motor and the spindle nut are also being sought for this system.

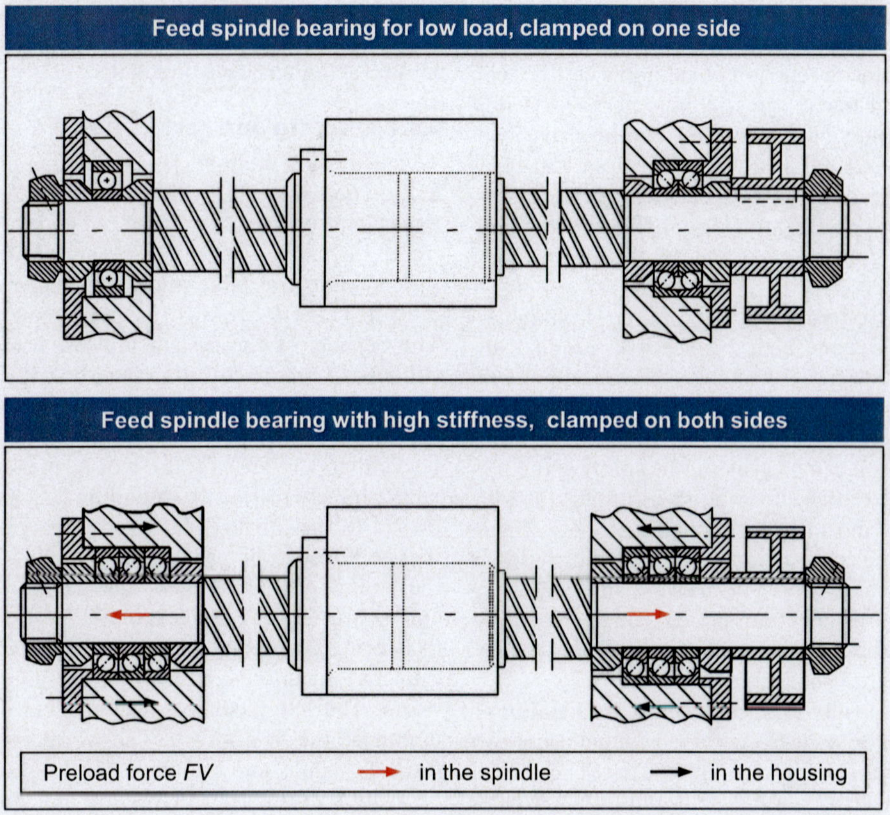


**Fig. 2.88** Bearing arrangement examples for ball screw spindles with axial angular contact ball bearings according to SKF

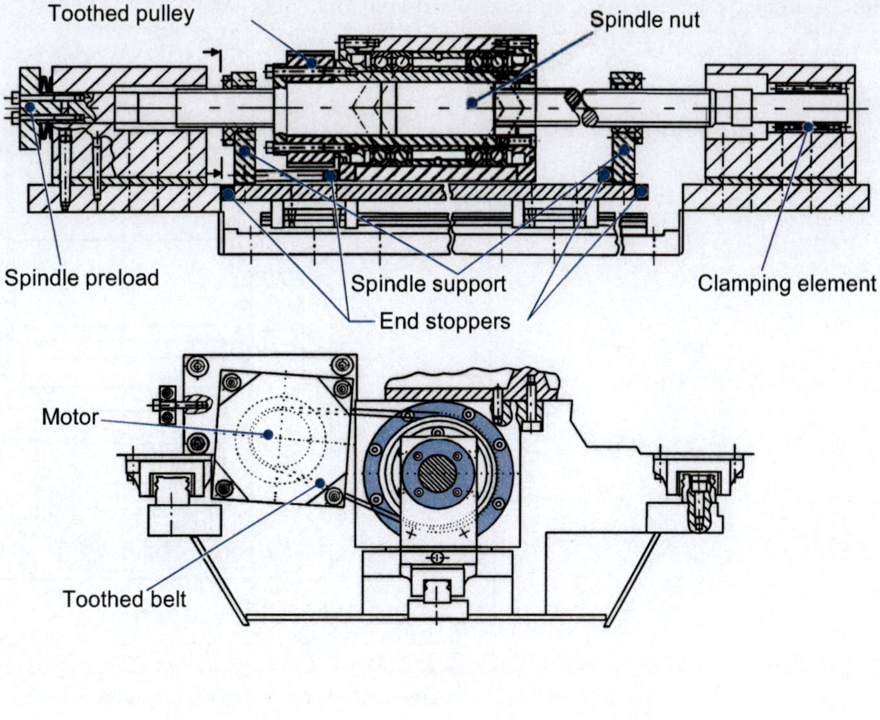


**Fig. 2.89** Spindle drive with static ball screw and spindle supports according to CHIRON Werke

#### 2.4.1.2 Rack and pinion drives

Using a rack and pinion drive is recommended for long travel distances. Any long feed paths can be realized by grouping rack segments together. The resulting overall stiffness of the rack and pinion drive is always independent of the travel distance length. This stiffness is made up mainly of the components of torsional stiffnesses of the auxiliary shaft gear and pinion shaft and the contact stiffness of the rack and pinion pairing.

The transfer of power at the pinion is characterized by low speeds and high torques. This requires additional gear unit stages. The entire drive train should be torsionally stiff and clearance-free. Clearance-free feed transfer between the pinion and rack in both directions of motion can be achieved by splitting the pinion. For the feed drive shown in ◘ Fig. 2.90, the play between pinion and rack is eliminated by the cogging of two helically geared pinions with a rack. The lower pinion is moved axially on a spline shaft shoulder using spring force. This is how both pinions, one on each opposite flank of the rack, make contact with the system. Gearing errors are also compensated.

Another possibility for clearance-free realization of rack and pinion drives is electrical tensioning using two motors. The motors drive one pinion each. The overall moment required, $M_{set}$, is not delivered in equal parts by the motors. Instead, the torque lines of the motors are displaced such that, at the zero-load moment required ($M_{set}=0$), the motors work in reverse directions with tension moment $M_V$. ◘ Figure 2.91 shows the tension characteristics. At the torque zero-crossing, the pinions have contact with the system at opposite rack flanks. The play is eliminated. At higher torques, the motors act in the same direction.

#### 2.4.1.3 Worm and rack drives

To avoid using multi-stage gears as required for a rack and pinion system, hydrostatic worm and rack systems (Johnson drives) are also commonly used for long travel distances. Typical applications for worm and rack systems are table and support drives for heavy-duty machine tools such as portal milling machines. The benefits of the system are low friction and high stiffness. ◘ Figure 2.92 illustrates how it works.

The rack has pressure oil pockets. Oil is supplied by hollow pistons pressed onto the feed lines of the oil pockets. In the area of the worm, pressure is applied by the hollow pistons to the oil pockets of the rack attached to the carriage. The back pressure generated at the series resistor integrated in the hollow piston causes the hollow pistons to press against the rack. Leaking and return oil runs freely into a collection channel via the worm. Gear teeth in the outer cylinder of the worm gear and an intermediate wheel are used to drive the worm. The intermediate wheel engages directly in the pinion of the feed motor. The worm is seated hydrostatically axially and radially in needles.

Worms/worm gear drives are in common use in machine tools to generate rotatory feed motions. ◘ Figure 2.93 shows the table drive of a hobbing machine with twin worm drive design. The preload is applied and play eliminated by a worm gear with hydraulic, axial adjustment.

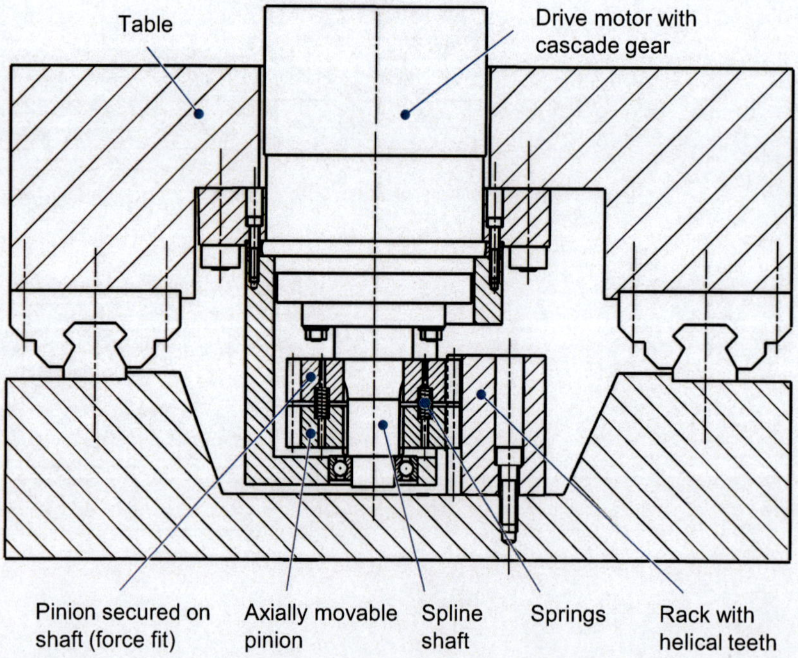


◘ **Fig. 2.90** Clearance-free pinion-rack system with split pinion

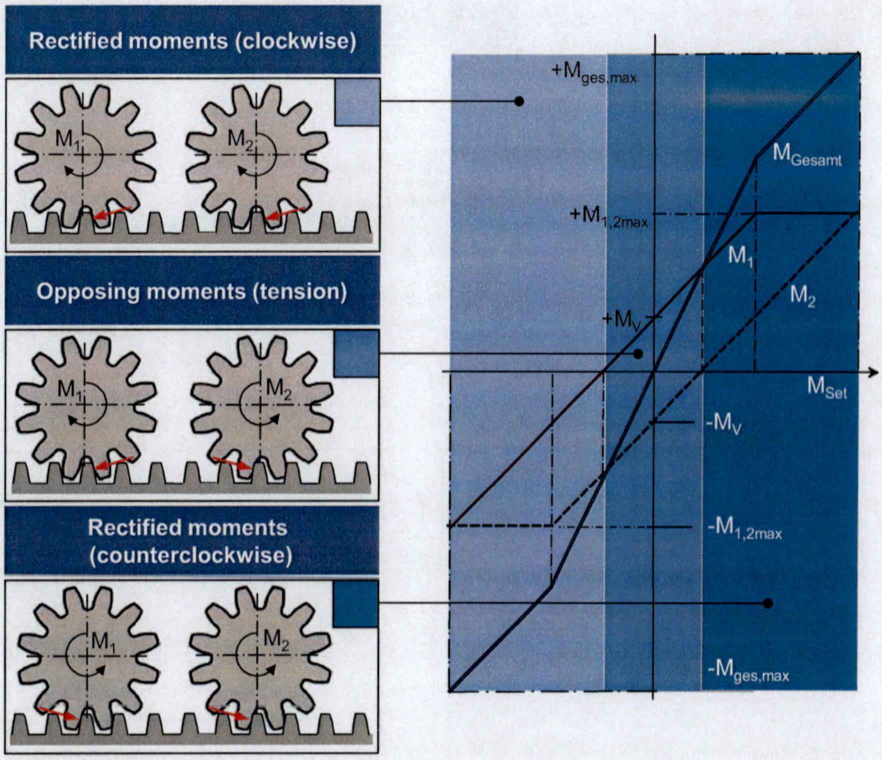


**Fig. 2.91** Clearance-free pinion-rack system with electric tensioning

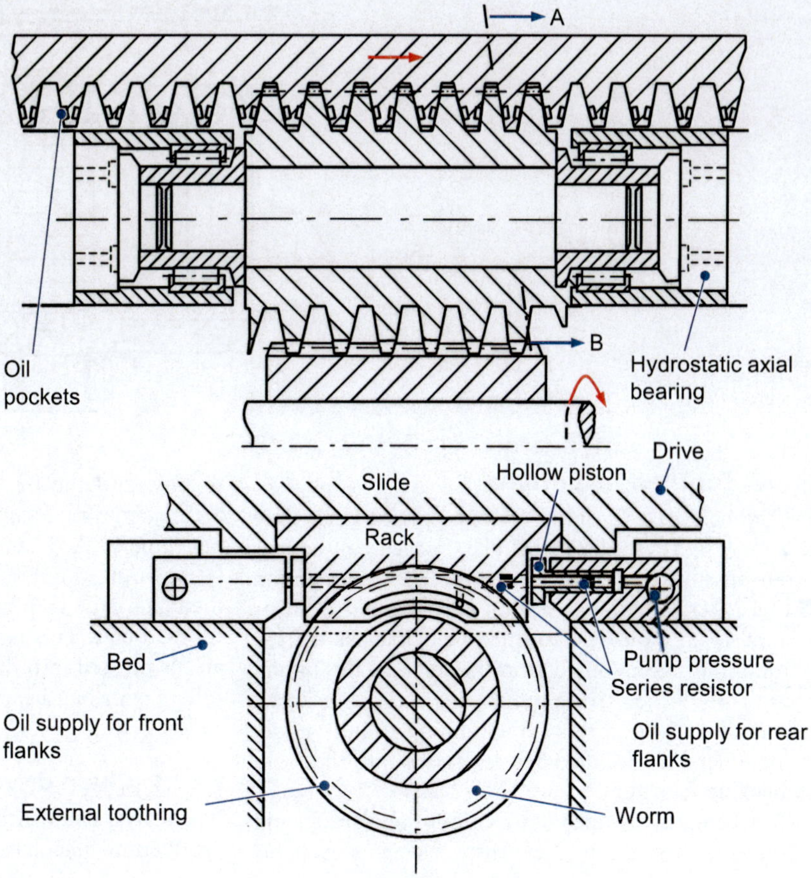


**Fig. 2.92** Feed drive with hydrostatic worm and rack system—Johnson drive according to Waldrich Siegen

2

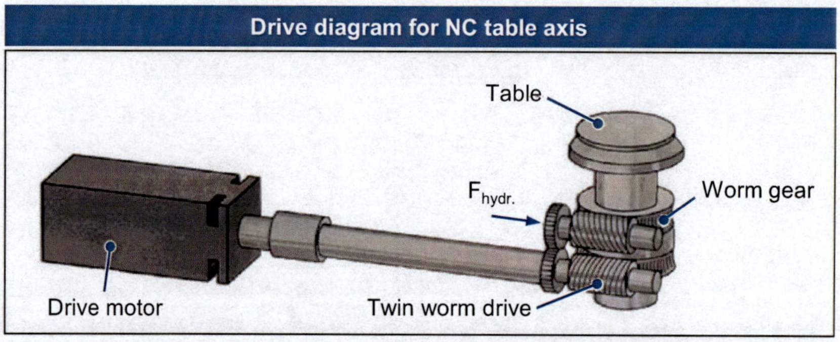


**Fig. 2.93** Table drive of a hobbing machine using hydraulically preloaded twin worm drive. *Source:* Gleason-Pfauter Maschinenfabrik

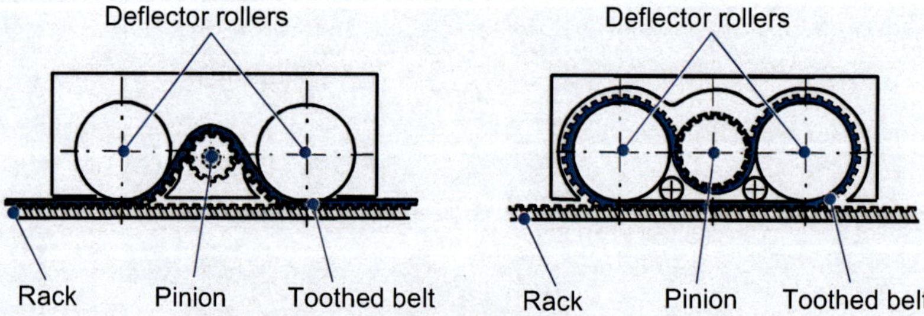


**Fig. 2.94** Functional principle of toothed belt drives with supporting toothed belt bar according to Zarian

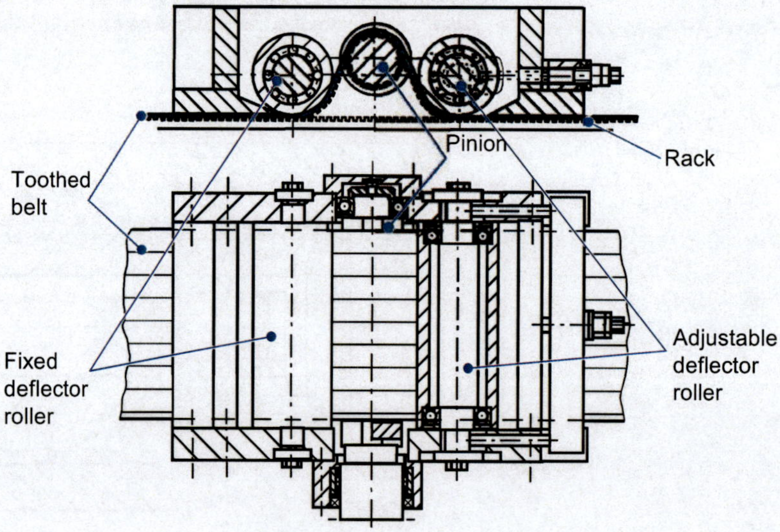


**Fig. 2.95** Toothed belt linear drive with stationary toothed belt according to CHIRON Werke

### 2.4.1.4 Toothed belt drives

Toothed belt drives, just like rack and pinion drives, are used for long travel distances. The toothed belt acts as a transmission element between the rack and pinion (Fig. 2.94). Rotational motion is converted to linear or rotatory motion by a toothed belt that has a fixed position and is only lifted from the rack at the pinion location. With the driving pinion, the belt forms a belt transmission. After a short free-running strand via deflector rollers, the toothed belt is pressed into the rack, establishing a form fit.

The benefits, compared to conventional rack and pinion solutions, are the inexpensive elements available to buy, and the freedom from clearance of the drive arising from pressing the elastic belt teeth into the rack and pinion. A detailed example of using such a drive is shown in Fig. 2.95. The motor, which moves with the table, drives a pinion having a rolling motion on the toothed belt. This is preloaded by two deflector rollers and is pressed into the static rack. One of the deflector rollers with ball bearings can be moved so the preload can be set.

### 2.4.1.5 Chain drive

Nowadays, chain drives perform various functions in production machines because of the benefits such as

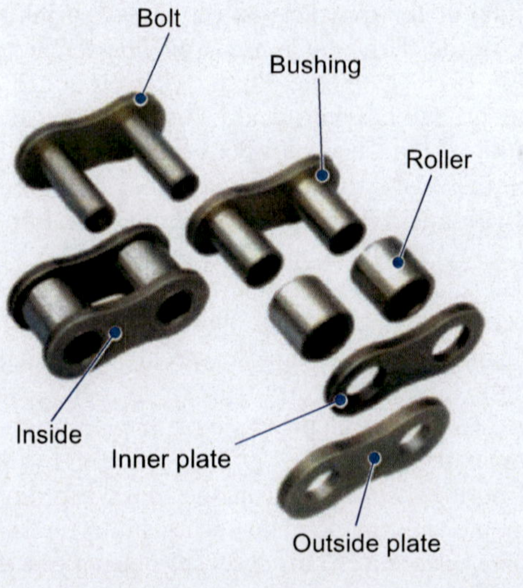


**Fig. 2.96** Elements of the chain link of a simple roller chain

slip-free power transmission, constant gear ratio and insensitivity to ambient factors such as temperature, humidity and dirt. As the selection of chain used is specific to the application, different chain designs have been developed. The chain type used most often is the roller chain shown in Fig. 2.96.

The rollers moving on the bushings roll on the tooth flanks of the sprocket with low friction such that a different point on the circumference of a roller keeps making contact. A pad of grease between the rollers and bushings dampens noise and vibration. The chain elements are pulled by sprockets using polygons. The number of corners they have corresponds to the number of teeth on the sprocket. In contrast to other gear types, it is simple to bridge larger center distances using chain drives. The downsides of chains are, in particular, the polygon effect (that causes a fluctuating chain speed), the maintenance overhead and susceptibility to vibration.

As shown in Fig. 2.97, the chain assumes a polygon shape around the sprocket wheel. The chain force and speed fluctuate due to the unwind radius changing over the rotation angle. The consequence of this is that, for constant circumferential speed $v_u$ of the driving sprocket, chain speed $v_K$ periodically fluctuates between thresholds $v_{K,min}$ and $v_{K,max}$. The thresholds are calculated as follows:

$$v_{K,\min} = v_u \cdot \cos\frac{\tau}{2} \tag{2.118}$$

$$v_{K,\max} = v_u \tag{2.119}$$

Alternating acceleration and deceleration of the chain subjects it to increased loading. The chain links are also twisted against each other during the up and down movement on the sprocket. This causes friction and wear at the joints and causes elongation of the chain. In addition, the chain elements and sprockets are subjected to impact loading when the chain travels upwards onto the sprocket. This effect, known as the polygon effect, is shown in Fig. 2.97.

## 2.4.2 Feed gear units

In addition to the gear units introduced in the previous chapter for converting rotational motion to translational carriage or table motion, additional feed gear units are often used in feed drives between the motor and spindle or pinion shaft. They are commonly used for gearing the high motor speeds to a speed suitable for the feed drive, and for torque adjustment. Furthermore, they offer the benefits of further reduction of the mass moments of inertia applied to the motor shaft. Feed gear units should be torsionally stiff, have low inertia and be free of torsional backlash.

### 2.4.2.1 Gear drives

Gear drives for feed drives are usually realized as cylindrical gears and, for higher gear ratios, also as planetary

**Fig. 2.97** Polygon effect for chain drives

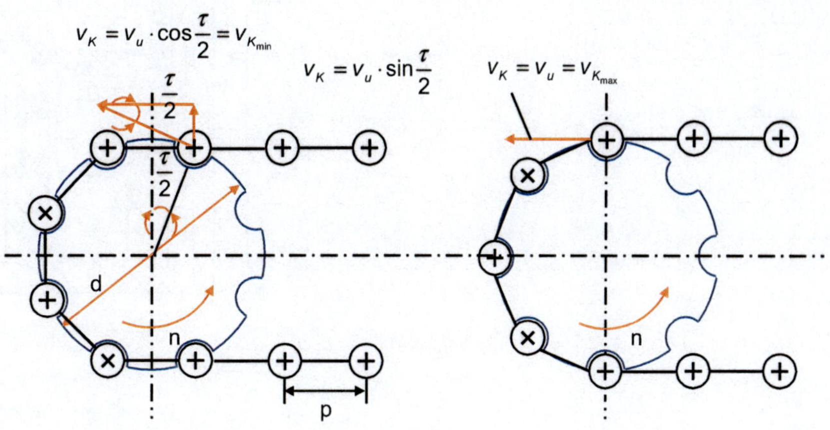

gears. In light of space restrictions, positioning the feed motor at an angle to the driving pinion or spindle is a sensible option in a lot of cases. Angular gears with bevel wheel pairs are used for this. Feed gears are commonly realized as cascade gears. A direct connection of the gear unit to the motor is realized when they are installed together. That a compensating element otherwise necessary for shaft skew (compensating coupling) can be eliminated is a benefit for cascade gears.

The moment of inertia of feed gears should be as low as possible for good acceleration behavior of the overall drive. For this reason, gear wheels should have small diameters (because these are included in the mass moment of inertia to the power of 4). This often requires multi-stage gear drives to be used instead of single-stage. To realize the absolute requirement of freedom from clearance, high structural effort is usually the consequence.

One alternative solution is tangential tensioning of two sprockets, pivot-mounted on a shaft, that cog with a counter wheel as wide as the two sprockets. This variant usually assists in reducing design and manufacturing overhead.

For the feed gear unit shown in ◘ Fig. 2.98, tangential preloading is solved in terms of design for both stages by a sprocket with a twistable sprocket ring flanged on. Other stiffness-improving measures can also be seen such as conical seats (oil interference fit assembly) of the sprockets on the shafts, shrink fits between motor shaft and pinion (oil interference fit assembly), bearing of the shafts with tensioned angular contact ball bearings, and bearing of the ball screw with stiff, preloaded axial roller bearings.

Another way to realize clearance-free feed gear units without dividing the wheel (i.e., shorter wheel width, lower mass moment of inertia) is to seat the sprocket shafts in adjustable eccentric bushings. The distance between axes can be changed to eliminate tooth flank play by twisting the eccentric bushings.

Using the planetary gear as an example, ◘ Fig. 2.99 shows the principle of play adjustment using conical sprockets, i.e., ground sprockets with continual profile displacement in the tooth gap direction. Axially pushing the planetary wheels into the sun gear eliminates clearance. This sprocket pairing can also be adjusted to be free of clearance with additional displacement of the ring gear. Unlike most other ways of designing feed gear units to be free of clearance, the standout features of this solution are that the gear unit can be loaded in both directions with the same forces and rotational frequencies, and that no intermediate gears are necessary.

The higher production effort compared to that of conventional sprockets is a downside. This is because in the production of the gear teeth, it is not possible to put restraints on the wheels in packages [WITT88].

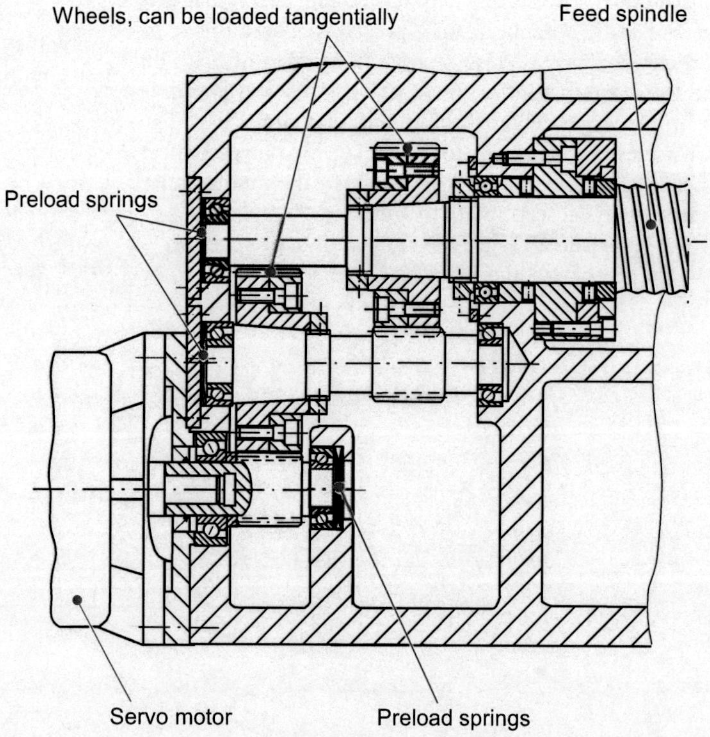


◘ **Fig. 2.98** Clearance-free cylindrical gear unit for feed spindles

**Fig. 2.99** Planetary gear with play adjustment using conically beveled sprockets according to Wittenstein alpha

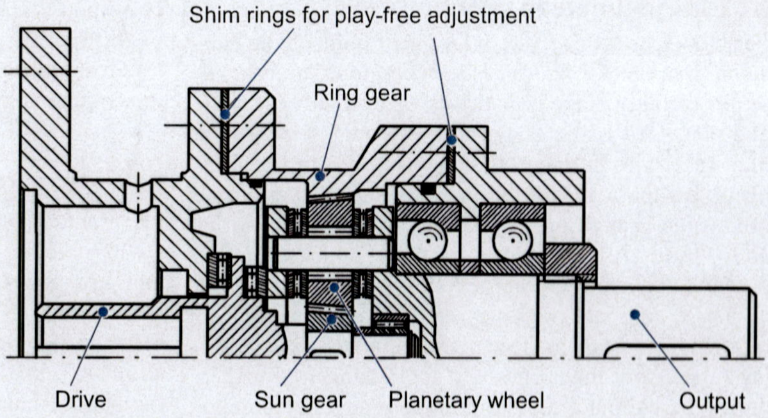


### 2.4.2.2 Toothed belt drives

In light of the disadvantages mentioned, toothed belt drives are frequently used as feed drives in place of gear drives when design reasons mean that an additional gear stage is necessary (**Fig. 2.100**). The toothed belt drive satisfies the requirements placed on feed gears used in NC machine tools in a particularly affordable way as regards stiffness, zero clearance and accuracy. High tensile strength, good level of compliance and low expansion can be attained using tensile members made of glass fibers, Kevlar fibers or steel cables.

The toothed belt is preloaded to increase stiffness and prevent play. This can be achieved easily with radial movability of the drive motor or with tensioning rollers. The tensioning rollers press against the outside of the belt and should not be spring-mounted. They also dampen the strand vibrations that occur, especially on longer, free belt lengths, and bring about a larger angle of contact so greater torque can be transferred. Furthermore, preloading also compensates the manufacturing tolerances over the toothed belt length.

The toothed pulleys are made of aluminum to improve dynamic behavior. The high level of material damping of the toothed belt material transfers the motor actuation with low vibration.

Also, the greater center distance means the toothed belt gear offers considerably better structural design options for installation of the motor. This results in feed drive concepts with a small installation space leading to small machine dimensions.

**Fig. 2.100** Feed drive of a bed-type milling machine with integrated toothed belt gear according to DMG MORI

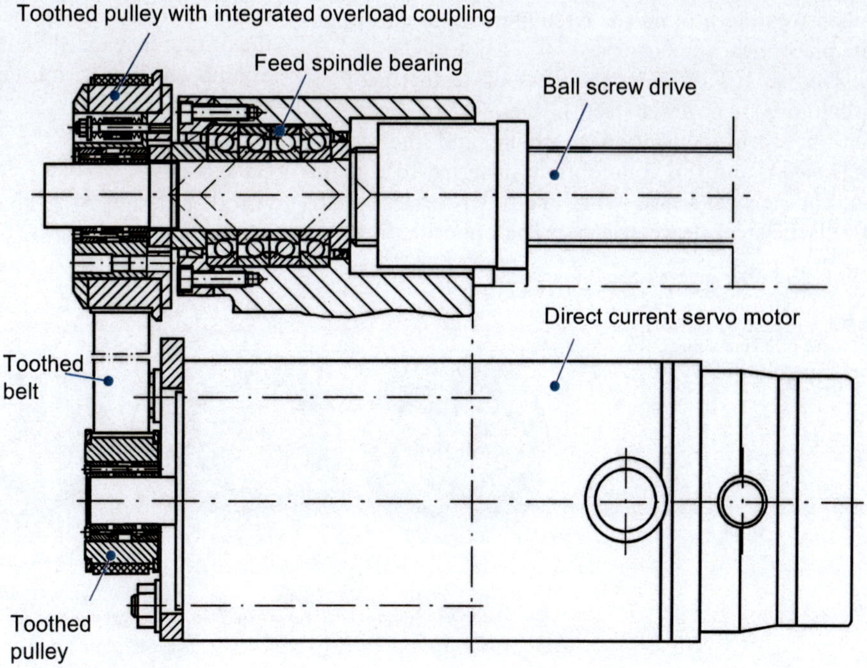

2

### 2.4.2.3 Custom feed gear units

For high gear ratios, custom feed gear units of the *Harmonic Drive* and *Cyclo* types are commonly used in addition to multi-stage gear drives with cylindrical gears and planetary gears. They satisfy the requirement of a high gear ratio in one gear stage, with a compact footprint, low mass and low moment of inertia, yet, at the same time, a high level of stiffness with coaxial drive and output. A downside is the lower efficiency compared to conventional gear drives.

These compact, high gear ratio drives are used as intermediate gears in feed drives and to drive rotating tables, tool magazines and tool changers. Furthermore, the compact gear units represent a key component in the joint drives of robots. Particular importance is attached to the following criteria: high gear ratio in the minimum footprint, coaxiality, high dynamic range, low clearance, high torsional stiffness and high overload capacity to realize highly dynamic drives with an extremely low level of clearance and with high positioning and repeat accuracies.

#### Harmonic Drive

Figure 2.101 shows the functional principle of a coaxial feed gear (Harmonic Drive) [LAGE13]. A rigid circular spline with internal toothing is connected to the housing so as to remain stationary. In this rigid spline is an elastic, cup-shaped steel sleeve (Flexspline) with external toothing that usually has two fewer teeth than the internal toothing. The Flexspline with external toothing is deformed by an elliptic disk joined to the drive, with ball bearing fitted (Wave Generator), and pressed into the internal toothing of the circular spline rotating around the circumference at two opposite places near the big axis.

Due to the different numbers of teeth (the Flexspline has two fewer teeth), there is relative rotation between the stationary spline and the flexible steel sleeve and this rotation is transferred to the output via the steel sleeve. When the roller carrier rotates, the elastic steel sleeve rotates in the opposite direction, corresponding to the difference between the numbers of teeth for the internal and external toothing.

Gear ratio $i$ is given by the numbers of teeth $z$ for the circular spline (CS) and Flexspline (FL) as

$$i = \frac{z_{FL}}{z_{FL} - z_{CS}} \qquad (2.120)$$

Harmonic Drive gears permit gear ratios of $i = 30 - 320$ for output torques of $T = 0.5 - 10{,}000$ Nm, and are free from clearance. The torsional stiffness is very high given the high percentage contact area of the dual-sided tooth contact (the tooth contact area is about 15% of the total number of teeth). The low mass moment of inertia of the gear unit continues to be particularly beneficial.

Figure 2.102 shows two different designs of the Harmonic Drive gear, both using an elliptical inner ring as Wave Generator. The cup design comprising the three gear components, circular spline, Flexspline and Wave Generator, is shown in the left part. An refinement of this gear type is the flat design and this realizes the benefits of the gear unit with compact external dimensions. Unlike the cup design, a fourth base element, the Dynamic Spline, serves here as the output. In conjunction with the elliptically-shaped Flexspline, this represents the gear ratio level by means of the low difference between teeth numbers, whereas the Flexspline and circular spline have the same number of teeth.

Figure 2.103 shows the use of the flat design as the feed drive of a milling machine. The benefits as regards high gear ratio within the smallest footprint, low gear play and straightforward installation options are apparent here. The drive motor drives the Wave Generator. The circular spline is secured in the housing and the output is realized via the Flexspline to the Dynamic Spline, which in turn is connected rigidly to the output shaft.

#### Galaxie drive

Another design of a coaxial hollow shaft gear drive is the Galaxie drive from Wittenstein, shown in

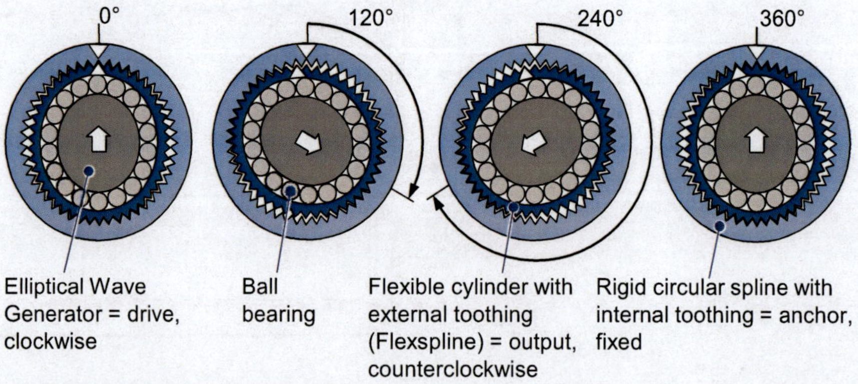


**Fig. 2.101** Functional principle of a Harmonic Drive according to Harmonic Drive AG

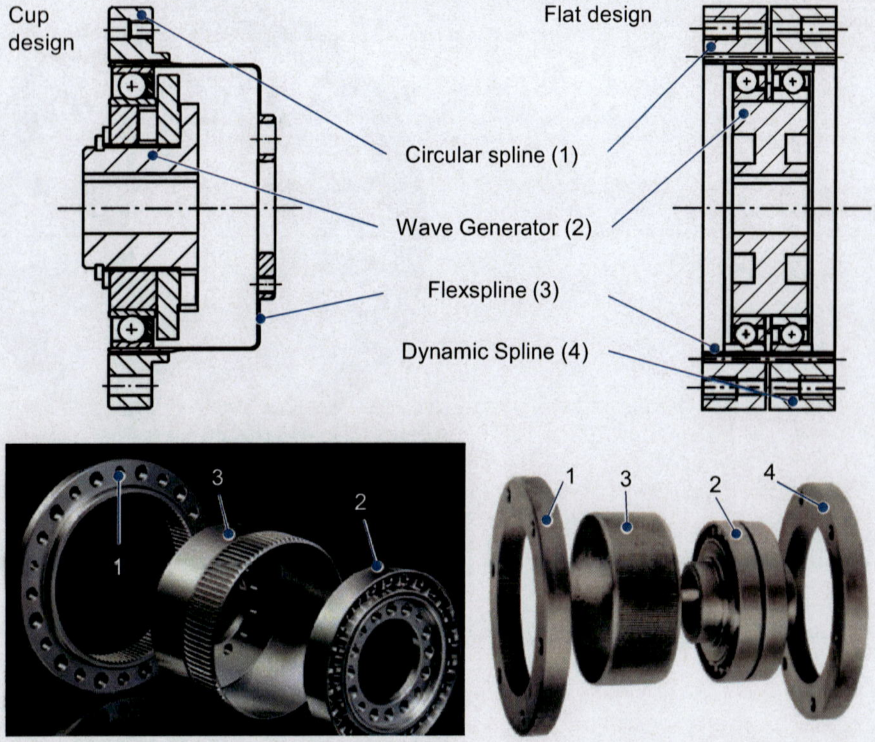


**Fig. 2.102** Designs of Harmonic Drive gears. *Source:* Harmonic Drive AG

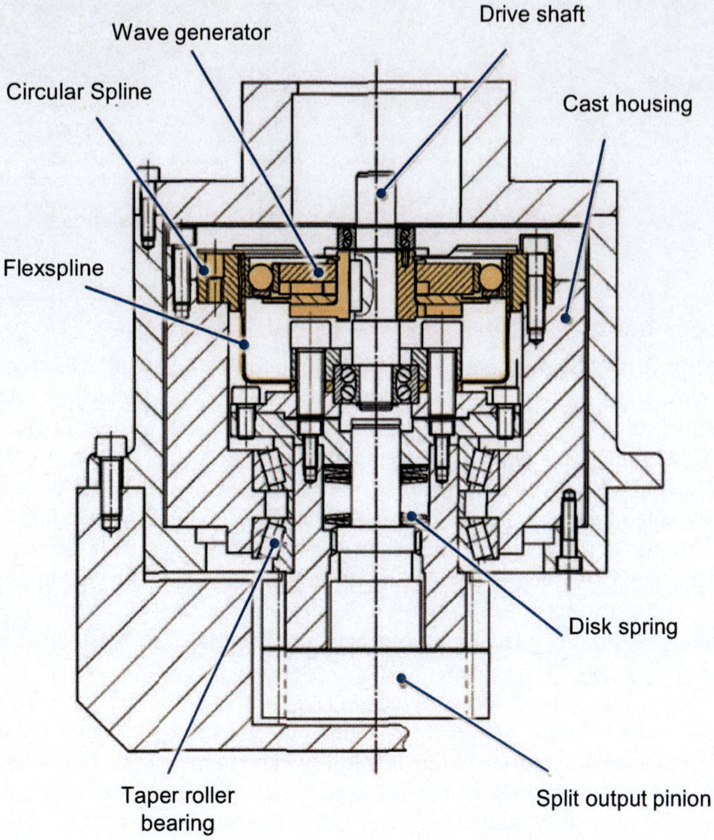


**Fig. 2.103** Feed drive for the milling head of a milling machine according to Harmonic Drive AG

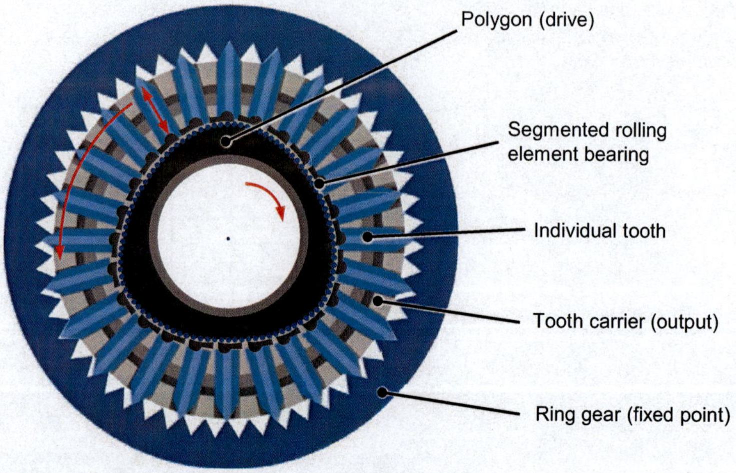


**Fig. 2.104** Design and function of a Galaxie gear drive according to Wittenstein alpha

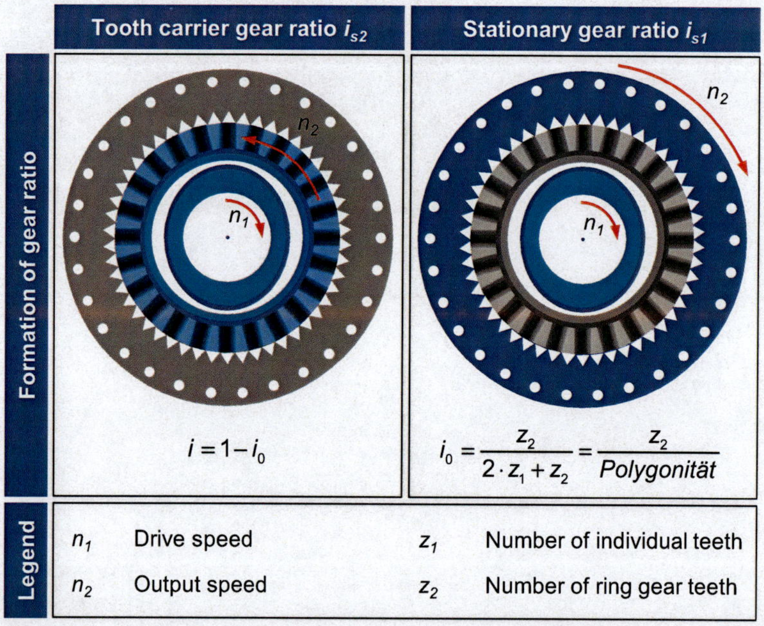


**Fig. 2.105** Gear ratio of a Galaxie drive according to Wittenstein alpha

Fig. 2.104. Its design and function differ from other hollow shaft gear drives. Unlike planetary gears for example, this gear unit has no gears with external teeth. Instead, the drive train is implemented as a polygon hollow shaft on which individual, pencil-shaped teeth are seated using a needle bearing. Every individual tooth sits on a segment of the needle bearing outer ring and is guided while radially sliding using the tooth carrier that also serves as the output. To prevent the segments twisting, two teeth are positioned one behind the other on every swivel segment, giving rise to two rows of teeth next to each other. The contour of the drive shaft means the individual teeth undergoes one stroke in the radial direction over one revolution. The number of strokes corresponds to the polygonity. The circumferential, generally static ring gear, has corresponding internal toothing via which teeth in the contact slip off causing rotation of the tooth carrier. In every stroke, the individual teeth in the planetary wheel move on by one tooth. Different rotational directions of the output relative to the drive rotational direction can be realized depending on the drive variant. The direction of rotation of the tooth carrier when the ring gear is stationary is opposite to the rotation of the polygon. If the stationary gear ratio is used, the drive and output directions of rotation are the same [o.V.15].

Gear ratio $i$ is given, as shown in Fig. 2.105, by the number of teeth on the ring gear and the number of polygons.

The contours of the tooth flanks of the individual teeth and of the ring gear follow the contour of a logarithmic spiral. The radial infeed of the individual tooth gives rise to a tangential contact, over the entire surface in the ideal case, at a spiral piece having the same pitch as the ring gear toothing. This surface contact allows a converging lubrication gap to build forming a hydrodynamic lubrication film and contributing to a friction reduction.

As there are several facing teeth in contact at any one time, there is a large contact area enabling high torques to be transferred at a high rotational speed. This also means the torsional stiffness of the gear unit is very high. Another benefit of the gear unit is the potentially small size while having a relatively big hollow shaft at the same time. Clearance-free toothing can be achieved with corresponding preloading of the gear unit [o. V.15].

Use of the gear unit is being investigated. On area for this is handling (e.g., for rotating tables) requiring high positioning accuracies and torsional stiffnesses. Another application for the gear unit is the drive train of lathes.

### Cyclo drive

Another design of these compact gear units is the eccentric drive. The most well-known of these is the Cyclo drive (◘ Fig. 2.106). The functional principle of the eccentric drive can be described as follows: A circular disk is driven by a cam and is subjected to a rolling motion within a stationary ring. Every point on the disk follows a cycloidal curve. Rotational motion develops on the disk at a considerably lower speed in the opposite direction. The actual ratio depends on the ratio of ring to disk diameters.

To prevent slippage during the rolling motion, the disk on the Cyclo drive has a closed cycloidal contour as the external shape and the ring is replaced by bolts arranged in a circle. Every cam plate has one fewer cam sections than there are bolts in the bolt ring. The locuses of the disk engage in the rollers of the static outer ring in a form-fit manner and roll on it. The reduced rotational motion of the cam plate is transferred to the output shaft via bolts engaged into holes on the plate. The gear ratio is determined by the number of cam sections on the cam plate.

Rollers are placed on the bolts of the outer ring and output shaft. These rollers produce rolling-only power transfer between the cam plate and bolt ring and the cam plate and driving pins of the output shaft. This reduces friction losses, noise and wear to a minimum. There are needles between the eccentric shaft and cam plate, also to keep friction low here [LORE26].

The Cyclo drive has four main components: drive shaft with cam, cam plate, outer ring with bolts and rollers, and output shaft with bolts and rollers. To bring about a balancing of masses, the gear units have two cam plates offset by 180°. These plates are driven by a double eccentric. This allows the output torque to be increased at the same time.

### Eccentric drive

Another design of the eccentric drive is shown in ◘ Fig. 2.107. The cycloids of the Cyclo drive are replaced by gear teeth. Contact of the two sets of teeth always takes place where the cam is nearest to the pitch circle of the ring gear.

When the cam is rotated by half a revolution from $K$ to $L$, the smaller planetary wheel is subjected to rolling motion in the larger ring gear. This means the contact point moves from $M$ to $N$ and by the same angle as the cam. As the smaller planetary wheel mounted on the cam has a correspondingly shorter circumference than the ring gear, it needs to rotate through a greater angle than the cam. So the path from $M$ to $N$ on the ring gear corresponds to path $M$ to $O$ on the small wheel. The proper motion of the planetary wheel is in

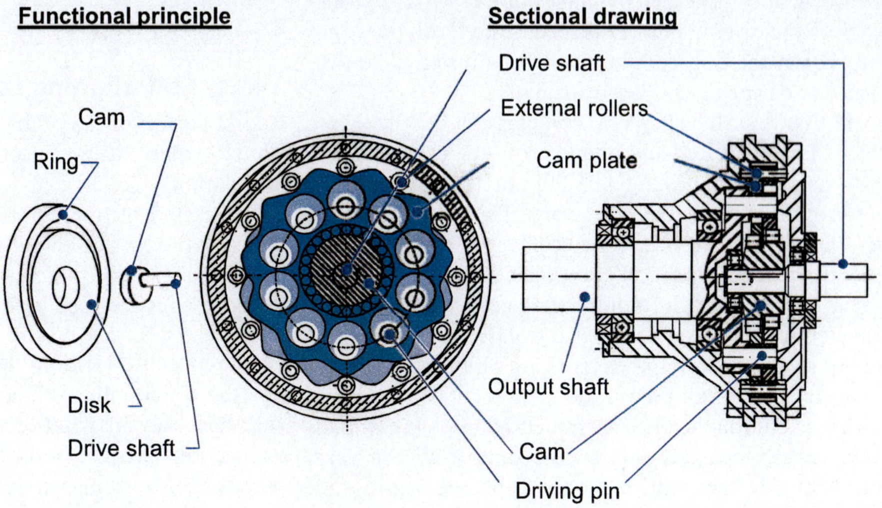


◘ **Fig. 2.106** Cyclo drive. *Source:* Sumitomo Drive Technologies

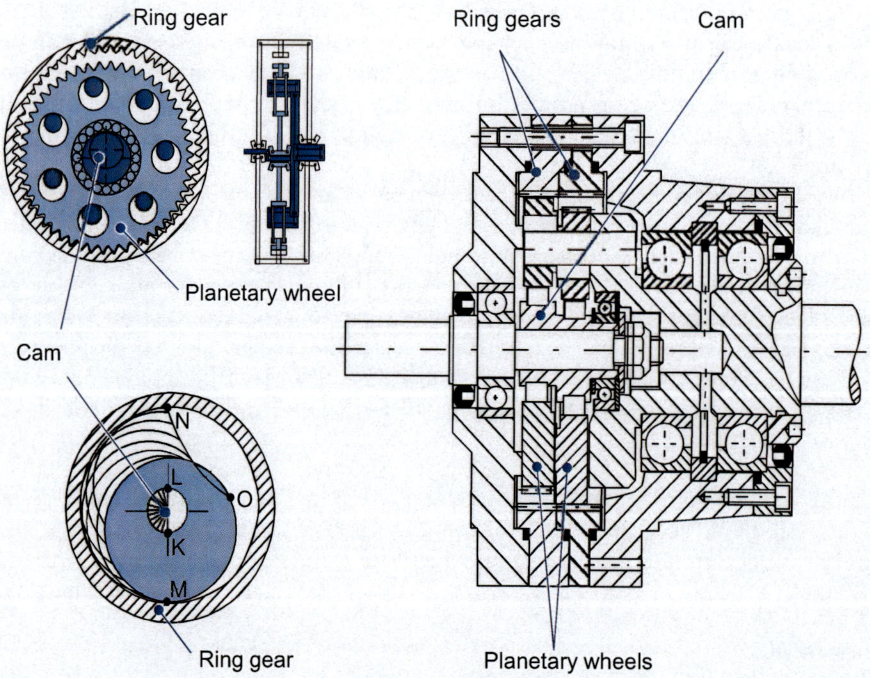


**Fig. 2.107** Eccentric drive according to AKIM.swiss Getriebetechnik

the opposite direction to the cam and corresponds to the difference between the circumferences of the two wheels.

To convert cycloidal motion to rotational motion, the toothed washer has holes engaged (as for the Cyclo drive) by corresponding driving pins attached in the output flange shaft. These holes and engaging bolts can be compared to a shaft coupling. The gear ratio is only determined by the pitch circle diameter and the number of teeth. Fitting two toothed washers, offset by 180° to each other, produces a weight counterbalance of the cam imbalance and creates a tensioning option.

With mutual twisting of the two ring gears during installation, the toothed washers are rotated around their own axes, and thereby make contact with opposite sides of the driving pins. This means the tooth play and the play in the bolt coupling can be adjusted. The two cams are designed as roller bearings.

Figure 2.108 shows a comparison of the Harmonic Drive, cycloid and planetary gears, part of comprehensive bench testing [o.V.90]. In the comparison of gear units, note the sizes vary. The measured values must therefore be referenced to the nominal torques that extend from 250 Nm (Gear unit 3) to 340 Nm (Gear unit 2). The Harmonic Drive gear has very good transfer properties at the lowest rotational speeds. At higher speeds however, the errors are on the same order of magnitude for all gear units.

Static compliance is determined by applying a load with an increasing torque to the output of the gear unit with the input drive blocked and measuring the resulting torsion angle. A linear relationship between load torque and torsion angle, with the characteristic profile as flat as possible, would be desirable here. The cycloid gear, having a stiffness of 118.5 Nm/arc min. at the nominal torque, performed particularly well here. Hysteresis caused by play and friction is low for all three gears albeit slightly different.

Even for dynamic loads, the cycloid gear achieved the best results, namely the lowest dynamic compliance at the highest resonance frequency. The efficiency of the planetary gear is far higher than the values for the other two. It improves for all types under load [ROSE94].

### 2.4.3 Couplings

#### 2.4.3.1 Self-aligning couplings

Radial, axial or angular shaft skew cannot be prevented when fitting mechanical components in the drive train. Under load, deformations that must be compensated also occur between components. This is the reason why self-aligning couplings are used as connecting elements. Torsionally stiff and elastic self-aligning couplings are used in feed drives.

**Torsionally stiff, bending-flexible couplings**

For the torsionally stiff connection of two shaft ends with accuracy of rotation angle (for direct drives, the connection between motor and ball screw spindle in particular), couplings are used in feed drives. They only

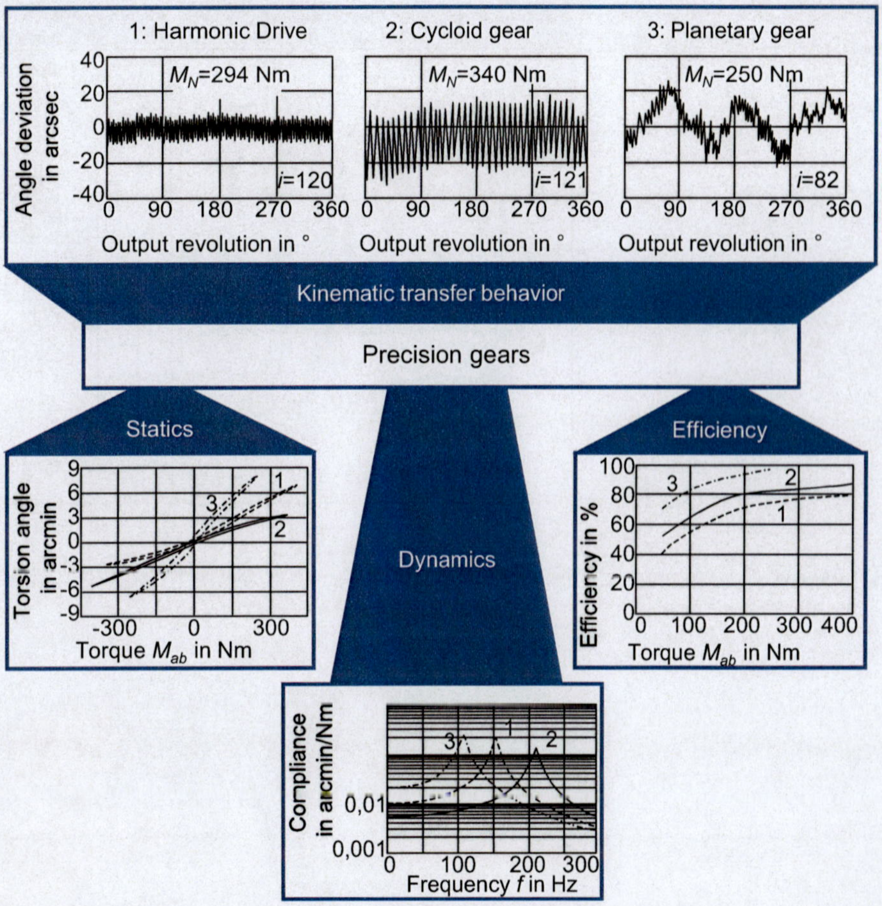


**Fig. 2.108** Operating characteristics of custom, high gear reducing feed drives

exhibit high stiffness in the circumferential direction. This means the rotational motion is transferred very accurately in the circumferential direction. All other load directions have a lower spring rigidity so that, within certain limits, radial, axial and angular alignment errors between the two shaft ends can be compensated.

Force-fit couplings are normally used for high precision feed drives. They satisfy best the strict requirements placed on torsional stiffness, zero clearance and low mass moment of inertia. Their design engineering is based upon torque to transfer, shaft diameter and torsional stiffness.

Figure 2.109 shows the most common variant of a torsionally stiff, flexible coupling—the bellows coupling. It comprises two ring-shaped bellows wheels connected by bellows that can be screwed to the corresponding shaft ends of the motor and spindle. The individual shafts of the bellows enable the aforementioned errors to be compensated [JAKO83]. For a larger angular misalignment, unusual for machine tools, other designs can be used, such as spiral toothed couplings.

### Elastic couplings with damping

Elastic couplings with damping are used for discontinuous machining processes with impact-like process forces. Rotational vibration and impacts occurring during operation can be dampened and reduced using elastic materials with high inner damping in the force flow. This protects the drive against dynamic overloading. In addition to rotational vibration damping, elastic couplings can usually also compensate the axial, radial and angle displacements of shafts to connect.

When using gear rings made of synthetic material having a high Young's modulus, the coupling becomes more torsionally stiff and commonly meets the requirements for machine tools. Their straightforward and compact design, together with low cost, means they are used frequently.

Figure 2.110 shows a torsionally flexible claw coupling. Two congruent coupling halves having concave claws inside are positioned, offset in relation to each other by half a division in the circumferential direction. Their design is such that an elastic gear ring made of synthetic material can be inserted in the space between them.

2

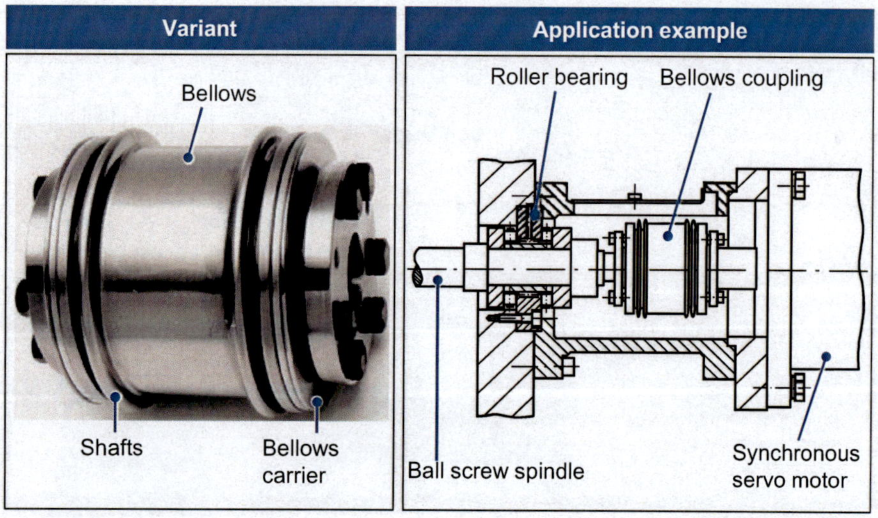


**Fig. 2.109** Torsionally stiff, flexible bellows coupling. *Source:* Jakob

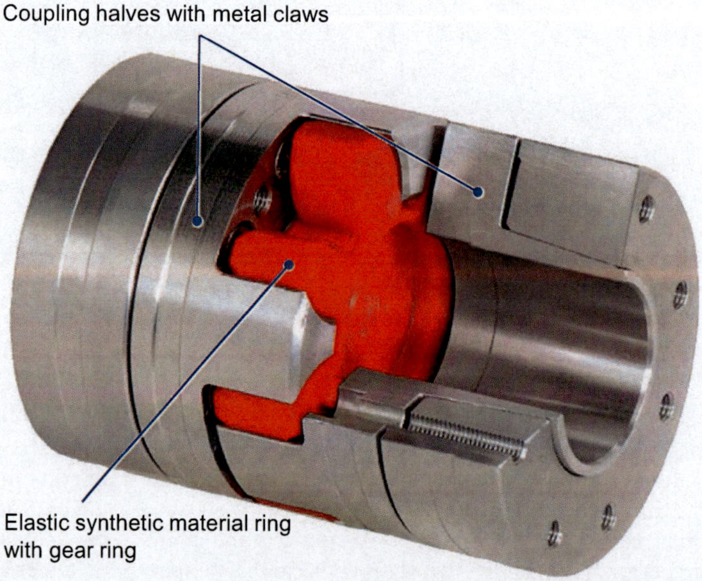


**Fig. 2.110** Elastic claw coupling for damping rotational vibration. *Source:* KTR Systems

## Safety couplings for overload protection

To effectively safeguard NC machine tools against overload and collision damage resulting from tool fractures as well as programming and operating errors, safety couplings are used. These limit the effective torque in a drive train to a maximum value. When this value is exceeded, the force flow is interrupted to reliably protect components at risk from damage [JAKO79].

The arrangement of safety couplings in the force flow depends on the positions of the components to protect and the positions of the machine components causing the high collision forces. These forces are essentially caused by two mechanisms. First, in the event of a collision with the sudden deceleration of a machine axis, high inertia forces need to be absorbed. These forces are determined by the kinetic energy of the moving machine components (such as carriage, workpiece, spindle and motor). All the kinetic energy is converted to deformation energy. Second, the motor torque in a collision increases momentarily to 3 to 10 times the nominal torque. Inertia forces and peak torque of the motor add up to the resulting overall collision force that leads to elastic deformations and, in the worst case, also to permanent deformations and breakages of machine components located in the force flow. Major significance is attached to the perfect position of a safety coupling to effectively avert collision damage to machine tools. Of interest in the analysis are the energies occurring between two machine components the instant they collide.

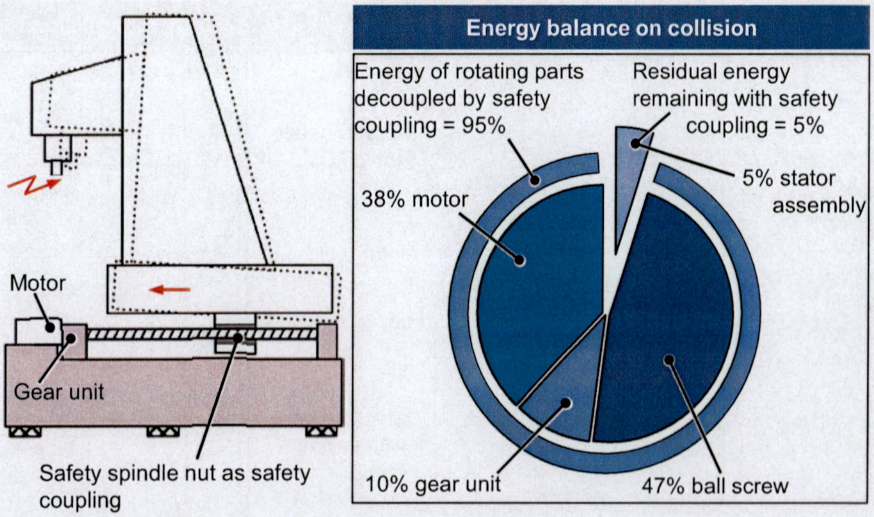


**Fig. 2.111** Energy balance for collision of a milling machine of moving column design according to Bohle

Figure 2.111 shows, for a milling machine with moving column design, the energy balance of the components in question when a collision occurs. At 5%, the mass moved translationally is by far the lowest. An effective reduction of collision damage can only be achieved here when all rotating masses are decoupled from the translational motion when a collision occurs. The transition point between the spindle nut and feed slide thus seems the best place to attach a safety coupling. In view of the practical considerations however, the coupling is usually attached between the motor and spindle.

The effectiveness of safety couplings in the force flow of feed drives has been verified in practical collision tests [ULRI82]. Figure 2.112 shows the feed drive of the cross slide for a slant-bed lathe with machine-specific data. Also, on the right there are two measured curve profiles for the collision force. These were recorded with and without the safety coupling when the carriage impacted the spindle housing in the x-direction. It was possible to reduce the impact by 64%, from 110 to 40 kN, using a safety coupling on the motor shaft.

Safety couplings should be designed such that the two coupling halves, after being disengaged by a mechanical mechanism, can re-engage exactly in relation to each other to prevent displacement of the reference point for the control system.

Safety couplings are designed as frictional slipping clutches and increasingly as spring-loaded form-fit couplings.

Figure 2.113 shows a variant of a form-fit overload coupling. While many of the familiar form-fit overload couplings transfer torque using balls or rollers, this coupling uses the principle of spur gearing that is kept in the contact with spring force. The faces of the spur gearing are helical so that, at the force transfer points, there is contact over a large area during the switch-off process as well as during the normal transfer of torque. This prevents excessively high Hertzian stress.

Furthermore, the coupling ring that moves axially is split into several segments. The cone shape of the spur gearing causes the springs to push the individual segments outwards into the internal toothing of the coupling housing. This means the rotational motion between coupling ring and housing is transferred completely clearance-free.

Figure 2.114 shows a safety coupling that makes it possible to limit the effect of inertia forces on the parts of the translationally moving masses of the feed drive when a collision occurs. For this, the sectioning point was placed between the carriage and spindle nut so that, when a malfunction occurs, there is no moment of inertia of the spindle leaving only the kinetic energy of the carriage itself.

The safety spindle nut is a hydraulically loaded force limiter acting on both sides. When it is engaged, the force flow passes from the spindle via the spindle nut to a flange bolted to it. An annular piston is used to secure this to another flange with axial precision. This flange is bolted directly to the carriage. Pressure oil is applied to the ring cylinder using the P-connector. This causes the annular piston to press against the precision-machined plane surfaces of the two flanges, and the ring cylinder to seal against the depressurized tank return (T). The force with which the annular piston joins the two flanges together depends on the annular piston surface and the hydraulic pressure. If the external force engaging at the carriage exceeds the annular piston force, the annular piston is raised from the plane surface of one of the two flanges. The annular gap between the annular piston and cylinder thus enables flow to the depressurized tank return. The hydraulic

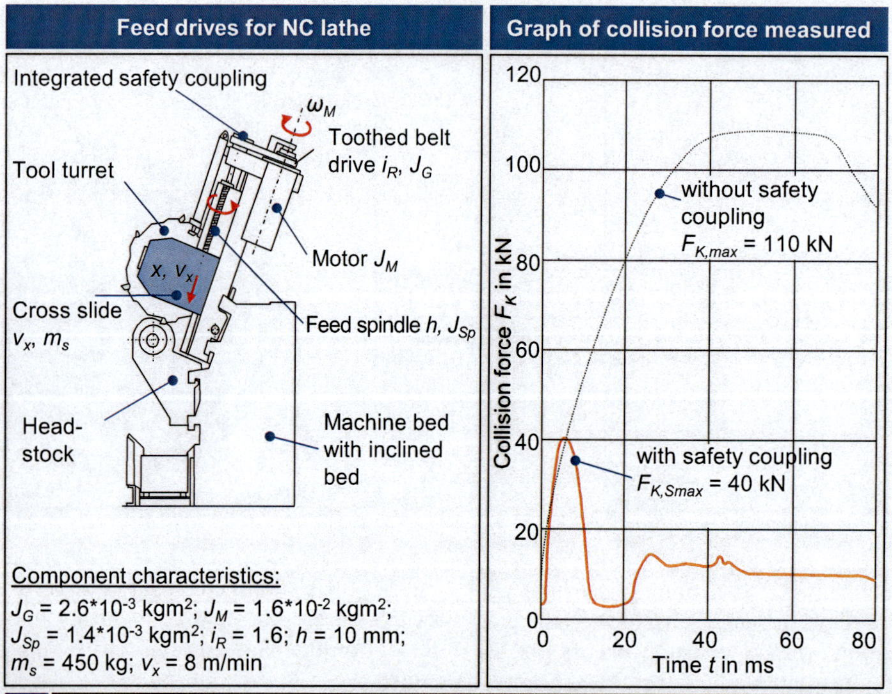


**Calculation**

Reduced masses:

$m_S$ = 450 kg; $m_{Sp,red}$ = 553 kg; $m_{G,red}$ = 3705 kg; $m_{M,red}$ = 22803 kg

| Collision forces calculated: | without coupling | with coupling |
|---|---|---|
| Case 1: $k$ = 1 kN/µm | $F_{K1}$= 699.4kN | $F_{K,S1}$ = 133.5 kN |
| Case 2: $k$ = 0.1 kN/µm | $F_{K2}$= 221.2kN | $F_{K,S2}$ = 42.2 kN |

**Fig. 2.112** Collision force profiles measured on an NC lathe

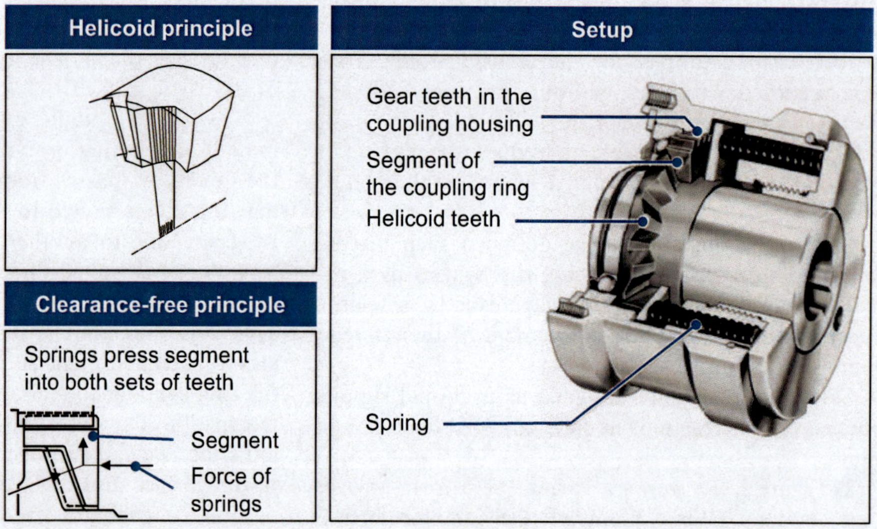


**Fig. 2.113** Coupling for limiting torque in feed gears according to RINGSPANN GmbH

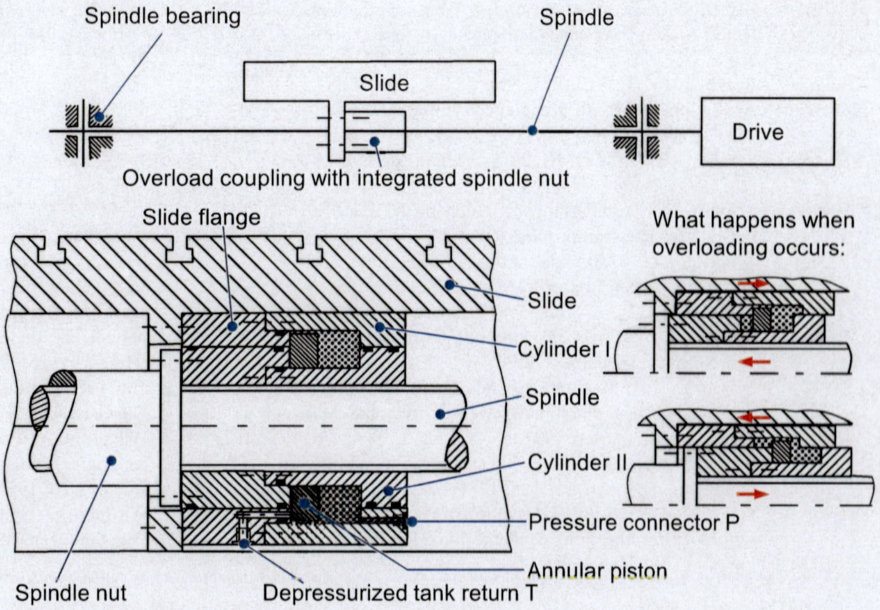


**Fig. 2.114** Hydraulic safety spindle nut according to Jakob

pressure at the ring cylinder drops immediately, and the force-fit joint between the two flanges or the spindle nut and carriage is disconnected.

Switches are used to detect the start of the disengage stroke. Then, the servo motor is braked. The coupling can be preloaded and returned to its fully functional condition after a collision by simply reversing the direction of the drive spindle.

## References

[AULM70] Aulmann, A.: Bausteine und Anwendungsbeispiele für die Übertragung und Verarbeitung von Winkelinformationen. In: Sonderdruck "Funk und Technik", 1970, 1–4

[BRAA00] Braasch, J.; Fa. Heidenhain: Genauigkeit von Vorschubantrieben. Traunreut, Juni 2000. Firmenschrift

[BRAA97] Braasch, J.: Genügen Drehgeber zur Positionierung von Werkzeugmaschinen? In: MSR Magazin, 1997, 7–8

[BREC02] Brecher, C.: Vergleichende Analyse von Vorschubantrieben für Werkzeugmaschinen. (Reihe: Berichte aus der Produktionstechnik. Bd. Bd. 2002,17). Aachen: Shaker, 2002

[BREC12] Brecher, C. (ed.): Effizienzsteigerung von Werkzeugmaschinen durch Optimierung der Technologien zum Komponentenbetrieb—EWOTeK. Apprimus Verlag, Aachen (2012)

[DIN71] Norm DIN 1319 (1971). Grundbegriffe der Meßtechnik

[DIN88] Norm DIN 32876 (1988). Elektrische Längenmessung

[ERNS98] Ernst, A.: Digitale Längen- und Winkelmesstechnik. (Reihe: Die Bibliothek der Technik. Bd. 165). 3. Aufl. Landsberg/Lech: Moderne Industrie, 1998

[FORS10] Forstmann, J.: Kugelgewindetriebe im Einsatz an Kunststoffspritzgießmaschinen. Duisburg, Essen, Univ., Diss., 2010, 01.01.2010

[GROS06] Gross, H., Hamann, J., Wiegärtner, G.: Technik elektrischer Vorschubantriebe in der Fertigungs- und Automatisierungstechnik. Publicis Corporate Pub, Erlagen (2006)

[HAME06] Hameyer, K.: Elektrische Maschinen 2. VorlesungsskriptRWTH Aachen, 2006

[HEID00] Heidenhain: Positionsmeßsysteme für elektrische Antriebe. Traunreut, März 2000. Firmenschrift

[HEID89] Heidenreich, W.: Aufbautechniken für Halbleiter-Magnetfeldsensoren. In: Technisches Messen. 56, 1989, 436–443

[HENN99] Henneberger, G.: Elektrische Maschinen I—Grundlagen, Wirkungsweise, Aufbau, Betriebsverhalten. VorlesungsskriptRWTH Aachen, 1999

[HEWL76] Hewlett-Packard: Laser-Transducer-System. System operating and service booklet. Santa Clara, CA, 1976. Firmenschrift

[HEWL92] Hewlett-Packard: Magnetisches Längenmeßsystem ML01, Band 30–05. Kirchzarten, 1992. Firmenschrift

[ISER99] Isermann, R.: Mechatronische Systeme. Springer, Berlin, Heidelberg, New York, Barcelona, Hongkong, London, Mailand, Paris, Singapur, Tokio (1999)

[JAKO79] Jakob, L.: Sicherheitskupplungen in Achsantrieben von NC-Maschinen. Konstruktion, Elemente, Methoden. Stuttgart: Konradin-Verlag, 1979

[JAKO83] Jakob, L.: Anwendungen drehstarrer, flexibler Kupplungen. In: Der Zuliefermarkt, Sept. 1983

[LAGE13] Lageder, F.: Entwicklung von Harmonic Drive® Getriebe für Raumfahrtantriebe, 10.–12.09.2013

[LARS03] Larsen, R.M.: Fieldbus Technology. Springer, Berlin Heidelberg (2003)

[LORE26] Schutzrecht USRE17811 E (30.11.1926)Lorenz Konrad braren: Gear Transmission

[MENZ73] Menzel; Mirande; Weingärtner: Fourier-Optik und Holographie. Springer, Berlin, Heidelberg, New York (1973)

[MILL90] Miller, W.: Bezeichnungen für Weg- und Winkelmeßsysteme, Verwirrende Begriffe. In: Industrieanzeiger 78, 1990

[ROSE94] Rosenbauer, T.: Getriebe für Industrieroboter—Beurteilungskriterien, Kenndaten, EinsatzhinweiseR-WTH Aachen, 1994

[SCHL15] Schlecht, B. (ed.): DMK 2015. TUDpress, Dresden (2015)

[SCHM09] Schmitt, G.: Ansteuerung von Hochvolt-IGBTs über optimierte Gatestromprofile, 2009

[SCHÖ09] Schöder, D.: Elektrische Antriebe—Regelung von Antriebssystemen, 1st edn. Springer, Berlin Heidelberg (2009)

[SCHÖ13] Schöder, D.: Elektrische Antriebe—Grundlagen, 5th edn. Springer, Heidelberg (2013)

[SESS91] Sesselmann, T.: Maßstäbe für interferentielles Abtasten ermöglichen Nanometer-Meßschritte. In: WB Werkstatt und Betrieb, 1991, 124-2

[SIMO71] Simon, W.: Die numerische Steuerung von Werkzeugmaschinen. Hanser, München (1971)

[SPIE90] Spies, A.: Längen in der Ultrapräzisionstechnik messen. In: Feinwerktechnik & Meßtechnik. 98, 1990

[TR14] Tränkler, H.-R.; Reindl, L. M.: Sensortechnik. Berlin: Springer, 2014

[ULRI82] Ulrich, H.: Verhütung von Kollisionsschäden an NC-Drehmaschinen. In: Industrie-Anzeiger 104. 45. 1982

[VDI73] VDI/VDE-Richtlinie VDI/VDE 2600–2 (November 1973). Metrologie (Meßtechnik)—Grundbegriffe

[WALC74] Walcher, H.: Digitale Lagemeßtechnik. VDI, Düsseldorf (1974)

[WECK90] Weck, M.; Eversheim, W.; König, W.; Pfeifer, T. (Hrsg.): Wettbewerbsfaktor Produktionstechnik: VDI-Verlag, 1990

[WECK99] Weck, M.; Krüger, P.; Brecher, C.; Wahner, U.: Components of the HSC-Machine. In: Schulz, H.; u. a. (Hrsg.): Proceedings of the 2nd International German and French Conference on High Speed Machining. Darmstadt, 10.-11. März 1999, S. 269–278

[WILD59] Wildermuth, E.: Der "Resolver", ein moderner Analogrechenbaustein. In: Sonderdruck "Feinwerktechnik". 63. 1959, S. 9

[WITT88] Wittman, G.: Spieleinstellbare Getriebe mit konischen Verzahnungen. In: Zuliefermarkt. 36, 1988

# Dynamic Behavior of Feed Axes

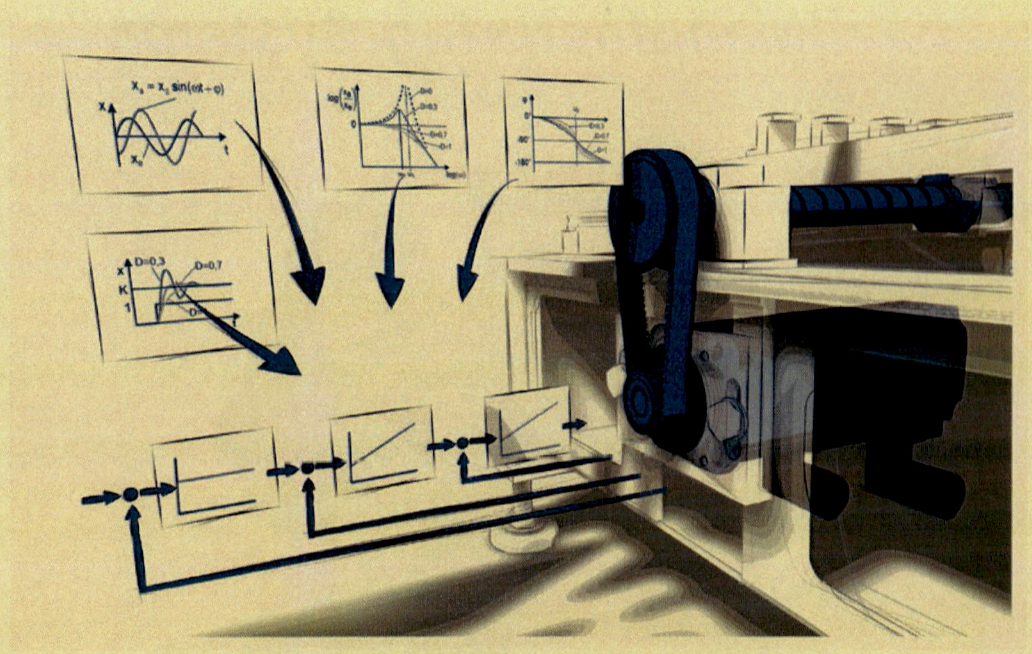

I thank my staff members Christian Fimmers M. Sc., Aleksandra Müller M. Sc. And Simon Storms M. Sc. for their contribution in revising this chapter.


C. Brecher and M. Weck, *Machine Tools Production Systems 3*,
Lecture Notes in Production Engineering, https://doi.org/10.1007/978-3-658-34622-5_3

Contents

The high workpiece accuracy required means the feed drives must be realized such that the position and velocity values specified by the control must be converted to the relative motion between tool and workpiece with the highest accuracy and without delay if possible. This requirement applies in particular to multi-axis, path-controlled machine tools where the individual axis motions are functionally dependent and every axis has its own feed drive. In addition, all drives here must have the same dynamic transmission behavior because otherwise errors occur when flat and spatial contours are generated.

The requirements on feed drives can be summarized as follows:

- *Distortion-free signal transmission:* The transition from one position to another must be without vibration; mechanical transmission elements must be free of reversal errors and play, with adequate stiffness.
- *High dynamic range:* Quick changes to the reference variable must be made with a short delay of the machine component to be moved.
- *Settling of disturbances:* Friction and cutting forces, and their fluctuations, require high static and dynamic stiffness of the drive or overall closed-loop control.
- *Adaptation of the individual axes:* For multi-axis machines, the transmission behavior of all feed drives must be the same, i.e., the drives having the faster reactions must be adapted to the slower drives.

Introduced first in this chapter are control-related principles—pertaining to linear, continuous and discrete-time transfer systems. After that, advanced methods for improving dynamic behavior are explained. These are the feedforward method and state control.

Also, control theory on a feed axis is illustrated. This pertains to the depiction of position control loops and using the most important parameters. In addition to closed-loop control, the transmission behavior of the mechanics also has a major bearing on the motion of the feed axis. Linear and non-linear behavior of the mechanical and electrical components are also described.

## 3.1 Control Engineering Principles

One of the tasks of machine tool numerical control (NC) is to output suitable reference variables to the feed drives using information on the paths to travel and speeds.

These reference variables are converted to position and velocity values either using a constant drive within a master position control loop or using a step motor in an open control chain, see ◘ Fig. 3.1. Every axis has its own position control loop or control chain that is normally independent of the other axes.

In control theory, terms "control" and "feedback control" are differentiated, with the course between input and output signal flows being analyzed. There is no clear notational distinction between control and feedback control.

One characteristic of a control chain is the open action flow in which there is no feedback of measured values. For a feed drive with position control, this means the actual position value is not compared to the reference variable, i.e., the reference position value. For position control, step motors are used as drive motors because the rotor of this motor type executes a defined angular rotation for one control pulse and the number of pulses is directly proportional to a certain rotation angle or feed path. The angular velocity can be

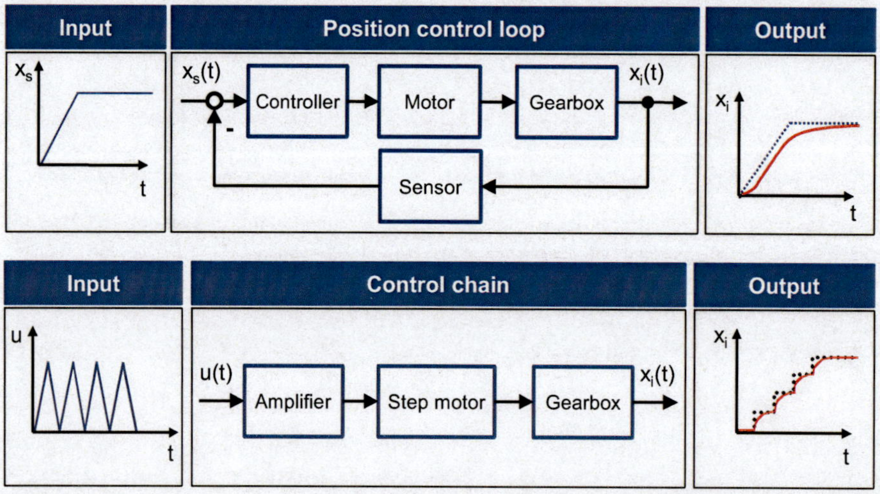


◘ **Fig. 3.1** Comparison of a control loop and a control chain

controlled using the frequency of the control pulses, ◘ Fig. 3.1 below.

A position control loop is required for constant drives, ◘ Fig. 3.1 above. ◘ Figure 3.2 shows the details of a constant feed control loop. A speed control loop and a current control loop are realized here as secondary loops for the position control loop to increase drive dynamics.

The position control loop enables the rotation and angular velocity of the constant drive motor to be regulated in line with the reference variable. The position control loop comprises the following components:

- Control system
- Drive system (amplifier, motor and mechanical transmission elements)
- Measuring system

In the control system, the actual rotation angle of the motor or actual position $x_i$ of the machine table is subtracted from the reference variable $x_s$, i.e., the reference position value. This yields the control deviation $x_w$.

$$x_w = x_s - x_i \qquad (3.1)$$

This value is passed to the position controller. In the controller, a time-varying control variable for the drive motor is calculated from the difference between the actual and reference values to correct this deviation. If control deviation $x_w$ is zero, the drive has reached its reference position.

The drive system comprises the power amplifier, the servo motor and the mechanical transmission elements. The power amplifier provides the power necessary for the drive. Nowadays, synchronous motors are mostly used as constant drive motors. The rotary motion of the motor is normally converted to the translational motion of the table using a spindle nut system (see Sects. 3.3). When strict requirements are placed on axis dynamics, linear motor drives are also used. In this case, there are no mechanical transmission elements.

The measuring system closes the control loop with position feedback. Slide position $x_i$ is read using an angle or position measuring system (see ▶ Sect. 2.4), and passed to the control equipment to calculate a reference value/actual value comparison. Using the position control loop, the actual position of the table tracks the reference position and the influence of disturbing forces is eliminated.

The requirement that the actual position value of the machine table follows the reference position as free of errors and delays as possible can be realized with:

- High control loop gain
- High level of damping to prevent instabilities and overshooting
- Small drive time constants
- Low mass moment of inertia of the rotating parts
- High mechanical natural frequency
- Great stiffness of the mechanical elements that are within the flux of forces
- Freedom from play for the mechanical transmission elements
- A ratio of the cutoff frequencies of mechanical transmission system and control loop of $\omega_{0,Mechanik}/\omega_{0,Regelkreis} > 2$

The secondary speed and current feedback paths also shown in ◘ Fig. 3.2 are used to improve the dynamic behavior of the drive, i.e., loop gain for simultaneous

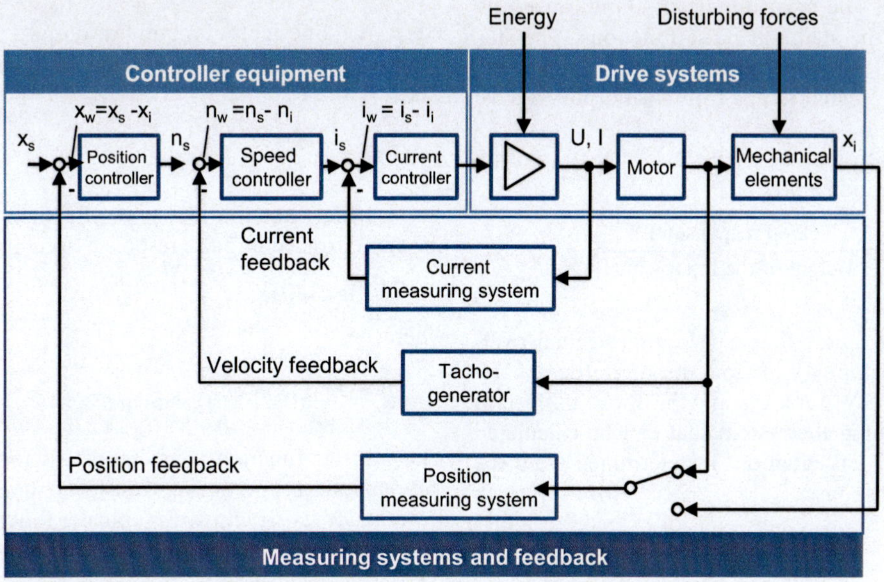


◘ **Fig. 3.2** Components of a position control loop with secondary speed and current feedback

3

improvement of damping behavior. This is addressed in detail in ▶ Sect. 3.2.

To define the dynamic behavior of the closed-loop position control, the transmission behavior of the entire system and the individual systems must be known in addition to the signal flow. An accurate analysis of the dynamic behavior of the closed position control loops requires some basic knowledge of control theory—summarized in brief below.

### 3.1.1 Linear, Continuous Transfer Systems

#### 3.1.1.1 Time Behavior of Control Loop Elements

The behavior of control loop elements, i.e., the dependency over time of the output parameter $x_a$ on the input parameter $x_e$, can be defined using a differential equation. Instantaneous values $x_e$ and $x_a$ play a part as well as their time derivatives, $\dot{x}_e$ and $\dot{x}_a$, $\ddot{x}_e$, $\ddot{x}_a$, etc.

$$a_n x_a^{(n)}(t) \ldots + a_2\ddot{x}_a(t) + a_1\dot{x}_a(t) + a_0 x_a(t) = b_0 x_e(t) + b_1\dot{x}_e(t) + b_2\ddot{x}_e(t) + \ldots + b_m x_e^{(m)}(t) \tag{3.2}$$

In Eq. 3.2, $x_e$ and $x_a$ are any physical variables, such as travels, forces or electrical voltages. Values $a_0$, $a_1$, $a_2$, $b_0$, $b_1$ and $b_2$ are system-defining constant coefficients. These coefficients need to be calculated or measured on a case-by-case basis depending on the hardware arrangement of the control loop elements [ABEL20, LUNZ08].

One method of investigating the time behavior of a linear control loop element is to solve the differential equation. To assess and compare control loop elements, special input signals are used and the resulting output signals assessed. The most important standardized input signals are the step and pulse functions. The step function is used more frequently; here, at time $t=0$, the system input signal jumps from an output to a set value.

Normalizing to the step height $\Delta x_e$ of the input signal yields step response $h(t)$:

$$h(t) = \frac{x_a(t)}{\Delta x_e} = \frac{\text{Step response}(t)}{\text{Height of the input signal}} \tag{3.3}$$

Step height, gain factor, time constant, dead time, overshoot range (damping), period duration (decay frequency) and decay duration are the most important parameters defining the system that can be calculated from the differential equation or determined from the measured step response.

◘ Figure 3.3 shows an example of the step response for a single-mass oscillator with different damping values.

Linear differential equations with constant coefficients for specified initial conditions can be solved easily using a Laplace transform. The Laplace transform uniquely and reversibly assigns the function $X(s)$ in the Laplace domain to the function $x(t)$ in the time domain—Eq. (3.4). The benefit of the transformation is that differentiation and integration of the time function, and the serialization and interconnection of system blocks, can be substituted by solving simple algebraic equations in the Laplace domain (◘ Fig. 3.4).

To solve a differential equation, the time function is first moved to the Laplace domain using a Laplace transform (◘ Fig. 3.3) [KUPF68].

$$X(s) = \mathsf{L}\{x(t)\} \tag{3.4}$$

Definition: $s$ is a complex frequency ($s=\sigma+j\omega$).

The transformation can be realized in most cases using correspondence tables. This gives an algebraic equation that can be solved easily. The correspondence tables are again usually adequate for the reverse transformation; a partial fraction decomposition might need to be performed beforehand (Eq. (3.5).

$$x(t) = \mathsf{L}^{-1}\{X(s)\} \tag{3.5}$$

Transforming a differential equation into the Laplace domain gives an algebraic equation of the following form:

$$X_a(s) = \underbrace{G(s)\cdot X_e(s)}_{A} + \underbrace{K_1 \cdot X_{a,t=0}}_{B} \tag{3.6}$$

Term $A$ is determined by the complex input signal $X_e(s)$; term $B$ depends on the initial conditions $X_{a,t=0}$. Setting the initial conditions for this representation to zero (i.e., $X_{a,t=0}=0$) yields the following equation:

$$X_a(s) = G(s)\cdot X_e(s) \tag{3.7}$$

Here, the quotient of complex output parameter to complex input parameter is called the transfer function $G(s)$:

$$G(s) = \frac{X_a(s)}{X_e(s)} \tag{3.8}$$

The transfer function $G(s)$ is the Laplace transform of the standardized impulse response (weight function):

$$G(s) = \mathsf{L}\{g(t)\} \tag{3.9}$$

where:

$$g(t) = \frac{d}{dt}h(t); \quad h(t) = \text{Step response} \tag{3.10}$$

The transfer function is an important tool in defining the dynamic properties of transfer systems [ABEL20]. In terms of its structure, the transfer function is identical to the system definition form used more commonly, the frequency response.

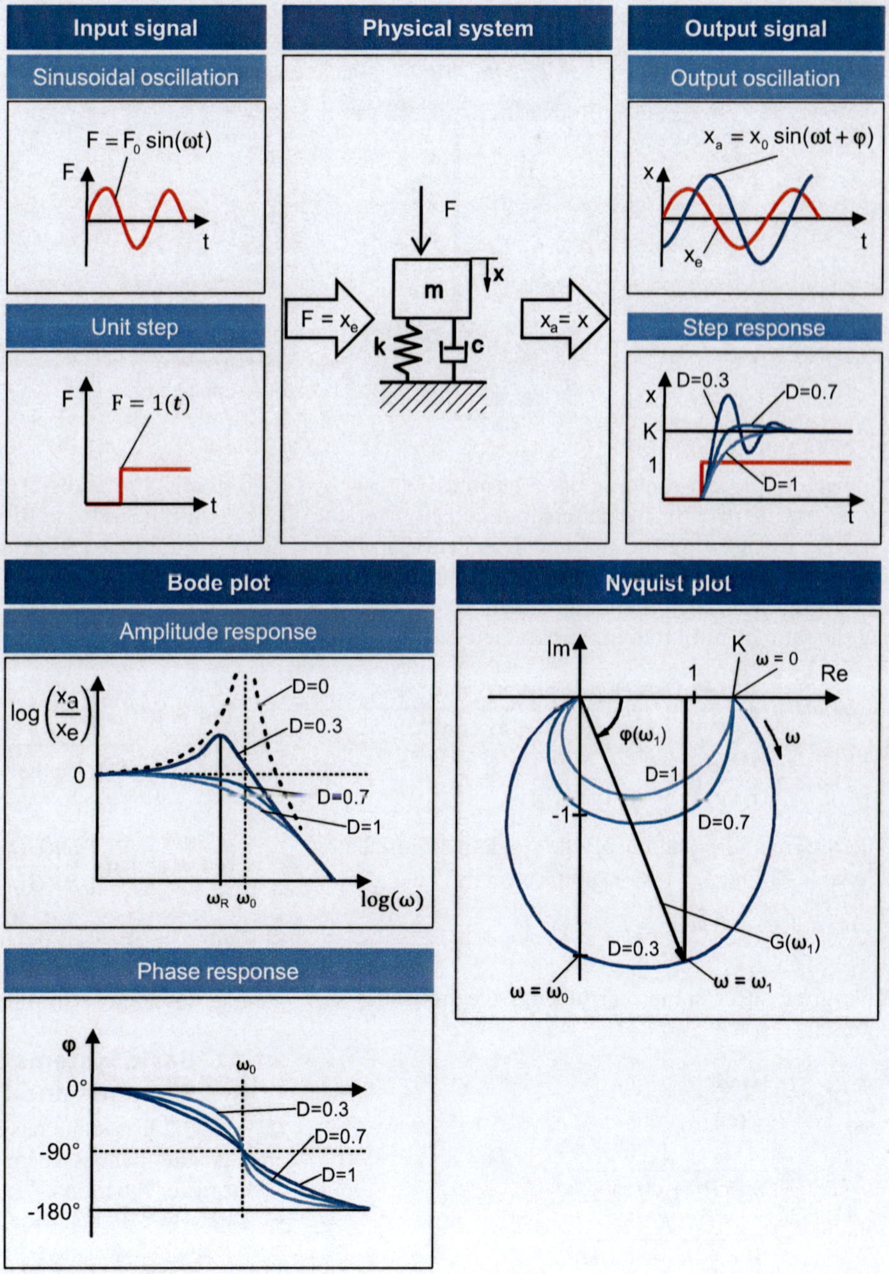


**Fig. 3.3** Different representations of a single-mass oscillator

The transfer function $G(s)$ fully defines the dynamic properties of the corresponding transmission element. However, as a function of the complex frequency $s$, the fact that it is less descriptive and barely measurable are downsides. Thus, complex frequency "s" is often substituted by "$j\omega$", corresponding to the real part $\sigma$ being neglected.

The frequency response $G(j\omega)$, however, is a complex function of real frequency values. Used as system input functions in the complex variable and time domains are easy-to-represent signals, namely harmonic signals having pure sinusoidal profiles, $x_e = \hat{x}_e \sin(\omega t)$. If the input parameter is varied sinusoidally, the output parameter also has a sinusoidal profile after a transient process, $x_a = \hat{x}_a \sin(\omega t + \varphi)$. After this transient process, the frequency of the output parameter is the same as that of the input parameter. The ratio of the amplitudes of input and output oscillation, and the phase shift between the two oscillations, are a function of the frequency in the steady state, Fig. 3.3 (top).

The frequency response of a transfer system is the function defining the amplitude ratio of input and output parameters (amplitude response) and the phase shift (phase response) as a function of the frequency.

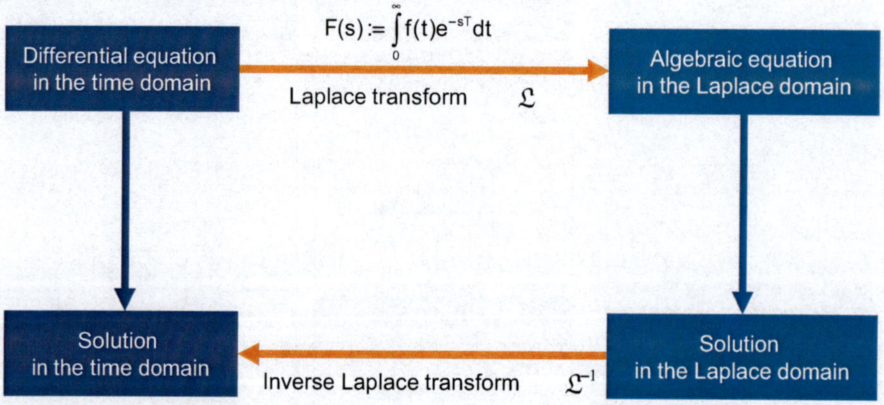


**Fig. 3.4** Methods for solving differential equations

The frequency response is determined either by measurement or with the differential equation for the system—by inserting the sinusoidal oscillations of the input and output parameters and their time derivatives into the appropriate differential equation and calculating the ratio of output to input parameter:

$$G(j\omega) = \frac{X_a(j\omega)}{X_e(j\omega)} = \frac{b_0 + b_1 \cdot j\omega + b_2 \cdot (j\omega)^2 + \ldots}{a_0 + a_1 \cdot j\omega + a_2 \cdot (j\omega)^2 + \ldots} = \frac{\hat{x}_a(\omega)}{\hat{x}_e(\omega)} e^{j\varphi(\omega)} \tag{3.11}$$

The differential equation of the single-mass oscillator shown in Fig. 3.3 is for example:

$$m \cdot \ddot{x} + c \cdot \dot{x} + k \cdot x = F \tag{3.12}$$

This gives, after some rearranging, the frequency response:

$$G(\omega) = \frac{\hat{x}(\omega)}{\hat{F}(\omega)} e^{j\varphi(\omega)} = \frac{1}{m(j\omega)^2 + c(j\omega) + k} = \frac{K}{1 + \frac{2D}{\omega_0} j\omega + \frac{1}{\omega_0^2}(j\omega)^2}$$
$$\text{mit } K = \frac{1}{k}, \omega_0 = \sqrt{\frac{k}{m}} \text{ und } D = \frac{c}{2 \cdot m \cdot \omega_0} \tag{3.13}$$

### Nyquist and Bode Plots

The frequency response is a complex function and can be shown on the complex plane. The graphical depiction of the real part as a function of the imaginary part of the frequency response is called the Nyquist plot. The frequency here is varied between $\omega = 0$ and $\omega \to \infty$. It is a parameter on the Nyquist plot.

The Bode plot is the graphical depiction of the frequency response in a rectangular coordinate system. Two graphs are necessary, one to show the amplitude or magnitude and one to show the phase. The coordinates are plotted logarithmically except for the phase plot.

Amplitude and phase can be determined as follows from Eq. 3.11:

$$\hat{G}(\omega) = |G(j\omega)| = \sqrt{(Re\{G(j\omega)\})^2 + (Im\{G(j\omega)\})^2} \tag{3.14}$$

$$\varphi(\omega) = \arctan \left\{ \frac{Im\{G(j\omega)\}}{Re\{G(j\omega)\}} \right\} \tag{3.15}$$

Figure 3.3 shows different depiction forms for a 2nd order oscillatory proportional system with delay (single-mass oscillator) for different damping ratios $D$.

#### 3.1.1.2 Basic Systems of Control Loop Elements and Their Representation

Table 3.2 lists some basic systems of control loop elements and defines them with differential equations and transfer functions:

- Delay-free P behavior (proportional): The output parameter is always proportional to the input parameter. There are no frequency-dependent amplitude or phase changes.
- Delay-free I behavior (integral): The output parameter corresponds to the integral of the input parameter over time.
- Delay-free D behavior (differential): The output parameter corresponds to the differential quotient over time of the input parameter.
- $PT_1$ behavior (proportional with 1st order delay)
- $PT_2$ behavior (proportional with 2nd order delay), highly damped (connection of two $PT_1$ systems in series)
- $PT_2$ behavior, weak damping

The corresponding Bode plots and step responses are shown in ◘ Table 3.1.

The output parameter is exactly proportional to the input parameter for $\omega = 0$ or $t \rightarrow \infty$, i.e., in the steady state. More time delays occur with increasing frequency, something that is accompanied by a phase shift and an amplitude change between input and output parameters (◘ Table 3.2).

### 3.1.1.3 Structure of a Control Loop

The function of a control loop is to keep a physical parameter in a system constant, or to change it in line with a prescribed program, thereby compensating disturbance effects. A control loop generally has the structure shown in ◘ Fig. 3.5 and the following control-related technical terms are used [ABEL20]:

**◘ Table 3.1** Step response and Bode plot of control loop elements

| Name (example) | Step response | Bode plot |
|---|---|---|
| P element<br>Transfer function:<br>$K$ | 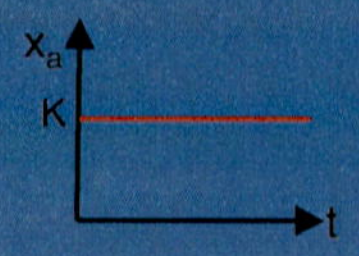 | 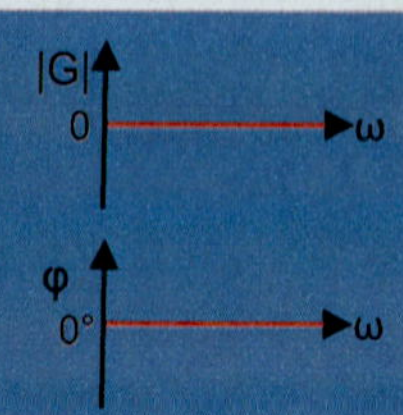 |
| I element<br>Transfer function:<br>$\frac{K}{(j\omega)}$ | 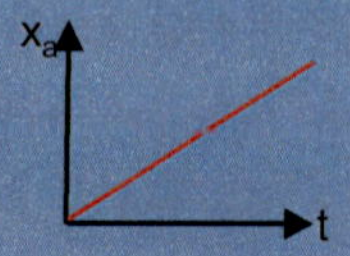 | 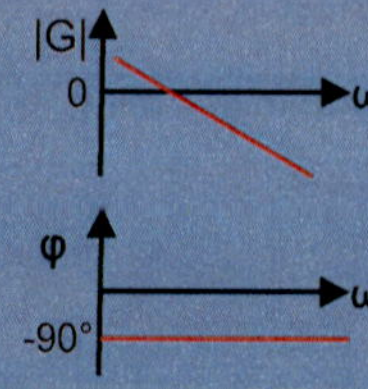 |
| D element<br>Transfer function:<br>$K(j\omega)$ | 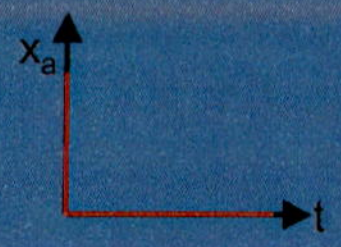 | 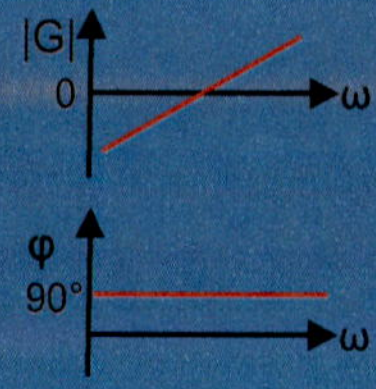 |
| $PT_1$ element<br>Transfer function:<br>$\frac{K}{1+T(j\omega)}$ | 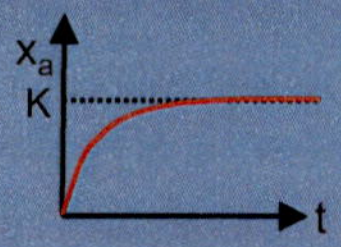 | 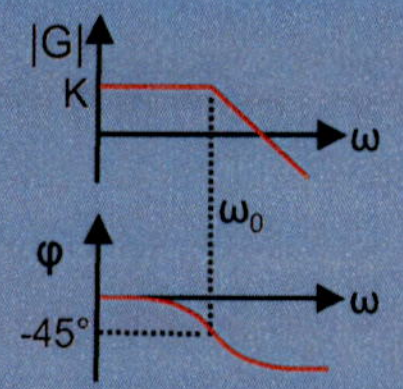 |
| $PT_2$ element, D>1<br>Transfer function:<br>$\frac{K}{1+(T_1+T_2)j\omega+T_1T_2(j\omega)^2}$ | 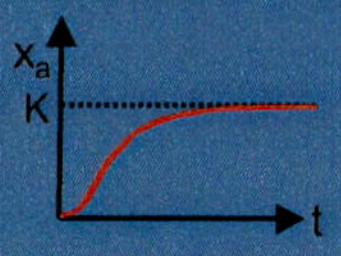 | 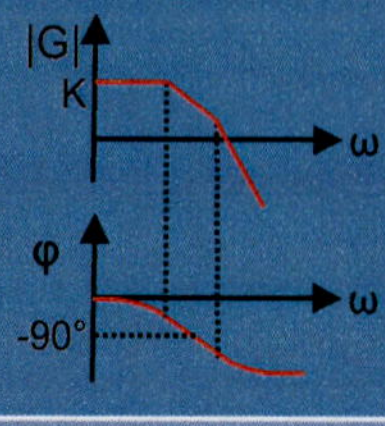 |
| $PT_2$ element, D<1<br>Transfer function:<br>$\frac{K\omega_0^2}{(j\omega)^2+2D\omega_0(j\omega)+\omega_0^2}$ | 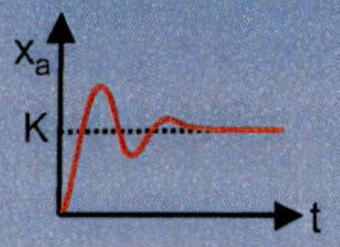 | 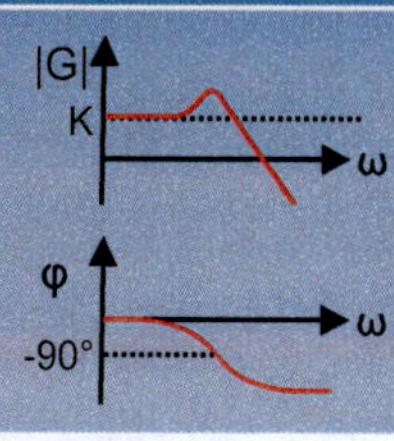 |

3

**Table 3.2** Basic systems of control loop elements and their transfer function

| Name (example) | Differential equation | Transfer function |
|---|---|---|
| P element (lever) | $x_a = Kx_e$ | $K$ |
| I element (relation between velocity and path) | $x_a = K\int x_e dt$ | $\frac{K}{(j\omega)}$ |
| D element (reversal of the I system) | $x_a = K\dot{x}_e$ | $K(j\omega)$ |
| $PT_1$ element (simplified electric motor) | $T\dot{x}_a + x_a = Kx_e$ | $\frac{K}{1+T(j\omega)}$ |
| $PT_2$ element, well dampened (single-mass oscillator) | $T_1T_2\ddot{x}_a + (T_1+T_2)\dot{x}_a + x_a = Kx_e$ | $\frac{K}{1+(T_1+T_2)j\omega + T_1T_2(j\omega)^2}$ |
| $PT_2$ system, weakly dampened (single-mass oscillator) | $\ddot{x}_a + 2D\omega_0\dot{x}_a + \omega_0^2 x_a = K\omega_0^2 x_e$ | $\frac{K\omega_0^2}{(j\omega)^2 + 2D\omega_0(j\omega) + \omega_0^2}$ |

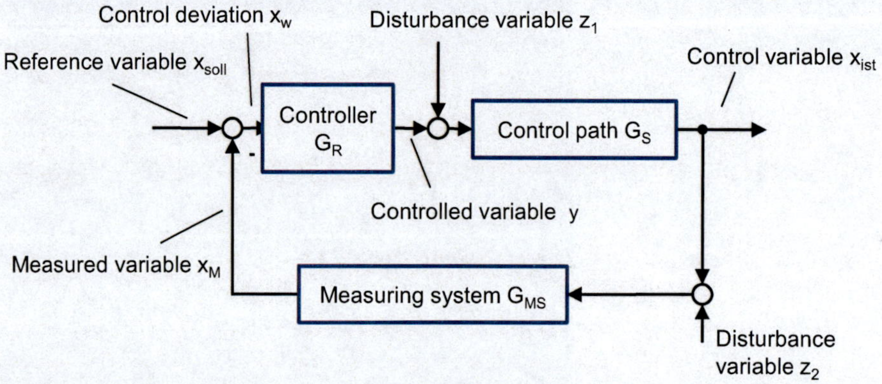


**Fig. 3.5** Structure of a control loop

*Control path:*
The system to be controlled in which one or more parameters (control variables) are kept as constant as possible against exterior factors or are to follow specified reference values in a manner that is as error-free as possible.

*Controller:*
Equipment for calculating the difference between actual and reference values, i.e., the control deviation, and for commanding the actuator in line with a specified dynamic internal behavior (such as P, I or PI behavior).

*Control loop:*
Connection of the control path with the controller.

*Controlled variable:*
Variable; to be kept constant or changed specifically for intended operation of the control path.

*Actual value:*
Instantaneous value of the controlled variable.

*Control deviation:*
Difference between the reference and controlled variables.

*Reference variable:*
Specified reference value of the controlled variable and this value is constant or varies over time.

*Control variable:*
Variable for influencing the energy flow of the path, and so the controlled variable. The control variable is the controller output.

*Actuator:*
Element fitted in the control path for controlling the energy flow (e.g., a valve).

*Disturbance variable*:
Every influencing factor that attempts to change the specified controlled variable (e.g., the load).

### 3.1.1.4 Signal Flow (Block) Diagram

The cycle of a control process is determined by the time behavior of the individual control loop elements and how they are connected (structure of the closed-loop control system).

In schematic form, the signal flow diagram shows the linking of the individual loop elements (◘ Fig. 3.6). For identification of the time behavior for the individual elements, the associated equations or individual step responses are also shown in the corresponding blocks. The signal lines can branch off or merge within a control loop. This creates branch and addition points in the signal flow diagram.

At a branch point, the line of action in the signal flow diagram divides and continues on to different control loop elements simultaneously. Every path receives the same signal.

At addition points, different signals meet and are superimposed on each other with the correct sign. The outgoing variable comes from subtracting or adding the incoming variables. The signs for incoming variables must always be specified at addition points.

In a control loop, multiple individual transfer systems are connected to each other with the time behavior of each being defined by its frequency response. A few calculation rules are required to determine the overall frequency response (◘ Fig. 3.7). The benefit of the

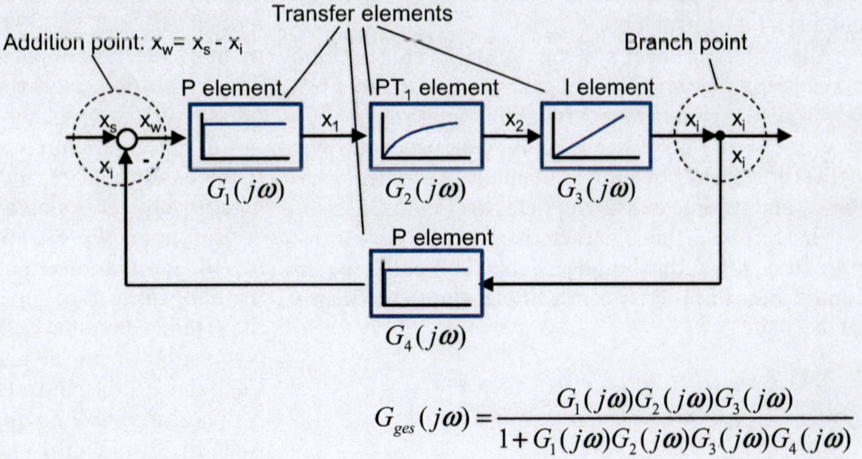


◘ **Fig. 3.6** A control loop in the signal flow diagram

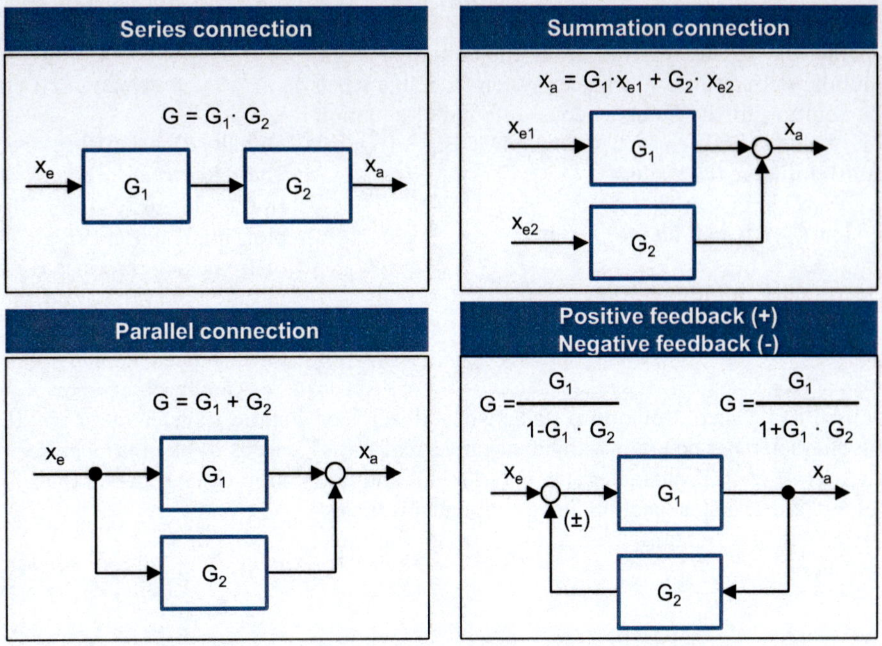


◘ **Fig. 3.7** The ways control loop elements can be connected

frequency response view is that connections in series and negative feedback occurring in the control loops can be calculated using simple additions and multiplications of complex individual frequency responses.



#### 3.1.1.5 Stability of Control Loops

Ascertaining whether a control loop is stable or not is one of the most important tasks in the mathematical investigation of control loops. Stability in the control theory sense means that a system "initiated" externally continues to oscillate with a decaying amplitude (and comes to a quiescent state) once the excitation is removed. An unstable system responds to an input signal with a monotonically or oscillating increasing output signal. This transitions the overall system to a not fully functional state, causing in many cases damage beyond repair of the system. A number of stability criteria have been developed to assess the stability or instability of a control loop.

Applicable to linear systems: A stable control loop is stable for any input signal and an unstable control loop is also unstable for every input signal. The stability of a system only depends on the homogeneous part of the differential equation containing the control variable $x_a$ and its time derivatives (left part of Eq. 3.2).

The zeros of the "characteristic equation" are required to solve the homogeneous differential equation of an n-th order system with constant coefficients [LUNZ08]:

$$\alpha_n\lambda^n + \alpha_{n-1}\lambda^{n-1} + \ldots + \alpha_1\lambda + \alpha_0 = 0$$

$$x_h(t) = \sum_{i=1}^{n} C_i e^{\lambda_i t}$$

**(Solution of a homogenous differential equation)** (3.16)

Coefficients $C_i$ are derived from the initial and the boundary conditions. A transfer system is stable when the solution to the homogeneous differential equation for $t$ tends to $\infty$, i.e., each summand of Eq. 3.16 must vanish for large time values:

$$\lim_{t\to\infty} C_i \cdot e^{\lambda_i \cdot t} = 0 \text{ für i} = 1,\ldots,\text{n} \quad (3.17)$$

The stability condition is thus:

$$Re\{\lambda_i\} < 0 \text{ für i} = 1,\ldots,\text{n} \quad (3.18)$$

Consequently, a control loop is stable when all zeros of the characteristic equation exhibit negative real parts. The zeros of the characteristic polynomial and the poles of the transfer function being equal means the assertion can also be transferred to the poles of the transfer function.

Algebraic stability criteria can only be applied when the differential equation or an analytic expression for the frequency response or transfer function of the closed control loop is known. Such expressions are not available in many cases, however. The Nyquist stability criterion offers a key simplification.

This stability criterion starts with the frequency response $G_0(j\omega)$ of the open control loop. Then, the control loop is disconnected at some point, in front of the controller, for example (◘ Fig. 3.8) such that the controller and path are connected in series. Multiplying the individual frequency responses then gives the frequency response $G_0(j\omega)$ of the open loop.

##### Stability Check Using the Nyquist Plot

For the stability check, frequency response $G_0(j\omega)$ of the open control loop is shown as a Nyquist plot. A check is performed to find where the Nyquist plot intersects the real axis. The location of the intersection point is the criterion for system stability or instability. For all points on the positive real axis, input and output oscillations have the same phase positions. If, in this case, the amplitude of the output oscillation is smaller than that of the input oscillation, the control loop is stable. If the Nyquist plot $G_0(j\omega)$ passes through critical point $P_k$ $(Re\{G_0(j\omega)\} = -1;\ Im\{G_0(j\omega)\} = 0)$, i.e., the input and output oscillations of the open control loop have the same amplitude, the system is on the stability boundary. If the intersection point of the Nyquist plot with the real axis is to the left of the critical point $P_k$, the control loop is unstable (◘ Fig. 3.8).

The simplified Nyquist stability condition applies when the open control loop $G_0(j\omega)$ is stable or exhibits integrating behavior. The condition is [ABEL20]:

$$\begin{aligned} &\text{Stability}: Re\{G_0(j\omega)\} \succ 1 \\ &\text{Stability threshold}: Re\{G_0(j\omega)\} = -1 \text{ at } Im\{G_0(j\omega)\} = 0 \\ &\text{Instability}: Re\{G_0(j\omega)\} < -1 \end{aligned} \quad (3.19)$$

##### Stability Check Using the Bode Plot, Amplitude and Phase Reserve

In the same way as stability analysis using the Nyquist plot, information can also be taken from the Bode plot of frequency response $G_0(j\omega)$, ◘ Fig. 3.8 on right. Amplitude reserve $A_R$ and phase reserve $\alpha_R$ can be introduced to identify the oscillation behavior. They define the distance of the Nyquist plot from the critical point.

Amplitude reserve $A_R$ is the factor by which the static transfer factor of the open control loop $G_0(j\omega)$ needs to be multiplied for the associated closed control loop $G(j\omega)$ to be on the stability boundary (Eq. 3.20).

$$A_R = \frac{1}{|G_0(j\omega_\pi)|} \quad \text{with } \varphi_0(\omega_\pi) = -\pi \quad (3.20)$$

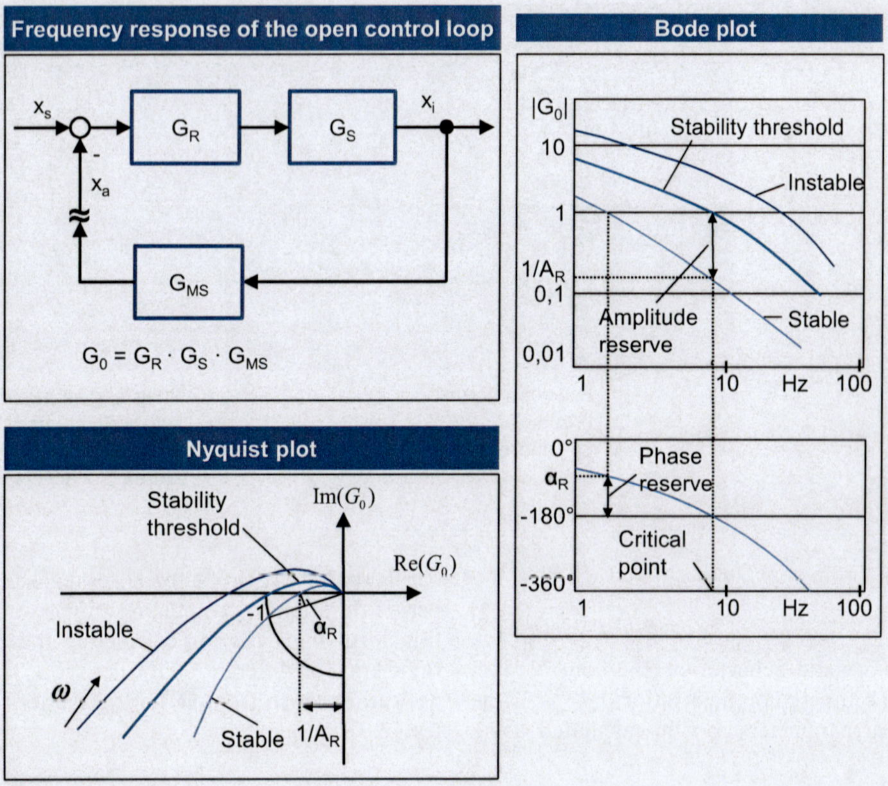


**Fig. 3.8** Nyquist stability criterion

Phase reserve $\alpha_R$ is the angle that a pointer (to the point at which the Nyquist plot intersects a circle around the origin with radius 1) forms with the negative real axis (Eq. 3.21, Fig. 3.8) [ABEL20].

$$\alpha_R = \varphi_0(\omega_d) + \pi \ \text{with} |G_0(j\omega_d)| = 1 \tag{3.21}$$

$A_R = 1$ and $\alpha_R = 0$ apply to a control loop on the stability boundary. Stable control loops have an amplitude reserve $A_R > 1$ and a positive phase reserve $\alpha_R$. This criterion is satisfied for the closed control loop $G_1$ with its Nyquist plot of the open control loop $G_{0,1}$ in Fig. 3.9.

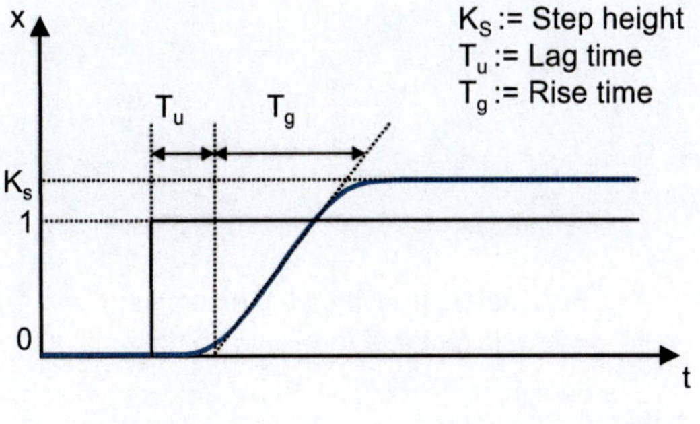


**Fig. 3.9** Measuring method for identifying the step response

### 3.1.1.6 Rules for Setting Analog Controllers

The dynamic properties of the controller are configured to optimize the dynamic behavior of the control loop. Table 3.3 shows a summary of the most common controller types and their defining parameters.

The challenge when structuring control loops is mainly to dimension the entire control loop such that the overall system is operable, i.e., it is sufficiently fast without excessively high overshooting or instability. The controller can be set using the analytic methods described in ▶ Sect. 3.1.1.5. Another method involves determining the optimum setting parameters for the controller using the characteristic values to be measured for the path of the system to be controlled.

One method starts from the control path step response measured or calculated. The principle of the measurement process is shown in Fig. 3.9. Actuators and measuring transducers are also included in the measurement process. The step is created at the point of disconnection after the controller output and the step response of the control path is recorded before the input into the controller.

As Fig. 3.9 shows, constructing the tangent of inflexion yields equivalent variables $T_u$ (lag time) and $T_g$ (compensation time) that, together with the transfer factor of the path, define

3

**Table 3.3** Summary of common controller structures and their parameters

| Controller | Differential equation | Transfer function |
|---|---|---|
| P | $x_a(t) = K_R x_e(t)$ | $\frac{X_a(j\omega)}{X_e(j\omega)} = K_R$ |
| I | $x_a(t) = K_I \int x_e(t)dt$ | $\frac{X_a(j\omega)}{X_e(j\omega)} = K_I \frac{1}{j\omega}$ |
| PI | $x_a(t) = K_R \left[ x_e(t) + \frac{1}{T_n} \int x_e(t)dt \right]$ | $\frac{X_a(j\omega)}{X_e(j\omega)} = K_R \left[ 1 + \frac{1}{T_n \cdot (j\omega)} \right]$ |
| PID | $x_a(t) = K_R \left[ x_e(t) + \frac{1}{T_n} \int x_e(t)dt + T_v \frac{dx_e(t)}{dt} \right]$ | $\frac{X_a(j\omega)}{X_e(j\omega)} = K_R \left[ 1 + \frac{1}{T_n \cdot (j\omega)} + T_v \cdot (j\omega) \right]$ |

P: Proportional
I: Integral
D: Differential

$K_I$: Integral gain factor
$K_R$: Proportional gain factor
$T_n$: Resetting time
$T_v$: Derivative action time

$$K_s = \frac{\Delta y}{\Delta x} \tag{3.22}$$

the dynamic and static behavior of the control path with sufficient accuracy. For time ratios $T_g/T_u > 3$, the controller setting parameters can be calculated using Table 3.4.

Another way to determine the controller parameters is based upon practical analyses of a closed control loop. Regardless of the type of controller to be implemented, the system to be controlled is first operated in a control loop with a P controller. Starting from stable operation, the gain factor $K_R$ of the P controller is increased until the control loop just starts to sustain oscillation. The gain factor $K_{Rkrit}$ reached here and the period duration $T_{krit}$ of the resulting continuous vibration form the basis for calculating the controller parameters. For different controller types, the setting values can be taken from Table 3.5 after the oscillation test [ABEL20].

### 3.1.2 Linear Discrete-Time Transfer Systems

Position controllers, closed-loop speed controllers and (for some years now) also closed-loop current controllers have been realized in modern control systems and drive amplifiers using software running on

**Table 3.4** Values for setting a controller in accordance with the step response of the control path for $T_g/T_u \geq 3$ (according to Chien, Hrones and Reswick)

| Controller | Parameter | Aperiodic controller profile | Controller profile with 20% overshoot |
|---|---|---|---|
| P | $K_R$ | $\frac{0{,}3}{K_S} \cdot \frac{T_g}{T_u}$ | $\frac{0{,}7}{K_S} \cdot \frac{T_g}{T_u}$ |
| PI | $K_R$ | $\frac{0{,}35}{K_S} \cdot \frac{T_g}{T_u}$ | $\frac{0{,}6}{K_S} \cdot \frac{T_g}{T_u}$ |
| | $T_n$ | $1{,}2 \cdot T_g$ | $1 \cdot T_g$ |
| PID | $K_R$ | $\frac{0{,}6}{K_S} \cdot \frac{T_g}{T_u}$ | $\frac{0{,}95}{K_S} \cdot \frac{T_g}{T_u}$ |
| | $T_n$ | $1 \cdot T_g$ | $1{,}35 \cdot T_g$ |
| | $T_v$ | $0{,}5 \cdot T_u$ | $0{,}47 \cdot T_u$ |

Controller parameters:
$K_R$: Gain factor
$T_n$: Resetting time
$T_v$: Derivative action time

Parameters from the step response:
$K_S$: Gain factor of the control path
$T_g$: Compensation time
$T_u$: Lag time

**Table 3.5** The values for setting a controller in accordance with a vibration test at the stability boundary (according to Ziegler and Nichols)

| Controller | Gain factor | Resetting time | Derivative action time |
|---|---|---|---|
| P | $0{,}5 \cdot K_{Rkrit}$ | - | - |
| PI | $0{,}45 \cdot K_{Rkrit}$ | $0{,}85 \cdot T_{krit}$ | - |
| PID | $0{,}6 \cdot K_{Rkrit}$ | $0{,}5 \cdot T_{krit}$ | $0{,}12 \cdot T_{krit}$ |

Controller parameters:
$K_R$: Gain factor
$T_n$: Resetting time
$T_v$: Derivative action time

Parameters from oscillation test:
$K_{Rkrit}$: Gain factor on stability boundary
$T_{krit}$: Oscillation period duration

microprocessors or microcontrollers. The sequential execution of corresponding routines means these controllers work according to a set timing cycle and not continuously. Systems of this type are known as discrete-time systems.

### 3.1.2.1 Representation of Discrete-Time Systems

As the majority of systems to be controlled (control paths) work in a continuous manner, it is appropriate to define for discrete-time transfer systems an equivalent representation in which the input and output parameters are also treated as continuous signals (Fig. 3.10).

In contrast to analog systems upon which the continuous input signal acts continuously, for discrete-time systems, the input signal $f_e(t)$ is scanned at discrete, equidistant intervals ($T_0$). This discretization is called "sampling". This yields a sequence of signal values $f_e(k \cdot T_0)$. A hold element converts the signal back if the downstream transfer system again requires continuous signals. From the input pulse $f_e(k \cdot T_0)$, the hold element generates an output parameter $f_a(t)$ that is equal in magnitude. It keeps this output parameter constant until the arrival of next input pulse $f_e((k+1) \cdot T_0)$. This provides a step-like progression of output parameters. Its equivalent continual progression has a time offset of approximately $T_0/2$ as compared to the input function.

A practical distinction is made between two control cases of discrete-time systems in regard to the analysis of dynamic transmission behavior and for stability considerations:

$$T_0 \leq T_g/10 \quad \textbf{quasi-continuous} \tag{3.23}$$

$$T_0 > T_g/10 \quad \textbf{discrete-time} \tag{3.24}$$

If the compensation time of the control path, $T_g$ (Fig. 3.9), is longer than 10 times the sampling time $T_0$ ($T_g \geq 10 T_0$), the sampling process has no major effect on the overall system dynamics. Consequently, these "quasi-continuous sampling systems" can be treated as continuous systems with the rules from ► Sect. 3.1.1. A series-connected dead-time element with a dead time $T_t = T_0/2$ need only be taken into account for the functioning of the sampling and hold elements.

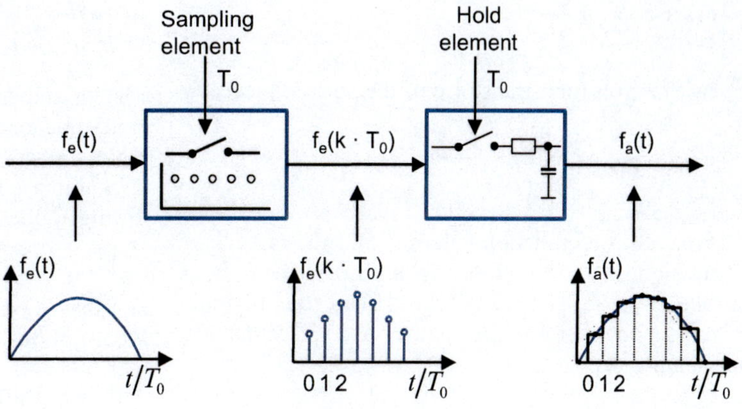


**Fig. 3.10** Operation of a sample and hold system

If the relationship between sampling time $T_0$ and the compensation time $T_g$ is different ($T_g < 10T_0$), the sampling process is a key factor in overall system dynamics. This influence can thus not be ignored. In this case, other mathematical tools must be used, the z-transform in particular. This is discussed in more detail in the next chapter.

### 3.1.2.2 Z-Transform

The z-transform is used to analyze the dynamic behavior of discrete-time systems where $T_g < 10T_0$ [ABEL20]. As with the Laplace transform, this permits a simplified mathematical definition of systems working discretely.

A constant, continuous function $f^*(t)$ is digitized by the sampling element into pulse values (weighting function), and converted to stepped characteristic $f(t)$ in conjunction with the hold element. The definition of the stepped characteristic is made possible by adding and subtracting step functions $\sigma(t)$ with amplitudes determined by function values $f_k$. So, this follows for $f(t)$:

$$f(t) = \sum_{k=0}^{\infty} f_k(\sigma(t - k \cdot T_0) - \sigma(t - (k+1) \cdot T_0)) \tag{3.25}$$

This yields the following when transformed to the Laplace domain:

$$\begin{aligned} \mathbf{L}\{f(t)\} &= \sum_{k=0}^{\infty} \frac{1}{s} f_k \left(e^{-k \cdot T_0 \cdot s} - e^{-(k+1) \cdot T_0 \cdot s}\right) \\ &= \frac{1 - e^{-T_0 \cdots}}{s} \cdot \sum_{k=0}^{\infty} f_k \cdot e^{-k \cdot T_0 \cdot s} \end{aligned} \tag{3.26}$$

$$\mathbf{L}\{f(t)\} = G_H(s) \cdot F_k(s) \tag{3.27}$$

$G_H(s)$ represents the frequency response of the hold function, and $F_k(s)$ is the sampled sequence of values in the Laplace domain.

Substituting in expression

$$F_k(s) = \sum_{k=0}^{\infty} f_k \cdot e^{-k \cdot T_0 \cdot s} \tag{3.28}$$

$e^{T_0 \cdot s}$ with $z$ gives the $z$-transform of the pulse sequence $f_k$.

$$F(z) = \mathbf{Z}\{f_k\} = \sum_{k=0}^{\infty} f_k \cdot z^{-k} \tag{3.29}$$

It is made up of function values $f_k$ and the factor $z^{-k}$. Expression $z^{-k}$ corresponds to a displacement by $k$ sample steps, so by time $k \cdot T_0$. $z^{-1}$ can thus be interpreted as an operator that causes a displacement by one sample step.

**Table 3.6** z-transform calculation rules

| Property | Calculation rule |
|---|---|
| Definition | $F(z) = \mathfrak{Z}\{f_k\} = \sum_{k=0}^{\infty} f_k \cdot z^{-k}$ |
| Linearity | $\mathfrak{Z}\{a_1 \cdot f_k + a_2 \cdot g_k\} = a_1 \cdot F(z) + a_2 \cdot G(z)$ |
| Shift to the left | $\mathfrak{Z}\{f_{k+m}\} = z^{+m} \cdot \left[F(z) - \sum_{i=0}^{m-1} f_i \cdot z^{-i}\right]$ |
| Shift to the right | $\mathfrak{Z}\{f_{k-m}\} = z^{-m} \cdot \left[F(z) + \sum_{i=1}^{m} f_{-i} \cdot z^{i}\right]$ |
| Summation rule | $\mathfrak{Z}\left\{\sum_{i=0}^{k} f_i\right\} = \frac{z}{z-1} \cdot F(z)$ |
| Convolution | $\mathfrak{Z}\left\{\sum_{i=0}^{k} f_i \cdot g_{k-i}\right\} = F(z) \cdot G(z)$ |
| Initial value | $f_0 = \lim_{z \to \infty} F(z)$ |
| End value | $\lim_{k \to \infty} f_k = \lim_{z \to 1}((z-1) \cdot F(z))$ |

Table 3.6 shows a summary of the computing operations with discrete functions in the time domain and the associated operations in the z domain.

### 3.1.2.3 Linear Difference Equations

In the same way as the definition of continuous systems using a differential equation (Eq. (3.2), the behavior of a continuous system at time k can be defined by a difference equation with the input sequence $u_k$ and the output sequence $y_k$.

$$\begin{aligned} & a_m y_{k-m} + a_{m-1} y_{k-m+1} + \ldots + a_1 y_{k-1} + a_0 y_k = \\ & b_n u_{k-n} + b_{n-1} u_{k-n+1} + \ldots + b_1 u_{k-1} + b_0 u_k \end{aligned} \tag{3.30}$$

System stability must also be guaranteed in the discrete-time case. Stability is, like in continuous analysis, a given when all poles of the characteristic polynomial lie within a unit circle around the origin. Thus, the following applies a stable system (see ▶ Sect. 3.1.1.5):

$$\|\lambda_i\|_2 \leq 1 \tag{3.31}$$

Table 3.7 depicts the various representations using a continuous and a discrete-time integrator as examples.

**Table 3.7** Continuous and discrete-time integrators

| Continuous | Discrete-time |
|---|---|
| 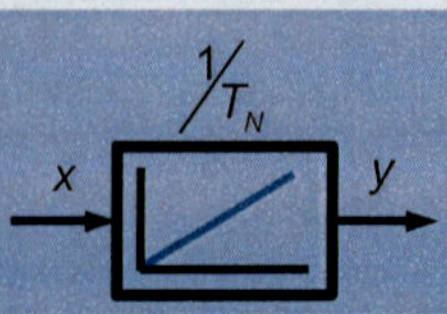 | 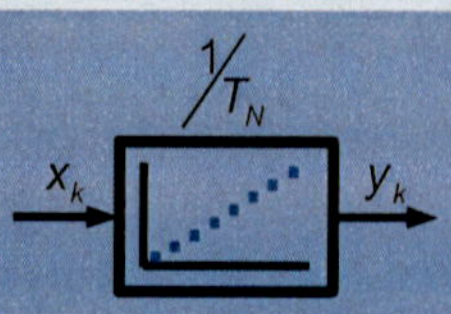 |
| $y(t) = \frac{1}{T_N} \cdot \int_0^t x(t)dt$ Resetting time: $T_N$ | $y_k = \frac{T_0}{T_N} \cdot \sum_{i=0}^{k} x_i = y_{k-1} + \frac{T_0}{T_N} \cdot x_k$ Resetting time: $T_N$ Sampling time: $T_0$ |
| Differential equation: $\dot{y} = \frac{1}{T_N} \cdot x$ Laplace transform: $s \cdot Y(s) = \frac{1}{T_N} \cdot X(s)$ | Difference equation: $\frac{y_k - y_{k-1}}{T_0} = \frac{1}{T_N} \cdot x_k$ z-transform: $\frac{Y(z) - Y(z)z^{-1}}{T_0} = \frac{1}{T_N} \cdot X(z)$ |
| Transfer function: $G(s) = \frac{Y(s)}{X(s)} = \frac{1}{T_N} \cdot \frac{1}{s}$ | z transfer function: $G(z) = \frac{Y(z)}{X(z)} = \frac{T_0}{T_N} \cdot \frac{1}{1-z^{-1}} = \frac{T_0}{T_N} \cdot \frac{z}{z-1}$ |

Solving the difference equation of a discrete-time system for $y_k$ directly gives the algorithm that can be implemented (for example) in software.

This then gives the following difference equation for the discrete-time PID controller with the input function $x_k$ and the output sequence $y_k$:

$$y_k = y_{k-1} + K_R \left( x_k \left( 1 + \frac{T_v}{T_0} + \frac{T_0}{T_n} \right) - x_{k-1} \left( 1 + 2\frac{T_v}{T_0} \right) + x_{k-2} \frac{T_v}{T_0} \right) \tag{3.32}$$

In the case of shorter sampling times ($T_0 < T_g/10$), the controller parameters can be calculated from the criteria for the $K_R$, $T_n$ and $T_v$ values of analog PID controllers. This means the calculation methods for continuous control loops can be applied (▶ Sect. 3.1.1.6).

The computer as a controller uses Eq. 3.31 to calculate the control variable using the PID algorithm. The current control deviation at the sampling time $x_k$, the values of control variable $y_{k-1}$ before a sampling time and control deviations $x_{k-1}$ and $x_{k-2}$ before one or two sampling times are factored in.

Using this procedure enables the behavior of continuous PID controllers to be simulated on the computer.

### 3.1.2.4 Rules for Setting Discrete-Time Controllers

Just as there are rules for setting continuous controllers (discussed in ▶ Sect. 3.1.1.6), there are also corresponding rules for discrete-time controllers. The setting parameters necessary ($K_R$, $K_I$ and $K_D$) are derived from the parameters of the control path step response (◘ Table 3.8). For this, the step response of the control path is recorded in a test (◘ Fig. 3.9), and the transfer factor $K_S$, lag time $T_u$ and compensation time $T_g$ are determined from this.

The parameters to be set for the discrete controller can also be taken from an oscillation test. Here, the critical controller transfer factor $K_{krit}$ and associated period duration $T_{krit}$ where the loop just starts to become unstable are determined (◘ Table 3.9).

### 3.1.2.5 Transfer Function in the Z-Domain

Just like the conversion of the differential equation of a transfer system to the frequency domain $G(s)$ using the Laplace transform, the difference equation of a system can be converted to a transfer function $G(z)$ using the z-transform. Transfer function $G(z)$ is defined as follows:

3

**Table 3.8** Rules for setting discrete-time controllers in accordance with the step response of the control path (for $T_u/T_g > 0.25$; according to Takahashi)

| Controller | Gain factor | Resetting time | Derivative action time |
|---|---|---|---|
| P | $\frac{1}{K_S}\frac{T_g}{T_u+T_0}$ | - | - |
| PI | $\frac{0,9}{K_S}\frac{T_g}{T_u+0,5 \cdot T_0} - 0,5 \cdot K_I$ | $\frac{0,27}{K_S}\frac{T_0 \cdot T_g}{(T_u+0,5 \cdot T_0)^2}$ | - |
| PID | $\frac{1,2}{K_S}\frac{T_g}{T_u+T_0} - 0,5 \cdot K_I$ | $\frac{0,6}{K_S}\frac{T_0 \cdot T_g}{(T_u+0,5 \cdot T_0)^2}$ | $\frac{0,5}{K_S}\frac{T_g}{T_0}$ |

Controller parameters:
$K_R$: Gain factor
$T_n$: Resetting time
$T_v$: Derivative action time
$T_0$: Sampling time

Parameters from oscillation test:
$K_S$: Gain factor
$T_g$: Compensation time
$K_I$: Integral gain factor
$T_u$: Lag time

**Table 3.9** Rules for setting discrete-time controllers in accordance with an oscillation test at the stability boundary (for $T_u/T_g > 0.25$; according to Takahashi)

| Controller | Gain factor | Resetting time | Derivative action time |
|---|---|---|---|
| P | $0,5 \cdot K_{krit}$ | - | - |
| PI | $0,45 \cdot K_{krit} - 0,5 \cdot K_I$ | $0,54 \cdot \frac{K_{krit}}{T_{krit}}$ | - |
| PID | $0,6 \cdot K_{krit} - 0,5 \cdot K_I$ | $1,2 \cdot \frac{K_{krit}}{T_{krit}}$ | $0,075 \cdot K_{krit}\frac{T_{krit}}{T_0}$ |

Controller parameters:
$K_R$: Gain factor
$T_n$: Resetting time
$T_v$: Derivative action time
$T_0$: Sampling time

Parameters from oscillation test:
$K_{krit}$: Gain factor
$T_{krit}$: Period duration
$K_I$: Integral gain factor

$$G(z) = \frac{X_a(z)}{X_e(z)} = \frac{b_0 + b_1 z^{-1} + b_2 z^{-2} + \ldots + b_m z^{-m}}{a_0 + a_1 z^{-1} + a_2 z^{-2} + \ldots + a_n z^{-n}} \quad (3.33)$$

with $X_a(z)$ as output parameter of the system
$X_e(z)$ as input parameter of the system
$a_i, b_i$ as system parameter

Parameters $a_i$ and $b_i$ of this transfer function can be determined, for example, from an available input and output signal profile. Arranging an equation system with $m+n+2$ difference equations as in Eq. 3.30 for different sampling times with the associated input and output parameters yields a linear equation system from which the $m+n+2$ unknown parameters ($a_i$ and $b_i$) can be calculated. The prerequisite is that the equations are not linearly dependent, as is the case, for example, for constant input and output signals.

If the output signal values are from a measurement, directly arranging and solving the equation system is not generally recommended because unavoidable measurement inaccuracies strongly skew the result. Parameter identification methods (from [ISER83], for example) return better results here. They determine the required parameters from a much greater number of measured values, thus reducing the effect of measurement inaccuracies and disturbances. The least squares method, for example, minimizes the square of the error between the output value calculated using estimated parameters and the output value actually measured.

The connections of discrete transfer systems can be calculated using z transfer functions G(z) using the same rules as for continuous systems—see Fig. 3.7. For the z transfer function of a sampling system comprising sampling element, the hold element and continuous control path, the following z transfer function results

$$G_z(z) = \frac{z-1}{z}\mathbf{Z}\left\{\frac{G_s(s)}{s}\right\} = \frac{z-1}{z}\mathbf{Z}\{H_s(s)\} \quad (3.34)$$

where $H_s(s)$ is the step response of the path after the Laplace transform. In most cases, z-transforms can be performed easily using correspondence tables (just like Laplace transforms)—see Table 3.10.

### 3.1.3 Feedforward Controller for Tracking Error Correction

Popular methods for compensating the tracking error are mostly based on the feedforward controller method. The difference between feedback controllers (included here are all methods known to date such as P controllers, and other state controllers as in ► Sect. 3.1.4) and feedforward controllers is that, instead of the deviation, the reference input value is processed using a second controller in the forward branch. The actual controller (feedback controller) remains. The function of the second controller is to offset the delay of the path or the entire position control loop using an inverted model control value. First, this assumes precise knowledge of the system, and, second, digital signal processing must be available because reversing a system model is virtually not practical in an analog circuit.

Known machine data or experimental measurements on the machine can be used to create the model of the closed control loop. The control computer does the signal processing in this case. Defining the digital position control loop in the form of a 2nd order z transfer function using a model usually suffices:

$$G(z) = \frac{Y(z^{-1})}{U(z^{-1})} = \frac{b_0 + b_1 z^{-1} + \ldots}{a_0 + a_1 z^{-1} + a_2 z^{-1}} \quad (3.35)$$

Figure 3.11 shows two elementary circuits for correction of the path error using the feedforward controller principle. In the top part of the figure, a parallel feedforward element is added to the control loop. The purpose of this element is to exactly compensate the dynamic behavior of the path. This is precisely the case

Table 3.10 z-transform and associated Laplace transform

| f(t) | F(s) | f(k) | F(z) |
|---|---|---|---|
| $1$ | $\frac{1}{s}$ | $1$ | $\frac{z}{z-1}$ |
| $t$ | $\frac{1}{s^2}$ | $k \cdot T_0$ | $\frac{T_0 \cdot z}{(z-1)^2}$ |
| $t^2$ | $\frac{2}{s^3}$ | $k^2 \cdot T_0^2$ | $\frac{T_0^2 \cdot z(z+1)}{(z-1)^3}$ |
| $e^{-at}$ | $\frac{1}{s+a}$ | $e^{-akT_0}$ | $\frac{z}{z-e^{-aT_0}}$ |
| $t \cdot e^{-at}$ | $\frac{1}{(s+a)^2}$ | $k \cdot T_0 \cdot e^{-akT_0}$ | $\frac{T_0 \cdot z \cdot e^{-aT_0}}{(z-e^{-aT_0})^2}$ |
| $\sin(\omega t)$ | $\frac{\omega}{s^2+\omega^2}$ | $\sin(\omega k T_0)$ | $\frac{z \cdot \sin(\omega T_0)}{z^2 - 2z\cos(\omega T_0)+1}$ |
| $\cos(\omega t)$ | $\frac{s}{s^2+\omega^2}$ | $\cos(\omega k T_0)$ | $\frac{z(z-\cos(\omega_0))}{z^2 - 2z\cos(\omega T_0)+1}$ |
| $e^{-at}\sin(\omega t)$ | $\frac{\omega}{(s+a)^2+\omega^2}$ | $e^{-akT_0}\sin(\omega k T_0)$ | $\frac{z \cdot e^{-aT_0}\sin(\omega T_0)}{z^2 - 2ze^{-aT_0}\cos(\omega T_0)+e^{-2aT_0}}$ |
| $e^{-at}\cos(\omega t)$ | $\frac{s+a}{(s+a)^2+\omega^2}$ | $e^{-akT_0}\cos(\omega k T_0)$ | $\frac{z^2 - ze^{-aT_0}\cos(\omega T_0)}{z^2 - 2ze^{-aT_0}\cos(\omega T_0)+e^{-2aT_0}}$ |

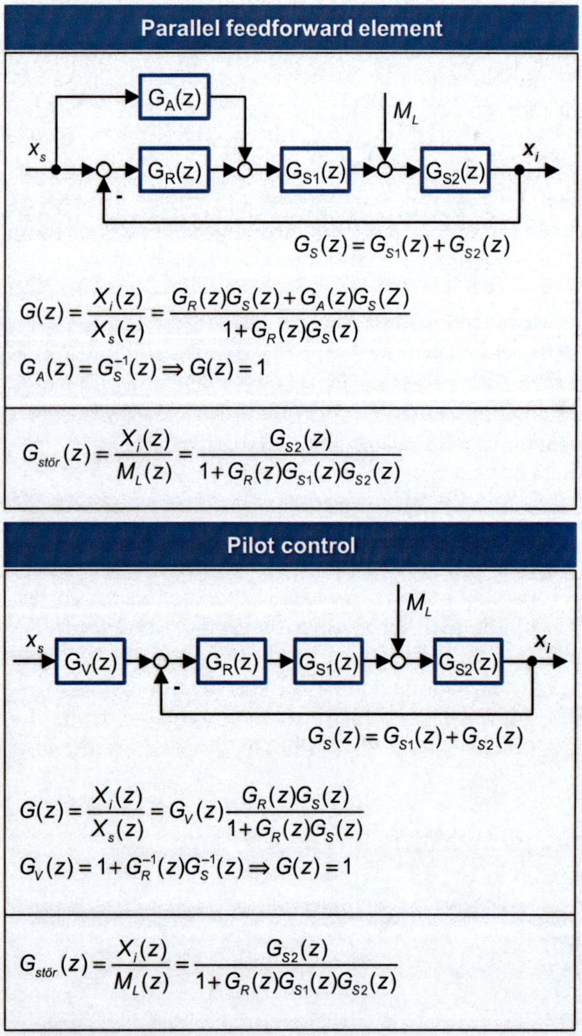


**Fig. 3.11** Elementary circuits for path error correction using the feedforward controller principle

when the transmission behavior of the parallel feedforward element is the inverse of the transmission behavior of the path ($G_A(z) = G^{-1}{}_S(z)$). Similarly, the approach on the right aims to compensate the inertia of the position control loop when $G_v(z) = 1 + G_R{}^{-1}(z) \cdot G_S{}^{-1}(z)$. Full compensation is theoretically possible in both cases if control variable limits are ignored. In practice, however, improvement of path behavior is determined by how accurate the model is. Given that it is not possible to create a model with perfect accuracy, an improvement in path error correction can only be as good as the model. Furthermore, problems such as instability occur on inverting the system model. These can only be resolved using special mathematical methods.

As Fig. 3.11 shows, it is possible to improve the tracking behavior with the feedforward controller. The disturbance behavior is not changed by doing so, however.

The positions of the zeros of $G(z)$ are key for the stability of the inverted model. "Inverse stability" is a given if they lie within the unit circle. It is therefore possible to design a feedforward controller as on the right of Fig. 3.11. The recommendation is to connect an additional low-pass filter to the pilot control. This is because its limited bandwidth means the position control loop can only follow the high-frequency output signals from the pilot control to compensate system inertia up to a certain frequency.

### 3.1.4 State Control

A transfer system or complete control loop can be defined by a system of general equations and differential equations, most of a higher order, in the time domain, that can be combined into a single differential equation defining the entire system behavior. In addition to the representation of this differential equation in the form of a transfer function (Eq. (3.11)), the state space representation is also used for control-related analyses. Here, the relationship between the output parameters of the system and the internal state variables also exerting an influence over system dynamics is made clear.

For this, the differential equations are converted to a system of coupled 1st order differential equations and general equations. This is accomplished by introducing intermediate variables, the state variables [LUNZ08]. One characteristic of state variables is that their profile versus time depends only on the input parameters of the system and the initial values of the state variables, i.e., the initial conditions. This means that the individual derivations of the output parameters occurring in the system-defining differential equation (Eq. 3.2) can be used as state variables.

Below, the state space representation is initially represented using a SISO system (single input–single output).

#### 3.1.4.1 Representation in the State Space

For the sake of simplicity, the representation is for a system with one n-th order differential equation, one input parameter with no time derivatives. Using a suitable transformation, every system with input polynomial order $\leq n$ can be expressed in this form, meaning the representation is generally applicable. Representation of control of the input parameters over the output parameters is omitted here for the same reason.

Determined from the system-defining n-th order differential equation with input parameter $u(t)$, output parameter $y(t)$ and its derivation over time $y^{(n)}(t)$

$$a_n y^{(n)}(t) + a_{n-1} y^{(n-1)}(t) + \ldots + a_1 \dot{y}(t) + a_0 y(t) = b_0 u(t) \tag{3.36}$$

is, by inserting intermediate variables $x_1(t)$, $x_2(t)$,..., $x_n(t)$

$$\begin{aligned} x_1(t) &= y(t) \\ x_2(t) &= \dot{y}(t) \\ x_3(t) &= \ddot{y}(t) \\ &\vdots \\ x_n(t) &= \overset{(n-1)}{y}(t) \\ \dot{x}_n(t) &= \overset{(n)}{y}(t) \end{aligned} \quad (3.37)$$

a system of $n-1$ 1st order differential equations

$$\begin{aligned} \dot{x}_1(t) &= x_2(t) \\ \dot{x}_2(t) &= x_3(t) \\ &\vdots \\ \dot{x}_{n-1}(t) &= x_n(t) \end{aligned} \quad (3.38)$$

and a 1st order differential equation

$$\begin{aligned} \dot{x}_n(t) = &-\frac{a_0}{a_n}x_1(t) - \frac{a_1}{a_n}x_2(t) - \dots \\ &- \frac{a_{n-1}}{a_n}x_n(t) + \frac{b_0}{a_n}u(t) \end{aligned} \quad (3.39)$$

of the form:

The system-defining equation system from now on comprises differential equations (Eqs. 3.37 and 3.38) that contain the derivatives of intermediate variables $x_i(t)$ and the input parameter $u(t)$. The output parameter $y(t)$ does not occur in the differential equations; it is expressed by an intermediate variable:

$$y(t) = x_1(t) \quad (3.40)$$

Grouping state variables $x_1$... $x_n$ into vector **x** enables Eqs. 3.37 to 3.39 to be expressed in vector notation (controllable canonical form):

$$\begin{bmatrix} \dot{x}_1(t) \\ \dot{x}_2(t) \\ \vdots \\ \dot{x}_n(t) \end{bmatrix} = \begin{bmatrix} 0 & 1 & 0 & 0 & \cdots & 0 \\ 0 & 0 & 1 & 0 & & 0 \\ & & \vdots & & \ddots & \vdots \\ -\frac{a_0}{a_n} & -\frac{a_1}{a_n} & -\frac{a_2}{a_n} & -\frac{a_3}{a_n} & \cdots & -\frac{a_{n-1}}{a_n} \end{bmatrix} \cdot \begin{bmatrix} x_1(t) \\ x_2(t) \\ \vdots \\ x_n(t) \end{bmatrix} + \begin{bmatrix} 0 \\ 0 \\ \vdots \\ \frac{b_0}{a_n} \end{bmatrix} \cdot u(t) \quad (3.41)$$

$$y(t) = [1\,0\,0 \cdots 0] \cdot \begin{bmatrix} x_1(t) \\ x_2(t) \\ \vdots \\ x_n(t) \end{bmatrix} = x_1(t) \quad (3.42)$$

or in compressed matrix notation:

$$\begin{aligned} \dot{\mathbf{x}}(t) &= \mathbf{A} \cdot \mathbf{x}(t) + \mathbf{b} \cdot u(t) \\ y(t) &= \mathbf{c} \cdot \mathbf{x}(t) \end{aligned} \quad (3.43)$$

The vector differential equation (Eq. (3.41) is known as the state differential equation, Eq. 3.42 as the output equation and both together as the state equations (Eq. (3.43). The components of state vector $x_i(t)$ determine the state space for every time $t$ the system in the $n$-dimensional space. To represent a multi-variable system in the state space, the derivatives of $n$ state variables $\dot{x}_i(t)$ are shown as functions of the $n$ state variables $x_i(t)$ and $p$ input parameters $u_i(t)$:

$$\begin{aligned} \dot{x}_1(t) &= f_1\left(x_1(t), \dots, x_n(t); u_1(t), \dots, u_p(t)\right) \\ &\vdots \\ \dot{x}_n(t) &= f_n\left(x_1(t), \dots, x_n(t); u_1(t), \dots, u_p(t)\right) \end{aligned} \quad (3.44)$$

The $q$ output parameters $y_i(t)$ form an equation system of general equations as functions of the state variables:

$$\begin{aligned} y_1(t) &= g_1(x_1(t), \dots, x_n(t)) \\ &\vdots \\ y_q(t) &= g_q(x_1(t), \dots, x_n(t)) \end{aligned} \quad (3.45)$$

This yields the following for the state representation in matrix notation:

$$\begin{aligned} \dot{\mathbf{x}}(t) &= \mathbf{A} \cdot \mathbf{x}(t) + \mathbf{B} \cdot \mathbf{u}(t) \\ y(t) &= \mathbf{C} \cdot \mathbf{x}(t) \end{aligned} \quad (3.46)$$

where
$\dot{x}(t)$ is $n \times 1$ derivative vector of the state variables
$x(t)$ is $n \times 1$ vector of the state variables
$u(t)$ is $p \times 1$ vector of the input parameters
$y(t)$ is $q \times 1$ vector of the output parameters
$A$ is $n \times n$ system matrix
$B$ is $n \times p$ input matrix
$C$ is $q \times n$ output matrix.

▫ Figure 3.12 shows the block diagram representation of Eq. 3.46. Matrix **D** shown in addition here (scalar $d$ for SISO systems) represents direct access of the input parameters to the output parameters. Such system behavior only occurs in exceptional circumstances for real systems. Matrix **D** is therefore ignored below.

As an example, the state space representation of a single-mass oscillator with damping $D$ and natural angular frequency $\omega_0$ is determined below.

Starting from the 2nd order differential equation of the system

$$\ddot{y} + 2D\omega_0\dot{y} + \omega_0^2 y = \omega_0^2 u \quad (3.47)$$

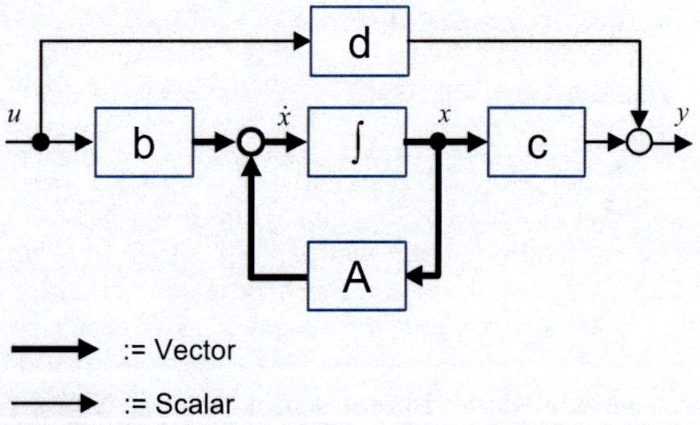


**Fig. 3.12** State space representation of a single variable system

gives, by introducing state variables,

$$\begin{aligned} x_1 &= y \\ x_2 &= \dot{y} \end{aligned} \tag{3.48}$$

the two coupled 1st order differential equations

$$\begin{aligned} \dot{x}_1 &= x_2 \\ \dot{x}_2 &= -\omega_0^2 x_1 - 2D\omega_0 x_2 + \omega_0^2 u \end{aligned} \tag{3.49}$$

or in matrix notation:

$$\begin{aligned} \dot{\mathbf{x}} &= \underbrace{\begin{bmatrix} 0 & 1 \\ -\omega_0^2 & -2D\omega_0 \end{bmatrix}}_{\mathbf{A}} \cdot \mathbf{x} + \underbrace{\begin{bmatrix} 0 \\ \omega_0^2 \end{bmatrix}}_{\mathbf{b}} \cdot \mathbf{u} \\ y &= \underbrace{\begin{bmatrix} 1 \\ 0 \end{bmatrix}}_{\mathbf{c}} \cdot \mathbf{x} \end{aligned} \tag{3.50}$$

### 3.1.4.2 Design of a State Controller

A state controller is a controller that uses the controller matrix **R** to feed the state variables of the system to be controlled back to the system input. Figure 3.13 shows the structure of the closed control loop with controller. In addition to the controller matrix **R**, the pre-filter matrix **S** is used. This also determines the behavior of the overall system.

For the dimensioning of controller matrix **R**, a number of different methods have been developed that differ in terms of design approach, number of state variables returned and design goal. Three methods for single variable (SISO) systems based on the return of all state variables are introduced briefly here.

#### Design with Pole Placement

For a controller design with pole placement, the controller vector **r** is specified such that the poles of the control path transfer function are shifted by the controller to specified values. For SISO systems, this is possible without restriction, i.e., any pole position in the overall system can (theoretically) be reached. Thus, the starting point of the design task is specification of the poles and formulation of the differential equation of the system behavior required.

The poles of the control path transfer function are derived from the zeros of the denominator polynomial:

$$a_n \cdot y^{(n)}(t) + a_{n-1} \cdot y^{(n-1)}(t) + \ldots + a_1 \cdot \dot{y}(t) + a_0 \cdot y(t) \tag{3.51}$$

If the denominator polynomial of the transfer function of control path and controller has the form

$$p_n \cdot y^{(n)}(t) + p_{n-1} \cdot y^{(n-1)}(t) + \ldots + p_1 \cdot \dot{y}(t) + p_0 \cdot y(t) \tag{3.52}$$

the following applies to the controller vector **r** according to [ACKE83]:

$$\mathbf{r}_R = [(p_0 - a_0) \quad (p_1 - a_1) \quad \ldots \quad (p_{n-1} - a_{n-1})] \tag{3.53}$$

One problem with this method is that, while a controller vector can always be calculated for a specified system behavior, the design method does not include, for example, any way to take into account control variable limitations that are always present in real-world systems.

Another characteristic of this design method is illustrated using the example of the single-mass oscillator with state space representation as in Eq. 3.46. If, for the overall system, the positions of the poles are selected such that the resulting system has a damping

**Fig. 3.13** Control path with state controller

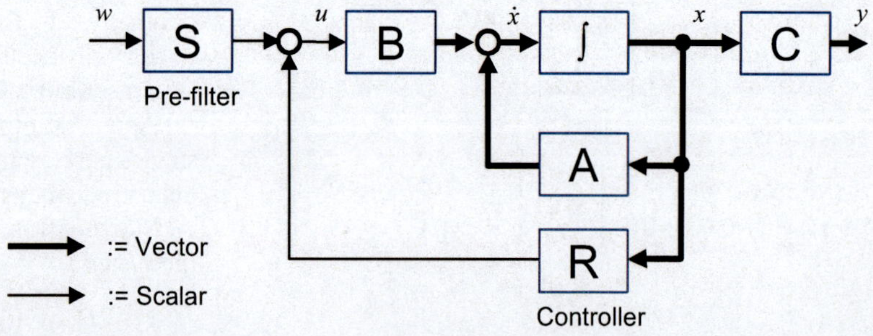

value of 1 and a natural angular frequency $\omega_1$, the differential equation for the overall system is:

$$\ddot{y} + 2\omega_1\dot{y} + \omega_1^2 y = \omega_1^2 u \tag{3.54}$$

So the Eqs. 3.52 and 3.46 yield the following for the controller:

$$\mathbf{r}_R = [(\omega_1^2 - \omega_0^2) \quad (2(\omega_1 - D\omega_0))] \tag{3.55}$$

The control law is as follows when expanded:

$$u_r = (\omega_1^2 - \omega_0^2)y + 2(\omega_1 - D\omega_0)\dot{y} \tag{3.56}$$

This control law defines a PD controller because the controller output parameter comes from the weighted controller input parameter and the weighted derivative of the controller input parameter. Steady-state accuracy cannot always be attained with this controller type. The pre-filter "S" or an integrating controller part is also required.

#### Design with Optimization of a Quadratic Quality Criterion

If the goal of the design task is not directly a specific dynamic behavior of the overall system but minimization of the control deviation, for example, the controller matrix **R** can be determined by solving an optimization problem. Instead of system dynamics, a suitable cost function must be specified here that is minimized using the optimization method. Typical cost functions assess, for example, the square of the control deviation and the square of the input parameter $u$ to also minimize the actuating energy required. Solving this optimization problem gives the quadratic vector equation, the Riccati equation, that can be used to determine the controller matrix [ACKE83].

#### Design for Finite Setting Time

If the controller is designed as a sampling system (i.e., discrete-time), it can be shown that it is always possible to transition the control path from any initial state to any final state using a finite number of sampling steps. The task in this dead-beat design is first to determine optimum input sequence $u_k$ that satisfies this requirement. For this sequence, a controller is then designed that generates the required sequence from the state variables. Theoretically, an $n$-th order system can be transitioned to a new final state in $n$ sampling steps. The physical limitations of the system can, of course, not be overcome here by the controller. If the optimum control variable sequence is expanded to more than $n$ steps, however, additional degrees of freedom are created in the design. This means that additional boundary conditions can be satisfied (such as inclusion of an control variable limitation). A downside of the dead-beat design is the controller's sensitivity to parameters. If the actual behavior of the control path deviates only slightly from the model used for the design, the controller quality reduces considerably in most cases. This is why dead-beat control is rarely used for machine tool feed drives. Varying the positions of slides, columns, etc. and the different masses due to the changing weights of jig and workpiece alter the path behavior noticeably.

## 3.2 Closed-Loop Control of Feed Drives

### 3.2.1 Feed Drive as Control Loop

#### Layout and Function of Closed-Loop Position Control

Figure 3.14 shows schematically the hardware setup of an individual feed axis. The motor powers a ball screw via a coupling and toothed belt. The ball screw nut converts the rotatory motion of the ball screw to the translational motion of the machine slide.

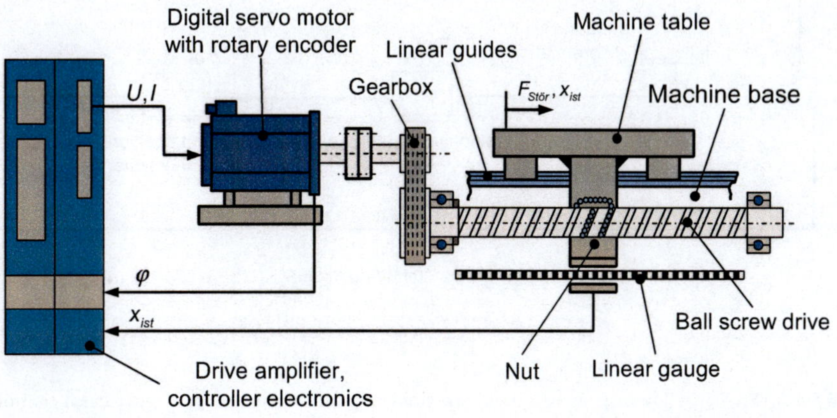


**Fig. 3.14** Schematic structure of a feed axis

The angular position of the motor is read by a rotary encoder. From this signal, the position of the machine slide can also be calculated indirectly using the gear ratio (e.g., the gear ratio $i$ of the toothed belt or the pitch $h$ of the spindle). Digital drive amplifiers available today also derive the rotation speed of the motor from the position signal. This makes another tachogenerator unnecessary.

Determining the position of the machine table using a direct linear length measuring system is commonplace nowadays. This increases the accuracy of motion because factors such as friction, mechanical clearance and geometrical errors in the drive train can be measured by the measuring system. For the purposes of improved differentiation, the position signal measured directly is called $x_m$ and the signal calculated from the rotary encoder of the drive is $x_a$. For the feed systems common nowadays in machine tools, either the direct or indirect measuring system can be used for the closed-loop position control.

■ Figure 3.15 shows the controller structure as used in most feed axes in machine tools today. It depicts the cause and effect relationships in the drive. The mechanical structure is of secondary importance here. The devices and electrical drive components are also in the controller structure, however.

In the analysis of the control loops, it is not the dynamic behavior of the mechanics, but that of the closed-loop control, that is paramount. Thus, the mechanics are assumed to be ideally rigid here as a simplification. Here, only the mass moment of inertia $J_{ges}$, reduced onto the motor shaft, needs to be considered. Also, the gear ratio of the mechanical transfer elements (spindle pitch $h$ and ratio $i$) are present in the constant $K_G$.

The reference variable $x_{soll}$ of the position control loop is first compared to the actual position of the machine slide, $x_{ist}$. The difference between these two signals is amplified by the proportional position controller. The control variable of the position controller is forwarded to the secondary velocity control loop, which in turn has a secondary current control loop. The velocity feedback and the closed-loop current controller mean the drive dynamics are increased considerably and overshooting is prevented by an improved damping ratio.

The velocity controller in the example shown has a PI character to be able to compensate the most minor of control deviations with no remaining errors. The closed-loop current controller, downstream of the velocity controller, permits the motor to be acceleration controlled. This prevents overshooting on acceleration and deceleration. The control variable for the motor is calculated from the difference between the reference value of the velocity controller $v_{soll}$ and the actual speed value $v_{ist}$ linked to the actual motor load (closed-loop current control). The power amplifier converts the control variable calculated this way to the correspond-

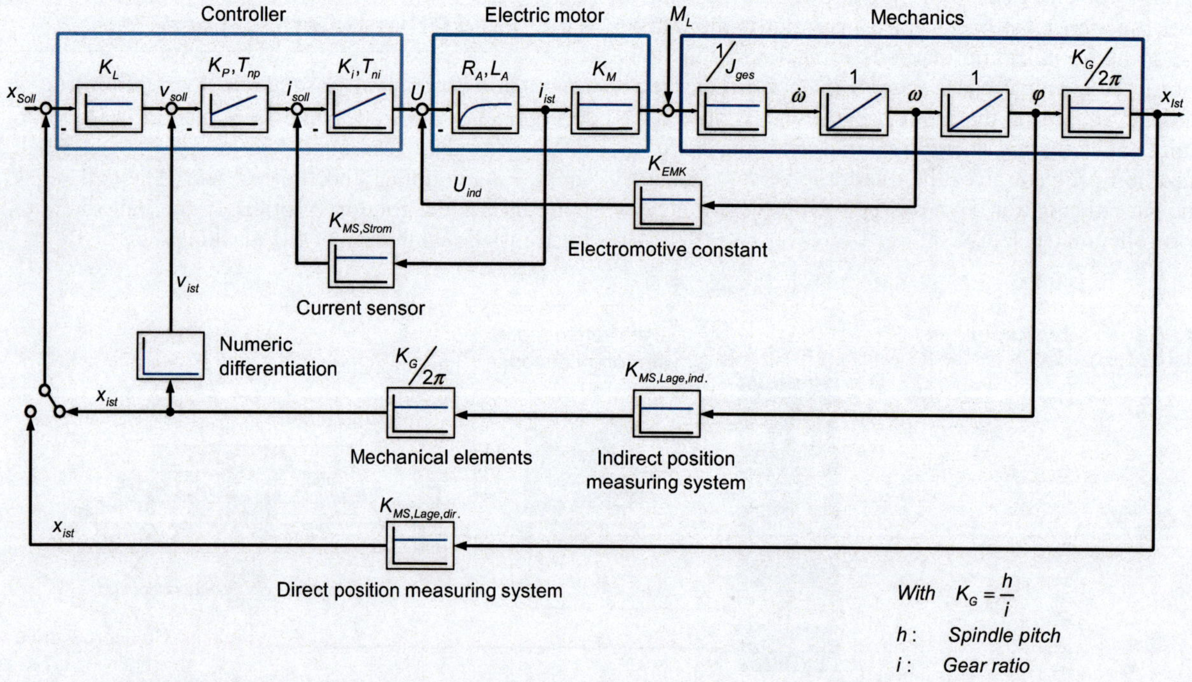


■ **Fig. 3.15** Signal flow diagram of a position control loop with secondary speed and current control loops ing motor voltage $U$ (see ◘ Fig. 3.14) that is applied to the motor coils.

A current builds up in line with the electrical parameters of the motor coil (resistance $R_A$ and inductance $L_A$). Proportional to this (by factor $K_M$), the motor torque is also generated that, together with the load torque, acts upon the inertia $J_{ges}$ (see ◘ Fig. 3.14) of the overall feed axis. According to the principle of conservation of angular momentum, rotational acceleration is generated that influences the motor speed $n$ via formal integration and relevant conversion. Another formal integration gives the angular position of the motor shaft $\varphi_a$. Multiplying with the gear ratio gives the position $x_{ist}$ (see ◘ Fig. 3.14) of the machine slide.

The feed drive as a whole is a higher-order system. Calculating systems of an order greater than 2 is difficult, however. This is why simplifications in the model of the control loops are often made when certain properties can be ignored for the analyses. ◘ Figure 3.16 shows one possible simplification of the control loop from ◘ Fig. 3.15. For simplicity, only direct position measurement is examined. The measuring systems normally exhibit no significant errors. They can thus be assumed to be P elements with a gain factor of 1 or ignored. At low speeds, the countervoltage induced is low and so feedback via the electromotive constant can be ignored (◘ Fig. 3.16). So the current control loop can generally be simplified to a 1st order delay element ($PT_1$ element). For modern drives, about 1 kHz can be assumed for the cutoff frequency. The dynamic behavior of the control loop can normally be neglected at frequencies this high. In this case, the current control loop can also be assumed ideal ($G_{SRK}=1$, corresponding to a P element where $K=1$). For approximate analyses of control response, the integral part of the velocity controller is often also ignored. The velocity controller thus becomes a P controller (like the position controller).

When examining the closed speed control loop analytically in its greatly simplified form, it can be summarized easily in line with the calculation rules in ◘ Fig. 3.7 and using ◘ Table 3.1. Assuming the current control loop to be a $PT_1$ element gives a 2nd order delay element for the closed speed control loop ($PT_2$ element). For a current control loop assumed to be ideal, the closed speed control loop can be combined into a $PT_1$ element.

### 3.2.2 Calculation of Continuous Position Control Loops

One or more simplifications are often used for qualitative handling of the position control loop. Individual transfer systems can be defined in the form of frequency responses using ◘ Table 3.1. The comparisons between reference and actual position values and between the position controllers are realized digitally in modern NC controllers and so, strictly speaking, need to be considered discrete-time controllers. If only the lower frequency range is analyzed, however, closed-loop position control can be assumed to be quasi-continuous. The descriptions of the control loop elements in ◘ Table 3.1 can then be applied.

Remember the calculation rules in ◘ Fig. 3.7 when formulating the tracking frequency response $G(j\omega)$. The frequency response of a control loop $G_{ges}(j\omega)$, comprising the forward branch $G_V(j\omega)$ and the return branch $G_R(j\omega)$, can be formulated using Eq. 3.57.

$$G_{ges}(j\omega) = \frac{G_V(j\omega)}{1 + G_V(j\omega) \cdot G_R(j\omega)} \tag{3.57}$$

To demonstrate the function of the secondary closed velocity control loop, the calculations for a position control loop with and without velocity control are performed in ◘ Table 3.11. For this, the control loop is simplified as in ◘ Fig. 3.16. The behavior of the motor with a coupled mass is approximated using a $PT_1$ element. To convert the rotational speed $n_s$ to the set point voltage $U_s$, the same constant ($K_p$) as for the closed-loop speed controller is used.

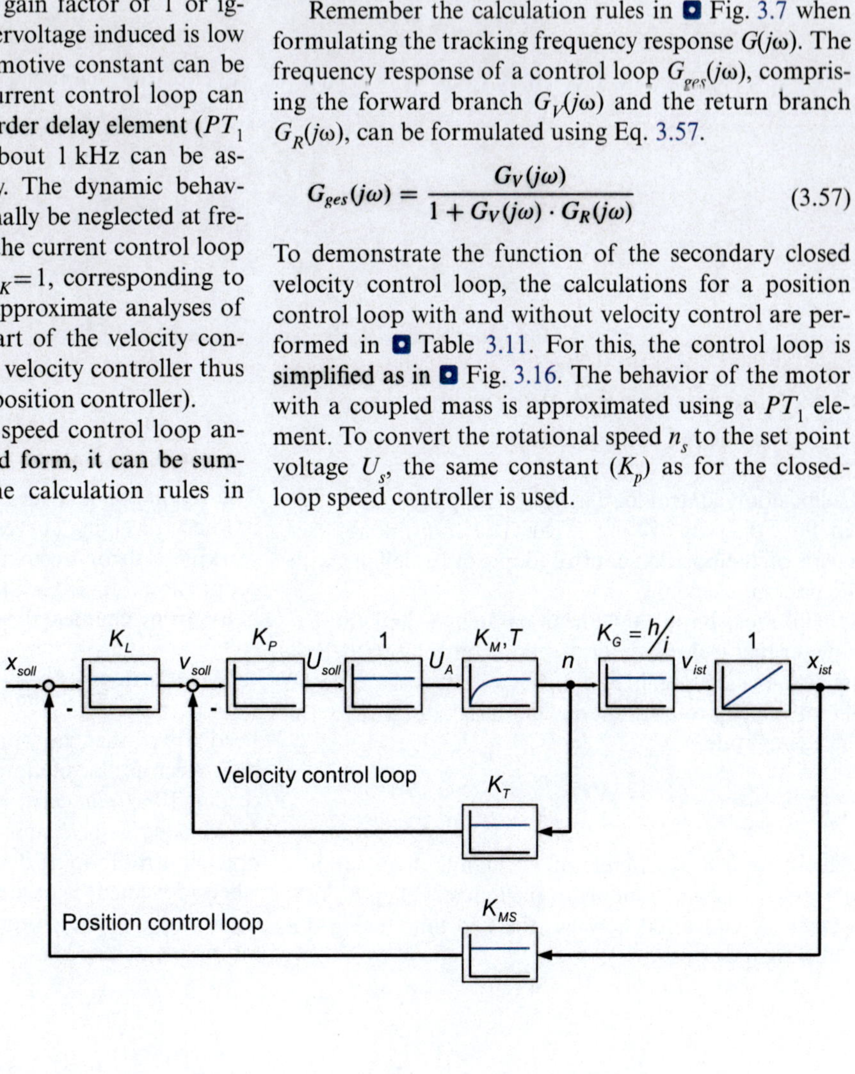


◘ **Fig. 3.16** Simplified position control loop with secondary velocity control loop

**Table 3.11** Calculation of a position control loop with and without velocity control

| Position control loop without velocity control | Position control loop with velocity control |
|---|---|
| Velocity gain | |
| $K_V = K_L K_P K_M K_G$ | $K_{V,G} = \frac{K_L K_P K_M K_G}{1+K_P K_M K_T}$ |
| Frequency response of the velocity control loop | |
| - | $G_{GS}(j\omega) = \frac{K_P \frac{K_M}{1+T\cdot(j\omega)}}{1+K_T K_P \frac{K_M}{1+T\cdot(j\omega)}}$ |
| Transfer function of the position control loop | |
| $G(j\omega) = \frac{K_L K_P K_G \frac{K_M}{1+T\cdot(j\omega)} \frac{1}{(j\omega)}}{1+K_{MS} K_L K_P K_G \frac{K_M}{1+T\cdot(j\omega)} \frac{1}{(j\omega)}}$ | $G_{GS}(j\omega) = \frac{K_L G_{GS}(j\omega) K_G \frac{1}{(j\omega)}}{1+K_{MS} K_L G_{GS}(j\omega) K_G \frac{1}{(j\omega)}}$ |
| Comparison with proportional element and 2nd order delay<br>$G_{PT2}(j\omega) = \frac{K}{1+\frac{2D}{\omega_0}\cdot(j\omega)+\frac{1}{\omega_0^2}(j\omega)^2}$<br>$\omega_0 = \sqrt{\frac{K_L K_P K_M K_G K_{MS}}{T}}$ | |
| $D = \frac{1}{2}\frac{1}{\sqrt{K_L K_P K_M K_G K_{MS} T}}$ | $D = \frac{1}{2}\frac{1+K_P K_M K_T}{\sqrt{K_L K_P K_M K_G K_{MS} T}}$ |

### 3.2.3 Transmission Behavior of the Linear Position Control Loop

The position control loop is shown simplified in the signal flow diagram (Fig. 3.15). The transmission behavior of the position control loop can be defined with the parameters below.

The most important linear parameter, and the parameter that is decisive for position control loop behavior, is the velocity gain $K_V$ ($K_V$ factor). It is the ratio of actual velocity to position deviation (lag distance) $x_w$ in the steady state,

$$K_V = \frac{\dot{x}}{x_w} \tag{3.58}$$

Velocity gain $K_V$ has units of $s^{-1}$. Units of $m\ /\ (min \cdot mm)$ is also used commonly in the industry. The $K_V$ factor can be converted between the two units using the following relationship:

$$1\left[\frac{m}{min \cdot mm}\right] = \frac{1000}{60}\left[s^{-1}\right] = 16{,}\overline{6}\left[s^{-1}\right] \tag{3.59}$$

The $K_V$ factor is a measure for how true the machine is when traversing curves on the plane or spatially. The workpiece error occurring due to the lag distance is proportional to $\dot{x}/K_V$. Standard values for velocity gain in electromechanical drive systems are in the range

$$K_V = 0{,}6\ldots4{,}8\frac{m}{min \cdot mm}\text{(equivalent to 10 to 80 s}^{-1}\text{)}$$

Feed drives that are optimized for high dynamics and have electric, linear motor drives exhibit $K_V$ values exceeding $10m/(min \cdot mm) = 166{,}\overline{6}\,s^{-1}$ in exceptional cases.

Velocity gain $K_V$ must be set to the loop gain of the open control loop. It determines the position deviation when movement is at a constant speed and is of major significance in the design of position control loops for multiple machine axes.

Besides the $K_V$ factor, the cutoff frequency $\omega_0$ and system damping $D$ are also key factors in position control loop behavior. The influence of the damping and cutoff frequency is shown in ◘ Fig. 3.17.

The cutoff frequency $\omega_0$ (or $f_0$) is a measure of the signal transfer quality. A high cutoff frequency means low signal distortion by the system. Generally, $f_0 = \omega_0/2\pi$ at 10 to 30 $Hz$.

The damping ratio $D$ defines the ability of the system to oscillate. A high damping ratio means a low tendency to oscillate. To approach the target position of a movement without significant overshooting, damping should be as high as possible ($0.8 \leq D \leq 1$) [STUT81].

The cutoff frequency $f_g$ of a position control loop is the frequency at which, for a sinusoidal reference position value change, the ratio of actual amplitude to the reference amplitude reaches a value of $1/\sqrt{2}$.

$$f_g \approx \frac{1}{2\pi} K_V \tag{3.60}$$

Mathematically, the relationships can be expressed with the following equation:

$$K_V = \frac{\omega_0}{2D} = \frac{2\pi f_0}{2D} = \frac{\pi f_0}{D} \tag{3.61}$$

Since, as mentioned, damping should be between

$$0{,}8 < D < 1{,}0 \tag{3.62}$$

the following dependency arises:

$$K_V = \frac{\pi}{0{,}8 \ldots 1{,}0} \cdot f_0 = (3{,}14 \ldots 3{,}93) f_0 \tag{3.63}$$

◘ Figure 3.17 shows this relationship graphically using the step response. Increasing the proportional component in the position controller, given constant masses to accelerate and equal drive torque, increases the $K_V$ factor and cutoff frequency but the damping is reduced.

The slide path $x$ is plotted versus time in ◘ Fig. 3.18. The profile of carriage speed $\dot{x}$ is depicted in the lower part. The parameter is velocity gain $K_V$. With increasing gain, the control deviation lessens while the reference velocity is attained quicker. For large gains, stronger oscillation occurs due to reducing damping on starting and stopping. Gain is thus restricted to a value at which the carriage reaches its position as fast as possible but without any major overshooting. The damping ratio $D$ is a measure of overshooting. Step-like speed changes make very good testing and rating of the dynamic behavior of feed drives possible. In practice, speed changes are realized with targeted speed control of the controller, meaning the drives can easily comply with the specifications and vibration excitation of the machine structure due to impermissible jerks is prevented (Jerk = Acceleration change $\equiv \dddot{x}$).

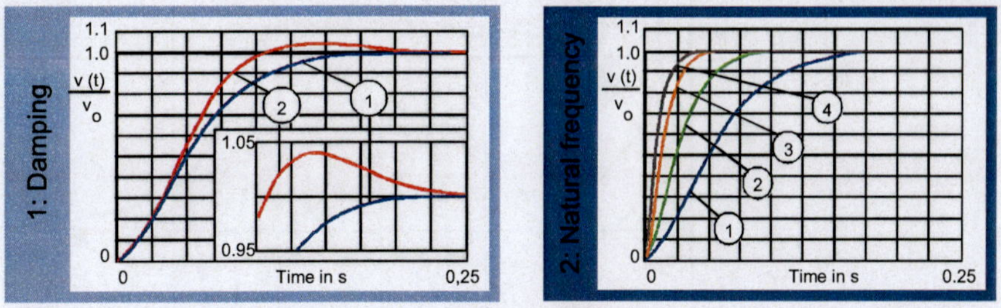


| 1: Influence of damping: System properties | | | | |
|---|---|---|---|---|
| System | | 1 | 2 | |
| Damping | D: | 0.9 | 0.7 | |
| Cutoff frequency (= constant) | $f_0$: | 5 | 5 | Hz |
| Velocity gain | $K_V$: | 1.05 | 1.35 | m/(min*mm) |

$v_0 = 0, t < 0$
$v_0 = const.$
$t \geq 0$

$$K_V = \frac{2\pi f_0}{2D}$$

| 2: Influence of natural frequency: System properties | | | | | | |
|---|---|---|---|---|---|---|
| System | | 1 | 2 | 3 | 4 | |
| Damping (= constant) | D: | 0.9 | 0.9 | 0.9 | 0.9 | |
| Cutoff frequency | $f_0$: | 5 | 10 | 20 | 30 | Hz |
| Velocity gain | $K_V$: | 1.05 | 2.09 | 4.19 | 6.28 | m/(min*mm) |

◘ **Fig. 3.17** Velocity profile: step response – influence of damping and cutoff frequency

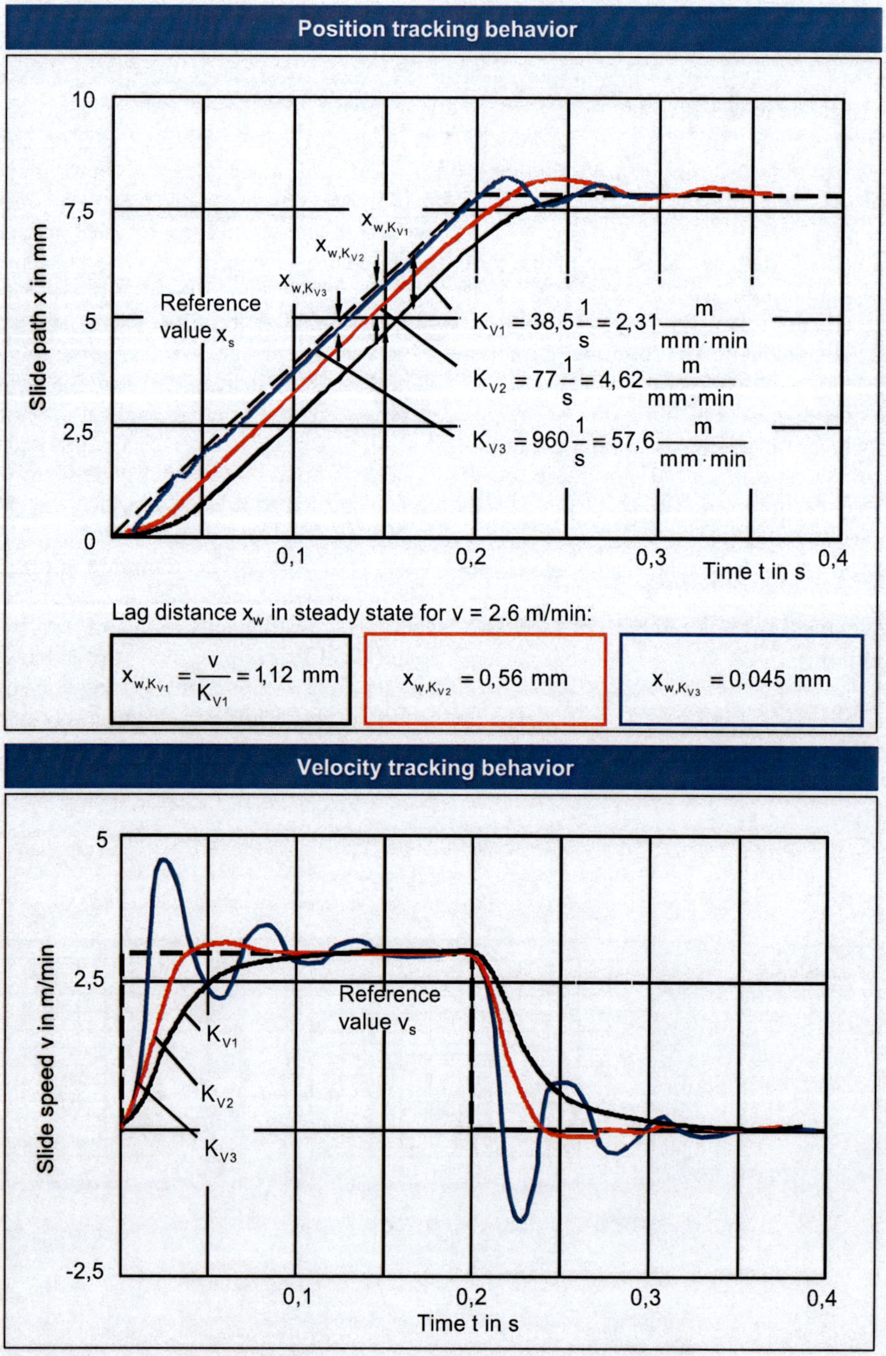


**Fig. 3.18** Tracking behavior of a position control loop (with velocity feedback)

### 3.2.4 Simulation of Feed Drives

Computer-aided simulation systems are an excellent tool in the design and analysis of dynamic systems. These simulation systems are based on libraries with functions for linear algebra, numerical integration, for solving differential equations and for performing matrix operations. All calculations required for simulation and analysis can be traced back to these functions. Any non-linearities and special characteristic functions can also be used in a signal flow diagram. It is mainly the user interface that differs. The users are able to formulate and solve their tasks using this interface.

Generally applicable simulation programs (such as MATLAB) require a mathematical formulation of the simulation task. Starting from a definition of the feed axis, a control block diagram must first be entered with exact formulation of the individual blocks in the form of numerator and denominator coefficients.

The diversity of functions available enables analyses to be performed in the time and frequency domains. The results can be shown numerically, graphically and be printed.

To simplify the step of the mathematical description of the simulation problem for control-related applications, input editors (such as Simulink) have been developed that permit direct graphical input of the block diagram. From different libraries, the user is able to assemble and configure the control loop elements required. The transfer function coefficients are then calculated automatically.

▫ Figure 3.19 shows the graphic modeling of a position control loop of a feed axis with simulated ramp response. For improved structuring of the model, use has been made of the ability to form self-defined elements. Internally, these are made up of more model elements. Both the drive and the feed mechanics are modeled here as an oscillatory 2nd order system. They are coupled together using the rotation angle of the shaft and the feedback torque of the mechanics acting on the drive.

User support extends even further for systems developed specially for the simulation of drive-related problems [ITI13]. Here, the element library no longer contains general control blocks but models of drive-related components (guide systems, spindle nut systems, servo drives, etc.) that users are able to combine as they wish. This means a user is able to concentrate on solving the actual simulation problem without having to know the underlying, internal structure of the models.

## 3.3 Transmission Behavior of the Mechanics

In addition to the linear transmission behavior described (and the associated tracking error for example), attributable mainly to the closed-loop control, non-linear transmission elements also influence the behavior of feed axes in drive amplifiers.

Also, the mechanical part of the control path has a feedback effect on the controller. The behavior of mechanical transmission elements can be classified into linear and non-linear behavior. Linear behavior is predominant in most cases, meaning non-linear components can be ignored in the design and calculations. Also related to this, however, is the fact that the non-linear behavior of a partial path can only be described or calculated accurately with difficulty using mathematical methods. It is mostly interfering factors that are in play here, those resulting from the physical boundary conditions and that thus cannot be prevented entirely.

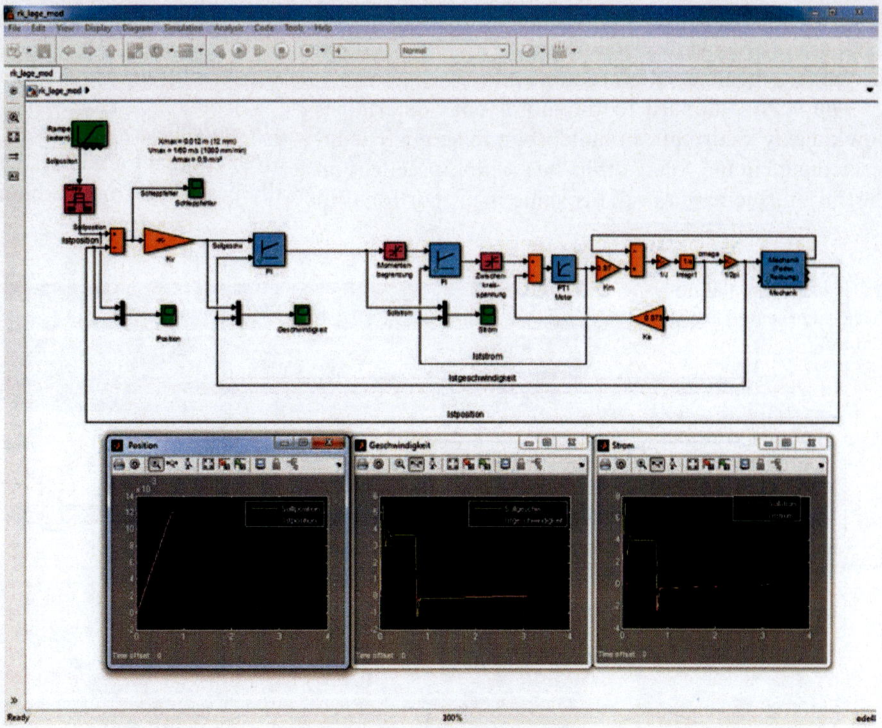


▫ **Fig. 3.19** Modeling of a position control loop in the MATLAB/SIMULINK simulation program

Non-linear transmission elements can be found at many places on a feed axis (◘ Fig. 3.20). They represent performance limits of the electrical system and non-linearities in the area of the mechanical transmission elements. Non-linearities include quantization of the measuring systems and digital control, limitation of motor current, backlash and reversal errors in mechanical components such as bearings, ball nut/spindle, toothed belt, spindle pitch errors, Stribeck friction in guide systems and the ball screw and signal noise. The following sections detail the most important influences on feed generation in machine axes.

### 3.3.1 Physical Boundaries of the Mechanical and Electrical Systems

At many places in electrical and mechanical systems, physical conditions have a limiting effect on impact parameters in the position control loop. These limits must be taken into account by the drive control so as not to overload the electrical or mechanical system. The most important limiting values are given and described briefly below.

*Intermediate circuit voltage*
The DC voltage generated in the intermediate circuit after rectification of the auxiliary current forms the energy source for the inverters (see ▶ Sect. 2.2). The voltage applied to the motor coils (see $U_A$ in ◘ Fig. 3.15) is thus limited to the magnitude of the intermediate circuit voltage. In certain circumstances, the internal resistances of the switch elements in the inverter (IGBTs) also have to be taken into account.

The voltage applied to the motor coils determines how quickly a current can build up in the coils. The intermediate circuit voltage thus has a limiting effect on the current increase rate. This value is proportional to the realizable jerk of the motor, i.e., to the change in acceleration of the drive.

In addition, the induced countervoltage of the motor must be overcome by the power source. So the realizable dynamic decreases more and more with increasing speeds because the voltage reserve is used up.

*Motor current, motor torque*
The current flowing through the motor coils is limited for two reasons. First, the motor coil can be overloaded by too high a current, and second, for synchronous motors, there is a risk of the large electromagnetic flux demagnetizing the magnet material of the rotor or secondary section.
Motors used today in feed drives are usually designed to handle 4 to 5 times the overload capacity for a short time.

*Motor speed*
The maximum motor speed is limited to the properties of the rotor. Centrifugal forces become so high at high speeds that, for synchronous motors, for example, the usually bonded rotor magnet plates can come loose. The rotors of asynchronous motors are generally more sturdy and, consequently, far higher speeds can be attained.

For linear motors, there is no limitation of maximum speed due to centrifugal force. In this case, the induced countervoltage limits the speed attainable.

*Critical whirling speed of the ball screw*
For long ball screws, the traversing speed of the carriage is limited by the critical whirling speed of the ball screw. The ball return system in the nuts also represents a speed-limiting factor.

*Maximum speed of the carriages*
Most compact guide systems used nowadays are generally permitted for maximum speeds of 300 m/*min*. For

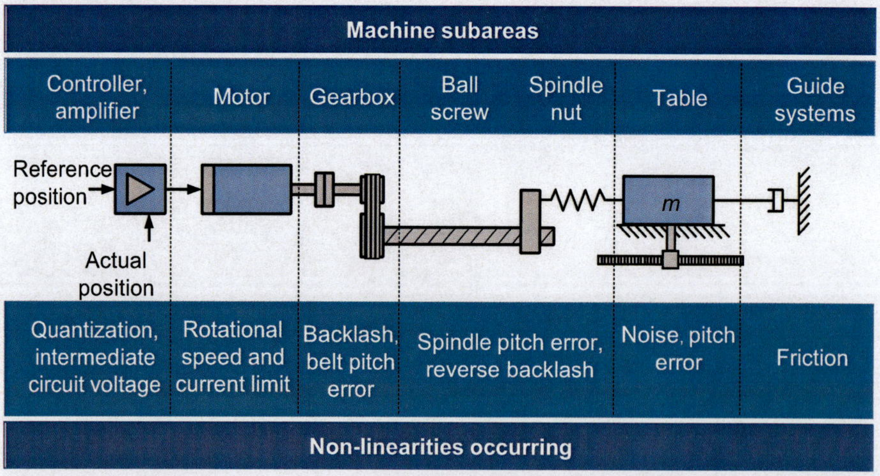


◘ **Fig. 3.20** Non-linearities in the position control loop of a feed axis high traversing speeds, the high centrifugal force loads of the rolling element diverters have a limiting effect.

*Threshold speed of the measuring system*
At high speeds, very high encoder frequencies are reached depending on the division period of the measuring system. They mean, first, that the signal amplitudes are attenuated in the amplifier electronics, and, second, that the maximum measurement input frequency of the drive controller can be exceeded.

## 3.3.2 Transmission Behavior of Electromechanical Drive Systems

The transmission behavior of the controller dynamics and of the mechanical components is a key contributing factor in the operational accuracy of machine tools. Every axis forms a complex, dynamic system that is deformed statically and excited dynamically by the reference and disturbance variables specified.

Other deviations between the workpiece shape realized by the tool and the reference workpiece contour output by the open-loop control are caused by defects in individual component fabrication (such as spindle pitch errors) or incorrect assembly.

The deviations resulting from the mechanics can be classified into errors in kinematic, static and dynamic transmission behavior.

### 3.3.2.1 Kinematic Transmission Behavior

Kinematic errors manifest themselves in geometric deviations of the feed system and are broadly independent of speeds, accelerations and forces. Kinematic transmission errors of mechanical components arise from defective production or assembly and as a consequence of wear. The following geometric component defects have a dominating impact on kinematic transmission errors:

- Spindle pitch, wobble and eccentricity errors
- Gearing defects,
- Circularity, eccentricity, wobble and alignment errors of bearing arrangements
- Straightness, angular, parallelism and alignment errors of guide systems

Kinematic transmission errors caused by geometric component defects can give rise to geometric machining errors as well as alternating forces that excite vibration, such as contact spacing errors for gear drives.

Adequate functional play between the components moving relative to each other is often required for the relative displacement of the mechanical transmission elements. It contributes proportionally to the kinematic transmission error in consideration of any increases or decreases in speed given alternating load directions. In particular, it is mechanical transmission elements directly in front of the machining point (such as gears and spindle nut systems) that should be clearance-free because this play has a direct impact in the form of machining errors.

The geometric deviations of individual mechanical transmission elements can be determined with individual error measurements (such as those for pitch errors on spindles). In most cases, however, it is not possible to determine the overall kinematic transmission error from multiple, individual errors having a combined effect.

The real, overall kinematic transmission behavior can be determined with transmission error measurement. The components contributing significantly towards the resultant kinematic transmission error due to their individual geometric errors can be detected by analyzing the error frequencies (such as the rotational frequencies of spindles and gear meshing frequencies). It is possible to reduce the kinematic transmission error with specific reworking of the defective components.

### 3.3.2.2 Static Transmission Behavior

Static transmission errors are caused by static deformations of the transmission elements as a result of static frictional forces, process forces and weight forces caused by the workpieces and by the machine components themselves. It is acceleration forces by the drive, however, that represent the most significant excitation. Accelerations of 10 to 20 $m/s^2$ at speeds of 100 to 120 m/*min* are the current requirement.

◘ Figure 3.21 shows a compliance analysis performed by measurement of the individual components of a spindle nut system. With a compliance value of 58%, the spindle is the weakest element in this case. If the static stiffness of the spindle is increased by selecting a longer diameter, static deformation reduces considerably and the resonance frequency increases. This can be a desired effect given that the natural frequency range of the mechanical structure should lie above the natural frequency of the control loop by a factor of at least 2.

Enlarging the spindle diameter increases the stiffness by the power of 2. However, the mass moment of inertia increases by the power of 4 at the same time.

- Stiffness $\sim d_{Sp}^2$,
- Mass moment of inertia $\sim d_{Sp}^4$.

The acceleration behavior of the feed drive can be negatively impacted by increasing the spindle diameter. This means that selecting the correct spindle diameter, and its pitch, is a compromise between sufficiently static stiffness yet a mass moment of inertia that is just still acceptable.

3

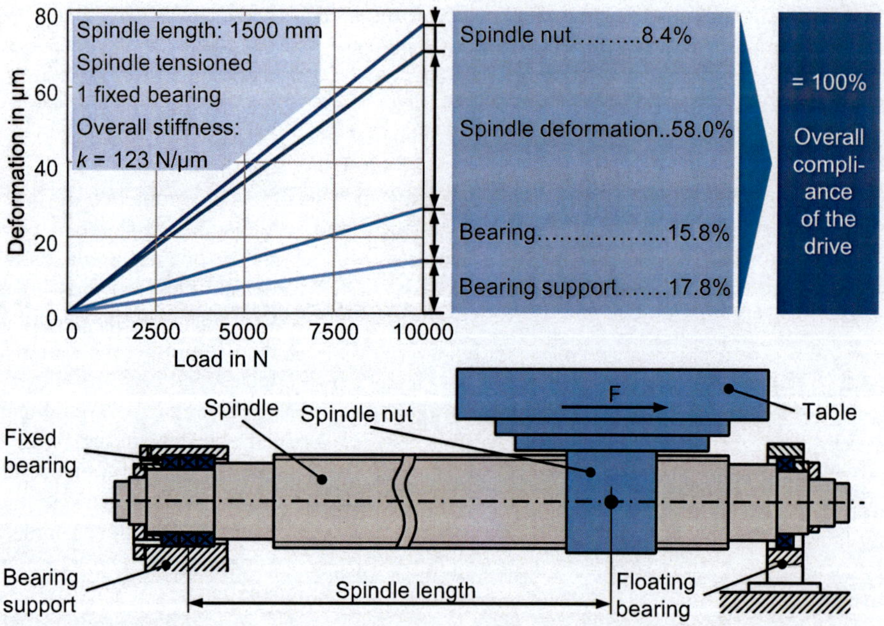


**Fig. 3.21** Static compliance behavior of a spindle nut system

### 3.3.2.3 Dynamic Transmission Behavior

Relative vibrations between the workpiece and tool entail geometric deviations and poor surface quality of the workpiece being machined. Vibrations can also cause increased levels of tool loading.

The dynamic transmission behavior of the mechanical transmission elements is defined by the frequency-dependent, functional relationship between the output parameter and the input parameter of the system. The actual output parameter here is referenced to the amplitude of the reference output value. The deviation of the actual from the reference output value is called transmission error. Figure 3.22 shows the dependency of the table path of a machining center (output parameter) on the rotation angle of the motor shaft (input parameter).

A high table mass in conjunction with relatively high compliance values for spindle nut and spindle is responsible for the lowermost mechanical resonance point on most feed drives with spindle. Although the system examined here does not show any dominating weak points in the measurement of compliance frequency responses and in the modal analysis, an increase of the amplitude ratio by a factor of 4 at 183 *Hz* can be seen.

Similarly, the same measurement on a grinding machine illustrates that the impact of the mechanical transmission elements has high significance for the dynamic behavior of the machine (Fig. 3.23).

The amplitude ratio shows clear peaks at 82 *Hz* and 145 *Hz* by factors of 2 and 9, respectively. It was possible using a modal analysis to confirm that these peaks are attributable to the shifting of the feed table. The modal analysis was performed with spatial excitation at the cutting point and includes the examination of the whole machine, meaning other machine components also affect the dynamic behavior here (Fig. 3.24).

The vibration amplitude is at its highest in the range of the resonance frequencies of the mechanical transmission system. Possible excitation frequencies should therefore have a sufficiently high frequency separation from the resonance frequencies. The following rule of thumb applies in this context for subcritical systems: The frequency of the first resonance point of the mechanical system should, as a minimum, be twice the natural frequency of the position control loop.

The force flow for converting the motor movement into a feed motion is comprised of many individual elements that, like vibration systems, exhibit mass, spring and damping characteristics. The feed drive as a whole can thus be considered a system of coupled spring-mass systems (multiple-mass oscillator). The system properties mean that, for dynamic excitation, the feed drive elements can perform vibrations. Knowledge of the dynamic tracking and disturbance variable behaviors is of primary importance here.

The dynamic transmission behavior is defined by a 2nd order differential equation:

$$[M] \cdot \{\ddot{x}\} + [C] \cdot \{\dot{x}\} + [K] \cdot \{x\} = \{f\} \tag{3.64}$$

with: $[M]$ Mass matrix (diagonal),
$[C]$ Damping matrix (tridiagonal),
$[K]$ Stiffness matrix (tridiagonal),
$\{x\}$ Displacement vector,
$\{\dot{x}\}$ Velocity vector,
$\{\ddot{x}\}$ Acceleration vector and
$\{f\}$ vector for external forces.

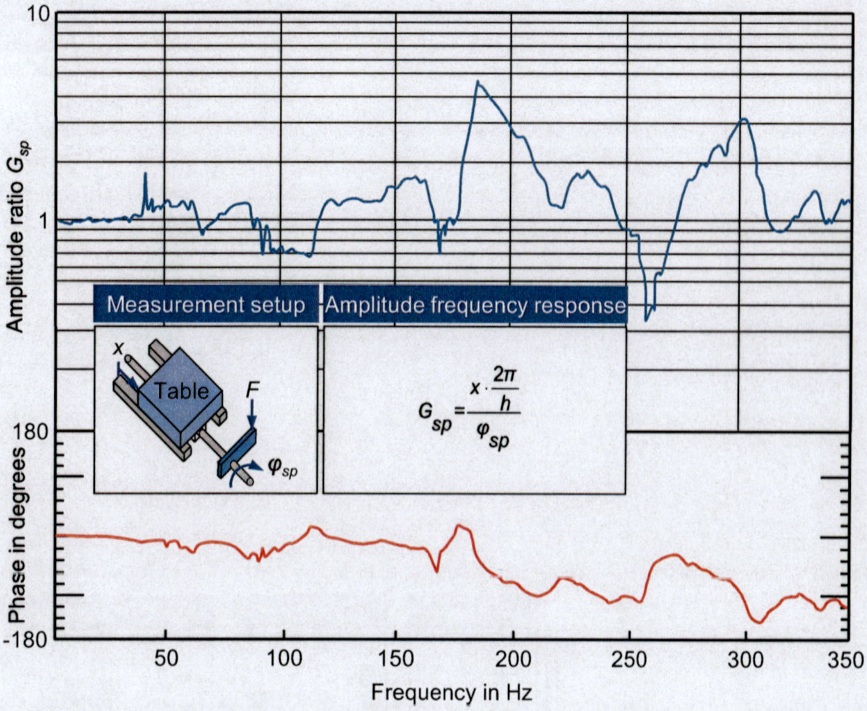


**Fig. 3.22** Frequency response of the mechanical transmission elements of a machining center feed drive

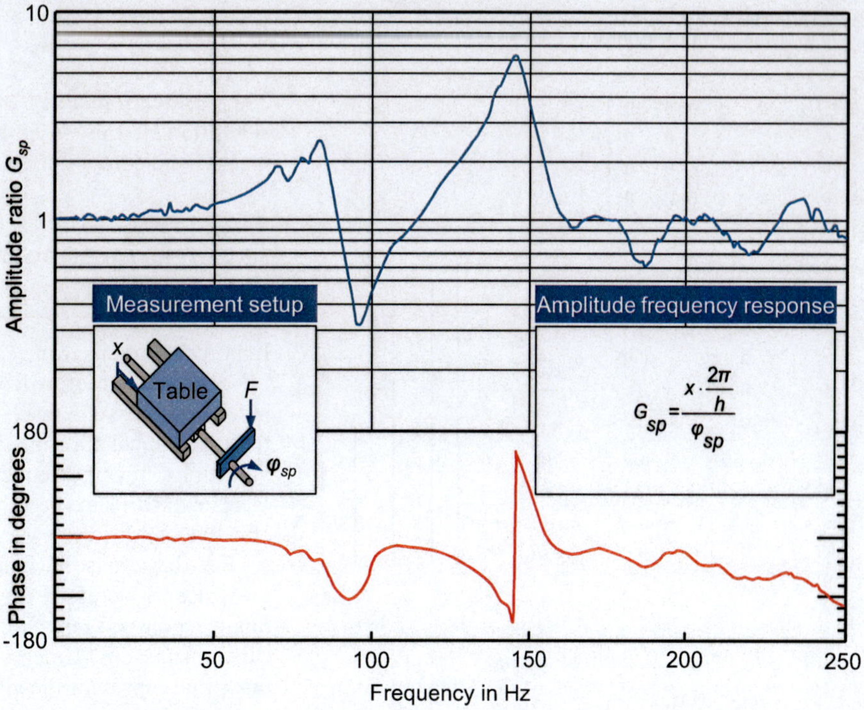


**Fig. 3.23** Frequency response of the mechanical transmission elements of a grinding machine feed drive

An electromechanical feed system, comprising a motor, transmission, ball screw and workpiece table, is depicted as a multiple-mass oscillator in Fig. 3.25. The spindle can be regarded here as two spring/damper elements connected in series. The reduced values, referenced to the drive-side variables φ3 and $x_4$, must be used for the mass and moment of inertia of the spindle. The following equation of motion is given for the feed system:

**Fig. 3.24** Vibration mode of a grinding machine at $f_{0.2} = 145$ Hz

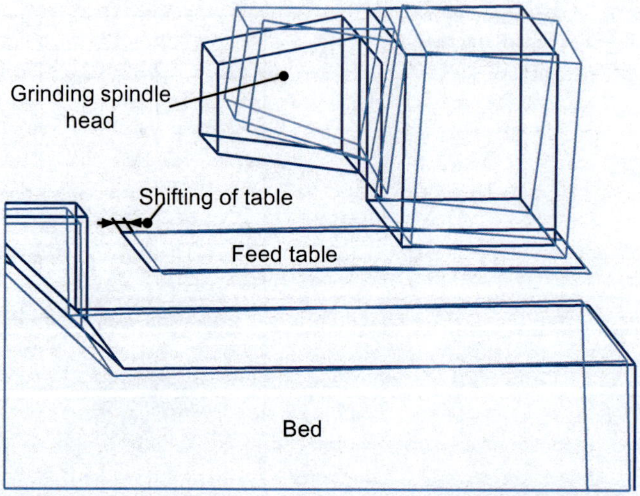


**Fig. 3.25** Depiction of a mechanical feed system as a multiple-mass oscillator

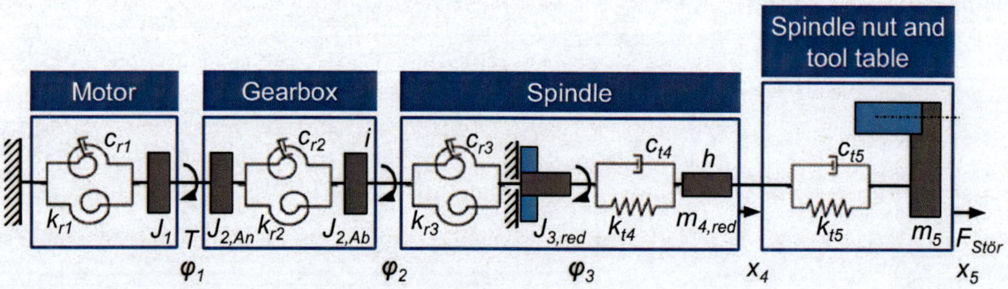


$$\begin{bmatrix} J_1 + J_{2,\mathrm{An}} & 0 & 0 & 0 & 0 \\ 0 & J_{2,\mathrm{Ab}} & 0 & 0 & 0 \\ 0 & 0 & J_{3,red} & 0 & 0 \\ 0 & 0 & 0 & m_{4,red} & 0 \\ 0 & 0 & 0 & 0 & m_5 \end{bmatrix} \cdot \begin{bmatrix} \ddot{\varphi}_2 \\ \ddot{\varphi}_2 \\ \ddot{\varphi}_3 \\ \ddot{x}_4 \\ \ddot{x}_5 \end{bmatrix} +$$
$$\begin{bmatrix} c_{r1} + c_{r2} & -c_{r2} & 0 & 0 & 0 \\ -c_{r2} & i \cdot c_{r2} + c_{r3} & -c_{r3} & 0 & 0 \\ 0 & -c_{r3} & c_{r3} + c_{t4} \cdot \frac{h^2}{4\pi^2} & -c_{t4} \cdot \frac{h}{2\pi} & 0 \\ 0 & 0 & -c_{t4} \cdot \frac{h}{2\pi} & c_{t4} + c_{t5} & -c_{t5} \\ 0 & 0 & 0 & -c_{t5} & c_{t5} \end{bmatrix} \cdot \begin{bmatrix} \dot{\varphi}_1 \\ \dot{\varphi}_2 \\ \dot{\varphi}_3 \\ \dot{x}_4 \\ \dot{x}_5 \end{bmatrix} +$$
$$\begin{bmatrix} k_{r1} + k_{r2} & -k_{r2} & 0 & 0 & 0 \\ -k_{r2} & i \cdot k_{r2} + k_{r3} & -k_{r3} & 0 & 0 \\ 0 & -k_{r3} & k_{r3} + k_{t4} \cdot \frac{h^2}{4\pi^2} & -k_{t4} \cdot \frac{h}{2\pi} & 0 \\ 0 & 0 & -k_{t4} \cdot \frac{h}{2\pi} & k_{t4} + k_{t5} & -k_{t5} \\ 0 & 0 & 0 & -k_{t5} & k_{t5} \end{bmatrix} \cdot \begin{bmatrix} \varphi_1 \\ \varphi_2 \\ \varphi_3 \\ x_4 \\ x_5 \end{bmatrix} = \begin{bmatrix} T \\ 0 \\ 0 \\ 0 \\ F \end{bmatrix} \quad (3.65)$$

with: $J_i$ Mass moment of inertia,
$m_i$ Mass,
$c_{r+t,i}$ Damping:rotary and translational,
$k_{r+t,i}$ Stiffness: rotary and translational,
$T_i$ Torque and
$F_i$ Force

The observations put forward here apply without consideration of the closed-loop feed controller. Only the mechanical components are included.

### 3.3.3 Transmission Behavior of Linear Motor Drives

In contrast to electromechanical drive systems, the motor of a linear motor drive is flanged directly onto the machine slide. As a result, there is no low-pass effect of the mechanical transmission elements. Tests show that vibrations having frequencies of many hundreds of Hertz can form in the velocity control loop because its effective bandwidth is very wide [WECK99].

For these reasons, increasing the controller settings gives rise to increased excitation of the mechanical resonance points. Here, however, it is not the mechanical transmission elements in conjunction with the slide mass that represent the vibratory system. Rather, as can be verified, the carriages must be interpreted as spring/damper elements given a very rigid machine slide design. This spring-mass system, comprising carriage and slide mass, is then excited to vibration by the linear motor drive in conjunction with the controller.

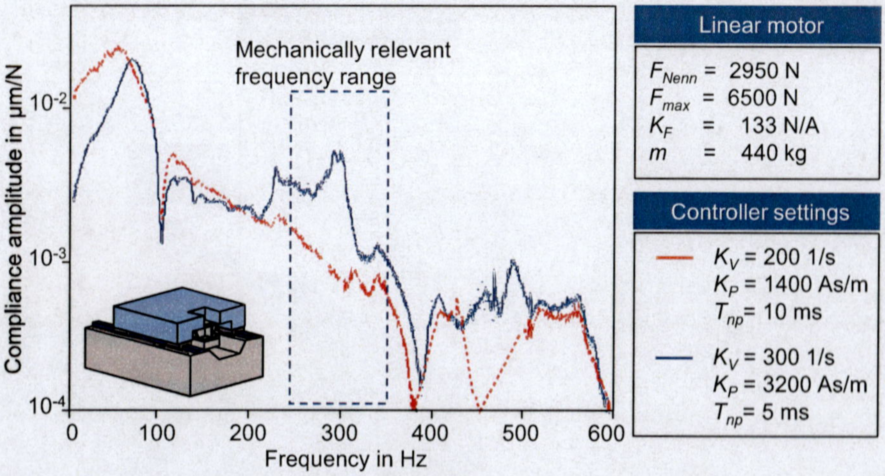


**Fig. 3.26** Disturbance frequency responses of the linear motor axis for different controller settings

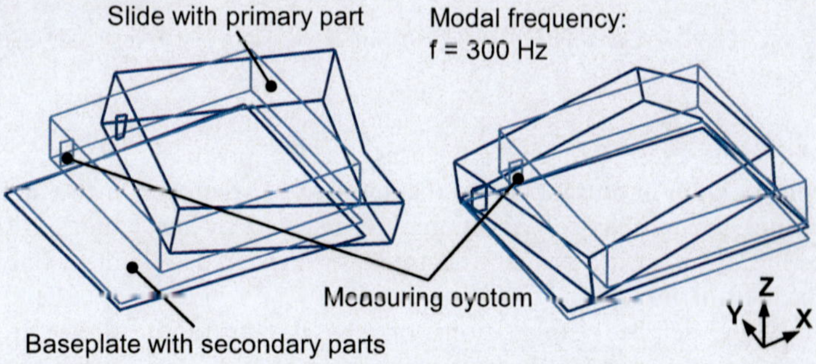


**Fig. 3.27** Measured vibration mode of the mechanical natural frequency

Figure 3.26 shows two disturbance frequency responses of a linear motor-driven axis for low and increased controller settings in the position and velocity control loops. It can be seen clearly that a mode shape is excited at a frequency of 300 *Hz* for higher controller settings and those at low settings are barely noticeable.

Figure 3.27 shows the vibration mode measured with a modal analysis. The slide structure hardly deforms and instead, the mass is cushioned into the carriage. The wobbling vibration of the slide means the dynamic movements are detected by the measuring system attached on the side and are relayed to the controller as deviation, resulting in instability.

These deliberations demonstrate that the controller settings in the highly dynamic velocity control loop are limited by the vibratory coupling of the slide mass. Given that damping in the position control loop is influenced greatly by the velocity control loop, the limited controller settings in the velocity control loop also have a limiting effect on the $K_V$ factor in the position control loop.

These interrelations can, by applying suitable simplifications in the control loops, also be calculated analytically. A stability assessment of the control loops can be performed with vibratory mechanics according to Nyquist, for example.

A field of the controller parameters can be calculated in the parameter space as the result (Fig. 3.28). In this field, the control loops set with parameter triple ($K_p$,$T_{np}$,$K_V$) are stable. $K_V$ here is the velocity gain in the position control loop, and $K_p$ and $T_{np}$ is the gain factor and resetting time in the velocity control loop. The properties of the electrical and mechanical systems are inputs into the graph. An accurate derivation of the method is in [WECK99].

## 3.4 Influences of the Measuring System on the Closed-Loop Feed Control

The vibratory properties of the mechanical transmission elements have a bearing on the dynamic behavior of the position control loop as a whole. The analyses described below show that the nature and position of the distance measuring system are key influences over overall system behavior. The mechanical structural

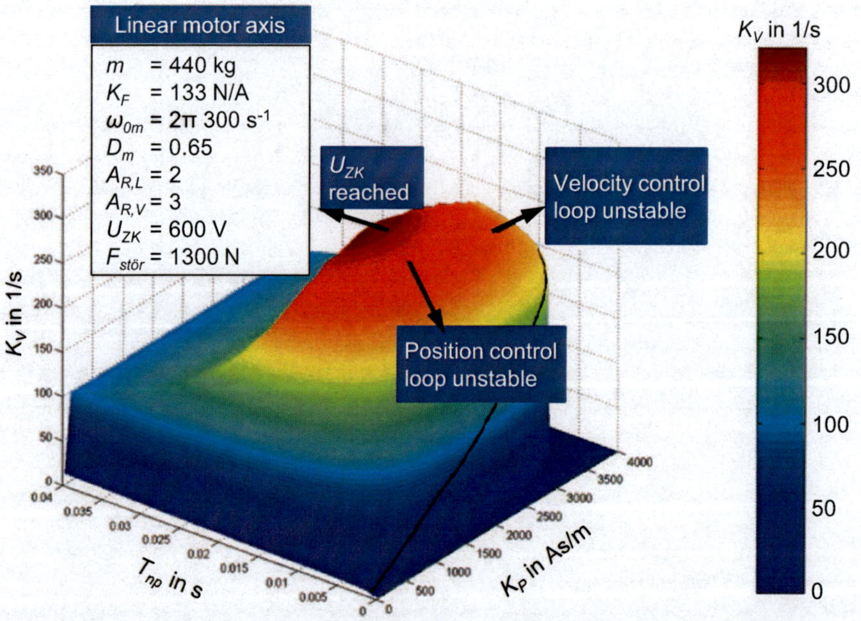


**Fig. 3.28** Setting range for the controller factors with regard to system stability and physical limitations

vibrations exert different influences on the position control loop. Important here is the question of whether or not the mechanical vibrations are detected by the position measuring system and thus are registered by the control loop.

More detailed explanations on the designs and working principles of the measuring systems used are in ► Sect. 2.4.

### 3.4.1 Behavior of Electromechanical Axes for Closed-Loop Control via Indirect and Direct Measuring Systems

The position measuring systems usual today for determining the positions of numerically controlled machine components are linear gauges that read the table position directly and rotary encoders mounted on the motor that read the angular position of the motor. Knowing the gear ratio in the transmission and the spindle pitch indirectly gives the travel distance of the slide. The benefits and downsides of both systems are shown in ► Sect. 2.4.

In the example here (see Fig. 3.29), the table mass with ball screw, nut and bearing arrangement forms the mechanical vibration system. As the measurement results of the transmission behavior show, a sizable resonance peak occurs for the drive with direct measuring system that can result in excessively high workpiece dimension deviations and render impossible stabilization of the drive control loop. The mechanical vibrations excited are input directly into the position determination measurement result.

In comparison, the indirect position measurement method has benefits here. The vibration amplitudes of the table vibration affect the position measurement only in the form of torsion excitation of the feed spindle. This can be ignored. The measurement result shows an acceptable change to the natural frequency range of the vibrating structure of the indirect position measuring system.

Simulation as in Figs. 3.30 and 3.31 illustrates the influence of the path feedback location on the dynamic behavior of the feed drive. Applying a step-like change of the set path signal to the control loop results, for the indirect position measuring system, in a ramp-like output signal of the feed motor that is converted to an actual path signal only changed marginally by the mechanical transmission elements. The same control loop becomes unstable when the direct measuring system is used. A step-like change to the set path signal gives rise to an oscillating angular position of the motor shaft and a dynamically unstable actual position (see Fig. 3.31).

### 3.4.2 Effect of the Measuring System for Linear Motor Drives

Particular importance is attached to the position of the measuring system when the requirement is to control linear motor drives. Only directly measuring gauges are used for linear motor drives (see Fig. 3.15).

Two effects can be seen when the gain factors in the position control loop, but also in the velocity control loop in particular, are increased to improve

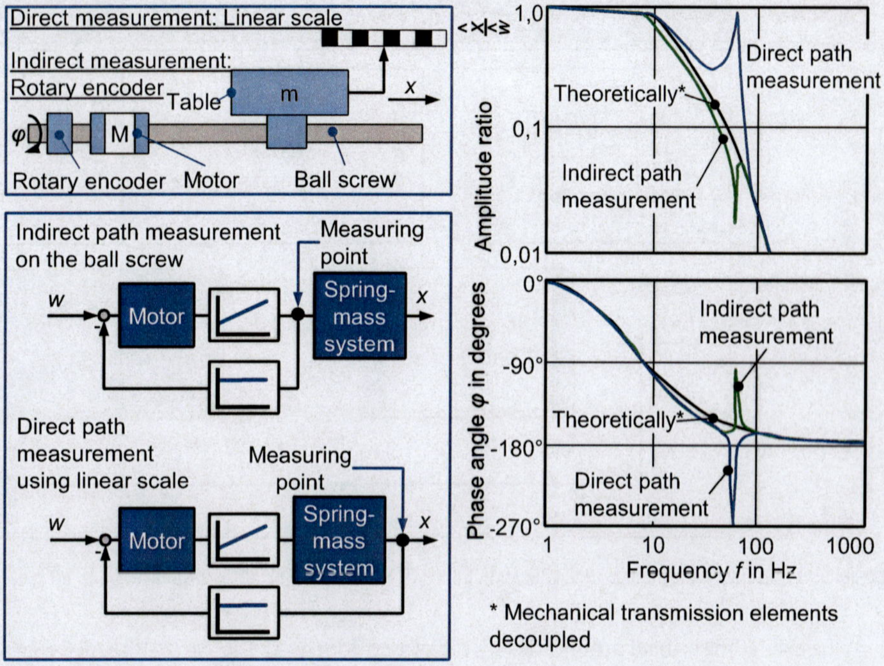


**Fig. 3.29** Effects of the transmission elements onto the position control loop of the feed drive for different position determinations

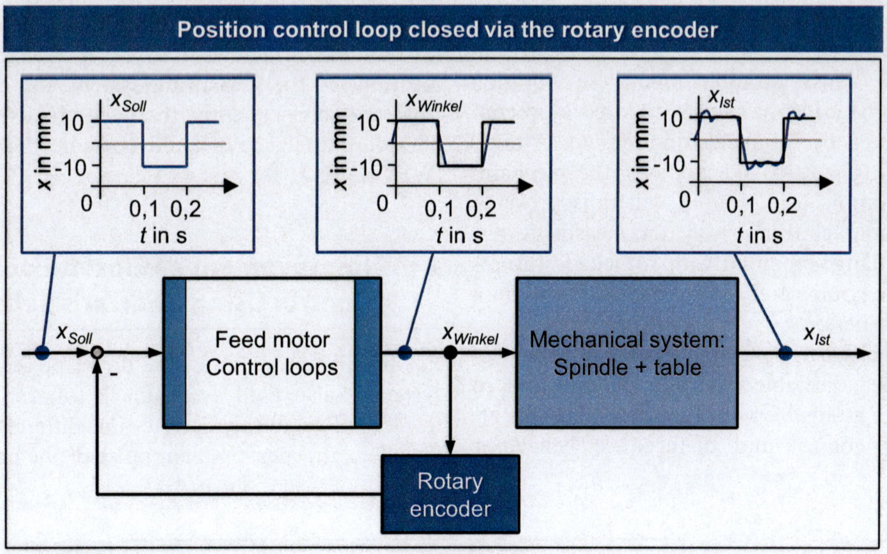


**Fig. 3.30** Effect of the path feedback location on the dynamic behavior of the feed drive; path measurement on the spindle with rotary encoder

transmission behavior: The drive generates an audible noise from quick flux changes in the motor and, when increased further, a high-frequency natural vibration of the machine slide can occur.

The drive noise can be explained by the fact that the signal noise in the measuring system is amplified by the controllers, in particular by the broad-band velocity control loop. And also by the fact that the position signal for generating the velocity signal is derived in a discrete-time manner. Signal noise rises steeply depending on the clock timing of the derivation.

One possible natural vibration is described in ► Sect. 3.3. It is not only the frequency and damping of the mechanical natural vibration that are of interest, however. In fact, position, orientation and attachment of the measuring system also influence the nature and natural mode shape of the system. The reason for this is that the natural mode shape of the mechanics might

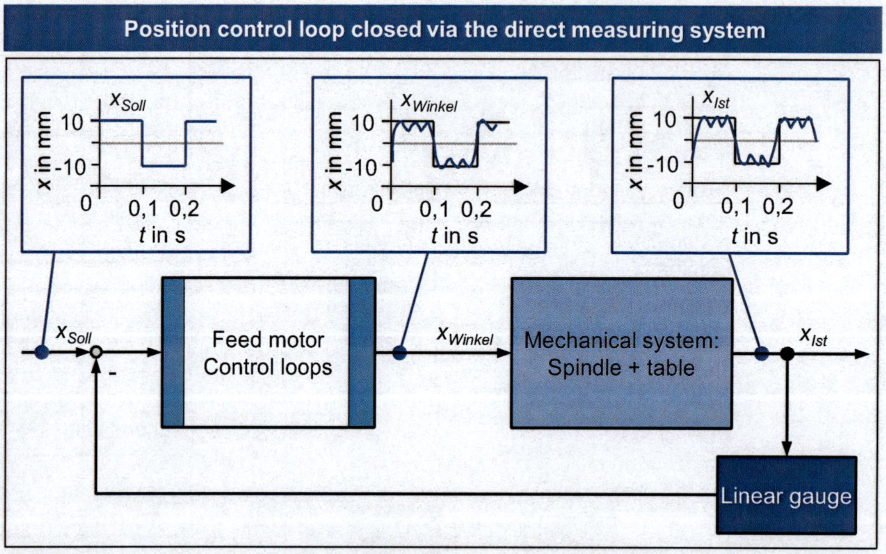


**Fig. 3.31** Effect of the path feedback location on the dynamic behavior of the feed drive; path measurement on the slide with linear gauge.

only be excited to critical amplitudes by the closed-loop control. The vibrations are recorded and amplified by the measuring system of the closed-loop control.

Given that the motor forces have an effect in the feed direction as well as, albeit to a lesser degree, in the normal and transverse directions, these vibration modes can also be critical as regards closed-loop control when detected by the measuring system. Attaching the measuring system at a node of the mechanical natural vibration, for example, can thus prove to be a sensible approach. Either way, the measuring system should be flange-mounted with sufficient stiffness to the mechanical frame and slide elements to prevent a vibratory system here.

Where the measuring system is attached has a great influence on the adjustable controller factors and so on the dynamics attainable with the feed axis. This affects both the control and disturbance behaviors.

Figure 3.32 shows the results of a wire cutting test. The position of the machine slide is shown after a step-like load (disturbing force of 981 *N*). The controller settings were set at the stability threshold of the system for both tests. It shows that, just by selecting a different location for the measuring system, the controller settings and, consequently, the dynamic disturbance stiffness of the feed axis could be more than quadrupled [WECK99].

### 3.4.3 Improvement of Closed-Loop Feed Control Using a Ferraris Sensor

For linear motor drives, the direct position measuring system is also used as feedback for the velocity control loop. Due to the discrete-time differentiation necessary here, the measurement noise of the position sensor

**Fig. 3.32** Measurement results for dynamic disturbance stiffness, cutting test

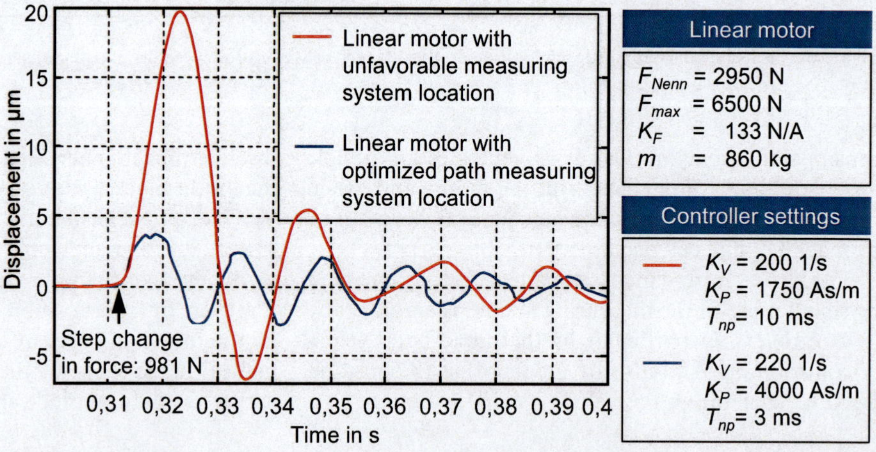

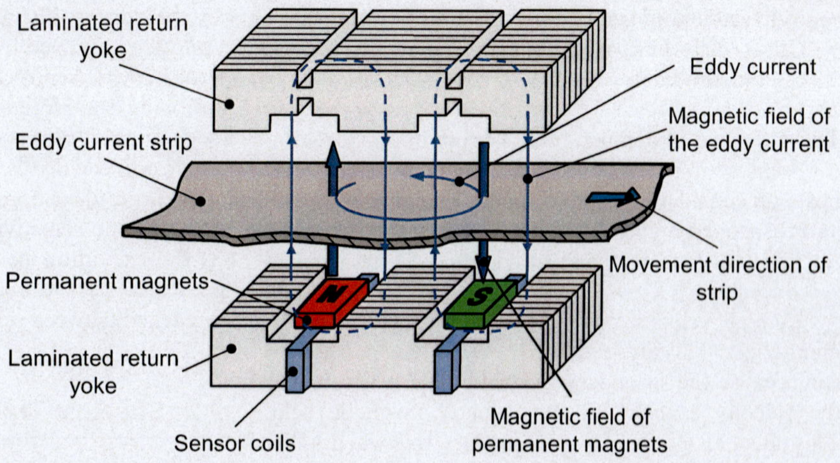


**Fig. 3.33** Basic design of a ferraris sensor (according to Hübner)

exerts a disturbing effect on the actual velocity calculated and so on the control quality of the overall system. Using an additional measuring system that can register the acceleration of the feed slide can be a viable option at this point. Suitable for this, in particular, are relative acceleration sensors that generate their signals directly between feed slide and the surroundings. These signals are thus comparable to those of the corresponding position measuring system.

The structure of a Ferraris relative acceleration sensor is detailed below for the linear case. This principle can, however, also be readily applied to the rotatory case. In this case, the eddy current strip of the sensor must be replaced by an eddy current disk. The term "Ferraris sensor" dates back to a test conducted by Ferraris in 1885.

The sensor comprises two parts—an eddy current strip (normally a narrow aluminum plate) and a sensor head (normally a slit) that contains the sensor system. The head grips the eddy current strip in a U-shape so that the strip is pulled backwards and forwards through the sensor when there is movement (Fig. 3.33). It is not relevant whether the plate or sensor head is moved because it is the relative motion between the components that is detected.

The sensor head comprises two laminated iron yokes that contain permanent magnets and generate a constant magnetic field penetrated by the eddy current strip. When there is relative motion between the plate and the magnetic field, the field induces a voltage in the plate causing eddy currents to build up (conductor moving in a magnetic field). These eddy currents, in turn, generate a magnetic field with a strength proportional to the relative velocity between the sensor head and plate.

In addition to the permanent magnets, the sensor head contains sensor coils that are positioned such that the magnetic field of the eddy currents flows through them. When the magnetic field generated by the eddy currents changes, a measurable voltage is induced in these coils.

A change to the magnetic field of the eddy currents is caused primarily by a change of the relative velocity between sensor plate and sensor head, i.e., when there is a relative acceleration between these components. The voltage induced in the coils of the sensor head is proportional to the relative acceleration between the plate and sensor head. This voltage magnitude is different depending on the sensor design. As a result, it may need to be amplified in line with the application desired before further processing. For this, some manufacturers of sensors offer electronics integrated in the sensor head (see Fig. 3.34).

In principle, a high quality velocity signal can be generated from the acceleration signal of the Ferraris sensor with integration because the noise in a measurement signal is reduced by integration. The signal of a Ferraris sensor does, however, also contain some disturbance factors that need to be filtered out. These are:
- Temperature influence
- Field displacement
- Geometric deviations of the eddy current strip

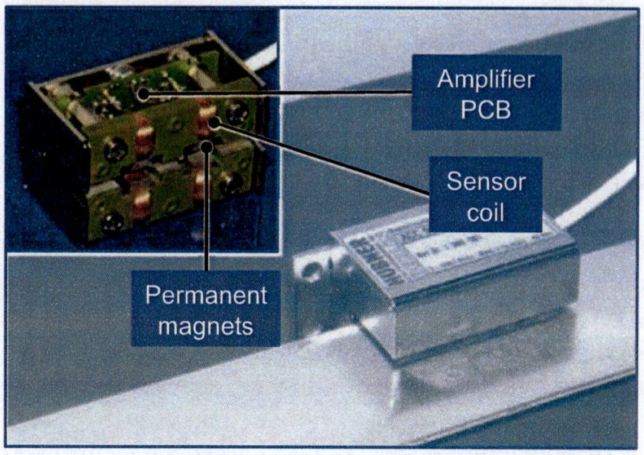


**Fig. 3.34** Linear Ferraris sensor with integrated amplifier PCB

- Material inhomogeneities of the eddy current strip
- Offset, drift, linearity and noise of the A/D converters and amplifiers



The influence of temperature is pronounced for rotatory sensors in particular because the eddy current disk heats up considerably due to the high relative velocities and associated high eddy currents. The increased material resistance means the sensitivity of the sensor signal is reduced.

At high velocities, there is also the field displacement effect. This also reduces the sensitivity of the sensor because the secondary magnetic field generated by the velocity cancels out the primary magnetic field. This effect also only plays a secondary role when linear sensors are used.

Since the fields of the exciter magnets are not homogeneous, geometric deviations of the eddy current strip also have a disruptive effect on the measurement signal. These are generally position-dependent (e.g., deformation in the metal sheet). Also however, the traversing speed has an influence over the amplitude of this disturbance.

Material inhomogeneities have an effect similar to geometric deviations. They are largely avoided when aluminum is used for the eddy current strips.

The amplification and analysis electronics also affect the signal of the sensors. A/D converters currently available generally have low levels of drift and offset. Sometimes, only errors in signal linearity occur. These errors become more significant when the requirement is to integrate the signal for reconstructing the speed because even minor errors add up over time.

To be able to use the signal in closed-loop feed control, these disturbance factors must first be compensated. Some of this is performed directly in the amplifier electronics (e.g., compensation of the temperature influence). Also, however, the signal usually needs to be filtered by a suitable observer structure. As the reference, it uses the position read by a parallel position measuring system.

The noise and thus associated excitation of the system can be reduced significantly by using a Ferraris sensor in closed-loop feed control [HILL98]. As a consequence, higher controller settings can be used for most systems. This gives rise to increased dynamic stiffness and improved control response of the drive.

As an alternative to improving the controller characteristics, a Ferraris sensor can also be used to "upgrade" an inexpensive length measuring system. For example, for measuring systems working on the reluctance principle that are recessed into the guide system, the resolution for differentiation to provide feedback to the velocity controller is not adequate. In this case, the signal quality necessary can be generated using a Ferraris sensor. The accuracy of the measuring system cannot be improved by doing so, however.

A Ferraris sensor can in principle also be used for improving the closed-loop control of rotatory drives. Much higher plate passage speeds occur here compared to linear drives. This has a negative effect on the temperature development (the rotation means the disk heats up) and sensitivity (the big secondary field weakens the excitation field). Consequently, the maximum speed of the sensors is limited and a comprise needs to be reached between sensitivity and permitted speed.

### 3.4.4 Shortest Traversable Increment

For precision machine tools, just how accurately the feed drives are able to realize the smallest of infeeds is often of interest. This makes strict demands on the measuring systems and closed-loop control of the system. The behavior of a feed axis in a step response test is clarified first below and then the cause of the deviations is explained.

To investigate these properties, step-like reference value changes are specified by the NC controller. These are to be reproduced by the feed drives. The resultant motions are read by a laser interferometer, for example, and depicted versus time (see ◘ Fig. 3.35).

In the tests, the drive must be given enough time to reach the individual positions. The path-time behavior for the smallest of control signals, the remaining path errors and the reversal behavior can be determined from the measurement plots. In the micrometer and submicrometer ranges in particular, most conventional drives behave completely differently from the behavior in the millimeter range.

The smallest increment is reached when the remaining path deviation is just as big (or even bigger) than the reference value specified. For conventional feed drives, noise is often superimposed on actual values, meaning the curve needs to be smoothed here as shown below in ◘ Fig. 3.35.

The reason the noise is present in the actual signal is that the actual position of the axis can only be detected by the measuring systems with a certain level of background noise. The control loops interpret this noise as an erroneous position and attempt to compensate this error. Control variables build up as a result causing minor movement of the motor. So a controlled axis always moves marginally backwards and forwards around its reference value.

These undesired movements are greater the higher the controller factors are set. On the other hand, the control loops also cannot be adjusted to a weak setting because otherwise the cutoff frequency of the feed system becomes too low and no more dynamic movements

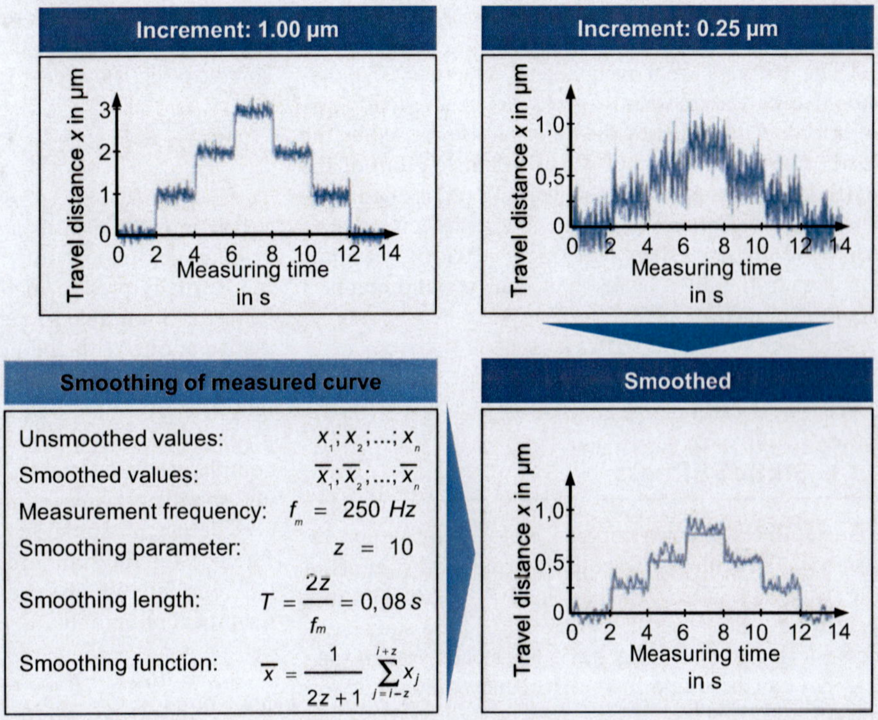


**Fig. 3.35** Determination of the smallest traversable increment of feed drives

are possible. Consequently, a comprise needs to be reached here depending on the application.

## 3.5 Static and Dynamic Stiffnesses of Feed Axes

In addition to the properties of the machine structure, the open-loop and closed-loop control of feed drives are exerting greater and greater influence on the behavior of machine tools.

The computing power of microprocessors, growing with every new generation, means drive controllers are becoming increasingly powerful. Higher control loop sampling rates can be realized as a result. The bandwidth of control loops is increasing and closed-loop control can still influence the dynamic behavior of the machine even at high frequencies. Also, compensation functions can be integrated into the controllers that, for example, filter out certain frequencies from the machine vibrations making the controller insensitive [ALTI11].

The requirements placed on machining speed have become very strict. For this reason, ball screws nowadays have much greater spindle pitches to attain high feed rates. On the other hand, process forces have a greater influence on the controller due to more reduced self-locking of the systems. For linear motor drives, mechanical transmission elements within the flux of forces are omitted completely. This is why all disturbing forces from the process and friction (in the guide systems) have a direct effect on the controller. To define the behavior of machine tools, the closed-loop control must be included in the analyses [HOFF08; OERT08].

The contour accuracy of feed motions of machine tools (i.e., measure of the actual path deviation of a feed slide from the path specified) is determined by a series of system properties and influencing factors. In the assessment of feed motion deviations, a distinction must be made between mechanical structural deformations caused by inner acceleration forces or external forces (process forces), and control-related reference variable deviations such as tracking, velocity and acceleration errors. Reference variable deviations can be reduced almost at will with control-related measures and viable velocity control (limited jerk/continual jerk change) in due consideration of the maximum possible drive acceleration and speed.

The mechanical structural deformations in the drive train are caused by acceleration forces and exterior process force components in the feed direction. Both forces have influence over the static and dynamic slide displacements in the same way. The increased use of electric linear motor drives inevitably raises the issue of a definition of the static and dynamic stiffness of these drive types because they lack the mechanical gear components that absorb forces and represent the basis for calculations using rotatory drives.

The definition of static and dynamic stiffnesses/compliances of feed drives is explained below. A distinction

3

is made here between systems driven electromechanically and with a linear motor.

The stiffness of a mechanical structure is understood to be the quotient ($k_{xx} = F_x/x$) of the load (e.g., of a force $F_x$) divided by the deformation caused by the load ($x$, in the direction of $F_x$) at a defined point of the machine structure, mostly the working point of the machine. The reciprocal of stiffness is called compliance ($d_{xx} = x/F_x = 1/k_{xx}$). It is used more often for dynamic analyses in the form of resonance curves and complex Nyquist plots (Eq. 3.65).

$$d_{xx}(jf) = \frac{X(jf)}{F(jf)} = \frac{\hat{x}(f)}{\hat{F}(f)} \cdot e^{-j\varphi(f)} \tag{3.66}$$

### 3.5.1 Static Stiffness

Static stiffness is understood to be the deformation resistance a structure of a static, external load exhibits in the steady state. If a step static load $F_{stör}$ is applied to a system and displacement x materializes, the static stiffness is calculated as in Eq. 3.65. The steady system state ($t \rightarrow \infty$) can be found in the frequency domain at frequency $\omega = 2\pi f = 0\ s^{-1}$. Consequently, static stiffness can be seen in the amplitude response at a frequency of 0.

$$k_{xx,stat} = \frac{F_{stör}}{x(t \rightarrow \infty)} = \frac{F(jf = 0)}{X(jf = 0)} \tag{3.67}$$

#### 3.5.1.1 Static Stiffness of Electromechanical Drives (Ball Screw Drives)

For a feed axis with electromechanical drive train, all mechanical elements lying within the flux of forces affect the stiffness magnitude. There is a differentiation here between the individual elements connected in series and parallel (see ◘ Fig. 3.36).

As only the influence of the drive train (and not the guide systems) is to be considered here, the process force is assumed to be effective in the feed direction. The lack of self-locking of the ball screw drive means these drives are not reactionless, i.e., the motor must apply a reverse torque (due to force $F_x$) when the friction forces are ignored (Eq. 3.66). The process force, spindle pitch $h$ and gear ratio $i$ are factored in.

$$M_{d,Rück} = F_{stör} \cdot \frac{h}{2\pi i} \tag{3.68}$$

When indirect measuring systems are used, the integrating component of the closed-loop control system (usually present in the closed-loop speed controller) is also not able to detect or compensate the elastic deformation. With the help of the controller, the motor only keeps to the specified angular position of the motor shaft or the ball screw spindle, respectively, applying reverse torque as necessary. Overall stiffness or compliance is determined mainly by the ball screw, nut and spindle bearing arrangement elements.

The overall static drive compliance is derived as in Eq. 3.68 for the standard series connection of individual compliances. The overall static drive stiffness is the reciprocal of compliance (Eq. 3.69).

$$\begin{aligned} d_{ges} &= d_{Mutter} + d_{Spindel} + d_{Lager} \\ &= 1/k_{Mutter} + 1/k_{Spindel} + 1/k_{Lager} \end{aligned} \tag{3.69}$$

$$k_{ges} = \frac{k_{Mutter} \cdot k_{Spindel} \cdot k_{Lager}}{k_{Mutter} \cdot k_{Spindel} + k_{Mutter} \cdot k_{Lager} + k_{Spindel} \cdot k_{Lager}} \tag{3.70}$$

During the constant acceleration/delay phase of the slide, the same mechanical system is compressed or stretched by the acceleration/delay forces. These deformations can also not be compensated by indirect measuring systems and so manifest themselves fully on the workpiece surface in the way described.

When a linear, direct position measuring system is used, the static, elastic deformation is compensated completely by the controller with integrating component such that the static stiffness of the drive effective outwards takes on a value of $\infty$ despite the inner, elastic structural deformations.

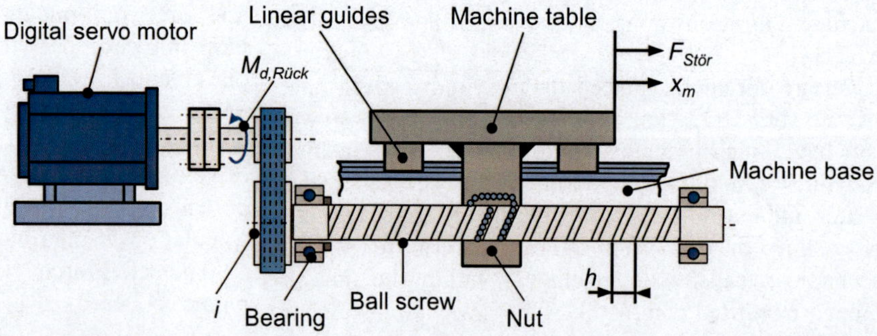


◘ **Fig. 3.36** Conventional drive axis

◘ **Fig. 3.37** Linear motor drive

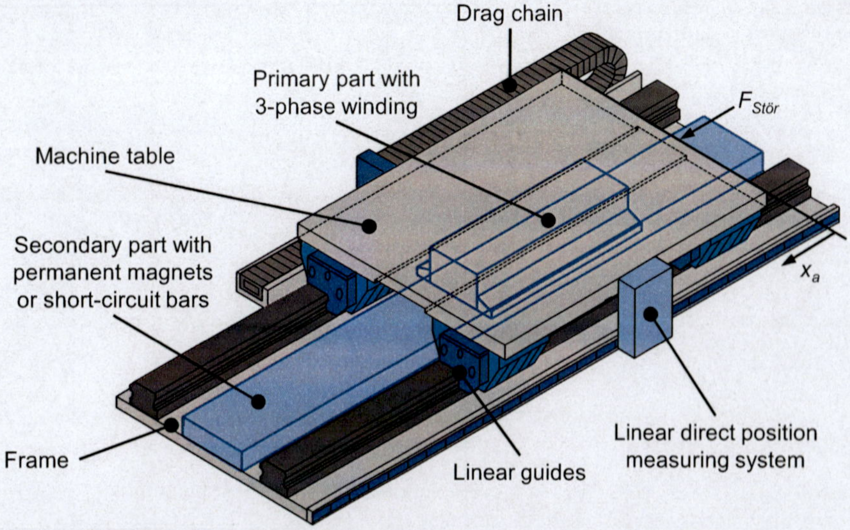


#### 3.5.1.2 Static Stiffness for an Electric Linear Motor Drive

The difference between electric linear motor drives and ball screw drives is that there are no mechanical drive components in the flux of forces and only direct position measuring systems can be used (◘ Fig. 3.37).

So the static stiffness is determined solely by the force generation in conjunction with the closed-loop control of the motor. The integral component of the controller ensures that the slide reaches the specified reference position under static, external loading. Its static stiffness is, as with electromechanical drives with direct position measuring system, infinite provided the external forces do not exceed the maximum force of the motor.

### 3.5.2 Dynamic Stiffness

Dynamic stiffness is understood to be the elastic deformation resistance of a structure to an external dynamic force. In practice, the reciprocal, dynamic compliance, is used. The frequency dependence of this parameter means a single variable cannot be used. Rather, it is a function (curve progression) that determines dynamic behavior versus frequency. The compliance frequency response or compliance Nyquist plot are informational forms for depicting the linear system behavior of this variable.

As with the analysis of static drive properties of electric linear motor drives, the dynamic compliance pertaining to the process disturbance force or acceleration processes is also determined solely by the control properties of the drive system. External forces act directly upon the motor and are no longer reduced by a ball screw or transmission. Only the table mass (including motor mass) that has a reducing effect on the vibration amplitudes with increasing excitation frequency of the disturbing force.

An acting disturbing force must then be compensated by the force of the drive. This drive force is built up by the controller as a reaction to a measured deviation from the reference parameters.

The consequence of this means the dynamic disturbance stiffness can be defined in the time domain as the quotient of the acting disturbing force $F_{Stör}$ and resultant maximum position deviation $\hat{x}_a$ of the machine slide (Eq. 3.70).

$$k_{dyn} = \frac{F_{stör}}{\hat{x}_a(t)} \tag{3.71}$$

◘ Figure 3.38 shows the reactions of two different drive systems to a step change in the force. It shows the profiles of the displacement of a ball screw drive (BSD) with a free ball screw (of 200 and 1100 $mm$) and a linear motor drive. The controllers were set at the boundary of each system.

It can be seen that the stiffness of the ball screw drive is dependent on the free length of the ball screw. The linear motor drive, on the other hand, has the same dynamic disturbance stiffness over the entire traverse range.

#### 3.5.2.1 Dynamic Stiffness of Electromechanical Feed Axes

Multiple influences must be factored in for conventional drives, as already set out for static loading. This is the purely mechanical vibration system, comprising table and workpiece masses, the nut, spindle and bearing elements acting as springs, and the damping, determined predominantly by the friction forces of the guide systems. The drive is not included here, i.e., it is assumed to be stiff (Eq. 3.71).

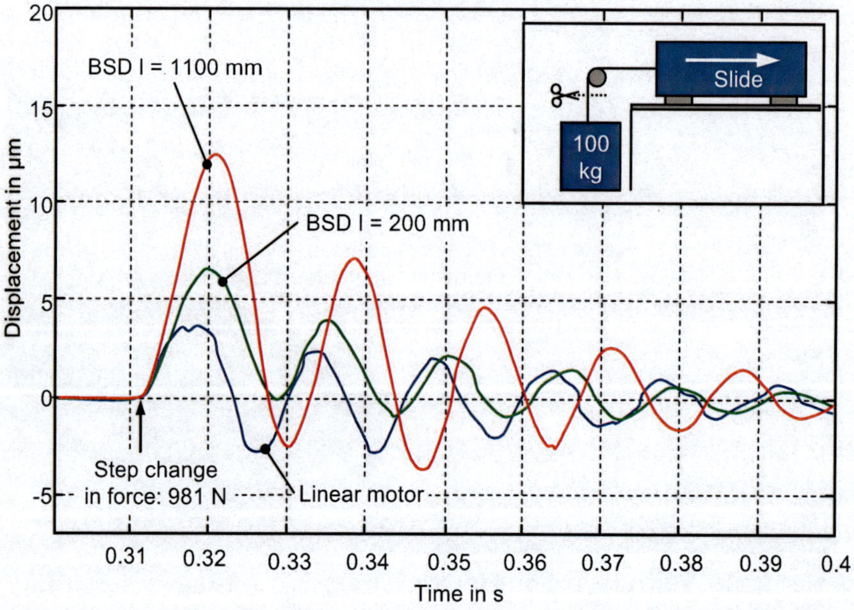


**Fig. 3.38** Comparison of the reactions of different feed systems to a step change in the force

$$G_{Mechanik}(j\omega) = \frac{x_m(j\omega)}{F_x(j\omega)} = \frac{\frac{1}{k_{ges}}}{1 + \frac{2D}{\omega_0} j\omega + \frac{1}{\omega_0^2}(j\omega)^2} \quad (3.72)$$

$$\text{with } D = \frac{c}{2m\omega_0} \quad \text{and} \quad \omega_0 = \sqrt{\frac{k_{ges}}{m}} \quad (3.73)$$

In addition, the controller dynamics of the drive are factored in as a reaction to disturbances (process load here), Eq. 3.73.

$$G_{Antrieb}(j\omega) = \frac{X_a(j\omega)}{F_x(j\omega)} \cdot e^{-j\omega t} \quad (3.74)$$

For small machines, the relatively low masses for high spindle stiffnesses mean the natural frequencies of the mechanics are generally higher than those of the drive control loop. So the drive can only excite vibrations of the mechanics to a small amount or not at all. The bandwidth of the machine axis is thus limited primarily by the controller dynamics of the drive train.

If the characteristic angular frequencies of the mechanics and the drive are right next to each other, the axis damping is reduced because the resonance points of the two systems are superimposed.

For large machines, it is essentially the stiffness of the mechanics that limits the dynamics of the axis. This is because heavy masses need to be accelerated and the slide moves a vibratory assembly, such is the case for portal milling machines. For the drive train, it is essentially possible to actively dampen vibrations occurring or suppress them using suitable velocity control.

Mechanical natural vibrations and vibrations excited by disturbances are not factored in when indirect measuring systems are used and thus cannot be compensated even with highly dynamic controllers. For direct measuring systems, the influences of both systems are coupled. This is why natural vibrations for poorly dampened and compliant, mechanical components can only be compensated with high controller overhead. For the active damping of mechanical vibrations, complex controller structures such as state control with observer structures must be used.

### 3.5.2.2 Electric Linear Motor Drive

The control loop of the linear motor drive corresponds to that of a rotatory servo drive comprising a cascade control system with current, speed and position control (Fig. 3.39). The drive speed is derived from the signal of the direct position measuring system.

In contrast to the electromechanical drive, it is no longer the resonance points of the mechanical transmission elements that have a detrimental effect on the control loops (especially the velocity and position control loops). Instead, it is mainly the properties of the electrical system such as sampling and dead times within the control loop, the current rise time of the motor and the behavior of the guide system that have a limiting effect (also refer to ▸ Sect. 3.2.3).

#### Influences of the Position and Velocity Control Loops

As the block diagram in Fig. 3.39 shows, the disturbance behavior of the linear motor drive is determined for ideal behavior of the current control loop by the control parameters of the velocity controller, gain $K_p$, resetting time $T_{np}$ and position controller gain $K_L$. For the linear motor drive, the gain factor of the position

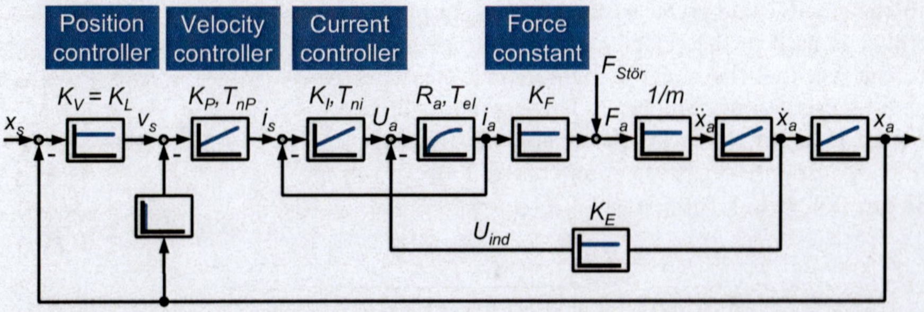


**Fig. 3.39** Control loops of a linear motor drive

controller must be set to the velocity gain $K_V$ due to the lack of a gear ratio, unlike electromechanical drives. It is the most important linear parameter, and the parameter that most determines the behavior of the position control loop. It also specifies the ratio of actual velocity to position deviation in the steady state (refer to ▶ Sect. 3.2.1).

While the load stiffness of the linear motor drive is specified primarily by the gains of the two controllers in the higher frequency range, the integrating component of the velocity controller acts on the stationary accuracy. Note here that the two controller gains cannot be changed independently. This can be illustrated in the signal flow diagram shown in **Fig. 3.40**. The behavior of the current control loop is assumed ideal here and the integrating component of the velocity controller is neglected (see ▶ Sect. 3.2.1).

The tracking frequency response of the position control loop is calculated as in Eq. 3.74 for this control loop.

$$G_{x,Regel}(j\omega) = \frac{x_a(j\omega)}{x_s(j\omega)} = \frac{1}{1 + \frac{1}{K_V}(j\omega) + \frac{m}{K_V K_p K_F}(j\omega)^2} \tag{3.75}$$

This tracking frequency response defines the behavior of a system with 2nd order delay behavior. Comparing coefficients gives a cutoff frequency pg $\omega_{0,\,Regel}$ and damping of $D_{Regel}$ for the position-controlled drive as follows:

$$\omega_{0,Regel} = \sqrt{\frac{K_V K_P K_F}{m}} \quad \text{and} \quad D_{Regel} = \frac{1}{2}\sqrt{\frac{K_P K_F}{K_V m}} \tag{3.76}$$

It is clear from Eq. 3.75 that the cutoff frequency of $\omega_{0,\,Regel}$ of the drive with a velocity gain of $K_V$ can only be increased meaningfully by simultaneously changing the gain of velocity controller $K_p$. Just increasing $K_V$ entails a lowering of the damping ratio $D_{Regel}$, something that (in theory) can be compensated by increasing $K_p$. This interrelation can also be found in the dependency of control parameters on one another as shown in **Fig. 3.28**.

### Calculation of the Dynamic Stiffness of Linear Motor Drives

In addition to the possibility of determining the dynamic stiffness by measurement, two methods are described below that permit mathematical estimation of the stiffness using the control parameters and mass.

First, a rule of thumb should be derived that is based upon neglecting the integrating component of the velocity controller and idealization of the current control loop. Derived from this is 2nd order system behavior ($G_{stör,\,P}$) whose final deviation is a measure of the stiffness of the original system.

Described as the second option is an improved method including the integral component of the closed-loop vibration controller and the slide mass. The current control loop is also assumed ideal here. First, this gives rise to a 3rd order system ($G_{stör,\,PI}$) that is simplified to a 2nd order equivalent system ($G_{stör,\,PI,\,ers}$). Its maximum vibration amplitude is used to determine the stiffness of the original system.

### Approximate Estimation of Dynamic Stiffness

The maximum deflection of a linear drive as a reaction to a disturbing force corresponds approximately

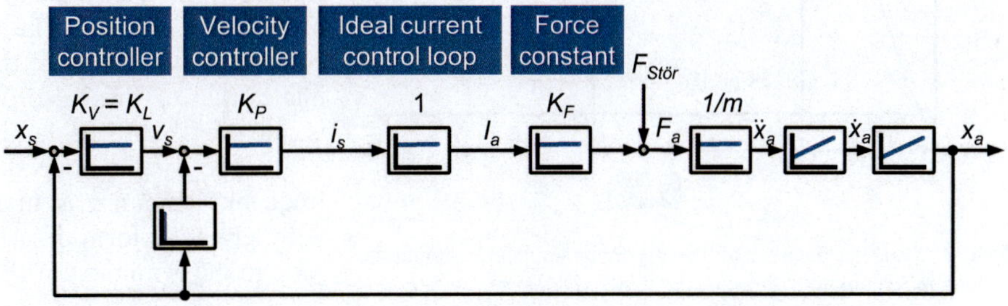


**Fig. 3.40** Simplified block diagram of the control loops, with a P velocity controller

to the steady-state error arising when a simple P controller is used in the velocity control loop. This is due to the fact that the integral component is not effective at high excitation frequencies. This error can be determined in a simplified manner using the tracking frequency response of the position control loop shown in ◘ Fig. 3.40 (Eq. 3.76).

$$G_{stör,P}(j\omega) = \frac{X_a(j\omega)}{F_{stör}(j\omega)} = \frac{\frac{1}{K_V K_p K_F}}{1 + \frac{1}{K_V}(j\omega) + \frac{m}{K_V K_p K_F}(j\omega)^2} \tag{3.77}$$

In line with the final value set of the Laplace transform, the remaining deflection of the P-controlled system, corresponding to the maximum deflection (as in Eq. 3.70 in this case), can be calculated as follows (Eq. 3.77).

$$x_{a,max} = \lim_{j\omega \to 0} G_{stör}(j\omega) \cdot F_{stör} = \frac{1}{K_V K_p K_F} F_{stör} \tag{3.78}$$

This yields the dynamic equivalent stiffness as in Eq. 3.78.

$$k_{stat} = \frac{F_{stör}}{x_{a,max}} = K_V K_p K_F. \tag{3.79}$$

To assess the dynamic behavior, the frequency response Eq. (3.76) provides extensive information with the derivable parameters $\omega_{0,\,Regel}$ and $D_{Regel}$ (Eq. 3.75). The maximum overshoot amplitude of the step response (i.e., of the response function to a force step) lends itself if the requirement is to restrict the behavior to one representative parameter.

According to ◘ Fig. 3.41, the maximum amplitude at time $t = \tau/2$ is:

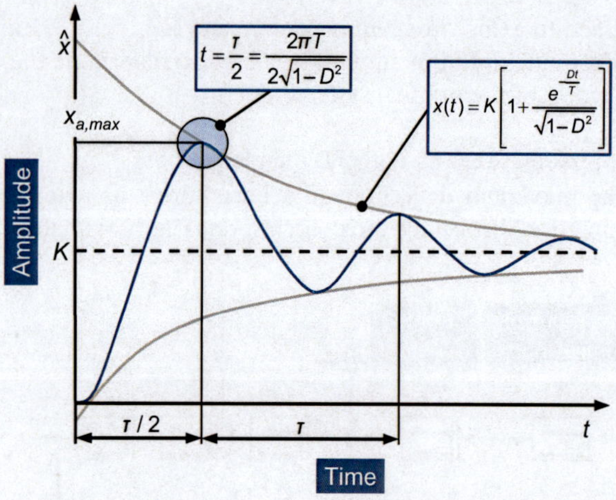


◘ **Fig. 3.41** Determination of the maximum overshoot amplitude of the step response

$$\frac{x_{max}}{F_{stör}} = K\left[1 + \frac{e^{-\frac{D\pi}{\sqrt{1-D^2}}}}{\sqrt{1-D^2}}\right] \tag{3.80}$$

$$\text{with} \quad D = \frac{1}{2}\sqrt{\frac{K_P K_F}{K_V m}} \quad \text{and} \quad K = \frac{1}{K_V K_P K_F} \tag{3.81}$$

The representative stiffness value is then the reciprocal value of Eq. 3.79.

$$k_{dyn} = \frac{F_{stör}}{x_{max}} = K_V K_p K_F \frac{1}{\left[1 + \frac{e^{-\frac{D\pi}{\sqrt{1-D^2}}}}{\sqrt{1-D^2}}\right]} \tag{3.82}$$

The disturbance stiffness can be estimated approximately using the formula derived here. Used as the basis here is the permanent deviation arising when the integral component of the velocity controller is ignored, the transmission behavior of the closed-loop current controller is assumed to be ideal and the influence of mass is compensated using the gain factor setting of the velocity controller. It is ignoring the integral component of the velocity controller in particular that results in an error however. This comes from the fact that it still exerts an influence in the range of frequencies analyzed (30–80 Hz).

## Improved Estimation of Dynamic Stiffness

Dynamic disturbance stiffness can be determined more accurately when the velocity control loop with PI controller is included in the analysis. The current control loop continues to be viewed as ideal. This yields the following model, ◘ Fig. 3.42. The disturbance frequency response of this system is derived as in Eq. 3.82.

$$G_{stör,PI}(j\omega) = \frac{X_a(j\omega)}{F_{stör}(j\omega)} = \frac{\frac{T_{np}}{K_V K_p K_F}(j\omega)}{\frac{mT_{np}}{K_V K_p K_F}(j\omega)^3 + \frac{T_{np}}{K_V}(j\omega)^2 + (\frac{1}{K_V} + T_{np})(j\omega) + 1} \tag{3.83}$$

The properties of this frequency response are extracted from a direct calculation because it is a 3rd order delay element. The goal is to find a 2nd order equivalent system where the dynamic stiffness approximates that of the 3rd order system. One such system is characterized by its damping and natural frequency.

To specify the equivalent system, the disturbance frequency response is first reformulated. This essentially gives the form of a $PT_2$ element, expanded by a term in the denominator (Eq. 3.83).

**Fig. 3.42** Simplified block diagram of the control loops with a PI velocity controller

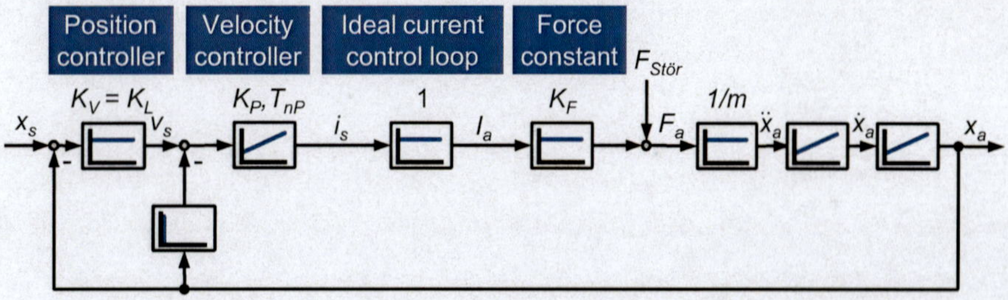


$$G_{stör,PI}(j\omega) = \frac{\frac{1}{K_V K_p K_F}}{\frac{m}{K_V K_p K_F}(j\omega)^2 + \frac{1}{K_V} j\omega + 1 + \frac{1}{K_V T_{np}} + \frac{1}{T_{np} j\omega}} \tag{3.84}$$

It is assumed below that the last denominator term can be ignored for the subsequent analyses because it tends towards a low value compared to the other terms in the relevant frequency range. The fact that this estimation generates errors at low frequencies still needs to be stated, however. As linear drives are generally designed such that natural frequencies greater than 50 Hz result, this error can be ignored in practice.

The disturbance frequency response of the equivalent system has the form of a $PT_2$ element (Eq. 3.84).

$$G_{stör,PI,ers}(j\omega) = \frac{\frac{T_{np}}{K_p K_F (1 + T_{np} K_V)}}{\frac{T_{np} \cdot m}{K_p K_F (1 + T_{np} K_V)}(j\omega)^2 + \frac{T_{np}}{(1 + T_{np} K_V)}(j\omega) + 1} \tag{3.85}$$

Its cutoff frequency, damping and gain can be determined easily with a coefficient comparison with a $PT_2$ element (Eq. 3.85).

$$\begin{aligned} \omega_0 &= \sqrt{\frac{K_p K_F (1 + K_V T_{np})}{m T_{np}}} \\ D &= \frac{1}{2}\sqrt{\frac{K_p K_F T_{np}}{m(1 + K_V T_{np})}} \\ K &= \frac{T_{np}}{K_p K_F (1 + K_V T_{np})} \end{aligned} \tag{3.86}$$

This means the parameters for the equivalent system are determined. To calculate the maximum deviation, the decay curve of the equivalent system as a result of an input unit step are again used (Fig. 3.41). The system oscillation is derived from a sinusoidal oscillation superimposed on an exponentially decaying curve. The maximum system deviation is given for the maximum value of the first deflection amplitude. This is reached after half an oscillation time (as a good approximation).

Calculating the maximum value of the first deflection period using the envelopes gives a relatively accurate value for the system deflection in relation to the disturbing force. This value corresponds to the dynamic compliance of the system. The reciprocal of this can be used to calculate the dynamic stiffness $k_{dyn}$.

The more accurate approximation of dynamic stiffness for linear motor drives is as follows:

$$k_{dyn} = \frac{F_{stör}}{\hat{x}_a} = \frac{K_P K_F (1 + K_V T_{np})}{T_{np}\left(1 + \frac{e^{\frac{-D\pi}{\sqrt{1-D^2}}}}{\sqrt{1-D^2}}\right)} \quad \text{with}$$
$$D = \frac{1}{2}\sqrt{\frac{K_P K_F T_{np}}{m(1 + K_V T_{np})}} \tag{3.87}$$

The values calculated for dynamic stiffness are exact, compared to the simple approximation in Eq. 3.78, over a far greater setting range of control parameters and masses moved.

## References

[ABEL20] Abel. D.: Regelungstechnik und Ergänzungen (Höhere Regelungstechnik). Mainz (2020)
[ACKE83] Ackermann, H.: Abtastregelung, 3. Aufl. Springer, Berlin (1983)
[ALTI11] Altintas, Y., Verl, A., Brecher, C., Uriarte, L., Pritschow, G.: Machine tool feed drives. CIRP Ann. – Manufact. Technol. 60(2), 779–796 (2011)
[HILL98] Hiller, B.: Relativbeschleunigungssensor – Potential und Einsatzmöglichkeiten in der Servo-Antriebstechnik. Stuttgart (1998)
[HOFF08] Hoffmann, F.: Optimierung der dynamischen Bahngenauigkeit von Werkzeugmaschinen mit der Mehrkörpersimulation. Apprimus-Verlag, Aachen (2008)
[ISER83] Isermann, R.: Systemidentifikation. Springer, Berlin (1983)
[ITI13] ITI: SimulationX Benutzerhandbuch. Dresden, (2013)
[KUPF68] Kupfmüller, K.: Einführung in die theoretische Elektrotechnik. Springer, Berlin (1968)
[LUNZ08] Lunze, J.: Regelungstechnik 1 - Systemtheoretische Grundlagen, Analyse und Entwurf einschleifiger Regelungen. Springer, Berlin (2008)
[OERT08] Oertli, T.: Strukturmechanische Berechnung und Regelungssimulation von Werkzeugmaschinen mit elektromechanischen Vorschubantrieben. Utz, München (2008)
[STUT81] Stute, G.: Regelung an Werkzeugmaschinen. Hanser, München (1981)
[WECK99] Weck, M., Krüger, P., Brecher, C.: Grenzen für die Reglereinstellung bei elektrischen Lineardirektantrieben. Antriebstechnik. 38 (2+3), 55–58 and 71–76 (1999)

# Feed Drives for Path Generation

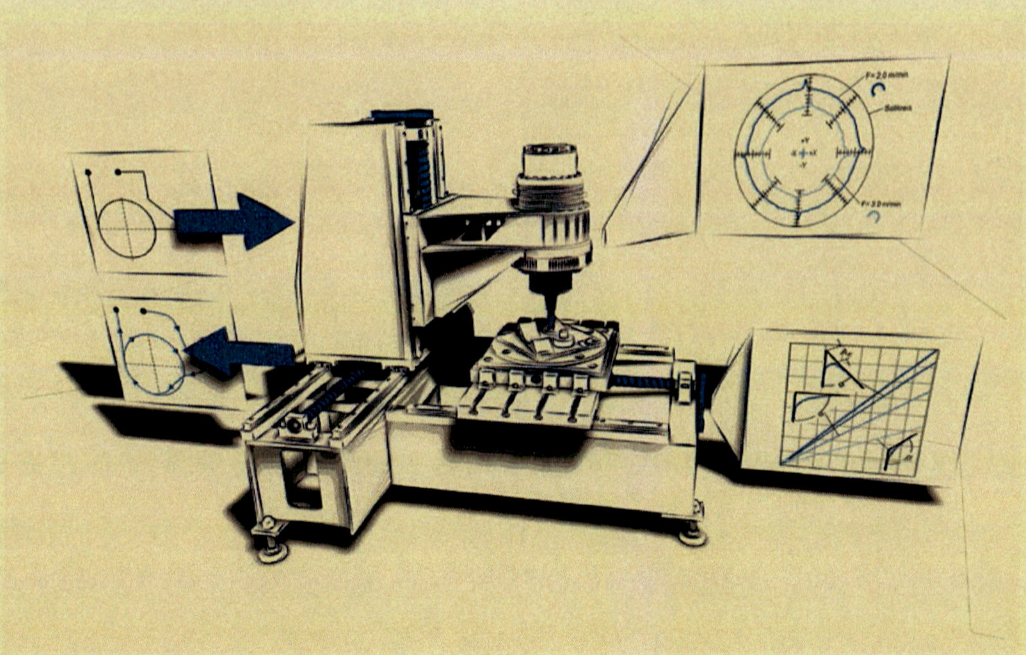

I thank my staff members Dr.-Ing. Marcel Fey, Christian Fimmers M. Sc., Matthäus Loba M. Sc. and Simon Storms M. Sc. for their contribution in revising this chapter.


C. Brecher and M. Weck, *Machine Tools Production Systems 3*,
Lecture Notes in Production Engineering, https://doi.org/10.1007/978-3-658-34622-5_4

MITSUBISHI ELECTRIC
Changes for the Better
Making complex tasks easy
A CNC always has to be a little more intelligent than the challenge it faces. Of course, it's useful to have the world's fastest dedicated processor installed so that both precision and speed benefit. See for yourself how much more you can get out of your machine tools – with the Mitsubishi Electric CNCs that think ahead.
www.Mitsubishi-CNC.de

Contents

4

## 4.1 Design of Path Controllers

In terms of design, a modern NC controller corresponds to a microcomputer using one or more processors. The different internal functional areas such as NC line processing, interpolation and position control, and also display and user interfaces, are implemented as software modules running on the computer under a customized operating system.

◘ Figure 4.1 shows in simplified form the key modules of an NC controller. NC lines, stored in the program memory in text form, are first preprocessed sequentially in the NC line processing phase. Here, geometry and control information, for example, are separated and forwarded to the appropriate functional modules. From the geometry data, the interpolator calculates the reference positions in a fixed timing cycle, factoring in the type of interpolation (such as linear, circular or spline interpolation) and the path velocity specified. These are the set point values for the position controllers of the individual axes, using these values and the actual measured positions to form the reference velocities for the feed drives (◘ Fig. 4.1). Chapters 11 and 13 in this volume contain detailed descriptions of the functions and designs of modern NC and RC controls.

## 4.2 Path Errors on Machine Tools

Geometric and kinematic errors have an influence on the working precision of machine tools. Kinematic deviations dominate in the traversing of machine axes. Potential sources of error can be found in the interpolator, the position control loops and the mechanical components of the feed drives.

### 4.2.1 Path Errors in the Interpolator

An erroneous contour calculation in the interpolator inevitably entails an erroneous path. When the interpolator is implemented as a software module on a microcomputer, the geometric accuracy only depends on the algorithm used and the internal computational accuracy. Dividing a movement across multiple axes, and given a very low path velocity on one of the two axes (such as when a flat bevel is traversed), the output granularity of the interpolator has a bearing on the path curve. On the axis with the slow velocity, reference value steps on the order of magnitude of the smallest measurable displacement increment are output. These individual steps can no longer be smoothed by the position control loops and cause discontinuous feed motion (◘ Fig. 4.2). This effect is overlaid by friction and elasticity factors.

Comparing geometrical errors caused by the interpolator with errors caused by the downstream position control loop shows that interpolation errors at permitted path velocities are usually on the order of magnitude of a single measurement increment and so form an insignificant part of the total error.

### 4.2.2 Typical Path Errors of Position Control

◘ Figure 4.1 shows the design of a 3-axis path controller for a machine tool. The input data for distances and velocities are passed to the interpolator by entering the NC datasets and are forwarded to the position control loops as set point values. When signals are transferred from NC data input to the tool, transfer errors occur that, from a control perspective, depend mainly on the

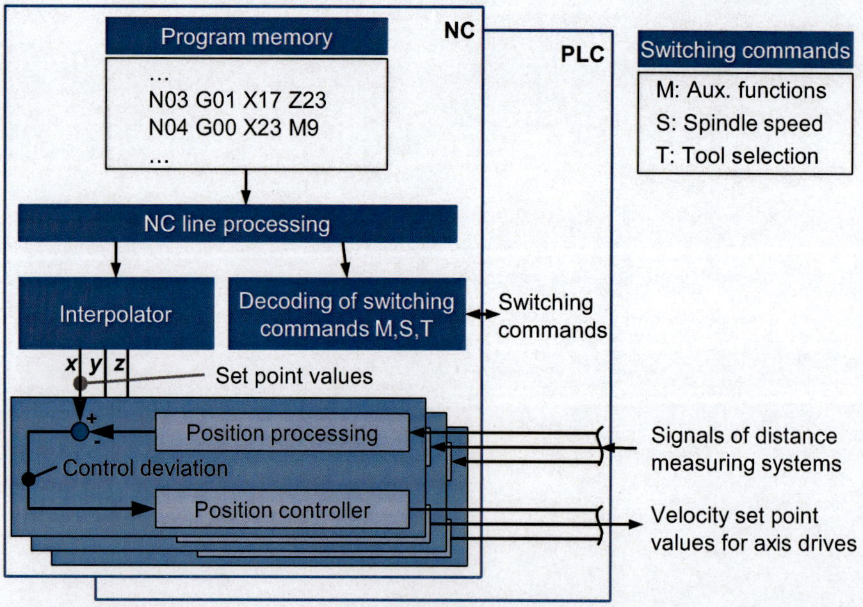


◘ Fig. 4.1 Internal structure of an NC controller

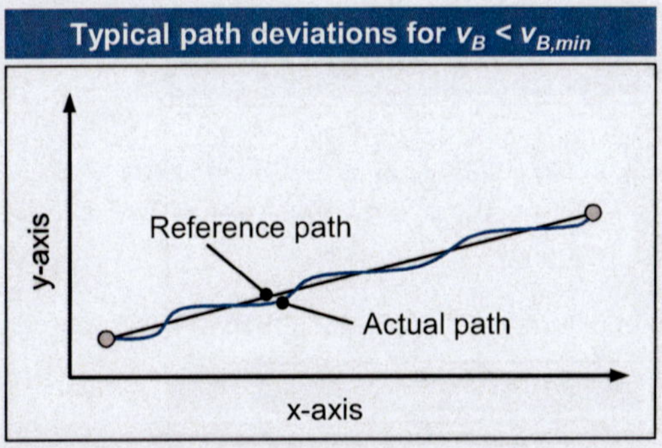


**Fig. 4.2** Typical path deviations for a very low path velocity

interpolation type (see ▶ Sect. 12.1) and the behavior of the position control loops. The causes of path deviations arising this way:

- Dynamic path deviations
- Different velocity gains in the individual position control loops
- Persistent control deviations due to the proportional action of the position controllers
- Unequal dynamic behavior of the drive units
- Non-linear signal transfers (reversal errors)

The resultant workpiece error results from superposition of the subsequent errors in the individual axes. The higher the speed and its change over time, the greater the resulting subsequent error. Shown below and then explained in more detail are typical deviations using a path with different geometrical elements.

Figure 4.3 shows an overview of the path deviations generated for different geometries when the path is traversed at a constant speed and without stops at the transitions and without compensation measures. The corners are rounded and the circle diameters change. With greater damping ($D>0.7$), they become smaller with increasing speed.

Figure 4.4 shows which dynamic axis deviations are likely when the same path is traversed. It can be seen in the lower part of the figure that the velocity and acceleration profiles deviate greatly from the target specifications. The dynamic properties of the feed drives are not adequate to conform to the path behavior required.

**Fig. 4.3** Path deviations for different geometrical elements

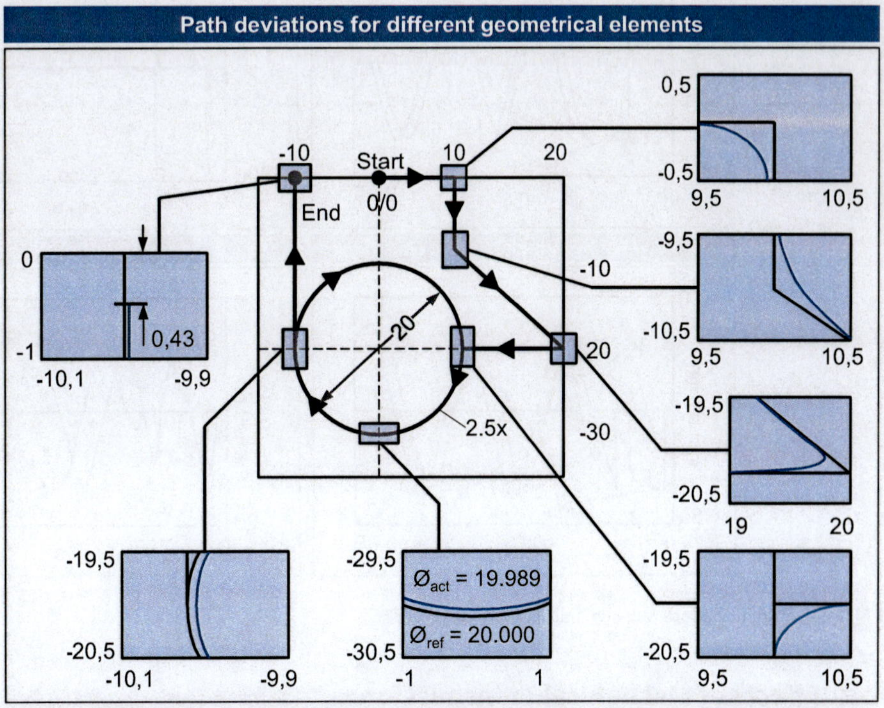


| Technological data | |
|---|---|
| Feed rate: | $v$ = 1 m/min |
| Radius: | $R$ = 10 mm |

| System data | |
|---|---|
| Natural frequency: | $f_0$ = 10 Hz |
| Damping: | $D$ = 0.8 |
| Velocity gain: | $K_V$= 2.34 m/min/mm |
| Circular frequency: | $f_u$ = 0.27 Hz |
| Tracking error: | $\delta_s$= 0.425 mm |
| Diameter difference: | $\delta(2R)$ = -0.011 mm |

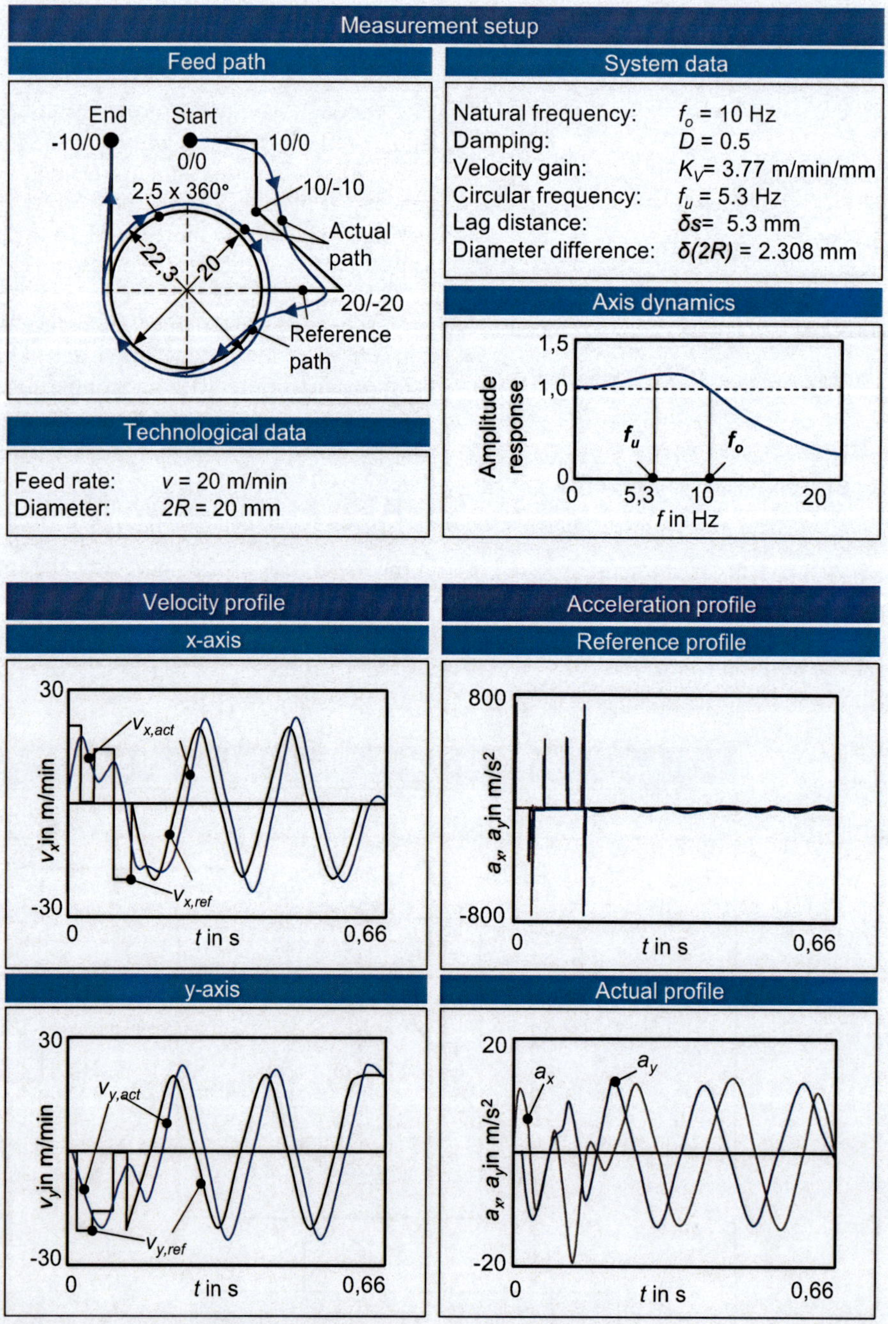


**Fig. 4.4** Path deviations, velocity and acceleration profiles

### 4.2.3 Effects of Mechanical Transmission Elements

Non-linearities in the position control loop that are caused by the mechanical transmission elements can also greatly influence the tool path [ALTI11]. Of particular importance here is the zero-crossing of the velocity on one axis. At these points, the movement and force directions of the feed drive reverse. This means the springs in the drive train (e.g., the ball screw) are first subjected to pressure relief and then loading in the opposite direction. The motor must also pass through any backlash present.

The consequence is that the axis at the zero-crossing of the velocity remains stationary until sufficient force has built up again to move the axis in the new

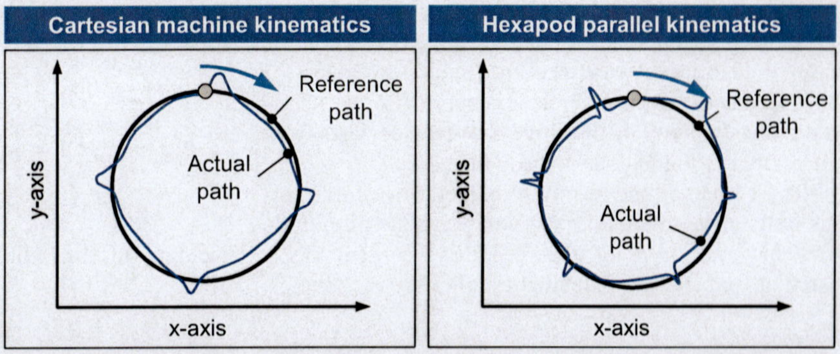


■ **Fig. 4.5** Circularity errors due to non-linearities in the position control loop

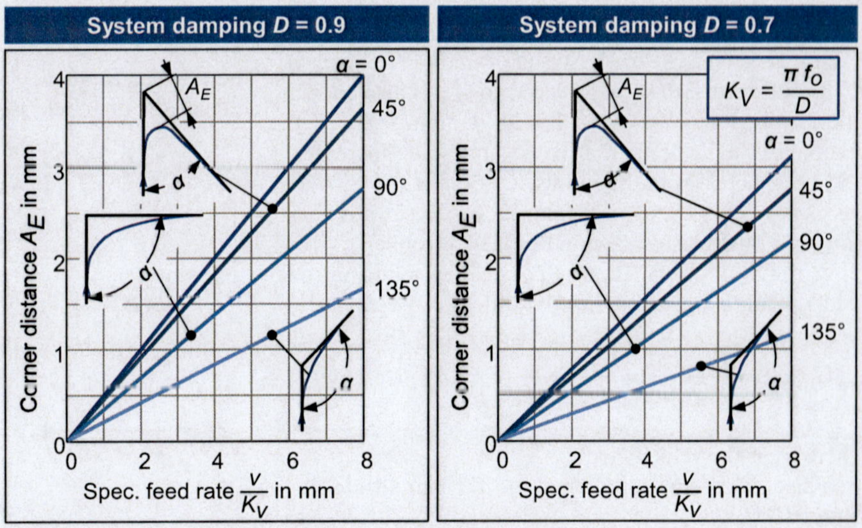


■ **Fig. 4.6** Corner distance as a function of angle, velocity and $K_V$ factor for different damping values

feed direction. First, the somewhat higher static friction component also needs to be overcome here. Machines with Cartesian axes thus exhibit quadrant transition errors when traversing a circle (■ Fig. 4.5, left).

For machines with parallel kinematics, the axes are not necessarily located vertically above each other. So the velocity zero-crossing of the axes also does not necessarily occur at the quadrant transition. For a machine with hexapod kinematics for example, that moves the tool with six drive axes, the circularity error shown on the right in ■ Fig. 4.5 is generated [POWE98].

### 4.2.4 Determination of Dynamic Path Deviations

Different factors such as geometrical machine errors, tool errors and temperature changes have a bearing on the working precision of a machine tool. The proportional action of the position controllers and the associated deviation from the reference contour exert the greatest influence on path generation (■ Fig. 4.4).

#### 4.2.4.1 Corner Smoothing

As ■ Figures 4.3 and 4.4 show, corner smoothing occurs when the feed rate is constant and the geometrical contour is discontinuous. Due to the tracking error the corner coordinate of the first axis of motion of the feed drive has not yet been reached while the interpolator is already heading in the other direction of motion.

■ Figure 4.6 shows the results of simulation computations. Corner distance $A_E$ is shown graphically as a function of $v/K_V$ for different corner angles ($0° < \alpha < 135°$). The graph on the left shows the deviations for a system with good damping ($D = 0.9$) and the graph on the right for a system with slightly less damping ($D = 0.7$). As can be seen, the deviations increase linearly with increasing value of $v/K_V$. These deviations are slightly smaller for the system with weak damping. Slight overshooting can occur here, however.

#### 4.2.4.2 Circularity and Diameter Deviations

■ Figures 4.3 and 4.4 convey a picture of the possible circularity and diameter deviations when the path is traversed at a constant speed with no stopping.

Large circularity deviations occur in particular when the circular path is entered and exited. For the case of identical $K_V$ factors and damping values for both feed axes, an exact circle is generated in the steady state. The diameter of the circle, however, deviates from the reference diameter as velocity increases.

The course of the resulting diameter or radius versus frequency is here an exact image of the amplitude frequency responses of the feed drive control loops. Circular frequency $f_u$ is defined as follows:

$$f_u = \frac{v}{2\pi R} \tag{4.1}$$

Figure 4.7 shows amplitude response $R_{act}/R_{ref}$ for a system having the value $D=0.9$. Circular frequency $f_u$ here is referenced to the system's natural frequency, $f_0$. Radius error $\delta R/R_{ref}$ referenced to $R_{ref}$ is also shown graphically. For systems with weak damping ($D<0.7$), resonance peaks occur that cause amplitude increases, and thus increases in the radius, as soon as the circular frequency nears the natural frequency. A descriptive example of this is also shown in Fig. 4.4 (top left).

A rough approximation is possible by reducing the feed system to a proportional 1st order delay element:

$$G_v(j\omega) = \frac{R_{act}(j\omega)}{R_{ref}(j\omega)} = \frac{1}{1+\frac{1}{K_V}j\omega} \tag{4.2}$$

where $K_V$ is the velocity gain.

This yields an approximation for amplitude response $|G_v(j\omega)|$:

$$|G_v(j\omega)| = \frac{|R_{act}(j\omega)|}{|R_{ref}(j\omega)|} = \frac{1}{\sqrt{1+\left(\frac{\omega_u}{K_V}\right)^2}} \tag{4.3}$$

Truncation after the 2nd order term of the Taylor series approximates the actual radius $R_{act}$ of the circle as follows:

$$R_{act} \approx R_{ref}\left(1-\frac{1}{2}\left(\frac{\omega_u}{K_V}\right)^2\right) \tag{4.4}$$

Yielding this for radius difference $\delta R$:

$$\delta R = R_{ref} - R_{act} \approx R_{ref} - R_{ref}\left(1-\frac{1}{2}\left(\frac{\omega_u}{K_V}\right)^2\right) = \frac{1}{2}R_{ref}\left(\frac{\omega_u}{K_V}\right)^2 \tag{4.5}$$

The angular frequency at which the radius is circled is calculated from circle radius $R_{ref}$ and path velocity $v_B$

$$\omega_u = 2\pi f_u = 2\pi \frac{v_B}{2\pi R_{ref}} = \frac{v_B}{R_{ref}} \tag{4.6}$$

where

$$f_u = \frac{v_B}{2\pi R_{ref}} \tag{4.7}$$

Resulting finally for radius difference $\delta R$:

$$\delta R = \frac{1}{2R_{ref}}\left(\frac{v_B}{K_V}\right)^2 \tag{4.8}$$

Large path deviation s when a circle is traversed occur in particular for small circle radii and high path velocities. Velocity $v_B = 5$ m/min, $K_V$ factor $K_V = 1$ m/min/mm and reference radius $R_{ref} = 10$ mm give, for example, a radius difference of 1.25 mm.

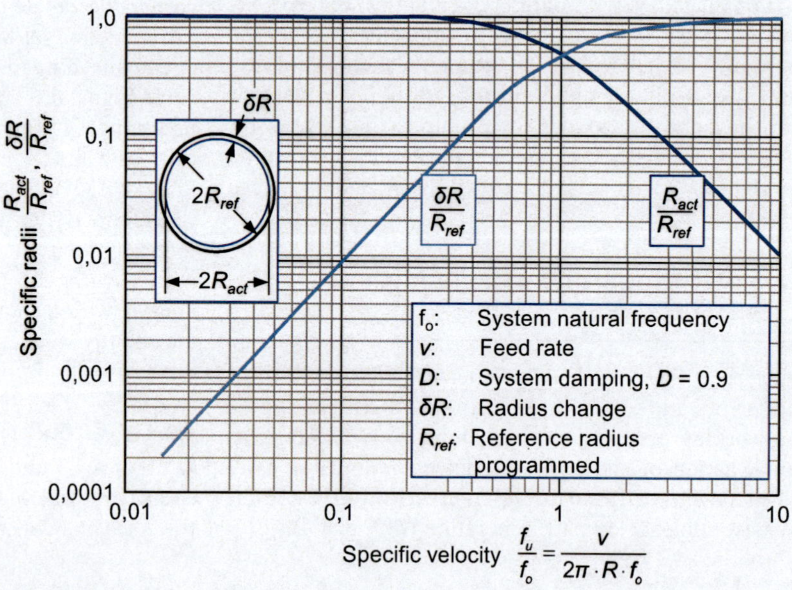


**Fig. 4.7** Circular path deviation as a function of feed rate, system's natural frequency and radius

## Measuring circularity

The circular test has proven its worth for machines with linear axes. It permits a geometrical and drive-related accuracy assessment that is straightforward to perform and of informational value. A more accurate description of the circular test is in Volume 2, ▶ Sect. 3.2, of this book series.

Figure 4.8 shows the results of a circular test with a grid measuring system, measured on a knee-type milling machine. At the return points of the direction of motion, the measurements show clear deviation peaks caused by static friction in the axis mechanics. Besides the direction of rotation dependency of these path deviations, the radius deviations increasing with feed rate can also be seen. These deviations are associated with the control of the axis with the limited $K_V$ factor causing a tracking error on both axes.

### 4.2.5 Influence of the $K_V$ Factor on the Path Deviations

The dynamic path deviation is closely dependent on the gain factor set for every position control loop. The interrelations derived and described in ▶ Sect. 4.2.2 for one axis also apply in a superpositioned manner in the axis group with multiple axes. Figure 4.9 shows the path trajectory following a right-angled corner without stops for one low and one high velocity gain $K_V$.

The velocity errors (lag distance) are calculated from the transmission behavior of the overall position control loop, i.e., from the loop gain $K_V$ or the velocity gain:

$$\Delta x \approx \frac{\dot{x}}{K_{Vx}} \tag{4.9}$$

$$\Delta y \approx \frac{\dot{y}}{K_{Vy}} \tag{4.10}$$

The equations apply in exactly the same way in the stationary state (t→ ∞ for constant axis velocities).

Figure 4.10 compares three different machining types as regards the machining error occurring. When a bevel is machined, a workpiece error does not occur when the velocity gains of the control loops are identical for all axes ($K_{Vx}=K_{Vy}$). Given that the lag distances behave just like the partial velocities, and are constant during the machining process, the actual value follows the reference value on the reference contour with a time delay.

$$\frac{\dot{x}}{\dot{y}} = \frac{\Delta x}{\Delta y} \tag{4.11}$$

A circular contour is generated by overlaying one sinusoidal and one cosinusoidal movement when the machine table can be traversed on two axes offset by 90°. The deviation shown in Fig. 4.10 is given by the start and stop processes in the movements of the x- and y-axes, and by the different lag distances caused by the alternating partial velocities.

A machining error also occurs when a 90° corner, for example, is followed at a constant, specified velocity. The error is greater, the poorer the dynamics of the drive systems.

In addition to machining errors caused by the system that can be reduced, but never eliminated

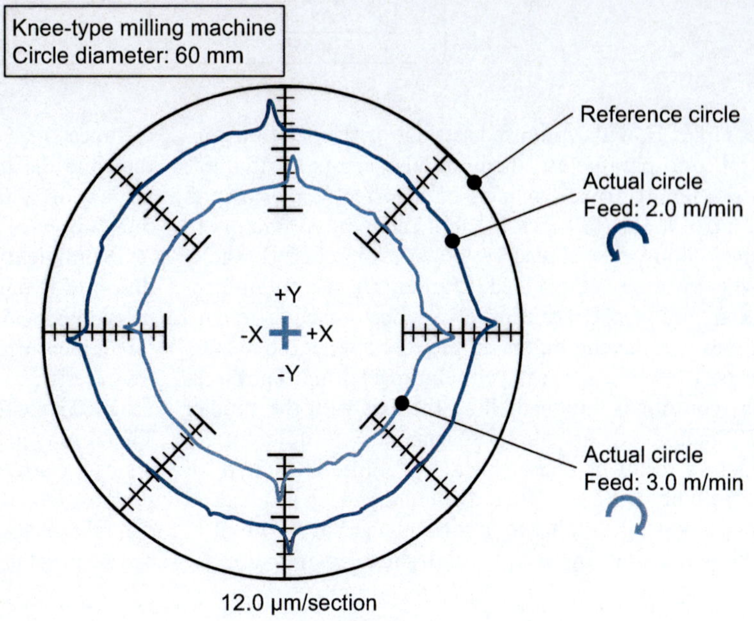


**Fig. 4.8** Circular traversing motion when the feed is varied

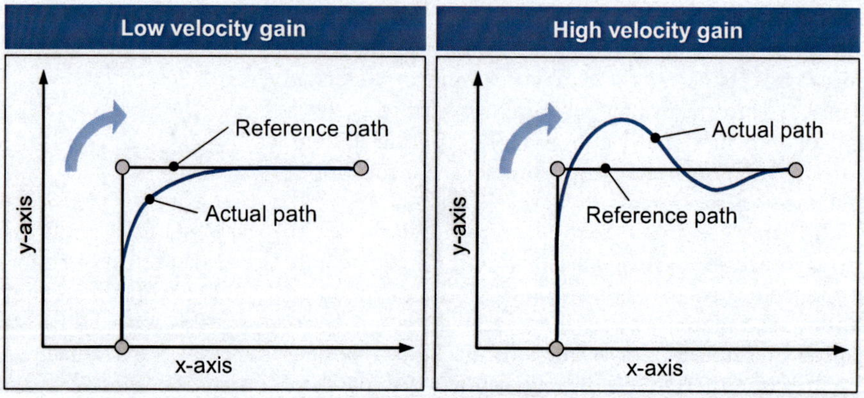


**Fig. 4.9** Path deviations for following a right-angled corner without stops

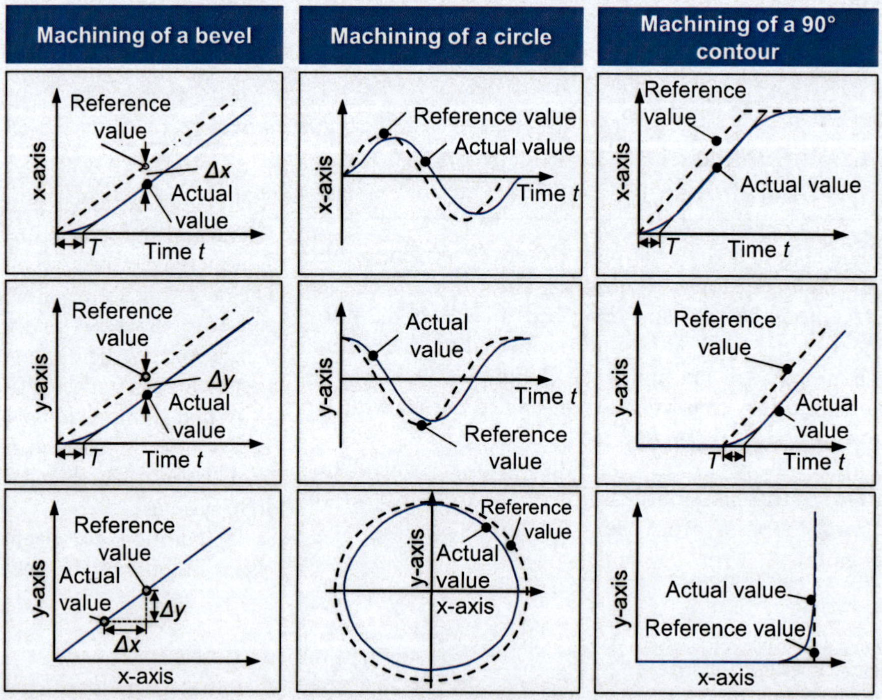


**Fig. 4.10** Machining errors caused by the system for a 2-axis path controller (two equal 1st order systems)

completely, with optimized settings of the position control loop parameters, there are also errors attributable to different drive dynamics or incorrectly set position control loops (Fig. 4.11). For different $K_V$ factors in the x- and y-axes, the target point is approached exactly when a bevel is machined. During this machining process, however, contouring errors occur having magnitudes that depend on the difference between the $K_V$ factors. The same applies to circular machining. The circular contour is flattened along the axis with the smaller $K_V$ factor, i.e., an ellipse is generated. If different $K_V$ factors result in phase shifts, the ellipse is rotated on the plane. Different drive dynamics can be established by traversing parallel to a ruler aligned at 45°, and at the same time by using a dial gauge to measure the distance from the moving machine component. From the time the machine is stationary until it is next stationary after traversing at 45°, the distance to the ruler must not have changed.

Non-linear signal transmission errors occur in addition to dynamic errors as a consequence of the transmission behavior of the control loop. These occur on indirect measuring systems as a result of the reversal error (hysteresis), caused by play, friction and compliance of the mechanical transmission elements when under load. This results in a path error having the magnitude of the reversal error. Contour distortion can be eliminated by the position control loop by selecting the position of the distance measuring system such that the reversal error can be measured (directly on the machine table).

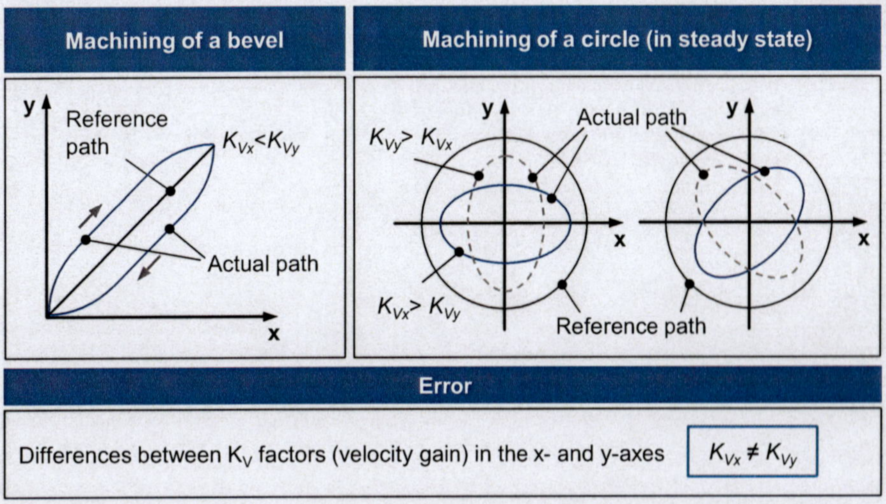


**Fig. 4.11** Machining errors for a 2-axis path controller caused by different drive dynamics of the individual systems

## 4.3 Measures to Reduce Path Deviations

Controllers popular today prevent the path deviations described in ▶ Sect. 4.2 with targeted control-related measures or velocity adaptations.

Given that large radius deviations are generally not acceptable for adhering to a standard manufacturing tolerance, various methods (some of which are intricate) for improving this behavior are used by controller manufacturers. Increasing the $K_V$ factor for circular paths is one of the simpler methods. In theory, the minimum radius deviation is given at about 1.4 times the maximum $K_V$ factor—to be set for overshoot-free positioning at corners.

Another possibility is path-dependent limitation of the feed rate by the controller. For this, Eq. 4.8 is used, for example, to determine, for a specified, maximum radius difference permitted, the path velocity associated with the radius at which the circle is then traversed regardless of the path velocity actually programmed.

One downside of both methods is that they only work for circular paths also programmed as such. If the circular path to traverse is made up of short linear segments, this is not detected by the controller and thus also not corrected. Such contours comprising short linear segments are generated, for example, by most CAM systems from CAD data.

Another more complex group of methods reduces dynamic deviations (which also include the radius difference) by using control-related measures. Feedforward velocity and acceleration is worthy of particular mention here. It is available from multiple controller manufacturers under different names [DOOM91, METZ89].

Dynamic compensation filters and higher-quality position control structures (state controllers) also continue to be used [GREF90, STOF77, WECK89]. These methods are discussed in detail in Sects. 3.1.3 and 3.1.4. Also, current scientific approaches are exploring the possibility of predictively adapting paths on the basis of an approximated error [ALTI10, KHOS14].

The corner error can also be reduced by using dynamic corrective measures. Traversing a corner at a constant path velocity without any errors at all is impossible, however. Even when optimum position control structures and optimized feed mechanics are used, a compromise must always be made between path velocity and corner error. Influencing the set point values as described at the start of the chapter can greatly reduce the corner error. If the highest priority is placed on an error-free contour, this contour cannot be generated using a constant velocity set value. In such a case, the path velocity at the corner must be reduced down to stationary or the specifications for the straight-line orientations are changed such that the contour traversed by the machine corresponds to the specification.

## References

[ALTI10] Altintas, Y., Sencer, B.: High speed contouring control strategy for five-axis machine tools. CIRP Ann. Manuf. Technol. **59**(1), 417–420 (2010)

[ALTI11] Altintas, Y., Verl, A., Brecher, C., Uriarte, L., Pritschow, G.: Machine tool feed drives. CIRP Ann. Manuf. Technol. **60**(2), 779–796 (2011)

[DOOM91] Dooms, K., Machielsen, K.: Intelligenter Bewegungscontroller. NC-Fertigung (6) (1991)

[GREF90] Greff, S.: Für Werkzeugmaschinen und Industrieroboter bestimmt. Industrie-Anzeiger (6) (1990)

[KHOS14] Khoshdarregi, M.R., Tappe, S., Altintas, Y.: Integrated five-axis trajectory shaping and contour error compensation for high-speed CNC machine tools. IEEE/ASME Trans. Mechatron. **19**(6), 1859–1871 (2014)

[METZ89] Metzger, K., Hullinger, P.: Schleppfehlerfreie adaptive Regelung mit PED-Algorithmus für NC-Achsen. Antriebstechnik **28**(8) (1989)

[POWE99] Powell, N.P., Whittingham, B.D., Gindy, N.N.Z.: Parallel Link Mechanism Machine Tools: Acceptance Testing and Performance Analysis. In: Boër C.R., Molinari-Tosatti L., Smith K.S. (eds) Parallel Kinematic Machines. Advanced Manufacturing. Springer, London (1999) ▶ https://doi.org/10.1007/978-1-4471-0885-6_23

[STOF77] Stof, P.: Untersuchung über die Reduzierung dynamischer Bahnabweichungen bei numerisch gesteuerten Werkzeugmaschinen. Springer, Berlin (1977)

[WECK89] Weck, M., Ye, G.: Bahnsteuerungskonzept zum hochgenauen Abfahren komplexer Kurven. Technische Rundschau (1) (1989)

4

# Design of Feed Drives

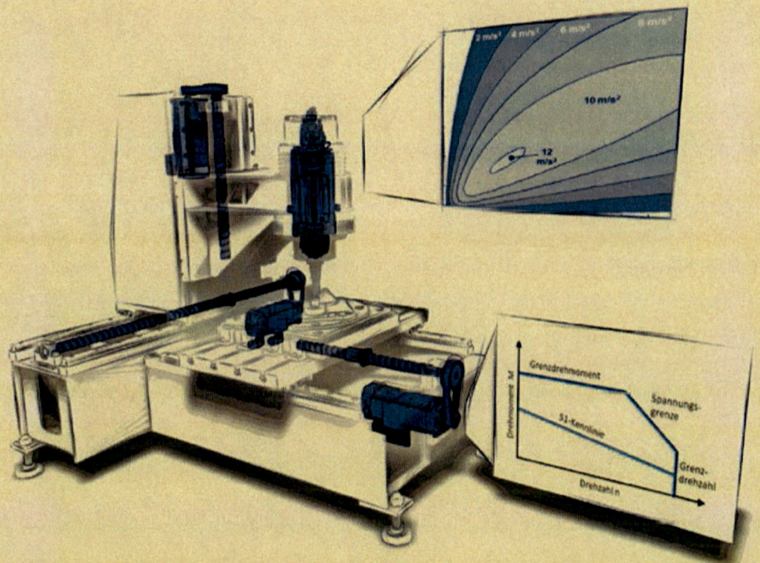


I thank my staff members Dr.-Ing. Marcel Fey, Christian Fimmers M. Sc., Simon Storms M. Sc. and Tiandong Xi M. Sc. for their contribution in revising this chapter.


C. Brecher and M. Weck, *Machine Tools Production Systems 3*,
Lecture Notes in Production Engineering, https://doi.org/10.1007/978-3-658-34622-5_5

Contents

5

## 5.1 Design of the Motor and Mechanical Components

In the machine tool industry, feed axes are driven electrically. A distinction can be made here between direct and indirect drives. Direct drives generate the required kinematic momentum directly by means of torque or linear motors (see ▶ Sect. 2.1). For indirect drive concepts, mechanical transmission elements (such as a ball screw spindle) are used to convert rotation to translation.

Key challenges in the design of feed drives are the selection and sizing of the drive motor as regards the required key performance indicators. The motor is specified by defining the gear ratio of the mechanical gear components, such as gear drives, toothed belt drives, ball screw spindles, and rack and pinion drives. Attention must thus always be paid to the interplay between the motor and the mechanical system in feed drive design.

◘ Figure 5.1 shows an overview of the method for designing a feed drive. The fundamental requirements and the drive principle are specified first. Included here, for example, are the motor type, the mechanical system, the masses moved, forces occurring, and the required speeds and accelerations (see ▶ Sect. 5.1.1). The next step entails selection of the mechanical components—because the parameters of these components are required for selecting the motor (see ▶ Sect. 5.1.2). Then, the motors can be selected on the basis of the load torque at the motor shaft (see ▶ Sect. 5.1.3.1). The dynamic design is different for each of these motors because the moment of inertia of the rotor in each case must be factored into the overall inertia of the mechanical system (see ▶ Sect. 5.1.3.2). An optimum gear ratio as regards attainable acceleration is linked to the ratio of moments of inertia on both sides of a transmission or feed element. The gear ratio is thus adjusted as required later on (see ▶ Sect. 5.1.3.3). The selection of the position measuring system has a considerable bearing on the possible accuracy. Direct position measuring systems are used for machine tools with strict accuracy requirements. In the design of the measuring system, the division period must be selected such that the resolution that satisfies requirements is attained but that the input cutoff frequency of the drive controller is not exceeded (see ▶ Sect. 5.2). The sizing of the motor's power modules is given from the motor current stipulated. Usually, drive manufacturers offer power modules appropriate for the motors, so that selecting the power modules goes hand in hand with specifying the motor. This is not discussed in any more detail here. To help in the design steps described here, many drive manufacturers offer special software that includes design equations and the corresponding technical characteristics of the product range.

A carefully designed drive is a prerequisite for achieving the dynamic behavior of the position control loops that meets requirements during commissioning. During commissioning, selecting suitable control parameters enables the frequency response behavior to be optimized (see ▶ Sect. 5.3).

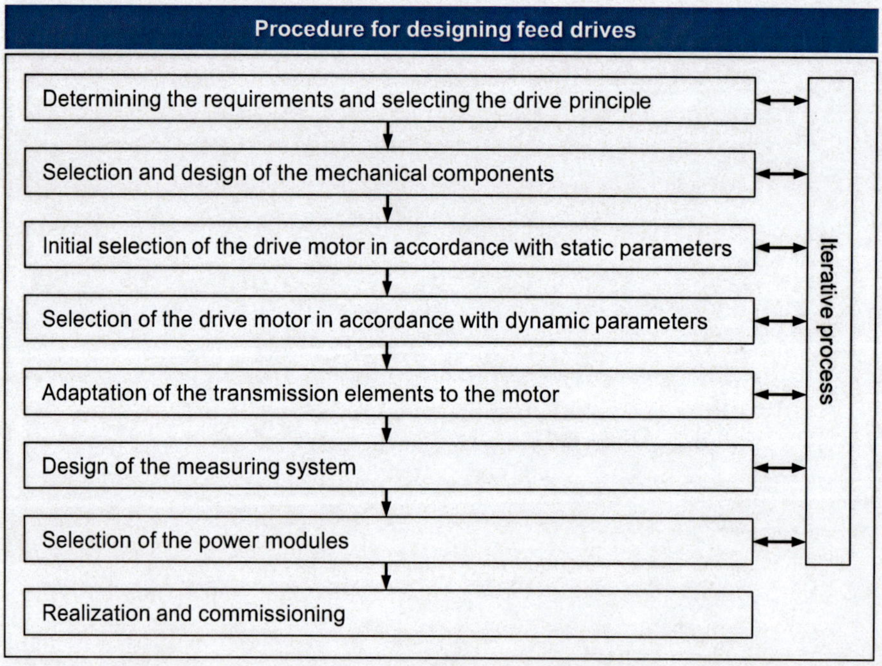


◘ Fig. 5.1. Procedure for designing feed drives

### 5.1.1 Determining the Requirements and Selecting the Drive Principle

The function of feed drives is to generate the relative motion between workpiece and tool and, in this process, to apply the feed forces, speeds and accelerations required. Feed drives are designed in consideration of the following:
- Attaining the positioning and path accuracies specified
- Realizing high accelerations

The following requirements can be derived from these:
- free of backlash
- Low static and dynamic compliance, yet high enough damping
- High first-order resonance frequency of the mechanical transmission elements
- Transfer behavior that is as linear as possible for the design elements of the feed drive
- High geometric and kinematic accuracies of the components and the installed drive train
- System runs smoothly, even at low speeds (low torque ripple of the motor at low speeds, no stick–slip effect of guide systems)
- Broad feed rate range due to the much different axis velocities when a path is traversed
- Capable of fast acceleration due to adequate motor torque, high momentary overload capacity, and low friction, masses and moments of inertia of the components accelerated
- Low masses of the overall feed system for axes positioned above each other (cross carriage)
- Excellent response and disturbance behavior of the position control loop

Due to their high accelerations, modern drive motors and control loops are able to excite the mechanics at the resonance frequency. This means that sometimes the full dynamic properties of the motors cannot be utilized (jerk limitation). Thus, in the design of highly dynamic feed axes, it is important to ensure there is sufficient separation between possible excitation frequencies of the position control loop and the mechanical resonance frequencies. The following rule of thumb applies in this context: The frequency of the first resonance point of the mechanical system should be at least twice the excitation frequency of the mechanics in the closed position control loop.

Electric linear motor drives have a direct effect on the machine components to be moved. They therefore require no mechanical transmission elements. This enables greater machining speeds to be attained for very high static and dynamic load stiffness.

Given the iterative nature of the design process, initially there is usually no reliable data on the masses to be moved or the static/dynamic compliances of the axes or their individual components. Information such as process forces, accelerations required, rapid motion rate, average machining speed, masses to move and cost form the design framework. Initially, it is a question of selecting the drive principle for the feed drive. A distinction must be made between the different ways to convert rotation to translation (spindle/nut, pinion/rack or rotatory drive with/without transmission) or a linear motor drive. Which final design is best suited for any given feed scenario can often only be determined by almost fully designing the components and then performing a subsequent recalculation. Key criteria in selecting the most suitable drive principle are positioning accuracy, travel distance/speed/acceleration, static and dynamic stiffnesses stipulated, production and installation overhead, and cost.

### 5.1.2 Selection and Design of the Mechanical Components

The masses to be moved and accelerated, and the moments of inertia, must be determined for sizing the motor. In addition to the machine table and workpiece to be moved, the motor inertia, couplings, transmission, spindle, guideway covers and cable drag chains must also be taken into account.

The transmission errors permitted are crucial in the design of mechanical transmission elements. The transmission error (kinematic, static or dynamic) exerting the greatest influence on the attainable machining result depends on the design of the feed system, the nature of the application and the load.

The mechanical components in the drive train must be selected and designed such that, in consideration of the maximum cutting and acceleration forces and the maximum workpiece weight, the deformations and natural frequencies of the mechanics lie within specified tolerances.

For ball screw drives, this especially includes determining the spindle diameter as regards stiffness, buckling and the critical natural frequency of bending vibration. Increasing the spindle diameter increases the stiffness by a power of 2. However, the mass moment of inertia increases simultaneously by a power of 4. The acceleration behavior of the feed drive is negatively impacted by increasing the spindle diameter. This means that, in selecting spindle diameter and pitch, a comprise must be made between adequate stiffness and mass moment of inertia.

In the system design, it is important to ensure there is sufficient separation between possible excitation frequencies and the resonance frequencies. For the computed design, it is enough here in many cases to determine the lowest or first resonance point using an approximation of the machine behavior by means of a 2nd order delay element. The following must be checked:

1. The natural frequency of the feed system, comprising moving components as masses and gear elements as springs, should be at least twice the excitation frequency in the closed position control loop.
2. The natural frequency of bending vibration for the feed spindle should be at least twice the maximum rotational frequency of the spindle at the maximum feed rate.

If the frequencies are not separated enough, the natural frequencies of the mechanical transmission elements must be shifted by varying the moments of inertia and spring constants. Changing the design might also be required (in regard to the ball screw bearing, for example). Using simulation methods is a sensible approach for the specific design of dynamic transfer behavior for mechanical elements.

A lower overall mass moment of inertia can be attained with optimized structural design of the individual feed drive components. This is explained using a gear drive as an example. ◘ Figure 5.2 shows two feed drives that are identical apart from the transmission gearing. To determine the necessary drive torques on the basis of the accelerations required, the equivalent moments of inertia referred to the drive shaft must be calculated. ◘ Figure 5.3 shows step by step the calculation required for this. The reduced moment of inertia applied to the drive shaft is calculated for the single-speed gearbox as in Eq. (5.1) and for the two-speed gearbox as in Eq. (5.2). The fact that the diameter is entered into the moment of inertia to the power of 4 is factored into these equations.

Single-speed gearbox:

$$J_{\text{gear, red}} = J_1 \cdot \left(1 + \frac{i^4}{i^2}\right) \tag{5.1}$$

Two-speed gearbox:

$$J_{\text{gear, red}} = J_1 \cdot \left(1 + \frac{i_{12}^4}{i_{12}^2} + \frac{1}{i_{12}^2} + \frac{i_{12}^4}{(i_{12} \cdot i_{34})^2}\right) \tag{5.2}$$

Assuming a gear ratio of 4 for the single-speed gearbox, the moment of inertia of the transmission is 17 times the moment of inertia of the gear on the drive shaft. If a second speed is added to the gearbox and both gear ratios are assumed to be 2, the total moment of inertia is only 6.25 times. This means the single-speed gearbox, for the same overall gear ratio and same gear widths, has a reduced moment of inertia applied to the motor shaft that is 2.7 times that of the two-speed gearbox. This results from the high moment of inertia of the drive wheel of the single-speed gearbox. Despite the lower moment of inertia of the two-speed gearbox, the single-speed gearbox is preferred here because, in addition to the lower compliance and backlash values, it costs less to manufacture. So, cost considerations ultimately also have a bearing on selecting the solution.

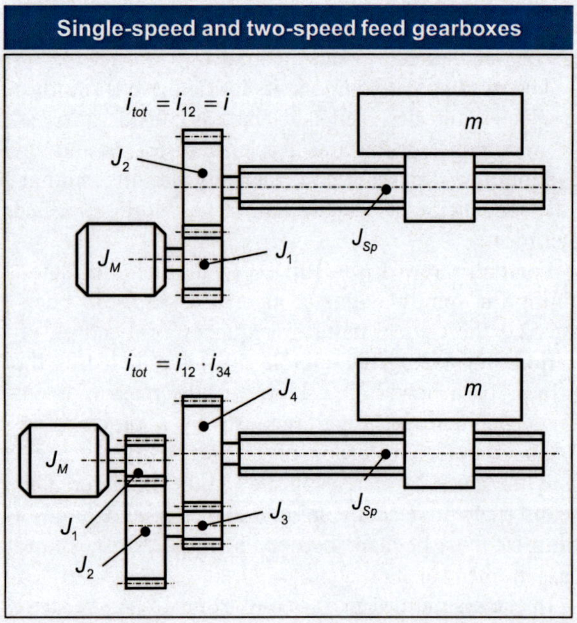


◘ **Fig. 5.2.** Spindle drives with single-speed and two-speed feed gearboxes

### 5.1.3 Selection and Design of the Drive Motor

The feed motor design must be viewed in conjunction with determination of the gear ratios for the mechanical components (feed gearbox, spindle pitch and pinion diameter). They permit the motor torque and speed to be adapted to the requirements regarding the feed force, rate and acceleration. There is no such adaptability for linear motor drives.

So the selection of the feed motor manifests itself as a design problem with parameters of gear ratio $i$ and spindle pitch $h$ on the one hand, and the motor characteristics on the other. The mutual influence exerted by the parameters means using an incremental process comprising the selection of a motor-transmission configuration followed by calculation of the static and dynamic properties. If the design is to be the best one for the requirements, multiple iterations (comprising

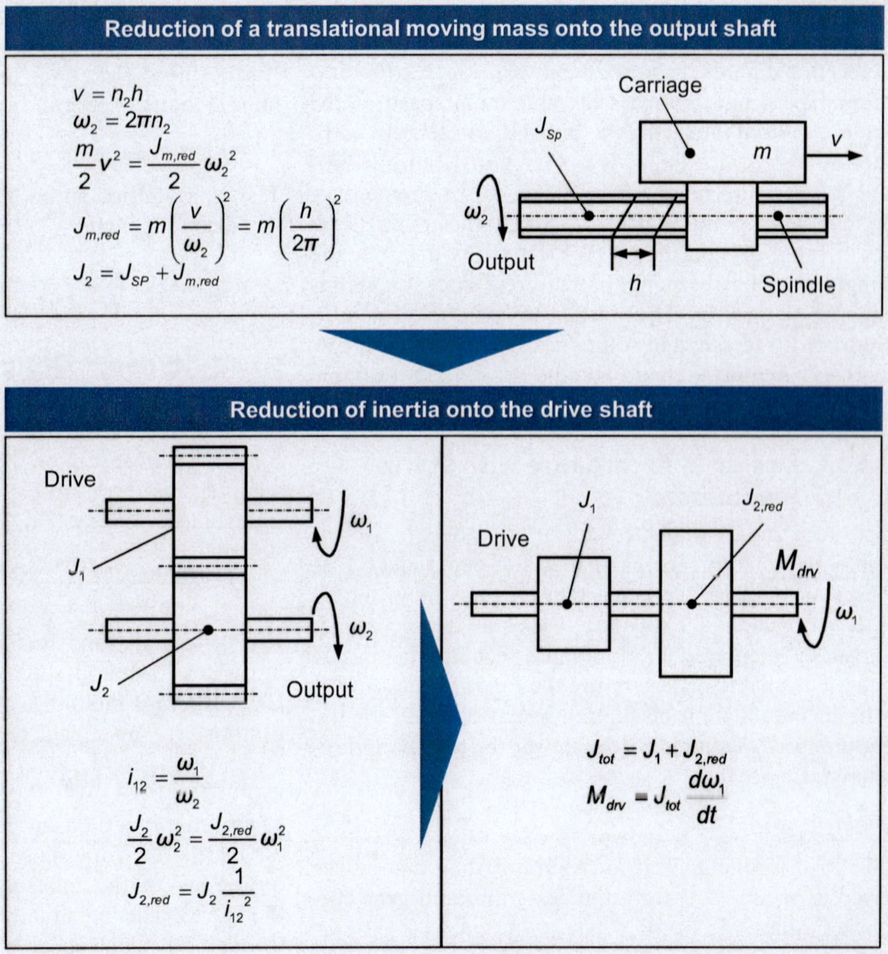


**Fig. 5.3.** Reduction of mass moments of inertia onto the drive shaft of a feed drive

selection of modified components and recalculations) are usually necessary.

Normally, only the extreme values of the machining and acceleration scenarios occurring within the machining cycle are used as the basis for the design. Of relevance are the maximum rapid motion speed, maximum acceleration, load duration, and standstill times after machining. The requirements put on the machining cycle must be contrasted with the capabilities of the drive motor selected (usually a permanent-magnet synchronous motor). The information necessary is available in the relevant torque-speed characteristic curves for the motors. A distinction is made between a characteristic for continuous operation (S1 characteristic) that reaches the maximum permitted motor temperature under constant load, and a characteristic for momentary loading that includes the maximum torque-speed range (Fig. 5.4). Staying below the S1 characteristic ensures that the maximum permitted motor temperature will not be exceeded. The simplified characteristic for momentary loading comprises the limits for torque, voltage and rotational speed. The torque limit specifies the maximum motor torque without the motor shaft being damaged. The voltage limit results from the maximum supply voltage of the motor that needs to compensate the countervoltage induced in the stator/armature winding by rotation. As the countervoltage

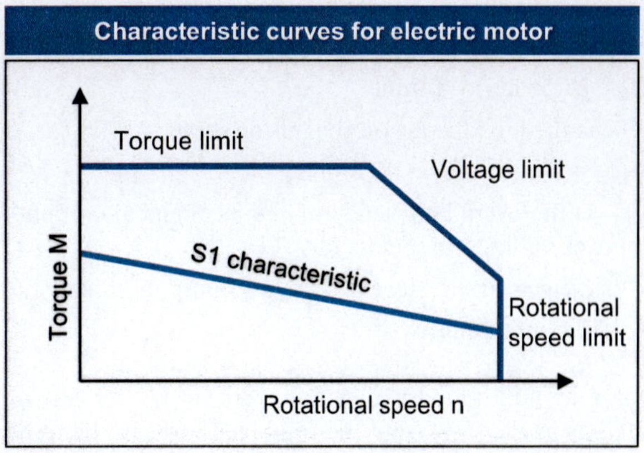


**Fig. 5.4.** Maximum torque-speed limits for continuous and short-term use

5

is dependent on the speed, the voltage limit is also dependent on the speed (with the exact profiles being different for different models of motor). Lastly, the rotational speed limit specifies the maximum speed of the motor without damaging the bearings or rotor. In addition to these three limits, the maximum permitted motor temperature must not be exceeded for momentary loading and going above the S1 characteristic (such as in rapid motion mode) [WEID08]. As a result, the length of time the power is required, and any standstill times, must be taken into account in the design in addition to torque and rotational speed. Shown below is the procedure taking a spindle drive as an example (◘ Fig. 5.2).

#### 5.1.3.1 Design in Accordance with Static Parameters

The feed motor and mechanical components are initially designed on the basis of static parameters. It is necessary first to guarantee that the maximum motor speed is high enough to attain the required travel speed, normally the rapid motion rate. Second, the motor must be capable of generating the load torques occurring in the different machining scenarios, both for the motor speeds required here and over the time period required, Eq. (5.3).

$$M_L \leq M_M \tag{5.3}$$

The load torque applied to the motor shaft can be derived as in Eq. (5.4) from the gear ratio and gear efficiency values.

$$M_L = \frac{1}{\eta_{tot}} \cdot \frac{1}{i} M_{Sp} \tag{5.4}$$

where:

$M_M$ – is the motor torque

$M_L$ – is the load torque on the motor shaft

$M_{Sp}$ – is the torque on the spindle

$\eta_{tot}$ – is the overall efficiency of the mechanical components excluding the spindle nut

$i$ – is the gear ratio between drive and outputs shafts

$h$ – is the spindle pitch

For a spindle nut drive, spindle torque $M_{Sp}$ is calculated as in Eq. (5.5) from the total feed force in the feed direction, the spindle pitch and the efficiency of the spindle nut:

$$M_{Sp} = F_{tot} \cdot \frac{h}{2\pi} \cdot \frac{1}{\eta_{Spindlenut}} \tag{5.5}$$

The total feed force is calculated in turn from the process, weight and frictional forces in the feed direction as in Eq. (5.6):

$$F_{tot} = F_{Feed,\,process} + F_{weight} + F_{friction} \tag{5.6}$$

Furthermore, the motor must be able to reach the maximum required speed in rapid motion mode, Eq. (5.7).

$$n_{max} \geq n_{rapid} \tag{5.7}$$

The rapid motion speed is calculated as in Eq. (5.8) for spindle nut systems.

$$n_{rapid} = i \cdot \frac{v_{rapid}}{h} \tag{5.8}$$

#### 5.1.3.2 Dynamic Design

For modern machines, the acceleration processes normally represent the strictest requirements on motor torque. The measure of the acceleration capability of a feed drive is the startup time constant of the (initially unregulated) feed system. It is calculated as in Eq. (5.9):

$$T = \frac{J_{tot,\,red.} \cdot 2\pi \cdot n_{max}}{M_{acc.,\,max}} \tag{5.9}$$

For the calculation, the individual moments of inertia must be known and applied to the motor shaft ($J_{tot,red}$, Eq. 5.10), and the motor torque available for acceleration $M_{acc.,max}$ must be determined.

$$J_{tot,red} = J_{Gear} + J_{Motor} + J_{Spindle,red} + J_{table,red} \tag{5.10}$$

The primary method for converting the mass moments of inertia of rotating or translating masses, based on the law of the conservation of energy, is shown in ◘ Fig. 5.3. The maximum motor torque minus the load torque is available as acceleration torque $M_{acc.}$. As the motor torque, and possibly the load torque, is/are dependent on the rotational speed, the resulting acceleration torque is also a function of the speed, Eq. (5.11).

$$M_{acc.}(n) = M_M(n) - M_L(n) \tag{5.11}$$

For acceleration, electric motors can generally be overloaded momentarily with 3 to 10 times the standstill torque $M_{M,max}$. The torque equilibrium considering the efficiency and friction in the guide systems as in Eq. (5.12) yields the motor torque available for acceleration, Eq. (5.13).

$$M_{M,max} = \frac{1}{\eta_{tot}} \cdot (M_{acc.} + M_{friction}) \tag{5.12}$$

$$M_{acc.} = \eta_{tot} \cdot M_{M,max} - M_{friction} \tag{5.13}$$

The maximum possible acceleration (Eq. 5.16) is calculated using the differential equation of motion (Eq. 5.14) and the relationship between translational acceleration and angular acceleration for ball screws as in Eq. (5.15).

$$J_{tot} \cdot \dot{\omega}_M = M_{acc.} \tag{5.14}$$

$$\text{where: } \omega_M = i \cdot \omega_{Sp} = i \cdot \frac{v \cdot 2\pi}{h} \tag{5.15}$$

$$\ddot{x} = \frac{h}{2\pi} \cdot \frac{1}{i} \cdot \frac{M_{acc.}}{J_{tot.}} \tag{5.16}$$

### 5.1.3.3 Optimum Gear Ratio

To increase the acceleration of a mass to be moved linearly, the drive system can be optimized (starting from the motor to the ball screw via any intermediate gears). Besides using different motors $M_{M,\max}$, possible variation parameters are the gear ratio of the feed gearbox $i$, the spindle pitch $h$, and the pinion diameter $d$.

For any given motor, there are combinations of gear ratio and spindle pitch, or gear ratio and pinion diameter, to maximize carriage acceleration, ◘ Fig. 5.5. The optimum parameter values for $i$, $h$ and $d$ are derived as solutions of the necessary condition for an extreme value problem, Eq. (5.17).

$$\frac{d\ddot{x}}{di} = 0, \frac{d\ddot{x}}{dh} = 0 \text{ or } \frac{d\ddot{x}}{dd} = 0 \tag{5.17}$$

The machine table acceleration is known from Eq. (5.16). The acceleration for the case of a spindle nut drive is determined by applying all of the reduced moments of inertia to the motor shaft, Eq. (5.18).

$$\ddot{x}_m = \frac{M_{drv.}}{J_{tot}} \frac{1}{i} \frac{h}{2\pi} = \frac{M_{drv.}}{2\pi} \frac{h \cdot 1/i}{J_M + J_1 + 1/i^2 \cdot (J_2 + J_{Sp} + J_m)}$$
$$\Leftrightarrow \ddot{x}_m = \frac{M_{drv.}}{2\pi} \frac{h \cdot i}{i^2(J_M + J_1) + J_2 + J_{Sp} + h^2 m/4\pi^2} \tag{5.18}$$

Thus, the analytical optimum can be specified by Eqs. (5.19) and (5.20) for a spindle nut system, and by Eqs. (5.21) and (5.22) for a rack and pinion system.

$$\Rightarrow i_{opt,acc.} = \sqrt{\frac{J_2 + J_{Sp} + mh^2/4\pi^2}{J_M + J_1}} = \sqrt{\frac{J_{output}}{J_{drive}}} \tag{5.19}$$

Feed drive with rack and pinion system

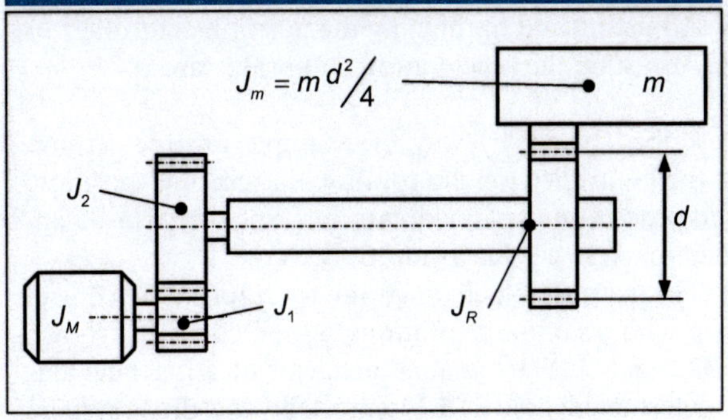


◘ **Fig. 5.5.** Feed drive with rack and pinion systems

$$h_{opt,acc.} = \sqrt{\frac{i^2(J_M + J_1) + J_2 + J_{Sp}}{m/4\pi^2}} \tag{5.20}$$

For the optimum parameter values of a rack and pinion system, note that the reduced moment of inertia of the machine table mass changes compared to the spindle nut system.

$$\Rightarrow i_{opt,acc.} = \sqrt{\frac{J_2 + J_R + md^2/4}{J_M + J_1}} = \sqrt{\frac{J_{output}}{J_{drive}}} \tag{5.21}$$

$$d_{opt,acc.} = \sqrt{\frac{i^2(J_M + J_1) + J_2 + J_R}{m/4}} \tag{5.22}$$

◘ Figure 5.6, for example, depicts the relationship of attainable acceleration and the spindle pitch and gear ratio parameters for a spindle nut system. The effect of suitable parameter selection on acceleration can be seen clearly—an area with optimum acceleration can also be determined. From a practical viewpoint, however, it must be stated that the possible gear ratios and spindle pitches are limited to minimum and maximum values by the inherent design.

## 5.2 Design of the Measuring System

Axes driven electromechanically generally have a measuring system in the motor, which can be used to close the rotational speed and position control loops. However, as the mechanical transmission elements affecting accuracy cannot be measured with this measuring system, the indirect motor measuring system is usually supplemented with a direct one in machines having strict accuracy requirements.

Two contrasting aspects must be kept in mind in the selection of the measuring system used. First, the measuring system must have the accuracy and resolution required. This can be achieved for example by selecting a scale with a low division period. Next, the maximum speed of the measuring system (reduction of the sensor amplitude) and the timing signal of the drive controller must not be exceeded during rapid machine movements.

Usual input cutoff frequencies are 250 kHz. For a measuring system with a signal period of 20 µm, for example, this corresponds to a maximum speed of 5 m/s = 300 m/min. If a measuring system with a division period of 4 µm is used, however, the maximum speed is only 1 m/s, so 60 m/min. This value is far exceeded in High Speed Cutting (HSC) machines [BREC02, ZIRN06, AUCH12]. It follows that, for HSC machines, either a measuring system with a lower division period or an analysis unit with a higher cutoff frequency needs to be used.

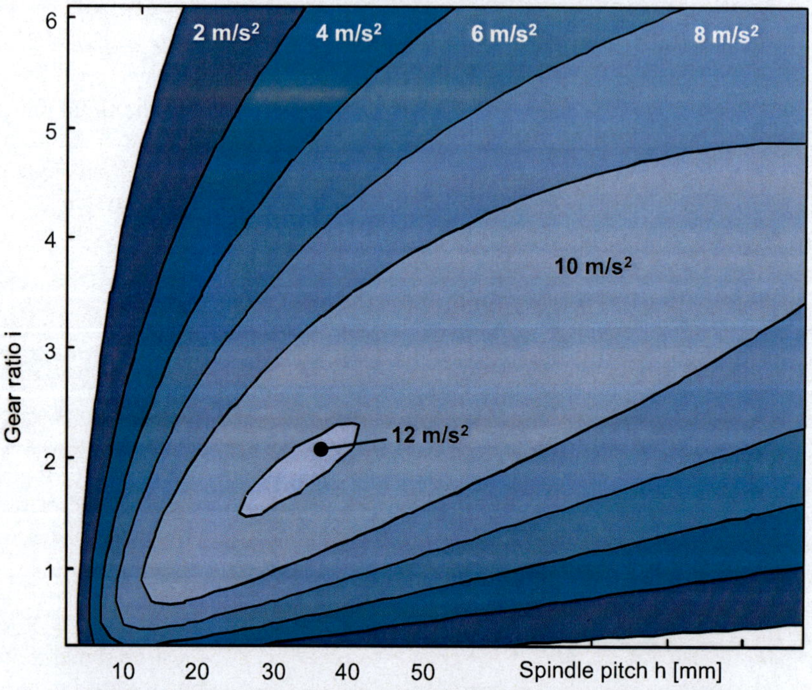


**Fig. 5.6.** Variation of gear ratios for screw drives

| Boundary condition | |
|---|---|
| Moved masses m = 935 kg | Motor torque $M_{max}$ = 65 Nm |
| Motor inertia $J_M$ = 0.0048 kgm$^2$ | Spindle length l = 1000 mm |
| Spindle diameter d = 40 mm | h, i → variable |

## 5.3 Commissioning of the Controller

Commissioning the servo drive and optimum adjustment of all control loops require several steps. The user is assisted by commissioning aids for the drive amplifier and controller (depending on the type and make of the motor). The commissioning procedure for a drive where all commissioning steps need to be performed by the user generally requires the steps described below. It is assumed here that the electrical system is fully installed and tested.

### 5.3.1 Manual Commissioning

The work steps below are usually performed for the manual commissioning of a feed axis:

- Setting of the current controller in line with the manufacturer instructions
- Drift compensation of the rotational speed for analog systems
- Setting of the rotational speed controller
- Connecting the reference value specification by the controller
- Checking the correct control direction
- Configuring the current or rotational speed reference value filters as required
- Adjusting and optimizing the position controller
- Adjusting the acceleration and brake ramps

Described below by way of example are the settings of the controller for the rotational speed and position. The procedure explained can, in principle, also be applied to axes with linear motor drive.

The general block diagram for a position control loop with secondary rotational speed control is shown in Fig. 5.7. The dynamic behavior of a machine axis is determined first and foremost by the drive system.

Due to this, it is advisable to first optimize the rotational speed and velocity control loops before adjusting the position controller [HEIN98].

#### 5.3.1.1 Adjusting the Rotational Speed Controller

The rotational speed control loop is first disconnected from the master position control loop. The dynamic behavior is determined using excitation with defined test signals (e.g., a step function, noise, sine function) as a rotational speed reference value and by recording the actual rotational speed.

For analog drives, the rotational speed reference value can be specified very easily using a signal generator on the reference value input of the amplifier. Similarly, the tachometer signal can be tapped and recorded externally.

As part of a network with corresponding software modules on the controller, modern digital drives have powerful functions able to both generate reference values and record the internal actual values. Furthermore, functions are generally available that help technicians in the correlation and evaluation of signals.

**Step-like Excitation**

The rotational speed reference value is often subjected to a step-like change for recording the transfer function. To eliminate non-linearities due to the current-limiting effect of the drive amplifier, the step magnitude must be set such that the drive does not reach its maximum current. This rotational speed is generally only 5% to 15% of the maximum motor speed. ◘ Figure 5.8 shows as an example the step response and current profile of a feed drive for a step from 0 to 80 $\min^{-1}$ for which the current limit of 40 A is not quite reached. The maximum motor speed is 2000 $\min^{-1}$.

Under the assumption that the natural angular frequency of the feed mechanics is far higher than that of the drive control loop with motor and coupled load, the behavior of the drive system versus time can be approximated by a 2nd order delay system with the transfer function defined in Eq. (5.23). This is the behavior for most machine tools. For robots, the natural angular frequency of the mechanics is commonly of the same order of magnitude or even lower than that of the drives, meaning the following equations cannot be applied here.

$$G_A(j\omega) = \frac{1}{1 + \frac{2D_A}{\omega_{0A}} j\omega + \frac{1}{\omega_{0A}^2} (j\omega)^2} \tag{5.23}$$

The following parameters can be determined graphically from the step response (see ◘ Fig. 5.7)

$T_r$: Rise time
$T_o$: Overshoot time
$T_s$: Settling time
$e$: End deviation
$A_o$: Overshoot magnitude

This means it is possible to calculate the parameters of damping $D_A$ and cutoff frequency $\omega_{0A}$ for the 2nd order system. The following applies:

$$D_A = \frac{1}{\sqrt{1 + \left(\frac{\pi}{\ln a_o}\right)^2}} \tag{5.24}$$

$$\omega_{0A} = \frac{\pi}{T_o \cdot \sqrt{1 - D_A^2}} \tag{5.25}$$

Where $a_0$: overshoot magnitude $A_0$ related to static end value

Also, the step response recorded and Eqs. (5.24) and (5.25) can be used to check the correct setting of the rotational speed controller. The drive should behave approximately like a 2nd order delay element with a damping value of 0.5 and a natural angular frequency that is as high as possible [GROS06]. If time constants $T_{\text{mech}}$ and $T_{el}$ of the motor with a coupled load are known, the setting parameters of the rotational speed PI controller can be determined computationally. Adjustment using the symmetric optimum, yielding the gain factor of the rotational speed controller $K_p$ and the integral action time $T_p$ as in Eqs. (5.26) and (5.27), is a much-proven method [STUT81].

$$K_p = \frac{T_{\text{mech}}}{8 \cdot T_{el}} \tag{5.26}$$

where $T_{mech} = \frac{J_{tot} \cdot \omega_{\text{max,motor}}}{M_{St,A}}$

$M_{St,A}$: Standstill torque of the motor

$$T_p = 4 \cdot T_{el} \tag{5.27}$$

**Frequency Response Measurement**

Another method to determine the dynamic behavior of control loops is to record the transfer function (frequency response) and show it as a Nyquist or Bode plot. The behavior of the drive within a certain frequency range is determined and shown here. For the measurement, an excitation signal is superimposed on the rotational speed reference value and the actual speed value is measured. Fourier analysis is then used to calculate the transfer function from the reference and actual rotational speed values.

**Fig. 5.7.** Determining the transfer behavior of drives

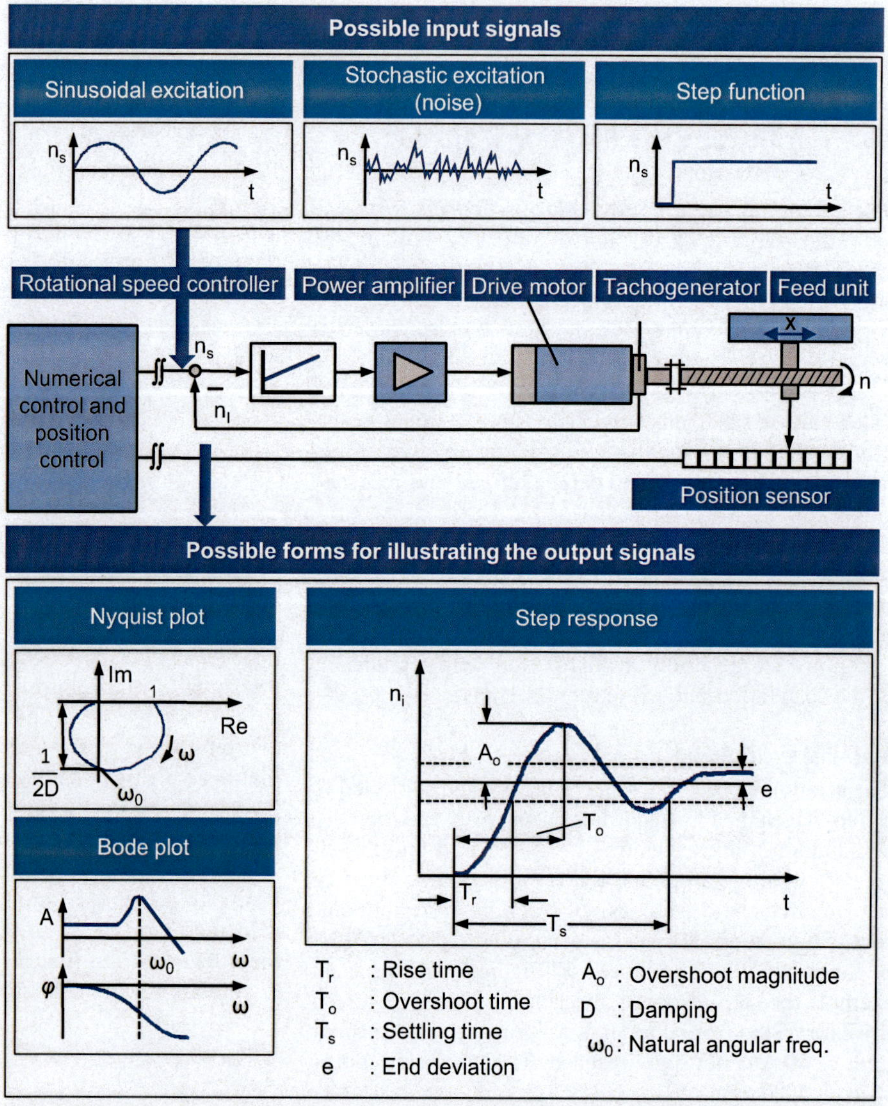


The excitation signal for the rotational speed reference value has a major influence on the measurement accuracy. Stochastic and deterministic signals can generally be used. Here, the excitation signal must contain the entire frequency spectrum to be analyzed by the Fourier analyzer. A sinusoidal signal, with a frequency that changes linearly over time between a minimum and a maximum value (*sweep*) during the measurement, returns good results.

For this measurement too, the excitation (amplitude) must be limited such that the current limit of the drive is not reached. Otherwise, the assumption of linear transfer behavior to calculate damping and natural angular frequency no longer applies.

Figure 5.9 shows the effect of current limitation on the frequency response of a feed drive. With sinusoidal excitation at about 70 Hz, this drive reaches the current limit at an amplitude $\hat{n}$ of the rotational speed vibration of 300 $\text{min}^{-1}$. The non-linear effect of current limitation is evident in the decrease of the natural angular frequencies at higher excitation amplitudes. The reasoning behind this decrease is that the maximum acceleration to be generated increases linearly with the vibration amplitude. The acceleration, in turn, is proportional to the motor current as a first approximation, meaning the current limit for large vibration amplitudes is already reached at lower frequencies.

In this measurement, a swept sinusoidal signal over the frequency range of 2 Hz to 100 Hz was used as an excitation signal. Drive behavior for frequencies below 2 Hz could therefore not be determined from the measurement data. This also means that the amplitude and phase profiles shown in the Bode plot do not reflect the actual drive behavior for frequencies lower than 2 Hz.

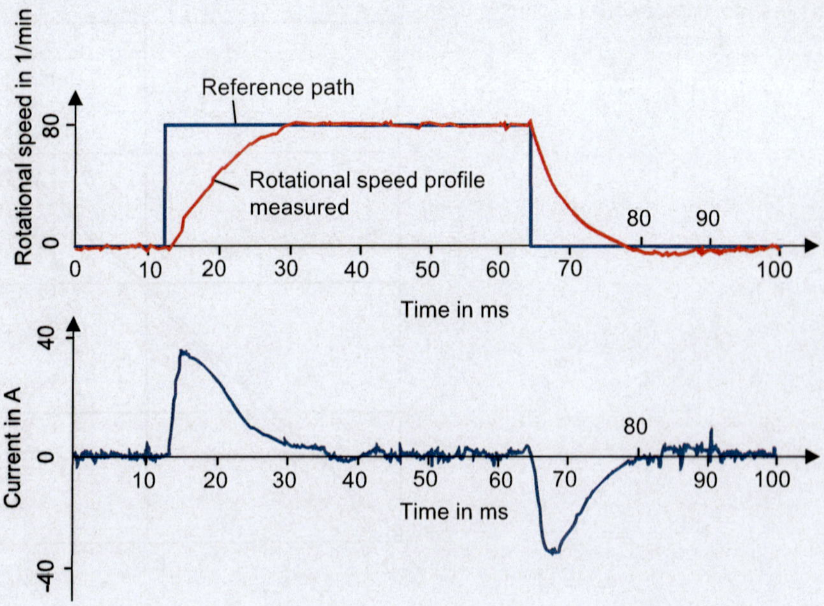


**Fig. 5.8.** Rotational speed step from 0 to 80 $min^{-1}$ and associated current profile

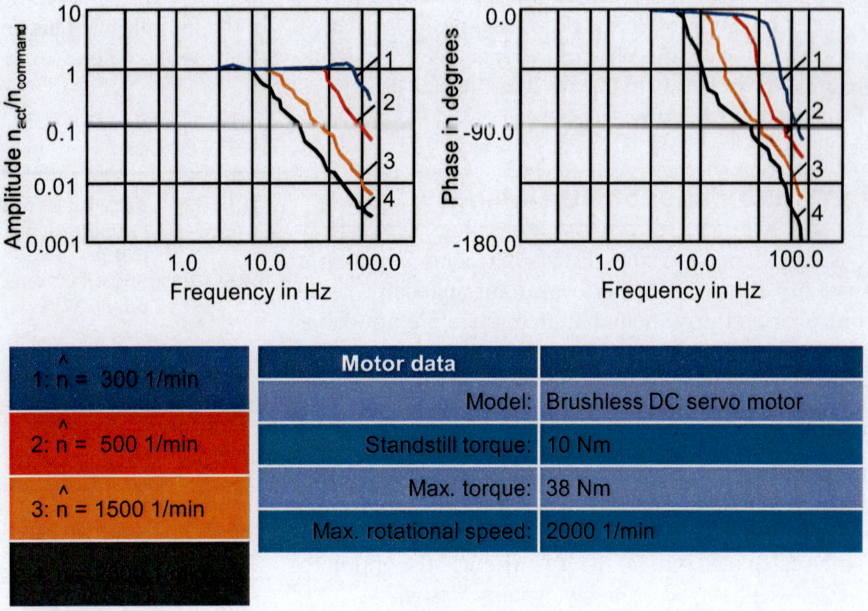


**Fig. 5.9.** Frequency responses nact/ncommand of a feed drive for different excitation amplitudes $\hat{n}$

### 5.3.1.2 Adjustment of the Position Controller

To increase the dynamics of the feed drives and to improve path behavior, the $K_V$ factor must be set as high as possible. Too high a $K_V$ factor, however, results in instabilities and overshooting during positioning. As a compromise, a value can be determined where there is virtually no overshooting when the normal machining speed is used for moving into a position, Fig. 5.10.

Well-established values for the $K_V$ factor are in the range:

$$K_V = 0.6 \ldots 4.8 \frac{\text{m/ min}}{\text{mm}} \text{ (elektromechanical system)}$$

$$K_V = 15 \ldots 25 \frac{\text{m/ min}}{\text{mm}} \text{ (linear motor)} \tag{5.28}$$

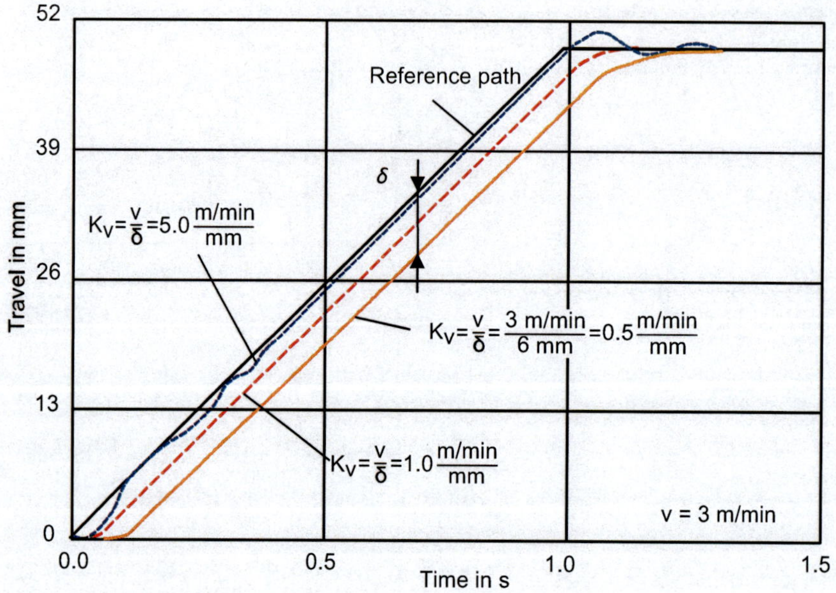


**Fig. 5.10.** Influence of the $K_V$ factor on a speed ramp

5

The higher $K_V$ factors for linear motor drives are linked to the elimination of compliant mechanical transfer elements. For electromechanical drives, they inevitably ensure a lower cutoff frequency and thus lower dynamics of the control loops.

### 5.3.2 Automatic Commissioning

It is clear from the steps listed that commissioning a servo drive can also be performed automatically. Some controller and drive manufacturers therefore now offer commissioning aids to assist the user.

The controller ensures that the steps described in the previous chapter are run automatically in sequence. For this, the drives are excited repeatedly with noise in the individual control loops, and the corresponding transfer function is determined. From it, the parameters for the PI controllers of the cascaded control loops can be calculated. Often, the user also has the option to select a well damped or highly dynamic design.

In addition, the tools for automated commissioning are often able to configure any reference value filters required in the current control or rotational speed control loops (so as to not excite resonance frequencies in the first place). This enables the tracking behavior of the drives to be improved further still.

## References

[AUCH12] Auch, S.: Fertigungszentren für die sechsseitige Komplettbearbeitung. WB Werkstatt und Betrieb **4**, 50–52 (2012)

[BREC02] Brecher, C.: Vergleichende Analyse von Vorschubantrieben für Werkzeugmaschinen. Berichte aus der Produktionstechnik, Bd. 2002,17. Shaker, Aachen (2002)

[GROS06] Gross, H., Hamann, J., Wiegärtner, G.: Technik elektrischer Vorschubantriebe in der Fertigungs- und Automatisierungstechnik. Publicis Corporate Pub., Erlagen (2006)

[HEIN98] Heinemann, G., Papiernick, W.: Hochdynamische Vorschubantriebe mit Linearmotoren. VDI-Z Special Antriebstechnik, 26–34 (1998)

[STUT81] Stute, G.: Regelung an Werkzeugmaschinen. Hanser, München (1981)

[WEID08] Weidauer, J.: Elektrische Antriebstechnik. Publicis Corporate Publ., Erlangen (2008)

[ZIRN06] Zirn, O., Weikert, S.: Modellbildung und Simulation hochdynamischer Fertigungssysteme. Springer, Berlin (2006)

# Process Monitoring

I thank my staff members Christian Fimmers M. Sc., Simon Storms M. Sc., Minh Trinh M. Sc. and Stephan Wein M. Sc. for their contribution in revising this chapter.


C. Brecher and M. Weck, *Machine Tools Production Systems 3*,
Lecture Notes in Production Engineering, https://doi.org/10.1007/978-3-658-34622-5_6

## Contents

## 6.1 Introduction

In addition to machining speed and accuracy, the utilization time of a manufacturing system is of particular economic importance. In order to ensure high utilization times, quality assurance tasks are carried out under specific timed conditions (see Sects. 6.1.1 and 6.1.2).

The influences on the functionality of the means of production and the product quality (see ▶ Sect. 6.1.3) determine whether continuous monitoring is required.

▶ Section 6.1.4 provides a description of the defining parameters in monitoring tasks and the structure of the monitoring system.



The aim of process control is to automatically optimize process management through constant adaptation to the behavior of the machine, tool, workpiece and material (see ▶ Sect. 6.1.5).

### 6.1.1 Background, Terminology and Objectives

Current manufacturing technology is characterized by high machining speed and accuracy. This leads to high wear or load on the tool and the machine. It also requires a wide range of process monitoring and control measures, diagnostics and maintenance to avoid disruptions and malfunctions as far as this is possible.

In order to achieve the objective of high productivity, the machine downtime must be kept to a minimum and high process reliability must be achieved during monitored and unmonitored machining. The actual utilization time $T_U$ of a manufacturing system, i.e., the plant's productive time, is influenced by various downtimes (◘ Fig. 6.1) defined in the VDI standard 3423 [FIRM95]. The organizational downtime $T_O$ is defined here as the total of all downtimes attributed to poor organization such as a lack of material. For comparison, the technical downtime $T_D$ is the total downtime caused by the technical disruptions experienced by the plant in question. A further distinction is made here between downtime caused by the manufacturer ($T_{DM}$) or the user ($T_{DU}$). Further downtimes are maintenance time $T_M$ and testing time $T_T$ but these are only valued as downtime when the plant is not able to manufacture products. Testing time is to be included in the utilization time if production is being upgraded. The total of these downtimes reduces the occupancy time $T_B$ to the actual utilization time $T_U$. Regardless of the shift pattern and the complexity of the plant, it is possible to define comparative criteria such as the total degree of utilization and these allow conclusions to be drawn on the weak points and corrective measures.

The items to be monitored can be either on the process or the machine side. For this reason, both the process, i.e., the tool and workpiece, and all machine components must be monitored.

**Utilization time and technical degree of utilization**

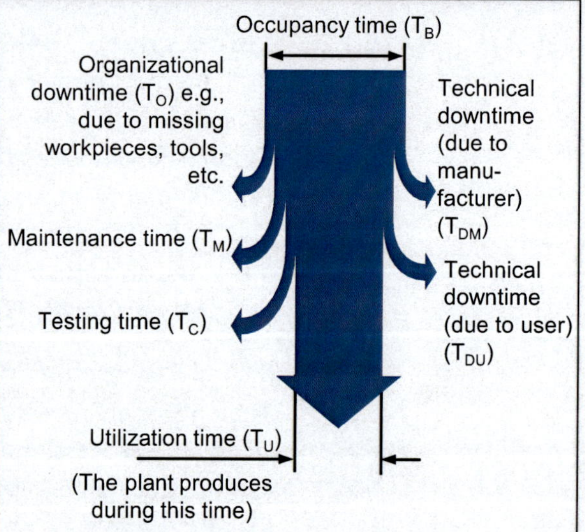


Total degree of utilization $U_{total}$:

$$U_{total} = \frac{T_U}{T_B} * 100\,\%$$

$$= \left(1 - \frac{T_O + T_{DM} + T_{DU} + T_M + T_C}{T_B}\right)$$

Technical degree of utilization $U_T$:

$$U_T = 1 - \frac{T_{DM}}{T_B} * 100\,\%$$

◘ **Fig. 6.1** Determining the utilization time and defining the total and technical degree of utilization according to VDI3423

The objective when monitoring the process is to ensure maximum process reliability. Process reliability means maintaining a given product quality and preventing or eliminating disruptions to processes (e.g. tool fractures).

The objective of machine monitoring is to check the actual condition of individual machine components and to identify erroneous functions of individual components or assemblies.

A more detailed definition of these two types of monitoring is given in ▶ Sect. 6.1.4.3.

To maintain process reliability and machine availability or to increase it, automated monitoring and diagnostic systems are being increasingly used to supplement organizational measures, such as qualifications of personnel, timely provision of materials, optimized machine scheduling, or regular maintenance intervals. These arrangements must.

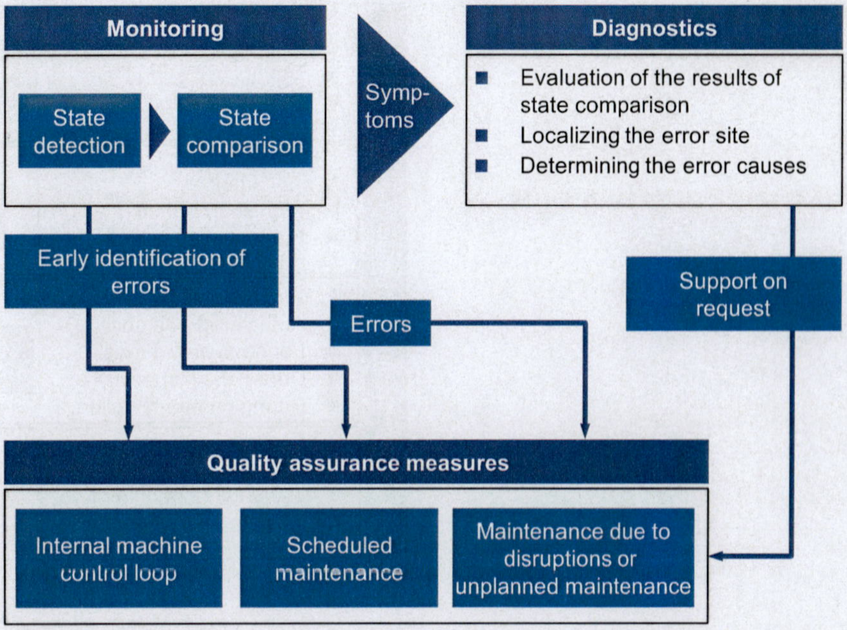


**Fig. 6.2** Tasks and measures in manufacturing from a quality assurance perspective

- monitor the process functions and the product quality for deviations or errors (disruptions),
- control the process in the event of deviations or
- diagnose the disruption on request (i.e., determine the type, location and cause of the fault) and
- provide the user wherever possible with assistance in error management.

The term "monitoring," which actually has its own functional hierarchy from a technical perspective (see ► Sect. 6.1.4.2), is used in the field of quality assurance for the two different tasks of state detection and state comparison (◘ Fig. 6.2).

Detecting the state of the plant or process includes recording process and machine parameters that reflect the current state of the means of production. State comparison means comparing the actual state with a given nominal state. Depending on the specific parameter, this nominal condition is defined by the machine data established at machine acceptance, taking environmental variations into consideration, or by given specific process, workpiece and tool variables.

This state comparison has two main purposes. These are checking limiting values and trends as well as inspecting timed processes. Checking limiting values can be monitoring the maintenance of minimum or maximum values or monitoring the violation of tolerances—so-called permitted ranges. There are many operating parameters on a machine that can be considered limiting values such as temperatures, forces, pressures, distances, workpiece dimensions and switching times. Violating these values provides information on errors. The symptoms resulting from the state comparison provide input parameters for the actual diagnosis. This diagnosis assures the function and quality of the relevant components in the manufacturing process. The diagnosis evaluates the results of the state comparison, localizes the error site and determines the cause of the error. The result of the diagnosis provides specific information about the location and cause of the error. Systematic diagnosis supported by appropriate tools such as error trees can reduce the cost of identifying errors. In addition to this, diagnosis allows early identification of an error as well as its cause. The deviations that occur can then be compensated by an internal machine control loop. Similarly, necessary repairs (i.e., maintenance or repair due to disruption) that have been identified early can be specifically addressed. Then, replacement parts orders and other organizational measures can be introduced in good time. Downtime due to errors and the associated costs can thus be reduced. Trend evaluation also allows errors to be identified and mitigated during their development phase.

These quality assurance tasks, identified in [SSBA99] as quality control and error management in quality assurance, can be carried out directly or indirectly. This means that machine and monitoring parameters are evaluated in terms of time and place either directly or indirectly through the effect of measurable parameters [INDR97]. An example of the latter is monitoring wear through force signals. Qualitatively safe operation based on the philosophy of Total Quality Management (TQM) as described in [SSBA99, SIEM98] includes having additional maintenance and inspection measures scheduled that, for example, can be carried out during breaks (changes of shift,

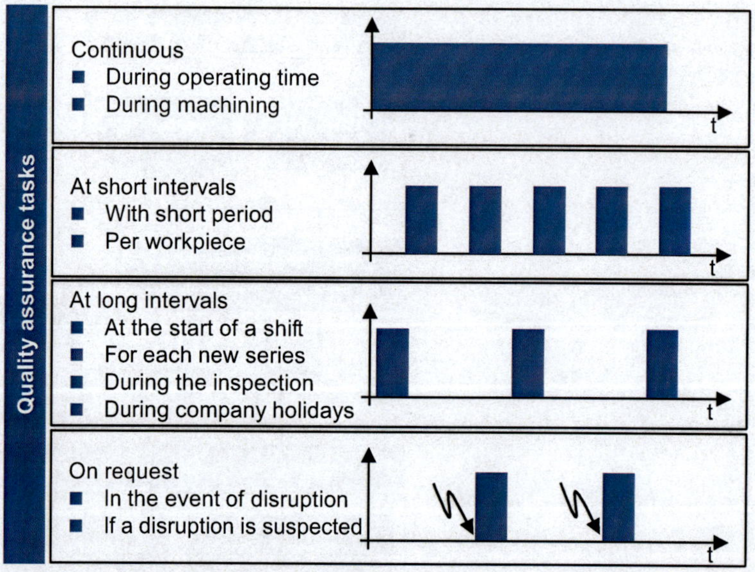


**Fig. 6.3** Chronological arrangement of quality assurance tasks and measures

holidays, etc.). The reliability of machines and manufacturing processes is influenced not only by quality assurance tasks and measures but also, decisively, by their chronological inspection criteria, frequency and time of implementation. A schedule must therefore be established and maintained during the manufacturing process. This schedule is dependent on the required product and process quality and the quality assurance equipment used prior to the start of production. A typical such arrangement is shown in Fig. 6.3.

Some quality assurance tasks and measures have to be completed on a continuous basis, e.g., throughout the machine's total operating time or during a machining process. For example, the working spindle systems that run at high speeds and are susceptible to temperature, have to be continuously monitored while the machine is operating. Contrasted with this, the cutting force and the wear on tools only have to be monitored and regulated if necessary during machining.

However, there are also parameters that have to be monitored on a periodic basis. A simple example of this is a central lubrication system. A lubrication pulse occurs as required at time intervals of a few minutes. The resulting pressure and the lubricant flow have to be monitored during this lubrication pulse. Other monitoring tasks, such as measuring tasks, are carried out only once per workpiece. For example, monitoring the clamping travel takes place only once when clamping each workpiece. In addition to these short interval quality assurance tasks, there are also long interval tasks which are only necessary once per shift or before starting a new series of workpieces. Long interval tasks and measures include all tasks that are only necessary during an inspection or company holidays (maintenance). These include, primarily, inspection of geometrical, kinematic and dynamic machine properties that change only very slowly due to wear and which require a great deal of measuring resources to determine. Finally, there are quality assurance tasks occurring on an intermittent basis such as manual or semi-automated measures to deal with errors triggered by a disruption to a machine. In this respect, disruptions also include damage to machines caused, for example, by a crash or the use of the wrong tools as well as more significant repair work on machines.

### 6.1.2 Economic Significance of Process Monitoring, Process Control, Diagnostics, and Maintenance Measures

When it comes to increasing manufacturing productivity, including the use of production facilities in shifts with fewer operators, it is important to take into consideration not only what can be achieved in technical terms but also the cost-effectiveness of the solution in question when designing measures to increase quality.

A systematic development in monitoring, control, diagnostic and maintenance systems requires detailed knowledge of the deviations and disruptions as well as their causes and possible effects. It is therefore important, as far as increases in productivity are concerned, to concentrate on those disruptions and their causes that result in the most significant manufacturing downtime and the greatest subsequent costs.

### 6.1.3 Factors Affecting the Function of the Means of Production and Product Quality

As with possible design or organizational solutions, all solutions for monitoring function and quality assume a broad knowledge of the potential factors affecting the function of the means of production and product quality. Potential influencing factors are shown in ◘ Fig. 6.4.

Disruptions to the function of the means of production and a lack of quality on the manufactured workpiece are caused by errors that can occur on various components involved in the manufacturing process. These include the electrical components such as the feedforward control, the drives, including the feedback control, and the measuring systems. Potential errors here are component failures, reading errors or, on the drive side, changes in the controller settings that can lead to a deterioration in the drive dynamics. On measuring systems, the error may be due to contamination or maladjustment of the scales. Errors on the mechanical components of the means of production are also to be taken into consideration. Errors on transmission elements manifest themselves, for example, as uneven clamping of the tool or the workpiece, inaccurate alignment of the spindle housing after a collision, or too large a reverse backlash in a machining axis [BEUC89].

People are a significant cause of error: They may enter incorrect cutting data due to inattentiveness or cause a collision due to a general operating error. The damage this causes and the associated machine downtimes and costs are often so considerable that automatic monitoring equipment is put in place to help recognize the position of the tool and the workpiece and avoid collisions (see ▶ Sect. 6.3.6). Suitable process monitoring equipment also reduces or avoids error effects. These effects come, for example, from tool wear or fracture, overdimensioning and incorrect material composition of the workpiece (see ▶ Sect. 6.3).

### 6.1.4 Strategies and Structure of Monitoring Systems

The complexity of processes and machines leads to many monitoring tasks and problems. Many applications have therefore been developed for specific technologies or tasks (see Sect. 6.3).

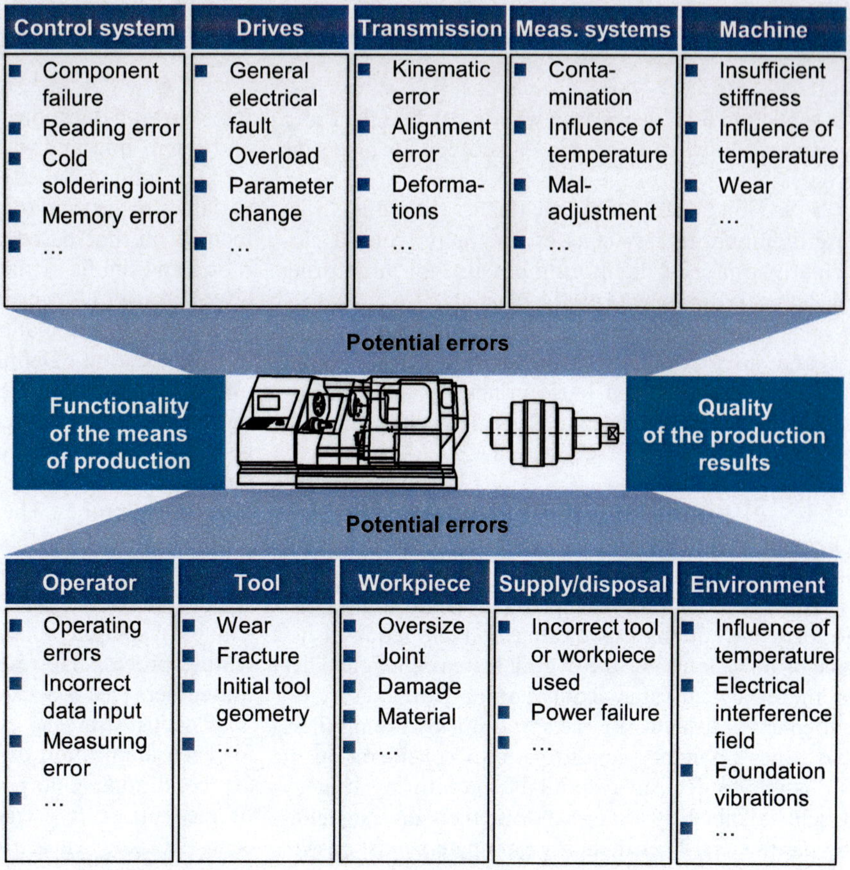


◘ **Fig. 6.4** Factors affecting the function of the means of production and product quality

6

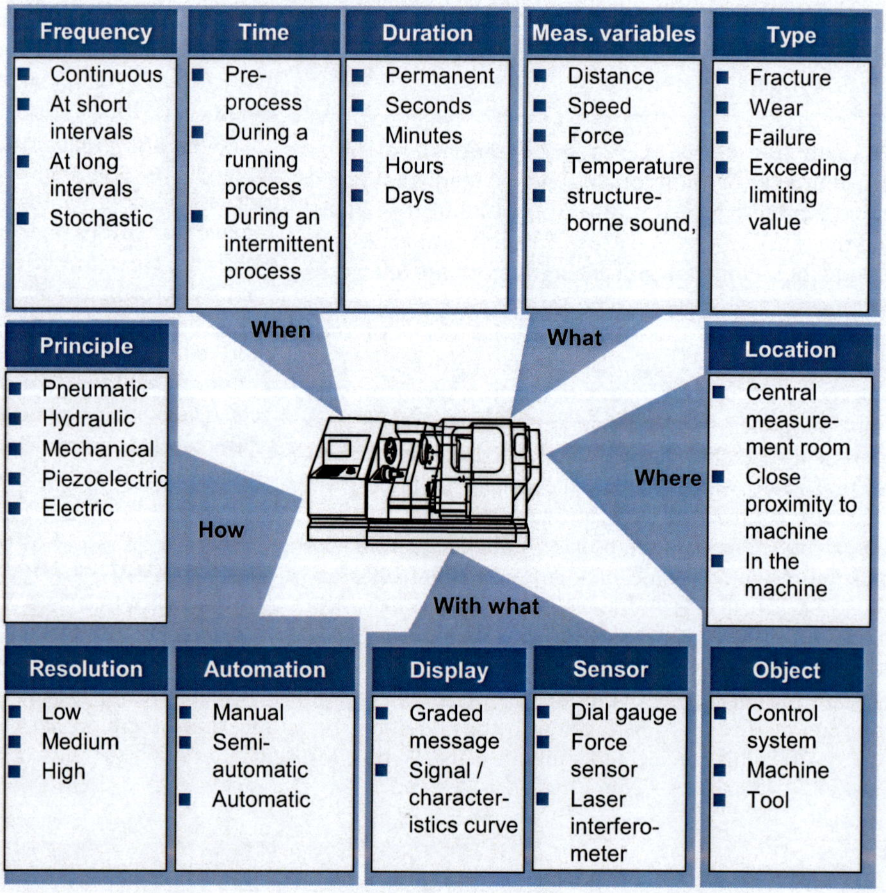


**Fig. 6.5** Defining parameters for creating a monitoring strategy (in this instance, tool fracture)

▸ Section 6.1.4.1 explains what must be taken into consideration when developing a strategy for a specific problem.

▸ Section 6.1.4.2 builds on this to show the structure of monitoring systems as they may be used particularly in process monitoring but also for monitoring mechanical components on the machine.

The third section discusses the functional similarities and differences between process and machine monitoring. The subsequent basic principles are then discussed in general and technological specifics are discussed separately.

### 6.1.4.1 Strategies for Monitoring Systems

There are many alternative monitoring strategies available for the technical/organizational implementation of each monitoring task. Which strategy is most appropriate depends, among other things, on the required degree of automation of the overall system, the reliability of the measured-value acquisition and the quality requirements of the workpiece spectrum. By taking these and other boundary conditions into consideration, it is possible to determine a suitable monitoring strategy which is defined by a range of extremely diverse defining parameters. These defining parameters must answer the following question: "What is to be monitored when and where, how and with what?" (Fig. 6.5).

Depending on the type of disruption, the measured variables to be monitored and the type of capture, the focus is on time-based variables such as frequency, duration and timing of the monitoring. The cost-effectiveness of the monitoring strategy is largely determined by the degree of automation, the resolution and the monitoring location. Defining parameters, which should primarily be chosen from a technical perspective, are the object, the measured variable, and the monitoring principle. The geometry of the object can thus be determined visually using camera systems or with simple mechanical probes. The last step in producing a monitoring strategy is establishing the sensor to be used and the relevant means of display on the evaluation system. Graded messages, which describe a disruption event, can be displayed in addition to the simple signal or feature process. However, conclusion strategies are required here (see ▸ Sect. 6.2.4).

Various strategies are available for consideration for each application depending on the specific boundary conditions. One possible solution to the problem of recognizing tool fractures, which is regarded as a central task when setting objectives for unsupervised

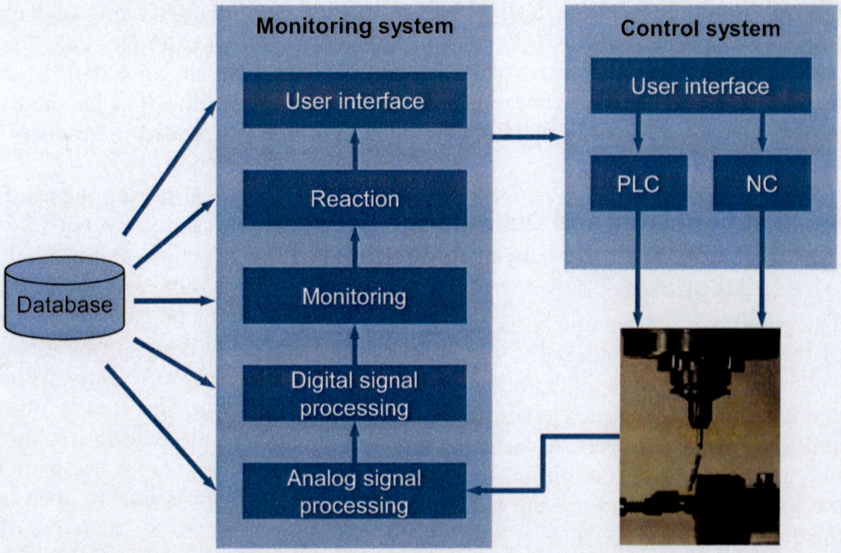


**Fig. 6.6** Information flow and components of monitoring systems

continuous manufacturing processes, is shown in Fig. 6.5. It is largely characterized by:
- permanent monitoring during continuous machining,
- continuous recording of measured values,
- an associated negligibly short capture time per measured value [GEOR98] and
- a definite correlation between measured variables and wear values.

Other systems that use probes, for example, can only capture the condition of the tool during process breaks and sometimes require a few seconds to acquire and process the measured value [HENN99].

General, technical preconditions for monitoring strategies are described in ▶ Sect. 6.2 using the available sensors, signal processing and pattern recognition. ▶ Section 6.3 then explains the various procedures for process monitoring and control for selected manufacturing processes while collision monitoring is explained in ▶ Sect. 6.3.6.

### 6.1.4.2 The Structure of Monitoring Systems

The first step is to classify a monitoring system according to its functional tasks (Fig. 6.6).

The machining process or the machine behavior is mirrored by the physical variables measured by sensors (see ▶ Sect. 6.1.4.4), processed as analog signals and then digitally after conversion (see ▶ Sect. 6.2).

The information can be classified based on these digital signals in order to assess the behavior of the process or the machine (▶ Sect. 6.2.4). A suitable reaction can then be introduced based on this classification.

There are a range of different procedures and processes for the individual steps in this signal processing chain. These are discussed in greater detail in the following chapters. However, the actual selection of suitable processes is directly dependent on the specific application. The various approaches taken in the field of process monitoring are discussed in ▶ Sect. 6.3. Other processes for machine monitoring are discussed in ▶ Sect. 6.5.

Figure 6.6 shows the modules required for the structure of a monitoring system. In accordance with this, a monitoring system can be subdivided into three modules:
- sensor module,
- monitoring module and
- control unit.

The basic objective is always to mount the sensor as close as possible to the object to be monitored (process or machine component). However, this is usually very complicated meaning that, in practice, a compromise has to be found between signal quality and the cost of integration. Individual sensor systems are discussed in detail in ▶ Chap. 9, Volume 2.

Preprocessed signals are evaluated in the monitoring modules. The exchange of information between the monitoring system and the machine control system usually takes place via the PLC. The monitoring system is synchronized with the machining process in that, for example, trigger signals are transmitted for the start and end of machining and tool-specific monitoring parameters. Disruption messages from the monitoring system are transmitted to the machine control system in the opposite direction via inputs to the PLC. Monitoring can also be implemented in the real-time portion of the NC controller if it is appropriately open [ECKH93]. The internal signals can be processed here with the NC cycle time. In this instance, evaluating the signals puts a load on the NC. On machines with many axes this can, for example, lead to having to increase the NC cycle time.

Due to the wide popularity of PC-based user interfaces on machine controls, the control unit of the monitoring system is currently increasingly integrated into the control system.

#### 6.1.4.3 Connection and Delimitation Between Process Monitoring and Machine Diagnosis

The general aim of monitoring is to assess the behavior of the object to be monitored.



Process monitoring therefore focuses on the machining process. This can relate to both geometrical and technological aspects. A particularly good application example here is tool monitoring where damage to the tool is identified to prevent subsequent damage in the downstream process flow.

The aim of machine monitoring is to protect the machine and its individual components. This assures the manufacturing quality produced by the machine as well as machine availability.

Tools and strategies in both application fields have a fundamental similarity in many areas. This means that similar sensor principles can be used for both areas. The approaches and problems involved in signal processing and the monitoring strategies are also similar. For this reason, the basic principles are dealt with in the following chapters (► Sect. 6.2).

However, there are quantitative differences in the specific requirements for sensor systems, electronics, and in the algorithmic domain.

For example, the sensor location for measuring significant variables is a decisive factor in sensor systems. The objective is always to place the sensor systems close to the object to be monitored, i.e., the process or the machine component, as the components that lie between the target object and the sensor represent a transmission path.

In principle, it is possible to identify the characteristics of the transmission path with a specific accuracy and to carry out appropriate corrections to the measured signal. In this way, the process monitoring can infer information about the process variable of interest from the sensor signal using the accuracy of this signal correction. However, this transmission path represents a physical limit on the quality of the signal to the point that not every event can be monitored. Critical applications are, for example, fracture monitoring on small drill bits as the process signal is too small in relation to the signal damping over the transmission path.

In machine monitoring, the behavior of the machine structure is the source of the signal. In process monitoring, this transmission path between the process signal and the sensor represents a disruption. If the characteristics of the machine structure change due to, for example, wear or damaged components, these represent an event to be detected as far as machine monitoring is concerned. In process monitoring, however, they lead to an error in the previously fixed signal correction.

However, the recognizable changes in machine monitoring are often so small that it is difficult to achieve reliable monitoring of critical occurrences.

The excitation of the object to be monitored must now be considered in more detail. In process monitoring, chip formation or deformation in the workpiece essentially causes excitation in the tool/machine system. The signals from this excitation can be captured by the sensor system. In contrast, in machine monitoring, these process-oriented signals are treated as disturbances due to their signal dominance and stochastic properties therefore, the machine structure outside the process has to be stimulated for process identification. A significant part of the signal quality is based in the appropriate excitation of the components or structures being monitored.

Finally, distinctions are to be made between sensor systems depending on the actual installation. Demands on the signal-to-noise ratio, sampling rate, etc., that characterize the sensor system differ between specific applications. They are highly dependent on the chosen monitoring strategy. Examples of this are provided in ► Sect. 6.3.

This is summarized in ◘ Fig. 6.7.

#### 6.1.4.4 Mechanical and Optical Sensors

Probes are generally used as mechanical sensors in process and tool monitoring. These probes contact an object and, when combined with the distance measuring system, allow the tool geometry to be determined. Highly accurate probes are also used to check workpiece geometry in the machine workspace. The measuring probe replaces a tool allowing, for example, the diameter of a hole to be probed. Communication from the probe to the machine control system can, for example, use an infrared link.

Photoelectric sensor systems are used as optical sensors in a similar way to switching measuring probes. However, the switch point is less accurate than with corresponding mechanical probes. A one or two dimensional image can be formed with the aid of photodiode arrays distributed along a line or over a surface. Several application examples of the use of mechanical and optical sensors in tool monitoring are given in ► Sect. 6.3.3.1.

#### 6.1.4.5 Control-Internal Information

The torque or current values in three-phase AC feed motors and mainly on spindle drives are available

**Fig. 6.7** Similarities and delimitation between process and machine monitoring

| Distinguishing criterion | Process monitoring | Machine monitoring |
|---|---|---|
| Sensor principle | Similar principles can be used in a wide range of areas | |
| Sensor parameters | Different sampling rate, signal-to-noise ratio, etc. | |
| Monitoring strategy | Strategies may be the same but they are usually application-specific | |
| Transfer behavior of the machine structure | Disruptive influence on the signal quality during monitoring, may be compensated | Object to be monitored |
| System trigger | Handled automatically by the process | Specific system trigger can be used to generate appropriate signals |
| Machining process | Object to be monitored | Disrupts the monitoring |

**Fig. 6.8** Internal control information

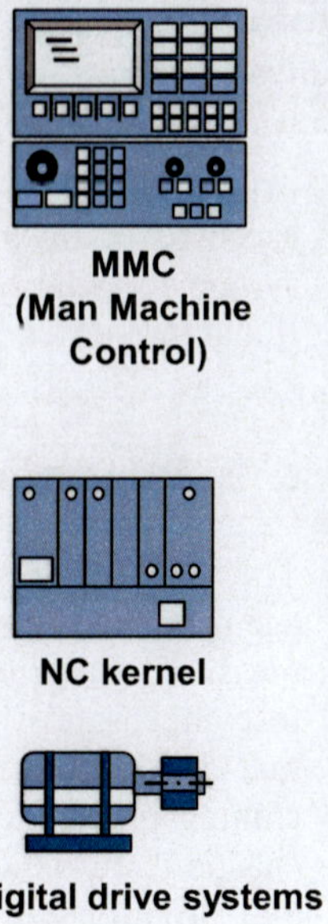


**Machining process signals**

- Motion signals, e.g., position, rotational speed, linear speed
- Performance indicators, e.g., active power, drive utilization
- Other controller variables, e.g., controller difference, lag distance

**Signals from the NC program procedure**

- Auxiliary functions and M-codes
- Information on the current tool
- The current NC operating mode (manual, semi-automatic, or automatic operation)
- Events during program procedure or during operation e.g., NC start, stop, reset
- Reference values for position, speed and acceleration

internally to the control system as they are managed by an underlying torque control loop (▶ Sect. 6.3). If there is access to the internal values, for example, through the use of open controls (Sect. 6.2.6, volume 4), the sensors required to measure these values are unnecessary. This is particularly beneficial if all the axes in a system are to be monitored as there is no longer a cost for multiple sensor systems. However, there is a high dependency on the openness of the control system and the signal quality produced by the control manufacturer at the open interface. Standard diagnostic and programming interfaces available are generally limited to the necessary functions.

Other information on the machining process and signals from the NC programming procedure is available in the control system in addition to these drive signals (◘ Fig. 6.8).

All information on synchronizing monitoring with the machining procedure, such as trigger signals for starting and ending the machining and tool and cutter information, are available in the control system without having to change the NC or PLC programs. The reference and actual values for the position, velocity and acceleration are also available and these can be used to offset the effect of the mechanical transmission path (▶ Sect. 6.2).

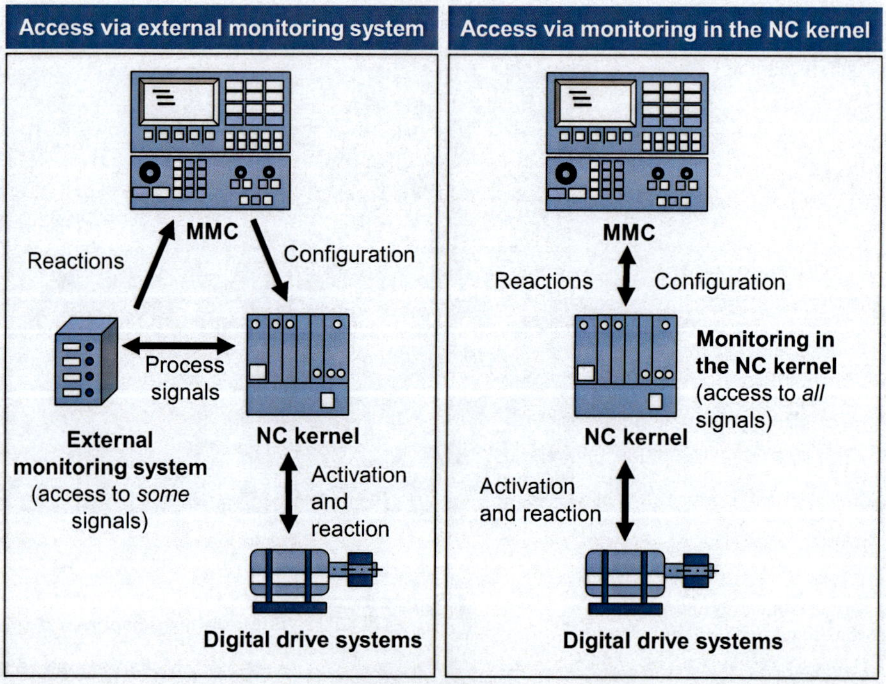


**Fig. 6.9** Access options to internal control functions

6

Two systems are briefly described below. They are based on the Siemens Sinumerik 840D control system. User-specific extensions have been integrated into this control in the real-time portion of the control (NC kernel, NCK) and the operator area of the control.

A system has been developed through a collaboration between Artis and Siemens that allows the current and torque values to be externally evaluated [ETEL95], Fig. 6.9, left. This system makes the current and torque values available to the external monitoring system via the PLC or on the PLC bus. The resolution of the signals is defined by the NC accuracy. The bus cycle (approx. 12 ms) and therefore the signals' frequency comply with the possible additional PLC load.

A monitoring system which is completely integrated into the control has been developed at the Laboratory for Machine Tools and Production Engineering (WZL) at the RWTH Aachen University in collaboration with Siemens and the machine tool manufacturer Chiron. In this system, all functions of the monitoring system are integrated into the machine tool control system, see Fig. 6.9, right. The sensor system comprises only control-internal signals. The signal processing, i.e., compensation of acceleration and friction effects, and the monitoring algorithms are executed in the control's NC kernel. Sensor signals and algorithms are computed during the position controller cycle of the control system (approx. 2–4 ms). The control's own user interface was extended to configure and parameterize the monitoring system [ECKH93]. However, the NC is still loaded by the control's real-time calculations. Complex, comprehensive calculations, therefore, have to be executed in the non-real-time portion of the control system to avoid having a negative impact on the functional properties of the NC.

### 6.1.5 Principles of Process Control

The aim of process control is to optimize process management. In Statistical Process Control, workpieces are measured at periodic intervals and the results are used to compensate systematic errors due to wear and thermal drift in the machine (see ▶ Sect. 6.4). Most machining processes are subject to the influence of stochastic disturbance factors such as material deviations and variations in tool quality. However, it is not possible to predict how a process will run, if tools or the machine will be overloaded, or whether tools will wear as anticipated. To not overload the machine and the tools, machine settings are selected with a safety margin so that, even in the worst case, the process will continue to run properly. However, basing processes on safety with respect to overloading machines and tools normally means that the performance of a machine (minimum machining or idle times) and tool (maximum cutting capacity) is not fully utilized. As a result, machining does not reach the desired economical optimal point.

This can be improved by running a control loop within the machine (see ▶ Sect. 6.3). A suitable control loop is used in machining centers to offset the relative displacement between the spindle and the workpiece caused by temperature, based on the bearing temperature of the milling spindle and the isotherm

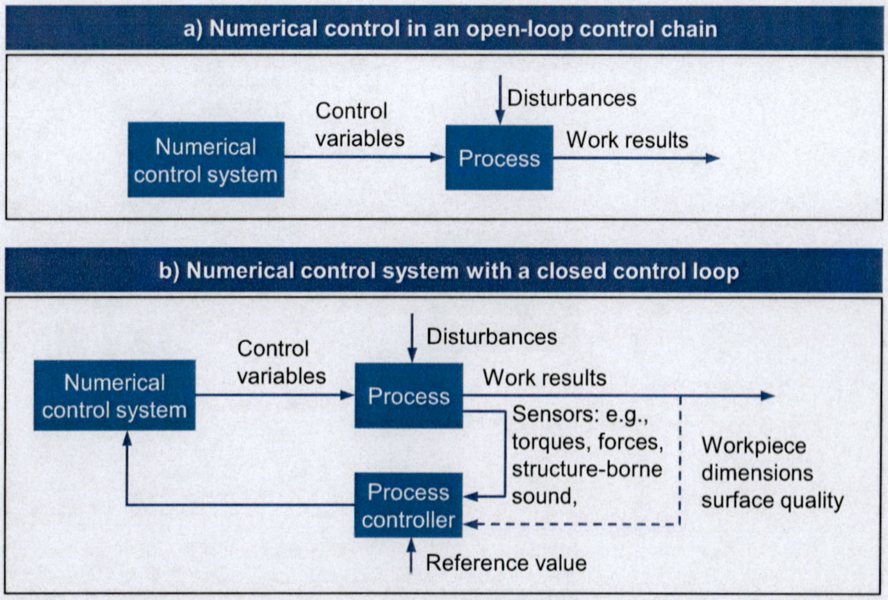


**Fig. 6.10** Comparison: conventional open-loop control and closed-loop control

field. In this respect, of equal interest are all the controls which have been developed in the field of Adaptive Control systems (AC). By measuring disturbances and feeding back the process variables, i.e., by closing the open-loop control chain as shown in Fig. 6.10a with a process controller to a closed control loop as shown in Fig. 6.10b, it is possible to improve the machining quality and safety. Relevant parameters for the process sequence are also measured and compared with given reference values. The required control variables to maintain the controlled variables are then calculated in accordance with the control deviation.

In the 1970s, many systems were developed in the field of Adaptive Control [SCHU93, KLÖC95; RUMM92] that pursued the various aims of process management and optimization. A distinction is drawn here between technological and geometrical AC systems. In the former, the technological process parameters are adjusted in such a way that, for example, the machine performance is fully used or the degree of tool wear remains within given tolerances. In geometrical AC systems, the controlled variables are the workpiece dimensions or machine deformations.

Another decisive feature is the method of specifying the controlled variable. This is known as Adaptive Control Constraint (ACC) if the reference values are based on fixed limits such as performance or stability thresholds. Adaptive Control Optimization (ACO) is understood as process controllers used to minimize manufacturing times or manufacturing costs.

Figure 6.11 shows the aims of process control with the appropriate parameters and required sensors.

## 6.2 Signal Processing and Pattern Recognition

The process or machine state variables to be measured such as tool wear, bearing wear, pitting on tooth flanks or imminent tooth breakage in gear drives, cannot be measured directly during operation or during the process. In most cases, indirect or process-related measuring methods are used. Measurable impacts on the changing state variable are captured in this instance, such as the change in force, acceleration or vibration signals, or variables such as current or structure-borne sound.

The individual process steps of these physical variables through a pattern recognition system for monitoring and diagnostics are shown in Fig. 6.12 [NIEM74, NIEM83] and are discussed in detail below.

Once the signal is captured by the sensors, (see ▸ Chap. 9, Volume 2), signal pre-processing commences. The aim of signal pre-processing is to extract specific features from the signal sequences in order to be able to make a secure assessment of the process or machine condition at that moment in time.

As there is usually a transmission path between the signal source and signal capture and as the captured signals are always the sum of various signal sources, both analog and digital signal pre-processing is necessary. While analog signal pre-processing essentially processes the signals in such a way that no significant errors occur during their digitization, digital signal processing is used to carry out signal correction in order to extract the relevant quantities for monitoring from the

6

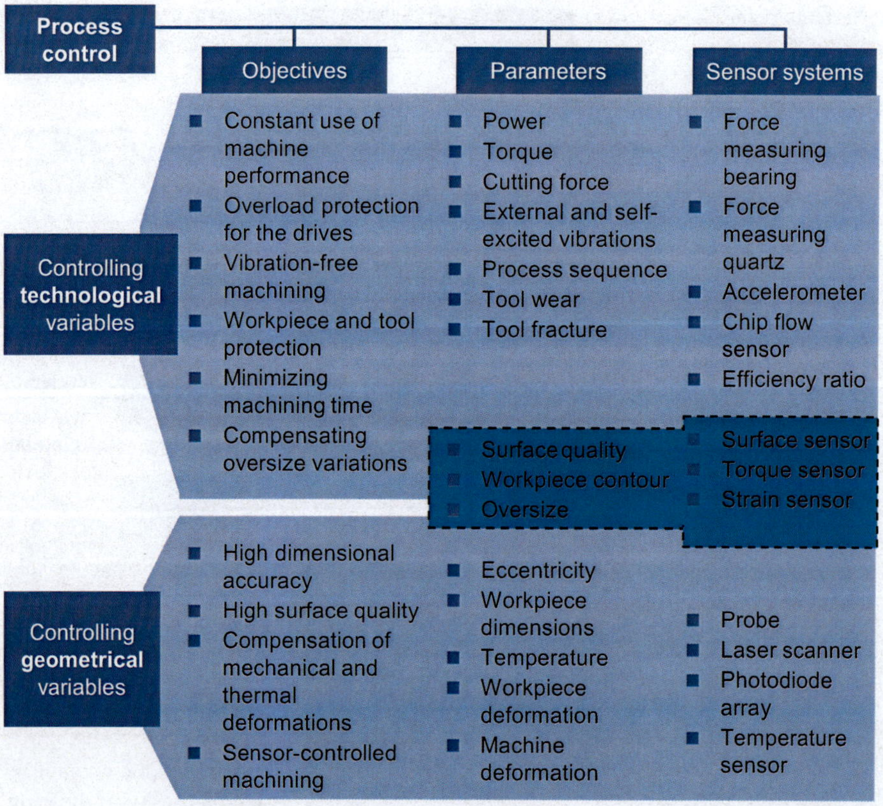


**Fig. 6.11** Additional process control options

captured signals. This usually takes place using suitable models, which, for example, simulate the drive train as a transmission path such that motor current signals are transferred into suitable feed forces. The models used here can have varying degrees of complexity depending on the specific tasks (see ▶ Sect. 6.2.2).

Feature extraction determines specific variables from the prepared signals that describe the event to be examined. Finding these features that describe the state follows detailed investigation of the degree of validity of these features with respect to characterizing the errors to be detected. For this reason, the determination of appropriate features and combining them into meaningful patterns for the various monitoring tasks has been the subject of many research projects.

If features having a significant correlation with the state to be monitored have been found, appropriate classification methods can be determined and prescribed for the automatic classification phase. The decision-making process classifies the features occurring in such a way that they model the required monitoring result.

There is a fundamental distinction between the learning or parameterization phase and the automatic classification phase. The learning phase is necessary to determine appropriate limiting values and other parameters for the specific task. In this phase, the pattern recognition system is set to provide reliable information about the machine or process state variables during operation (classification phase). In the system's learning phase, the state sample vectors (feature vectors) are determined from random feature samples and correlated with the process or plant conditions. In principle, all significant process and plant signals such as force, torque and vibrations can be used as input variables.

The parameterization of the classifier often requires a protracted, iterative process. Due to the difficulty in encompassing the variety of parameters, current practice is generally to use individual features and simple classification procedures, for example, comparing one feature with one or more limiting values.

The learning phase is complete once the feature has been selected and the classifier parameterized.

The selected feature vectors are allocated to classes using the processes designed for this purpose during the classification phase, i.e., they are classified. Possible classes can be, for example, "sharp tool " or "worn tool".

### 6.2.1 Analog Signal Processing

Disruptions are eliminated by analog signal processing and signal components which carry information

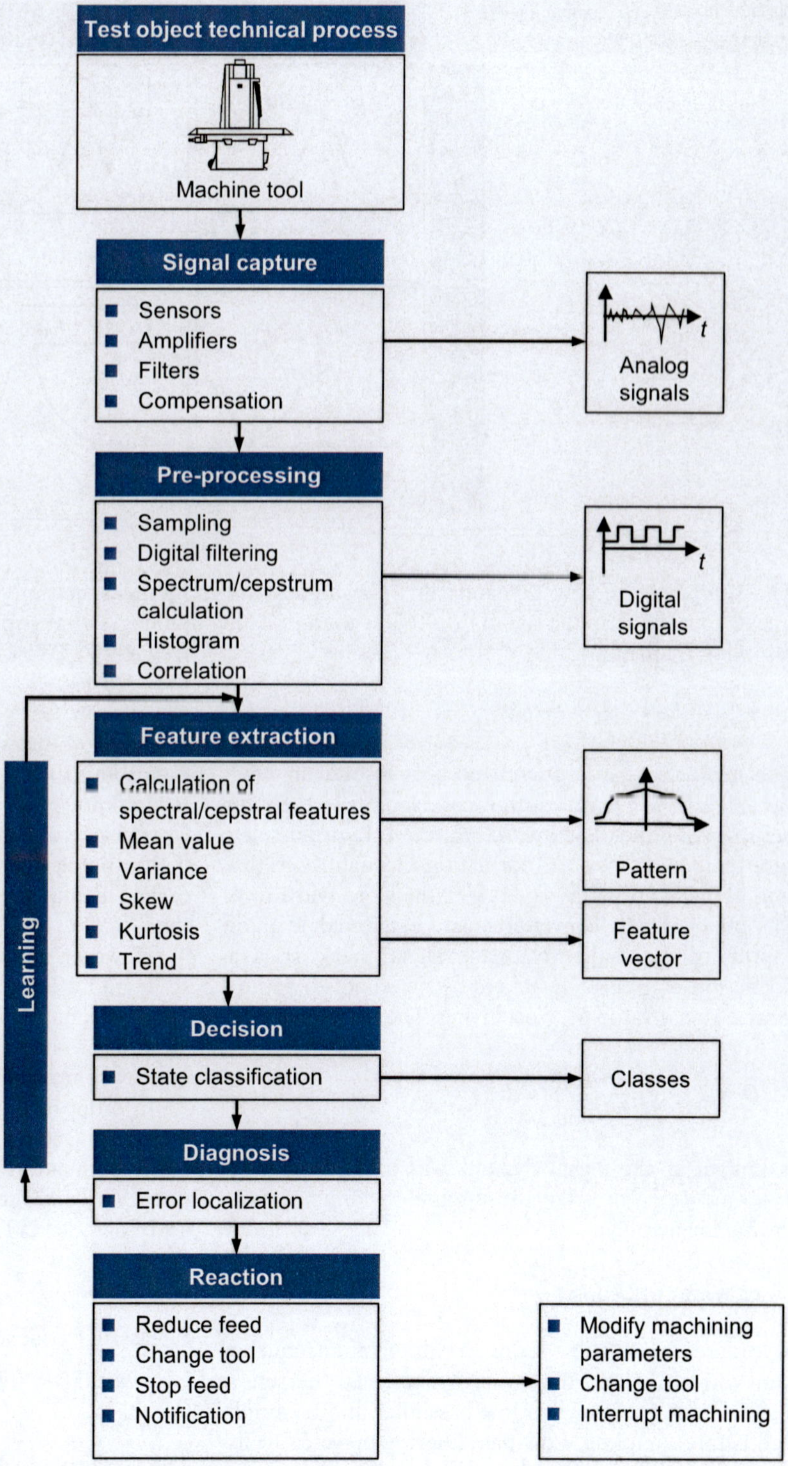


**Fig. 6.12** Monitoring and diagnosis using pattern recognition terms

are processed such that they can be digitized as cleanly as possible and with maximum resolution. The sensor signal continuous over time and amplitude $u(t)$ is converted into a series of discrete values $u_i$ through the analog/digital conversion process (A/D conversion) while the signal is sampled at regular time intervals. In order to ensure no information is lost during digitization, Shannon's sampling theorem must be met.

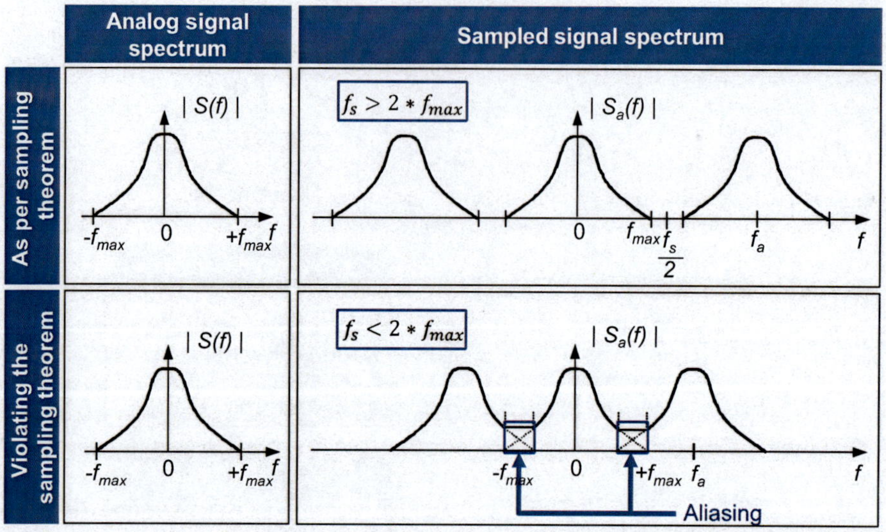


**Fig. 6.13** Effect of sampling on the signal spectrum

The sampling theorem demands that the highest frequency occurring in the signal $f_{max}$ be less than half the sampling frequency $f_s$:

$$f_{\max} < \frac{1}{2} \cdot f_a \tag{6.1}$$

The analog signal is discretized at a resolution of $2^w$ value levels. The word width $w$ corresponds to the number of bits in the digital word produced. In order to ensure there is no loss of information through discretization of the continuous signal $u(t)$, the word width $w$ of the analog/digital converter must be selected to allow capture of the smallest change in signal that carries information $U_{min}$ (usually limited by signal noise) and the largest signal value $U_{max}$ occurring. The relationship

$$D = 20 \cdot lg \left[ \frac{U_{\max}}{\Delta U_{\min}} \right] \quad [dB] \tag{6.2}$$

is known as the signal dynamic. In order to capture $U_{min}$ and $U_{max}$, thus the entire signal dynamic, the following must apply:

$$w \geq \log_2 \left( 10^{\frac{D}{20}} \right) \approx \frac{D}{6dB} \tag{6.3}$$

In order to meet these requirements on the frequency bandwidth $f_{max}$ and the signal dynamic $D$, the sensor signals must go through a low-pass filter and be amplified before sampling. Low-pass filtering prevents transgression of the sampling theorem. Amplification is necessary to make full use of the input voltage range of the A/D converter.

### Low-Pass Filtering

Sampling the signal changes the frequency spectrum of the signal. This produces new frequency components in the spectrum due to a periodic extension of the analog signal's spectrum with a period of $f_s$(sampling frequency) [WECK99]. The relationship is shown in Fig. 6.13.

As can be seen from Fig. 6.13, sampling at $f_s < 2f_{max}$ produces an overlap in the mirrored frequency bands and therefore a distortion of the signal. This effect is known as aliasing. The loss of information due to transgressing the sampling theorem can be avoided if the analog signal is put through a low-pass filter producing a maximum residual signal frequency of less than $f_s/2$.

However, actual analog low-pass filters have a finite slope and a limited damping $\delta_s$ in the stopband (stopband attenuation), Fig. 6.14b. The stopband attenuation $\delta_s$ produces the damping level in dB for the frequency components in the stopband of the filter.

In order to keep signal distortion due to aliasing in the useful frequency range as low as possible, the stopband attenuation $\delta_s$, signal dynamics $D$, sampling frequency $f_s$, and limiting frequency $f_{lim}$ must be selected as follows, see Fig. 6.14:

$$f_s > f_{\lim} + f_u \tag{6.4}$$

$f_u$ – useful frequency range limit,

$f_{lim}$ – limiting frequency of low-pass filter,

$$\delta_s > D \tag{6.5}$$

The condition in Eq. (6.4) prevents signal distortion in the useful frequency range due to the finite slope of the filter, see Fig. 6.14c.

The condition in Eq. (6.5) ensures that the aliasing effect of the finite stopband attenuation of the filter is kept sufficiently low such that it remains within the order of magnitude of the signal noise.

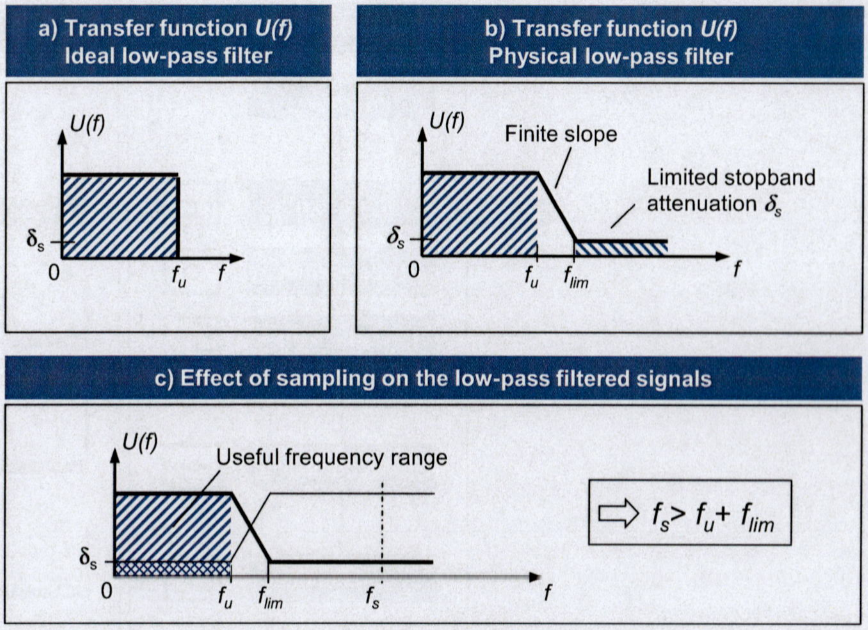


**Fig. 6.14** Limiting the aliasing effect with low-pass filtering

### Amplification

The output signals from the sensor elements, particularly piezoelectric sensors, are at a very low power and are usually sampled at a very high electrical impedance. These sensor signals are therefore prone to interference. Amplifying the signal reduces its interference susceptibility and allows optimum adjustment to the voltage range of the A/D converter and therefore good resolution of the signal.

### Analog Signal Processing Chain

Analog signal processing requires an amplifier and an anti-aliasing low-pass filter as a minimum. For force, strain, and current signals, signal amplification followed by a low-pass filter is sufficient.

In this instance, the useful bandwidth of the signal depends on the measured variable and the monitoring task.

For wear detection, typically, force and strain signals up to 50 Hz have to be evaluated. For fracture detection, on the other hand, evaluations of up to several kilohertz (kHz) are required because the signal peaks typical of fracture cannot otherwise be captured.

### Analog Signal Processing for Structure-Borne Sound Signals

A more expensive signal conditioning system is required for analog processing of structure-borne sound signals as the relevant information is contained in higher frequency signal components. The signal is initially filtered through a high-pass filter, see Fig. 6.15 [KÖNI92]. This allows low-frequency interference such as network hum, direct voltage drift or machine vibrations, for example, due to imbalance of rotating shafts, to be suppressed. This interference could otherwise lead to over-control of downstream components (e.g., amplifiers or A/D converters). The signal is then amplified for the reasons given above.

Band-pass filtering is used to select informative frequency bands for the specific application. Depending on the application, they lie in a range from 1 kHz to over 1 MHz [MEYE91, SCHE91]. For direct digitization, these high-frequency signals would require complicated A/D conversion (sampling frequencies up to 2 MHz). Furthermore, direct sampling would produce a data quantity which would be extremely difficult to process (up to $2 \cdot 10^6$ values/s).

The signal is thus converted via rectification and then low-pass filtering into a low-frequency envelope signal (50 Hz to 4 kHz). This type of signal processing corresponds to demodulation of the band-pass filtered signal as used in communications technology. The frequency band is transformed by rectification into a frequency range starting at 0 Hz. The curve of envelope signals produced in this way includes the relevant monitoring information and is digitized for further processing.

Structure-borne sound signals processed in this way are used in both plant and process monitoring. It is therefore possible to use meaningful, high-frequency signal components in spite of the uncomplicated A/D conversion and digital signal processing. By transforming the relevant frequency band to the frequency range starting from 0 Hz, the required sampling rate and volume of data produced by the sampling process is significantly reduced.

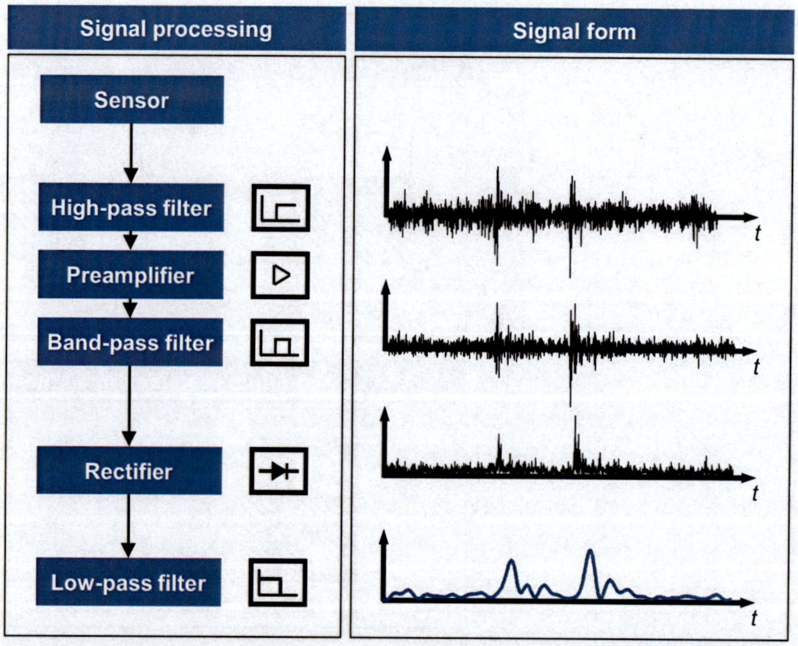


**Fig. 6.15** Analog signal pre-processing for structure-borne sound signals according to König

### 6.2.2 Digital Pre-Processing

Digital signal pre-processing prepares the captured signals in such a way that the signal features which describe the object under observation can be obtained using feature extraction.

The first step is to correct the signal in order to compensate the disruptive influences on the signal from its source to the sensor as well as potential influences from the sensor technology.

Typical examples of this signal correction are:

- digital filtering to suppress known parasitic frequencies [EX-C93],
- averaging to reduce stochastic influences,
- spectral analysis, cepstral analysis [STEA79]
- determining the frequency distribution (histogram) [NIEM83]

In addition to this, models that simulate the behavior of the transmission path can be used for signal correction. A typical application of this is simulating the drive mechanism between the point of process action and the motors.

Models based on spring damper masses are usually used to simulate these mechanical properties. The electrical components (motor, power converter and controller) are simulated using known control-theory approaches. However, such a high level of detail is very expensive due to the modeling and the subsequent parameterization of the model. This level of detail can be useful in machine monitoring as the behavior of individual mechanical components in the drive train is of particular interest here and the stochastic influences are less than in process monitoring.

For process monitoring, it can make sense to use more straightforward models based on transfer elements with less and simple control technology (e.g., $PT_1$). The disadvantage is the lower accuracy of the model, particularly at higher frequencies. However, this type of model can be set up and parameterized more quickly.

Other control-theory approaches to signal correction, such as the monitoring principle, are currently under investigation with respect to their suitability [FRIT99].

### 6.2.3 Feature Extraction

Features (parameters) which describe the process or machine state to be monitored are extracted from pre-processed signals.

The following methods of evaluating signals are typically considered when extracting features from process and machine signals [KÜHN85]:

- absolute and weighted values,
- trends, derivations of the mean value curves,
- statistical parameters (mean value, distribution, frequency distribution, variance, slope, kurtosis, etc.) [DIN71],
- spectra parameters,
- cepstra parameters,
- deviations from reference values (residuals).

**Fig. 6.16** Forming the moving average. *Source* König

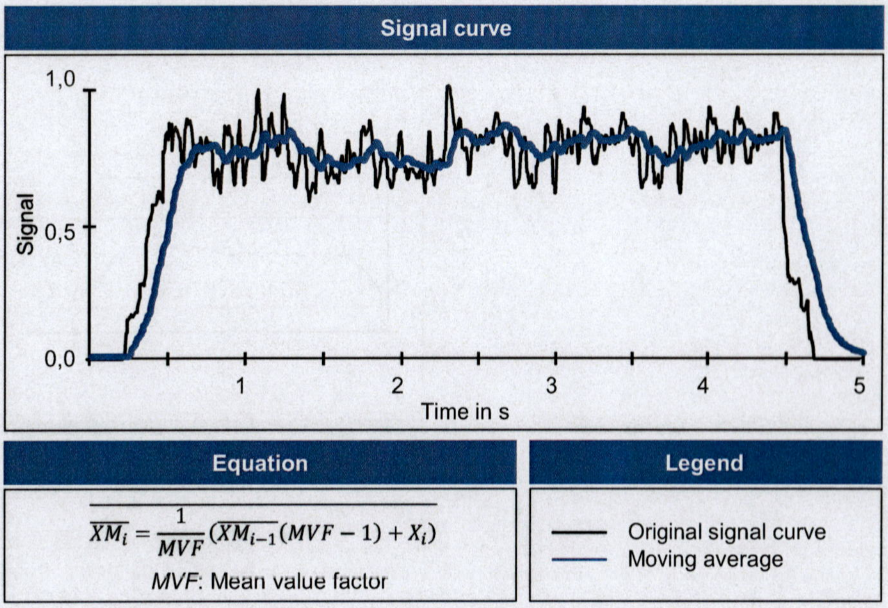


One feature often used in process monitoring is described in more detail below.

### Moving Average

The moving average is used, for example, to identify stages in the dressing process when grinding or for eliminating stochastic signal components in structure-borne sound signals when monitoring bearings. A typical example of a moving average for a measurement signal is shown in Fig. 6.16 [DIN88] and corresponds to a low-pass filter with a $PT_1$ element. It is calculated as follows:

$$\overline{XM_i} = \frac{1}{MVF}(\overline{XM}_{i-1}(MVF - 1) + X_i) \tag{6.6}$$

where

$\overline{XM}_i$ – new moving average,

*MVF* – mean value factor,

$X_i$ – current average,

$\overline{XM}_{i-1}$ – previous moving average.

This produces the adjustable mean value factor *MVF* used to calculate the new average $XM_i$ by weighting the relevant measured value. A high mean value factor allows the moving average to follow the original signal correspondingly slowly. In other words, the greater the mean value factor, the greater the low-pass effect in the formation of the mean value.

Other important features are static parameters such as variance or slope [WALC74] as well as spectral features such as partial outputs from the frequency spectrum and parameters produced by cepstral analysis.

## 6.2.4 Classification

In the next processing stage, the features extracted from the measured values have to be evaluated to produce information on the state of the process or machine. The reactions resulting from the state detection differ depending on the application. Obvious examples are notifications of imminent damage to the machine, which is to be corrected in good time where possible, or reaching tool wear limits, which results in changing a tool.

Constant changes, such as flank wear, which cause a dimensional change on the workpiece, can be offset through intermittent or constantly acting process control. The pre-processed sensor signals are processed further in application-specific controllers. Examples are provided in ▶ Sect. 6.3 describing the functionality of process controllers. In process and plant monitoring, the calculated parameters are allocated to classes using the classification process described below (e.g., "sharp tool" or "worn tool"). Effective classification is extremely reliant on the intuition and experience of the relevant experts.

### 6.2.4.1 Fixed Limits

The most straightforward option for classification is comparing a feature with a fixed limiting value. Being above or below this limiting value highlights a process disruption or a defective machine element depending on the monitoring task.

An example of the principle is shown in Fig. 6.17a for tool monitoring. Four limiting values have been established for "missing tool", "wear", "tool fracture" and "collision" [BRAA97]. Exceeding

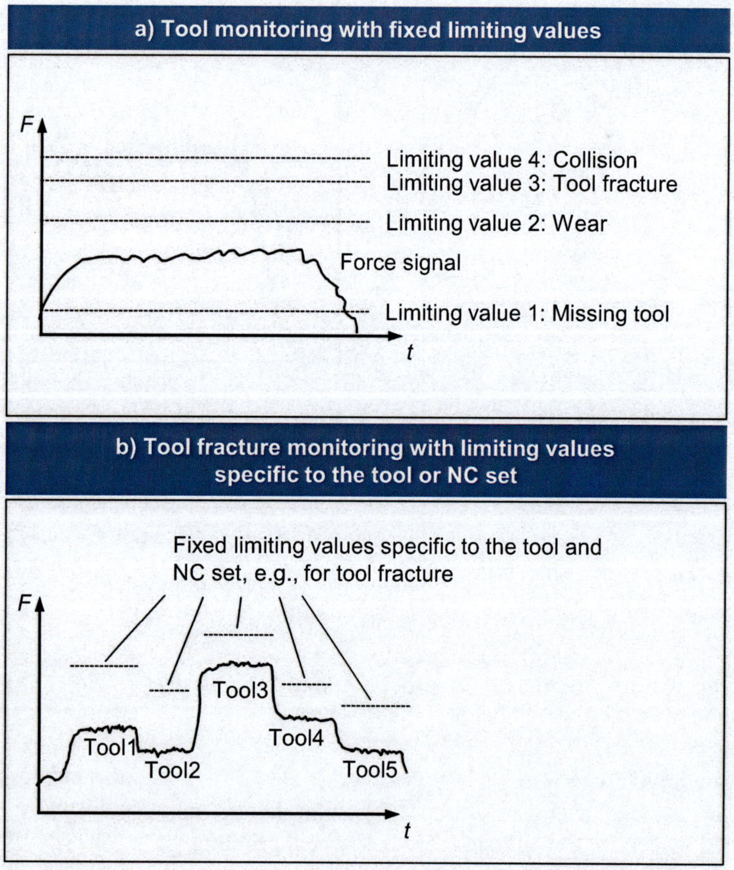


**Fig. 6.17** Tool monitoring with fixed limits



the collision limiting value leads to the feed drive being switched off immediately. In less critical cases (e.g., exceeding the wear limit), exceeding the limit simply results in a notification to the user or an automatic tool change. The limiting values are generally set automatically by the system as a measured value relative to the tool's new condition but they have to be adjusted subsequently by the user.

In process monitoring, the limiting value frequently has to be adapted to changing process conditions. In these instances, there is an option of defining, for example, limiting values specific to an NC set or a tool. Figure 6.17b shows a typical example of a force curve for processing a workpiece using five NC sets with five different tools.

Fixed limiting values are also used for machine and plant monitoring. For example, rolling bearing damage can be identified when a fixed limiting value for the overall structure-borne sound level is exceeded.

#### 6.2.4.2 Moving Thresholds

The moving thresholds are used to identify sudden signal changes caused by damage such as tool fracture or collision. Moving thresholds are calculated during operation and are based on the current moving average (see ▶ Sect. 6.2.3). A typical example of the principle is shown in Fig. 6.18 for monitoring tool fracture.

A specific value is added to the moving average in order to calculate the upper threshold. The lower threshold is calculated by subtracting the same value from the moving average. As both thresholds are based on the moving average, which follows the signal with a delay, this leads to values falling above or below the threshold if the original signal changes quickly. The width of the tolerance range is usually set proportionally to the absolute value of the quantity to be monitored. The advantage of tracking the thresholds dynamically as opposed to the fixed limits described above is being able to adjust the tolerance range to the signal curve as process parameters change, e.g., oversize variations, batch adjustments or tool wear during tool fracture monitoring.

#### 6.2.4.3 Multi-Dimensional Classification

As mentioned above, evaluating a single feature is often not enough to complete the monitoring task. In this instance, either several different physical variables or different parameters, which are extracted from a signal curve, are monitored; decision making is based on their variations and relationships to each other.

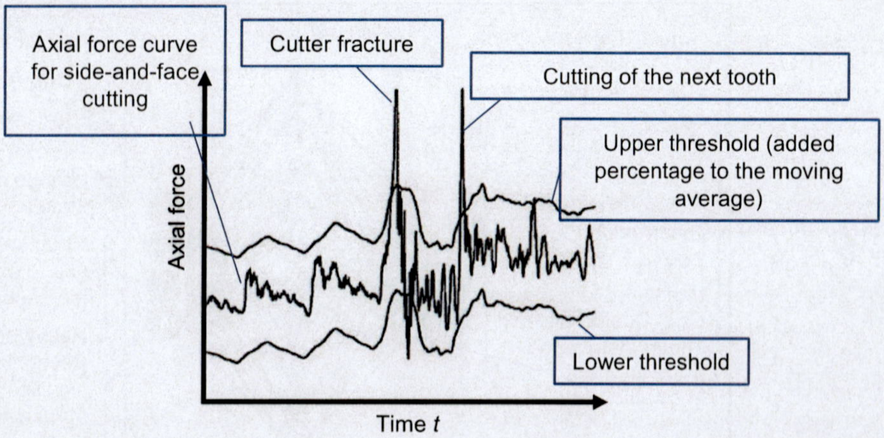


**Fig. 6.18** Tool fracture monitoring with moving threshold

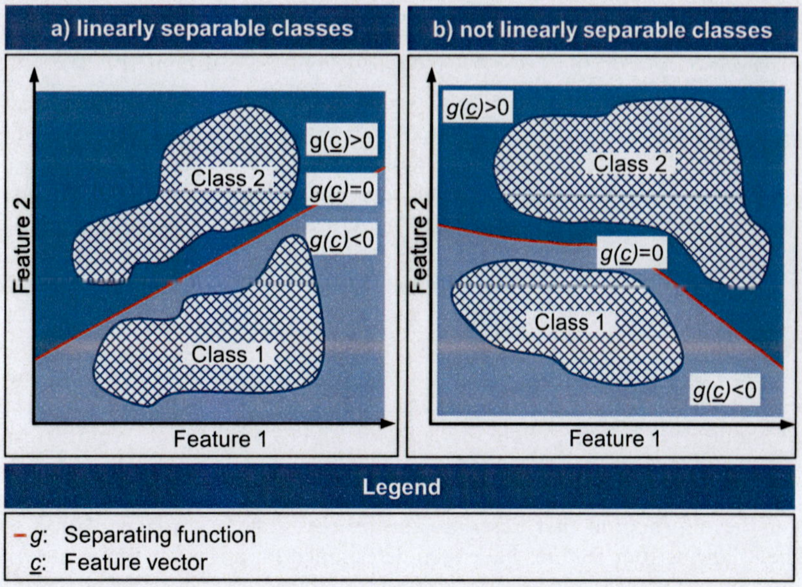


**Fig. 6.19** Classification rule for a two-class problem

These features form the components of the so-called n-dimensional feature vector c (n = number of features). This feature vector could, for example, contain the following components:

$$\underline{c} = \begin{bmatrix} L_{0-1000} \\ L_{\text{total}} \\ \mu \\ \sigma \end{bmatrix} \quad \begin{array}{ll} L_{0-1000} & \text{overall output level of the structure-borne sound signal in frequency range of 0 to 1000 Hz,} \\ L_{\text{total}} & \text{overall output level of the total structure-borne sound signal recorded,} \\ \mu & \text{mean value of the feed force signal,} \\ \sigma & \text{standard deviation of the frequency distribution of the structure-borne sound signal.} \end{array} \tag{6.7}$$

If the features have been properly selected, the states to be classified form separate point sets in the relevant state space. This is illustrated in Fig. 6.19 using an example in two-dimensional space. The features $M_{m1}$ and $M_{m2}$ are entered on the axes of the coordinate system. The so-called separating functions $g(\underline{c})$ must be used as the decomposition provision that allows the various process states to be clearly allocated to a defined class (= point set). The algorithms for determining the separating functions are usually iterative processes [LUDA91]. The example shown in Fig. 6.19a illustrates this for a two-dimensional case with two classes which can be separated on a linear basis [VDI73]. The separating function is a straight line. For example, for the tool monitoring problem in question, the classes are: Class 1: "Tool OK" and Class 2: "End of service life".

A random sample with feature vectors of typical process states is determined in the learning phase. This random sample is used to calculate the separating function. New suitable features, as well as the parameters of

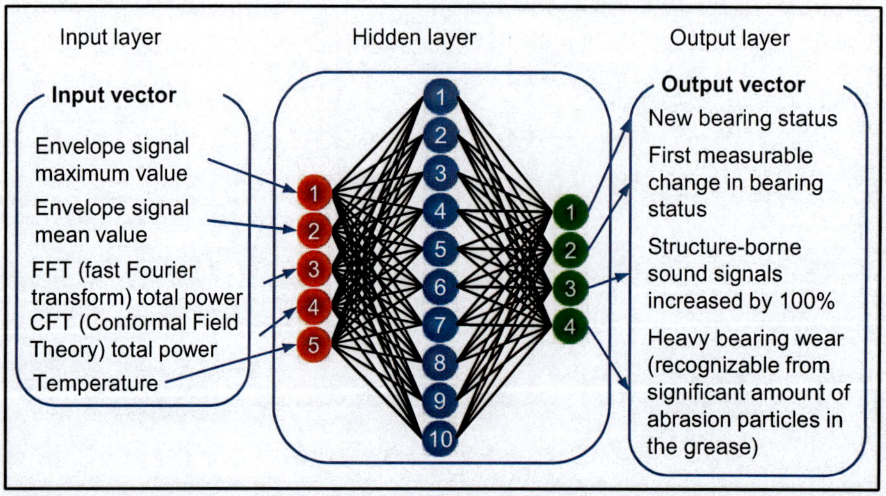


**Fig. 6.20** Structure of a neural network using bearing monitoring as an example

the separating function, must be determined for each monitoring task.

The end of service life of drilling tools on a machining center (drill bit diameters: 4 mm and 6 mm) has been determined using this process of pattern recognition.

The feed force signal, the structure-borne sound signal of an accelerometer mounted on the workpiece clamp and two signals, one signal proportional to the spindle speed and the other signal proportional to the feed rate in the z-direction, are used as input signals for the monitoring system. While the last two signals are necessary for automatic detection of a new machining cycle, the feed force signal and the vibration signal contain the information that describes the process state. The following four features have been determined for a four-dimensional sample vector as a result of the analysis and learning phase (see ► Sect. 6.3.3):

- overall output level of the structure-borne sound signal,
- standard deviation of the structure-borne sound histogram,
- overall output level of the feed force signal,
- standard deviation of the feed force histogram.

The number of features and the error quota in the classification are optimized here such that a worn tool can be detected before fracture with a 99 percent probability.

The linear separating function was used as the classifier [LUDA91]. The linear separating function is a hyperplane in the feature space separating the two different classes from each other. This means that Class 1 features lie on one side of the hyperplane in the four-dimensional feature space while Class 2 features lie on the other.

A further classification process also able to separate classes that cannot be separated on a non-linear basis as shown in Fig. 6.19b is neural networks [BRAA00]. Artificial neural networks are systems built from simple processing units, known as neurons. These units are connected to a network using weighted connecting lines, see Fig. 6.20. Neural networks can either be built with hardware using special processors or simulated with software. Neural networks are able to allocate an output vector that, for example, determines the classification of the feature vector to a given input vector (feature vector). The parameterization of this neural network classifier is a result of learning from examples. The network learns automatically in accordance with a given set of learning rules. There are many different types of networks which differ in terms of their network structure (topology), neural structure (Fig. 6.21) and the learning rules used. ► Section 6.5.3.4 includes an example of neural networks in practice.

## 6.3 Technological Process Monitoring and Process Control in Different Manufacturing Processes

### 6.3.1 Turning

#### 6.3.1.1 Sensor Systems for Measuring Torque and Cutting Force

Torques and cutting forces are used to indirectly measure the process state (e.g., chatter) and the tool condition (e.g., sharpness), particularly in roughing. When finishing, current sensors are generally insufficiently sensitive for measuring wear.

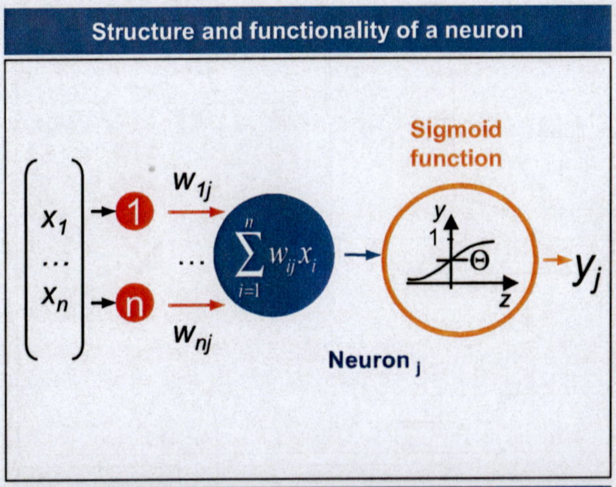


**Legend**

| | | |
|---|---|---|
| $x_i$ | : | Input values for the neuron |
| $w_{ij}$ | : | Weighting factors for the input values $x_i$ |
| $y_j$ | : | Output values for the neuron |
| $\theta$ | : | Transfer function parameters |
| $y_{ij}$; $z_{ij}$ | : | Auxiliary variables |

**Fig. 6.21** Structure and functionality of a neuron

Torque is rarely measured directly on lathes. The disadvantages here are the signal transmission away from the rotating spindle that is generally required and the inertia of the workpiece mass which damps the signal.

On lathes, it is often more appropriate to measure the cutting force directly in the area of the bit holder than to measure the torque. This is characterized by low disturbance sensitivity and particularly by low delay times. Figure 6.22 shows one possible structure of a cutting force sensor [MILL90]. The square turret supporting the tools is in turn supported on four measuring rings that are under high preload (approx. 80 $kN$). The individual measuring rings each have eight active and two passive strain gauges attached. The elastic deformation of the deformation element, which is surrounded by protective rings, is a measure of the actual load that is recorded as a change in resistance in the active strain gauge. The two passive strain gauges not directly in the flux of forces are used to compensate for the effect of temperature. The design ensures that the flux of forces can only flow through the measuring rings. The cutting force $F_c$ is determined by totaling all the changes in load measured at the points of

**Fig. 6.22** Configuration and structure of the cutting force measuring rings

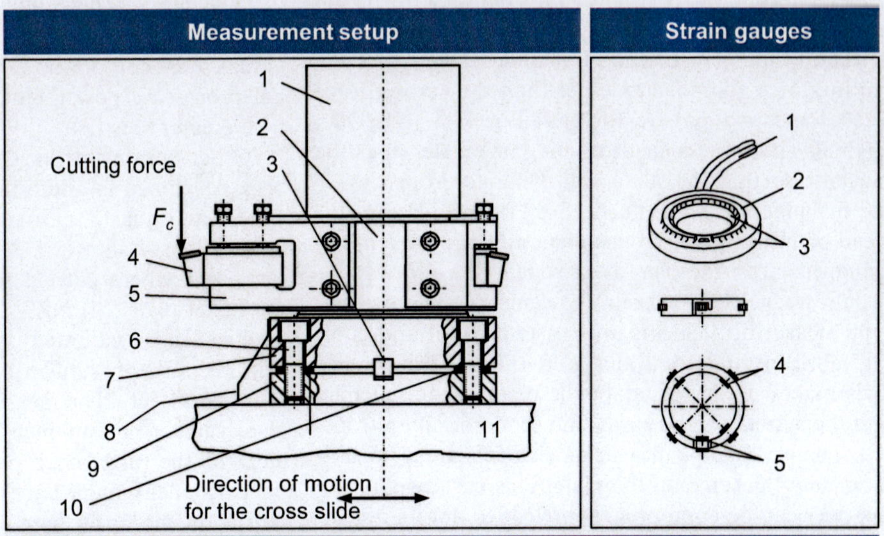


**Legend**

| Measurement setup | | Strain gauges | |
|---|---|---|---|
| 1: | *Hydraulic control* | 1: | *Connecting cable* |
| 2: | *Square turret* | 2: | *Deformation element* |
| 3: | *Key* | 3: | *Protective rings* |
| 4: | *Tool holder* | 4: | *Active strain gauge* |
| 5: | *Tool carrier* | 5: | *Passive strain gauge* |
| 6: | *Fixing screw* | | |
| 7: | *Fixed lower component* | | |
| 8: | *Protective cover* | | |
| 9: | *Baseplate* | | |
| 10: | *Measuring rings* | | |
| 11: | *Cross slide* | | |

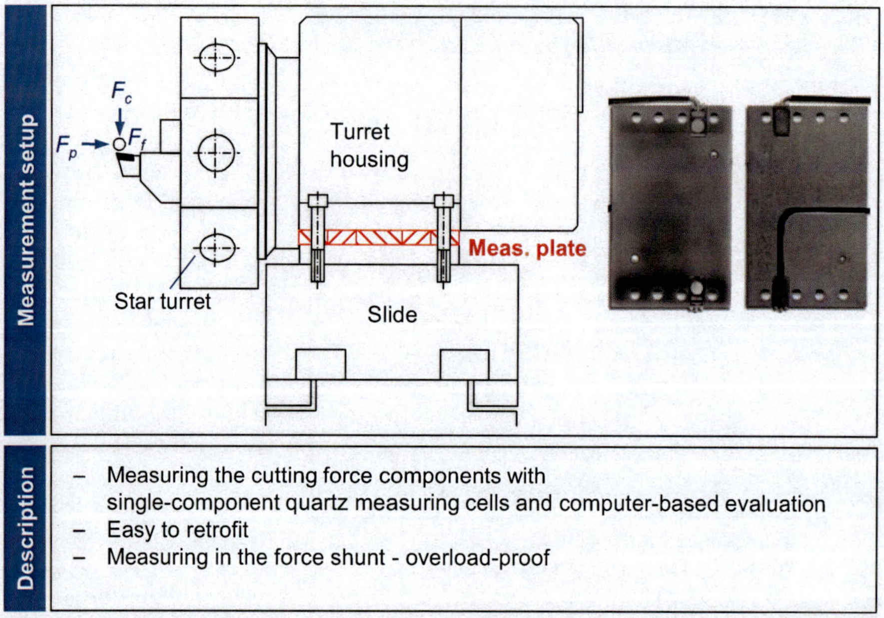


**Fig. 6.23** Flexible force measuring plate for process monitoring on lathes. *Source* PROMETEC

support. The cutting force measurement for a tool is therefore not dependent on the cutter geometry, the adjustment length, or the position of the tool bit with respect to the tool holder.

An appropriately dimensioned measuring plate mounted between the machine slides and the lathe's turret housing performs additional measurement of the cutting force components other than the cutting force (feed force and passive force) Fig. 6.23 [HEID00, ERNS]. This plate is made of the same material as the machine so that the static and dynamic properties of the machine are not influenced by its installation. Instead of utilizing a fixed installation of the force measuring quartzes, the plate is provided with grooves for fitting wedge-type measuring elements. The wedge-type measuring elements are resistant to oil and cooling lubricant and are available in 1D and 3D designs. This makes it easy to adapt the plate to the machine and the system can be easily and economically retrofitted. The wedge-type measuring elements are a few micrometers thicker than the plate. This ensures that the quartzes can be appropriately preloaded during assembly so that tensile, compressive and shearing forces can all be measured. The quartz sensors are located in the force flux bypass and support only a very small component of the overall preload and process forces in this arrangement. This process is also used to protect the sensor itself from overloading and destruction if the sensor is sufficiently sensitive. The measuring elements can also be fitted between the Hirth ring and the turret housing. This ensures that the sensors are closer to the machining process and the signals are more sensitive and less frequently distorted due to the shorter transmission path. However, the design modifications to the turret housing required for this have kept this assembly location from being used as yet.

By using the force and moment equations, it is possible to use the forces measured by the measuring plate to calculate the machining forces $F_c$, $F_p$ and $F_f$ that stack up vertically [SESS91]. The calculation of individual force components demands a great deal of signal processing power but allows wear to be detected with greater sensitivity.

Other options for measuring machining forces have been developed in addition to the systems using strain gauges and quartz sensors described above. The deformation sensor shown in Fig. 6.24 is used directly at the point where cutting forces are generated behind the turret plate [MENZ73]. The actual sensor is an inductive sensor mounted on the end face of the cylindrical shank of standardized VDI tool holders, see Fig. 6.24, left. The measuring principle is based on a displacement measurement, i.e., deformation measurement of the turret head plate. Changes in the field of the permanent magnet are measured using an inductive sensor on the turret base element. The deformation is proportional to the feed force $F_f$ in the z-direction. The system can be easily retrofitted as only one measured value sensor is required per machine (Fig. 6.24 right) The additional cost for fitting the magnet to each tool holder is relatively low.

#### 6.3.1.2 Controlling Force, Torque and Power During Turning

To ensure processes run correctly, it is reasonable to monitor tools and machines to ensure specific given cutting threshold values and torques are not exceeded. This also prevents overloads from occurring.

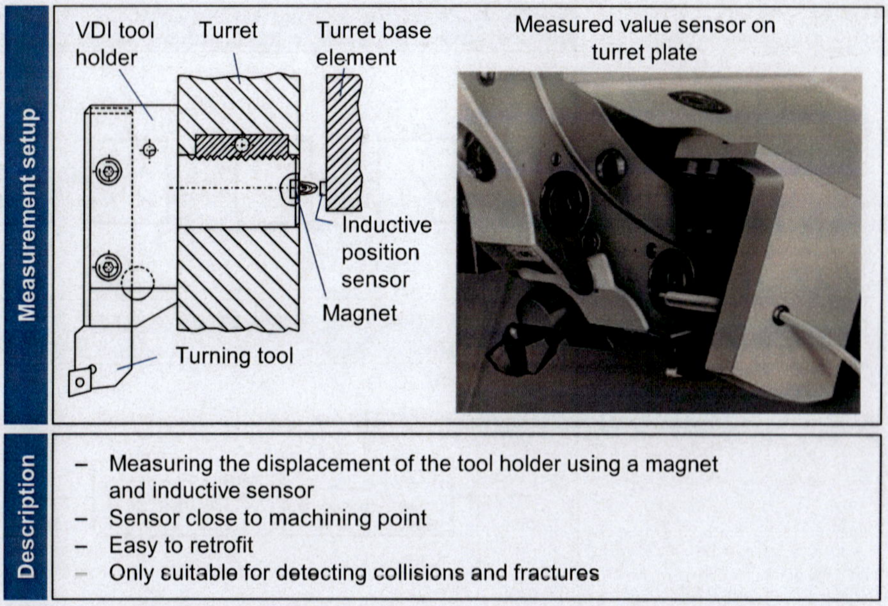


**Fig. 6.24** Magnetic strain sensor. *Source* Kennametal

It is also advisable to run the machine and the tool at the limits of their power, force, or torque without exceeding these values, particularly during roughing operations.

The objectives during turning are therefore:

- protecting the machine by limiting the torque $M$ and the power $P$:

$$M \leq M_{\max} \tag{6.8}$$

$$P \leq P_{\max} \tag{6.9}$$

- protecting the tool by limiting the cutting force $F_c$:

$$F_c \leq F_{c,\max} \tag{6.10}$$

- fully utilizing the machine to minimize the machining time:

$$P = P_{\max} \tag{6.11}$$

By measuring one of these three quantities $M$, $P$ or $F_c$, it is possible to calculate the other two using the following formulas as the NC control system already knows the spindle speed $n$ and workpiece radius $r$.

$$P = M \cdot n \cdot 2 \cdot \pi = F \cdot r \cdot n \cdot 2 \cdot \pi \tag{6.12}$$

$$M = \frac{P}{2 \cdot \pi \cdot n} \tag{6.13}$$

$$F_c = \frac{M}{r} \tag{6.14}$$

Figure 6.25 shows the principle of a simple technological threshold control for utilization of available machine power during turning. The controlled variable here is therefore the machine power. The torque or the cutting force could also be used as the controlled variable. In this example, the motor current is measured as the parameter proportional to the electrical power consumed. The feed rate (control variable) required to maintain the reference power is output by the controller after comparing the reference and actual values.

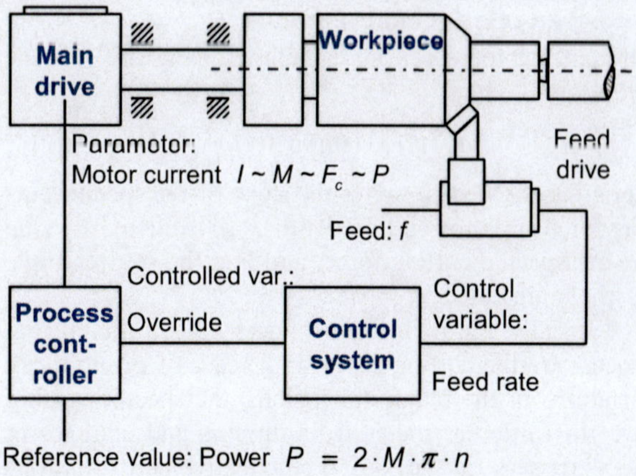


**Fig. 6.25** Functional diagram of a simple technological threshold control (power regulation) during turning

A significant disadvantage of measuring the power via the power consumption of the main drive is that the measuring point is a relatively long way away from the cutting point and changes in the cutting power with substantial delay times affect the power consumption of the main drive. It is therefore not possible to react quickly to variations in the disturbance input (see ▶ Sect. 6.3).

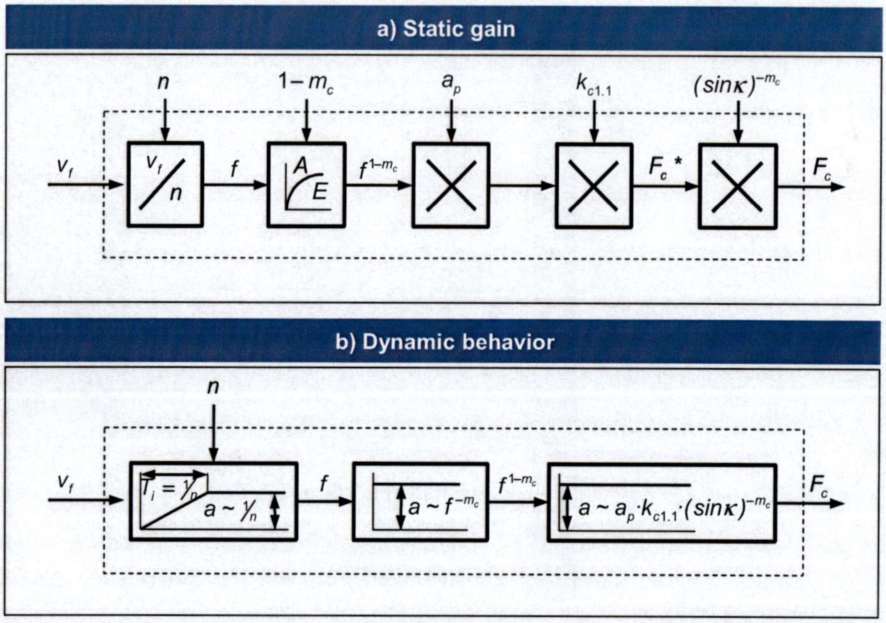


**Fig. 6.26** Cutting force control path during turning

### Cutting Process as Control Path

The cutting force can be calculated using the Kienzle equation:

$$F_c = f^{1-m_c} \cdot a_p \cdot k_{c1,1} \cdot (\sin\kappa)^{-m_c} \tag{6.15}$$

Here, $f$ is the feed, $1-m_c$ is the slope of the specific cutting force, $a_p$ is the cutting depth, $k_{c1.1}$ is the main value for the specific cutting force, and κ is the setting angle for the main cutter.

Typically, the main disturbances during the cutting process are the cutting depth in the case of pronounced variations in the blank dimensions, the specific cutting force for different material hardnesses and changes in the sharpness of the tool. The effective control variable for regulating the cutting force is therefore the feed $f$ (tool path per spindle revolution) and this can be changed with the feed rate $v_f$. The feed and feed rate are related as follows:

$$f = \frac{v_f}{n} \text{ [mm / rev.]} \tag{6.16}$$

Figure 6.26 illustrates the static and dynamic transfer behavior of the cutting force control path [SPIE90]. This shows that the disruptive influences act exclusively on the control path gain.

The behavior of the cutting force control path over time, i.e., the course of the force versus time as the feed rate changes, is influenced by the rotational speed only.

Figure 6.27 illustrates the influence of the spindle speed on the control path versus time. Figure 6.27a shows the cutting force curve as the cutting depth jumps from zero to a value of $a_0$ at a constant feed rate (controlled operation).

Once the tool touches the workpiece, it takes precisely one spindle revolution until the cutting force $F_c$ moves from zero to a constant value. The cutting edge requires one revolution to leave the material. During this time, the cutting thickness decreases continuously and the cutting force returns to zero.

Figure 6.27b illustrates the cutting force profile for an abrupt change in the feed rate at a constant cutting depth. However, in this case, the effects of the jump in the control variable are only completed after one spindle revolution.

### Power Regulation Loop

Figure 6.28 shows a simplified block diagram for a power regulation loop for a turning operation in which the torque is captured as a load parameter [NELL91]. The control variable is the feed rate $v_f$ set by a NC in this instance.

The disturbances acting on the process and the machining result should be compensated by the actions of the control device. A PI controller is used for this. In order to generate the control difference, the torque $M$, which is captured with an appropriate sensor, is multiplied by the actual main spindle speed $n$ and the result is compared to the reference value for the given cutting power $P_n$. Division is used to normalize the results to achieve favorable modulation of the controller for all given output values $P_n$.

As changes in the rotational speed influence the dynamic behavior of the control path, their variable time behavior is compensated using a multiplying function of the rotational speed connected to the I component of the PI controller.

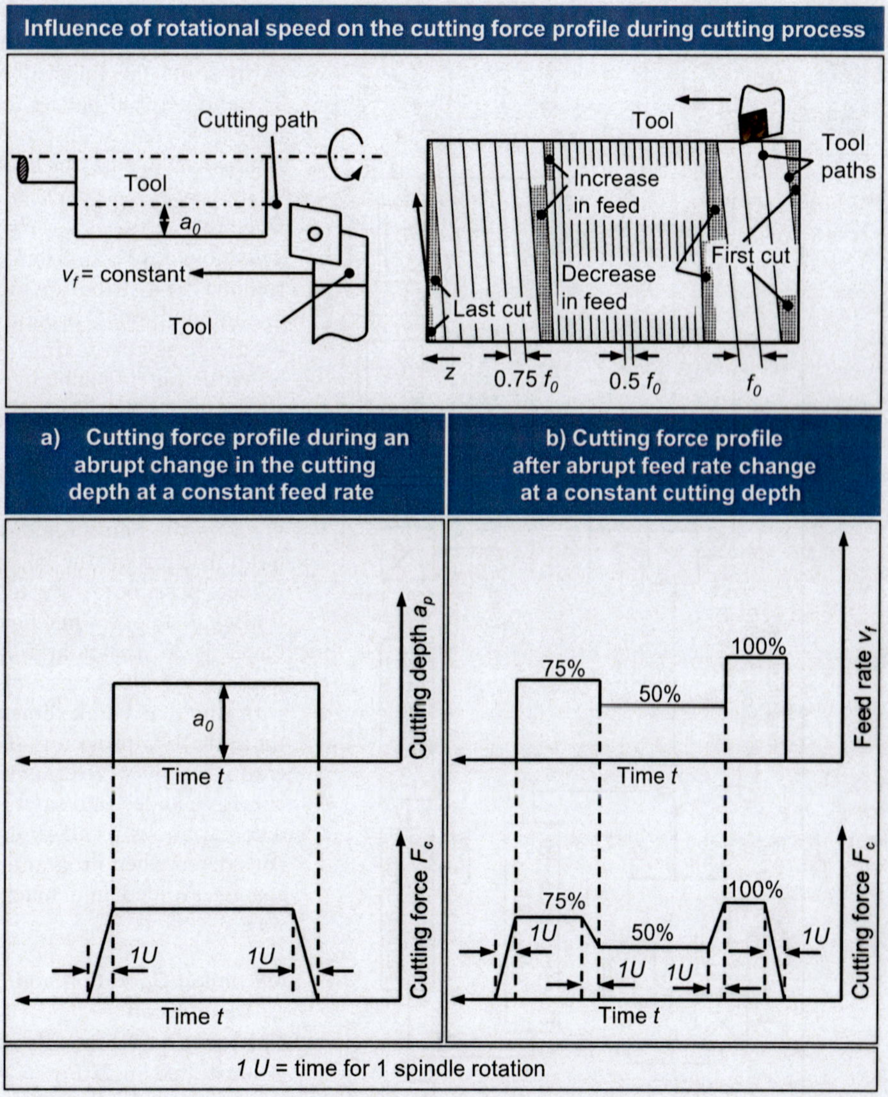


**Fig. 6.27** Influence of rotational speed on the cutting force profile during the cutting process

The controller output parameter if fed into a multiplying function with the reference feed $f_0$ and then with the actual rotational speed $n_{mom}$. Multiplying by the rotational speed $n_{mom}$ offsets the change in the control path gain related to the rotational speed and calculates the feed rate control variable. Through this measure, "self-adjustment" is, strictly speaking, understood as the adaptation of the controller to the rotational speed process parameter. The control variable $v_f$ is fed to the interpolation module of the numerical control influencing its output speed in this way. This closes the control loop.

Periodic disturbances frequently occur in turning. Typical examples of this are out-of-round or square workpieces (e.g., forged blanks) where machining generates a pulsating torque. After a short period of time, this generates a steady-state oscillation of the controlled variable around the reference value without changing the usual control characteristics or without special measures. Maximum value memory is used to avoid periodic overshooting of the controlled variable. This piece of equipment feeds the maximum torque of the last relevant revolution to the controller as an actual control value instead of the periodically varying actual torque. If an even greater torque occurs during the current revolution, this is fed directly to the controller, ◘ Fig. 6.28.

If idle travel occurs during the execution of NC sets (approaching the workpiece from the starting point or bridging recesses), these distances can be traversed at higher feed speeds in order to achieve the shortest possible machining times. The module Power identification ($M_{max} = 0$) is added to the circuit shown in ◘ Fig. 6.28 for this purpose.

6

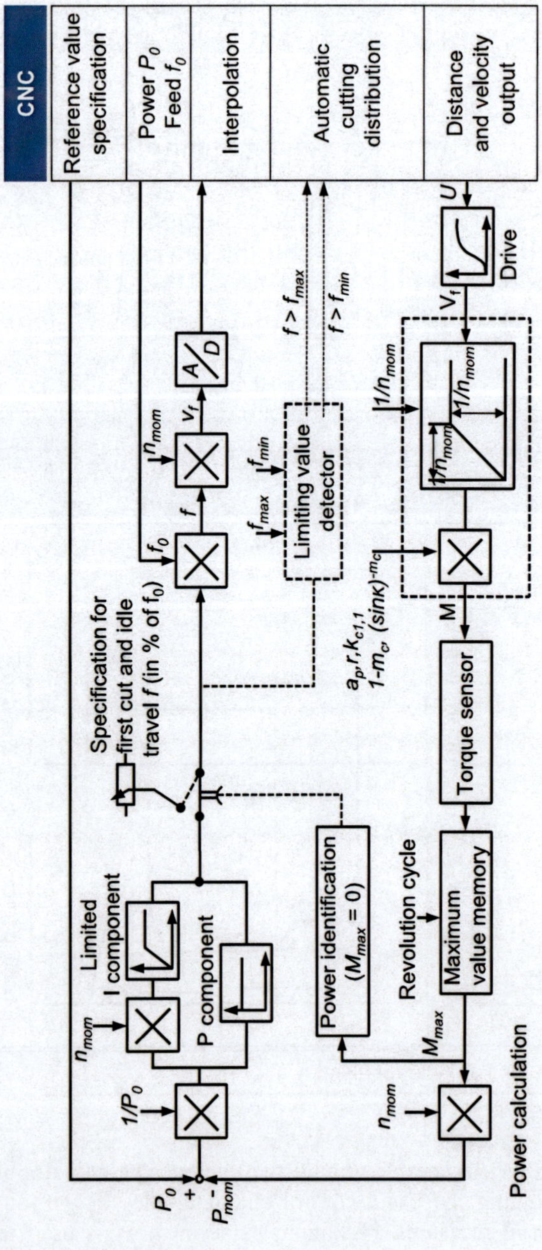


**Fig. 6.28** Integration of the turning process control

### 6.3.1.3 Automatic Cut Distribution for Turning

An NC program always has to take the dimensions of the blank into consideration. Cutting depths and the required cut distributions are determined by the geometry of the blank. If a blank deviates from the dimensions provided during programming, e.g., the machine operator selects a blank with a diameter of 120 mm instead of the specified 100 mm diameter, the tool and the machine are overloaded at the first cut and this leads to tool fracture and subsequent collision between the tool shank and the workpiece. This frequently leads to subsequent machine repairs and expensive downtimes. It would be desirable for the machine to be able to machine the finished part regardless of the blank size and without having to modify the NC program. A fully developed solution for automatic cut distribution is already currently available for turning [SPIE90].

#### Objective

The most important objectives to be achieved with automatic cut distribution include:

- simplifying NC programming,
- use of any blank dimensions,
- reduction in machining times,
- securing machines and tools against overloading.

Simplifying NC programming is achieved by only having to program the contours of the finished workpiece. Data on the blank dimensions are not necessary. This not only shortens the NC machining programs but also reduces possible error sources when programming.

Being able to machine without reference to the blank is a particularly important factor for industrial use of NC technology as this allows workpieces with different blank dimensions to be machined using the same NC programs. This is also beneficial for operational safety. Overloading of the tool and the machine is avoided and safety tolerances which accommodate variations in blank dimensions do not have to be considered when programming. Unnecessary idle travel can be avoided and machining times reduced in this way.

#### Machining Operation and Strategies for Automatic Cut Distribution

The functional principle of automatic cut distribution is illustrated in Fig. 6.29 for a straightforward finished part (cylinder). The only programmed items are the finished part contour, the reference values $P_0$ (reference power), $f_0$ (reference value for the feed) and $a_0$ (reference cutting depth); there is no need to describe the contour of the blank.

The starting point for machining (start of set) is outside the maximum possible overdimension. Once machining has started, the tool approaches the workpiece to bridge the idle travel at an increased feed along the programmed contour, i.e., at the finished part diameter. Once there is contact between the workpiece and the tool, the power regulation switches itself on and this attempts to maintain the reference power value by changing the feed rate. Due to the fact that the cutting depth is too great, at $a_0 = a_{max}$ the feed drops below the lower feed limit $f = f_{min}$, triggering the $f \geq f_{min}$ alarm. This alarm immediately switches off the current NC set and saves the actual position of the tool. With the output of the next NC block, which the automatic cutting division generates automatically, the cutting depth

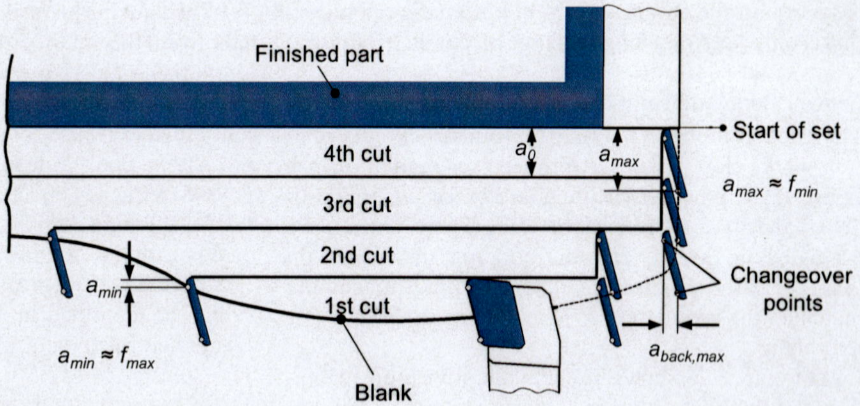


**Fig. 6.29** Functional principle of an automatic cut distribution for turning

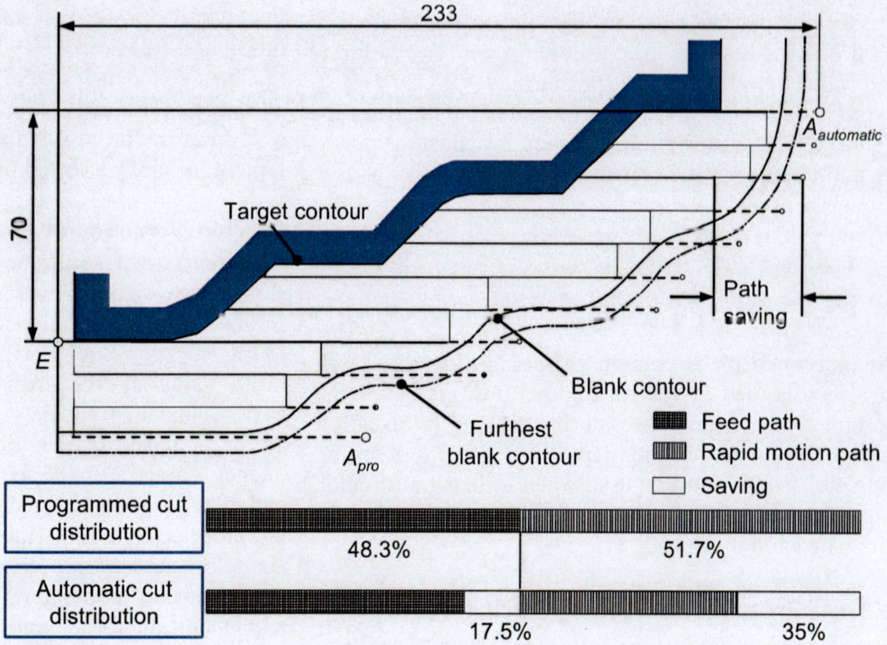


**Fig. 6.30** Path saving when using automatic cut distribution for turning

is reduced by the programmed cutting depth $a_0$ and the tool is retracted by this amount $a_0$. After the reduction in the cutting depth, the cut distribution program calculates the next linear set along the axis that reaches the intersection point with the finished part contour.

In this example, after the first reduction in cutting depth, the remaining cutting depth ($f \leq f_{min}$) is still too great meaning that the cutting depth needs to be reduced further still by $a_0$. This cutting depth is now within the permissible cutting depth range and machining continues with longitudinal feed until a new alarm $f \leq f_{min}$ occurs. After a further reduction, caused by the increase in the blank contour, the cutting depth finally falls below the minimum cutting depth (as shown on the left-hand side of Fig. 6.29): $a_p \leq a_{min}$, $f \geq f_{max}$. Machining is now stopped and the machine returns to the last release point. If there are no further alarms, the remaining cuts 2 to 4 are completed one after the other.

Figure 6.30 shows a working example of producing a turned part from a forged blank with programmed and automatic cut distribution. Here, the total distance in rapid motion and at working speed for the programmed cut distribution is set at 100%. In this example, the automatic cut distribution saves 17.5% of the feed path and 35% of the rapid motion path. These values may be lower or higher in other examples.

Automatic cut distribution is always of benefit if oversized blanks with more than one cutting depth (e.g., when turning from bar stock) or blanks with overdimensions that vary greatly (e.g., forged or cast components) are used. However, suppressing the chatter phenomena, which occur frequently, requires a reduction in the cutting depth. In automatic operation, this is only possible using this type of cut distribution.

#### Control, Regulation and Tool Requirements

The programming system for automatic cut distribution works closely with adaptive power regulation, see ◘ Fig. 6.28. When given limits are exceeded, this produces alarms for the current feed. These alarms are evaluated in the cut distribution program. Based on these alarms, NC sets are produced during machining matching the geometrical machining process to the current blank.

6

◘ Figure 6.28 shows how the adaptive limit regulation and the NC control system work together with automatic cut distribution. The cut distribution program is also stored in the control computer in this instance.

By comparing the controller output parameters (f as a % of $f_0$) to the limiting values $f_{min}$ and $f_{max}$, a limiting value detector triggers the alarms $f \geq f_{max}$ or $f \leq f_{min}$ when the limiting values are reached. These alarms are fed to the cut distribution program. As constant power is maintained by the control loop, the following relationships apply:

$$f \leq f_{\min} = a_p \geq a_{\max} \tag{6.17}$$

$$f \geq f_{\max} = a_p \leq a_{\min} \tag{6.18}$$

so that when the maximum permissible cutting depth $a_{max}$ is exceeded or the cutting depth drops below the minimum $a_{min}$, the above alarms provide information about the current cutting depth. A tool with a setting angle of $\kappa > 90°$ must be used when utilizing automatic cut distribution so that plane surfaces also can be approached and machined.

### 6.3.1.4 Process Monitoring During Turning

Process disruptions during turning are tool or machine related disruptions such as tool wear or fracture or a missing tool or machine or tool overloading. They may also come from disrupted process sequences due to chatter or unsuitable chip shapes. Monitoring methods, which accompany the process, are used for quicker identification of these disruptions. They evaluate process forces, structure-borne sound signals, the effective power of the main drive or the signal from the digital drives. These signals are also used for collision monitoring.

#### Effective Power and Signals from Digital Drives

The effective power and the signals from the digital drives are used for tool fracture monitoring. However, the changes in the signal from the feed drive are too small to identify tool wear because friction in the guide systems and accelerations obscure the relatively small changes in the machining forces [GEOR98].

When facing at a constant cutting speed, the signals from the main drive are once again not suitable for monitoring as the accelerations which occur here are also too large. The limited dynamics of the signals cause further restrictions:

- chatter does not generally manifest itself in the low-frequency effective power signals,
- identifying a tool fracture or a collision takes 20 to 60 ms due to the low-pass behavior of the transmission path. In contrast, collisions and tool fractures can be identified in 3 to 5 ms by using direct measurement of the machining forces.

As the consequential damage from collisions increases progressively the longer the time required to identify the event, there is only a very limited application for using effective power measurement to monitor tool condition and collisions.

The dependency of the effective power on the overall cutting force reduces its sensitivity when measuring wear because the cutting force $F_c$, the largest force component, is barely affected by wear [PFEI76].

#### Structure-Borne Sound

Research into structure-borne sound measurement has shown that signal slopes specifically caused by wear can be measured in frequency ranges above 30 kHz, see ◘ Fig. 6.31 top [FAHR93]. The variance and the static mean value of the structure-borne sound envelope curve signal in frequency ranges greater than 30 kHz are particularly suitable for this. The most suitable frequency bands and the associated slope values to the end of service life are specific to machines and machining processes and must therefore always be determined again for each new process.

The signal slopes due to wear are caused by dynamic cutting forces and changes in contact in the shearing zone and are two to three times greater than the corresponding force slopes [FAHR93]. However, structure-borne sound signals display greater scatter than the cutting forces despite averaging and thus provide comparable identification security.

Tool fractures manifest themselves through signal pulses in the structure-borne sound envelope signal, similar to when measuring cutting forces ($f > 30$ kHz), see ◘ Fig. 6.31, bottom. The second, somewhat smaller, pulse occurred when the broken cutter struck the workpiece a second time after two workpiece revolutions. It is not always possible to distinguish between normal signal scatter or cavities and hard spots in the workpiece.

#### Cutting Forces

The passive force $F_p$ and the feed force $F_f$ are excellent means of monitoring wear during turning. However, at cutting depths in the order of magnitude of the corner

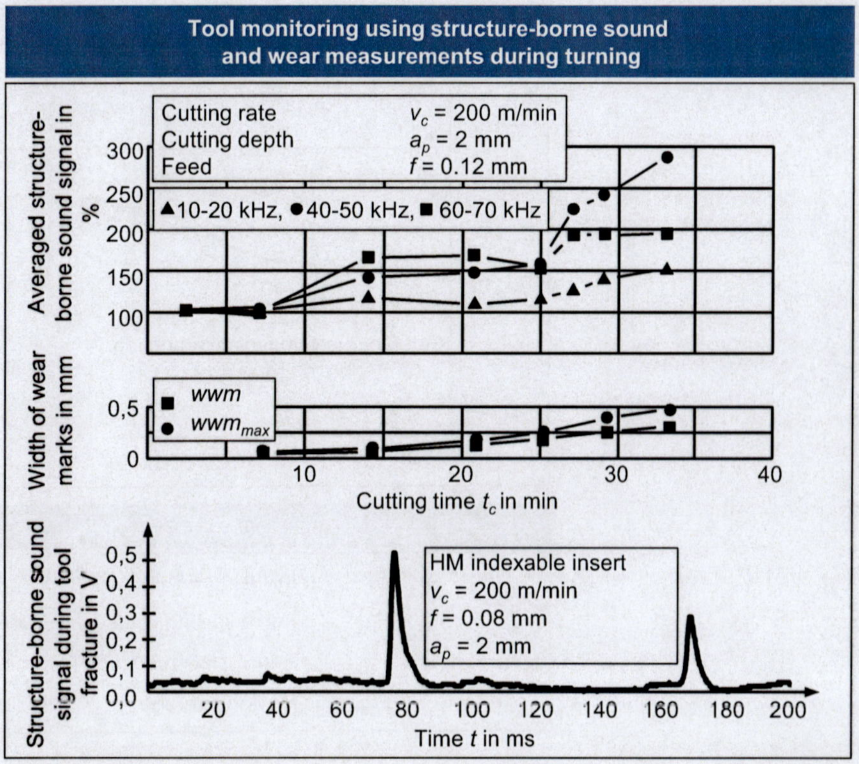


**Fig. 6.31** Tool monitoring using structure-borne sound and wear measurements during turning

radius and with a setting angle of $\kappa > 90°$, changes in the passive force are virtually no longer dependent on wear.

Changes in the cutting force component $F_c$ caused by wear are very low. For this reason, measurement of the overall force is less sensitive and less suited to monitoring wear than is measuring the feed force [FAHR93, PFEI76].

Wear measurement should always be performed under the same cutting conditions to measure the force slope caused by the wear independently of other influences from the cutting condition. In general, continuous wear monitoring is not necessary as the wear is a constant and slowly growing quantity. Measuring points are therefore defined on the workpiece with cutting conditions as constant as possible for each tool used. The changes in the force caused by the wear are captured during the process [PFEI76].

The end of service life is detected using the fixed limiting values shown in Fig. 6.32. At first, the moving average $F_{Mn}$ is generated. The first value for $F_{Mn}$ corresponds to the initial force value $F_{in}$ for a sharp cutter. The force limiting value at the end of service life $F_{eol}$ is determined either through learning cuts or is provided by the user. As soon as this limiting value is exceeded, the user receives an end of service life message.

The permissible reduction in force $F_{drop}$ is defined in order to identify crater wear which can lead to a reduction in the machining forces.

The cutting forces are also suitable for fracture monitoring. Total tool fractures are easy to detect in the cutting force components due to pulse-like slopes (<1 ms) and a subsequent drop in signal.

Tool fracture is usually monitored through moving thresholds so that variations in the force caused by wear and oversizing do not have a detrimental influence on fracture identification. Changes in the force at initial and final cuts or when drilling holes progress significantly slower than in tool fractures. The delayed moving thresholds do not follow the rapid signal changes in tool fractures. The values fall either above or below the moving thresholds in tool fractures, see Fig. 6.33. If the characteristic signal curves shown in Fig. 6.33 occur, tool fracture is detected and a message is sent to the PLC and the user. If a rapid response is required, the controller enable for the feed drives is disabled. This prevents further damage to the tool, machine and workpiece.

**Fig. 6.32** Strategy for identifying end of service life according to Kluft

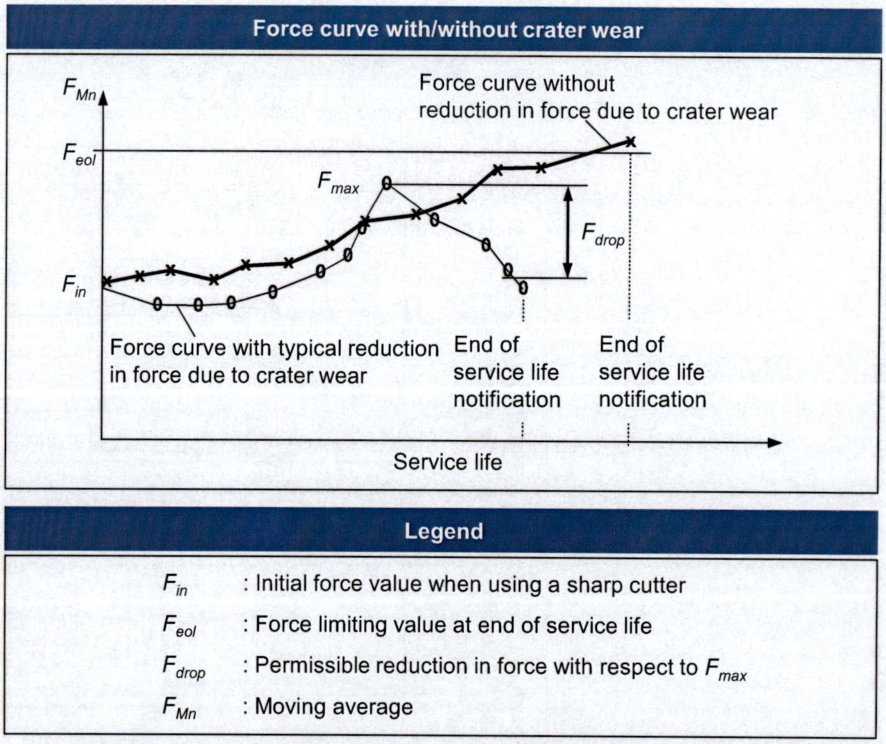


**Fig. 6.33** Strategy for detecting fracture and total fracture a) with and b) without exceeding the upper threshold

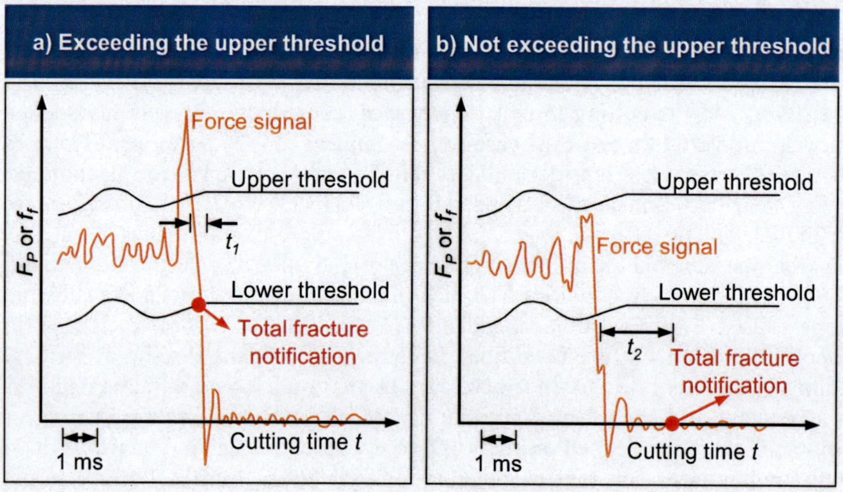


Before the monitoring process can start, the monitoring parameters must be entered by experts or determined through a teach-in process during the service life of the first tool. Important monitoring parameters include:

- cutting force for sharp tool as a reference value,
- limit for wear, fracture and collision monitoring,
- mean value factor for forming the moving average (▶ Sect. 6.2.3),
- width of the tolerance range for moving thresholds,
- permissible cutting interruption time when turning over holes,
- suitable cutting points on the workpiece for wear monitoring during the process,
- measuring range and gain factors.

Depending on the monitoring system, these parameters are either provided by the user, set at the factory, or determined by the system during a learning (teach-in) phase. System reliability relies on the correct entry of these parameters. As a large number of monitoring parameters have to be set for each process, lengthy testing and adjustment phases are often required.

## 6.3.2 Milling

### 6.3.2.1 Sensor Systems and Processes for Process Monitoring During Milling

Milling machines with automatic tool and workpiece change have been developed for automated production on machining centers. The machining operations on these centers range from milling and drilling with extremely small tool diameters, thread tapping, reaming, drill-finishing fitting holes to heavy-duty machining with large inserted-tooth cutters. This must be taken into consideration when equipping a machine with a sensor system for process monitoring. For a description of the sensor functionality, please refer to ► Sect. 6.2.

The application of sensor systems in process monitoring on milling machines or machining centers is significantly more complex and problematic than on lathes. It is difficult to attach sensors on the workpiece side as the force has to be measured behind or underneath the pallet due to automatic workpiece changes together with the vise pallet.

More favorable relationships for force-based sensor systems are available on the tool side as the spindle is a much better option for measuring forces and torques. As several cutting edges are usually engaged simultaneously during grinding, it is not possible to determine the force components of one individual cutting edge precisely as is possible in turning operations.

Using external sensors is difficult so current or power measurement or internal control information is used provided the process forces are sufficiently large to be measured downstream of the mechanical transmission path.

A system for process monitoring integrated into the control system and based on the signals from digital drives and a torque, force and acceleration sensor system integrated into the spindle is described below.

#### Process Monitoring Integrated into the Control System

Modern, open control systems allow process monitoring functionality through access to the current or torque signals from digital drives as well as connection to data for the NC set and the control system. The main benefit of these types of process monitoring systems integrated into the control system compared to previous commercial solutions is that, in many cases, monitoring is possible without having to use sensors or other additional hardware [ECKH93].

Reliable process monitoring requires high-quality information on the monitoring signals being used. The drive signals for the actual current value and reference torque value do not equal zero even when there is no cutting process, i.e., when idling. The following typical effects should therefore be taken into consideration:

- acceleration effect from inertia forces,
- friction effect when moving axes/spindles, e.g., friction from guidance systems or rolling friction,
- signal offsets at standstill caused by static friction. These effects basically allow any standstill current to occur at the drive which produces a force less than the static friction limit.

If the influences of idling behavior are of a similar order of magnitude or even greater than the useful signals produced by the process during the cutting process, it is no longer possible to analyze the process state reliably. ◘ Figure 6.34 provides an overview of the corrective measures used to eliminate disruptive signal influences and to extract a qualitatively high-value useful signal. Using open-loop controls provides a particularly efficient option for correcting signals as additional information such as the position, rate, acceleration and, for example, drive parameters, and gear unit stages are directly available from the software interface of the control system. In this project, it was possible to reduce disruptive influences by up to 90% by using comparatively simple models.

If the excitation signals from the process are too low so that they are too heavily damped by the mechanical transmission path of the feed drive, the drive signals can no longer be used for reliable monitoring. In this instance, the system has to be supplemented with additional sensor technology. As part of the project discussed above, the external sensor technology was connected to the control system using additional signal-processor-based hardware. The structure of this system is explained in the next chapter.

#### Machine-Integrated Sensor Technology

The aim of machine-integrated sensor technology is to capture different process forces and torques during milling such as those required to monitor tools with small diameters.

◘ Figure 6.35 shows a suitable design for a simple external sensor system to capture the resultant torque on an NC bed-type milling machine. The torsional deformation of the spindle is measured using strain gauges on a weakened spindle component between the front bearing arrangement and the ground wheel [PHIL91]. The disadvantage of this torque sensor is the weakening of the spindle necessary to achieve the desired sensitivity as well as the signal transmission of the rotating machine tool spindle.

If the aim is to avoid weakening the spindle, then a sensor arrangement with higher sensitivity must be used such as quartz sensors as shown in ◘ Figs. 6.36 and 6.37 [HEWL 76b]. A torque sensor is applied to the coupling sleeve between the main spindle and the

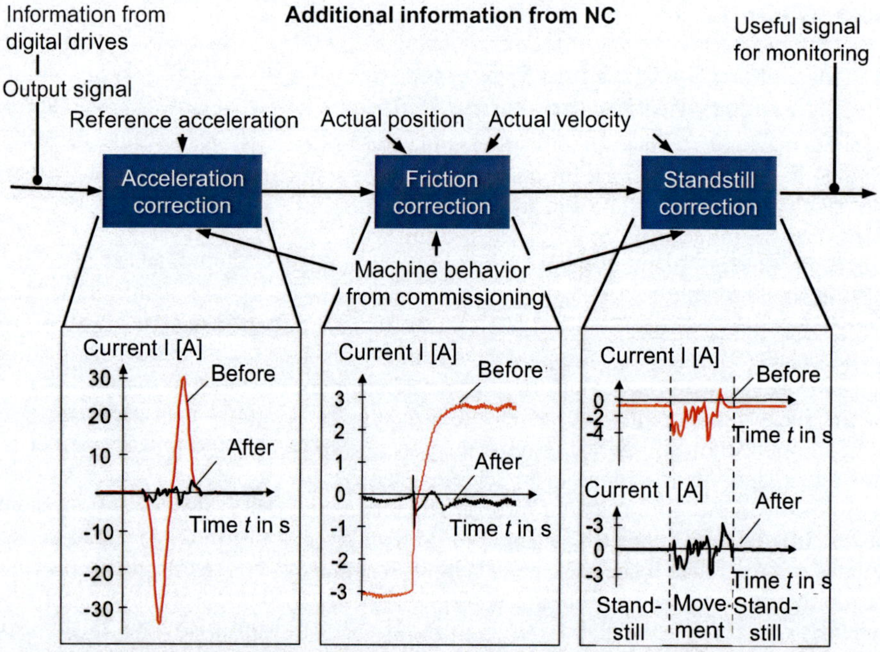


**Fig. 6.34** Eliminating non-process-related influences on the signals from direct drives

6

motor. Torsion on the coupling sleeve caused by torque leads to a relative displacement of the two clamping rings. The resultant forces are measured with piezoelectric force quartzes, Fig. 6.36.

An acceleration sensor system is also located on the spindle. This is integrated into two components, which run on the spindle and which replace the original bearing spacer sleeve. A ring contains the sensor and the transmission loop while the necessary electronics are located in a sleeve in suitably milled pockets. A piezoelectric accelerometer captures frequencies up to 20,000 Hz, Fig. 6.37.

The sensor signals are captured and pre-processed by rotor electronics that run on the spindle and integrate the charge amplifier, the A/D converter, and data transfer functionality. The simultaneous, contactless transmission of energy and sensor signals between the stationary and rotating components of the evaluation electronics system uses inductively linked conductor loops. Special measures are required to protect the machine and the sensors due to the potentially high rotational speeds on the main spindle (15,000 rpm). This includes, specifically, casting all endangered components in an electrically insulating but thermally conductive epoxy resin compound. In addition to carrying out the HF power feed and signal recovery, the stationary stator electronics also performs the D/A conversion to prepare the control connection. This evaluation electronics system is based on the process described in [BRYA79, HEWL 76a]. Its layout and technical implementation have been optimized for use on the high-speed spindle.

Four piezoelectric force sensors, which measure forces in the z-direction, are placed in the relevant holes in the headstock to supplement the spindle sensor technology. Ceramic rods act as strain transducers to increase the sensitivity. By positioning the sensors on the x or y coordinate axes, it is possible to achieve a simplified multi-sensor evaluation to obtain a three-dimensional cutting force vector.

The evaluation of the external sensor signals can be optimized by using internal control data inasmuch as, for example, the signal quality is improved by speed or position-dependent correction.

The sensor system shown in Figs. 6.36 and 6.37 is particularly suitable for capturing dynamic processes in the machining process. The capture of purely static forces and torques in contrast is problematical due to the drifting which occurs. The drift in the measurement signals is primarily produced by thermoelastic deformation which is captured by the force measuring cells.

### 6.3.2.2 Process Monitoring for Milling

#### Tool and Tool Class Related Monitoring

As part of the "Control-integral process monitoring" project (see ▸ Sect. 6.3.2.1), a completely new approach was developed for detecting overload and fracture on the tool. The advantage of this process is that the monitoring is largely independent of the NC partial program and learning cycles are not necessary.

Tool-based monitoring means that the monitoring methods are dependent on the specific tool as the maximum torque or the maximum force a tool can support

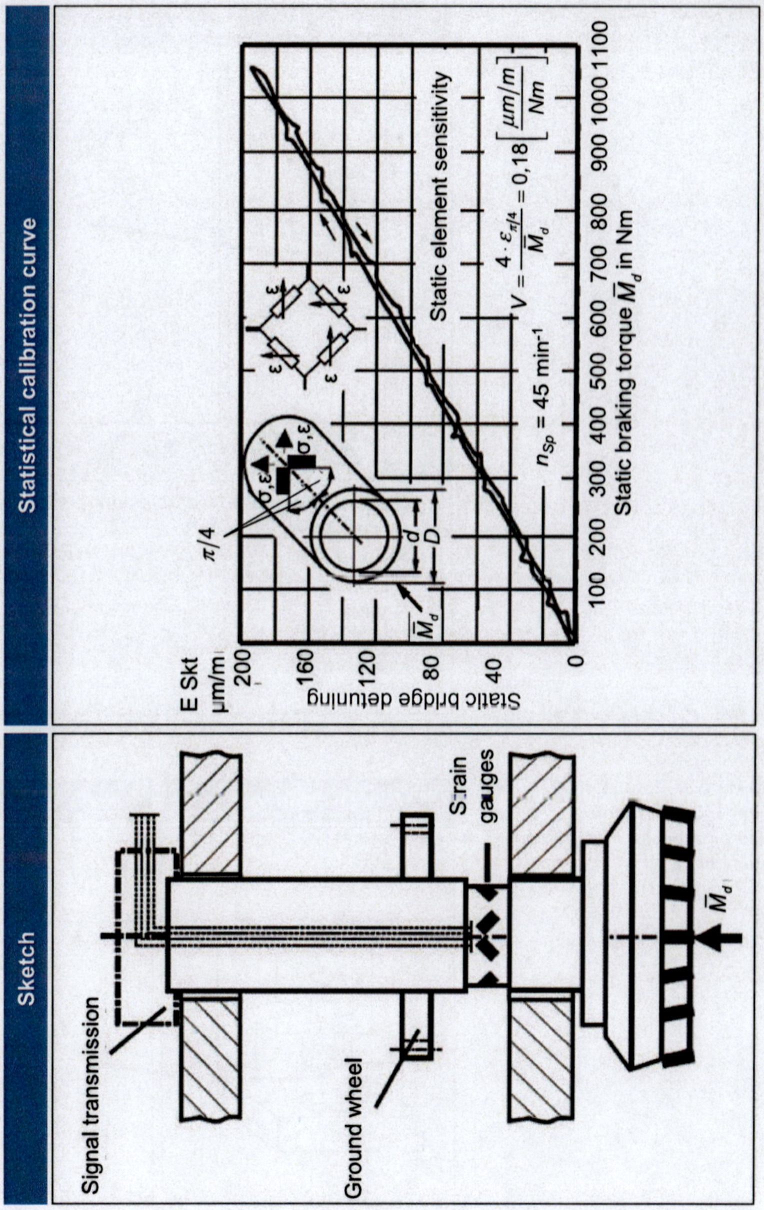


■ **Fig. 6.35** Torque sensor on a bed-type milling machine, design arrangement and statistical calibration curve

is defined by the tool itself and not by the cutting conditions or the machined material.

However, as it is not possible to have the parameters available for every possible tool, the tools are divided into appropriate classes to which the necessary parameters are allocated. If the interpreter encounters a command to change a tool while processing the NC program, the first step is to determine whether a defined method exists for the specific tool. If this is the case, this method is activated and loaded into the NC kernel. This is then used as a basis for monitoring the tool. Otherwise, the monitoring method allocated to the relevant tool class is activated. Obviously, class-based monitoring is not as precise as tool-based monitoring. However, this method provides relatively inexpensive basic monitoring starting with the first cut.

### Total Fracture Monitoring for End Mills

In the above project, the main spindle current is monitored via the method of moving thresholds to detect end mill fractures (see ▶ Sect. 6.2.4.2). The fixed threshold process is used to monitor overloading (see ▶ Sect. 6.2.4.1).

■ Figure 6.38 shows the online monitoring of internal main spindle current signals during milling with a 5 mm end mill using the method of moving thresholds.

6

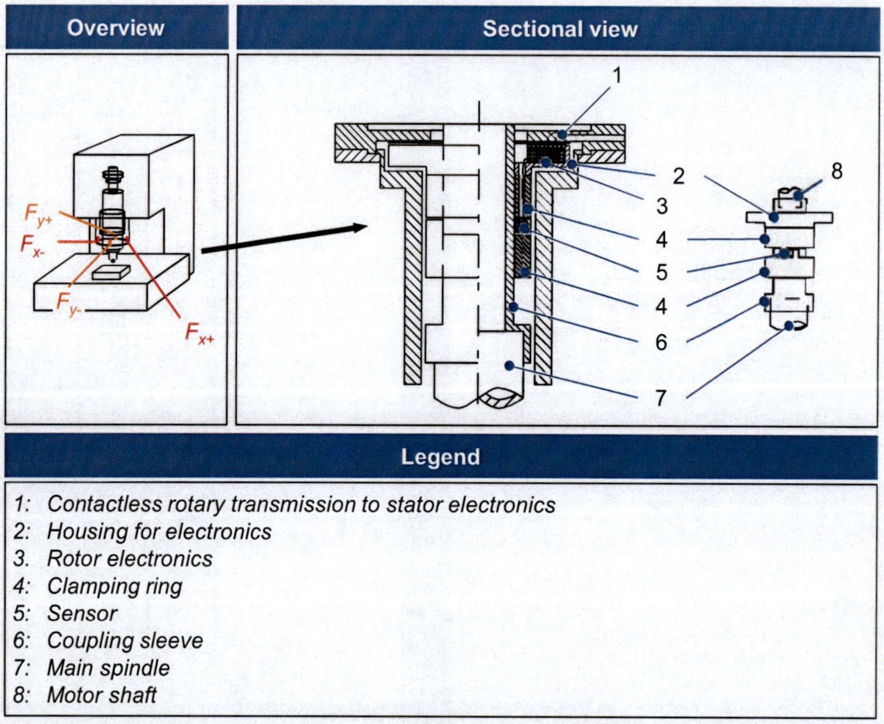


**Fig. 6.36** Torque sensor technology for process monitoring

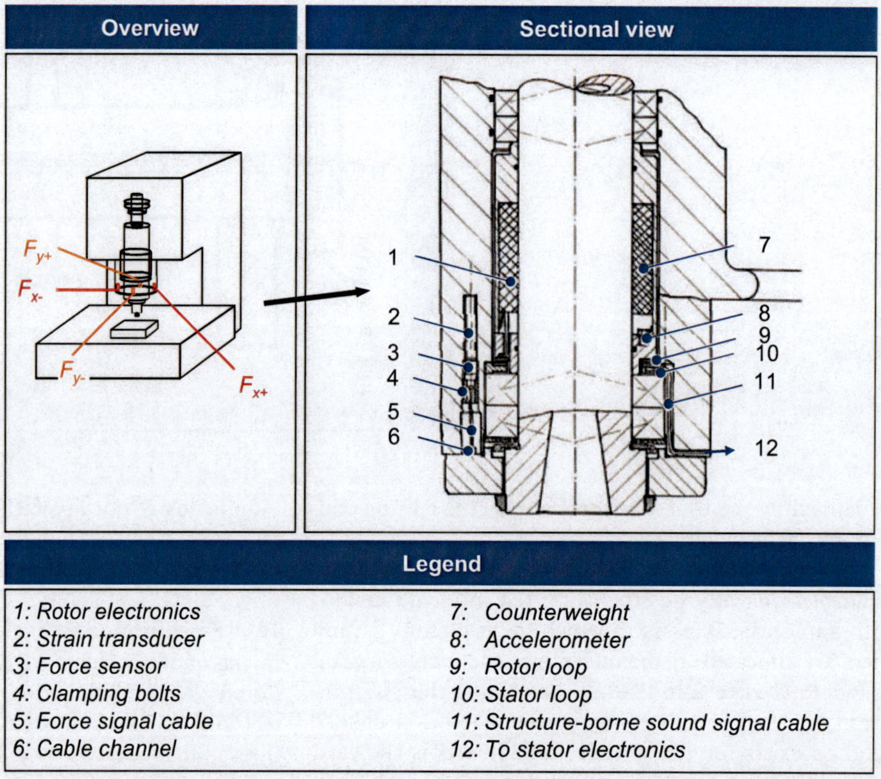


**Fig. 6.37** Structure-borne sound and force sensor technology for process monitoring

Total fracture of the tool occurs during machining. The sharp drop in the cutting torque leads to a correspondingly sharp drop in the main spindle current. This results in the upper threshold being exceeded and this triggers a fracture alarm and the feed is stopped immediately.

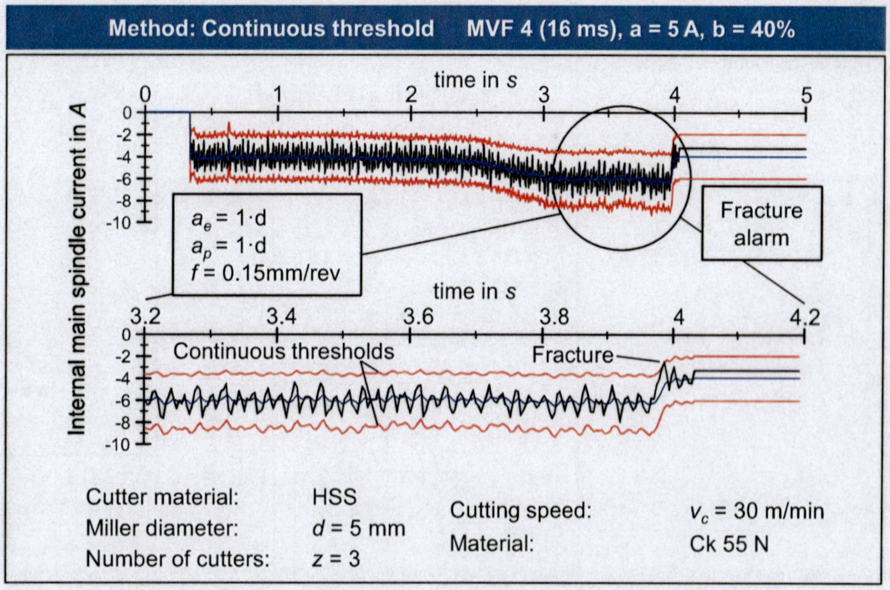


**Fig. 6.38** Total fracture monitoring on an end mill

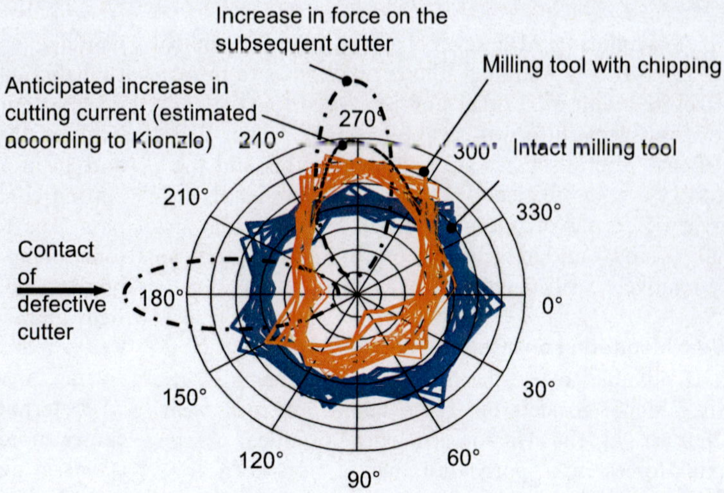


**Fig. 6.39** Comparison of the main spindle current in intact and defective milling tools

### Cutting Edge Chipping on End Mills

A further example of the use of internal control signals can be found in monitoring a mill for cutting edge chipping. This process disruption usually causes a deterioration in the workpiece surface and can be the first stage in total fracture of the milling tool.

The problem when monitoring this process disruption is that the effect occurs only locally (at one cutting edge) and the monitoring signal, for example the main spindle motor current, is therefore influenced only locally. The overall course of the signal however does not display typical disruption characteristics. This means the process disruption cannot be identified using classification processes. These work with only one signal and the local effect in the signal cannot be reliably separated from the stochastic influences.

A characteristic feature of this disruption is the periodic occurrence of imperfect cutting in the workpiece (Fig. 6.39). This effect, which is synchronized with the angular position of the main spindle and occurs every 360°, can be detected if the angular position is used in addition to the monitoring signal from the main spindle current. It is possible in this way to implement a simple pattern recognition algorithm in the time domain. This algorithm uses the principle of moving thresholds but with significantly narrower thresholds than in conventional tool fracture monitoring. As a result of the pattern recognition synchronized with the spindle position, stochastic influences within the time signal, which would exceed a moving threshold, can be largely eliminated and the significance of the signal verified as the cutting edge chipping occurs.

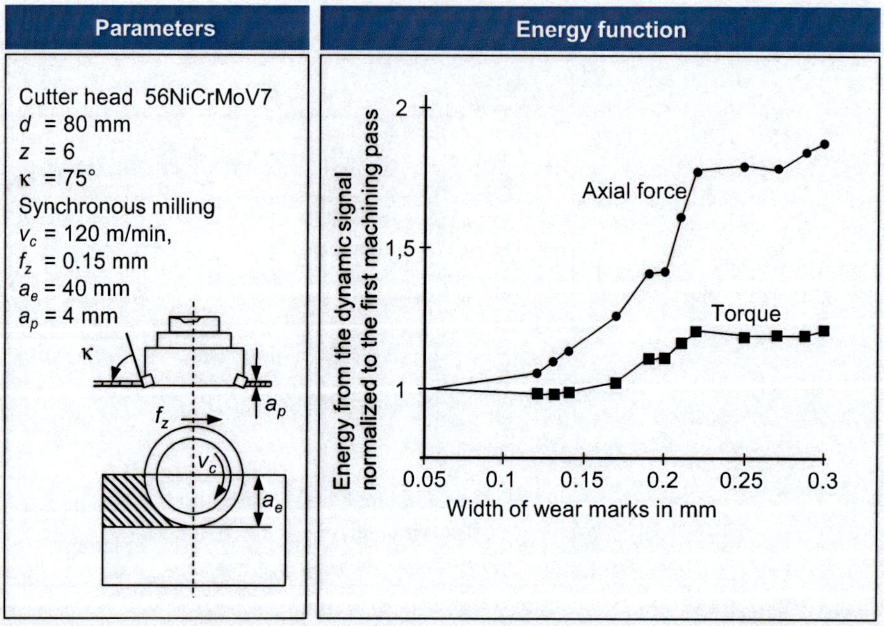


**Fig. 6.40** Energy of the dynamic axial force signal and torque as a function of the cutter wear on side-and-face cutters

A significant advantage of using internal control signals is that this strategy allows two signals to be synchronized with each other and processed together in a relatively straightforward way. In classic sensor-based systems, several sensors are required for this and there may be even more analog pre-processing of signals necessary. In terms of internal control, both different signals can be found in the same control interface and can be relatively easily combined.

### Wear Monitoring on Inserted-Tooth Cutters

It is possible to use the dynamic components in the force signal to determine the degree of tool wear. Changes in the friction behavior produce greater axial forces in a worn tool than a non-worn tool. Figure 6.40 shows the energy from the dynamic component of the axial force when roughing steel with an inserted-tooth cutter as a function of the width of the wear marks on the main cutting edges. The axial force on the tool is formed by the passive forces on the individual cutting edges. In terms of the absolute value, this passive force is the least of the forces acting on the cutter. However, it has the greatest dependency on wear when compared to the other force components [WILD59]. As the wear on the cutting edges increases, the waviness of the force curves also increases, i.e., the alternating power of the signal rises. If the alternating power is integrated over a fixed period of time, e.g., the duration of one machining pass, this produces the energy value. If only the alternating component is considered, errors present in the static measuring signal due, for example, to offset values and temperature drift, are not taken into account. Integration would cause these sources of error to have an extremely significant effect.

For comparison, Fig. 6.40 shows the relationship between the integrated power in the dynamic torque and the average width of wear marks on the hard metal cutting edges on the cutter head over one machining pass. The torque is proportional to the main cutting force component on a single cutting edge. This component is only slightly dependent on the width of the wear marks. Research in turning has shown that crater wear on a cutting edge causes a reduction in the main cutting force [PFEI76].

Determining tool wear during milling is significantly more difficult than fracture detection. Fluctuations in oversize on the blank or differing workpiece material strengths due to different batches can produce significantly greater variations in the cutting force components than those caused by tool wear. A certain independence from workpiece parameters can be achieved if the relationships between the force components are evaluated instead of the individual force components or their proportional values [AULM70].

In finishing machining, as with milling using small tools, the machining forces are very low. The initial developments in this application area are described in [HEWL92]. A piezoelectric force measuring ring has been developed for this system. It is located in the main spindle casing. The measuring elements are positioned within the main flux of forces. However, the forces during finishing machining can only be clearly measured in the laboratory using expensive systems such as force measuring plates. Here, evaluating the structure-borne sound signal is often the only option for monitoring the tool.

### Tool Monitoring with Structure-Borne Sound Sensors

The fracture of a hard metal or HSS tool is characterized by the generation of a broad-band structure-borne sound signal that propagates through the machine structure. A substantial energy component of this signal pulse lies in the ultrasound range up to frequencies of a few hundred kilohertz. This occurrence is often described as acoustic emission. This acoustic emission has various causes, including the separation of the material structure during the actual fracture [FAHR93]. Structure-borne sound sensors allow the sound pulses occurring during tool fracture to be captured. To do this, the sensor must be mounted on the machine at a location where the sound signal is not too heavily damped by the transmission path. A location as close as possible to the process point is usually selected.

Significant cutter wear on small milling tools means that the milling cutters can no longer perfectly shear the material. Sometimes, they only scrape over the workpiece resulting in tool deformation. This scraping can be identified in the structure-borne sound as a signal burst. ◘ Figure 6.41 shows the structure-borne sound signal when machining a workpiece with a sharp HSS milling tool and a heavily worn HSS milling tool with a diameter of 6 mm. The signal was captured with a structure-borne sound sensor on the rotating spindle of a machining center.

Each of the three worn cutters on the milling tool produces the typical sound pulses. These pulses increase disproportionately in terms of frequency and amplitude together with the wear that occurs. There is also a dependency between the tool's contact cross section and the size of the tool itself. At best, this characteristic provides the process with a qualitative assessment of the wear condition of the cutters.

### 6.3.2.3 Process Control for Milling

The general aim of milling process control is to use the installed machine power or to achieve the maximum cutting performance within the limits of the maximum permissible tool loading while still remaining without chatter. This yields virtually the minimum manufacturing costs for roughing work.

Due to the use of multi-cutter tools and the changing contact conditions during milling, the curve for the spindle torque is more complicated than during turning due to the large number of parameters.

First and foremost, the tangential forces acting on the individual milling cutter edges, these forces adding to produce the torque on the working spindle via the cutter diameter, are decisive for the cutting work during milling. The cutting force, the cutting torque and the cutting power in this case depend on the contact conditions of the milling tool.

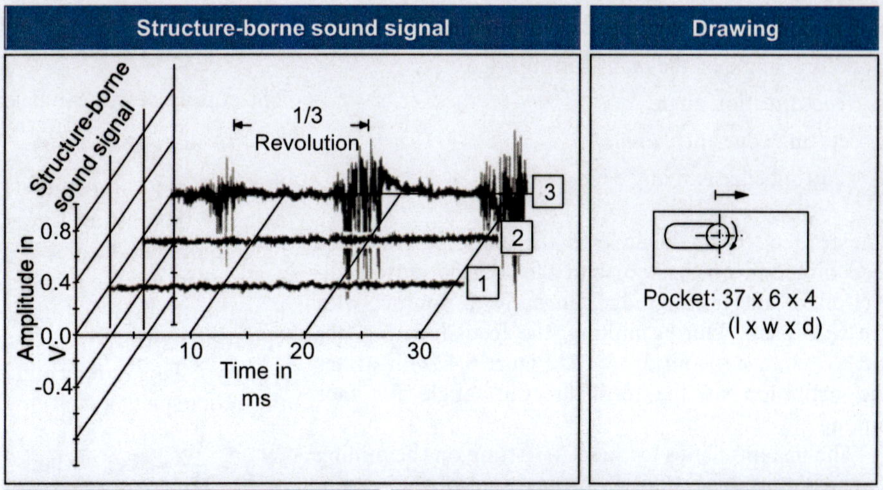


| Parameters | |
|---|---|
| Tool: | HSS end mill cutter<br>6 mm<br>3 cutters |
| Material: | 42CrMo4V |
| Cutting speed: | 30 m/min |
| Rot. speed: | 1591 rpm |
| Feed: | 0.04 mm/rev |
| Coolant: | emulsion |

| Legend | |
|---|---|
| 1 | Idle mode |
| 2 | Sharp milling tool in operation wwm = 0 mm, tool life distance = 0 m |
| 3 | Blunt milling tool in operation wwm = 0.15 mm, tool life dist. = 2.3 m |
| | (wwm = width of wear marks) |

◘ Fig. 6.41 Structure-borne sound pulses from worn cutters on a small HSS milling tool

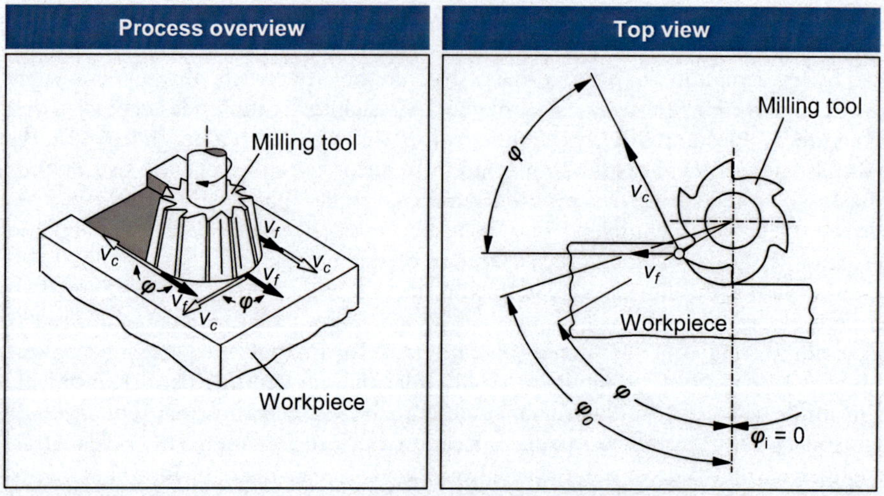


**Fig. 6.42** Feed direction angle when using a face milling tool according to DIN 6580

The instantaneous value for the cutting force per cutting edge is:

$$F_c(\varphi) = a_p \cdot f_z^{1-m_c} \cdot (\sin\kappa)^{-m_c} \cdot k_{c1,1} \cdot (\sin\varphi)^{1-m_c} \quad \text{für } \varphi_i \leq \varphi \leq \varphi_0 \tag{6.19}$$

where:

$a_p$ – cutting depth,
$f_z$ – feed per cutting edge and
$Z$ – number of milling teeth,
$1 - m_c$ – slope of the specific cutting force,
$k_{c1,1}$ – main value of the specific cutting force,
$\kappa$ – setting angle of the main cutting edge,
$\varphi$ – feed direction angle
$\varphi_i$ – cutting edge entry angle,
$\varphi_0$ – cutting edge exit angle

The feed direction angle φ is the angle between the feed direction of the tool and the specific cutting direction of the cutting edge currently in contact with the workpiece. During milling, the feed direction angle φ changes constantly. ◘ Figure 6.42 illustrates the definition of the feed direction angle for face milling.

The instantaneous torque $M(t)$ acting on the milling spindle comes from the arithmetical sum of the cutting forces $F_c(\varphi)$ on the individual cutting edges $j$ in contact on the cutting arc $\varphi_s$ at the instantaneous feed direction angle $\varphi_j$:

$$M(t) = \frac{D}{2}\sum_{j=1}^{j=z_E} F_c(\varphi_j(t)) \tag{6.20}$$

with a cutter diameter of $D$, the number of cutting edges in contact $z_M$ and the cutting edge index $j$.

The feed direction angle $\varphi_j$ and the number of cutting edges in contact $z_M$ are time-dependent variables changing periodically as the milling tool rotates.

The number of cutting edges in contact $z_M$ is only constant when the cutting arc $\varphi_s$ is an integer multiple of the cutter pitch angle $\varphi_T$. Assuming chatter-free machining, the time characteristics for the torque can be described using Eq. (6.19):

$$M(t) = \frac{D}{2} \cdot a_p \cdot f_z^{1-m_c} \cdot (\sin\kappa)^{-m_c} \cdot k_{c1,1} \cdot \sum_{j=1}^{j=z_E(t)} \left(\sin\varphi_j(t)\right)^{1-m_c} \tag{6.21}$$

The cutting power can be calculated as follows:

$$P(t) = 2 \cdot \pi \cdot n \cdot M(t) \tag{6.22}$$

◘ Figure 6.43 illustrates the various time characteristics of the cutting forces or torques for four specific machining conditions [HEID89]. Depending on the milling arc and the number of cutting edges, this results in various pulsating torques. The frequency of the base vibration corresponds to the tooth engagement frequency $f$. This is determined by the cutting speed $v_c$ and the cutter spacing measured on the circumference $b_z$:

$$f = \frac{v_c}{b_z} = n \cdot z \quad \text{where} \quad b_z = \varphi_T \cdot \frac{D}{2} \tag{6.23}$$

Both the cutting depth $a_p$ and the cutting arc $\varphi_s$ during machining can vary due to the variables influencing the torque during milling (see Eq. (6.21)) and can disturb the cutting process. For this reason, it is not possible to gather any information on the current cutting depth using only the magnitude of the measured torque as can be done with turning.

◘ Figure 6.44 shows the basic structure of threshold controls as they can be applied to milling to utilize

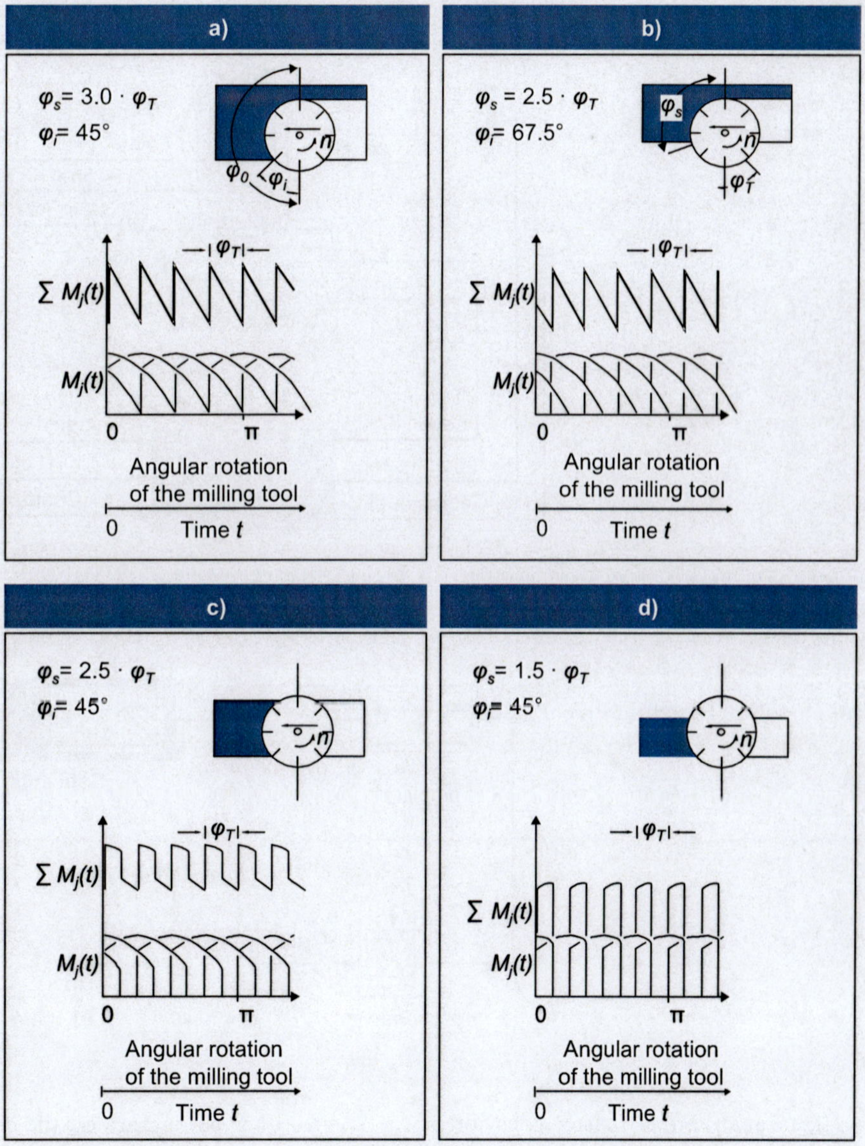


**Fig. 6.43** Time characteristics for individual torques on each cutting edge $M_j(t)$ and the total torque ($t$) when milling for various contact conditions

the available machine power. As with turning, utilization of the machine or the tool is maintained at a given constant value by a control device. The motor current, **spindle torque or deflection force on the front spindle** bearing can be used as the measuring variable and the feed or feed rate, spindle speed or cutting depth can be used as the control variable. However, the latter requires automated cut distribution in the machine. This cannot be achieved with today's commercial control systems. As already explained in ► Sect. 6.3.1 on turning, the torque, bearing load or expansions, rather than the motor current, are currently captured as the measuring variable due to the response behavior in the listed threshold controls.

When milling, the cutting power of the machine is often limited not by the power of the installed motor but by self-excited vibrations known as chatter. Stable machining is only possible if the chatter vibrations are captured by special chatter recognition logic and the process controller is able to automatically guide the milling process to a stable operating point when chatter vibrations occur. As explained in more detail in ► Sect. 6.3.2.5, a **speed modification and reduction in** the cutting depth are suitable measures for eliminating chatter vibrations on milling machines.

Unlike turning, the control loop of a torque control **for milling is characterized by another influencing fac**tor, the cutting arc $\varphi_s$.

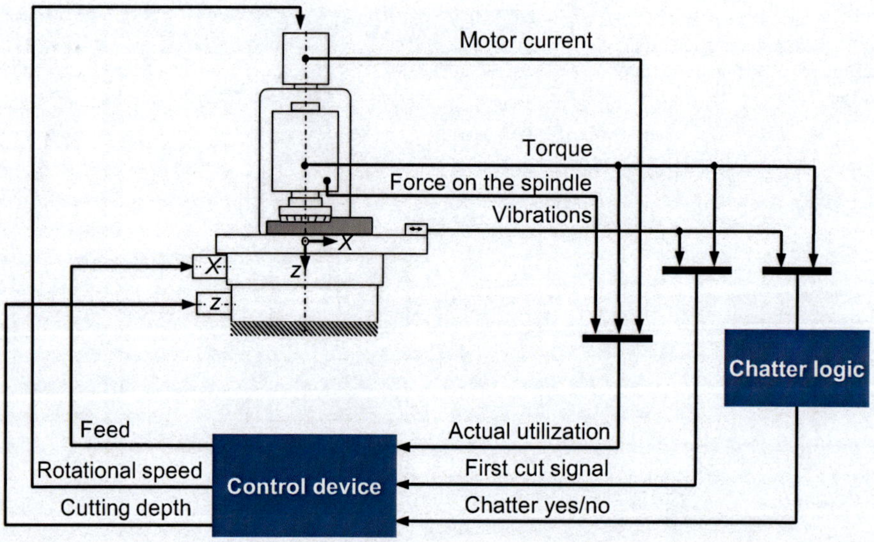


**Fig. 6.44** Process control structure for milling

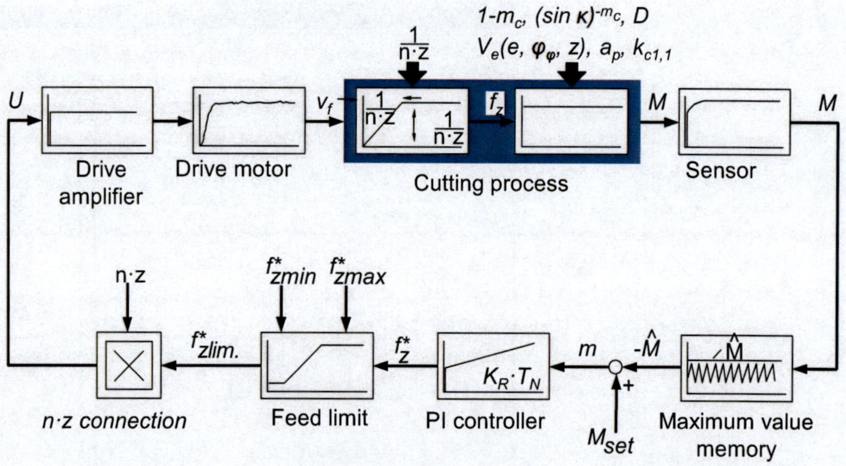


**Fig. 6.45** Control loop for a torque control for milling

If, to analyze the transfer behavior of the control path, the peak value of the torque $\hat{M}$ within the time interval of one single tool rotation

$$\tau = \frac{1}{n}$$

is considered instead of the pulsating torque over time M(t), see Eq. (6.21), this produces:

$$\hat{M} = \frac{D}{2} \cdot a_p \cdot f_z^{1-m_c} \cdot (\sin\kappa)^{1-m_c} \cdot k_{c1,1} \cdot V_e \tag{6.24}$$

where

$$V_e = \left\{ \sum_{j=1}^{j=z_E(\varphi_s,t)} \left(\sin\varphi_j(t)\right)^{1-m_c} \Bigg|_{\varphi_j=\varphi_0}^{\varphi_j=\varphi_0} \right\} \tag{6.25}$$

The maximum value for the total expression $V_e$ is part of the gain factor for the control path determined by the cutting arc $\varphi_s$, the cutting edge entry angle $\varphi_i$ and the number of cutting edges $z$.

■ Figure 6.45 shows the entire control loop for torque control for milling [HEID89]. As the tool speed during machining generally remains constant when milling, the power is also maintained constant by controlling the torque.

With this control, the cutting process is part of the control path with the feed rate $v_f$ as the input variable and the torque $M$ as the output variable. The cutting depth $a_p$, the material strength $k_{c1.1}$ and the gain factor $V_e$ are the main disturbances to the cutting process. These disruptive influences affect only the control path gain while the speed $n$ and the number of teeth $z$ affect the time behavior of the control path.

Machine utilization is captured by a torque sensor. As the spindle torque shows a sawtooth profile when milling due to the tooth contact impacts and the variation over time of the number of teeth in contact, a

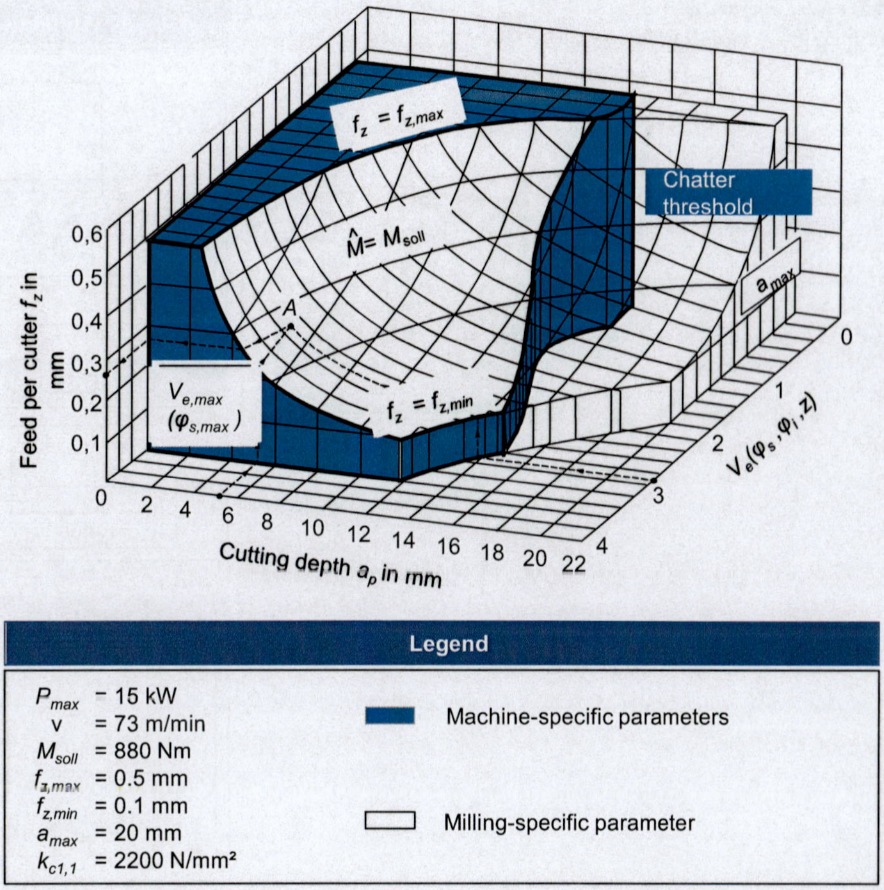


**Fig. 6.46** Control range for torque control

peak value memory (maximum value memory) has been added downstream of the sensor to regulate the maximum torque occurring during a single rotation of the milling tool. This avoids control loop oscillation at the tooth engagement frequency as well as overloading the individual milling cutting edges. A PI controller is used. This adapts automatically to the changing milling speeds and therefore to the variable time behavior of the path as well as to variation in the number of teeth in different tools. The output variable of the controller, the feed per tooth $f_z$, has an upper and a lower limit.

At small cutting depths $a_p$, the corresponding torque is very low. In this case, the controller attempts to achieve the reference torque by increasing the feed rate $v_f$. This can lead to too great a feed per tooth. To stop this, the feed is limited to a specific value $f_{zmax}$. Conversely, if the cutting depth $a_p$ is too great, the given value of $f_{zmin}$ occurs. The process would now have to be stopped and a new cut distribution set by the machine operator. In the future, it is conceivable that machine controls, as described in ► Sect. 6.3.1for turning, would automatically perform the cut distribution in these instances.

The control range for the torque control is achieved by considering the limits for the control variables $f_{zmin} \leq f_z \leq f_{zmax}$ and the chatter range as shown in ◘ Fig. 6.46 for a practical example. The influencing factors of cutting depth $a_p$ and gain factor $V_e(\varphi_s, \varphi_i, z)$ are entered as the abscissa and ordinate values.

The control variable for the control—the feed per cutter $f_z$—is shown on the ordinate. This varies as a function of the influencing factors. The hyperbolic arched surface is that of the constant torque, in this instance, the reference torque $M_{set}$. The surface $\hat{M} = M_{set} = const.$ describes the control range for the torque control. This is limited by $f_{zmin}$ and $f_{zmax}$ explained above as well as the maximum possible value $V_{emax}$ and the maximum possible cutting depth $a_{max}$. In most practical cases, the chatter threshold follows a similar course to that shown in ◘ Fig. 6.46 [HEID89] through the middle of the control range and limits it considerably. The chatter threshold is influenced by a large number of variables so that its position in the control range diagram can vary significantly. It is possible that the torque control may not be able to correct all the variations in the disturbances due to the

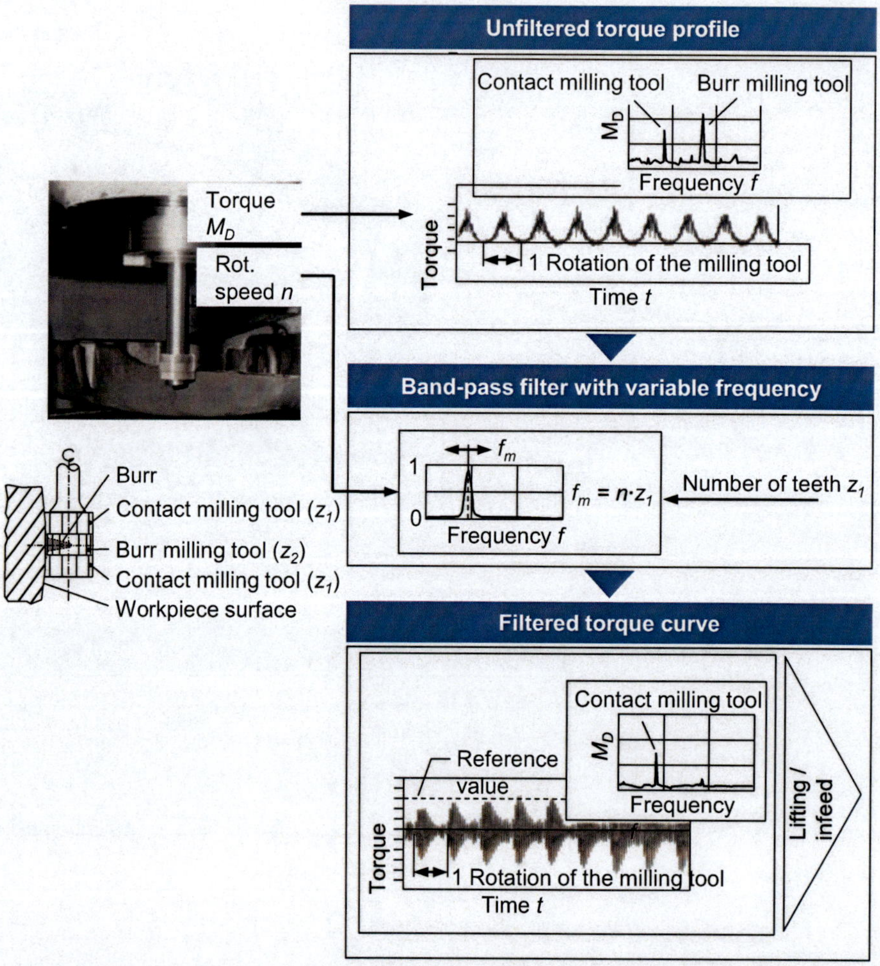


**Fig. 6.47** Functional principle of the sensor system for controlling the fettling quality during peripheral milling

technological, geometric and vibration-related limits on the control range. If the remaining limits are reached, the chatter elimination strategies described in ► Sect. 6.3.2.5 must be implemented and the machining parameters are to be modified so that the torque control can operate in the safe control range and the reference value is maintained.

### 6.3.2.4 Process Control When Fettling

Research work has been carried out as part of process control on the processes and tools for deburring cast workpieces [LENO93]. The same problems always occur here when programming the blank contour, regardless of whether robots or machine tools are involved. In one case, the precise geometry of the cast part is frequently not known and then there are often process-related tolerances on the blank part of up to 3 mm. Machining the flash using NC-controlled machines can only be completed with online monitoring of the tool position with respect to the surface of the cast part using appropriate sensors and an effective motion strategy.

Figure 6.47 shows a sensor system which can be used for deburring during peripheral milling. The milling tool consists of three standard side mills assembled on a single arbor. The side mills have a different number of cutting edges on the same diameter.

The positioning of the tool with respect to the burr is such that the burr is only processed by the middle milling tool. The milling tool is introduced such that the upper or lower contact milling tool barely touches the surface of the cast part. This condition is detected when the tooth engagement frequency of these milling tools is maintained in the cutting force, torque or structure-borne sound signal. If this frequency is missing in the signal, the tool is moved towards the workpiece. If the signal is too large, the tool is retracted as required.

A similar process can be used for face milling if the axial force is measured on a suitably prepared face milling tool and evaluated as described above.

This means that the NC program only has to approximate the deburring path. The correct position of the milling tool is achieved via the superimposed sensor

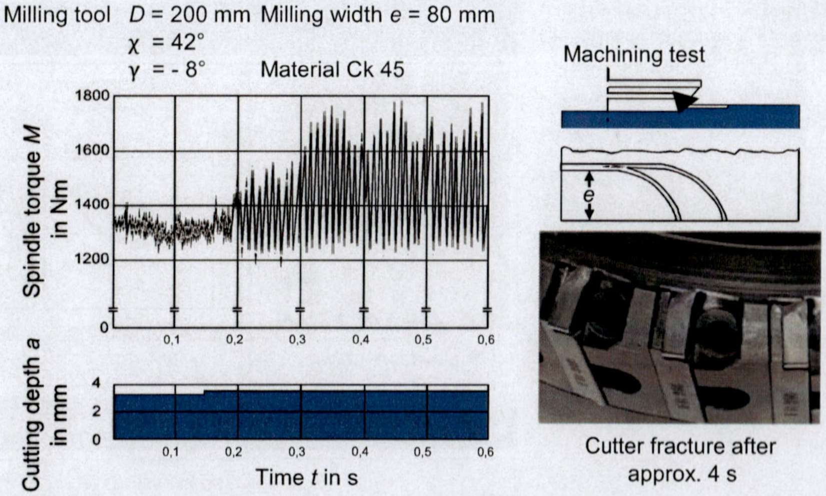


**Fig. 6.48** Effects of chatter vibrations on milling machines

guidance. However, the NC controller must have the interfaces required for this.

### 6.3.2.5 Automatic Chatter Elimination

Under certain conditions (see ▶ Chap. 3, Volume 2), cutting machine tools tend towards unstable behavior, i.e., self-excited vibrations. Vibrations are always excited in a machine when machining. The frequency occurring is determined by the compliance behavior of the machine system. The vibrations cause a displacement between the workpiece and the tool. A cutter on a milling tool therefore produces waviness on the workpiece surface. If the following cutting action meets an unfavorable phase relationship in this waviness, it produces a chip thickness modulation and the vibration builds up further.

Self-excited vibrations in metal-cutting machine tools are characterized by sudden increases in the vibration amplitude when a specific threshold cutting depth $a_{cr}$ is exceeded. Figure 6.48 shows the curve for the spindle torque during the transition from stable to unstable cutting. The vibrations started after a step change in the cutting depth of only 0.5 mm and the dynamic component of the torque increased by 600 Nm after 0.2 s. This high level of dynamic loading puts the workpiece, tool and machine in significant danger. In this instance, all the cutting edges on the milling tool were broken within approximately 4 s.

To guide the milling process automatically to a stable operating point through appropriate measures, it is necessary to identify the self-excited vibrations of the chatter while the process is running, i.e., to distinguish these vibrations from externally excited vibrations (e.g., from cutting edge contact impacts). Such identification can be performed in both the frequency domain and the time domain.

Figure 6.49 shows the basic sequence used by a sensor system for identifying chatter vibrations in the

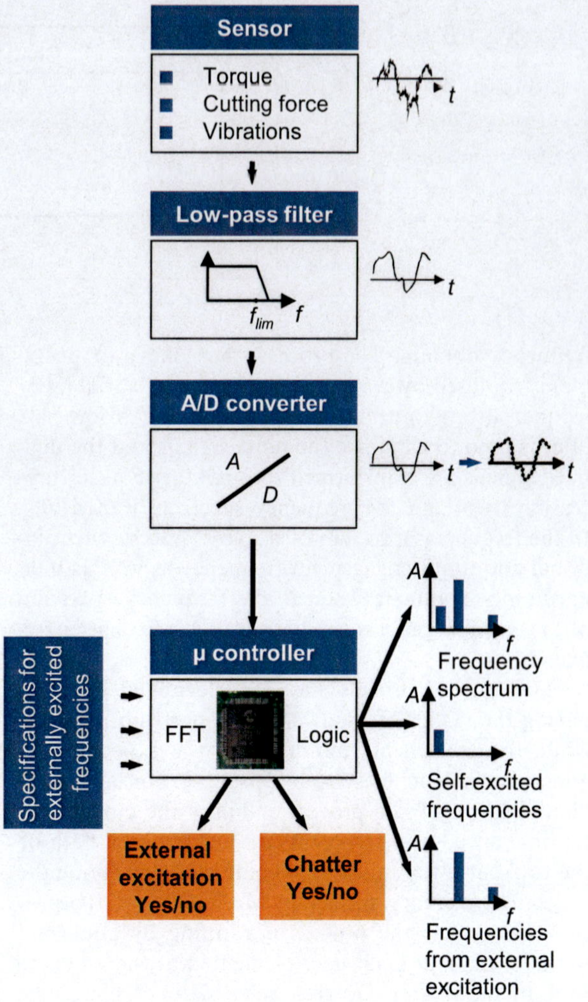


**Fig. 6.49** Sequence used by a sensor system for chatter identification

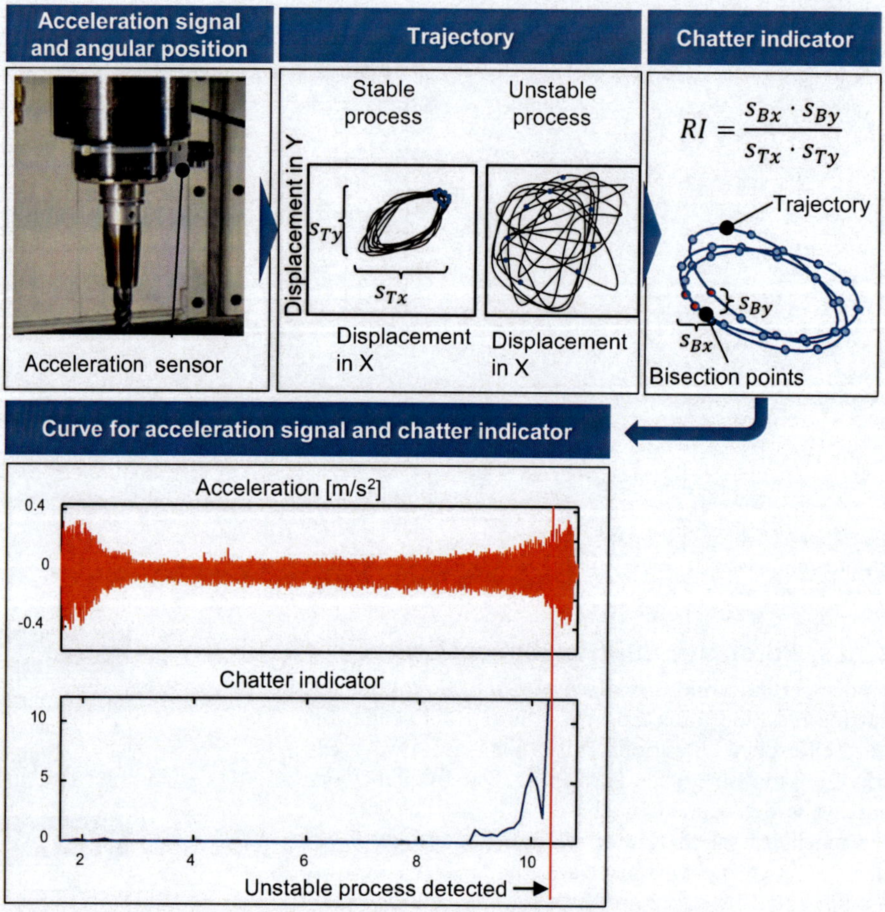


**Fig. 6.50** Process stability assessment using measurement of TCP displacement

frequency domain. The process state (stable or unstable) is captured by evaluating the measured variables of torque, cutting force or machine vibrations. A low-pass filter is used to eliminate the noise signals and the digitized signals are transformed through fast Fourier processing to obtain the frequency spectrum [ÖZME00]. In the frequency spectrum of a stable process, the rotational and tooth engagement frequency as well as their harmonics dominate whereas the frequency spectrum of an unstable process shows a peak at the chatter frequency.

The current process state is determined by comparing the values of the frequency spectrum recorded while the process is running with the expected frequency spectrum of a stable process. For a reliable identification of the process stability, the current rotational speed and the contextual technology data of the tool currently in use are required to determine the expected frequency spectrum. Monitoring of the process state while the process is running by analyzing the frequency domain is implemented at the machine tool manufacturer Dörries Scharmann Technologie (DST). By means of integrated acceleration sensors in the main spindle, the current frequency spectrum can be determined online in an evaluation electronics system. In addition, the rotational speed and the contextual technology data from the machine control system are transmitted to the evaluation electronics system to calculate the expected frequency spectrum. If the recorded values of the frequency spectrum deviate from the calculated values, the machine operator receives a warning, or other actions, such as a feed stop, are triggered.

The procedure for a process stability assessment in the time domain is shown in Fig. 6.50 using the procedure developed at Laboratory for Machine Tools and Production Engineering (WZL). Here, the process stability assessment measures the TCP displacement and the angle position of the spindle using the Poincaré section. The displacements in the feed direction and its normal are plotted in two dimensions over a defined time interval. The resulting trajectory describes the displacement of the TCP for one or more spindle revolutions. This trajectory is used to determine bisection points, describing the TCP displacement depending on a defined spindle rotation angle. The process stability is determined using a chatter indicator, which describes the relation of the standard deviation of the total

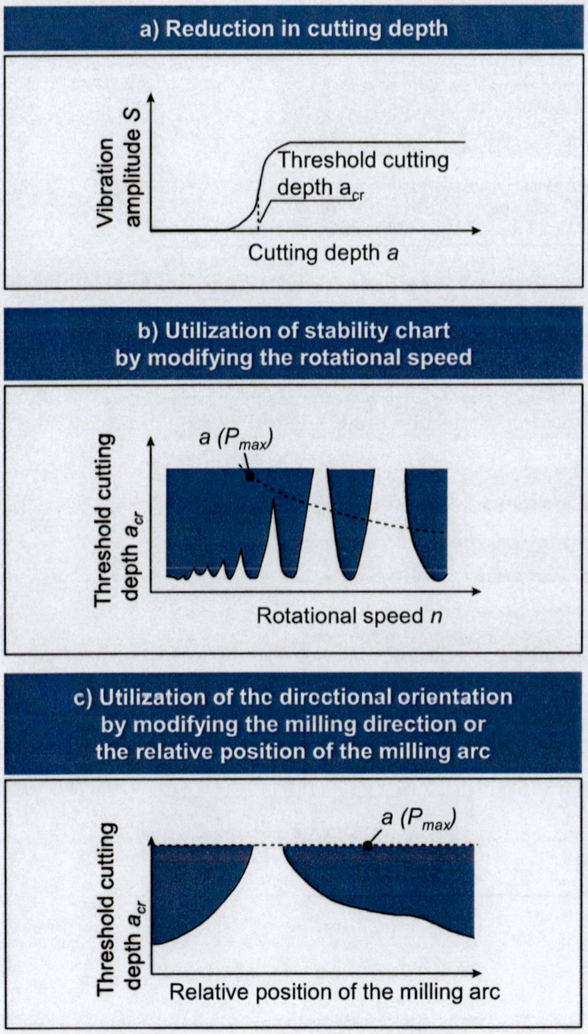


**Fig. 6.51** Possible measures for eliminating chatter vibrations

trajectory to the standard deviation of the bisection point cloud. Stable processes show a clearly defined set of curves and a small scatter of the bisection points within the sensitive plane so that the chatter indicator is close to zero. Unstable processes, on the other hand, result in a chaotic set of curves and pronounced scatter of the bisection points. The chatter indicator increases considerably when instabilities occur.

Suitable strategic measures for eliminating chatter vibrations, which can be implemented on milling machines during machining, can be derived from the typical characteristics already known for milling machines, see Fig. 6.51. The reduction in cutting depth is the safest and most effective measure for avoiding chatter vibrations as this always leads to stable machining. The disadvantage of this measure is that additional cuts are required. Here, the control system must implement automatic cut distribution. Commercial controls are currently not able to do this. The other measures described in Fig. 6.51, i.e., speed modification or utilizing directional orientation, scarcely reduce the cutting power due to their stabilizing effect but their scope is limited.

The strategy of speed modification is based on using the stable areas in the stability chart where the curve for the threshold cutting depth $a_{cr}$ is shown versus the milling tool speed $n$ [WECK77] (Fig. 6.51b). To achieve this, the area of the current stability chart containing the machine's working point for the current rotational speed must be known. If the rotational speeds from the stability chart associated with the lowest points of the unstable areas are known, it is then possible to decide whether a reduction or increase in the rotational speed leads to a stable area between two "instability valleys."

If an oscillatory system approximates single-mass oscillator behavior, it is possible to determine the critical rotational speeds $n_{cr}$ at which the minimum stability values occur (i.e., the minimum threshold cutting depth $a_{crmin}$):

$$n_{cr}(m) \approx \frac{f_{chatter}}{z(m - 0,25)} \qquad (6.26)$$

where $m = 1, 2, 3\ldots$ ; $f_{chatter}$ = chatter frequency

Here, $m$ is the number of waves cut on the surface at the given chatter frequency between two cutting edge contacts.

There are areas of greater stability between the minimum stability values whose maximum values do not lie at the very center between the minimum stability values within the stability chart due to the asymmetrical shape of the unstable areas. The positions of the minimum stability values can be described using the following empirically derived approximation formula:

$$n_{st} = n_{cr}(m+1) + 0,6[n_{cr}(m) - n_{cr}(m+1)] \qquad (6.27)$$

The flow chart in Fig. 6.52 shows the sequence used for the chatter elimination strategy requiring this formula for the calculations. A set from the NC program used to program the rotational speed $n$, is transferred to the control program before machining begins. If chatter occurs during machining, the process computer detects the chatter frequency in an alarm and uses the current chatter frequency to determine the minimum stability values in the rotational speed area as well as the adjacent target rotational speeds. The restrictions on the permissible area for the change in the rotational speed then have to be taken into consideration [HEID89]. The area for the change in the rotational speed is defined by the permissible cutting rates for the milling process. If a minimum stability value has to be exceeded in order to move from a chatter range to a stable area in the stability chart, the feed is briefly stopped while the speed is changed. This avoids potential increases in the chatter vibrations.

6

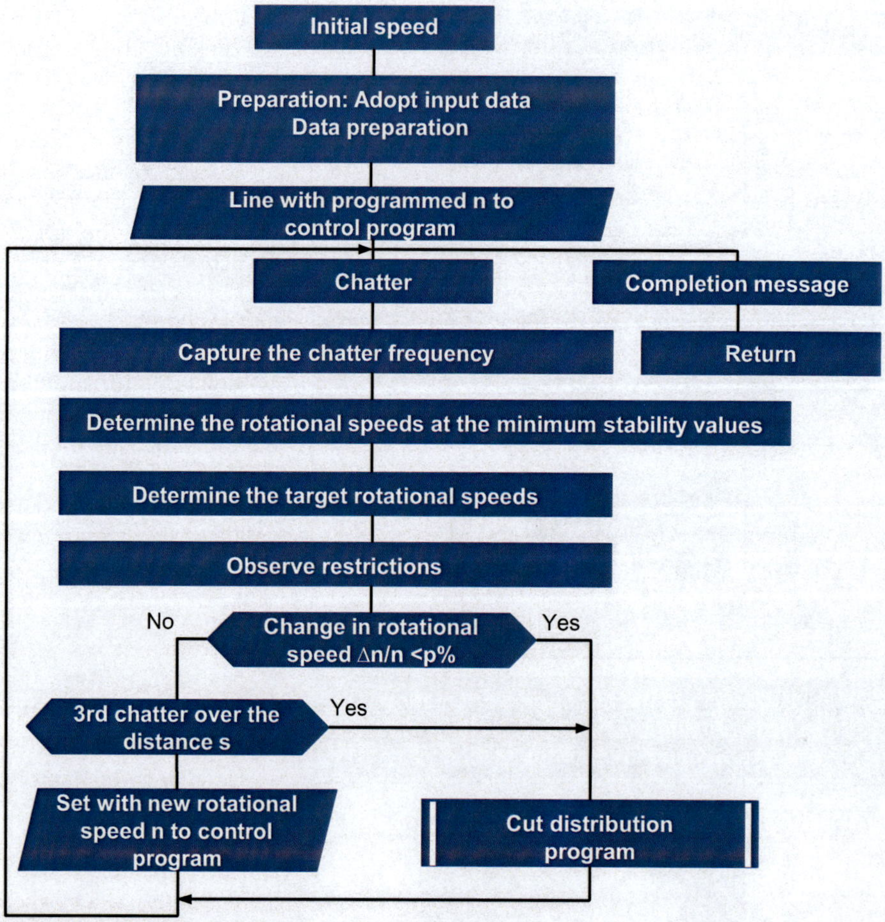


**Fig. 6.52** Basic flow chart for the strategy for changing the rotational speed

**Figure 6.53** shows an example of chatter elimination using speed modification. There is no maximum stability value in the permitted area for the change in the rotational speed. The rotational speed is therefore controlled up to the limit of the permissible area. The milling process is still stable at this speed. **Figure 6.53** shows the spindle torque, the amplitude spectrum of the torque signal, the chatter alarm and the rotational speed of the main motor, one above the other.

In this example, an FFT from 0 to 250 Hz is completed within approximately 100 ms (see **Fig. 6.49**). Stabilization of the machining process by speed modification can be clearly seen. The position of the rotational speeds before and after contact is shown in the upper section of the stability chart.

Another strategy for avoiding chatter is based on the elementary interactions during self-excited vibrations. In this case, the spindle speed is modified on a step-by-step basis when chatter occurs until the milling process returns to a stable condition [WITT88].

The greatest permissible axial cutting depth where the milling process using an inserted-tooth cutter still operates in a stable manner is described as follows:

$$a_{cr} = \frac{-1}{2 \cdot k_{cb} \cdot \mathrm{Re}\{G_g(jf)\}_{\mathrm{neg}}} \tag{6.28}$$

where

$a_{cr}$ – greatest permissible axial cutting depth for machining,

$k_{cb}$ – specific dynamic cutting force coefficient of the material,

$\mathrm{Re}\{G_b\{if\}_{neg}\}$ – negative real part of the directional compliance function between workpiece and tool}

The relationship between the chatter frequency $f_{chatter}$ and the tooth engagement frequency $n \cdot z$ on the milling tool is:

$$f_{\mathrm{chatter}} = n \cdot z\left(m + \frac{\varepsilon}{2\pi}\right) \tag{6.29}$$

where

$f_{chatter}$ – chatter frequency occurring,

$n$ – spindle feed,

$z$ – number of cutting teeth,

$m = 1, 2..$ – number of waves between cutter contacts

**Fig. 6.53** Stabilizing the milling process through automatic speed modification

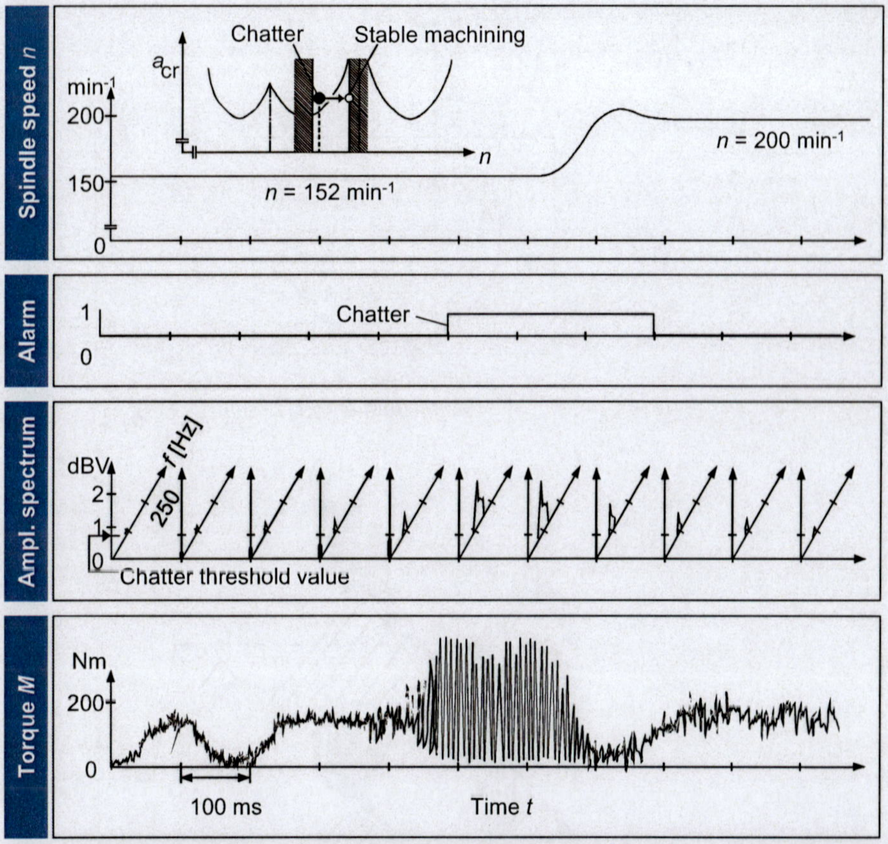


| Milling data | Machining data |
|---|---|
| $d$ = 200 mm | $a_p$ = 3 mm |
| $z$ = 28 | e = 60 mm |
| $\kappa$ = 42° | $f_x$ = 0.2 mm |
| $\gamma$ = -8° | $v$ = 90 – 125 m/min |

The phase shift ε between the vibration of the cutting edge and the waviness caused by the previous cut is described as follows for the stability boundary:

$$\varepsilon = 2\pi - \arctan\left(\frac{\text{Re}\{G_g(jf)\}}{\text{Im}\{G_g(jf)\}}\right) \tag{6.30}$$

Figure 6.54 shows graphically the relationships based on the directed frequency response of a simplified machine system in the form of a single-mass oscillator. Figure 6.54a depicts the stability chart for the process. Figure 6.54b shows the chatter frequency. The straight lines for the tooth engagement frequency $n \cdot z$ and their 1/m multiples ($m$=2, 3) $2 \cdot n \cdot z$ and $3 \cdot n \cdot z$ are also shown as a function of the rotational speed $n$. As the rotational speed decreases, with the chatter frequency constant, several waves ($m$) are shown between the cuts on the surface.

It should be noted that at the points of maximum stability, the phase shift ε between the vibration of the cutting tooth and the waviness on the workpiece surface caused by the previous tooth is precisely $\varepsilon = m \cdot 2\pi$. In this instance, there is no chip thickness modulation which leads to a build-up of vibrations. With $\varepsilon = 0$ or $\varepsilon = m \cdot 2\pi$, Eq. (6.29) becomes:

$$n \cdot z = \frac{f_{\text{chatter}}}{m} \tag{6.31}$$

First of all, the cutting depth $a$ and the milling tool rotational speed $n_A$ is given for a typical milling process, point A in Fig. 6.54a. The tooth engagement frequency occurring at this speed or its 1/m = 1/2 multiple is illustrated in Fig. 6.54b by points A' and A" on the straight lines $n \cdot z$ or $2 \cdot n \cdot z$. This working point lies within the unstable area. Chatter starts at a frequency of $f_1$, Fig. 6.54b. This frequency is measured

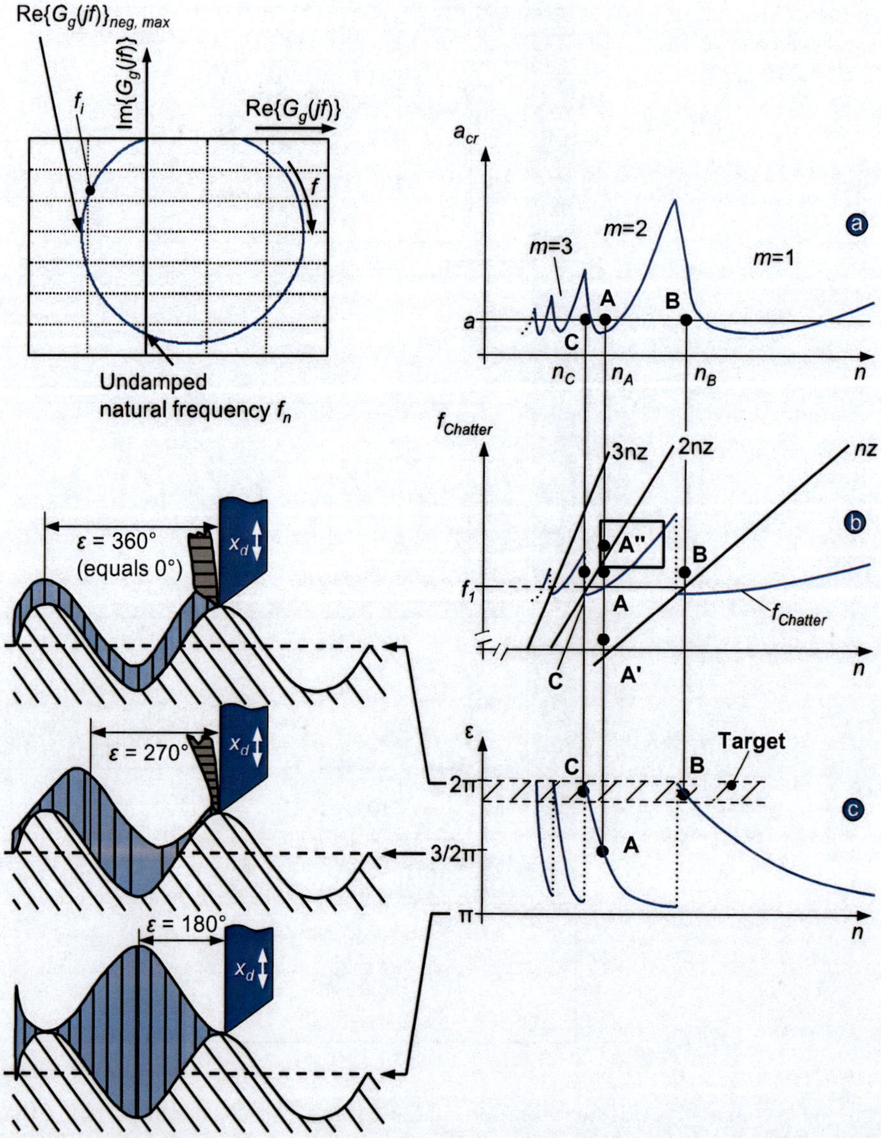


**Fig. 6.54** Displacement of the working point into stable machining areas according to Tlusty

using an appropriate sensor system (see Fig. 6.49). The control system, which must know the number of teeth on the milling tool, calculates a new spindle speed so that the tooth engagement frequency $n \cdot z$ corresponds precisely to the measured chatter frequency. In Fig. 6.54b, point A' moves along the straight line $n \cdot z$ to point B. The process repeats itself if necessary if the new selected rotational speed is still not in the stable area. In this way, the working point is moved step-by-step into the stable area.

The process effects a displacement of the phase relationship ε between the vibrating tool cutter and the waviness on the workpiece surface towards $m \cdot 2\pi$. With this phase shift, practically no chip thickness modulation occurs.

If, for example, an upper rotational speed limit is specified due to the performance of the machine or for technological boundary conditions, the process can also be stabilized by reducing the rotational speed. In this case, the spindle speed is reduced until the next 1/m multiple (m=2) tooth engagement frequency, $2{\cdot}n \cdot z$, corresponds exactly to the measured chatter frequency. In Fig. 6.54b, point A" moves towards point C. As can be seen in the stability chart, this point is also within the working area without chatter but one instability valley lower (m = 2).

As shown in the stability chart in Fig. 6.54a, there is a distinct possibility of locating an appropriate rotational speed between the instability valleys $m = 1$ and $m = 2$. The stable areas become notably smaller at lower rotational speeds, i.e., as the m values of the instability valleys increase. Looking for a stable area when m values are greater than 6 is largely pointless.

When using this control strategy for a range of different machining operations, experience has shown that a maximum of iteration steps are required to stabilize the process after the occurrence of chatter [WITT88]. However, if there are no areas in the stability chart where stabilization is possible for a given cutting depth, the only option is to reduce the cutting depth.

## 6.3.3 Drilling

### 6.3.3.1 Process Monitoring When Drilling and Deep Drilling

Drilling using twist drills is a frequently used machining method. Monitoring the drilling process is particularly important in relative terms. The potential of reducing costs by monitoring the tool has less to do with saving tools because twist drills are relatively inexpensive items. However, the subsequent damage which can result from drill bit fracture has a much greater impact. A piece of the broken drill bit usually remains in the material and so workpieces that have already undergone significant machining, and are therefore expensive, are useless. Furthermore, finishing tools such as tap drills and reaming tools will also be destroyed when they encounter the broken fragment.

Probes are used as mechanical sensors to monitor drill bit fracture. These probes make contact and, in conjunction with the position measuring systems, allow the tool length to be determined. Photoelectric sensor systems are used as optical sensors. A one or two dimensional image can be formed with the aid of photodiode arrays distributed along a line or over a surface.

Figures 6.55 and 6.56 show several application examples of the mechanical and optical sensors used

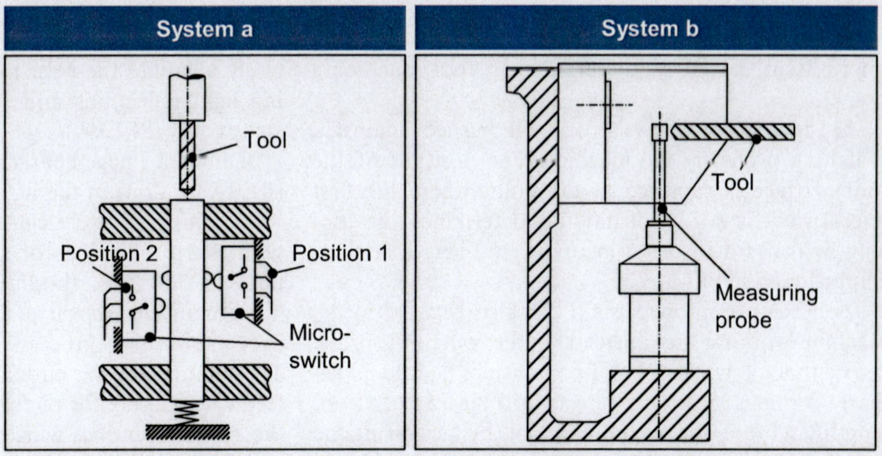


**Fig. 6.55** Systems for detecting drill bit fracture with mechanical sensors. According to Hüller Hille

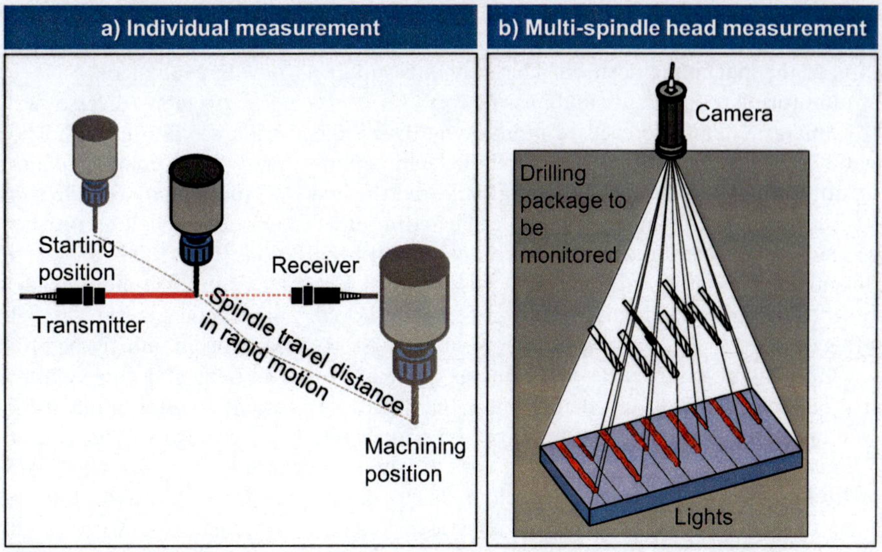


**Fig. 6.56** Systems for detecting drill bit fracture with optical sensors. *Source* Artis

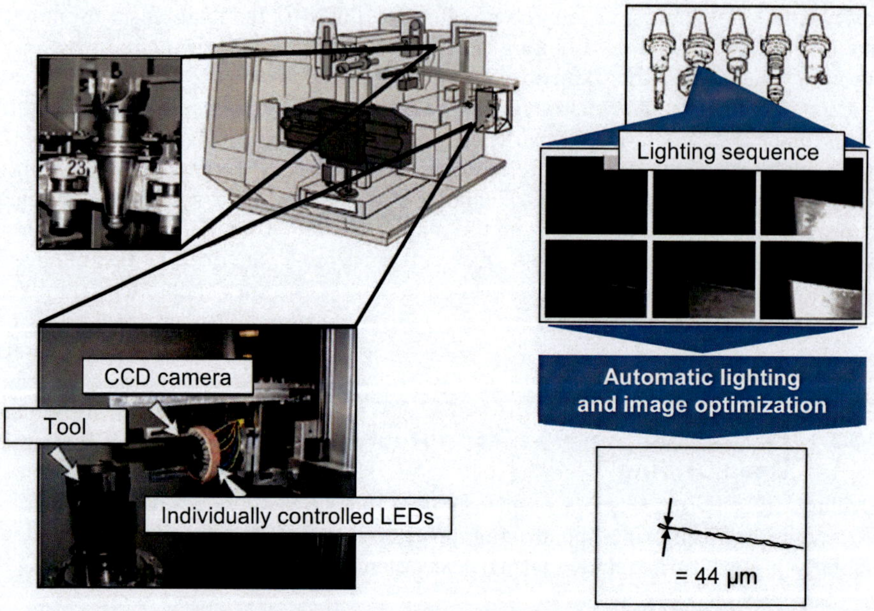


**Fig. 6.57** Use of a CCD camera with connected image processing system for measuring wear on drill bits. *Source* Pfeifer

6

in tool monitoring systems. The processes used here are intermittent, i.e., they operate during tool machining breaks.

Figure 6.55a shows a drill bit pushed against a switching probe by the machine. The position of the tool carriage is captured at the point where the first microswitch is switched and this determines the tool length. Tool fracture and incorrect tool length settings can be found in this way.

A measuring probe system can also be used in the machine working area instead of the switching probe device that allows sampling from all sides (axial and radial) (Fig. 6.55b). This detects drill bit fracture (tool length) and tool wear on milling tools by measuring the tool diameter in increments in the range of 1/100 mm.

The design of an optical inspection device for detecting drill bit fracture is shown in Fig. 6.56a. The drill tip is scanned while moving from the starting position to the machining position. This system is suitable for monitoring tool fracture and missing tools.

Camera systems are used to monitor multi-spindle heads, see Fig. 6.56b. The multi-spindle head is photographed by the camera in a learning cycle and the image is saved in the monitoring module. The contour of the learned reference image is then compared with the contour of the actual image taken. Missing drill bits and the fracture of individual drill bits are monitored in this way.

CCD (charge coupled device) cameras are particularly suitable for precisely determining the width of wear marks on the main cutting edges on a drill bit. Figure 6.57 shows an arrangement with a camera in the tool magazine on a machining center. The drill bits in the magazine are checked by the camera during the machining process. An image processing system precisely identifies the wear marks on the cutting edge using light reflections and shadows and measures their size exactly [PFEI99].

Different image characteristics easily occur through slight variations in the lighting, the tool or the type of wear. This effectively causes the image to be overlaid with sharp reflections or shadows. Information in the image is thus lost (brightness override) or additional 'false' contours appear in the image. A widely used image optimization process has been developed for this. The lighting on the cutter is varied using a fixed pattern when taking the picture. By comparing the images, the system identifies which contours in the image have changed position or moved, i.e., as a function of the lighting. These are 'false' contours from reflections or shadows and are eliminated in the contour image. What remains is an image of the 'true' wear contour that can now be evaluated.

Extremely diverse wear patterns occur on twist drills, see Fig. 6.58. Thorough investigations have already been completed on the effects of different sorts of individual wear to static and dynamic forces and torques as well as on the production of sound vibrations [HIMB88].

A suitable monitoring process has been established in practice based on detecting vibrations that typically occur shortly before drill bit fracture. Heavy corner wear and guide chamfer wear will cause a drill bit briefly to seize in the drilling hole, to twist and to rip itself out again. This causes vibrations which can be determined in the axial force and the torque. These vibrations in structure-borne sound can be very clearly detected even with small drill bits, see Fig. 6.58.

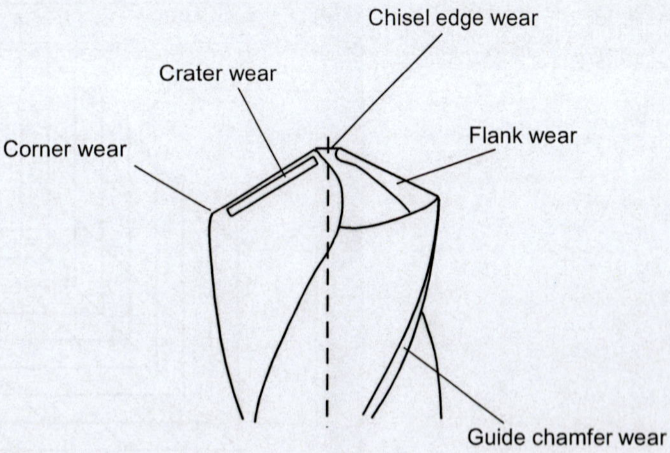


Fig. 6.58 Types of wear on a twist drill

The drill bit breaks shortly after these vibrations occur due to the heavy static and dynamic loading.

The vibration frequencies occurring can have a very wide spread. These frequencies depend on the vibration characteristics of the overall system consisting of the machine, the workpiece, the tool, and the clamping devices. Generally, the broad-band measurement of the overall structure-borne sound level is sufficient for monitoring the drilling process. The occurrence of drilling vibrations and the violent acoustic pulse found when a tool spontaneously fractures are detected when a fixed limiting value is exceeded.

A further very clear criterion is the spread in the curve of the histogram (amplitude density distribution) of the structure-borne sound time signal. During undisturbed machining, the drilling process and the machine produce a noise with an amplitude density distribution similar to a Gaussian function, see Fig. 6.59, bottom. The spread of this distribution increases as additional vibration amplitudes occur in the time signal.

In practice, a range of sensor systems based on measuring the feed force are used to monitor drilling. As the drill bit becomes increasingly blunt, the feed force increases with the same feed rate. Figure 6.60 shows a sensor where the feed force on the axial bearing of the ball screw is measured using a specially designed deformation element acting as the bearing support and fitted with strain gauges [EVER93]. The bottom of the Figure shows typical signal curves of the sensor with two limiting values.

When using this sensor, there must be the option of installing an additional force measuring sleeve that can accommodate the axial bearing. If this is not possible for design reasons, then force measuring bearings can be used, see Fig. 6.61.

Here, standard bearings are provided with a circumferential groove in the outer ring. The force transmitted from the bearing can then be measured using appropriately attached strain gauges. These bearing forces are modulated with the overrolling frequency of the rolling elements so that the signal initially has to be put through a low-pass filter. This solution is not suitable for dynamic measurements, e.g., capturing chatter vibrations. In principle, these bearings can be prepared for measuring radial and axial forces.

Monitoring the drill bit using the current of the main or feed motor is also frequently used in practice. However, additional errors occur here due to frictional losses in the drive train (belt drive, gearbox). Dynamic processes can only be captured in the very low frequency range. The inertia in mechanical components acts as an extreme low-pass filter. For larger drill bit diameters, monitoring the motor current is, however, a functional process in practice.

A significant improvement can be achieved by using the control-internal monitoring system described in ► Sect. 6.3.2.2. In this system, the frictional and acceleration effects are offset by using information on the position, rate and acceleration accessible from the control system.

Producing deeper holes using a deep drilling process requires a constant coolant circuit for cooling and lubricating the cutting edge and support strips. It is also responsible for the continuous flow of chips away from the hole. Transporting the chips away through the long boring rod requires short swarf produced by the specific design of the cutting edge (cutting edge distribution, chip groove) and suitable machining parameters (cutting speed, feed).

Identifying the flow of swarf and chip shape is a necessary condition for automating the deep drilling process. One option for chip shape identification is to use the chip breaking frequencies. Figure 6.62 shows the conceptual arrangement of an inductive ring sensor used for this purpose [ROSE94]. The nozzle in the coolant return which has been specially designed for the drilling spindle is required to ensure there is a

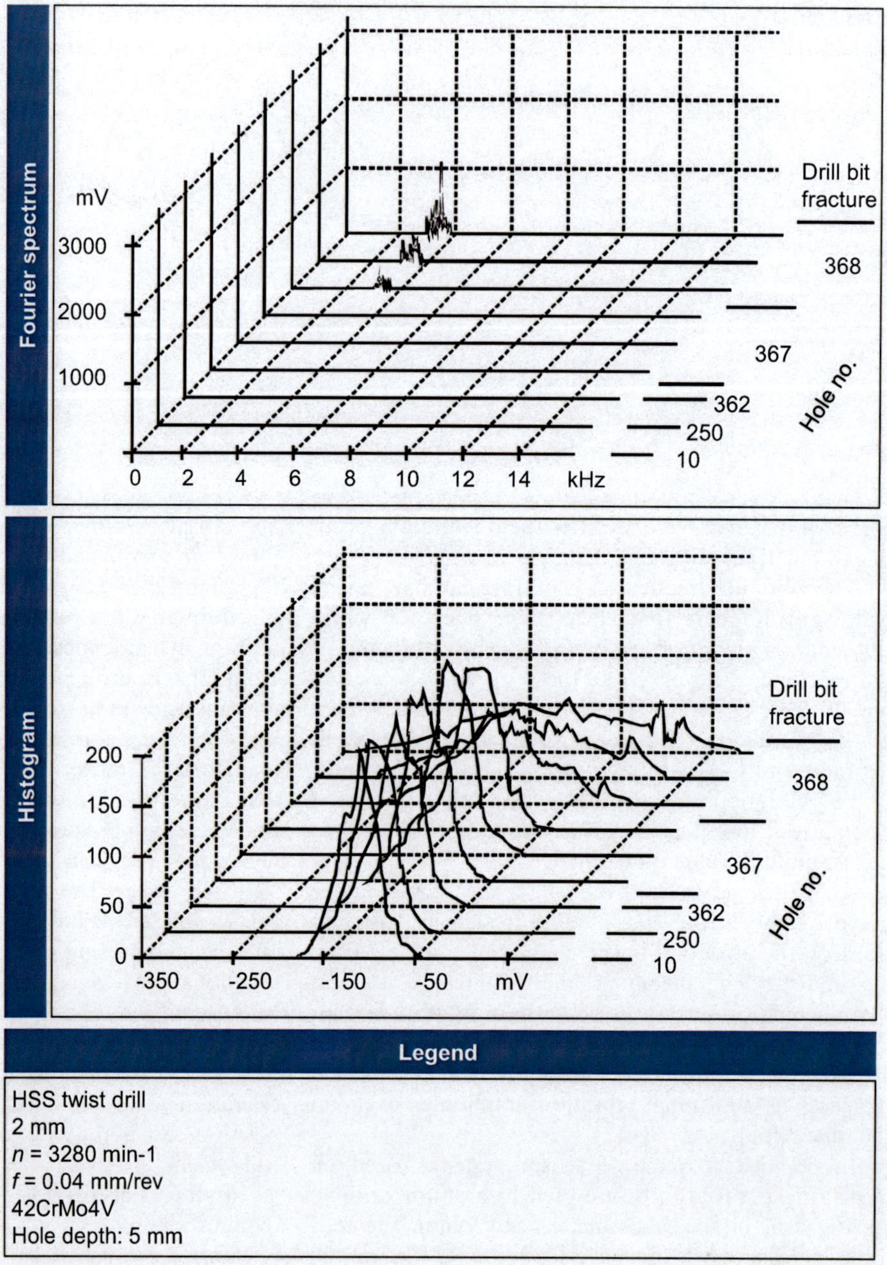


**Fig. 6.59** Variations in structure-borne sound shortly before fracture of a twist drill

6

sufficiently large flow velocity to produce a high resolution in the area of the measurement path.

The output signal is dependent on the number (= pulse frequency) and size (= pulse width) of the chips so that the actual process state can be described using an adapted evaluation electronics system and the system can react appropriately using a programmed strategy in the event of chip shape deviations.

### 6.3.3.2 Process Control for Deep Drilling

The deep drilling manufacturing process is very important in specialized areas of application. The high drilling quality achieved today with this process is possible due to specialized tools. The achievable drilling depth / diameter ratio can be up to approximately 400. The conditions for a perfect hole are a hole that is as straight as possible and a perfect flow of chips away from the hole.

Setting technological limits and, above all, monitoring the process flow are currently often left to operator's instincts and urgent action can be required to prevent potential subsequent damage to expensive tools and workpieces. Process control using the measurement of all relevant process state variables is an absolute

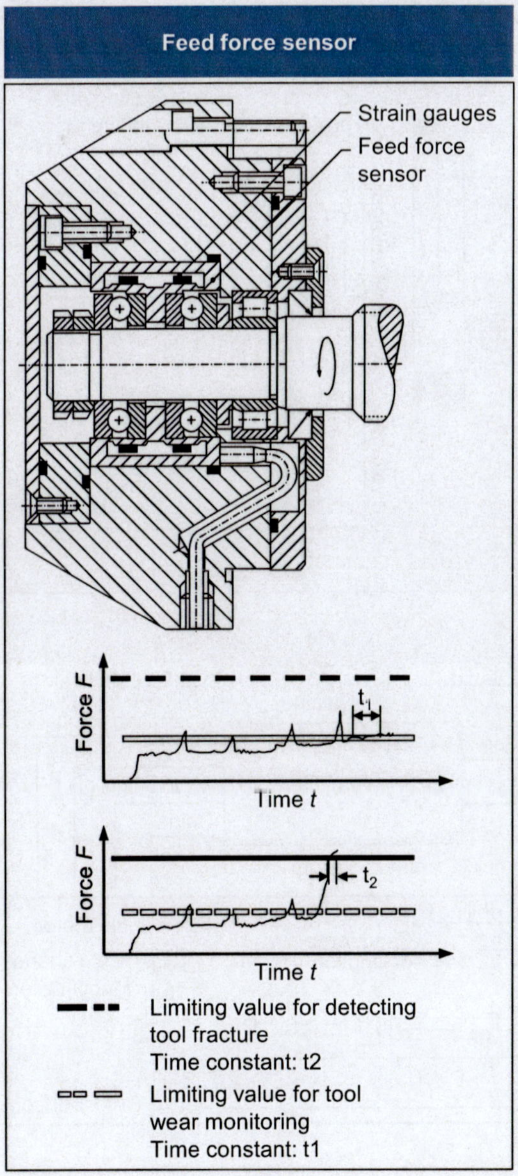


**Fig. 6.60** Feed force sensor for process control and monitoring. According to Sandvik

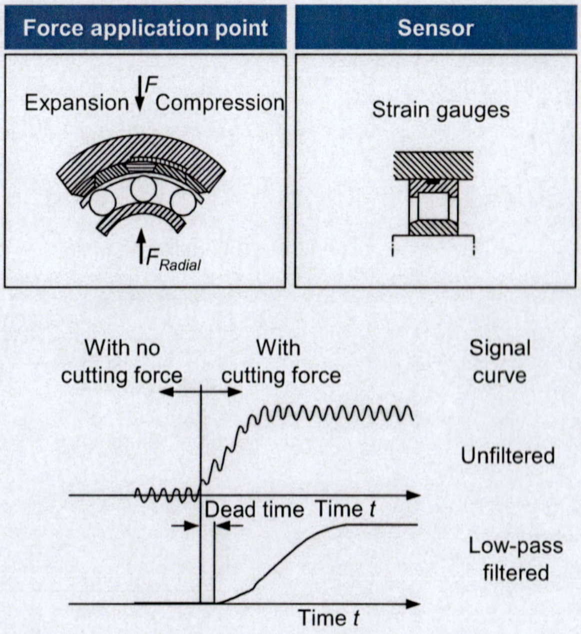


**Fig. 6.61** Structure and functionality of a force measuring bearing according to Promess

essential if the intended degree of automation is to be achieved.

A closed control loop for the deep drilling process is shown in Fig. 6.63 [ROSE94]. Input parameters for the process controller are feed force, drilling torque, machine vibrations as well as information on the chip shape and feed path. These signals are obtained through appropriate sensors during the process flow. The two parameters of drilling torque and feed force are suitable basic controlled variables for controlling and monitoring the deep drilling process. Variations in the tool cutting edge are clearly visible in the feed force signal while variations in the support strips are easier to detect in the drilling torque signal [ROSE94].

The process-dependent control variables of spindle speed and feed rate are generated by the controller for the machine control system from the sensor signals and reference values. To achieve this, the reference values, such as the tool-dependent feed forces or drilling torques and information on the chip shape must either be specified or learned. The system shown here also controls the drill head setting using information from the rotary encoder to monitor the approach, entry and exit positions.

## 6.3.4 Grinding

### 6.3.4.1 Process Control

In the sequence of various machining steps, grinding is usually last, after hardening of the workpiece. This is a particularly important process as it defines the required dimensions and surface qualities.

There are considerable problems here in measuring the quality characteristics mentioned above directly and as part of the process. Measuring regulations, which capture the change in the workpiece diameter during machining with NC-controlled, mechanical measuring heads and provide the switchover points of the feed rate for roughing, finishing and sparking out to the machine control system, are regarded as the

6

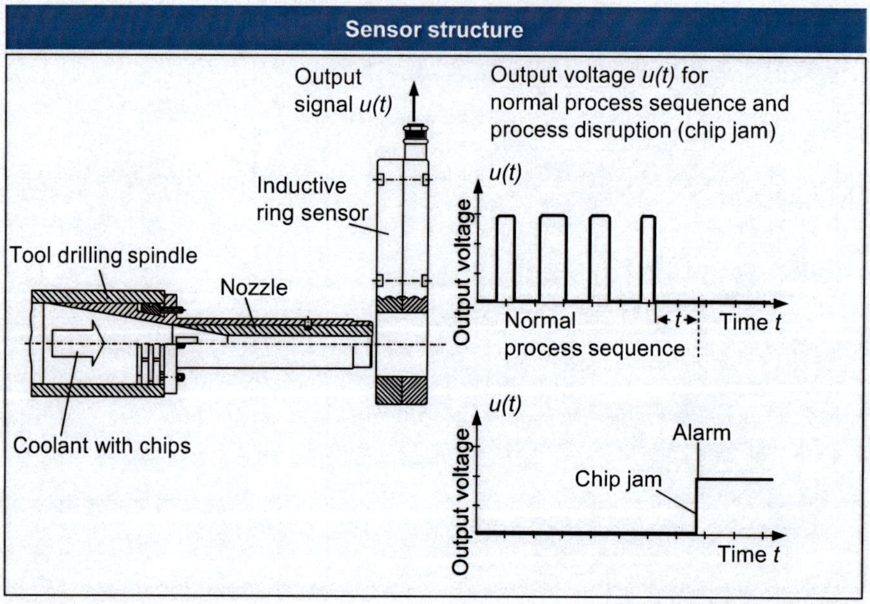


**Fig. 6.62** Inductive ring sensor for measuring swarf flow rate frequencies in deep drilling according to Baier

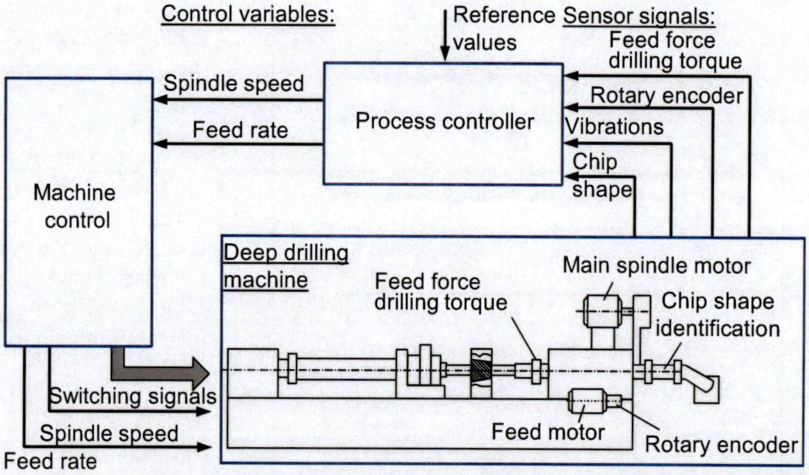


**Fig. 6.63** Hardware configuration for deep drilling process control

standard solutions today. The machining process is stopped once the nominal diameter is achieved.

However, there are proven systems that reduce the machining time by avoiding or reducing gap elimination before the first cut [JAKO83]. The grinding wheel feed rate is normally changed from rapid motion to roughing speed at a given maximum workpiece oversize. In this system, the wheel infeed takes place at a reduced rapid motion rate until the grinding wheel contacts the workpiece. This first cut is detected with the aid of structure-borne sound sensors—for safety's sake, usually with a redundant design.

In a standard three-stage grinding process, machine bending caused by the cutting forces is largely dissipated in the finishing phase by the decrease in the workpiece diameter. Such systems, which deliberately extend the roughing phase, shorten the machining time to reduce the amount of machine bending also due to rough machining [JAKO79], Fig. 6.64.

By extending the time of the roughing phase, the workpiece is machined with greater cutting power almost to the point of the final dimension while the finishing phase is reduced. However, this increased infeed during the extended roughing phase would produce a corresponding undersize in the workpiece during subsequent finishing. For this reason, the headstock is withdrawn just before reaching the final dimension. The precise finished dimension is thus achieved during subsequent finishing work, see Fig. 6.65.

The machining parameter settings for the grinding process are based on engineering empirical values and the elastic machine properties. The setting

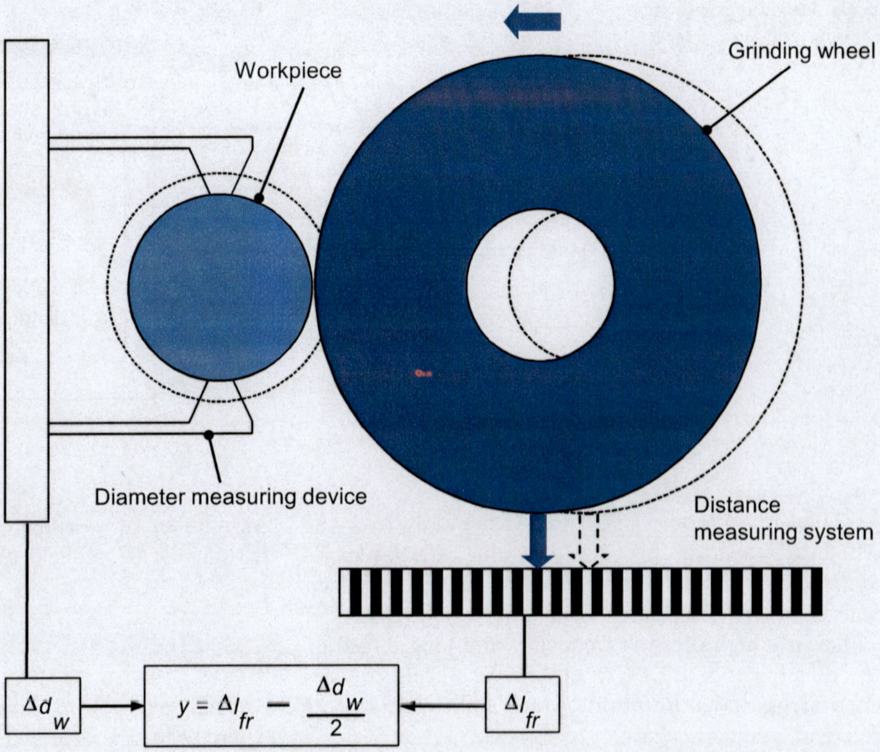


**Fig. 6.64** Calculating the machine bending y from the feed $\Delta l_{fr}$ and the decrease in diameter $\Delta d_w$

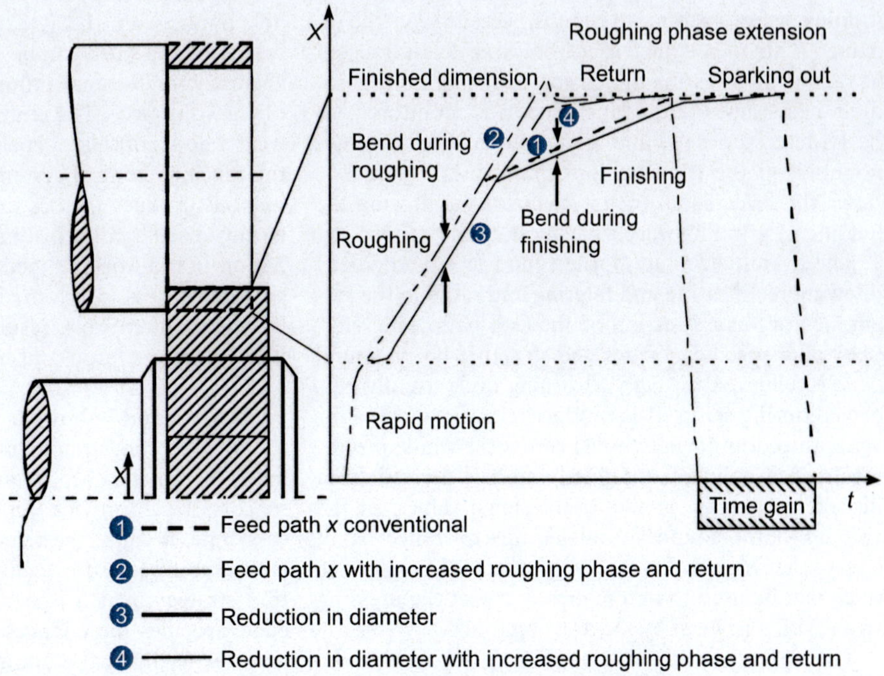


**Fig. 6.65** External cylindrical grinding with head stock withdrawal

values are prepared in the form of value tables or databases. The optimum machine setting frequently has to be established by the user through quality checks on workpieces and targeted variation of the setting parameters.

#### 6.3.4.2 Dressing Monitoring

The aim of the dressing process is to produce the required geometry and cutting capability of the grinding wheel to ensure workpiece quality in the following grinding process.

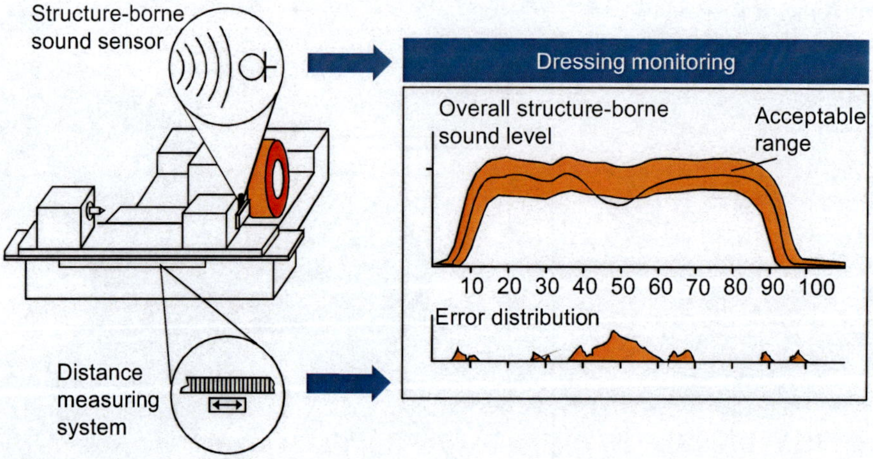


**Fig. 6.66** Principle of a dressing monitoring system

The grinding wheels do not wear evenly over the whole edge width. In particular, the external edges and the sharp, protruding profile elements on profiled wheels experience greater wear. For safety reasons, a sufficiently high dressing frequency and long infeed of the dressing tool are set to ensure the wheel is sharpened across its entire profile width and it achieves its reference geometry again.

As the amount of wear is not a precise quantity, too much rather than too little is delivered so that the grinding wheel loses more than is necessary on the diameter. This means increased tool and dressing costs. To reduce the dressing dimension to the amount absolutely necessary, the dressing process is monitored via the structure-borne sound signal curve. The sensor is mounted on the dressing tool bracket. Figure 6.66 shows the basic setup of a system for monitoring the dressing of grinding wheels having stationary tools.

The overall structure-borne sound level is digitized following rectification and filtering and saved in the system memory as a function of the feed path. The reference value specification for the structure-borne sound curve is achieved through a learning cycle usually performed during setup. This produces the signal curve for a guaranteed uniform removal across the whole profile. A tolerance band of a defined width is derived from this reference value profile. If the signal values for the structure-borne sound lie outside this tolerance band, an error message is sent via the system PLC interface which can be used to automatically repeat the dressing stroke with a new cut by the dressing tool.

The screen also shows a frequency distribution for the dressing errors over the last 32 dressing cycles. This is shown below the current structure-borne sound signal curve. The erroneous actual value profile depicted in Fig. 6.66 shows that there is increased wear in the wheel center. Due to this deviation, the wheel has to be dressed again. In order to optimize this process, the infeed can be increased as a preventive measure and/or the number of workpieces to be ground between two dressing cycles can be decreased.

### 6.3.5 Electrical Discharge Machining (EDM)

Correct operation of equipment for EDM requires an experienced operator with appropriate knowledge of the technological interactions between the many setting options on the equipment and the results achievable by the work [ULRI82]. Assessment criteria for the results of the work can be: surface quality, dimensional accuracy, or thermal influence on the peripheral zone of the workpiece. The removal rate (workpiece) and the wear rate (tool) also depend on the chosen setting for the machine control system and are influenced by various disturbance factors such as the rinsing conditions at any specific point in time or the degree of contamination in the working medium (dielectric). The control system's task is to use the best machine control setting for the required work results regardless of the capabilities of the operator, and to offset the disturbance factors.

Figure 6.67 shows a schematic for the design of a controller on equipment for EDM. Firstly, appropriate parameters are obtained from the process using sensors, which provide information on the current process state, because the removal rate or wear cannot be measured directly for technical reasons. The optimization strategy, which is based on technological knowledge, specifies the changes in the control variables required to achieve the optimum working point depending on the target variable (e.g., short machining time, low tool electrode wear).

Figure 6.68a shows possible characteristic discharge voltage curves at the discharge gap. The curve described by the spark discharge indicates a discharge with intense erosion while curves of the no-load and short form do not generate erosion and the abnormal

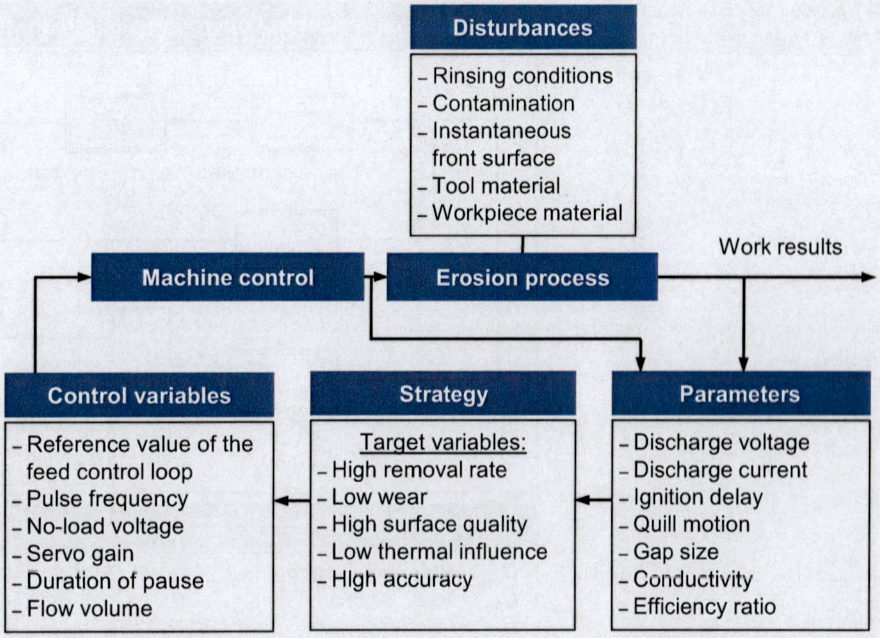


**Fig. 6.67** Flow chart of a process control system for electrical discharge machining with descriptive parameters

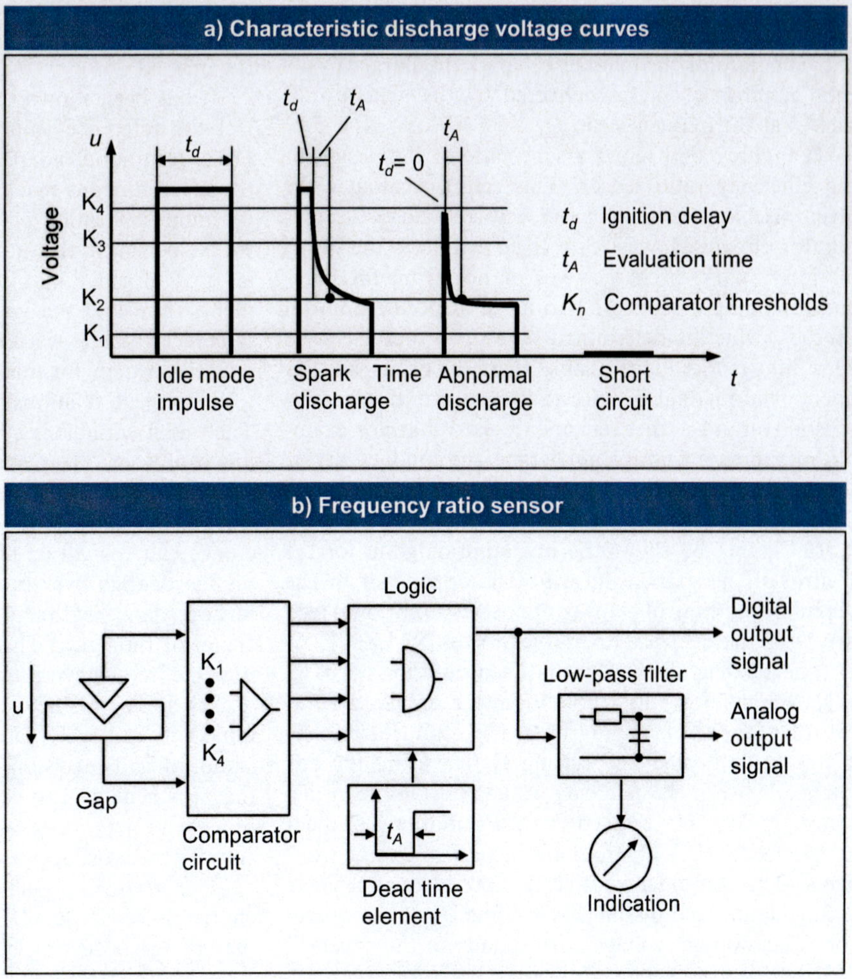


**Fig. 6.68** Characteristic discharge voltage curves and schematic diagram of the frequency ratio sensor

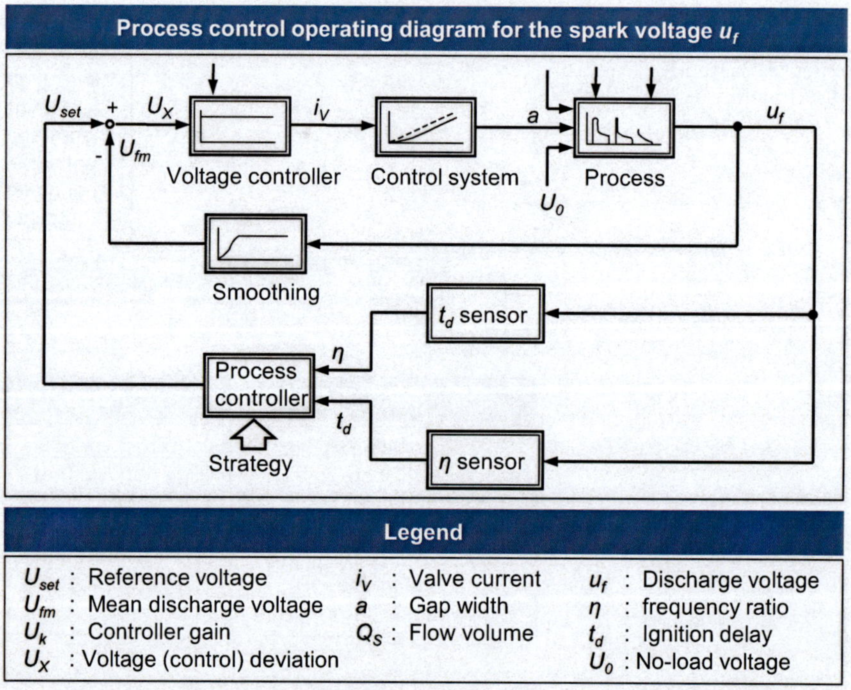


**Fig. 6.69** Process control operating diagram for the spark voltage $u_f$

discharge indicate the tendency toward unwanted arcing. The ratio of the number of spark discharges to the total number of pulses delivered by the generator is known as the frequency ratio η.

Figure 6.68b shows a schematic of the design of the efficiency ratio sensor. This sensor evaluates the given discharge voltage curves with the aid of a comparator circuit. Downstream logic generates the information "Spark discharge—yes or no" from the comparator's output signals *K*1 to *K*4 at a specific point in time, $t_A$, which is determined by a dead time element. This information is available digitally or as an frequency ratio η analog signal (averaged over time) at the sensor's output. Other sensors are used that, for example, measure the ignition delay time, the conductivity of the gap, and the valve current of the hydraulic feed system.

Figure 6.69 shows the operating diagram for the control of the spark voltage $u_f$. The upper half of the diagram (the control path) contains a conventional feed drive and the electrical discharge process [SALZ97].

The reference value for the feed control loop $U_{set}$ is set manually using a potentiometer on the generator used for the machining task. The quill is driven by a voltage controller and setting system using the difference between the reference value and the mean discharge voltage $U_{fm}$ derived from a smoothing element in the feedback path from the actual discharge voltage $u_f$. This sets a gap width *a* that produces an average voltage influenced by the flow volume $Q_S$ of the dielectric. This average voltage corresponds to the specified value $U_{set}$. The disturbances given in Fig. 6.67 act on the process.

It has been shown that it is difficult to correctly select the reference value depending on the machining task. Continuous adaptation of this control variable to the instantaneous machining conditions contributes to optimum utilization of the machining plant. The controller performs this adaptation. For this, the controller uses information on the process state from measurements of η and the ignition delay $t_d$ and changes the reference value $U_{set}$ according to a given strategy.

The strategy for this simple controller and the basic technological relationships are shown in Fig. 6.70. The target variable is a high erosion rate with relatively low tool wear. The technology has shown that the erosion rate $V_W$ is proportional to the frequency ratio η as can be seen in the lower section of Fig. 6.70. The reference value is varied using a inquiry procedure based on the change over time of the parameters η and $t_d$ in accordance with the decision matrix shown in the center of the figure. This procedure achieves the largest possible frequency ratio at the maximum erosion rate.

For some time now, there have been efforts to determine how to develop adaptive control devices for both standard and precision machining. It has been shown that it is beneficial to define the burn delay time as another parameter besides the ignition delay time to characterize the erosion process [STRO98], Fig. 6.71.

This makes it possible to describe the abnormal discharge pulses in detail. These are characterized in that the voltage occurring before discharge lies in the area

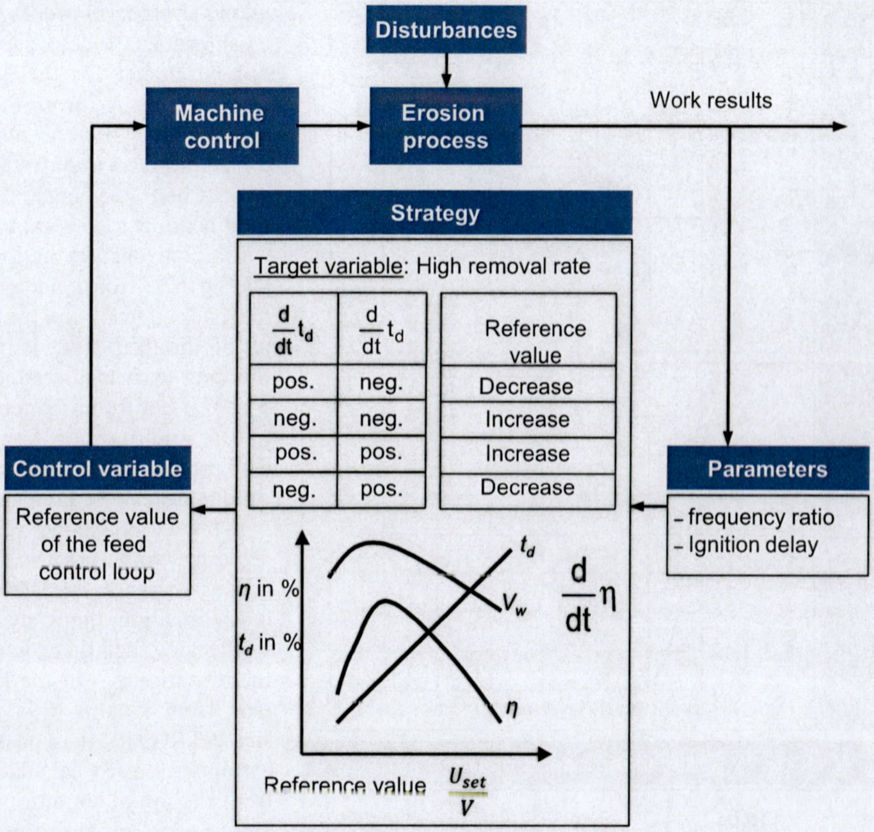


**Fig. 6.70** Example of a partial process control system

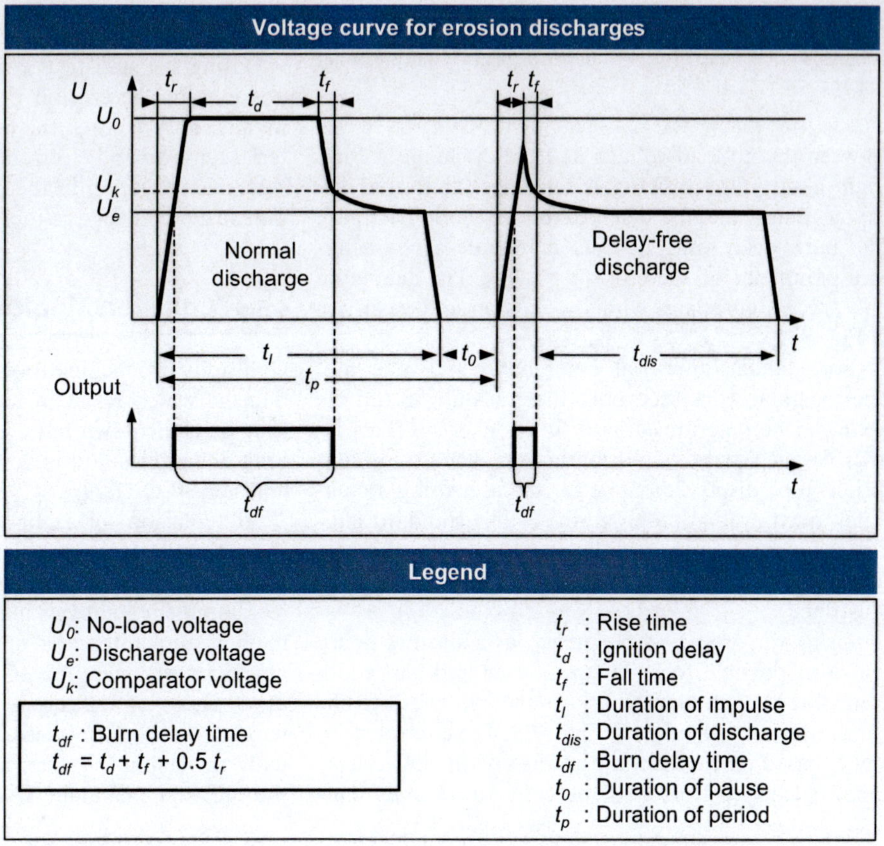


**Fig. 6.71** Extended illustration of the parameters of an erosion discharge

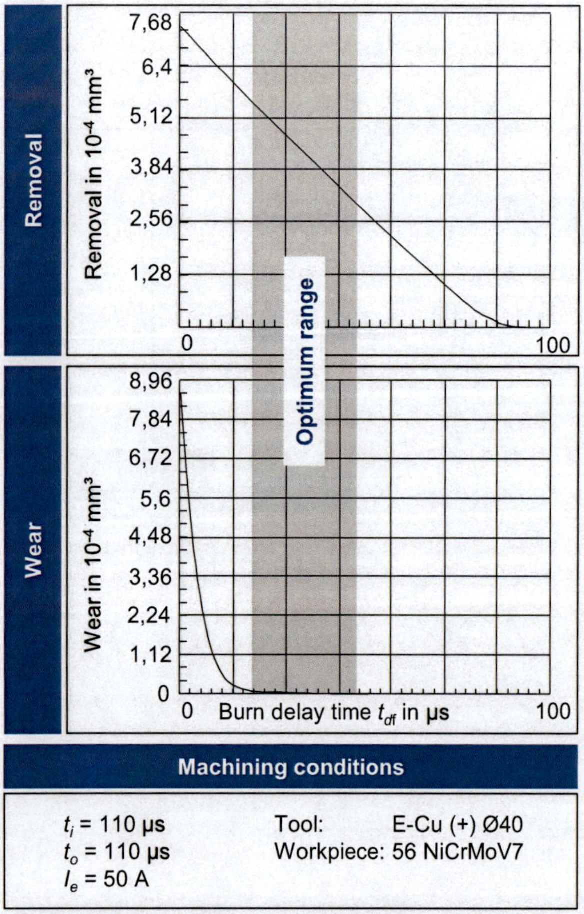


**Fig. 6.72** Relationship between burn delay time $t_{df}$ and erosion or tool wear

between the no-load voltage and the discharge voltage (right-hand pulse in Fig. 6.70). These abnormal discharge pulses can be detected by measuring ($t_{df}$>0). The burn delay time $t_{df}$ thus represents a characteristic parameter of the erosion process. The change in this process correlates with the erosion and wear, see Fig. 6.72.

For standard material combinations of the tool electrode and workpiece, both the erosion rate and the wear can be determined as a function of $t_{df}$. This allows detailed process information on the current condition to be displayed online for the user using a computer-aided evaluation program. Unlike standard indicators for the frequency ratio, this allows the effect of the ignited discharges on the erosion and the wear to be illustrated.

Using the process parameter $t_{df}$ as a starting point, it is also possible to build up a structured gap width control that continuously adjusts the gap width to the current process state, see Fig. 6.73. The control strategy is based on the relevant average of the burn delay times calculated from 30 consecutive discharges. This allows short-term process fluctuations to be determined as shown in the lower section of Fig. 6.73. The input parameter for the adapted proportional controller $t_{df,s}$ is formed from the deviation of the short-term mean value $t_{df,s}$ of the given reference value $t_{df,\mathrm{set}}$. However, long-term trends should be taken into consideration when assessing the whole process. They cause a modification to the gain factors $V_r$ and $V_v$ of the gap width controller in the form of a superordinate system (Fig. 6.73, top). The gain ratio from the forward to feedback gain of the quill drive, and thus the dynamics of the feed drive, is influenced by the deviation of the long-term mean value $t_{df,l}$ from the given value $t_{df,\mathrm{set}}$. The controller generates a signal of variable amplitude and duration $L_v$ or $L_r$ that is transmitted to the feed motor's speed controller. In practice, it has proven prudent to choose high gain over a long duration.

This action allows very short control cycles to be achieved. And yet, it is no longer necessary to distinguish between $L_v$ and $L_r$ for reasons of temporal resolution meaning that only the gains have to be adapted. If the fact that there is a linear relationship between the deviation $t_{df,l}$ in the long-term mean and the quotient from $V_v$ and $V_r$ is also taken into consideration (see [STRO98]), it is possible, using interpolation and extrapolation, to calculate the best gain ratio $V_v = V_r$ from the initial conditions and another working point. The specific initial setting of the parameters $L_i$, $V_v$, $V_r$ and $t_{df,s}$ held in the control loop is thus based on empirical values and the results of experimental investigations.

Using the statistical evaluation presented here, it is possible for the erosion process to operate in a stable manner and to reach the best state with respect to material removal and electrode wear. This results in a considerable reduction in the machining time while simultaneously reducing the tool wear.

### 6.3.6 Collision Monitoring

Collisions in machine tools usually lead to substantial damage which result in high repair costs and lead to long machine downtimes. Of the machine failures with long downtimes and high repair costs, 70% are due to collisions [LORE99].

If collision protection systems are used, these should at best prevent the collision or at least minimize subsequent damage. They should be able to detect imminent collisions in all operating conditions, i.e., manual and automatic operation. Statistically speaking, incorrect operation of the machine is the most frequent cause of collisions. This includes chucking the wrong tool or workpiece, inattentiveness when manually traversing the axes, as well as errors in entering the NC sets manually or when setting the correction

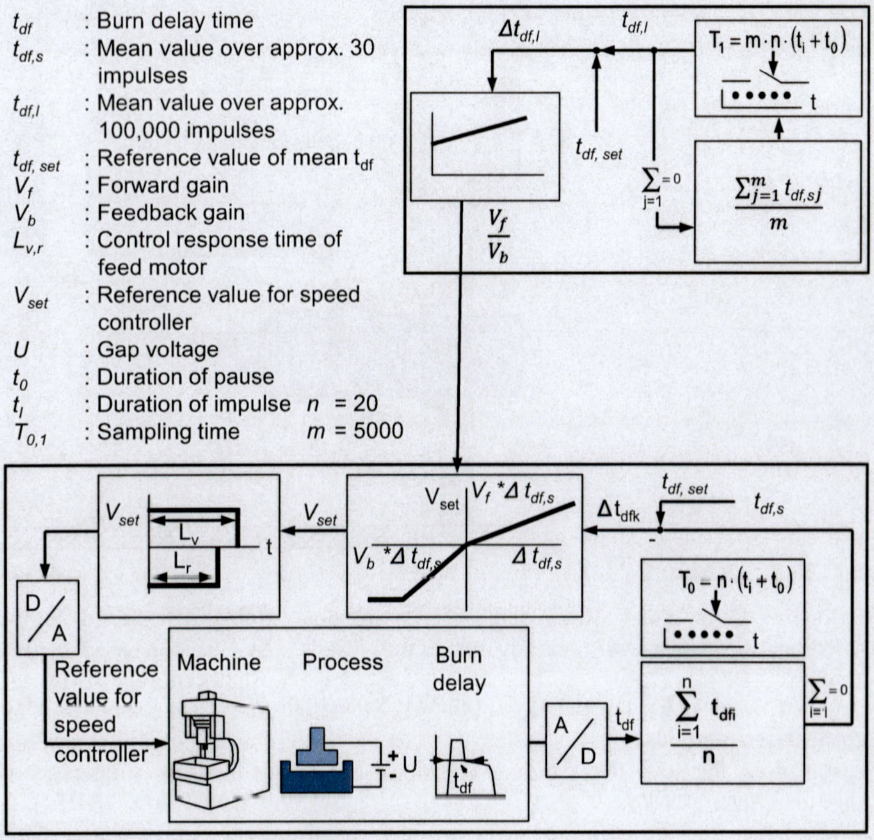


**Fig. 6.73** Structure of gap width control

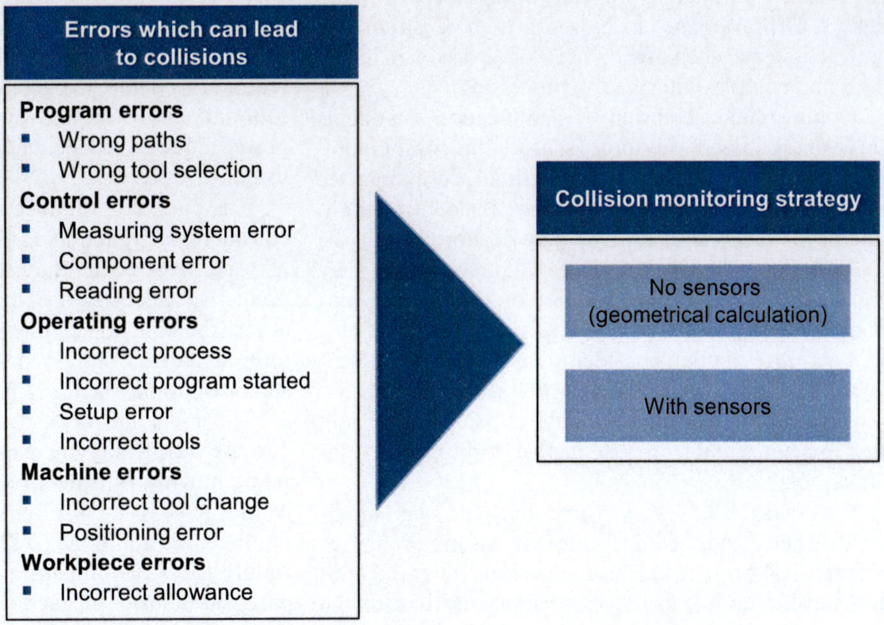


**Fig. 6.74** Causes of collisions and associated monitoring strategies

values. Other causes of collisions are control and programming errors that cannot always be detected despite monitoring and tests, see Fig. 6.74.

A collision protection system should capture all sources of error that can lead to collisions if at all possible. There are basically two different methods here.

6

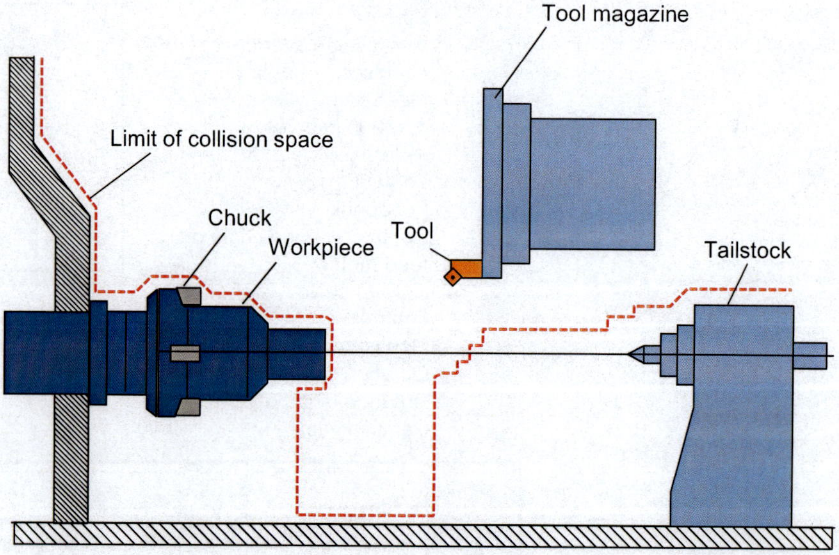


Fig. 6.75 Defining the collision space

Sensor-based and non-sensor-based systems are discussed below together with their advantages and disadvantages.

### Sensor-Based Methods

Sensor-based methods for avoiding collisions use a range of different sensors.

These may include optical sensors in the form of cameras. The vast majority of potential collisions can be prevented through appropriate evaluation of camera signals. Disadvantages include the high sensitivity of optical systems when used in the workspace of machine tools and complex information processing.

Commercial collision identification systems are usually based on force sensors in combination with targeted vulnerabilities (overload coupling, see ► Sect. 2.4.3) or a device for blocking the controller enable. Add to this the fact that they do not prevent the collision but only detect it or keep damage minor, they often do not capture every point of the moving machine components that can be involved in the collision.

There are similar considerations for acoustic sensors. They only capture data in one direction and are easily disrupted by swarf, coolants, etc. For this reason, they are not suitable for use in the workspace of machine tools.

All axes, thus every direction, can be monitored when using control-internal information (see ► Sect. 6.1.4.5). The collision threshold in rapid motion can be set at a low level by compensating for the acceleration effects. It is usually possible to reduce the damage, at least in tangential collisions, in combination with a rapid reaction using the same timing as the position controller.

### Non-Sensor-Based Methods

Non-sensor-based methods for avoiding collisions are usually purely software-based. The geometric collision space, including tools, workpiece, clamping elements and surrounding parts for all moving axes, is defined by simple geometrical elements (primitives) [INDR 92a; FÖRD97].

It is possible to identify the danger of a collision at any time mathematically by using the position and speed of the moving machine components (tools, tool holder) and the known geometries. The problem with this geometrical calculation is the high volume of data required to define the geometries of the workspace, the tool and the workpiece as well as the high computing power required to calculate the potential danger of collision.

It is possible to reduce the online computing power considerably by reducing the number of points which may be involved in the collision to a single point by predictive calculation. This single point of consideration (the collision point) should be that point whose motion path is described by the machining program, so, for example, the tool tip.

The areas in a workspace where a collision may occur are artificially increased such that the collision point moving outside these artificially enlarged areas cannot lead to a collision with any point of the moving machine component. Figure 6.75 shows the procedure for a two-dimensional instance. The artificially enlarged items or surfaces which are obstacles as far as the tool tip is concerned are marked by the dashed line. The collision calculation now only needs to consider the motion path of the tool tip with respect to the enlarged obstacle surfaces.

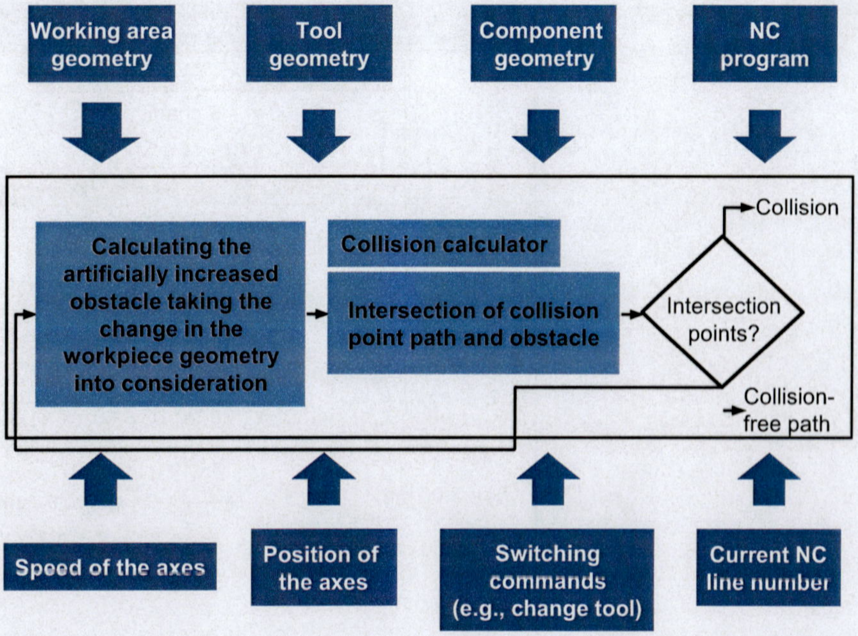


**Fig. 6.76** Information flow in a collision protection system

If the path of the collision point in the movement under investigation intersects the artificially enlarged obstacles, a collision signal is generated and the movement is canceled (Fig. 6.76). Otherwise, the path does not indicate a collision and can be completed. The collision computer then checks the next movement.

The collision space is enlarged further by considering the stop distances of the feed axes with respect to their speed, or the speeds are automatically reduced when the collision space is approached. This guarantees that the feed motion can be stopped before the collision space is entered.

This procedure assumes that all geometrical data have been entered correctly and that the variable geometrical values (tool, workpiece, etc.) match what is actually in place. Substitution of an incorrect, e.g., too large, tool or workpiece or entering an incorrect workpiece program or incorrect operation will inevitably lead to a collision with this system.

## 6.4 Statistical Process Control

Another option for systematically compensating the causes of error on a finished workpiece is the Statistical Process Control method (SPC). In SPC, workpieces are taken out of production at periodic intervals and their critical dimensions (for example bearing seat diameters with particularly tight tolerances) are checked using appropriate measuring equipment or with standard measuring devices such as micrometers or vernier calipers. The benefit of the appropriate measuring equipment is evident in volume production as many features are generally captured with a single measuring process and the result of the check can be automatically documented in electronic data carriers.

The purpose of process control is to offset systematic error components, i.e., a trend, caused by tool wear or thermal drift in the machine. It is necessary to be able to distinguish between systematic and random errors (see ▶ Chap. 9, Volume 2). Random errors are found with a machine capability study. Usually, 50 consecutive parts are manufactured and measured. Systematic influences are to be kept as low as possible during the capability study. This means maintaining a constant ambient temperature and using a tool cutter that has been used a few times in order to prevent progressive initial wear.

The standard deviation $s_x$ and the given tolerance from the drawing $T$ with respect to the machine capability $C_m$ are calculated for the critical features using the scatter of the manufactured workpiece dimensions:

$$C_m = \frac{T}{6s_x} \tag{6.32}$$

where:

$s_x = \sqrt{\frac{\sum_i (x_i - \overline{x})^2}{n-1}}$ – standard deviation,

$x_i$ – measured value,

$n$ – sample size,

$\overline{x} = \frac{1}{n}\sum_i x_i$ – mean value

To be able to control the process, the random scatter that cannot be controlled must be small with respect to the ratio of the width of the tolerance range. The capability indices required for this are between 1.3 and 1.67.

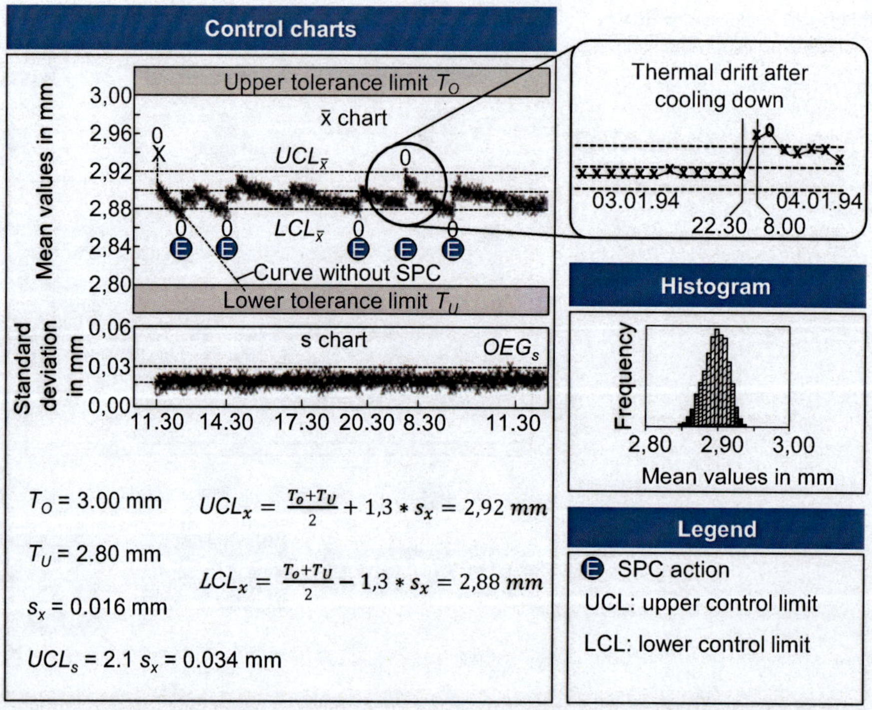


**Fig. 6.77** Control charts for SPC

In statistical process control, a random sample of up to five parts is removed from production during the actual machining process at periodic intervals and measured. The standard deviation and the mean value are calculated for each random sample. The measurement results are entered in process control charts, see Fig. 6.77. In the first instance, the upper and lower tolerance limit for the controlled feature—in this case the diameter of a synchronizer ring for an automotive gearbox—are entered in the control chart.

Control limits are then established for the mean value chart ($\bar{x}$ chart) and the standard deviation chart ($s$ chart). The control limits are calculated such that values outside the limits clearly indicate a trend. The standard deviation $s_x$ for the individual values is known from the machine capability study. Statistically speaking, systematic deviation is where a measured value lies outside a range of $\pm 3\ s_x$ where all measured values are within a completely random distribution of 99.73%. The curve for the measured values is smoothed by formation of the mean value. A trend can be more easily identified because the mean values of a random sample do not spread as widely as individual values. If the individual values in a random sample are distributed with a standard deviation of $s_x$, the standard deviation $s_{\bar{x}}$ for the mean values of $n$ values between the random sample results in

$$s_{\bar{x},n} = \frac{1}{\sqrt{n}} s_x \qquad (6.33)$$

The natural spread containing 99.73% of all mean values from randomly sampling five pieces parts is therefore $\pm 3/\sqrt{5} \cdot s_x$.

The control limits of the mean value charts are calculated as follows:

$$UCL_{\bar{x}} = \frac{T_o + T_u}{2} + \frac{3}{\sqrt{5}} s_x \approx \frac{T_o + T_u}{2} + 1,3 s_x$$
$$LCL_{\bar{x}} = \frac{T_o + T_u}{2} - \frac{3}{\sqrt{5}} s_x \approx \frac{T_o + T_u}{2} - 1,3 s_x \qquad (6.34)$$

The natural spread of the standard deviation of a random sample of five parts can be rendered as:

$$UCL_s = 2,1 * s_x \qquad (6.35)$$

As machine users require capability indices of between 1.3 and 1.67 [KYNA99; KIEL92; ARNO93; INDR 92b; MURR 99b], this guarantees that the individual values of the random samples do not exceed the tolerance limit if a mean value approaches its control limit.

The standard deviations and mean values determined are entered in the prepared control charts during the actual machining process. In modern SPC stations for volume production, measurement, evaluation, and transfer of data to the control charts is automatic. Figure 6.77 shows an excerpt from a control chart of this kind. Due to the wear on the conical reamer, smaller and smaller diameters are manufactured over time. If a mean value reaches the lower control limit, the reamer is axially adjusted and abrades larger diameters again.

6

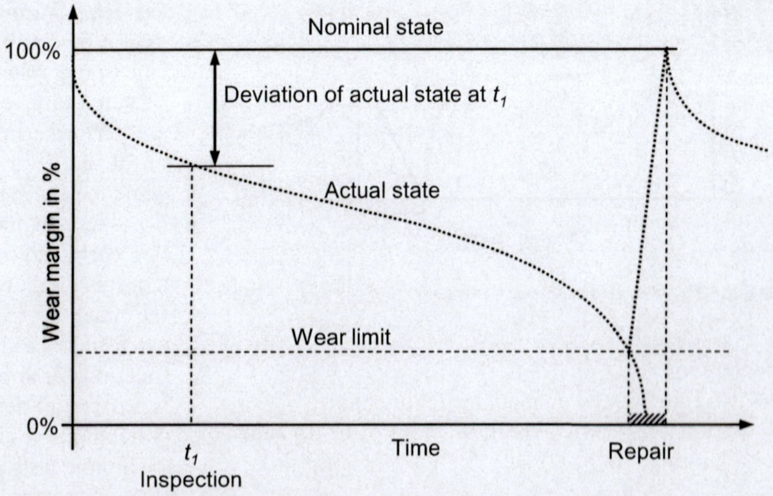


**Fig. 6.78** Curve of the wear margin $t_s$ according to [DIN12]

The second effect which can be clearly seen on the control chart is the machine's thermal drift. It shows that the first random sample, taken at the beginning of the new shift, was manufactured above the upper control limit. This required the process to be reset. The histogram shows the frequency distribution of 4785 individual values produced during this time period.

It shows that, on average, the diameters were manufactured slightly below the mid-tolerance but all synchronizer rings were within the tolerance. Without statistical process control, the process would have exceeded the tolerance after three hours.

## 6.5 Machine Condition Monitoring

Current manufacturing systems have an extremely complex structure in the tradeoff space between availability, efficiency, costs and cost-effectiveness. In particular, availability has a high added value, which is why enormous demands are made of it. At times, an Overall Equipment Effectiveness of 98% [OPHE98] or more is demanded. Condition monitoring is particularly important to avoid unplanned downtimes due to maintenance or repairs.

Various methods and algorithms for condition monitoring are presented and illustrated with practical application examples in the following chapters.

### 6.5.1 Service and Maintenance Procedures

High utilization of the manufacturing capacity is required to amortize the acquisition costs of production equipment [HAGE99]. This degree of utilization requires the machines to have high reliability and availability. The overarching objective of maintenance work is to maximize the reliability and availability of the production facility and to counteract wear on the machine and its components [MURR 99a]. During the utilization phase, the wear margin, which defines the maintenance and repair intervals for the machine, is depleted (Fig. 6.78).

Maintenance measures are divided in [DIN12] into:

- **Maintenance**: Measures for preserving the nominal condition of technical systems,
- **Inspection**: Measures for establishing and assessing the actual condition,
- **Repair**: Measures for producing a required (nominal) machine condition.

Inspection, which is largely understood as machine condition monitoring, is examined in greater detail in ► Sect. 6.5.2.

Another significant aspect of maintenance is the maintenance strategy. This establishes when and which different maintenance measures are performed and what event initiates them. Typical strategies are [MURR 99a]:

- Maintenance triggered by failure
- Time-dependent maintenance
- State-oriented maintenance

#### 6.5.1.1 Reactive Maintenance

Maintenance measures are carried out as required, i.e., following machine or component failure. There is no maintenance plan. The machine is regarded as functioning until it fails (Fig. 6.79).

**Disadvantages**

- High warehouse costs as the failure cannot be scheduled and replacement parts must be stocked.
- If the necessary replacement parts are not available, the result is extended downtimes.
- The secondary damage to the machine caused by the component failure cannot be predicted.

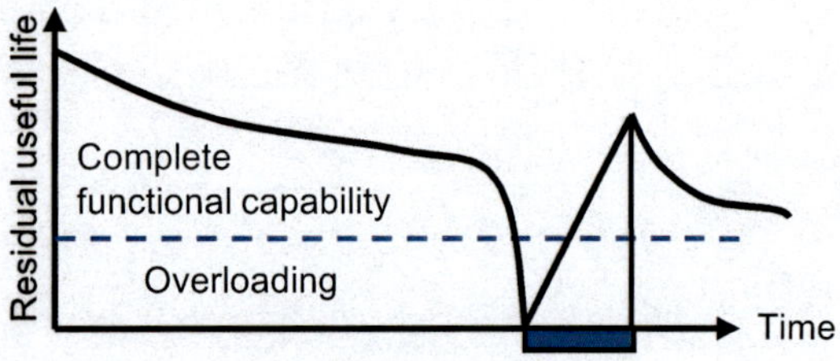


**Fig. 6.79** Reactive maintenance strategy

6

**Advantages**
- No time wasted on machine maintenance.
- No on-costs for condition monitoring systems or planning.

### 6.5.1.2 Time-Dependent (Preventive) Maintenance

This is the maintenance strategy currently most frequently used by manufacturing companies. The machine manufacturer or operator defines, based on previous experience, the intervals at which wear components are replaced. The actual condition of the component is not considered. The estimated intervals are often very conservative for safety reasons and therefore only take standard component wear during operation into consideration. This prevents those production failures that are foreseeable and can be scheduled (Fig. 6.80).

**Disadvantages**
- Those failures caused by poor quality components or by overloading cannot be prevented.
- Components that are not worn are replaced.

**Advantages**
- Fewer unplanned downtimes than with failure-based maintenance.
- Better scheduling of the need for replacement components.
- The maintenance schedule can be planned in advance allowing the economic downside of machine downtime to be kept to a minimum.

**Example** Maintenance and inspection are generally driven by company guidelines as well as standards specific to the system and machine.

For example, for machine tools, regular inspections are performed on the lubrication or friction conditions on all guideways. By traversing the axes without load while measuring the motor current or the feed forces, it is possible to detect deterioration in the guideway condition early and to initiate remedial measures. To this end, a corresponding trend evaluation over the service life of the machine is required. The functionality of switches can also be easily tested. By triggering various switching functions or approaching the limit switches in all axes, it is possible to determine the switches' condition from the switching time and magnitude of the contact bounce.

Simple tests are carried out at the beginning of each shift or at short intervals for a relatively quick check of the machine accuracy. In this instance, a measuring probe, instead of a tool, is used and various reference positions are approached on the machine table. The geometrical accuracy should definitely be checked in more detail at longer intervals, either quarterly or every six months. A range of procedures is available to measure the linear positional deviation as well as the straightness and angle deviations of the machine as described in [YE92]. Additional tools can be used, such as a measuring ruler combined with measuring probes or a laser interferometer.

One example of a resource for manual monitoring of machine condition is the machine tool data logger. Its principle is shown in Fig. 6.82. The data logger's job is to store important information and data during machine operation over a specific time period. These include (Fig. 6.81):
- current NC data,
- operational interventions,
- process parameters (e.g., torques, forces and temperatures),
- machine parameters (e.g., distances, rotational speeds and feeds) and
- switching signals.

This is shown in Fig. 6.81 using a window which runs along the time axis. The window provides the

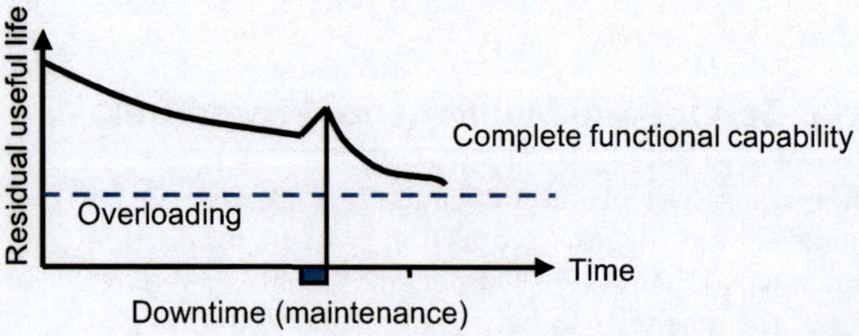


**Fig. 6.80** Preventive maintenance strategy

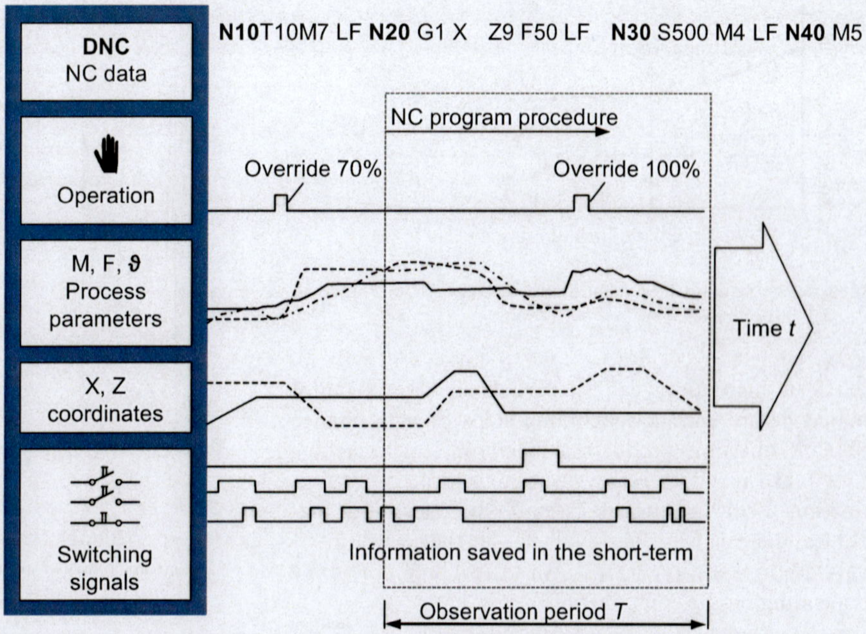


**Fig. 6.81** Principle of a machine tool data logger

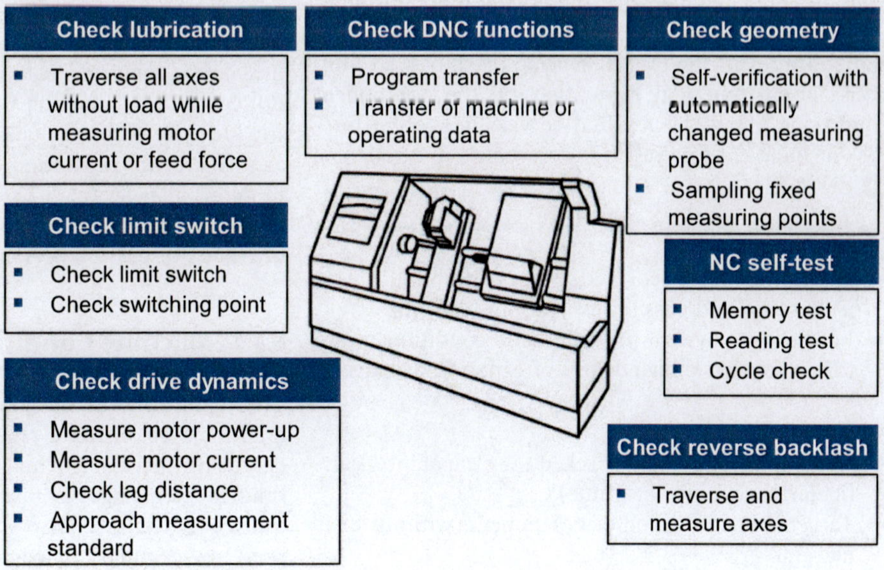


**Fig. 6.82** Tasks for general functional tests

observation period. All information within this window is saved and can be recalled at any time. Under normal circumstances, the window moves forward in time t, i.e., as new information is recorded, old information is lost at the same rate. If the machine is disrupted, the window is stopped. The information and data for the observation period T before the disruption and up to the occurrence of the disruption are retained. An accurate analysis of this information can often reveal the cause of the disruption.

Many controls have additional integral routines for automatic inspection. NC controls offer integral functionality to check characteristic machine parameters and the drive dynamics, e.g., determining the transfer behavior of the axes [ISER83]. The latter is achieved by evaluating various measured values recorded as the feed and main drive motors power up without load. Characteristic values here include the lag distance, the rotational speed profile over time, and the motor current.

In addition to this, essential internal functions, such as the timer and machine tool functionality such as hydraulics and pneumatic pressure, can be checked and logged accordingly in the event of malfunction.

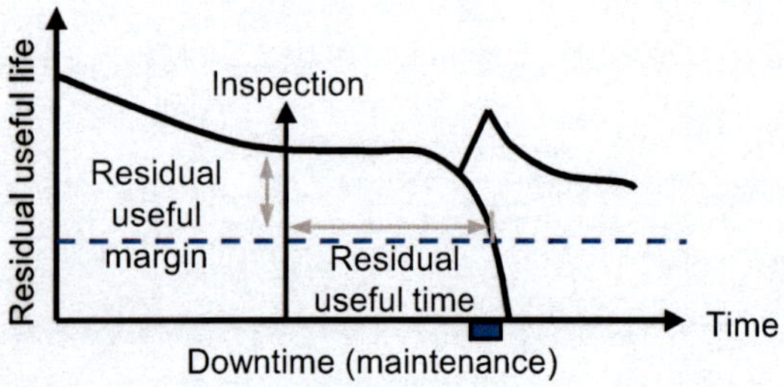


Fig. 6.83 State-oriented maintenance strategy (in this instance, through inspection)



Basic inspections can be performed or even automated before specific machining steps or at a change of shift on some production machines (e.g., machine tools) (Fig. 6.79). However, in high-throughput production facilities, such as assembly or packaging machines, there is a further challenge in that every interval-based inspection causes a short machine downtime depending on the component.

#### 6.5.1.3 State-Oriented Maintenance

State-oriented maintenance represents another significant approach to further reducing the number of machine failures in the overall scheme of things. Continuous monitoring and evaluation of the component conditions allow the residual service life to be forecast or incipient failures to be prevented in good time (Fig. 6.83).

**Disadvantages**

- High investment costs due to additional sensor technology and possible evaluation electronics.
- Producing a system for automatic condition monitoring requires a high degree of expert knowledge.

**Advantages**

- Maximum maintenance scheduling capability (particularly machine downtime).
- Targeted identification of components without wear margin.
- Replacement parts do not have to be stocked or warehoused and can be ordered as required.

**Example** State-based maintenance is suitable for a wide range of components that are prone to wear and that can be monitored by sensors. Figure 6.84 shows two torque profiles versus time. By defining the reference profile with a moving threshold, it is easy to determine whether the components (in this case, a belt) are still functioning properly or not.

The individual approaches to maintenance were expanded in the 1980s to form the strategic concept of Total Productive Maintenance (TPM) [BACK87]. This is a proactive and preventative maintenance strategy

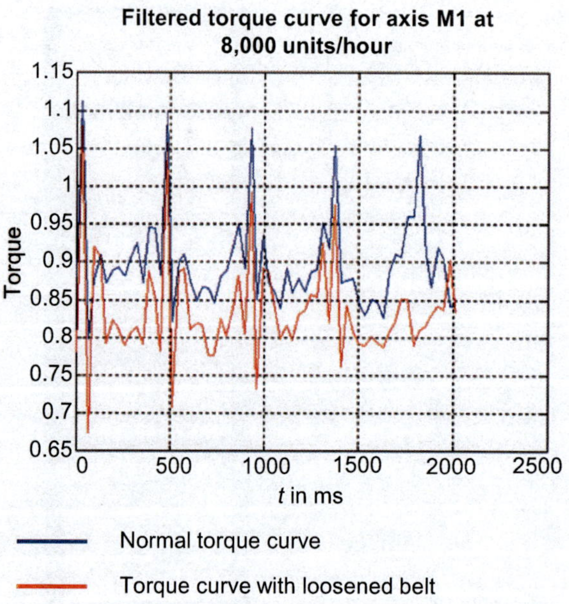


Fig. 6.84 Torque curve for different belt tensions (standard, loose belt)

incorporating the entire business into the maintenance measures to increase equipment effectiveness and does not cover just the action phase of the plant but its entire life cycle. "Losses due to plant failure" are regarded as a significant disruption to efficiency. This is covered in detail in ► Sect. 6.5.2. However, the use and implementation options of this strategy provoke a great deal of heated discussion [RAKE93].

### 6.5.2 Machine Condition Monitoring

The significance of machine monitoring, or more generally, inspection, is a result of having the option of capturing the current machine condition. This avoids failures, reducing the machine downtime or the frequency of failure. Also, changing components early prevents secondary damage to machine components or the overall machine that may occur as a result of failure.

#### 6.5.2.1 Parameters

The essential parameters for describing availability resulting from downtime and productive operation are defined in the following subsections.

**Overall Equipment Effectiveness**

Overall equipment effectiveness (OEE) is often used as a measure of the quality of maintenance [BACK87, BACK88, GROß81]. This parameter describes the level of effectiveness of the available plant. It is calculated according to [GROß81] as follows:

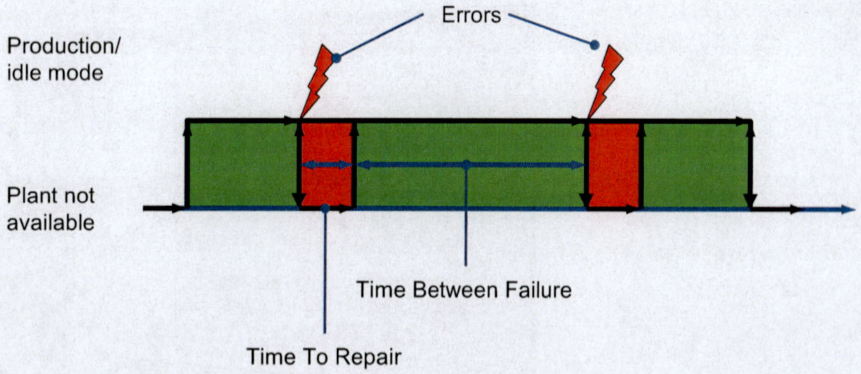


**Fig. 6.85** Time To Repair and Time Between Failures with respect to overall production time

$$O.E.E. = \text{Availability} \cdot \text{effectiveness} \cdot \text{quality factor } [\%] \quad (6.36)$$

The essential part, which can be influenced by maintenance strategies, is, in this instance, the availability.

#### Reliability R(t)

The precise definition of reliability varies depending on the source of information [DOET85, MEYR92, KUPF68]. It basically means that a machine "… has to meet the demands of its intended use placed on the behavior of its characteristics over a given time period" [MEYR92]. Empirically, it is defined as:

$$R(t) = \frac{\text{number of available units}}{\text{number of observed units}} \quad (6.37)$$

#### Failure Rate λ

The failure rate is derived as a statistical measure from the relationship between the number of failed units and the number of non-failed units over the observation period $\Delta t$:

$$\lambda(t) = \frac{n(t) - n(t + \Delta t)}{n(t) \cdot \Delta t} = -\frac{dR(t)}{dt} \cdot \frac{1}{R(t)} \quad (6.38)$$

#### Mean Time Between Failures (MTBF)

The MTBF defines the anticipated value of the operating period between two consecutive failures (Fig. 6.85). In the event of constant failure rates λ, the mean time between failures is defined as:

$$\text{MTBF} = \frac{1}{\lambda} \quad (6.39)$$

#### Mean Time to Repair (MTTR)

The Mean Time To Repair (MTTR) is the mean time required for a repair (Fig. 6.85). This time is regarded as technical downtime during which the machine tool cannot operate productively. If the machine fails unexpectedly, this time has to be extended by the amount of time required for diagnosis and replacement part procurement. Significant use of machine monitoring is based on this fact.

MTBF and MTTR can thus be used to define availability as follows:

$$A = \frac{\text{MTBF}}{\text{MTBF} + \text{MTTR}} \quad (6.40)$$

#### Availability

The availability value is based on operational factors. It indicates the productive time for the plant. The cost efficiency of a production system is therefore directly dependent on the availability. The availability of a plant can also be defined as follows as an alternative to (6.40):

$$\text{Availability} = \frac{\text{total time - total downtime}}{\text{total time}} = \frac{I + B}{P}$$

and $P = I + B + R + M$

where:

*P* – planned operation,

*I* – idle state,

*B* – production,

*R* – repair,

*W* – maintenance

There is no explicit distinction drawn between planned and unplanned maintenance.

### 6.5.2.2 Sensor-Based State Detection

When monitoring the condition of mechanical components, this usually involves a monitoring system covering the whole service life of the machine over long periods of time (see Fig. 6.3). This is based on the fact that wear effects and damage to mechanical components frequently extend over longer periods of time. These components are therefore usually monitored by sensor-based systems.

Using this data set as a starting point, procedures such as vibration analysis, thermography, vibro-acoustic

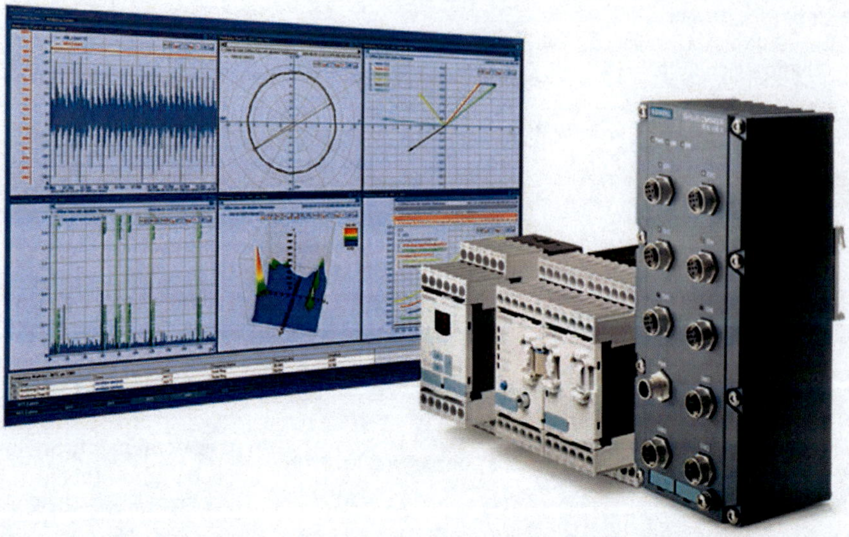

**Fig. 6.86** SIPLUS CMS from Siemens. *Source* Siemens

evaluation or oil analysis can then be carried out [STRO98]. A fundamental distinction is drawn here between control-internal and control-external condition monitoring (sensors and evaluation).

### Control-External Condition Monitoring

Control-external condition monitoring is often used in production engineering plants. This can be motivated by a range of factors. Often, only specific components or their wear are monitored or the performance of the existing control system is not sufficient for condition monitoring. Subsequent retrofitting of solutions for condition monitoring can be one reason for deciding to use external systems.

Solutions available on the market offer not only the necessary external sensor technology to achieve this. They also provide the corresponding hardware containing evaluation algorithms, visualization and the interface to management systems and directly to the internet. Examples of these systems are "FAG DTECT X1 s" from Schaeffler,[1] "Ω-Guard" from Bachmann,[2] "@ptitude Asset Management System" from SKF[3] or "SIPLUS CMS" from Siemens.[4] The latter offers a scalable system. At its most basic level, it allows bearing and vibration monitoring without any necessary parameterization (CMS1000) and, at its highest performance level, provides data from application-specific parameters as well as project-based implementation of the users' own monitoring algorithms (see Fig. 6.86).

The ability to scale products allows inexperienced users to overcome problems when using more complex and expensive condition monitoring systems.

A few advantages and disadvantages of using control-external systems are given in the following table.

There are three reasons why systems with exclusively control-external monitoring are suitable for continuous machine condition monitoring of closed systems such as machine tools:

1. Under difficult ambient conditions in the workspace of machine tools (cooling lubricant vapors, vibrations, etc.), these sensor systems become less reliable over the necessarily long monitoring period.
2. Machine tools are subject to constantly fluctuating loads meaning the signals to be monitored can vary significantly and have a wide distribution.
3. Furthermore, connecting temporary sensors to the machine is difficult in terms of contact quality and reproducibility of the sensor location.

### Control-Internal Condition Monitoring

Existing control-internal values, such as drive data in the form of currents, torques and rotational speeds, are frequently used for condition monitoring on production machines as an alternative to external capture and, particularly, for evaluating sensor data. This often means there is no need for additional sensors and pre-calculated system variables can also be used.

Table 6.2 compares a few advantages and disadvantages when using control-internal systems, as with Table 6.1.

Components that can typically be monitored by this sort of integral control approach are shown in Fig. 6.87.

The optimum system is always based on specific components, machines or applications. Today, a

1 For example, Schaeffler: ► http://www.schaeffler.de/.
2 For example, Bachmann: ► http://www.bachmann.info.
3 For example, SKF:► http://www.skf.com.
4 For example, Siemens (Automation):► http://www.automation.siemens.com.

**Table 6.1** Advantages and disadvantages of control-external condition monitoring

| Advantages of control-external monitoring | Disadvantages of control-external monitoring |
|---|---|
| Potentially very high sampling rates (regardless of control cycle) | High costs due to expensive measuring hardware (e.g., CompactRio®, dSpace, Norma) |
| No delays due to, for example, bus hardware in signal capture | Additional visualization hardware required |
| Any process variables can be monitored with appropriate sensors | Sensors represent additional sources of errors |
| High flexibility: Can be retrofitted, range of application locations | Cost of calibration |

**Table 6.2** Advantages and disadvantages of control-internal condition monitoring

| Advantages of control-internal monitoring | Disadvantages of control-internal monitoring |
|---|---|
| Economical as no additional measuring or processing solutions are required | Sampling rate fixed to the control system's cycle time |
| No new sources of errors from additional sensors | Delays due to bus systems |
| Evaluation results can be visualized directly at the machine interface | Only estimates are available for specific process values (e.g., torque and acceleration) or the quality is dependent on the calculated values from the motor model |
| | Limited performance of the calculation hardware |
| | Access to the control project essential |

combination of external senors (e.g., vibration monitors in packaging machines) and internal data (e.g., drive values) is used to reap the benefits of both systems (Fig. 6.88).

In addition to just the evaluation solutions, there are other service providers which provide advice and support for users, from selecting the correct system to integration into the automation pyramid [SCHI90], i.e., from the actuator and sensor level to the ERP and beyond. A typical such provider is ifm consulting with its concept of "Real Time Maintenance®".[5]

Another product, which is characterized by providing a large number of evaluation algorithms, is the "TwinCAT Condition Monitoring Library" from Beckhoff.[6] The user can combine and use application-specific, scalable solutions from a set of "modular mathematical algorithms." The application range is restricted to the platforms that can be configured with TwinCAT.

5 Ifm consulting: ▶ http://www.ifm-consulting.com.
6 Beckhoff: ▶ http://www.beckhoff.de.

### 6.5.3 Measured Value Analysis

The measured values required for an inspection depend on the complexity of the object in question. According to [PRES64], the actions can basically be divided into.

- measuring and
- testing.

Capturing data using sensors (measuring) has been dealt with in detail in ▶ Sect. 6.5.2.2. This section deals with testing.

Testing is part of the diagnostic process and means using measurement to establish compliance with specific quality criteria. As a result, testing always includes a decision about the condition. Quantifiable values do not necessarily have to be captured to achieve this. For example, the usual visual inspections, reading available displays, etc., are also part of testing as is the machine operator being aware of irregularities in a machine's operation, such as unusual noises or components overheating. The challenge facing testing is to precisely

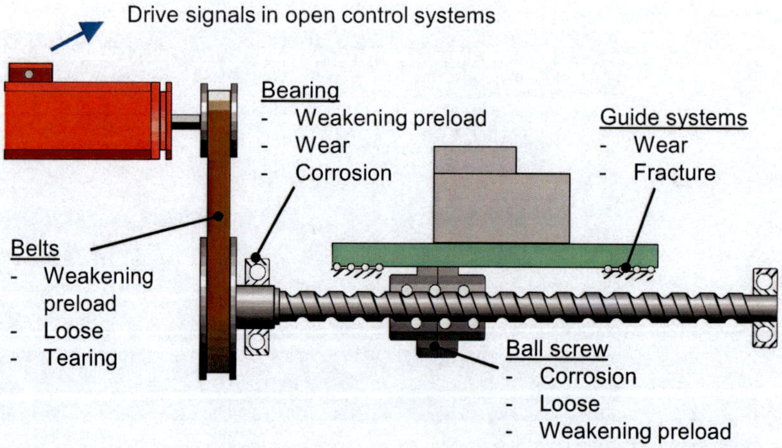


**Fig. 6.87** Typical components for control-internal machine monitoring

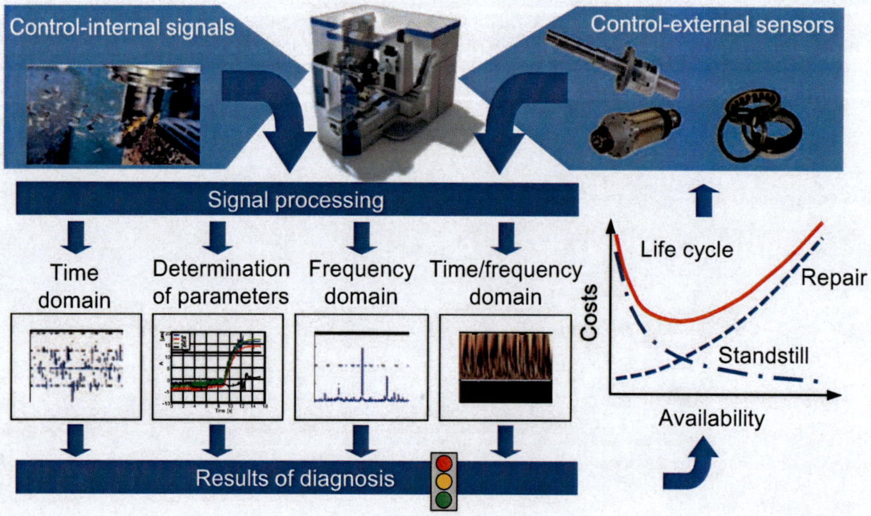


**Fig. 6.88** Combination of control-internal and control-external signals for condition monitoring according to DS-Technologie, JenaTec, Schaeffler, Weiss

localize and determine the cause of a disruption once it has been detected. A significant task here is the classification of the signal patterns that have been captured and prepared (symptoms) in such a way as to allow allocation of the disruption that occurred to various components. Some of the applicable functional principles and algorithms of the various applicable tools have already been discussed in ▶ Sect. 6.2.

Various transformations, algorithms and methods are discussed in this section that allow automatic analysis and evaluation to be performed based on captured data from sensors and control data.

#### 6.5.3.1 Conventional Diagnostic Functions

The diagnosis is usually performed in various stages using a range of functions, algorithms and methods depending on the degree of complexity of the diagnostic task. The machine operator can be the first link in the diagnostic chain here. However, the analytical possibilities available to an operator are limited to visual diagnosis and the quality is heavily dependent on the degree of expertise available. For this reason, the rest of this section will focus on automated diagnostic systems.

A fundamental distinction is made between signal-based, model-based and behavior-based analytical methods (**Fig. 6.89**) [RUDO14]. *Signal-based*

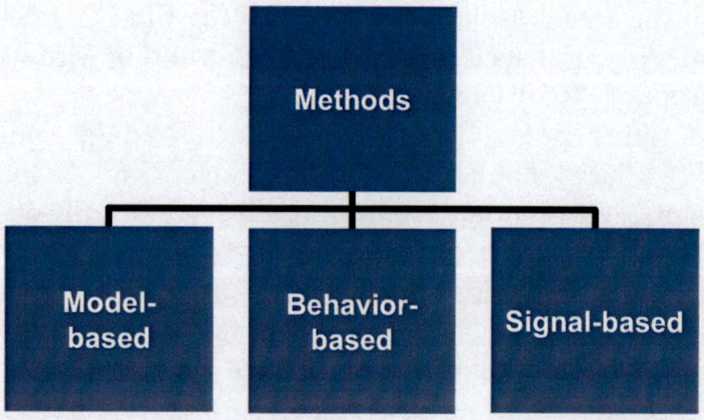


**Fig. 6.89** Categorization of current methods for monitoring and diagnosis. *Source* [RUDO14]

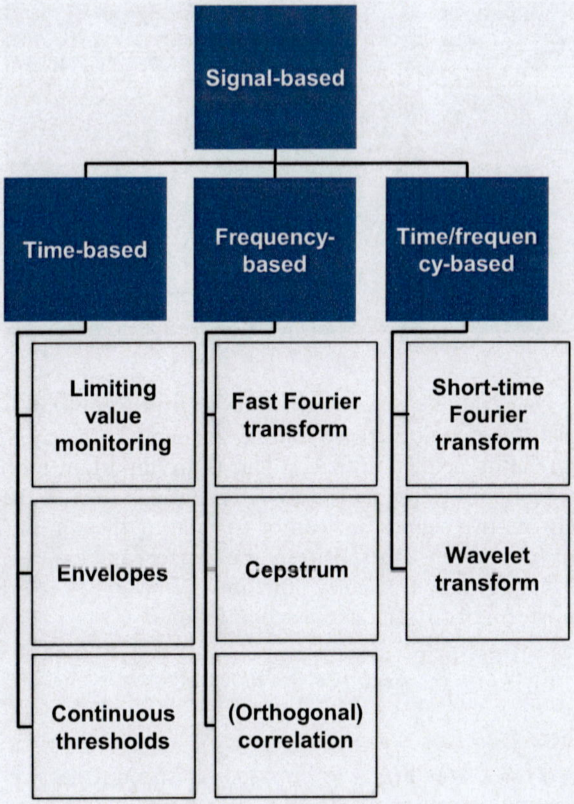


**Fig. 6.90** Overview of signal-based analytical methods

methods monitor the signals provided by sensors and controls and are further divided into time-based (e.g., envelopes), frequency-based (e.g., Fourier transform) and mixed time/frequency-based functions. *Model-based* methods do not transform the signals directly but extract further information from these signals or calculate indirect signals, which cannot be measured directly, by monitoring several signals at the same time. These can be statistical processes for estimating parameters (e.g., least squares) where estimation of the state space (e.g., state observer) is derived from the correlation of various signal states that can then be transferred to parity equations in state space models. The third category is *behavior-based* methods. These include, for example, artificial neural networks that learn parameters for the specific application. Figure 6.89 shows a differentiated overview of this categorization structure.

## Signal-Based

In signal-based signal analysis, time-based, frequency-based or, as a mixed form, time/frequency-based methods are used for diagnosis. Figure 6.90 provides an overview of current analytical methods.

### Time-Based

In time-based signal analysis, subtraction is the simplest way of gaining information on the similarity of signals. By monitoring the difference signal, it is possible establish the points at which the signals differ from each other. Analyzing the length of the intervals in which the deviations occur provides information on the type of signal difference. If two binary signals are being assessed, a simple bit comparison is also possible by using the AND logical operation. If all the deviations of the difference signal are added together, it is possible to make a quantitative assessment of the degree to which two signals deviate from each other. Otherwise, it is not possible to obtain local information on the differences between the signals.

Another algorithm to compare or analyze two signals is cross correlation. This is a measure of the linear dependency between two signals. By correlating different time shifts τ, it is possible to obtain information on the time shifts between two signals. For real-value functions x(t) and y(t), the cross correlation is therefore defined as:

$$R_{xy}(\tau) = \lim_{T\to\infty} \frac{1}{T} \int_{-T/2}^{T/2} x(t) \cdot y(t+\tau)dt \tag{6.42}$$

where

$\tau$ = time shift
$T$ = time frame considered
$x(t), y(t)$ = signals for cross correlation

If discrete signals are to be analyzed, the sum from— to + are considered in specific instances instead of the integral of various time shifts.

In addition to this, envelope analysis or limiting value monitoring is frequently performed, as shown in Fig. 6.91. In this case, upper and lower, fixed threshold values are provided (see Fig. 6.91a) and the motor current may not cross these thresholds for collision monitoring. As greatly fluctuating process values frequently occur during a machining process, static upper and lower limits are generally not adequate. Envelopes of an anticipated curve are then generated as an alternative. These allow specific limiting values to be defined for each discrete point in time (see Fig. 6.91b).

### Frequency-Based

An alternative to analyzing a signal in the time domain is frequency analysis. The aim here is to “describe and analytically investigate processes in a mathematical and systematic manner “ [KOLE08, p. 342]. To achieve this, continuous signals are transformed from the time domain to the frequency domain. The mathematical relationship between the representation of the signal in the

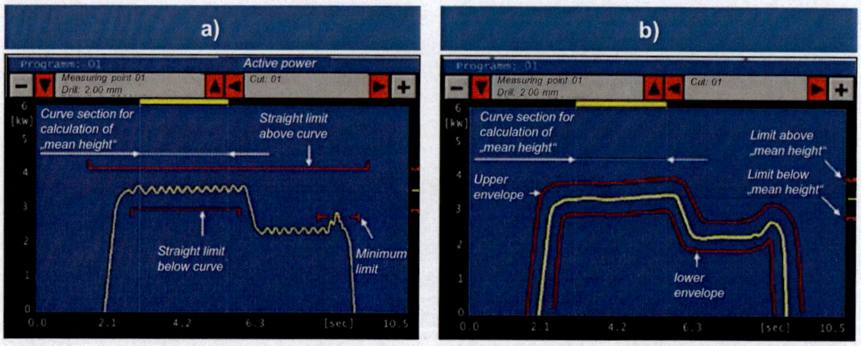


**Fig. 6.91** Collision monitoring on a machine tool by monitoring limits (**a**) and upper and lower envelopes (**b**). *Source* Nordmann

6

time domain and its representation in the frequency domain is described by the Fourier transform [KOLE08, p. 42]. According to [KOLE08, p. 345] the transformations can be described as follows:

$$X(f) = F\{x(t)\} = \int_{-\infty}^{+\infty} x(t)e^{-j2\pi ft}dt \tag{6.43}$$

$$x(t) = F^{-1}\{X(f)\} = \int_{-\infty}^{+\infty} X(f)e^{+j2\pi ft}df \tag{6.44}$$

where

$X(F)$ – signal in frequency domain

$x(t)$ – signal in time domain

It is clear from the reversibility between the transformation (6.43) and the inverse transformation (6.44) that it is neither possible to lose information nor gain new information in the Fourier transform [KOLE08, p. 345]. Furthermore, analysis in the time domain considers **when** certain signal characteristics occur, while frequency analysis is concerned with **how often** the characteristics occur [KOLE08, p. 38].

As frequency analysis is best suited to investigating periodic signals, it is used in, for example, vibration diagnostics. As Kolerus states in [KOLE08, p. 37], this process allows vibrations to be allocated to individual components in a machine and this would not be possible in the time domain in this form. The fast Fourier Transform (FFT) is currently frequently used for this. This is a memory-efficient and rapid form of the discrete Fourier transform and can also be used on embedded systems for frequency analysis.

A further benefit of the Fourier transform is that convolutions in the time domain become multiplications after the transformation and vice versa. However, the fact that transformation into the frequency domain and back takes time must be taken into consideration as this time is vital for process monitoring (e.g., collision monitoring).

One example of using the Fourier transform to calculate what are otherwise complex formulas is the cross correlation defined in (6.42). This is an important tool for comparing two signals to determine the time shifts between two signals. Instead of calculating these in the time domain by superposition, the signals can be transformed into the frequency domain. The inverse Fourier transform of (6.42) is expressed as follows:

$$R_{xy}(\tau) = \int_{-\infty}^{\infty} S_{XY}(f)\cdot e^{i2\pi f\tau} df) \tag{6.45}$$

where

$S_{XY}(f) = X^*(f)\cdot Y(f)$ – as cross-power spectral density from multiplication of the transformed signals

If the inverse Fourier transform is generated from the cross-power spectral density, which represents a multiplication of the transformed signals in the frequency domain, this again results in the cross correlation in the spatial domain. This conversion is described in formula (6.45).

As has been shown before, the Fourier transform is particularly suitable for investigating periodic signals. (Figure 6.92a) shows a Wolf tubular bag machine. A film take-off attached to the formation tube continuously takes the film off a roller and transports the tubular bag film. During operation, the contact pressure on the film take-off is to be monitored. For just condition monitoring the wear of the belts is to be monitored. A vibration sensor is attached for this purpose to the film take-off (Fig. 6.92b). As the teeth on the film take-off belt mesh during the drive phase, vibrations occur that are modulated by the vibrations on the film take-off. If incorrect operation occurs or if there is imminent wear, this results in amplitude variations in the spectrum (Fig. 6.92c), red areas, blue circles) [ZuPack].

The second example concerns a cross seal unit on the packaging machine (Fig. 6.93a and b), where filled bags are sealed and separated at the same time by a knife. Over a period of extended operation, contact

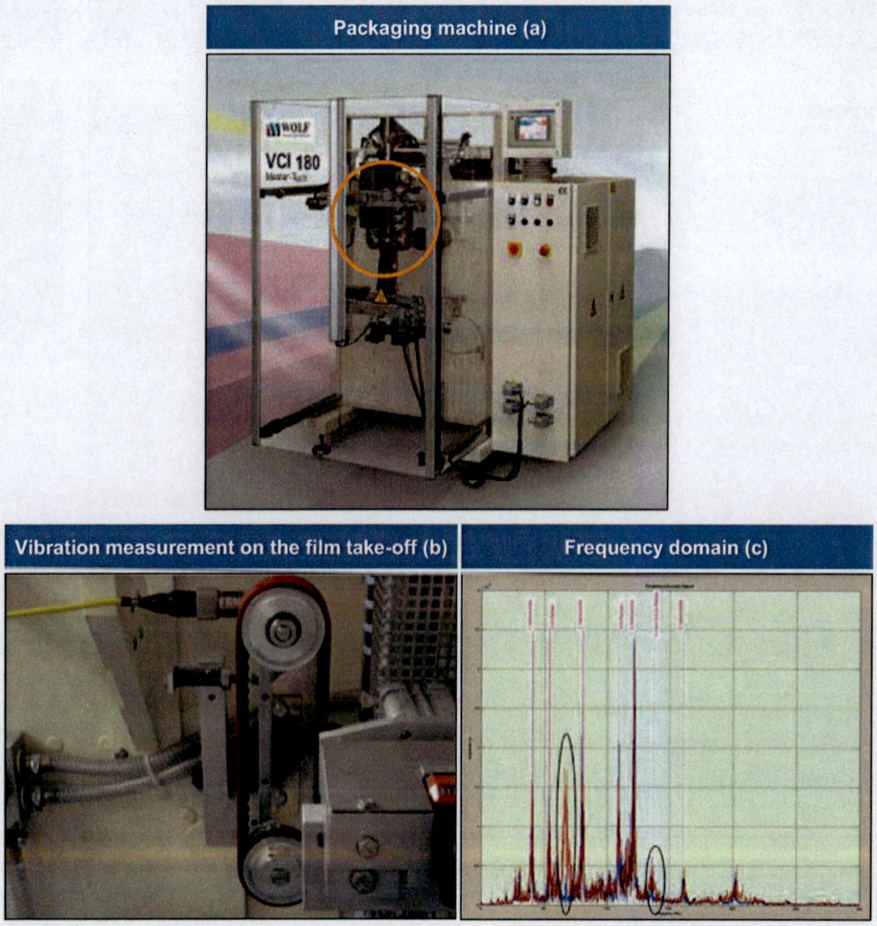


**Fig. 6.92** Vibration analysis of the film take-off of a tubular bag machine (packaging machine). *Source* Wolf

can occur with the sealing table due to incorrect initial parameterization or variations in the mechanical system. To detect this, a piezoelectric sensor was attached to the acceleration pickup. While evidence on contact with the sealing table in the time domain (Fig. 6.93c) provides little information, a significant deviation between the reference condition (red) and the damage pattern during contact (blue) can be detected in the frequency domain (Fig. 6.93d). Although the vibration phenomena are very complex, a decision can be made based on, for example, FFT and envelope analysis.

#### 6.5.3.2 Model-Based

Model-based methods extract patterns and parameters such as parameters of statistical distribution from measurement signals to calculate further information from this data. Furthermore, indirect signals, which cannot be measured directly, can also be specifically generated. An overview of frequently used methods is shown in Fig. 6.94.

As an example, state observers can be used to reconstruct unknown state variables from captured measured values. In this case, a monitor is a model connected in parallel that uses monitoring to approximate the actual system through an iterative process. However, this assumes that all state variables can be observed. Further information on state observers and the observability of a system can be found in ▶ Sect. 3.4.3 (Fig. 6.94).

#### 6.5.3.3 Behavior-Based

The third category, behavior-based methods (Fig. 6.95), attempts to learn the system behavior. For example, artificial neural networks or neural fuzzy systems are created and their parameters such as weighting, transfer and activation functions, as well as threshold values are learned for the specific application. A comprehensive example of feature classification using neural networks can be found in ▶ Sect. 6.5.3.4. Figure 6.89 shows a differentiated overview of this categorization structure (Fig. 6.95).

##### Expert systems and Artificial Intelligence

One approach often found in machine diagnostics is knowledge-based diagnostics. It is used in what are

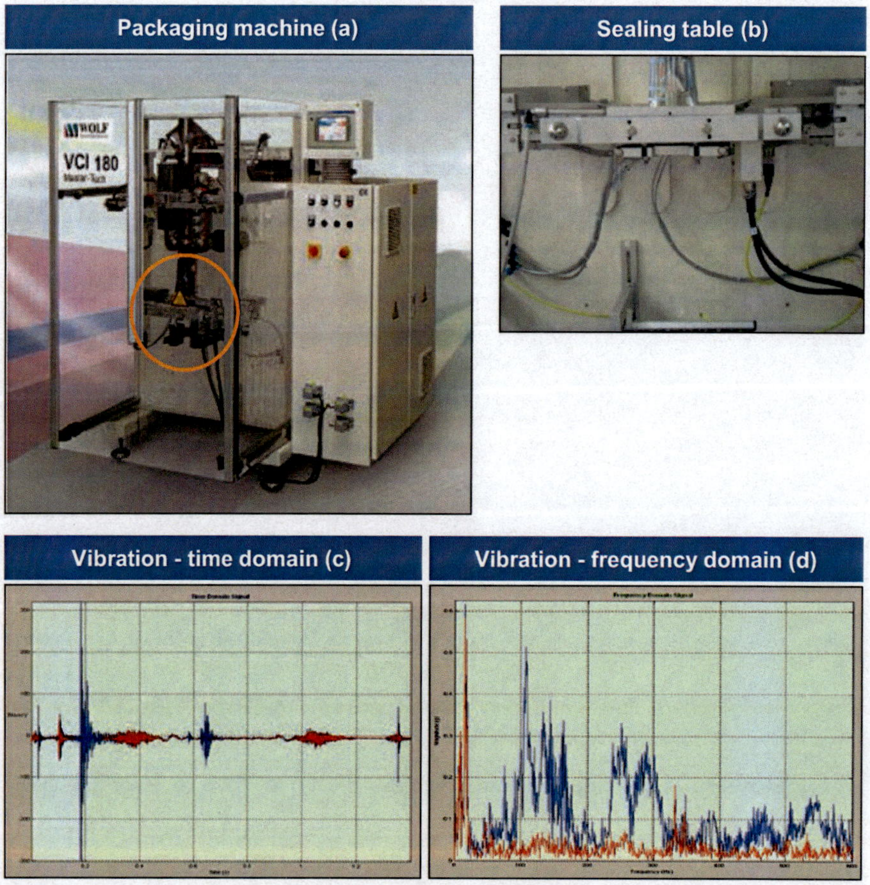


**Fig. 6.93** (Acceleration) measurements on the sealing table of a packaging machine in the time domain (c) and the frequency domain (d). *Source* Wolf

known as expert systems. This type of system is able to answer queries on precise and complete problem definitions using an internal knowledge base and to contribute to implementing the answer through support functions [WECK01]. This field of artificial intelligence (AI) attempts to emulate the knowledge of human experts. Such an expert system consists of (Fig. 6.96):

- a knowledge acquisition component,
- a knowledge base (with static and dynamic knowledge),
- an explanation component and
- an inference engine that arrives at a conclusion for the specific case from the knowledge base.

The inference engine can be implemented using various inference processes such as neural networks or logic-based programming languages such as PROLOG.

Scientific research was performed during the 1980s in particular on the use of these expert systems for machine diagnosis.

However, these highly complex technical systems have not been very successful in industrial applications. The key reasons for this are:

- The knowledge acquisition by these systems is too broad to maintain an adequate knowledge base for industrial applications. This problem of knowledge acquisition was approached using research [PHIL92] but a usable system was not found for the complex field of machine diagnosis.
- The error rate of these systems when reaching a conclusion is directly dependent on the quality of the knowledge base. If a good match between the query and the knowledge base is not found, it is easily possible that the expert system will provide incorrect statements that are incomprehensible to the user. The direct consequence of this is that it leads to the user losing trust in the system.

#### 6.5.3.4 Areas of Application

One area of application for diagnostics is monitoring gearboxes (Fig. 6.97). In this application, the structure-borne sound forms the physical base variable for assessing the condition of the gear teeth. A range of individual parameters can be compiled from the structure-borne sound signal's three processing stages of time periods, spectra and cepstra either by statistical calculation procedures (mean values and frequency distributions) or by calculation and compilation of individual levels (e.g., basic system frequencies, their harmonic series and level differences). An associated,

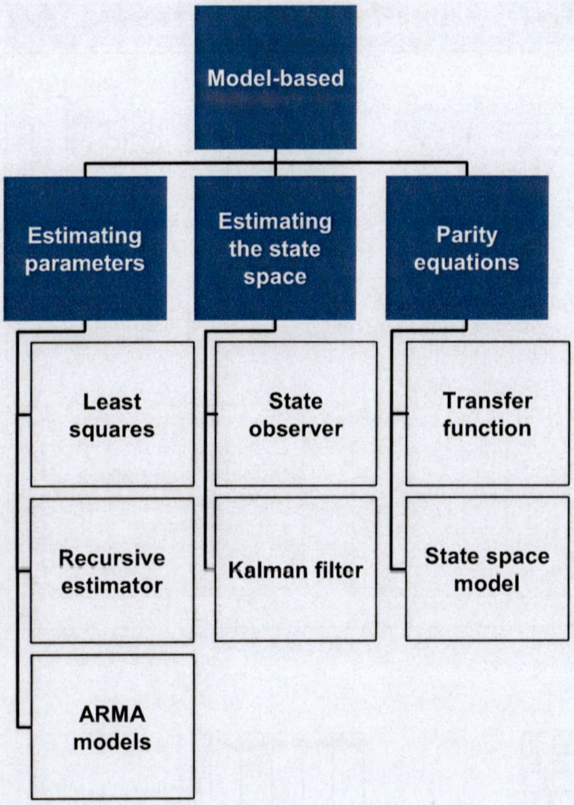


**Fig. 6.94** Overview of model-based analytical methods

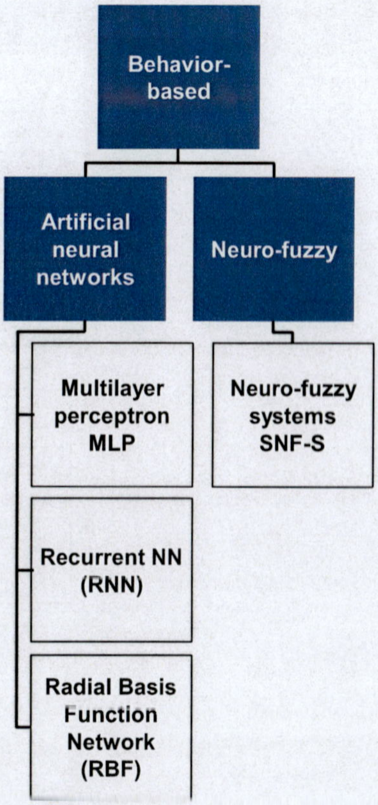


**Fig. 6.95** Overview of behavior-based analytical methods

vector-based consideration of these features and their variations $K_i(t)$ thus allows changes in the components to be detected at an early stage by drawing comparisons with known damage patterns that have occurred. To process different types of damage as well as different applications, a range of comparative patterns with corresponding characteristics must be available (examples shown in Fig. 6.97 for $K_{1i}(t)$, $K_{2i}(t)$).

Figure 6.98 illustrates the typical curve of several significant features for the formation of pitting damage with subsequent tooth fracture at the end of the test. The "mean amplitude" feature represents the mean value for the amplitude components within a specific frequency band. "Autocorrelation" describes the change in a signal with respect to itself after a time shift while "skew" represents an evaluation of the amplitude frequency distribution of the spectrum within a specific frequency range. The "amplitude difference" is understood as the distance of a system's frequency (e.g., tooth engagement frequency or rotational frequency) to a background noise component that can be determined around this frequency.

A gentle increase in the characteristic "mean amplitude" with a simultaneous decrease in the statistical characteristic "skew" can be clearly seen after approximately the twelfth operating hour. The changes in these variables are frequently associated with changes in the microgeometry. The first signs of

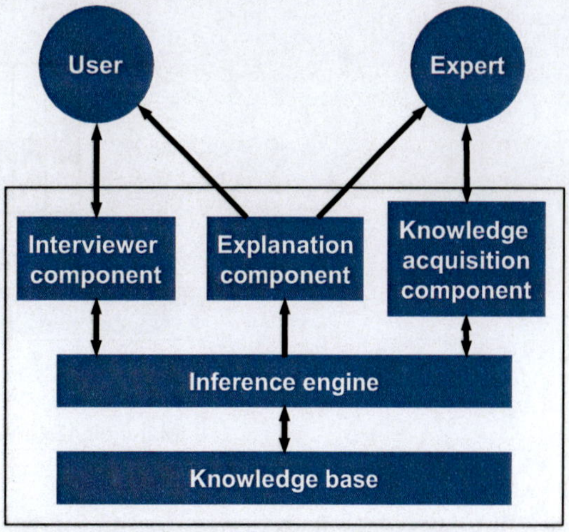


**Fig. 6.96** Architecture of an expert system according to [MILB92]

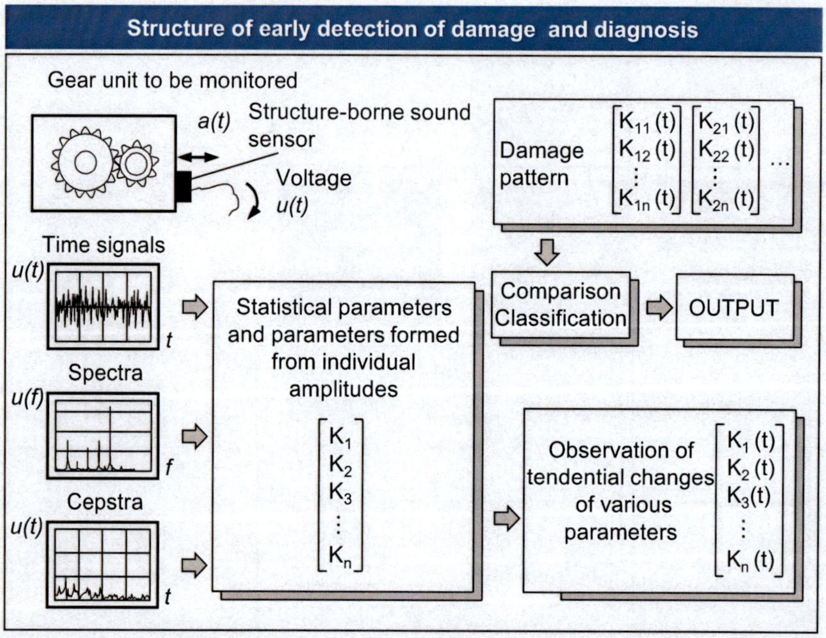


**Fig. 6.97** Structure of early detection of damage and diagnosis according to [MENZ73, WECK77]

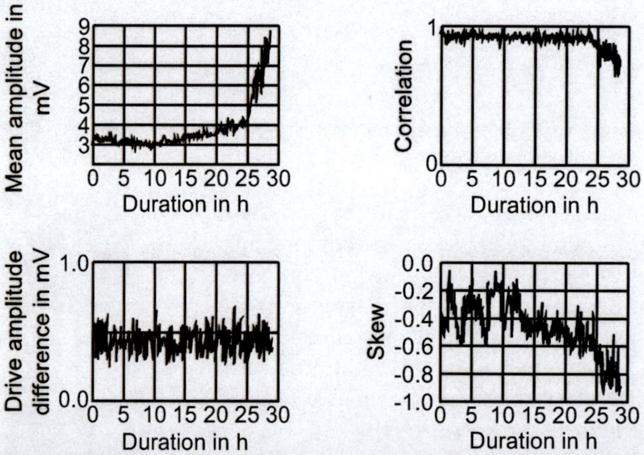


| Geometry data | Operating data | Test run |
|---|---|---|
| $m_n$ = 5 mm  $b$ = 16 mm | $T$ = 800 Nm | – Heavy pitting |
| $z_1$ = 21  $z_2$ = 23 | $n$ = 3,000 min$^{-1}$ | – Tooth root fracture |
| Material: 42 CrMo4 V | $\vartheta_m$ (oil) = 75 °C | after 29 hours of operation |

**Fig. 6.98** Development of characteristics on a test rig with tooth root fracture

damage (pitting) on the flanks can be seen at $t = 25$ h, which, in this case, can be seen in sharp jumps in the features mentioned above. The damage increases rapidly over the following time. This is once again reflected in the sharp rise or fall in the features. The tooth root fractured after 29 h.

Classification on the basis of these specific features can be completed using different mechanisms, see ▶ Chap. 3. One example of classification is allocation using neural networks. This is illustrated below using the example of a rolling bearing [WECK92].

Features from the structure-borne sound and temperature signals were selected as elements of the feature vector. Their curves correlate to bearing damage:

- Maximum value of the structure-borne sound envelope curve signal (30 to 100 kHz)
- Mean value of the structure-borne sound envelope curve signal (30 to 100 kHz)

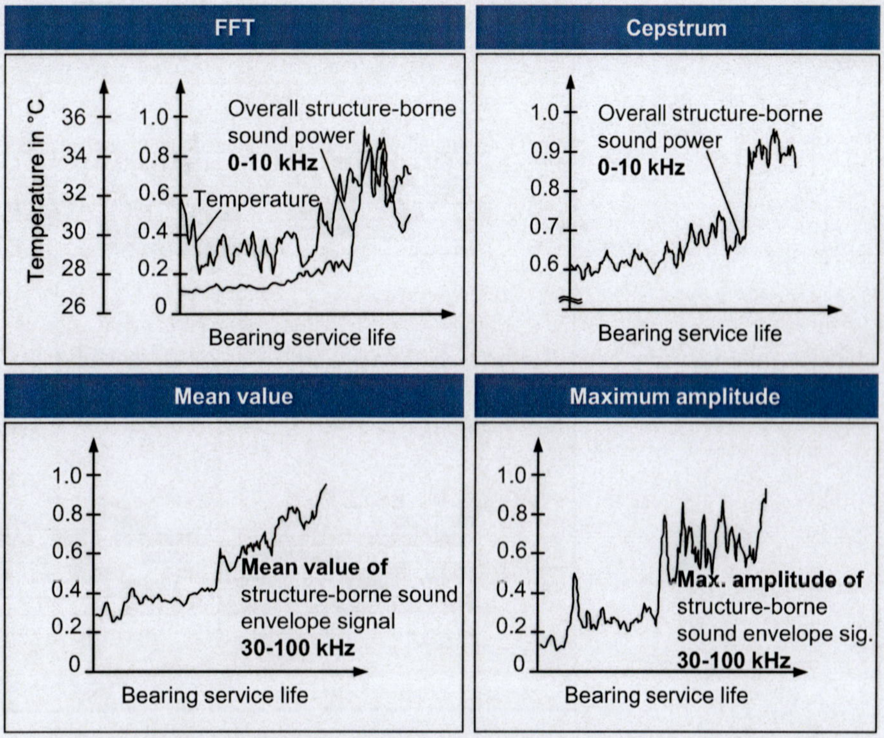


**Fig. 6.99** Curve for various signal features over the service life of the bearing

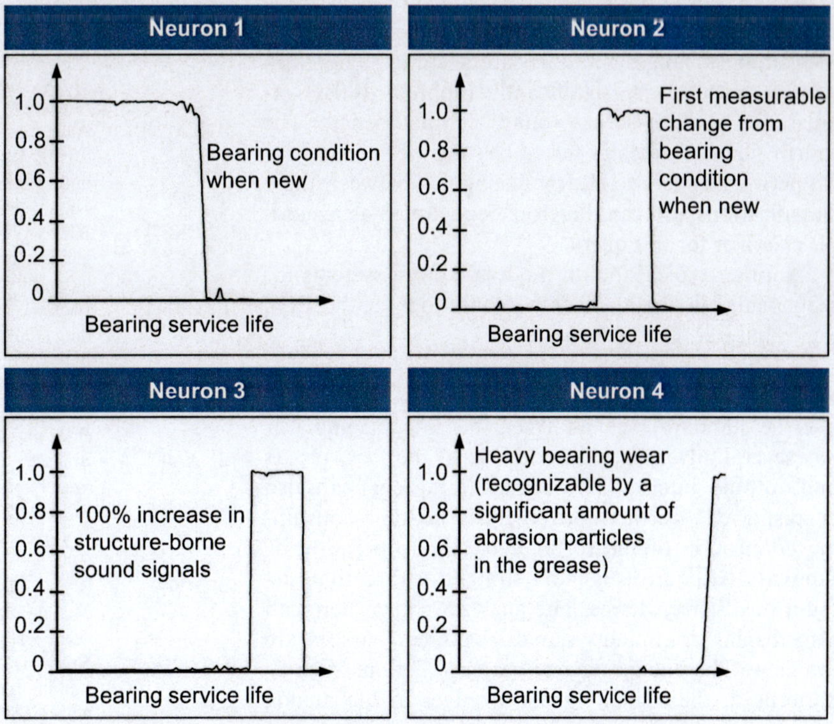


**Fig. 6.100** Classification of the bearing status by a neural network

- Spectral overall output level of the structure-borne sound signal (0 to 8 kHz)
- Cepstral overall output level of the structure-borne sound signal (0 to 8 kHz)
- Temperature of the bearing outer ring.

Figure 6.99 shows the signal curves over the service life of the bearing.

Figure 6.100 shows the classification of the bearing status using the activation of neurons in the output layer over the service life of the bearing. The first

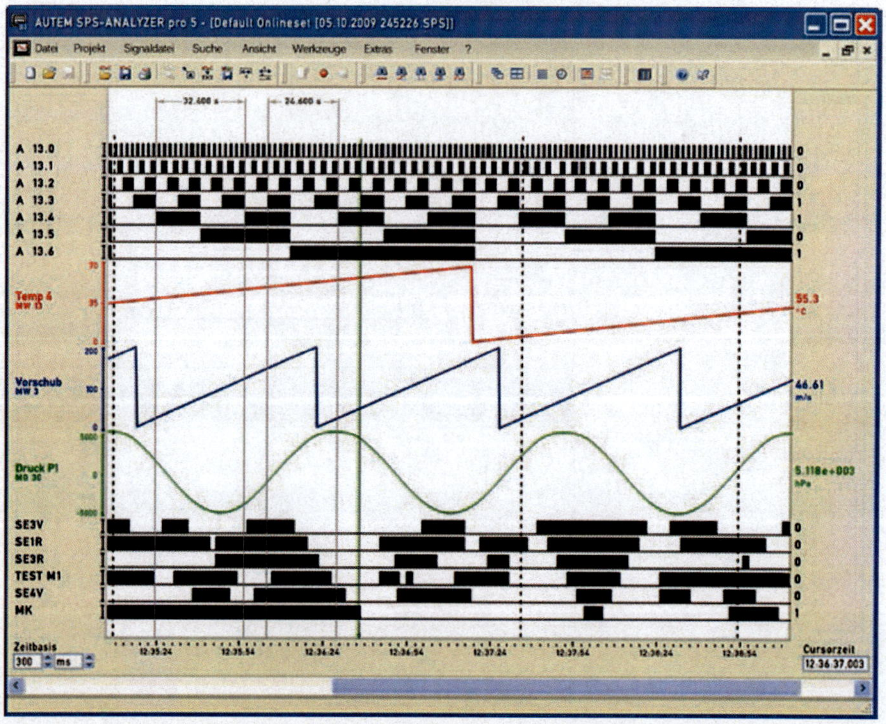

Fig. 6.101 Time display on a PLC logic analyzer. *Source* Autem

6

neuron is active even when there is no measurable bearing damage. The second neuron indicates the first measurable changes in the bearing status. The third neuron responds at significantly (approx. 100%) increased structure-borne sound characteristics. The fourth neuron indicates heavy bearing wear. The neural network is able to clearly distinguish between these classifications and can therefore be regarded as a suitable classifier for this query.

Another typical tool in modern control systems for diagnosing disrupted machine conditions is the PLC logic analyzer (Fig. 6.101).

A PLC logic analyzer can show PLC sequences over time and at the interfaces on the NC screen. Unlike the standard logic analyzer, this integral analyzer has several advantages. In addition to the PLC inputs and outputs, internal PLC signals (e.g., flags) can also be displayed. Furthermore, there is no time-consuming connection of measuring cables and operation of familiar NC controls is more straightforward than using a third-party device. The analyzer shown here can also display continuous signals. Different cursors are available. There is a range of search functions such as patterns in the digital signals or edges in the analog signals.

## References


[ARNO93] Arnold, M.: Feldbussysteme, Teil 1: Grundlagen der offenen Kommunikation; Teil 2: Konzepte und Ausführungen realisierter Systeme. ELRAD (1993)

[AULM70] Aulmann, A.: Bausteine und Anwendungsbeispiele für die Übertragung und Verarbeitung von Winkelinformationen. Sonderdruck "Funk und Technik", Nr. 1–4 (1970)

[BACK87] Backé, W.: Fluidtechnische Realisierung ungleichmäßiger periodischer Bewegungen. Ölhydraulik und Pneumatik 5, 22–38 (1987)

[BACK88] Backé, W.: Neue Möglichkeiten der Verdrängerregelung (1988)

[BEUC89] Beuck, W.: Qualitätssicherung bei der Planung und Realisierung sowie der Störungsanalyse von flexiblen Fertigungssystemen und CNC-Koordinaten-Meßgeräten

[BRAA00] Braasch, J.; Fa. Heidenhain: Genauigkeit von Vorschubantrieben. Firmenschrift, Traunreut (2000)

[BRAA97] Braasch, J.: Genügen Drehgeber zur Positionierung von Werkzeugmaschinen? MSR Magazin, Nr. 7–8 (1997)

[BRYA79] Bryan, J. B.: Design and Construction of a precision 84 inch Diamond Turning Machine. Precision Engineering 1, 13–17 (1979)

[DIN12] Norm DIN 31051:2012-09: Grundlagen der Instandhaltung (2012)

[DIN71] Norm DIN 1319: Grundbegriffe der Meßtechnik (1971)

[DIN88] Norm DIN 32876: Elektrische Längenmessung (1988)

[DOET85] Doetsch, G.: Anleitung zum praktischen Gebrauch der Laplace-Transformation und der z-Transformation. Oldenbourg, München (1985)

[ECKH93] Eckhardt, V., et al.: Die Grenzen liegen jetzt bei der Mechanik. Elektronik 18 (1993)

[ERNS] Ernst, A.: Digitale Längen- und Winkelmeßtechnik, Bd. 3. Verlag Moderne Industrie, Landsberg/Lech (1998)

[ETEL95] ETEL S.A.: Lineare kollektorlose Motoren großer Kraft. Môtiers, Schweiz (1995)

[EVER93] Eversheim, W., König, W., Pfeiffer, T., Weck, M.: Wettbewerbsfaktor Produktionstechnik. VDI-Verlag, Aachener Werkzeugmaschinen Kolloquium. Düsseldorf (1993)

[EX-C93] EX-CELL-O GmbH: Hochgeschwindigkeits-Bearbeitungszentrum XHC 240 (1993)

[FAHR93] Fahrbach, C.; Hammann, G.: Ohne Umwege – Servoantriebe in Vorschubachsen erhöhen die Bearbeitungsgeschwindigkeit. Maschinenmarkt **99**(22) (1993)

[FIRM95] Firmenschrift der ETEL S.A.: Lineare kollektorlose Motoren großer Kraft. Môtiers, Schweiz (1995)

[FÖRD97] Fördergemeinschaft SERCOS interface e. V.: IEC 1491 SERCOS interface – Digitale Schnittstelle zwischen numerischen Steuerungen und Antrieben an numerisch gesteuerten Maschinen. Fördergemeinschaft SERCOS interface e. V., Frankfurt (1997)

[FRIT99] Fritz, S.: Process Monitoring in Turning, Milling, Drilling and Grinding Processes. Proceedings of the 8th International Imeko Symposium on Technical Diagnostics (1999)

[GEOR98] GEORGII KOBOLD GmbH & Co. KG: Dreh-strom-Servomotoren. Leinfelden-Echterdingen (1998)

[GROß81] Groß, M.: Elektrische Vorschubantriebe an Werkzeugmaschinen (1981)

[HAGE99] Hagemeister, W.: Auslegung von hochdynamischen servohydraulischen Antrieben für eine aktive. Frässpindellagerung RWTH, Aachen (1999)

[HEID00] Heidenhain: Positionsmeßsysteme für elektrische Antriebe. Firmenschrift, Traunreut (2000)

[HEID89] Heidenreich, W.: Aufbautechniken für Halbleiter-Magnetfeldsensoren. Technisches Messen **56**(11), 436–443

[HENN99] Henneberger, G.: Elektrische Maschinen I – Grundlagen, Wirkungsweise, Aufbau, Betriebsverhalten. Vorlesungsskript RWTH, Aachen (1999)

[HEWL 76a] Hewlett-Packard: Laser-Transducer-System. System operating and service booklet. Firmenschrift, Santa Clara (1976)

[HEWL 76b] Hewlett-Packard: Laser Transducer Systems Computer Interface Electronics. Santa Clara, California (1976)

[HEWL92] Hewlett-Packard: Magnetisches Längenmeßsystem ML01, Bd. 30-05. Firmenschrift, Kirchzarten (1992)

[HIMB88] Himburger, T.: Die Dynamik des Bohrvorgangs als Grundlage eines Überwachungskonzeptes. TU Hamburg, Hamburg (1988)

[INDR 92a] Indramat GmbH: Hochpräzise CNC-Bearbeitung mit digitalen, intelligenten Antrieben. Lohr a. M (1992)

[INDR 92b] Indramat GmbH: Digitale, intelligente AC-Servoantriebe. Lohr a. M (1992)

[INDR97] Indramat GmbH: CNC-Steuerungen und AC-Antriebe – Produktprogramm. Lohr a. M (1997)

[ISER83] Isermann, R.: Systemidentifikation. Springer, Berlin (1983)

[JAKO79] Jakob, L.: Sicherheitskupplungen in Achsantrieben von NC-Maschinen. Konstruktion, Elemente, Methoden. Konradin-Verlag, Stuttgart (1979)

[JAKO83] Jakob, L.: Anwendungen drehstarrer, flexibler Kupplungen. Der Zuliefermarkt (1983)

[KIEL92] Kiel, E., Schierenberg, O.: Einchip-Controller für das SERCOS-Interface. Elektronik **6**, 50–59 (1992)

[KLÖC95] Klöckner-Moeller GmbH: Automatisieren – Katalog Antriebstechnik. Bonn (1995)

[KOLE08] Kolerus, J.: Zustandsüberwachung von Maschinen (Bd. 79). expert-Verlag (2008)

[KÖNI92] König, W., Ketteler, G., Klumpen, T.: Process Monitoring in Turning, Milling, Drilling and Grinding Processes. Proceedings of the 8th International Imeko Symposium on Technical Diagnostics. Dresden, 23–25th September (1992)

[KÜHN85] Kühne, L.: Entwicklung eines universellen Überwachungssystems für Fertigungseinrichtungen. RWTH, Aachen (1985)

[KUPF68] Kupfmüller, K.: Einführung in die theoretische Elektrotechnik. Springer, Berlin (1968)

[KYNA99] Kynast, R.: SERCOS interface – Technical Overview, 15–18 März (1999)

[LENO93] Lenord+Bauer: Minicoder GEL 243/244. Oberhausen (1993)

[LORE99] Lorenz, L., Kanelis, K.: Antriebslösungen optimiert – Signalverarbeitung und Leistungschalter vereint. Components, 13–16 (1999)

[LUDA91] Luda, G.: Drehstrom-Asynchron-Linarantriebe: Grundlagen und praktische Anwendungen für industrielle Zwecke. Vogel Verlag, Würzburg (1991)

[MENZ73] Menzel, E., Mirandé, W., Weingärtner, I.: Fourier-Optik und Holographie. Springer, Vienna (1973)

[MEYE91] Meyen, H. P.: Acoustic Emission (AE) – Mikroseismik im Schleifprozeß. RWTH, Aachen (1991)

[MEYR92] Meyr, H.: Regelungstechnik I+II (1992)

[MILB92] Milberg, J.: Werkzeugmaschinen – Grundlagen. Springer, Berlin (1992)

[MILL90] Miller, W.: Bezeichnungen für Weg- und Winkelmeßsysteme, Verwirrende Begriffe. Industrieanzeiger **78** (1990)

[MURR 99a] Murrenhoff, H.: Servohydraulik (1999)

[MURR 99b] Murrenhoff, H.: Grundlagen der Ölhydraulik (1999)

[N. N] N.N.: MatrixX user's guide

[NELL91] Nelle, G.: Optoelektronische Meßverfahren zur digitalen Längenmessung. Konstruktion **43** (1991)

[NIEM74] Niemann, H.: Methoden der Mustererkennung. Akademische , Frankfurt a. M (1974)

[NIEM83] Niemann, H.: Klassifikation von Mustern. Springer, Berlin (1983)

[OPHE98] Ophey, L.: Leistungssteigerung mit Direktantrieben? (1998)

[ÖZME00] Özmeral, H.: Elektromagnetisches Fast Tool Servo System für den Einsatz in der Ultrapräzisionstechnik. Berichte aus der Produktionstechnik, Bd. 12). Shaker, Aachen (2000)

[PFEI76] Pfeifer, T., et al.: Verfahren und Methoden zur Beurteilung der Arbeitsgenauigkeit von Werkzeugmaschinen (1976)

[PFEI99] Pfeifer, T., Wiegers. L.: Reliable tool wear monitoring by means of image optimization in machine vision. Darmstadt, 10–11 March (1999)

[PHIL91] Philipp, W.: Linear-Direktantriebe hoher Genauigkeit und Dynamik (1991)

[PHIL92] Philipp, W.: Regelung mechanisch steifer Direktantriebe für Werkzeugmaschinen. Springer, Berlin (1992)

[PRES64] Pressler, G.: BI-Hochschultaschenbuch: Regelungstechnik. Bibliographisches Institut, Mannheim (1964)

[RAKE93] Rake, H.: Regelungstechnik A+B (1993)

[ROSE94] Rosenbauer, T.: Getriebe für Industrieroboter – Beurteilungskriterien, Kenndaten, EinsatzhinweiseRWTH. Aachen (1994)

[RUDO14] Rudolf, T.: Adaptierbare Parametrierung von Diagnosesystemen durch Verwendung digitaler Antriebssignale in der Prozessüberwachung. Aachen (2014)

[RUMM92] Rummich, E., Gfrörer, R.: Elektrische Schrittmotoren und -antriebe, Funktionsprinzip – Betriebseigenschaften – Meßtechnik. Expert-Verlag Ehningen, Kontakt & Studium (1992)

[SALZ97] Salzmann, T., Klug, R.: Mehr Qualität für Antriebe im Megawattbereich. Eng. Autom. **1–2**, 38–40 (1997)

[SCHE91] Schehl, U.: Werkzeugüberwachung mit Acoustic Emission beim Drehen, Fräsen und BohrenRWTH. Aachen (1991)

[SCHI90] Schierling, H.: Selbstinbetriebnahme – eine neue Eigenschaft moderner Drehstromantriebe. atp. **32** (1990)

[SCHU93] Schuhmacher, W., Kiel, E.: Elektrische Antriebe in den neunziger Jahren. Elektronik **7**, 94–98 (1993)

[SESS91] Sesselmann, T.: Maßstäbe für interferentielles Abtasten ermöglichen Nanometer-Meßschritte. WB Werkstatt und Betrieb, Nr. 124-2 (1991)

[SIEM98] Siemens AG Automation & Drives: Gleichstrommotoren für drehzahlveränderbare Antriebe von 0,7-1550 kW. Nürnberg (1998)

[SPIE90] Spies, A.: Längen in der Ultrapräzisionstechnik messen. Feinwerktechnik & Meßtechnik. **98** (1990)

[SSBA99] SSB-Antriebstechnik GmbH: Bürstenlose SSB-Servomotoren, SSB-Hohlwellenservomotoren. Salzbergen (1999)

[STEA79] Stearns, S.D.: Digitale Verarbeitung analoger Signale. Oldenbourg, München (1979)

[STRO98] Stromag Elektronik GmbH: Frequenzumrichter UDM…3. für DS-Asynchronmotoren und Kompaktverstärkermodul CDC…1/2. für Synchron-Servomotoren. Unna (1998)

[ULRI82] Ulrich, H.: Verhütung von Kollisionsschäden an NC-Drehmaschinen. Industrie-Anzeiger **104**(45) (1982)

[VDI73] VDI VDI/VDE 2600-2 (November 1973). Metrologie (Meßtechnik) – Grundbegriffe

[WALC74] Walcher, H.: Digitale Lagemeßtechnik. VDI, Düsseldorf (1974)

[WECK01] Weck, M.: Werkzeugmaschinen Bd. 5: Messtechnische Untersuchung und Beurteilung. Springer, Berlin (2001)

[WECK77] Weck, M., Teipel, K.: Dynamisches Verhalten spanender Werkzeugmaschinen. Springer, Berlin (1977)

[WECK92] Weck, M., Fauser, M.: Lagerüberwachung mit neuronalen Netzen. In: wt Produktion und Management. Nr. 12. Jg., 1992, S. 34

[WECK99] Weck, M., Krüger, P., Brecher, C.: Grenzen für die Reglereinstellung bei elektrischen Lineardirektantrieben. Antriebstechnik **38**(2+3), 55 (1999)

[WILD59] Wildermuth, E.: Der "Resolver", ein moderner Analogrechenbaustein. Sonderdruck "Feinwerktechnik" **63**, 9 (1959)

[WITT88] Wittman, G.: Spieleinstellbare Getriebe mit konischen Verzahnungen. Zuliefermarkt **36** (1988)

[YE92] Ye, G.: Erhöhung der Bahngenauigkeit NC-gesteuerter Vorschubachsen mit Hilfe eines Kompensationsfilters. VDI, Fortschritt-Berichte VDI, Düsseldorf (1992)

6

# Automation of Machines and Plants

I thank my staff members Christian Fimmers M. Sc., Oliver Petrovic M. Sc. and Simon Storms M. Sc. for their contribution in revising this chapter.


C. Brecher and M. Weck, *Machine Tools Production Systems 3*,
Lecture Notes in Production Engineering, https://doi.org/10.1007/978-3-658-34622-5_7

Contents

## 7.1 Definitions

Automation encompasses all measures for the complete or partial automation of processes; the processes are controlled by either a program, which has been created earlier, or an artificially produced intelligence (AI) independent of human beings. Preconditions for this are mechanization and control technology, see ◘ Fig. 7.1. *Mechanization* replaces human power and effort with devices and motor drives on machines. *Control technology* encompasses the storage, logical processing, and transmission of path and switching information as well as the control of actuators.

An *automatic machine* is thus a device or machine that carries out defined tasks, either partially or completely, autonomously once it has completed a one-time setup or programming procedure and is loaded wth primary and auxiliary materials.



A *semi*-automatic or *partially automatic machine* in the machine tool industry is a machine which autonomously carries out all the main and feed motions for the tool and workpiece. Functions such as material feed, workpiece clamping, and speed selection are performed manually.

Fully automatic machines acting as machine tools complete all motions and functions autonomously; there is no need for human involvement in the workflow. However, people still carry out control functions and sometimes have to insert raw material into the magazine or blanks in the chuck.

## 7.2 Historical Developments and Reasons for Automating Machine Tools

The first devices with properties similar to machine tools can be found as far back as the Middle Ages. For example, on pole lathes from the fourteenth century, the workpiece is guided on a wooden structure between two adjustable centers and is worked exclusively by muscle power with a gouge. The cutting action, i.e. the rotation of the workpiece, is produced via a rope drive by means of a treadle; the feed motion is produced manually [WITT60].

The advent of steam-powered machines towards the end of the eighteenth century saw the development of motor-driven machine tools. While, initially, the tool still had to be guided manually on these machines, metal-working machines were developed around 1790 and these incorporated guided tool slides (supports) and frame components made from cast iron instead of wood. In 1830, *Maudslay* built a lathe in England with an autonomous linear feed. This was the first stage in mechanization and, at the same time, the precondition for automation [BRUI66]. There has been a continuous transition from mechanization to automation in the last 150 years of industrial development. Control technology has developed from a purely mechanical system using tappets, levers and cam control, to hydraulic, electrical and electrohydraulic positioning and post-forming controls, to electrical control.

The development in the field of manufacturing process automation in recent decades has been largely determined by the huge improvement in the performance of electronic components and the associated software. Since around 1960, numerical controls and, since 1970, programmable and computer-aided controls have been used in the broad application field of control technology. These controls, and the opportunities for electronic data processing and communication, are now the basis for integrating consistent data models into the overall flow of information in a manufacturing operation. The greatest degree of automation is currently achieved on transfer lines for mass production as well as flexible manufacturing systems. These systems were developed for the automatic manufacturing of a range of different workpieces within specific families of parts. This includes automatic machine retooling. Here,

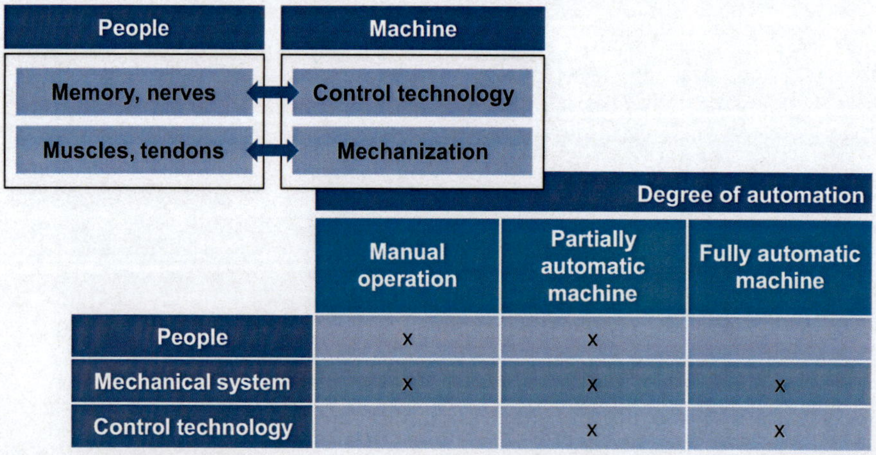


◘ **Fig. 7.1** Relationships between human activity, mechanization and automation

several numerically controlled machines are linked to each other via automatic material transport devices or robots so that the manufacturing sequence for the workpieces can be set as desired. Control systems are used to control the machines and the material flow [BEYE07]. Due to the complexity and high investment costs of these types of flexible, interlinked systems, they are not easy to make economically viable. High disruption rates have led to the creation of smaller systems, so-called manufacturing cells, which are easier to oversee and easier for the user to operate. Here, more and more of the control system's functionality is integrated into the control leading to the increased creation of mixed forms comprising centralized approaches with elements organized in a decentralized way.

This development will, in the future, allow the unappealing, monotonous and physically demanding work undertaken by people to be replaced by creative tasks such as planning, design, monitoring, and optimization.

Figure 7.2 shows a general overview of the individual development stages in automation using turning as an example. In the first stage, all control and auxiliary functions (operating the machine, transporting the workpiece, and measuring) are performed manually. In the second stage, the semi-automatic machine relieves the user of some work. However, due to the greater quantities created by faster processing, there is more manual transportation and measuring work. This leads to the third stage of full automation of machining and workpiece transportation as well as handling the workpieces and tools by handling devices such as industrial robots. Measurement is still a manual task here but, in the fourth stage, this process is also carried out by an automatic measuring device. People are only involved if the measurement result requires an adjustment to the automatic lathe. Finally, the fifth stage represents a fully automated manufacturing process. The workpiece measuring system integrated into the machine allows autonomous correction of the tool infeed rate. People are only involved in the event of a disruption.

The foundations for automation at the start of industrialization lay mainly in the demand for higher productivity, i.e., quicker and easier manufacturing of greater volumes of workpieces. In the past, Taylor's

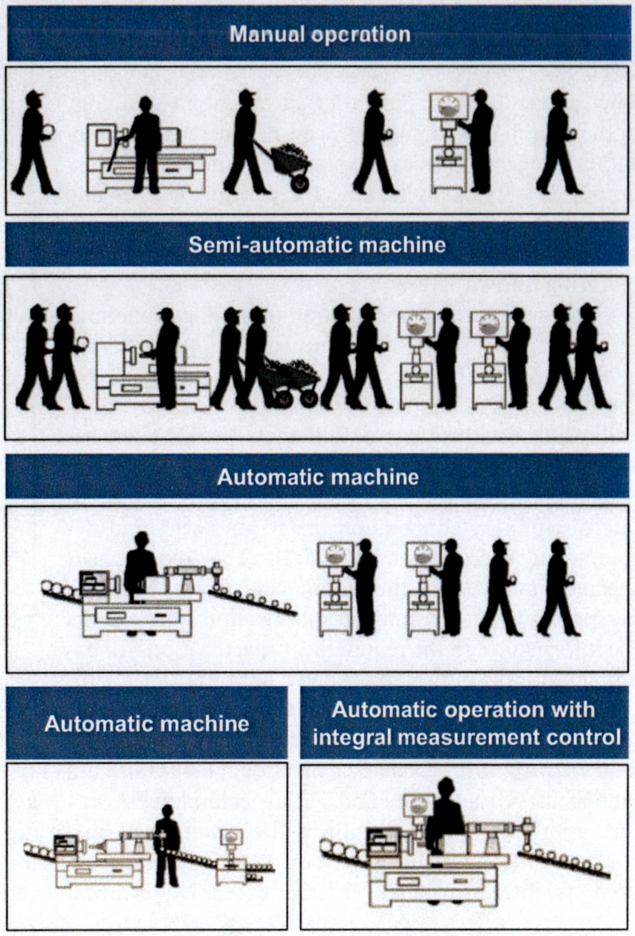


Fig. 7.2 Stages of automation according to Saljé

approach to the division of labor allowed enormous increases in productivity to be achieved. Right up to the present, the predominant technical solutions in the field of automation technology have been guided by the principles of large volume production. However, more recent developments indicate a trend away from large volume production towards small-lot and single-component production. This means that qualified specialist work combined with an integrated approach will become increasingly important.

Moving to the field of machine tools, the job of specialists in this area is now characterized by wide-ranging job content, from programming, optimization, to monitoring manufacturing processes. The basic precondition for supporting people as they work with new, innovative technical solutions is to be found in the ergonomic design of the machine user interface. Modern control systems generally provide programming interfaces that allow individualized adjustment to the user's requirements. At the same time, the control systems can be extended with additional functionalities. This includes, for example, remote diagnosis and connection to the intranet and internet.

## 7.3 Control and Automation Technology as a Subtask of Machine Development

The value-added component of automation technology in the overall development process has increased significantly in recent decades. The tasks of automation technology now typically include the control of axes and peripheral components, integration of safety technology, and IT links between the machine and the manufacturing infrastructure.

The integration of automation technology functions produces interactions at the mechanical, electrical engineering, and IT level. The complexity this generates can only be handled through cross-departmental cooperation between a range of different specialist departments.

Standard tools to help with interdepartmental and interdisciplinary cooperation are requirement and development specifications [VDI12]. The requirement specification contains the requirements from the user's perspective. The development specification supplements the information in the requirement specification, for example, with the information shown in ◘ Fig. 7.3.

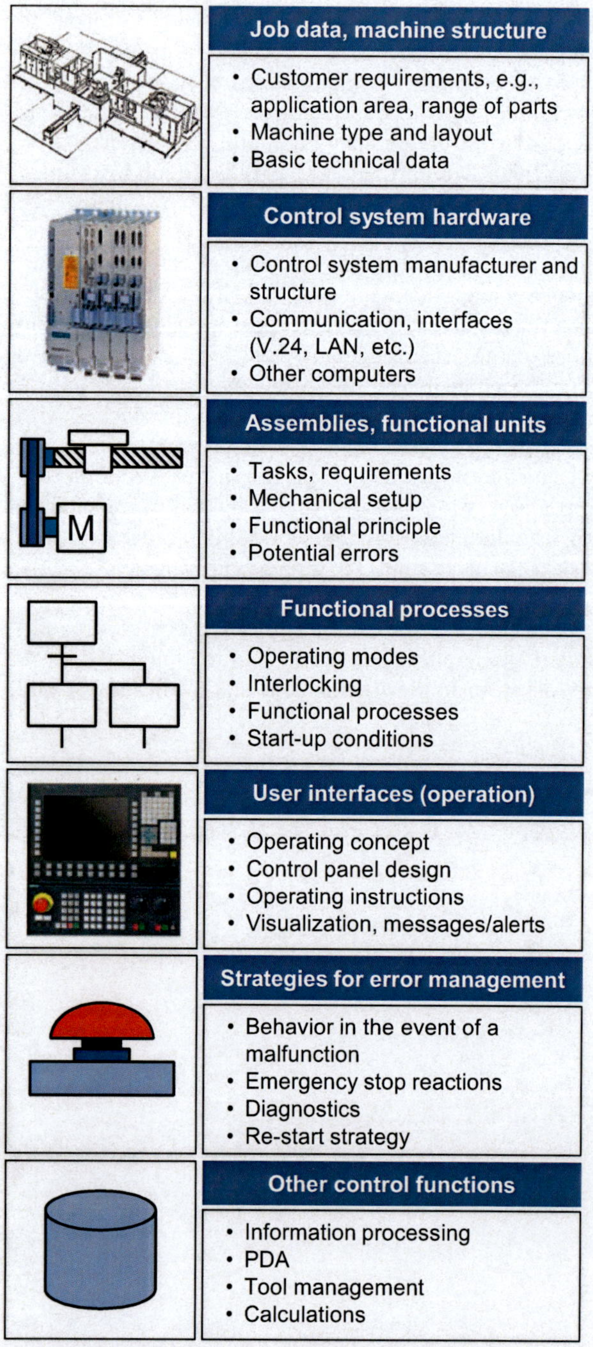


◘ **Fig. 7.3** Typical structure of a development specification

Requirement and development specifications are often supplemented by further specification methods. Examples of this are schematics, functional diagrams, and tabular lists of actuators and sensor technology. A further approach to specification is the structure of mechatronic systems with functions. Here, functions are used as interdisciplinary descriptive tools; technical elements are allocated to these to implement them (see [VDI04]).

Standard *Product Lifecycle Management (PLM) systems* are being increasingly used in automotive engineering as well as in mechanical and plant engineering to manage the interdisciplinary product data. The management of product structures, CAD data, access rights, and projects is a central function of today's

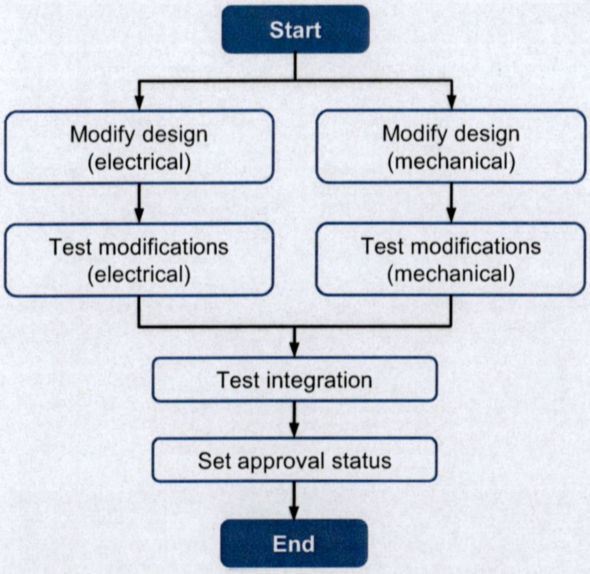


**Fig. 7.4** Example of a design workflow in the PLM system

PLM systems. Other central components of a PLM system are to map the design workflows and to monitor their progress, see Fig. 7.4. In addition to this, they also provide opportunities to manage data on suppliers, simulations, and maintenance. PLM systems generally have an interface to the Enterprise Resource Planning (ERP) system, which coordinates the manufacture of individual product components and the allocation of manufacturing resources.

## 7.4 Control of Operating Sequences

### 7.4.1 Production Facility Functions that Can be Automated

There are many opportunities in the field of production engineering to automate manual functions as well as human thought processes for operating and controlling machines. The individual production processes and tasks usually require specially adapted solutions.

The current state of the art of mechanical manufacturing is characterized by the high level of development of machine tools capable of solving complicated machining tasks with correspondingly high technological requirements. In addition to this, fully-automated machines (e.g., automatic lathes), as well as linked plants (e.g., flexible manufacturing systems), are now being extensively used in mechanical production facilities. In recent years, the focus in automation has been to increase the flexibility of production facilities for small-volume and single-component production. The autonomous flow of alternating machining, handling and clamping tasks for a large number of different workpieces requires simple and rapid reprogramming of the functional sequences and the adaptation of gripping and clamping devices.

The basic functions that can be performed on production facilities are allocated to machines, tools, devices, and workpieces in Fig. 7.5. These basic functions are the main focus of automation.

Electronic data processing in the form of programmable logic controllers (PLCs, ▸ Chap. 10) and Numerical Controls (NCs, ▸ Chap. 11) are a decisive precondition for the automation of the functions shown in Fig. 7.5 (transportation, handling, organizing workpieces, machine setup, etc.). As the complexity and the investment value of the plants increase, the reliability of the components and an appropriate diagnostic system become increasingly significant so that errors in the plant can be identified and localized early. This is particularly important with respect to resulting damage and expensive downtimes as a consequence.

The subfunctions required for the material and resource flow in a machine tool are illustrated in principle in Fig. 7.6. The functions shown are initiated, executed and monitored by an NC with a PLC. As automation of the various functions in a plant or machine is always associated with a high level of investment, it is generally the economic analysis that decides which subset of the functions shown in Fig. 7.8 is automated.

When considering any automation, it has been shown in practice that the frequency of disruption increases overproportional with the complexity and the degree of automation. Each application should therefore be analyzed to establish whether the automation system can approximately achieve the process security which is achieved with human thought processes.

Deviations from the planning data often occur in the complex interaction between organization (material and tool management, programming and upgrading the machine, production control) and optimization of the technological process flow. Automating the identification of this deviation and subsequent initiation of a suitable measure to alleviate this should make the plant self-optimizing, a characteristic currently in development for individual sections of the plant or machining centers but not yet available for universal implementation in networked production facilities. Cost-effective solutions to resolve these deviations are usually only to be found by qualified employees and not by automated systems.

### 7.4.2 Functional Sequences

One precondition for an automated production process is the coordination of the subfunctions in time and space. Machine control performs this task. It consists of storing the precise sequence of all subfunctions,

**Fig. 7.5** Automatable functions

| Function | Item: Machine | Tool | Device | Workpiece |
|---|---|---|---|---|
| Transporting, handling, organizing | | X | X | X |
| Setup | X | X | X | |
| Storage | X | X | X | X |
| Clamp, release | (X) | X | X | X |
| Machining | X | (X) | (X) | X |
| Prepare | | X | X | X |
| Inspect (measure and evaluate) | (X) | X | (X) | X |
| Diagnostics (machine and tool condition) | X | X | X | |

X - item performs the function
(X) - item indirectly involved in the function

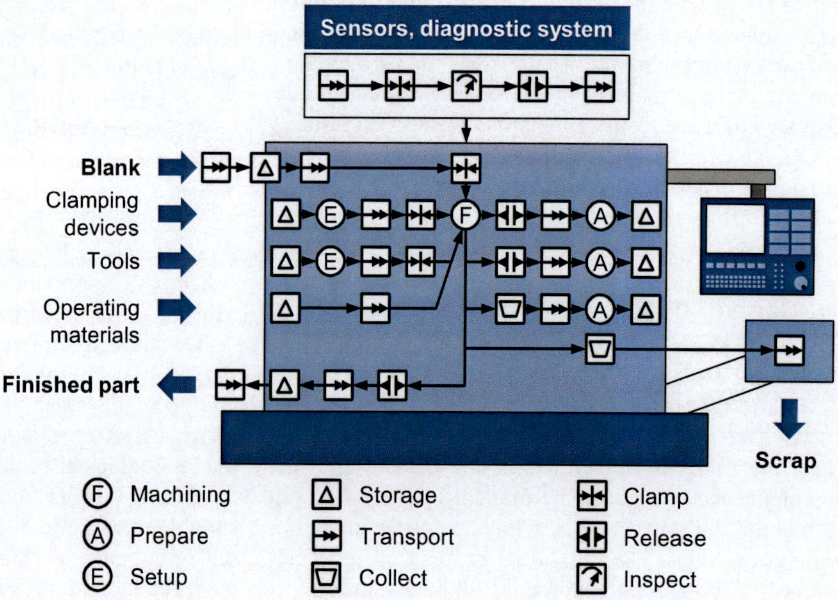


**Fig. 7.6** Basic structure of the subfunctions in a machine tool

initiating the processes in the correct order at the relevant time, monitoring the execution, if necessary, regulating the manufacturing process (measurement control, adaptive control) and continuously diagnosing the condition of the overall system.

Apart from the central task of machine control, i.e., controlling the manufacturing process, the machining trajectories and the required tools, as well as the relevant resources and clamping devices, must be saved, transported, clamped and unclamped. Manufacturing quality is checked during and after machining, i.e., the workpiece dimensions are measured and, if necessary, appropriate corrective measures are initiated. Monitoring and diagnosis of nearly all machine components, as well as the actual production process, takes place over the entire operating time. The complete realization of this all-encompassing control task can only be achieved with very expensive equipment such as flexible manufacturing systems.

Above all, it is still extremely difficult to monitor the objects involved in the process, e.g., the tool condition, due to a lack of sensors that can measure the wear

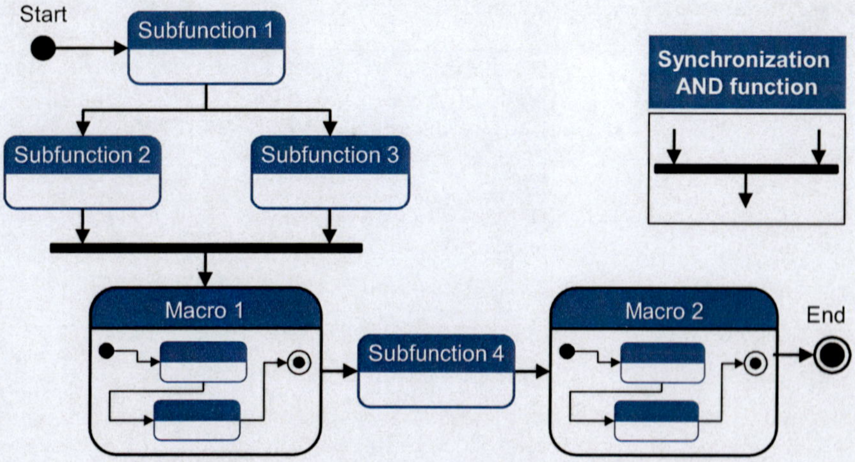


**Fig. 7.7** Flow of a functional sequence in Unified Modeling Language (UML)

variables directly and that can be used in unfavorable environmental conditions (contamination, temperature, etc.).

The program or process control in a plant must be reprogrammed for each different machining task. Figure 7.7 illustrates the general flow of a functional sequence comprising subfunctions and macros. Macros are a reusable, complete concatenation of several subfunctions. They make it easier for users to create programming instructions for complicated, comprehensive operating sequences. The subfunctions and macros of a working cycle (machining a workpiece) are initiated and executed in accordance with the programmed functional sequence. The functions to be executed run either consecutively, i.e., once the previous function(s) have been completed, or in parallel if this is possible or required. The completion of the last function to be executed (e.g., measuring the workpiece) ends the execution of the functional sequence and simultaneously initiates the first subfunction for the next workpiece machining operation or brings the machine to a halt.

### 7.4.3 Control, Programming and Storage Elements

The control system is constructed using mechanical, hydraulic, pneumatic, electrical or electronic elements or a combination of these. The complexity of the elements for executing the subfunctions depends on the type of control. Simple elements are manual or limit switches where the latter can generally be used to complete the previous path function.

In mechanical control systems, the function calls frequently require complicated switching mechanisms. End stops and cams are added to the appropriate carrier systems which, for their part, move with the machine slides and determine the feed path using their location with respect to the mechanical switching probes. They initiate switching and control functions that are mechanically executed via the cam disk (see ▶ Chap. 8).

In electrical and electronic controls, electrical switches (manual and limit switches) are used to initiate and complete functions. In electrical controls, the logical signal processing takes place with the aid of relays; in electronic controls, depending on the scope of the task, this takes place with the aid of integral circuits, microprocessors and process computers. In electronic controls, amplifiers or servo components are used to convert the control commands into executable functions; these control commands usually have very low energy levels. Contactors, transistor and thyristor elements, or valves and couplings are usually used (see ▶ Chap. 10).

Electric and hydraulic motors are used to produce the rotational or translational motion for the main working motions (e.g., linear motion of press rams and planing slides, rotational motion of main spindles on turning and milling machines) and feed motions (e.g., cross tables, quills, and roll and gripper feed motions when processing sheet metal). Depending on the application, these are either speed-controlled or force-controlled. Suitable converters frequently have to be included (e.g., gear units or spindle nut systems) to achieve the required speed or torque range, or to convert the rotational motion of the motor into a linear drive motion.

Switches, sensors and measuring systems are used to monitor the machine functions and the manufacturing process. The choice of the correct control is always a compromise between comfort and functional scope and the price the customer is prepared to pay.

An important precondition for reusing programmed workpiece machining sequences is to store the information required to execute the action. This includes the sequence of the individual functions, i.e., the sequential provisions of the switching and path information that, as a whole, produces the machining program. The switching information is used to execute subfunctions

■ **Fig. 7.8** Storage media for switching and path information

| | | Storage medium | | |
|---|---|---|---|---|
| **Control technology** | **Programming** | **Path function** | **Switching function** | **Execution of the logic and program flow control** |
| Cam-controlled machine | Production of mechanical storage media and setting of cams | - Fixed stops<br>- Cam plates<br>- Curve template | - Cams on auxiliary control shaft and moving slides | - Coupling, gearbox<br>- Shift linkage<br>- Cams |
| Postforming (copying) machine | Production of mechanical storage media and setting of cams | - Templates<br>- Curve template<br>- 3D models | - Limit switches<br>- Cams on strips or drums | - PLC<br>- Relays |
| Electronically and cam-controlled machines | Production of mechanical storage media using display and keyboard | - Fixed stops<br>- Cam plates | - Limit switches<br>- Cams on strips or drums and moving slides | - PLC<br>- Relays (also some NC) |
| Electrically/electronically controlled machines (PLC) | Using crossbar distributors, plug boards or displays and keyboards | - Gauges | - Plug boards<br>- Limit switches<br>- Cams with switches | - PLC<br>- Relays |
| Numerically controlled machine (micro-processor-based) | Online using displays and keyboards, offline using programming languages or graphics | NC programs<br>- ROM and RAM memory<br>- Magnetic and optical storage (network, USB, hard disk) | | - NC<br>- PLC<br>- Relays |

(e.g., changing speeds and feeds, switching over the tool turret, clamping the machine slide, clamping of tools and workpieces). Path information contains all the data for the motion sequences in the individual axes of the various assemblies such as slides, quills, etc. required to produce the workpiece geometry. ■ Figure 7.8 is a summary of the various storage media for axis motions (geometry) and switching commands depending on the control technology used.

## 7.5 Examples of Automated Functions

### 7.5.1 Path and Switching Information

Defined, sequential motions in time of machine slides, spindle(s), etc. within specific path limits must be specified when machining workpiece contours such as bores, turning recesses or milling rectangular components. A precondition for controlling the motions and the switching processes from one function to the next is having information on the path limit from either linear displacement transducers (see ► Sect. 7.5.1.1) and cam strips, switch panels (see ► Sect. 7.5.1.2), or absolute or incremental rotary encoders (see ► Sect. 7.5.1.3). The physical principles belonging to the measuring procedure are explained in ► Sect. 2.3.

#### 7.5.1.1 Linear Displacement Transducers

The position of machine components with linear guidance, such as main axes, is generally determined with optical, linear length measuring devices, see ■ Fig. 7.9.

Incremental optical measuring devices use the functional principle shown in ■ Fig. 7.10. The measuring unit, consisting of a light source, condenser, scanning plate, and photoelements, is fixed to the moving machine axis and is guided parallel over a stationary, fixed glass scale. The glass gauge has a code track which is

**Fig. 7.9** Optical-electronic displacement transducer for determining the position of a machine axis

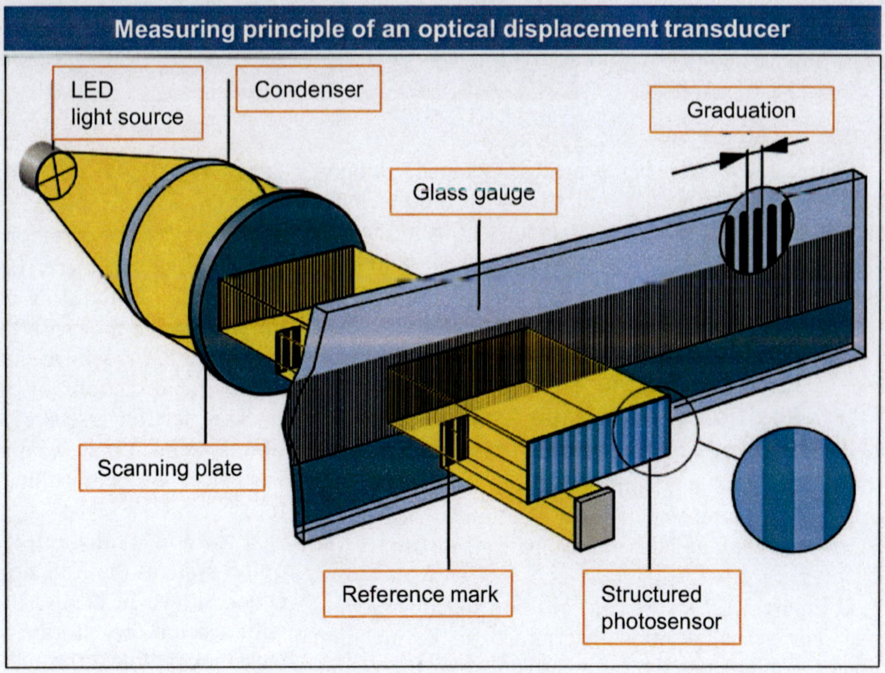


**Fig. 7.10** Principle of operation of an optical incremental displacement transducer. *Source* Heidenhain

alternately translucent and opaque [SCHÖ97]. As the measuring unit passes over the gauge, the number of light impulses (recorded by the photosensitive elements) and the graduation of the code track allow the travel distance, and thus the position of the axis, to be calculated, see ► Sect. 2.3.2.

The position information can then be used in the machine control system to control motion sequences and switching procedures.

### 7.5.1.2 Cam Strips and Switch Panels

The easiest way to specify and store path information is with cam strips, which are arranged as required either next to or behind each other on slideways or rolls and are adjustable. The cams can be detected mechanically with a lever or with electrical, pneumatic or hydraulic switches. By storing this displacement distance, switches, which stop and start running functions, can be activated when this position is reached. ◘ Figure 7.11 shows the cam strips on a cross table in a bed-type milling machine. The cams establish that a specific distance has been reached and thus initiate the subsequent functions such as changing the feed rate and direction or the spindle speed.

The result of various switching functions can also be stored on cam drums. The drum continues to rotate as the work progresses. After one revolution, the work cycle is complete and can be started again. The absolute switching accuracy of the cam is approximately 0.1 mm.

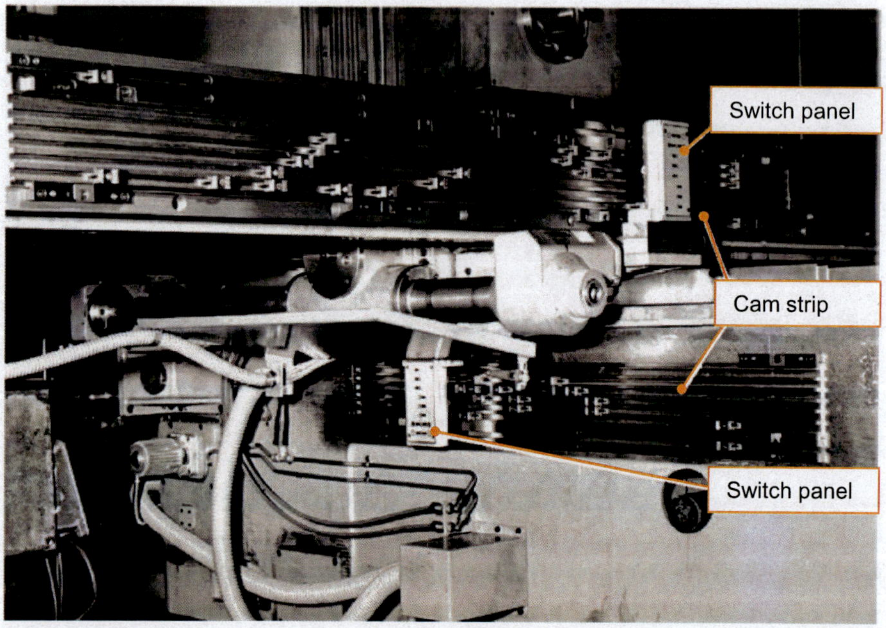


**Fig. 7.11** Cam strips for limiting travel and initiating switching procedures

7

Where a workpiece is repeatedly machined, it makes sense to reuse the whole cam strip with the mechanical cams. To this end, the strip is removed from the machine and refitted when the production task comes up again.

#### 7.5.1.3 Absolute and Incremental Rotary Encoders for Measuring the Actual Position of a Machine Assembly and for Controlling Rotational Speed

Rotary encoders are combined with ball screws for (in)direct measurement of table positions or for direct measurement of angular positions, e.g., on rotary tables.

Figure 7.12 shows an NC-controlled machine slide. The actual position determined in the machine control system from the output signal from the rotary encoder via the pitch of the ball screw is compared to the corresponding reference position value.

The reference value for the motor speed is determined from the difference between the actual and reference values and transmitted to the drive amplifier. The speed is controlled by the drive amplifier, which supplies a rotor and field voltage as a function of the reference rotational speed, and this is corrected in accordance with the actual rotational speed determined by the tachogenerator. Switching and path functions can also be initiated depending on the actual value for the position of the rotary encoder (e.g., changes in speed, feeding other axes, etc.).

### 7.5.2 Workpiece Transportation and Handling

Automating workpiece transportation is an important contribution to reducing non-productive times. Providing a machine tool with raw material generally means providing bar stock (round, square or hexagonal), ring material, bar sections, or preformed blanks.

This section discusses devices used in transportation systems or as auxiliary equipment moving the workpiece to be machined to the relevant machine tool.

Providing a fully-automated lathe with bar stock requires systems that can store and feed the material. The chuck shown in Fig. 7.13 which is fitted in a lathe with a drum tool storage system, is one solution. The bar stock is drawn through the hollow working spindle by the opposing chuck. This makes it possible to provide a flexible material feed due to the distance the tool carriage can travel. This chuck also facilitates machining of both sides of the workpiece.

An alternative option for feeding bar stock for a lathe is shown in Fig. 7.14. In this instance, the material is transported using a two-piece material feed (telescopic feed) consisting of an external component and an internal feed. The external component of the telescopic feed remains at the rear end of the working spindle as the rod decreases in length while the internal feed, which has a smaller diameter, provides the feed for the rest of the rod in the spindle area. The internal

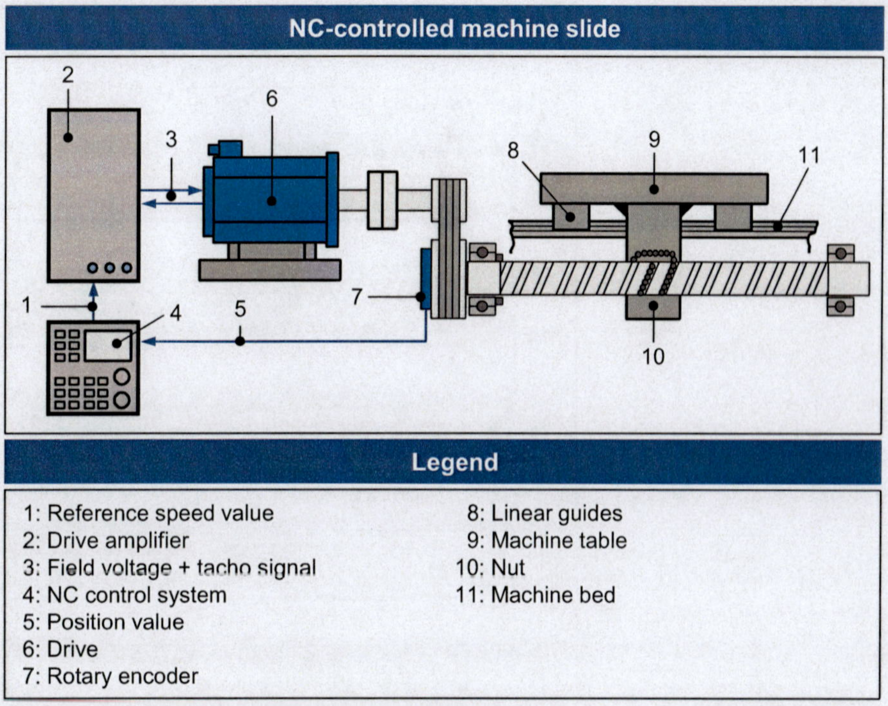


**Fig. 7.12** NC-controlled machine slide. *Source* Boehringer

**Fig. 7.13** Tool storage system with chuck for implementing bar feed

feed is driven through the slot in the external component via the chain drive and the dog.

As the whole bar rotates during machining, a suitable bearing arrangement must be provided in the area of the bar loading magazine. This is usually achieved with rollers or as an oil guide in a closed tube which generates a hydrodynamic oil film due to the rotation of the bar.

In fully automated, flexible manufacturing systems (Volume 1, ► Chap. 9), the blanks to be machined are prepared in a manufacturing cell (e.g., saw cells or manual workstations) and are then machined in other manufacturing cells using appropriate machine tools. Transportation between the individual cells is usually via handling devices, industrial trucks or conveyor systems as shown in Fig. 7.15. The flexibility of

7

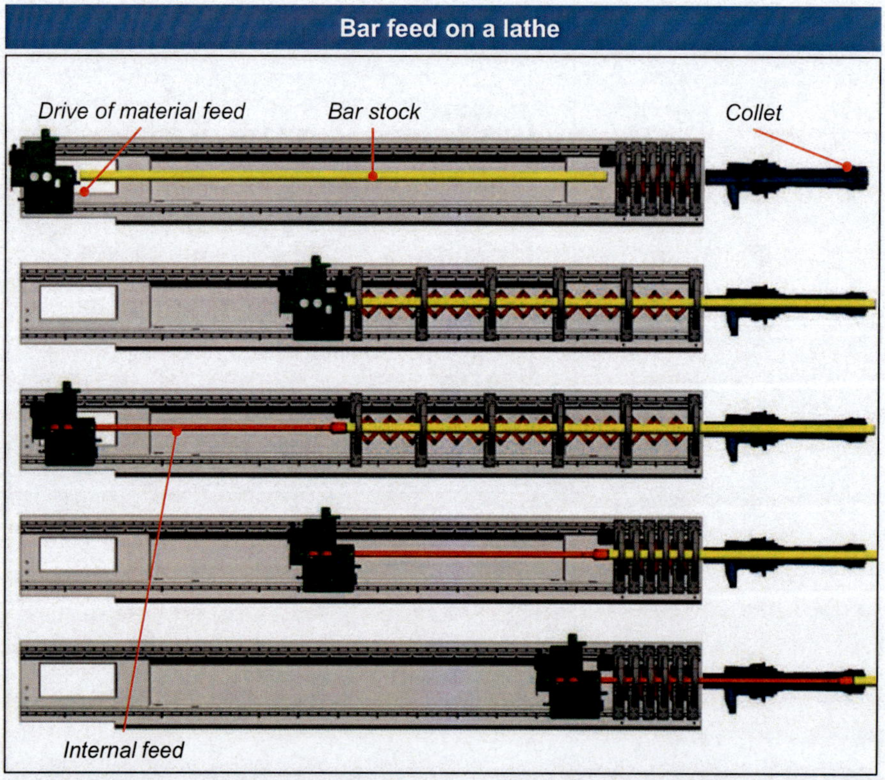


**Fig. 7.14** Material storage system for bar feed on a lathe. *Source* Index

**Fig. 7.15** Fully automated manufacturing line for milled components. *Source* GROB-WERKE

programmable handling devices such as industrial robots or modular gripper and transportation systems allows the retooling times, and therefore the costs when changing to a new workpiece, to be kept to a minimum.

Using automatic pallet changing equipment increases the degree of utilization of the machine significantly as workpieces can be clamped or removed automatically or manually at the same time as machining is in process. Suitable pallet transportation equipment allows several machines to be linked into one flexible manufacturing system.

In computer-aided manufacturing, the material and information flow are inseparable. This is particularly true of flexible, automated manufacturing. Electronic identification systems are used to identify the pallets with the relevant clamped workpieces. They consist of a code carrier and a reading unit. The code carrier can be designed to suit the specific requirements, e.g., as a transponder, barcode or RFID (*radio-frequency identification*). The carrier is built into the pallet and has a unique identifier. The pallet is read and thus identified by a reading unit connected to the computer (master

**Fig. 7.16** Precision swivel arm for transferring workpieces to a machining module on a CNC multi-spindle lathe. *Source* Alfred H. Schütte

computer, cell controller, machine control system). This information allows the transport system to control the corresponding machine tool following the routing plan and the appropriate NC program to be sent to the machine.

Automatic loading systems can therefore automatically change and transfer the workpieces on the lathes. Figure 7.16 shows a precision swivel arm for transferring between different machining units on a lathe. The swivel arm has three gripper modules that are mechanically identical. Once a machining cycle has been completed, the swivel arm picks up the workpiece and rotates through 120° to transfer it to the next machining station. This type of design allows simultaneous machining of several workpieces.

### 7.5.3 Tool Handling and Storage

Fully-automated machining cannot be achieved without automating the tool change and tool handling processes. This requires appropriate interfaces between machines, tools, handling devices, tool magazines, identification systems, controls systems, etc. Typical interfaces for milling tools are the SK (DIN ISO 7388) [DIN14] steep-angle taper and the HSK (DIN 69893, [DIN11] hollow taper shank). The cylindrical shank with taper key (DIN ISO 10889) [DIN06] as well as other interfaces are used on lathes.

Figure 7.17 shows an automatic quick-clamping device with a hollow taper shank as used on lathes and in drilling, milling, and grinding machines. The unclamped and clamped positions are shown in the upper and lower halves of the figure. The tool is essentially linked mechanically to the working spindle by the drawbar and the collet. By moving the drawbar in the direction of the spring force, the cone forces the collet apart and thus clamps the tool. The retaining force for the drawbar is applied via the disk spring assembly integrated in the spacer. This ensures that the tool does not fall out if the power supply fails. The clamping device can be opened electrically, pneumatically or hydraulically. To achieve this, a piston is put under pressure producing a force in the opposite direction to the spring force compressing the disk spring assembly. The drawbar is simultaneously displaced to the left which releases the cone from the collet. As soon as the control edge has moved past the collet, the tool can be removed.

Electronic identification systems, similar to those used in material flow monitoring, consisting of a reading head and an individualized code carrier, are used to monitor the tool location, i.e., to uniquely identify the tool, see Fig. 7.18. A computer (master computer,

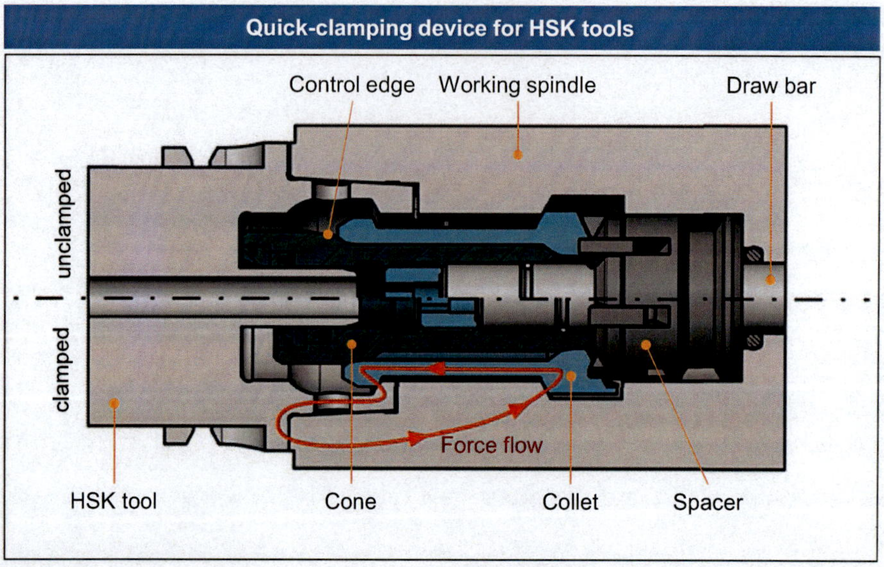


Fig. 7.17 Automatic quick-clamping device for hollow taper shank tools. *Source* Ott-Jakob

Fig. 7.18 Tool with electronic identification system. *Source* GROB-WERKE

cell controller, machine control) can read the code with a reading head and identify this tool. It is thus possible to check, for example, that the correct tool has been inserted when changing the tool.

Using the code also provides the further benefit of automatic measurement of the tools' service life.

The tools required in a lathe are stored in either a turret head, Fig. 7.13 right, or similar systems, and are switched to the working position in one after another. Tools for milling machines are frequently located in various types of chain storage systems. Figure 7.19 shows a tool chain storage system integrated into a machine tool. The machine operator can manually add more tools or remove those no longer required using a tool change station.

The tool required for machining is removed from the magazine using a handling device and put into the transfer position for the tool change; the tool changer then replaces the tool in the working spindle with this tool. The old tool is then returned to its location in the chain storage system by the manual handling system.

Figure 7.20 shows the tool change procedure with a double gripper in a machining center.

To change the tool, the gantry portal with the tool held in the gripper moves to the tool transfer position. The double gripper in the workspace moves forward, Fig. 7.20a), and pivots through 90°, Fig. 7.20b). The tools are gripped and locked in this position and then unclamped in the spindle and gripper of the gantry portal. The double gripper then moves to the swivel position, pivots through 180°, Fig. 7.20b to d, and moves back to the tool transfer position. The tools are unlocked from the double gripper and clamped in the spindle and gripper of the gantry portal. The double gripper then pivots back through 90° and moves back to its starting position.

An alternative to a chain storage system is to prepare the tools in separate storage cassettes containing the tools for a specific task for a defined period of time, e.g., three shifts. In this solution, the tools are aligned in the magazine using asymmetrical drive grooves or an orientation groove.

### 7.5.4 Process Monitoring, Process Control, Diagnostics and Safety

Monitoring and the direct correction of the production process have an important role to play in the field of automated manufacturing. Production costs can be reduced by running diagnostics on the machines and the tools as, for example, downtimes can be avoided by early identification of machine faults. In addition to this, po-

Fig. 7.19 Manual tool change station of an integral tool chain storage system

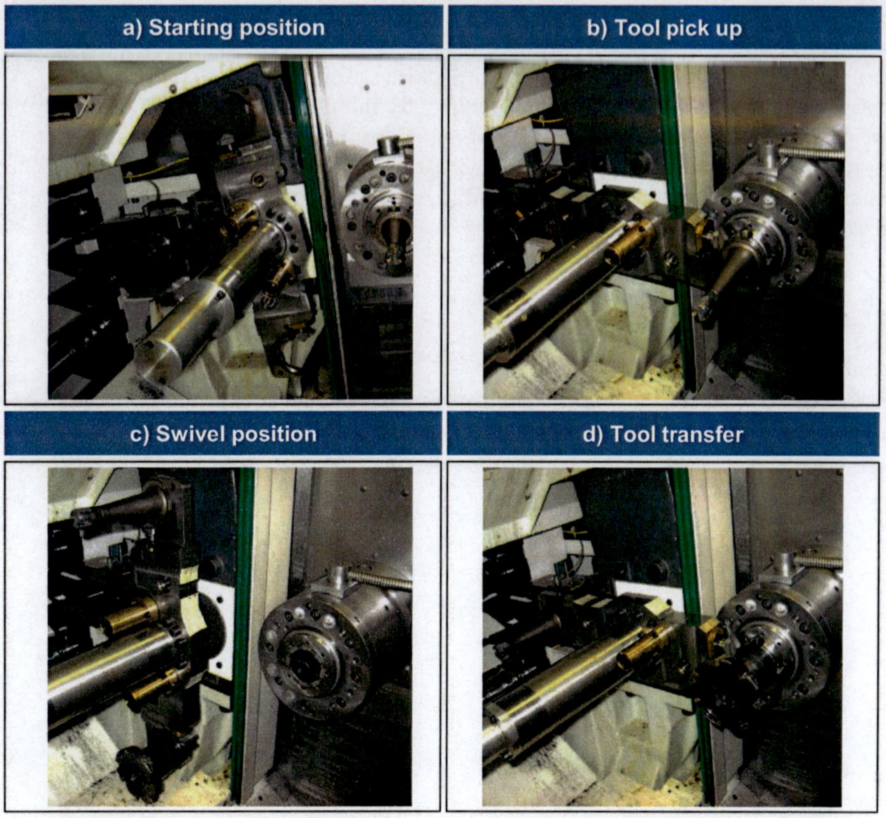


Fig. 7.20 Tool change in a machining center

sition deviations can be detected during the machining process by suitable sensor systems and the manufacturing process can be corrected thus increasing the quality of the manufactured components. Also, compensation of tool wear or thermodynamic deformation can lead to lower dimensional deviations, see ▶ Chap. 12.

| | |
|---|---|
| **Planning level** | - Production planning<br>- Design<br>- Work preparation |
| **Control level** | - Detailed task planning<br>- Coordinating implementation of the task<br>- Material and tool flow control across cells |
| **Cell level** | - Management of all cell tasks<br>- Process control by performing a cell task<br>- Material and tool flow control within cells |
| **Control level** | - Controlling and monitoring processes |
| **Sensor/ actuator level** | - Controlling and monitoring components |

**Fig. 7.21** Hierarchy levels in a control system

Unlike small and medium-sized CNC machine tools, heavy-duty machine tools frequently operate without cladding around the working area. This presents a hazard to the operator and third parties who may enter the workspace during machining. By monitoring the workspace with optical sensors or ultrasonic systems, it is possible to stop the machine, for example, when someone enters the workspace and to avoid danger to people entering the workspace.

### 7.5.5 Control Technology

The term control technology encompasses the control and monitoring of multi-machine systems. The data processing in a manufacturing system (for example, in a flexible manufacturing system) is carried out by a range of specialized software modules that are part of the hierarchy levels of the control system, see Fig. 7.21, ▶ Chap. 14.

The *planning level* includes functions such as design or production planning. The *control level* is responsible for converting tasks from the planning level in the manufacturing system by issuing appropriate manufacturing, assembly or transportation tasks to various units on the design level (*cells*).

The tasks are implemented by the cell controllers in control instructions for the relevant plant elements and then transferred to the controls for the machines, transportation systems, warehouses, etc. These then convert the control instructions into the required actions in the unit (machine, robot, transportation system, warehouse, etc.) and capture error messages as well as machine and operating data that are transmitted to the higher levels and evaluated there.

### 7.5.6 Disposal

When automating machine tools, the disposal of coolants and chips is particularly important in terms of environmental sustainability. In addition to handling tools, clamping devices and workpieces, the transportation of the increasing volumes of chips created by

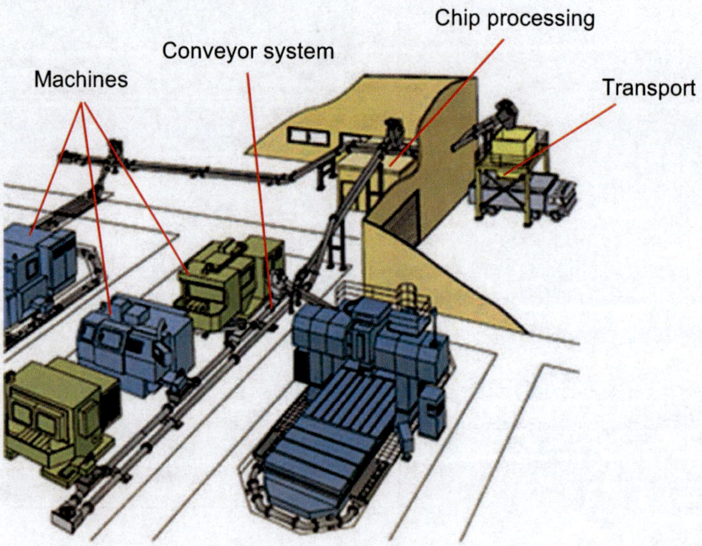


**Fig. 7.22** Automatic conveyor system for centralized chip capture from automatic lathes. *Source* Ins-europe

higher machine productivity has to be taken into consideration when establishing the automation criteria. Fully automated machine tools are designed such that the chips can fall unobstructed into the free space below the machine frame. Appropriate cutter geometry and relevant, optimized process parameters allow short, uncoiled chips to be produced. Moving protective covers prevent the chips that are flying about from being able to enter the workspace. Under these conditions, the chips can be collected and then removed from the machine area using conveyor belts or other means of conveying systems. Larger manufacturing systems or machine configurations have integral conveyor systems that consolidate the chips centrally, see ◘ Fig. 7.22. If the machining process includes coolants, these are collected in containers under the chip conveyors and separated from the chips in centrifuges.

## References

[BEY07] Beyerer, J.: visIT Leittechnik. Karlsruhe: Fraunhofer IITB, 2007

[BRUI66] Bruins, D.H.: Werkzeuge und Werkzeugmaschinen. Hanser, München (1966)

[DIN06] Norm DIN ISO 10889: Werkzeughalter mit Zylinderschaft (2006)

[DIN11] Norm DIN 69893: Kegel-Hohlschäfte mit Plananlagen (2011)

[DIN14] Norm DIN ISO 7388: Werkzeugschäfte mit Kegel 7/24 für automatischen Werkzeugwechsel (2014)

[OPIT70] Opitz, H.: Moderne Produktionstechnik. Stand und Tendenzen. Girardet, Essen (1970)

[SCHÖ97] Schöne, A.: Meßtechnik. Zweite, überarbeitete. Aufl. Springer, Berlin (1997)

[VDI04] Richtlinie VDI 2206: Entwicklungsmethodik für mechatronische Systeme (2004)

[VDI12] Richtlinie VDI 2519 Blatt 1: Vorgehensweise bei der Erstellung von Lasten-/ Pflichtenheften (2012)

[WITT60] Wittmann, K.: Die Entwicklung der Drehbank. VDI, Düsseldorf (1960)

# Mechanical Control Systems

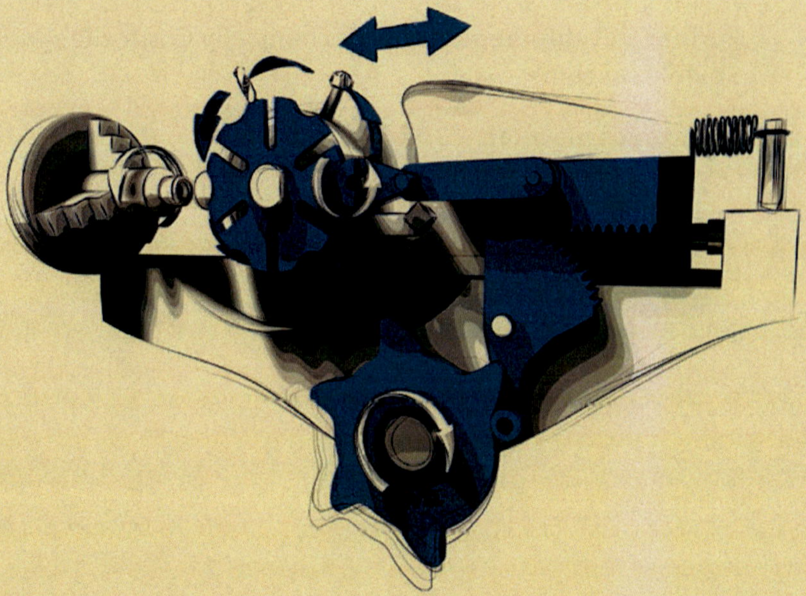

I thank my staff members Dr.-Ing. Marcel Fey, Christian Fimmers M. Sc., Nils Frenkel M. Sc., Simon Storms M. Sc. and Stephan Wein M. Sc. for their contribution in revising this chapter.


C. Brecher and M. Weck, *Machine Tools Production Systems 3*,
Lecture Notes in Production Engineering, https://doi.org/10.1007/978-3-658-34622-5_8

Contents

The aim of this chapter is to provide an understanding of how movement and control processes were carried out before the introduction of NC technology. Due to their considerable robustness, mechanically controlled motion sequences are still used for frequently recurring unchanging functions.

In mechanical control systems, mechanical components such as switching cams, motion control cams and levers are used to store, to initiate and to carry out the workflow. Despite the rapid development of electronic or NC-controlled machines, mechanical control systems are still used commercially today, mainly in automatic lathes where the shortest cycle times are required due to the large number of workpieces and average to longer set-up times can be accepted. In addition to high productivity, other advantages of mechanically controlled automatic lathes are their high degree of operational safety and their low purchase, operational and maintenance costs. Due to the relatively long set-up times and the fact that it is more complicated to carry out production processes on mechanically controlled machines, there is less flexibility with these machines compared to numerically controlled machines.

In the case of automatic lathes, which will be used in the following to illustrate the operating characteristics of mechanical control systems, a distinction is made between single and multi-spindle automatic lathes, ◘ Fig. 8.1.

With a single-spindle automatic lathe, the turret tools are used one after another. There is a longer tool idle time after the short working time of the turret tool during its work phase. Other tools are fitted to cross slides. These are either used one after another or parallel to the turret tools depending on the machining step.

Mechanically controlled single-spindle machines were built into the 1980s although many functions were gradually being replaced by new NC-controlled machine concepts. Today, mechanically controlled multi-spindle machines are no longer manufactured [HEIS14]. These machines are available on the second-hand market after being retrofitted and, despite their age, they are still used due to their considerable robustness.

◘ **Fig. 8.1** Automatic lathes with mechanical control; above: single-spindle machine "Index 1;" below: six-spindle machine "SG 18" manufactured by Schütte. *Source* Alfred H. Schütte

With the multi-spindle automatic lathe, between four and eight spindles can machine one workpiece each at the same time. A tool set that carries out the feed and cutting motions is fitted to each spindle position. Once the machining of a workpiece in a spindle position is finished, then the spindle can be moved to the next position by turning the spindle drum. The workpiece can then be further machined with the tool set of the next spindle position. By machining several parts on the different spindles of a multi-spindle machine at the same time, job times of only a few seconds can be achieved depending on the machine type and workpiece. Compared to the single-spindle machine, the multi-spindle machine is many times more productive [WECK13].

The functions of mechanically controlled multi-spindle machines are being replaced more and more with NC-based technology. On new machines, only auxiliary functions are still mechanically controlled. The machines available on the market range from mechanically controlled machines and hybrids to machines whose axes of motion are exclusively NC-controlled. In contrast to single-spindle machines, mechanically controlled multi-spindle machines are still being further developed and are being manufactured in greater numbers. Despite the limited flexibility of these machines, adapted tool sets are providing the opportunity to manufacture a wide product range of high-precision components in large quantities.

An overview of the history of the single and multi-spindle automatic lathes can be found in [HEIS14].



## 8.1 Single-Spindle Automatic Lathe

The various components of a mechanical control system are shown using a single-spindle automatic lathe as an example (◘ Fig. 8.2).

In ◘ Fig. 8.3, the front view of the automatic turret lathe from ◘ Fig. 8.2 is sketched in cross-section with the different components of the mechanical control system. The main drive on the bed base has two gear-change stages, which can be changed manually. One of the stages allows the spindle speed to be varied. The other stage changes the rotational direction and sets the difference in speed between the clockwise rotation and the counterclockwise rotation for the feed and return when thread cutting. Chains are used to transfer the motion for the clockwise rotation and the counterclockwise rotation to the working spindle via the clutches.

◘ Figure 8.4 shows the plan view of the gearbox structure from ◘ Fig. 8.2. Cam disks and switching cam disks are attached to the main control shaft. The former control the motion of the longitudinal and cross slide of the tool while the latter use switching cams to operate the high speed clutches (see ◘ Fig. 8.5) that initiate the fast-running auxiliary functions. The motion of the slow-running main control shaft is derived from the fast-running auxiliary control shaft using change gears. These are designed such that one revolution of the main control shaft corresponds to the machining time of one workpiece.

The initiation point of the various control and switching functions is determined by the position of

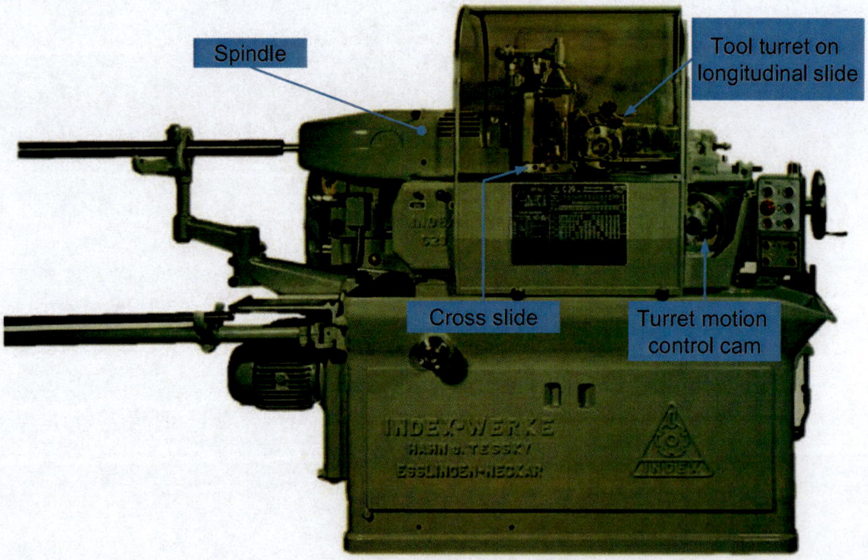


◘ **Fig. 8.2** Single-spindle automatic lathe "Index C29" *Source* INDEX-Werke

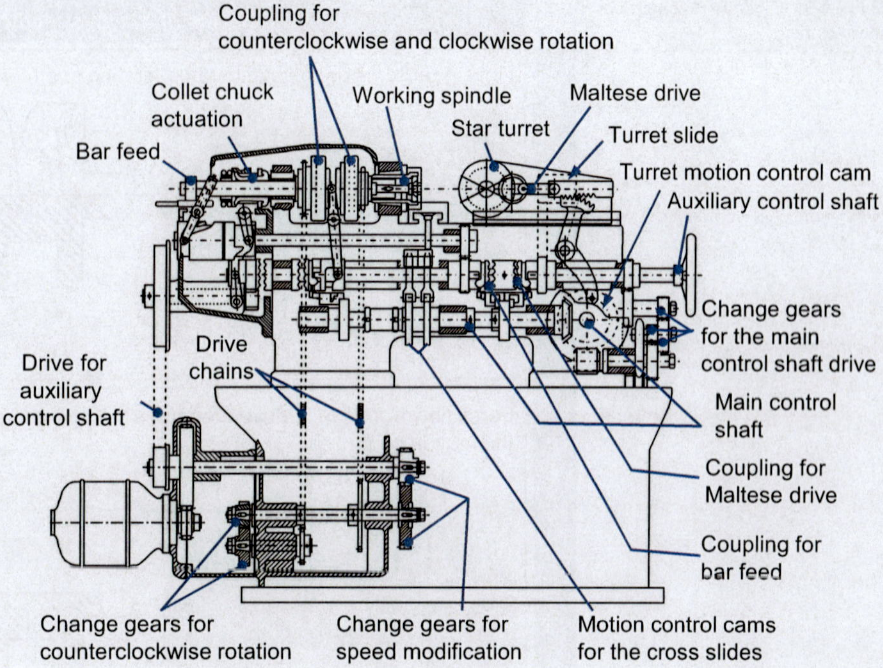


**Fig. 8.3** Drive and mechanical control of the automatic turret lathe (side view). *Source* INDEX-Werke

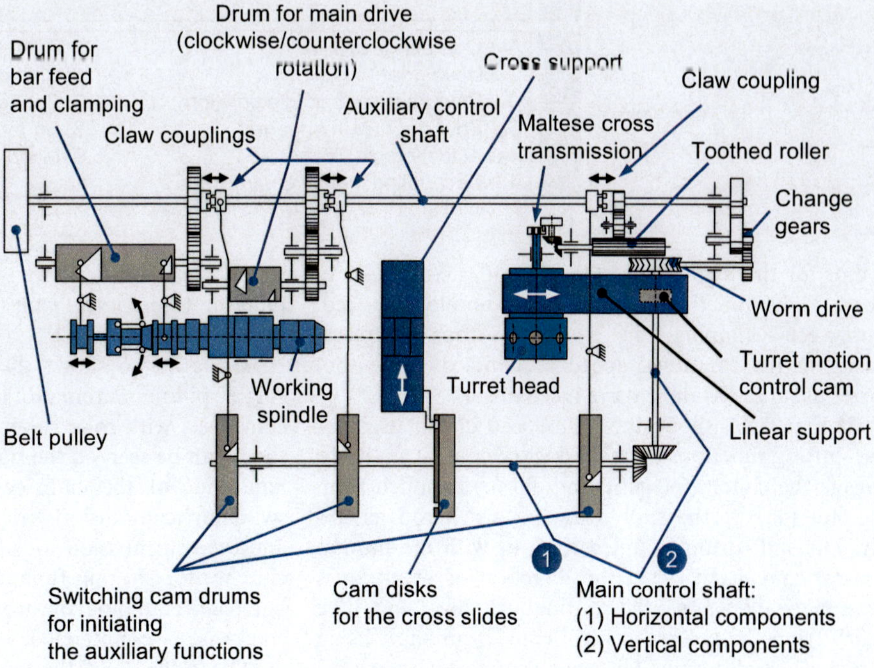


**Fig. 8.4** Gearbox diagram of a mechanically controlled automatic lathe with auxiliary and main control shafts (plan view), according to INDEX-Werke

the switching cams on the switching cam drums attached to the main control shaft. The main control shaft consists of two shaft lines which are positioned vertically on top of one another and coupled via a bevel gear. The vertically drawn component (2) in Fig. 8.4 controls the longitudinal feed of the turret via the turret motion control cam while the horizontal component (1) controls the cross feed via the cam disk for the cross slide.

The fast-running auxiliary control shaft is driven by the main drive via a pulley, Figs. 8.3 and 8.4. The high speed claw couplings already mentioned above are fitted to the auxiliary control shaft. When these couplings are engaged, they drive the motion control cam

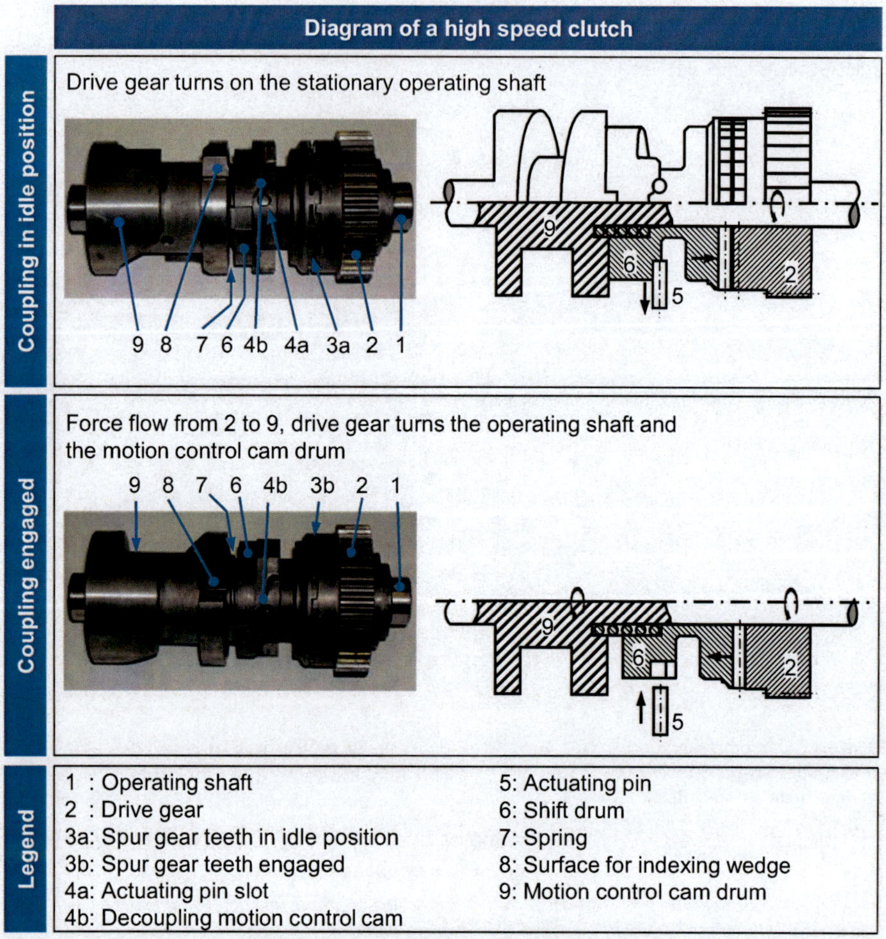


**Fig. 8.5** Diagram of a high speed clutch

drums for the high speed indexing and auxiliary functions. These functions include, for example, bar feed, workpiece clamping, the swiveling stops in place, changing the rotational speed and indexing the tool turret using the Maltese drive mechanism.

Figure 8.5 shows the high speed clutch used for the shifting functions of the auxiliary control shaft. To engage the clutch, the actuating pin (5) is pulled from the slot (4a) of the shift drum (6) via a rod assembly. The shift drum, coupled form-fit with the motion control cam drum (9) in the direction of rotation, is pressed axially against the continuously revolving drive gear (2) on the motion control cam drum shaft using a pretensioned spring. The spur gear teeth on the revolving gear and shift drum produce a form fit between the two components. The torque required to drive the motion control cam drum is transferred from the drive gear (2) to the motion control cam drum (9) via the shift drum (6). Shortly after the initiation of the shifting function, the actuating pin (5) is again extended radially. Once the shifting function has been performed, that is, after one complete or one half revolution of the auxiliary control shaft, the shift drum of the index bolt is pulled away axially from the drive gear via the decoupling motion control cam (4b). The force flow is again interrupted.

Figure 8.6 shows the main control shaft of the single-spindle automatic lathe from Fig. 8.2. The cam disks with roller levers for moving the four cross slides can be seen on the right of the figure. On the left-hand side of the main control shaft, there are three switching cam disks with the initial switching cams and the transmission rods for actuating the high speed clutches for shifting functions such as turret indexing, bar feed, clamping the workpiece and selecting either clockwise or counterclockwise rotation of the spindle.

The indexing of the tool turret head (lengthwise) is shown in the following figures as an example of a secondary movement driven by the auxiliary control shaft. Figures 8.7 and 8.8 show the turret multi-speed gearbox and the stages of a high speed turret indexing sequence.

Figure 8.7 illustrates the indexing sequence of the turret head. A switching cam on the switching cam disk of the main control shaft engages the claw coupling for the Maltese cross transmission on the auxiliary control

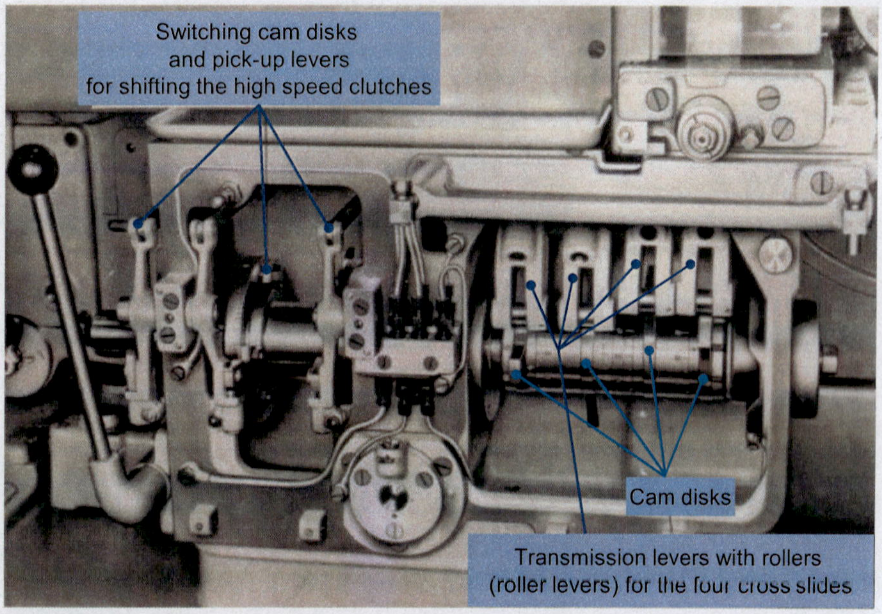


**Fig. 8.6** Main control shaft of an automatic lathe with cross slide motion control cams and switching cam disks. *Source* INDEX-Werke

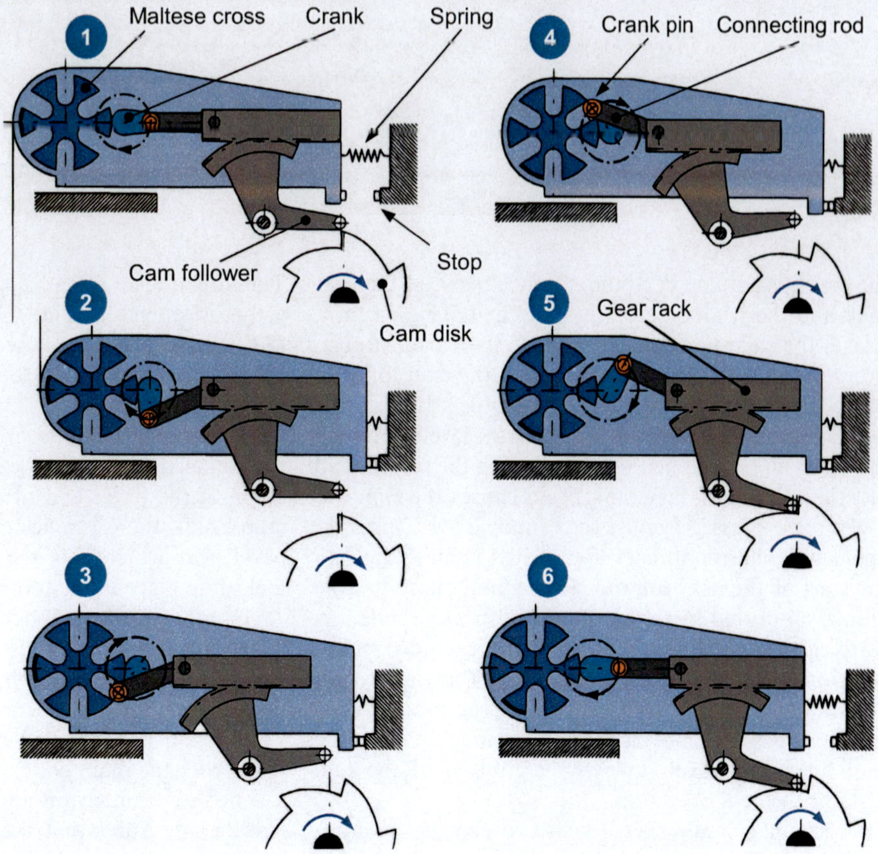


**Fig. 8.7** High speed turret indexing of the automatic lathe turret head from **Fig. 8.2**. According to: INDEX-Werke

shaft via levers which in turn causes the auxiliary control shaft to turn exactly one revolution. Before the crank is actuated (1), the gear train is still in the extended position and the turret head is indexed so that the slide moved by the cam follower and the gear rack follow the contour of the cam disk. A strong spring pulls the slide back. This ensures that the cam follower roller makes precise contact with the cam disk. When

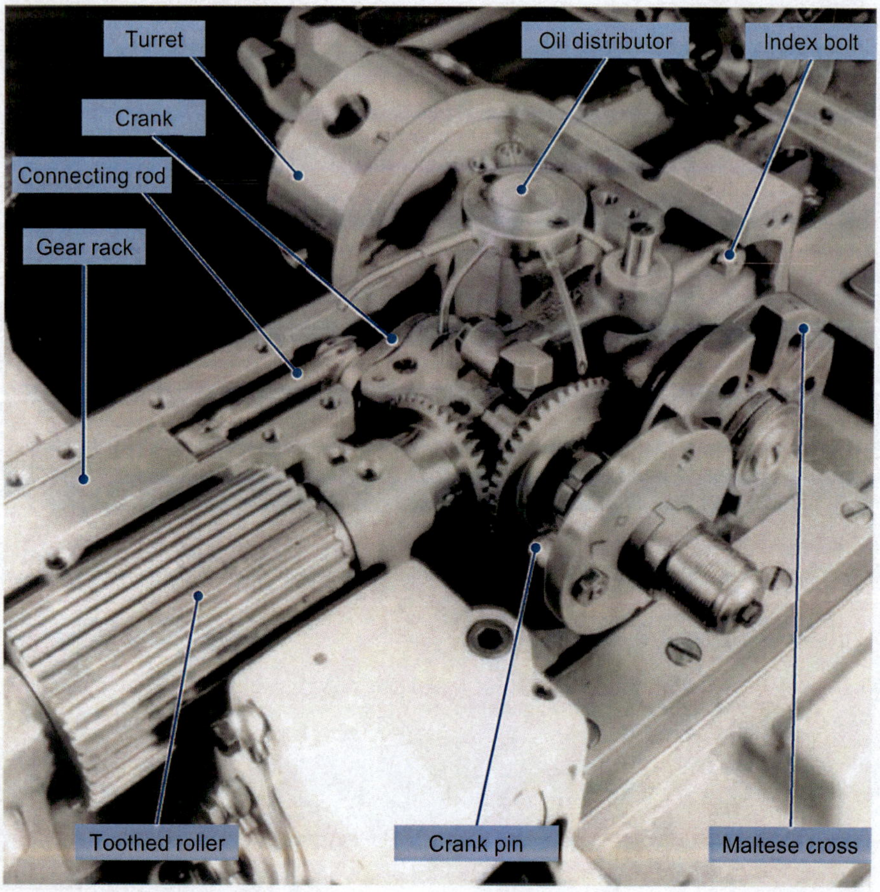


**Fig. 8.8** The turret multi-speed gearbox of the automatic lathe from Fig. 8.2. *Source* INDEX-Werke

8

the crank is actuated (2), the distance between the gear rack and the Maltese cross is shortened. This, in turn, causes the slide to be pulled back to its rearmost position against a stop by the spring and the cam follower to be lifted. In this position, the crank advances the Maltese cross one slot (3). In this way, level jumps on the disk can be spanned. After indexing the turret head (4), the motion of the crank again causes the cam follower to be pressed against the cam disk (5). This disk, meanwhile, has continued to turn by a small angle to the start of the next contour. In the final phase (6), the crank, supported by the cam disk, pushes the slide forward against the spring force. In the extended crank position, the turret head is blocked by an index bolt (see Fig. 8.8), which is shifted via the motion control cams beside the bevel gear. In the indexed state, the slide again exclusively follows the contour of the cam disk for the next operation.

The open multi-speed gearbox can be seen in Fig. 8.8. The disk with the crank pin that meshes with the Maltese cross can be seen in the front part of the figure. The rotational motion for the crankshaft is transferred from the auxiliary control shaft via a claw coupling and an intermediate gear to the toothed roller. The roller also remains meshed with the intermediate gear when the slide is moving longitudinally (see Fig. 8.4). The force flow runs from the toothed roller via the bevel gear drive to the crank and the crank pin of the Maltese drive.

The crank, which is connected to the gear rack via the connecting rod, can be seen in the same angular position at the other end of the crankshaft. Also on the crankshaft, there is a switching cam located beside the bevel gear which lifts the index bolt via a lever and locks it in place in the tool turret.

In more modern machine constructions, the auxiliary functions that must run fast are either partly electromotive or electrohydraulic meaning that the fast-running auxiliary control shaft can be omitted. The system is controlled by a control drum synchronized with the main control shaft. With the help of positioned switching cams and electrical switches, the control drum ensures that the control functions are initiated on time.

Another way of initiating the control functions in accordance with the work progress is to use a rotary encoder attached to the main control shaft (see ► Sect. 2.3). Upon reaching certain angles of the main control shaft, the individual functions required by the appropriate machining plan will be performed by either electrical or other actuators.

### Calculation for the Cam Disk

The working precision of a cam-controlled machine essentially depends on the exact calculation of the cam disk contour. The turret slide motion control cam is calculated in the following for a single-spindle automatic lathe (◘ Fig. 8.10) with a six-part turret using the example of the turned part in ◘ Fig. 8.2 (top left) that is machined from the bar. The following work steps are to be performed based on the workpiece drawing in ◘ Fig. 8.10.

- *Preparation of the tool sequence plan*: The work sequence is to be specified such that the workpiece is completely machined using the six available turret positions and a maximum of four cross slides. To reduce the single part production time, as many tools as possible are to be engaged in the multiple tool holders on the turret head and the cross slides. Taking these considerations into account, the tool sequence plan is to be drawn with the corresponding slide end positions as per ◘ Fig. 8.9.
- *Preparation of a calculation sheet for a motion control cam*: If the tool sequence is specified after performing a collision check, then the sequence of the work operations and indexing functions will also be defined. The sequence is entered separately for the turret and cross slides into the lines of a prepared calculation sheet as in ◘ Fig. 8.10. The columns are used to enter all the motion control cam calculation data that are dependent on the work operations.
- *Determination of the working spindle speeds, distances and feeds*: The required spindle speed $n$ is determined from the cutting speed $v$, which is recommended in material tables, and the largest turning diameter $d_{max}$ as per Eq. (8.1) because the speed remains constant during machining.

$$n = \frac{v}{\pi \cdot d_{max}} \tag{8.1}$$

The entry is made in the header of the calculation sheet, ◘ Fig. 8.10. The working distances are taken from the workpiece drawing. To prevent the tools from cutting when moving the slide up quickly, a safety margin of 0.5 mm is added and the resulting values entered in column 1.

In column 2, the required feed values $F_i$ are entered in mm/revolution. These values are dependent on the required surface quality, the machinability of the material, the thread pitch, the stiffness of the turned part and the tool holder, the desired service life, the cooling of the tool and the power of the drive motor. These values are taken from reference value tables.

- *Calculation of the working spindle revolutions of a work operation:* The cam disk used to manufacture a workpiece rotates once. As this rotation is derived from the working spindle, this means that it is necessary to calculate the number of spindle revolutions required to determine a motion control cam section of a work operation. The revolutions $U_h$ required for the other calculations also allotted to the machining time $t_h$ consist of the allotments $U_{hi}$ for the individual work operations i (column 3). These are calculated using the working distances $l_i$ and the corresponding feeds $F_i$, Eq. (8.2).

$$U_h = \sum U_{hi} = \sum \frac{l_i}{F_i} \tag{8.2}$$

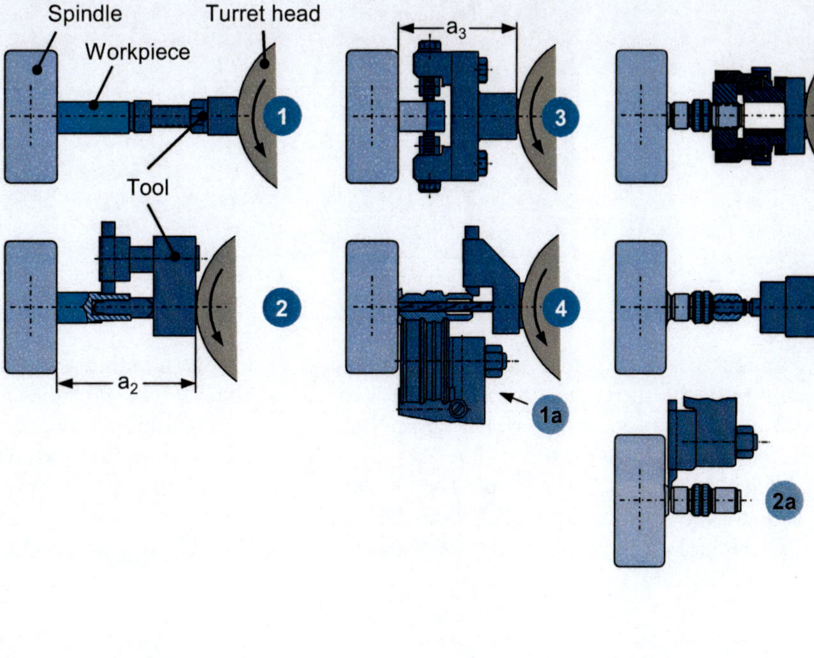


◘ **Fig. 8.9** Tool sequence plan with the corresponding slide positions

**Fig. 8.10** Calculation sheet for a motion control cam

thr.: threads
(..): values not used in the calculation

**Revolutions for a workpiece:**
(545 * 100) / 73 = 746

**Single part production time:**
(60 * 746) / 3000 = 15 sec

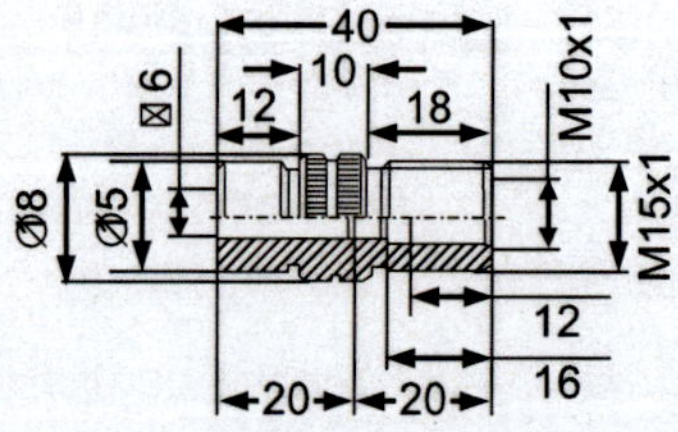


| Material: round brass 18 DIN 1756 Ms 58 | | |
|---|---|---|
| Spindle rpm | Turning | 3000 |
| | Thread cutting, outside | 1500 |
| | Thread cutting, inside | 1500 |
| Cutting speed in m/min | Turning | 170 |
| | Thread cutting, outside | 70 |
| | Thread cutting, inside | 47 |
| Required revolutions for one workpiece | | 746 |
| Single part production time, seconds | | 15 |

| Work operation in accordance with the tool sequence plan | | | Working distance | Feed during one revolution of the working spindle | Working spindle revolution: for respective working distance | Working spindle revolution: to be taken into account | Cam disk hundredths: for idle travel | Cam disk hundredths: for working dist. | Cam disk hundredths: from | Cam disk hundredths: to |
|---|---|---|---|---|---|---|---|---|---|---|
| | | Column | 1 | 2 | 3 | 4 | 5 | 6 | 7 | 8 |
| Turret head | 1 | Material stop | | | | | 4 | | 0 | 4 |
| Turret head | | Indexing the turret head | | | | | 4 | | 4 | 8 |
| Turret head | 2 | Drilling and stripping the threaded end | 23.5 | 0.15 | 157 | 157 | | 21 | 8 | 29 |
| Turret head | | Indexing the turret head | | | | | 4 | | 29 | 33 |
| Turret head | 3 | Knurling – Forward | 12 | 0.4 | 30 | 30 | | 4 | 33 | 37 |
| Turret head | 3 | Knurling – Back | 12 | 0.6 | 20 | 20 | | 2.5 | 37 | 39.5 |
| Turret head | | Indexing the turret head | | | | | 4 | | 39.5 | 43.5 |
| Turret head | 4 | Through-drilling and shoulder turning | 22 | 0.125 | 175 | 175 | | 23.5 | 43.5 | 67 |
| Turret head | | Indexing the turret head | | | | | 4 | | 67 | 71 |
| Turret head | 5 | External thread – Cutting | 17 thr. | | 34 | 34 | | 4.5 | 71 | 75.5 |
| Turret head | 5 | External thread – Return | 17 thr. | | 17 | 17 | | 2.5 | 75.5 | 78 |
| Turret head | | Indexing the turret head | | | | | 4 | | 78 | 82 |
| Turret head | 6 | Internal thread – Cutting | 14 thr. | | 28 | 28 | | 3.5 | 82 | 85.5 |
| Turret head | 6 | Internal thread – Return | 14 thr. | | 14 | 14 | | 2 | 85.5 | 87.5 |
| Cross slide | 1a | Front cross slide Form turning | 2.5 | 0.025 | 100 | (100) | | (13) | 54 | 67 |
| Cross slide | 2a | Rear cross slide Parting | 4.5 | 0.075 | 60 | 60 | | 8 | 87.5 | 95.5 |
| Cross slide | 2a | Rear cross slide Parting | 1 | 0.1 | 10 | 10 | | 1.5 | 95.5 | 97 |
| | | Allowance after parting | | | | | 3 | | 97 | 100 |
| | | | | | | 545 | 27 | 73 | | |

The components of the cross slides are not taken into account with the turret slide motion control cam (column 4) as long as they are performed simultaneously with the machining of the turret slide.

- *Distribution of the motion control cam sections*: The cam disk is divided into 360°/100 = 3.6° sections. The angles on the disk are therefore specified in one-hundredth divisions. The total circumference of the cam disk corresponds to the sum of all the hundredths for the machining times $H_h$ and the non-productive times $H_n$. The number of hundredths for the non-productive times are stipulated in the manufacturers' operating manuals e.g., 4 hundredths for indexing the turret head or 3 hundredths as an addition after parting (column 5). The number of remaining hundredths for the machining time can be calculated directly from Eq. (8.3).

$$H_h = 100 - \sum H_{ni} \quad (8.3)$$

As there is a direct proportionality between the hundredths of the cam disk and the number of working spindle revolutions, the hundredths $H_{hi}$ for the individual machining time components (column 6) are determined in accordance with equation (8.4).

$$H_{hi} = H_h \cdot \frac{U_{hi}}{U_h}, e.g. H_{h2} = 73 \cdot \frac{157}{545} = 21.09 \quad (8.4)$$

From this data, the distribution sequence can then be determined in columns 7 and 8.

- *Calculation of the job time:* The total time $t_w$ required for either the production of a workpiece or for a revolution of a cam disk is included in the master data in the header of the calculation sheet. The total time can be calculated from the total number of working spindle revolutions $U_{tot}$ as per Eqs. (8.5) and (8.6).

$$t_w = \frac{U_{tot}}{n} \quad (8.5)$$

$$U_{tot} = U_h \cdot \frac{100}{H_h} \quad (8.6)$$

- *Design of the cam disks:* Pre-printed motion control cam sheets containing a diagram of the unmachined cam disk on a scale of 1:1 are usually used for drawing. The division of the circumference of the circular disk in hundredths is represented by arcs with the radius of the cam follower, Fig. 8.11.

In addition, the point corresponding to the most forward position of the turret head in the tool sequence plan ($a_3 = a_{min}$ in Fig. 8.9) is placed on the outside diameter of the unmachined disk. Taking the roller radius into account, the motion control cams are drawn according to the turret head positions shown in the tool sequence plan, the working distances from column 1 and the calculated one-hundredth divisions in columns 7 and 8, Fig. 8.10.

Previously, cam disks for longitudinal and cross slides were largely designed and produced by hand in accordance with the contour sketch. Today, numerically controlled milling machines are used for this purpose. With the help of computer programs, both the NC data records for controlling the milling machine as well as the corresponding operating plan data for the workpiece, which is to be produced on the automatic lathe, are automatically generated.

## 8.2 Multi-Spindle Automatic Lathe

With a multi-spindle automatic lathe (four-, six- or eight-spindle machine) the workpieces are machined on the various spindles of the machine at the same time, Fig. 8.12. The tool sets are distributed to the individual spindle positions according to the work sequence. Several machining steps can be performed on the workpieces at the same time. The maximum number of workpieces that can be machined on the lathe at the same time corresponds to the number of spindles. If the work operations are distributed between several spindle positions, one workpiece will be fully machined

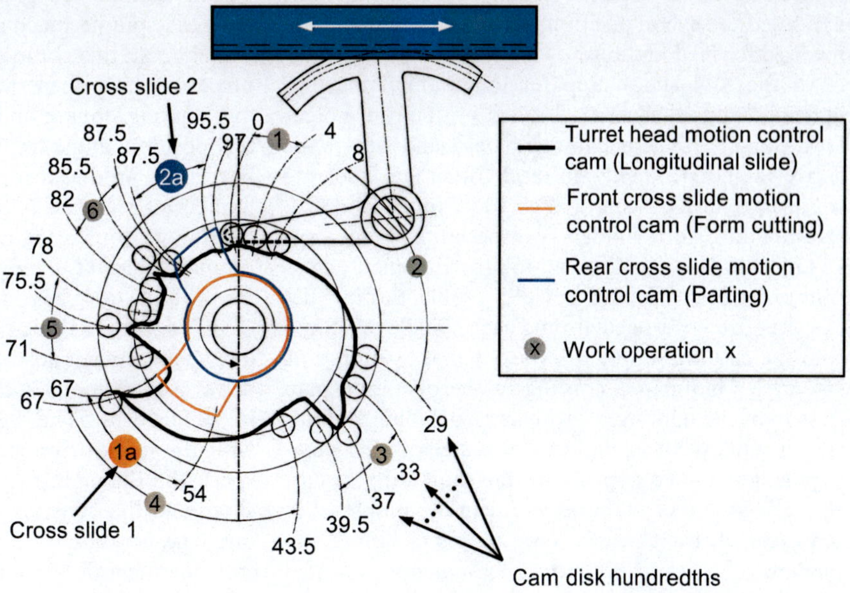


Fig. 8.11 Cam disks for longitudinal and cross slides

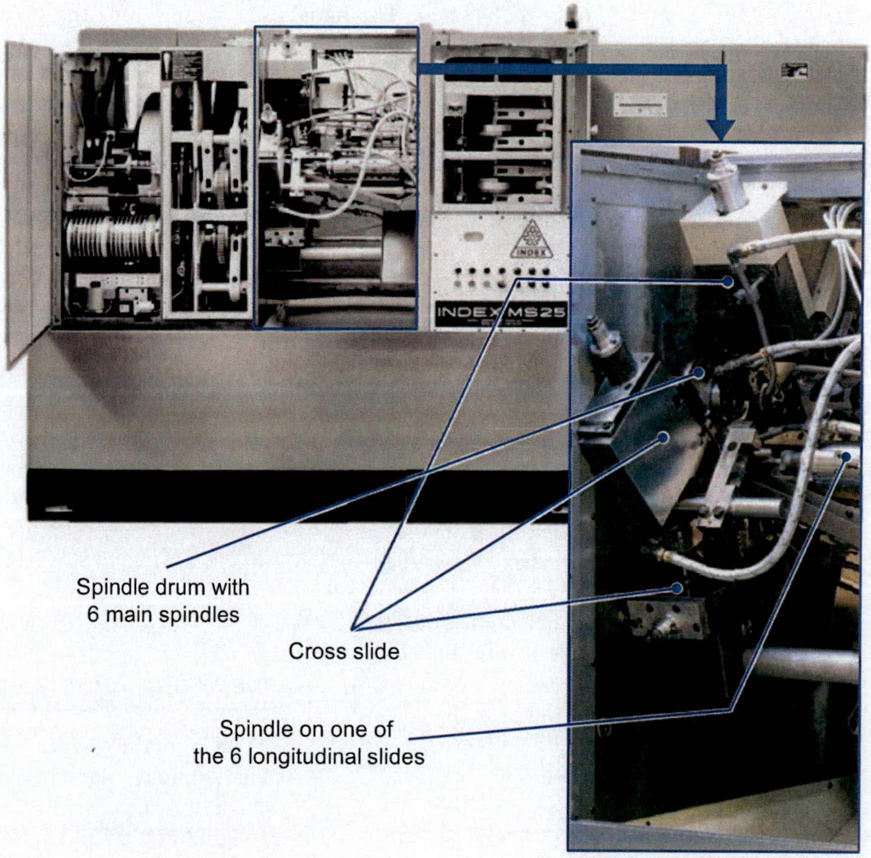


**Fig. 8.12** Mechanically controlled six-spindle machine, according to INDEX-Werke

after each work cycle. A new blank can then be inserted in the lathe for processing or bar stock can be introduced. In contrast to single-spindle automatic lathes, which require the tool turret to be indexed after each work operation, in multi-spindle machines, the spindles with the workpieces clamped in the spindle drum are indexed one slot to the next spindle position. The tools distributed to several positions remain in place and are only used for feed motions.

All the tool slides used for longitudinal machining (longitudinal slides) are located on the tool holder block (longitudinal machining block) and are to be controlled independently of each other via separate cam disks. During facing, one or more tool slides with corresponding tool holders or special-purpose tools like tangential or radial thread rolling equipment, cross drilling devices, gripping devices, profile turning devices, etc., are positioned around each spindle position.

In the case of simple workpieces, if the number of work cycles required according to the operating plan is less than the number of available machining stations then, if multiple tool holders allocated to the spindle positions are equipped with the same tools, a corresponding number of workpieces can be completed in each cycle. If, for example, only three machining stations are required to produce a component then, if a six-spindle machine is equipped with double tooling, two components can be manufactured in each work cycle. In addition, the same tools are mounted at spindle positions 1 to 3 and spindle positions 4 to 6. As a result, the same two production steps can be carried out in parallel for both workpieces.

An example of producing a complex component on a six-spindle machine can be seen in Fig. 8.13. To fabricate this component, the following production processes must be performed: form turning and drilling, tapping, square milling, cross drilling into the correct position, angle slotting and parting. The adjustable bearing is produced in six steps corresponding to the number of stations. There are a total of ten machining operations, four using cross slides and six using longitudinal slides, Table 8.1. The total machining time for the adjustable bearing is 2.1 s.

A section through a multi-spindle machine (six-spindle machine) with gear trains for the spindle drive, feed motions and control functions is sketched in Fig. 8.14. The figure shows the spindle drum with the main drive shaft running through the workspace and the middle of the spindle drum, three facing slides and the units for bar feed and workpiece clamping. The gearbox construction for the drive and control functions of the automatic lathe is also depicted.

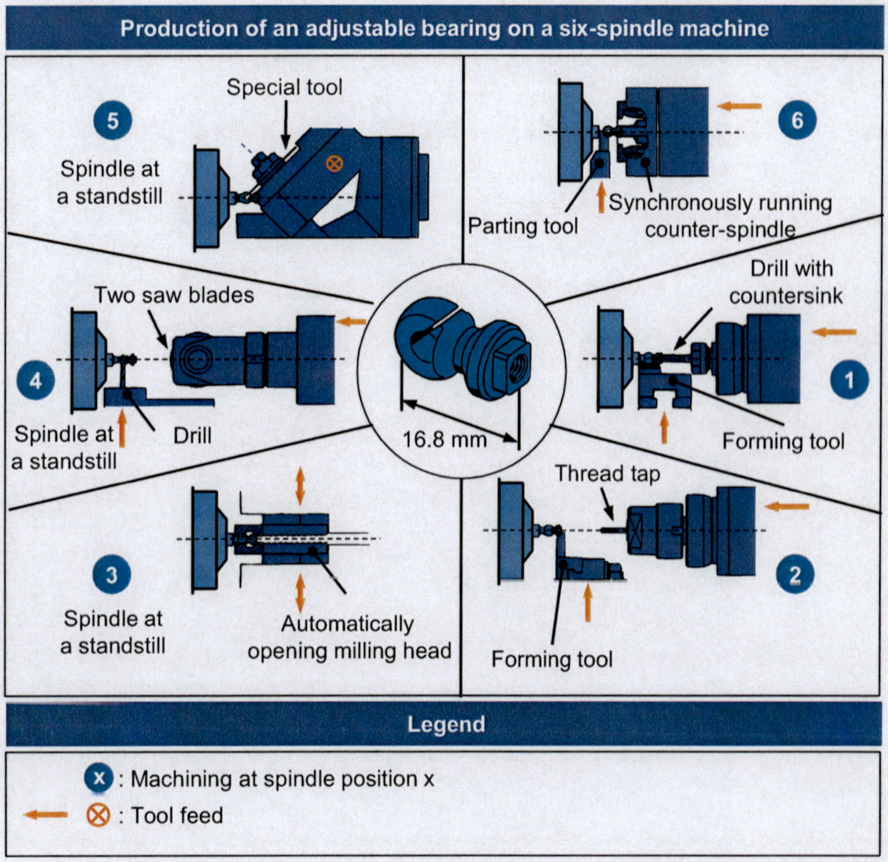


**Fig. 8.13** Production of an adjustable bearing on a six-spindle machine, according to INDEX-Werke

The longitudinal machining slides, which are positioned around the central shaft in the workspace, are not shown in Fig. 8.14 for clarity.

All motions for the working spindles (main and additional spindles), feed slides and auxiliary equipment are derived from the central main drive motor. All six of the main spindles in the spindle drum are driven by the pulley using detachable change gears and a planetarily arranged spur gear drive. Both control shafts are driven by the pulley via a change gear drive, clutches, safety clutches and worm gears. The control shafts perform the functions of the main and auxiliary control shafts. In each work cycle, they perform one full revolution. This is composed of a slow-running component for the machining feed motions and a fast-running component used for the auxiliary motions and to return the slides. For the work operation, the change gear drive is engaged by the work operation clutch. The fast-running auxiliary functions like material feed, spindle drum indexing and actuation of the clamping jaws are initiated by engaging the quick motion clutch. Approximately 160° of the control shaft rotation is available for the actual working range and approximately 200° for the auxiliary functions.

The rotational motions for the additional spindles transmitted to the longitudinal slide via change gears, e.g., for driven spindles for high speed drilling, reaming and thread cutting, are also derived from the central shaft (9), Fig. 8.15. The operating principle is explained here using the example of a thread cutting spindle. The required differences in speed between the main spindle and the thread cutting spindle (6) are produced by the gear train shown. When carrying out thread cutting, two electromagnetic clutches (EMC) on the axes of gears (1) and (3) can be engaged and disengaged using the proximity switch (8) to control the advance and return of the tool. At the same time, the rotation of the central shaft is transferred via various gears to the two electromagnetic clutches and from there using additional gears to the thread cutting spindle (6). The tool is advanced via change gears (1) and (2) using an electromagnetic clutch a while the tool is returned using change gears (3) and (4) with electromagnetic clutch b. When cutting a right-hand thread (advance), the thread cutting spindle rotates according to the pitch of the thread being produced, which is slower than the speed of the main spindle. In contrast, when the tool is returned, the thread cutting spindle

**Table 8.1** Work operations performed to machine the adjustable bearing

| Machining at spindle position | Active slide | Work operation |
|---|---|---|
| 1 | Cross slide 1 | Machining outer contour with the first forming tool |
| 1 | Longitudinal slide 1 | Drilling pilot hole for thread and countersinking |
| 2 | Cross slide 2 | Machining outer contour with the second forming tool |
| 2 | Longitudinal slide 2 | Thread tapping |
| 3 (Spindle at a standstill) | Cross slide 3 | Two-sided plane surface machining of the bearing head and the square end with an automatically opening milling head |
| 4 (Spindle at a standstill) | Cross slide 4 | Cross drilling of the bearing head |
| 4 (Spindle at a standstill) | Longitudinal slide 3 | Completing the square end using two power-driven saw blades located at a distance |
| 5 (Spindle at a standstill) | Cross slide 5 | Fabricating the angled slot using a special tool (angular milling head) |
| 6 | Longitudinal slide 4 | Clamping of the workpiece in the synchronization device |
| 6 | Cross slide 6 | Finish turning of the sphere with simultaneous parting of the adjustable bearing using the forming tool in the counter-spindle of slide 6 and subsequent ejection |



rotates faster than the main spindle. The thread cutting spindle is driven by an idler gear (7) on the central shaft (9). This gear can drive several thread cutting mechanisms. The force flow from the central shaft (9) to the thread cutting spindle (6) is shown at the bottom of Fig. 8.15 on the left-hand side.

In Fig. 8.16, the bar feed and clamping device is shown in detail in accordance with the section shown in Fig. 8.14. This device is only provided at those spindle positions where the finished workpieces are cut off and material feed is required. With double tooling, two components are produced in each work cycle. Then there will also be two material feed devices as two workpieces are required in each work cycle.

Through the indexing motion of the spindle drum, the spindles involved are swiveled into the clamping jaw (9) using the clamping sleeve (4) and into the feed connector (16) using the bearing on the guide ring (1). After a part has been cut off, the bar stop is swiveled into the workspace to determine the length of the blank. The clamping sleeve (4) and the clamping jaw (9) are then positioned above the clamping slide (10). The slide meshes with the motion control cam of the motion control cam drum (13) in the direction of the collet chucks (7).

This motion engages the clamping lever (12) with the clamping sleeve. At the same time, the tension of the disk spring assembly (11) and the initial tension between the spindle and the collet chuck (7), which is produced as a result, are reduced and the collet chuck is pushed out of the clamping cone. As the collet chuck is sprung, it opens automatically and releases the bar stock. The exact position of the collet chuck, as well as the initial tension produced by the disk spring assembly, can be set using the notched nut and the clamping nut ((2) and (3)).

When the bar is being fed, the feed collet (8), i.e., the inside pipe, is moved to the right via the second motion control cam of the motion control cam drum. It is shown here in its forward end position. The supplied

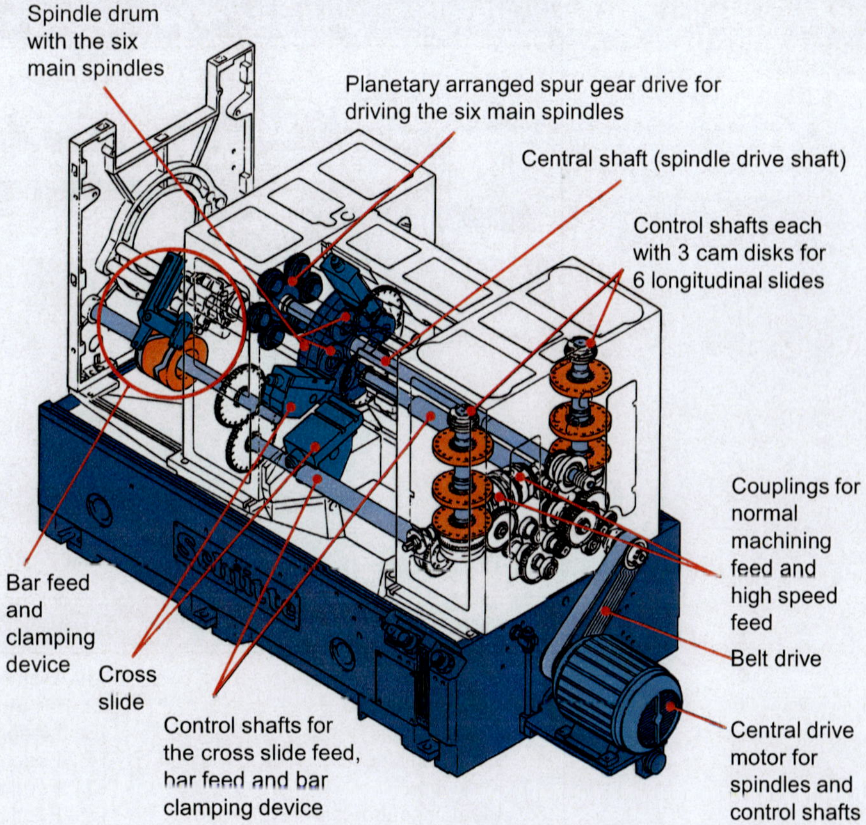


**Fig. 8.14** Section through a multi-spindle machine with gear trains for the spindle drive, feed motions and control functions, according to Alfred H. Schütte

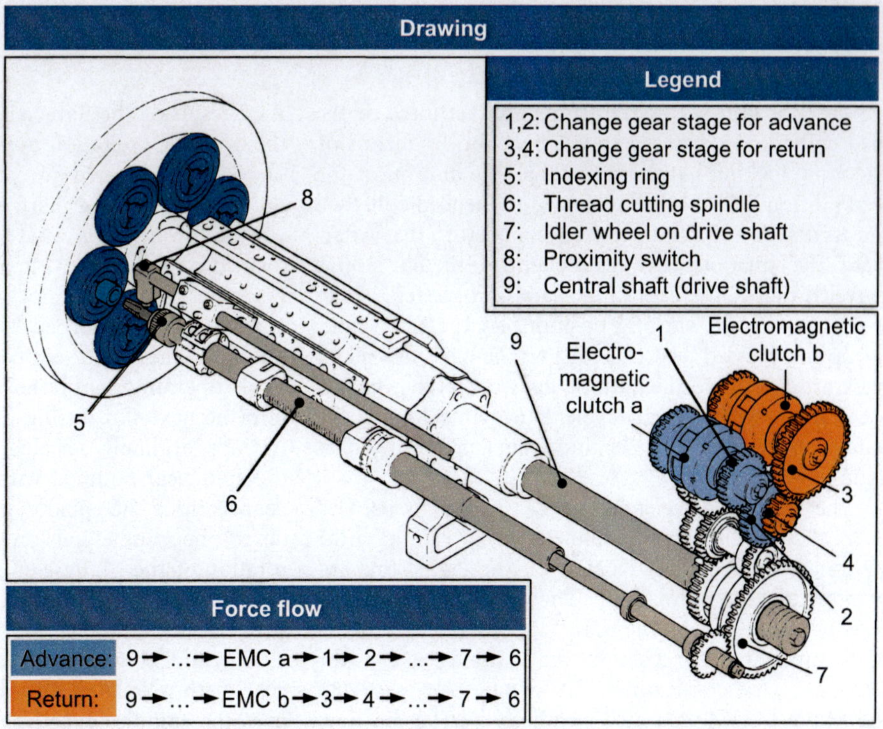


**Fig. 8.15** Gear train for additional spindles on the longitudinal slides (thread cutting in this case), according to Alfred H. Schütte

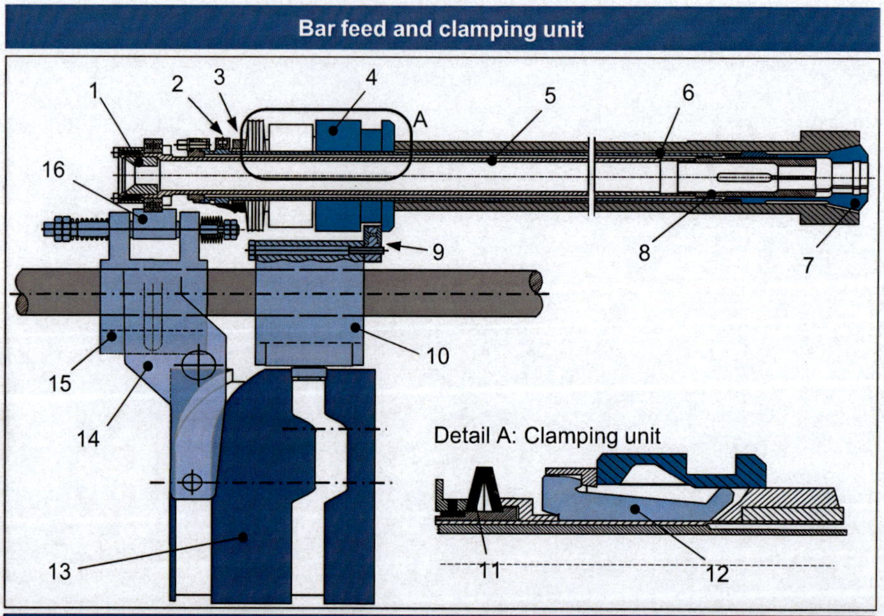


**Legend**

1: Guide ring
2: Notched nut
3: Clamping nut
4: Clamping sleeve
5: Feed tube
6: Clamping tube (left-hand thread)
7: Collet chuck (left-hand thread)
8: Feed collet (right-hand thread)
9: Clamping jaw
10: Clamping slide
11: Disk spring assembly
12: Clamping lever
13: Motion control cam drum
14: Feed motion link
15: Feed slide
16: Feed connector

**Fig. 8.16** Bar feed and clamping unit, according to Alfred H. Schütte

bar stock is grasped by the spring contact force of the feed collet and advanced to the bar stop by means of frictional locking. At the same time, the motion of the feed motion link (14) is transferred to the feed collet via the feed slide (15) and the feed connector (16). Afterward, the collet chuck is again clamped, the bar stop is swiveled out of the operating space and the feed collet is returned to the starting position. At the same time, the spring-pressed feed collet (8) on the bar stock slides back around the feed length of the workpiece, which is again required in the next cycle. During machining, the bar makes contact with the guide ring (1) and the collet chuck.

The Maltese drive shown in Fig. 8.17 is used to index the spindle drum. A planetary gear riding on the stationary spur gear is located on the crank that rotates with the control shaft. The pin, which is eccentrically positioned on the planetary gear, runs tangentially into and out of a slot on the Maltese cross. At the same time, the crank turns 120° counterclockwise and the Maltese cross is rotated another 90°. The pin travels along the sketched tripod-shaped motion path K. As a result of the reduced jerk, this movement reduces the dynamic excitation by 50% when indexing the spindle drum compared to a conventional Maltese drive. So that the moving inert mass is not only decelerated via the pin, there is what is called the locking curve S formed by the concentric area of the crank. When the pin comes out of a slot, the motion of the Maltese cross is gently absorbed by the locking curve. The rotation of the Maltese cross is then locked during the work operation until the pin again runs tangentially into the next slot during the next revolution. Through the appropriately selected gear ratio $z = 2/3$ between the spur gear coupled with the Maltese cross and the gear teeth of the spindle drum, the spindle drum is indexed one spindle division (360°/6 = 60°) during the 90° rotation of the Maltese cross. A detailed description of gearboxes with intermittent movements can be found in [LICH53].

The indexing device in the form of three-part spur gear teeth is located on the spindle drum. This device fixes the spindles exactly into the next machining position once the indexing has been performed by the

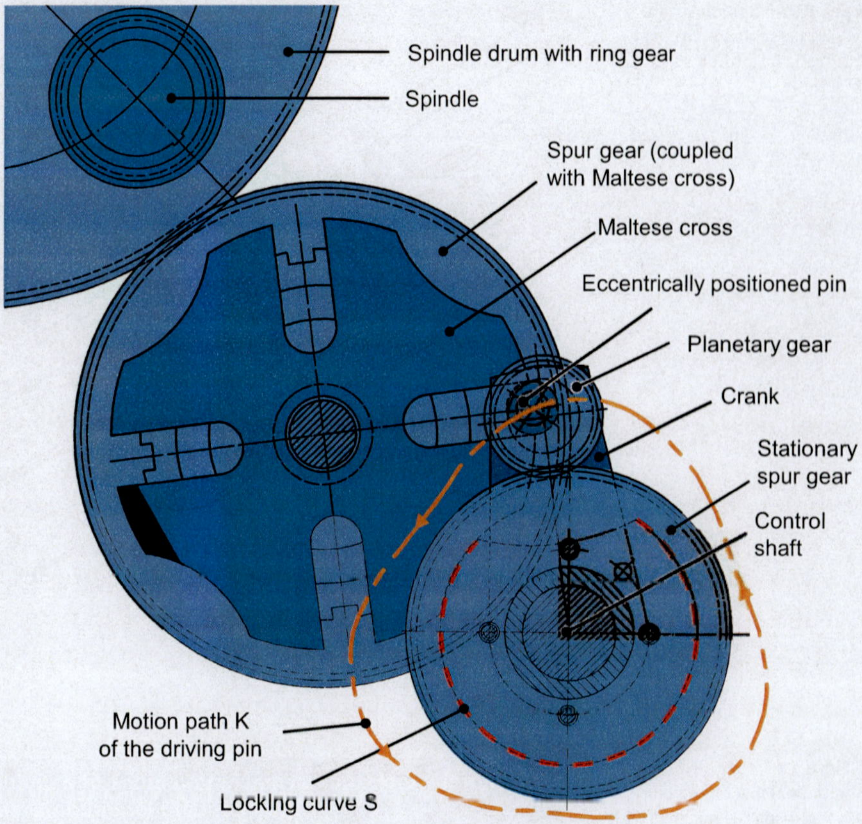


**Fig. 8.17** Diagram showing the spindle drum being indexed using a modified Maltese drive, according to Alfred H. Schütte

Maltese drive, Fig. 8.18. The broad spur-toothed ring is pressed into the two narrow rings for positioning purposes and fits them tightly together. To index the spindle drum, the broad spur-toothed ring, pressed on by a large number of springs, is lifted out of the gear teeth axially against the spring force by a lever mechanism. The narrow interior ring attached to the spindle drum can now twist relative to the outer ring attached to the spindle housing. If the spindle drum is rotated another spindle division, then the ring, which can move axially, will move back and position the two narrow rings such that they fit tightly together again.

The six slides used for longitudinal machining are located on a central block as shown in Fig. 8.19. The slides move in cylinder segments whose axes align exactly with those of the rotating spindles. Cover fillets on the block prevent torsion of the slides. The central block is fixed to either the drive box or the longitudinal beam using a retaining arm. Threaded holes are provided in the longitudinal slides so that the tool holders and special-purpose devices can be installed.

A longitudinal slide drive, which machines the workpiece in the spindle (1) using a tool, is mounted on the longitudinal slide (2) shown in Fig. 8.20. The rotational motion of the cam disk (15) is transferred from the motion control cam roller (14) of the motion link lever (12) to the connecting rod (11) and then on to the guide tube (8). The cam disk has both an outside and an inside motion control cam so that the motion control cam roller can be guided on both sides. This also means that both the machining and return movements can be transferred. The stroke is limited by a stop piece (10) on the guide tube and a stop screw (9) in the machine column. The operating position is set roughly using the two clamping rings (5) and (7) and then accurately adjusted using the setting screw (6), which is similar to a micrometer screw. The feed motion is transferred via the driving rod (4) onto the clamping ring (3), which is used to change the different tools. The stroke length is set using the position of the coupling point of the connecting rod (11) in the motion link slot (13). In this way, the feed length of the longitudinal slide is determined.

Figure 8.21 shows the drive for the cross slide. As with the longitudinal slides, the motion of the cross slide (7) is transferred from the control shaft (6) via the machining and return motion control cam of the cam disk (3), the motion link lever (2), the pull and connecting rod (1), the toothed segment (5) and the gear rack (4). As with the longitudinal slides, the feed length can be varied for each slide using the adjustable motion link levers.

As the stroke can be adjusted using transmission levers, there is a broad field of application for a cam disk. It also means that lengthy setup and changeover times

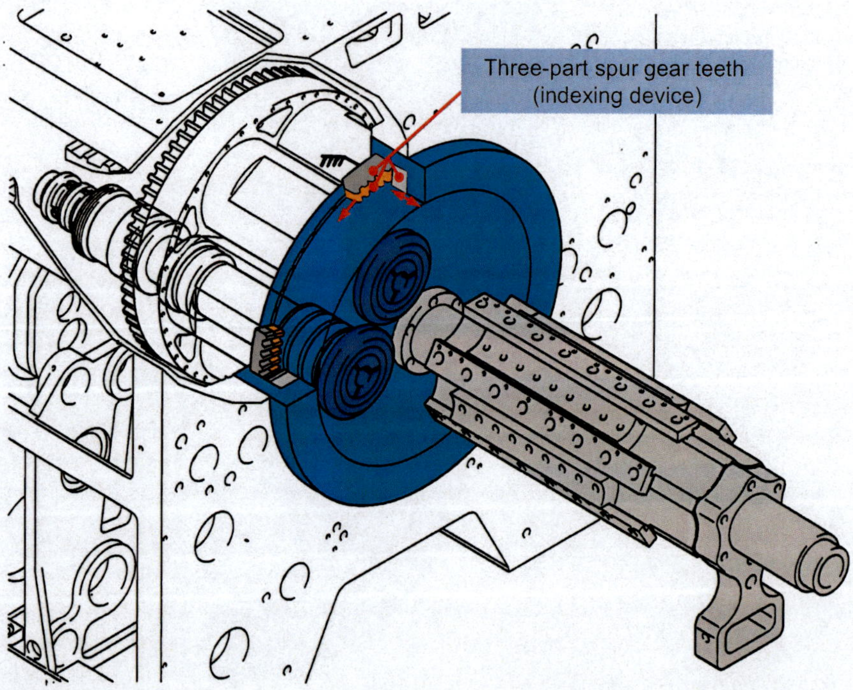


**Fig. 8.18** Spindle drum locking using three-part spur gear teeth, according to Alfred H. Schütte

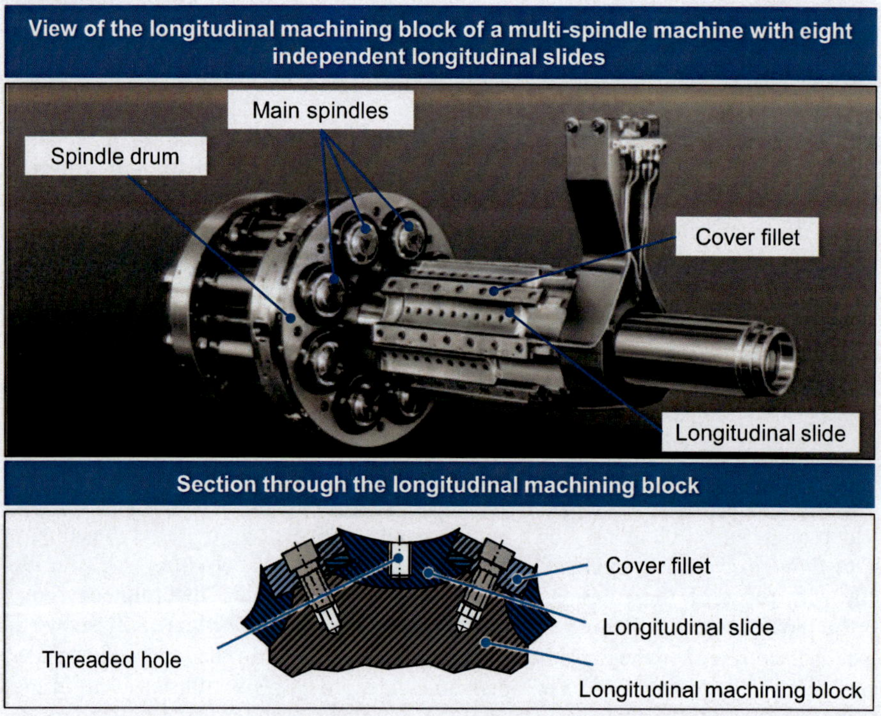


**Fig. 8.19** View of the longitudinal machining block of a multi-spindle machine with eight independent longitudinal slides, according to Alfred H. Schütte

are avoided. The feed rate of the slide comes from the angular velocity of the disk, i.e., from the control shafts and the slope of the motion control cams. The greatest feed force is limited by the maximum torque of the control shafts, the maximum transmission forces of the cam disk drive and the transmission ratio of the levers.

The work sequence diagram shown in Fig. 8.22 illustrates the chronological assignment of the control functions of a multi-spindle machine. Partial rotation of the control shaft from 0° to 200° at high speed controls all the auxiliary functions that do not have any effect on the immediate progress of the machining operation.

**Fig. 8.20** Schematic diagram of the longitudinal slide drive, according to Alfred H. Schütte

**Schematic diagram of the longitudinal slide drive**

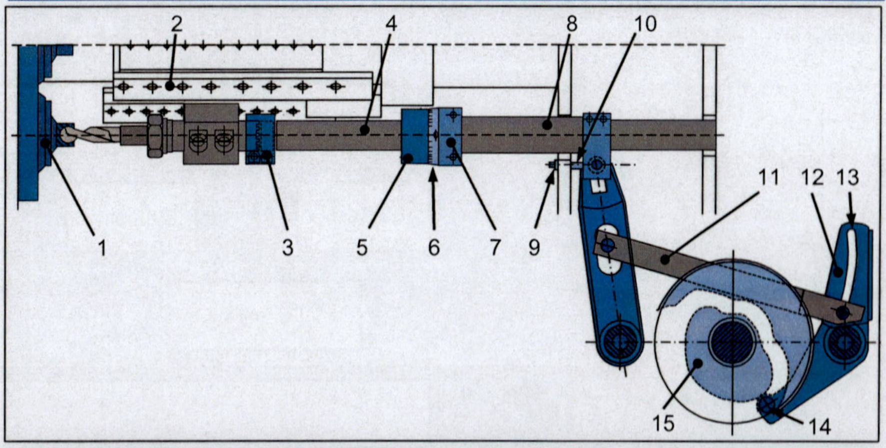


**Legend**

| | | | |
|---|---|---|---|
| 1: | Rotating spindle | 9: | Stop screw |
| 2: | Longitudinal slide | 10: | Stop piece |
| 3: | Clamping ring | 11: | Connecting rod |
| 4: | Driving rod | 12: | Motion link lever |
| 5: | Clamping ring (rough adjustment) | 13: | Motion link slot |
| 6: | Setting screw with scale (fine adjustment) | 14: | Motion control cam roller |
| 7: | Clamping ring | 15: | Cam disk with machining and return motion control cam |
| 8: | Guide tube | | |

**Fig. 8.21** Schematic diagram of the cross slide drive, according to Alfred H. Schütte

**Schematic diagram of the cross slide drive**

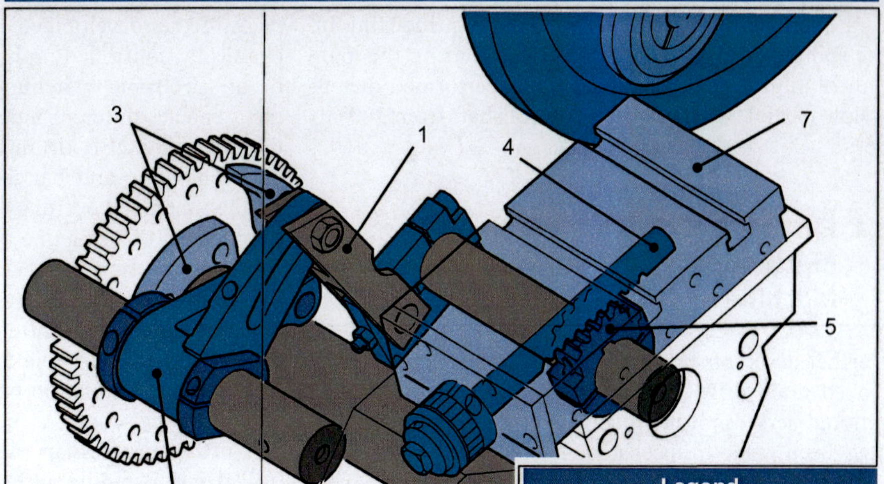


**Legend**

1: Pull and connecting rod
2: Motion link lever
3: Machining and return motion control cam
4: Gear rack
5: Toothed segment
6: Control shaft
7: Cross slide

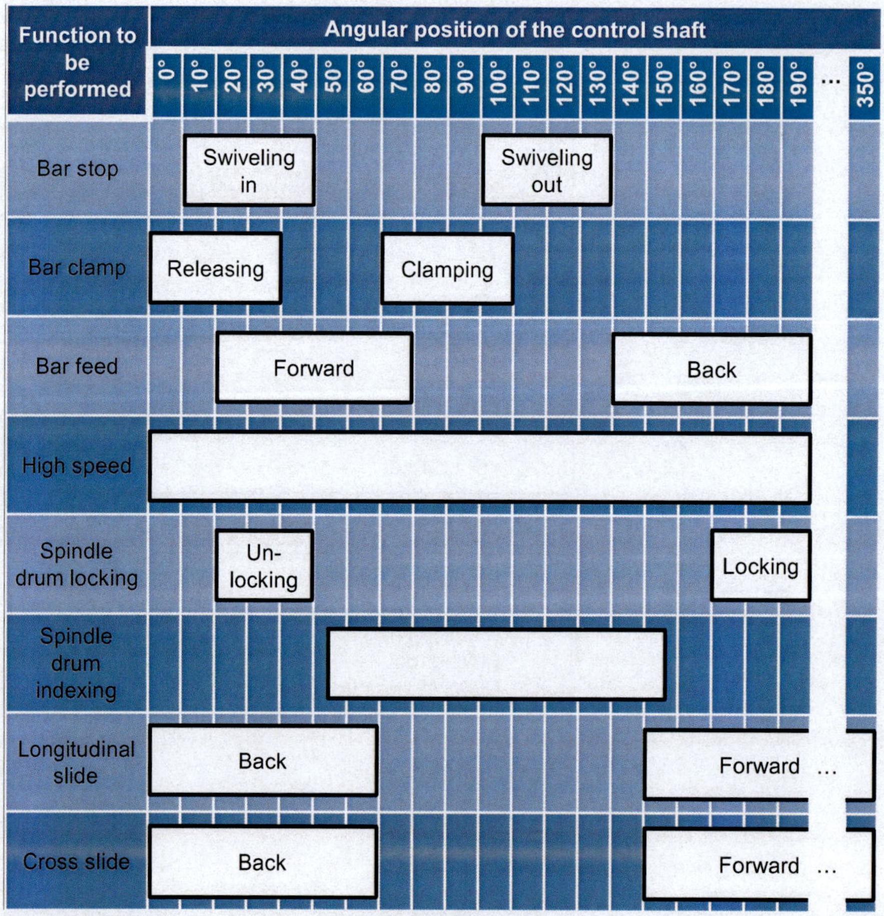


**Fig. 8.22** Work sequence diagram of a mechanically controlled multi-spindle machine, according to Alfred H. Schütte

Functions like bar feed and clamping, indexing of the spindle drum, etc., are carried out during this partial rotation. The machining itself is performed during a slow partial rotation of the control shaft from 200° to 360°.

## 8.3 Further Development of the Mechanically Controlled Multi-Spindle Machine

Further developments based on the use of a speed controlled main drive motor for the spindles and an additional servo motor, which drives the control shafts, have been made to mechanically controlled multi-spindle machines so that functions like feeding, bar feeding, workpiece clamping, etc., can be handled, Fig. 8.23. The advantage of this principle is the variable setting of the spindle speed and the feed rate. Another enhancement made on some variants is the addition of a drive motor for each spindle. This allows the speed to be set at each spindle separately and independently from the other spindles and that, in turn, allows optimal machining.

A further development which has been made to mechanically controlled multi-spindle machines is the use of an electronic switching cam group. This development enables different indexing functions such as spindle drive ON/OFF, drum locking and indexing, workpiece clamping and bar feed to be performed directly by the machine control system and various servo motors instead of using one or more control shafts. These subfunctions are started depending on the position of the control shafts (measurement via rotary encoder) and, consequently, as a function of the production progress, Fig. 8.24. Depending on the drive of the control shafts, the initiation points can be steplessly varied and optimally set.

Further progress in the field of mechanically controlled multi-spindle automatic lathes is based on the use of NC-controlled tool slides for facing. This technology combines the high speed of a mechanically controlled machine with the flexibility of NC-controlled slides, e.g., it can be used to fabricate complicated outside and inside contours and it also minimizes the retooling time (Fig. 8.25).

Nowadays, most multi-spindle machines no longer just have purely mechanically controlled functions but

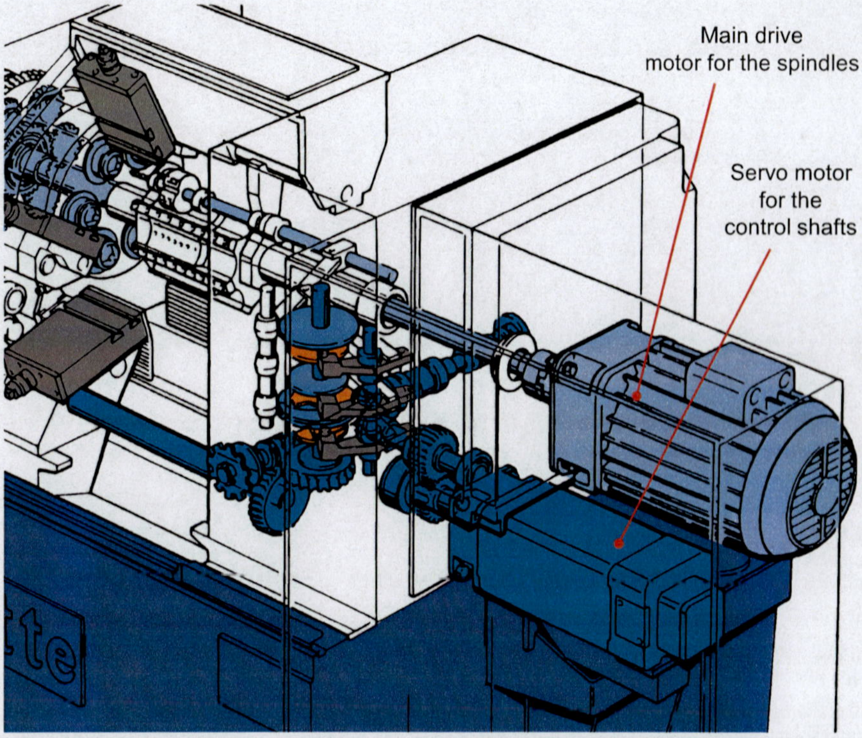


**Fig. 8.23** Mechanically controlled multi-spindle machine with a main drive motor and a servo motor to drive the control shafts, according to Alfred H. Schütte

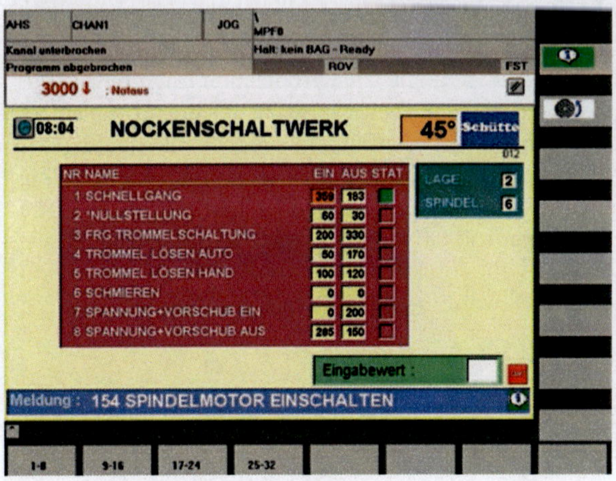


**Fig. 8.24** Switching cam group display on the NC control system of a multi-spindle automatic lathe. *Source* Alfred H. Schütte

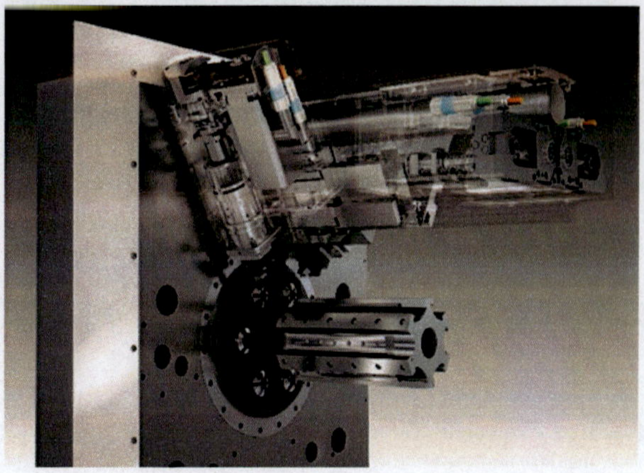

**Fig. 8.25** Operating space of a multi-spindle machine with an additional NC-controlled cross slide. *Source* Alfred H. Schütte

also NC-controlled functions. The number of mechanically controlled functions has decreased more and more over time. With these machines, it is often only the auxiliary functions that are still mechanically controlled. Therefore, an exact division between mechanically controlled and NC-controlled multi-spindle machines is no longer possible as the machines combine both control concepts (hybrids). These hybrids are still being built today although they are no longer being further developed. Nowadays, with multi-spindle machines, new developments are, without exception, based on NC-controlled machine concepts, Fig. 8.26.

## 8.4 Electronic Line Shaft

The ability to guide tools on complex paths and dynamically control several—partially synchronized—axes of motion is in many machines crucial for the completion of the application-specific machining task. The classic mechanical solution for such drive

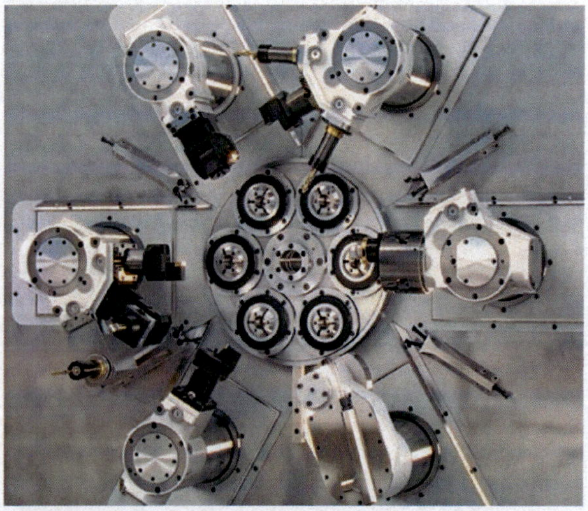

■ Fig. 8.26 Operating space of a multi-spindle automatic lathe with only NC-controlled functions. *Source* Alfred H. Schütte

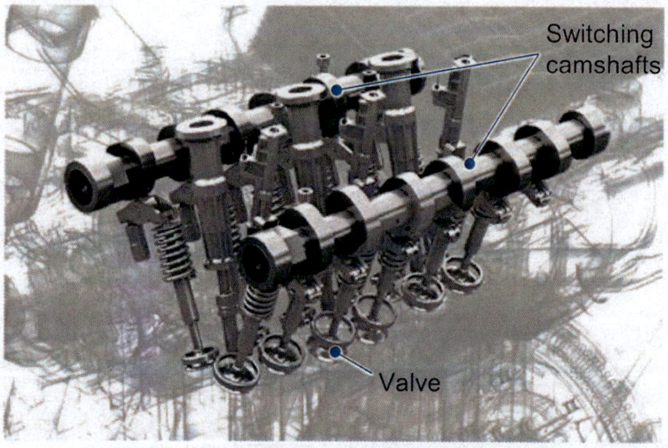


■ Fig. 8.27 Example of a mechanical machine construction with switching camshafts. *Source* MAHLE

8

tasks—in particular with cyclically operating machines—is to use a single motor with a central drive shaft, i.e., the line shaft. It drives the other axes of motion and synchronizes the individual movements, e.g., by using cam disks or bevel gears. The switching camshafts shown in ■ Fig. 8.27 perform a similar task.

Motivated by the requirement to design more flexible machines, electrical concepts in the field of mechanical engineering are supplementing and replacing mechanical concepts. For the drive structure, this usually means changing from one common motor to a servo drive for each axis of motion. With the electronic line shaft, machine developers are being given the opportunity to draw on conventional engineering methods for the new design of the drive concept.

### 8.4.1 Operating Principle

In the electrical drive concept, the functions of the mechanical line shaft, i.e., drive control and synchronization, are performed by the motion control software (see ▶ Sect. 10.5) that provides the individual synchronization controls implemented there together with servo motors, ■ Fig. 8.28.

The motion tasks resulting from using the line shaft in the mechanical drive concept are assigned to the motion control function "Angle synchronization," ■ Fig. 8.29. In angle synchronization, in contrast to speed synchronization, it is possible for the movement of the leading axis to continually adapt to the fluctuations of the sequence movements. The mechanical coupling better reflects this using the line shaft [WEID08]. The motion control software contains the virtual line shaft. Using this, the software generates the guide value used to link the following axes in the synchronous operation.

■ Fig. 8.28 Comparison of the mechanical (**a**) and the electrical (**b**) line shaft

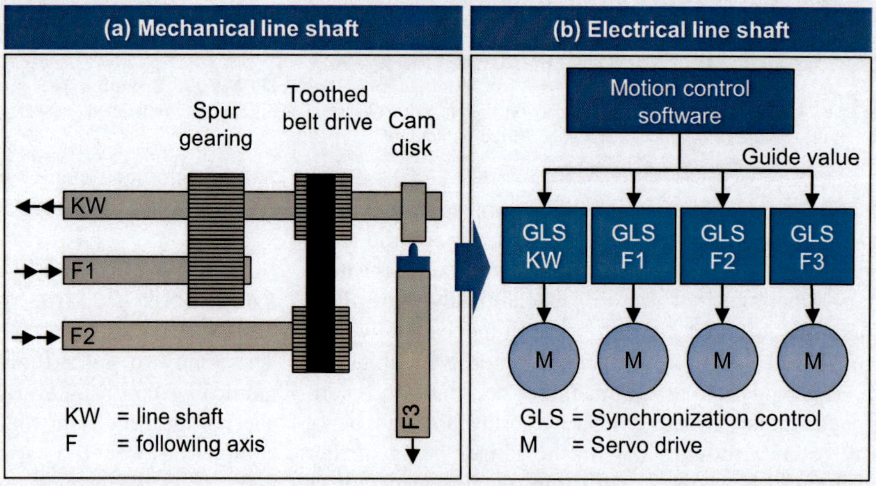

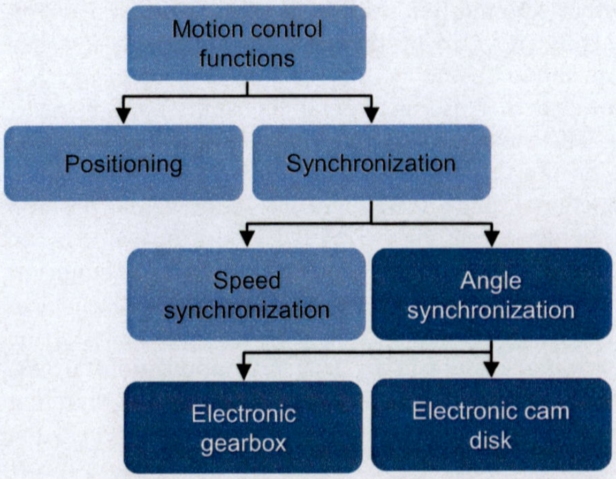


**Fig. 8.29** Motion control functions, according to SIEMENS

In the synchronization controls of the following axes, the reference position values for the position controllers of the follower drives are derived from the guide value and transmitted to the controllers. This creates a hierarchical relationship between the movement of the leading axis (referred to as the master) and the following axes (slaves). The functional structure of a synchronization control is described in detail in the following, see Fig. 8.30 [WEID08]:

The first step specifies how the guide value is generated. In principle, the guide value of a real master or a virtual master axis can be used. If an electronic line shaft is used, then the guide value will be generated by a virtual leading axis. The guide value then usually specifies the angular position of the leading axis.

In the second step, the guide value is processed for further use by normalizing it to the axis cycle

**Fig. 8.30** Functional components of a synchronization control, according to SIEMENS

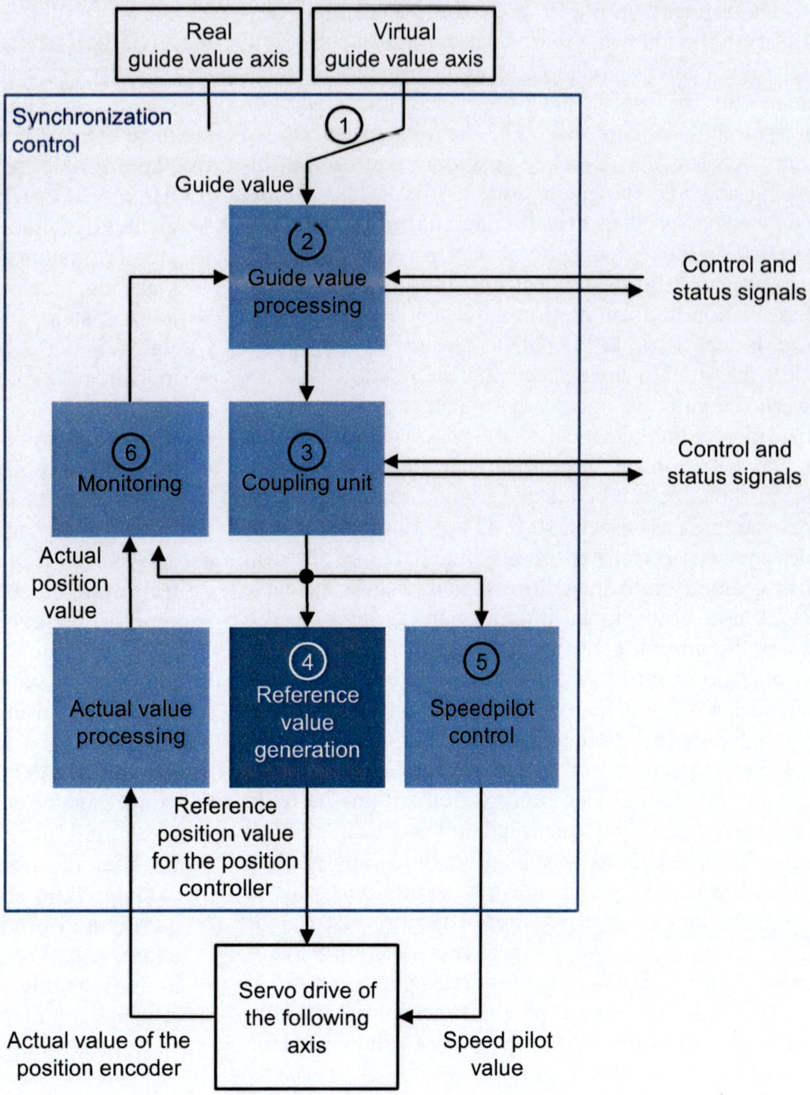

(corresponds to one revolution in the case of a line shaft). In addition, the guide value processing step processes the results of the referencing to the mechanical system if necessary while, during synchronous running, this happens cyclically. However, with the electronic line shaft, referencing is omitted in many cases as the following axes are frequently provided with absolute value encoders. The third task of the guide value processing step is the addition of a possible offset between the guide value and the mechanical system. This offset can be adjusted manually by the operator so that the system can later be optimally set and small errors corrected.

A coupling unit, which can carry out the engage and disengage functions, then follows. It ensures that the machining processes can be interrupted, e.g., if there is a defective product. The safe engaged and disengaged positions of the following axis are also to be observed.



The reference position value for the position controller of the following axis is determined in the fourth step. The reference value generation forms the core element for the replacement of a mechanical gearbox or a mechanical cam disk. The mathematical relationships between the reference position value of the following axis and the guide value of the virtual leading axis can be specified here for the control system. Depending on the design, either an electronic gearbox with a fixed gear ratio or an electronic motion control cam disk for non-uniform motion sequences is implemented (see ► Sect. 10.5; ◘ Fig. 10.46 Electronic cam disk). With the latter, a table specifying the relationships between the guide value and the reference position value instead of a mathematical relationship is usually stored in the synchronization control. The interpolation values in the table can be created using a cam disk editor, for example (see ► Sect. 10.5; ◘ Fig. 10.44). This enables the user to transfer the machine data and the cam disk geometry into the editor. It makes them available to the user both graphically and numerically and also stores the motion control cam spreadsheet in the synchronization control. Multiple motion control cam disk spreadsheets can always be stored there to withstand a power failure in case the product is changed [EIDA97]. It is necessary to interpolate the guide values between the support points. The interpolation within the reference value generation functional unit is usually handled using linear functions or spline functions. The result is a two-dimensional graph showing the motion profile of the following axis as a function of the processed guide value. From this graph, the reference position value is determined in relation to the processed guide value.

The fifth component of the synchronization control is the calculation of the speed pilot value. However, with electronic line shafts, the pilot value is also frequently provided by the virtual leading axis and also transferred to the synchronization control of the following axis. This functional unit is omitted in this case.

The movement of the following axis is monitored to ensure that it is synchronized with the guide value, functional unit 6 in ◘ Fig. 8.30. If the actual position value deviates from the reference position value by an amount that is too large, this will result in a significant tracking error signifying an overload of the mechanical drive train. The end positions are also monitored if the following axis is a linear axis. If these positions are exceeded, a check is made to see if the axis is vibrating unacceptably. Unacceptable vibrations can occur if the load is too high or if the controller has been set incorrectly. If this is not the case, errors can be expected either in the actual-value acquisition or in the determination of the reference point.

### 8.4.2 Application

As well as close alignment to the already familiar mechanical structure, the use of a virtual line shaft and the appropriate software brings further advantages [LOHS94, EIDA97]:

- greater dynamics
- greater system flexibility with regard to project planning, machine construction and operation
- lower system costs due to a simpler mechanical design
- increases in productivity by up to 30 to 40% [BACH98]
- higher quality of the end product if applicable
- minimization of the setup times as only the machine data have to be changed in the editor
- more complex machine concepts are made easier
- easy monitoring of the end positions identifying defects more quickly
- simple integration of safety functions

Besides these advantages, there are disadvantages associated with using a virtual line shaft compared with a mechanical concept and these must be solved [LOHS94, EIDA97]:

- safety must be ensured if there is a power failure
- greater risk of position errors
- the user of the machine must possess technological know-how and be open to the technology as the quality and productivity of the systems depends to a large extent on the controller settings
- limitations due to the resolution of the measuring systems, the dynamics of the servo motors and the speed of the hardware

## References

[BACH98] Bachmann, G.: Virtuelle Königswelle bietet mehr Flexibilität. In: Konstruktion, Nr. 50/1, S. 28 (1998)

[EIDA97] Eidam, J.: Beurteilung und Simulation des Betriebsverhaltens von lagegeregelten elektrischen Direktantrieben als "Elektronische Kurvenscheibe". Dissertation (1997)

[HEIS14] Heisel, U.: Handbuch Spanen. (Reihe: Ed. Handbuch der Fertigungstechnik). Hanser, München (2014)

[LICH53] Lichtwitz, O.: Getriebe für aussetzende Bewegung. Springer, Berlin (1953)

[LOHS94] Lohse, G.: Konstruktion von Kurvengetrieben (Reihe: Kontakt & Studium. Bd. 454: Maschinenbau). Expert-Verlag, Renningen-Malmsheim (1994)

[WECK13] Weck, M., Brecher, C.: Maschinenarten und Anwendungsbereiche (Reihe: VDI-Buch, eds. Weck, M., Brecher, C.). 6. Aufl. Springer Vieweg, Berlin (2013)

[WEID08] Weidauer, J.: Elektrische Antriebstechnik. Publicis Corporate Publication, Erlangen (2008)

# Basics of Information Processing

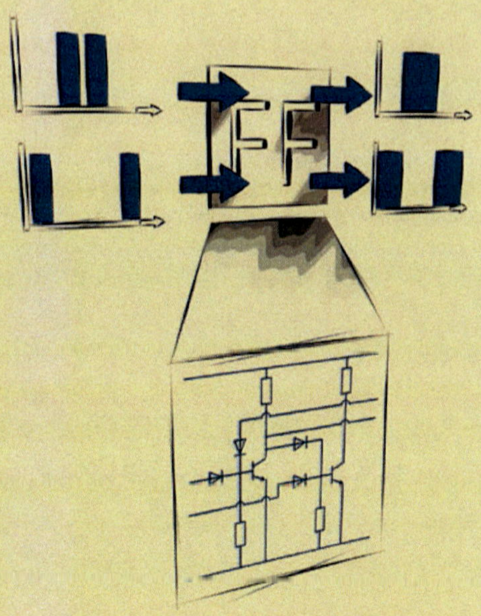


I thank my staff members Christian Fimmers M. Sc., Christoph Pallasch M. Sc., Simon Storms M. Sc. and Christoph Susen M. Sc. for their contribution in revising this chapter.


C. Brecher and M. Weck, *Machine Tools Production Systems 3*,
Lecture Notes in Production Engineering, https://doi.org/10.1007/978-3-658-34622-5_9

## Contents

Stricter requirements are being made on information processing as production processes become increasingly automated. Electronic control technology has been developed on different levels. First, it has had an effect on the low-level machine and process level using powerful, miniaturized control components, and second, the development of powerful data processing systems has provided a way of incorporating the control of production processes into the overall information flow of a company. The information flow of higher-level integrated manufacturing systems includes the transfer of control data to the machining and transport units. From there, production data are recorded, collected and analyzed to monitor the manufacturing process and to control it in line with specified strategies.

Common to all digital information processing equipment is that the data to process must be defined in a particular form. The display form of data is centered around how the devices work—based on electrical "On" and "Off" states. These parameters, that can only assume two values, are called binary variables.



Binary information can be processed in logic networks. The logic networks are made of switching elements that implement logical functions (Boolean operations and basic functions according to DIN 19226). The following chapters cover, for example, the conversion of information data to different display forms and their processing in logic networks [BAIL74; BORU89; DWOR86; FLEG93; KOCH77; KORE81; BUES94]. Then, the communication technology used in automation is described. This technology is very important against the backdrop of stricter requirements placed on integrating workflows, on organizational flexibility and on the independence of sites for manufacturing companies [BUES94].

## 9.1 Basics

### 9.1.1 Number Systems

The number system in general use is the decimal system. Practical requirements mean other number systems are better suited for control systems.

In every number system, a number can be represented as a sum of powers of a base number. The general form of such a number representation is specified in the top part of Eq. (9.1).

$$Z = b_n \cdot B^n + b_{n-1} \cdot B^{n-1} + \ldots + b_1 \cdot B^1 + b_0 \cdot B^1 + b_{-1} \cdot B^{-1} + \ldots + b_m \cdot B^m$$

Short form:

$$Z = \sum_{i=m}^{n} b_i B^i \quad 0 \leq b_i \leq B \quad \begin{matrix} B{:}Base \\ b_i{:}Value\ of\ the\ i\text{−}th\ digit \end{matrix} \tag{9.1}$$

The value of a position within a multi-digit number is determined from the product of the digit and the significance of the digit position. The significance is an integer power of the base used to name the number system. The associated exponent corresponds to the number of digits away from the decimal point. The first position to the left of the decimal point is position zero. From there, the count increases to the left and decreases to the right. Example significances in the hexadecimal system:

$$0.01_{16} = 16^{-2}$$
$$0.1_{16} = 16^{-1}$$
$$1_{16} = 16^{0}$$
$$10_{16} = 10^{1}$$
$$100_{16} = 16^{2}$$

The binary system is especially significant because logic networks can easily process binary numbers. In the binary system, the base is 2 and the digits 0 and 1 can be differentiated. As representing large numbers requires a very large number of digits, however, the octal (8) and hexadecimal (16) systems are common in digital computing. Numbers in these number systems are very easy to convert to binary numbers while also being easy to understand. ◻ Figure 9.1 shows a comparison of the most common number systems. The different digits are 0, 1, 2, 3, 4, 5, 6, 7, 8 and 9 in the decimal system, 0, 1, 2, 3, 4, 5, 6 and 7 in the octal system and 0, 1, 2, 3, 4, 5, 6, 7, 8, 9, A, B, C, D, E and F in the hexadecimal system.

If the requirement is for logic networks to process decimal numbers, they must first be converted to binary numbers. Also, the special importance of the binary system becomes evident when decimal numbers are converted to octal or hexadecimal numbers since it serves as an intermediate step.

For a conversion, a decimal number is split into integers and fractions, Integer division and a modulo operation convert both parts to the corresponding parts of the binary number (◻ Fig. 9.2). To differentiate the number representation in different systems, the base is specified as the index.

The subsequent conversion of the binary number to an octal number is as follows: Starting at the decimal point of the binary number, the number is sectioned into three-digit positions on each side. Zeros are added to the remaining digits to make three digits. Each of these blocks is, as shown in ◻ Fig. 9.1, converted to an octal digit. Taken together, they represent an octal number. Conversion to a decimal number uses the power syntax.

Converting from binary numbers to hexadecimal numbers and back is similar to converting to octal numbers. Only in this case, the digit blocks comprise four instead of three digits.

**Fig. 9.1** Number systems

**Power syntax**

$$Z = b_n \cdot B^n + b_{n-1} \cdot B^{n-1} + \ldots b_1 \cdot B^1 + b_0 \cdot B^0 + b_{-1} \cdot B^{-1} + \ldots b_m B^m$$

**Short form**

$$Z = \sum_{i=m}^{n} b_i B^i \quad 0 \le b_i \le B$$

$B$: Base

$b_i$ : Value of the i-th digit

| Decimal | | | Binary | | | | | | | Octal | | | Hexadecimal | |
|---|---|---|---|---|---|---|---|---|---|---|---|---|---|---|
| $10^2$ | $10^1$ | $10^0$ | $2^6$ | $2^5$ | $2^4$ | $2^3$ | $2^2$ | $2^1$ | $2^0$ | $8^2$ | $8^1$ | $8^0$ | $16^1$ | $16^0$ |
| | | 0 | | | | | | | 0 | | | 0 | | 0 |
| | | 1 | | | | | | | 1 | | | 1 | | 1 |
| | | 2 | | | | | | 1 | 0 | | | 2 | | 2 |
| | | 3 | | | | | | 1 | 1 | | | 3 | | 3 |
| | | 4 | | | | | 1 | 0 | 0 | | | 4 | | 4 |
| | | 5 | | | | | 1 | 0 | 1 | | | 5 | | 5 |
| | | 6 | | | | | 1 | 1 | 0 | | | 6 | | 6 |
| | | 7 | | | | | 1 | 1 | 1 | | | 7 | | 7 |
| | | 8 | | | | 1 | 0 | 0 | 0 | | 1 | 0 | | 8 |
| | | 9 | | | | 1 | 0 | 0 | 1 | | 1 | 1 | | 9 |
| | 1 | 0 | | | | 1 | 0 | 1 | 0 | | 1 | 2 | | A |
| | 1 | 1 | | | | 1 | 0 | 1 | 1 | | 1 | 3 | | B |
| | 1 | 2 | | | | 1 | 1 | 0 | 0 | | 1 | 4 | | C |
| | 1 | 3 | | | | 1 | 1 | 0 | 1 | | 1 | 5 | | D |
| | 1 | 4 | | | | 1 | 1 | 1 | 0 | | 1 | 6 | | E |
| | 1 | 5 | | | | 1 | 1 | 1 | 1 | | 1 | 7 | | F |
| | 1 | 6 | | | 1 | 0 | 0 | 0 | 0 | | 2 | 0 | 1 | 0 |
| | 1 | 7 | | | 1 | 0 | 0 | 0 | 1 | | 2 | 1 | 1 | 1 |
| | | … | | | | | | | … | | | … | | … |
| | 2 | 7 | | | 1 | 1 | 0 | 1 | 1 | | 3 | 3 | 1 | B |
| | | … | | | | | | | … | | | … | | … |
| 1 | 0 | 0 | 1 | 1 | 0 | 0 | 1 | 0 | 0 | 1 | 4 | 4 | 6 | 4 |

The reasoning behind the central status of the binary system is that binary numbers can be added using digital modules. All other arithmetic operations (such as subtraction and multiplication) can be derived from addition. This means that machine calculation is possible to a large extent using binary numbers.

Figure 9.3 shows how the four basic arithmetic operations with binary numbers are performed. To represent negative binary numbers, the "two's complement" is formed. Now, subtraction can be converted to addition. The overflow (left digit) is ignored.

The two's complement is formed, as shown in Fig. 9.3 on the right: All digits of the binary number are inverted and a "1" is added to the resultant binary number. The number of digits is selected to be large enough so there is a "0" in the first digit for the positive binary number. Forming the two's complement fills the first digit with a "1" thereby denoting the binary number as negative.

Apart from this exception, calculations with binary numbers are the same as with decimal numbers. A division involving two integer binary numbers (integer number) can result in a fractional binary number (real number). Fractional binary numbers can be represented as two types - fixed point numbers and floating point numbers. In contrast to the fixed point representation, the floating point representation also lends itself to the processing of numbers of different magnitudes (such as a multiplication of a large number by a small number). Computing with floating point numbers is more complex because every number is split into a mantissa (coefficient) and an exponent to base two.

**Fig. 9.2** Converting numbers for representation in different number systems.

**Splitting of the decimal number**

$Z_{10} = 13{,}6875 = 13 + 0{,}6875$

**Conversion of the decimal number to a binary number**

| | | | |
|---|---|---|---|
| 13 | $:2 = 6$ | Remainder 1 | Dual number 1101 |
| 6 | $:2 = 3$ | Remainder 0 | |
| 3 | $:2 = 1$ | Remainder 1 | |
| 1 | $:2 = 0$ | Remainder 1 | |
| 0,6875 | $\cdot 2 = 1{,}375$ | $= 0{,}375 + 1$ | Dual remainder 0,1011 |
| 0,375 | $\cdot 2 = 0{,}75$ | $= 0{,}75 + 0$ | |
| 0,75 | $\cdot 2 = 1{,}5$ | $= 0{,}5 + 1$ | |
| 0,5 | $\cdot 2 = 1{,}0$ | $= 0 + 1$ | |
| 0 | $\cdot 2 = 0$ | | |

$Z_2 = 1101{,}1011$

**Conversion of the binary number to a hexadecimal number**

$Z_2 =$ 1101, 1011
$Z_{16} =$ *D* , *B*

**Conversion of the binary number to an octal number**

$Z_2 =$ 001 101, 101 100
$Z_8 =$ 1 5, 5 4

**Conversion rule for decimal numbers**

$$Z_{10} = \sum_{i=m}^{n} b_i \cdot B^i$$

$B$: Base of the number system
$b_i$: Digit with significance i; $0 < b_i < B$
$n+1$: Number of digits to the left of the decimal point
$m$: Number of digits to the right of the decimal point

## 9.1.2 Data Codes

For data to be entered into and processed in digital devices, the data must be represented in a suitable manner. The term used is coding data.

BCD (binary-coded decimal) codes lend themselves to the digit-by-digit representation of decimal numbers as binary numbers. These codes combine the usual decimal number format with the binary notation.

Four-digit binary numbers (tetrads) are required to represent a decimal digit. Tetrads provide $2^4 = 16$ possible binary combinations, only ten of which are required here. The individual decimal digits are assigned to the individual tetrads depending on the technical requirement. The four most common codes are shown in Fig. 9.4.

In the 8421 BCD code, the decimal digit is shown as the equivalent binary number. For the 8421 hexadecimal code, all 16 possibilities of a tetrad are used.

The Gray code assigns the decimal digits to the tetrads such that the tetrads only change in one place from one decimal digit to the next. This is the same for representing hexadecimal digits. In terms of structure, there are any number of assignments that conform to the method of calculation of the Gray code. A binary-reflected Gray code is shown in Fig. 9.4. For a binary-reflected Gray code, there is also only one change in the tetrad on a transition from the most significant digit ("9" or "F") to "0". The Gray code is used, for example, to encode absolute linear scales. This is done to get only one information change in one digit when transitioning from one value to the next. This lends the code the benefit that dirt on or damage to the scale can be detected due to the disruption of the pattern, preventing reading errors.

Letters, punctuation marks and special characters can be coded just like digits. These symbols can then be processed in logic networks. About 50 to 60 characters can be represented in this way. To define such a range of characters, 6-digit binary combinations are required ($2^6 = 64$; DIN 66024). In general data processing, in the PC sector in particular, the ASCII code is popular. It comprises 8 bits and has 128 different characters (letters, numbers and special characters). The 7th

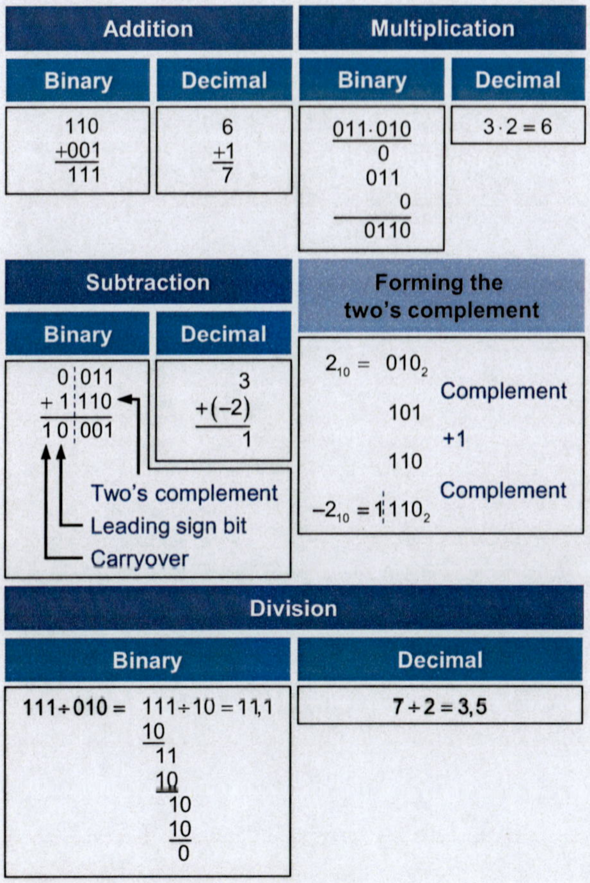


**Fig. 9.3** Arithmetic using binary numbers

bit is the parity bit. This code is shown on the right in Fig. 9.4.

The increasing level of international data exchange requires the coding of national special characters (such as umlauts). The coding of all special characters in use across the globe requires far more characters than the ASCII code contains. As a result, the International Standards Organization (ISO) assigned almost every character and every symbol in every language a number from 0 to 65,535 ($2^{16} - 1$). This character set is known as Unicode [FREY97] and has 16-bit coding.

### 9.1.3 Boolean Algebra

Binary information can be processed in digital devices known as logic networks. They contain switching elements that implement logic functions. The most important of these are the AND and OR functions. They are discussed in more detail below. Logic networks connect the states of the input variables (signals at the inputs) to output variables (signals at the outputs). Figure 9.6 shows the general form of a logic network. The outputs can be switched to inputs to create special functions.

A typical feature of logic networks is that the input and output states and the internal mode of operation use two values (binary). The possible states of signals occurring are defined as follows:

- "1": Signal is present (voltage being applied, current flowing, pressure being applied).
- "0": Signal is not present (voltage, current or pressure is zero).

The logic function of a logic network can be defined and simplified mathematically using Boolean algebra. It provides all the rules for calculating with logic variables.

The fundamental mathematical operations in Boolean algebra are contained in five postulates (see Fig. 9.5). Postulate 1 defines the binary character of logical state variable X. Postulate 2 defines the inversion (negation) and stipulates that state "0" is the complement of state "1", and vice versa. A bar above the variable is used as a symbol for the inversion.

Postulates 3 to 5 define the results of all combinations possible with two input variables for the AND and OR operations for which mathematical notation symbols are standardized.

The international symbols for representation are used below (DIN 66000), see also Fig. 9.7.

The five most important basic functions of logic networks are specified in Fig. 9.7. The Boolean notation, the value table and the graphic symbol are shown for every function. All three representation variants provide the same information.

The value table shows all possible variations of the input states and assigns to them the associated output states. Every row in the value table is a link from the states of the input variables shown – here A and B – to the specified state of output variable F.

The n variables yield $2^n$ rows for the value table. This means the value table, for two input variables, contains four rows and the output variable can be defined by a 4-digit combination of states "0" and "1".

The variation with a repeat of this 4-digit number combination for the output variables results in 16 different expressions. They are shown in Fig. 9.8. The top table shows the combinations of input states and their logical operation yielding the 16 possible output functions. The bottom table explains these logic operations. Four of the logic operations shown are the AND, OR, NAND and NOR functions already mentioned ($F_2$, $F_8$, $F_{15}$ and $F_9$). Apart from these, the names of those logic operations available as function modules are also specified: XOR (antivalence, exclusive OR) and XNOR (equality).

**Fig. 9.4** Tetrad codes and 8-bit ASCII code

| Tetrads | 8421 BCD code | 4-bit binary code | Gray BCD code | 4-bit Gray code |
|---|---|---|---|---|
| 0000 | 0 | 0 | 0 | 0 |
| 0001 | 1 | 1 | | F |
| 0010 | 2 | 2 | 9 | 9 |
| 0011 | 3 | 3 | | E |
| 0100 | 4 | 4 | 1 | 1 |
| 0101 | 5 | 5 | 2 | 2 |
| 0110 | 6 | 6 | 8 | 8 |
| 0111 | 7 | 7 | 3 | 3 |
| 1000 | 8 | 8 | | B |
| 1001 | 9 | 9 | | C |
| 1010 | | A | | A |
| 1011 | | B | | D |
| 1100 | | C | 6 | 6 |
| 1101 | | D | 5 | 5 |
| 1110 | | E | 7 | 7 |
| 1111 | | F | 4 | 4 |

| 8-bit ASCII (bit 7: parity bit) | | | | | | | | |
|---|---|---|---|---|---|---|---|---|
| Bits 654 / 3210 | 000 0 | 001 1 | 010 2 | 011 3 | 100 4 | 101 5 | 110 6 | 111 7 |
| 0000 0 | NULL | DC0 | | 0 | @ | P | ` | p |
| 0001 1 | SOM | DC1 | ! | 1 | A | Q | a | q |
| 0010 2 | EOA | DC2 | " | 2 | B | R | b | r |
| 0011 3 | EOM | DC3 | # | 3 | C | S | c | s |
| 0100 4 | EOT | DC4 | $ | 4 | D | T | d | t |
| 0101 5 | WRU | ERR | | 5 | E | U | e | u |
| 0110 6 | RU | SYNC | & | 6 | F | V | f | v |
| 0111 7 | BELL | LEM | ' | 7 | G | W | g | w |
| 1000 8 | FE | S0 | ( | 8 | H | X | h | x |
| 1001 9 | HTAB | S1 | ) | 9 | I | Y | i | y |
| 1010 A | LF | S2 | * | : | J | Z | j | z |
| 1011 B | VTAB | S3 | + | ; | K | [ | k | |
| 1100 C | FF | S4 | , | < | L | / | l | ACK |
| 1101 D | CR | S5 | - | = | M | ] | m | UC |
| 1110 E | SO | S6 | . | > | N | ↑ | n | ESC |
| 1111 F | SI | S7 | / | ? | O | ← | o | DEL |

| Postulate 1 | | Postulate 2 | |
|---|---|---|---|
| 1 | $X = 0$, if $X \neq 1$ | 2 | $\overline{0} = 1$ |
| 1' | $X = 1$, if $X \neq 0$ | 2' | $\overline{1} = 0$ |

| | Postulate 3 | Postulate 4 | Postulate 5 |
|---|---|---|---|
| AND | $0 \cdot 0 = 0$ | $1 \cdot 1 = 1$ | $0 \cdot 1 = 1 \cdot 0 = 0$ |
| OR | $0 + 0 = 0$ | $1 + 1 = 1$ | $0 + 1 = 1 + 0 = 1$ |

**Fig. 9.5** Fundamental mathematical operations of Boolean algebra

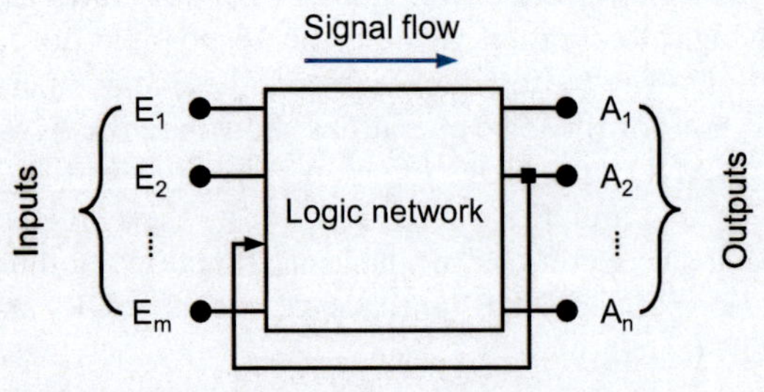


**Fig. 9.6** General representation of a logic network

The Boolean expressions for functions $F_1$ to $F_{16}$ are specified in the third column of the list. Each of these functions can be expressed as an equation, so for $F_7$ for example:

$$F_7 = \overline{A} \cdot B + A \cdot \overline{B} \tag{9.2}$$

This representation of a logical function is called Boolean equation. A Boolean equation contains all input variables used to derive the states of the output variables using logic operations. However, only those logic operations on the input states are included where the output variable assumes the value "1". Boolean equations are only made up of ORed AND functions, including the negation function.

Advantageously, they can be created from the verbal formulation of logical conditions. The operational state of a machine tool can be defined, for example, as follows with the states of machine functions A, B and C:

The machine is in operational state F when:

- Functions A and B have state "1", or
- Function B has state "1" and Function C has state "0".

**Fig. 9.7** Basic functions and graphic symbols

| Syntax (DIN 66000) | AND<br>A·B=F | OR<br>A+B=F | NOT<br>$\overline{A}$=F | *NAND*<br>$\overline{A\cdot B}=F$ | *NOR*<br>$\overline{A+B}=F$ |
|---|---|---|---|---|---|
| Value table | A B \| F<br>0 0 \| 0<br>0 1 \| 0<br>1 0 \| 0<br>1 1 \| 1 | A B \| F<br>0 0 \| 0<br>0 1 \| 1<br>1 0 \| 1<br>1 1 \| 1 | A \| $\overline{A}$ \| F<br>0 \| 1 \| 1<br>1 \| 0 \| 0 | A B \| F<br>0 0 \| 1<br>0 1 \| 1<br>1 0 \| 1<br>1 1 \| 0 | A B \| F<br>0 0 \| 1<br>0 1 \| 0<br>1 0 \| 0<br>1 1 \| 0 |
| Graphic symbol | 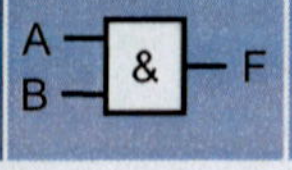 | 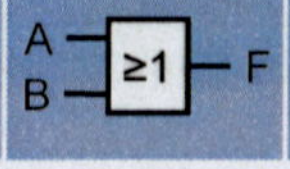 | 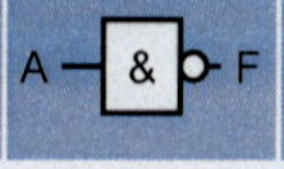 | 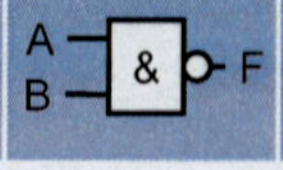 | 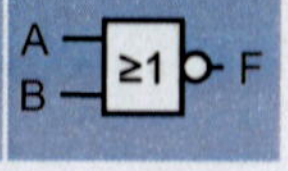 |

**Fig. 9.8** Possible operations with two input variables A and B

| Inputs | | Possible output functions $F_i$ | | | | | | | | | | | | | | | |
|---|---|---|---|---|---|---|---|---|---|---|---|---|---|---|---|---|---|
| A | B | $F_1$ | $F_2$ | $F_3$ | $F_4$ | $F_5$ | $F_6$ | $F_7$ | $F_8$ | $F_9$ | $F_{10}$ | $F_{11}$ | $F_{12}$ | $F_{13}$ | $F_{14}$ | $F_{15}$ | $F_{16}$ |
| 0 | 0 | 0 | 0 | 0 | 0 | 0 | 0 | 0 | 0 | 1 | 1 | 1 | 1 | 1 | 1 | 1 | 1 |
| 0 | 1 | 0 | 0 | 0 | 0 | 1 | 1 | 1 | 1 | 0 | 0 | 0 | 0 | 1 | 1 | 1 | 1 |
| 1 | 0 | 0 | 0 | 1 | 1 | 0 | 0 | 1 | 1 | 0 | 0 | 1 | 1 | 0 | 0 | 1 | 1 |
| 1 | 1 | 0 | 1 | 0 | 1 | 0 | 1 | 0 | 1 | 0 | 1 | 0 | 1 | 0 | 1 | 0 | 1 |

| Output function | Statement | Boolean expression $F_i$ | Logic operation name |
|---|---|---|---|
| $F_1$ | Never | 0 | |
| $F_2$ | A and B | $A\cdot B$ | AND |
| $F_3$ | A and not B | $A\cdot\overline{B}$ | |
| $F_4$ | Equal to A | $A$ | |
| $F_5$ | B and not A | $\overline{A}\cdot B$ | |
| $F_6$ | Equal to B | $B$ | |
| $F_7$ | A not equal to B | $\overline{A}\cdot B+A\cdot\overline{B}$ | XOR |
| $F_8$ | At least one | $A+B$ | OR |
| $F_9$ | Neither A nor B | $\overline{A}\cdot\overline{B}=\overline{A+B}$ | NOR |
| $F_{10}$ | A equal to B | $\overline{A}\cdot\overline{B}+A\cdot B$ | XNOR |
| $F_{11}$ | Not B | $\overline{B}$ | |
| $F_{12}$ | A or not B | $A+\overline{B}$ | |
| $F_{13}$ | Not A | $\overline{A}$ | |
| $F_{14}$ | B or not A | $\overline{A}+B$ | |
| $F_{15}$ | One is not | $\overline{A\cdot B}=\overline{A}+\overline{B}$ | NAND |
| $F_{16}$ | Always | 1 | |

Logic operation

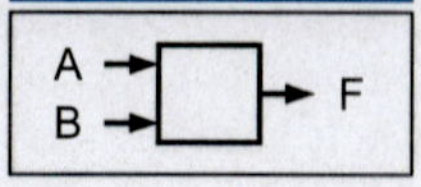

The Boolean equation is then:

$$F = A \cdot B + B \cdot \overline{C} \tag{9.3}$$

Logical expressions given in the form of Boolean equations can often be simplified. This is especially the case when the functions contain three or more input variables. The theorems of Boolean algebra shown in ◘ Fig. 9.9 are required for this simplification. The simplification is viable when the logic circuit is realized using discrete elements (relays, logic gates). The number of components can be reduced in this way.

The Boolean equation of a logical function cannot always be derived directly from the verbal formulation – particularly not for complex logical conditions. To carefully consider and check all input states, a value table is used as an aid. In this, all combinations of the input variable states are listed row by row. With only two possible states per input, the table has $2^n$ rows for n input variables. The state to be formed by combining the input signals is entered in a column for the output variable. The logical conditions for the value "1" of the output variables are also specified in the value table alongside those for the value "0".

◘ Figure 9.10 shows an example of how the Boolean expression for a logical function can be determined from the value table and then simplified using the theorems. Only output variable $F_1$ is considered initially. In this case, its state is a function of three input variables (A, B and C), so the value table has $2^3 = 8$ rows. It specifies the signal combinations of A, B and C where the output signal is or is not present. To arrive at the Boolean expression for the function of variable $F_1$, the input variable states of all rows with an output state "1" are expressed as AND functions. Logical OR functions (disjunctions) are applied to these expressions (rows). This Boolean expression is called "full disjunctive normal form". This expression can be minimized to the "simplified disjunctive normal form" by applying the theorems. To implement function $F_1$ using a logic network, the overhead in terms of logic elements can be reduced considerably via minimization, as explained above.

If the value table contains more rows with state "1" for the output variable than rows with state "0",

| | Theorem | | Continuation |
|---|---|---|---|
| 1 | $X + 1 = 1$ | 2 | $X + 0 = X$ |
| 1' | $X \cdot 0 = 0$ | 2' | $X \cdot 1 = X$ |
| 3 | $X + X = X$ | 4 | $\overline{(X)} = \overline{X}$ |
| 3' | $X \cdot X = X$ | 4' | $\overline{\overline{(X)}} = X$ |
| 5 | $X + Y = Y + X$ | 6 | $X + \overline{X} = 1$ |
| 5' | $X \cdot Y = Y \cdot X$ | 6' | $X \cdot \overline{X} = 0$ |
| 7 | $X + XY = X$ | 8 | $(X + \overline{Y})Y = XY$ |
| 7' | $X(X + Y) = X$ | 8' | $X\overline{Y} + Y = X + Y$ |

| | Continuation | Name |
|---|---|---|
| 9 | $X + Y + Z = (X + Y) + Z = X + (Y + Z)$ | Associative laws |
| 9' | $XYZ = (XY)Z = X(YZ)$ | |
| 10 | $XY + XZ = X(Y + Z)$ | Distributive laws |
| 10' | $(X + Y)(X + Z) = X + YZ$ | |
| 11 | $(X + Y)(Y + Z)(Z + \overline{X}) = (X + Y)(Z + \overline{X})$ | |
| 11' | $XY + YZ + Z\overline{X} = XY + \overline{X}Z$ | |
| 12 | $(X + Y)(\overline{X} + Z) = XZ + \overline{X}Y$ | |
| 13 | $\overline{(X + Y + Z + \ldots)} = \overline{X} \cdot \overline{Y} \cdot \overline{Z} \cdot \ldots$ | De Morgan |
| 13' | $\overline{X \cdot Y \cdot Z \cdot \ldots} = \overline{X} + \overline{Y} + \overline{Z} + \ldots$ | |

◘ Fig. 9.9 Theorems of Boolean algebra

| Inputs | | | Outputs | |
|---|---|---|---|---|
| A | B | C | $F_1$ | $F_2$ |
| 0 | 0 | 0 | 0 | 1 |
| 0 | 0 | 1 | 0 | 0 |
| 0 | 1 | 0 | 0 | 1 |
| 0 | 1 | 1 | 1 | 1 |
| 1 | 0 | 0 | 1 | 0 |
| 1 | 0 | 1 | 1 | 0 |
| 1 | 1 | 0 | 1 | 0 |
| 1 | 1 | 1 | 1 | 0 |

**Full disjunctive normal form for $F_1$**

$$F_1 = \overline{A}BC + A\overline{BC} + A\overline{B}C + AB\overline{C} + ABC$$

**Forming the simplified disjunctive normal form:**

$$\begin{aligned}
F_1 &= \overline{A}BC + A\overline{BC} + A\overline{B}C + AB\overline{C} + ABC && \text{Theorem:}\\
&= \overline{A}BC + A(\overline{BC} + \overline{B}C + B\overline{C} + BC) && (10)\\
&= \overline{A}BC + A\left[\overline{B}(\overline{C} + C) + B(\overline{C} + C)\right] && (10)\\
&= \overline{A}BC + A(\overline{B} \cdot 1 + B \cdot 1) && (6)\\
&= \overline{A}BC + A(\overline{B} + B) && (2')\\
&= \overline{A}BC + A\\
F_1 &= BC + A && (8')\\
\\
F_2 &= \overline{A}(\overline{C} + B)
\end{aligned}$$

◘ Fig. 9.10 Value table and disjunctive normal form

considerably more terms must be provided to yield the Boolean expression. The procedure can be different in this case—a Boolean expression is derived in disjunctive normal form only on the basis of rows having output variable "0". Then, the complete expression is negated to arrive at the final Boolean expression. The second output combination, $F_2$, is shown directly without further derivation.

### 9.1.4 Karnaugh Map

Another way to minimize Boolean expressions is to use a Karnaugh map (K-map). This is a graphical minimization method in which the major benefit is that applying the theorems (often very complex) is unnecessary.

A K-map is made up of fields positioned like a chessboard. Just like the value table contains $2^n$ rows for n input variables, the K-map contains $2^n$ fields. ◘ Figure 9.11 and ◘ Fig. 9.12 show K-maps for two and three input variables. Doubling the number of fields enables the map to be set up in each case for one input variable more. The number of input variables is in theory unlimited. K-maps with more than five input variables become too difficult to manage.

The input variables are spread around the edge of the matrix such that all fields are occupied by different combinations of all input variables, i.e., each input variable covers half of the $2^n$ fields directly and the other half with its negated value. To simplify understanding, the row numbers of the value table are entered in the associated fields of the K-map. So every field in the K-map corresponds to one row in the value table.

The assignment of input variables to the rows and columns of the map can now be condensed with clever grouping. Adjacent fields in rows and columns in a K-map differ by just one variable. This also applies when several K-maps are arranged next to each other in practical applications, something that is possible in all directions. ◘ Figure 9.13 shows a variant of the K-map for four input variables.

The graphical minimization of the function in ◘ Fig. 9.10 is shown as an example of applying the K-map. The value table is specified on the right in ◘ Fig. 9.12. Shown on the left is the associated K-map for the three input variables, A, B and C. First, the

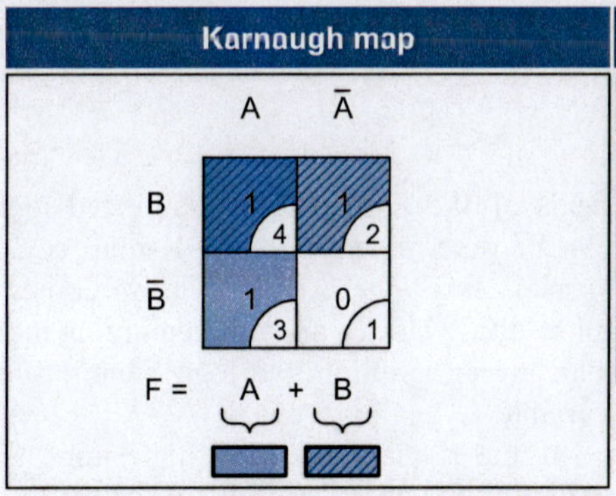


Value table

| No. | A | B | F=A+B |
|---|---|---|---|
| 1 | 0 | 0 | 0 |
| 2 | 0 | 1 | 1 |
| 3 | 1 | 0 | 1 |
| 4 | 1 | 1 | 1 |

◘ **Fig. 9.11** Karnaugh map for two input variables, A and B (using the OR function as an example)

◘ **Fig. 9.12** Karnaugh map for three input variables, A, B and C (using the $F_1$ function as an example)

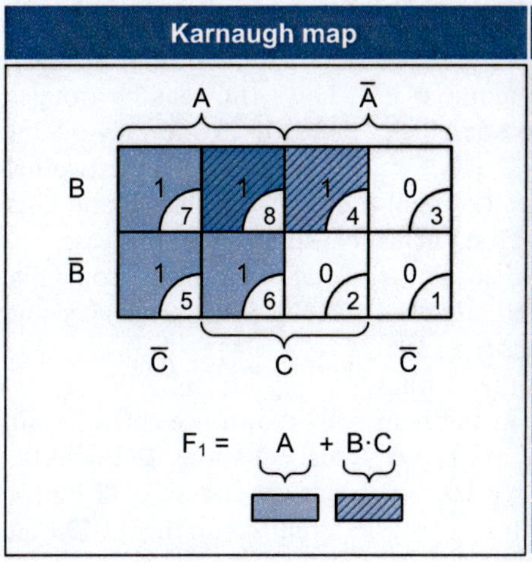


Value table

| No. | A | B | C | $F_1$ |
|---|---|---|---|---|
| 1 | 0 | 0 | 0 | 0 |
| 2 | 0 | 0 | 1 | 0 |
| 3 | 0 | 1 | 0 | 0 |
| 4 | 0 | 1 | 1 | 1 |
| 5 | 1 | 0 | 0 | 1 |
| 6 | 1 | 0 | 1 | 1 |
| 7 | 1 | 1 | 0 | 1 |
| 8 | 1 | 1 | 1 | 1 |

**Fig. 9.13** Karnaugh map for four input variables, A, B, C and D (using the $F_2$ function as an example)

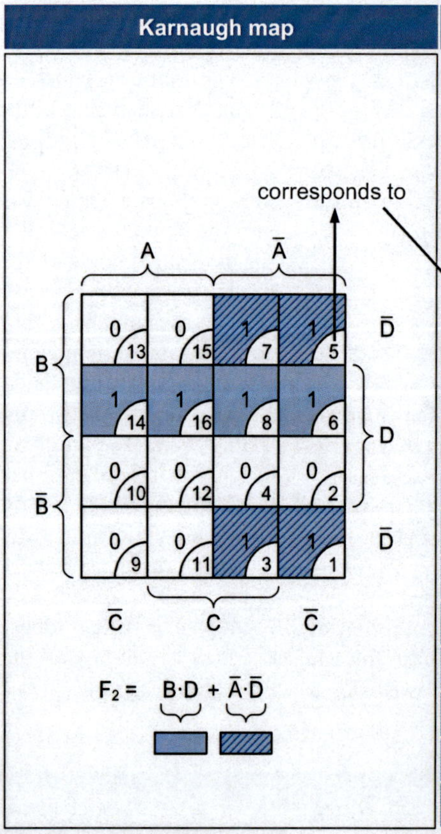


Value table

| No. | A | B | C | D | $F_2$ |
|---|---|---|---|---|---|
| 1 | 0 | 0 | 0 | 0 | 1 |
| 2 | 0 | 0 | 0 | 1 | 0 |
| 3 | 0 | 0 | 1 | 0 | 1 |
| 4 | 0 | 0 | 1 | 1 | 0 |
| 5 | 0 | 1 | 0 | 0 | 1 |
| 6 | 0 | 1 | 0 | 1 | 1 |
| 7 | 0 | 1 | 1 | 0 | 1 |
| 8 | 0 | 1 | 1 | 1 | 1 |
| 9 | 1 | 0 | 0 | 0 | 0 |
| 10 | 1 | 0 | 0 | 1 | 0 |
| 11 | 1 | 0 | 1 | 0 | 0 |
| 12 | 1 | 0 | 1 | 1 | 0 |
| 13 | 1 | 1 | 0 | 0 | 0 |
| 14 | 1 | 1 | 0 | 1 | 1 |
| 15 | 1 | 1 | 1 | 0 | 0 |
| 16 | 1 | 1 | 1 | 1 | 1 |

states of output variable $F_1$ ("0" or "1") from every row in the values table are transferred to the relevant field. As a result, the K-map has the same information as the value table.

For the minimization, join as many adjacent fields containing "1" as possible to make rectangular blocks containing 2, 4, 8..., i.e., $2^m$ (m = 1, 2, ..., n-1) fields. The location of every individual field is defined by the combination of n variables. If a block, for example, contains half of all the fields in the K-map (having the same information content), it can be defined by a single variable. If two adjacent fields having the same information content (1 or X = "Don't care") can be grouped into one block, only n-1 combinations are required to define the field positions; for four fields into one block, only n-2 variables are needed, etc.

Two basic rules must be observed to get the minimized logical function with the K-map. First, fields may be included in more than one block when blocks are formed. Overlapping blocks are permitted and even desired, for the minimization. Second, "concatenated blocks" need not necessarily border each other within a map. The four outer fields 1, 3, 5 and 7 in the map form one concatenated block, for example. As shown in Fig. 9.13, these fields can be grouped by $\overline{AD}$. The same applies to the outer columns and the four corner fields of the K-map. The minimized logical equation can be determined with the K-map considering these special cases. The overall solution comes from ORing all resulting blocks, and any remaining individual fields. This gives the minimized logic function for the output variable.

If the K-map is used to determine the minimized definition for output variable F for the two examples in Fig. 9.11. First, one block containing the left half of the fields can be formed. This means that this block is defined solely by variable A.

One field in the example below (Fig. 9.12, top row, third field from the left) has yet to be included. This ($\overline{A}BC$) can be grouped with its adjacent field, used already in the previous block formation, to form another block. These two fields are defined by the ANDing of variables B and C, meaning a definition with BC is sufficient in this case.

So all fields containing a "1" in the K-map in Fig. 9.12 are included and the logical function for $F_1$ is:

$$F_1 = A + B \cdot C \quad (9.4)$$

To show once again the significance of minimizing logical equations, Fig. 9.14 shows the operation for output variable $F_1$ in Fig. 9.10 as a logic circuit with different components. Figure 9.14 shows, on the right,

■ **Fig. 9.14** Logic circuit in different technical realizations

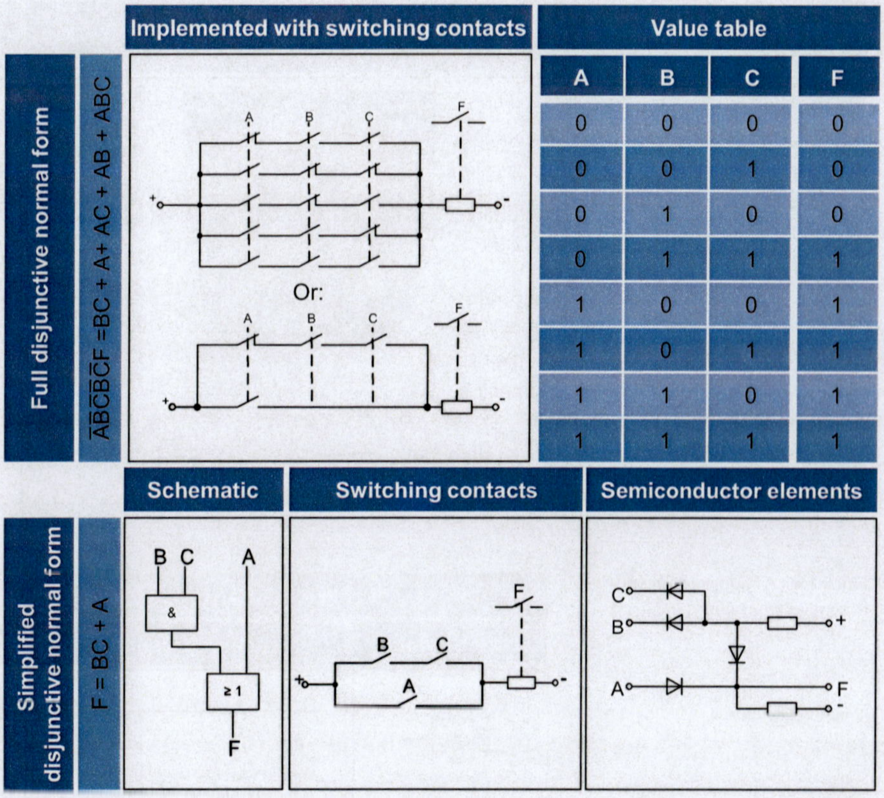


| A | B | C | F |
|---|---|---|---|
| 0 | 0 | 0 | 0 |
| 0 | 0 | 1 | 0 |
| 0 | 1 | 0 | 0 |
| 0 | 1 | 1 | 1 |
| 1 | 0 | 0 | 1 |
| 1 | 0 | 1 | 1 |
| 1 | 1 | 0 | 1 |
| 1 | 1 | 1 | 1 |

the value table and, on the left, the full disjunctive normal form as can be taken directly from the value table. The direct implementation of this equation using switching contacts and simplification by consolidating parallel switching contacts is shown to the left of the value table. The lower part shows the simplified disjunctive normal form with the associated wiring diagram and the corresponding implementations using switching contacts and semiconductor elements.

The minimization achieved can be seen most clearly from the circuit using switching contacts. Without minimization, 10 normally open contacts and five normally closed contacts (i.e., a total of 15 switching contacts, or 10 switches after the removal of evidently redundant contacts), instead of the three normally open contacts actually required, would be necessary to arrive at function $F_1$ from the value table.

## 9.2 Modules

### 9.2.1 Implementing the Basic Functions

The graphic symbols for the logic functions allow the creation of schematic diagrams for logic networks regardless of the nature of the technical implementation of the basic functions. The basic functions are available as modules (depending on the technology used) and can be interconnected to form complex logic networks as per the schematic diagrams. Different technologies can be used to implement logic networks depending on application area.

The simplest way to implement logic functions is using switching contacts. They represent binary signals, with appropriate electrical wiring, by the "Voltage present" and "No voltage present" states. Switching contacts are divided into make and break contacts. A break contact, that is, a contact closed in the idle state, represents inversion and is denoted by a bar above the variable name.

If the switching contacts are actuated by a solenoid coil, they are known as a relay. Relays enable one or more contacts to be actuated by an electrical control current over larger distances and at inaccessible places. With the right design, far greater currents can be switched by a small exciter current in the solenoid coil (power gain) using the contacts. ■ Figure 9.15 shows a picture of a relay and a schematic. The moving armature sits on the solenoid coil. When excitation current flows through the coil, a magnetic field is generated and the armature moves. This movement is transferred to the switching contacts via an actuating lever. The relay (photo) shown in ■ Fig. 9.15 has four changeover contacts, i.e., four contacts are opened and four contacts closed simultaneously. Interconnecting relays allow creation of logic networks of any size.

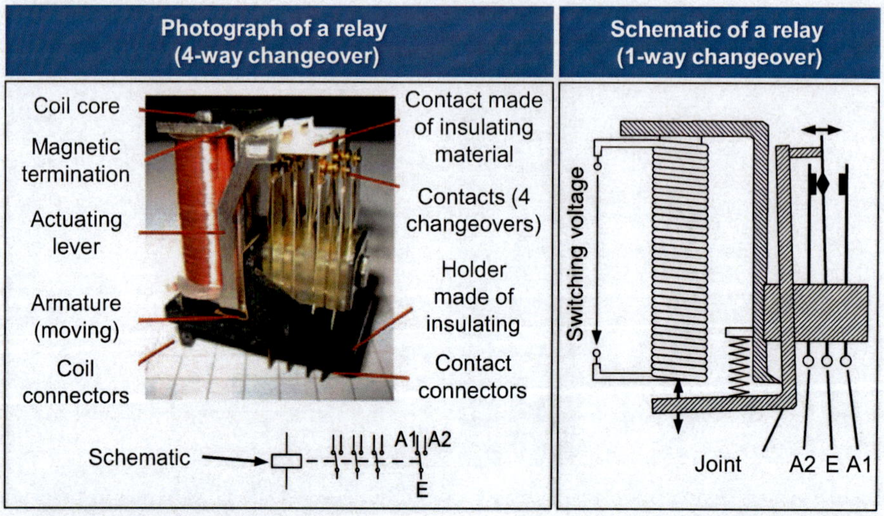


**Fig. 9.15** Picture and function of a relay



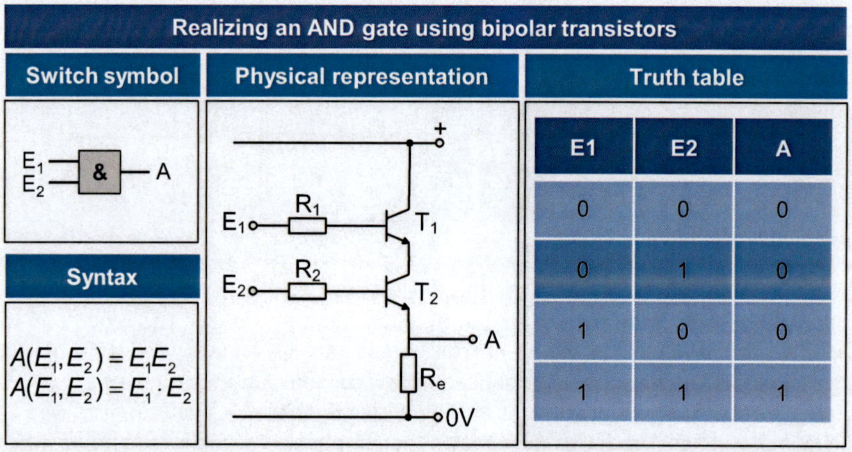


**Fig. 9.16** AND gate realized with bipolar transistors

Given the availability of smaller, faster and lower-cost electronic components for logic circuits, relays are rarely used nowadays. Relays for activating higher-power devices (motors, heating coils, etc.) are called contactors. They can have thermal overload protection, i.e., a contactor deactivates once the power limit is exceeded.

Logic devices using semiconductor technology were unavailable until the development of solid-state amplifier elements known as transistors (= transfer resistor). Like a relay, a transistor can also switch currents—using much lower control currents [DURC83]. While transistors may be analog components, i.e., they can continuously amplify voltages and currents, connecting multiple transistors appropriately means virtually digital behavior can be attained. These transistor circuits are excellently suited for use in logic arrays. For example, using suitable connections, logic gates AND, OR and NOT can be implemented in the simplest case and all other logical circuits can be realized using these.

Figure 9.17 shows a simple AND gate. Here, two transistors are connected in series and the output of the digital signal is realized as a voltage drop with emitter resistor $R_e$. If neither transistor, or only one of $T_1$ and $T_2$, is switched by applying a base current, no current flows through the emitter resistor. As a result, there is no voltage drop. Output A is connected to chassis ground and so virtually to 0 V or logic false. By applying a voltage at the two inputs, $E_1$ and $E_2$, the base current generated ensures that current flows through both transistors. Current consequently flows through the emitter resistor and at output A, there is a voltage drop of virtually the magnitude of the supply voltage. Logic true appears at the output.

In the same way as this functional principle, the two other logic gates can be explained as a basic illustration of the ways to implement digital or logic circuits. There are even more techniques.

Some examples of the technologies for interconnecting transistors and other components to form basic

**Fig. 9.17** Circuitry variants for realizing NOT and NAND elements

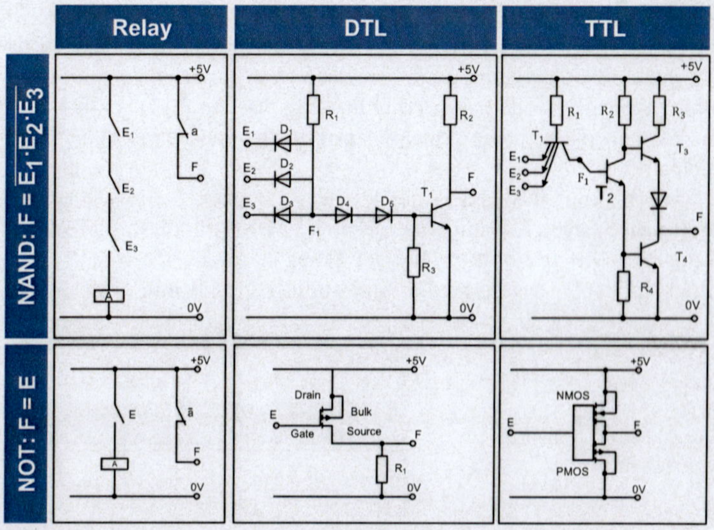


logic functions - DTL (Diode-Transistor Logic), TTL (Transistor-Transistor Logic) for bipolar circuits, MOSFET (Metal Oxide Semiconductor-Field Effect Transistor, NMOS, PMOS) and CMOS (Complementary MOS) for unipolar circuits. The difference between bipolar and unipolar is that, for bipolar elements, electrons and holes are involved in charge transfer, while for unipolar (field effect) elements, only one charge carrier type is responsible for current conduction. ◘ Figure 9.17 shows NOT and NAND elements in the different circuit technologies.

For Diode-Transistor Logic, shown in the middle of ◘ Fig. 9.17 at the top, transistor $T_1$ acts as a voltage switch, i.e., it is driven in the conductive state. Upstream of this is a diode AND element ($D_1$, $D_2$, $D_3$, $R_1$) and a diode OR element ($D_5$, $R_3$). Exactly 0 V is present at output $F_1$ of the diode AND element if one of the inputs $E_1$, $E_2$ or $E_3$ is at 0 V because, in this case, the relevant diode, $D_1$, $D_2$ or $D_3$, conducts while the other two diodes do not. Diode $D_4$ increases the blocking margin, so that a negligibly small current can flow through $D_4$ but cannot initiate transistor $T_1$ conducting. So signal $F_1 = E_1 \cdot E_2 \cdot E_3$ is applied to $F_1$. Output signal F is generated by inverting $F_1$ using the transistor. For Transistor-Transistor Logic, shown in the top right part of ◘ Fig. 9.17, a multiple-emitter transistor ($T_1$) with comparable function is used in place of the diodes $D_1$ to $D_4$. The transistor conducts and pulls the base from $T_2$ to 0 V as soon as at least one of the emitters is at 0 V. So, here too, the signal at point $F_1$ is $E_1 \cdot E_2 \cdot E_3$. While $T_2$ replaces diode $D_5$, $T_3$ and $T_4$ are used as transistors switching alternately to produce a higher output current.

A MOSFET comprises a semiconductor channel (source-drain) whose conductivity can be controlled with almost no power by a voltage applied between the gate and bulk (semiconductor substrate) connectors. A distinction is made between normally On (i.e., conductive with no voltage applied) and normally Off (i.e., only conductive when voltage is applied), and N-channel and P-channel types (polarity of the control voltage; specified by doping). Although MOSFETs can only seitch low currents in comparison to relays, they can do it at extremely high switching speeds. CMOS comprises two transistors (one P channel and one N-channel) that are connected such that the circuit output has either a conductive connection to the supply voltage or to ground.

## 9.2.2 Extended Functions

Control engineering tasks can often not be resolved directly using the basic functions. These are fabricated as function modules in the form of special circuitry [DURC83; WOLF77].

### 9.2.2.1 Flip-flop

The flip-flop (bistable multivibrator) is a component used to store a binary signal. It can be seen from the graphic symbol in ◘ Fig. 9.18 that a flip-flop has two inputs and two outputs.

Applying a signal to the set input S, the direct output Q is set as per the value table and held until the flip-flop is reset from the reset input R, regardless of what happens to the input S. The output $\overline{Q}$ has the complementary, i.e., the opposite, signal state of output $Q$ at all times. This circuit is also called an RS flip-flop from the input signal names.

The circuit diagram in ◘ Fig. 9.18 shows how a flip-flop can be made from two NOR function elements. The output of one logic element is connected to the input of the other logic element, thereby creating the memory behavior.

### 9.2.2.2 Edge-triggered Flip-flops

Switching technology often needs storage elements that sample and hold the input information at a particular time. A positive edge-triggered D flip-flop (◘ Fig. 9.19) is explained here as one example of such an element.

When input T (clock) changes its state from 0 to 1 (positive edge), the information at input D (data) is transferred to the output. For all other states of the clock input T (negative edge and static states 0 and 1), output $Q$ or $\overline{Q}$ retains its logical value regardless of any change at input D. Refer to the value table in ◘ Fig. 9.19.

### 9.2.2.3 1:2 Divider

The D flip-flop forms the basis for many clock-controlled circuits. One common application is the 1:2 divider implemented by feeding back the inverted output Q to the D input (◘ Fig. 9.20).

The pulse graph shows that the output is set/reset on every positive clock edge. Consequently, the frequency of the input clock is divided in a ratio of 1:2.

9

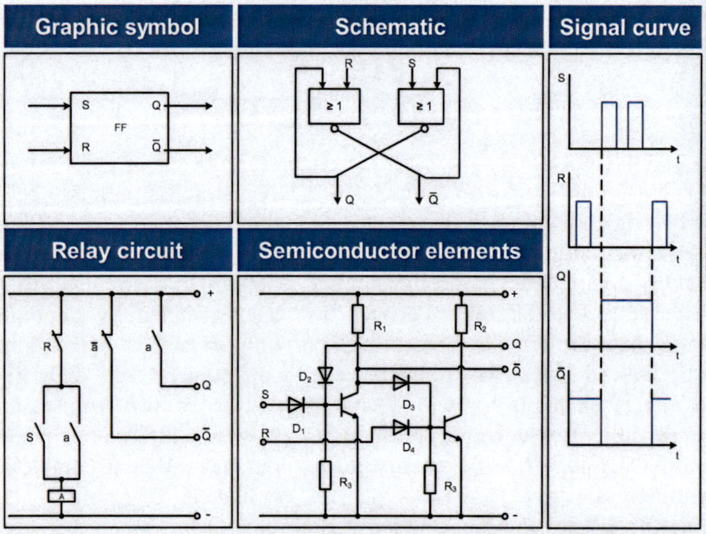


Value table

| S | R | Q | $\overline{Q}$ |
|---|---|---|---|
| 1 | 0 | 1 | 0 |
| 0 | 1 | 0 | 1 |
| 0 | 0 | Q | $\overline{Q}$ |
| 1 | 1 | Not def. | |

◘ Fig. 9.18 Flip-flop memory using different technology implementations

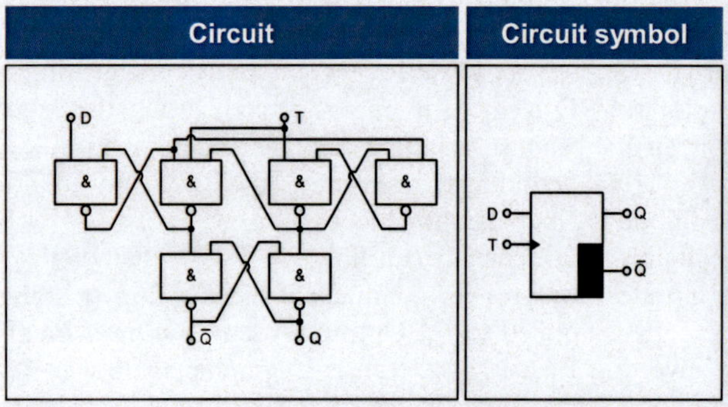


Value table

| T | D | Q | $\overline{Q}$ |
|---|---|---|---|
| 0 | X | Q | $\overline{Q}$ |
| ⎍ (rising edge) | 0 | 0 | 1 |
| ⎍ (rising edge) | 1 | 1 | 0 |
| 1 | X | Q | $\overline{Q}$ |
| ⎍ (falling edge) | X | Q | $\overline{Q}$ |

X = Don't care

◘ Fig. 9.19 D flip-flop

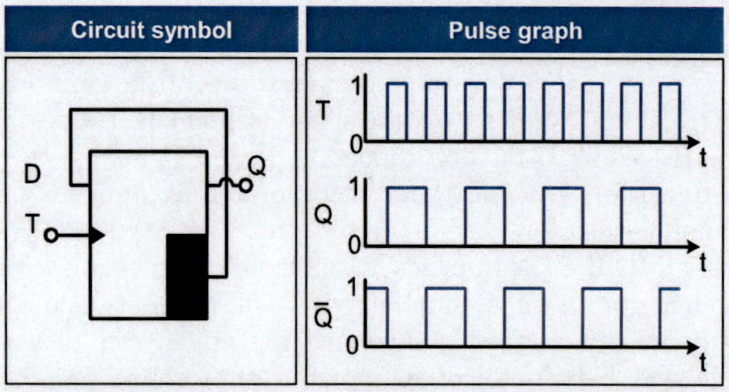


◘ Fig. 9.20 1:2 divider

#### 9.2.2.4 Binary Counters

When multiple trailing-edge-controlled flip-flops are coupled (◘ Fig. 9.21), every flip-flip is clocked by the preceding one except the first one, which uses the counting input clock. At the output, every stage halves the frequency of its input clock as the pulse graph shows. At outputs $Q_0$ to $Q_3$, the number of input clocks can be read as a binary number. When n = 4 stages, $2^4 = 16$ different binary numbers can be realized, corresponding to decimal numbers 0 to 15.

Using additional static set and reset inputs, the counter stages can be set to a defined "0" or "1" so that the count process can begin at a preselectable start value. Increasing the number of stages enables the counting range to be extended as needed.

When complementary outputs $\overline{Q}_0$ to $\overline{Q}_3$ are used for the count display, the arrangement can be used as a decrementer with the same counting capacity.

The functionality of information processing devices based on microprocessors or PCs can be defined by programming (software). Accordingly, it is possible to simulate a counter with software functions. Hardware-based counters can normally process a higher clock rate and they are preferred for use. The same applies to arithmetic modules.

#### 9.2.2.5 Half Adder

Besides storage and counting modules, modules that perform mathematical operations are also required in controllers. ◘ Figure 9.22 shows a circuit used to add

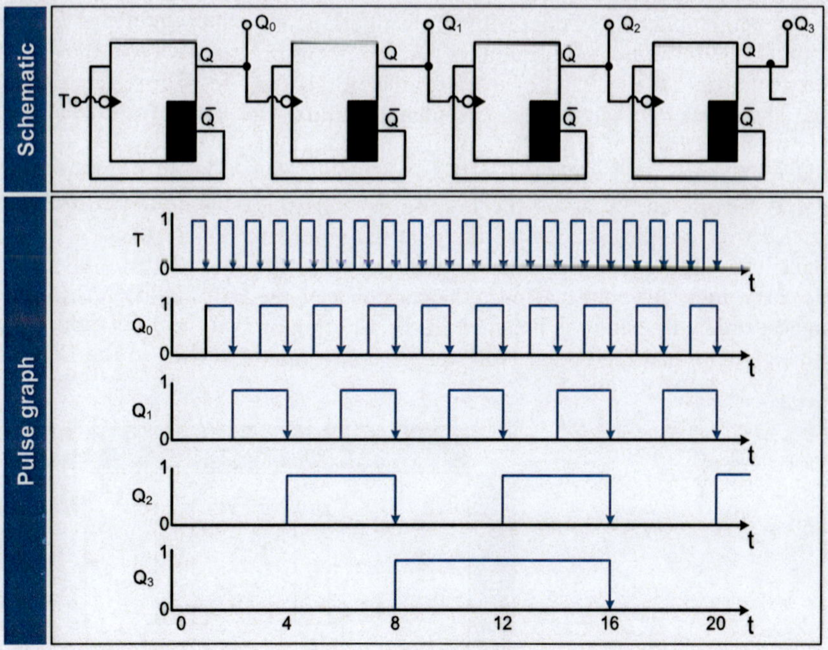


◘ **Fig. 9.21** Trailing-edge-controlled binary counter

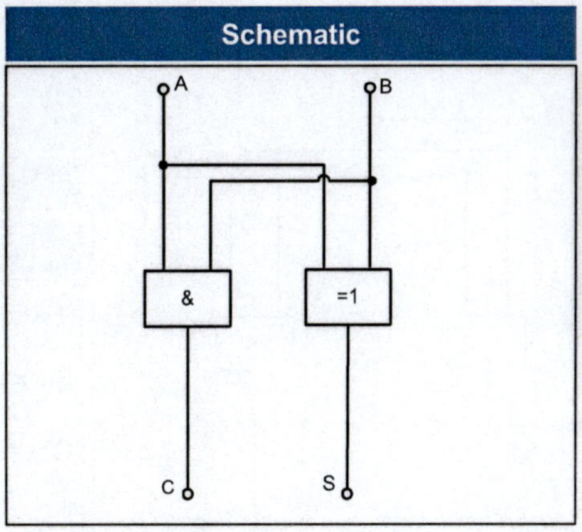


**Value table**

| A | B | S | C |
|---|---|---|---|
| 0 | 0 | 0 | 0 |
| 1 | 0 | 1 | 0 |
| 0 | 1 | 1 | 0 |
| 1 | 1 | 0 | 1 |

**Disjunctive normal form**

$S = A \cdot \overline{B} + \overline{A} \cdot B$

$C = A \cdot B$

◘ **Fig. 9.22** Design of a simple adder (half adder)

two 2-digit binary numbers. The value table specifies how this occurs and when there is a carryover to the next highest digit position. The circuit diagram is based upon the disjunctive normal form (► Sect. 9.1.3). This logic network is only suitable for the addition of one binary digit position. For multi-digit binary numbers, the carryover from the lower-ordered position must be included.

### 9.2.2.6 Full Adders and Accumulators

A full adder is used to add the nth digits, $A_n$ and $B_n$, of two binary numbers, A and B, taking into account carryover $C_{n-1}$ from the lower-ordered digit. The value table in ◘ Fig. 9.23 shows the link of inputs $A_n$, $B_n$ and $C_{n-1}$ to outputs $S_n$ and $C_n$.

The full adder is made up of two half adders. The first half adder totals digit positions $A_n$ and $B_n$, the second factors in carryover $C_{n-1}$.

By combining a half adder with n full adders, an accumulator can be created for adding two (n + 1)-bit binary numbers (◘ Fig. 9.37). The half adder totals the two lower-ordered digits of the two numbers and carries over the result to input $C_{n-1}$ of the following full adder. One such accumulator is called the ripple carry adder because the total of the lower-order digits is carried over for the addition of the next higher-order digit. The addition of two binary numbers by a ripple carry adder is completed once the carryover of the highest-ordered position is included in the addition of the highest-valued full adder. However, an unfavorable combination of two binary numbers might mean the total of the lower-ordered position has a bearing on the following sub-additions, and the calculation of the overall result of the addition is delayed because carryover $C_{n-1}$ needs to be factored in again for every sub-addition.

However, as calculations must be deterministic in computer architectures to guarantee correctness, using ripple carry adders is unsuitable because the constellation of the binary number digits has a crucial effect on the duration of an addition operation. The serial adder in ◘ Fig. 9.37 takes a deterministic approach. A serial adder essentially comprises three n-bit long shift registers, a full adder and 1 bit of storage, realized as a positive-edge-triggered D flip-flop. A shift register is a memory, n bits in length, the contents of which can be shifted bit by bit left or right as required. Two shift registers are used as operand registers to load the two binary numbers. The subsequent result of the addition is written to a third shift register.

Before the actual addition starts, the operand registers are loaded with the relevant operands and the result register is cleared (so it contains a value of 0). The D flip-flop contains the value 0. The serial adder now performs a clocked 1-bit addition from the low-ordered to the highest-ordered bit. On every clock cycle, the total of the addition is written to the result register, the contents of the two operand registers are shifted right by one position and the carryover is cached in the D flip-flop. For the next addition in the

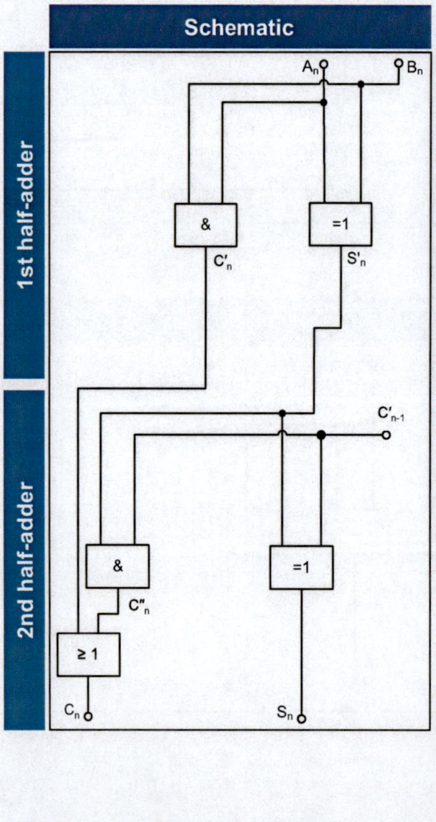


**Value table**

| $A_n$ | $B_n$ | $C_{n-1}$ | $S_n$ | $C_n$ |
|---|---|---|---|---|
| 0 | 0 | 0 | 0 | 0 |
| 1 | 0 | 0 | 1 | 0 |
| 0 | 1 | 0 | 1 | 0 |
| 1 | 1 | 0 | 0 | 1 |
| 0 | 0 | 1 | 1 | 0 |
| 1 | 0 | 1 | 0 | 1 |
| 0 | 1 | 1 | 0 | 1 |
| 1 | 1 | 1 | 1 | 1 |

**Formulas**

$$S'_n = A_n \cdot \overline{B_n} + \overline{A_n} \cdot B_n \qquad C'_n = A_n \cdot B_n$$

$$S_n = \overline{C_{n-1}} \cdot S'_n + C_{n-1} \cdot \overline{S'_n}$$

$$C''_n = S'_n \cdot C_{n-1}$$

$$C_n = C'_n + C''_n$$

◘ **Fig. 9.23** Full adder made of two half adders

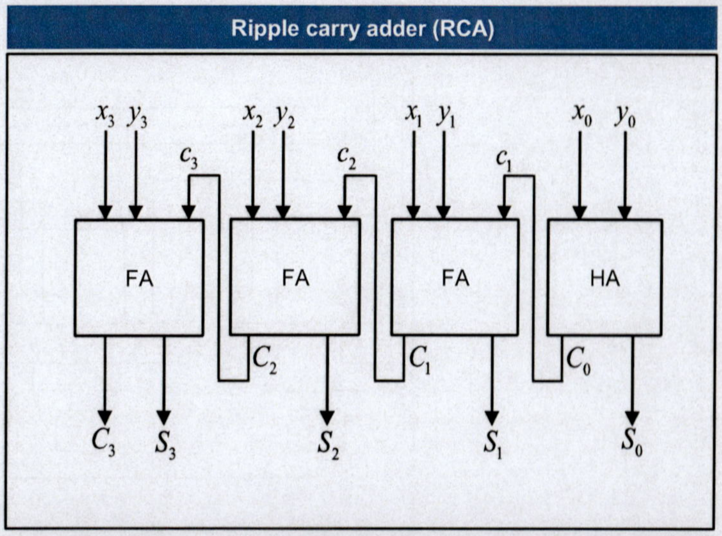


**Worst case of an addition using RCA**

| | | | | | |
|---|---|---|---|---|---|
| $x$ | 0 | 1 | 1 | 1 | 1 |
| $y$ | 0 | 0 | 0 | 0 | 1 |
| $C_3S$ | 1 | 0 | 0 | 0 | 0 |

**Fig. 9.24** Simplified ripple carry adder

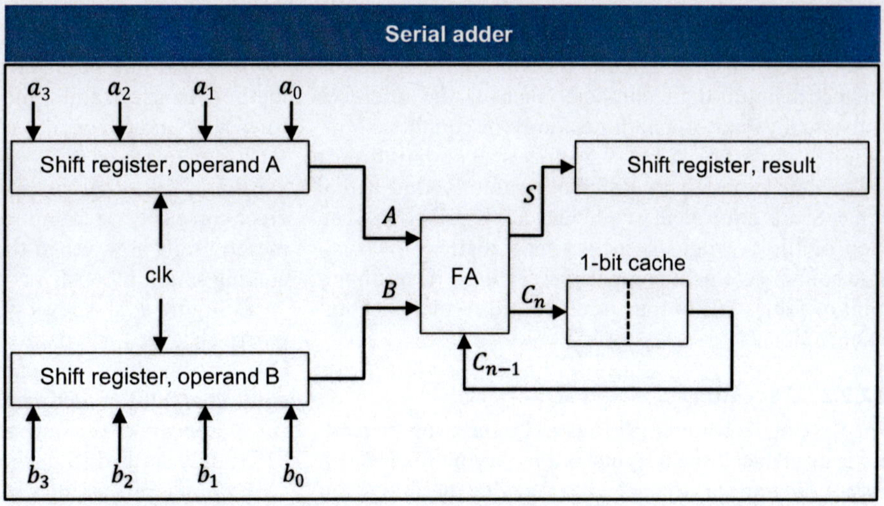


**Fig. 9.25** Simplified serial accumulator

next clock cycle, the current value in the D flip-flop is factored into the addition as carryover $C_{n-1}$. This process repeats until all bits of both operands have been processed. Because the whole process is deterministic using clock cycles, a result is available after n + 1 steps for an addition of two n-bit binary numbers – one clock cycle for loading the operands and n cycles for the actual addition.

### 9.2.2.7 Comparator

When processing information, it is often required to compare the numerical values of two signals. Figure 9.26 shows the circuit diagram and value table for the "Equivalence" (EXNOR) logic function. Output signal F has state "1" when output signals A and B simultaneously have state "1" or "0", i.e., when the signals are equivalent.

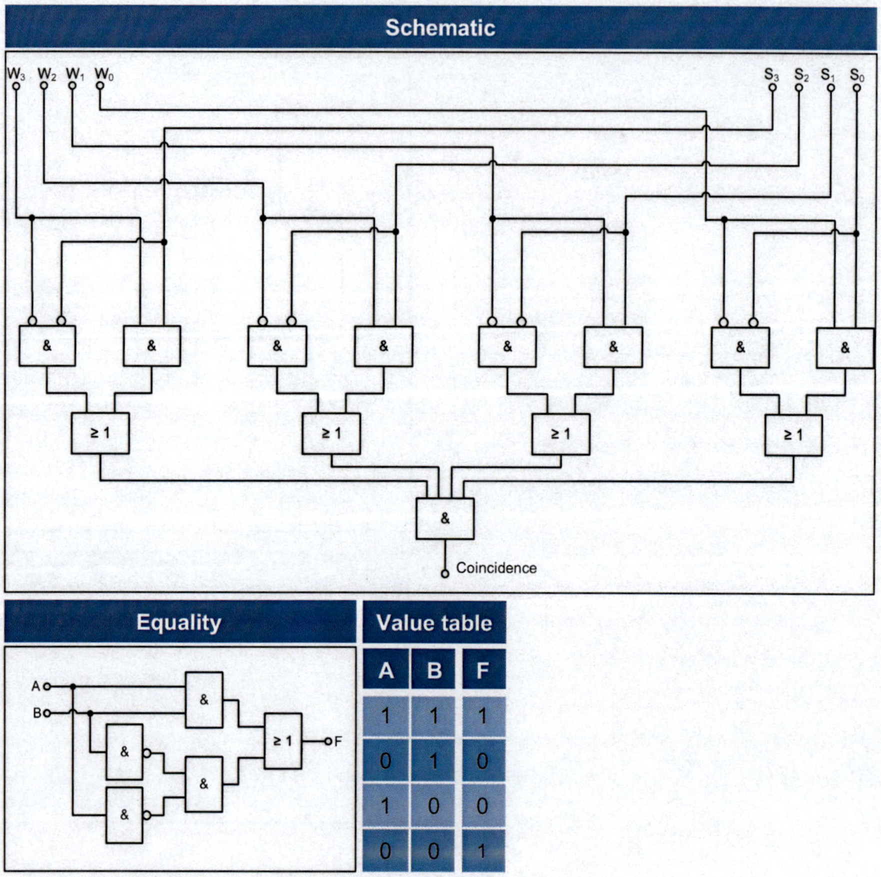


| A | B | F |
|---|---|---|
| 1 | 1 | 1 |
| 0 | 1 | 0 |
| 1 | 0 | 0 |
| 0 | 0 | 1 |

**Fig. 9.26** Binary comparison of reference and actual values

Coupling multiple circuits gives a comparator that can check multi-digit numbers (such as the reference and actual values of a path position) for equality.

The top part of Fig. 9.26 shows a comparator for 4-digit binary numbers. Reference value W and actual value S are equivalent (coincidence signal "1") when each of the $W_i$ digits coincides with digit $S_i$, $i=0\ldots3$. One equality element compares every digit position i (a total of four). The outputs are ANDed to make a coincidence signal.

#### 9.2.2.8 Decoder

For the binary counter (Fig. 9.21), the counter reading is available at the outputs as a binary number. Figure 9.27 shows a decoder, that decodes the 4-digit binary numbers and can drive a decimal display. The value table (Fig. 9.27) compares decimal digits 0 to 9 with the corresponding binary numbers.

#### 9.2.2.9 Parity Checker

Increasing the level of data integrity is the aim for every form of data transfer, be it between different computers or within one computer. Detecting errors in transmissions is an important step. A very simple method to ascertain some transmission errors is also to specify and check a parity bit. The parity bit is then set for example if an even number of bits in the data word are "1", otherwise it is "0". After the transfer, an error can easily be found by testing this condition. This method fails however in the rare case of two errors occurring simultaneously.

Figure 9.28 shows the simplified diagram of a parity checker for 8 data bits. This circuit is also available as a module. The parity checker comprises interconnected XOR functions (exclusive OR, antivalence). A "1" appears at the output of an XOR module when a "1" is only applied to one input.

Example processing of a data byte (8 bits) by a parity checker is shown in Fig. 9.28. A toggle can be used to select when an even or uneven number of "1" bits are checked for every data word. The parity bit is 0 in the example shown (assuming error-free transmission) because even parity is selected and there is already an even number of "1" bits in the data word. Accordingly, 0 is present at the error output, i.e., no errors have been detected.

**Fig. 9.27** Decoding decimal digits from the 8 4 2 1 code

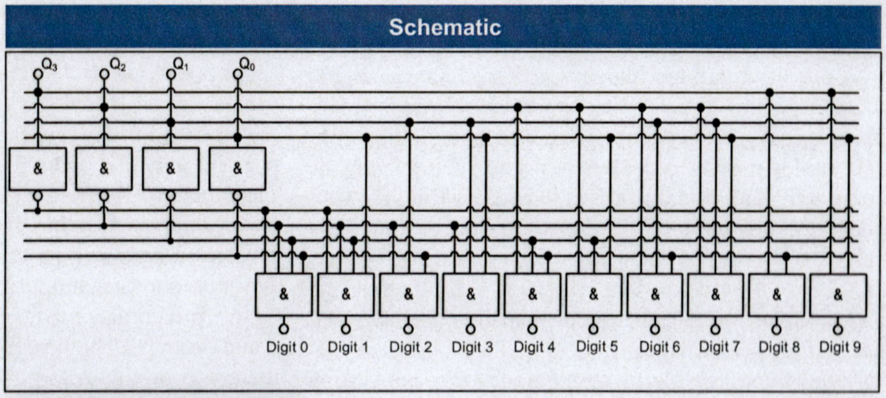


**Value table**

| $Q_3$ | $Q_2$ | $Q_1$ | $Q_0$ | Digit |
|---|---|---|---|---|
| 0 | 0 | 0 | 0 | 0 |
| 0 | 0 | 0 | 1 | 1 |
| 0 | 0 | 1 | 0 | 2 |
| 0 | 0 | 1 | 1 | 3 |
| 0 | 1 | 0 | 0 | 4 |
| 0 | 1 | 0 | 1 | 5 |
| 0 | 1 | 1 | 0 | 6 |
| 0 | 1 | 1 | 1 | 7 |
| 1 | 0 | 0 | 0 | 8 |
| 1 | 0 | 0 | 1 | 9 |

**Fig. 9.28** Simplified diagram of a parity checker

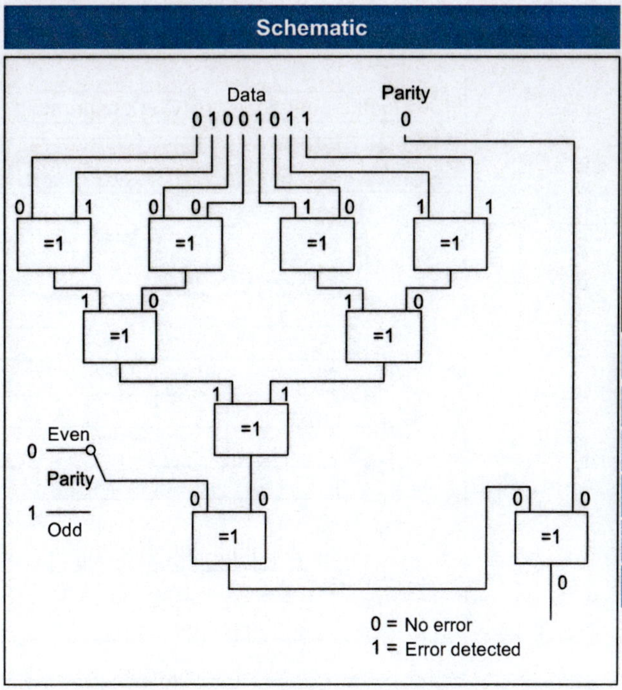


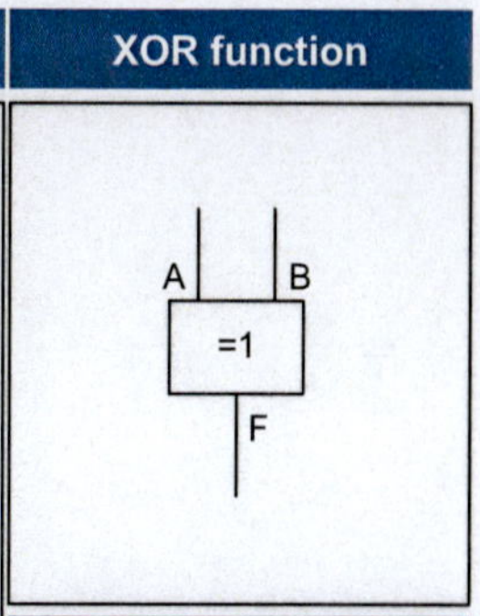


**Value table**

| A | B | F |
|---|---|---|
| 0 | 0 | 0 |
| 0 | 1 | 1 |
| 1 | 0 | 1 |
| 1 | 1 | 0 |

### 9.2.2.10 A/D Converter

For it to be possible for measured values to be handled by a process computer, they must be available in digital form. However, as many sensors provide analog signals, process computers generally also have circuits that convert analog signals to the representation suitable for the computer. These circuits are called analog/digital (A/D) converters. In addition to the resolution of these circuits, determined by the width of the result word generated (e.g., 14-bit converter: $2^{14} = 16{,}384$ steps across the measurement range), the circuit realization is also important for performance.

The most well-known methods for converting analog to digital signals are [ECKL90]:

- Voltage/frequency conversion
- Ramp method
- Successive approximation
- Parallel conversion
- Semi-parallel conversion
- Sigma-delta converter

9

A voltage/frequency converter provides a frequency corresponding to the input signal at its output. This is realized with charge/discharge processes. A capacitor is charged by the voltage to be measured until the voltage has exceeded the threshold voltage on a comparator, and activates it. This activation starts discharging the capacitor and the output of the comparator keeps switching to "1" until the voltage of the capacitor has fallen below the threshold voltage. The higher the measuring voltage, the quicker the capacitor is recharged and the faster the pulses follow each other at the capacitor output. Since computers cannot process frequencies directly but rather must analyze them with prolonged counting processes, this form of conversion is not used in the computing field.

A distinction is made in the ramp method between 1, 2 and 4 ramps. The benefits of the last two methods are mainly in attainable accuracy. ◘ Figure 9.29 shows the circuit diagram of an A/D converter using the 1-ramp method. In this method, the voltage of a sawtooth generator is continually compared in the converter to the measuring voltage at comparator K1. K1 and the additional comparison result K2 of the sawtooth voltage with the zero potential are connected to an AND gate. Output Y here is processed by another AND gate with the clock signal of an internal clock generator. The higher the input voltage, the longer the clock pulses of the clock generator are connected to the counter input via the AND gate. This means, after every voltage ramp generated, a digital value proportional to the input voltage is available at the counter output.

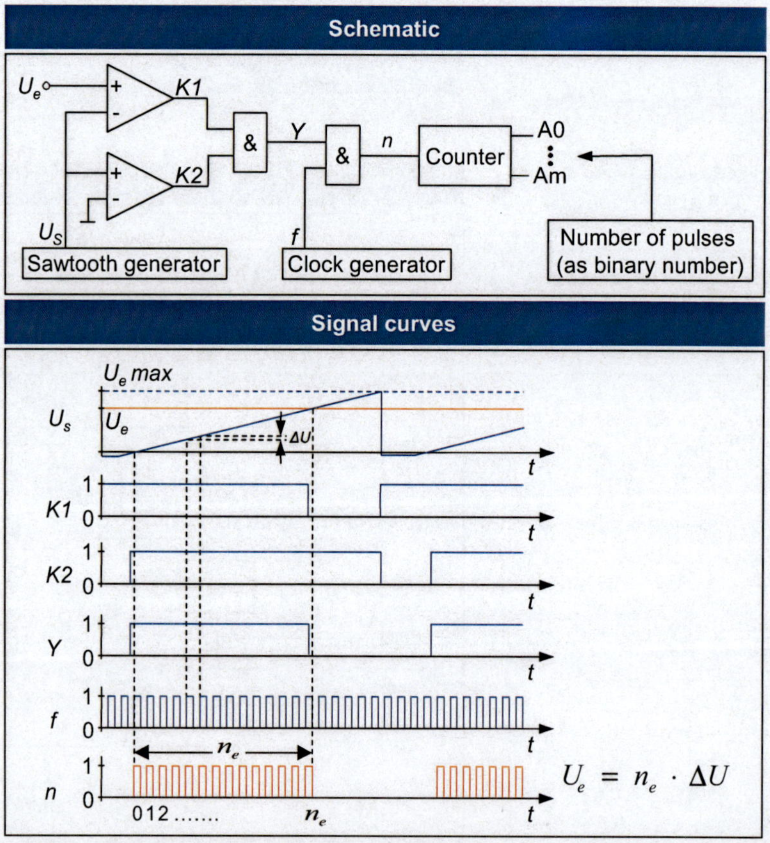


◘ **Fig. 9.29** A/D converter using the 1-ramp method

In the 2-ramp method, a reference voltage and the input voltage are integrated. From this, the result becomes more independent of fluctuations in both the clock frequency and voltage ramp. Greater improvement can be achieved using the 4-ramp method. This is essentially two 2-ramp measurements in sequence.

In the successive approximation method, an attempt is made to generate a voltage equal to the input voltage using a successive approximation process. A comparator and a digital/analog converter are required for this. Using this method with a 14-bit A/D converter, exactly 12 measurement steps are required for the result. Starting from half the maximum input voltage, an attempt is made to generate a voltage identical to the input voltage by increasing the reference voltage by smaller and smaller amounts. A check is performed after every step on whether the reference voltage is more or less than the input voltage. If it is more, the last addition is reset and the next-smallest voltage amount is used to continue. A bit pattern that is proportional to the magnitude of the input voltage is derived from the additions performed and reset. This method is extremely quick, with measuring times of only 3 µs per measurement for 14-bit resolution, for example.

Parallel converters attain the highest conversion rate. In terms of circuitry, they are considerably more complex than the methods already mentioned. For parallel converters, a comparator with a corresponding reference voltage is required for every possible voltage state. The measurement result is available almost immediately after the voltage is applied but the complexity doubles with every bit of additional resolution (8 bits $=2^8-1=255$ comparators, 14 bits $=16{,}383$ comparators). To reduce this complexity, two parallel converters for n/2 bits are used as semi-parallel converters for n bits. In the first step, the input signal is routed via the first parallel converter. The most significant bits represent the result of the conversion. A digital to analog converter converts this result to an analog voltage that is subtracted from the input voltage. The remaining differential voltage is amplified by a factor of $2^{n/2}$ and routed to the second parallel converter. Its result is the n/2 least significant bits of the result. This means $2 \cdot 2^{n/2}$ instead of $2^n$ comparators are required for an n-bit converter.

Sigma-delta A/D converters have been known since the 1960s. The availability of appropriate semiconductor technology since the 1990s with high integration density and speed means they are enjoying greater practical significance. Sigma-delta converters use the oversampling method to attain high resolution with simple components. However, before the function of these converters is presented in detail, the principle is explained using an example [BLAC99].

The main factor influencing the resolution of an A/D converter is quantization noise. If voltage level U is sampled multiple times in sequence, the measured values differ due to this quantization noise. The effective resolution of an A/D converter in bits is thus defined such that quantization noise is below 1/2 the LSB (least significant bit).

As shown in ◘ Fig. 9.30 at the bottom, quantization noise for multiple samples gives rise to a distribution of the measured values around the actual value. By then averaging, the actual value can be determined more accurately despite this noise. This means the final result is more accurate than the resolution of the converter used.

The top part of ◘ Fig. 9.30 shows how this principle is realized for sigma-delta converters. It essentially comprises an integrator, a comparator (corresponds to a 1-bit A/D converter), and a 1-bit D/A converter (DAC). The output signal of the DAC is subtracted from the input signal. The remaining signal is integrated and converted to a 1-bit data word (0 or 1) by the comparator. The resulting bit is fed back to the input via the DAC where it is again subtracted from the current input signal.

The process with the feedback loop is performed at a very high oversampling rate. The resulting bit sequence is then filtered digitally and decimated, i.e., converted to a signal with a higher bit resolution but a lower data rate. The density of the logical ones in the output signal is proportional to the input signal.

The noise energy is distributed over a broader frequency spectrum by oversampling (◘ Fig. 9.30 on right). The subsequent filtering and decimation mean however that only a small part of the noise is in the usable frequency band. So the resolution of the converter in the usable frequency band is far higher. Only a simple anti-aliasing filter is required (as compared to other A/D converter architectures).

The downside of a sigma-delta converter is the throughput delay. This is considerably longer than for other converter types. But it does represent an affordable, energy-efficient and highly accurate solution in the low-frequency range (such as audio technology). The transient response of the digital filter means it is more suited to continuous measurements than for multiplexed signals. This is because, for multiplexed signals (two inputs are read alternately by an A/D converter), there must be a delay time of at least one decimation period between two conversions.

#### 9.2.2.11 D/A Converter

Generating control variables with process computers and implementing certain analog–digital converters requires converting digital signals to analog signals. There are also a number of methods here:

- Parallel method
- Successive approximation
- R-2R network
- Counting method

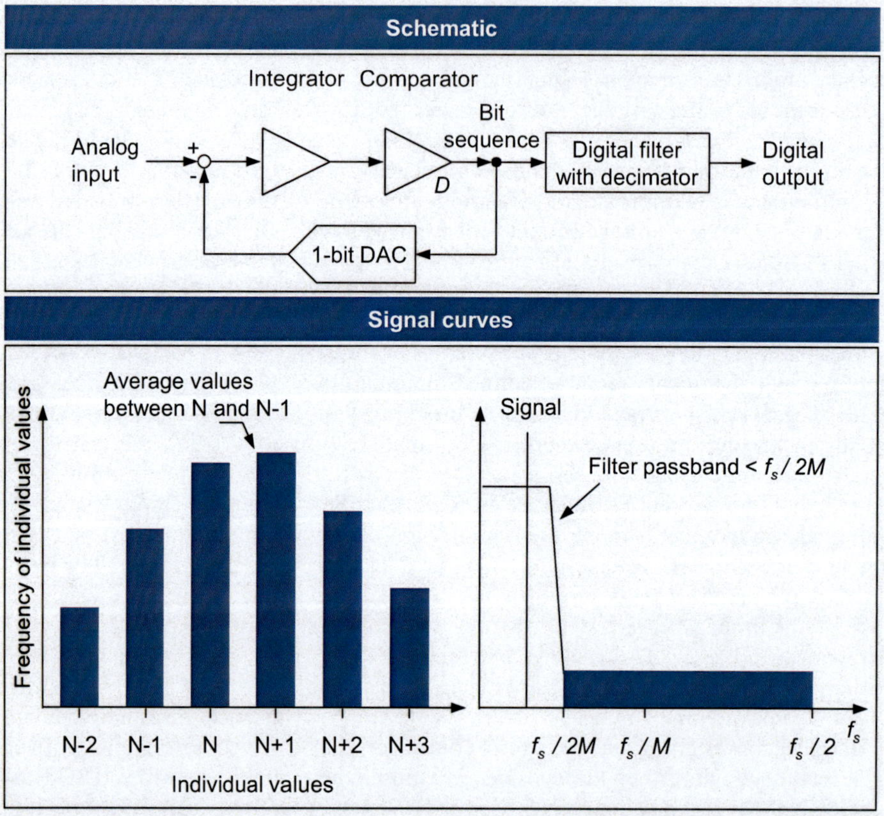


■ **Fig. 9.30** Sigma-delta converter

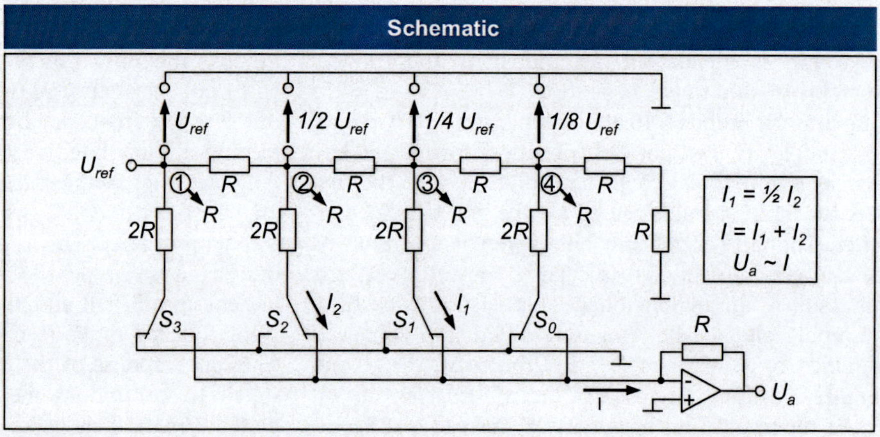


■ **Fig. 9.31** Digital/analog converter in the form of a standard R-2R network

The parallel method entails the greatest circuitry complexity here, too. Using a resistor network, every possible output voltage can be applied to the output of the converter using a separate switch. Which voltage corresponds to the digital information applied, and so which switch should connect, are determined by a "1-of-n decoder". This type of digital network has n inputs and $2^n$ outputs. Only that output corresponding to the input bit pattern is connected.

In the successive approximation method, one switch is required for every bit. Appropriately weighted resistors are connected to the output via this switch. So, a voltage corresponding to the significance of the respective bit is applied to the output by way of each resistor. The sum of the different voltages then corresponds to the bit pattern created. An R-2R network realizes the weighting of the levels using a continuous voltage divider. As shown in ■ Fig. 9.31, a voltage that is one half the preceding voltage is present at every network node (here, 1 to 4) because the total resistance of the network is exactly R as seen from every node. This means the voltage from one node to the next splits exactly in a ratio of R/2R. The operational amplifier is configured such that it converts current I to voltage $U_a$ that is proportional to the current. Assuming an ideal op-amp, the negative input is virtually at ground potential meaning the current can always flow to chassis ground regardless of the settings of switches $S_0$ to $S_4$. Consequently, the currents flowing through the switches are divided by 2, just like the voltages at the nodes.

The counting method is used to convert a frequency to an output voltage. For this, a switch is closed for a certain time on every rising edge at the input slowly charging a capacitor. The voltage at the output changes by how often this charge process occurs, thus as a function of the input frequency. More frequent switching results in higher output voltages. The downside of this method is that there is never a constant output voltage. Instead, it generates a voltage curve oscillating around the desired mean value. Suppressing this ripple requires additional circuitry.

### 9.2.3 Integrated Circuits

Semiconductor circuits are generally implemented as integrated circuits (ICs). Over an area of just a few square millimeters, they contain a large number of components and are manufactured for circuits required in large volumes. State-of-the-art semiconductor technology allows the placement of very many components on the smallest footprint in ICs. In VLSI circuits (Very Large Scale Integration), MOSFETs mostly serve as switch elements.

The heat losses that occur represent one obstacle to high integration densities. If the requirement is to place millions of switch elements on a chip, each element must have only a low power consumption. This explains the particular suitability of MOSFETs for very large-scale integrated circuits.

The number of function elements realizable on one chip has increased dramatically over recent decades. In addition to increasing the chip surface itself, this stems mainly from developments in the production technology for silicon chips. While, in 1960, the smallest structures to be fabricated were about 20 µm in width, now values of 25 nm or less are being achieved. Processors today, for example, are using 14 to 20 nm technology. Experts predict that processors with feature sizes even down to 7 nm are possible [STIL15]. With this technology, individual processor cores reach internal clock frequencies of up to 4 GHz.

Embedded systems are used alongside generally available processors. In the context of integrated circuits, an embedded system is essentially a microchip comprising a collection of different electronic assemblies performing various functions and contributing towards overall system functionality. Microcontrollers commonly used in this context comprise, in addition to a CPU and RAM, D/A converters, flash memory, ROM, bus interfaces, digital inputs and outputs and other special microchips. CPUs used in microcontrollers have a data word width from 8 to 32 bits and a clock rate between 1 and 32 MHz (using quartz oscillators). RAM is typically 128 bytes to 8 megabytes in size. In addition to RAM, ROM is also used to correctly perform important steps necessary for analyzing sensor information and activating actuators.

In contrast to volatile "Random Access Memory", data in "Read-Only Memory" remains stored permanently even after the voltage supply is switched off. The contents of ROM are hard-wired, and so can only be read and not written to after the fabrication process. In contrast, data can be rewritten to RAM dynamically. ROM is barely used nowadays because it has been replaced by EEPROM (Electrically Erasable Programmable Read-Only Memory). This can also be rewritten, just like RAM. One difference lies in the ability to permanently store data after power is switched off. This applies to EEPROM, but not to RAM. Another difference is the speed of write operations. While RAM is designed for fast read and write cycles, EEPROM needs more time to store data.

Application-specific integrated circuits (ASICs) are also seeing increasing use in the development of new systems and devices. In IC fabrication, the electronic functions designed by the circuit designer are converted to geometric structures on the semiconductor material. Several high-precision photolithographic steps are required for this. To reduce fabrication costs, components (ASICs) are offered where not all fabrication steps are performed in accordance with customer specifications, instead, only wiring of the base elements, for example. It is possible to put together an IC from ready-to-use library elements [HAGE93; PALO13]. ASICs are divided into design forms *Full-Custom, Semi-Custom* and *programmable logic* [PALO13].

For a *Full-Custom* ASIC design, the entire chip design, the integration density, the power loss, etc. are optimized for the respective application. This design form is associated with the greatest cost in view of the high development overhead. A *Semi-Custom* design allows the use of standard ICs having custom connectivity to implement the application. In contrast, *programmable logic* ASICs are completely standardized chips. FPGAs (Field Programmable Gate Arrays) and PLDs (Programmable Logic Devices) with internal logic circuits that can be programmed are used for the custom configuration. Chips of this type only differ in configuration and speed [PALO13].

A *programmable logic* ASIC essentially comprises a grid structure, similar to a 2-dimensional matrix with n horizontal and m vertical inputs and outputs. The grid fields each contain one vertical and one horizontal input, x and y, and one vertical and one horizontal input, v and u. Each grid field can be written to such that interconnection of the horizontal with the vertical input can be realized by means of predefined switching functions at a grid point for every output – $u = f_1(x, y)$ and $v = f_2(x, y)$.

Programmable logic ASICs are divided into writable and rewritable programmable logic arrays. A Programmable Array Logic (PAL) device is a programmable logic device that can be written to once. The grid points of this logic comprise AND and OR gates with the AND gates being programmable and the OR gates being hard-wired. All connections are available in the unwritten state. This kind of ASIC is programmed by the targeted destruction of connections to AND gates using overvoltage. Connections cannot be restored once destroyed. This is the reason programming is possible only once.

A programmable logic array (PLA) can be rewritten. Programming here uses selective writing of individual grid points to establish fixed states so that a function in the respective grid point is realized with well-defined basic modules.

The following four basic modules are used strictly for circuit realization, as shown in ◘ Fig. 9.32:

1. Identity: $u = y \quad v = x$
2. Adder: $u = x + y \quad v = x$
3. Multiplier: $u = y \quad v = x \cdot y$
4. Negative multiplier: $u = y \quad v = x \cdot \overline{y}$

The modules are designed such that an input signal is always routed to an output downwards or to the right or that it is connected with the other input signal. The identity function connects the input signals to the relevant output signals. In the unprogrammed state, a programmable logic array is made up only of identity functions. The adder performs an OR operation on both input signals and relays the signal to the right. The multiplier performs an AND operation on both input signals and relays the result downwards. The negative multiplier performs the same operation as the multiplier, only here the horizontal input y is negated before the AND operation. The basic modules are set up such that switching functions can be implemented in conjunctive normal form (CNF), as in ◘ Fig. 9.33. The number of columns corresponds to the number of terms in all switching functions to be realized. The number of rows corresponds to the number of different variables in all switching functions and the number of switching functions to realize. Binary value 0 is applied to the rows at whose outputs the switching functions are realized (result rows). The terms are now formed from top to bottom using multipliers and negative multipliers. The terms are finally totaled in the result rows using adders.

## 9.2.4 Controls and Displays

Logic operations can be realized with different technology (e.g., electronic or electric) and be used in control devices (refer to ▸ Sect. 9.2.1). The input signals of logic networks are generated by the controls and signal transducers while output signals actuate displays or drives (valves, couplings, motors, etc.). Just as it was mainly electrical controllers that were used in the machine tool sector previously, so the machine tools of

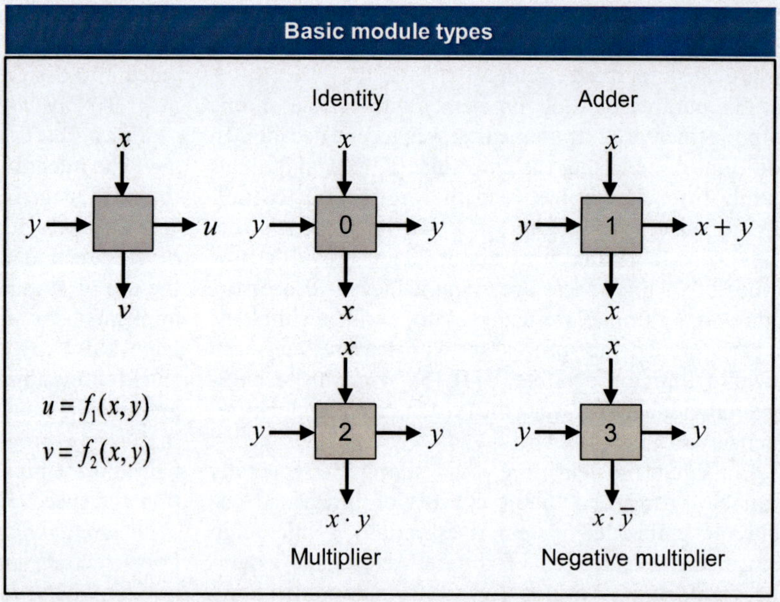


◘ Fig. 9.32 Basic module types of a programmable logic array

**Fig. 9.33** Implementing functions in programmable logic

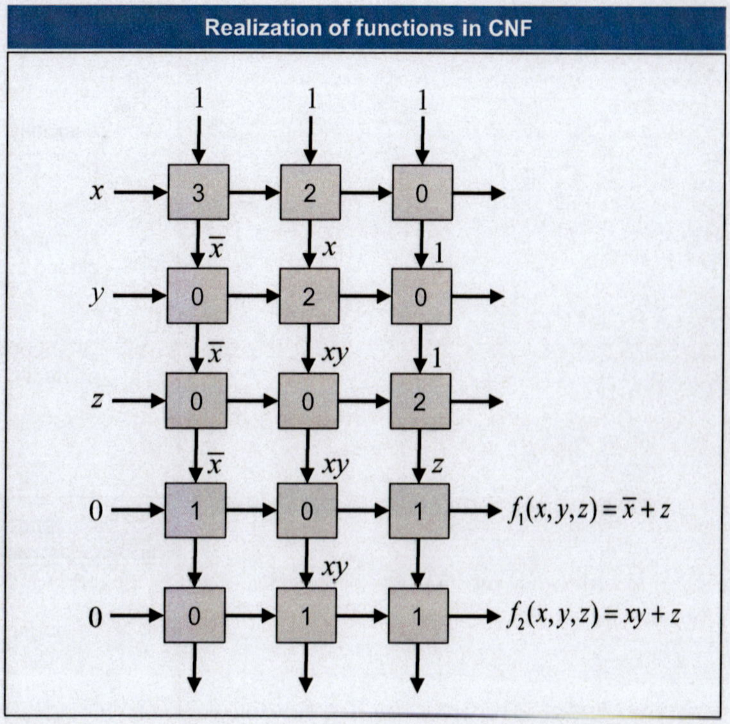


today have electronic controllers. Described below are simple, fundamental components for input and output devices in electronic circuits [BAIL74; FLEI85; FRIS78; HÄUS80; HILB92; TIET93; LEHM74; LEHM83].

■ Figure 9.34 shows the graphic symbols for the most common switching elements and displays as per DIN EN 60617 [DIN97]. The switching elements can be used as signal transmitters. Pressure and temperature sensors can be realized, for example, depending on the type of switch drive. The signal transmitters can be designed as push-buttons or latching switches that only differ in operating function. For push-buttons, the contacts remain activated as long as it is pushed. Upon release, the contacts return to their rest positions under spring force. Latching switches, on the other hand, must be returned to their initial position by an actuating force. Push-buttons and latching switches can be fitted with multiple contacts (normally closed contacts and/or normally open contacts).

It is also necessary, for NC machines, to specify signal values as digital numbers and possibly also to add explanatory text. Different displays have been developed for these.

■ Figure 9.35 shows the basic layouts for 7-segment, 16-segment, 5/7 matrix displays and a LCD. The segments activated to show letters and digits are also depicted. Whole text lines can be shown by combining multiple display elements. A matrix display allows any characters and, given the corresponding number of pixels, even complete graphics to be shown (LCD or electroluminescence flat screens).

Individual segments or pixels are activated by decoding circuits that receive their information from logic networks or programmable devices.

### 9.2.5 Computers

The high integration density of electronic circuits has led to a steady growth in digital electronics and the trend towards digital computers that were, in essence, an integration of digital electronic components. Depending on requirements and objectives, different architectures arose over the course of the years. The computers and other computing machines of today are based on these. The architectures essentially differ in regard to type of switching concept for command and data memories. Harvard and Von Neumann are two important and well-known computer architectures.

The Harvard architecture is a switching concept in which the command memory and data memory are separated logically and physically. Historically, the name has it origins in the development of the Mark I electromechanical computer developed jointly by IBM and Harvard University. The logical division arises from different address spaces and machine commands defined specially for access to the two memory areas. The physical separation is implemented using two separate memories connected independently using two

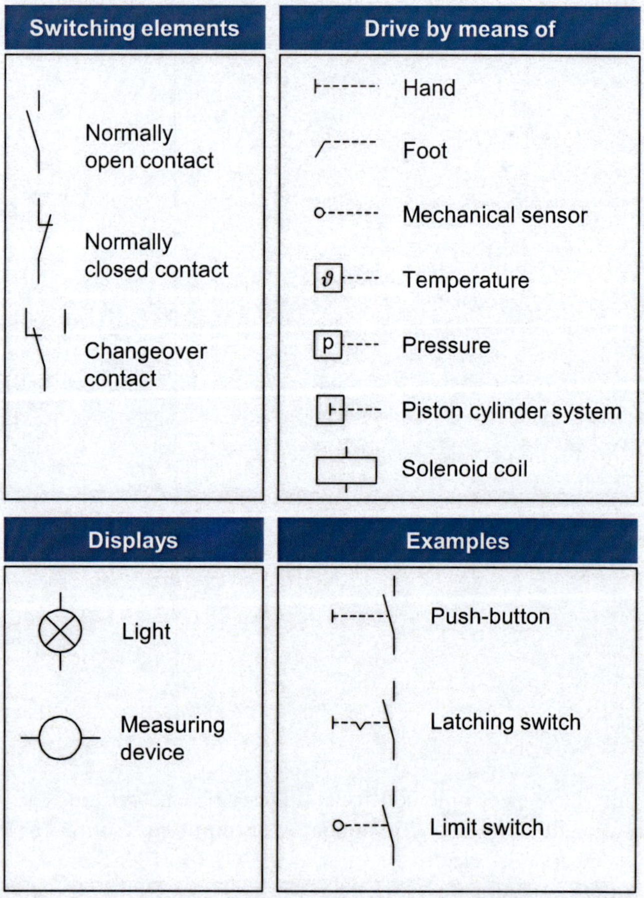


**Fig. 9.34** Controls and displays

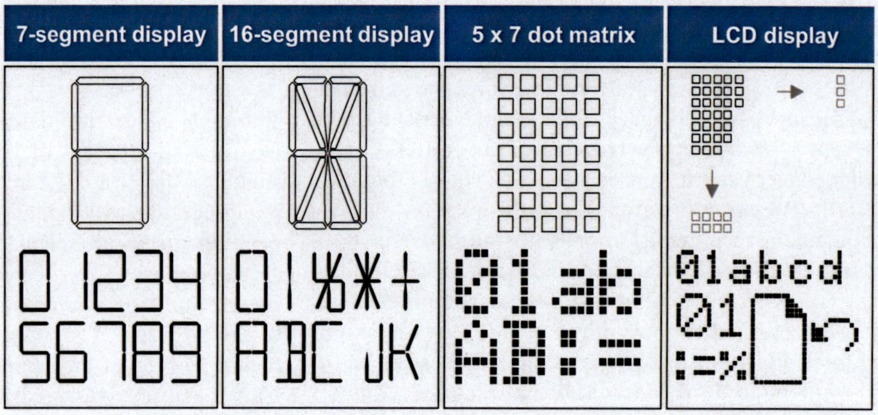


**Fig. 9.35** Display of digits and letters using different display elements

different data buses. The benefit of this architecture is that data and commands can be loaded and written simultaneously in one cycle. The physical separation of command and data memory means access right separation and memory protection are easy to realize. This prevents program code from being overwritten accidentally or by intentional or unintentional software bugs. A downside of this architecture is that additional electronic components must be used in the implementation.

The Von Neumann architecture, named after mathematician John von Neumann, is a switching concept that defines the principle of a universal computer able to run a variable program using a minimum of hardware. Program and data are in shared memory. Also, the concept defines a systematic split into corresponding function groups using specialized circuits such that operations can be structured flexibly and more efficiently. Most computers nowadays are based on the basic principle of this architecture and have been modified over time depending on requirements.

The most important components of the Von Neumann architecture are the arithmetic logic unit (ALU), control unit, memory and the I/O unit. The Von Neumann architecture also exhibits a division between different memory categories. These start with low-level CPU memories such as registers and caches that themselves are on the CPU, RAM, extending to hard drives and optical/electronic media. This gives rise to a downside of this architecture, namely the "Von Neumann bottleneck". This denotes the connection between memory and processor, which carries both the data and program code.

The relatively high clock rate of the processor, and the low clock rate of the memory relative to it, mean the processor is not able to process the data at the speed intended. This is compensated however by intermediate storage, or caches, directly on the processor so that data blocks are always directly reloaded instead of individual instructions. One key component of the Von Neumann architecture is the ALU. This is a hardware component providing the standard arithmetical and logic operations for the processor. Standard operations are addition, subtraction, multiplication, division, logic operations of Boolean algebra and shift operations. Also, an ALU has an accumulator, and operand and result registers. It also has a status register in which certain bits are set when a particular situation occurred in the last operation – such as when there is a carryover or overflow, or the result was zero.

In both architectures, the processor must process instructions from the memory in three phases. First, the instruction is loaded from the memory, the fetch phase. In the second phase, the decode phase, the instruction is interpreted. In the subsequent execute phase, the instruction is run, and more data or the next instruction fetched. To now enable more efficient processing of commands, a "pipelining" mechanism is used. This allows even more instructions to be processed within a short time. Here, as soon as an instruction has passed the fetch phase and transitioned to the decode phase, the next instruction is loaded directly in the fetch phase. In the next cycle, the instruction loaded first is already in the execute phase, the second instruction in the decode phase and the next new instruction in the fetch phase. This means that, while an instruction is being executed, two more instructions can be prepared to increase throughput.

## 9.3 Communication in Automation Technology

### 9.3.1 Communication Technology Requirements

On the different levels of the automation pyramid, specific requirements are made on the communication technology used (see Table 9.1). On the upper levels, the data volume to be transferred is large. On field or sensor/actuator level, stricter requirements are made on the real-time capability and reliability of data transmissions when exposed to various interference factors such as dirt, vibration, electromagnetic waves and liquids. There are also requirements about averting hazards to people, machinery and the environment (see ► Sect. 10.4.1). Integration options between applications on the automated system on individual levels of the automation pyramid (horizontal integration), and also cross-level (vertical integration), are a key prerequisite for comprehensive information availability and controllability of the system and put additional requirements on the communication technology used. This pertains first to the hardware components used, such as connectors, cables and DIN rails, and second the communication protocols used must be compatible with one another. The broad availability and comprehensive applicability of diagnostics and project planning tools are an absolute necessity in the planning and parallel operation of different communication technologies within the system to automate. Physical interdependencies between the technologies used must be identified early on and the resultant negative effects averted. This applies in particular to wireless equipment.

### 9.3.2 OSI Reference Model

Previously, communication interfaces for different local networks and fieldbuses always had a manufacturer-specific design and so, in general, were mostly incompatible with each other. However, the lack of compatibility between communication interfaces leads to high costs in integrating automation equipment in multi-vendor environments. This was the reason that, between 1978 and 1983 under the name Open Systems Interconnection (OSI), a reference model for implementing open, vendor-independent communication interfaces was defined by the ISO [DAY83].

The OSI reference model, ISO 7498 [ISO/94], is a framework guideline defining an architectural concept. Using this concept, the activities of various

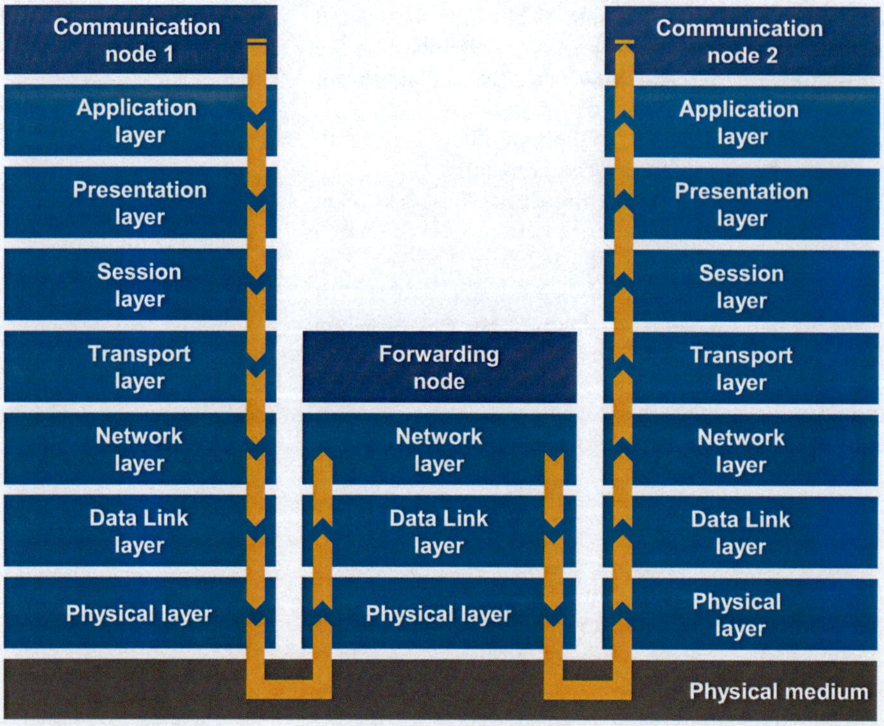


**Fig. 9.36** OSI/ISO layer model as per ISO/IEC 7498–1 according to [DAY83; ISO/94]

9

standardization committees, working independently of each other on the creation of open communication interfaces, are to be coordinated and put on a shared platform. The OSI reference model defines open systems. That is, compatible systems for this kind of communication. It also defines the interaction of these systems in an OSI environment and a layer model. To better understand the OSI reference model, these terms are explained in more detail in the following sections.

The OSI layer model splits communication between two partner processes abstractly into seven levels (Fig. 9.36). Specific communication functions are assigned to each of these layers. They communicate with each other over the interfaces defined by the OSI. The specific implementations of the layers can be exchanged while complying with the interface definitions. The OSI layer model thus defines a framework based on which communication systems can be assembled from components from different vendors. Layers 1 to 4 contain common functions or offer services used for secure transport of information between application processes. They are known jointly as transport-oriented layers. In contrast, Layers 5 to 7 are known as application-oriented layers because of the services and functions they make available to the application processes. Information density and service granularity increase from lower to higher layers [SCHN12].

The functions of the seven layers of the OSI reference model are described below [ZIMM80; ISO/94]:

- Layer 7 (Application Layer) represents the interface to the application of the communication process. A web browser, for example, uses Layer 7 to access the Internet.
- Layer 6 (Presentation Layer) is responsible for data presentation. Depending on the computer and lower levels, data need to be converted. Strings, for example, can be presented in different ways using the familiar ASCII characters, say.
- Layer 5 (Session Layer) covers dialog control. This is used to specify which communication node can send and when. Token management is also on this layer. Tokens are exchanged and certain operations may only be executed when in possession of the token. Restart points are also provided. If an error occurs during an operation, such as when sending files, Layer 5 ensures that the session is continued at the point where it was interrupted.
- Layer 4 (Transport Layer) establishes end-to-end connections between applications. For this, the continuous data flows are split up into packets. The layer sets up new connections between communication nodes. Application processes are addressed for this. The layer assumes, as needed, the handling of errors such as lost data and incorrect packet sequence. Layer 4 also performs flow control to adjust the data rate.
- Layer 3 (Network Layer) assumes the transfer function between homogeneous and heterogeneous

networks. It routes data packets, i.e., sends data packets on their intended path, or addresses packets to unique computer addresses. Layer 3 also performs multiplexing of multiple logical connections over one physical link.

- As a service for the Network Layer, Layer 2 (Data Link Layer) provides error-free transmission of data between two communication nodes. It checks that individual packets are error-free by adding a checksum on the transmit side. For local networks, overloading of the physical medium is detected and prevented.
- Layer 1 (Physical Layer) transports individual bits over a particular physical medium. It encodes individual or groups of bits for transmission.

Communication in the OSI layer model is always between two layers on the same level. Information passes through all the lower layers (in descending order) at the sending communication node and is processed and forwarded by the appropriate implementation of the layer in line with the interface definitions of the OSI. On Layer 1, the data are then processed and sent onto the physical transmission medium. The data then pass through all of the layers at the receiving communication node, in ascending order. It is possible that not all the layers in the OSI layer model are available. This depends on the communication system [HERF13; SCHN12].

### 9.3.3 Bus Architectures and Access Methods

Different bus architectures represent the basis for the communication technologies used in automation technology. Figure 9.37 shows diagrams of the most common architectures. They describe the communication nodes of the network with direct communication. They also show the connections and other stations that messages pass during transmission between two communication nodes.

In the star architecture, also called the parallel wiring arrangement, all communication processes pass through a central point. If there is a fault at this point, no further communication is possible. If one cable connection fails, only communication to or from this node is down. If there is a fault with one node or a connection in the ring architecture, the whole network is down. One node missing in a bus architecture is not critical for the network. If there is a cable break or the connecting communication channel fails, subsequent communication is not possible.

The nodes of the network use a shared transmission medium for communication regardless of the bus architecture selected. Communication processes must be coordinated to prevent collisions in communication, i.e., simultaneous use of the transmission medium by more than one sending node, and the accompanying failure to send telegrams. This is achieved on the basis of bus access methods. The most prevalent methods are described below.

In the Token Passing method, a token is forwarded in a cyclic sequence between all communication nodes. The communication node currently in possession of the token is authorized to send data. This method is used for ring architectures (Token Ring) in which the token is always sent to the next physical neighbor. It is also used for bus architectures where the token is forwarded on the basis of a permanently defined logical sequence. This method generally provides good real-time behavior, i.e., response times are predictable (deterministic communication).

In the Master-Slave method, communication is coordinated by a single communication node, the Master. The Master is the only node that can use the communication medium and regularly polls all other nodes, the Slaves, for data. The Slaves only send messages once the

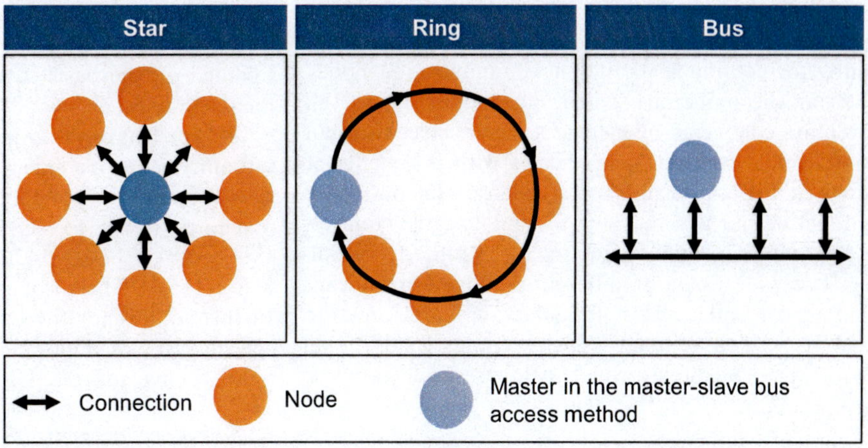


Fig. 9.37 Bus architectures in automation technology

**Fig. 9.38** Properties of bus architectures

| | Bus architecture | | |
|---|---|---|---|
| Fault | Star | Ring | Bus |
| Failure of one connection | Non-critical | Critical | Critical |
| Failure of one node | Critical for central node, otherwise non-critical | Critical | Non-critical |
| Cabling expense | High | Moderate | Low |

Master has sent a corresponding request. This method is straightforward to implement but is highly susceptible to failure because no communication is possible should the Master fail. This is why, in practical applications, this method is often combined with a Token Passing method. Here, the role of Master switches between multiple nodes on the basis of the Token Passing method.



The Carrier Sense Multiple Access method (CSMA) regulates asynchronous access by multiple stations to one shared medium. Every station monitors the medium. Before attempting to send, a station checks whether the medium is free. Depending on the method variant selected (CSMA/CD—CSMA with Collision Detection or CSMA/CA—CSMA with Collision Avoidance), a transfer is started immediately and started again after a waiting time in the event of a collision or communication is started after a random waiting time should the medium still be free, thereby averting collisions for the most part. CSMA/CD is usually used for hard-wired transmission, CSMA/CA usually for wireless transmission. CSMA results in non-deterministic transmission. Response times also increase with an increasing number of nodes.

### 9.3.4 Bus Systems

It is primarily on the lower levels of the automation pyramid that mainly bus systems are being used currently for communication. The communication nodes, such as the numerous sensors and actuators on this level, are connected by means of short cables to a bus cable routed centrally. This brings with it benefits of lower cabling and maintenance expense—in contrast to parallel or star wiring where a separate cable connects every communication node to a centralized controller. The wide range of different fieldbus types means that sensor and actuator manufacturers are forced to support several of these systems entailing additional costs. Also, predicting which fieldbuses will gain/lose significance is difficult. There is a risk here of misdirected investment in the development of fieldbus interfaces.

The German Commission for Electrical, Electronic & Information Technologies of DIN and VDE (DKE for short) is responsible in Germany for fieldbus standardizations. DKE standards are used as drafts for international standards. Also, the draft standards of various user organizations, whose aim is to support certain bus systems, flow into the standardization process. The International Electrotechnical Commission (IEC) manages international fieldbus standards efforts while the Commité Européen de Normalisation Electrotechnique (CENELEC) handles the European fieldbus standards efforts. Contractual regulations lay down that IEC standards are also put to the vote as European drafts. The work of both organizations is performed by employees working in industry as well as at public institutions.

The IEC 61158 standard (DIN EN 61158 of CENELEC) defines existing fieldbuses and structures them on the basis of the OSI/ISO reference model. The IEC 61784 standard (DIN EN 61784 of CENELEC) defines, building upon IEC 61158, additional structuring of fieldbuses by protocol families. Examples of protocol families are, for Ethernet-based fieldbuses, "CIP," "Profibus-Profinet" and "P-Net". For classic fieldbuses, these are "Foundation Fieldbus," "CIP" and "Profibus & Profinet". Safety bus systems are also included in the standards. The standards are being continuously updated on the basis of up-to-date technical developments [SCHN12].

To avert hazards to people, machinery and the environment, European law stipulates that machine manufacturers must conduct hazard analyses for their machines (refer to standards EN ISO 14121, EN ISO 13849, EN IEC 62061 and 2006/42/EC, and see ▶ Sect. 15.1). This must cover all components installed. In this context for the installed bus systems, there are specific risks of data being corrupted and lost or telegrams being addressed incorrectly. There is increased

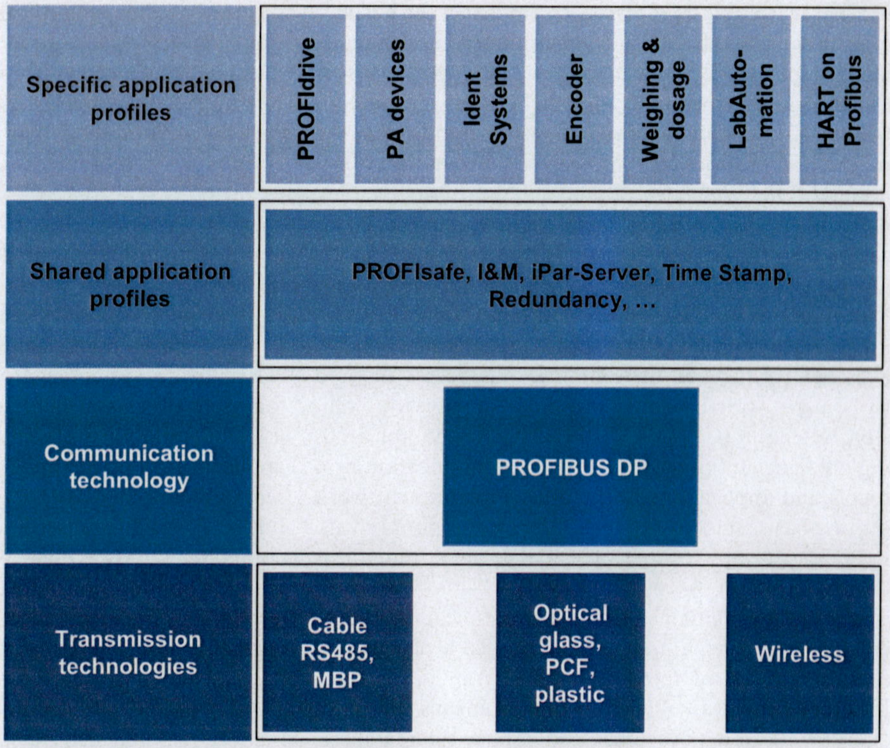


■ Fig. 9.39 Profibus system modules

hazard potential for potentially explosive zones, such as gas distributors and storage tanks. In these zones, "intrinsically safe bus systems" are used whose maximum energy absorbed is limited as regards potential sparking. Potting plugs and sand filling of connectors are additional measures for the safe operation of bus systems in potentially explosive areas. The latter describes the design of embedding components that are potential ignition sources in sand inside the housings of connections so that no explosion can occur inside the housings. Safety buses available on the market include AS-I Safety at work, Profisafe on Profibus and INTERBUS Safety [SCHN12; HERF13].

The prevalent Profibus standard comprises specifications of Layers 1, 2 and 7 of the OSI/ISO reference model. It also includes application profiles, defining certain properties, features and behaviors of devices and systems between manufacturers and users (■ Fig. 9.39). Profibus comprises specifications for data transmission over copper cable, fiber optics and wireless technologies on Layer 1 of the OSI/ISO reference model. On Layer 2, Profibus defines the Master–Slave method in combination with the Token method for coordination of multiple masters. For Layer 7, PROFIBUS defines the PROFIBUS DP communication protocol, available in three performance versions (DP-V0, DP-V1 and DP-V2). The versions comprise capabilities for cyclic and acyclic data exchange between master and slave, e.g., to connect distributed input and output modules to a PLC, and for isochronous data exchange (for motion control applications, for example).

Application profiles serve the purposes of interchangeability and integration of different devices and systems. They define, depending on the purpose, the properties of certain device classes (such as of robots and drives), but are in part also industry-specific and application-related. Application profiles define the integration of subsystems of other standards and manufacturers, such as the HART Profibus application profile. A distinction is made between general application profiles used in different applications and industries, and specific application profiles whose uses are associated more strongly with specific applications and industries. The "PA Devices" profile specifies the properties of process technology devices in process automation as regards Profibus. This also includes safety to protect from explosions in potentially explosive areas. The Profibus (PROcess FIeld BUS) fieldbus standard was developed by several companies and institutes as part of a joint project. It is updated and enhanced by the PROFIBUS user organization and its international umbrella organization PROFIBUS & PROFINET [SCHN12; PNO10].

### 9.3.5 Industrial Ethernet

The use of Ethernet technologies for automation technology was being discussed back in the '90s. Classic fieldbuses were to be replaced by the Ethernet bus system, familiar from the office environment and already being deployed to some degree on the upper levels of the automation pyramid. These fieldbuses were to contribute towards more comprehensive standardization of communication technologies in automation technology. The standardization of connectors, cables, network topologies and transmission media (so in areas of Layers 1 and 2 of the OSI/ISO reference model) for Ethernet means the proven, and affordable hardware components as regards sourcing and operation, are readily available. The Ethernet standard (IEEE 802) defines the non-deterministic bus access method to be used—Carrier Sense Multiple Access with Collision Detection (CSMA/CD)—and a protocol for sending data packets. Some established communication protocols and applications in the office environment work on this basis, such as the TCP/IP protocol family with its important application area of internet communication [SCHN12].



Hardware components and communication protocols based on the Ethernet standard see broad use on the upper levels of the automation pyramid. For this, Ethernet provides sufficiently high transmission rates (such as 10 Gbit/s over fiber and copper cabling specified in IEEE 802.3). The use of Ethernet technology at the field or sensor/actuator level, the real-time applications to be supported, such as motion control, and the necessity for reliable communication in industrial environments where interference factors are great, give rise to stricter requirements. The increasing use of image processing applications in which large data volumes are transferred for evaluation is benefiting from the upsides of high data rates realizable. Also, web technologies based on the TCP/IP protocol family are being used for diagnostics and direct control of field devices, such as by the installation of web servers on controllers ([EISE12; SCHN12]).

The coupling of quick, synchronized drives requires a short and equidistant cycle time of the communication equipment used (sometimes under 1 microsecond). The Ethernet standard is only suitable up to a point for this application. The underlying bus access method, CSMA/CD, leads, with increasing use of the communication channel (i.e., usage of the channel by many nodes), to growing delays of the communication processes. Although taking various countermeasures, such as selecting a star topology to connect field devices (parallel wiring arrangement) and restricting the communication processes as regards frequency and transferable data volume for the individual nodes, can counter this effect, the solutions implemented this way are not always compatible, to each other and between applications and vendors [SCHN12; HMS14].

Communication components used in automation technology must feature a high level of reliability as regards sending of information under field conditions. They must, for example, work at different temperatures and under the effect of liquids, dirt and vibration. Furthermore, increased electromagnetic compatibility (EMC) is required. For some potentially explosive environments, requirements are also placed on the maximum energy absorbed by the component to prevent sparking (intrinsic safety). Ethernet components used in office environments do not satisfy these requirements [SCHN12].

The necessity of integration capabilities between Ethernet-based bus systems from different vendors and classic fieldbuses comes from protecting the investment for users of classic fieldbuses and from the increased level of suitability of classic fieldbuses for particular applications. For the integration of different fieldbuses, requirements, such as compatibility of connectors and DIN rails used for installation, are put on the hardware components. Also, there is a necessity to extend and couple configuration and diagnostics software already available. This is intended to provide integrated management of bus systems and automation components, particularly those based on Ethernet.

The following Ethernet-based solutions for automation technology have established themselves on the market—Sercos III, SafetyNET p, VARAN, Profinet, EtherNet/IP, Ethernet Powerlink and EtherCAT. Cables and connectors familiar from the office environment, such as the RJ45 connector, are being used in variants suited to industrial use, as are newly developed connectors such as the M12. It is possible to interface between different connectors and buses using adapters and terminals [EISE12].

### 9.3.6 Wireless Communication

Wireless technologies used in automation equipment differ in the transferable data rates, frequency ranges used, real-time capability and susceptibility to failure. The benefits of wireless technologies over hard-wired communication technologies arise in the applications of long-range systems and systems with mobile production equipment (such as driverless transport systems). To select and use wireless technologies in automation equipment, the application scenario and associated requirements must always be considered. ◘ Figure 9.40 contrasts some example wireless technologies in automation equipment using range and data rate.

In addition to these parameters, also important in the selection and operation of conventional bus systems, additional issues of mutual interference between wireless technologies used and the impact of geometric and structural properties of the operational environment are of key significance in selecting wireless technologies. Radio waves can be reflected, and so distorted

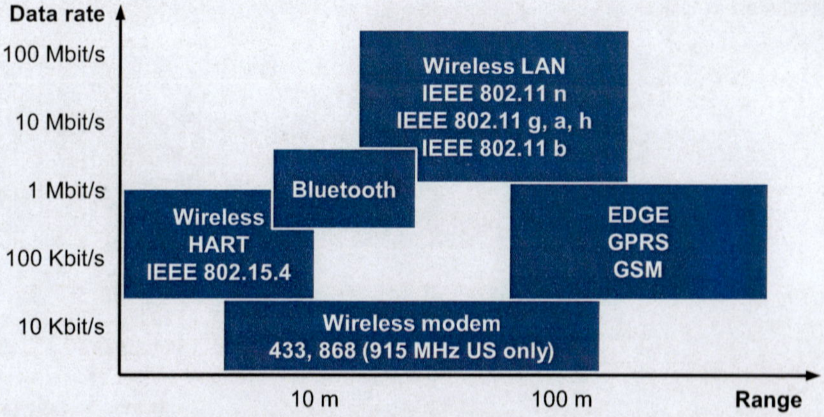


**Fig. 9.40** Selected wireless technologies in automation equipment, in order of range and data rate according to [WOLF10]

and superimposed, by metallic objects. The requirement in selecting and designing wireless systems is to avoid this effect, called multipath scattering. Wireless technologies operated in parallel can cause mutual interference. The goal of "coexistence management" is to prevent interference such as by decoupling all wireless systems used in parallel in at least one of the following traits: location, frequency and time [WOLF10].

VDI guideline 2185 defines the requirements placed on wireless technologies in automation equipment. They are assigned to the individual levels of the automation pyramid. The requirements are also structured in terms of factors such as availability stipulated, necessity of redundant transmission paths, cyclic/acyclic data traffic, real-time communication, number of networking stations, transmission bandwidths, use of stationary/mobile peripherals, distance to span and electromagnetic compatibility necessary. Application-specific requirements are also defined. These arise, for example, from the use of mobile peripherals and safety–critical sites of operation. Requirements on integration capabilities in existing systems are also described. Furthermore, VDI Standard 2185 explains the processes of coexistence management. Also, technical features of wireless technologies and their application-compatible selection on the basis of assessment criteria are presented in the form of a proposal for an application-specific assessment matrix. The standard also introduces a recommendation for characterizing wireless solutions. This provides for measuring and control of parameters and influencing variables defined in the standard as part of the proposed assessment scenarios [VDI/12a; VDI/12b; VDI/13].

The Institute of Electrical and Electronics Engineers (IEEE) manages some standards of the two lower layers of the OSI/ISO reference model and these form a basis for wireless technologies in automation equipment. They include standards for Wireless Metropolitan Area Networks (WMAN—IEEE 802.16 and IEEE 802.20), Wireless Personal Area Networks (WPAN – IEEE 802.15), Wireless Local Area Networks (WLAN – 802.11) and Ultra Wide Band (UWB – 802.15.3a and 802.15.4a).

The wireless technologies used in automation equipment are, in particular, Wi-Fi, Bluetooth, Wireless Interface for Sensors and Actuators (WISA), WirelessHART, ZigBee, MeshScape and nanoLOC. All of these wireless technologies use a license-free wireless frequency band (ISM band) in the 2.4 GHz range. Provider-based wireless technologies such as GSM/GPRS and UMTS are also used in automation equipment. Their range and virtually global network coverage can be deemed benefits. Dependencies of network operators are becoming increasingly negligible for selection decisions in view of the steadily reducing data transmission costs.

### 9.3.7 Near Field Communication

Near Field Communication (NFC) is an international transmission standard for contactless exchange of data using wireless technology over short distances (of just a few centimeters up to about 20 cm) in the 13.56 MHz frequency band. Data are exchanged between two NFC devices and communication is possible between an NFC device and either RFID transponders or RFID readers. Consequently, NFC devices combine the functions of an RFID reader and an RFID transponder. In industrial applications, highly integrated NFC transceiver modules are available. These contain all the components required for NFC communication, such as transmitter, receiver and load modulator, that are integrated into electronic devices such as cell phones. NFC is defined in the international ISO/IEC 18092 NFCIP-1 and ISO/IEC 21481 NFCIP-2 standards. The prerequisite for communication with RFID transponders and readers is compliance with the ISO 14443 standard.

In the communication of NFC devices, a distinction is made between "Active" and "Passive" modes. Active

**Table 9.1** Communication technology requirements

| Level on the automation pyramid | Applications | Requirements | | | | | |
|---|---|---|---|---|---|---|---|
| | | Data volume per second | Transmission reliability | Response times | Interference | Distance to span | Deterministic transmission |
| Management level | Internet, office | MB | Low | Minutes | Low | Kilometers | No |
| Process level | SCADA, HMI, engineering, diagnostics | KB | Moderate | Seconds | Moderate | Meters | No |
| Control level | PLC, soft PLC, motion control | Bytes | Raised | Milliseconds | Raised | Meters | No |
| Field level | Decentralized I/O, drives with changeable speeds, servo drives | Bits | High | Microseconds | High | Meters | Yes |

9

mode is used to send data between two NFC devices. Passive mode enables communication between an NFC device and RFID transponders in "Reader Emulation" mode, or RFID readers in "Card Emulation" mode. In Active mode, communication between the devices is started by one NFC device, the NFC initiator, broadcasting a signal. The signal induces a voltage in the attached antenna loop of the second NFC device, the NFC target. The data are sent using amplitude modulation of the alternating magnetic field transmitted (ASK modulation). The transmit direction is reversed for data transmission from the NFC target to the NFC initiator. In Passive mode and Reader Emulation mode, the connection to the RFID transponder is also established by the NFC device by transmitting a signal that induces a voltage. The signal is maintained unmodulated. Load modulation is used to send data from the RFID transponder to the NFC device. Load modulation is also used for communication between the NFC device and RFID readers in Card Emulation mode [ECMA13].

RFID technologies lend themselves to the realization of decentralized control concepts. The readable and writable RFID transponders contain information on workpieces and tools, for example, that can be read and changed in the process independent of a centralized master computer. RFID technologies are used in the machine tool industry, for tool, workpiece and workpiece carrier identification, for example. Tools and tool holders here are fitted with RFID transponders containing information such as usage recommendations and state variables, preventing tool mix-ups [FINK12].

### 9.3.8 Middleware Protocols

In VDI/VDE Standard 2657 [VDIV13], middleware in automation is defined as a software layer to connect different components, such as systems and machines, of an automation technology application. Middleware in this case can be run on a field device and a computer. Key requirements specifically put on middleware in automation are abstraction of communication and mapping of a data and information space.

The OPC Unified Architecture (OPC UA) is a middleware solution for automation, already standardized, that satisfies the requirements in the aforementioned standard. OPC UA is standardized in DIN EN 62541 and also includes several companion standards, that are based upon OPC UA and link in corresponding fields of automation (such as configuration of drive systems and sensor integration) using OPC UA. OPC UA is the most up-to-date of a series of specifications published by the OPC Foundation on cross-vendor data exchange in automation. Platform-independent information exchange based on cross-equipment and cross-vendor client–server communication by different types of systems and devices is the core of OPC UA. Security and authentication measures on the Transport layer safeguard the identities of clients and servers as well as data integrity. OPC UA defines a fixed set of services provided by a server. Information is represented using an information model, also called a namespace, that can be written to, read from and changed by a client using the server services. Since the nature of information transfer is specified by OPC UA, different manufacturers can generate appropriate information models for their own devices. These models can then be used by the clients of other vendors. In addition to accessing current and historic data as well as transmitting alarms, clients can also be informed of important changes by the server.

Overall, OPC UA is structured in layers. This separates core concepts of communication from underlying computer technology-specific and network transport-specific implementations. The mapping of standards to future technologies is thus guaranteed. Communication processes are currently possible using binary and XML formats as well as with TCP or SOAP web services over HTTP. Earlier specifications from the OPC Foundation, such as OPC Data Access (OPC DA), are still enjoying widespread use but are proprietary compared to the OPC UA Standard – (COM/DCOM from Microsoft). The goals of the OPC UA are to combine the functionality of earlier specifications and to offer the benefits of technology independence described [DIN12]. In the process, it can form the foundation for integrating IT systems into automation technology [APPE10; ROHJ10]. Direct and seamless feedback of process data to Development and Engineering is thus possible.

## References

[APPE10] Appelrath, H.-J., Nieße, A., Sonnenschein, M., Tröschel, M.: Energieinformatik 2010. Oldenburg, OFFIS (2010)

[BAIL74] Bailey, F.J.: Halbleiter-Schaltungen. Oldenburg, Wien (1974)

[BLAC99] Black, B., Ranzenberger, R.: A/D-Umsetzung - aber richtig. In: Elektronik (24) (1999)

[BORU89] Borucki, L.: Digitaltechnik. Teubner, Stuttgart (1989)

[BUES94] Bues, M.: Offene Systeme. Berlin, Heidelberg: Springer Berlin Heidelberg (1994)

[DAY83] Day, J.-D., Zimmermann, H.: The OSI-Reference Model. In: Proceedings of the IEEE (71), 1334–1340 (1983)

[DIN12] Fachbericht SIN CLC/TR 62541–1: OPC Unified Architecture - Teil 1: Übersicht und Konzepte (08/2012)

[DIN97] DIN EN 60617: Graphische Symbole für Schaltpläne (1997)

[DURC83] Durcansky, G.: Digitaltechnik. Physik-Verlag, Weinheim (1983)

[DWOR86] Dworatschek, S.: Grundlagen der Datenverarbeitung. Walter de Gruyter, Berlin (1986)

[ECKL90] Eckl, R., Pütgens, L., Walter, J.: A-D- und D-A-Wandler. Franzis, München (1990)

[ECMA13] ECMA International: Standart ECMA-340. 3. Ed. (2013)

[EISE12] Eisenmann, U.: Vom Sensor ins Rechenzentrum oder vielleicht doch gleich in die Cloud. In: Fach.Journal, 130–131 (2012)

[FINK12] Finkenzeller, K.: RFID-Handbuch. Hanser, Carl, München (2012)

[FLEG93] Flegel, G., Birnstiel, K., Nerreter, W.: Elektrotechnik für den Maschinenbauer. Hanser, München, Wien (1993)

[FLEI85] Fleischer, D.: Digitale Schaltglieder. Berlin, München: Siemens-Aktiengesellschaft, [Abt. Verl.] (1985)

[FREY97] Freytag, A.: Zeichen setzen - Unicode in der Praxis (14) (1997)

[FRIS78] Frisch, H.: Elektronik. Grafenau/Württ., Düsseldorf: Lexika-Verlag; VDI-Verlag (1978)

[HAGE93] Hagelauer, R., Ronge, K.: ASIC-Entwurf für die Mikrosystemtechnik - Varianten, Verfahren, Systeme. Conference Paper (1993)

[HÄUS80] Häusler, L.: MOS-Technologie. Berlin, München: Siemens-Aktiengesellschaft, [Abt. Verl.] (1980)

[HERF13] Herfs, W.: Automatisierungstechnik für Produktionsanlagen. VorlesungWerkzeugmaschinenlabor RWTH Aachen (2013)

[HILB92] Hilberg, W.: Grundlagen elektronischer Schaltungen. München, Wien: Oldenbourg (1992)

[HMS14] HMS Industrial Networks GmbH: Feldbusse.de. URL: ▶ http://www.feldbusse.de. Accessed: 27. Febr. 2014

[ISO/94] [ISO/94]ISO/IEC: 7498–1. Genf (1994)

[KOCH77] Koch, G.: Einführung in die Informatik für Ingenieure und Naturwissenschaftler. Hanser, München (1977)

[KORE81] Koren, J., Thaller, W.: EDV für jedermann. Mannheim (1981)

[LEHM74] Lehmann, J.: Feldeffekttransistoren. Vogel-Verlag, Würzburg (1974)

[LEHM83] Lehmann, J.: Dioden und Transistoren. Würzburg: Vogel (1983)

[PALO13] Palotas, L.: Elektronik für Ingenieure. 1. Ed. Wiesbaden: Vieweg & Teubner (2013)

[PNO10] PNO: POFIBUS Systembeschreibung

[ROHJ10] Rohjahns, S.: Eine standardkonforme und semantikbasierte Kommunikatiosnarchitektur für die Automationsebene in Smart Grids (2010)

[SCHN12] Schnell, G.: Bussysteme in der Automatisierungs- und Prozesstechnik. Vieweg & Teubner, Wiesbaden (2012)

[STIL15] Stiller, A.: Prozessorgeflüster. In: c't (7) (2015)

[TIET93] Tietze, U., Schenk, C.: Halbleiter-Schaltungstechnik. Springer-Verlag, Berlin (1993)

[VDI/12a] Richtlinie VDI/VDE 2185 Blatt 1: Funkgestützte Kommunikation in der Automatisierungstechnik (08/2012)

[VDI/12b] Richtlinie VDI/VDE 2185 Blatt 2: Funkgestützte Kommunikation in der Automatisierungstechnik - Koexistenzmanagement von Funklösungen (08/2012)

[VDI/13] [VDI/13]Richtlinie VDI/VDE 2185 Blatt 3: Funkgestützte Kommunikation in der Automatisierungstechnik - Anforderungen und Spezifikationen an die Energieversorgung basierend auf Batterien und Energy Harvesting (10/2013)

[VDIV13] 2657: Middleware in der Automatisierungstechnik Grundlagen (01/2013)

[WOLF10] Wolfgang Bölderl-Ermel: Funktechnologien in der Industrie. In: Fertigungs- und Maschinenautomation (6), 2–5 (2010)

[WOLF77] Wolf, G.: Digitale Elektronik. Franzis-Verlag, München (1977)

[ZIMM80] Zimmermann, H.: OSI Reference Model. In: IEEE Transactions On Communications (4), 425–432 (1980)

# Electrical Controls

I thank my staff members Philipp Blanke M. Sc., Christian Fimmers M. Sc. and Simon Storms M. Sc. for their contribution in revising this chapter.


C. Brecher and M. Weck, *Machine Tools Production Systems 3*,
Lecture Notes in Production Engineering, https://doi.org/10.1007/978-3-658-34622-5_10

Contents

In light of the complexity of production processes nowadays and the high degree of automation of modern production systems, electrical controls are being used as standard to guarantee high levels of quality and efficiency. The rapid development in the microelectronics sector, for microprocessors in particular, means powerful controller systems are on the market that are able to reliably process several thousand process signals, even under strict real-time conditions. The functions of a controller here are to read the multitude of signals from the process (position, speed, rotational speed, temperature, fill level, pressure signals, etc.), to determine the necessary control signals in accordance with the logical operations and to output these signals to the actuators. The process to control can be a single machine component, such as a feed drive for a machine table, or a complete production system, such as a machining center or transfer line.

As the logic part of large control tasks can become very complex and confusing, using Boolean algebra and the associated graphical tools can provide effective support in the planning and design of control logic. One of the goals here is to minimize the number of Boolean operations for the definition of the control task, to keep as low as possible the device costs for relay controls, for example. Programmable logic controllers available today, however, are so powerful that minimizing Boolean operations for the sake of improved transparency is often unnecessary. When strict real-time requirements are made of the controller, however, using Boolean algebra can shorten the cycle time of a program.

## 10.1 Design and Categorization of Electrical Controls

■ Figure 10.1 shows a schematic of the structure and signal flow of an electrical control system.

Process signals are read on the *input level*. Signals here can be sensor signals internal to the process for example, or external operator inputs. Internal signal transducers are, for example, sensors for reading machine states or process feedback signals (temperature/pressure/fill level sensors, displacement transducers, angle encoders, etc.). Operator inputs can be read from manual controls on a machine tool, such as manually operated buttons and switches.

Before the signals can be processed in the logic part of the controller, there is generally galvanic isolation and an alignment to the voltage level of the controller. Optocouplers, for example, or relays, are used for decoupling and signal matching. The much faster optocouplers work photoelectrically (as opposed to electromechanical relays). When actuated electrically, an LED generates a light signal that is converted back to

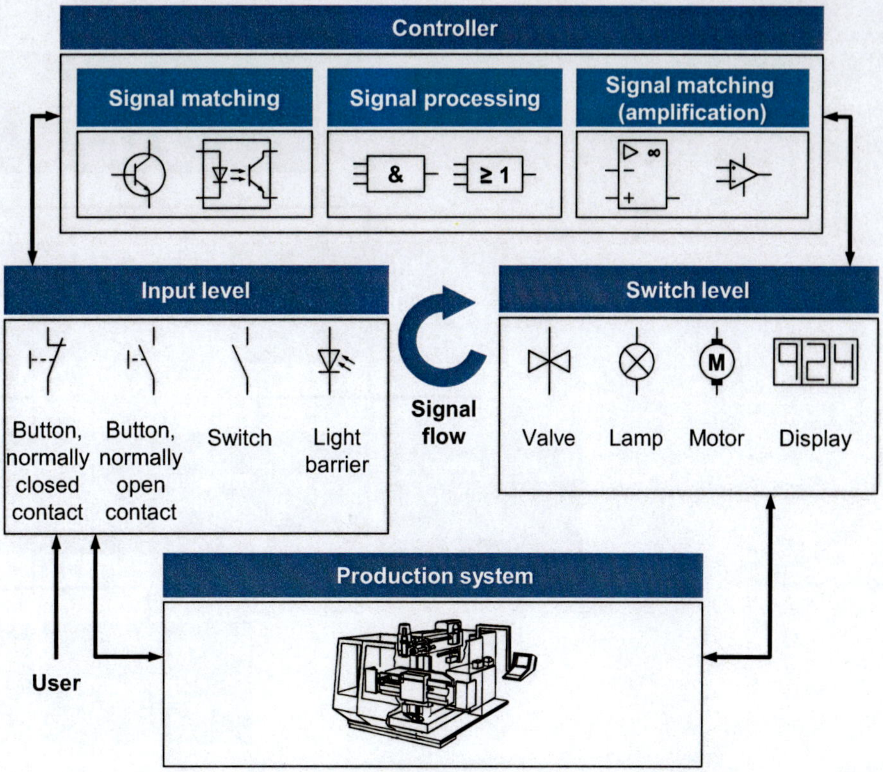


■ **Fig. 10.1** Schematic of the structure and signal flow of an electrical control system

an electrical signal by a light-sensitive photo semiconductor.

Signal processing represents the actual core of the controller. The output signals are determined by logical operations applied to the input signals. This logic is specified by the process and developer. The technical implementation of logical operations can be realized either electromechanically with relays and contactors, electronically using semiconductor elements or with a program and computer.

To activate the elements on the switch level (motors, couplings, displays, etc.), the signals must often be amplified by electronic circuitry. For relay controllers, on the other hand, the maximum switching capacity of the relay is usually sufficient. If the requirement is for higher powers, contactors are used.

As mentioned above, the logical operations can be implemented in different technical ways representing, at the same time, a key distinguishing criterion from electrical controllers. As shown in ◘ Fig. 10.2, electrical controllers can essentially be differentiated into hard-wired controllers and programmable logic controllers (PLCs).



For hard-wired controllers, the controller function is determined by the connection of the components (relays, contactors and semiconductor modules). If this uses terminals, soldering or screws, the controller function can only be changed with mechanical intervention. That is why these controllers are called "hard-wired controllers."

The program logic for hard-wired controllers can be changed with pluggable line links, patchboards and diode matrices. These types of controllers were commonly used in the past to enable, for example, *reprogramming* when changes are made to the workpiece-related control process. These types of controllers, such as crossbar switch and patchboard controllers, have almost been completely replaced by PLCs, however.

In contrast to hard-wired controllers, the control function for a PLC is determined by a program (software) stored in memory. The controller hardware is thus independent of the application and is always the same for a device type in the base configuration. Subsequent subdivision of this controller type is based on the memory type used.

For programmable controllers, the content of program memory (RAM, battery-backed CMOS RAM, EEPROM) can be changed without mechanical interventions in the control unit. If ROM (read-only memory) is used, the controllers are PLCs with interchangeable memory, i.e., the software can only be modified by swapping out the memory components.

PLCs with interchangeable memory are subdivided again into controllers with unchangeable (ROM, PROM) and changeable (EPROM) memory—see ► Sect. 10.3.2.

In addition to categorizing controllers by how the control program is realized (◘ Fig. 10.2), DIN 19226, Part 3 (Control technology) [DIN94] essentially differentiates again between logic controls and sequential controls (◘ Fig. 10.3).

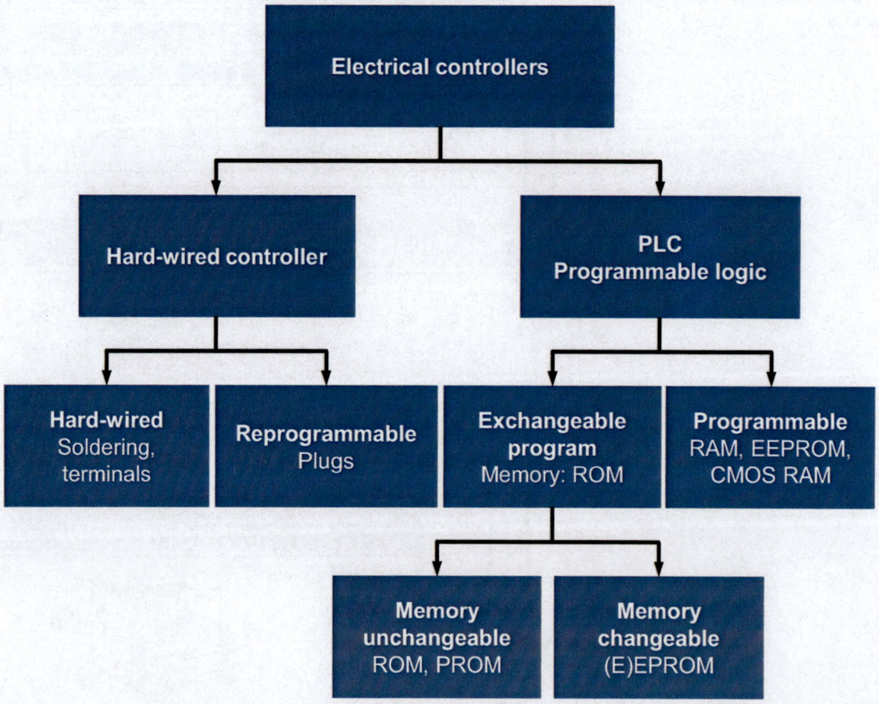


◘ **Fig. 10.2** Classification of electrical controls by implementation

**Fig. 10.3** Classification of electrical controls by control principle

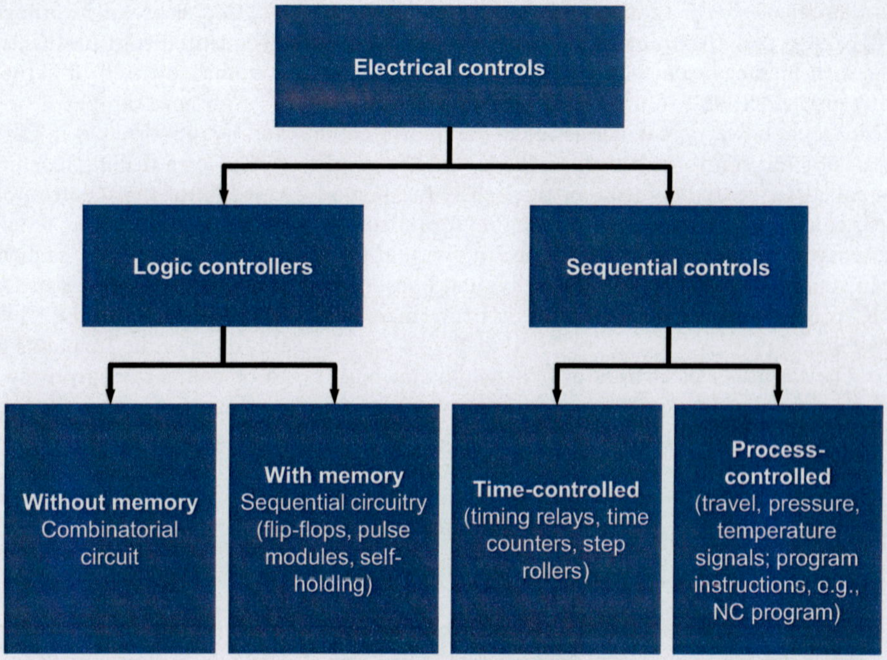


### 10.1.1 Logic Controllers

The logic controller term is defined in DIN 19226–3 as "a controller" that assigns to the output signals certain signal states resulting from Boolean operations performed on the signal states of the input signals. Typical Boolean operations include AND, OR, NOT, NOR and NAND functions—refer to ▶ Sect. 9.1.3. Crane and excavator controllers are practical examples of logic controllers where the sequence of individual actions is specified solely by the operator. The controller for selecting the rotation direction of a motor or to actuate a feed drive, with all of the interlocks required, is also broadly a logic controller, see ▶ Sect. 10.2.1.

Logic controllers can again be subdivided into those with and without memory (◘ Fig. 10.3).

For logic controllers *without memory*, also called combinatorial circuits, the output signals at any point in time depend solely on the states of the input signals. So the behavior of a controller can be defined in full by the relationships between input and output signals in the form of a value table, see ▶ Sect. 9.1.3.

For logic controllers *with memory*, also called sequential circuits, the output signals also depend on the internal memory states and the history of the system. If the requirement, for example, is to turn a motor on by briefly pressing a button and to turn it back off by pressing the button again, the behavior of the controller is dependent on the history, namely whether the motor is running or not.

### 10.1.2 Sequential Controls

Sequential controls are of major importance alongside logic controllers. In the machine tools sector, the number of sequential controls used far exceeds that for pure logic controllers [OEST86]. Sequential controls are controllers with mandatory step-by-step cycles. Here, advancing to the next step in the program depends on step enabling conditions. The most important property of a sequential control is the unique functional and time-based assignment of individual steps to technological processes. While, for a logic controller, all operations can be run at any point in time, for sequential controls, control instructions belonging to a step are only run when the relevant step enabling condition of the previous step is satisfied.

If the step enabling conditions only depend on time, the term used is *time-controlled* sequential control. Timing relays, time counters or controller cylinders with constant speed, for example, can be used to generate the step enabling conditions. If the step enabling conditions only depend on the signals of the system controlled or the process (such as reaching the end position of a feed motion with subsequent advance to another axis motion), the term used is *process-controlled* sequential control system (◘ Fig. 10.3).

Although there is a strict delineation between the terms logic control and sequential control in DIN 19226–3, it is important to note that more complex machine controllers in particular can include a combination of the control systems mentioned [BIEL90], and the transitions from a logic controller with memory to a sequential control are straightforward [ZAST95]. If the states of the system's process steps, however, can be assigned uniquely to the states of the controller program, the control principle is called sequential control. In this sense, the numerical control (NC) system of a machine tool also represents a sequential control because individual steps can be assigned uniquely to the individual lines of the NC program when the workpiece is being machined (► Chap. 11).

The controller of an automatic assembly machine is one practical example of a process-controlled sequential control system[PLAG90]. The controller must rotate, lower, grip, lift, open, etc., the handling device in line with the step-by-step work cycle of the automatic machine – feed, clamp, screw, declamp, transport, etc.

Also, a broadly automatic cycle is characteristic of a sequential control. It is triggered by a start signal and continues by itself until the end.

Another example of how a sequential control works and its properties is in ◘ Fig. 10.4. It is part of a controller for a drilling machine. The control process is divided into steps corresponding to the operation sequence of the drilling process.

To depict the sequence, the Sequential Function Chart according to DIN EN 60848 and DIN EN 61131-3 [DIN14b; DIN14c] has been selected because this enables the functions and behavior of a controller to be defined descriptively and independent of the implementation. Also, the Sequential Function Chart can serve as a means of communication between different specialists involved in the project planning for the controller, such as design engineers and control engineers (► Sect. 10.3.4).

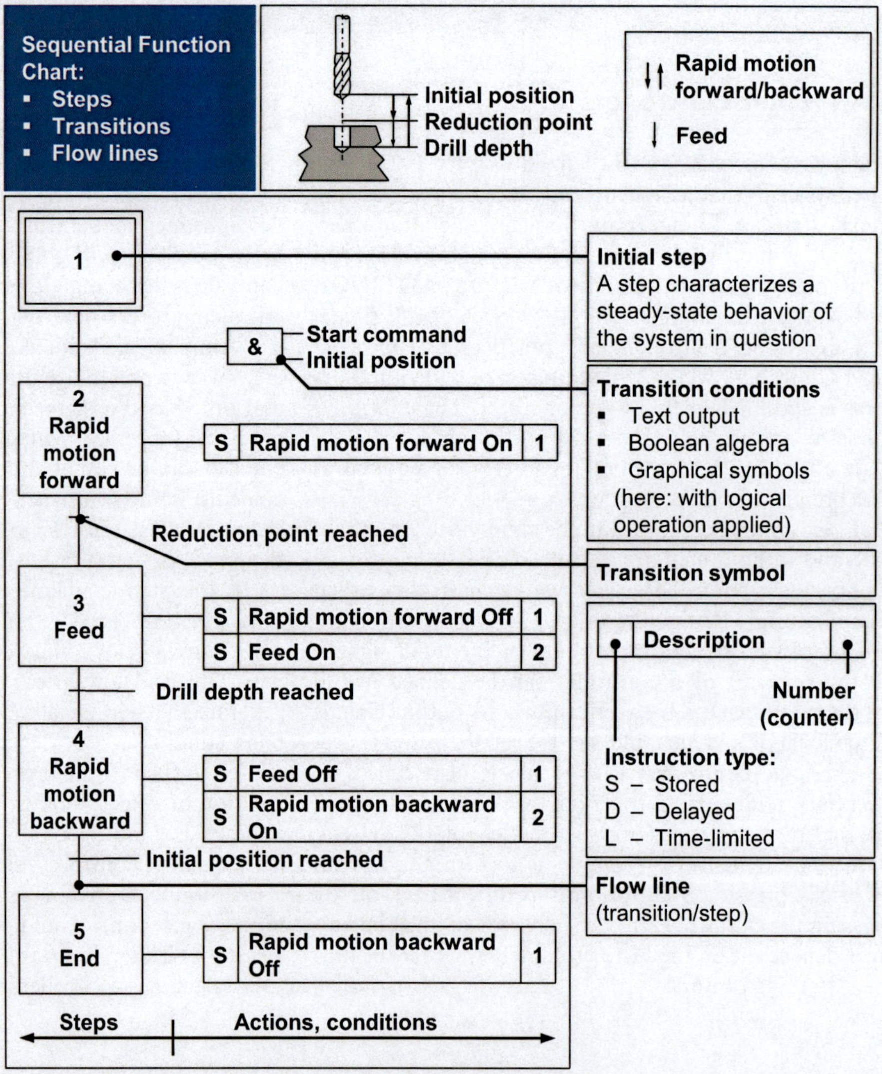


◘ **Fig. 10.4** Sequential control of a drilling machine with Sequential Function Chart, according to DIN EN 60848 and DIN EN 61131-3

The Sequential Function Chart is essentially defined by three symbols—steps, transition conditions and flow lines that connect steps and transitions (◘ Fig. 10.4). The steps define individual states (such as Step 2, *Rapid motion forward*) to which one or more commands (e.g., *Rapid motion forward On*) can be assigned. Transition conditions (such as *Reduction point reached*) define the condition to satisfy to switch from one step to the next. The syntax of the function chart becomes clear in ◘ Fig. 10.4 in the example of the drilling machine.

The initial state, defining the initial behavior of the controller, is shown by the *initial step* (Step 1). The drilling cycle is triggered by a start command such as manually by the operator or automatically by production equipment connected. If the drill is in the initial position, the drilling cycle starts when rapid motion is enabled (Step 2). When the reduction point is reached, a switching cam is activated (transition condition from Step 2 to Step 3) and the speed is reduced to normal feed. The same applies to the cycle from Step 3 to 5. Finally, Step 5 transitions back to initial state 1 and enables triggering of the next drilling cycle.

This chapter has covered the fundamental categorization of electrical controls as regards implementation and control principle. The next two chapters aim to discuss in more detail the two most important implementations: hard-wired controls (► Sect. 10.2) and programmable logic controls (► Sect. 10.3).

## 10.2 Hard-wired Controls

### 10.2.1 Application Areas and Tasks

Although hard-wired controllers have been displaced in many application areas by more flexible and more powerful programmable logic controllers, their use is still common for relatively straightforward control tasks and for safety–critical applications, in particular (► Sect. 10.4). Rapid progress in the field of microelectronics and the associated price drop for electronic components, and for microprocessors in particular, have meant PLCs offer cost benefits starting at controller configurations of just 5 to 10 relays/contactors. However, the use of contactors cannot be eliminated entirely because higher electric current, from main drives and larger solenoid valves for example, cannot be switched directly by the PLC. Only main or load contactors can handle this.

When safety requirements are more stringent, for press control systems, for example, or for the monitoring of emergency stop circuits on machine tools, hard-wired controllers also have the benefit of lower complexity and they enable malfunctions to be detected easily (► Sect. 10.4.3).

In addition to simple basic logic circuits, time delays and pulses of any length can also be generated with relay circuits (e.g., as clock generators). Time relays with defined switch-on and switch-off delays are used for this. Some typical applications of hard-wired controls are presented below using machine tool examples.

### 10.2.2 Application Examples

As already mentioned, logic controllers and sequential controls are dependent on the implementation and can also be implemented as hard-wired controllers. The *Self-holding* and *Interlock* functions used in virtually all controllers are explained below using a simple logic controller. ◘ Figure 10.5 shows for this a circuit diagram for the rotational direction control of a three-phase a.c. motor. Both the control circuit, usually operated with 12–24 VDC, and the load circuit with the three-phase a.c. motor to control, connected directly to the 380 V three-phase network (U, V, W), are shown.

Taking the left-hand current path in the control circuit yields the *Self-holding* function as follows. When button S2 is pressed, the current flows via contacts S1, S2 and a2 through contactor A1. Contactor A1 is energized and all auxiliary contactors of A1 (denoted a1) are moved to the operating position. The drive motor shown in the load circuit then switches to counterclockwise.

Closing auxiliary contact a1 in parallel with Switch S2 maintains the current flow through A1 even after S2 is opened. A1 is thus self-holding. Parallel contact a1 is called a "self-holding contact" and the circuit "self-holding." Button S1 is used to disable counterclockwise rotation. When pressed, the circuit is interrupted, de-energizing contactor A1.

Analogously, the right-hand side of the circuit in the control circuit is used to switch the drive motor to clockwise rotation via contactor A2. The rotational direction is reversed by simply swapping two conductor phases (V and W). To prevent simultaneous switching into the clockwise and counterclockwise directions, one auxiliary contact of contactor a1 (a normally closed contact) is provided in front of contactor A2. If contactor A1 is activated, this contact a1 is open and A2 cannot be energized. In the same way, the current path of contactor A1 is protected against contactor A2 being energized at the same time. This circuit is called an interlock and the associated contacts are interlocking contacts.

The controller of a feed drive shown in ◘ Fig. 10.6 is discussed as an example of a somewhat more complicated logic controller.

The feed drive permits three motions—rapid motion forward, rapid motion backward and working

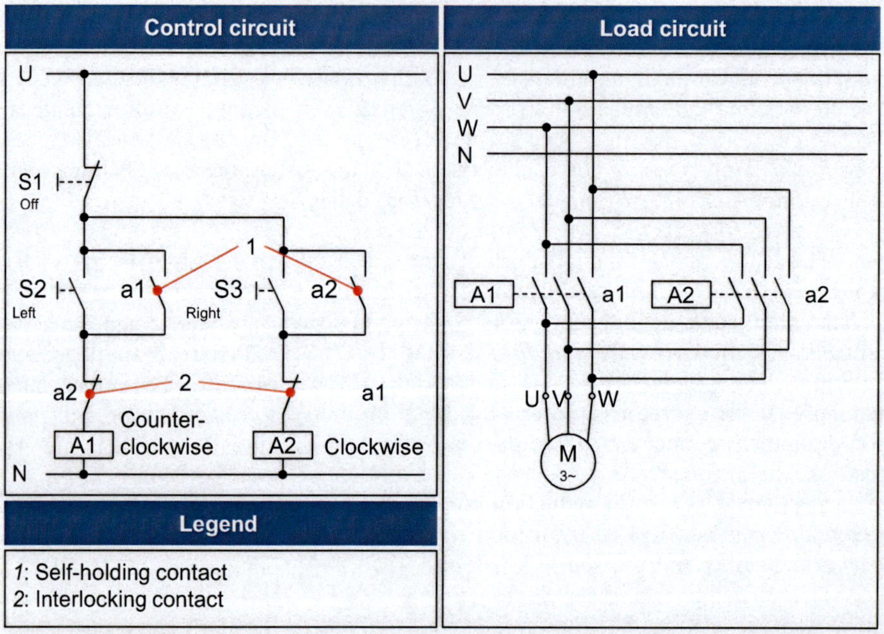


**Fig. 10.5** Circuit for selecting the rotational direction of a three-phase a.c. motor



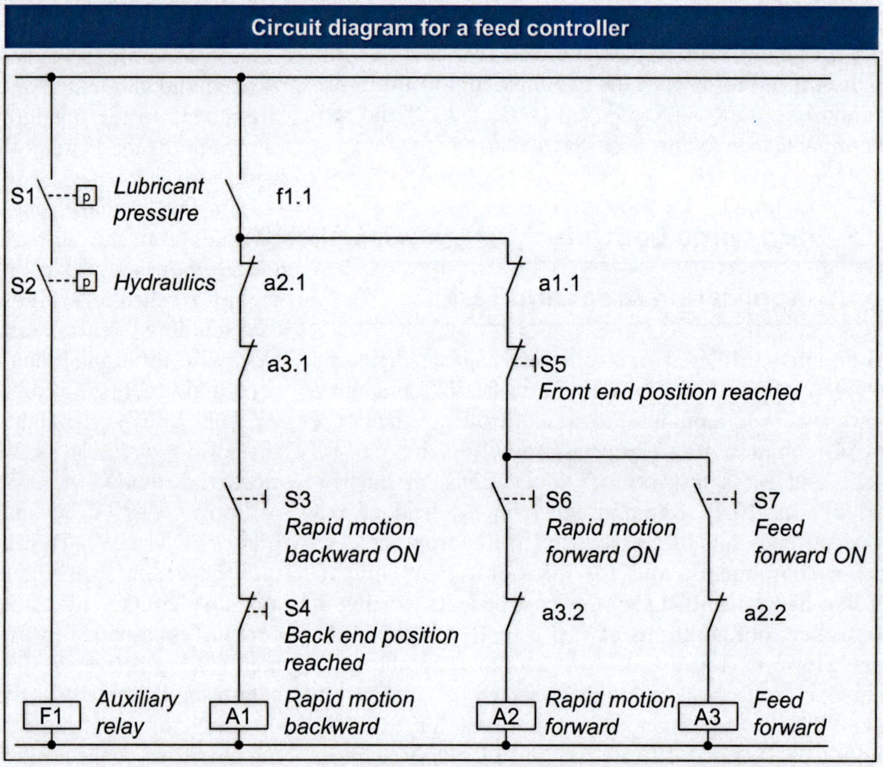


**Fig. 10.6** Implementation of a feed controller

feed. All of these table motions can be initiated via a button and they continue until the button is released. However, a button press only works if neither of the two other movements is initiated (interlock). The motions stop as soon as a limit switch is reached. Pressing a button also has no effect when the hydraulic or lubrication system is faulty.

Semiconductor technology can also be used to implement the logic controller described. Usage of hardwired semiconductor controllers, however, has all but

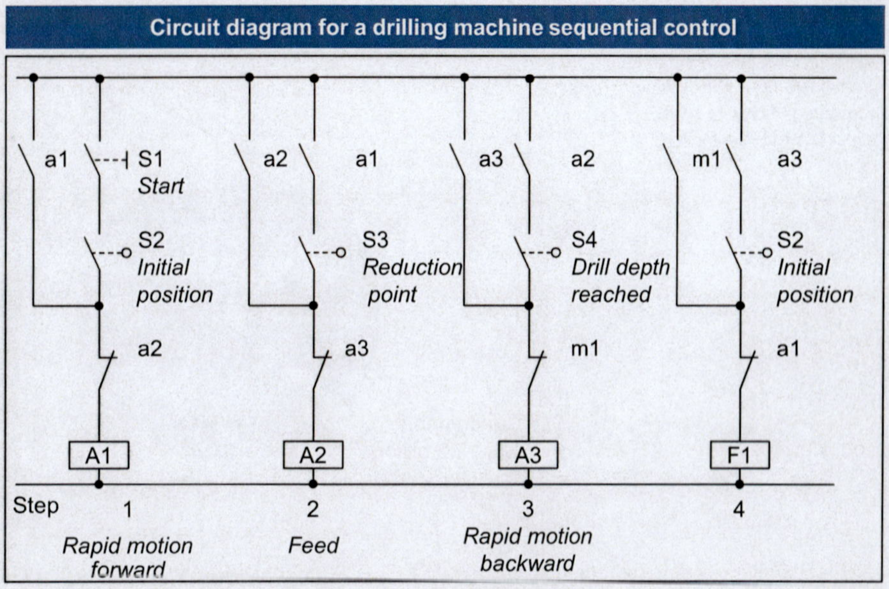


**Fig. 10.7** Circuit diagram for the sequential control of a drilling machine as in Fig. 10.4

died out because they have been displaced by more flexible programmable logic controllers (► Sect. 10.3).

The same components used for logic controllers (contactors and semiconductor modules) can implement *sequential controls*. Figure 10.7 shows the implementation of the control task in Fig. 10.4 as an example of a sequential control in the hard-wired programmed logic form. Given that the step enabling conditions are essentially identical in this example, the circuit is only explained for the first step. If the drill is in the initial position and start button S1 is pressed, contactor A1 activates and rapid motion forward is initiated. Cam-operated switch S2 for the initial position opens, but contactor A1 initially stays in the energized state via its self-holding contact a1. With active contactor A1, its auxiliary contact a1 is also closed in the current path for A2. As soon as the cam-operated switch S3 closes for the reduction point, contactor A2 energizes and opens interlocking contact a2 in the current path of A1 so that A1 de-energizes again. The next step keeps clearing the previous until the drilling cycle has ended.

## 10.3 Programmable Logic Controllers (PLCs)

### 10.3.1 Application Areas and Functions

Since their market launch at the start of the 1970s, programmable logic controllers (PLCs) have been deployed in virtually all fields of industry. They are currently one of the most important modules in automation technology. While PLCs initially served as low-cost, flexible alternatives to the relay controllers exclusively used previously, their functional scope today extends well beyond the logic operations of binary signals. Hardware performance has risen sharply and has resulted in ever more complex controller functions being implemented cost-effectively with a PLC. In addition, the process computers used previously have broadly been superseded by the development of faster multiprocessor systems. The application areas of PLCs are consequently very diverse and are mainly in the automation sector. In Germany, more than a quarter (28%) of German electronics market revenues come from this sector, making up a good third (35%) of the sector for industrial goods [ZVEI19]—see Fig. 10.8. In terms of global production, Germany also holds a considerable position in the automation sector (Fig. 10.9). The market volume of the German electronics industry was EUR 122 billion in 2017, corresponding to 3% of the world's electronics market.

Figure 10.10 shows an overview of the tasks PLCs can perform for machine tools. The main functions of PLCs are the control and monitoring of mechanical functional units. These include the time and plausibility monitoring of individual modules in addition to logical operations and interlocking functions (such as swivel time monitoring of the tool changer and pair monitoring of limit switches). Many functions of the HMI (Human Machine Interface) are also controlled by the PLC. These include switching the drives, the coolant supply, and the chip conveyors on and off.

Furthermore, PLCs today are being used to collect, format and relay process data and to manage the tools available in the machines. Larger data processing tasks, however, are often being performed by Industrial Personal Computers (IPC) connected to a PLC—refer to ► Sect. 10.3.2.1. For machine tools, PLCs must

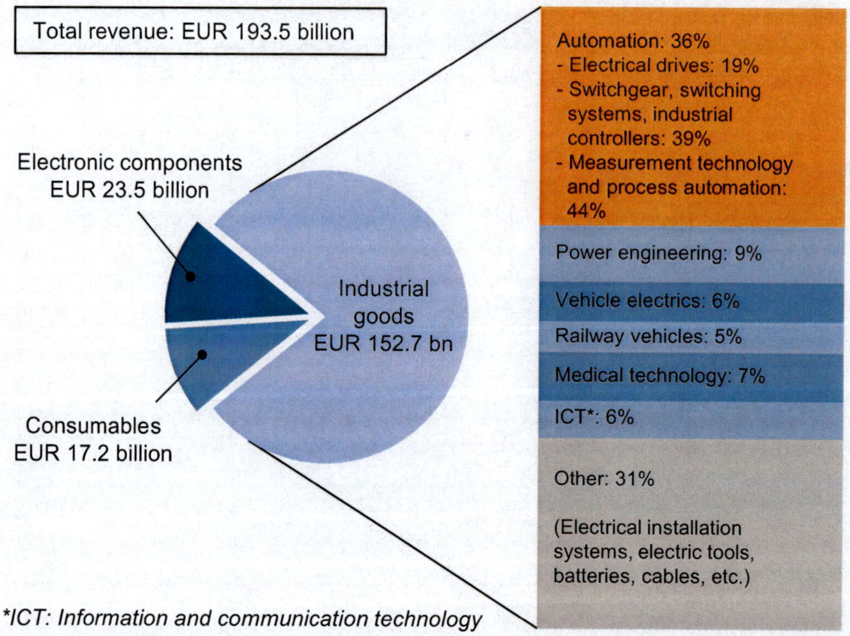


**Fig. 10.8** Market share of automation equipment as a percentage of sector revenues for industrial goods in the German electronics industry in 2018. (*Source* [ZVEI19])



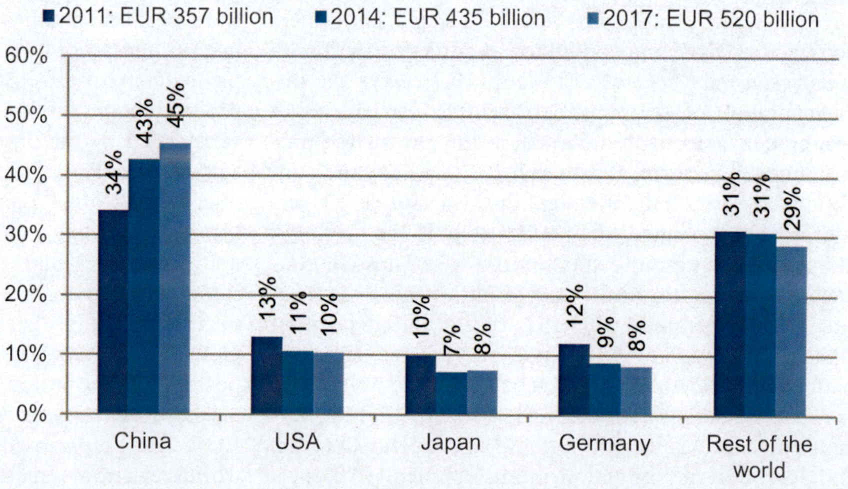


**Fig. 10.9** Global production and market share by country in the automation sector. (*Source* [KEGE14;KEGE16;KEGE19])

also manage information interchange with the numerical controller (NC)—see ▶ Chap. 11. While the NC extracts and processes the geometric information for traversing the axes (coordinates, feed rates) from the underlying NC program, the switch instructions in the NC blocks (such as M and T commands) are relayed to the PLC and run there. Additional signals between the NC and PLC are used for synchronizing the two systems and interchanging status messages and other information (NC program ended, error messages, etc.). In addition to data interchange with the NC, a PLC can also communicate with a master control computer or with the PLCs of other machines. A typical example of this is an interlinked transfer line where every machining station is controlled by its own PLC.

For PLCs, the control task is not realized as for hard-wired controllers with suitable circuitry comprising relays and contactors but by a controller program. The control task can thus be changed without additional wiring overhead by simply changing the PLC program. The controller hardware is designed to be independent of the task at hand. It has the function of running the process signal logic operations as specified in the controller program and of providing the logic results to the process. Since a PLC is clock-controlled, i.e., one instruction is run after another (serial operation) and only a single signal state can be processed at any one time, there is another difference to hard-wired controllers that can process any number of signal states at the same time (parallel operation). However, since

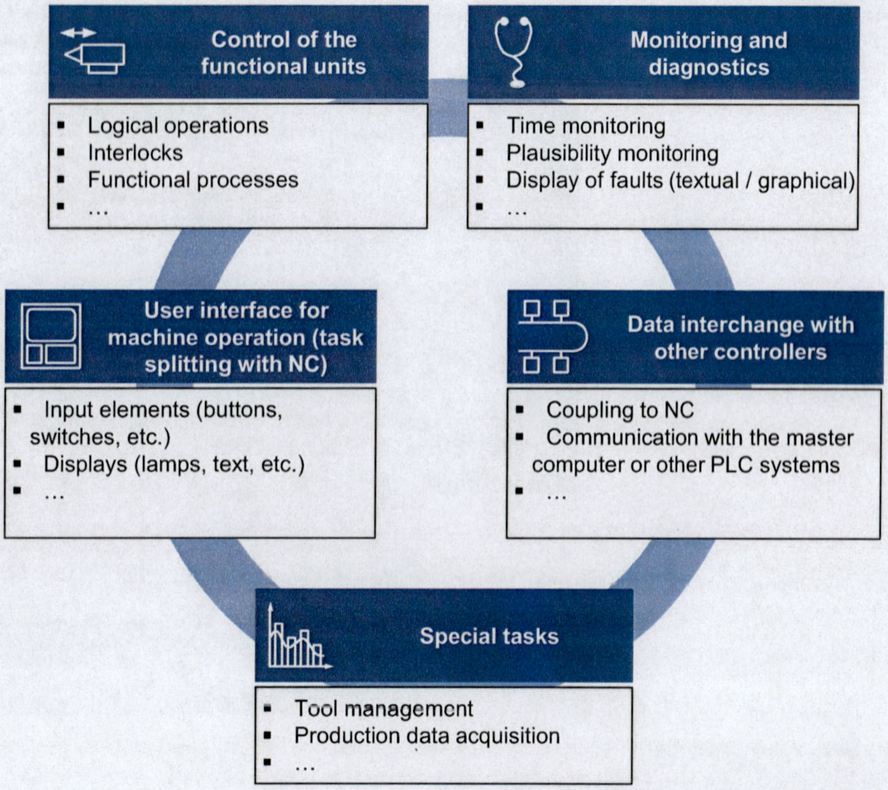


**Fig. 10.10** Tasks of PLCs for machine tools

**Fig. 10.11** Comparison of hard-wired controllers and programmable logic controllers

| | Hard-wired controller | Programmable logic controller |
|---|---|---|
| Programming | Hardware (wiring) | Software (instructions) |
| Mode | Parallel | Serial |
| Technology | Electromechanical (relays) or contactless (semiconductor elements) | Contactless semiconductor elements (Microprocessor) |

the time for running an instruction is in the order of microseconds, we can speak of a "quasi-parallel" operation. The key features of both controller types are summarized in Fig. 10.11.

## 10.3.2 Design and Function

### 10.3.2.1 Design

Figure 10.12 shows the basic design of a PLC. It can be divided schematically into input, output and processing parts. Terminals or plug connectors directly connect the signal transmitters and actuators for the process to be controlled to the input and output modules of a PLC.

The input modules are tasked with matching the voltage levels of the actuators (e.g., 24 V or 220 V) to the system voltage of the PLC (5 V), eliminating interference from the signals (e.g., using RC filters to filter out interference voltage peaks) and galvanically isolating them from the PLC (using optocouplers, for example). Built-in LEDs display the signal states.

Besides the elements for galvanic isolation and the LEDs for indicating the signal states, the output modules have an output amplifier that amplifies the signals internal to the PLC and provides the output power necessary to drive the actuators. The output modules are normally designed for a signal level of 24 VDC, but other voltages are also standard (such as 48 VDC or 220 VAC). The currents provided by the output modules (0.5 to 2 A) are usually high enough for driving the actuators directly. Power contactors are usually connected in series to deliver higher output powers for actuating drive motors, for example.

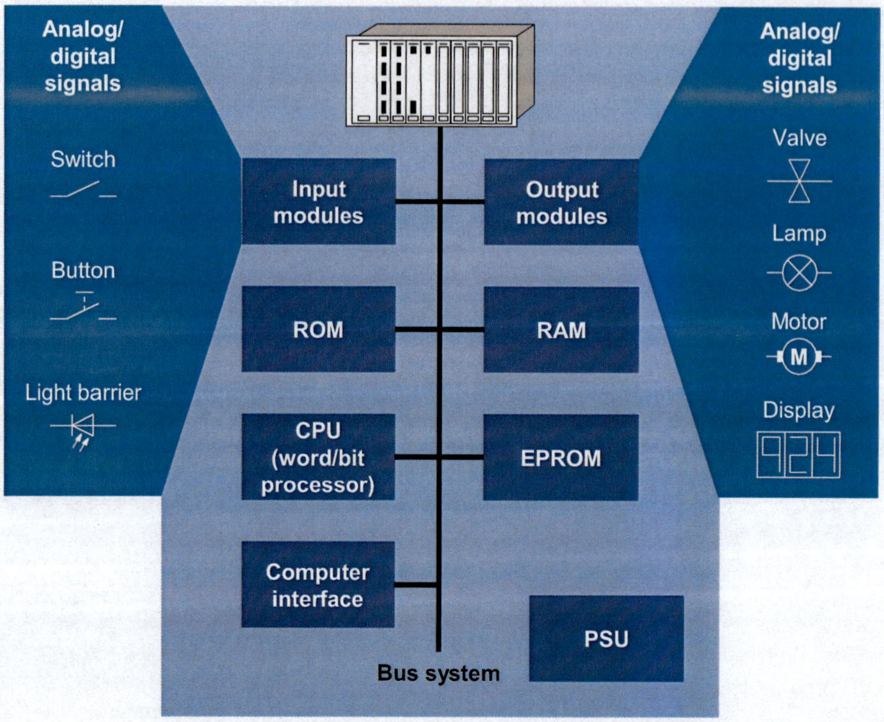


**Fig. 10.12** Hardware components of a programmable logic controller

10

The core of every PLC is the processing part, comprising one or more processors (CPUs) and different memories (RAM, ROM, EPROM). As modern microprocessors are byte-oriented or word-oriented (8, 16 and 32 bits) but typical applications in control engineering represent only bit operations (such as open/close chuck), special bit processors are commonly used for PLCs in addition to a word-oriented processor. Besides the bit processor, responsible for purely binary signals, the word-oriented processor handles analog values and text and performs arithmetic calculations. Bit and word operations can be combined as required in programming because task splitting is internal to the controller and has no impact on programming. The performance of today's processors means leading edge PLCs are usually single processor systems. When very strict timing requirements are made of PLCs, those featuring real multi-processor capabilities are also used. Here, one controller program is split across different processors.

The memory of a PLC can be divided into system, program and data storage. The controller operating system, responsible, for example, for reading and correct processing of user programs and for monitoring the cycle time, is stored in the *system memory* (read-only memory, ROM) and cannot be manipulated by users. The user program is normally created on an external programming device and can be loaded via an interface into the *program memory*, the RAM (random access memory) of the PLC. This memory usually has battery backup. The program can also be stored in EPROM (Erasable Programmable ROM) for more reliable storage. The program stays in this memory after the PLC is switched off, even without battery backup. Every time the controller is switched on, the program is copied from EPROM to RAM. When configured to do so, the program boots automatically when a PLC is switched on. Intermediate results occurring while the program is running (such as flags) are stored temporarily in *data memory* (RAM). Interchangeable memory cards (CF cards) can be used to save, port and access machine data.

One example design of a modular PLC is in Fig. 10.13. Since the individual hardware components of the controller are built as separate modules, a PLC can be configured specific to the system depending on the number of inputs and outputs required (digital and analog), the memory capacity necessary and external interfaces. As the main module with the CPU (central processing unit) is also a separate module, the processing speed can be increased when requirements on the control system are stricter by installing a more powerful CPU module.

PLCs can also have positioning modules. These modules enable drive systems to be controlled by a PLC, too. Typical applications here are handling equipment as well as textile, packaging and printing machines. In contrast, numerical controls (NC) are generally used to control axes for machine tools. Controller manufacturers offer integrated solutions for this. That is, they provide control hardware with PLC, NC

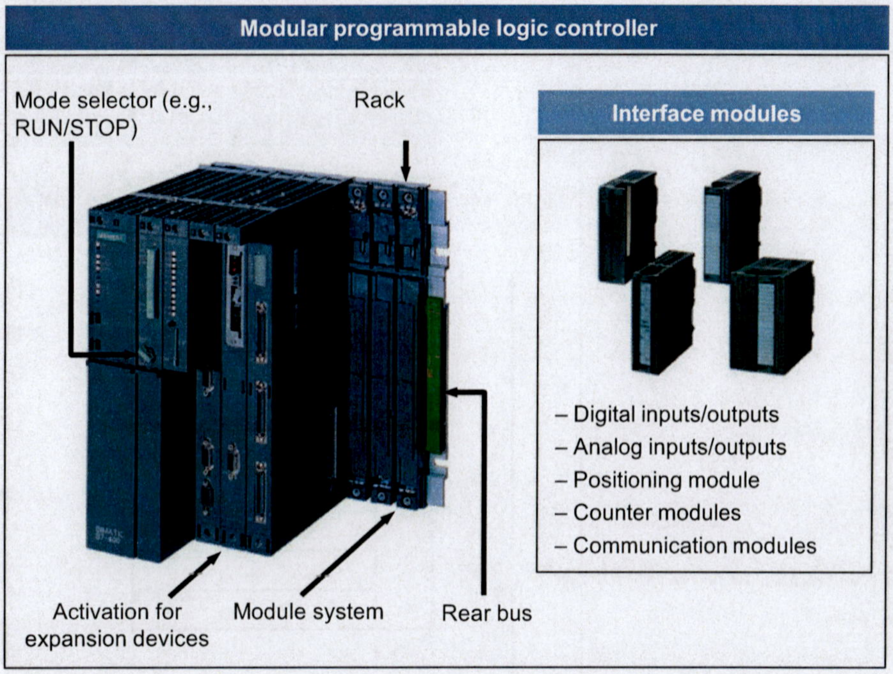


**Fig. 10.13** Modular programmable logic controller. (*Source* Siemens)

and HMC (Human Machine Control) components, see ▶ Chap. 11.

When the requirement is for a PLC to process larger data volumes, it is also possible for an industrial personal computer (IPC) to assume this function. An industrial PC is a PC "suited for industrial use" that, like a PLC, satisfies requirements made of the hardware. These requirements include temperature, vibration and electromagnetic compatibility (EMC). Data interchange between the PLC and IPC means that complex data processing tasks, such as tool data management and operational data collection, can be handled by the IPC. Together with a screen and operator panel as the HMI (Human Machine Interface), the IPC can also be used to visualize data and operate the system to control. The benefits of an IPC are the ability to use existing software applications developed for PCs and to easily connect the controller to a host or operational computer. An IPC can also be used by the programmer to create PLC programs, see ▶ Sect. 10.3.3.

In addition to connecting PLCs to external IPCs and IPC modules as extensions to modular PLCs, using IPC-only solutions is expanding more and more. In terms of hardware, the PLC is integrated in the form of a card plugged into the IPC. In terms of software, the PLC is implemented as a "Soft PLC", i.e., without its own PLC processor, on the IPC [INTE15]. Actuators and sensors are normally connected via external input/output modules and a fieldbus system—refer to Sect. 10.3.5.

#### 10.3.2.2 Operating Principle

A PLC works in cycles, reading the signal states present, running the controller program, and providing controller commands to the output terminals. The controller program is stored as a sequence of instructions in program memory and is run once to completion in every cycle. Jump instructions enable forward and backward jumps to be realized within a program—see Fig. 10.14.

In addition to normal, cyclic program processing, bigger PLCs nowadays also offer event-driven, interrupt-driven and time-driven processing. In *event-driven* processing, a signal arriving from the process (such as the pressing of an Emergency stop button) can interrupt cyclic program processing to run a specific program such as the Emergency stop routine to safely switch off the machine. If the main program is generally not interrupted, such as with the Emergency stop, the processor, after running this program, continues in the cyclic program at the point it was interrupted. In *time-driven* processing, cyclic program processing can be interrupted by an internal time signal (wake-up alarm) to run a specific program. The cyclic program is then resumed at the point it was interrupted, as with event-driven processing. Figure 10.15 shows an example program cycle with the aforementioned processing types.

While the signals of the input and output modules for process computers are commonly accessed directly, PLCs normally work with the "process image."

**Fig. 10.14** Cyclic program processing in a PLC

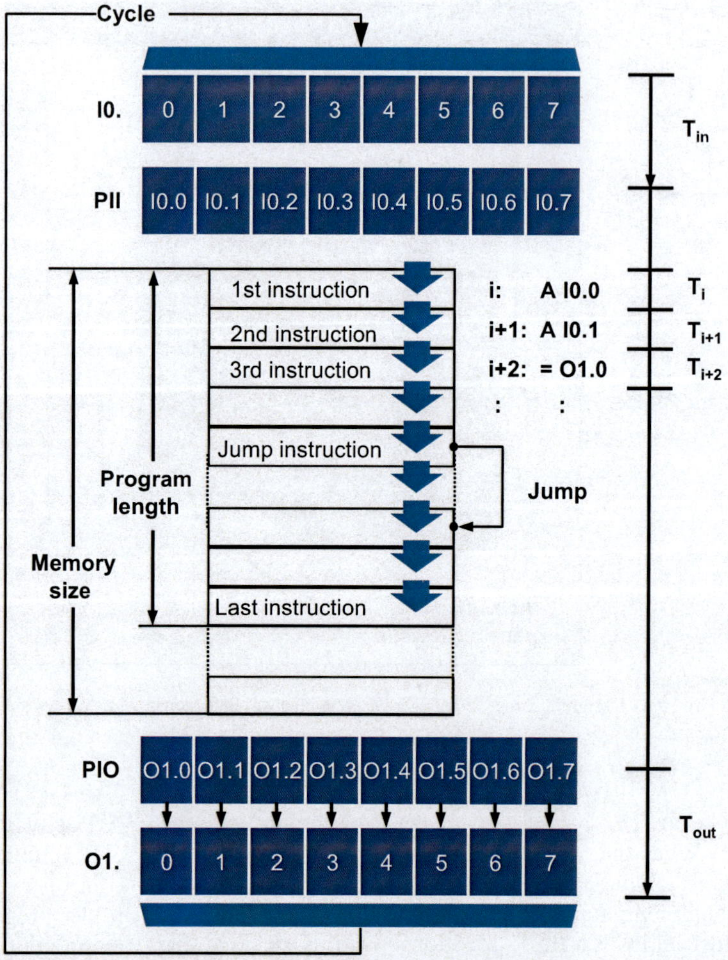


PII: Process image input, PIO: Process image output

There are, however, also PLCs having direct access to the input and output modules. When the process image is used, the signal states of all inputs are retrieved at the start of a cycle and transferred to the input image memory. This state image remains constant for one entire cycle, meaning state changes occurring at the inputs within a processing cycle are ignored by the controller. In the processing cycle, logic operations as in the user program are applied to the stored values. The values are then cached in the output image memory and relayed collectively to the output terminals. Using the process image, the input states are read in parallel and, after program processing, the outputs are driven simultaneously. Given that such a cycle repeats at intervals of a few milliseconds ("cycle time") depending on program length and device, quasi real-time behavior is realized that guarantees sufficiently quick responses to all process changes. To determine a comparative value for the processing speeds of different PLCs, the processing of 1 K (1024) instructions is compared. Standard values here are 0.1 to 5 ms.

The instructions necessary for programming the control process are predominately a subset of Boolean logic (such as AND and OR operations). Furthermore, time conditions can also be programmed for example (see ▶ Sect. 10.3.3.4). These correspond to the functions of time relays for electromechanical controllers. Other special instructions are used for programming counters and jumps, both conditional and unconditional, within a program. Unconditional jumps are always performed provided the previous logical condition is satisfied (branching). Many arithmetic instructions are also available for analog value processing. If the relevant hardware is used (modules for analog input and output, path and speed measurements, and step motor control) with these instructions, complex controller functions can also be realized, traverse axes can be positioned, and control loops can be designed.

For complex controller programs, the PLC program is no longer contained in a single block. Instead, it is divided into multiple self-contained routines (modules or blocks) for improved structuring. A module here is

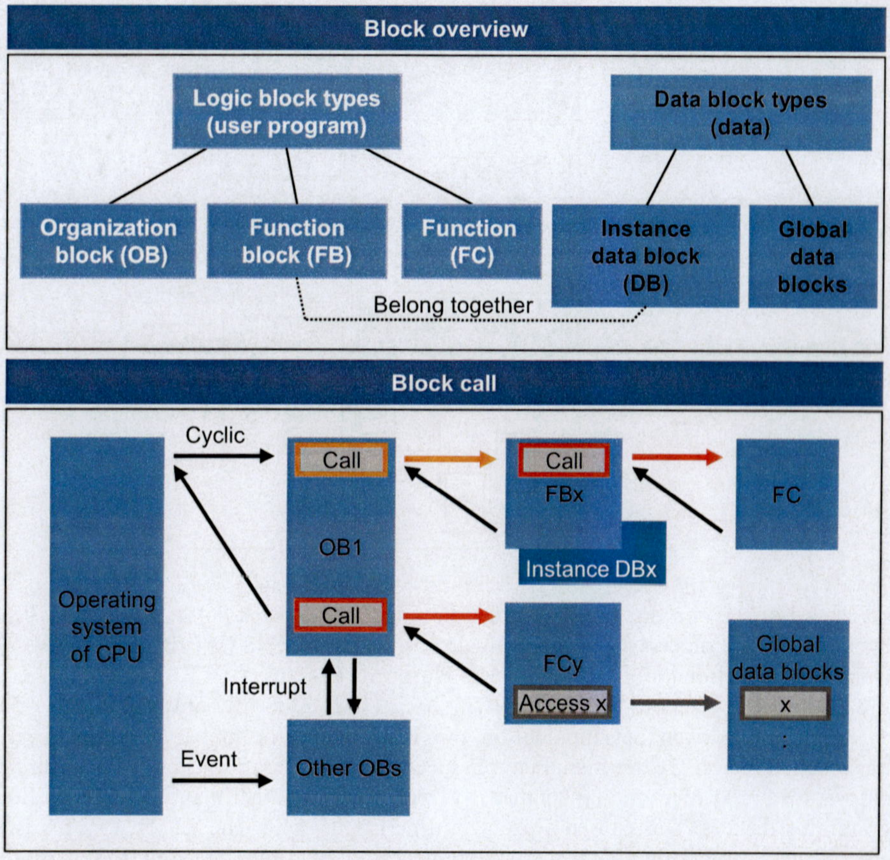


**Fig. 10.15** PLC configuration per IEC 61,131–3

part of the program delineated by function, structure or purpose. Benefits of the modular approach are, for example, the simpler and more orderly programming of complex program procedures, the ability to standardize and reuse frequently required program modules (subroutine method) and the simplification of program testing and startup (also refer to ► Sect. 10.3.4). The international IEC 61,131–3 standard reduces the various block types of individual PLC manufacturers to three standardized basic types (program, function and function block) [DIN14c]. As shown in ◘ Fig. 10.15, they are linked to runtime programs with call attributes (cyclic or time/event-driven and call priority) and assigned to an arithmetic unit (resource) of the hardware configuration [JOHN09].

### 10.3.3 PLC Programming

For programming PLCs, several basic programming languages have established themselves. These can be divided into mnemonic and graphic programming methods depending on the representation type (◘ Fig. 10.16).

While programming using *Instruction List (IL)* and *Structured Text (ST)* fall under mnemonic programming, the *Ladder Diagram (LD)* and *Continuous Function Chart.*

*(CFC)* are graphical programming methods (refer to Sects. 10.3.3.1 to 10.3.3.3). Structured Text, based strongly upon high-level language programming and including typical high-level language elements such as WHILE loops, IF and ELSE-IF statements, is discussed separately in 10.3.3.5. There is also another method in the standard that lends itself particularly well to the program structuring of more complex sequential controls. The *Sequential Function Chart (SFC)* used here is described in ► Sect. 10.3.3.6.

To generate programs, the actual PLC and a separate programming device are required for all programming methods. Programming devices range from specialist devices associated with the PLC to extensive programming systems for normal PCs. The benefit of the latter is that no manufacturer-specific programming device hardware is required.

In addition to offline programming on the programming device, online connection is also possible with direct data interchange with the PLC. In this mode, the system state of the PLC can be shown directly on the programming device so that, for example, the incremental testing of programs is possible and the responses of the controller program to external process signals can

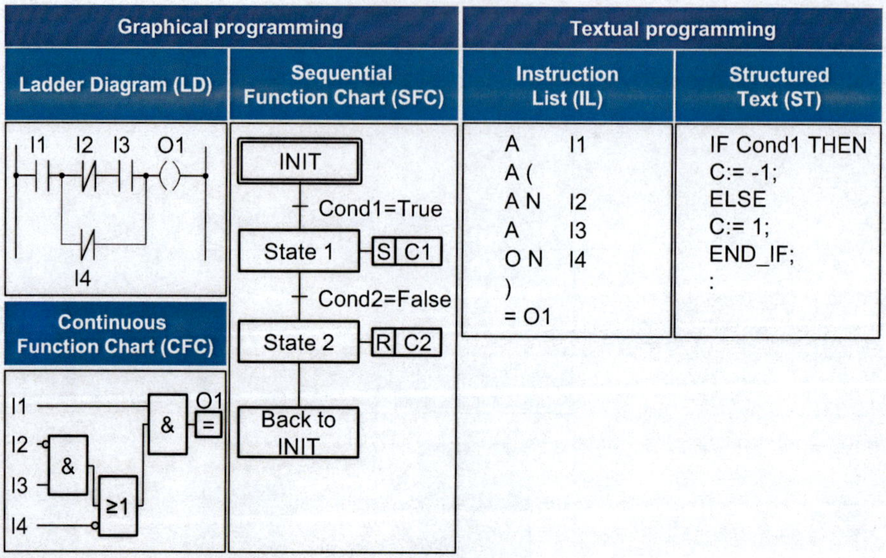


**Fig. 10.16** Conventional programming languages for PLC programming as per programming standard IEC 61,131, Part 3



be viewed directly on the screen. Simulation-aided feedback of external process signals to evaluate correct control-related functionality is also known as Hardware-in-the-Loop (HiL)—refer to ▶ Sect. 10.3.4.3.

Application-driven development in the PLC field has resulted in ever greater manufacturer-specific differences—for both programming languages as well as programming devices. This caused a problem for PLC users. Changing to another controller system necessitated time-consuming adaptation of user programs and additional financial overhead for acquiring new programming devices.

In the light of this situation, efforts were made to establish greater international standardization. The outcome of these endeavors is international PLC programming standard IEC 61,131, Part 3 [DIN14c]. The programming languages and Sequential Function Chart in graphical and text form, shown in Fig. 10.16, represent parts of the standard. The German version of IEC 61131–3 (previously IEC 1131–3) was published as Europe standard DIN EN 61131–3. It superseded the previous DIN 19239 standard. This meant other standards applicable for PLC programming were also rendered redundant by this, such as VDI Guideline 2880, DIN 40700 and DIN 40719.

On the basis of IEC 61,131–3, different manufacturers and users of controller and programming systems have joined forces to form the international PLCopen organization. This group is pushing for usage and dissemination of programming in accordance with IEC 61,131–3. PLCopen is pursuing the objective of simplifying the portability of user programs on devices from different manufacturers. For the certification of programming systems, four different levels of compliance to the standard have been defined by PLCopen. Programming systems currently available on the market, such as Step 7 [SIEM10], TwinCat 3 [BECK14], WinSPS [WEIß16] and cross-vendor systems such as Codesys [WELL14] and Accon ProSys 1131 [TRÖS11], mostly only conform to the IEC 61,131–3 standard on the first level (Base Level). This means the PLC programs generated are still not portable to the PLCs from other manufacturers without modifications.

The reason for this is that, while the standard defines precisely the attributes of the programming languages, it does not exclude vendors from integrating special, proprietary language elements. This is required, however, because it is precisely the special functionality that is often the feature distinguishing PLCs from different manufacturers. Even though user programs generated with IEC-compliant programming systems cannot be interchanged between the PLC systems of different manufacturers without additional effort, the basic language elements and program structure are still compliant (see ▶ Sect. 10.3.2.2). This simplifies the porting of PLC programs to third-party systems [JOHN09].

Prior to the introduction of the IEC 61,131–3 standard, a quasi-standard from the market leader established itself in Europe [SIEM93]. Despite the introduction of the standard, programming systems, as described above, still only permit broadly manufacturer-specific programming, however, the old quasi-standard is still well established. The program examples below in the Ladder Diagram, Continuous Function Chart and Instruction List are shown in accordance with the original syntax.

#### 10.3.3.1 Ladder Diagram Programming

Ladder Diagram programming or circuit diagram programming is the most widely established programming method in the US, accommodating PLC users who have previously developed controllers using relays.

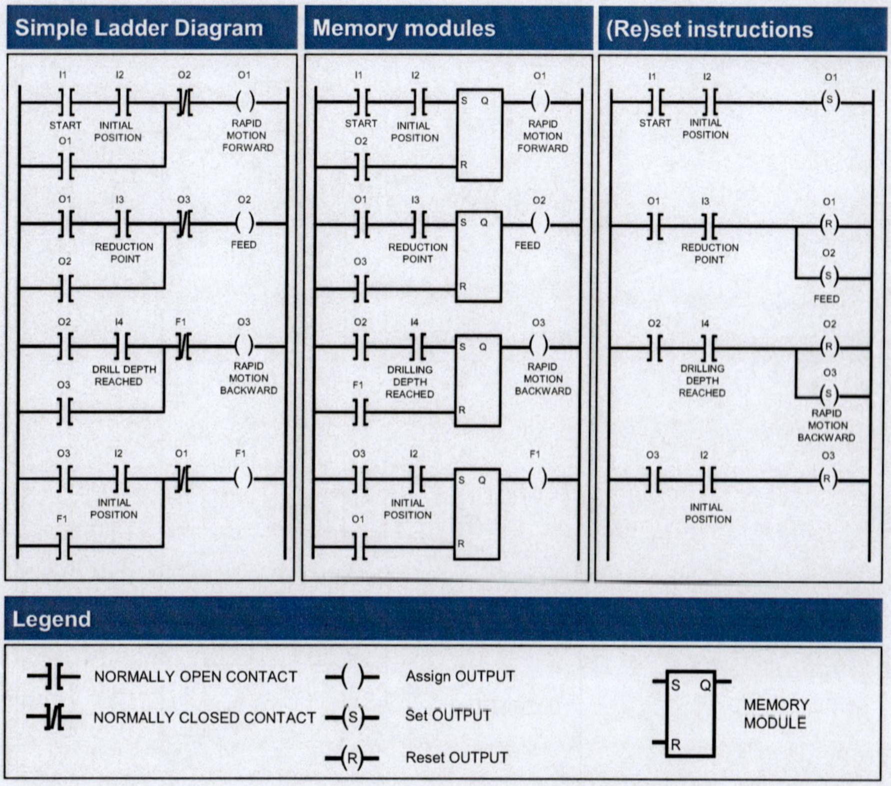


**Fig. 10.17** Ladder Diagram (LD) for the drilling machine of Fig. 10.4

An existing relay circuit diagram can be realized by substituting symbols, usually directly into the relevant PLC program. Figure 10.17 shows the implementation of the drilling machine sequential control from Fig. 10.4.

Inputs (Start button, cam-operated switches) are denoted I, outputs (rapid motion forward/backward, feed) are denoted O, suffixed with a differentiating digit. Start button S1 and cam-operated switches S2 to S4 are each connected individually to one PLC input, I1 to I4; the series connection of S1 and S2 in Fig. 10.7 is already part of the PLC Ladder Diagram. Flags (F) are used in the Ladder Diagram for relays not activating an output and that only perform a logical function (auxiliary relay). Since a PLC has only a limited number of outputs, flags can be used to store intermediate results. These can be queried again elsewhere in the program. In addition to relays with make and break contacts, the Ladder Diagram programmer has timing relays, clock generators, counters and other complex function blocks that simplify the creation of PLC programs (see ▶ Sect. 10.3.3.4).

Shown in the middle of Fig. 10.17 as an example is the Ladder Diagram described above using memory modules. They perform the function of a self-holding circuit. Output Q is set via Input S and reset via R. In the example, block outputs Q are connected to PLC outputs O1 to O3, or Flag F1. It is also possible to connect a block output to the input of the next block. If additional gate functions (such as AND and OR operations) and an inverter (providing the inverse value of the input signal) are used as function blocks, the Ladder Diagram becomes increasingly similar to a Continuous Function Chart—refer to ▶ Sect. 10.3.3.2.

Illustrated on the right of Fig. 10.17 is how Set (S) and Reset (R) instructions can simplify and shorten this Ladder Diagram even more: In contrast to the simple assignment that continually assigns the result of the logical operation to the outputs (Q) or the flags, using the memory modules (also called storage elements and flip-flops) means the values of the outputs/flags are set continuously using set/reset conditions. They also retain their values when the conditions are no longer satisfied. A set/reset condition is satisfied once the result of the logical operation at the input (S or R) is True. If the truth value is applied simultaneously to both inputs of the memory module, the type of storage element decides whether the preference is to set (set-dominant) or reset (reset-dominant) its output.

### 10.3.3.2 Continuous Function Chart Programming

A PLC program written as a Continuous Function Chart (CFC) corresponds to a wiring diagram made up of logic block types (AND, OR, INVERTER). Figure 10.18 shows the Continuous Function Chart

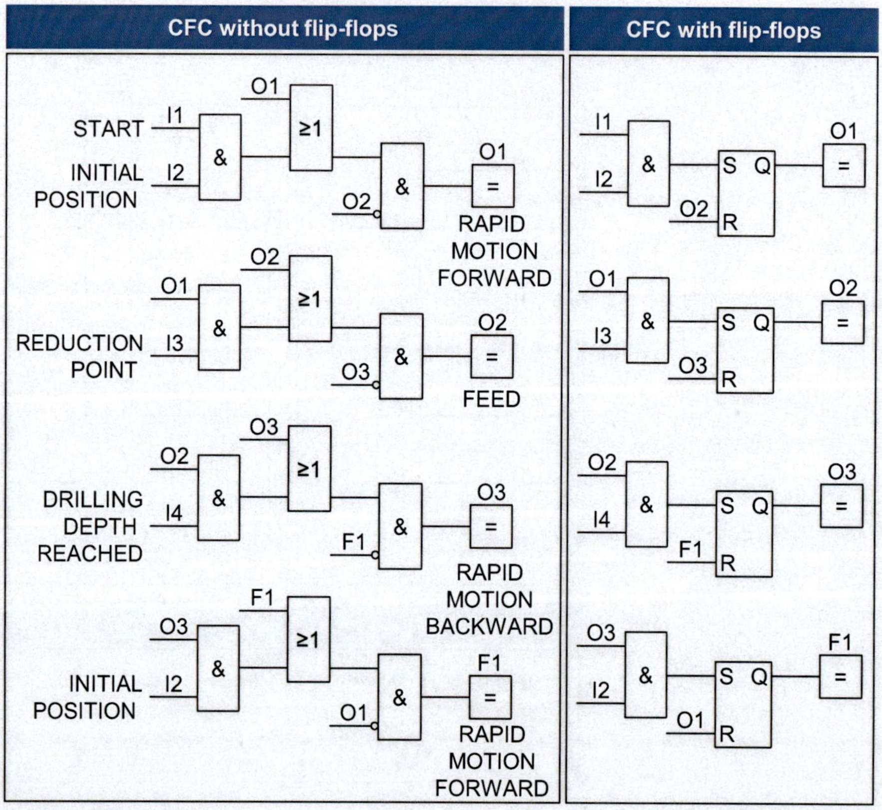


**Fig. 10.18** Continuous Function Chart for the drilling machine of Fig. 10.4



of the drilling machine from Fig. 10.4. The layout of the CFC can be simplified using flip-flops. The design of these elements is similar to the memory modules of the Ladder Diagram. In the CFC too, more function blocks are available to the user, such as delay elements, clock generators, counters and comparators (refer to ▶ Sect. 10.3.3.4).

### 10.3.3.3 Programming with Instruction List

The Instruction List (IL) can be created directly from the Boolean equations of the controller function as the Continuous Function Chart. The equations can be converted directly to an Instruction List using the mnemonic instructions. The "AND before OR" rule applies here just like in Boolean algebra. This can be bypassed using brackets. Figure 10.19 shows the Instruction List for the drilling machine with added comments. The numbering for the inputs and outputs corresponds to the previous examples.

As with the Ladder Diagram, the Instruction List can be simplified and shortened even more using set and reset instructions. Figure 10.20 shows the way to structure a program by using jump instructions. They cause the PLC, depending on logical conditions, to skip program sections, thus shortening the cycle time. This is shown here using a "step splitter". Using a step splitter shortens the cycle time of the program because only the instructions of one step are processed, not the entire PLC program.

### 10.3.3.4 Examples of More Complex Program Instructions

The previous chapters have explained the basic Ladder Diagram, Continuous Function Chart and Instruction List programming methods, using the drilling machine of Fig. 10.4 as an example. Now some more complex program instructions are introduced.

#### Time Functions

One function required very often in PLC programs is the timing relay or timer. Switch-on and off delays of any duration, wait and monitoring times as well as pulse modules and clock generators can be realized, for example.

Figure 10.21 shows an example use of a timing relay to implement a switch-on delay. In addition to the circuit and timing diagrams for explaining the mode of operation, programming using the Ladder Diagram, CFC and Instruction List is also shown for comparison purposes. In the Ladder Diagram and CFC graphical programming languages, the time function is indicated by a rectangular symbol, with the inputs on its left and the outputs on its right.

**Fig. 10.19** Instruction List for the drilling machine of Fig. 10.4

| Program code | Instructions |
|---|---|
| ( | AND operations |
| A I1 START | A Ix Input x |
| A I2 INITIAL POSITION | A Ox Output x |
| O O1 SELF-HOLDING STATE | A Fx Flag x |
| ) | |
| AN O2 RESET BY FEED | OR operations |
| = O1 RAPID MOTION FORWARD | O Ix, O Ox, O Fx |
| | |
| ( | Inversions |
| A O1 RAPID MOTION FORWARD | AN Ix, AN Ox, AN Fx NOT AND |
| A I3 REDUCTION POINT | ON Ix, ON Ox, ON Fx NOT OR |
| O O2 SELF-HOLDING STATE | |
| ) | = Ox Assign Output x |
| AN O3 RESET OF RAPID MOTION BACKWARD | = Fx Assign Flag x |
| = O2 FEED | S Ox SET Output x |
| | S Fx SET Flag x |
| ( | |
| A O2 FEED | R Ox RESET Output x |
| A I4 DRILL DEPTH REACHED | R Fx RESET Flag x |
| O O3 SELF-HOLDING STATE | |
| ) | JP x JUMP to Line x |
| AN F1 RESET OF FINISH | CJP x CONDITIONAL JUMP i.e., depending on the previous operation) to Line x |
| = O3 RAPID MOTION BACKWARD | |
| ( | |
| A O3 RAPID MOTION BACKWARD | |
| A I2 INITIAL POSITION | S Cx SET COUNTER x |
| O F1 SELF-HOLDING STATE | R Cx RESET COUNTER x |
| ) | CU Cx COUNT UP COUNTER x |
| AN O1 RESET OF START | CD Cx COUNT DOWN COUNTER x |
| = F1 FINISHED | |
| | SET x SET SWITCH-ON DELAY with TIMER x |

Input I 5.3 is used to start the timing relay and Input I 5.2 to reset it. The delay time is loaded as a constant at the input with ID "TW". The duration is composed by a factor and by the time base (factor · time base). Values 0.0 (= 10 ms), 0.1 (= 100 ms), 0.2 (= 1 s), and 0.3 (= 10 s) are available for the time base. Constant CT 500.0 (500 · 10 ms) therefore gives a time of 5 s. Outputs "DU" and "DE" show the current duration (DU in dual coding, DE in BCD coding). Output O 5.2 is the actual "set output", that is set or reset in line with the programmed time profile.

This means the function of the timing relay is such that O 5.2 is set with a delay of 5 s after I 5.3. The input only remains set for as long as the signal "1" is applied to I 5.3. The output signal (O 5.2) remains in state "0" if the start signal (I 5.3) is reset within the duration programmed (refer to the timing diagram in Fig. 10.21). The reset signal (I 5.2) causes the output signal to be reset regardless of the state of the timing relay.

## Counters

Counters are used for PLCs, for example, for measuring sets (such as quantities, liquid amounts and weights) and for forming very long delay times by connecting several counters in series. Actual counting is performed directly by the CPU of the PLC. PLCs have multiple counters, that can be used independently of each other and can count up and down. A counter counting up allows, for example, the number of workpieces on an assembly line to be counted and displayed while counters counting down start from an initial (programmed) state and count down to zero, issuing a signal (such as "Pallet fully loaded") when the counter has reached zero. The function and programming for a down counter are explained in Fig. 10.22.

Counters are available in the Ladder Diagram and CFC as separate symbols as for the timing relay. Every counter has connections for counting up (*CU*) and counting down (*CD*), setting (*S*) and resetting (*R*), loading a counter value (*CV*) and polling a counter read-

**Fig. 10.20** Instruction List with set and reset conditions (a) and jump instructions (b)

**Reset condition (a)**

```
A   I1   START ?
A   I2   INITIAL POSITION ?
S   O1   RAPID MOTION
         FORWARD ON

A   O1   RAPID MOTION FORWARD ?
A   I3   REDUCTION POINT ?
R   O1   RAPID MOTION FORWARD
         OFF
S   O2   FEED ON

A   O2   FEED ?
A   I4   DRILL DEPTH REACHED ?
R   O2   FEED OFF
S   O3   RAPID MOTION
         BACKWARD ON

A   O3   RAPID MOTION
         BACKWARD ?
A   I2   INITIAL POSITION
R   O3   RAPID MOTION
         BACKWARD OFF
```

**Jump instruction (b)**

```
STEP SPLITTER:
1   A     F1   STEP 1 ACTIVE ?
2   CJP   12   JUMP TO STEP 2
3   A     F2   STEP 2 ACTIVE ?
4   CJP   18   JUMP TO STEP 3
5   A     F3   STEP 3 ACTIVE ?
6   CJP   24   JUMP TO STEP 4

               STEP 1:
7   A     I1   START ?
8   A     I2   INITIAL POSITION ?
9   S     O1   RAPID MOTION
               FORWARD ON
10  S     F1   SET STEP FLAG 1
11  JP    28   TO PROGRAM END

               STEP 2:
12  A     I3   REDUCTION POINT ?
13  R     O1   RAPID MOTION
               FORWARD OFF
14  S     O2   FEED ON
15  R     F1   CLEAR STEP FLAG 1
16  S     F2   SET STEP FLAG 2
17  JP    28   TO PROGRAM END

               STEP 3:
18  A     I4   DRILL DEPTH REACHED ?
19  R     O2   FEED OFF
20  S     O3   RAPID MOTION
               BACKWARD ON
21  R     F2   CLEAR STEP FLAG 2
22  S     F3   SET STEP FLAG 3
23  JP    28   TO PROGRAM END

               STEP 4:
24  A     I2   INITIAL POSITION ?
25  R     O3   RAPID MOTION
               BACKWARD OFF
26  R     F3   CLEAR STEP FLAG 3
27  JP    28   TO PROGRAM END

28  PE         PROGRAM END
```



ing (*DU* – dual counter reading, *DE* – counter reading in BCD format). In the example in Fig. 10.22, Counter 1 (*C*1) is set to counter value 4 when Input 4.1 (Set) is activated. "CCV" stands for constant counter value. Signal "1" now appears at Output 2.5. Every time Input 4.0 is activated (counting down), the counter value decrements by 1. Output 2.5 is set to "0" when the counter reading is "0." Setting Input 4.1 again reloads counter value 4, even before the counter has reached zero (refer to the timing diagram in Fig. 10.22).

### Word-oriented Processing

PLCs today can not only perform simple binary operations such as logical operations, but they also allow comprehensive byte (8-bit), word (16-bit) and double word (32-bit) operations. This means calculations, comparisons and (closed-loop) control are also possible with PLCs. Similarly, this makes possible the processing of analog signals, although an analog/digital (A/D) or digital/analog (D/A) conversion is necessary here since the PLC can only process numbers.

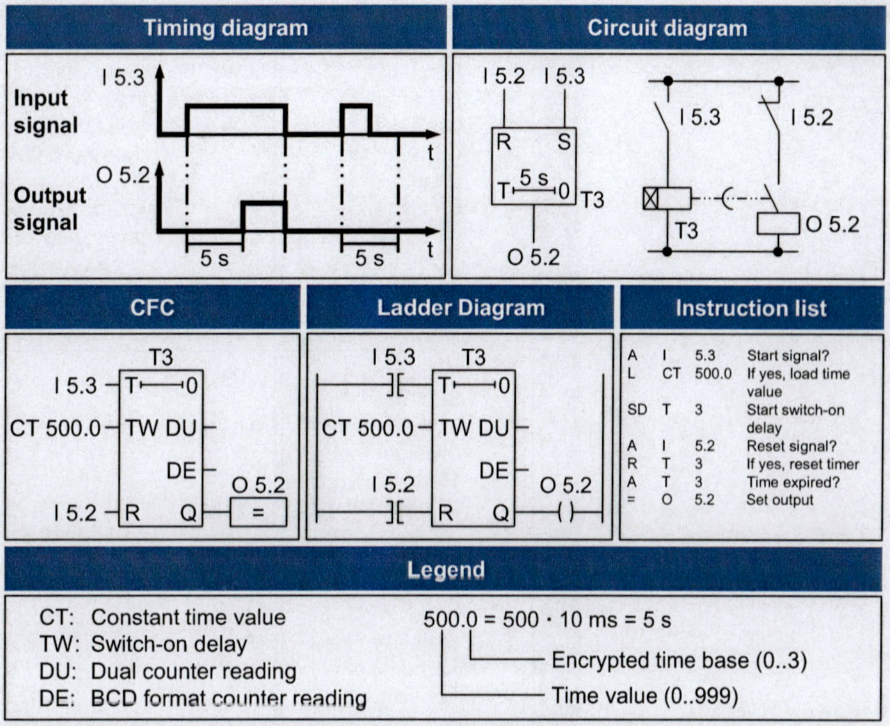


**Fig. 10.21** Sample implementation of a switch-on delay

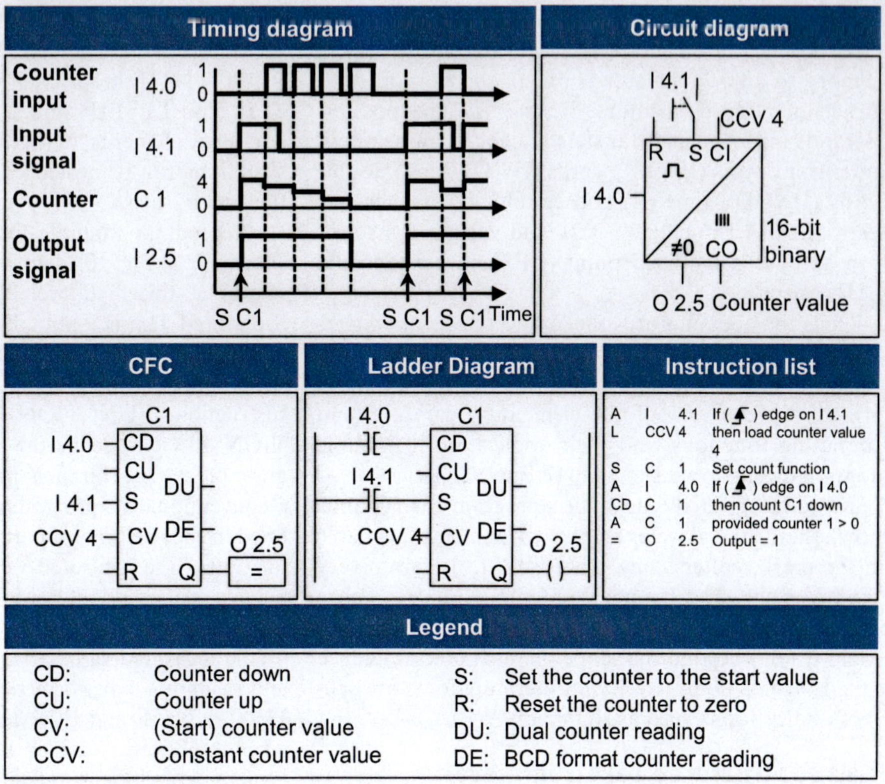


**Fig. 10.22** Sample implementation of a down counter

Important functions in byte and word processing are load and transfer functions. These allow information interchange on a digital basis between peripheral modules and the PLC and between different data areas inside the PLC. This information interchange is not "direct" as such, and instead is always associated with

**Fig. 10.23** Command overview and program example for word-oriented processing (compare and computing functions)

**Compare functions**

| Command overview | | Program example | |
|---|---|---|---|
| != | Equal to | L IW 10 | Load value of input word IW 10 (16-bit value) |
| >< | Not equal to | L DW 22 | Load value of data word DW 22 (32-bit value) |
| > | Greater than | ! = F | IW 10 equal to DW 22? (16-bit fixed point comparison) |
| >= | Greater than or equal to | = F 170.0 | If yes, flag 170.0 = "1" (1-bit value) |
| < | Smaller than | > F | IW 10 greater than DW 22? |
| F | 16-bit fixed point | = F 170.1 | If yes, flag 170.1 = "1" (1-bit value) |
| D | 32-bit fixed point | | |
| G | 32-bit floating point ① | | |

**Arithmetic functions**

| Command overview | |
|---|---|
| + | Addition |
| - | Subtraction |
| X | Multiplication |
| : | Division |
| F | 16-bit fixed point |
| G | 32-bit floating point |

① Possibilities vary depending on processor type

Program example:

Problem:

<DW 30> = <DW 28> + 3 x <FW 196>

(3: Constant factor; <FW 196>: Contents of flag word 196)

Solution:

| | |
|---|---|
| L CB 3 | Load Factor 3 as a constant (byte) |
| L FW 196 | Load the contents of flag word FW 196 |
| XF | Multiplication (16-bit fixed point multiplication) |
| L DW 28 | Load the contents of data word DW 28 |
| +F | Addition of DW 28 to the subtotal |
| T DW 30 | Transfer the end result to data word DW 30 |

a "detour" via special intermediate registers of the processor known as accumulators. Information flow from memory to an accumulator is called *Load L* and from an accumulator to memory *Transfer T*. The possible operands for loads and transfers are, for example, input and output bytes (IB/OB), words (IW/OW) and double words (ID/OD), time (T) and counter (C) values (see Sects. 10.3.3.4.1 and 10.3.3.4.2), and various constants such as byte (CB), fixed point (CF) and hexadecimal (CH) constants.

Once the operands are loaded into the accumulators of the PLC processor using the load functions, comprehensive compare and computing functions can be performed. An overview of the elementary compare and computing functions, and their application in a short program example, are shown in ◘ Fig. 10.23.

In addition to the advanced program instructions shown here, the development environments available on the market offer many other ways to generate very comprehensive and complex programs. Since a more extensive discussion of the programming capabilities would extend beyond the scope of this book, reference is made at this point to relevant literature for more specifics on this topic, such as [JOHN09; WELL14].

#### 10.3.3.5 Structured Text (High-level Language Programming)

As indicated in ► Sect. 10.3.3, a high-level language has been defined in the IEC 61,131–3 PLC programming standard with "Structured Text" (ST). Its development was given impetus by PASCAL as well as ADA, MODULA and C [MITT91]. WHILE and REPEAT UNTIL loops, and other control structures such as IF and ELSEIF, and CASE statements, are typical language elements of Structured Text. Individual instructions can be nested in any way in compliance with the syntax rules.

A program example for sorting measured values is shown in ◘ Fig. 10.24 to explain the syntax and capabilities of this high-level PLC language. A program in Structured Text is essentially split into a declaration part in which all variables are defined symbolically and an instruction part containing the actual program instructions. Instructions FUNCTION_BLOCK and END_FUNCTION_BLOCK define the start and end of the ST program block that can then, for example, be integrated into a conventional IL program as a separate subroutine. Defined in the declaration part of the example program are, for saving the measured values, a field with 10 components of data type Integer and the required constants and local program variables. In the instruction part, sorting of the measured value array with continuous comparing and swapping is performed in the two nested loops (REPEAT UNTIL and FOR loops). The process is stopped when the Boolean variable SWAP is logical "1", i.e., there is no more swapping, and all values are sorted.

As the example shows, more challenging tasks can also be programmed compactly using Structured Text. A corresponding program in the Instruction List would be considerably longer and thus also unwieldy.

Fig. 10.24 Example program for sorting measured values in Structured Text (ST)

Declaration part

```
FUNCTION_BLOCK

CONST
   VALUES        := 10;
   MAX           := VALUES - 1;
END_CONST;
VAR_IN_OUT
   VALUE_FIELD :ARRAY[0..MAX] OF INT;
END_VAR;
VAR
   HELP          :WORD;
   SWAP          :BOOL;
   I             :INT;
END_VAR;
```

- Block start
- Declaration of the constants
- Declaration of in-out parameters (one-dimensional measured value array)
- Declaration of local program variables

Instruction part

```
BEGIN
REPEAT
 SWAP := TRUE;
 FOR I := MAX TO 1 BY -1 DO
  IF VALUE_FIELD[I]<VALUE_FIELD
             [I-1]THEN
   HELP     := VALUE_FIELD [I];
   VALUE_FIELD[I]:= VALUE_FIELD[I-1];
   VALUE_FIELD [I-1]:= HELP;
   SWAP            := FALSE;
  END_IF;
 END_FOR;
UNTIL SWAP;
END_REPEAT;
END_FUNCTION_BLOCK;
```

- Start of the instruction part
- Start of the repeat loop
- Comparison of the measured values in pairs and swapping as required for sorting
- Termination cond. for the loop
- End of the repeat loop
- Block end

Another approach for high-level PLC language programming is using a generally adopted high-level language with IT standardization such as C. Some PLC manufacturers have already developed special C-code compilers for this, enabling PLC-specific target code to be generated automatically from program code (such as ANSI-C) written by the user. It is also possible to combine high-level language programming with other familiar PLC languages. For example, a program can be created with a Ladder Diagram, CFC or IL and only contain individual program modules programmed in C, such as complicated control algorithms.

#### 10.3.3.6 Sequential Function Chart

The SFC (Sequential Function Chart) defined in the IEC 61,131–3 standard [BRUG99]) does not in itself represent any new programming language. Rather, it is a convenient way of logically structuring and graphically programming sequential controls (Fig. 10.25, see ▶ Sect. 10.1.2). The Sequential Function Chart offers benefits, in particular for complicated controller sequences with multiple parallel processes and branches. Programming using the programming methods described thus far is very complex and confusing here.

Steps and transitions are the basic elements of the Sequential Function Chart. Every step comprises a program section that is run until a condition specified in the transition is satisfied. One example of a simple sequential control with only one process chain, with no branches or parallel processes, is the drilling machine of Fig. 10.4. The system corresponds exactly to the process chain with steps (such as "Rapid motion forward On") and transitions ("Until reduction point is reached"). If the drilling machine is considered part of a bigger system, also including, for example, sawing, turning and milling machines, a process as in Fig. 10.25, defining the overall system, would be conceivable.

After the blanks are sawn in Step 3, either turning (S4 and S5) or milling (S6) is performed depending on the workpiece (Transitions T4 and T6). This is followed by drilling (S7) ("OR" branch). The outer and inner contours are turned at the same time in the turning process ("AND" branch). Then the parts are installed in Step 8.

To program this example using the Sequential Function Chart, the higher-level program procedure is specified on the overview level by entering the

**Fig. 10.25** Example of a complex flowchart with overview and detailed views

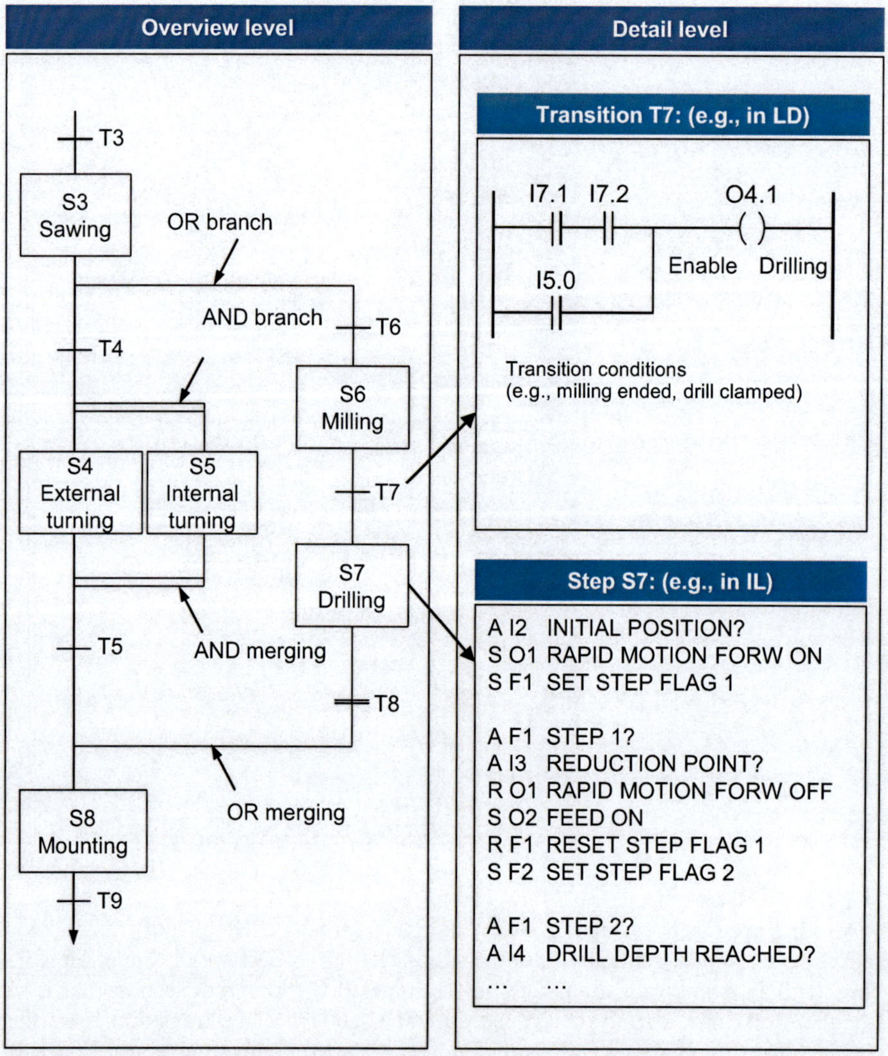




steps and transitions graphically/interactively. Then, in the detailed view, the logic for the individual steps and transitions is programmed. The user is able to select from the different programming methods available. In Fig. 10.25 for example, transition condition T7 is shown in the Ladder Diagram view, and Step S7, the familiar example of the drilling machine, in the Instruction List (IL). The program for the drilling machine represents a separate sequencer and could equally have been programmed again as a separate process chain as in the overview.

Programming of sequential controls is greatly simplified using the Sequential Function Chart because the higher-level program logic has already been specified with the creation of the overview level. The explicit, normally complicated, programming of the program logic is unnecessary, enabling the programmer to concentrate on programming the modules for steps and transitions. The Sequential Function Chart supports structuring of the control task and the systematic top-down procedure in program development.

### 10.3.4 Procedure for Systematic Development of Complex PLC Programs

The increased performance and the scope of tasks performed by today's PLCs have greatly expanded both the complexity and the scope of PLC programs. Control systems with over 1,000 inputs and outputs and comprehensive programs with high storage requirements are not uncommon in the machine tool industry. Software development costs as a percentage of overall development costs are correspondingly high for machine tools. Given the reduction in hardware prices, the cost for a simple PLC today corresponds roughly to the daily rate of a qualified PLC programmer.

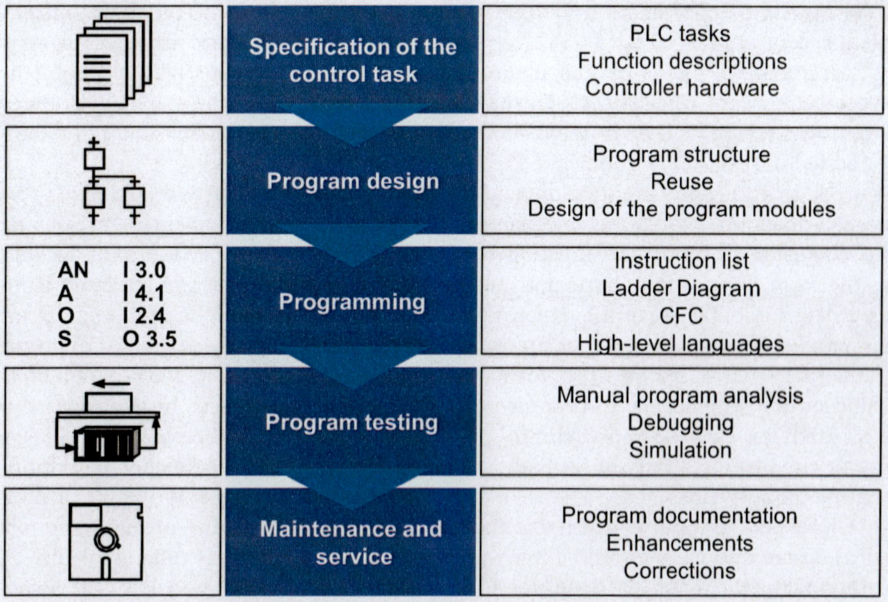


**Fig. 10.26** Phases of PLC software development

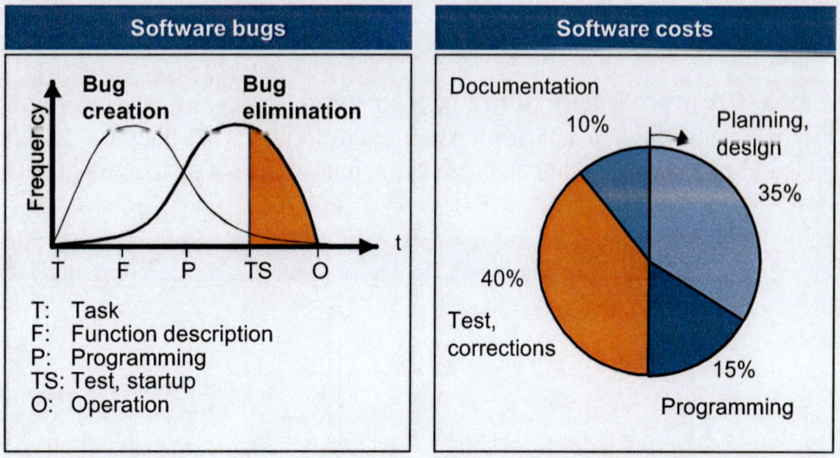


**Fig. 10.27** Bugs and costs in PLC software development, according to [OEST86]

This illustrates that PLC software development has become a key cost factor. The systematic organization of the PLC software development effort, starting from the **specification of the control task through to program testing, is thus of major significance.** This is especially true for machine tool manufacturers with high custom programming overhead (for special-purpose machines, for example). Figure 10.26 shows an overview of the basic procedure and individual phases of a systematic PLC software development effort.

#### 10.3.4.1 Specification of the Control Task

Indicative for the industrial development of PLC software is the fact that most software bugs do not occur when programs are written but in early development phases, during specification of the control task. Most bugs, however, are not fixed until the testing and startup phases (refer to Software bugs in Fig. 10.27). The cost for the testing phase as a portion of the overall development costs is correspondingly high—refer to the software costs in Fig. 10.27.

So, the most important condition for developing complex PLC programs and avoiding costly consequential errors is **the systematic specification of the control** task, i.e., the collection and documentation, in a de**velopment specification, of all machine/system information relevant to control. The development specifi**cation, however, should not only be the basis for PLC software development. Rather, it should serve as a central, continually updated work basis for all departments involved over the entire order processing period—from Sales, Mechanical engineering and Electrical design through to testing and startup. Within a company, Engineering Data Management (EDM) systems can be used instead of coordination by one or more development specifications.

Although the merits of informational development specifications are beyond dispute, they are rarely created to an adequate level in the machine tool industry. In actual industrial environments, the key information sources for PLC software development are preliminary design documents from the mechanical design, such as schematic diagrams and function descriptions, e.g., in the form of function charts and function diagrams, as well as a tabular list of all actuators (motors, valves, etc.) and sensors (limit switches, transducers, light barriers, etc.) used. Based on this information, the control engineer first specifies the PLC hardware (CPU, peripheral modules, power supply, controls, switch cabinet, safety circuits, etc.), but can in parallel also start with writing the software.

Close cooperation between mechanical and electrical design and early coordination with regard to the solution are the basic prerequisites for creating PLC programs that are as fault-free as possible. Actual industrial environments however are showing that this is not happening to a great enough extent and is a major reason behind misunderstandings and bugs.



#### 10.3.4.2 Program Design and Programming

Once the control task is sufficiently known and documented, the software generation process can commence. Some of the main objectives in industrial practices are to standardize software structures and programming methods within a company, to improve the readability and transparency of the programs and to increase the level of reusability of the software modules already developed.

The first step for the Software department is to analyze, and supplement as required, the control task specified. On the basis of this information, the fundamental program structure is first designed and then a check is made to find program modules from previous projects, already developed and tested, that can be reused. In the next step, the program modules to be developed anew are designed. In this process, additional specifications (such as the operating panel layout) need to be analyzed alongside the function descriptions from mechanical design.

A key principle in program design is the unique assignment of program modules to corresponding mechanical functional units. If standardized modules are used, this allows the associated software modules to be reused, often without modification. A prerequisite for this is a standardized design of the modules, based on defined interfaces, such that the internal functionality of the modules can remain unknown provided there is knowledge of the interfaces. This procedure is generally known as "information hiding" in software engineering.

Figure 10.28 illustrates, using a program module for controlling a tool changer as an example, how

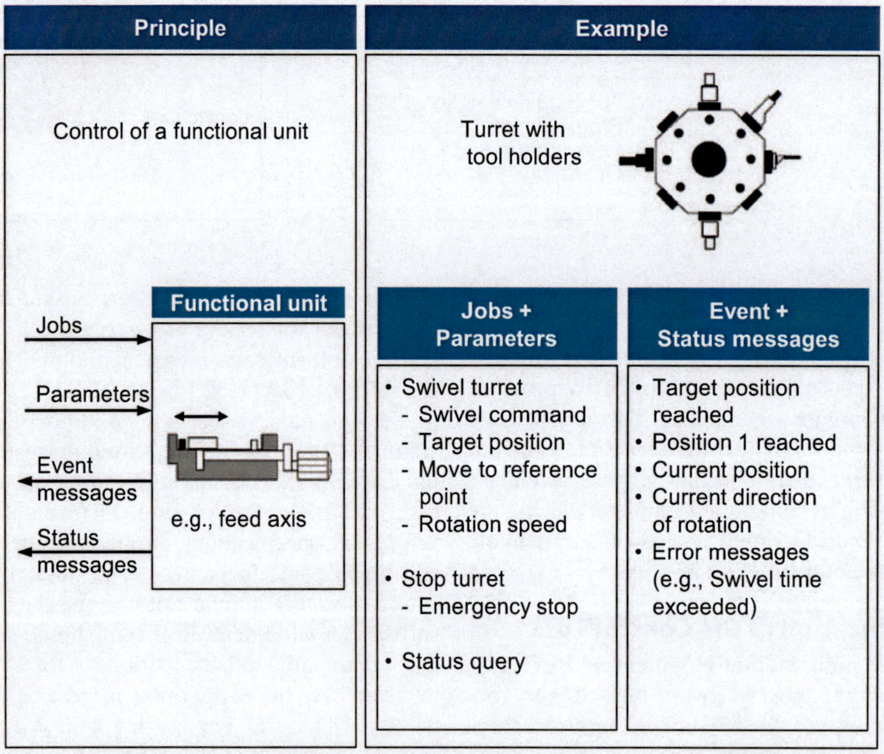


Fig. 10.28 Structuring of PLC software modules aligned to the mechanical functional units

such an arrangement of PLC software modules can be realized on the mechanical modules. The interfaces to a master program module are reduced to defined commands (such as "Swivel turret") that are defined more precisely with corresponding parameters (such as "To target position 7"). The feedback signals from the program module are event messages (such as "Target position reached" and "Fault occurred") and status messages (such as exact error code number and current position or direction of rotation).

Following the design of the program modules is the actual implementation using a programming device. Here, all the programming methods described in ▶ Sect. 10.3.3 can be used depending on the task at hand and the programmer's individual preferences. The next important step is testing the programs developed.

#### 10.3.4.3 Program Testing

The key objective in testing PLC programs is to provide, at the beginning of the startup process, programs that have undergone extensive testing. The goal of this testing is to reduce the generally high software test overhead as part of the overall startup process [WECK90; SOSS89].

A control panel is a simple test aid for PLC programs in common use earlier at many companies (◘ Fig. 10.29). It comprises an input panel with switches and buttons that are connected to the PLC inputs and a series of lamps displaying the PLC outputs. There are also potentiometers and analog displays for analog signals. Particular input states of the machine can be set specifically using the switching elements. The controller reads these signals, processes them in line with the PLC program and sets the relevant output signals. If the PLC reaction displayed on the lamps does not correspond to expectations or requirements, the software must be modified accordingly. Some PLC manufacturers offer special simulation cards to save users from having separate test setups.

The key downside of this simple test method is that the user must manually emulate the process behavior of the machine. This procedure requires a high level of abstraction and only replaces the time behavior of the machine approximately, meaning it is virtually impossible to test dynamic processes. There is also a lack of clarity with this method, meaning complex controller processes are virtually incomprehensible. The frequent requirement, therefore, is to simulate the time behavior of the machine to generate the feedback signals provided by the sensors and limit switches at the correct time as a function of the PLC output signals. A more favorable solution is a computer-aided simulation of the logic and time-based process behavior of the machine.

◘ Figure 10.30 illustrates the basic principle of a test environment for PLC software. The ready-to-use PLC is connected to a simulation computer that simulates the control-related process behavior of the machine. From the viewpoint of the PLC, the simulation computer acts like the machine. It receives commands in the form of output signals arriving from the PLC and generates, based on the machine behavior specified in a simulation model, appropriate feedback signals (time-synchronized to each other) from the transducers and sensors. The corresponding test method where the actual controller electronics are tested using simulated input signals is called Hardware-in-the-Loop (HiL).

##### Hardware-in-the-Loop simulation

In Hardware-In-The-Loop simulation, the controller is linked into the simulation environment. Decentralized input and outputs are simulated using a plug-in PC card. This enables the mechatronic behavior of the machine to be planned and verified on the control computer. Original NC and PLC hardware is used for simulation. Simulation is performed on the computer that simulates the properties of the components connected. Connecting to the actual controller means there can be

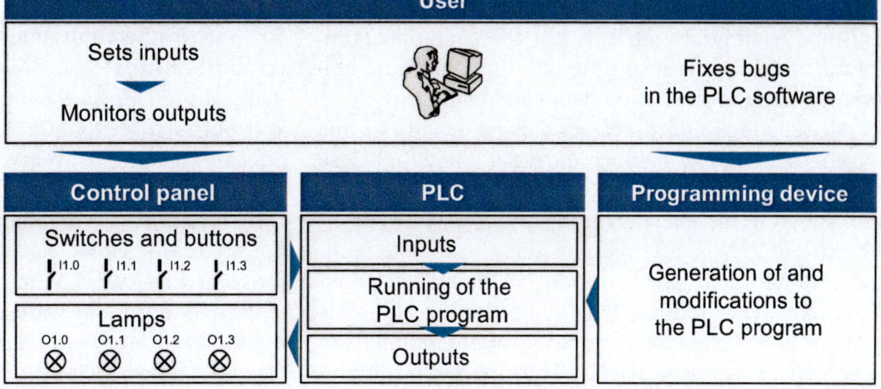


◘ Fig. 10.29 Simple test method for PLC programs (no longer used)

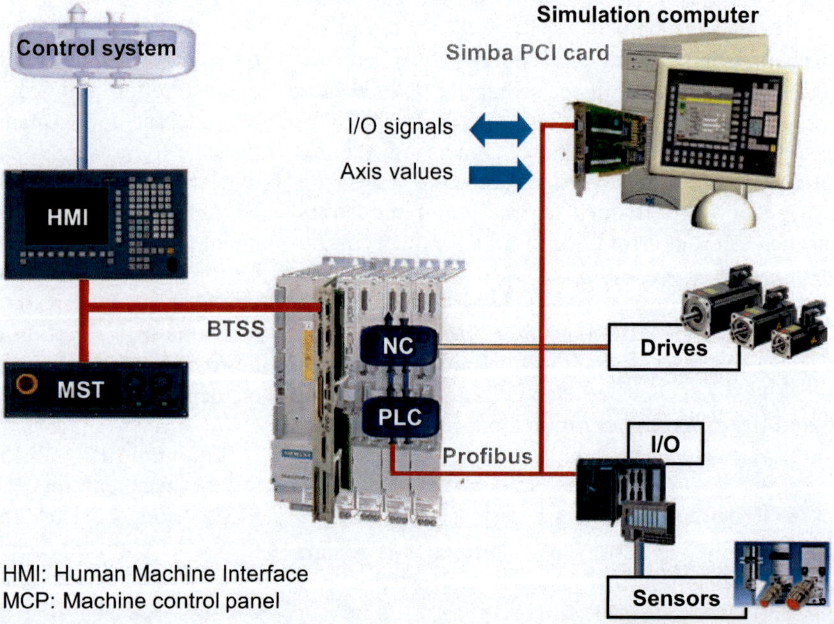


**Fig. 10.30** Setup of a Hardware-in-the-Loop test environment



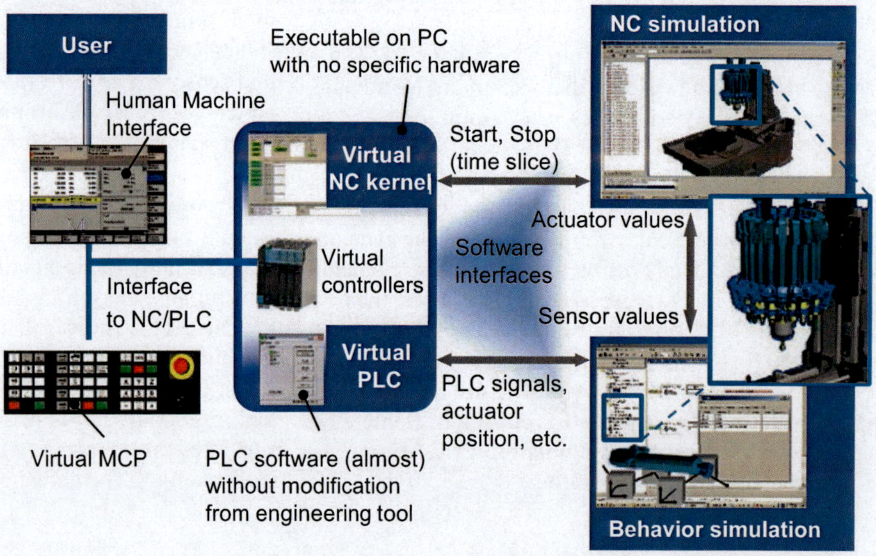


**Fig. 10.31** Software-in-the-Loop simulation: Full control simulation of machine tools

real-time communication between the components involved. It is important to note that this procedure limits somewhat the simulation depth and complexity.

All functions and processes can be visualized and logged in a suitable manner during a simulation test. Consequently, the user can check whether the results correspond to the machine behavior expected. By altering process signals, it is also possible to simulate faults and external factors and to monitor the associated effects. Experience tells us that fault detection and diagnostic routines often make up for far more than half of controller programs [SEIF92]. Thus, particular importance is attached to this capability.

### Software-in-the-Loop simulation

In addition to HiL testing, it is also possible to simulate the PLC hardware, together with the inputs and outputs [HERF10]. This enables inputs and outputs to be set and observed directly on the programming computer allowing a program to be tested in the early program development phase. The corresponding method is called Software-in-the-Loop (SiL)—refer to Fig. 10.31. When Software-in-the-Loop simulation is used, software can be tested without having the hardware used later for implementation. In the example shown, a machine tool is simulated in full. This enables NC programs, for example, to be run with the behavior of the tool changer emulated in the simulation.

### Opportunities and risks of simulation applications

Using simulation environments provides greater process understanding and production systems testing to be performed without damage, allowing more efficient usage of resources. The formalization of test processes enables controller tests to be automated easily. The coupling of simulation systems also permits the identification of interactions within the production system. Simulation has limitations as well as benefits. For example, the modeling overhead can sometimes be very high and so not cost-effective (depending on the level of detailing and scope in question). Also, models only ever represent just a part of the actual situation, meaning that significant effects might be overlooked. In general, the value added by a simulation is mainly determined by the underlying process understanding and technical and organizational realization. Coupling mechanisms require further analysis and research even in the future.

### Bus Systems

In production and assembly systems, input and output signals generally originate in several remote components, peripherals and machines that are connected to the PLC. To minimize the wiring overhead between the components involved in communication, the signals flow via bus systems. They combine multiple pieces of information and distribute them to the bus users. Bus systems differ, for example, in terms of real-time capability, fail-safe operation and intrinsic safety (such as in potentially explosive areas) and are selected depending on system requirements. A bus architecture defines how bus users are configured or networked (e.g., in a ring or star arrangement). Refer to Sects. 9.3.3 and 9.3.4 for a detailed description of the different bus systems and architectures together with their most important properties.

## 10.4 Safety Controllers

As control errors in automated machines and production systems can cause serious accidents entailing physical injury and material damage, fail-safe controllers are used in safety–critical applications. These guarantee defined behavior of the overall system in the event of internal controller faults or malfunctions at the process interfaces.

◘ Figure 10.32 shows an overview of possible measures (explained below) for increasing the levels of

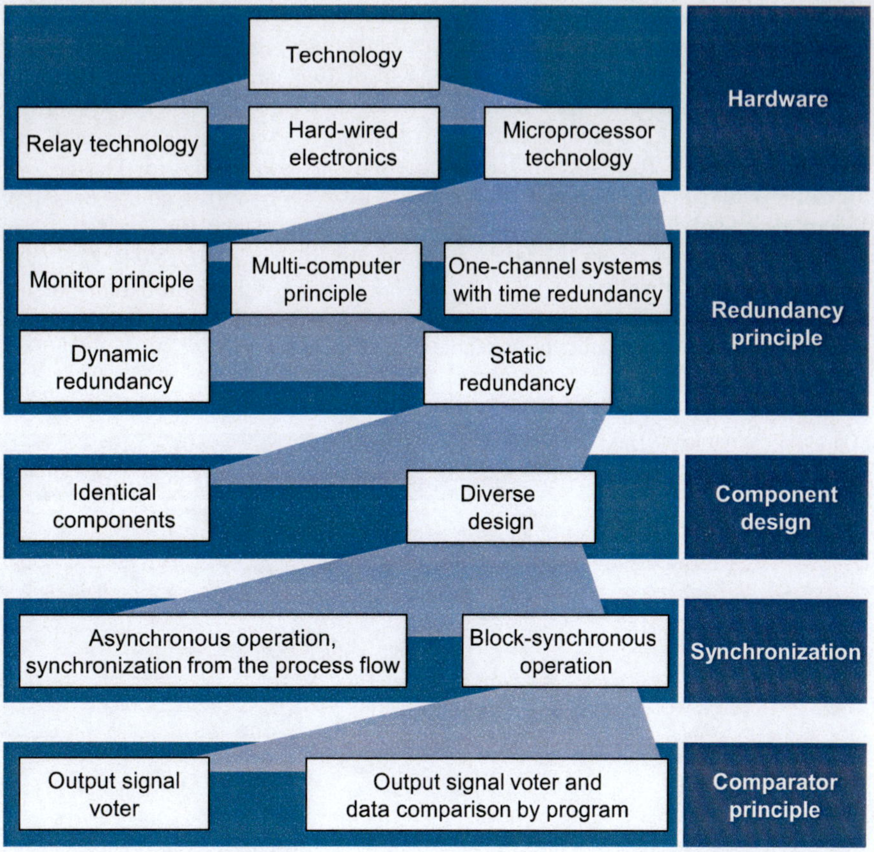


◘ **Fig. 10.32** Categorization of the measures for designing fail-safe controller systems

safety and availability of controller systems [FREN87]. In the manufacturing technology sector, for example, presses and machine tools are equipped with safety controllers protecting vertical shafts from falling. In addition to classic relay controllers with redundancy, programmable logic controllers and even fieldbus systems have prevailed for use in safety-critical application areas [PILZ13].

The main function of a fail-safe controller must be to enter a defined, safe state immediately after a fault is detected. For machine tools and industrial robots, this can mean disabling drives, for example. In contrast, increased importance is attached to tolerating faults for systems not able to immediately enter a safe state. An airplane in the air, for example, cannot immediately enter a safe state after the failure of a safety-relevant component. Such cases require fault-tolerant controllers that are able to correctly resume their original tasks after detecting faults. Whether simple fault detection is sufficient in the design of a fail-safe controller or whether fault tolerance is needed ultimately depends on the function in question.



### 10.4.1 Machinery Directive

December 29, 2009 saw the national implementations of the 2006/42/EC Machinery Directive take effect in the European Economic Area (EEA) [EURO06;EBER12]. The new version of the Machinery Directive was developed as part of the harmonization of statutory provisions within the EC to abolish trade constraints and to create standardized legal regulations pertaining to safety. To market machines and systems within the EEA, machine manufacturers must satisfy obligations as laid down by the Directive:

- The machine must satisfy the safety requirements in Annex 1 of the Directive.
- Technical documentation of the system, on risk assessment in particular, must be available.
- Documentation (operating instructions) required for operation must be available.
- A conformity assessment procedure must be performed.
- The EC declaration of conformity must be issued and included with the machine.
- The CE marking must be on the machine.

General information on the safe design of machines and systems is provided in Volume 2 of this book series. Further, more specific requirements also apply to control engineering. Specific requirements made of controllers and control devices, for example, are explicitly detailed in § 1.2 of Directive Annex 1. Some examples of these requirements:

- The machinery must not start unexpectedly.
- The machinery must not be prevented from stopping if the stop command has already been given.
- The emergency stop function must be available and operational at all times, regardless of the operating mode.
- Furthermore, electrical controls automatically fall under the Low Voltage and EMC Directives. The regulations of these Directives must also be satisfied.

### 10.4.2 Performance Levels

To assess the reliability of safety functions in a standardized manner, Performance Levels (PLs) are defined in EN ISO 13849-1, "Safety of Machinery—Safety-Related Parts of Control Systems—Part 1: General Principles for Design." A PL is a "discrete level specifying the ability of safety-related parts of a controller to perform a safety function under predictable conditions" [DIN08]. ◘ Figure 10.33 defines a range of the average probability of a dangerous failure of the controller system per hour for each of the five Performance Levels.

◘ **Fig. 10.33** Performance Levels and probabilities of failure, according to [DIN08]

| Level (PL) | Average probability of a hazardous failure per hour [1/h] |
|---|---|
| a | $\geq 10^{-5}$ to $< 10^{-4}$ |
| b | $\geq 3 \times 10^{-6}$ to $< 10^{-5}$ |
| c | $\geq 10^{-6}$ to $< 3 \times 10^{-6}$ |
| d | $\geq 10^{-7}$ to $< 10^{-6}$ |
| e | $\geq 10^{-8}$ to $< 10^{-7}$ |

To assess reliability based on the PL, the required PL for the safety function in question must be determined (similar to the risk assessment of EN 954–1) using the following criteria: extent of damage, frequency/duration of occurrence and the ability to prevent the danger. A decision tree simplifies the determination. The tree provides a unique path to the relevant PL using the criteria mentioned.

The required PL is compared to the actual PL and the latter must be greater than or equal to the required PL for a positive assessment of the safety function implemented. The actual PL is determined on the basis of the controller components and safety components. First, their Mean Time To Failure (MTTF) is determined. Second, the components are split into Categories B, 1, 2, 3 and 4, each representing the extent of integrated safety functions. The actual Performance Level can then be determined using the standard depending on combination of category and MTTF.

### 10.4.3 Two-channel, Fault-Detecting Control Structure

A process that is to enter a defined, safe state after a fault occurs in the controller or at its interfaces to the machine requires a two-channel control system design as in ◘ Fig. 10.34.

The two subsystems perform the same operations and compare, at specific times, their input and output signals and, as required, the intermediate results or system states. If a fault occurs in one of the two components, it is detected by the two-way comparison and a relevant fault can be indicated. For redundant control systems with identical hardware and software, there is a risk that faults due to the system go undetected because they can occur simultaneously in both controller computers and so possibly mimic a fault-free state. A diverse design of the entire system can help here [DIN90; HÖLS84]: The redundant components differ in terms of design (such as different microprocessors), although they should perform the same task. Only then can it be assumed that a fault in one subsystem cannot cause the same effects in the other subsystem. While selecting different microprocessors for electronic controllers results in diverse design of the hardware, diverse creation of software can ultimately only be guaranteed with the involvement of more than one independent programmer. This is because there is a risk that a single, undetected error in the design of the program structures is transferred to all individual systems.

### 10.4.4 Three-channel, Fault-tolerant Control Structure

The primary downside of two-channel control systems is that the safe state is entered immediately after a fault is detected. This inhibits continued operation at first. Furthermore, when a fault does occur, the two units usually cannot decide which subsystem has the fault.

A marked increase in system reliability, associated with a further increase in safety, can be attained with a three-channel variant (◘ Fig. 10.35).

After the failure of one subsystem, operation can be maintained and so the fault tolerated for a certain time. The 2-of-3 voter determines, once the controller outputs have been compared, the correct output variables by means of a "majority vote." The subsystem identified as faulty is excluded from the subsequent decision-making process while the two units still working switch over to two-channel, fail-safe mode. The overall

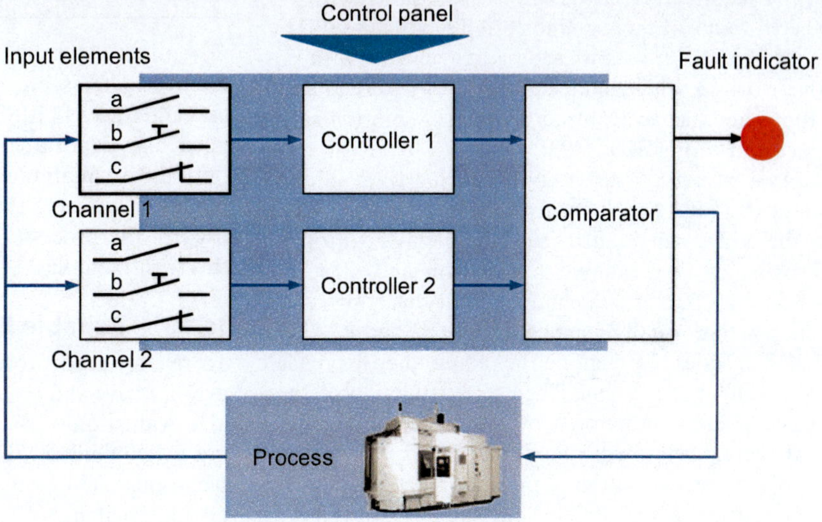


◘ Fig. 10.34 Two-channel, fault-detecting control structure

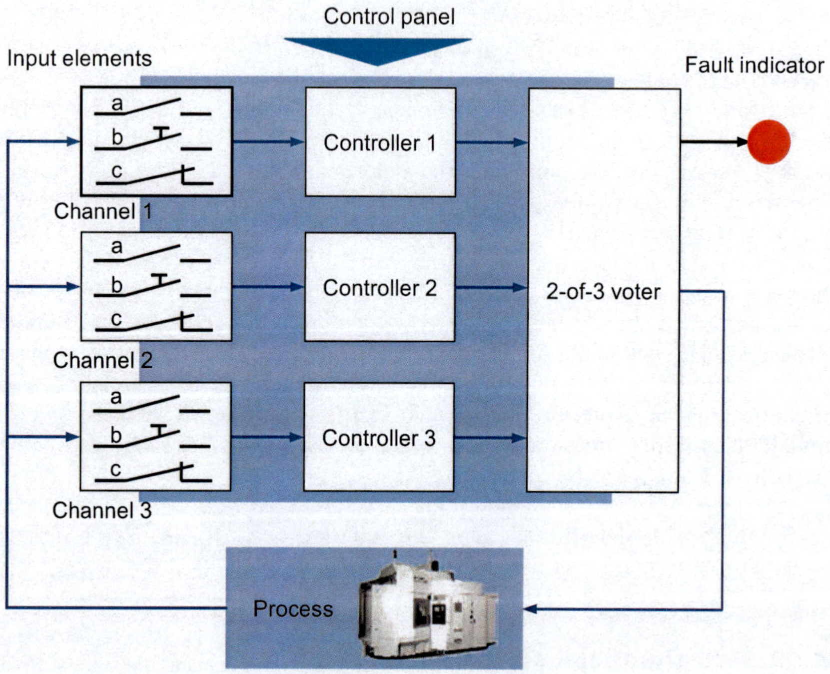


**Fig. 10.35** Three-channel, fault-tolerant control structure



system need only be transitioned to a safe state once another fault occurs. For three-channel control systems with fault-tolerant behavior, it is often possible to swap out and repair the faulty component during operation. This avoids expensive and unproductive downtimes.

### 10.4.5 Conventional Safety Circuits Using Relay Technology

Given their straightforward, easily monitorable design, safety-relevant circuits (such as the Emergency stop circuit for machine tools) are normally implemented with a redundant relay design in accordance with IEC 60,204–1 and VDI Guideline 2854 [DIN14a; VDI91].

Figure 10.36 shows such a relay circuit with positively driven relay contacts as used in practical applications and also available as a compact solution saving space and wiring [KOCH93].

As a rule, the system is only enabled when all three relays (K1, K2 and K3) are working correctly. Faults in the relays can be detected with regular, automatic checking of their ability to switch on and off. As in ► Sect. 10.4.3, it is two-channel, fault-detecting redundancy with normally open contacts k2.2 and k3.2 connected in series representing the redundancy for reliable switch-off of the system. The correct function of these two contacts is monitored by momentary opening of contact k1.5 before every energization.

When the On button is pressed, relay K1 activates and transitions to the self-hold state via the parallel normally closed contacts k2.7 and k3.7. At the same time, relays K2 and K3 are activated and also transition to the self-hold state. The self-hold state of K1 is canceled by opening k2.7 and k3.7. K1 can only de-energize once K2 and K3 have energized. When the Off button or Emergency stop switch is pressed, the K2 and K3 relays are de-energized again. If one of these relays does not de-energize when a fault occurs, energizing K1 again is prevented by the series-connected normally closed contacts k2.6 and k3.6. This prevents the system from switching back on.

### 10.4.6 Fail-Safe Process Coupling

Given that about 90% of all failures with conventional control systems are attributable to faults at the process interfaces [LEIS87], redundancy in this area is of particular importance. Fault detection here pertains to both the input and output channels of the control computers as well as to the signal transducers and actuators used in the process. The following chapters will show this with examples.

#### 10.4.6.1 Reliable Evaluation of Process Inputs

To reliably read process inputs, mechanical, positively driven relays and switches can be used that enable reliable evaluation of switch information using two opposing switch contacts. Figure 10.37 shows as an example a simple circuit using relay technology to evaluate a position switch.

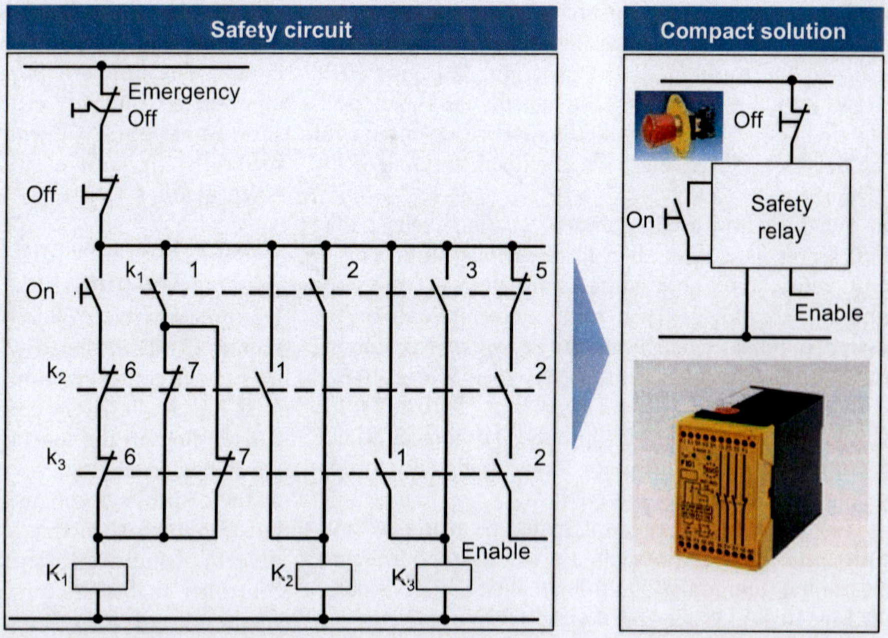


**Fig. 10.36** Emergency stop circuit using reliable relay technology, according to [KOCH93]

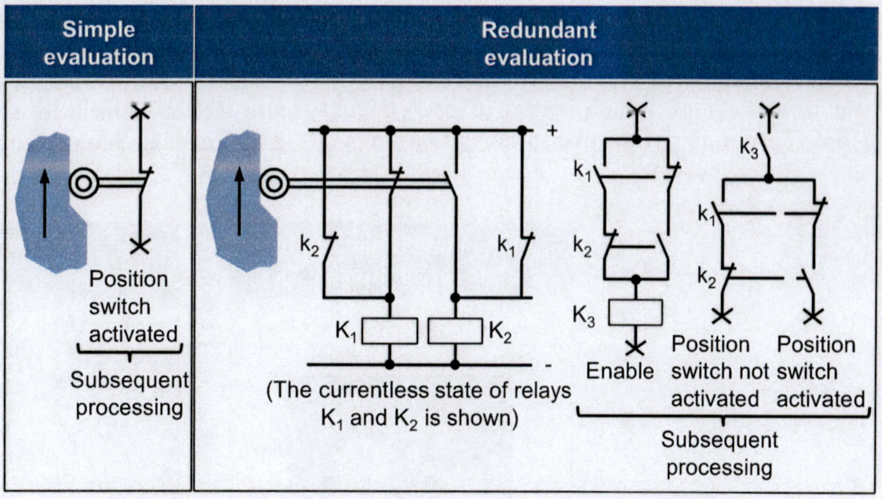


**Fig. 10.37** Switch evaluation circuit using relay technology

While only the switch contacts themselves are used for subsequent processing in the simple switch evaluation circuit (Fig. 10.37 on left), additional components are added to the redundant circuit that can be monitored, shown on the right in the figure. In the first case, a defective or "sticking" push-button cannot be detected. The situation is different for the second solution. Here, a position switch with normally closed and normally open contacts is used. The consequence of the mechanical, positively driven operation of these contacts is that, although both contacts are open momentarily when switching, they can never be closed at the same time. Two auxiliary relays, K1 and K2 are used to evaluate the contacts. Only one of these relays is live when there is no fault. If both relays are live due to short or open circuits, the enable contact is interrupted. The feedback of relays K1 and K2, in conjunction with the enable signal circuit, satisfy two demands made of the evaluation circuit. First, when switching, the short, simultaneous opening of the contacts may not create a fault in enabling the machine. Secondly, switching information should only be relayed once the switch contact is closed, that is, when switching has finished.

Besides the switch function, any short and open circuits at the switch contacts and controller connectors can also be detected. However, detecting mechanical defects at the process inputs (such as the breaking off of a stylus) is often only possible with a redundant design of the input elements.

For programmable controllers, the operations realized in relay controller wiring are defined as instructions in the form of a text program. ◘ Figure 10.38 shows a PLC program that implements the switch polling described. The controller requires two inputs for the normally closed and normally open contacts of the switch.

Flags are now used in place of auxiliary relays. Flag F10 serves as a status flag for a serious fault. This is only active when both contacts are closed at the same time (I1 = 1 and I2 = 1). If both contacts are open, the switch is polled again because the switch is changing state or there is a cable break (see also ◘ Fig. 10.39). Only for combination I1 = 1 / I2 = 0 or I1 = 0 / I2 = 1 is flag F1 set to logical ONE ("pressed") or logical NULL ("unpressed") for the state of the normally closed-normally open contact combination.

For evaluating redundant switch elements by multiple controller units (such as two microcomputers), additional boundary conditions must be considered (◘ Fig. 10.39). While there are opposing switch positions in Case 1 (fault-free), both contacts are open momentarily in the following Case 2 while the switch element undergoes a state change. To differentiate this from a longer-term cable break, the duration of the fault state is monitored over time T. While T is canceled again after the correct state is read (Case 3), exceeding the time duration (Case 6) also causes triggering of a fault (as in Case 4) when both contacts are closed at the same time or when this state is mimicked by a short circuit. The different program runtimes of individual computers with diverse designs can also mean Case 5 shown in ◘ Fig. 10.39 can occur where the subsystems read different input states. In this case, the read process is repeated.



### 10.4.6.2 Fail-safe and Fault-Tolerant Process Outputs

Faults at the controller outputs and downstream actuators can be detected with appropriate feedback signals to the control computers, as shown in the example in ◘ Fig. 10.40. Every control system can interrupt the current flow in the load by disabling the relevant output. The current state of the load circuit is transferred to the control systems using relay 3 via a feedback input (monitor) such that short-circuit detection is possible by a multi-stage energization process—only one controller enables its output at a time while the corresponding output of the other controller remains disabled. If a current flow in the load is signaled by the monitor feedback signals despite this, there is a short circuit in the relays and the process must immediately enter the safe state.

In the design of fault-tolerant controllers, there is also the requirement for configuring the actuators in a 2-of-3 configuration so that a fault in the output area

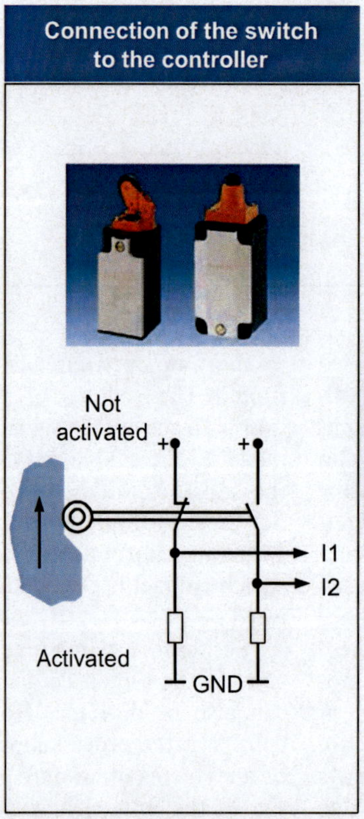


**PLC switch polling**

```
A    I1    Input 1
A    I2    Input 2
CJP  10    Jump to 10 if both
           switches closed
AN   I1    Input 1 inverted
AN   I2    Input 2 inverted
CJP  5     Jump to 5 if both
           switches open
A    I2    Input 2
AN   I1    Input 1 inverted
=    F1    State "Activated"
           |
Continued process
           |
5  Repeat switch polling
           |
10 S  F10  Both switches closed:
           Serious fault!
           |
Call of the safe state
```

◘ **Fig. 10.38** Switch polling program

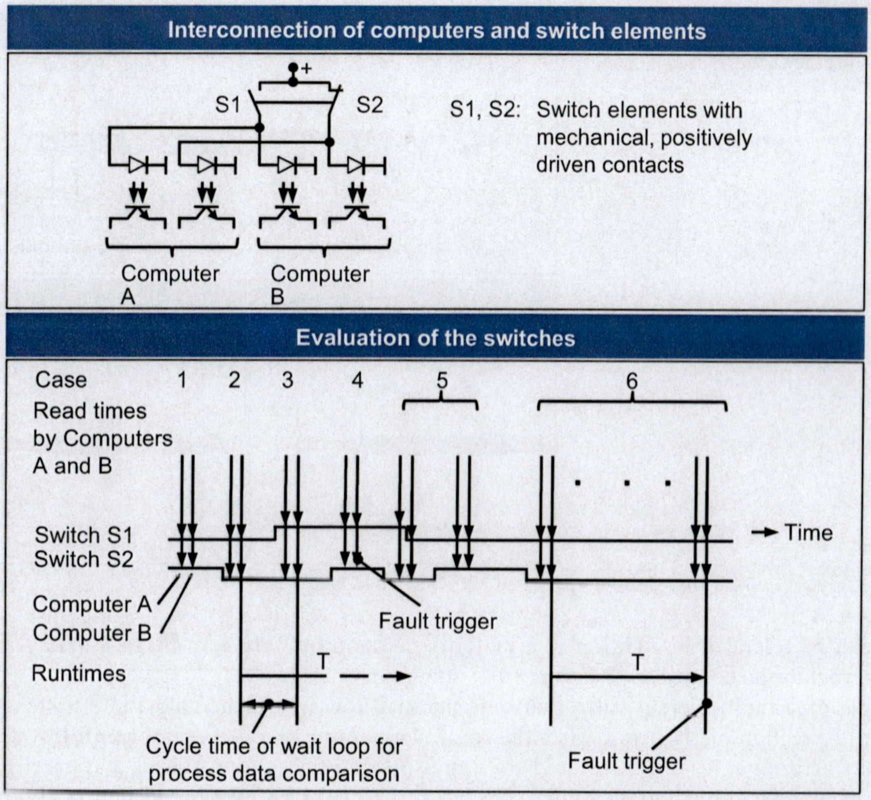


**Fig. 10.39** Connecting and evaluating process inputs

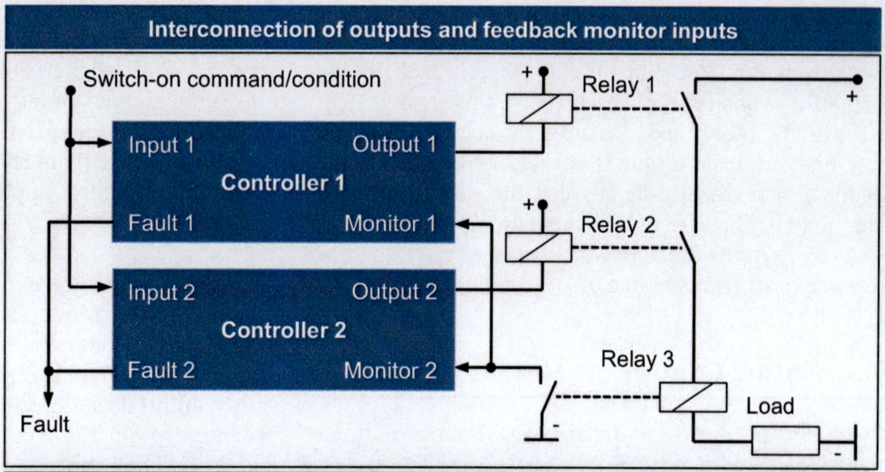


**Evaluation of the actual switching state (identification of controller and relay faults)**

**Safe (monitored) switch-on:**
Setting of Outputs 1 and 2 (load is activated),
Monitors 1 and 2 must become active.
**Safe switch-off:**
One or both outputs are switched off (load is switched off),
both monitors must become inactive.
**Short circuit test:**
Alternate setting of just Output 1 or 2.
Monitors 1 and 2 must remain inactive.

**Fig. 10.40** Principle of a monitored fail-safe output

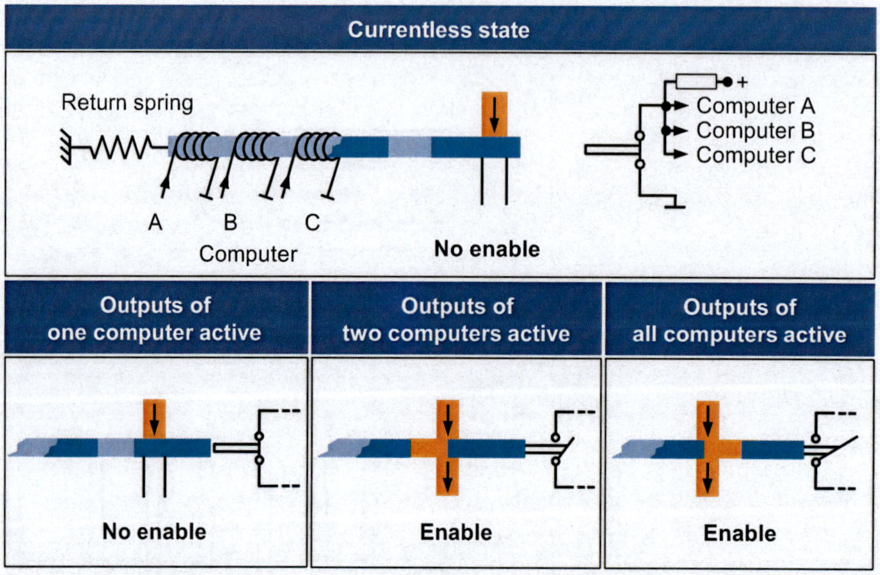


**Fig. 10.41** Example of a fault-tolerant output with one actuator in a 2-of-3 configuration

10

can be tolerated in addition to a majority decision internal to the controller. Figure 10.41 shows an example of a fault-tolerant valve that only permits flow in the direction of the arrow when the controller computers activate at least two solenoid coils, thus moving the opening far enough to the right.

The assumption here is that every coil is either fed a constant current in only one direction or is fed no current at all. A return spring ensures that, when the coils are not carrying current, the opening is far enough away from the flow pipe. Using the feedback in the form of a switch (Fig. 10.41, right), this arrangement can also be monitored: Before the valve is activated, only one output at a time is enabled alternately. If the feedback here already signals that the valve is enabled (that may be the case only if there are no faults and at least two outputs are active at the same time), there is an error short circuit at one of the outputs.

## 10.5 Motion Control

To satisfy the diverse and strict performance requirements for motion control in today's production systems, controller functions that extend beyond the feature set of conventional programmable and hard-wired controllers are required. Real-time capability and synchronization, as well as the high processing powers needed by interpolation tasks, require in particular the use of specific controller solutions. The design and the operating principle, as well as typical system features, of motion control systems are thus presented below.

### 10.5.1 Basics and Application

Generally, these systems initially included all forms of motion control systems. The evaluation of positions, speeds and accelerations and the presence of actuators and sensor systems are just as characteristic of such systems as real-time capability. Furthermore, motion control systems are characterized by the grouping of logic functions and technology functionality, such as pressure and temperature control, into one unit. For the control of precision drives, such as in machine tools, simultaneous monitoring of position, speed and acceleration is standard. This requires a cascaded control loop structure that features a large dynamic range.

In motion control applications, there is generally a differentiation between *closed-loop* and *open-loop* controllers.

*Closed-loop controllers* are characterized by feedback of the controlled variable and continuous comparison to the command variable. Given an appropriate controller design, this principle allows minimization of the control deviations resulting from unforeseen disturbances within the control path. Consequently, this structure is important for precision applications in particular. A number of configurable modules, e.g., in the form of servo drives representing one unit comprising sensor systems and actuators, support the implementation effort today. The real-time capability of systems was ensured by the introduction of fieldbuses to connect the feedback signals. The design of a conventional motion-controlled axial system with feedback of the actual value is shown in Fig. 10.42.

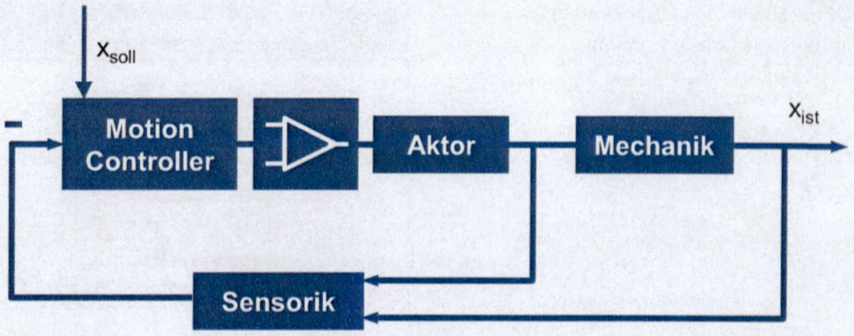


**Fig. 10.42** Design of a conventional motion control system

In contrast to closed-loop controllers, there is no feedback of the controlled variable for *open-loop controllers*. This principle is usually only used for simple applications, such as stepper motors, that are unable to compensate any control deviation caused by disturbance factors. Widespread usage and low-cost access to servo motors mean the principle of open-loop control on its own usually plays a secondary role in the industrial applications of today.

The difference between the actual position and reference position determined in a motion controller is first converted to a current equivalent to the relative motion and relayed to an actuator by a power amplifier. Using the current being applied as the starting point, the actuator then applies a corresponding force onto the mechanical transmission elements of the system. The resulting displacement is then read by a sensor and fed back to the motion controller. This continuous closed-loop control strategy minimizes the remaining control deviations. In particular, it is the strict requirements on the transmission speeds between individual system elements that separate motion control systems from simple closed-loop control systems.

## 10.5.2 Design and Operating Principle of MC Systems

Although the functional scopes defined in IEC 61,131 represent a basis for standardizing motion control systems, different basic architectures exist that are characterized by application-specific benefits and disadvantages. The different forms of the system architecture are thus presented below.

### 10.5.2.1 System Architecture of MC Systems

Systems with a central motion controller implementing only closed-loop control on a single axis are currently only used in applications requiring processing speeds in the sub-millisecond range. The high availability of computing power in current processors and the loss of flexibility accompanying the use of centralized systems require that this form of system architecture only be used in high-speed applications such as packaging machines, Fig. 10.43, a) [GEVA06].

In comparison, systems with a decentralized design exhibit far greater flexibility due to the parallel control of multiple axes. Individual axes are linked using real-time capable fieldbus systems. The demands of synchronizing such systems require, given the increased level of complexity, greater overheads in design and startup. These must be taken into account in the conceptual design of the system architecture. Today's CNC control systems with digital drive controllers that manage all closed-loop control parameters internal to the controller, can be cited as examples of systems with decentralized design—Fig. 10.43, b) [GEVA06].

Another option is the use of programmable logic controllers in combination with expanded, special modules. As already described, the computing power of conventional PLCs is not sufficient to perform comprehensive interpolation functions. This problem can be bypassed by using suitable expansions such as the Simotion C240. The higher costs and limited level of compatibility between manufacturer-specific components represent clear disadvantages with this architecture, Fig. 10.43, c) [GEVA06].

Furthermore, there are real-time capable PC-based solutions having comparably high processing powers and large storage capacities. So that the decentralized configuration does not create any disadvantages compared to conventional systems, use of a bus system capable of real-time operation is mandatory—Fig. 10.43, d) [GEVA06].

Figure 10.43 shows a summary of the different system architectures.

### 10.5.2.2 Fieldbus Systems for Motion Control

As described above, the benefits of a decentralized structure can only be leveraged using a real-time fieldbus. Since conventional fieldbus systems such as PROFIBUS and CANopen do not generally satisfy this requirement, specialized systems have moved into the mainstream for the drive sector. Using Ethernet-based bus systems, in particular, is a low-cost way to implement real-time connections. In turn, the various Ethernet fieldbus systems differ in the manufacturer-dependent operating principles. This means that compatibility between individual components must be addressed separately in the conceptual de-

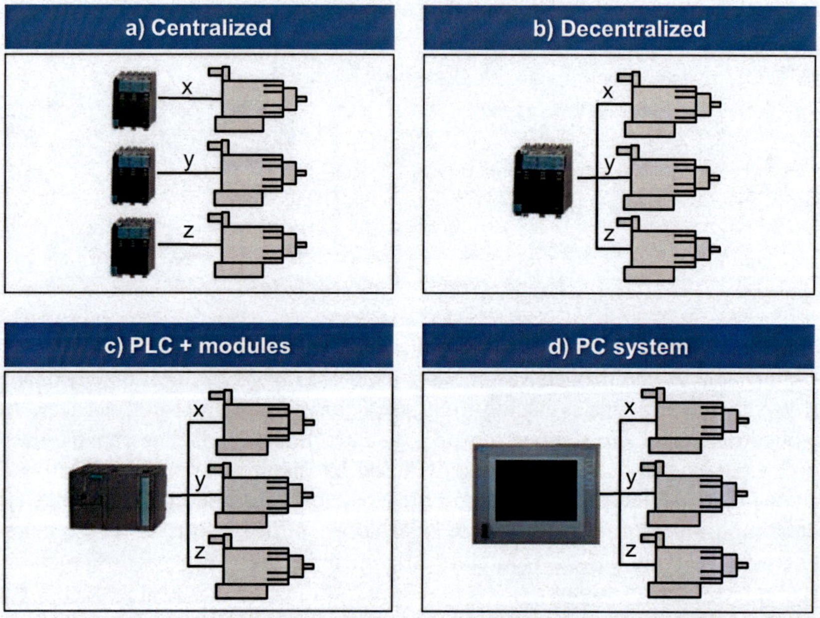


**Fig. 10.43** Different system architectures for MC systems, according to Siemens

10

sign of a motion control system. EtherCAT, SERCOS III, PROFINET, Ethernet/IP, Ethernet Powerlink and VARAN represent the most common system profiles.

In addition to Ethernet-based technologies, multimedia buses such as IEEE 1394 (Firewire) lend themselves to system communication [GEVA06].

### 10.5.3 Programming

The fundamentals of the application and system architecture of motion control systems were described above. Now the programming and project planning of the systems will be discussed.

#### IEC 61,131

The majority of standards specified in IEC 61,131–3 on programming controllers are used to implement motion control systems (see Sects. 10.2 and 10.3). The function blocks provided in the proprietary project planning environment usually exhibit platform-specific properties, however. Often, more programming languages are also offered within project planning software—with the aim of providing intuitive operation.

#### PLCopen

In contrast, the function toolkit of PLCopen represents an attempt to establish a fully standardized function library that can be used in a manufacturer-independent manner. General drive states were initially defined for this. These can be used as the basis for complex motion and control profiles (Fig. 10.44). A brief explanation of the state diagram is provided below.

The "Blocked" state corresponds to the initial state of the axis. This state can be exited by enabling and running the Reset command. If there is an axis error when the Reset command is run, the system transitions to "Emergency stop" (see 3). "Emergency stop" can be reached from any state (see 1) and can only be exited when the axial motion is error-free or the enable signal is retracted (see 2 and 4). When the system successfully switches to "Standstill" (see 4, 5 and 6), all motion states can be reached from here, and it is possible to switch between them. The routine for moving to the "home position" can also be accessed from Standstill. As with a discrete motion, reaching the reference position is confirmed (i.e., acknowledged) and a transition is made back to Standstill. All motion states can be transitioned to the Stop state at any time using a Stop command (yellow arrows). In this state, the axial motion is stopped in a controlled manner and is then switched to Standstill (see 6).

The state transitions numbered in Figure 10.44 can be described as follows [PARK13]:

1. Error has occurred in any state
2. Enable signal was retracted in any state
3. Enable signal was issued for an active axis error
4. Reset was confirmed with power available and error-free axial system
5. Enable signal was issued with power available and error-free axial system
6. Axial motion was successfully stopped and is transitioning to Standstill

Based on the motion and control profiles, a large number of modules and functions are defined in the PLCopen motion control specification that, because of the

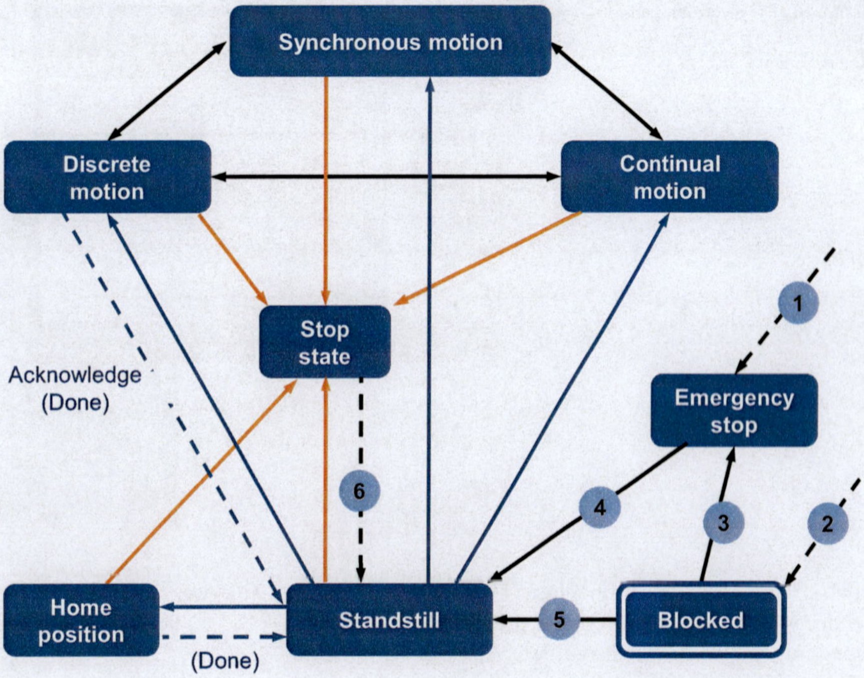


**Fig. 10.44** Defined motion states in the PLCopen motion control standard, according to PLCopen

standardized structure, are compatible with an array of products from different manufacturers.

#### 10.5.3.1 Project Planning of Motion Control Systems

In the SIEMENS Simotion Scout project planning software shown in Fig. 10.45, languages familiar from IEC 61,131–3 (ST, Ladder Diagram and CFC) and the Motion Control Chart can be used for programming.

In addition to classic programming, the project planning environments frequently allow a function test of the system and include advanced software tools for startup and system diagnostics. There are also additional applications, usually visualized graphically,

| SIMOTION SCOUT | | SCOUT option packages |
|---|---|---|
| **Project management** | **Programming** | **Drive Control Chart (DCC)** |
| ■ Workbench<br>■ Managing all equipment (controllers, drives, etc.)<br>■ Integrated library functions | ■ Ladder Diagram (LD)<br>■ Continuous Function Chart (CFC)<br>■ Structured Text (ST)<br>■ Motion Control Chart (MCC) | ■ Graphical project planning and enhancement of device functionality<br>■ Depiction of control engineering structures |
| **Project planning / configuration** | **Test and startup** | **CamTool** |
| ■ Hardware and network configuration<br>■ Creation of technological objects<br>■ Creation of cam plates (basis) | ■ Diagnostics for test and startup (trace) for axis control panel<br>■ STARTER drive and startup tool | ■ Graphical cam plate editor<br>■ Generation and optimization of cam plates |

**Fig. 10.45** SIEMENS Simotion Scout project planning software, according to Siemens

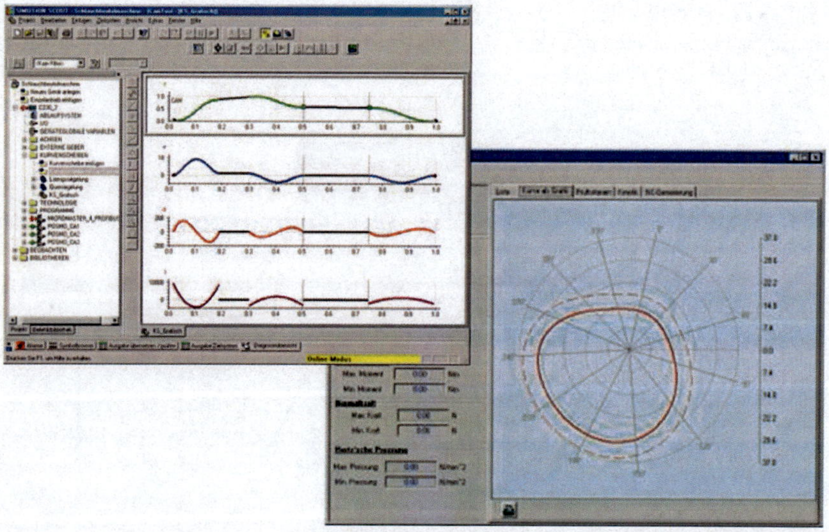

**Fig. 10.46** Electronic cam plate in SIEMENS Simotion Scout. (*Source* Siemens)

such as the electronic cam plate. These allow the user to access known structures and integrate existing motion control solutions into new system environments (Fig. 10.46, see ▶ Sect. 8.4).

10

## References


[BECK14] Beckhoff: SPS und Motion Control auf dem PC. Verl (2014)

[BIEL90] Bielefeld, H.: Speicherprogrammierbare Steuerungen in der Prozeßleittechnik. VDE-Verlag, Berlin (1990)

[BRUG99] Bruggink, B.: Programming XE "Programming" PLCs using Sequential XE "Sequential Function Chart" Function Chart XE "Sequential Function Chart." Universität Nijmegen, Diss (1999)

[DIN08] Norm DIN EN ISO 13849–1: Sicherheit von Maschinen – Sicherheitsbezogene Teile von Steuerungen – Teil 1: Allgemeine Gestaltungsleitsätze (2008)

[DIN14a] Norm EN 60204–1, VDE 0113–1: Sicherheit von Maschinen - Elektrische Ausrüstung von Maschinen - Teil 1: Allgemeine Anforderungen (2014)

[DIN14b] Norm DIN EN 60848: GRAFCET – Spezifikationssprache für Funktionspläne der Ablaufsteuerung (2014)

[DIN14c] Norm DIN EN 61131–3: Speicherprogrammierbare Steuerungen - Teil 3: Programmiersprachen (2014)

[DIN90] Norm DIN VDE 080: Grundsätze für Rechner in Systemen mit Sicherheitsaufgaben (1990)

[DIN94] Norm DIN 19226: Begriffe zum Verhalten von Schaltsystemen(1994)

[EBER12] Eberhardt, O.: Die EU-Maschinenrichtlinie. Renningen: expert (2012)

[EURO06] Europäische Union: Richtlinie 2006/42/EG des Europäischen Parlaments und des Rates über Maschinen und zur Änderung der Richtlinie 95/16/EG. Brüssel: Amtsblatt der Europäischen Union (2006)

[FREN87] Frentzen, B.: Entwurf redundanter Steuerungssysteme - Ein Beitrag zur Steigerung der Sicherheit programmierbarer Steuerungen. Diss. RWTH Aachen (1987)

[GEVA06] Gevatter, H.-J.: Handbuch der Mess- und Automatisierungstechnik in der Produktion. (Reihe: VDI). 2., vollst. bearb. Ed. Berlin: Springer (2006)

[GONT11] Gontermann, A.: Deutsche Elektroindustrie - innovativ und wachstumsstark. In ifo Schnelldienst **64** (2011)

[HERF10] Herfs, W. J.: Modellbasierte Software in the Loop Simulation von Werkzeugmaschinen. Diss. RWTH Aachen (2010)

[HÖLS84] Hölscher, H., Rader, J.: Mikrocomputer in der Sicherheitstechnik. Verlag TüV Rheinland, Köln (1984)

[INTE15] Intel: Collaboration Accelerates Collaboration Accelerates the Internet of Things and Industry 4.0. USA. Case Study Beckhoff, Firmenschrift (2015)

[JOHN09] John, K.-H., Tiegelkamp, M.: SPS-Programmierung mit IEC 61131–3. Konzepte und Programmiersprachen, Anforderungen an Programmiersysteme, Entscheidungshilfen. (Reihe: VDI). 4., neubearbeitete. Ed. Dordrecht: Springer (2009)

[KEGE14] Kegel, G.: Pressekonferenz des ZVEI-Fachverbands Automation zur Hannover Messe. Hannover (2014)

[KEGE16] Kegel, G.: Pressekonferenz des ZVEI-Fachverbands Automation zur Hannover Messe. Hannover (2016)

[KEGE19] Kegel, G.: Pressekonferenz des ZVEI-Fachverbands Automation zur Hannover Messe. Hannover (2019)

[KOCH93] Koch, T.: Zustimmungsschalter. (Reihe: Schriftenreihe der Bundesanstalt für Arbeitsschutz, Fb 681). Bremerhaven: Wirtschaftsverlag NW (1993)

[LEIS87] Leist, K.: Sicherheitsgerichtete speicherprogrammierbare Steuerungen. Automatisierungstechnische Praxis atp, Jg. 29, (6), 267–275 (1987)

[MITT91] Mittmann, R.: International standardisierte SPS-Programmierung nach IEC 1131 zum Vorteil für Anwender und Hersteller. (Reihe: VDI Berichte). Düsseldorf: Springer (1991)

[OEST86] Oestreicher, T.: Rechnergestützte Projektierung von Steuerungssystemen. Diss. TH Karlsruhe (1986)

[PARK13] Parker: PLCOpen State Machine. URL: ▶ https://www.parkermotion.com/dmxreadyv2/faqsmanager/faqsmanager.asp?question=plcopen-state-machine. Accessed: 07. Juni 2017

[PILZ13] Pilz GmbH: Das Sicherheitskompendium. Ostfildern (2013)

[PLAG90] Plagemann, B.: Speicherprogrammierbare Steuerungen. Hüthig Verlag, Heidelberg (1990)

[SEIF92] Seifert, H.-J.: Modellgestützte Diagnose komplexer Produktionssysteme - Ein Beitrag zur Erhöhung der Verfügbarkeit kapitalintensiver Fertigungsanlagen. Diss. Univ, Bochum (1992)

[SIEM10] Siemens AG: Simatic - Programmieren mit STEP 7 Handbuch. Nürnberg, Siemens-Bestell-Nr. 6ES7810-4CA10-8AW0 (2010)

[SIEM93] Siemens AG: STEP 5 Handbuch. Nürnberg, Siemens-Bestell-Nr. C79000-G8500-C800–01 (1993)

[SOSS89] Sossenheimer, K. H.: Entwickeln von Instrumentarien zur rationellen Planung und Steuerung der Inbetriebnahme komplexer Produkte des Werkzeugmaschinenbaus. Diss. RWTH Aachen (1989)

[TRÖS11] Tröster, F.: Steuerungs- und Regelungstechnik für Ingenieure. (Reihe: Oldenbourg Lehrbücher für Ingenieure). 3., überarb. und erw. Aufl. München: Oldenbourg (2011)

[VDI91] Richtlinie VDI 2854: Sicherheitstechnische Anforderungen an automatisierte Fertigungssysteme (1991)

[WECK90] Weck, M. et al: Wege zur Verkürzung der Inbetriebnahme- und Stillstandszeiten komplexer Produktionsanlagen. In: Weck, M., Eversheim, W., König, W., Pfeifer, T. (Ed.): Wettbewerbsfaktor Produktionstechnik: VDI-Verlag (1990)

[WEIß16] Weiß, T., Habermann, M.: STEP7-Workbook. Einführung in die STEP7-Programmiersprache : mit TIA Portal, STEP7 V5.x und WinSPS-S7. 4. ed. Bretten: MHJ-Software GmbH & Co. KG (2016)

[WELL14] Wellenreuther, G., Zastrow, D.: Automatisieren mit SPS - Theorie und Praxis. (Reihe: Studium). 5. korrigierte und erw. Aufl. Wiesbaden: Vieweg + Teubner (2014)

[ZAST95] Zastrow, D., Wellenreuther, G.: Steuerungstechnik mit SPS. Vieweg, Wiesbaden (1995)

[ZVEI19] ZVEI - Zentralverband Elektrotechnik- und Elektroindustrie e.V.: Elektroindustrie in Zahlen. Frankfurt am Main (2019)

# Numerical Controllers

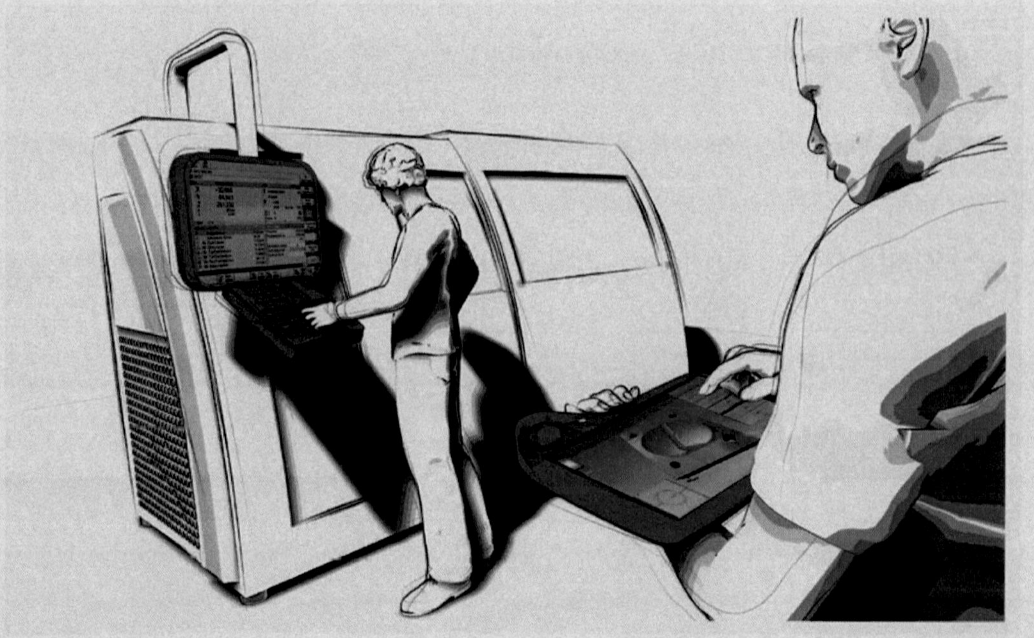

I thank my staff members Dr.-Ing. Marcel Fey, Christian Fimmers M. Sc., Ziqi Huang M. Sc. and Marian Wiesch M. Sc. for their contribution in revising this chapter.


C. Brecher and M. Weck, *Machine Tools Production Systems 3*,
Lecture Notes in Production Engineering, https://doi.org/10.1007/978-3-658-34622-5_11

## Contents

Numerical controllers (NC) offer far more flexibility than the sequential controls discussed in ▸ Chap. 10. For a workpiece, the type and sequence of production steps are defined in an NC program that contains the control information in the form of alphanumeric characters. So, from a global viewpoint, numerical controllers are part of the family of sequential controls, whereby the individual control steps are specified by NC lines (as alphanumeric characters). Modern-day numerical controllers must perform logical operations as well as complex computing operations [KAMP70][KIEF99][NN99][SAUT87; SIMO71][SPUR91][SPUR94]. Microprocessor technology forms the core of numerical controllers today as a result.

## 11.1 Historical Development of Numerical Controllers

Parsons drew up the initial concepts for a numerical machine tool controller in 1949. Contracted by the U.S. Air Force, he developed a computing machine to enable automatic processing of the integral airplane parts that were becoming increasingly complicated [SPUR93b]. Using Parsons' ideas, the first numerical controller developed for machine tools based on vacuum tubes and electromechanical relays was developed at the Massachusetts Institute of Technology (MIT) in 1952. Using binary encoded punched tape for data entry, a milling machine was able to perform a simultaneous movement over three axes [KIEF99].



From 1960 onwards, the expensive and sensitive relay and tube controllers, which had not enjoyed widespread use up to this point however, were being replaced by numerical controllers using transistor technology. These variants of numerical controllers (up to about 1975) were designed completely with hardwired, discrete semiconductor elements used in digital information processing (refer to ▸ Sect. 10.2) and were called hardwired logic controllers. Suitably tailored solutions had to be developed for every application (turning, milling, etc.). Changes of and extensions to the controller functions were only possible with intricate modifications to the controller hardware [NN99].

Controllers for simple machining tasks were sometimes also programmed manually with encoded connectors using a programming panel. Similar to a patchboard controller, the position and feed rate, as well as the machine functions, were entered numerically for every axis.

Outer wiring had to be plugged in elsewhere to make changes to routines (see ▸ Sect. 10.3.2). A machining task could be archived by placing a template in the form of a perforated plastic film onto the data module panel before the plugs were inserted. After programming, this film was removed with the plugs and kept.

With numerical controllers becoming increasingly widespread, requirements from users were becoming stricter in terms of availability, flexibility, ease of maintenance, functional scope and ease of use of controllers, and their simple adaptability to the machine types. These requirements could only be satisfied by hardwired controllers with great overhead. After 1968, a simpler and more cost-effective solution for realizing such complex controllers arose with the development of highly integrated electronic components (IC technology). Since 1972, one or more computers were used as the central component (first minicomputers up to about 1977, afterward microcomputers were more common). This gave rise to computerized numerical control (CNC).

Behind the conceptual idea of numerical controllers based on computers is the production of large volumes by standardized and reliable computer hardware. By implementing different system programs, they can then be adapted to production jobs as well as to special machines. This means the controller software can be upgraded at a later time with additional functions that are offered to the user as options. No major modifications to hardware are required here. The significance of numerical control for machine tools has risen continually due to its increased functionality. ◘ Figure 11.1 shows the requirements nowadays of numerical control inside a machine tool.

Since the end of the 1970s, all numerical controllers contain microprocessors. Use of the "CNC" term made sense in the transition period to differentiate between hardwired numerical control and computer control. Today, however, NC should always be used instead of CNC.

The competitiveness of machine tool manufacturers and users is determined to a high degree by the performance and price of the controller used. In the control technology market, so important for this very reason, European manufacturers were in danger of missing out at the start of the 1990s. Japanese products dominated the world market for numerical controllers in particular, both in terms of technology and volumes. Since then, the market (made up of over 60 European controller manufacturers represented primarily on the European market) has been consolidated substantially. In 2015, following numerous acquisitions and mergers, just a few high-profile European manufacturers are represented on the world market with numerical controllers. In addition, there are some smaller companies with more of a regional significance.

Control technology has also benefited from the rapid advance of personal computers for the office and

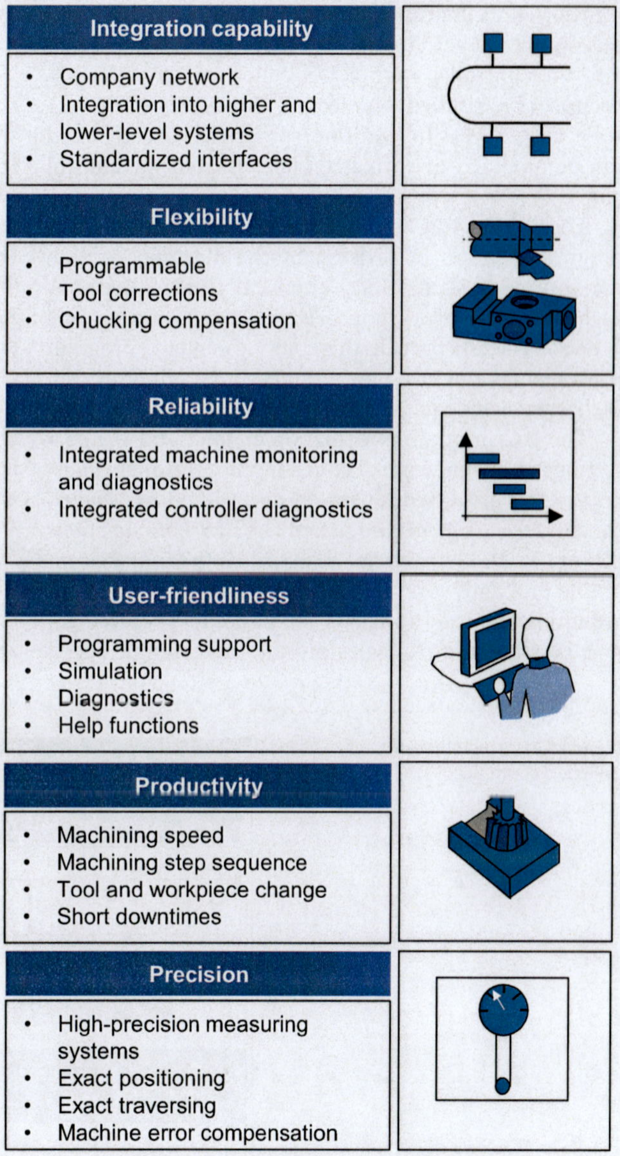


**Fig. 11.1** Requirements of numerical control inside a machine tool [DIN96]

home user markets. With commercially available PC-based hardware and software, controller manufacturers have been able to reduce considerably their investments for launching new products. This has meant the importance of software has increased significantly compared to hardware. Extending beyond their roles as providers of systems only, control manufacturers are increasingly using their expertise to offer tailored customer services.

Hardware and software standards, standardized interfaces, as well as open control systems that are expandable flexibly are increasingly in demand by users. In particular, the capability of integrating a controller into networks and master systems took on enormous significance at the turn of the millennium.

## 11.2 Design and Functional Description of Numerical Controllers

### 11.2.1 General Functional Description

The main task of NC is to control relative motion between the tool and workpiece. Path and velocity instructions are specified in the form of an NC program (parts program) and contain control information in the form of alphanumeric characters [DIN96].

An NC program can be created in two different ways depending on the organizational form. Firstly, NC programs are generated in production planning using automatic NC programming systems and are transferred to the controller (refer to ▶ Sect. 11.4.2). In this solution, the data of an NC program are entered into the controller from a USB stick or network connection (Distributed Numerical Control (DNC)—see ▶ Sect. 14.1.2.2). In addition, programs already read can be modified or corrected later on from the control panel.

The second way to enter simple machining tasks is program creation or revision by the machine user on the controller itself. Here, graphically aided methods such as WOP (workshop-oriented programming, see ▶ Sect. 11.4.2.4) provide the ability to enter a program conveniently and easily in production planning as well as at the machine itself.

The NC program information entered and the corrections to be made are decoded in the controller (*interpreter*) and processed separately according to *geometrical* and *technological data*, as well as *switching functions* (◘ Fig. 11.2). Geometrical data are all the information on the tool and workpiece paths to traverse. These data are ultimately used to create the desired workpiece geometry. Functions for selecting the tool, spindle speed and cutting speed are examples of technological information.

The switching functions arrive at the "*logic controller*" of the machine tool as switching commands. Here they are logically linked with feedback signals coming from the machine tool and converted to control commands for the units to be activated. Nowadays, the largest part of logic control is realized using a PLC (programmable logic controller, refer to ▶ Sect. 10.3). The logical operations comprise to a large extent locking and safety circuits, so that conflicting actions and actions that can endanger the machine user and machine (such as "Feed on" before "Spindle on") cannot be triggered. With the help of logic control, machine-independent functions of the numerical controller can be reproduced on (adapted to) machine-specific devices.

For a specified path segment, an *interpolator* calculates the motion sequence to be coordinated on the axes by direction and velocity. From this, it generates the *reference variables* for the drives of the axes [SAUT87]. The superposition of individual



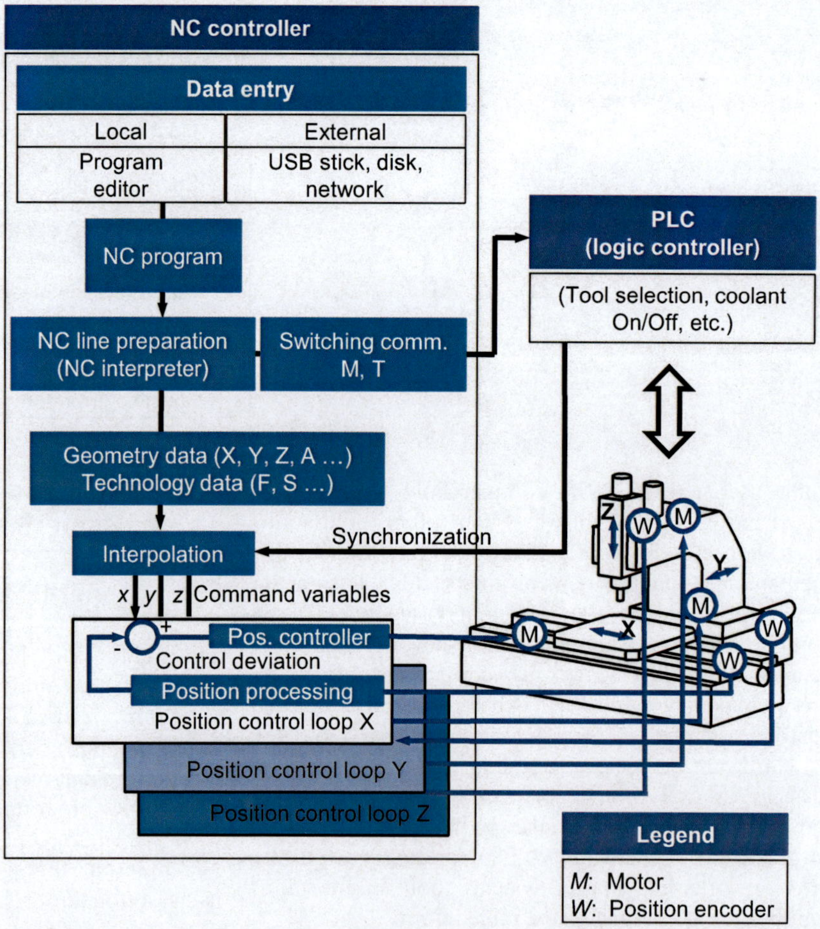


◘ **Fig. 11.2** Structure of a numerical controller

axis movements means a tool movement is then created along the programmed workpiece contour (see ► Sect. 12.1).

The contour to be generated is programmed by specifying the contour end point coordinates and type of connection between the respective start and end points as a path condition (e.g., straight line or arc). Controllers of varying complexity are suitable for realization here depending on the movement trajectory. In this regard, a distinction is made between point-to-point, line motion and path controls [DIN96] [KAMP70]. Point-to-point and line motion controls are no longer in use as autonomous controllers. Sometimes they are only used as modules inside PLCs.

*Point-to-point controllers* are used for simple positioning tasks such as drilling and spot welding. When they are used, the tool must not be in contact during the positioning process because the programmed end point of the tool movement is traversed on a non-defined path.

For *line motion control*, the end point of a machining section is traversed in a straight line and the tool may be in contact when traversing. Traversing is only possible on axially parallel straight lines with simple line motion controls and on any straight lines with advanced line motion controls. Line motion controls are suited, for example, to simple milling and turning processes and to controlling transfer lines.

Tool and table movements to generate arbitrary contours are only possible using *path controllers*. Every numerical controller nowadays has a path interpolator, making it a continuous path controller. These continuous path controllers can normally perform movements in straight lines and arcs and sometimes even spline segments. Movements on individual axes have a strict functional dependence on one another. Continuous path controllers are used in the machine tool industry for turning, milling and grinding machines, laser beam and oxyacetylene cutting systems, nibblers, die sinking and wire-electro discharge machines, water jet cutting machines, stereolithography machines and many more.

Before the controller sends the reference axis values to the drives, these values must be adapted to the current workpiece and tool position. For example, NC programming is independent of the current clamping position of the blank on the table of a milling machine. Thus, the current position of the workpiece zero point in relation to the machine zero point must also be entered and be taken into account by the controller using a coordinate transformation (see ► Sect. 12.2).

Similar correction calculations are performed when the tool length and diameter differ from the programmed data. The workpiece contour required is programmed free of the actual tool geometry, i.e. independent of the radius and length offsets. The current tool geometry data are not transferred to the controller until just before the machining process—manually or using DNC. The controller then calculates the Tool Center Point path and outputs it to the axis controllers in the interpolation cycle rate.

### 11.2.2 Hardware and Interfaces of a Numerical Controller

Increasing demands made of numerical controllers in regard to automation and precision of the process initially led to a distribution of control tasks across different controllers or modules. This was because initially the performances of individual processors were not high enough. Advances in the microprocessor sector facilitated this approach, with more and more processors with special features (numerical coprocessors and multi-processor systems for example) coming onto the market [SOSO90].

This situation however, brought about by rapid advances in semiconductor technology, has fundamentally changed, particularly in the area of personal computers. Nowadays, standard components from the PC sector are powerful and especially affordable due to the high volumes. Since the mid-1990s, they have also seen increased use in numerical controllers.

Given the special requirements in control technology, the hardware is generally developed and produced by controller manufacturers themselves. Standard processors and components are also used here, however. For special tasks, in particular interfaces to external hardware (SERCOS, PROFIBUS, etc.), ASICs (application specific integrated circuits) are used. These are cost-effective even for medium-sized volumes.

◘ Figure 11.3 shows the components of a numerical controller in the higher performance class [SIEM16]. The controller manages a maximum of 31 feed axes or spindles. These can be distributed over up to 10 parallel channels. The hardware components include:

- An MMC (Man Machine Control) module using an integrated industrial PC and a control panel with TFT color display and full NC keyboard as a central display and input unit
- The machine control panel for operating the machine by hand, for specifying the type of motion and override values and for defining, by the machine manufacturer, of individual keyboard assignments (e.g., running of particular NC line when a key is pressed)
- An NCU (Numerical Control Unit) with an integrated NC and PLC. The multi-processor NCU module is integrated directly into the digital converter system and connected to the drive modules. Profibus/Profinet can connect input/output modules for machine control directly to the NCU.

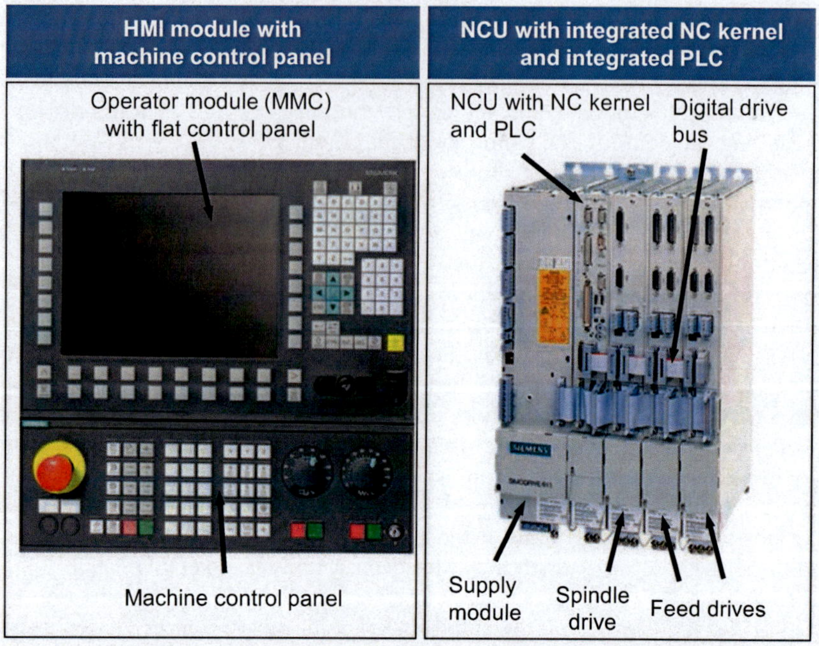


**Fig. 11.3** Components of a numerical controller, according to Siemens



All components are networked together using a serial bus system. Other controller components can also be connected to the bus.

In the upper performance category in particular, distributed hardware with a separate processor for operation, NC and PLC functionality, is used in many cases. The clear-cut separation between the real-time and operator areas is an argument in favor of this for reasons of safety and reliability. This division is sometimes also due to the historical lack of performance of individual processors. This is no longer an issue, however, for processors running at 500 MHz and up.

Figure 11.4 shows a SERCOS III Master PCI card allowing a standard office PC to be connected directly to the drive amplifiers. The split between the individual functions of operation, NC and PLC only takes place on software level here. A prerequisite is a powerful real-time operating system with priority-controlled multitasking. Using Linux derivatives with real-time kernels is a possibility.

### 11.2.2.1 Internal Structure

As already shown, the hardware of a numerical controller comprises different modules depending on variant. Historically, the selection of special hardware has been the basis for the controller performance. In many controllers, there is a classic split into three hardware components:

- *Operating area hardware:*
  for all operation and display functions; not real-time capable, normally PC-based hardware with Windows operating systems
- *NC kernel hardware:*
  for running geometric calculations and path control in line with the specifications of the NC program; real-time capable, normally special hardware with a real-time operating system
- *PLC* hardware:
  for machine adaptation of the controller (logic control); real-time capable, normally special hardware with firmware for cyclic sequencing.

In contrast to the two other hardware areas, the PLC was freely programmable from the very start for machine manufacturers. Experience has taught them to specialize in integrating their own software into the PLC. Consequently, in the absence of alternatives, management tasks such as tool management, pallet management or order processing, some of which were complex, were integrated into the PLC that was actually intended for logical operations and processes. It was not until open controller systems emerged that it was possible to correct these misguided developments (refer to ▶ Sect. 11.2.6). Complex management tasks can be realized in the operating area. This takes the workload off the PLC area (to the benefit of its original real-time tasks in machine control) and results in considerable performance enhancements.

With increasing acceptance and ever shorter innovation cycles for semiconductor technology, hardware is losing importance, however. Within the shortest of times, commercially available processors are doubling their performance. On the other hand, the performance and thus also the scope of software are increasing.

■ Fig. 11.4 SERCOS III Master PCI card, according to Bosch

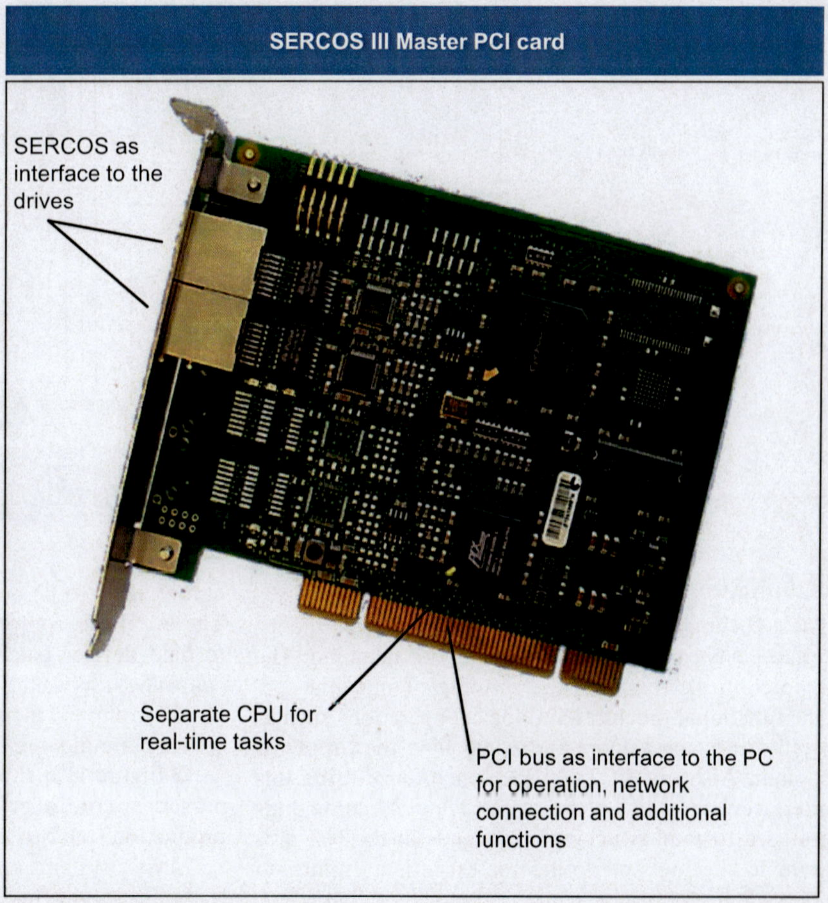


This increases the need to reuse expensively developed software and to use it independent of the hardware.

So, a logical and software-oriented split of the controller tasks by functional areas nowadays seems more viable than a classic split into hardware-oriented structures. Such a split by functional areas is called a reference architecture [OSAC00]. Modern-day controllers have modular designs in line with these architecture approaches and have defined interfaces between individual modules. This enables individual software modules to be deployed and reused in controllers of different performance classes. Controller manufacturers and users have more freedom to select the functionality (software) and the hardware used for implementation. The key functional areas of a numerical controller are explained below.

### HMC Functional Area

The HMC (Human Machine Control) functional area comprises all of the functions for operation and data management of the controller. This includes visualization of process and master data, programming of NC workpiece programs using a text editor or WOP system and managing different production data.

### MC Functional Area

The MC (Motion Control) functional area comprises NC data preparation and interpolation of the reference position values for every NC channel. The multi-channel design means different axis groups can perform movements independently of one another. Each channel represents an autonomous system with multiple axes and its own NC program as well as its own NC line preparation and interpolation. The programs on the individual channels can process in parallel or synchronized to each other. ■ Figure 11.5 shows a schematic of a controller structure with four channels for managing six axes. In this example, two channels each are combined into one machining group. Channels in one machining group are normally started at the same time and always in the same operating mode.

### AC and SC Functional Areas

The AC (Axes Control) and SC (Spindle Control) functional areas manage one or more groups of axes or spindles. In terms of the software architecture, they comprise all of the feedback control even if this is housed in the drive amplifier on separate hardware.

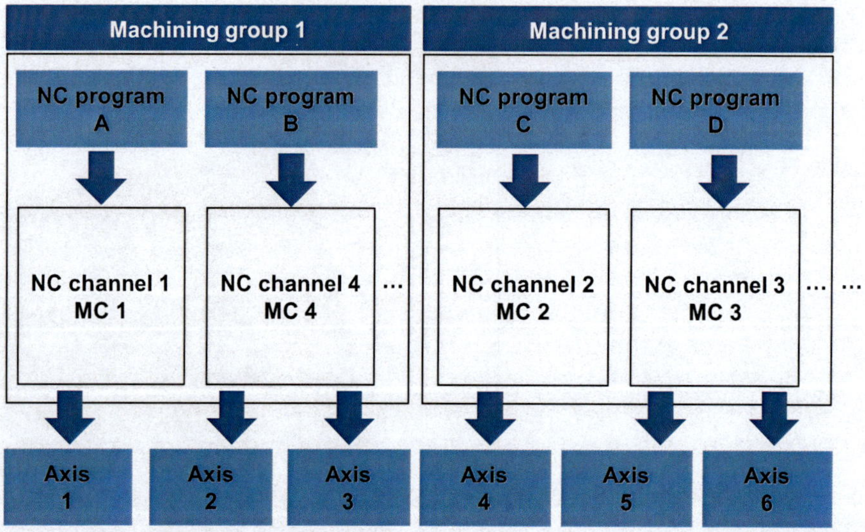


**Fig. 11.5** Multi-channel structure in the NC kernel, according to Siemens

11

### LC Functional Area

The LC (Logic Control) functional area forms the interface between the controller and the machine. The logic controller assumes the running of machine-specific functional processes and logical operations such as monitoring guard doors and controlling the supply of coolant. Also, the MC and LC synchronize using this interface. In NC line preparation, T and M commands are used to send switching commands in the NC program to the logic controller for execution. Inputs on the machine control panel are also processed here.

### Other Functional Areas

Additional functions such as the organizational and planning tasks of a numerical controller can be assigned to other functional areas (refer to ► Sect. 11.2.5.2):

- *Tool management:* for saving and managing all tool data (tool numbers, geometries, wear information, etc.)
- *Order management:* for managing production orders and the material flow of workpiece and tool.
- *Control of conveying equipment:* for performing tool conveying, tool changes, pallet conveying and pallet changes within one machine tool.

## 11.2.2.2 External Interfaces

An array of different communication interfaces are available to integrate a numerical controller into a production-related environment. A fundamental distinction can be made between interfaces to higher and lower-level systems and to the machine user.

The pronounced proliferation of network technologies in production means a numerical controller is hardly likely to be operated in a self-contained system. The interfaces comprise real-time capable connections to field devices (such as sensors, actuators and other controllers) as well as less real-time capable interfaces to the company network (diagnostics, order and resource planning—refer to ► Sect. 14.2).

Figure 11.6 shows the functional interfaces of a numerical controller in relation to the level model of production (refer to ► Sect. 14.2).

Master systems are computers on the cell or control level. These systems are to supply the controller with NC programs and tool data, as well as, if required, remote control (program start/stop). The term used for this mode is DNC, that is deployed for automated production systems as well as flexible manufacturing systems (FMS) and flexible manufacturing cells (FMC). Refer to ► Sect. 14.3.

The interface to the machine user is on the same level in the form of graphic screen, keyboard, handwheels, rotary switches, etc., enabling manual, direct operation of the controller and machine (see ► Sect. 11.5). The lower-level actuator/sensor level is controlled by reference position, velocity and current values for the drives (amplifiers) and switching functions. It returns the actual values (from encoders), process data (such as temperature changes) and any error messages.

Now that the functional interfaces of a numerical controller have been described, the physical interfaces are explained below. Figure 11.7 shows the typical peripherals and physical interfaces of a numerical controller [INDR99]. Ethernet and TCP/IP have established themselves as the standard for connections to cell/master computers. For older controllers, serial interfaces (RS-232 and RS-422) were commonly used as point-to-point connections. Ethernet is also the

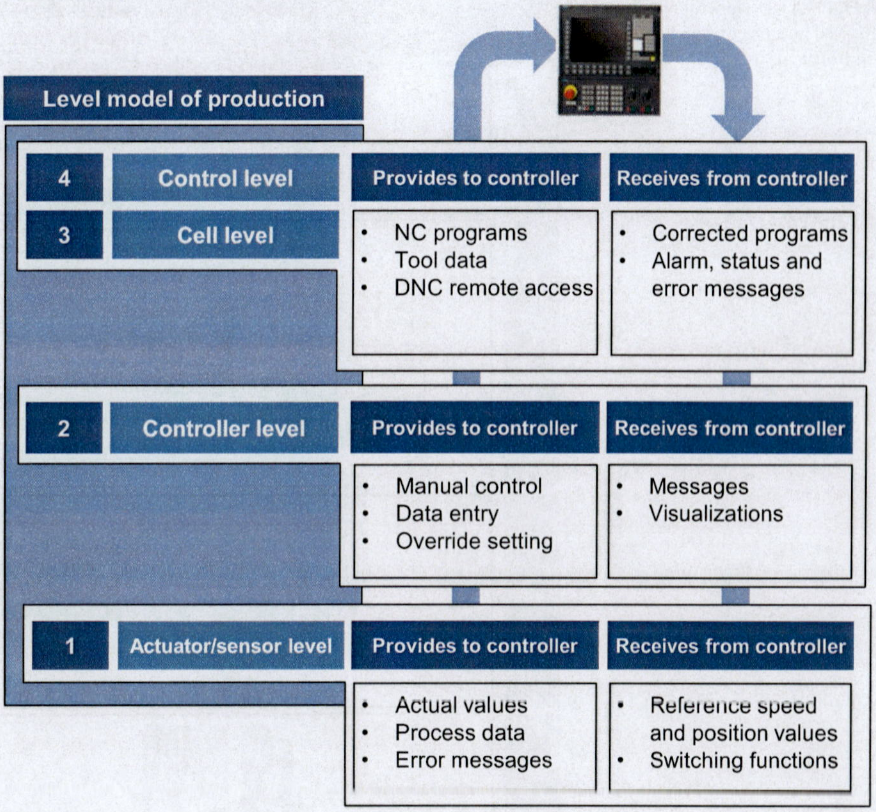


**Fig. 11.6** Functional interfaces of a numerical controller in the level model of production

preference now for networking numerical controllers. Fieldbus systems such as INTERBUS and PROFIBUS-DP are deployed when there are special security or real-time requirements.

Fieldbus systems are also used for digital inputs and outputs to connect to actuators and sensors. Handwheels and other controls also use this bus.

The drives to move the machine axes and spindles are normally connected over digital interfaces such as SERCOS. For low-cost controllers, the analog 10 V interface common previously continues to be used (▶ Sect. 2.2.3).

### 11.2.3 Numerical Controller Software

The prime tasks of a numerical controller are interpretation of the input information in the form of alphanumeric characters (NC program) and preparation of the corresponding path/switching information for every machine tool axis/actuator. The use of programmable microprocessor technology since the start of the 1970s has enabled the realization of individual numerical controller functions by means of manufacturer-specific NC functional programs (such as interpreter and interpolator). At the start, these were complicated to program in assembler and now are being developed with the convenience of a high-level language such as C or C++.

The increasing performance of NC hardware provided options in developing NC software more effectively. This has been due not least to the integration of standard PC cards in modern-day controllers. The low-cost and comprehensive software market for PCs has blossomed as a consequence. Besides creating software on the controller, however, there is normally development on host computers. Here, workstations or powerful PCs with high processing and memory capabilities provide far more convenient options in the development and test phases. Once the functional programs are complete, they can be transferred to the hardware of the numerical controller (target system). This method of programming is generally known as cross-development.

The *software structure* of a numerical controller is described below (■ Fig. 11.8). The main part of the software is made up of the NC functional modules that essentially embody the controller functionality. The functional modules here include for example the NC kernel programs such as the interpreter and interpolator, and additional functions such as fault monitoring functions, an NC program editor, a simulation system and a tool management system.

The central interface between the hardware and functional modules is the *operating system* of the

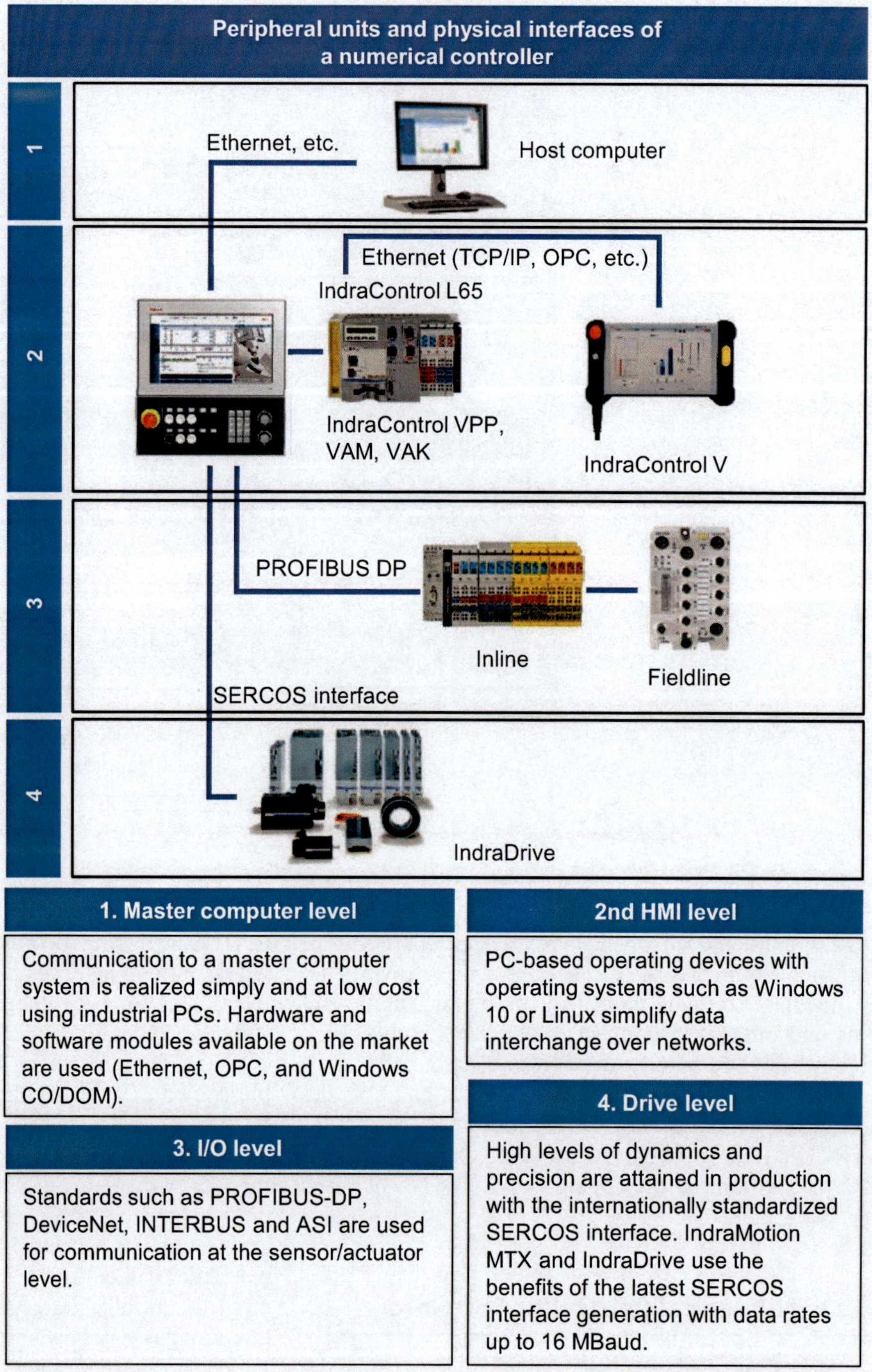


**Fig. 11.7** Peripheral units and physical interfaces of a numerical controller, according to Bosch



controller. The operating system handles memory management, input/output management and synchronization of the individual processes [KREB92]. The operating system can be split into hardware-dependent and hardware-independent parts (Fig. 11.8). A standard operating system suitable for numerical controllers must support multitasking—because multiple functional modules (tasks), such as interpreter and interpolator, must be active at the same time. Real-time capability is also a requirement, i.e. a task with higher priority runs without delay and replaces any other tasks with a lower priority. For example, a task for online collision monitoring must have a higher priority than a task for displaying information on the screen. Real-time capability guarantees deterministic responses, without delay, of certain processes to external

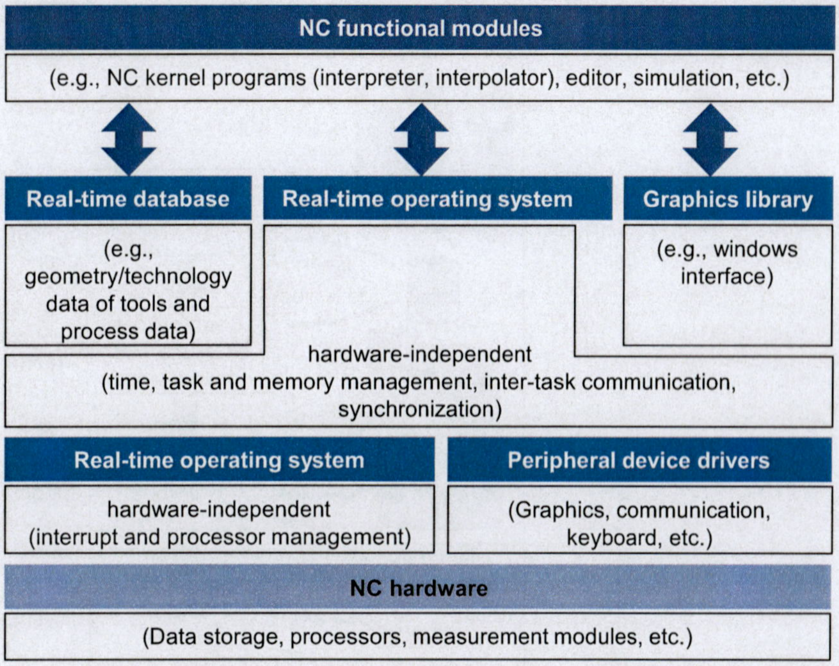


**Fig. 11.8** Software modules of a numerical controller

events within defined cycle times. Peripheral drivers are the software link to the peripheral units of the controller, such as drives and the screen.

Powerful hardware continues to integrate the system software that can be realized independently of the numerical controller functionality and is understood to be a platform for the NC functional programs. Graphic libraries and real-time databases are generally made available by the operating system. Real-time databases not only offer open interfaces (e.g., SQL as an ISO standard) for data management (e.g., tool data and technological data), they also guarantee secure access to time-critical data (such as correction values).

## 11.2.4 How a Numerical Controller Works

The information flow is used below to describe the NC core functionality—starting with interpretation of the NC program.

The way a numerical controller works is determined by the NC core functionality and by additional functions. The NC core functionality described in this chapter is an absolute requirement to control a machine tool. It is present in every controller in a comparable form. Additional functions, such as programming and tool management, are specific to the technology and/or the user and so are not found in every controller. The additional functions are described in ► Sect. 11.2.5.

Figure 11.9 shows by way of example the data flow in a numerical controller. Nowadays, all functions (such as interpolation and transformation) are integrated into the controller as software using a high-level language such as C or C++. The number and sequence of the different subfunctions are generally controller-specific and vary according to the manufacturer's strategy [KLEI95].

The input data for a controller are the NC program work cycles, tool data and operating data.

### 11.2.4.1 NC Interpreter

The function of the NC interpreter is one of syntax analyzer (parser) that translates the different formats of the NC programs and input data to a standardized, internally readable form. This means different data formats such as simple DIN 66025 NC programs [DIN83b], complex surface information such as per VDAFS (*Verein Deutscher Automobilhersteller Flächen Schnittstelle*, see ► Sect. 11.4.2) or STEP-NC (see ► Sect. 11.3.2), can be processed by one and the same controller.

The NC interpreter passes the position of the path corner points, the type of motion stipulated (such as straight line or circular motion) and the necessary feed rate to the *geometrical data processing* function block for every NC line.

Switching functions such as tool change and workpiece feed are passed later to the logic controller (usually realized by a PLC). Of course, these functions must be synchronized with the geometry data. It must be guaranteed for example that a tool change (as a switching command to the PLC) may only be performed once the last geometry block has run to completion with the previous tool.

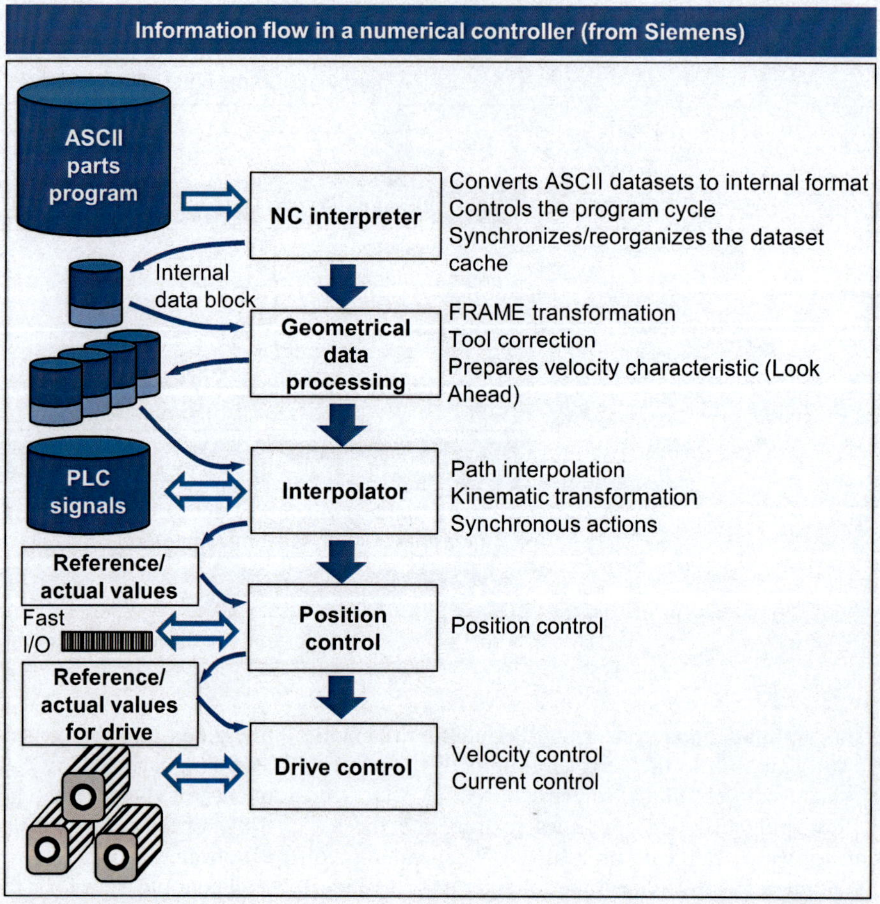


**Fig. 11.9** Information flow in a numerical controller, according to Siemens



### 11.2.4.2 Geometrical Data Processing

The functions Velocity control, Geometric transformations and Correction calculations are combined into one function block in Fig. 11.9.

Each individual function comprises in turn subfunctions where the order and implementation can be realized differently from controller to controller.

To enable NC programming independently of machine type and actual tool geometry, the controller makes *geometrical transformations* (see ▸ Sect. 12.2). First, a zero point offset is realized here that describes the position of the workpiece zero point relative to the machine zero point. This enables the further transformation between workpiece and machine coordinates. Second, the tool correction, generally comprising a tool length and tool radius correction, permits the calculation of the necessary equidistances to the path programmed as a function of the current tool data. This guarantees independence of NC programming from the exact length and diameter of the tool used later.

As part of *velocity control*, the velocities and accelerations of the programmed feed rate are aligned to the appropriate boundary conditions (see ▸ Sect. 12.1.1). For major velocity changes, such as when small radii are circumscribed, their limited dynamic range means the drives are not able to follow the motions specified. In these cases, the controller automatically decreases the path velocity to keep to the workpiece dimension tolerances. Modern-day controllers have a Look-ahead function to optimize further velocity control. Using this monitoring function, multiple NC lines (10 to 100) are analyzed in advance to determine whether the velocity profile can be followed or improved (see ▸ Sect. 12.1.2.2).

### 11.2.4.3 Interpolation

The task of *interpolation* is to calculate intermediate points that lie on a path section defined by the NC program. This splits the path movement arising from the simultaneous movement of multiple axes, piece-by-piece across the individual axes, such that the axes change their positions independently. Simple methods such as linear and circular interpolation are used, as well as more intricate types such as spline interpolations (see ▸ Sect. 12.1.4) [KIEF99]. The other tasks of interpolation are following the velocity profile specified by velocity control and, for non-Cartesian machines, *kinematic transformation* of the coordinates to the individual machine axes.

Furthermore, external factors, such as correction of the feed or spindle speed (override), as well as static and dynamic displacements, must be factored in or compensated with *compensation calculations* in the interpolation process. Using a rotary switch for the feed correction, the machine user is able to change the programmed path velocity (generally between 0 and 120%). The compensation calculation calculates the velocity change during the machining process without allowing geometric deviations to occur on the workpiece. In addition, any displacements must be compensated continually during the process. These displacements come from static loads or thermal distortion of the machine structure, such as thermal extension of the spindle, the reversal error of the spindle nut, the spindle pitch error or tool wear (see ► Sect. 12.3).

#### 11.2.4.4 Axis Control

Depending on the controller design, the individual feed drive units are supplied with reference position, velocity or current values by geometrical data processing. The control principle predominantly in use today is cascading position control. The motor current controller forms the inner control loop using a P or PI controller. It is superimposed by the speed control loop that receives its reference variables from the position controller. The controllers used are generally PI or PID controllers for the speed control loop and P controllers for the position control loop (see ► Sect. 3.2).

On some numerical controllers, *position control* takes place in digital form internal to the controller, i.e., reference velocity values are relayed to the drives. The velocity and current controllers and also the position controller depending on the design, work on the drive side in the servo modules (see ► Sect. 2.2.2). The described information flow in the numerical controller (i.e., from the interpretation of an NC program to output of the reference axis values) occurs in one NC channel at any one time. With every NC channel, multiple axes or spindles are operated asynchronously or synchronously in an axis group. The NC channels can be divided into operating mode groups (◘ Fig. 11.5). More NC axes can be defined, for example to control turret, tool magazine, workpiece/tool pallets, tailstock, quill, cutting-off slide or workpiece and tool loaders.

### 11.2.5 Functional Scope of Modern Numerical Controllers

Besides the basic numerical controller tasks described, namely interpretation and running of a parts program, additional functions have been integrated into the numerical controller arising from the stricter requirements coming from machine manufacturers and end users. These functions are technology or user-specific and broaden or increase the area and ease of use of the controller. This means these functions not only simplify operation and program entry at the machine, they also make possible data connections between machine tools or manufacturing cells and a production system. Standard functions are generally implemented by control manufacturers themselves, whereas additional functions that are dependent to a high degree on the user are integrated into the controller by the machine manufacturer. Described below are functions offered by both control and machine manufacturers.

#### 11.2.5.1 Standard Functions

◘ Figure 11.10 shows the possible functionality of numerical controllers. All of the functionality is not required for operation, however. This depends on the application and configuration of the controller.

##### Programming

For parts programming on the controller, different programming tools are available that cover a wide range as regards price and functionality—from a simple alphanumeric text editor, semi and fully automatic NC program generation systems, to a graphically oriented WOP system. The syntax for programming is based on DIN 66025 [DIN83b; DIN88] and the available language scope is frequently complemented by manufacturer-specific extensions. There is a detailed description of the different programming methods in ► Sect. 11.4.

##### Graphics-Aided Machining Simulation

An additional function is the graphic machining simulation on the screen for checking or optimizing NC programs on the controller without the machine having to actually perform the individual operations. This can reduce the setup time of the machine or the startup time of the NC program. In this respect, high importance is attached to graphical simulation. For workshop-oriented programming, this is usually used on the numerical controller or for automatic offline programming in production planning. Depending on the requirement, 2D or 3D models are used for modeling and depiction.

Simulation has two objectives. First, to check the NC program for possible collisions in the machine space when performing traversing movements, for tool changes and for limited checking of the dimensions and contours (such as detection of undercuts). Second, simulation offers a traverse path optimization capability. Here, unnecessary traversing movements (such as air cuts) are detected and eliminated, or bypassed with rapid machine

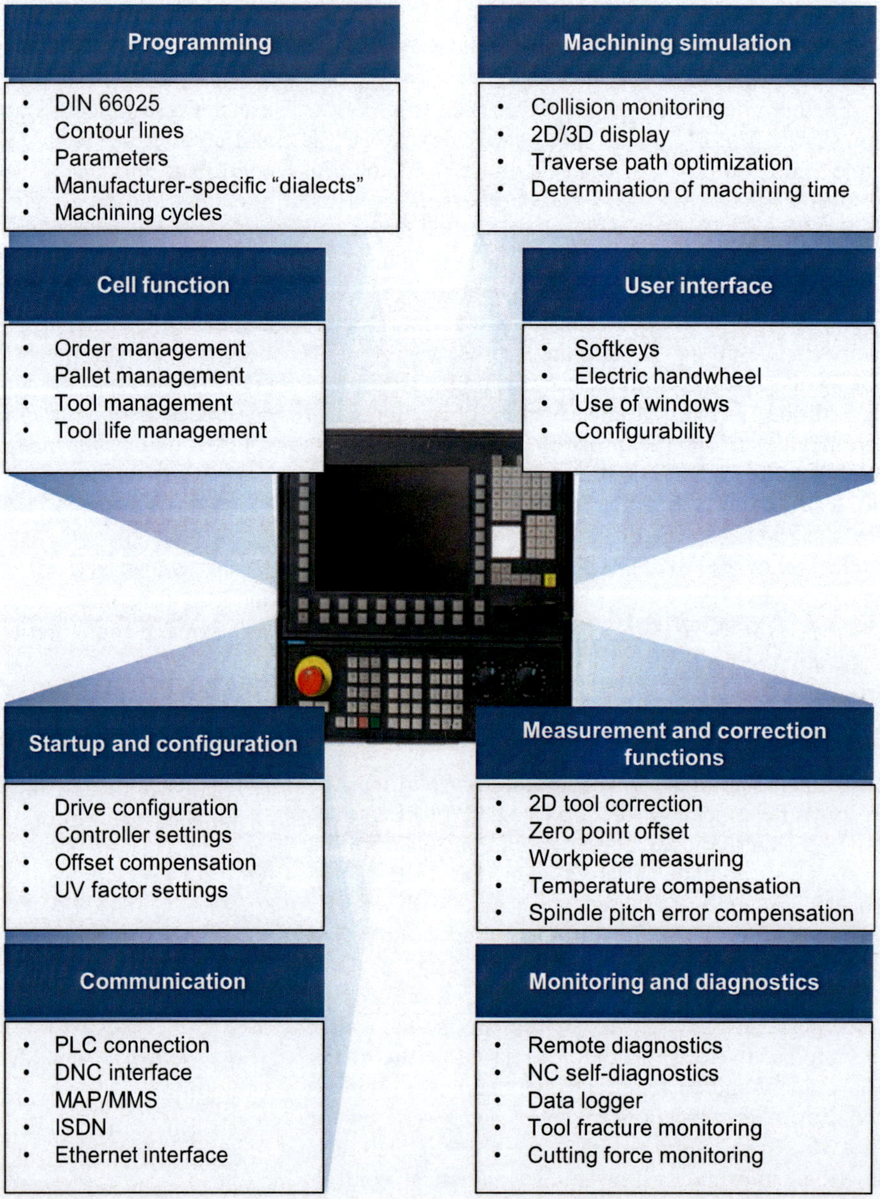


**Fig. 11.10** Functional scope of a numerical controller

11

movements, to shorten single part production times accordingly [PFEI93; SIEM99; WECK91].

Figure 11.11 shows by way of example the simulation of a roughing process for a turning operation. A magnifying glass function can be used on the view any time to zoom in and out. During and after a simulation, all current data are visualized, such as NC line number, tool in contact (T), reference and actual path information (X, Y), feed (F), spindle speed (S), cutting speed (v), rotational direction (M) as well as current and overall program runtimes. Also, the graphic contour can be shown at any time to check the geometry input data. In the 3D view, the workpiece can be rotated as required in space in every phase.

In manual mode, simulation in parallel with working with the handwheels makes possible a real image of the production process in real time. This is of particular benefit for difficult to see interior machining and when cooling lubricants make visibility poor.

Realistic emulation of the machining process requires modeling of the tools, tool holders and tool carriers that is as accurate as possible as well as correct movement of individual carriages over time during the entire machining process for the workpiece.

Graphic simulation represents an important tool for detecting collisions. Powerful simulation systems permit simulation of different machine types and machining processes by accessing a machine and tool file.

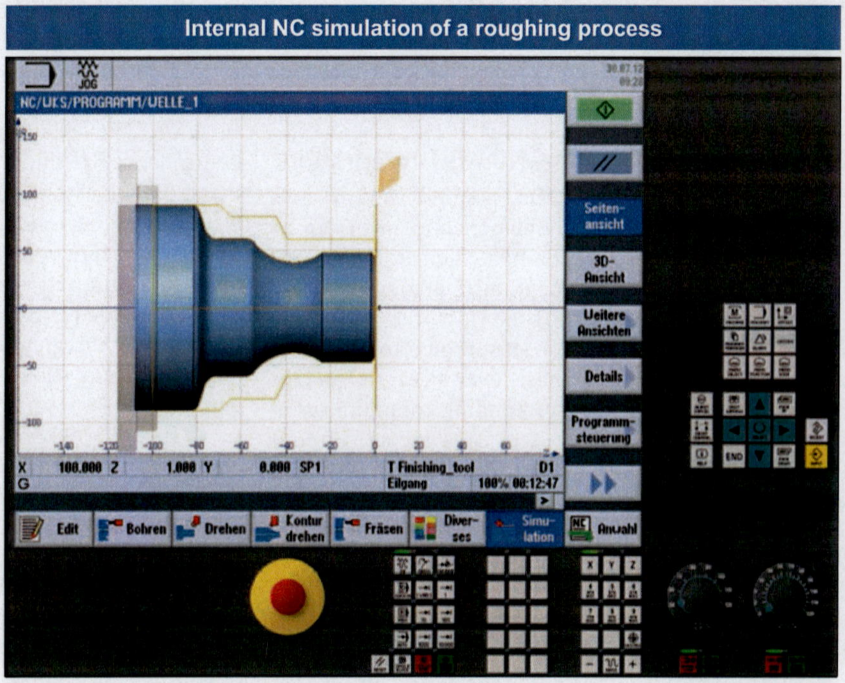


**Fig. 11.11** Screenshot of a roughing process simulation, according to Siemens

This contains the geometry data for tool turret, working area with spindle, clamping devices, tools, etc.

Machining simulations are commonly included already in commercially available offline programming systems or WOP systems (see ▶ Sect. 11.4).

## Monitoring and Diagnostics Functions

Monitoring and diagnostics functions are used to guarantee error-free machining processes and to prevent damage to the machine, workpiece and tool. These functions, which are implemented directly on the numerical controller or on additional devices, are divided into the areas for machine monitoring (e.g., monitoring of the drive dynamics, geometric accuracy and lubrication, ▶ Sect. 6.5.2), the installed hardware and software (e.g., numerical controller self-test, checking of lines and switches, see ▶ Sect. 6.5.2) and the machining process (e.g., tool fractures, tool wear, workpiece dimensions, surface quality, see ▶ Sect. 6.3).

The monitoring-only function is normally complemented by fault localization diagnostics functions that are more or less complicated. The different strategies and methods for monitoring and fault diagnostics are discussed in detail in ▶ Chap. 6.

## Measurement and Correction Functions

When strict quality requirements are in place, axis path corrections must be performed continuously during the machining process. Mechanical machine deficiencies, wear of the tools and temperature-related displacements of the machine structure are reasons for this. Static errors such as spindle pitch errors and reverse backlash are generally measured once on the machine during startup and stored in a correction memory.

Measurement cycles are used for parameters that vary constantly, such as the clamping position of the workpiece (clamping tolerances), tool wear and spindle expansion due to temperature. Using measuring devices, sensors and measuring probes, dimensional changes on the workpiece and tool as well as machine displacements are measured and corresponding correction values determined. These values can then be processed by the controller for dimensional compensation. There is a detailed discussion of this in ▶ Sect. 12.3.

## User Interface

The term user interface comprises all aids made available to the user of the numerical controller. It can be divided into two areas: the screen and devices for manual control (e.g., switch, keyboard and handwheel).

Simple adaptability of a user interface is an indication of the quality and openness of a controller. The term "open controller" indicates the trend, for example, towards granting the machine manufacturer as much freedom as possible in configuring the interface. In addition to the actual design of the interface, it is the ability to freely specify the data from the process or information flow to be visualized in the NC that is important. Modern-day interfaces are window-oriented, are based almost exclusively on the MS Windows operating system, have softkeys and assignable function keys and are implemented on flat screens. The user interface is discussed in detail in ► Sect. 11.5.

### External Communication—Network Integration

The functional scope of external communication has a major bearing on how easily numerical control can be integrated into the production-related environment. All interconnectivity options mentioned are used. These extend from the simple analog interface, through to the digital fieldbus and actuator/sensor bus, to the local network. Even remote diagnosis over a secure Internet connection is used. The corresponding communication protocols and mechanisms used for a numerical controller are described in separate Sects. (14.3.3, 14.3.2.2 and 2.2.3).

### Machine Startup and Configuration

During machine startup, the numerical controller is tailored to the features of the machine tool and the user requirements. For this, the controller is configured in regard to axes, spindles, handwheels, measuring systems, control loops, velocities, accelerations and interpolation modes.



Specific auxiliary functions with graphical evaluation are provided by some controller manufacturers for startup and machine diagnostics. To optimize the controller setting, the behavior of an axis control loop, for example, can be shown on the controller monitor for different control parameters, see ◘ Fig. 11.12 [ROBE99].

Offset compensation and Kv factor setting (unit for the gain in the control loop) are therefore easy to perform for axis optimization. The machine axis to check is activated by the controller using a defined path-time function. The response of the drive is displayed on the controller monitor in the form of the actual movement. For this, the position measuring systems of the feed axes are read. Documentation of the control loop setting as a function of the workpiece dimensions enables an optimized controller presetting in subsequent use. Given adequate resolution of the position measuring systems, the graphical display of the reference and actual contours shows dynamic deviations down to the $\mu$m range.

◘ Figure 11.13 shows the actual and reference traverse paths for a workpiece on the screen. Enlargement to any degree can be displayed (such as the contour area of the square).

The reference and actual positions as well as reference and actual velocities are shown for a defined point on the reference and actual contours. Furthermore, the reference and actual velocities, reference and actual accelerations and the lag distance can be shown graphically for the entire contour profile.

The quality of the controller settings, the contour accuracy and the feed rates can be adapted and checked in line with the accuracy requirements stipulated.

### Additional Functions of Numerical Controllers

The functional scope of numerical controllers can be expanded by integrating software modules for the

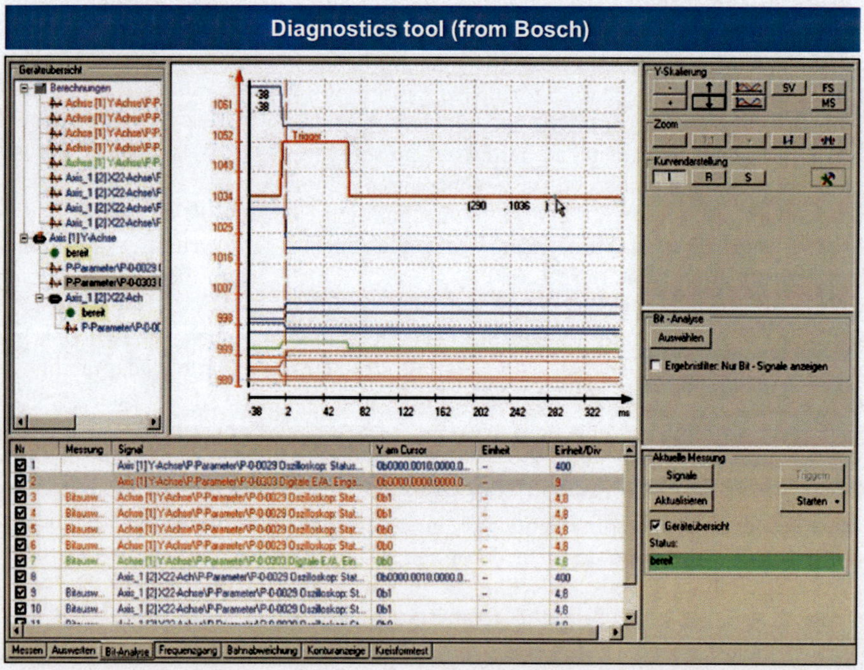


◘ **Fig. 11.12** Diagnostics tool, according to Bosch

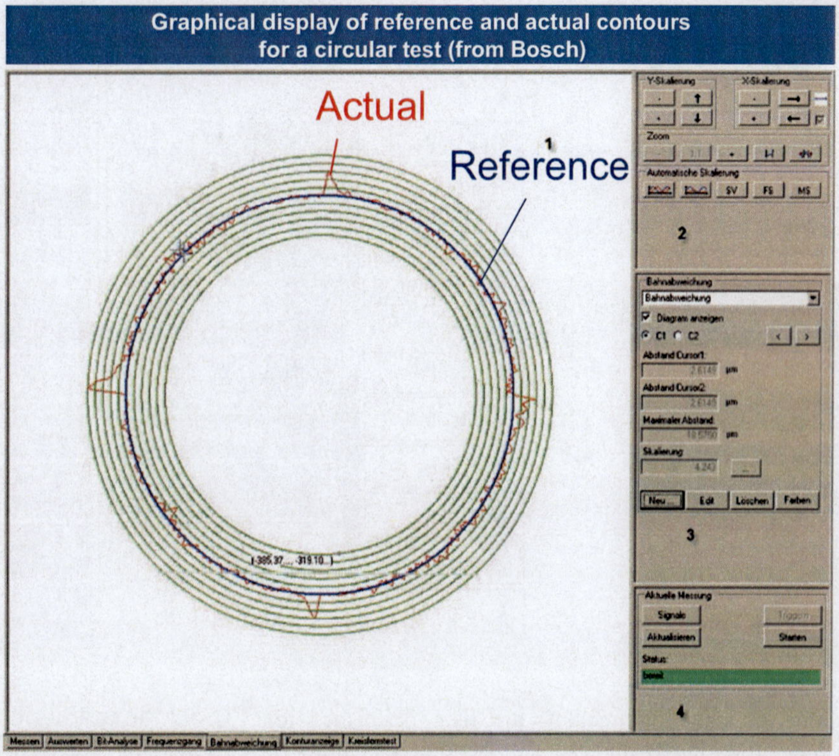


**Fig. 11.13** Graphical display of the reference and actual contours, according to Bosch

production data acquisition and machine data acquisition (PDA/MDA). Data acquired are divided into *order data* (order duration, setup time, quantities, scrap, etc.), *personnel data* (presence, working hours) and *machine data* (production time, downtime, fault causes, fault times, etc.). A weak-point analysis of the organization and machines can be conducted by statistically analyzing the data (refer to ► Sect. 6.5).

Additional functionality is provided by *NC program management*, that provides, on demand, the management of information on NC main and subprograms (e.g., article code, version no., status and runtime) and the necessary production equipment requirement (tools, equipment and measuring devices) for machining a workpiece.

Functions for collision and working area monitoring are available. The programmer can use these on certain feed paths to protect the machine from collisions. Also worthy of mention in conjunction with additional functions for numerical controllers are controlled gear shifting systems, modules for maintaining constant cutting speeds and exact spindle positioning in several defined positions for automatic workpiece, tool and clamping device changes.

### 11.2.5.2 Functions for Controlling Automated Production Cells

Viable linking of machines and appropriate automation of material handling for the individual machines are required to increase manufacturing productivity. The objective of a highly automated production system is to attain high utilization enabling 3-shift operations with low manning levels through high flexibility. To allow autonomous machining of a pool of orders for different parts over a certain time period without requiring user intervention, the standard functional scope of a numerical controller as offered by a controller manufacturer must be broadened. These machine-specific functions are generally developed and integrated into the controller by the machine manufacturer.

Extending beyond NC programs, the input information of a numerical controller for a manufacturing cell comprises data on tools, pallets and orders that must be managed by the controller. The various tasks can be divided into order, tool and pallet management They are described below in line with this division.

#### Order Management

In addition to actually managing the production tasks to perform in the manufacturing cell, order management also handles the execution of orders. Orders are either passed automatically from the control level to the controller or entered manually at the machine by the operator. After an order is accepted, order management is responsible for inventory control of the material and the workpieces in the cell. ◘ Figure 11.14 shows some example data for managing. Also, requirement calculations for NC programs and production

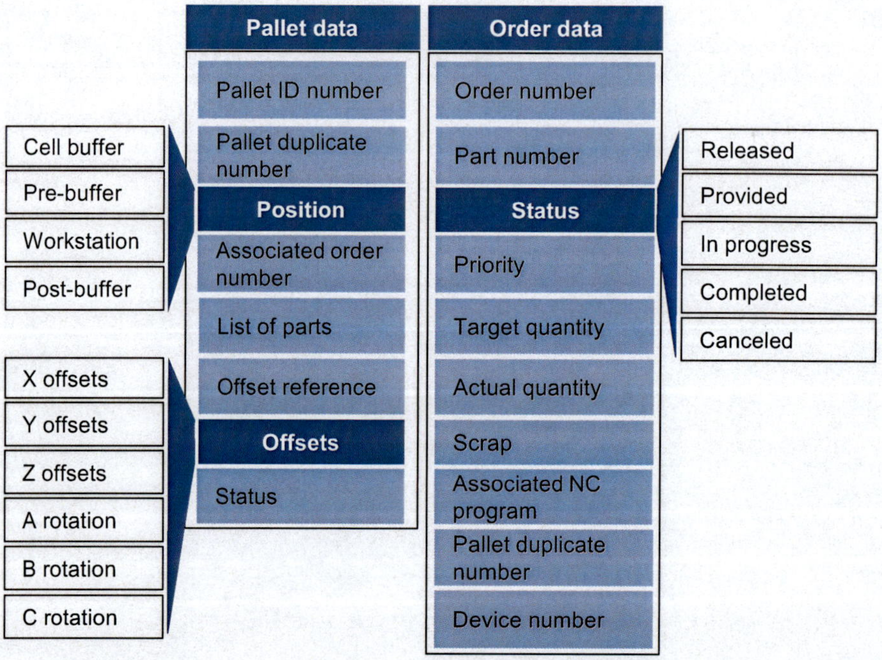


**Fig. 11.14** Order and pallet management data internal to the controller



aids (tools and equipment) are then performed for every order. Order management optionally changes or specifies the order sequence. It also handles the execution of individual tasks, such as load and start NC programs, transfer finished workpieces to storage and transfer blanks to local storage (see ▶ Chap. 14).

## Pallet Management

If a machine tool has its own pallet conveyor system or if it has multiple machining workplaces, there must be management of the pallets, equipment, buffer zones and machining workplaces. Furthermore, the material flow of pallets, equipment, materials and tools needs controlling.

Pallets are transport units able to accommodate and temporarily store one or more workpieces or tools at specific, stipulated locations. This means the workpieces and tools from and to a machine can be transported by a driverless transport system, roller conveyor or transport trolley on rails. It is also possible to use a handling system to move workpieces from a pallet to the machining position and back.

The task of pallet management is to coordinate material flow within the manufacturing cell. This management entails the entry and removal of pallets as well as data on pallet storage locations and pallet contents. Workpieces and tools on a pallet, workpiece displacements and status information are examples of pallet data (Fig. 11.14).

## Tool Management

To machine a workpiece (blank) in one operation, several different tools are commonly required. Allowing speedy tool changes between two machining steps uses tool magazines that have the appropriate tool change mechanisms and are located near the machine. Normally, position coding in the tool magazine, barcodes, and memory chips (such as RFID chips) uniquely identify the tools themselves. Actual information such as geometry, correction values and remaining service life for individual tools are kept in the numerical controller or in master tool management. Figure 11.15 shows some example tool management data internal to the controller. A distinction is made here between the basic data of a tool and the specific information of a tool cutter because a tool can comprise several cutters or layers.

The tasks of tool management are data management, correct tool selection and provision for an NC program given by the setup sheet or routing plan. Before the machining process, the tools necessary must be in the magazine or turret. During a machining process, a simple switching command in the NC program initiates a tool change. Tool management must execute this command once the last NC line using the previous tool is completed.

Furthermore, modern-day systems monitor tool service lives so that worn and defective tools can be replaced in good time by twin tools. Owing to the limited

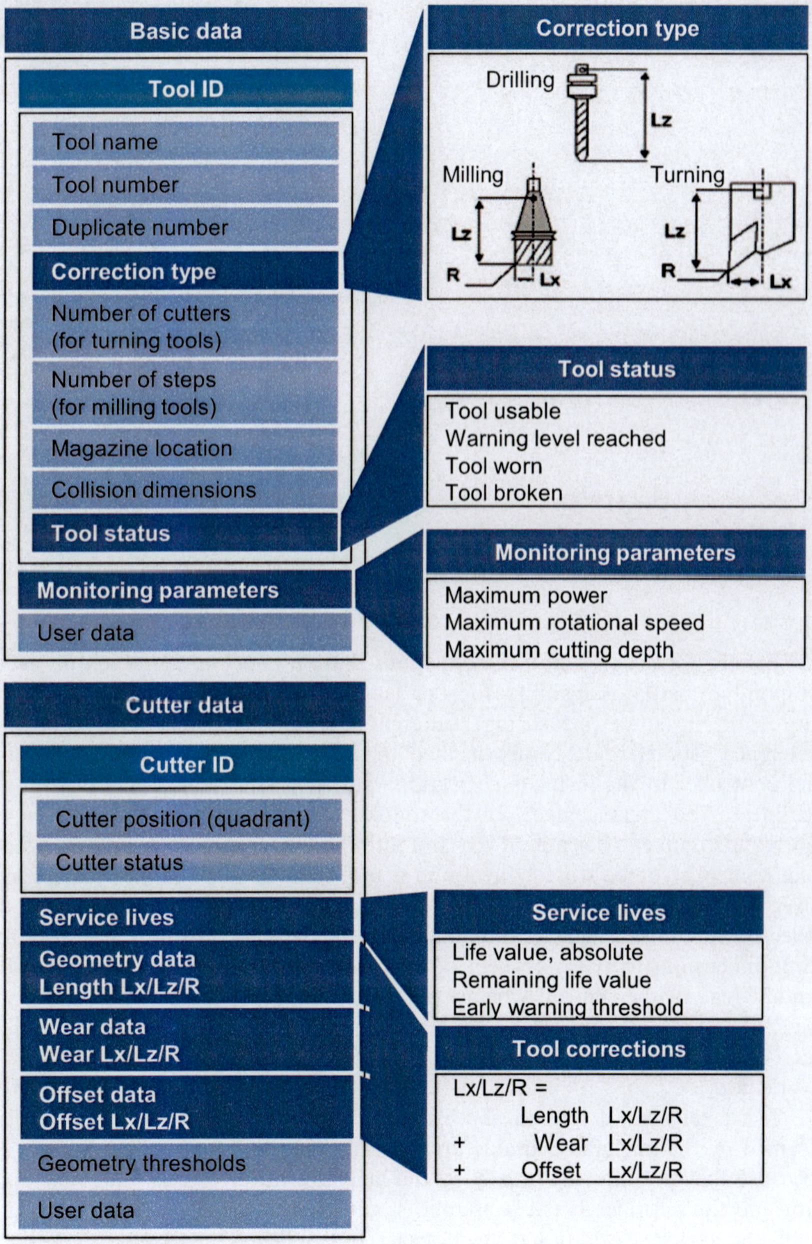


**Fig. 11.15** Tool management data internal to the controller, according to Bosch Rexroth

number of spaces in a tool magazine and the different space requirements for tools of different sizes, some systems provide tool magazine assignment planning that loads tools into a magazine so as to optimize space.

As tool management needs to initiate many switching functions for tool changes, the software is frequently also implemented in the PLC (see ▶ Sect. 10.3). Open controller systems allow more sensible solutions. For example, they permit a division between managing functions, which are far easier to implement on a PC with database access, and switching-only functions that are implemented in the logic controller.

Before an NC program is entered into the controller and production can start, some preparatory work needs to be performed on the NC machine. This work is called the machine tool setup. The setup sheet, in paper form or digital form on the computer, lists the tools required, among other things. Before the tools are loaded into the magazine or turret of the machine, they are calibrated on an optical tool presetter (Fig. 11.16). The tools are either set to a certain geometry (length and radius) during this *setting* process, or, when *measuring*, only the difference of the actual geometry from the reference geometry is recorded.

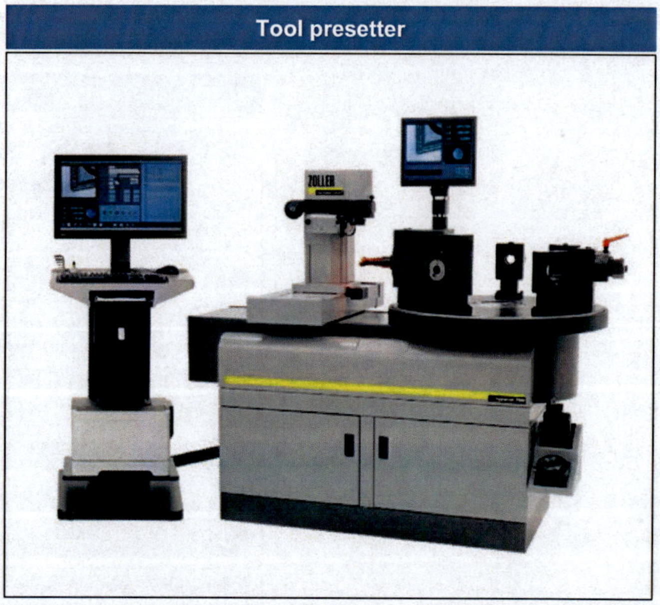


**Fig. 11.16** Tool presetter, according to Zoller

For the former (setting), the tools must have a corresponding setting capability for the length and radius. For the latter (measuring), differences from the reference geometry are communicated to the numerical controller in the form of correction values or offsets from tool management. In the simplest case, the correction data can be entered by hand into the numerical controller or be stored directly in a tool database over a network connection. If memory chips are affixed to the tools as identification data carriers, the setting dimensions can also be kept directly on the tools and be read by the controller before the machining process. ▶ Sect. 14.3.2.4 includes a supplementary description of the organization of production resources in the workshop.

The extensions described above are generally performed by the machine manufacturer for the end user because they are affected mainly by the machine variant and the peripherals (such as robots, and tool and pallet magazines). Additional functions to control flexible manufacturing cells, which extend beyond the functional scope of numerical controllers, are described in ▶ Chap. 14.

### 11.2.6 Openness of Controller Systems

Nowadays, the market is demanding machines and systems with continually growing functionality. The purpose of innovative concepts and rationalization measures is to reduce costs for all involved in the value-added chain. Associated with this is the users' demand for independence from individual suppliers. To increase the levels of flexibility and quality in production even more, new additional controller functions are necessary. Production technologies such as hard-fine turning instead of grinding or laser machining stipulate the integration of process-specific functionality into the numerical controller. Machine concepts such as parallel kinematics (see Volume 1, ▶ Sect. 4.3.3) also require additional functionality in numerical controllers. Machine end users demand standardized and technology-specific interfaces. Such features ensure simple integration of machines and controllers from different manufacturers into the user's production environment, and cross-machine planning, maintenance and diagnostic capabilities.

#### 11.2.6.1 Motivation and Objectives of Open Controller Systems

The diversity of the requirements described above means that they cannot be satisfied by control manufacturers themselves. Aiming for speedy responses to customer requirements, the traditional customer/supplier relationship is transforming into a mutual partnership for adding value. Customer are becoming increasingly involved in the development of controller products. For this reason, the development of open controller systems and the addition of custom NC functionality have gained in significance and established a foothold in the market [WECK93].

Of paramount importance here is providing open interfaces that allow the integration of additional software by the customer in the control cluster. This offers the customer flexible extension options to the standard functionality of a numerical controller, providing the freedom to design innovative solutions. This provides machine manufacturers with the option of easily integrating software, such as tool fracture monitoring, into their controllers.

#### 11.2.6.2 Variants of Open Controller Systems

A fundamental difference can be made between the "internal" and "external" openness of a controller, Fig. 11.17 [WECK93]. External openness pertains to a simplification or standardization of the external interfaces of a controller, openness such as harmonized NC programming, standardized communication interfaces, a user interface that is extendable and set up to the same basic rules, and a universal drive interface. Internal openness enables access to the software for internal controller functions and integration of additional modules on the basis of a fixed ruleset. It can be seen from this that internal openness is predominantly of importance to controller and machine manufacturers, while external openness benefits end users in particular.

A number of standards and de facto standards exist already for *external openness*. Many manufacturer-specific solutions as well as the MMS standard (production Message Specification) are being used for external communication. A number of different fieldbuses have established themselves as interfaces to sensors and actuators (see ► Sect. 14.3.3). Digital solutions are increasingly being deployed as drive interfaces (such as SERCOS, PROFIBUS-MC and Fast Ethernet, see ► Sect. 2.2.3). As regards programming, program structuring according to DIN 66025 is regarded as the standard input format for numerical controllers. The overriding opinion, however, is that it is no longer adequate for more complex machining operations and is in need of urgent rework. A new standard is on the horizon here in ISO 14649 (see ► Sect. 11.3.2).

As regards *internal openness*, some manufacturers are now offering their own solutions. These, however, differ greatly in terms of the extent of openness (Fig. 11.18). The term used in this context is *graduated openness*. This permits the customer (depending on the requirement profile) to access various levels having greater functionality but with increasing complexity.

Many manufacturers offer open interfaces in the operating area. This enables users to integrate non-time-critical software into the controller and access updated data of the NC kernel using defined interfaces (Fig. 11.18, top).

A few manufacturers go one step further and provide the customer with a function library to integrate their own software at specific places in the data flow of the NC kernel. This enables the NC kernel data to be accessed directly and even changed. An elaborate development environment is generally required to integrate proprietary software into the real-time area of the controller. Applications for external corrections (such as temperature compensation) can easily be integrated on this basis, see Fig. 11.18, center [ROBE99; SPUR93b].

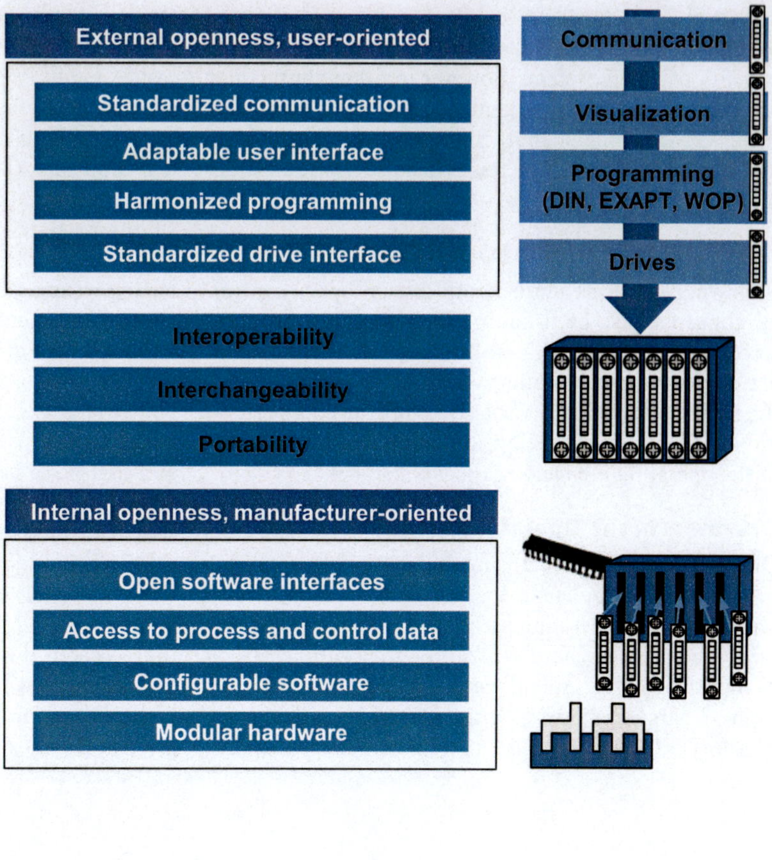


**Fig. 11.17** Internal and external openness of a numerical controller

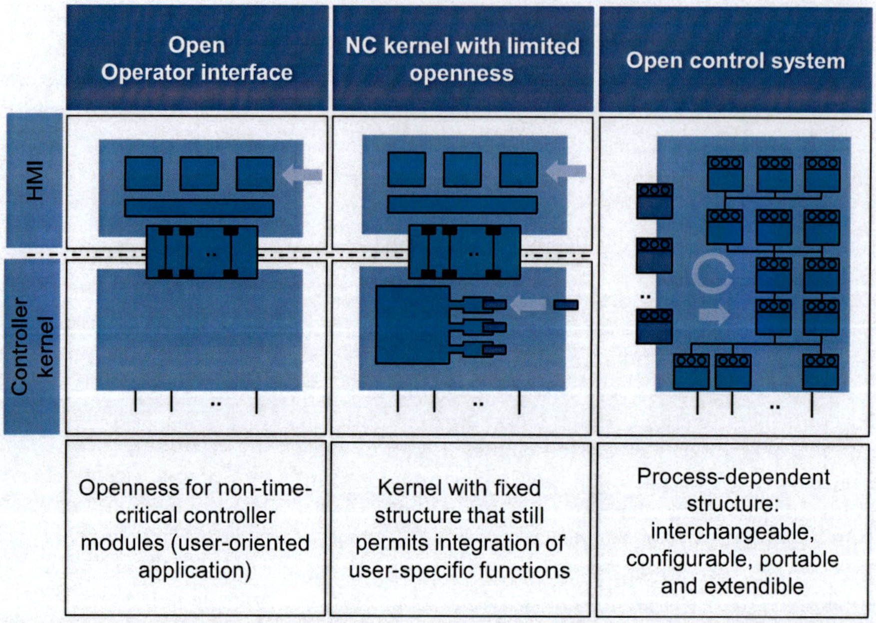


**Fig. 11.18** Variants of open control systems, according to [WECK96]

11

The highest level of openness enables integration of any software modules into the NC kernel. The data flow is configurable and software modules are interchangeable. This variant grants the user the greatest freedom but also requires extensive knowledge of the system structure of the controller. It is offered by few controller manufacturers and has seen little use (Fig. 11.18, bottom) [OSAC00; NN99].

All open controller systems available are restricted to manufacturer-specific solutions. To utilize all the benefits of open systems however, cross-manufacturer standardization is an urgent requirement. Not until then will machine manufacturers, software providers and end users be able to reuse their software and deploy it on other controllers.

### 11.2.6.3 Realization of Open Controllers

Following vocal requests from users, open controller concepts have been finding their way into products since the early 1990s, in parallel with extensive research activities, and these products have been launched onto the market as different variants. The concepts realized are introduced briefly below, differentiated by openness in the operating area and in the NC kernel.

#### Openness in the Operating Area

One of the variants of openness that is widespread for modern-day controllers is the use of PC-compatible assemblies in the operating area. This enables integration of PC-compatible hardware into the controller (such as network cards, sound cards and external devices such as barcode readers). In combination with the use of standard operating systems (almost exclusively MS Windows nowadays), the user is able to integrate programs from the office world (Office programs, editor, database, speech processing, etc.) into the controller.

Influence being exerted over the design of the user interface is the case now for almost all controllers. For modern-day controllers, a configuration tool is provided for designing and configuring the user interface. Using the standards of the Windows operating systems (Visual Basic, DDE, COM, OPC, etc.), any operator interfaces can be created and integrated into the controller. What is important to the user is that these tools are easy to use by IT specialists as well as technologists.

When the user develops proprietary software for integration into the operating area, access to NC kernel data is required in most cases, be it for displaying positions or states of the controller, for reading internal controller information such as tool data, or even for having write access to internal data. Almost all modern controllers offer this capability in the form of access to variables (Fig. 11.19). The variables are generally addressed using a name and can be read. Some can also be written.

From the user perspective, access to data is either in a high-level language (such as C, C# or C++), using an API (Application Programming Interface) or on the basis of standardized objects or components such as DDE (Dynamic Data Exchange), COM (Component Object Model) or OPC (OLE for Process Control). Components are normally based on standards from the office sector that are specified by Microsoft. This has the benefit that they can be used independent of a special programming language (Visual Basic, C#, C++ or Java) and can, to some extent, even be activated directly

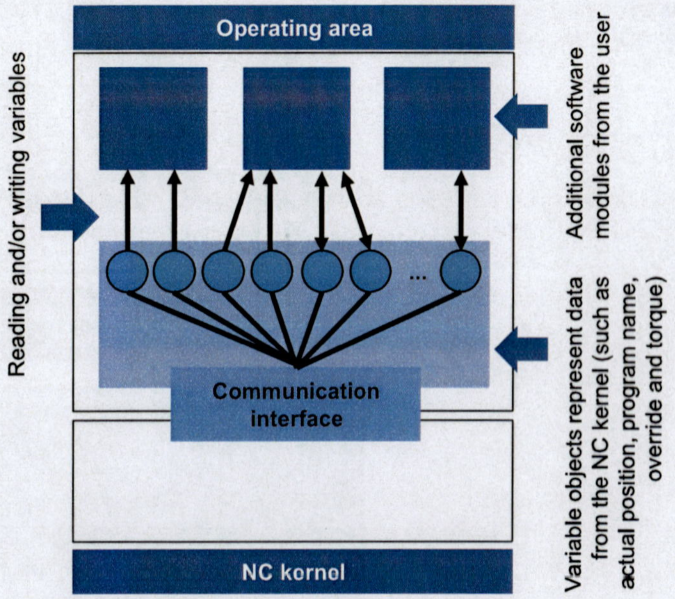


**Fig. 11.19** Openness by accessing NC kernel data

from Office applications such as Microsoft Excel. Also, the user is able to use the many development tools available on the market.

## Openness in the NC Kernel

Controllers offering openness in the NC kernel to integrate real-time-capable applications include those from Siemens (SINUMERIK 840D family) and Bosch (IndraMotion MTX), as well as those from some smaller companies that offer an NC kernel as a software kit in the source code. These controllers can be adapted flexibly to special requirements by integrating software that is specific to the machine or process. Users generally have access to all the standard functions of the controller and can use, adapt or replace them in part with new functionality. Application examples for using openness in the NC kernel:

- Transformations for specific kinematics (e.g., complex parallel kinematics)
- Online collision monitoring of the workspace
- Distance control for a laser machining head
- Process monitoring integrated in the controller (monitoring for tool fractures and wear)
- Prevention of chatter
- Thermal compensation
- Adaptive feed control depending on the process force when roughing
- Connecting a space mouse for 6-axis control

The requirements of giving users as much freedom as possible to integrate their software while guaranteeing the reliability and safety of a controller necessitate some restrictions. For example, the global system structure is generally specified and cannot be changed by a user. Two different system structures, compile cycles and job lists (described briefly below), have established themselves on the market.

## Compile Cycles

The *compile cycles* used by Siemens are applications written in C++ that are developed by the customer, linked into the NC kernel and activated using customer-specific data. Functionality is enabled by the controller manufacturer. Compile cycles in the NC kernel offer the same real-time capability as the base system. They are created in a powerful development environment for the C++ language. Here, the infrastructure of the NC kernel with communication between tasks, alarm mechanisms, PLC signals, machine data and internal NC lines can be used.

An object-oriented NC kernel in C++ forms the base system of the controller. At defined places in the NC kernel software system, there are exit points (or *events*) at which a customer can include its own software in the form of compile cycles (Fig. 11.20). Events reflect the internal function structure of the NC kernel. They provide over 50 exit points from the base system of the controller. Virtual access methods (or *bindings*) can be used to attain the necessary base system data. These methods are part of the open interface and thus guarantee compatibility with all future software versions. Bindings currently allow access to over 700 data items [SIEM97].

## Job Lists

Another way to integrate customer-specific software into the NC kernel is the concept of *job lists*, developed by Bosch. Controller modularity is achieved by

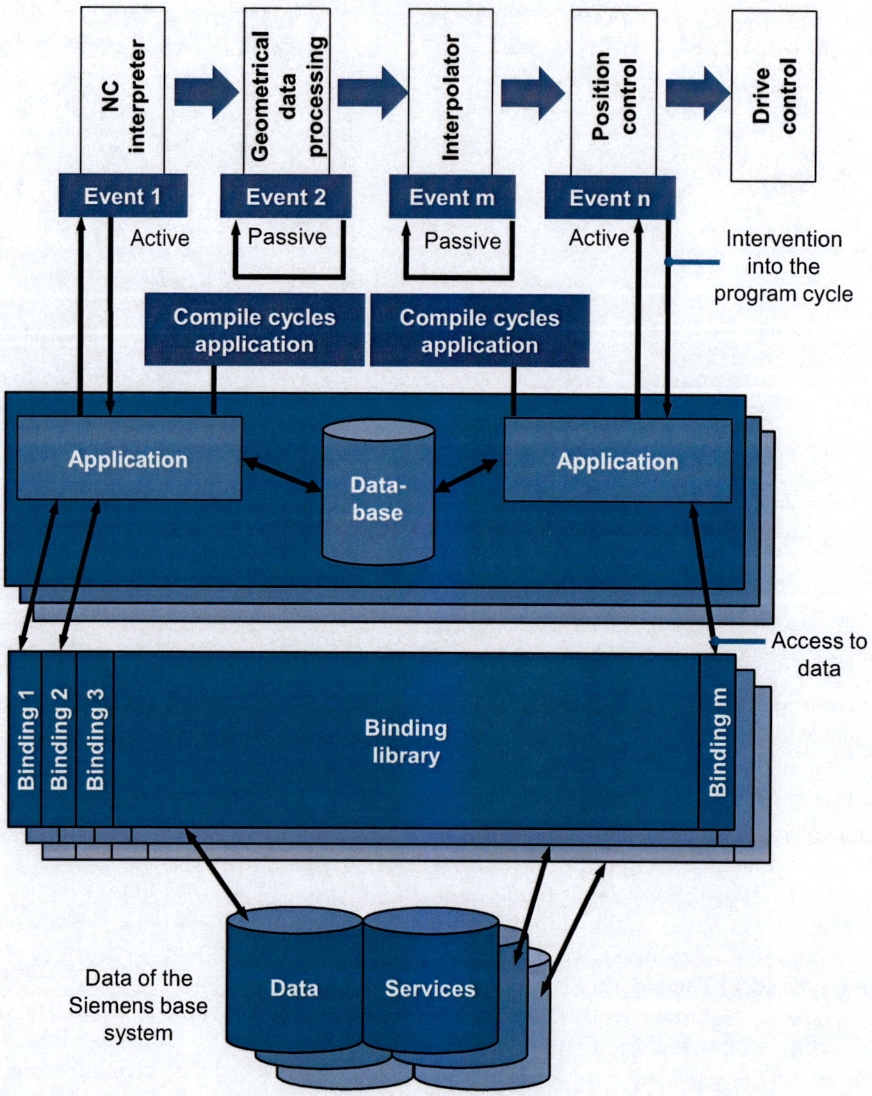


**Fig. 11.20** Openness using compile cycles, according to Siemens



splitting the functionality into small, self-contained units. *Subsystems* each comprise a self-contained area of the controller, such as NC line processing and movement generation. Communication between the subsystems uses defined interfaces. The functions of the subsystems are split further into individual functions (or *jobs*) that are programmable by the user as C functions (◘ Fig. 11.21).

The jobs can now be assigned to individual NC functions in configuration tables. An NC function corresponds to one instruction in the NC program. The functionality of this instruction is composed of one or more jobs in the *job list*. Every job is called at a particular place according to its priority while the NC line is being prepared or executed. An API can be used to access the controller data. By configuring the job lists, the user has a variety of ways to expand or modify the controller sequence and behavior.

#### 11.2.6.4 Cross-Manufacturer Standards for Open Controllers

The fundamental idea of open controllers is to define a modular, flexible controller architecture that enables a controller to be tailored (with suitable flexibility) to user requirements. On the basis of software modules, a controller can be reconfigured by the user depending on the needs by extending individual software modules, adding new ones or replacing them. However, the user, in particular, has an understandable interest in the standardization of open interfaces across manufacturers.

◘ Figure 11.22 illustrates the software modules needed to make up a controller. The functionality of the controller is first divided into individual, functionally related function areas. The Human Machine Interface (HMI and HMIServer), Motion Control (MC), Axis Control (HAC) and Machine Kinematics

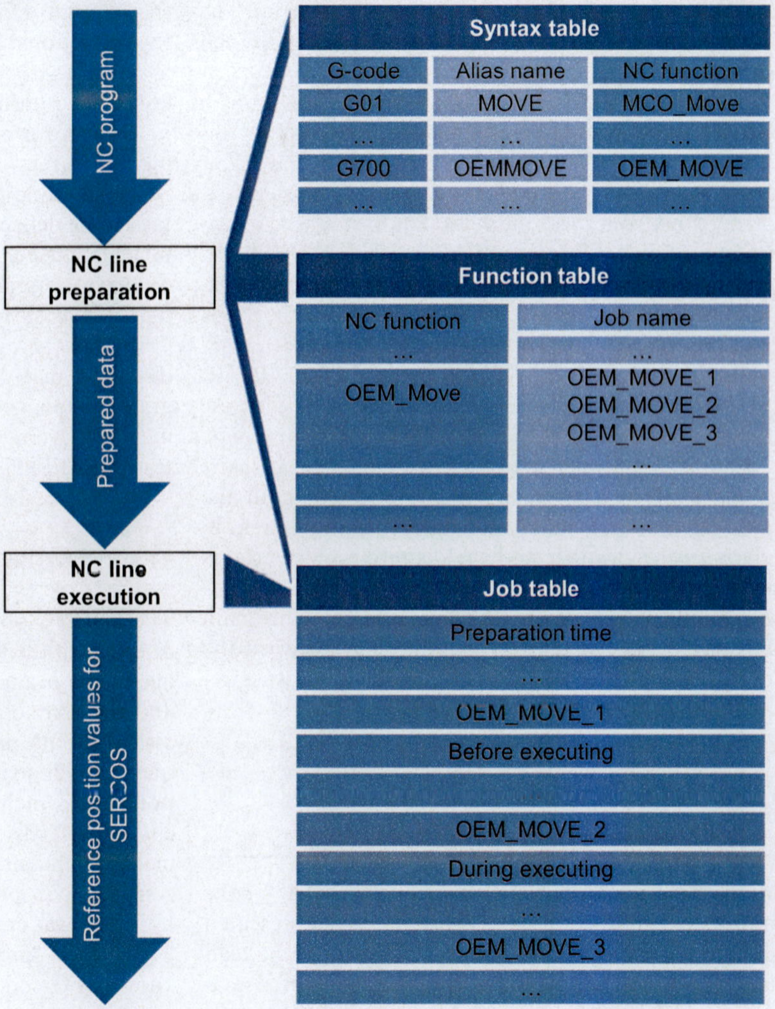


**Fig. 11.21** Openness using job lists, according to Bosch

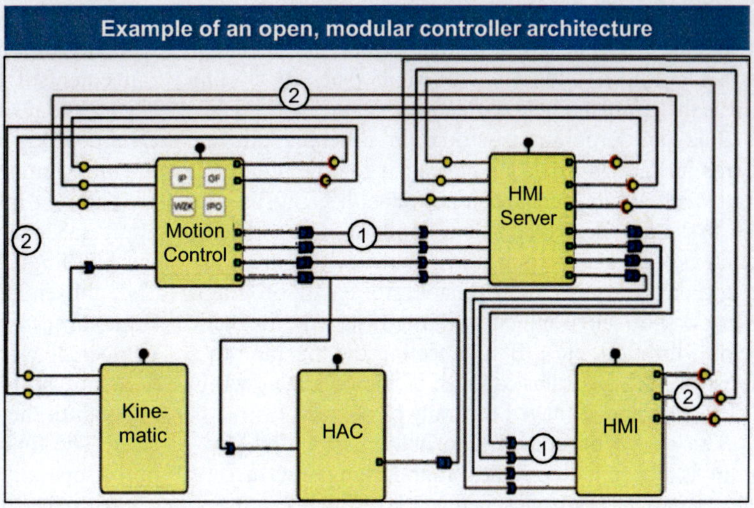


**Fig. 11.22** Example of an open, modular controller

(Kinematic) functional areas form the central part of the numerical controller. The MC functional area coordinates the geometric calculations and path planning inside the controller. The HAC functional area coordinates groups of individual machine axes and includes position control. The Kinematic functional area handles the tasks of kinematic transformation. The HMI comprises the user interface with its additional functions.

Each of these functional areas is implemented as a software module that is self-contained but broadly independent of other modules. For the overall control system to work however, information must be exchanged between these modules. For example, the program cycle in MC is controlled from the HMI (1). In the other direction, the MC and almost all other functional areas request and visualize data from the HMI (2). This example controller also has modules for actuating a tool changer (TC) and pallet changer (PCH). Information, such as tool change requests, is also exchanged between these modules. Thus, the structure of a modular controller requires data exchange at many places between the individual modules. So, both the reasonable delineation of individual functional areas in relation to each other and the definition of interfaces between the functional areas are prerequisites for implementing a cross-manufacturer solution. The main unresolved issue is the working interplay between the individual units, meaning also that liability issues remain unresolved.



## 11.3 Workpiece Programming in NC Production

In NC programming, all information necessary for the automated production of a workpiece is represented and stored in an NC program such that it can be read and interpreted by the numerical controller of a machine tool. This information includes geometric details for generating workpiece and tool movements, technological information (such as feed rate and spindle speed) as well as instructions that require a tool change or activation of additional functions (such as cooling lubricant pump and chip conveyor).

The NC program is stored in different physical forms in line with data management for the numerical controller. While punched tape was still being used in the first numerical controllers, modern storage media (such as USB sticks and external hard drives) are available today that also allow reliable storage of programs in workshop environments (with dust, oil, magnetic fields, vibration, etc.). By connecting the machine to a network, NC programs can also be interchanged within a company and managed centrally on servers.

The representation of information in an NC program is dependent on the programming interface of the numerical controller being used. Not all numerical controllers are based on the same data model nor do they process programs written in the same programming language. Furthermore, they differ in regard to the internal implementation of the NC programs and their functional scope. So, in addition to various standardized programming interfaces, there are manufacturer-specific interfaces used to implement the additional, non-standardized functions of the individual controller manufacturers in an NC program.

Corresponding to the complexity of the machining task to be defined in an NC program and the programming language, there are different, generally computer-aided, programming methods available during planning and at the machine.

Before the different NC programming methods are discussed in ▶ Sect. 11.4, the format of an NC program is explained below and the most important agreements regarding machine coordinate systems are illustrated [KIEF99].

### 11.3.1 Format of Line-Based NC Program

The first NC programs had a purely sequential design in line with the algorithms, computing power and storage media available. Instead of defining the workpiece and the technological/organizational information necessary for its production, they contained a fixed sequence of individual NC lines with traversing instructions and switching commands for one particular machine tool. This was thus the programming of an NC machine and not of a workpiece. Probably the most common NC programming language, DIN 66025 or ISO 6983 was created on this basis.

An NC program compliant with DIN 66025, known as G-code, comprises a number of NC lines that each correspond to one machining step. The machining instructions in a NC line are encoded in "words," combinations of code letters (address) and series of digits, with or without a leading sign. The meaning and arrangement of these words are defined in the program key of the respective numerical controller, upon which DIN 66025 is based. ◘ Table 11.1 lists the meanings of the most important words in DIN 66025 [DIN83b].

The first instruction in an NC program is the program start character, "%." Following on line by line are the NC lines, each starting with the NC line number "N" and ending with the end of line character "LF" (Line Feed). Each NC line may contain a different number of words. Most functions are modal, i.e., they need not be restarted in every NC line. Instead, they apply until they are overwritten in a new NC line or are reset. The functions applicable block by block are denoted separately by a dot in ◘ Table 11.1.

With the "G" path condition (geometry specifications), the programmer can select, for example, the interpolation type (linear or circular), type of dimension input (mm or inch, incremental or absolute), the tool

**Table 11.1** Excerpt from the program key of DIN 66025

| | ID | Function and meaning |
|---|---|---|
| % | | Program start |
| : | 1 to 9999 | Main NC line |
| N | | NC line number |
| /: | | Suppressible main NC line |
| /N | | Suppressible NC line |
| G | 00 | Rapid motion |
| | 01 | Linear interpolation |
| | 02 | Clockwise circular interpolation |
| | 03 | Counterclockwise circular interpolation |
| | 33 | Thread cutting |
| G | * 04 | Hold time preset at address X (in ms) |
| G | 17 | Level selection X – Y |
| | 15 | Level selection X – Z |
| | 19 | Level selection Y – Z |
| G | 39 | Corner correction for milling radius path compensation |
| G | 40 | Cancellation of the tool correction |
| | 41 | Tool path correction left, tool left of the workpiece |
| | 42 | Tool path correction right, tool right of the workpiece |
| G | 53 | No zero point offset |
| G | 54 | Zero point offset 1 |
| | 55 | Zero point offset 2 |
| G | 70 | Input system inch |
| | 71 | Input system metric |
| G | 80 | Cancelation of the work cycle |
| | 81 to 89 | Work cycles |
| G | 90 | Entry of absolute dimension |
| | 91 | Entry of chain dimension |
| G | 92 | Set actual value storage |
| G | 94 | Feed at address F in mm/min |
| | 95 | Feed at address F in mm/rev |
| D | 1 to 99 | Tool correction number |
| X | 0 to ±99999.999 | Path information in mm |
| | 1 to 99999.999 | Hold time in ms |
| Y | 0 to ±99,999.999 | Path information in mm |
| Z | 0 to ±99999.999 | Path information in mm |
| 4th axis | 0 to ±99999.999 | Path information in mm or degrees |
| | | Possible addresses A, B, C, U, V and W |
| R | 0 to n | Parameter (for subprograms for example) |
| I | 0 to ±99999.999 | Interpolation parameter for X-axis for circular interpolation |
| | or | Thread pitch in mm |
| | 1 to 2000.000 | |
| J | 0 to ±99999.999 | Interpolation parameter for Y-axis for circular interpolation |
| | or | Thread pitch in mm |
| | 1 to 2000.000 | |
| K | 0 to ±99999.999 | Interpolation parameter for Z-axis for circular interpolation |
| | or | Thread pitch in mm |
| | 1 to 2000.000 | |
| F | 0 to Fmax | Feed in mm/min or mm/rev |
| S | 0 to Smax | Spindle speed in min$^{-1}$ |
| T | 1 to 9999 | Tool number |
| H | 1 to 999 | Auxiliary functions |
| L | 01 to 99 | Number of the subprogram to call |
| M | * 00 | Programmed stop, compulsory |
| | 01 | Programmed stop, optional |
| | 02 | Program end without rewind, is in the last NC line of the program |
| | 30 | Program end with rewind to program start, is in |
| | | last NC line of the program |
| M | 03 | Direction of spindle rotation, clockwise |
| | 04 | Direction of spindle rotation, counterclockwise |
| | 05 | Spindle stop |
| M | 00 to 99 | Additional functions, some assignable |
| ( | | Start of comment |
| ) | | End of comment |
| LF | | Line end |
| | ID | Function and meaning |
| % | | Program start |

path correction, the zero point offset and work cycles (drilling cycles, cycles for clearing pockets, etc.). Movements along the individual machine axes are specified by the path information. This comprises a letter for selecting the axis in question (X, Y, Z for linear axes and A, B, C for rotation axes, see ◘ Fig. 11.24) and a series of digits for specifying the distance or rotation angle. When circular paths are traversed, the relative position of the arc center point in relation to the start point of the arc is specified by circular interpolation parameters I, J and K.

In addition to this geometric information describing the tool movements, an NC program contains technological instructions for workpiece machining:

Word F (Feed) specifies the feed rate in mm/min or the (revolution) feed in mm/rev. The selection is made by using command G94 or G95 (◘ Table 11.1). The spindle speed in $\text{min}^{-1}$ is programmed by Word S (spindle rotation).

For machines with tool change, command T (Tool) selects a tool. For every tool, the tool geometry data for a later tool correction is stored under the associated correction number "D" or in an internal controller database. With this information, the tool radius, length or diameter can be corrected when the program runs.

11

For milling, the milling radius correction represents a key aid in generating NC programs. Instead of the path of the tool center point, the workpiece contour to be generated is programmed when correction is active. G Words G41 and G42 communicate to the controller the relative position of the tool in relation to the workpiece in the direction of motion. This means that the tool diameter can be calculated perpendicular to the path direction in the controller when the axis motions are generated. If the material to cut is to the right of the tool path, G41 is set instead of G42.

In the programming of a turning process, such differentiation is unnecessary. A number of controllers use G46 for automatic correction of the cutting radius. The direction in which the correction is to take effect is derived from the position of the turning tool, to be stored internal to the controller (◘ Fig. 11.27).

The many additional functions, such as Spindle (counter)clockwise, Switch gear speed, Coolant On/Off and Program end, are specified using M codes (Miscellaneous functions). The L function calls subprograms. It saves on the programming of frequently recurring machining cycles, as they occur for cut distributions and hole patterns for example. The variable parameters must then simply be reset as parameters (Siemens: R parameters) and passed to the subprogram.

The end of the program is set by supplementary function M02. M30 is a function that, back when punched tape was being used, was necessary to rewind back to the start when the end of the program was reached. Originally, punched tape was the most common storage medium for NC programs because appropriate storage media were not yet available in the controller or were too susceptible to dirt and magnetic fields in the workshop area.

Nowadays, NC machines have internal memory and are able to import complete programs or routines and to run them at a later point in time.

There are different ways to transfer programs to the controller. In many cases, programs are read from disk or USB stick. Serial interfaces (such as workshop PCs) and DNC mode represent other viable ways to transfer programs. In DNC mode, comprising the tasks of NC program management and distribution, a computer with a large hard drive capacity is directly connected to several machine tools over a LAN (Local Area Network). The NC programs or routines required can then be requested over the network by the machine tool controllers or, for old controllers without their own permanent storage, from DNC terminals.

### 11.3.2 Structure of an Object-Oriented NC Program

In contrast to DIN 66025, the information required to perform a machining operation in an object-oriented NC program is not divided sequentially into individual NC lines, but stored in a combined and structured manner in larger logical units. This has the benefit that, at any time during the machining process, not only is a traversing instruction (e.g., G01 X...) or switching command (e.g., M06) present in the NC program, but the relationship to the current component characteristic or the workpiece as a whole can always be established. With higher-order program information, intelligent functions can be realized in numerical machine control while the controller is running. Examples of these are retraction strategies, collision monitoring and re-planning of machining operations for a new tool.

ISO 14649-FDIS, or STEP-NC for short, is one example of an object-oriented programming interface. This was enhanced as part of the international STEP-NC project. The data model of ISO 14649 is based on that of STEP (ISO 10303). In addition to the language definition (Part 11, Part 21), parts of the geometric description (IR 42/43) and feature definition (AP214/224) are also taken from STEP. These data elements are processed in modern CAD/CAM systems and used as an interchange format for geometry information. Using a standard data format from planning to production, product data created before can be used multiple times without having to be regenerated. This avoids errors and inaccuracies as they occur, for example, in the conversion of geometry information. This thought process is being driven forward mainly by large companies that manufacture at many locations and so need to exchange NC programs. ▫ Figure 11.23 shows, in simplified form, a small part of the data model of ISO 14649 in EXPRESS-G. EXPRESS is a language for depicting data structures. EXPRESS-G is the corresponding graphical display form.

ISO 14649 differentiates between geometrical and technologically operative data. Also, the geometrical data is split into 2½D and 3D elements in line with their degrees of freedom. For 2½D machining, interpolation uses only two axes, while fixed values are provided to the third axis. These include elements such as bores, pockets and grooves, or "machining features" (machining_feature). Freeform surfaces are described by special surface features.

The tool, cutting data, strategy, etc., are not necessarily applicable to only one component characteristic and so are described globally as operations (machining_operation) in a self-contained object. The attribute "toolpath" belongs to the "machining operation" for programming special or optimized path strategies. Tool paths defined in advance are stored here. This way, it is possible, for example, to provide the precise machining paths for a feature. Otherwise, the traverse paths are only calculated in the controller in line with the strategy selected.

It is not until a work step (machining_workingstep) that the controller is informed of which operation on a feature is to be run, i.e., how a geometry is to be machined. In the controller, relevant production features are then triggered, and traverse paths generated. These are tasks handled previously by CAD/CAM systems (with the exception of cycle programming) [ISO00].

The object-oriented design means there is a structured model of the entire machining task in the controller. This can be used for visual representation or for additional functionality on the part of the controller.

The extensive geometrical, technological, and organizational information is available as structured units

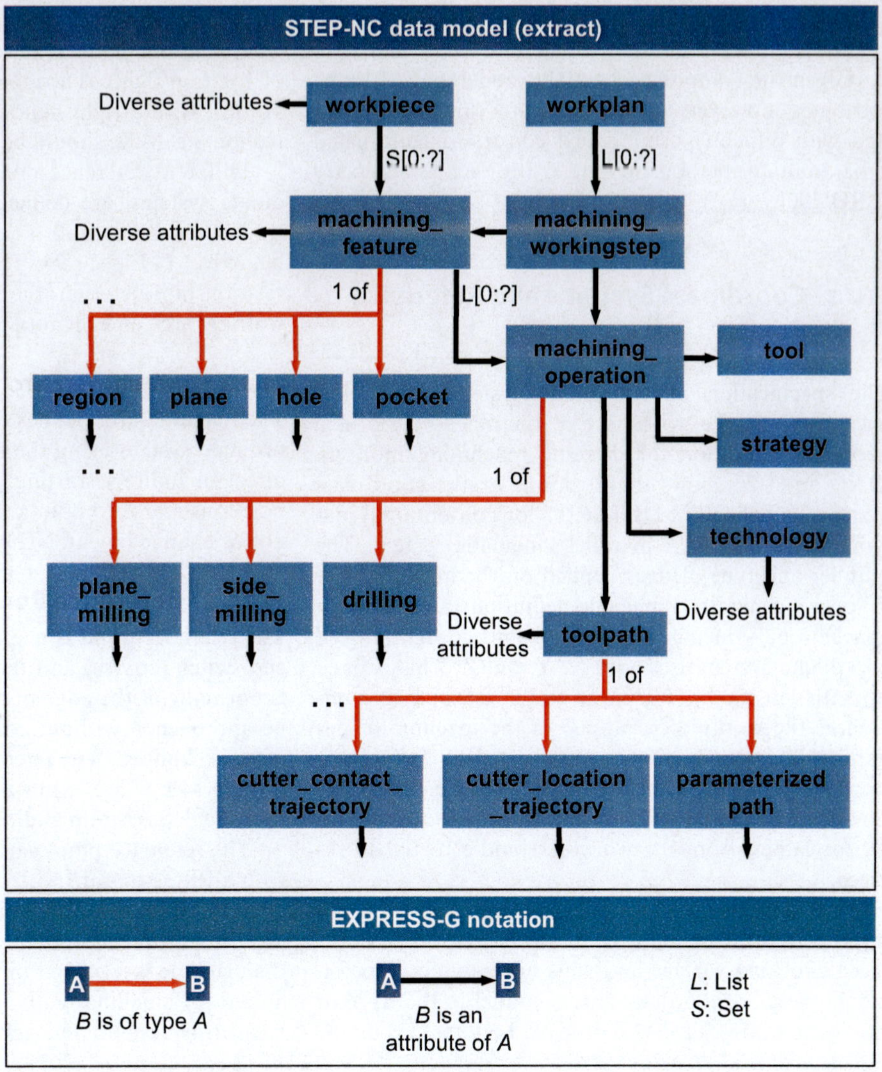


▫ **Fig. 11.23** Simplified ISO 14649 data model

in an object-oriented format. Information interchange between CAD/CAP/CAM systems and numerical controllers having an interface for such a comprehensive data format becomes more reliable because data conversions between the systems are unnecessary. Changes can be tracked better in all systems. If changes arise once an NC program is started for example, they can be made directly on the machine and returned to planning via the NC program [BROU00].

The description is as controller-neutral as possible in that the NC program separates geometrical and technological operative data. NC programs can be exchanged more easily between individual machine tools. Differences in the controller and machine functionality are realized quickly and easily by means of program changes at a few easy-to-find places in the operation definitions. For example, a changed cutting depth does not entail recalculation of all traversing movements, as in the case of programming to DIN 66025. Instead, the relevant attribute is changed. The controller (sometimes tailored to the machine better than any CAD/CAM system) does not generate the corresponding paths until runtime on the machine.



Given the comprehensive data model and its many attributes, however, an NC program can only be edited with difficulty using a text editor. An automated programming method or special browser is necessary [BROU00].

### 11.3.3 Coordinate Systems and Reference Points

The specification of coordinate systems and reference points in the workspace of the machine tool is a necessary condition for defining machining motions. ◘ Figure 11.24 shows the positions of the coordinate systems as defined in DIN 66217 for conventional machine tools and for a parallel kinematic system. The number and type of axes depend on the machine. The following generally applicable definitions are made for conventional kinematics. A right-handed, right-angled coordinate system uses axes X, Y and Z. This is lined up with the main guide paths of the machine and references the workpiece clamped in the machine (right-hand rule). The Z-axis lies parallel to the axis of the working spindle or coincides with it. The positive direction of the Z-axis runs from the workpiece to the tool or, for lathes, from the working spindle to the workpiece.

The X-axis is the main axis on the positioning plane. It essentially lies parallel to the workpiece chucking surface and runs horizontally if possible. If there are coordinate axes parallel to the X, Y or Z-axis, they are called U, V and W. Letters A, B and C denote rotational axes, such as rotary tables, assigned to the X, Y and Z coordinates. The positive direction of rotation is specified in line with the right-hand rule (◘ Fig. 11.24).

To generate a positioning motion, either the tool or workpiece can be moved. If the tool is moved, the positive direction of motion matches the positive axis direction and it is denoted +X, +Y, etc. When the workpiece is moved, the positive direction of motion is in the opposite direction to the axis direction, and it is denoted +X′, +Y′, etc. In the first case, term "workpiece coordinates" is also used—in the second case term "tool or machine coordinates." The same applies to the rotational axes A, B and C.

As the coordinate system pertains to the workpiece, programming is independent of whether the workpiece or tool is moved in the machining process. The programmer always assumes the tool moves relative to the coordinate system of the workpiece, which is thought to be stationary.

For programming the direction of spindle rotation for a lathe, miller or drilling machine, a rule of thumb is given at this point. The rule where a positive direction of rotation tightens a right-handed screw (tool) applies. To drill with a right-handed drill, spindle clockwise direction, i.e. M03, should be programmed for example.

Different reference points, in addition to the coordinate systems, are defined on every numerically controlled machine tool. The most important are highlighted in ◘ Fig. 11.24. They are used to establish the relationship between the workpiece position in the working area and the tool position.

#### 11.3.3.1 Machine Zero Point M

The manufacturer defines the origin of the machine coordinate system using the machine zero point. It is the direct or indirect starting point for all other reference positions and zero points. The machine zero point cannot be changed by an NC program.

#### 11.3.3.2 Reference Point R

The reference point is a specified point locatable by an end switch (coarse) and the measuring system (fine). It is normally at the edge of the workspace so that it can be approached without collision even when the workpiece is clamped. The reference point is set to fixed distance XMR, ZMR from the machine zero point when the machine is set up and then remains unchanged.

The reference point can be approached manually or with a PLC program with an accuracy of one displacement increment (normally in the range 0.1 to 1 μm) using the "Reference point run" command. Every time the machine is switched on, a reference point run is required for machines with absolute-cycle or incremental measuring system. The actual position counters of the

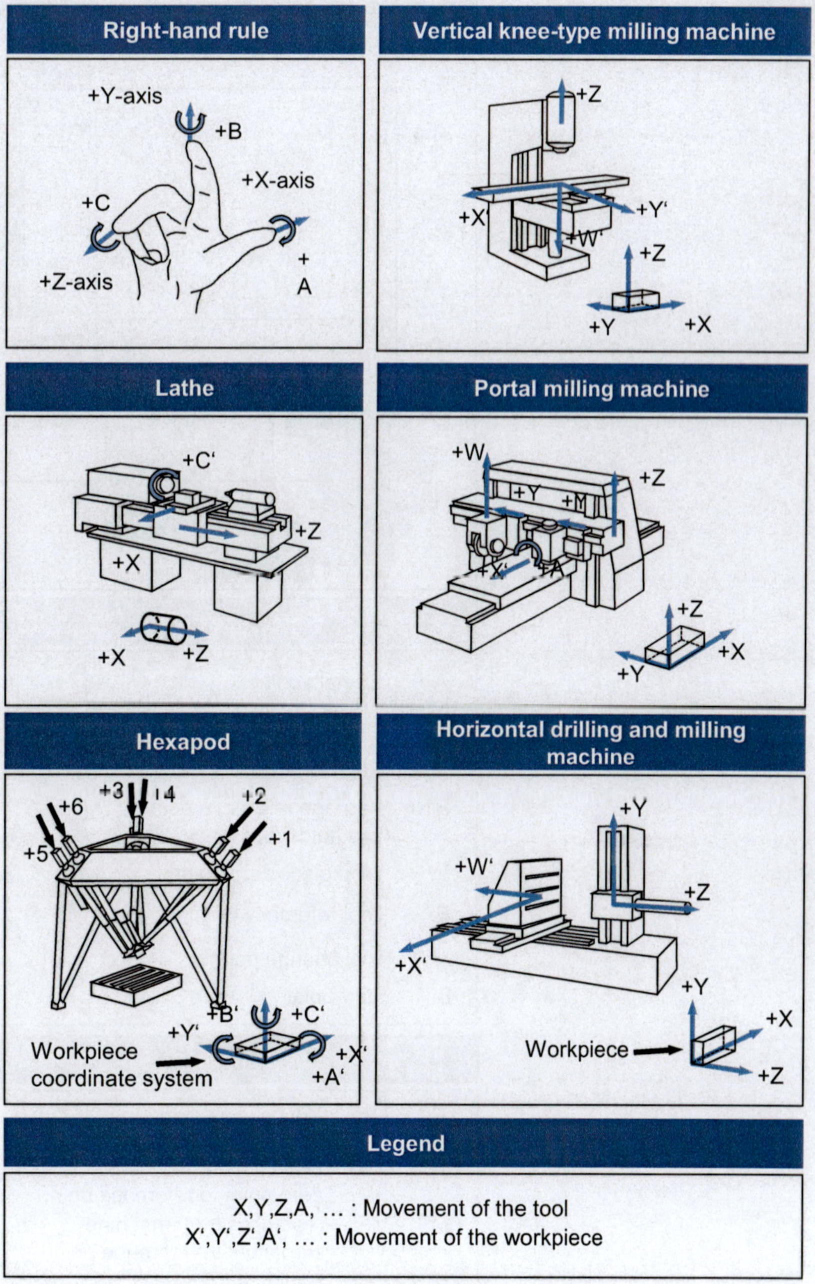


**Fig. 11.24** Coordinates of machine tools, according to DIN 66217

controller are then automatically set to zero or other values specified on startup, such as XMR and ZMR (Fig. 11.25).

### 11.3.3.3 Workpiece Zero Point

The workpiece zero point is the origin of the workpiece coordinate system referenced by parts programming. Its position is defined by the programmer and should be positioned where most dimensions start in the diagram. For turned parts, this is usually the intersection point of the rotation axis with the reference edge of the length dimension. For a machining scenario as in Fig. 11.25, the machine user would, for example, set the workpiece zero point relative to the machine zero point using dimension ZMW. This simplifies NC programming because the dimensions can be taken directly from the drawing.

### 11.3.3.4 Tool Reference Point E

The dimensions determined when the tool is measured and stored in the tool data memory for tool correction are referenced to the tool reference point.

**Fig. 11.25** Reference points for an NC lathe and on a milling tool

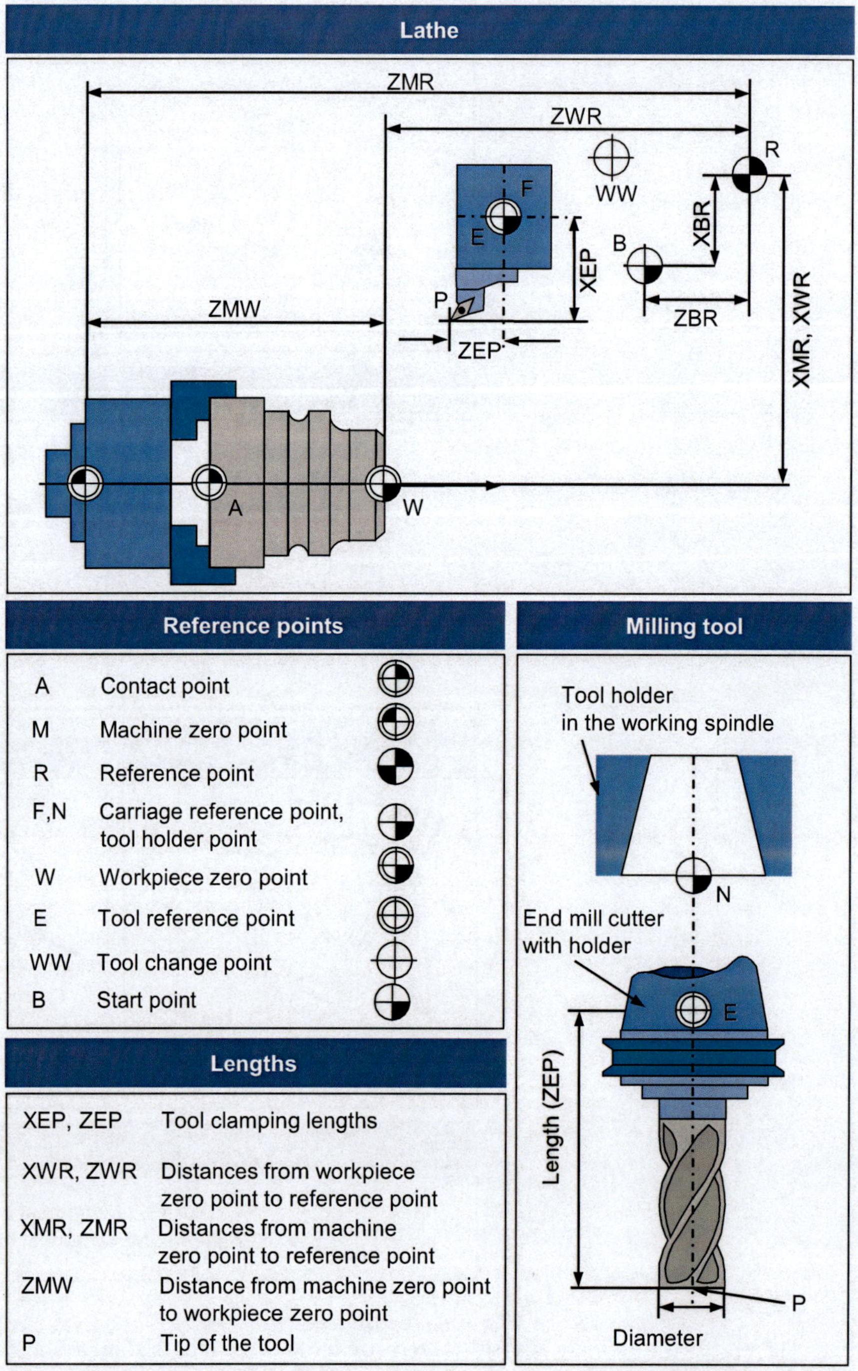


For milling, the tool reference point coincides with tool holder point N in the tool holder. For turning, it coincides with carriage reference point F.

The position of the tool reference point is not uniquely defined because the holder in the tool holder on the machine side does not make contact with a flange so as to form-fit but is housed in a cone. This point, necessary for the tool setting, must be determined depending on the machine and the holder. The further the taper enters the cone of the working spindle holder, the further point "E" is in the direction of the tool. This problem does not occur for interfaces with defined planar contact area, such as the HSK.

#### 11.3.3.5 Tool Holder N and Carriage Reference Point

For milling and turning, the terms used are tool holder point F and carriage reference point F, respectively. The points are the equivalents on the machine side to the tool reference point and are moved toward the reference point during the reference point run.

#### 11.3.3.6 Start Point

Start point B denotes the position of the tool reference point when the program starts. Before the program starts, B is approached from the reference point and again, once the programmed workpiece machining is complete, to start machining the next workpiece from there.

#### 11.3.3.7 Tool Geometry

When a workpiece is machined, several tools, such as roughing, finishing and drilling tools, are generally used. Distances XEP and ZEP (◘ Fig. 11.25) between the chip-forming cutter and tool setting point E are known as clamping lengths and have different values depending on the tool. With every tool exchanged, associated coordinate pair XFP, ZFP must be assigned to the path calculation. Minor deviations of the tool geometry, due to wear, inaccurate tool presetting or temperature-related machine displacements, can still be corrected by the machine operator before the machining process using the tool fine correction. The correction values are read in the NC program using the D instruction with the relevant storage number.

## 11.4 NC Programming Methods

When creating a program, the programmer must model the machining task required in the NC program such that the machine tool controller can interpret it unambiguously and such that the machining task can be performed on a particular machine with its restricted workspace and specific functionality. The programming language forms the basis here. It contains the command set and a corresponding modeling format to describe the machining task. Programs of different sizes and scopes are created depending on the form of the component to fabricate and the machining complexity. If the requirement is for one program to control multiple processes (machining centers), synchronization marks and change functions must also be added.

In addition to the appropriate NC programming language, suitable aids must be made available to the programmer to enable creating, changing, and simulating such a program reliably, quickly, and cost-effectively.

### 11.4.1 Manual NC Programming Methods

#### 11.4.1.1 Fundamentals and Procedure

In manual programming, the programmer splits the workpiece to be machined into individual machining features. On the basis of these features, the programmer then describes the geometry, the operations necessary and the technological boundary conditions. The programmer then uses them to specify the individual machining steps.

For conventional, sequential NC programming per DIN 66025 or ISO 6983, the programmer composes, together with the NC functions specified in the program key of the NC machine (◘ Table 11.1), NC lines that, when taken together, yield the NC workpiece program. The machine power output, workpiece and blank dimensions, material, tool geometry and workpiece surface qualities must be factored into the selection of cutting conditions and cut distributions (◘ Fig. 11.26).

The sequence of machining steps is defined in a routing plan. The tool and clamping plans to be created in addition inform the machine operator about the clamping position of the workpiece, the tools to use, by their code numbers and adjustment lengths (◘ Fig. 11.15). These plans also often contain the distance of start point B of the carriages from the reference point (distances XBR and ZBR in ◘ Fig. 11.25). The programmer used all this information when creating the program. If it is not possible to keep to individual specifications, because, for instance, the blank can only be clamped in an offset position, the NC program must be adapted (with a corresponding zero point offset, for example).

The longest time by far in programming is taken up with defining the tool movements. The entire movement generating the workpiece shape must be divided into individual movement segments. Blank allowances that cannot be cut in one operation must be broken down into multiple machining processes. This cut distribution requires the calculation of path points that are not usually dimensioned in the workpiece drawing. It can be very complex. Due to this, assistance is provided by geometry calculation programs and computers (as described later).

The shapes and lengths of the movement segments are derived from the interpolation functions of the appropriate NC programming language. In DIN 66025, only straight lines, arcs and parabolas are available. These must be used to approximate all paths to be traversed. In line with the complexity of the geometry to machine and the accuracy stipulated, very many NC lines are occasionally required to attain a sufficiently good approximation using straight-line and circle commands. This means the NC program is correspondingly

**Fig. 11.26** Data for creating the control information for an NC machine

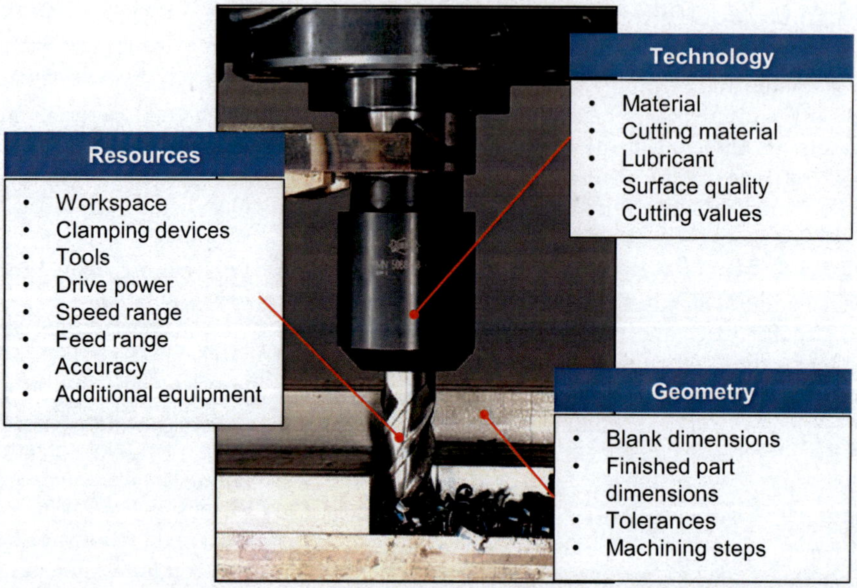




long and the number of NC lines increases with only short path segments. The time problems arising here during program execution are discussed separately in the chapters on interpolation. Modern-day programming languages and controllers also offer superior geometric descriptions such as splines. They enable complex, and at the same time longer, path segments to be described in only one NC line. There is however no standardization for these superior programming approaches like DIN 66025 provides for traditional NC programming.

Particular caution is called for when specifying the path characteristic to ensure the machining process is collision-free, also in regard to the resultant axis motions (refer to ► Sect. 12.2.2). There are particular hazards of collision with the workpiece when machining workpiece areas that are tight and angled or when multiple tools are in contact simultaneously on different feed slides.

There are two ways to specify traverse paths to DIN 66025. In the absolute specification of dimensions (absolute dimension programming, G90), the coordinate values of the end point—in relation to workpiece zero point W (Fig. 11.25) are programmed in the workpiece coordinate system. In the incremental specification of dimensions (chain or relative dimension programming, G91), the difference between the current location of the tool and the end point of the next path segment is specified.

Absolute programming is the predominant method in practical applications (G90). In this type of dimensioning, monitoring or correcting the program is often easier because dimensioning generally starts from individual reference edges in workshop drawings. Placing the zero point for absolute programming on these reference edges means the dimensions for NC programming can be taken directly from the drawing. Also, making changes to individual dimensions does not affect all of the following programs as is the case for relative dimensioning. In contrast to absolute dimensioning, in relative dimensioning there is a risk of chain dimension errors (the accumulation of tolerances for individual dimensions). Despite this, chain dimensioning offers benefits when the requirement is to use routines repeatedly. As dimensioning here pertains to the point previously approached, individual routines (such as subprograms) can be copied to the relevant place or be called from there as subprograms. This type of dimensioning is beneficial especially for the production of hole patterns where the same holes are required at different places. It is possible in workpiece programs to switch the dimension type between absolute and incremental dimensioning as required.

The position of the workpiece and that of the workpiece zero point in the workspace of the machine depend on the clamping device used, the position of the clamping device in the workspace (e.g., on the milling table) and the blank dimensions. So, in one of the first instructions of an NC program using absolute programming, the position of the workpiece zero point must be specified to the controller. This is possible by calling one of the G54, G55, …, G59 commands.

Example: N12 G54

The dimension of the displacement (ZMW in Figure 11.25) is taken from the setup sheet. Before the program starts, it is entered at the controller operating panel of the machine. Correction value memories are available as storage. They can be accessed from an NC program using a corresponding G command, in this case G54. Setting the actual

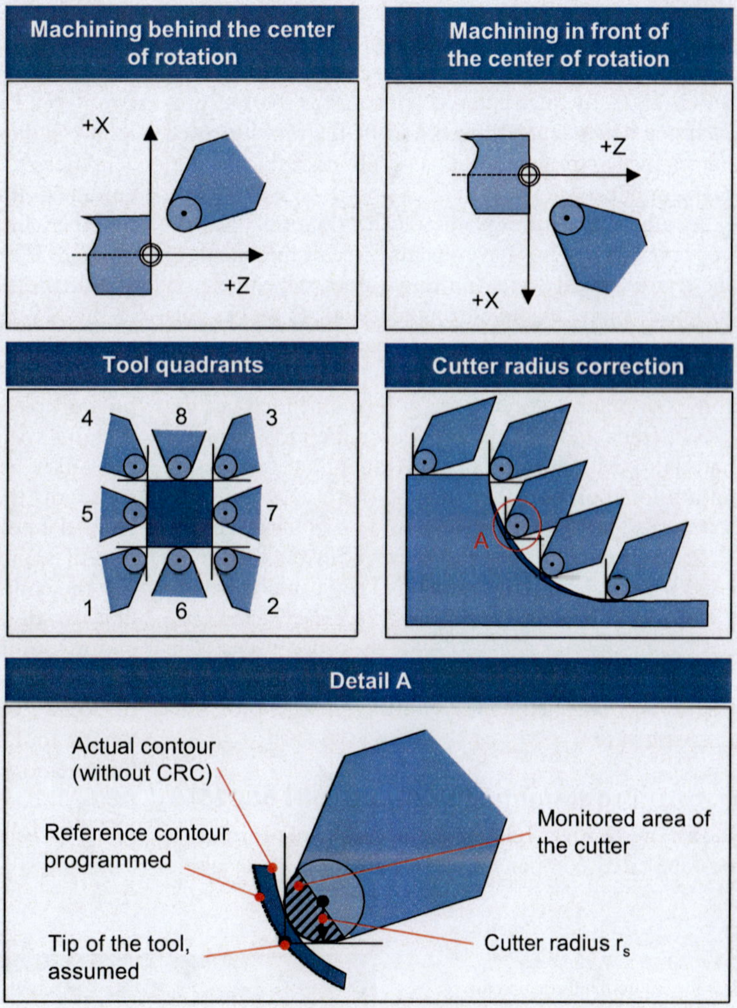


**Fig. 11.27** Orientation of the turning tool and possible shape errors when turning without cutter radius correction

value memory with command G92 is another way to change the zero point offset. The values of the current position are set in memory and so can specify the workpiece zero point indirectly.

Example: N9 G92 X100 Z200

If, for example, carriage reference point F is at the reference point and if the distance of reference point R from the workpiece zero point W (XWR, ZWR) along the individual axes is known (100 mm in X, 200 mm in Z-direction, Y = 0), these measurements can be set in the actual value memory and specify the relative position of the reference point in relation to the workpiece zero point.

This command is commonly also used in conjunction with "tool scratching" on the turned workpiece. For this, the programmer, before running the program, scratches the turned part on the end face and enters the Z-component of this position together with an X-value, which is away from the component, into the NC program. The X/Z positions can be approached in the program and the actual value memory be set here to Z = 0 using G92. This means that the programmer is able to define the workpiece zero point without having to take the machine zero point, clamping or component/clamping lengths into account in more detail.

When a workpiece geometry is programmed, the finished contour of the workpiece is always programmed without first factoring in the tool radii. For turning tools, the tool radius is understood to be the cutting radius $r_s$ of the indexable inserts used (described in Fig. 11.26), and, for milling tools, it is the milling radius. The controller, knowing the tool radii, then calculates the equidistant path for the workpiece contour for the cut rounding/miller center point. This calculation is also called cutting or milling radius correction. Without a path correction, shape errors can occur in the machining process, Fig. 11.27 (refer to ► Sect. 12.2.2).

The tool radius is not set in the NC program, Instead, it is entered into the controller before the start of the program. For turning work, the position of the tool in the tool quadrants, Fig. 11.27, and the dimensions for the tool length corrections (XEP, ZEP), Fig. 11.25, are also entered. This means the NC program is independent of the radius of the tool just used that changes continually (as a result of resharpening, for example).

In addition to the geometric information, the NC program must also have technological information. These are primarily the cutting values dependent on workpiece material and cutting material, as well as cutting depths, cutting speeds and feeds. Reference value tables (such as INFOS) [WERK79] and tool catalogs from tool manufacturers specify recommended cutting speeds, feeds and cutting depths for the most popular material combinations. The individual limit load information for tools and the machine (such as torque and power) also need to be factored into program creation.

Programming workplaces have been developed to assist with manual programming. They usually include a desktop computer to simplify necessary geometric and technological calculations for the programmer. The desktop computer is usually based on the input device of the NC machine or PCs with the relevant programming software.



### 11.4.1.2 Programming Example (DIN 66025)

Shown in Fig. 11.28 is an example of turning a pre-fabricated component. A lathe is used to machine this component. As already mentioned, many controllers have their own non-standardized programming commands to implement additional functions. The controller used in the lathe also offers commands different to those in DIN 66025 (Table 11.1).

After rough machining has been finished, a plunge-cut must first be made using two tools. This is followed by finishing machining using the contour points marked, $P_1$ to $P_6$.

As the first step, a zero point offset is performed in the NC program (Fig. 11.29) using the G59 command. This effects a zero point offset relative to the current workpiece reference point W. If, as in this case, no workpiece reference point is yet defined, the offset pertains to the machine zero point M. The values for the offset here are X=0 and Z=300. The rotational speed of the spindle is also defined. The next line, N02, defines the type of dimension input, the direction of spindle rotation and the tool to change, T1003. For the controller used as the basis in this example, the first two digits of the T instruction denote the position of the tool in the turret, that is, in the tool storage system of the lathe (Position "10" here). The second pair of digits, "03", specifies the storage location in the file for the tool fine correction. So, in this case, the correction memory is not selected by the D command but directly in conjunction with the tool change call. In the NC lines following on, first the right-hand and then the left-hand edge of the plunge-cut are produced.

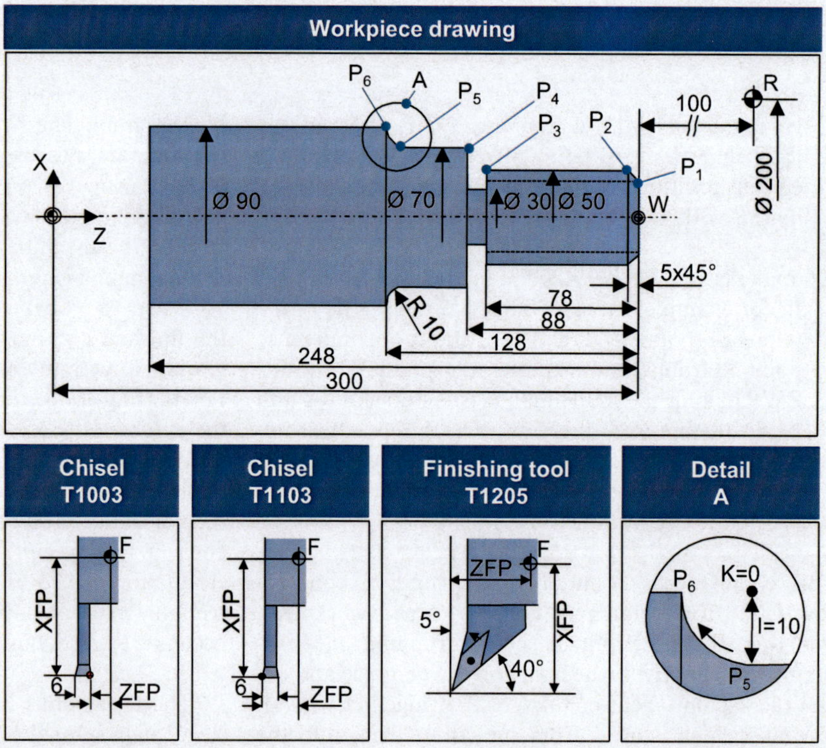


**Fig. 11.28** Workpiece drawing and tool information for the DIN 66025 example program. *Source* Traub

**Fig. 11.29** NC program example for the turning in **Fig.** 11.28. *Source* Traub

| Main program | |
|---|---|
| **Program:** | **Comment:** |
| % | Program start |
| N01G59 X0 Z300 S1300 | Zero point offset from M to W, 1300 min-1 |
| N02 G90 T1003 M04 | Entry of absolute dimension, tool change T1003, spindle "On" (CCW) |
| N03 G00 X54 Z-78 | Rapid motion via plunge-cut position |
| N04 G01 X30 F0.1 | Plunging, right-hand edge, feed 0.1 mm/rev |
| N05 X54 | Retract |
| N06 G00 X200 Z100 | Rapid motion to reference position |
| N07 T1104 | Tool change T1104 |
| N08 G00 X74 Z-88 | Rapid motion via plunge-cut position |
| N09 G01 X30 | Plunging, left-hand edge |
| N10 X74 | Retract |
| N11 G00 X200 Z100 | Rapid motion to reference position |
| N12 T1205 | Tool preselection T1205 |
| N13 G91 M06 | Chain dimensioning, tool change |
| N14 G46 | Cutter radius correction "On" |
| N15 G00 X-80 Z-98 | Rapid motion in front of P1 |
| N16 G96 V250 | Constant cutting speed 250 m/min |
| N17 G01 Z-2 F0.15 | Infeed towards P1, feed 0.15 mm/rev |
| N18 X5 Z-5 | P1 to P2 |
| N19 Z-75 | P2 behind P3 |
| N20 G00 X10 Z-6 | Rapid motion in front of P4 |
| N21 G01 Z-32 | To P5 |
| N22 G02 X10 Z-10 I10 K0 | P5 to P6 (circular interpolation) |
| N23 G01 X2 | Traverse away |
| N24 G40 | Cutter radius correction "Off" |
| N25 G90 | Entry of absolute dimension |
| N26 G00 X200 Z100 | Rapid motion to reference point |
| N27 M05 | Spindle "stops" |
| N28 M30 | Program end, reset to Program start |

Then, the finishing tool T1205 (line N12) can be changed and contour points $P_1$ to $P_6$ be followed. These movements are dimensioned incrementally and programmed. In contrast to absolute dimensioning where it is important to note that diameter values are specified in the X-direction, only the relative offsets of the start and target positions are included in chain dimensioning. Command G46 (line N14) enables the cutting radius compensation of the lathe.

The definition of circular programming is to be discussed specially again. First, starting from the start point of the circular motion ($P_5$, end of line N21), the center point of the circle is defined. For this, starting from the start point of the circular motion, the center point in the X- (Y-) and Z-directions is determined and the values are assigned to variables I (J) and K (line N22). If one of the variables is not specified, it is set to "0". The X-, Y- and Z-values are assigned the end coordinates of the circular motion. Command G02 signifies a clockwise circular interpolation.

At the end of the program (from line N25), a switch is made back to absolute dimensioning and the reference point is approached. Command M30 then ends the program and resets it to character "%" (program start) so that it can be run again by pressing "Start".

#### 11.4.1.3 Additional Commands for Program Entry

Nowadays, a broad range of different operating and programming aids is available to the programmers of modern numerical controllers. This way, all numerical controllers include NC program memory for multiple NC subprograms, correction memory for process optimization and tool correction memory for storing the current geometries of multiple tools.

Special machining cycles for thread cutting, plunging and boring and the ability to mirror, scale and mask individual NC lines, are just as much a part of the configuration of convenient numerical controllers as subroutines.

## Subroutines

Subroutines are commonly used to define frequently recurring machining sequences (such as drilling cycles, hole patterns or milling patterns) as subroutines, to call them using symbolic names and to pass the current parameters. The commands used, such as chain dimensioning, are (unlike high-level languages such as C and Pascal) not limited locally to the subroutine but valid for the entire NC program. ◘ Figure 11.30 shows an example for thread cutting.

The thread in question is to be made as a left-hand thread using a subroutine. For this, a supplementary program section to main program A (◘ Fig. 11.30) and the actual subroutine must be created.

The thread (M50 X 1.5) is to be made using the T 0906 tool. For this, the thread tool is to remove material in three passes to attain the required thread depth of 1.3 mm. The subroutine, denoted by the letter "O" and program number 8100, is called three times (N31, N33, N35) in the main program with command G22. As the transfer parameter (L1), it receives each of the machining depths that are separated by "/" signs from L1 in the main program. Feed F of the carriage is determined by the thread pitch.

Command G33 (line N103, thread cutting) finally synchronizes the feed and rotational motion of the spindle for this machining process.

◘ **Fig. 11.30** Programming example for creating a subroutine for the workpiece in ◘ Fig. 11.28

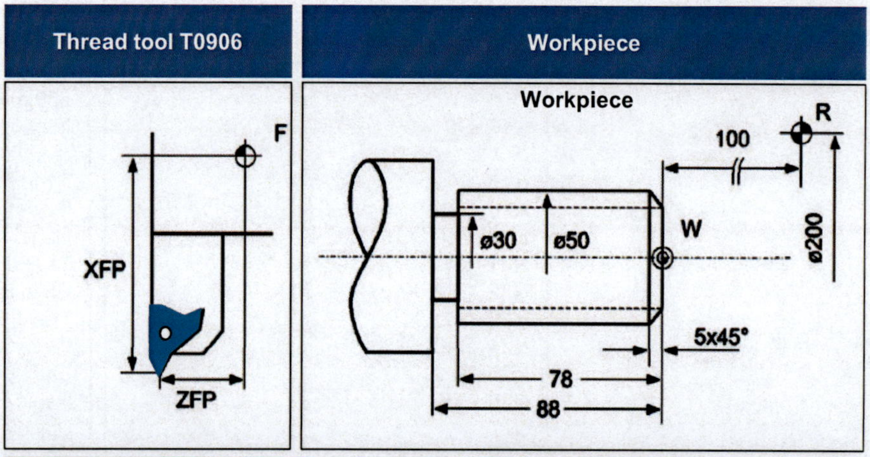


**Additions to main program A**

| | | | | |
|---|---|---|---|---|
| N28 | G97 | S500 | M04 | Constant speed of 500 min-1, spindle "On" (CCW) |
| N29 | | | T0906 | Tool change T0906 |
| N30 | | L1/49 | | Variable preassignment (L1 = 49 mm) |
| N31 | G22 | | A8100 | Call of subroutine |
| N32 | | L1/48 | | Variable preassignment (L1 = 48 mm) |
| N33 | G22 | | A8100 | Call of subroutine |
| N34 | | L1/47.4 | | Variable preassignment (L1 = 47.4 mm) |
| N35 | G22 | | A8100 | Call of subroutine |
| N36 | G00 | X200 Z100 | | Rapid motion to reference point |
| N37 | | | M05 | Spindle "stops" |

**Subroutine O8100**

| | | | | | |
|---|---|---|---|---|---|
| N101 | G00 | X54 | Z2 | | Rapid motion |
| N102 | | XL1 | | | Rapid motion in front of the workpiece |
| N103 | G33 | | Z-83 | F1,5 | Thread cutting (left-hand thread) |
| N104 | G00 | X54 | | | Rapid motion in X-direction |
| N105 | | | | M99 | Subroutine end |

## Contour Line Programming

In contour line programming, the finished part contour is broken down into single contour lines that extend over two or three points and are formulated as NC lines. In this process, the controller automatically calculates the start and end points of the traversing movements using the input parameters, thereby relieving the programmer of geometric calculations. The different contour lines are programmed by entering the end point coordinates and angles that uniquely define the contour lines and that come directly from the workpiece drawing. Standard bevels and radius transitions are not programmed separately but defined as boundary conditions of the contour lines using specified NC addresses. ◘ Figure 11.31 shows a contour line programming example.

Extending out from start point A, there are the following three predefined contour lines:

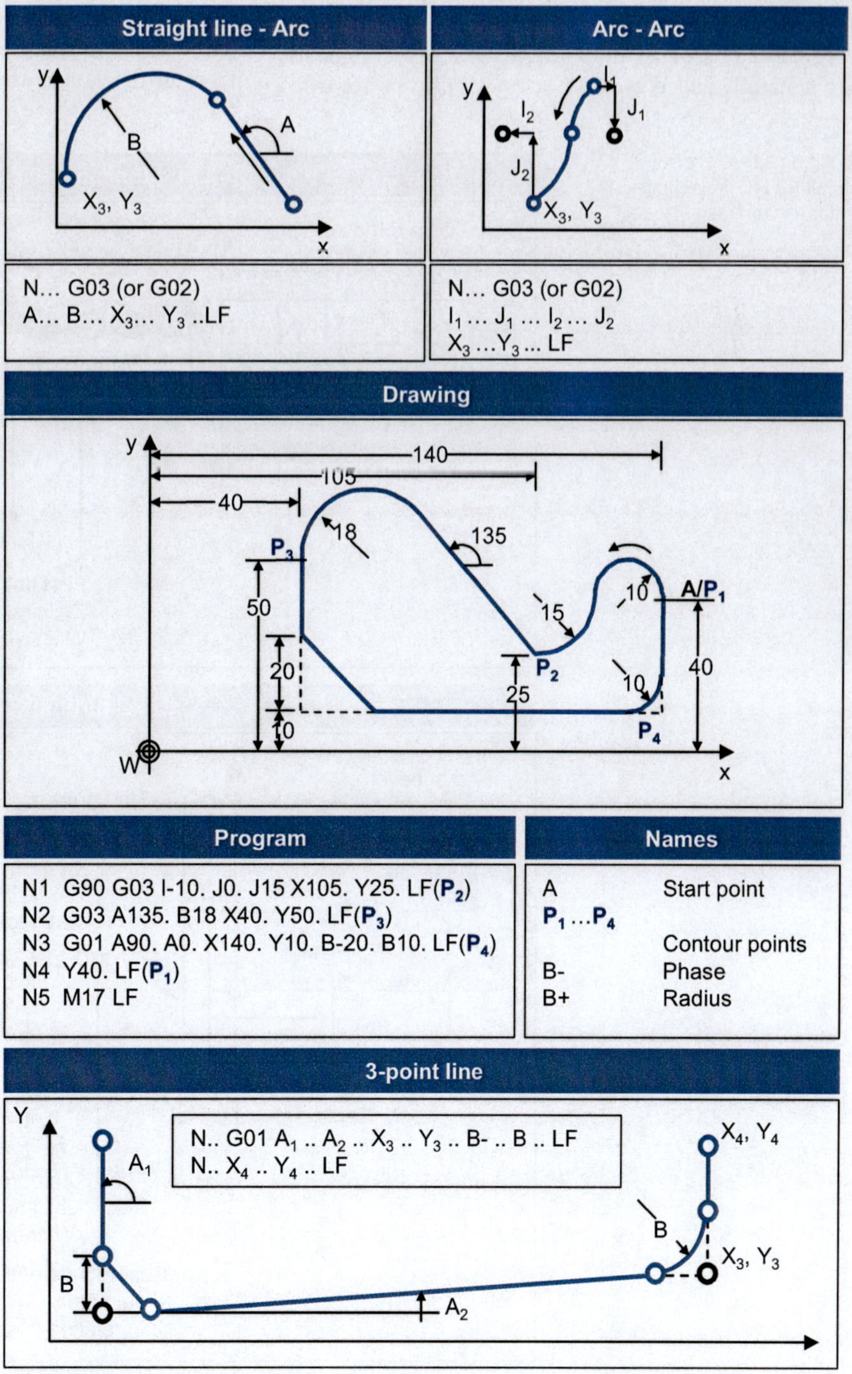


◘ **Fig. 11.31** Contour programming, according to Siemens

- Arc, arc (to $P_2$)
- Straight line, arc (to $P_3$) and
- 3-point line with bevel, radius (to $P_4$) and straight line (to $P_1$)

The saving compared to conventional NC programming is five NC lines and the often complicated calculation of intermediate points.

## Cycle Programming

To simplify manual programming, additional manufacturer-specific commands have been included in the command sets of many controllers. These commands are generally called cycles. Cycles can also be regarded as hard-coded subroutines in the controller. The transfer of parameters takes place in a defined form depending on the cycle.

Discussed below, as shown in Fig. 11.32, are two commonly used cycles—the drilling cycle and bevel turn cycle.

Cycles are generally called using G instructions. They can be assigned differently depending on the controller in use. In this example, a controller from TRAUB is selected [o.V.92. company publication].

The drilling cycle is called using command G83. This call can perform non-centric longitudinal bores. A condition for this, however, is that the machine has tool drives in the lathe turret. Execution of the command

11

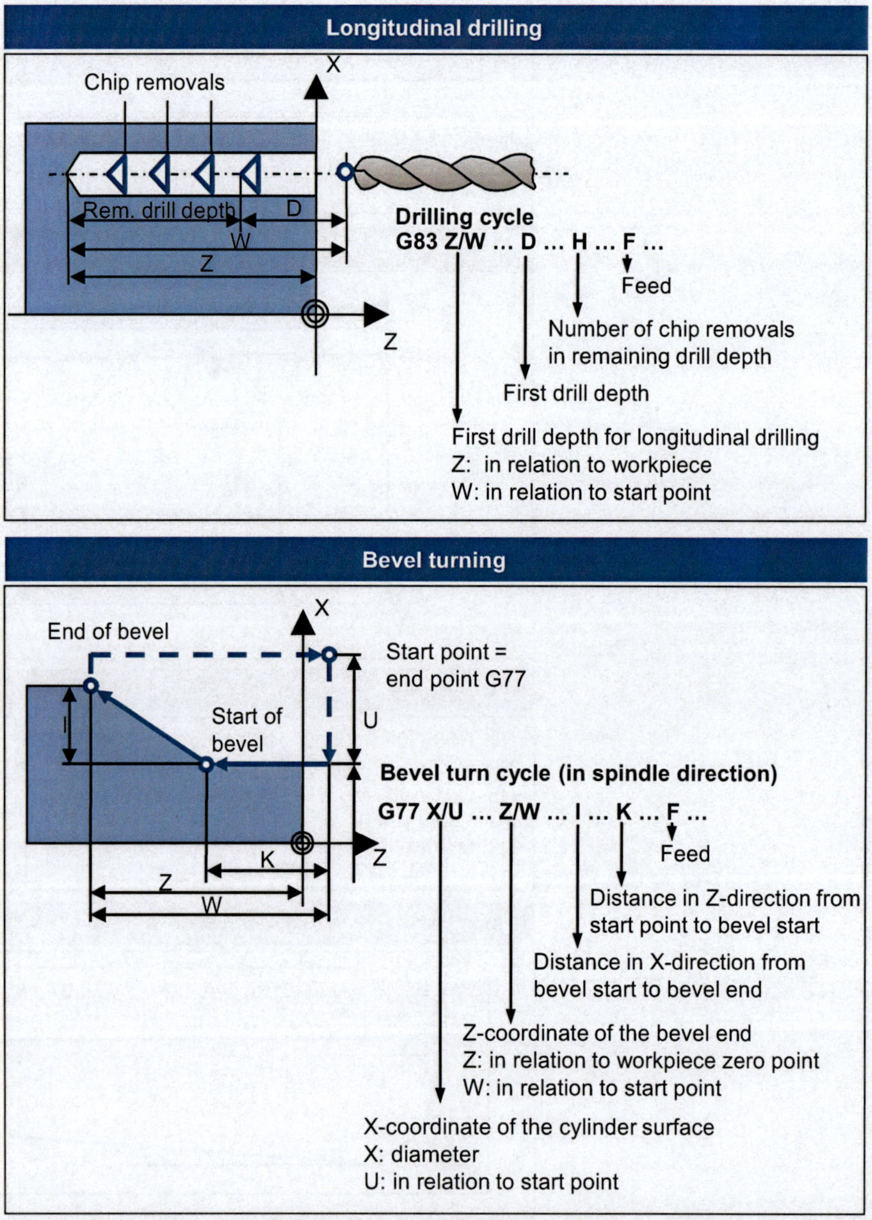


**Fig. 11.32** Cycles for simplified NC programming, according to Traub

starts on the axis of symmetry of the bore in front of the workpiece. This start point is also the end point of the movement. Refer to ◘ Fig. 11.32 for the individual transfer parameters.

The bevel turn cycle is called using G77. In this example, this is a machining process that first rotates a cylindrical piece in the spindle direction, and then produces the bevel with the required diameter. In this case too, the starting point is also the end point of the movement. Here however, unlike the drilling process described previously, machining is performed only once. The geometry of the bevel is specified by the parameters in ◘ Fig. 11.32.

#### 11.4.1.4 Limits of Programming in Accordance with DIN 66025

The descriptions and examples above for NC programming as per DIN 66025 illustrate how complex the language is and how little the command scope is tailored to users. The reason for this is the historical development of NC technology. Initially available computing power was limited. This meant that complex interpolation functions and traverse instructions over multiple axes could not be executed with these controllers. Programming techniques also did not correspond to the capabilities of programming languages today. The graphical display is limited by the disjointed single blocks and the capabilities of modern-day user interfaces cannot be used. Another problem is that, although the language is standardized, the meanings of individual commands can vary not just between different machining types (turning, milling, nibbling, etc.). Every controller manufacturer adds its own instructions to the language scope (specially tailored to the machine) or the existing meanings of commands are replaced by the manufacturer's own meanings [KIEF99].

The development of NC technology shows the command scope is no longer adequate and does not permit optimum utilization of all the functions of modern-day machine tools. The trend of numerically controlled machines is increasingly towards a greater degree of automation. This means that future controllers, in addition to existing functions, will be assuming tasks such as process monitoring, measurement cycles and simultaneously performing different processes. Efforts to integrate these capabilities, too, into the machining process include adding the standardized command set of a high-level language to the command scope of DIN 66025 [ISO89].

◘ Figure 11.33 explains the possible integration of such a new data format into a conventional NC program. The extended functions are bound by corresponding characters ("[,,"]") and inserted into the NC data format (refer to the flow diagram in ◘ Fig. 11.33). Available to the programmer within these characters are functions that are already familiar from other high-level languages, including some NC-specific functions. It is possible to exchange arbitrary variables between the data formats so that, in a high-level language level for example, calculations can be performed, and the conventional NC code initiates the resultant movements (◘ Fig. 11.33).

In such extended data formats, interpolation functions such as splines can also be used to machine free-form surfaces, for example. This format is also not object-oriented however, meaning closed information on individual machining features cannot be depicted and there is a risk that larger programs may become unmanageable. Also, these NC programs are programmed to run individual traverse and switch instructions on a machine with a special controller. The format is not an interchange format using machine-independent data describing the machining task. Thus, the benefits of manual programming are in the areas of simple workpiece geometries and short programs that can be created in a standard text editor. More complex tasks and programs that require the work of several programmers, can only be created reliably and cost-effectively with automated aids.

### 11.4.2 Automatic NC Programming Methods

Although a programmer workplace for manual program creation brings a sense of relief for the programmer, a series of often recurring activities, which are time-consuming and susceptible to errors, still remain. This includes checking material and tool files, taking into account machine data, calculating spindle speeds and feeds and determining coordinates in consideration of roughing allowances. To relieve the programmer of this work and to reduce its susceptibility to errors, computer-aided programming methods have been developed where a computer system can automatically handle these activities to a large degree.

In contrast to manual programming, workpiece machining is described in automatic programming using a problem-oriented language, an object-oriented model or graphically/interactively. Problem-oriented languages use mnemonic expressions, i.e., words and symbols that are easy to remember. In object-oriented programming interfaces, information is mapped to a specified hierarchical structure using appropriate software. Interactive graphic systems enable the user to generate NC programs in dialog and with graphic explanations.

One of the first and most powerful problem-oriented languages is the APT programming language (Automatically Programmed Tools), created by MIT (Massachusetts Institute of Technology) in 1958. From this language, comprising about 300 symbolic words, a

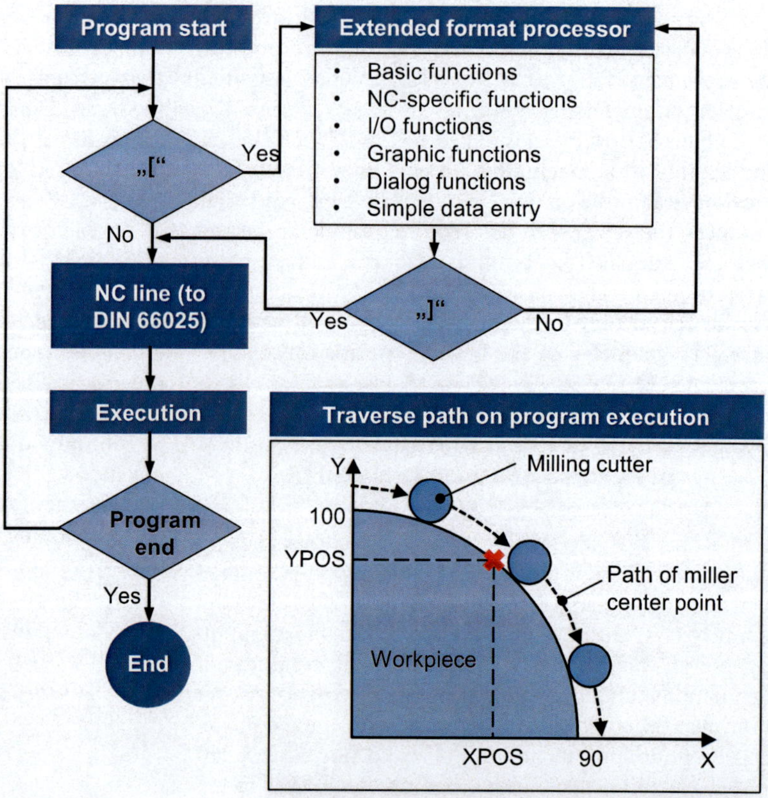


**Example NC program**

```
%
[ DATA
  LOCAL REAL; XPOS; YPOS; XPOSGRAD;
  ENDDATA;]
N10 G90 G41
N20 G00 X-2 Y100 Z110 F100
N30 G01
[ XPOS := 1; YPOS := 100;
  WHILE XPOS < 90 DO
     XPOS := XPOS + 1;
     XPOSGRAD := XPOS
     YPOS := 100 * COS( XPOSGRAD ) ]
N40 X[XPOS] Y[YPOS]
[ ENDWHILE;
  ENDPROG;]
N50 G01 X ...
```

**Fig. 11.33** Example of an extended data format for DIN 66025 (as per ISO/DIN 6132)

series of subsets have emerged (ADAPT, EXAPT, MINI-APT, etc.) [SHAH72]. The reasons for the development of these simplified languages, designed with limited applications and smaller computers in mind, were the strict (by the standards back then) requirements of the computing capacity of APT.

The source program created in an NC programming language like APT is generally known as a "parts program". It contains all data and instructions required to machine a workpiece. Table 11.2 shows an example of the main structure of an EXAPT parts program defining the machining of the turned part shown in Fig. 11.34. It starts first of all with general instructions. The blank and finished part contour is described from a start point, to be defined, by circumscribing the contour clockwise, and concatenating the individual contour elements. For turned parts, the description is restricted to one half of the symmetrical body cross section (that is automatically mirrored later). Technological instructions for the machining

**Table 11.2** Excerpt from an EXAPT parts program

| EXAPT 2 processor | Processor heading/message |
|---|---|
| $$-general instructions-<br>PARTNO/TURNED PART<br>...<br>MACHTDT/130,150,0.1,<br>10,5,3000,0.8 | Workpiece ID<br>Machine specifications (spindle power: 30 kW)<br>Torque: 150 mKp (meterKilopond), feed range from 0.1 to 10 mm/rev, speed range from 5 to 3000 $min^{-1}$<br>Correction factor for roughness depth: 0.8 |
| $$-blank description-<br>CONTUR/BLANCO<br>BEGIN/-5,0,YLARGE,<br>PLAN,-5<br>RGT/DIA, 70<br>...<br>TERMCO | Start of the blank description<br>Start at point (X = 5, Y = 5) in the positive Y-direction with plane surface at X = −5<br>To the right, cylinder with diameter 70 mm<br>End of contour description of the blank |
| $$-finished part description-<br>SURFIN/FINE<br>CONTUR/PARTCO<br>M1,BEGIN/0,0, AY-<br>LARGE,PLAN,0,BEV<br>EL,2<br>...<br>M2,RGT/<br>PLAN,70,ROUND,1.5<br>... | Surface quality (surface finish)<br>Start of the finished part contour description<br>Start at point M1 (X = 0, Y = 0) in positive Y-direction with plane surface at X = 0. At the end of the contour element is a chamfer (BEVEL) 2 mm wide<br>To the right at point M2; plane surface at X = 70 with radius: 1.5 mm |
| M3, RGT/DIA,0<br>TERMCO<br>... | To the right at point M3; cylinder with diameter 0, closing of the contour line<br>End of the contour description for the finished part |
| $$–Technological instructions–<br>PART/MATERL, 250<br>...<br>ROUGH = CONT/<br>SO,LONG,ROUGH,TOO<br>L, 123, SETANG,-90 | Material specified using code number of tool file<br>Definition of roughing for the contour (CONTOUR), single operation (SINGLE OPERATION), longitudinal machining (LONG), roughing (ROUGH) with tool (TOOL) 123 as per the tool change file, tilt angle (SETTING ANGLE) −90° |
| §§–Executive instructions–<br>COLLNT/ON<br>WORK/ROUGH<br>CUT/M3,RE,M2<br>...<br>CLAMP/160,INVERS<br>...<br>WORK/NOMORE<br>FINI | Coolant on (COOLANT ON)<br>Machining call (WORK) of the defined work conditions<br>Machining call: from M3 opposite to the description direction (REVERSE) until M2<br>Re-clamping on clamping level X = 160<br>End of machining<br>End of parts program |

type (e.g., roughing) or information on the clamping position of the workpiece come next. At the end, the machining processes and their sequence are defined by corresponding executive instructions.

This kind of parts program is converted to control information for the machine tool by two computer programs called "Processor" and "Postprocessor" (Fig. 11.35, Alternative 1).

In the Processor, the parts program instructions are first checked for formal errors. All arithmetical and geometric instructions and the path of the tool reference point are calculated. With the data from the material file and tool file, the Processor determines the cutting values such as cutting depth, cutting speed and feed rate. Any necessary cut distributions and collision calculations are also performed. The data created by the Processor signify a general solution to the machining problem and are independent, for the most part, of the machine tool used. They are called CLDATA (Cutter Location Data, DIN 66215). The benefit of generating such intermediate code is that this data can be used for a large number of NC machines [DIN74].

These machine-independent data are not tailored to a particular machine tool until the Postprocessor operation. The parameters calculated by the Processor, such as paths and speed/feed values, are assigned the coding stipulated by the machine controller in the NC language (to DIN 66025 for example) and are output as control information in the controller-specific format of the machine tool.

Figure 11.35 shows two more alternatives for machine-specific generation of an NC program. Alternative 2 assumes a machine-independent program and uses it to generate a machine-specific NC program using appropriate Postprocessors. The format of the neutral program can vary greatly. Many programming systems use machine-independent NC code that is based to a great extent on DIN 66025. Being based on DIN 66025 has the benefit that the strong similarity between the two NC programs means the Postprocessors can usually be configured easily by the users themselves [COSC93].

Other systems use their own problem-oriented languages and translate the program created directly (without an intermediate format such as a CLDATA file being generated) to the NC code for the relevant machine. Alternative 3 (in Fig. 11.35) describes such direct generation of a machine-specific NC program for a particular target machine. In most cases, these kinds of systems are developed by machine tool manufacturers for their own products.

Having a machine-bound, sequential NC program as a result is common to all three alternatives. Switching to a similar machine or machining with other tools is generally not possible. Also, changes by the specialist on the machine cannot be relayed back to planning

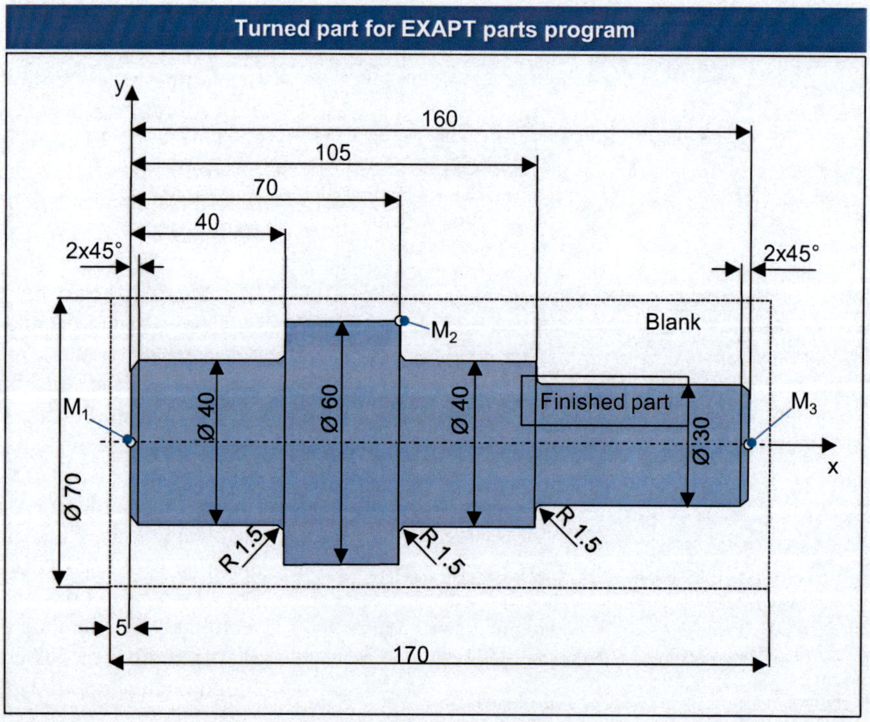


**Fig. 11.34** Turned part for an EXAPT parts program, according to Siemens

11

via the NC program. This is why any geometry changes and the optimal parameters are known only to the specialist and exist only in the NC program.

An NC program where the format and data model emerge from modern CAD and CAM systems without complex transformations and modifications offers several benefits compared to the programming alternatives introduced. These include that errors are prevented, that information on changes can be exchanged between all areas and, consequently, the workpiece data are always complete and up to date. Alternative 4 in Fig. 11.35 shows the use of such a data format (e.g., ISO 14649). A Postprocessor is not needed here.

Programming (Fig. 11.35, left) for the first three alternatives is either with text using an editor for the applicable programming system or graphically/interactively on modern programming systems. For Alternative 4 however, the data model is very large and an NC program would be too complex to edit it directly. The specialist or planner can only generate and change such an NC program using appropriate software.

The input information necessary for programming is the same for all four alternatives presented. The information to be entered by the programmer can essentially be divided into the following three types (see Fig. 11.26):

- General information
- Geometric information
- Technological information

General information is understood to be information that must be specified before the actual machining description, such as the workpiece number and type of machine used for machining.

Geometric information pertains to the geometric elements of the part. It includes contour descriptions and special geometries such as grooves and pockets.

Technological information generally contains the tools to use, machining instructions (roughing, finishing), etc.

For the automatic generation of the programmed machining processes, such as cut distribution and determination of feeds and speeds by the CLDATA Processor, technological data such as material and tool parameters are required. Material parameters define the relation between feed rate and cutting speed. This relation is given, for example, by the material—cutting material combination and by the machining type (turning, milling, etc.). Also, the geometry information, such as length correction values for the tools used, is available.

Finally, the machine file contains all machine-specific data such as spindle powers and speed ranges. These data permit the Postprocessor to generate an NC program tailored to the machine tool.

The previous discussions focused more on automatic programming using problem-oriented languages (APT, EXAPT) because of historical developments. Computer-aided systems and methods are explained

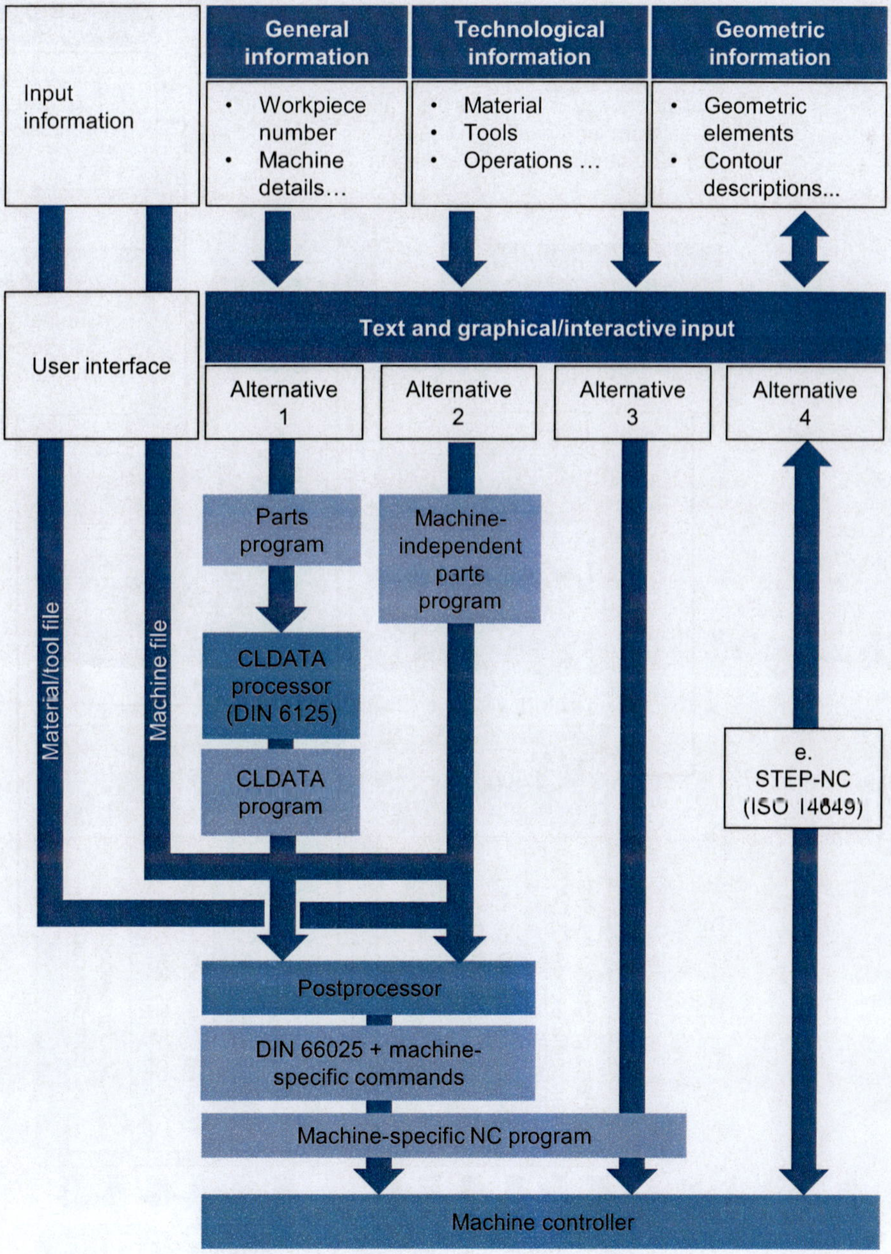


**Fig. 11.35** Alternatives for generating an NC program

below. These permit more effective program creation using the coupling alternatives to systems in other sectors (such as design engineering) and by using programming tools tailored specifically to the machining process and user.

### 11.4.2.1 CAD/CAP/CAM Coupling

Trends nowadays for implementing a continuous CAD/CAP/CAM process chain are tending towards integrating CAD systems more closely into the process of NC program generation. Factoring in this requirement essentially yields two different variants of integrating the CAD systems (Fig. 11.36):

- Usage of one single system that combines the CAD system and the NC programming system into one. In this case, all the functions of both systems have direct access to the identical internal data structures.
- Coupling of the CAD and NC systems using suitable data interfaces. The aims here are to use the geometry information created in the CAD system and

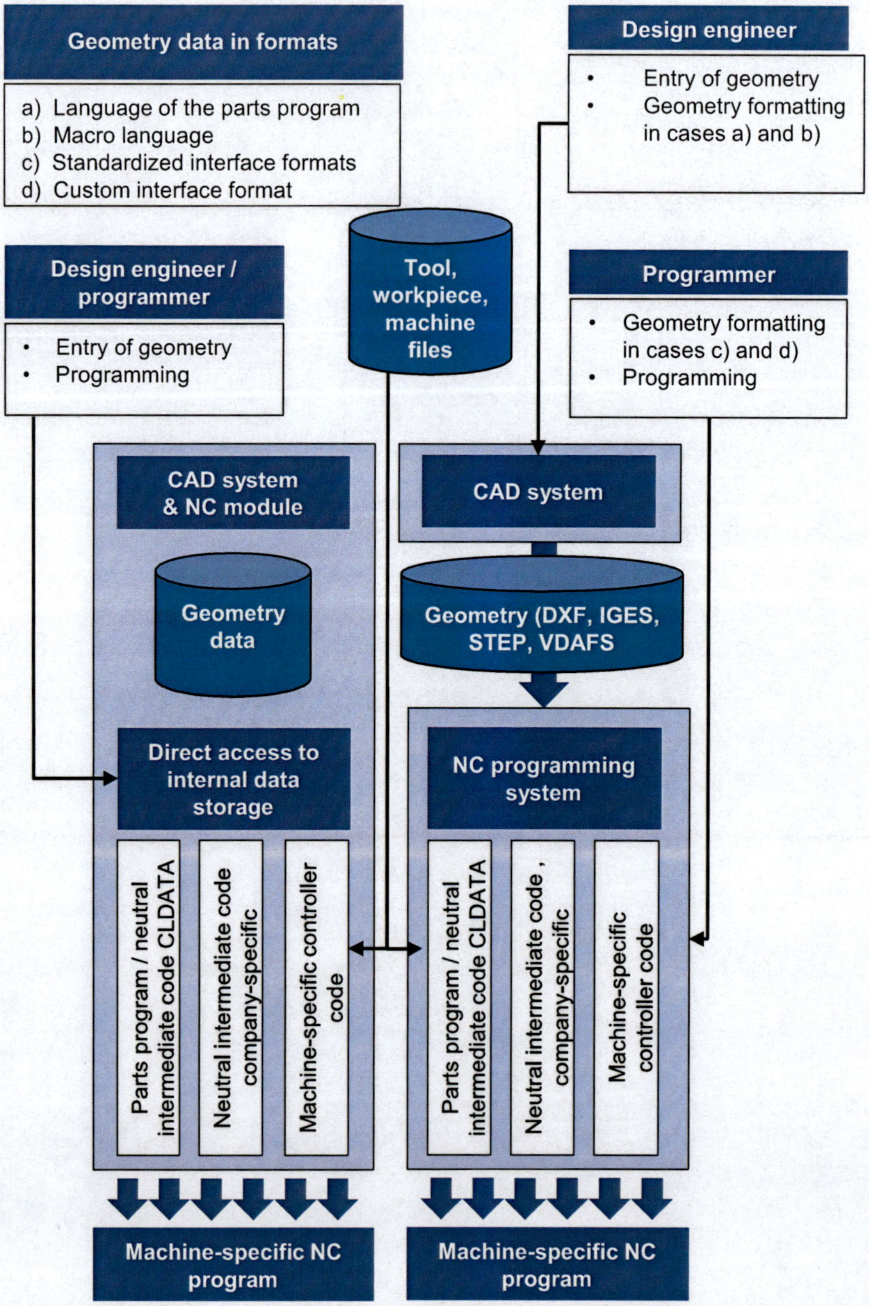


**Fig. 11.36** Methods for coupling CAD and NC programming systems



to prevent reentering the geometry data into the NC programming system. The different capabilities of this coupling, and the accompanying difficulties, are discussed in more detail in a later chapter.

Now, NC programs are generally created centrally in production planning.

The programmer receives the necessary workpiece information from the design department in the form of workshop drawings. The NC program created is transferred to the machine tool in production where it is optimized as required and run by the user on the machine.

There is a clear division of tasks here between design, planning and production. In contrast to this however, a clear trend can be seen in the generation of NC programs moving to production areas. This means that programs are becoming more and more programmable on a decentralized basis and broadly independent of their place of use. In this context, reference is made to workshop-oriented programming directly on the machine—discussed in more detail in ▶ Sect. 11.4.2.4.

The programming system to use and the place it is used are determined mainly by the range of parts to produce and the production structure [MILB89].

Considering the criteria for where it is used, programming for single-component or small-lot production for example would be advisable in the workshop because program generation in production planning would entail unjustifiable organizational overhead. For high volumes, however, where time-optimized machining is paramount, production planning would be better equipped to satisfy this criterion. Unlike the first case, workers in production here would not need extensive programming knowledge.

When a company has computer-aided design (CAD), the aim is to use the geometry and machining information already captured as data (such as roughness and tolerances) for NC program generation. CAD data, however, is generally available in a form that differs greatly from the format of NC programming systems. This means that the geometry information must always be prepared accordingly. The coupling capabilities of CAD/CAP/CAM systems can be classified by the different ways of this preparation and the division of tasks between CAD and NC programming systems. Different alternatives of this coupling are depicted in ◘ Fig. 11.36.

An integrated system, i.e. a CAD system with integrated NC module, is defined on the left. This alternative has the benefit that the NC module has direct access to the shared data memory [ANDE90]. All information required can be accessed and is available immediately in the required format. Also, changes to the component by the NC module directly affect the CAD model. Additional entries for any updating of the CAD data are thus no longer necessary. The accessing of just one shared model of the component by all modules is also called bidirectional associativity.

The motion paths programmed in such an integrated system, and the resultant workpiece contours, are easy to depict graphically using the CAD capabilities of the workstation.

After programming, the NC program (as detailed in ◘ Fig. 11.36) is available in the form of a parts program (often also directly in the CLDATA format) in neutral intermediate code or directly in machine-specific code. Then, running the Postprocessor represents subsequent processing for the first two alternatives. The result is an executable NC program (DIN 66025) for a particular machine tool.

Programming on an integrated system is mainly done by a programmer in production planning together with the design engineer as required. This presupposes, however, that the design engineer and programmer are familiar with the overall system. These integrated systems are used mainly for 3-axis and 5-axis machining of components with freeform surfaces, consequently surfaces with arbitrary bending, because the laborious preparation of complex geometries is automated by the software system.

The right side of ◘ Fig. 11.36 defines the coupling points of distributed systems. There are four different formats here for exchanging geometry data between the CAD and programming systems:

- Language of the parts program (a)
- Macro language (b)
- Standardized interface format (c)
- Custom interface format (d)

The first two data formats are specified with the NC programming system requirements very strongly in mind. Accordingly, the CAD system itself performs the geometry formating. ◘ Figure 11.37 shows examples of the parts program language (a) and use of the standardized interface (c) in Siemens NX.

### Language of the Parts Program

In the first case, the geometry data are transferred over a language interface, that is based on the syntax and semantics of the appropriate language of the parts program (APT or similar). The geometry information is available directly in the language of the NC programming system. In the subsequent work steps, the technology information necessary is entered in the NC programming system and the NC program is created by running the Processor or Postprocessor. One downside of this type of coupling is that the CAD system can only be connected to the programming system that also processes the corresponding language. One benefit is the ability to use a simpler programming system because only technology data need to be added to the parts program generated in the CAD system.

### Macro Language

In the second case, of sending geometry information in the form of macros, data exchange uses defined machining geometries. The individual geometry elements such as bores, pockets, undercuts and threads are described and configured in the CAD system as macros. The NC programming system is able to interpret these macros and adds relevant machining information. Prerequisites for this method however are that the range of parts is narrow and varies little as regards its geometry. Only then can an adequate scope of the macro language be realized with justifiable overhead.

### Standardized Interface Formats

The last two coupling alternatives are at the same time the most common. In both cases, geometry data are sent that is not intended specifically for subsequent processing on an NC programming system, meaning data preparation is necessary. Using these kinds of

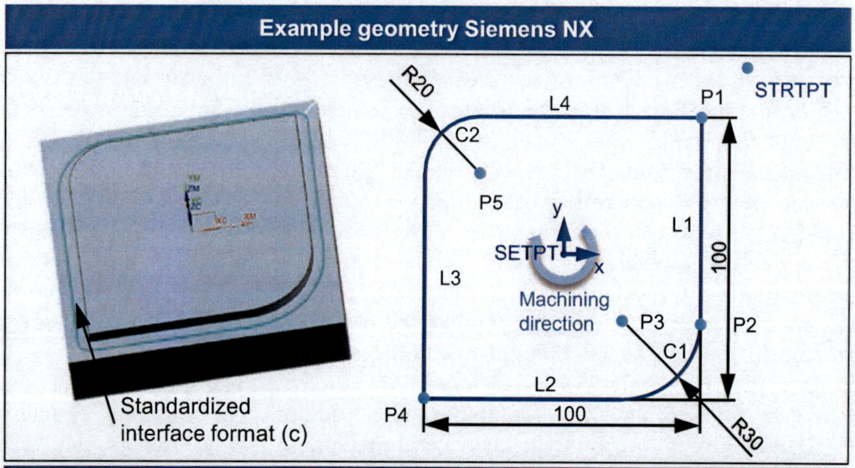


**Fig. 11.37** Examples of the different formats for exchanging geometry data, according to Siemens NX

**Description using APT (language of the parts program (a))**

```
PARTNO / APT-1
CLPRNT
UNITS / MM NOPOST
CUTTER / 10.0
$$GEOMETRY DEFINITION              // Creation of geometry elements
SETPT = POINT / 0.0, 0.0, 0.0
STRTPT = POINT / 70,70,0
P1 = POINT / 50, 50, 0
P2 = POINT / 50, -20, 0
P3 = POINT / 20, -20, 0
C1 = CIRCLE / CENTER, P3, RADIUS, 30
P4 = POINT / -50, -50, 0
P5 = POINT / -30, 30, 0
C2 = CIRCLE / CENTER, P5, RADIUS, 20
L1 = LINE / P1, P2                 // Creation of relationships
L2 = LINE / P4, PERPTO, L1
L3 = LINE / P4, PARLEL, L1
L4 = LINE / P1, PERPTO, L1
PLAN1 = PLANE / P1, P3, P4         // Definition of machining plane
PLAN2 = PLANE / PARLEL, PLAN1, ZSMALL, 16
$$MOTION COMMANDS                  // Programming of tool motion
SPINDL / 3000, CW
FEDRAT / 100, 0
FROM / STRTPT
GO/TO, L1, TO, PLAN2, TO, L4
TLLFT,                             // Radius compensation
GOFWD / L1, TANTO, C1              // Tool path
GOFWD / C1, TANTO, L2
GOFWD / L2, PAST, L3
GORGT / L3, TANTO, C2
GOFWD / C2, TANTO, L4
GOFWD / L4, PAST, L1
NOPS GOTO / STRTPT
FINI                               // Program end
```

standardized geometry interfaces has the benefit that this information can generally be used for a number of different programming systems. Nowadays, interface formats such as VDAFS (see Fig. 11.44), IGES (**I**nitial **G**raphics **E**xchange **S**pecification) and DXF (**D**ata E**x**change **F**ormat) are supported by most systems. One major drawback of these data formats however is that only data defining the shape are sent; information relevant to machining is not supported. They must be sent manually from the CAD workshop drawing to the NC programming system. This is not a satisfactory solution in terms of a continuous CAD/CAP/CAM process chain.

One approach to improve data interchange is the use of product data technology, and correspondingly more powerful interface formats, based on standards for product data definition and product data interchange instead of geometry-only interfaces. The ISO standard STEP (STandard for the Exchange of Product Model Data) was published as a series of ISO 10303 standards and forms the basis for such an interface.

In its current scope, the standard, with the application protocols built on it, has not yet been adequate to define all data elements necessary for NC programming. Continuous STEP-based NC programming was not possible until ISO 14649, which references parts of STEP (ISO 10303) and individual application protocols.

Compared to interface formats for geometry interchange such as VDAFS, DXF and IGES, STEP has the benefit that it is an interface for the comprehensive definition of product data. Consequently, it allows reliable interchange of data in all areas of integrated production, so CAD, CAP, CAM, ERP and CAQ. Additional criteria are the minimization of storage requirements, efficiency in converting the data stored internal to the computer (server solution), general validity and correctness of the models, and the mathematically correct and unambiguous specification of relationships [EVER94; [o.V.90].

The STEP standard comprises the actual models for describing product data from *description methods, implementation methods* and the *conformance testing methodology and framework*. The core of STEP is represented by "partial models" however. They contain the different product information that is implementation-independent and forms a standardized basis for different application protocols [PROS17]. They belong to the "Integrated Resources" (◘ Fig. 11.38). They are used to define, for example, geometry, topology, material, tolerance, shape element and structure information.

Also, differentiation is made between the models for special applications, the "Application Resources". To be able to use the partial models for certain applications, it is necessary to define such application protocols (APs). An interface for supporting all data relevant to production was not defined until this enhancement such as in the case of the CAD/CAP/CAM coupling. One AP, for example, specifies that, in a certain context, a straight line described in accordance with the Integrated Resources is a dimension line. So, the application protocol specifies the meaning and relation between individual base data elements.

### Custom Interface Formats

Unlike the standardized interfaces described above, the custom interface is tailored perfectly to the relevant NC programming system. Usually however, the processors that convert CAD data to the data format of the programming system are written by users themselves and are not supported by the CAD system vendor. This can be a downside when translation errors occur or there are new versions of the CAD software. The specific coupling of both systems means the connection of the programming system to other CAD systems is still greatly limited (◘ Fig. 11.39).

### Problems of CAD/CAP/CAM Coupling

Problems when coupling CAD/CAP/CAM systems, as already mentioned, arise mainly when information relevant to geometry and production are received from the upstream CAD systems. Information relevant to production, such as shape and positional tolerances, allowances, undercuts, roundings, bevels, surface quality and text additions, are available in the CAD systems in widely varying formats (◘ Fig. 11.40) and need to be prepared accordingly for NC programming. Other information, such as hachures and dimensioning, must be deleted from drawings because they are irrelevant for subsequent machining.

There are four main problems in the geometry transfer:

- Geometry adjustment
- Different accuracies of the two systems
- Incomplete geometry
- Use of asymmetrical tolerances.

For a geometry adjustment, the workpiece zero point, machining position and machine zero point must be specified because other reference points are usually selected in design engineering. Furthermore, contours sometimes need to be changed, or when different machining technologies or clampings are used, possibly segmented. Below, starting from a component model, individual geometry models are first generated that serve as the basis for subsequent programming.

The different accuracies of the coupled systems present problems of a different kind. If, for example, the programming system works with a higher accuracy than the CAD system, contours or general surfaces/bodies of the component might no longer be joined to each another. ◘ Figure 11.41 (top) shows the effect using a simple example.

Different requirements placed on the geometry by the design engineer and programmer are another reason for disjoint geometry elements. For the former, a graphical-only depiction of the component is usually adequate. The engineer rarely needs to work with the highest resolution on the screen of the CAD system. This resolution is often necessary however to detect gaps or overlaps between the elements.

The different types of tolerance dimensioning in the two systems often give rise to the greatest difficulties in the geometry transfer [FRAN91]. In CAD systems, the nominal dimension of the contour is specified, in NC programming systems, it is always the contour

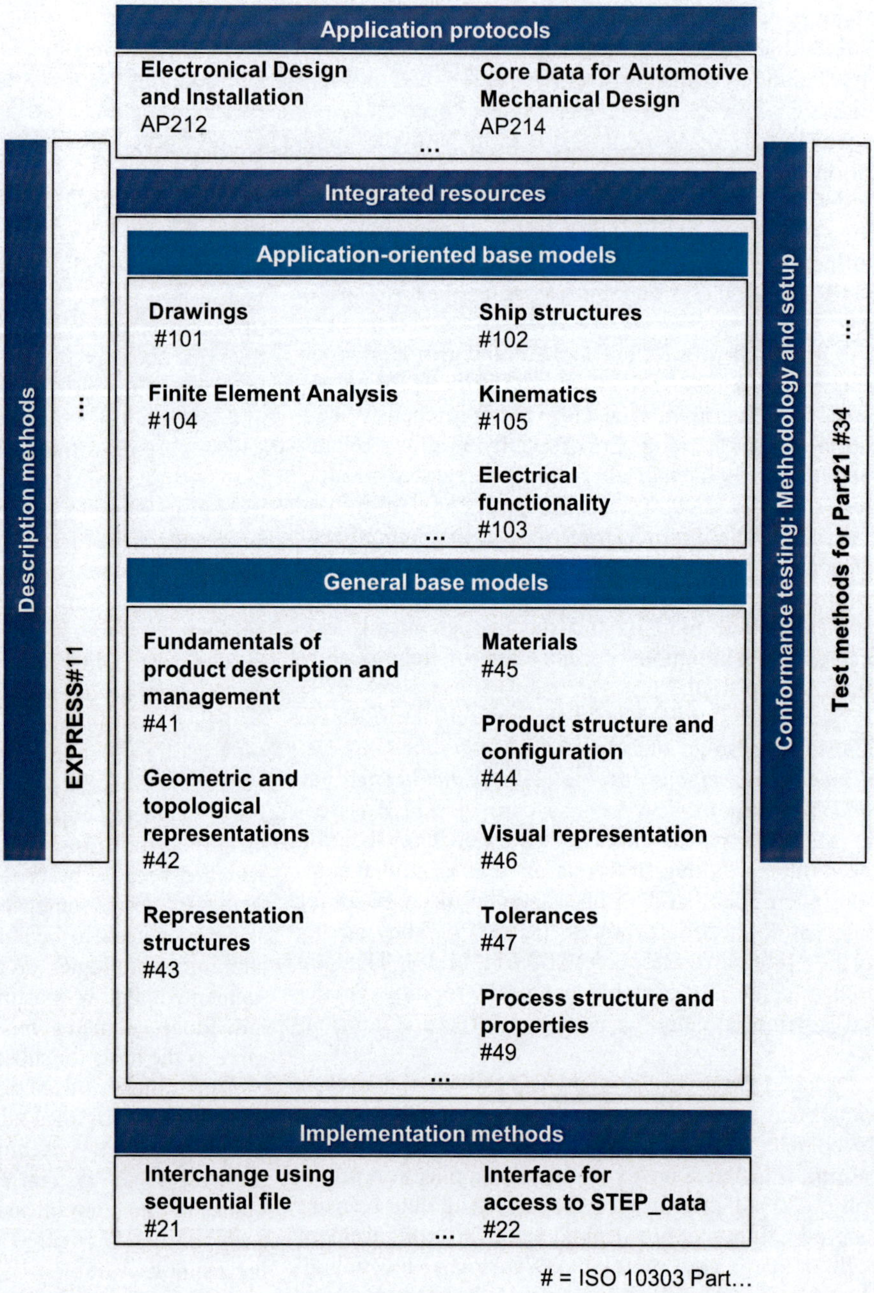


**Fig. 11.38** Structure of the STEP standard (according to NIST: "STEP on a page")

dimension oriented at the mid-tolerance. The latter is necessary because there are equally positive and negative machining deviations on the machine tool.

The asymmetric tolerances of CAD geometry data and their changes in the NC programming system have greater effects in part that also have a bearing on adjacent contour elements. Figure 11.41 (bottom) shows this problem by way of a turned part. The contour, here an undercut, is made up of a cylindrical and a conical section. The cylinder, or its diameter, is dimensioned at the start in the CAD system asymmetrically as 20 + 0.5/+0.1 mm. Offsetting the diameter to the mid-tolerance now causes the angle of the cone to also change (or to be changed by the programmer). Accordingly, for more complex components with a large number of asymmetric tolerances, manual geometry formating (that is therefore time-consuming and susceptible to errors) is always required.

### Freeform Surface Machining

Program creation for freeform surface machining is discussed separately once again below because this type

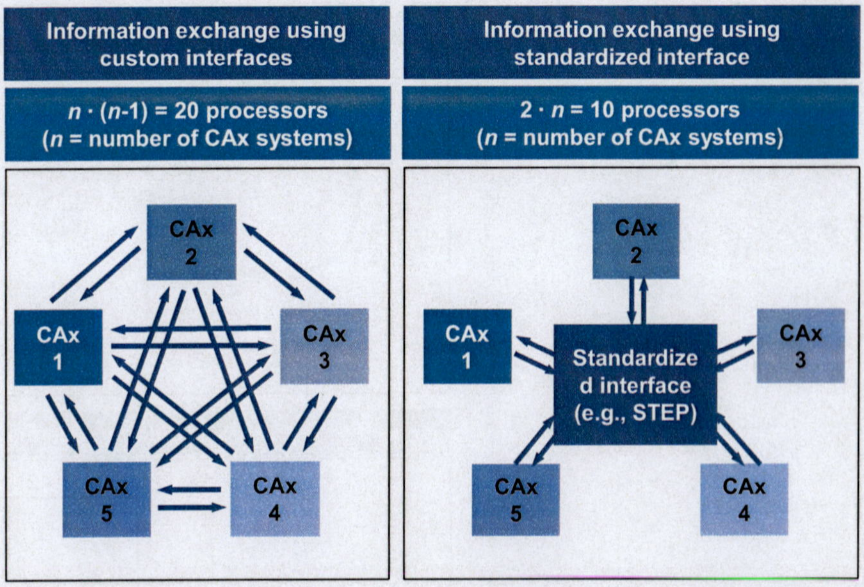


**Fig. 11.39** Benefits of standardized interfaces

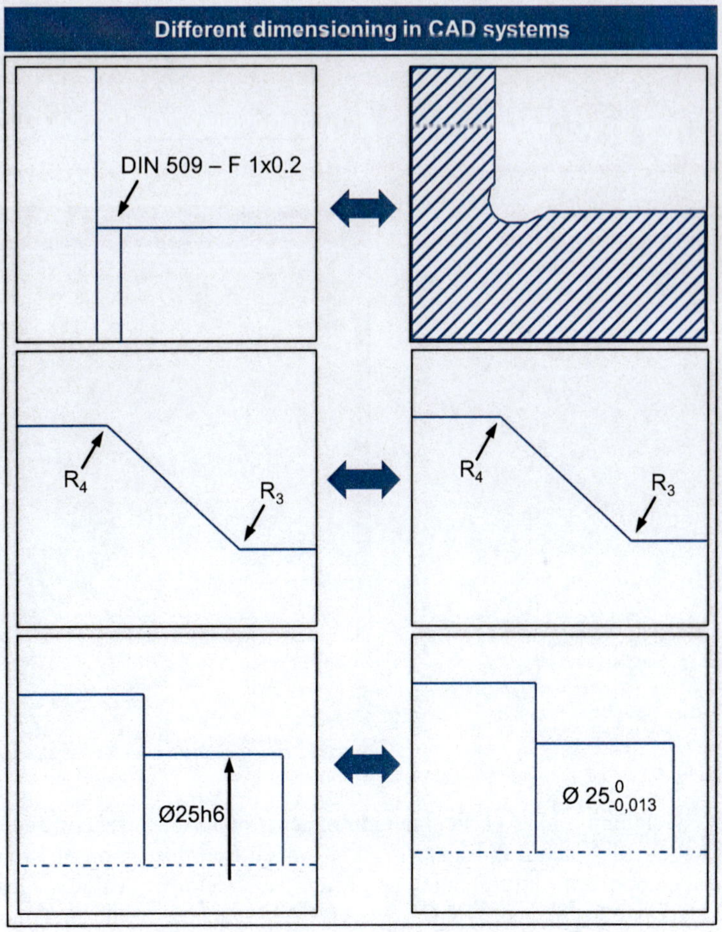


**Fig. 11.40** Different dimensioning in CAD systems, according to Pfeiffer

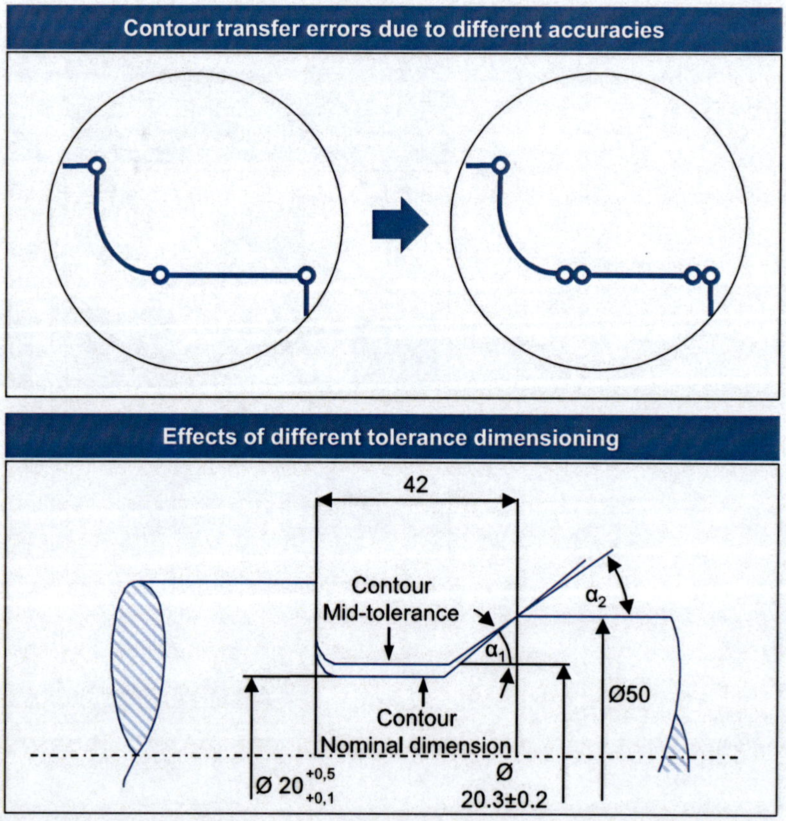


**Fig. 11.41** Problems in transferring CAD data to an NC programming system, according to Franz



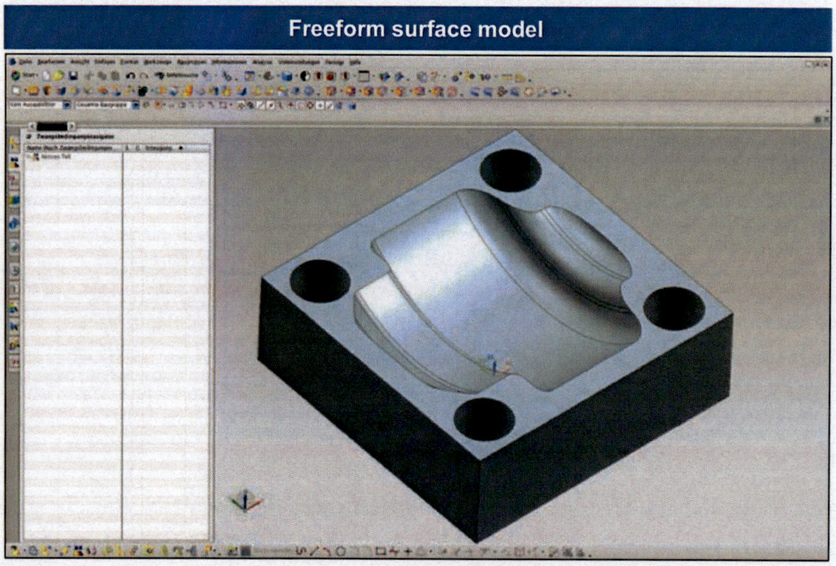


**Fig. 11.42** Example of a freeform surface model in Siemens NX, according to Siemens

of machining makes particular requirements of CAD/CAP/CAM systems. ◘ Figure 11.42 shows an example of a component (fixture) made up partially of freeform surfaces in the Siemens NX CAD/CAP/CAM system.

There are essentially three different ways to perform finishing machining of freeform surfaces as shown in ◘ Fig. 11.43. These are using 3, 3½ or 5 axis milling. In contrast to 2½ axis machining, that is, interpolation of two axes and advancing the third axis by a fixed value, at least three axes are moved synchronously in the first method mentioned. In 3 axis machining, the movements of exactly three axes are interpolated. The orientation of the milling tool stays constant here in the direction of the Z-axis of the machine coordinate

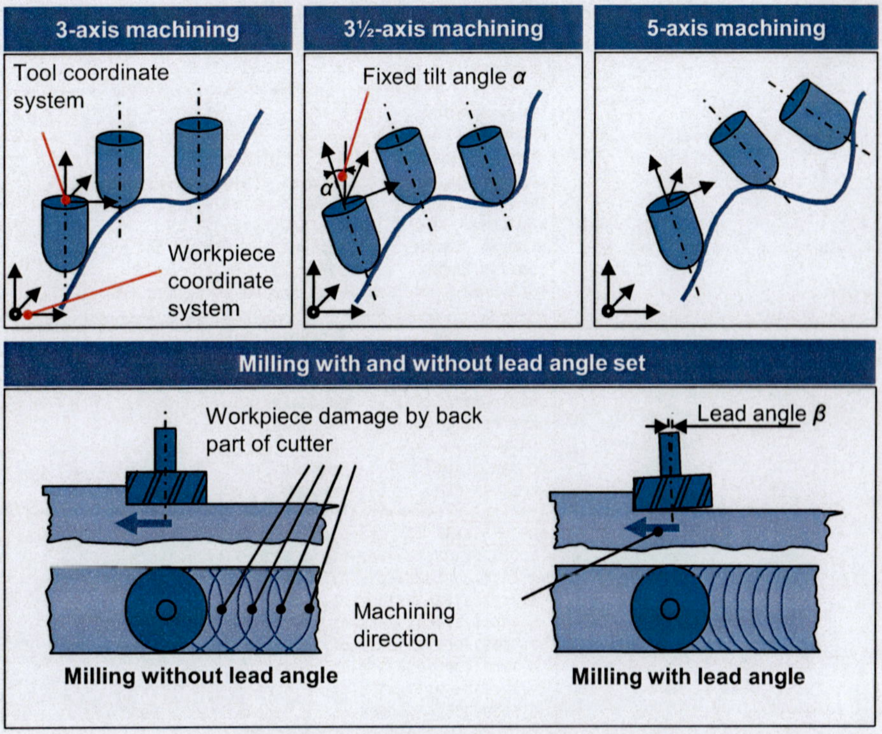


**Fig. 11.43** Freeform surface machining alternatives for different numbers of axes (top), and effects of milling with and without a set lead angle (bottom)

system. In 3½ axis machining, this orientation is changed by constant angles so that the rotation axis of the milling tool no longer coincides with the Z-axis of the workpiece coordinate system. For simplified graphical depiction, Fig. 11.43 shows the tool positioned at angle α. Finally, 5 axis machining permits the milling tool orientation to be aligned continuously. This guarantees the optimal position of the tool in relation to the workpiece surface at all times.

3 and 3½ axis milling have the benefit that, due to fewer axes, tool path calculations are less complicated and so less time-consuming. Collisions can be calculated more easily and singularities (non-unique assignment of Cartesian positions in relation to the corresponding axis positions) do not occur. Furthermore, since only linear axes are traversed for workpiece positioning, higher feeds, and thus higher machining speeds, can be attained. In contrast to this, machine running times are shorter for 5 axis machining. The optimal alignment of the tool to the workpiece contour in each case permits higher infeeds (greater material removal) thereby keeping the reworking overhead much lower due to the optimal tool orientation. Complex component geometries can be produced more accurately in one operation, in one clamping, and usually with no rework. This means re-clamping and repeated, time-consuming alignment of the workpiece are not necessary.

However, the kinematically more complex processes also require strategies designed for them. For example, slight tilting (dive) of the face mill by lead angle β in the machining direction is necessary so that only a defined area of the milling cutter circumference engages with the workpiece. This prevents component damage in the back area of the cutter.

CAD/CAP/CAM systems designed specially for this are available to satisfy the stricter requirements. This is because, in contrast to simpler turned part geometries, the complex geometry for freeform surface machining is normally generated using a CAD system. In this case, the CAD system is commonly integrated in an overall system or is connected to the programming system via a neutral interface (Fig. 11.36).

In the automotive industry for example, geometry information is often exchanged over the VDA surface interface (VDAFS). Figure 11.44 shows part of a VDAFS file. In the definition of the interface, the German Association of the Automotive Industry (VDA) is essentially limited to the definition of geometric elements. Dimensioning, definitions of component attributes (such as surface quality), text, etc. are omitted to allow definition of a less complicated interface format.

The command scope of VDAFS Version 2.0 is limited to 17 different types of commands. Of these, nine commands are for geometry definition and eight for file structuring. Geometry elements are, for example, single

**Fig. 11.44** Part of the VDAFS file for a freeform surface

**Taken from the VDAFS file for a freeform surface**

```
SLDWORKS = HEADER / 20                                                   00000001
************************************************************************ 00000002
VDAFS VERSION     :  2.0                                                 00000003
---------------------------- SENDER DATA ---------------------------- 00000004
SENDER COMPANY    :  WZL RWTH AACHEN                                     00000005
CONTACT PERSON    :  Michael Königs, Marian Wiesch                       00000006
TELEPHONE NO.     :  0241 80 27733 - 0241 80 28473                       00000007
ADDRESS           :  Steinbachstrasse 19                                 00000008
SENDING SYSTEM    :  SolidWorks-2014292                                  00000009
SENDING DATE      :  09/07/2017                                          00000010
FILE NAME         :  WZL_FREEFORM_SURFACE.vda                            00000011
----------------------------- PART DATA ----------------------------- 00000012
PROJECT NAME      :  FREEFORM_SURFACE                                    00000013
OBJECT CODE       :  001                                                 00000014
VARIANT           :  1                                                   00000015
CONFIDENTIALITY   :  UNCLASSIFIED                                        00000016
DATE EFFECTIVE    :  09/07/2017                                          00000017
-------------------------- RECEIVER DATA --------------------------- 00000018
COMPANY NAME      :  UNKNOWN                                             00000019
RECEIVER NAME     :  UNKNOWN                                             00000020
************************************************************************ 00000021
SR9 = SURF / 15, 18, 0.0000000000000000E+00, 5.3024374665930803E-02,    00000430
 1.0252231938902499E-01, 1.6226214122865501E-01,                         00000431
 2.4878200018264601E-01, 3.33611
 4.2183087485331300E-01, 4.86043
 5.7468618796396298E-01, 6.70443
 7.6435230667751397E-01, 8.24140
 8.7208443007856395E-01, 9.21920
 9.8311588024776497E-01, 1.00000
 0.0000000000000000E+00, 6.25000
 1.2500000000000000E-01, 1.87500
 2.5000000000000000E-01, 3.12500
 3.7500000000000000E-01, 4.37500
 5.0000000000000000E-01, 5.62500
 6.2500000000000000E-01, 6.87500
 7.1875000000000000E-01, 7.50000
 7.8125000000000000E-01, 8.12500
 8.7500000000000000E-01, 9.37500
 1.0000000000000000E+00,
4, 4, 0.0000000000000000E+00, -0
 0.0000000000000000E+00, -0.0000
 8.0912135832404406E+00, -0.0000
 -3.9031278209478160E-15, 2.6020
 -1.8790150988578060E+00, 0.0000
 7.8062556418956319E-15, -5.2041
 2.8487916460635115E-01, -3.3565
 4.1660363730381539E-05, -9.1175263953330310E-07,                        00000455
 8.8817845730693403E-16, -1.4329222014501055E+00,                        00000456
 3.1346596363263579E-01, 5.8301646808845822E-02,                         00000457
```

**Surface**

`name = SURF/nps,npt,((nps+1))*[pars],((npt+1))*[part]((nps*npt))*[iordu,iordv,((iordu*iordv))*[ax],((iordu*iordv))*[ay],((iordu*iordv))*[ax]]`

| | |
|---|---|
| • name | Name of the surface |
| • nps,npt | Number of surface segments in the u- and v-directions |
| • iordu,iordv | Polynomial degree of the current surface segment in the u- and v-directions |
| • pars,part | Definition of the surface segments in the s- and t-directions |
| • ax,ay,az | Corresponding coefficients of the current surface segment |

points, series of points and curves/surfaces shown as polynomials. Structuring commands are the start and end definitions of geometry groups and comment lines. The commands are called in an APT-like format:

- name = BEGINSET
- name = POINT/Parameter
- name = CURVE/Parameter
- name = SURF/Parameter
- name = ENDSET

Here, "name" stands for a variable name, the word after the equals sign is the relevant command. For geometry commands, the parameters describing the geometry are specified after the slash.

The example shown in Fig. 11.44 describes a freeform surface made up of 15 surface segments (patches) in the u-direction and 18 surface segments in the v-direction. These patches, each defined by exactly one polynomial, belong to one surface (SR9), the surface of the component.

$$x(u,v) = \sum_{k=1}^{iordv-1} \sum_{j=0}^{iordu-1} ax_{j,k} u^j v^k \tag{11.1}$$

$$y(u,v) = \sum_{k=1}^{iordv-1} \sum_{j=0}^{iordu-1} ay_{j,k} u^j v^k \tag{11.2}$$

$$z(u,v) = \sum_{k=1}^{iordv-1} \sum_{j=0}^{iordu-1} az_{j,k} u^j v^k \tag{11.3}$$

Parameters

$ax_{j,k}, ay_{j,k}, az_{j,k}$ – Coefficients of the surface segment

$u, v$ – Local parameters of the surface segment

*iordu, iordv* – Polynomial order of the surface segment in the u-direction or v-direction

Local parameters u and v, each set to one surface segment, run over these surfaces in the boundary areas from 0 to 1. To access positions across the surfaces, global parameters "s" and "t" are defined that correspond to local parameters u and v. Their definition range is specified in this line

*SRF*00002 = *SURF*/3,1,0.,1.,2.,3.,0.,1.,

after the "/" character and the first two digits (number of surface segments in $u=3$ and v-direction = 1). So parameters runs in the first surface segment from value 0 to 1, in the second from 1 to 2, etc. Accordingly, parameter t runs from 0 to 1 in each of the three patches.

The following line specifies the order of the polynomial (iordu, iordv) in the u-direction and v-directions. Finally, coefficients ax, ay and az are then specified. To clarify their order, the coefficients (ax, az) of the equations in the x-direction and z-directions are highlighted black. Command ENDSET finally ends the surface description.

In front of a sequence of geometry groups defined this way comes the file header. This specifies the ID and name of the file and the number of following text lines (comments). It also provides information on creation details such as company, date and recipient(s). A file is ended by calling command END (file end).

Freeform surfaces can be described straightforwardly this way and they can be exchanged between a large number of systems. Compliance with the simple VDAFS format however can only be ensured by limiting the set of geometry elements to a minimum. Information that is often desirable, such as machining data linked to the geometry elements using logical operations, or topology information, i.e., the linking of individual geometry elements, is not supported. This deficit is usually offset by transferring workpiece drawings with the VDAFS file. This means relevant information (surface quality, tolerances, neighboring surfaces, etc.) is then entered separately on the target system.

At the start of NC programming for freeform surface machining, the blank and finished part geometries are calculated on the basis of the geometrical input information (◘ Fig. 11.45). The missing topology data are generated automatically by most systems. This means the surface information required is available and planning of NC machining can start.

Starting from the blank and finished part geometries, roughing and finishing operations are defined first. For roughing, it is mainly about removing a lot of material in a short time. For this reason, axis-parallel traversing of the carriage is the predominant selection, resulting in a staircase appearance of the contour machined. Shown in ◘ Fig. 11.46 is the result of roughing on the model of the fixture in ◘ Fig. 11.42. The generation of milling tracks for roughing is automatic on most systems.

Afterwards, the machining strategies for finishing are selected. Individual surfaces are combined into surface groups and define the suboperations [SPUR93a]. Suboperations define the surface segments to be produced in one operation. Technological parameters such as milling strategies are assigned to the suboperations and are generally defined by control parameters such as tool, machining type, accuracy specifications and milling direction. On the basis of these values, the milling tracks are generated and checked for collisions.

The inputs described are made usually interactively using a graphical interface. In parallel to this, a program is created that enables the entire machining process to be visualized.

Finally, the machine-specific NC program is created in the last step. Information such as machine/controller type, workpiece material and position of the workpiece zero point are inputs into the postprocessor.

Given the limited interpolation types of DIN 66025, the geometries presented so far in the CAM system as splines are divided into discrete straight line segments (G01) in the Postprocessor. The result of this is that programs are created that can be up to 500 MB in size for large components. Machining times longer than 100 h are not unusual. Also, tool corrections can no longer be performed by the machine itself because only information on movement at the Tool Center Point (TCP) assumed is available with the DIN 66025 NC program. When tools are worn, the user must rerun the Postprocessor or have several identically configured tools ready at once. Either alternative is very cost-intensive.

Developments undertaken in the field of freeform surface machining have given rise first to a definition of a standardized interface permitting efficient transfer of freeform surface descriptions, such as in the form of splines. Second, there are numerical controllers currently on the market that are able to read these formats and perform corresponding calculations directly in the controller.

The programming result of freeform surfaces can be seen in ◘ Fig. 11.47. It shows both the roughing (left) and the finished workpiece (right).

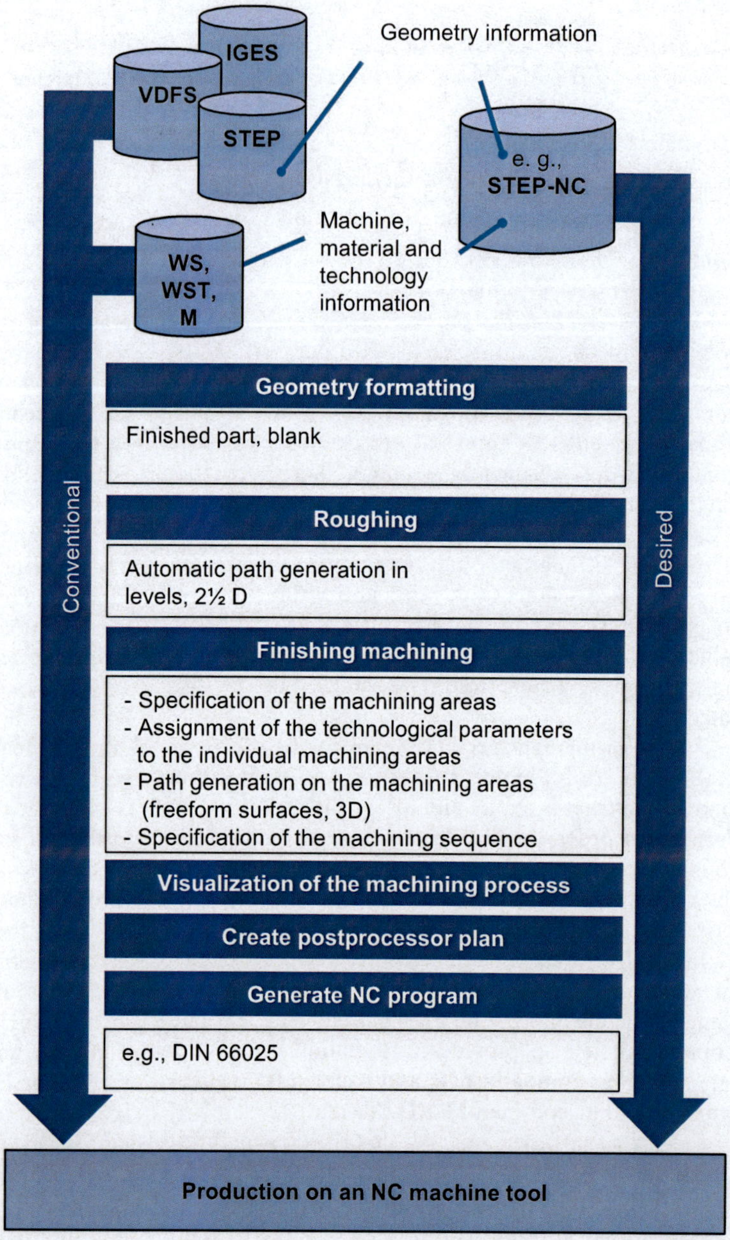


**Fig. 11.45** Work steps for NC programming of freeform surfaces

#### 11.4.2.2 Programming Example Using the EXAPT System

The exact process of programming for the different systems available on the market varies considerably in part and is dependent on the functionality included, applications, etc., of the respective products. But the different systems do have a series of common features that have come about due to defined requirements of users (see ▸ Sect. 11.4.2.4, workshop-oriented programming) and have been included in the developments of NC programming systems.

The EXAPT system is a graphical NC programming system with a broad application spectrum. It is used as an example below to outline the procedure for machine programming. The example is restricted to the turning machining type.

EXAPT is a graphically interactive, technology-oriented NC programming system that can be used as a standalone programming station or integrated within a CAD/CAP/CAM process chain using additional modules (Fig. 11.48). For NC planning and programming, a window-oriented base module is provided that

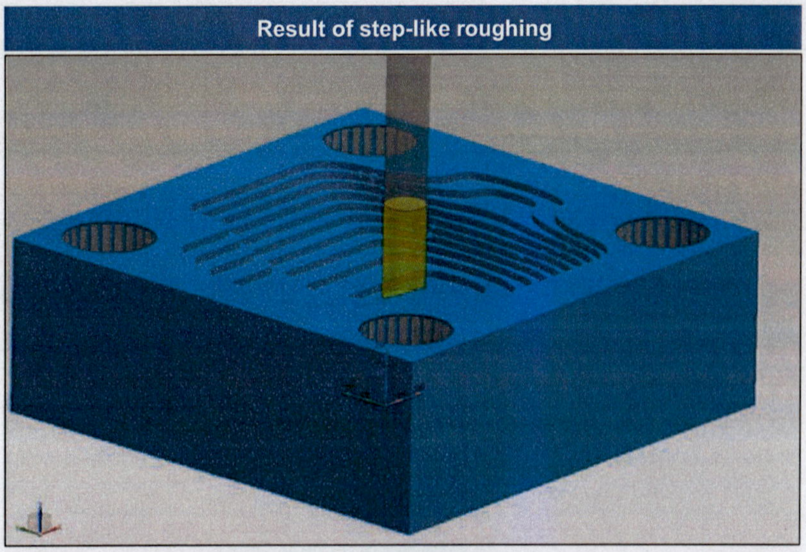


■ **Fig. 11.46** Illustration of a freeform surface component after roughing

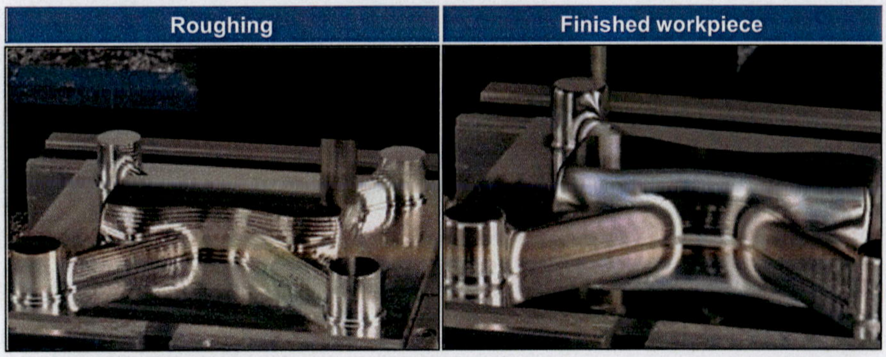


■ **Fig. 11.47** Workpiece with freeform surface. *Source* Datron

is upgradeable for the different production types. Also, it is possible to integrate technology databases and postprocessors tailored to the requirements at hand. Here are some of the features determining the programming environment:

- Identical programming with standard dialogs for all machining types
- Graphically interactive operation
- Graphical/dynamic simulation of the machining process
- Creation and optimization of programs in the same way
- Integration of tool and technology data
- Integration to CAD, ERP and shop floor control system
- DNC

Modules are integrated into this programming environment. These modules are aligned towards the different production processes (turning, milling, drilling, electrical discharge machining, oxyacetylene cutting, circular grinding, etc.). "EXAPTplus Turning" is one of these modules.

The first step in programming a turning process for the component shown in ■ Fig. 11.49 is generation of the blank or finished part contour. Both geometries can be created from scratch in the EXAPTplus system or be imported from a CAD system via interfaces. ■ Figure 11.50 shows the component geometry in the CAD module (window at bottom left) and the imported contour (window on right).

In the next step, the turrets of the machine tool are loaded with the tools present on the actual machine. The user opens the "Equipment management" module to gain access to a database and interactively selects the tools required for every work step. ■ Figure 11.51 shows the tool overview on the left. The turret assignment for 2-carriage machining is shown on the right.

There are essentially two different ways to actually program the tool paths or the machining process. One way is automatic generation of the motion sequences by the system itself, another is programming of the traverse instructions in single steps.

For the former, the user interactively selects the segment to machine and specifies the machining direction.

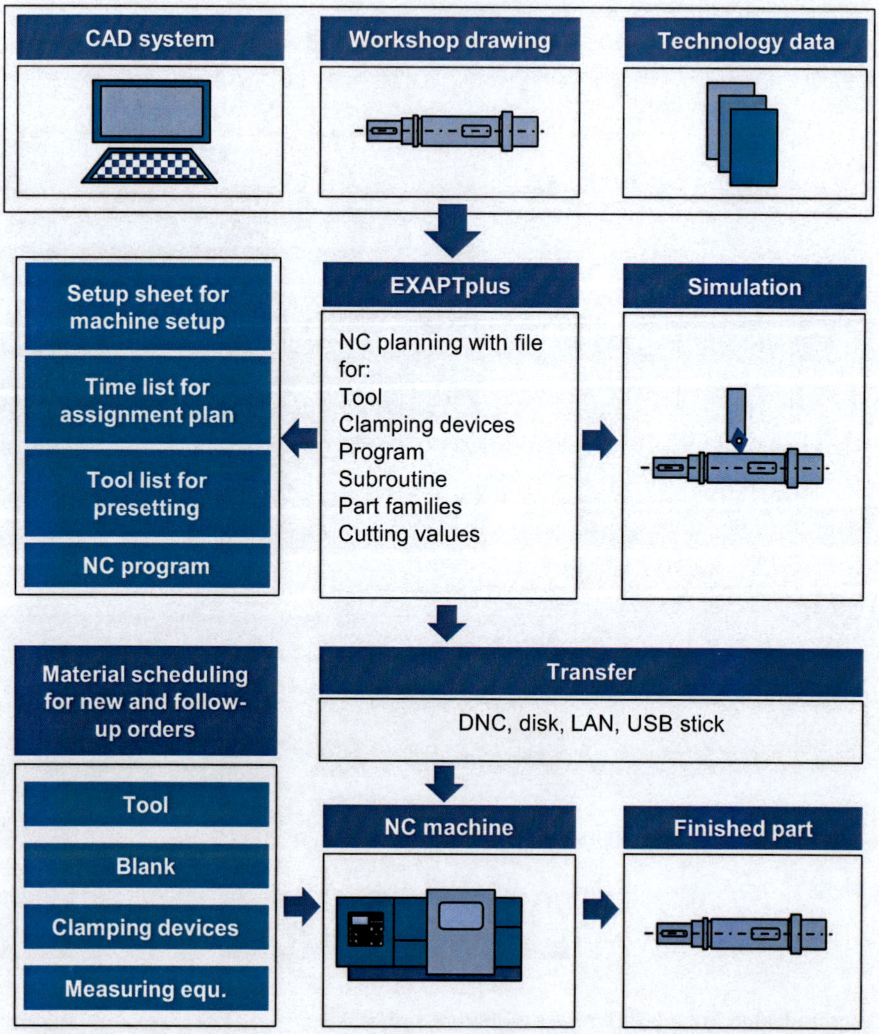


**Fig. 11.48** Flow of computer-aided NC programming using the EXAPTplus system, according to EXAPT



Specifying the machining direction defines the machining task. Infeed, approach and retract motions and cut distributions are then calculated automatically by the system. Figure 11.52 shows face turning of the blank as the first machining step. The quality of the machining operation (roughing here) is first set by selecting (5). The start and end points of this simple movement are denoted by (6).

Using single commands is an alternative to this programming aided by the system. It is step-by-step programming of the entire component contour. The possible machining variants are shown in Fig. 11.52 under (5). They can also be used for positioning the tool in the workspace.

Displayed in the bottom right of Fig. 11.52 is the time axis. It displays the current machining time in seconds.

If, instead of the standard settings mainly used, different technology data are required, the EXAPT module TDO (Technology Data Organization) from the EXAPTplus environment can be used. Figure 11.53 shows the transfer of technology data for the next machining step, "External turning". Listed in the TDO module are technological data for the different production processes, material pairings, machining parameters, etc. Users are able to take the required data from the database or add their own knowledge to this database and expand it as required.

In the programming process, the machining process can be checked by a simulation. Figure 11.54 shows the synchronous turning process with two carriages. The two tools are spatially offset when in use. The lower tool bit (carriage 1) orients itself to machining the upper tool bit (carriage 2) and is already removing the next layer of the component.

EXAPTplus creates a CLDATA file containing the results of programming for NC production. A postprocessor handles formatting of the data for the numerical controller and organization of accompany-

**Fig. 11.49** Turned part for the following example of graphically interactive programming. Source: EXAPT

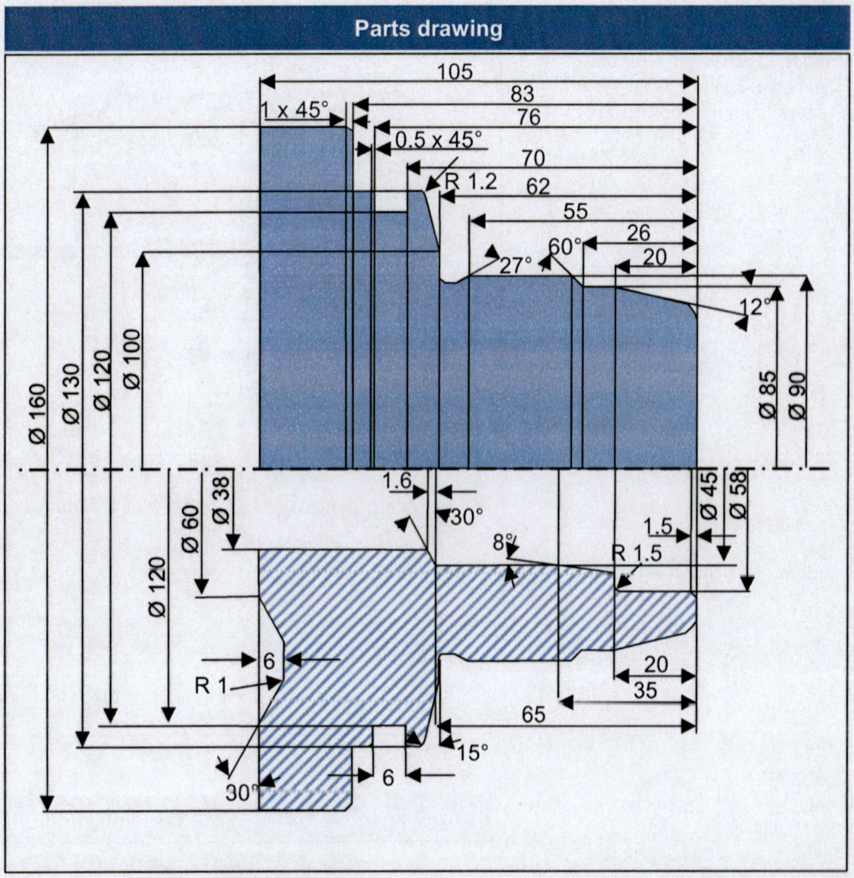


**Fig. 11.50** Geometry transfer: the geometry data created previously in a CAD system are imported using the EXAPTplus CAD coupling module. *Source* EXAPT

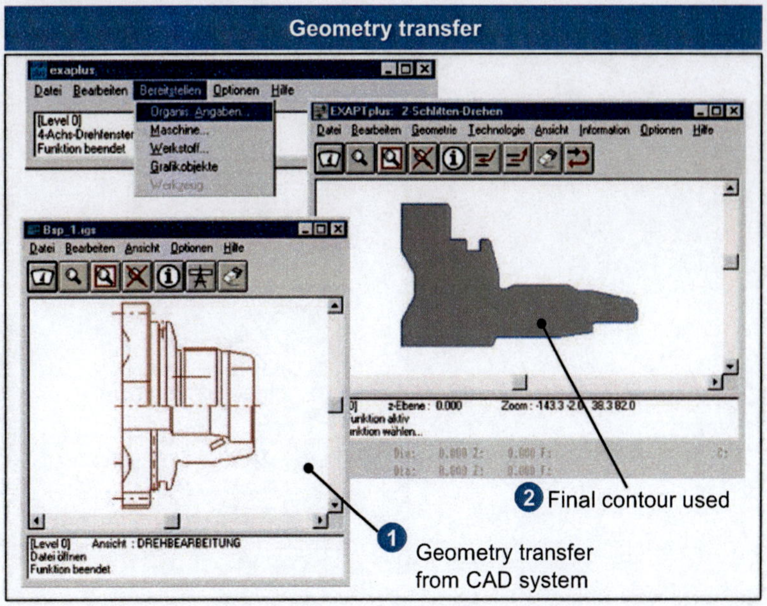


ing information (such as plain text for the control data, times list, setup information and tool table) in line with the organization and process.

Figures 11.55 and 11.56 show the NC program and the time sheet for machining. Machining is divided into its individual primary and secondary times on the time

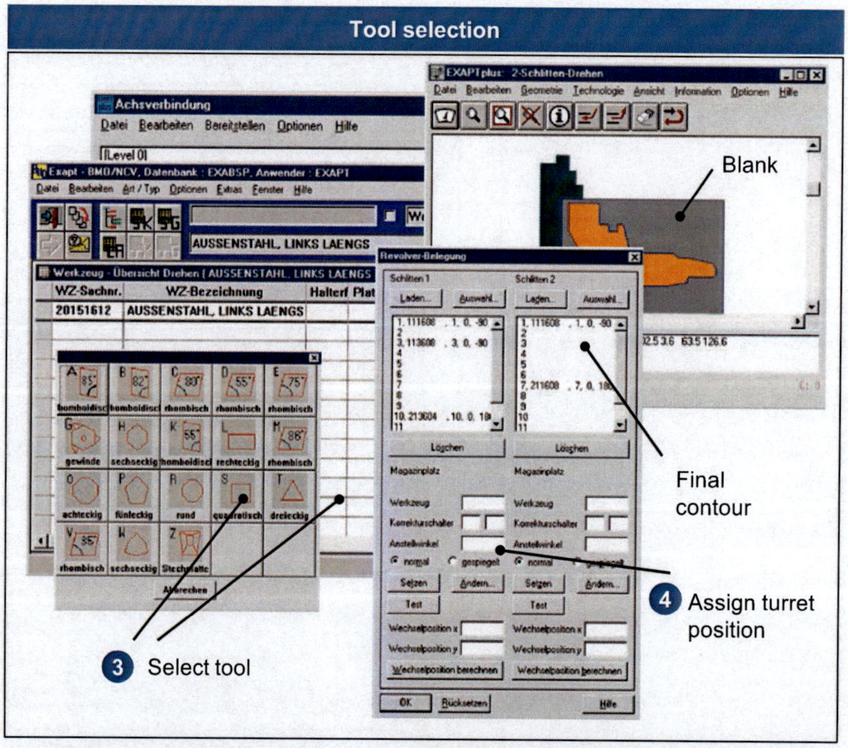


**Fig. 11.51** Tool selection: In the Equipment management menu, tools can be selected from a database (3) and be assigned to the individual turret positions (4)

11

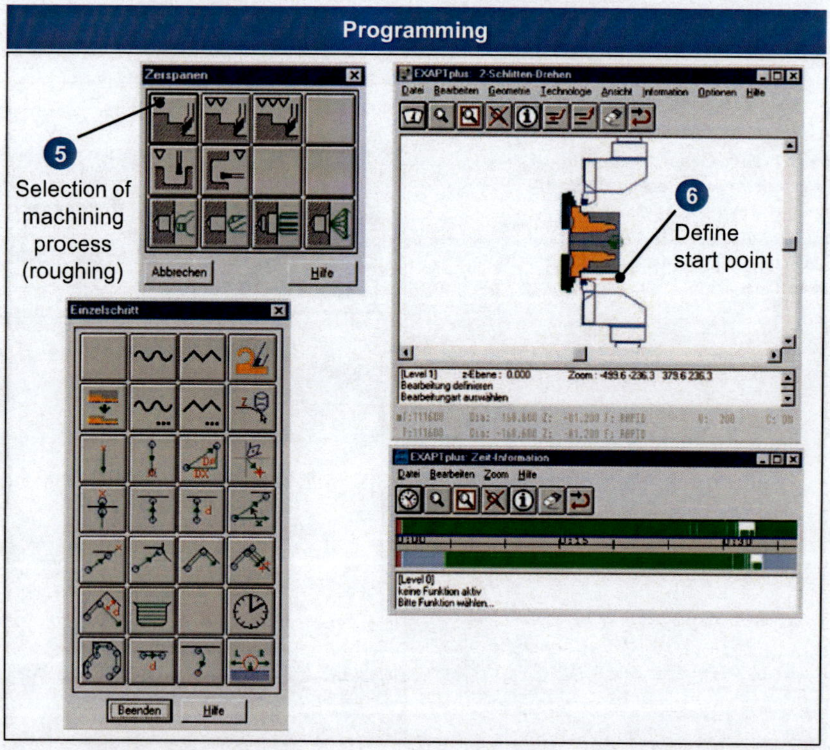


**Fig. 11.52** Programming: Once the machining type (roughing here) and relevant tool (5) are selected, the start and end points for machining are selected by clicking (6)

sheet. It contains information on any optimization possibilities for the program sequence. A tool sheet with a list of relevant correction values and the setup sheet shown in Fig. 11.57 are also generated. The setup sheet provides information on the machine type, clamping device used, clamping position and zero point offset performed.

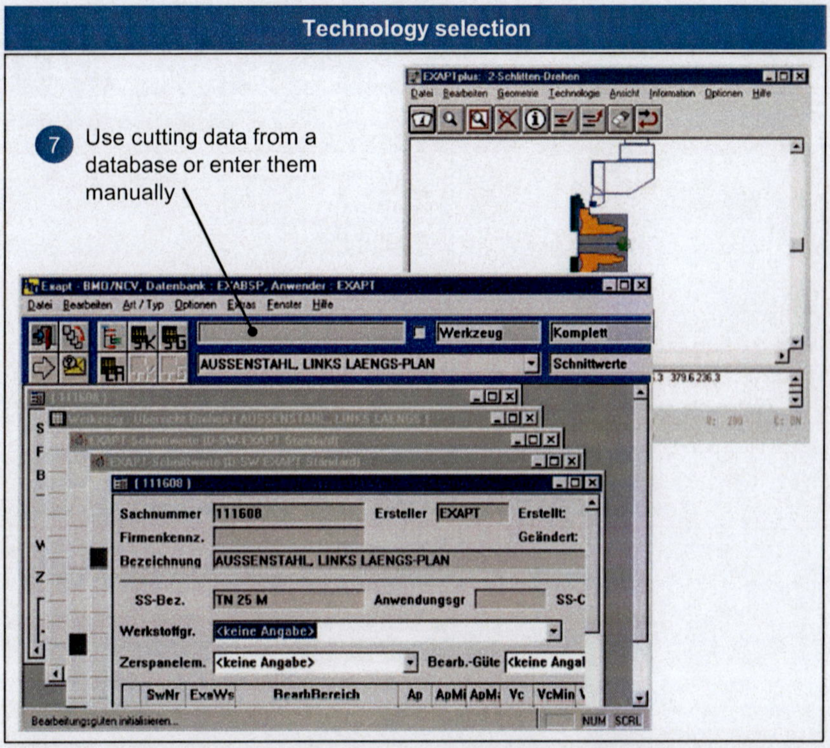


**Fig. 11.53** Technology selection: instead of using the default machining parameters, values can be taken from the TDO (Technology Data Organization)

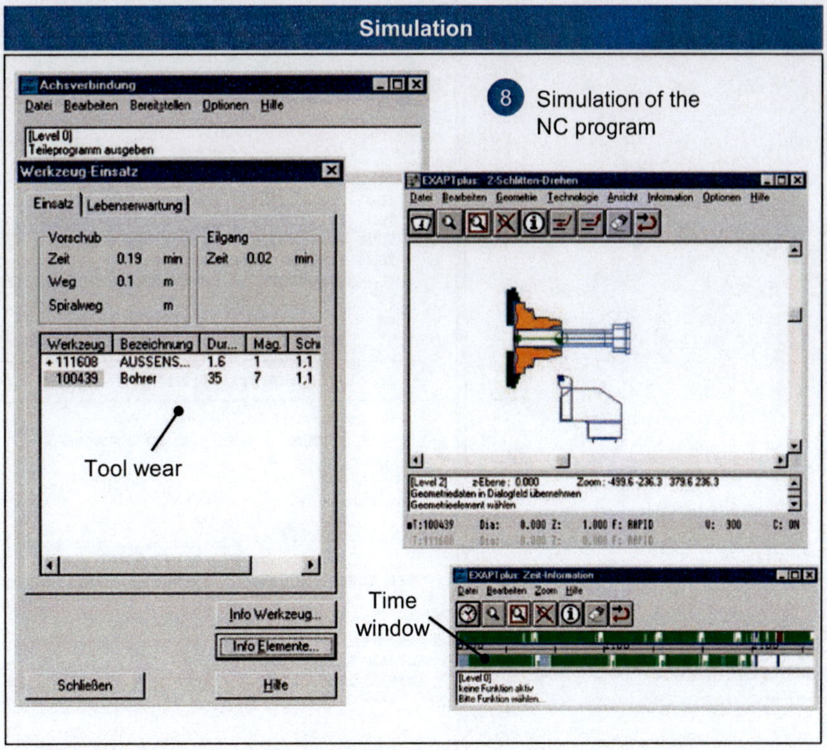


**Fig. 11.54** Simulation: After synchronizing the tool motions, the program sequence can be simulated. The time window provides information on the duration and sequence of the tool movements

In addition to the CLDATA file, there is also an EXAPT parts program keeping the results protocol of the NC programmer session. This parts program can be read as plain text and edited. This is so changes can also be made directly in the parts program. It is also the starting point for graphically interactive similarity

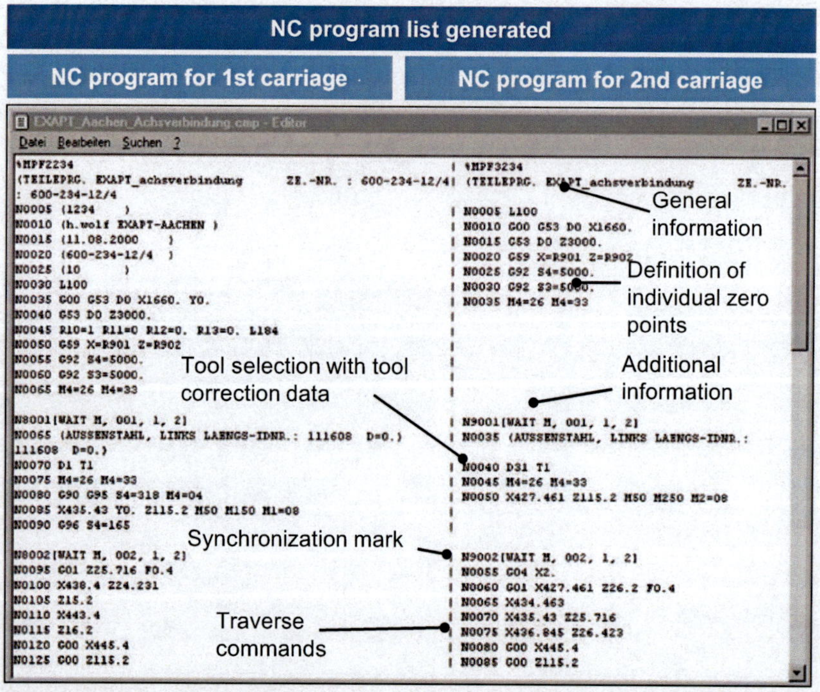


**Fig. 11.55** NC program list generated from the EXAPTplus session

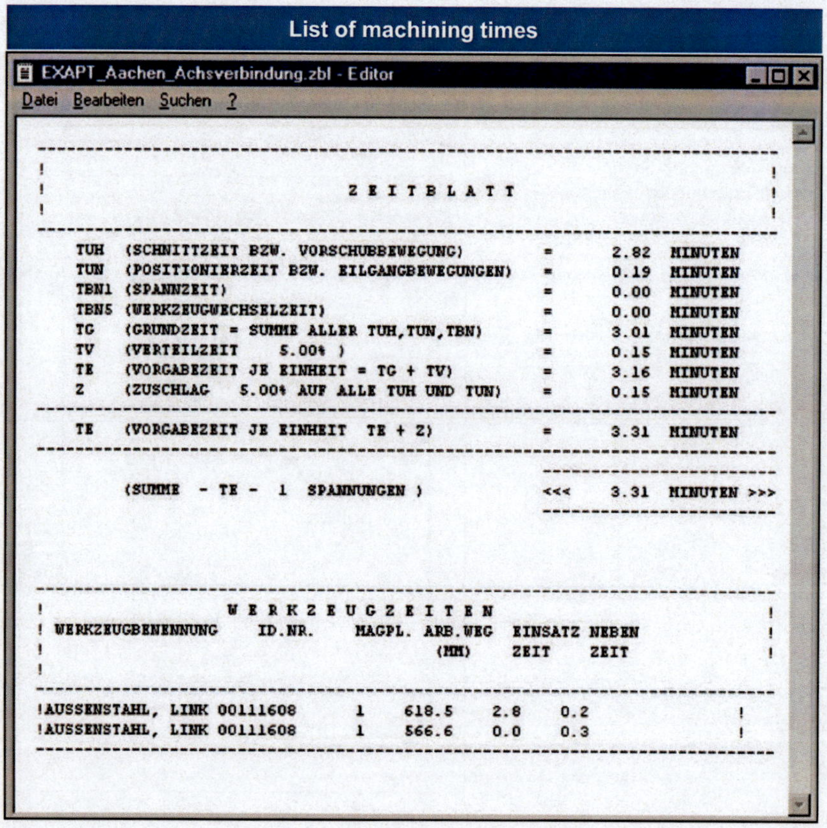


**Fig. 11.56** List of the different times for machining the programmed component in the first clamp

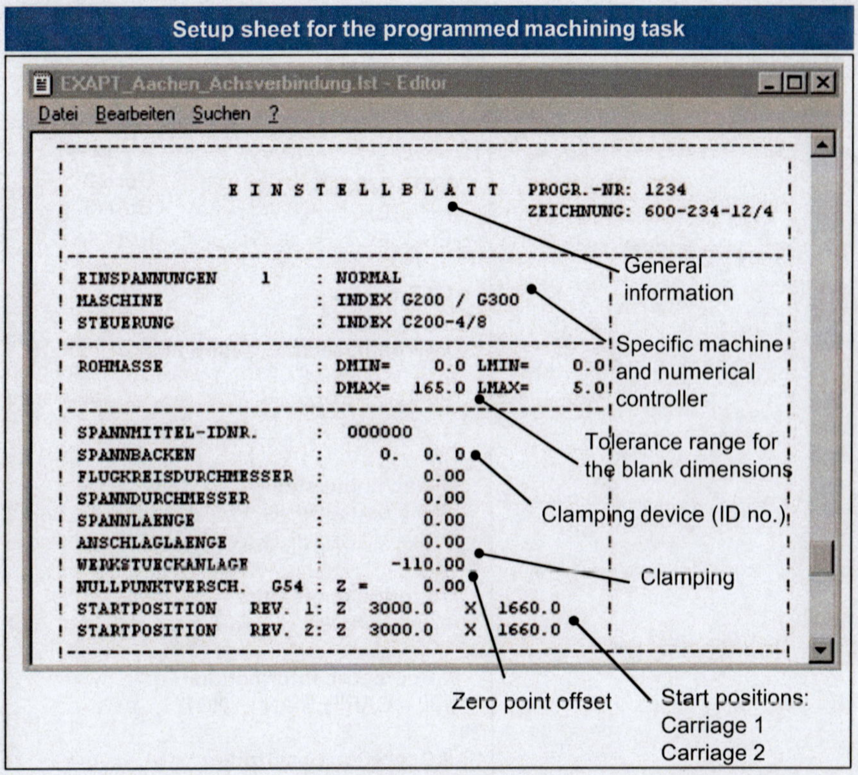


**Fig. 11.57** Setup sheet for the machining task programmed

programming. With this, programs for similar components can be created effectively on the basis of existing parts programs.

### 11.4.2.3 Programming Example for an Object-Oriented NC Program (STEP-NC)

ISO 14649 is a standard aiming to provide a standardized, end-to-end, object-oriented data format in planning and on the shop floor. This means geometry data available in a STEP database can be used directly in the NC program. Converting between different formats with the associated problems is eliminated, increasing data integrity.

Using a CAM system, the technological and organizational data are mapped to the ISO 14649 data model. The NC program is completed by combining geometrical with operative data in Workingsteps.

ISO 14649 is not an NC programming language that can be edited directly by the programmer. The data model is too complex for this. Part of such an object-oriented NC program is shown in ◘ Fig. 11.58. In light of the ongoing standardization activities, see ISO 14649 and ISO 10303-238 for further information [HARD13].

### 11.4.2.4 Workshop-Oriented NC Programming

The introduction of computer-aided systems in individual company divisions was previously been determined by division-specific tasks. This has given rise to standalone solutions, known today as CAD, CAQ, CAP, DNC and NC programming systems [HAAS93].

For the purposes of cost-effective production however, networking and the integration of individual systems is paramount. Rigid assignments of tasks presuppose inflexibility, that cannot be justified, especially in terms of ever smaller lot sizes. For this reason, in regard to more efficient NC program generation, efforts have been made to resolve the classic distribution of tasks between Design engineering, Production planning and Production. This means that programming, previously a central part of production planning, is being moved increasingly into design engineering and the workshop.

The trend back to the workshop and the accompanying upgrading of the workplace suits the specialist to a particular extent. This person broadens the nature of the tasks at hand from purely executing tasks to planning, control and management tasks[HEKE92]. The increased level of motivation of the specialist means in

**Fig. 11.58** Part of an NC program using 2.5D machining to ISO 14649 as an example

**Part of a STEP-NC program (ISO 14649)**

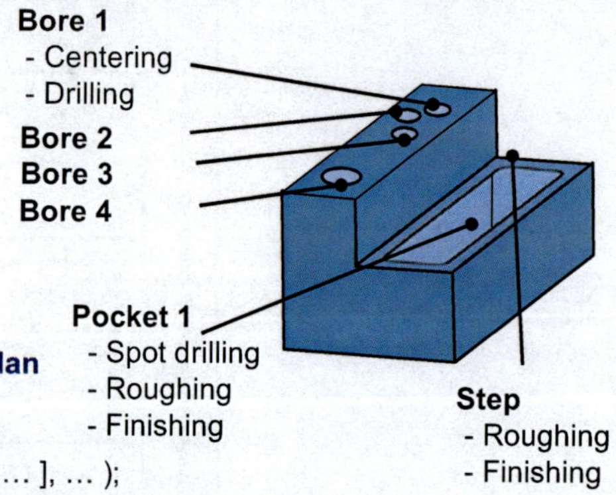


```
ISO-10303-21
HEADER;
ENDSEC;

DATA;

// Workpiece and routing plan
#1 = WORKPIECE( … );
#2 = MATERIAL( … );
#3 = WORKPLAN([#10,#11, … ], … );

// Machining steps
#10 = MACHINING_WORKINGSTEP('Bore1_Center',#20, … );
#11 = MACHINING_WORKINGSTEP('Pocket1_Spot_drilling',#20, … );
…
// Production feature
#20 = ROUND_HOLE( … , #1, #30, #40, … );
…
// Geometric information
#30 = CARTESIAN_POINT( … );
…
// Operation description
#40 = DRILLING( … , #50);
#41 = BOTTOM_AND_SIDE_ROUGH_MILLING( … )
…
// Tool data
#50 = CUTTING_TOOL(`spiral_drill_9mm', … );
```

turn that this person detects any problems, faults and malfunctions early—and that they can possibly be corrected quicker.

Programming on the shop floor or at the NC machine is essentially nothing new. Manual feed controllers have been in use for some time now. This is a numerical controller with an integrated, graphically interactive programming system. Familiar functions are supported including simple geometry entry for blanks and finished parts, cut distributions, collision monitoring and simulations. Being designed for a special machine type means, however, they are isolated from other systems—a downside of using manual feed controllers. When different machine tools are used, controller-specific NC programs, for example, can no longer be tailored to other machines. Also, every new manual feed controller requires the specialist to undergo training on the system.

Efforts have been made to develop a programming concept integrated into the production cycle that sidesteps the downsides of existing programming systems and is based on the requirements of specialist-oriented programming.

For the practical realization of these ideas, the "Workshop-oriented programming procedure (WOP)" joint project was launched in 1984. German machine and controller manufacturers, various software and system houses, university departments and users were involved in this project [HEKE92]. The aim of the project was, first of all, realization of a programming concept based on the following:

- a programming method based on the metal-cutting manufacturing processes, with standardized dialogs
- graphically interactive input without a programming language
- a graphic simulation of the machining process
- optimizing/changing programs using the same method as reprogramming
- effective use of a specialist for numerically controlled production equipment
- module for managing and transferring data for tools, clamping devices and programs
- standardized system for workshop and production planning

When these criteria are compared to the functionality of today's programming systems, it can be seen that a large number of providers have already integrated at least a subset of the points into their systems. The term "WOP" does not have any sort of protection, meaning it can be used arbitrarily. This is problematic in this context. So, the different "WOP systems" can sometimes differ greatly.

The CNC ShopMill and CNC ShopTurn systems from Siemens are WOP systems and discussed in more detail below. An array of functions, such as graphically interactive programming, simulation of the machining process and standardization of program creation in production planning and the workshop, are supported by many systems nowadays. Separate details on the technological information, the machining process and the geometry are required in the system. On the basis of these details, the machine is able to run the process directly. The process on the Siemens ShopTurn editor is detailed below. ◘ Figure 11.59 shows the specification of the blank geometry in line with the blank drawing using definition of line segments and arcs.

After specification of the blank geometry comes programming of the finished part contour as per the parts drawing in the same way as creation of the blank. For this, either standardized cycles such as contour turning or boring are used, or corresponding contours are specified (2)—Fig. 11.60. Once a new contour has been created and the start point defined, the individual contour elements that make up the contour can be defined (◘ Fig. 11.62). If no values are entered in some of the fields, the contour calculator assumes these values are unknown and attempts to calculate them from other parameters.

After planning the machining operations and specifying the routing plan in step (3), the complete machining operation is displayed using a simulation function (4) - ◘ Fig. 11.61. Accordingly, a similar input process with corresponding graphical visualization is provided for ShopMill.

Taking into account the Design engineering department in addition to Production planning and the shop floor gives the following option of including this department into the workshop-oriented CAD/CAP/CAM group. Geometry information (in DXF or DWG format, for example) is loaded into the Shop programming system using a neutral interface. The information is modified here for subsequent processing (◘ Fig. 11.62).

The tool management systems, the zero point offset or preset table and the R-parameter list of the controller are linked directly to the editors besides the general information on machining, geometry and technology.

The functionality described should serve as an example to illustrate the way a specialist-oriented WOP system can be realized. Examining the high number of WOP systems in existence today for the original criteria, it is clear that a large number of requirements have actually been met. How effective these implementations are only manifests itself in how flexibly such systems respond in practice to unforeseen modifications and

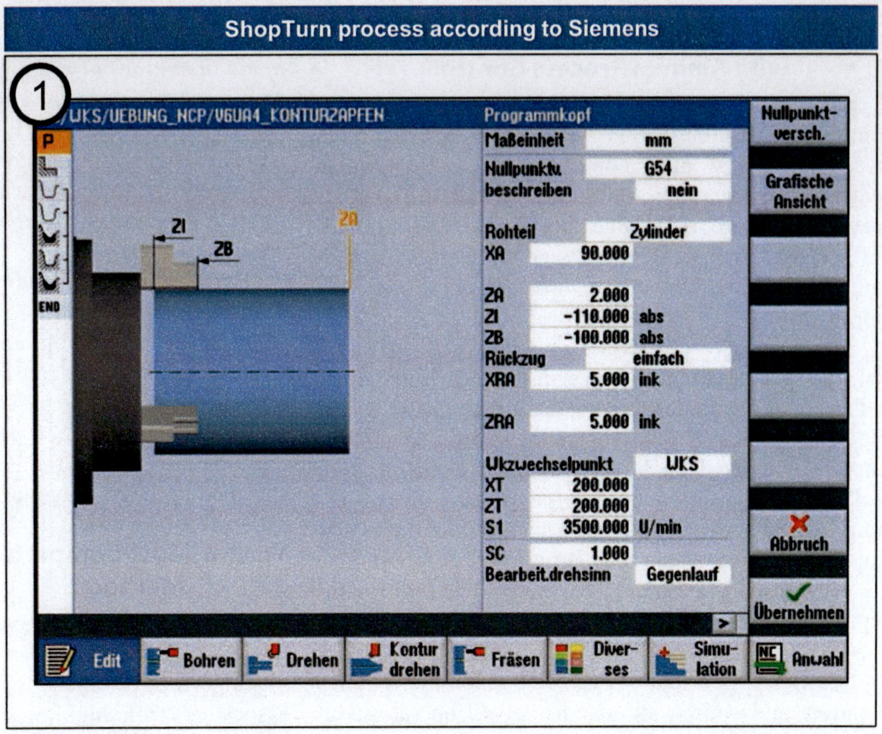


◘ **Fig. 11.59** ShopTurn process (entry of blank geometry), according to Siemens

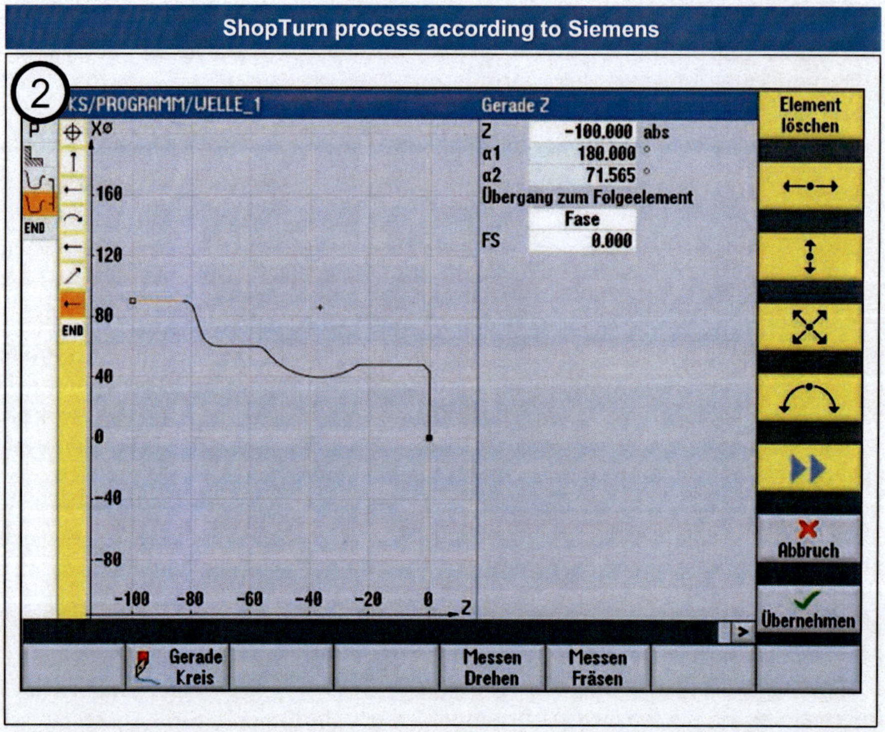


**Fig. 11.60** ShopTurn process (entry of finished part contour), according to Siemens

faults. That is to say, speedy intervention is only possible when there is a high level of integration of all systems involved in the production process, and when, first and foremost, the specialist at the machine is able to contribute to minimizing the effects with the appropriate options.

#### 11.4.2.5 Workshop-Level Programming with Manual Process Control

A key downside of the workshop-oriented programming methods described is the fact that a division between technology and geometry is a given for informatics-oriented programming. In contrast, the past has seen different programming concepts and, as a methodical enhancement, novel machine concepts that are tailored to the requirements of workshop-level production.

The purpose of these concepts in general is to assist to a specific extent the specialist work with NC machine tools. Characteristic for this requirement is that the knowledge of technological relationships in production is aided in a targeted manner by using controller software. This means the process-related actions of the specialist at the machine are paramount, not the creation of the geometric parts program. The key differentiating features of the programming concepts are illustrated by means of the following examples [HEKE92; KROM94].

Creation of a parts program generally requires planning and coding, in advance, of all operations in the sequence of the machining steps to be performed. Also associated with this conventional procedure is that the coded geometric motion sequences are specified before machining starts. On the basis of the system introduced in [KELL08] that has a special geometry model, it is possible to change the parts program dynamically at runtime. This means a change can be made between different operating modes (such as manual and automatic operation) at program runtime. The manual machining steps performed are recorded equally by the controller software and, if required, included in calculating the remaining material of the workpiece blank. In the event of a new switch to Automatic mode, the operations performed manually in regard to program cycle are stored.

Furthermore, flexible organization of the machining sequence of individual program and work steps is possible. This enables program modules to be started individually as parts of a routing plan or for them to be interchanged within the sequence with low overhead. The selection of a simulation parallel to or prior to machining aids in checking the geometric traverse commands specified in the routing plan.

#### 11.4.2.6 Cost Comparison of Programming Methods

It must be possible to measure programming methods using the same criteria as for numerical controller or machine tool products, for example. Complex systems associated with high procurement costs that require

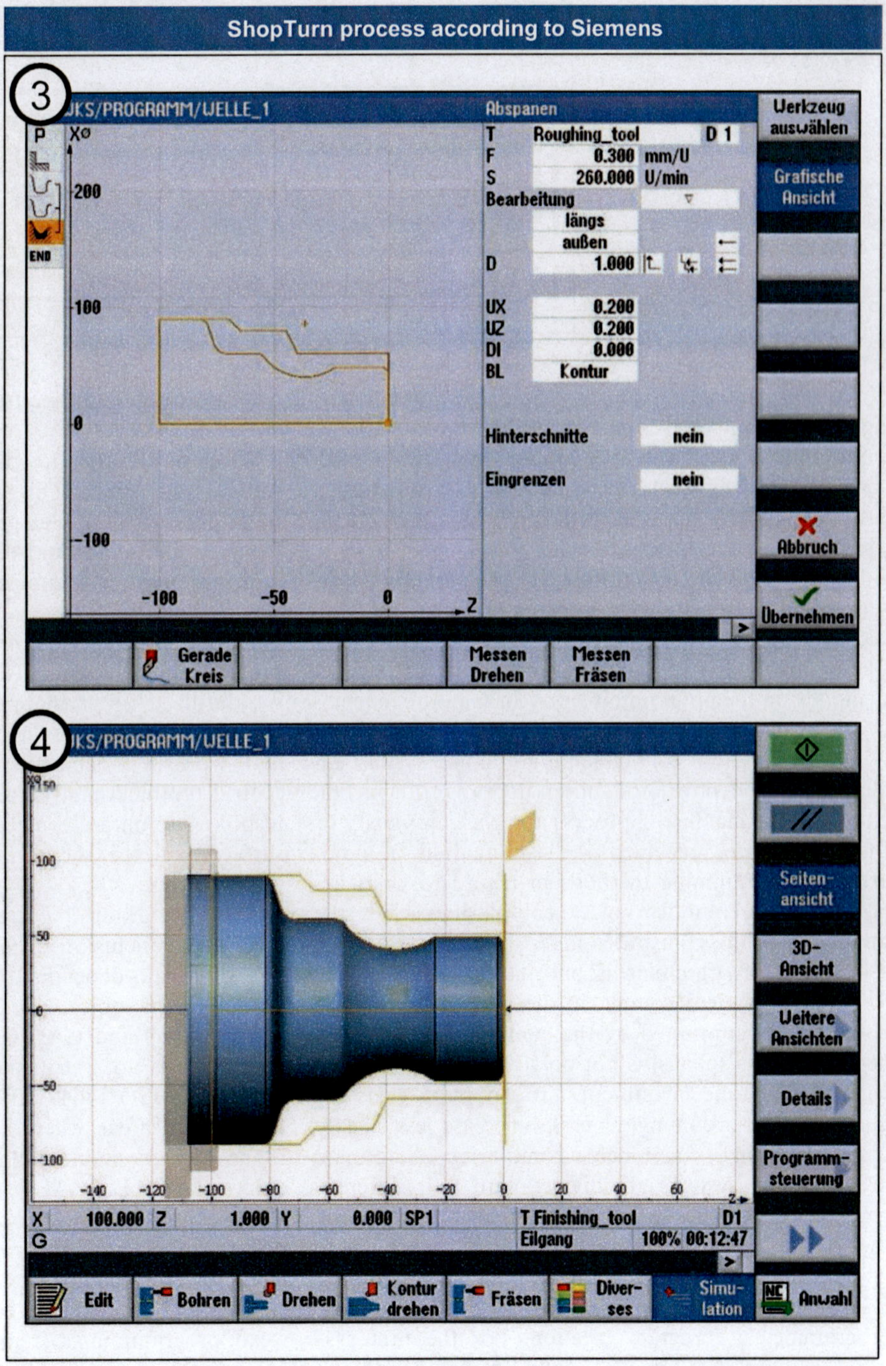


**Fig. 11.61** ShopTurn process (planning and simulation of the machining operations), according to Siemens

extensive training of operating personnel are only worthwhile when they yield correspondingly high financial savings or qualitative improvements in the generation of NC programs.

The costs for programming numerically controlled machine tools play a considerable role, particularly in the production of low volumes. The cost-effective use of NC machines often depends on the programming costs (related to a single workpiece). This means the programming method used is crucial in the cost listing. However, considering the large number of factors influencing the selection of a suitable programming system, it is not possible to tell which programming method is the most favorable overall.

In addition to labor and non-wage labor costs, the workplace costs (machine hourly rate of the system including equipment) for the programming workplace are generally used as inputs for determining programming

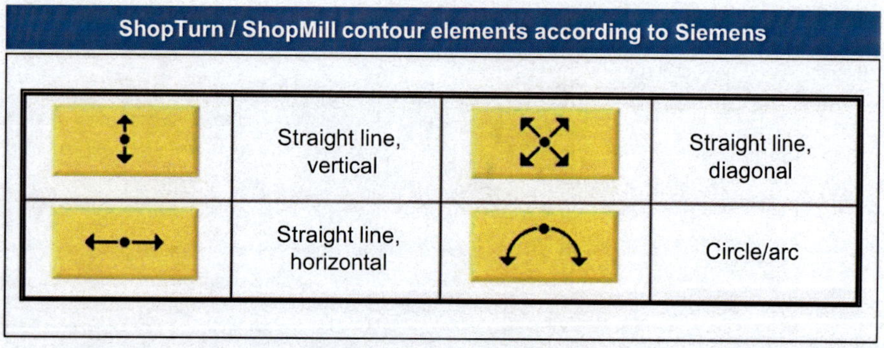


**Fig. 11.62** ShopTurn/ShopMill contour elements, according to Siemens

costs. For manual programming however, they are negligibly low compared to the workplace costs for computer-aided methods that essentially comprise hardware and software costs. These additional costs must be offset by higher utilization of the programming systems used.

The costs for an automatic programming system can differ greatly. They are determined predominantly by the machining types to program. For example, 5-axis milling of freeform surfaces makes stricter demands on the programming system than the machining of turned parts.

The selection of a suitable programming system, including cost-effective production, depends to a pivotal extent on the machining process and the complexity of the components machined. A basic analysis of modern-day programming methods in regard to costs accrued for different workpiece complexities gives the qualitative profiles illustrated in Fig. 11.63.

Manual programming is only used cost-effectively for low component complexity because programming contours not supported by the command set of DIN 66025 (such as splines) is not possible or only possible with a great time investment. Programming costs for automatic programming generally increase less steeply with increasing component complexity. Automatic programming systems are divided into four different groups based on the functionality and integration of the systems into a system cluster.

The first two groups represent NC programming systems that differ in terms of system support for the programmer in program creation. For the first system (2), the user is forced to enter, in dialog with the system, all information defining the production process. Aids such as automatic tool selection, technology data determination and automatic generation of the machining sequence are not provided. The second group (3) of programming systems has these functions. Here, only the entry of blank and finished part geometries and some general information (target machine, tools available, workpiece material, etc.) are necessary to generate the program largely automatically, checked by the user. This fully automatic support in this scope has so far only been realized for simple machining processes, however.

Group (4) represents the systems that are integrated into a system cluster in terms of a continuous computer concept. Complex geometry data can be accessed from CAD systems and be processed. Using geometry information already captured as data allows lower-cost production.

The last group (5) contains the workshop-oriented programming systems. The benefits mentioned for a standardized programming system aligned to specialists for the Design engineering, Production planning and Workshop departments, and the termination of central production structures enable quick and flexible responses to unforeseen events. Machining complex 3D surfaces with this is currently only possible with limitations, however.

Another group is currently being tested in the industry. A conclusion on costs is fictitious at the current stage and dependent on the successful implementation of new, object-oriented programming interfaces in controllers and CAD/CAM systems. Consequently, this group is not yet included in Fig. 11.63. An interface like ISO 14649 offers saving potential first and foremost in areas where complex NC programs need to be created and exchanged between different machines and CAD/CAM systems. As the data model can be read directly by all systems, there is no overhead for postprocessors and NC program management is simplified. However, the format can no longer be created and changed by the machine operator in a text editor. Instead, programming systems are necessary. Whether the investment is worthwhile for smaller companies that make less complex parts and for whom no rescheduling overhead is generally involved will depend on the costs of such systems.

### 11.4.3 Digitization of Workpiece Geometries for NC Data Generation

One way to generate NC data is to derive path information from mechanical storage (template, master piece,

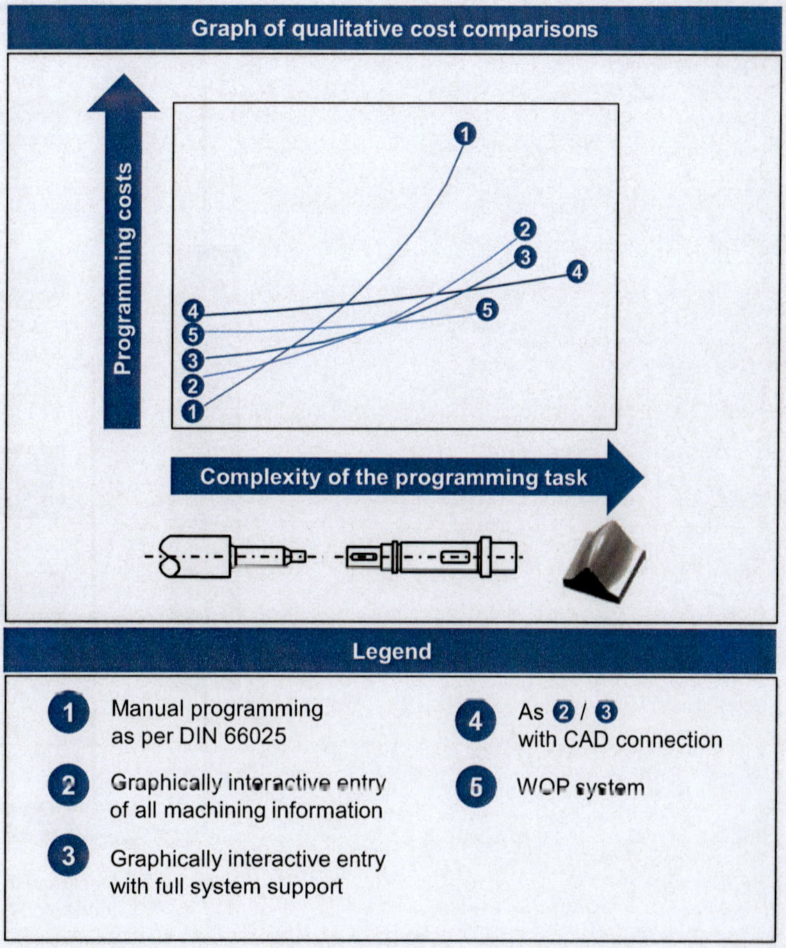


**Fig. 11.63** Qualitative cost comparison of different programming methods shown graphically

model) using "postforming systems". The development of such systems started back in the eighteenth century with the first, still hand-operated postforming machines. After the turn of the century, automatically operating servo postforming systems were developed that were used in niche areas. These are still being used today with a modified design, in mold and die production. The importance of these devices has declined considerably through the use of NC machine tools in conjunction with CAD/CAP systems and they are no longer used in new designs.

Characteristic for classic, directly working postforming systems is that the sampling and machining processes are linked to each other spatially as well as timewise. This means a workpiece is machined simultaneously with the sampling process, the sampling and machining systems normally representing one cohesive unit. Such a postforming system exhibits several downsides. The direct coupling between sampling and machining systems means there is hardly any way to alter sensor data to make changes to dimensions or smoothing surfaces. Also, machine vibrations caused by the machining process can have a negative effect on reading of the geometry by the probes. Furthermore, the storage costs for the sampled models must not be underestimated. The models must be kept as long as there is a necessity for re-machining or reworking a part. Periods of 10 years and longer are not uncommon.

In light of these downsides, postforming systems have been developed for which the sampling and machining processes are fully separated spatially and in terms of time [WOLL86] Fig. 11.64 shows the basic design of such a system. It essentially comprises a sampling unit, a digitization and data compression unit and an NC machine tool. The differences between systems available today are in the measuring devices and sampling systems used and in terms of computational processing and conversion into NC blocks of the geometry information obtained.

### 11.4.3.1 Measuring Devices for Digitizing Workpieces

In practical industrial applications, devices and measuring methods from the field of coordinate measuring technology have become established as extremely flexible and efficient tools for the geometric testing of

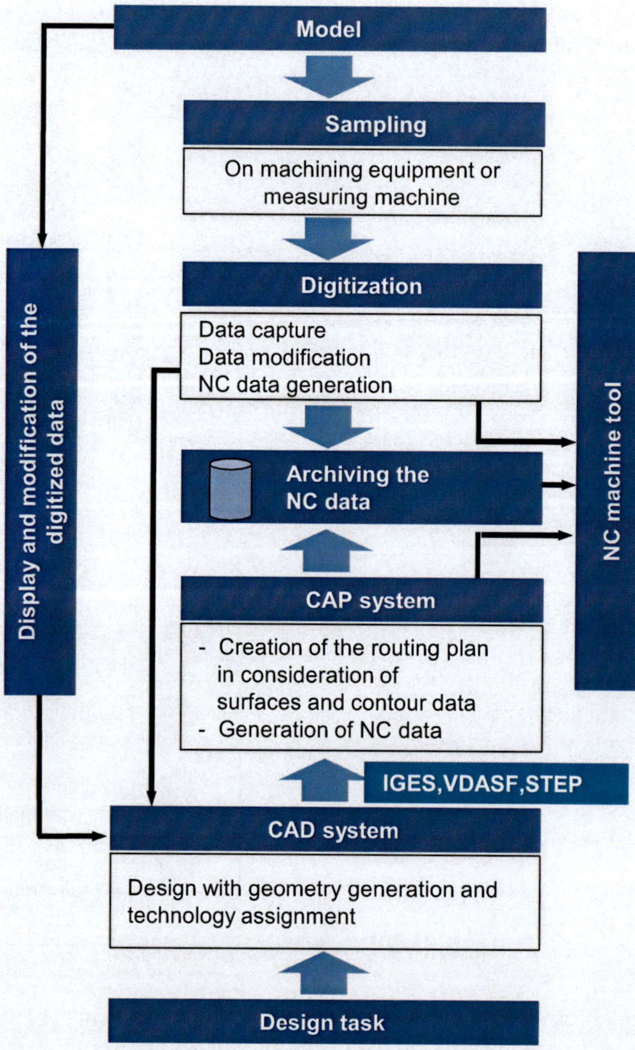


**Fig. 11.64** Design of an NC postforming system with CAD/CAP/CAM coupling



workpieces. In coordinate measuring technology, object geometries are defined by the mathematical linking of individual surface points in a common coordinate system. These points in space are measured using coordinate measuring devices [PFEI98]. The main application area of coordinate measuring devices is the reading of actual geometries on the part geometries relevant for the function of a component (bores, flanges and similar) to determine dimension, shape and position deviations by comparing them to the reference geometry. However, virtually all coordinate measuring devices available today are also suitable for digitization, i.e., for the full or partial measurement of object geometries to generate a digital image.

Figure 11.65 shows a dual-column coordinate measuring device for measuring large objects. A precision, Cartesian guide feed system and a tactile sampling system are the main parts. Other sampling systems can also be exchanged automatically depending on the task. This type is built for sampling areas up to 25 $m^3$ and larger. This allows full digitization of models the size of automobiles.

An array of specially designed samplers has been developed for low-cost digitization. Figure 11.66 shows an example of a 3-axis coordinate measuring device with tactile sampling system for workpieces up to 0.1 $m^3$ in size. The reduced dimensions means high scan speeds (to 140 measurement points/second) combined with high accuracy (<50 μm) are possible. The attainable, very low contact forces permit the use of very delicate sampling systems so that fine details can also be resolved.

Figure 11.67 shows another variant for fast, contactless digitization of very small, freely formed models, such as for dental applications. It is a contactless dental scanner. Using a highly accurate optoelectronic sensor can provide precision better than 10 μm.

Besides coordinate measuring devices, machine tools can also be used for digitizing workpieces. The

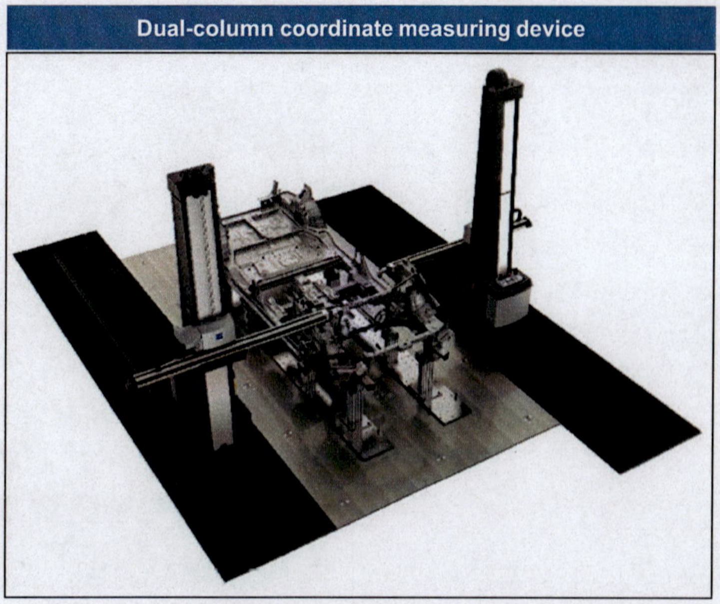


■ **Fig. 11.65** Dual-column coordinate measuring device for large workpieces. *Source* Zeiss

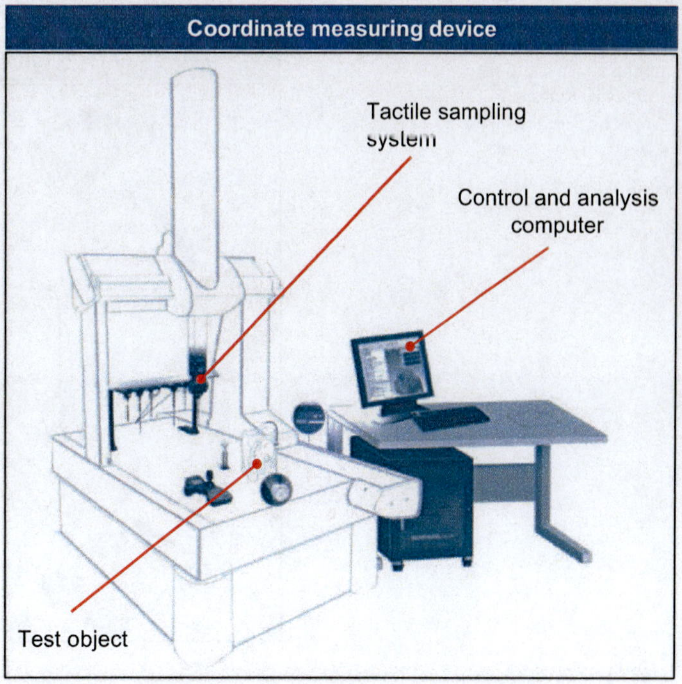


■ **Fig. 11.66** Coordinate measuring device for measuring smaller objects. *Source* Renishaw

use of existing machine tools and the associated reduced investment costs mean that digitization systems are also affordable for small and medium-sized businesses [BART99]. An example of such a system is in ■ Fig. 11.68. It essentially comprises a tactile sampling system that is connected to the standard tool holder of the machine tool by an appropriate adapter and to a computer and special software for controlling the measurement processes and for processing the measurement data. In subsequent upgrade steps, the system can be integrated more closely into the machine controller to achieve performance increases of the measurement processes.

#### 11.4.3.2 Sampling Strategies

For ruled geometries, the complete description of an object geometry is usually possible with few spatial points and the associated mathematical descriptions. For simple bodies in particular, digitization with manual probing of single points is beneficial. For more

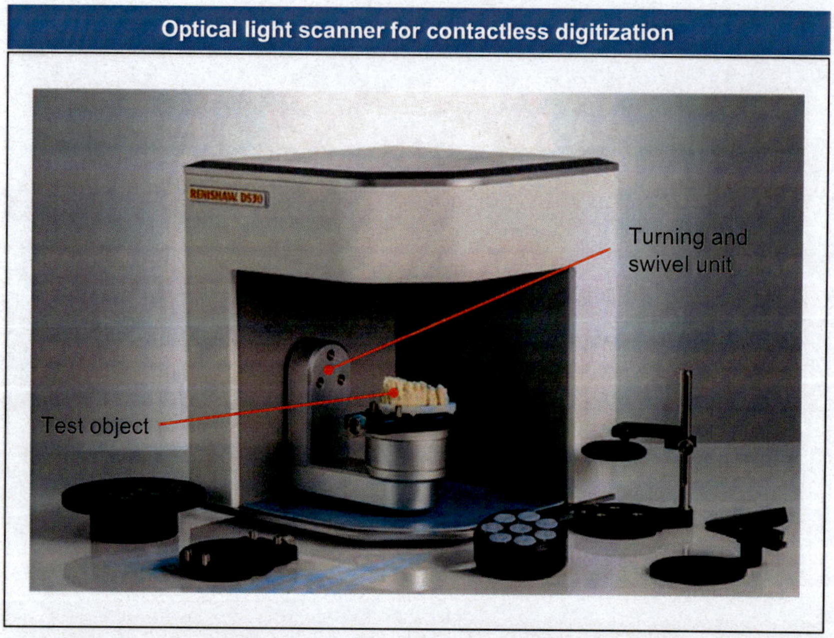


**Fig. 11.67** Optical coordinate measuring device for digitizing very small objects. *Source* Renishaw

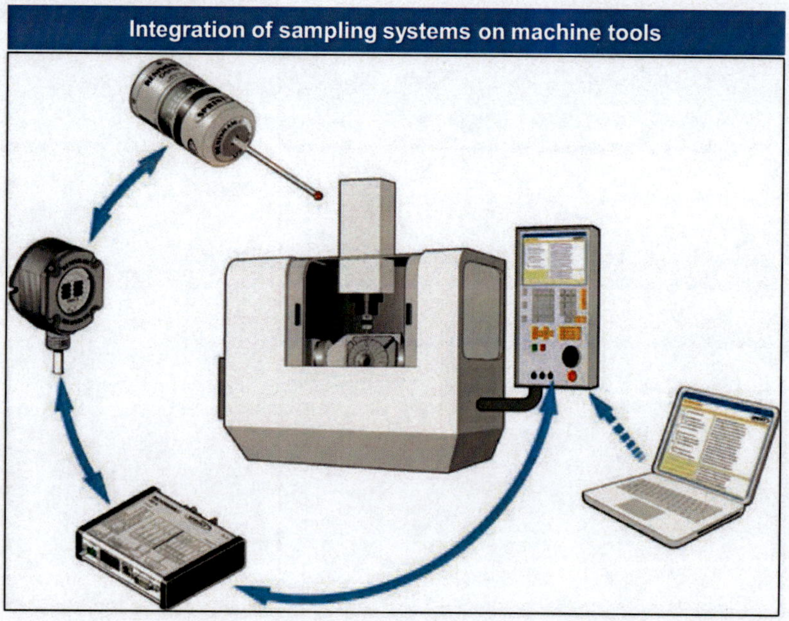


**Fig. 11.68** Integration of sampling systems into machine tools. *Source* Renishaw

11

complex geometries, in particular to describe free-form surfaces, considerably greater data volumes are required. The measurement points are, depending on the sampling system used, normally captured in scanning mode, i.e., with continuous measurement of spatial points along the workpiece geometry. Of key importance for the quality of digitization results is the use of a suitable sampling strategy. This must be selected in line with the measuring instrument and sampling system used, the available workpiece geometry, and/or depending on the accuracy to be attained, detailing, sampling duration, etc. Usual strategies here are linear sampling, sampling in sections in different directions and contour sampling (Fig. 11.69).

An alternative to scanning mode is digitization using iterative approximation of the object surface with single-point probing. The first step, usually with manual probing of individual boundary points, defines an initial geometry. Then, in an iterative process in line with a specified strategy, additionally single points are probed automatically, and they are added to the geometry description as new interpolation points if a

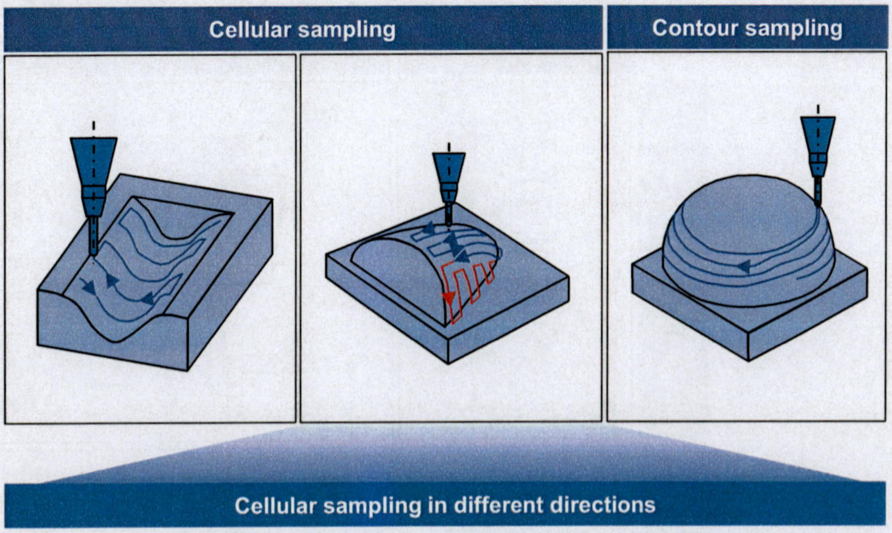


**Fig. 11.69** Examples of strategies for sampling different model shapes

**Fig. 11.70** Segmentation of an object geometry. *Source* Zeiss

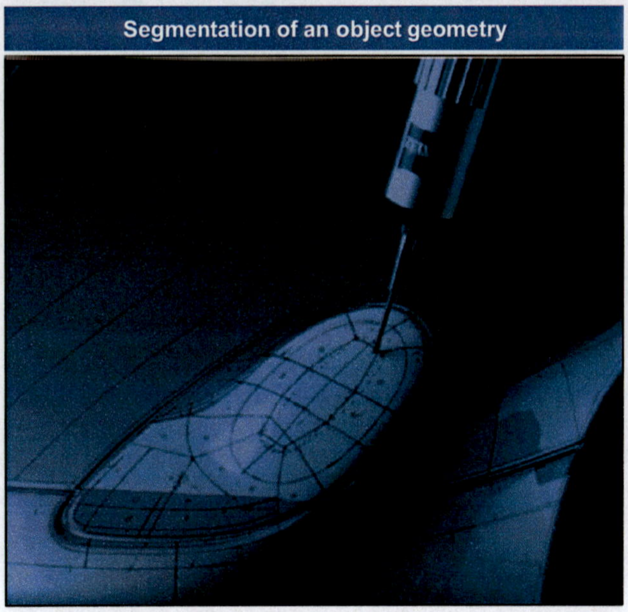


minimum geometry deviation is exceeded. This process is ended as soon as a specified minimum deviation is no longer exceeded by adding additional points.

To measure complex surfaces in full, segmentation of the object geometry into individual partial surfaces is commonly required. These partial surfaces are marked on the object and reported to the sampling controller using manual sampling of the boundary curves (Fig. 11.70). After this subdivision, the aforementioned sampling strategies can be applied to the individual segments.

#### 11.4.3.3 Sampling Systems

The sampling systems used in digitization systems for recording measurement points can essentially be divided into contact (tactile) and contactless (usually optoelectronic) systems.

##### Tactile Sampling Systems

Tactile sampling systems generally comprise a probe head, one or more styluses and a ruby ball for every stylus. The ruby ball is used to establish contact with the workpiece surface. A differentiation can

**Fig. 11.71** Tactile sampling systems. *Source* [PFEI98]

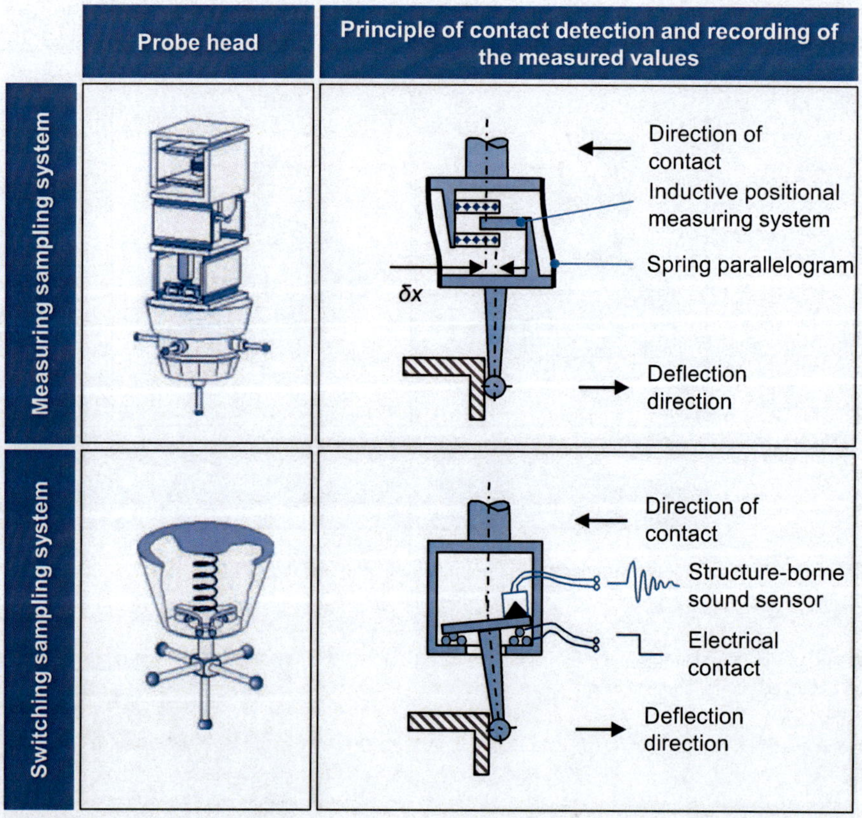




be made between switched and measuring variants. Figure 11.71 shows their basic designs.

Switched sampling systems have a kink. This is used to deflect the stylus when the workpiece is touched (Fig. 11.71, bottom). Deflecting the stylus triggers a switching signal. In turn, recording of the measured value reads the displacement transducers of the measuring device. When the stylus is deflected, the traverse axes of the measuring device are stopped by the controller and then moved back in the direction opposite to the direction of contact. So, with switched systems, only single-point probing is possible, albeit with very high accuracies. The slow speed of the process and the resulting small number of measurement points measurable in an acceptable time mean it is mainly suited to the digitization of ruled geometries that can be described mathematically using few points.

Figure 11.71 (top) shows the design of a measuring tactile sampling system. The stylus holder here is affixed at three spring parallelograms positioned at right angles to each other. The deflection of the holder due to a stylus deflection is measured by an integrated one-dimensional displacement transducer. Measuring sampling systems allow, in conjunction with a corresponding postforming controller, the stylus to maintain contact with the surface during digitization. The measurement point is then calculated from the measured deflection of the stylus and from the position of the machine axes using the position values of the displacement transducers. The continual mode of the process yields far shorter measuring times than for switched systems but with the same high accuracies.

### Optoelectronic Sampling Systems

Optoelectronic sampling systems are also used for digitizing workpiece geometries. Its contactless mode of operation means it is also suited particularly well to the quick digitization of complex geometries and soft/sensitive materials. Figures 11.72 and 11.73 show the method used in this context.

Their simple physical design and variably configurable measurement range mean triangulation sensors that measure pointwise are well suited to the optical sampling of bodies for digitization. This is why their use is becoming increasingly common in industry. To measure the distance between the sensor and workpiece, a laser diode emits a light pulse (Fig. 11.72, top left). This hits the workpiece and is reflected diffusely, totally or directed depending on the surface. The light energy reflected by the object is mapped onto a receiver element, usually a charge coupled device line or position-sensitive diode (PSD) by a lens positioned at a defined angle $\gamma$, the triangulation angle, with respect to the transmit beam. The position of the light point

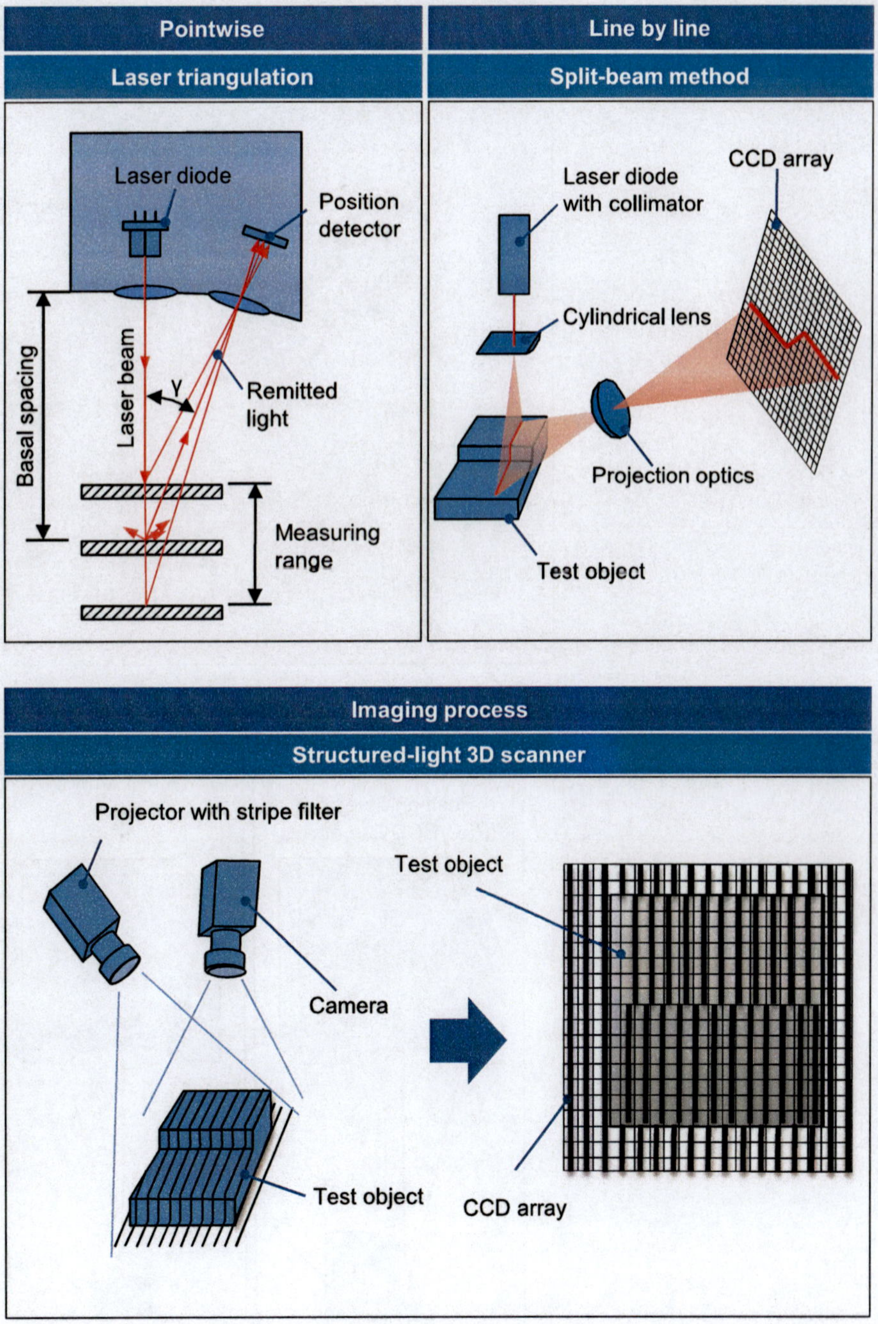


**Fig. 11.72** Functional principles of selected optoelectronic measuring methods for three-dimensional shape recording

on the receiver element can be used to determine the distance to the object. The sampling rate standard in the industry for sensors is 10 to 20 kHz and can be increased to 100 kHz.

The downside of all optical sensors using the triangulation principle is that the absolute measurement accuracy of the sensor depends on the reflection properties of the test object. It can only be specified in relation to a diffusely reflecting reference material. White paper is generally assumed here. Accuracies below 1/1000 of the measurement range can be attained using triangulation sensors measuring pointwise. On shiny bodies, and also metals, the light reflection is not diffuse, some of the light is totally reflected. This can lead to measurement errors particularly when there are height differences on the measurement object (Fig. 11.74). The surface must be covered with a matte spray paint in these cases. CCD receiver elements increasingly being used allow, in contrast to analog position-sensitive diodes, extensive compensation by

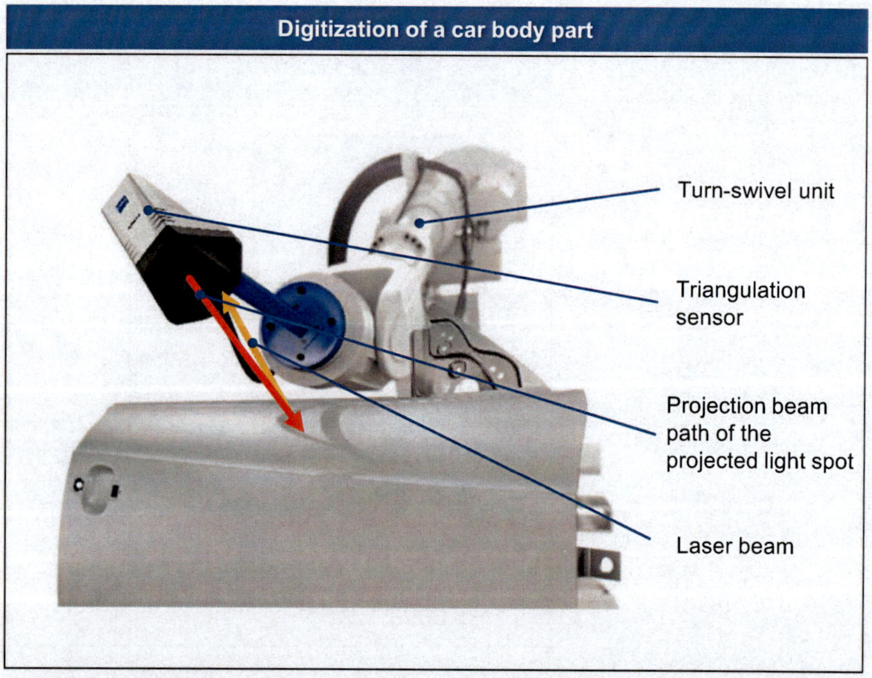


**Fig. 11.73** Digitization of a car body part using a triangulation sensor on the rotary swivel arm. *Source* Zeiss

11

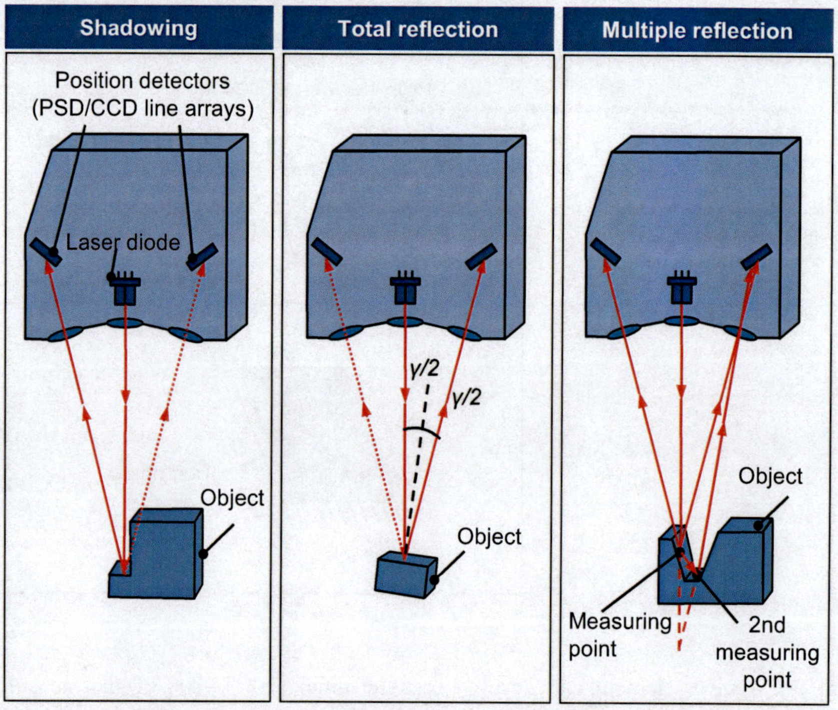


**Fig. 11.74** Typical sources of error in triangulation

computation of surface influences and inclination errors, meaning modern sensors are robust against measurement errors.

Although in principle one receiver element is enough to determine the distance, the trend is towards sensors with two receiver elements to reduce the aforementioned problems. Surface and inclination influences, as well as shadow errors, can be reduced by analyzing the measured values output by the two receiver elements. There are now also rotationally symmetric triangulation sensors especially for the digitization of complex freeform surfaces. With coaxial positioning of the transmitter and the receiver element arranged in a circle around the transmitter, these sensors exhibit higher levels of insensitivity to measurement errors especially when there is shadowing.

Optical sampling systems using the split-beam method also work using the triangulation principle (◘ Fig. 11.72, top right). Here, a light line is generated by expanding the laser beam using a cylindrical lens or oscillating mirror. A CCD array serves as the receiver element. This element is offset by a specified triangulation angle in relation to the laser. With the quasi-parallel evaluation of one complete light line for standard measurement frequencies of 5 to 50 Hz, these sensor systems can be used to significantly shorten, compared to sensors measuring pointwise, the required measuring times for digitizing workpiece geometries with lower attainable accuracies and resolutions.

Methods for 3D measuring over surfaces have made considerable advances in recent years with intensive research and development work at universities as well as in industry. Extremely quick digitization of even complex bodies is made possible by the ability to capture complete views of an object with a single measurement. This is why these methods are being used more in the field of digitization. Methods measuring over surfaces can be divided into active and passive methods. Active methods, such as the structured-light 3D scanner method, have one or more lighting devices used to project suitable lighting patterns onto the measurement objects that are then captured by one or more cameras. Passive methods, such as stereoscopic image processing, do not use these lighting patterns. Here, the object is measured by at least two cameras with different viewing directions. From the known positions and orientations of the cameras in relation to each other, the 3D coordinates of an object point can, using the triangulation principle, be calculated from the two-dimensional pictures of the camera images. Identification and assignment of object points in each camera image are necessary for this. They frequently cannot be done unambiguously, however. This is the reason passive methods are only used with limitations in the industry at the current time [PFEI98].

The structured-light 3D scanner method is used as an example for an active measuring method over surfaces for reading 3D geometries. In this method, stripes are projected onto an object by a projector using a lattice having a constant lattice spacing (◘ Fig. 11.72, bottom). These stripes are captured by one or more cameras and the lattice distortion is analyzed by an image processing system to describe the surface. If the basal spacing, the inclination angles of camera and projector, and the position of the lattice, are known, the shape of the object can be calculated using the intercept theorem. ◘ Figure 11.75 shows an example of the digitization of a PC mouse. The markings visible on the picture under the test object are used to calibrate the system.

◘ Fig. 11.75 Structured-light 3D scanner method for capturing geometries. *Source* FhG-IPT

So that no glancing intersections occur between the projection rays and the image rays and to attain good depth resolution, a relatively large basal spacing between the camera and projector is required. This could give rise to shadowing. Measuring discontinuous and steep object surfaces represent the limitations of use of these measuring systems, as ◘ Fig. 11.76 shows. Steep object surfaces are either hit by few projected lines or mapped onto the CCD array by a few image rays. As a result, the resolution and the measurement accuracy of the sensor system fall with increasing gradient. Even greater height differences give rise to shadowed areas that cannot be measured and result in imperfections in the digitized object surface. For complete and accurate digitization of the geometric shape of an object, measuring it from different directions of view is frequently required.

Measuring objects with discontinuous transitions, such as sharp edges, poses another difficulty. At the steps, the projected lines are disrupted and must be assigned to each other after the steps (something that is not readily possible unambiguously for complex bodies). This problem can be resolved with sequential lighting of the object using different, specially coded luminous patterns (Gray code method). When the camera images are analyzed, coding makes it possible to assign image points to the projected lines unambiguously.

For the structured-light 3D scanner, the camera images are analyzed by calculating the phase positions of the projected lines. Accuracies of approx. 1/20 of the

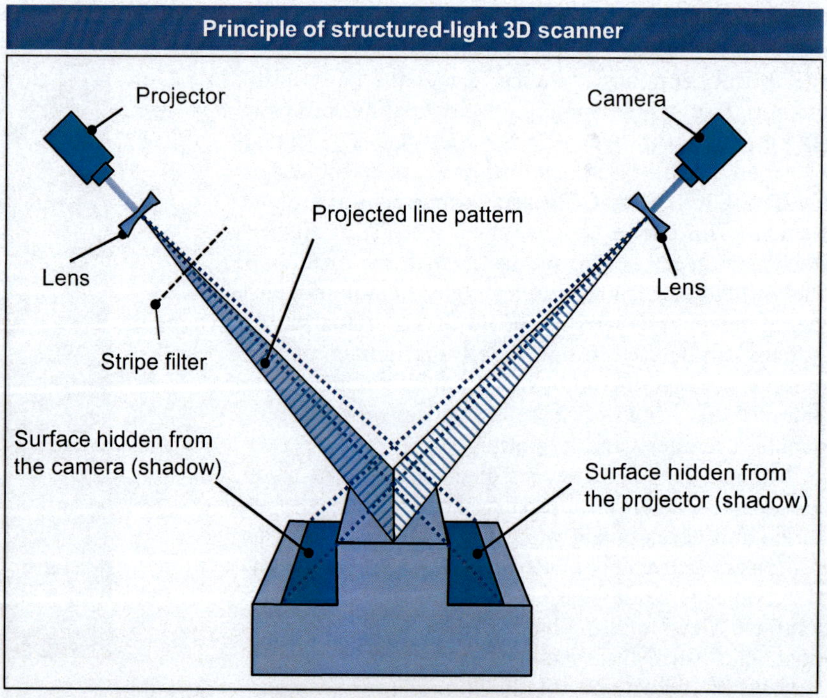


**Fig. 11.76** Sampling of a spatial model surface

11

line width can be attained in this method. The measurement accuracy and resolution can be increased by reducing the line width. However, this is limited by the camera resolution. In another method to increase accuracies, the structured-light 3D scanner method can be broadened by various methods. The phase shift and Moiré methods are worthy of particular mention here. In the phase shift method, the pattern of stripes is offset sequentially to increase the resolution in a stripe period. This means a phase accuracy to 1/100 of the stripe width can be attained. In the Moiré method, a lattice is introduced into the light path of the camera in addition to the lattice in the light path of the projector. The lattice in front of the camera means the mapping of stripes onto the object gives rise to dark and light-colored lines "beat" on the CCD array. These lines are analyzed to calculate the object geometry. Height resolutions in the nanometer range can be attained with this method, and so it is also used for machine investigations.

#### 11.4.3.4 Preparation and Processing of Measurement Data

Processing data gained by recording measurements, depending on the quality and degree of order of the point cloud measured, requires a number of additional processing steps. Some of these are very complex in terms of computation and can only be discussed briefly here.

If measurements are recorded by a tactile sampling system, the measurement data obtained describe the position of the center point of the probe tip at the respective measuring time. As the probe tip has only finite expansion, the point actually required, the point of contact of the probe tip with the object surface, differs from the position measured. Thus, as in Fig. 11.77, an equidistance to the actual workpiece contour is measured. This is given by the probe tip radius and deflection of stylus l. The true workpiece contour must then be determined by calculating these equidistances using a "probe tip radius correction".

A general problem in digitizing object geometries is the data volume involved. Sampling increases the memory requirement, which is not limited in theory, as the size of the models increases. As editing digitization data is considerably more efficient in prepared data structures, it does not seem viable to perform calculations with unnecessarily large data volumes.

Conventional CAD systems are not designed to process the large data volumes occurring in digitization. Geometries are usually described in CAD systems using volume information or body edges, that model equivalent geometries more efficiently. This yields a demand for effective methods of data reduction to minimize the data volumes involved without having to accept information losses for depicting the model sampled.

Standard methods of data reduction for sequentially sampling sensor systems check during sampling whether the path errors caused by data reduction are still within a permitted tolerance range. For this, tolerance bands are defined. The shape of these bands depends on the type of interpolation method selected

**Fig. 11.77** Sampling of a spatial model surface

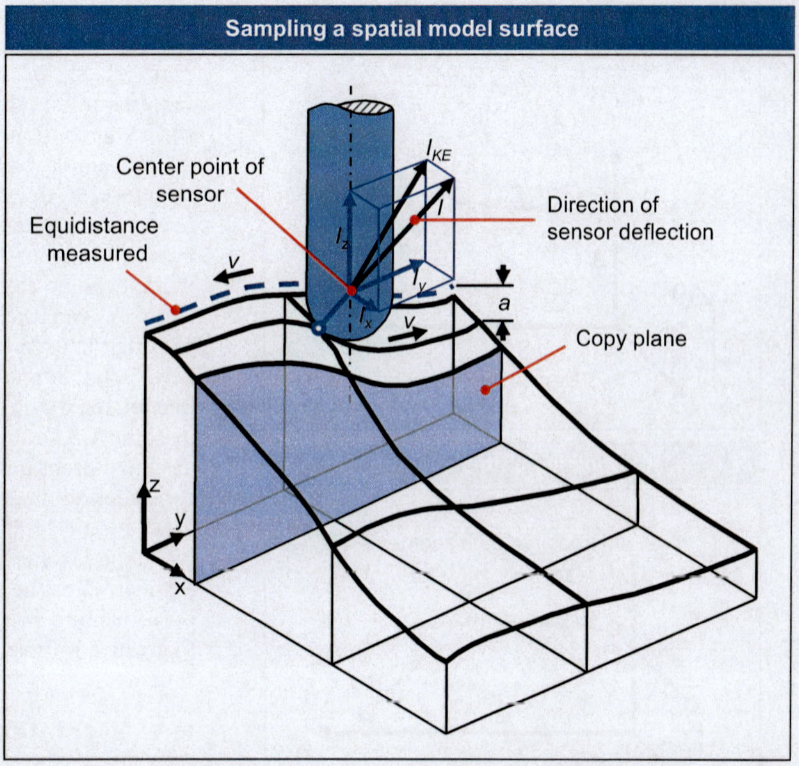


[o.V.86]. Figure 11.78 shows the procedure using an example of linear interpolation. A sequence of sampling points, $P_k$ to $P_n$, is replaced by a straight line segment extending from $P_k$ to the last sampling point determined, $P_n$ (Fig. 11.78, bottom). If intermediate points $P_{k+1}$ to $P_{n-1}$ are still within a tolerance band placed around the straight line segment, linear interpolation between points $P_k$ and $P_n$ is permitted and the next sample value, $P_{n+1}$, is included in the analyses. This is performed by placing a straight line segment from $P_k$ to $P_{n+1}$ and then checking whether points $P_{k+1}$ to $P_n$ are still within the tolerance band around the new straight line segment.

This procedure of gradual extension of the straight line segment is continued until, for the straight line segment in question, any intermediate point is no longer within the relevant tolerance band. In this case, it can be ascertained that the last extension, for example the transition from $P_n$ to $P_{n+1}$, was no longer permitted as a new end point and the previous end point $P_n$ must be registered as the new relevant interpolation point. At the same time, all intermediate points ($P_{k+1}$ to $P_{n-1}$) can be deleted because they are no longer required. The reduction algorithm then starts again, i.e., Point $P_n$ becomes the start point of a new straight line segment. Its end point results from the other sampling values.

Shown in Fig. 11.78 at the top is a contour section. It was possible to use the reduction method described to delete seven intermediate points from the 11 path points recorded for this section. Linear interpolation was assumed in the example described. But circular and spline interpolation methods are also possible. Using these reduce further the number of sample values required. The shape of the tolerance band must then be aligned accordingly here, meaning a circular band for circular interpolation, for example.

Particularly for the digitization of 3D objects using optical sampling systems measuring over surfaces, a number of processing steps, some of them interactive (that are not discussed in detail here however), are normally required before measurement data are processed further. Different views of an object need to be transformed into a common coordinate system with redundant information being eliminated from overlapping areas. If the position and direction relationships of the views relative to each other are not known, the parameters for the transformation can also be calculated with approximative methods on the overlapping areas of the individual views (matching methods). There is also the problem of object recognition from the camera images, i.e., clear-cut differentiation between the background and relevant image points describing the measurement object. Given that the data volume is extremely high sometimes, usually exceeding several megabytes per image for modern camera systems, the use of powerful algorithms for data reduction is also required. This data reduction is normally performed depending on the level of surface

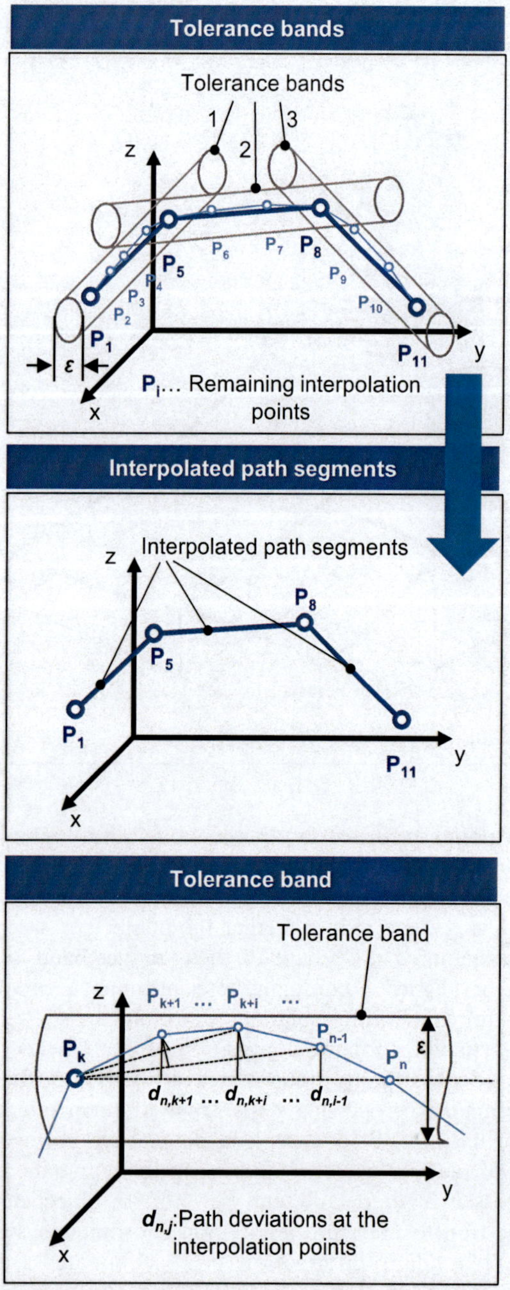


**Fig. 11.78** Data reduction for linear interpolation (according to Pritschow)

curvature. This means that major gradient changes in the surface, such as occur at edges, are represented by higher point densities than level surfaces.

NC data is generally created on the basis of measurement points using a parametric description (such as Nurbs surfaces) or using a polygonized depiction (e.g., triangulated triangle surfaces) of the object surface. Polygonization here can also serve as a preliminary stage for later calculation of the parametric representation (Fig. 11.79). This problem, known as surface reconstruction, can only be automated for relatively simple object geometries and depends on the structure of the measurement data. Complex components must, either during or after the measurement, be broken down interactively into individual segments using the point cloud measured, converted into partial surfaces, and then reassembled. The various systems provide the users with assistance in the form of a number of functions. The software necessary here is either part of the digitization system or available as a module in some CAD/CAM systems. There are also specially designed programs that have comprehensive surface reconstruction functionality.

Figure 11.80 shows an example visualization of NC data generation using a CAD system. A hand-held scanner scans the actual component surface in different positions and orientations (on left) and uses this data to create a reverse digital geometry (on right).

## 11.5 User Interfaces on Machine Tools

### 11.5.1 Control Panels on Machine Tools

A control panel is used as the interface between user and NC machine. This is used to manually activate the available functions. This control panel generally comprises several components required for operating machine-specific as well as controller-specific functions. Figure 11.81 shows an overview of the layout and functional scope of a controller and machine operating panel [SIEM12].

The controller panel, shown in the top part of the picture, comprises an alphanumeric keyboard (A), various softkeys (B) and a monitor (using an TFT flat display). Nowadays, a low-profile industrial PC running the Windows operating system is normally integrated in the controller panel. NC programs can be edited, correction values can be set manually and NC information, for example, can be accessed using the input interface. The softkeys are used to call numerical controller functions from a structured menu tree. Entering tool correction data, changing program parameters and calling a machining simulation are examples of these functions. More detailed information on the machine status can be accessed by the user (this is dependent on the mode selected).

In contrast, the machine operating panel contains function elements that are required for manual control

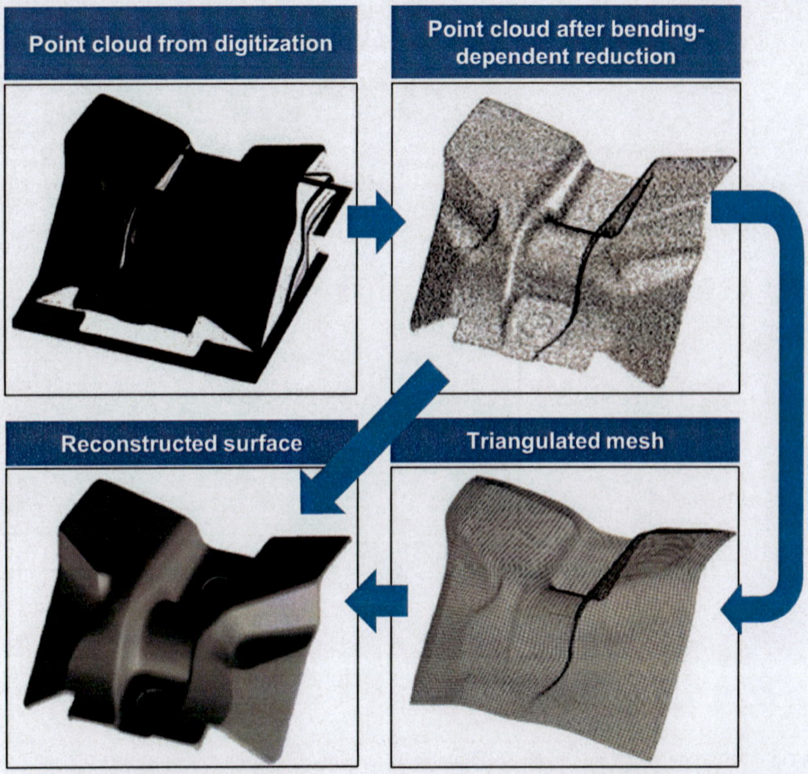


**Fig. 11.79** Surface reconstruction from point clouds. *Source* Zeiss

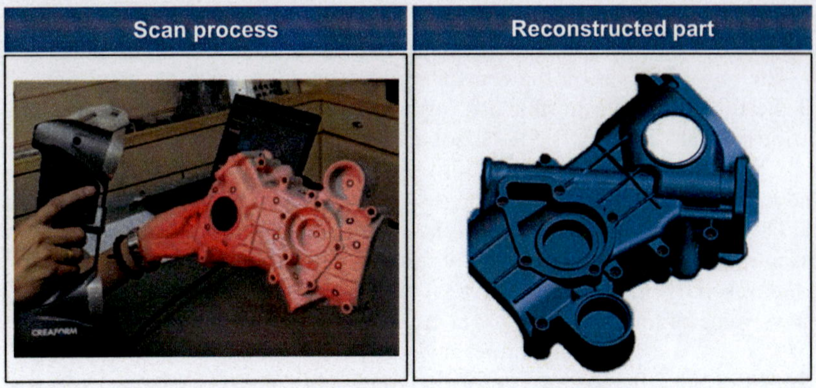


**Fig. 11.80** Digitized surface relief (point rendering) and calculated milling path. *Source* CreaForm

of the machine. To describe the functions available, the elements can be divided into different groups. The Group C buttons (see Fig. 11.81) can be used to

- select the mode (Setup, Manual entry, Automatic, Single block)
- reset the controller
- select referencing mode
- set the sensitivity for incremental mode (e.g., B. 0.01 mm/pulse).

Button group D can be assigned as needed by the machine tool manufacturer. These buttons are normally used to activate machine-specific devices (e.g., switch on the chip conveyor, activate the coolant supply, clamp and release the tool).

Group E buttons are required for manual control of the machine axes. For this, a corresponding selection of the axes (X, Y, Z, …) and the traverse direction desired (positive or negative axis direction) must be entered using the keypad.

The rotary switches (Group F) are used to change the programmed spindle speed and feed rate. The spindle speed can normally be changed manually in a range of 50 to 120% of the value programmed using the "Override" function. In most cases, the feed can be varied in a range of 0 to 150%. Other buttons are assigned

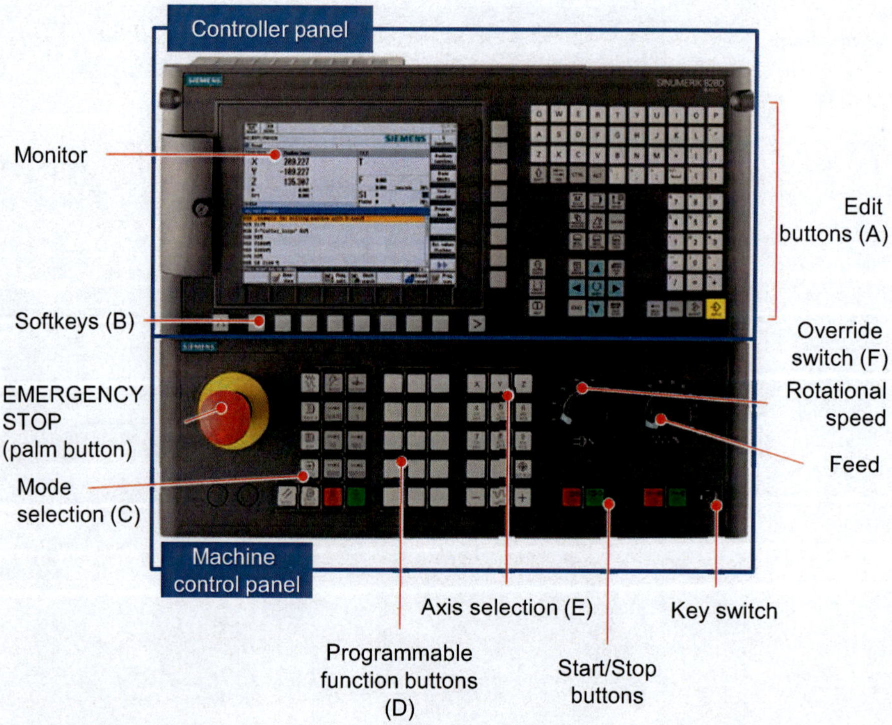




**Fig. 11.81** Machine control and operating panel for a machine tool. *Source* Siemens

to the rotary switches. These buttons can enable or disable the feed and speed for the entire system.

The set of buttons on the controller panel contains all the functions of a normal PC keyboard including various special characters for cursor and screen control. The only differences are in terms of positioning and icons on the buttons.

The icons on the function buttons of the controller and machine operating panel are described in DIN standards [DIN82; DIN81; DIN83a]. Figure 11.82 shows some of the most important function icons. In this context, a cross-manufacturer style guide was developed in the Hümnos project for designing and simplifying the user interfaces of machine controllers. The guideline contains a lexicon of terms with translations of important specialist terms, as well as rules for designing controller interfaces and operator dialogs [OSAC97].

## 11.5.2 Manual Process Control

### 11.5.2.1 General Overview

In regard to workshop-oriented use of machine tools, for single component and small-lot production in particular, central importance is attached to manual process control. Unlike series production where automated monitoring of the running processes is paramount, the broad spectrum of tasks here necessitates the use of qualified and experienced specialists. They require knowledge of the ongoing processes, from machine setup to manual machining, as well as programming and monitoring the production process. In addition to the workshop-oriented programming methods described in ► Sect. 11.4.2.4, setup, manual machining, process control in Automatic mode and the topic of user-oriented process analysis are described here.

### 11.5.2.2 Controls Required for Process Control

To perform operations as part of the work activities in question, the user has available different controls that enable manual control of the axis paths and the axis velocities in Setup and Automatic modes. Once all preparatory work is complete, such as the reference point run and clamping, supporting and aligning the workpiece, it is essentially only a matter of defining the exact position of the workpiece in the machine coordinate system. Manual positioning of the NC axes is required for this. For these operations, inching buttons are provided as standard on the controller panel. These enable approximate and fine positioning of the NC axes in conjunction with the axis selection and direction buttons and with the feed override switch.

For approximate positioning, an NC axis keeps moving while the relevant inching button remains

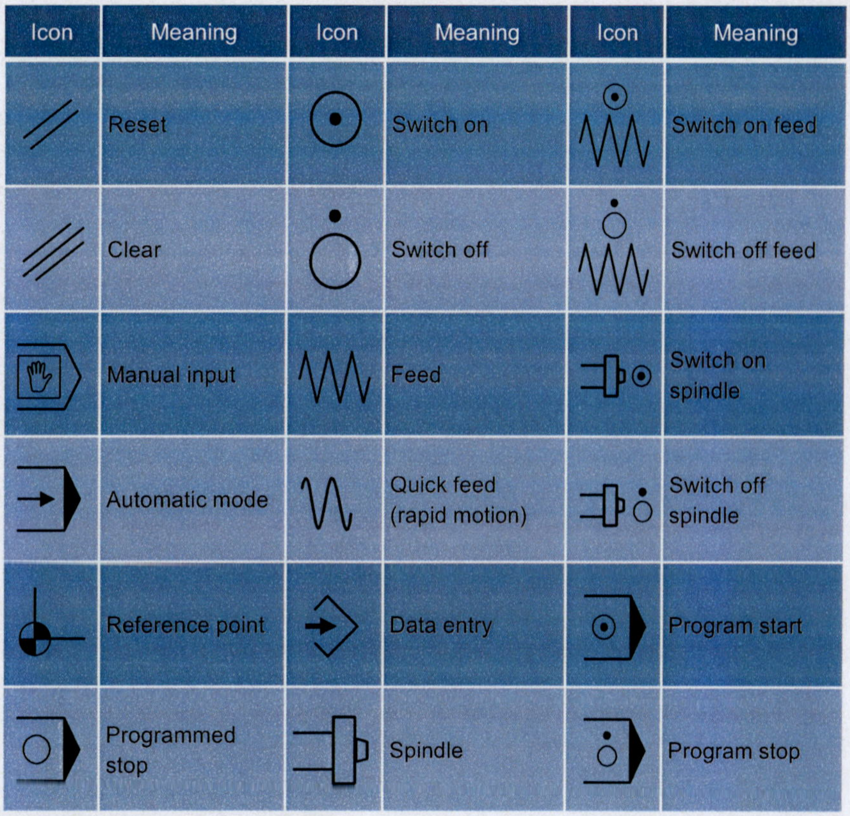


**Fig. 11.82** Function icons for use on machine tools. Taken from the DIN 24900, 30600 and 55003 standards

pressed (continuous movement). The velocity of the axis here is set using the feed override switch. In contrast, fine positioning of an NC axis enables a defined axis path to be initiated manually for every inch operation (incremental movement). The setting range for specifying an axis path is generally an adjustable value range that is specific to the controller. This enables the control panels to be used to move towards the required axis positions manually and within the machine accuracy range.

Electronic hand-held control units for Setup mode are available as part of an extended user interface (Fig. 11.83).

The user can pick up this control unit and walk towards where the process is running. It enables improved monitoring of critical work situations (such as fine positioning of the NC axes).

A handwheel presents an alternative to the inching buttons described for controlling the NC axes. This control is essentially identical in function to the mechanical handwheels on conventional machine tools. The key difference between mechanical and electrical conversion of the rotational motion is that, for the electronic component, there is a coupling on signaling level to the machine controller. This means that machining-dependent process information (such as forces occurring during a machining operation) cannot be measured by the user. Here however, the path parameters can be specified manually using the rotation angle, and the velocity can be controlled as a function of the rotational velocity using a single control. This enables approximate and fine positioning of the NC axes in Setup mode, and performing simple, manual machining operations without the aid of other controls. Several electronic handwheels can be integrated into control panels in line with the design of conventional machine tools. These input devices usually then have a fixed assignment to the axes in question, meaning there is no selection of the active axis from the axis selection buttons. This type of control panel design also permits manually controlled movement of multiple axes simultaneously.

For direct intervention into the program-controlled process, particular importance is attached to the override switches mentioned at the beginning. The variation of velocity parameters represents one way of optimizing the production process as regards technological boundary values. Also, when an NC program runs the first time, the traversing speeds of the axes can be reduced for test purposes using the feed overrides. Single block mode also proves useful in this context. When this mode is selected, machining is stopped after every NC line, and can be continued line by line by pressing the "NC Start" button again. A test run is possible by

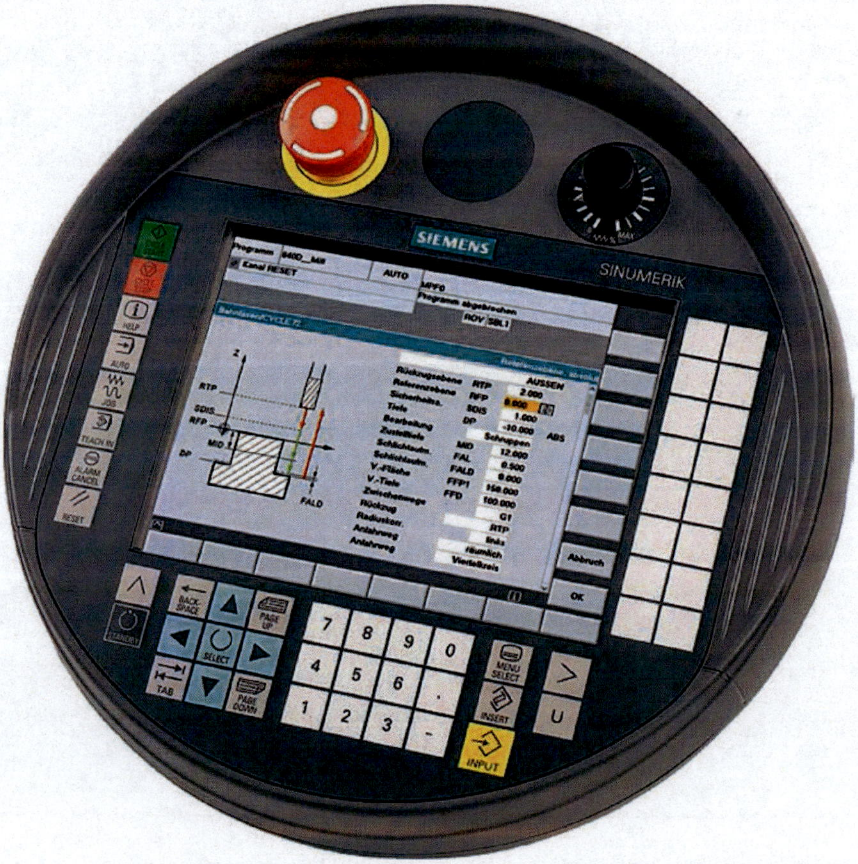

**Fig. 11.83** Hand-held control unit for a machine tool. *Source* Siemens



checking the line-related remaining path display and the remaining NC lines.

#### 11.5.2.3 Alternatives for Realizing a User-Oriented Process Control

The topic of previous and current research work is the design of application and user-friendly human–machine interaction forms on machine tools. In the process, suitable technical components are being developed that permit user-specific experience and knowledge to be channeled into the production process. It is also possible to collect new experience and knowledge by using these components. In conjunction with the elements discussed here for manual process control, it was possible, using numerous analyses in operational practice, to collate demands made on the design of these components from interdisciplinary viewpoints. Working the requirements described into a suitable technical concept can be explained using the depiction in Fig. 11.84.

As a basis, a multi-axis control in the form of a joystick was devised for the manual control of NC milling machines. Due to the limited adjustment range of the control (approx. ±20°), a velocity-proportional input option is provided, i.e., a corresponding velocity is set for the assigned NC axis depending on the deflection angle of the joystick axis.

To take into account user-specific interests and process-related requirements (e.g., approximate and fine positioning), specification of the reference velocity value as a function of the displacement movement must be configurable flexibly. This requirement can be implemented in that the relationship between the deflection angle of the control and the assigned reference velocity of the NC axis is realized with characteristic curves. These curves can be modified in line with the depiction in Fig. 11.84 for realizing joystick deflection at the traverse speed for the applications [NN99].

To also give the user a sense of process progress, the machining forces occurring for cutting, manual machining are fed back to the adjusting axes of the joystick. Depending on the machining scenario, tool-specific feedback characteristic curves can also be included in the analysis of machining forces. These curves would define an assignment between tool/machine loads occurring and the resulting feedback force (Fig. 11.84).

Implementing process transparency by feeding back the machining-dependent process forces is not

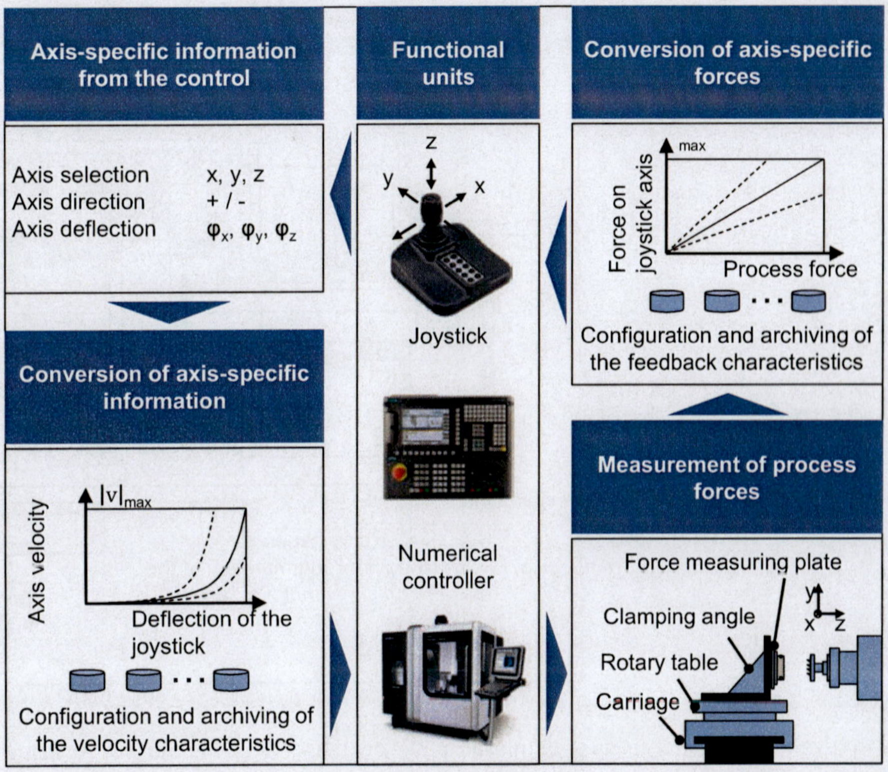


**Fig. 11.84** Technical concept for the controls required for manual process control

limited to the control introduced here. As shown in Fig. 11.85, a conventional, electronic handwheel can also be fitted with a suitable feedback mechanism [WECK97].

Regarding the component realized here, a direct drive electric motor is used to feed back the process forces. The linear relationship between current supplied and motor torque is utilized here. With this, a counter torque that is proportional to the force is generated easily as a function of the process force measured.

Initial practical analyses have shown that using the controls introduced here can yield methods to improve process control on machine tools. This conclusion applies to both technical components for performing manual machining operations on simple workpiece geometries. Worthy of particular note however is the joystick. It can be described as the ideal, user-oriented input device for simple positioning of NC axes and for cutting operations [WECK05].

Beyond actual process control, navigation within the user interface and control of the machine tool also represent key parts of human–machine interaction. Up to a few years ago, only physical controls were used for interaction (see Fig. 11.81). Touch-sensitive interfaces are also state-of-the-art now (Fig. 11.86).

Different touch technologies (introduced briefly below) are available depending on the application area and the resultant requirements:

a) Capacitive: Capacitive touchscreens can be divided into surface capacitive and projected capacitive.
In the **surface-capacitive** touch technology, a transparent conductive film (usually coated with metal oxide) is applied to a glass surface. Applying alternating current in the corners of the screen generates a homogeneous electrical field. When the panel is touched with a conducting object (such as a finger), a charge transfer takes place in its direction and changes the electrical field. The resulting currents in the screen corners are measured, and using the ratios to each other, the exact position of the point of contact is calculated.
In the **projected-capacitive** method, a mesh of conductive tracks comprising two conductive track levels on top of and isolated from each other, is applied. One of these tracks acts as a driver (signal transmitter) and the other as a sensor. If a conductive object is in the immediate proximity of an intersection point of two conductive tracks, the capacitance change can be detected. From this, the coordinates of the point of contact are calculated. The mesh is normally mounted between two glass panes, so it is protected from environmental effects. The covering glass layer can be up to 20 mm thick without appreciably impacting the touch functionality [SCHÖ08].
b) Resistive: non-conductive spacers are mounted in a dense pattern between an externally flexible

**Fig. 11.85** Electronic handwheel with integrated feedback unit

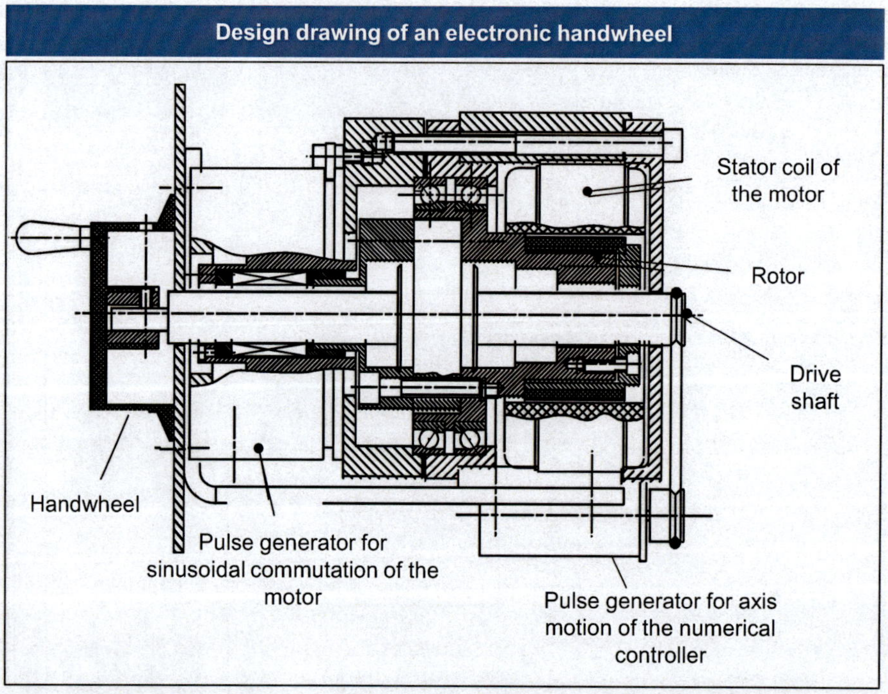


11

polyester layer and a glass or plastic pane on the back. The insides of the panes are coated with a semiconductor material (such as indium zinc oxide). Then, direct current is applied to one side each. Under mechanical pressure (by being touched), the two layers are connected, and the electrical circuit is closed. The position of the pressure point in the x- and y-directions can be determined using the voltage divider principle [SCHÖ08].

c) Optical: different approaches exist here that are all based on the same technology, the use of infrared light. It is either refracted through a glass pane on the side (Frustrated Total Internal Reflection (FTIR)) or projected onto the glass pane from the back (Diffused Illumination (DI)).
In the first case, the light inside the pane is interrupted when a finger is applied. This causes scattering at the point of contact. In the second case, the light is reflected directly by the finger. In both cases, an infrared camera directed at the glass from the back is used to determine the position in the camera image from the reflections of the infrared rays [SCHÖ08].
Another variant can be realized in the form of a 2-dimensional infrared light curtain. Any non-transparent objects interrupt the infrared rays in two directions. This provides the x/y-coordinates of the touch point [SPAT09].

d) Acoustic: There are also several variants of acoustic-based touch technologies.

In the **Surface Acoustic Wave method (SAW)**, signal transmitters introduce active horizontal and vertical ultrasound waves into a glass pane, where they are detected at the opposite end by sensors. Applying a finger or another vibration-damping medium changes the amplitudes of the waves. The energetic change caused by this is measured by the sensors and used to determine the position.
In contrast to the SAW technology, the **Dispersive Signal Technology (DST)** does not require an active signal source. Using piezo sensors installed at

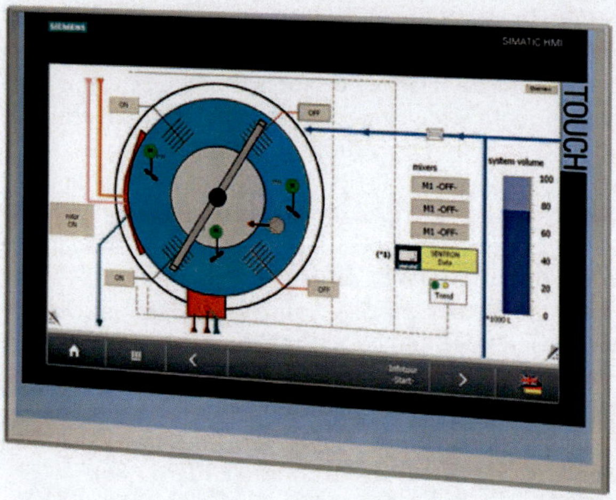


**Fig. 11.86** Touch-based HMI for process monitoring. *Source* Siemens

the edge of the screen, vibrations are detected that are generated when the screen is touched. The position of the point of contact can be calculated using the runtime differences of the signals. To evaluate the signal profiles, special algorithms must be applied that include or filter out application-specific influences such as glass dispersion effects and disruptive influences such as reflections.

Another method to analyze bending waves is **Acoustic Pulse Recognition (APR)** [NORT06]. Similar to the DST, screen touches are detected by piezo detectors, but there is intentionally no laborious correction and runtime analysis of the signal. Rather, the characteristic signal curve of every single pixel is considered to be a unique, acoustic fingerprint and is stored in the controller. The acoustic profile, created in this way and unique for every point of contact, is used to uniquely determine the position.

More detailed information on the design of the available touch technologies and how they work can be found in [SCHÖ08; NORT06] and [SPAT09] as well as other sources.

All touch technologies have in common the downside of there being no tactile feedback because of the planar touch surface. In contrast to mechanical controls, the operator does not receive any haptic feedback on inputs, something that can result in input errors and lack of user satisfaction. Optical and acoustic feedback is easy to generate in conjunction with software-based user interfaces, but in the context of blind operation of the machine on the shop floor, it is only perceptible with difficulty, or not at all, for the human user. This is why research is ongoing into feedback mechanisms for touch-based user input. In most cases, existing solutions are based on micro actuators that introduce amplitude and frequency modulated vibration signals in the touch surface to convey feedback on the input and any fault states to the operator, even for blind operation of a production machine [BREC12a].

#### 11.5.2.4 Development Trends

An analysis of the trends in modern production facilities shows that single component and small-lot production has gained in significance compared to large scale production. A prerequisite for efficient production, however, is that suitable forms of interaction are available for qualified operating personnel at the production equipment. This demand covers all specialist activities at the machine tool, ranging from NC programming to manual process control.

The trend described is being considered directly in the machine tool production industry at this time. For example, machine tools are being offered by many manufacturers (at least for "turning" production) that enable new forms of user-specific input using an optimized human–machine interface (see ► Sect. 11.5.2.3). In particular, the technical components and technologies described in regard to manual process control and basic HMI operation are playing a key role here.

The use of voice input and output is giving rise to new capabilities for direct human–machine interaction [HOYM05]. For a defined functional scope, action-oriented input operations on the NC keypad can be realized for certain parts with a direct verbal vocabulary. Exceptions here are safety functions (such as emergency stop), the effects of which must always be guaranteed regardless of the detection rate. Series of experiments using this technique have shown that the transformation towards other input modalities is of particular importance when the activities of the operator on the control panel of the controller can be more efficient, i.e., quicker and easier, with verbal input. In this context, access to different controller and machine functions is possible, such as:

- activating switches and buttons for additional machine-specific equipment (for tool change, pallet feed, etc.)
- direct selection of the information displays from the menu tree on the NC
- entry of numbers for programming.

Smart phones, and the associated advances in multi-touch technology that have increased by leaps and bounds, have greatly increased the acceptance levels for such interfaces. While only 2-point detection is available for dual multi-touch, true multi-touch enables the detection of virtually any number of touch points for touch-supported multi-person applications. Initial devices with (dual) multi-touch support are also already available for simple operating and monitoring tasks in production technology. The operator gestures supported are limited mainly to input patterns familiar from cell phone technology (zooming and wiping/scrolling of images displayed) and are generally restricted to two fingers.

Spatial gesture detection goes one step further and extends gesture detection to a third spatial dimension. Camera-aided systems, such as Microsoft Kinect®, generate at runtime 2D-only image information as well as 3D spatial data to track the working area and objects in it. Current research activities are in the field of validating automated assembly processes. Here, image attributes are extracted in multiple steps to perform object detection. ◘ Figure 11.87 illustrates this process. A reference object is captured optically in a camera image (a). From a point cloud (b), representing a rough approximation of the spatial geometry, filter algorithms can be used to extract the precise dimensions or individual attributes (c).

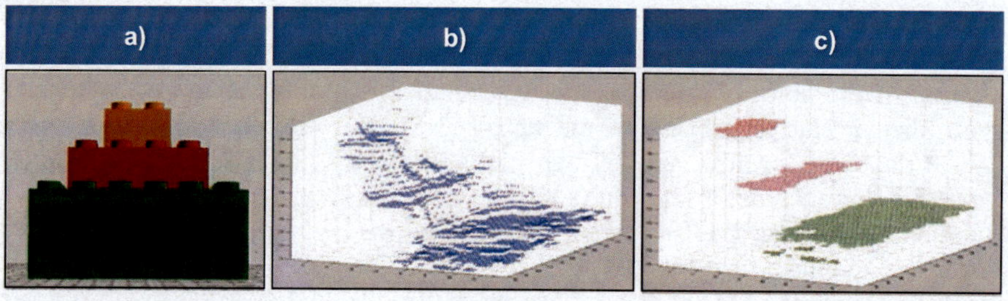


**Fig. 11.87** Depth-camera-supported object detection, according to [BREC12b]

In the future, these kinds of systems could be used to control production systems with physical gestures. Research here is focused on the actual hardware-based imaging processes, and the operating concepts required for users to intuitively access the machine HMI. In this context, one particular focus is avoiding operator input errors. Incorrect input from wrongly interpreted or unintentional operator gestures, and any resultant harm to people or machine damage, must be prevented by software.

### 11.5.3 User-Oriented Depiction of Process-Related and System-Related Parameters

#### 11.5.3.1 Initial Situation

The user needs various information items as the basis for the manual process control described in ► Sect. 11.5.2. These items must be determined from the machining process. Also, system-specific information can contribute to achieving a machining-dependent optimum with manual intervention by the operator.

Process analysis for the user proves problematic due to the design of modern production machines. Compared to the manual-only control and operation of machines, machine tools are operated encapsulated for safety reasons. This considerably restricts the perception of process changes [WECK92]. When heavy-duty machine tools are used, the physical distance to the machining point poses a similar problem for the user. This decoupling of operating personnel from the production process is being intensified on machine tools in that the complexity of machining processes has increased over the years and processes are running at far higher rates.

For user-oriented process monitoring and analysis, additional technical components must be applied in the form of sensors on NC machines that help humans process information. Also, improved representation of the process flow can be achieved by providing system-specific parameters.

#### 11.5.3.2 User-Oriented Communication of Parameters

To depict process-dependent parameters, a purely visual display of information, such as on the screen of the controller operating panel is usually used. A user-friendly communication method means, however, that the process information is made available on as many perceptual channels as possible in a form adapted to humans. In this context, the term "integrated sensual perception" plays a key role. Figure 11.88 shows an overview of the characteristic process parameters available to the user for state analysis and how these signals can be measured. The display forms that can be selected for the different information to guarantee improved process analysis is also shown.

On the basis of automated process monitoring, sensors (detailed in ► Sect. 6.4) can be used in part to capture useful signals. This applies in particular to capturing acoustic and tactile process information where accelerometers and piezoelements are used for example. The analysis of automatic signal classification is omitted here. At this point, there is in operator-oriented process monitoring, appropriate communication of the information for the user. The user is then able to perform a subsequent evaluation. Compared to the process parameters technically communicated by the sensors, additional components are also used for machine tools to improve visual process monitoring by the user. An example of these applications is the integrated camera system mentioned in Fig. 11.88. The process control loop is ultimately closed by the user by including the controls described in ► Sect. 6.4 as equipment supporting actions in manual process control on machine tools.

The latest technological advances mean additional options for intuitive, user-centric visualization are coming to the fore. The continued miniaturization of hardware has opened up a new segment in the projector market—pico projectors. These devices have very small dimensions and can be fitted in mobile devices. In combination with an initial projection of striped patterns,

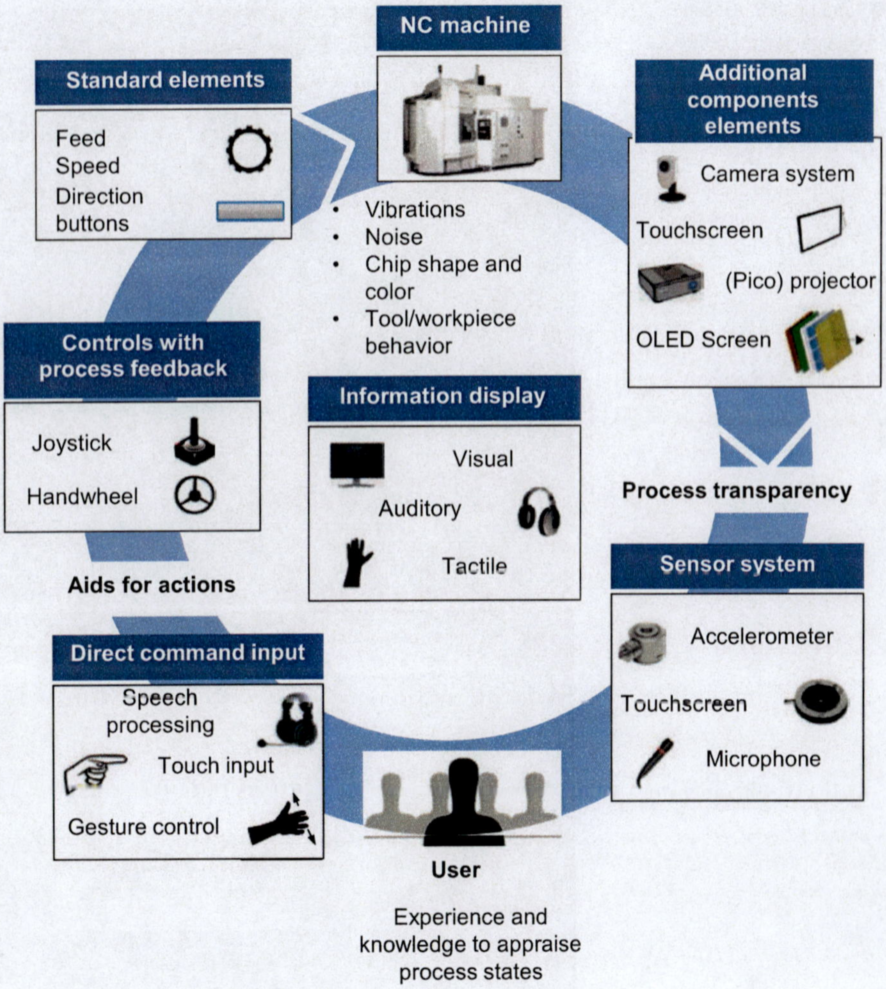


**Fig. 11.88** Process monitoring and control with help from the user

surface distortions and contrast differences can be offset. In addition, OLED screens (Organic Light Emitting Diode) have become cost-effective to use for the first time due to product series in large numbers. These technologies enable surfaces previously unused to become visualization and interaction surfaces at runtime. This opens new research fields in production technology for human–machine interaction.

#### 11.5.3.3 Technical Realization and Application Examples

In ▶ Sect. 11.5.2, it was shown how using a power feedback control for process control in manual mode can increase process transparency using tactile feedback. Added to this, more alternatives for process monitoring are explained for optical as well as acoustic applications using selected technical realizations.

One application example deals with the usage of Augmented Reality for process monitoring. Augmented Reality (AR) means that a viewer is shown virtual objects, that are embedded in the real environment or overlay it [AZUM97]. For this, the virtual objects are usually shown on a Head Worn Display (HWD) with see-through properties. This allows simultaneous viewing of the real world and virtual objects. Current research activities are grappling with the issue of how virtual information can viably be embedded into the environment of a machine tool. As the machine information to display always depends on the situation, the machining process is divided into scenarios. The greatest aid potential for the user is seen in the scenarios described below of Setup, Startup (including monitoring the running process) and Fault management.

When setting up workpieces with complex geometries, mounting the clamping devices is also often highly complex. When using AR techniques, the clamping devices specified in the production order are superimposed over the actual workpiece to clamp as a virtual display. In this way, the operator is provided a good aid to affix a clamping device quicker and easier than would be possible using a sequence described in text.

**Fig. 11.89** Augmented Reality for visual process monitoring

When a program is started, visual process monitoring and supervision are particularly important because here critical safety parameters are checked for correctness. However, visibility during the machining process is so heavily restricted due to flying swarf and coolant swirling around that assessing the process state is generally difficult. In the simulation of the machining process in the HWD, these problems can be surmounted using Augmented Reality. For this, the operator is shown the current tool movement in the ongoing production process, the features to machine (such as grooves and bores) and a few main workpiece reference edges. Figure 11.89 illustrates this scenario.

Fault management on machine tools is aided particularly effectively by mobile machine operation systems. For example, Augmented Reality can be used to show error messages in the field of vision of the operator. Optimization of this display method is desirable

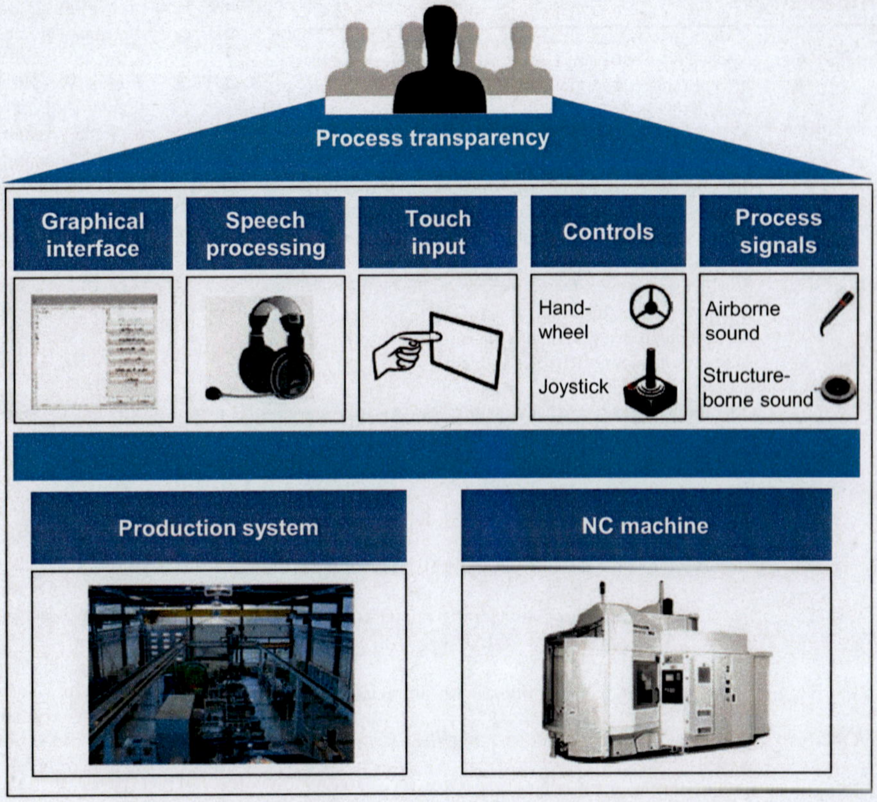


**Fig. 11.90** Design variants for human–machine interfaces

for a variety of reasons. Previous fault and diagnostics aids come up against their limitations in the often deficient text descriptions of situations and information. An AR display comprising images, symbols or markings (such as a virtual circle around an actual switch) considerably simplifies the work processes. Fault diagnostics can be made much more efficient by integrating component names and assembly instructions. For example, (virtual) animated repair instructions superimposed visually over a faulty machine component can contribute towards shortening repair times thus reducing costs [PLAP99].

Technically, the operator position must be detected exactly for superposition of the virtual information on the actual components of the machine tool. "Head trackers" are used for this. They detect the head position or viewing direction of the user. Tracking methods can be based on the most varied of physical working principles. Magnetic, optical, inertial and ultrasound trackers are very common. The highest accuracies however are attained by hybrid tracking methods that combine the benefits of the methods mentioned. Less suitable for use on machine tools are the common magnetic methods. This is because a lot of metal in the machine environment considerably reduces the accuracies attainable.

Note generally that, from a technical viewpoint to display process information, there are very many options to perform an information transformation to the respective perceptual channels of the user. So, for targeted provision of process information, the display form offering the best user assistance for the specific tasks at hand must be checked for every application. Figure 11.90 shows an overview of adapted technical solutions for human–machine interfaces. The realizations here are based on requirements coming from the different areas of computer-aided interfaces. First, graphically interactive and, where applicable, touch-based, user interfaces in combination with voice input have proven suitable for the control of complex production systems and machine tools (Fig. 11.90, left).

Second, it is a sensible approach to use combined input/output elements for the workplace on the machine tool. Furthermore, it should not just be the formatting and visual representation of information that is understood by the term human–machine interface. In the first step, the design of possible forms of interaction must include the systematic analysis of the user-specific working environment. It is not until the specific requirements are taken into account that appropriate technical concepts can be drawn up that form the basis of adequate technical solutions.

## References


[ANDE90] Anderl, R., Castro, P.: CAD/CAM – Auf dem Weg zu einer branchenübergreifenden Integration. Springer, Berlin (1990)

[AZUM97] Azuma, R.: A survey of augmented reality. Presence **6**(4), 355–385 (1997)

[BART99] Bartelt, V.: Konzeption und Entwicklung eines Systems zur Digitalisierung und anschließenden Bearbeitung unbekannter Freiformflächen. Dissertation RWTH Aachen (1999)

[BREC12a] Brecher, C., Kolster, D., Herfs, W.: Audio-tactile feedback mechanisms for multi-touch HMI panels of production engineering systems. International Journal of Automation Technology, Nr. 3 (2012)

[BREC12b] Brecher, C., Breitbach, T., Müller, S., Mayer, M. P., Odenthal, B., Schlick, C. M., Herfs, W.: 3D assembly group analysis for cognitive automation. J. Robot. **9**, 1–18 (2012)

[BROU00] Brouer, N.: NC-Steuerungskern mit Datenschnittstelle für eine Autonome Produktionszelle. Shaker, Aachen (2000)

[COSC93] COSCOM: CIM-Bausteine CAD-CAM-DNC. Ebersberg, 1993. Firmenschrift

[DIN74] Norm DIN 66215. Programmierung numerisch gesteuerter Arbeitsmaschinen. CLDATA. Allgemeiner Aufbau und Satztypen (1974)

[DIN81] Norm DIN 55003 Teil 3. Bildzeichen numerisch gesteuerter Werkzeugmaschinen (1981)

[DIN82] Norm DIN 24900 Teil 10. Bildzeichen für den Maschinenbau (1982)

[DIN83a] Norm DIN 30603 Teil 1. Graphische Symbole für Einrichtungen (1983)

[DIN83b] Norm DIN 66025 Teil 1. Programmaufbau für numerisch gesteuerte Arbeitsmaschinen (1983)

[DIN88] Norm DIN 66025 Teil 2 (1988). Industrielle Automation. Programmaufbau für numerisch gesteuerte Arbeitsmaschinen. Wegbedingungen und Zusatzfunktionen (1988)

[DIN96] Norm DIN ISO 2806. Industrielle Automatisierungssysteme – Numerische Steuerung von Maschinen – Begriffe (1996)

[EVER94] Eversheim, W., Müller G., Katsy, B.R.: NC-Verfahrenskette. Beuth, Berlin (1994)

[FRAN91] Franz, D.: CAD/CAM-Basisbetrachtungen. Berlin. Firmenschrift, Siemens AG (1991)

[HAAS93] Haasis, S., Zimmermann, R.: Effizienter Einsatz der CAD/NC-Kopplung. Expert, Ehningen (1993)

[HARD13] Hardwick, M., Zhao, Y. F., Proctor, F. M., Nassehi, A., Xu, X., Venkatesh, S., Odendahl, D., Xu, L., Hedlind, M., Lundgren, M., Maggiano, L., Loffredo, D., Fritz, J., Olsson, B., Garrido, J., Brail, A.: A roadmap for STEP-NC-enabled interoperable manufacturing. Int. J. Adv. Manuf. Technol. **68**(5–8), 1023–1037 (2013)

[HEKE92] Hekeler, M.: Werkstattorientierte Produktionsunterstützung. Esslingen. Firmenschrift, TRAUB AG (1992)

[HOYM05] Hoymann, H.: Entwicklung einer sprachlichen Benutzungsschnittstelle für Werkzeugmaschinen. Diss. RWTH Aachen (2005)

[INDR99] Indramat GmbH: Indramat MTC 200. Lohr. Firmenschrift (1999)

[ISO00] Norm ISO 14649-CD. Data model for CNC controllers, commitee draft (2000)

[ISO89] Norm ISO/DIS 6132 (1989). Industrial automation systems – Numerical control of machines – Extended format and data structure

[KAMP70] Kamp, W.: NC-Maschinen, Fachwörter und Definitionen. VDI Düsseldorf (1970)

[KELL08] Keller, S., Reuter, W.: CNCplus Production 5.0. Wuppertal. Firmenschrift (2008)

[KIEF99] Kief, H.B.: NC/CNC-Handbuch 2000. Hanser, München (1999)

[KLEI95] Klein, F.: NC-Steuerung für die 5-achsige Fräsbearbeitung auf der Basis von NURBS. Diss. RWTH Aachen (1995)

[KREB92] Krebser, G.: Betriebssystem für NC. Springer, Berlin (1992)

[KROM94] Kromberg, J.: Erfüllter Traum. NC-Fertigung, 2 (1994)

[MILB89] Milberg, J.: NC-Werkzeugmaschinen und rechnerintegrierte Produktion. Vorlesungsumdruck, Lehrstuhl für Werkzeugmaschinen und Betriebslehre. TU München (1989)

[NN99] N.N.: Der weite Weg zur Werkstatt-CNC. Maschine + werkzeug 10, 136–143 (1999)

[NORT06] North, K., D'Souza, H.: Acoustic pulse recognition enters touch-screen market. Inf. Disp. **12** (2006)

[OSAC00] OSACA Association: OSACA Handbuch 2.0. Stuttgart (2000)

[OSAC97] : Style guide Werkzeugmaschinen. Ein Handbuch zur Gestaltung von Benutzungsoberflächen für Werkzeugmaschinen. Fraunhofer IRB, Stuttgart (1997)

[PFEI93] Pfeiffer, H.-J.: Optimierung und Kontrolle von NC-Drehprogrammen durch Simulation. VDI-Z-Special CAD/CAM, **5**, 6–18 (1993)

[PFEI98] Pfeifer, T.: Fertigungsmesstechnik. Oldenbourg, München (1998)

[PLAP99] Plapper, V., Wenk, C., Weck, M.: Augmented Reality unterstützt den Teleservice. wt Werkstatttechnik **89**(6) (1999)

[PRIT86] Pritschow, G., Storr, A., Gruhler, G., Schuhmacher, H. (Hrsg.): Off-line programming system with geometrical data recording by Manually guided industrial robot. IFIP Working Conference on Off-Line-Programming of Industrial Robots, Amsterdam (1986)

[PROS17] PROSTEP AG: ► http://www.ProSTEP.de. [Stand: 2017]

[ROBE99] Robert Bosch GmbH: Bosch Typ 3 osa. Erbach. Firmenschrift (1999)

[SAUT87] Sauter, R.: Numerische Steuerungen im Werkzeugmaschinenbau – Funktionen, Programmierung und Betrieb. Vogel, Würzburg (1987)

[SCHÖ08] Schöning, J., Brandl, P., Daiber, F., Echtler, F. H. O., Hook, J., Löchtefeld, M., Motamedi, N., Muller, L., Olivier, P., Roth, T., Zadow, U.: Multi-touch surfaces : A technical guide. Technical Report TUM-10833, München (2008)

[SHAH72] Shah, R.: NC Guide. NCA, Zürich(1972)

[SIEM12] Siemens AG: Siemens Automation Webseite. ► http://www.automation.siemens.com. [Stand: 25.09.2012]

[SIEM16] Siemens AG: Sinumerik 840 D. Katalog CA 01. Erlangen. Firmenschrift (2016)

[SIEM97] Siemens AG: User's Manual, SINUMERIK 840D, OEM-Package NCK. Erlangen (1997)

[SIEM99] Siemens AG A&D: Schneller von der Zeichnung zum Werkstück: Einfacher Fräsen mit ShopMill. Erlangen. Firmenschrift (1999)

[SIMO71] Simon, W.: Die numerische Steuerung von Werkzeugmaschinen. Hanser, München (1971)


11

[SOSO90] Sosonkin, V.L.: Mikroprozessorsysteme für CNC-Steuerungen für Werkzeugmaschinen. Technik, Berlin (1990)

[SPAT09] Spath, D., Weisbecker, A. (Hrsg.), Laufs, U., Block, M., Link, J., Ardilio, A., Schuller, A., & Bierkandt, J.: Multi-Touch: Technologie, Hard-/Software und deren Anwendungsszenarien. Stuttgart. Firmenschrift (2009).

[SPUR91] Spur, G.: Vom Wandel der industriellen Welt durch Werkzeugmaschinen: eine kulturgeschichtliche Betrachtung der Fertigungstechnik. Hanser, München (1991)

[SPUR93a] Spur, G., Woijcik, L., Flexibel vom CAD-Modell zum NC-Programm. ZwF **2**, 71–74 (1993)

[SPUR93b] Spur, G.: Zusammenhang von Informationstechnik und Fertigungstechnik. In: FMS Flexible Manufacturing Systems; Past – Present – Future. 25th CIRP International Seminar on Manufacturing Systems. Bled (1993)

[SPUR94] Spur, G., Specht, D., Schröder, S.: Die numerische Steuerung. Cult. Tech. Innov. 621–735 (1994).

[TRAU92] TRAUB AG (Hrsg.): Drehen. (Reihe: CNC-Ausbildung für die betriebliche Praxis. Bd. 3). München: Hanser. Firmenschrift (1992)

[VDI90] VDI-Gemeinschaftsausschuß CIM; VDI-Gesellschaft Entwicklung, Konstruktion, Vertrieb (Hrsg.): Produktdatenverarbeitung. (Reihe: Rechnerintegrierte Konstruktion und Produktion. Bd. 2). VDI, Düsseldorf (1990)

[WECK05] Weck, M., Brecher, C., Wisniowski, J., Wolf, J.: Prozessgekoppelter Joystick zur Maschinenführung. wt – Werkstattstechnik online **95**(5), 314–319 (2005)

[WECK91] Weck, M.: Simulation in CIM. TüV Rheinland, Köln (1991)

[WECK92] Weck, M., Mertens, R., Schubert, I.: Sicherheit an Großmaschinen I und II. VDW-Forschungsbericht, Aachen (1992)

[WECK93] Weck, M.: Die offene Steuerung - Zentraler Baustein leistungsfähiger Produktionsanlagen AWK'93: Wettbewerbsfaktor Produktionstechnik, p. 4. VDI, Düsseldorf (1993)

[WECK96] Weck, M.: „Wie offen hätten Sie's denn gern?" – Offene Systeme in der Fertigung. AWK'96: Wettbewerbsfaktor Produktionstechnik, p. 2. VDI, Düsseldorf (1996)

[WECK97] Weck, M., Brouër, N., Daude, R.: Steuerung und Benutzungsschnittstelle einer Autonomen Produktionszelle. In: Weck, M., Spur, G., Pritschow, W. (Hrsg.) Zukunftsweisende Steuerungs- und Maschinenkonzepte für die Fertigung. (Reihe: 2). VDI, Düsseldorf, 1–17 (1997)

[WERK79] Werkzeugmaschinenlabor WZL der RWTH Aachen: Handbuch des Informationszentrums für Schnittwerte (INFOS). Zerspanseminar. Eigendruck WZL/RWTH Aachen (1979)

[WOLL86] Wollenberg, R.: Geometriedatenerfassung bei räumlichen Modellen. tz für Metallbearbeitung **80**, 8 (1986)

# Command Variable Generation and Interpolation

I thank my staff members Dr.-Ing. Marcel Fey, Christian Fimmers M. Sc., Ziqi Huang M. Sc. and Marian Wiesch M. Sc. for their contribution in revising this chapter.


C. Brecher and M. Weck, *Machine Tools Production Systems 3*,
Lecture Notes in Production Engineering, https://doi.org/10.1007/978-3-658-34622-5_12

## Contents

In an NC controller, the processing of the geometrical data plays a key role as this processing converts the traversing information, which is specified in the form of NC programs, to machining motions. As well as generating a path according to the programmed data, this task also involves monitoring functions and taking corrections into account. Corrections and monitoring functions must be performed online. It is not possible to carry out a precalculation offline.

In general, current NC controllers are implemented as digital computers. This also means that the entire function scope is created and implemented in the form of software. Hardware solutions for interpreters or interpolators belong to the past. The control software is created by the manufacturer using high-level languages. These languages have replaced machine-oriented programming in Assembler, which was common before. The reasons for this include that earlier restrictions caused by the limited performance of the processors have been eliminated. The manufacturers are also benefiting from the advantages of high-level programming languages. These include considerably shorter development times, independence from the computer platform used and easier maintainability, expandability, and reuse of the functionalities.

The variables to be processed can be represented in either floating point format or integer arithmetic. In the past, only integer arithmetic was used as numerical coprocessors, which enable calculations to be carried out quickly based on floating point numbers, were either not available or involved costs which were too high. This has changed in the recent past. Now, in newer, high-level performance NC controllers, geometrical data can be processed quickly using the floating point format.

The position of the Geometrical Data Processing function block within the whole control system is shown in ◘ Fig. 12.1. Tool data, correction values, and NC program data, which are usually available in accordance with DIN 66025 [DIN83] (cf. ► Chap. 11), serve as input. In general, the program data are converted line by line by the interpreter into a control-internal data format and then passed on to the Geometrical Data Processing function block. This essentially has the functions shown in ◘ Fig. 12.1: path corrections according to the current tool geometry, interpolation, position, and velocity influencing as well as transformation. Both the user input and correction values must be included in the online calculation due to either tool geometry deviations or position sensor signals. User input can either come from actuators such as rotary switches or it can come directly from the keyboard of the control board. These data primarily concern override values, changeable by the user, for the feed, spindle speed, and zero offsets. Besides the correction values specified by the user, the Geometrical Data Processing

12

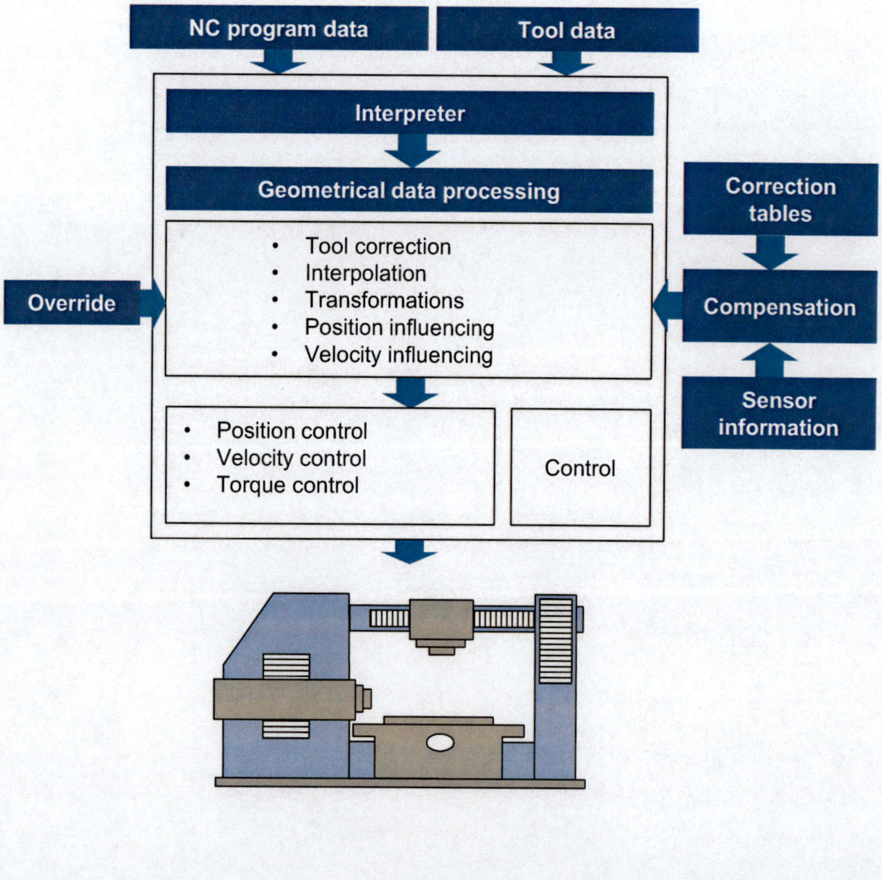


◘ **Figure 12.1** Position of the Geometrical Data Processing within the NC controller

function block must calculate by addition the control variables determined by a Compensation calculation function block. With the help of sensor signals, both the compensation of mechanical or thermal displacements (path influencing) and the adaptive control of tool loading and machine power can be performed by changing the feed rate.

Following the correction calculations, the position reference values determined are passed to the position control. Together with the underlying velocity and torque control, this position control is to control the machine axes so that, despite the high acceleration and frictional forces caused by the drive mechanism and other disturbance variables (process forces, thermoelastic displacements of the machine structure, etc.), any deviations from the desired path as kept are small as possible.

## 12.1 Interpolation

In the mathematical sense, interpolation is defined as the calculation of the intermediate values of a specified function using interpolation points in accordance with the specified calculation rules [MESC66]. However, for NC controllers, the term interpolation is usually taken to mean this function together with the generation of a desired velocity and acceleration profile along the path. So that the required path and velocity characteristics can be produced, the interpolation process must calculate the reference position values for the individual machine axes and then output them in a timely manner to the axes so that, taking the local path velocity into account, the specified path can be followed. The desired workpiece contour is then produced by the guided tool along this path. The essential requirements on an interpolation process are given below:

- The shape elements produced by the interpolation process (straight lines, circles, splines, etc.) are to approximate the desired workpiece contour as closely as possible.
- For technological reasons, it must frequently be possible to set the path velocity within broad limits and, taking the acceleration limits of the feed drives into account, this velocity must be followed almost constantly regardless of the contour shape.
- The numerically specified end points must be reached precisely by the interpolator as otherwise interpolation-related path errors (e.g., rounding errors) can add up.
- External corrections (e.g., position of the override) must be calculated online.
- If possible, the reference values of the velocities, accelerations and possibly the jerk are to be calculated or limited such that they can actually be achieved by the actual machine.

The geometry and the reference velocity of the path to be followed are passed to the NC controller by the information contained in the NC programs. Nowadays, the programs are usually executed according to DIN 66025 (cf. Sect 11.3) [DIN83]. The geometry elements contained in this standard are straight lines, circles, and parabolas lying in a plane. Most of the control system manufacturers circumvent the restrictions resulting from this by making their own enhancements to the standard programming interface. This means that, in general, the programming for different control systems is not standardized.

The interpolation process can make use of up to three individual modules. These are an upstream coarse interpolator (if desired or necessary), the actual interpolator and a downstream fine interpolator. The last one is part of the position control in most control systems, ▫ Fig. 12.2.

The coarse interpolator has the task of approximating complex programmed geometry elements at the specified accuracy using simple geometries like, for example, straight lines or circles. The interpolator calculates intermediate points along the paths it has received so that the desired path and velocity can be followed by having the individual axes approach these values at the correct time. Besides interpolation in the mathematical sense, the interpolator is also responsible for the velocity and acceleration control. Usually, the downstream fine interpolator calculates a specified number of equidistant intermediate values between the axis values generated by the interpolator. Most control systems have at least the interpolator and the fine interpolator.

The output data produced by the interpolation refer to the reference positions of the machine axes and the reference times when the axes should be in these positions. In general, they are passed on to the position controller as absolute values.

In the subsequent ions, the velocity and acceleration controls are initially described separately from the fundamental path. The interpolation of simple paths is then explained. How the creation of these paths relates to the general velocity and acceleration control is also shown.

### 12.1.1 Functions for the Velocity and Acceleration Control of Simple Paths Based on NC Lines

The *physical variables* position or *travel* $s(t)$, *velocity* $v(t)$, *acceleration* $a(t)$ and *jerk* $r(t)$ are of key importance for the velocity control. Through differentiation with respect to time, each variable in the given sequence results from the previous variable as can be seen below:

$$r(t) = \frac{da(t)}{dt} = \frac{d^2v(t)}{dt^2} = \frac{d^3s(t)}{dt^3} \qquad (12.1)$$

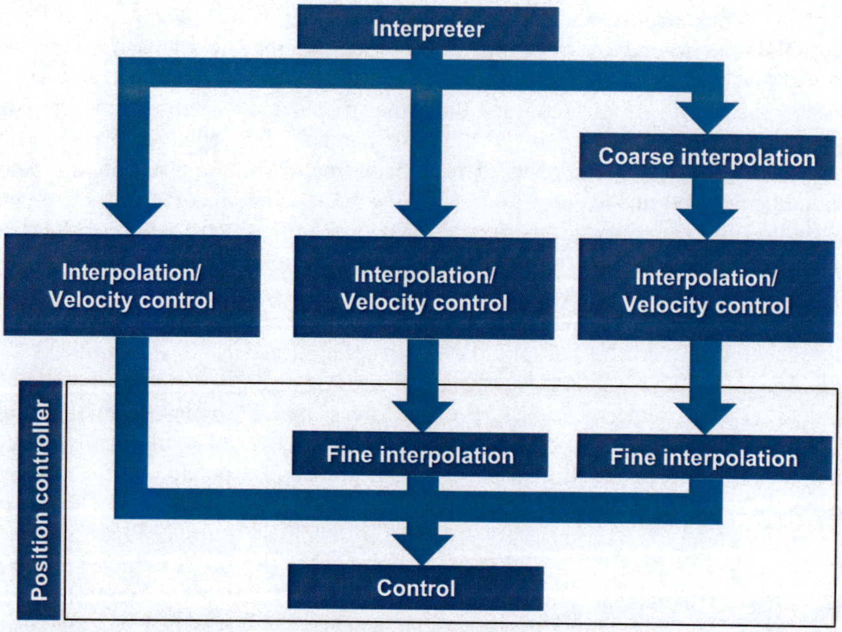


**Figure 12.2** Alternative interpolation structures



In most cases, from both an economical and technological point of view, it would be ideal to use the constant programmed feed rate as the surface would then have equal roughness values and the machining time would be kept to a minimum. However, this is not always possible with strongly curved or cornered paths due to the limited acceleration capacity of the feed drives. Sudden changes in direction, e.g., at workpiece corners, would require endless accelerations and decelerations at a finite constant velocity. The velocity control must then specify the feed rate for the axes of motion concerned so that the programmed contour can be followed as closely as possible taking into account the finite dynamic range of the drive systems. The motors and drive mechanisms of the machines must not be overloaded. The velocities and accelerations of all the axes must be taken into account when considering the dynamics. These, after all, generate the path velocity through vectorial superposition.

Disregarding the rise time of the current, which for modern motors is on the order of 1 to 10 ms, an initial approximation can assume a *sudden change in the acceleration* of the drives provided that this change is less than the possible limiting acceleration. This model representation leads to a controlled movement with a continuous velocity characteristic and has the advantage that relatively simple computing algorithms, which can also be processed using the processors available in older control systems, can be used, Fig. 12.3. The decisive parameters to be specified for this purpose in the control system are the maximum axis acceleration, which also represents the largest possible axis acceleration jump, and the maximum axis velocities.

However, this drive velocity control operating with a theoretically infinitely large jerk has distinct disadvantages. So that the load on the mechanical drive components arising from the sudden changes in acceleration can be kept within the permitted limits, the maximum possible drive acceleration jump cannot be used.

Modern processors can handle even more complex calculations. Therefore, in most cases, a *linear or higher-order acceleration change* model will be the basis for new control systems. This behavior avoids both velocity and acceleration jumps and has a finite jerk, Fig. 12.4.

The defining parameters for such a control system are the maximum axis velocity, the maximum axis acceleration and the maximum change of the axis acceleration over time, i.e., the maximum jerk. These values are permanently set in the control system and stored as machine parameters.

With this type of velocity control, the machine motion is quieter and smoother. Using higher maximum accelerations usually more than makes up for the time lost as a result of the smaller change in acceleration.

The *velocity control with acceleration jumps followed by low-pass filtering* is a third, lesser used, type of velocity control. The resulting velocity profile is usually referred to as bell-shaped or sine$^2$-shaped.

In principle, velocity control can be implemented in two different ways in numerical control systems. In each case, the control variable must be discretized as a function of the travel or time. The difference between the two ways is that the output of the traversing motions is carried out using either a constant travel grid

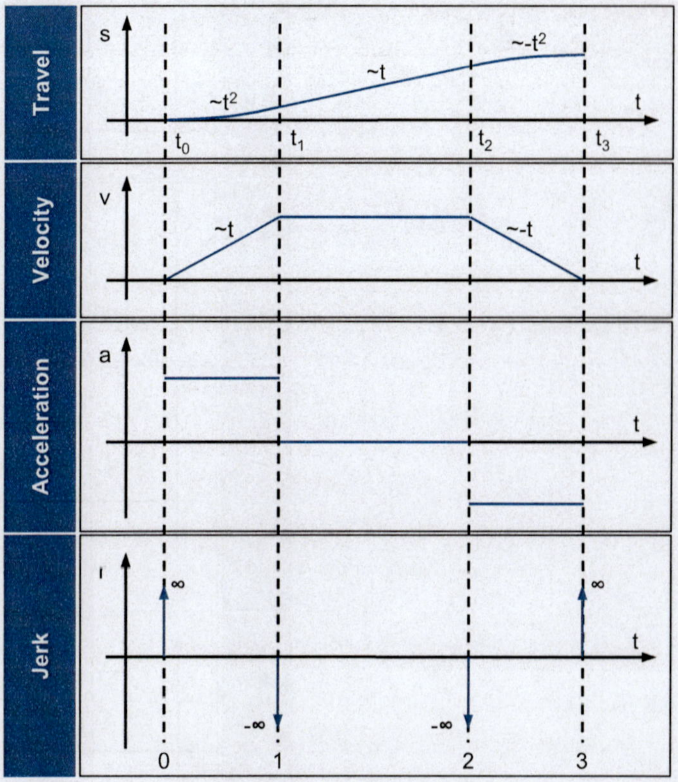


**Figure 12.3** Velocity control with sudden change in acceleration (infinite jerk)

or a constant time grid. A constant travel grid in this context means that the relative distance which is to be traversed for each output is constant whereas, with the constant time grid, the time period between two outputs must be constant, Fig. 12.5.

In both cases, the other parameter is used to model the velocity profile, i.e., with the *time grid method* the travel is used and, with the *travel grid method*, the time is used. The travel grid method is commonly used in the field of step motor technology for example. However, in practice, the time grid method is the much more frequently used variant. Therefore, in the subsequent sections, only this method will be considered. If a timed interpolation is assumed, the velocity control must control the output of the reference positions of the individual axes such that the reference difference in the axes to be traversed in each timing cycle models the specified path and velocity profile.

The timing cycle valid for the interpolation is called the *interpolation cycle.* All the calculations for the velocity control as well as the mathematical interpolation must be completed within this cycle. Otherwise, the machine would jerk or come to a complete stop. Therefore, the velocity control must execute the following subtasks:

- increase the velocity during the acceleration phase,
- implement the constant velocity phase,
- reduce the velocity during the deceleration phase,
- brake application point recognition,
- consider the feed override,
- proactive velocity control.

First of all, it is assumed that the path velocity at the beginning and end of each NC line is zero. The interpolation of simple paths is covered in ▶ Sect. 12.1.3. The feed override, as well as the processing of it in the control system, are covered in ▶ Sect. 12.3.2.

#### 12.1.1.1 Acceleration and Deceleration Phase

Acceleration is implemented either in one or three phases depending on which law of acceleration is used. If acceleration jumps are used, acceleration is constant in one single phase until either the programmed maximum path velocity or that velocity stored as a machine-specific value is reached or exceeded in the following time interval. Here, it is assumed that the maximum path velocity is specified as being equal to the smallest maximum velocity of all the axes. Making this assumption greatly simplifies the other calculations. The acceleration phase is completed when the axis velocity has been increased by the residual difference that is smaller than the maximum acceleration capacity, Fig. 12.6.

This residual difference results from the fact that an integral multiple of the interpolation cycles is not required to generate the programmed velocity difference.

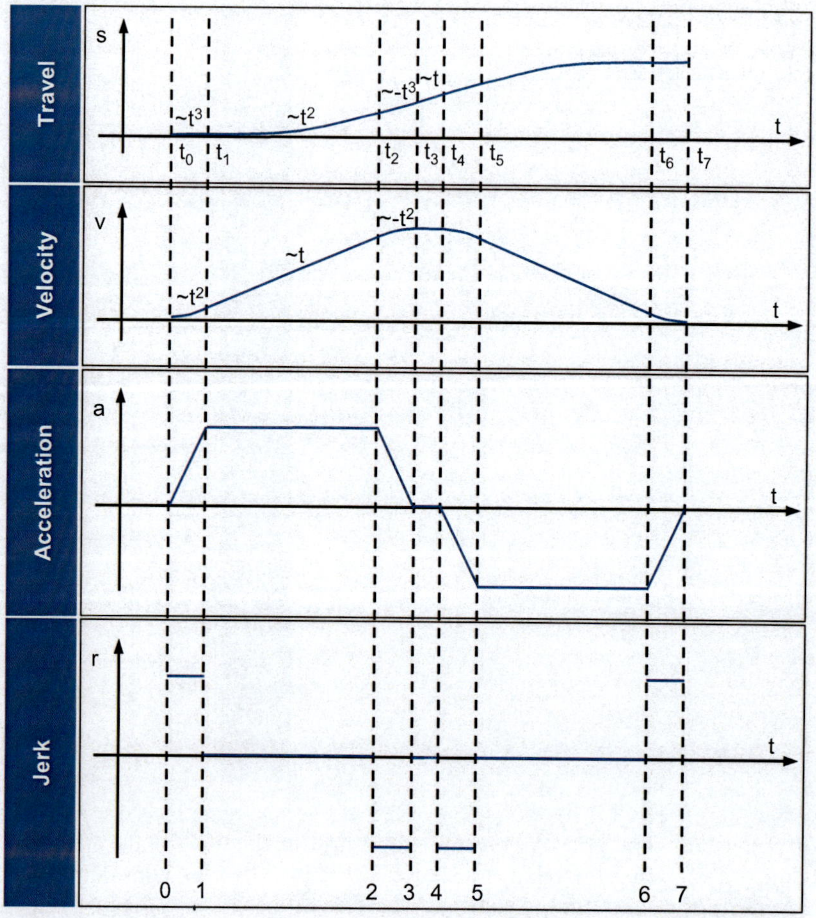


**Figure 12.4** Velocity control with continuous acceleration (finite jerk)

For example, the change in velocity within one interpolation cycle at a maximum acceleration of 1 m/s$^2$ and a cycle duration of 1 ms is 1 mm/s or 60 mm/min. With a programmed change in velocity of 126 mm/min or 2.1 mm/s the residual velocity difference, which is to be output in the third cycle after two cycles in which the full acceleration has been output, is 0.1 mm/s. The acceleration for the 3rd interpolation cycle is therefore reduced to a = (0.1 mm/s)/0.001 s = 0.1 m/s$^2$. Here it is important to note that the velocity achieved during the acceleration phase must be monitored in each interpolation cycle. The same applies to the brake application point recognition described in the following. The corresponding values are also to be monitored at the end of the current interpolation interval.

If acceleration jumps are used, then the following applies to the entire acceleration phase: 0 ⇒ 1 ($t_0 \leq t < t_1$) as per Fig. 12.3.

$$a(t) = a_{\max}$$

$$v(t) = \int_{t_0}^{t} a(t)dt + v(t_0) = a_{\max} \cdot (t - t_0) + v(t_0) \tag{12.2}$$

With $v(t_0) = 0$, it becomes

$$s(t) = \int_{t_0}^{t} v(t)dt + s(t_0) = \int_{t_0}^{t} (a_{\max} \cdot (t - t_0))dt + s(t_0)$$
$$= \frac{1}{2} \cdot a_{\max} \cdot (t - t_0)^2 + s(t_0) \tag{12.3}$$

As with the other variants, it is assumed here that, at the beginning and end of motion, both the velocity and acceleration are zero. If a ramp-shaped acceleration profile is used, the acceleration phase will be split into the three steps shown in Fig. 12.4. Comparison with Fig. 12.3 shows that the acceleration profile here has the same qualitative characteristic as the velocity characteristic produced when acceleration jumps are used. The task of the acceleration phase is now divided into the three phases in accordance with Fig. 12.4: a linear increase in acceleration to its maximum value with constant positive jerk, constant acceleration and a linear reduction of the acceleration to zero with constant negative jerk.

During each of these phases, residual values are output so that the desired final values can be achieved exactly. In the range of the acceleration rise 0 ⇒ 1, the following applies to $t_0 \leq t < t_1$ (Fig. 12.4):

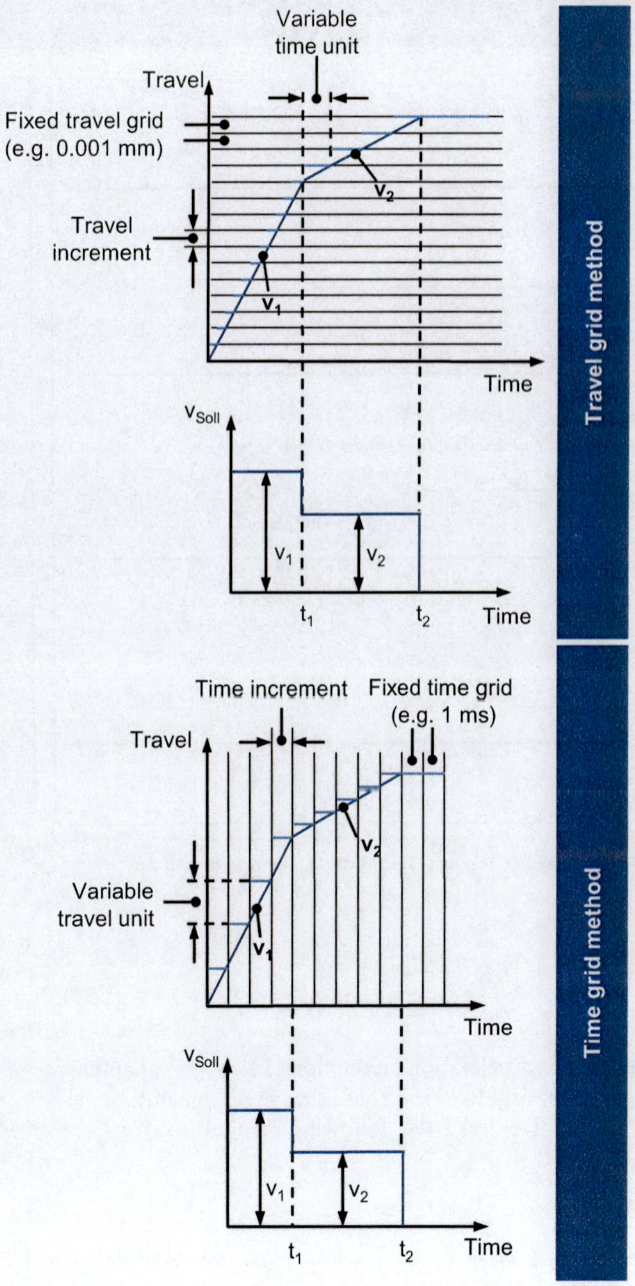


**Figure 12.5** Schematic diagram of the travel grid and time grid methods

$$a(t) = r_{\max} \cdot (t - t_0)$$

$$v(t) = \int_{t_0}^{t} a(t)dt + v(t_0) = \int_{t_0}^{t} (r_{\max} \cdot (t - t_0))dt + v(t_0)$$

$$= \tfrac{1}{2} r_{\max} \cdot (t - t_0)^2 + v(t_0)$$

With $v(t_0) = 0$, it becomes

$$s(t) = \int_{t_0}^{t} v(t)dt + s(t_0) = \int_{t_0}^{t} \left(\tfrac{1}{2} r_{\max} \cdot (t - t_0)^2\right)dt + s(t_0)$$

$$= \tfrac{1}{6} r_{\max} \cdot (t - t_0)^3 + s(t_0) \qquad (12.4)$$

For constant acceleration 1 ⇒ 2, the following applies to $t_1 \le t < t_2$ (Fig. 12.4):

$$a(t) = a_{\max}$$

$$v(t) = \int_{t_1}^{t} a_{\max} dt + v(t_1) = a_{\max} \cdot (t - t_1) + v(t_1)$$

$$s(t) = \int_{t_1}^{t} v(t)dt + s(t_1) \qquad (12.5)$$

$$= \int_{t_1}^{t} (a_{\max} \cdot (t - t_1) + v(t_1))dt + s(t_1)$$

$$= \tfrac{1}{2} a_{\max} \cdot (t - t_1)^2 + v(t_1) \cdot (t - t_1) + s(t_1)$$

where

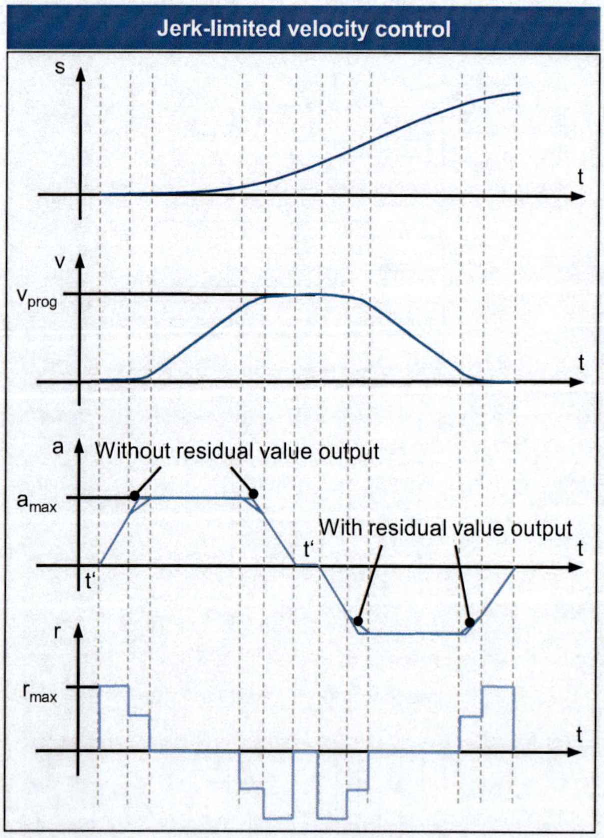


**Figure 12.6** Residual value output in the velocity control

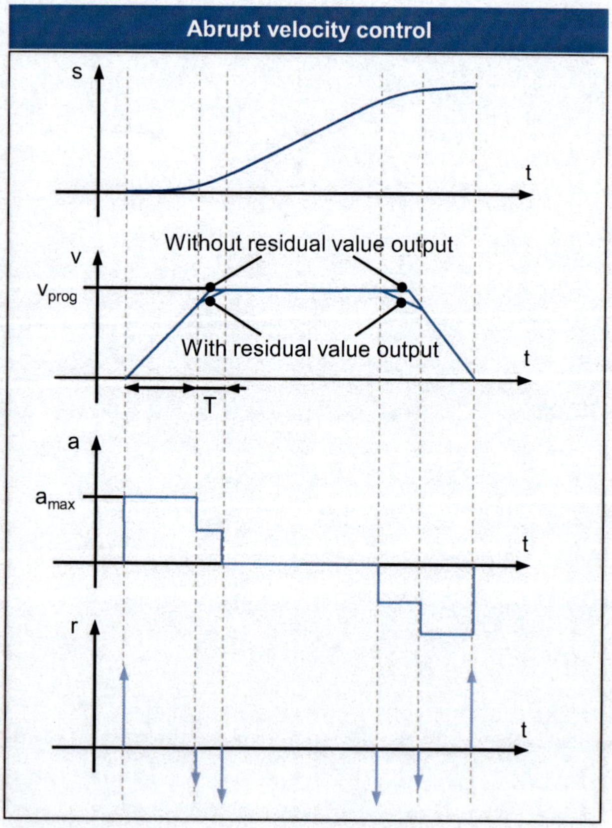


**Figure 12.6** (continued)

$$\begin{aligned} v(t_1) &= \tfrac{1}{2} r_{\max} \cdot (t_1 - t_0)^2 \\ s(t_1) &= \tfrac{1}{6} r_{\max} \cdot (t_1 - t_0)^3 + s(t_0) \end{aligned} \tag{12.6}$$

For the acceleration reduction 2 ⇒ 3, assuming that the negative jerk has the same magnitude as the positive jerk, the following applies to $t_2 \leq t < t_3$ (Fig. 12.4):

$$\begin{aligned} a(t) &= a_{\max} - r_{\max} \cdot (t - t_2) \\ v(t) &= \int_{t_2}^{t} a(t)dt + v(t_2) = \int_{t_2}^{t} (a_{\max} - r_{\max} \cdot (t - t_2))dt + v(t_2) \\ &= a_{\max} \cdot (t - t_2) - \tfrac{1}{2} r_{\max} \cdot (t - t_2)^2 + v(t_2) \\ s(t) &= \int_{t_2}^{t} v(t)dt + s(t_2) \\ &= \tfrac{1}{2} a_{\max} \cdot (t - t_2)^2 - \tfrac{1}{6} r_{\max} \cdot (t - t_2)^3 + v(t_2) \cdot (t - t_2) + s(t_2) \end{aligned} \tag{12.7}$$

where

$$\begin{aligned} v(t_2) &= a_{\max} \cdot (t_2 - t_1) + \tfrac{1}{2} r_{\max} \cdot (t_1 - t_0)^2 \\ s(t_2) &= \tfrac{1}{2} r_{\max} \cdot (t_2 - t_1)^2 \\ &+ \tfrac{1}{2} r_{\max} (t_1 - t_0)^2 \cdot \left(\tfrac{1}{3} \cdot (t_1 - t_0) + (t_2 - t_1)\right) + s(t_0) \end{aligned} \tag{12.8}$$

In general, for acceleration jumps and acceleration ramps, the deceleration phase takes place in much the same away as the acceleration phase. However, one difference is that the remaining travel or residual velocity (Fig. 12.6) is not output at the end as during the acceleration but instead is output at the start of each section. This is necessary to be able to reach the target point as quickly as possible.

#### 12.1.1.2 Constant Velocity Phase

The constant velocity phase follows the acceleration phase. As the name suggests, this phase ensures that the path velocity remains constant. It is important to note here that a constant path velocity or constant path acceleration is only equivalent to a constant axis velocity or acceleration in the special case of a linear motion, Figs. 12.7 and 12.4. The time/travel relationships resulting for this phase ($t_3 \leq t \leq t_4$) are as follows:

$$\begin{aligned} a(t) &= 0 \\ v(t) &= v(t_3) = \text{const.} \\ s(t) &= v(t_3) \cdot (t - t_3) + s(t_3) \end{aligned} \tag{12.9}$$

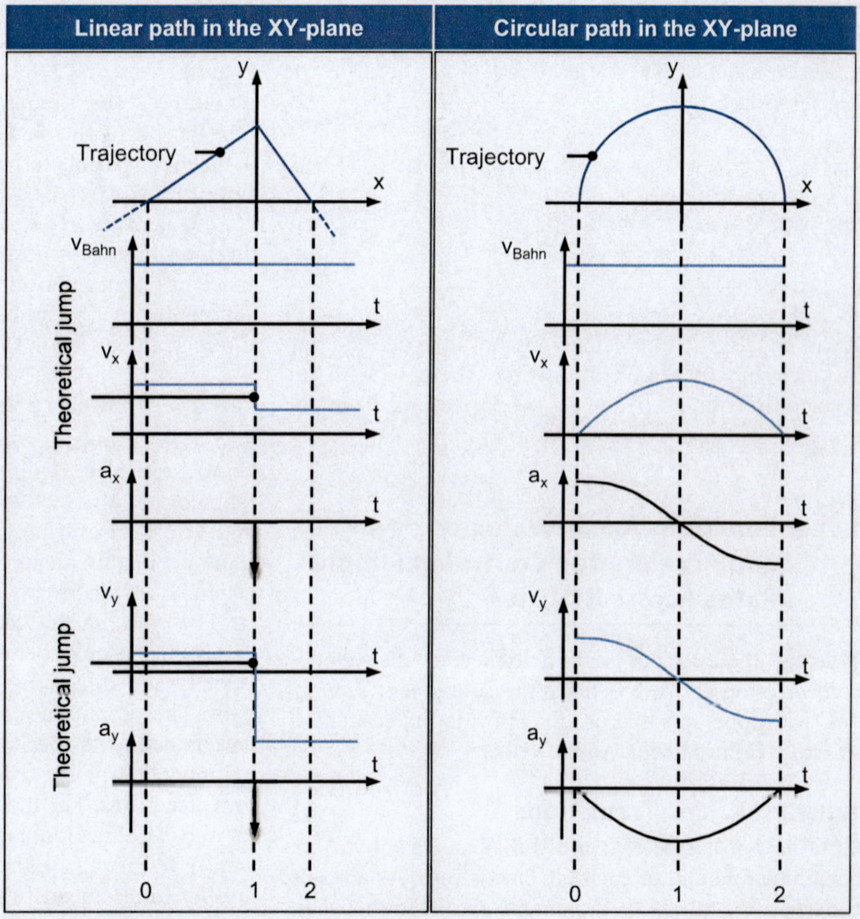


**Figure 12.7** Path and axis velocity for a linear and a circular motion with constant path velocity in the x/y plane

### 12.1.1.3 Brake Application Point Recognition

The brake application point recognition is to activate the deceleration phase in a timely manner so that the end point of the current NC line for traversing can always be reached exactly without any overshoot. Furthermore, to ensure the quickest machining times, the deceleration should not be started any earlier than required. In addition, it is also necessary to check the travel remaining at the end of the current interpolation cycle. This function must be active not only during the constant velocity phase but also during the acceleration phase. Within one interpolation cycle, this function is always executed before all other calculations. Depending on the result, it then branches into the functions for the stated phases, e.g., constant path velocity or deceleration. When there is a linear acceleration change, the times when a changeover is to occur between the three phases of the acceleration and deceleration must be monitored in a similar way.

If the velocity control is already operating in the constant phase, the minimum remaining travel, which must still be available from the end of the current travel interval, can be calculated as follows with specification of the acceleration jumps (Fig. 12.3):

$$s_{Rest} = \int v(t)dt = \int (v_{prog} - a_{max} \cdot t)dt$$
$$s_{Rest} = \frac{1}{2} \cdot \frac{v_{prog}^2}{a_{max}} \tag{12.10}$$

where $v_{prog}$ is the programmed reference velocity. In the case of a linear acceleration change assuming that $a_{max}$ is reached in the deceleration phase, the calculation then becomes a bit more complex. Having $v_{prog}$ as the programmed reference velocity yields the following calculation as per Fig. 12.4:

$$s_{Rest,4\Rightarrow 5} = \frac{a_{max}}{r_{max}}\left(v_{prog} - \frac{1}{6}\frac{a_{max}^2}{r_{max}}\right)$$
$$s_{Rest,5\Rightarrow 6} = \frac{1}{2} \cdot \frac{v_{prog}(r_{max} \cdot v_{prog} - a_{max}^2)}{r_{max} \cdot a_{max}}$$
$$s_{Rest,6\Rightarrow 7} = \frac{1}{6}\frac{a_{max}^3}{r_{max}^2} \tag{12.11}$$
$$s_{Rest} = s_{Rest,4\Rightarrow 5} + s_{Rest,5\Rightarrow 6} + s_{Rest,6\Rightarrow 7}$$
$$s_{Rest} = \frac{1}{2}\left(\frac{v_{prog}^2}{a_{max}} + \frac{a_{max} \cdot v_{prog}}{r_{max}}\right)$$

As can be seen, the required theoretical deceleration distance is dependent on the maximum jerk. It is always longer than the resulting deceleration distance for a velocity control with acceleration jumps and the same limiting values. Having an acceleration jump yields the following calculation for the time $t_{verz}$ required for the deceleration:

$$t_{verz} = \frac{v_{prog}}{a_{max}} \quad (12.12)$$

and with an acceleration ramp:

$$t_{verz} = \frac{a_{max}^2 + v_{prog} \cdot r_{max}}{a_{max} \cdot r_{max}} \quad (12.13)$$

It can be seen that, for calculating both the deceleration distance and the deceleration time, the case with the acceleration jumps (from $r_{max} \to \infty$) results from the more general case with acceleration ramps.

## 12.1.2 Functions for the Velocity and Acceleration Control of Simple Paths Across NC Lines

Whereas the previous section looked at the velocity control using single NC lines, the subsequent sections deal with functions that use the information content of several NC lines to control the velocity.

12

### 12.1.2.1 NC Line Transitions

Until now, it has been assumed that the velocity at the beginning and end of each NC line is zero. As a result, consideration of the velocity and acceleration behavior could be limited to single NC lines. However, control systems offer the possibility of selecting between exact positioning and an approximate velocity control via an NC program command. This means that, in the first case, the path velocity is reduced to zero towards the end of the NC line and, in the second case, it abruptly changes to the velocity of the next NC line. In addition, some manufacturers offer the option of specifying a velocity as a deceleration target at NC line transitions, ◘ Fig. 12.8. This represents a compromise between the two previous solutions.

For optimal production, it is important to keep the path velocity as constant as possible even at NC line transitions. However, at corners, the close link between the velocity control, the controller and the dynamic machine behavior is particularly apparent. A machine is an elastic, physical system with a mass and is unable to execute velocity or acceleration jumps. As a result, deviations arise in the actual values of the travel, velocity and acceleration behavior compared to the reference values.

The consequences of these deviations are shape defects that become all the bigger, the higher the velocity and acceleration jumps or their changes are. Even in the case of a reference velocity that has been reduced to zero at the corner point for an infinitely small period of time, the resulting path will deviate from the path coming from the reference velocity because a certain amount of time is required to reduce the tracking error.

The size of the deviation at the corner can be set as a machine parameter for the exact positioning function. In the extreme case of a permitted deviation from zero, it is necessary for the reference position value to dwell on this point (exact positioning) until the lag distance has been reduced. Some examples of the path resulting during a procedure with and without exact positioning can be seen in ◘ Fig. 12.9.

### 12.1.2.2 Proactive Velocity Control

The stated problems, which can result from the velocity characteristic at NC line transitions, have led to an anticipatory analysis of several NC lines being performed and velocities and accelerations being adapted to the accuracy requirements at the NC line transitions. In general, this function is known as Look-ahead.

The Look-ahead function analyzes several NC lines in advance, usually between around 10 and 100, for any velocity and directional changes occurring in the feed path. The permitted feed rates or accelerations can be determined for a specified permitted path deviation using modeling of the dynamic behavior of the machine or its feed axes. The modified reference velocity profiles for the NC lines under analysis are calculated in this way. In ◘ Fig. 12.10, the velocity profile determined using the Look-ahead function is compared to a profile which has been determined without it. The same maximum jump height for the velocity at the NC line transition is specified for both profiles.

As can be seen in ◘ Fig. 12.10, the Look-ahead function complies with the guaranteed accuracy and also permits use of the maximum allowed feed rate. This maximum rate can also be reached for a row of NC lines with a very small NC line length if the directional changes of the single NC lines are not too big. This is typically the case when machining freeform surfaces. A conventional velocity control system, together with its brake application point check, would ensure that the machine can stop within the assessable area. As this area only stretches to the end of the current block or, in special cases to the end of the next NC line, the travel length of, e.g., 0.05 or 0.1 mm is therefore not sufficient for the acceleration to deceleration or a complete stop of the programmed path velocity. On the other hand, a considerably larger section is analyzed in advance by the Look-ahead function. If it is determined that it is not necessary to stop an axis and the occurring velocity jumps or deviations do not exceed the specified limiting value at the NC line transitions, then the procedure can continue at a considerably higher feed rate.

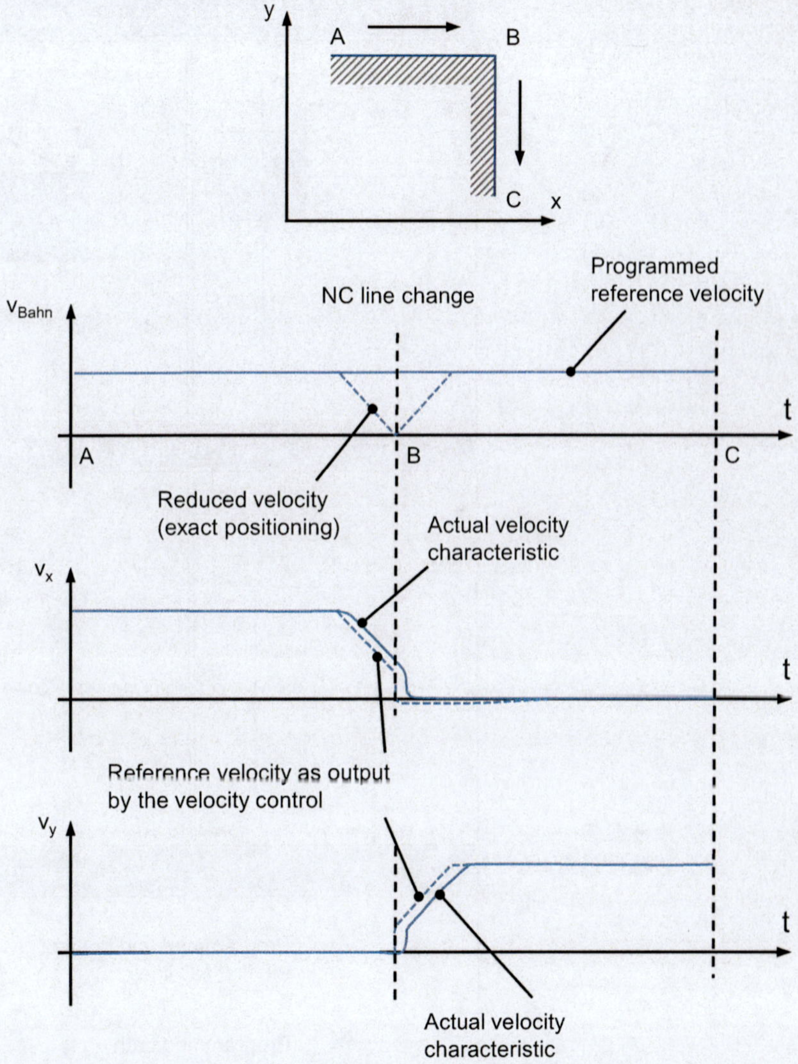


**Figure 12.8** NC line change with programmed reducing velocity (by Siemens)

For example, suppose a modern machine tool has an axis acceleration capability of 2 m/s$^2$. For the purposes of simplification, it is also assumed that the path acceleration is equal to the smallest axis acceleration. Then a machine could, within one NC line having a length of 0.1 mm or two NC lines with a combined length of 0.2 mm, accelerate to a maximum of 848.5 mm/min or 1200 mm/min, respectively. This is because it must be ensured that the machine can come to a complete stop at any time inside the area being considered. However, for modern high-speed machining operating with feed rates of up to 15 m/min, such restrictions are not acceptable.

In particular, in the area of high-speed freeform surface machining, the Look-ahead function allows considerably increased feed values while achieving a constant high level of shape accuracy compared with conventional control systems. It offers the possibility of working at the highest possible feed rate despite having to adhere to a defined accuracy. This applies in particular to consecutive NC lines that are short in length and only have a small amount of directional change.

### 12.1.3 Interpolation of Simple Paths

The task of the interpolation is to calculate the reference position values for the individual machine axes so that the desired path can be formed through their superposition. This creates the task of dividing a programmed path (e.g., a circular path) into intermediate points for the individual axes. A variety of interpolation methods have been developed to determine the individual axis-dependent commanded position variables. In general, derivation assumes the exact mathematical equation that can be represented for a Cartesian coordinate system in the following basic forms [BRON84]:

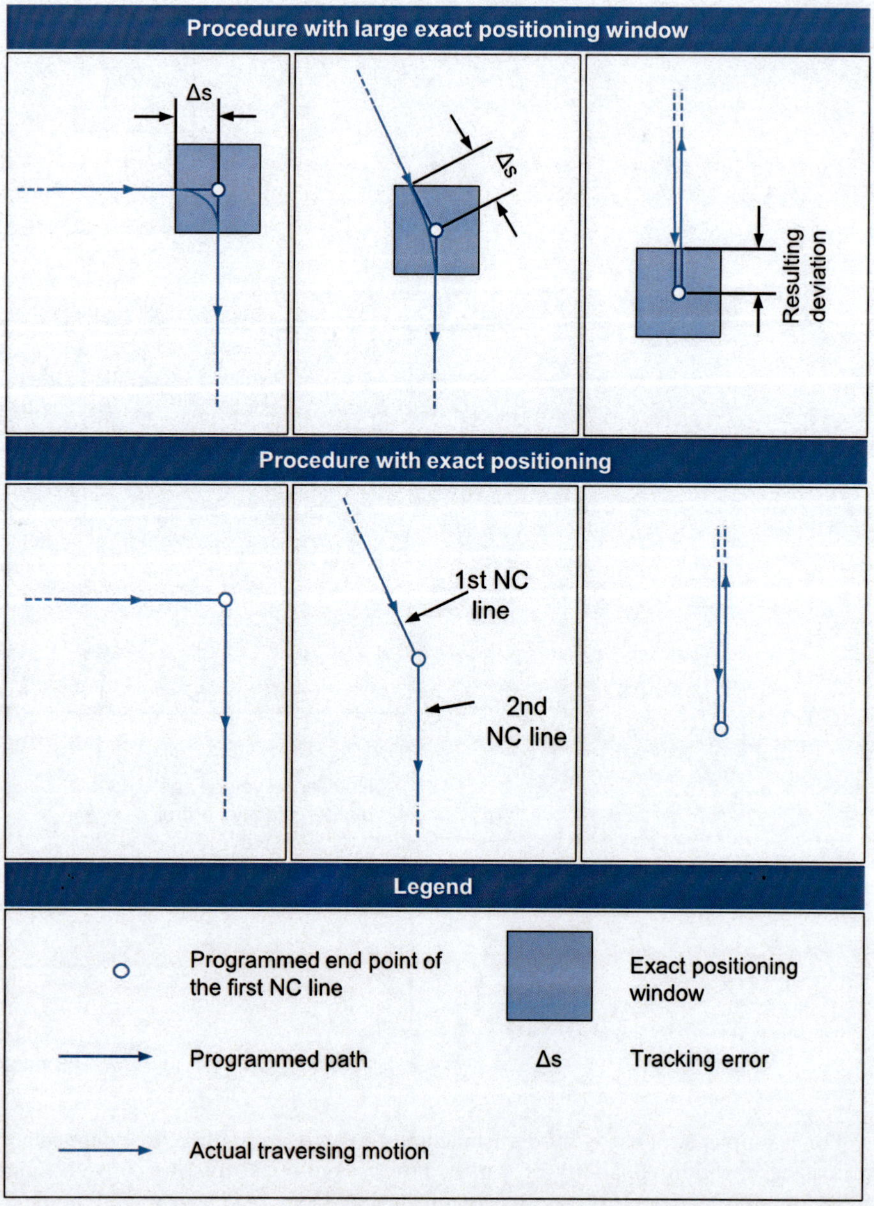


**Figure 12.9** Traversing motion with large exact positioning window and exact positioning



Implicit representation :

$$F(x, y, z) = 0 \tag{12.14}$$

Explicit representation :

$$\begin{aligned} x &= F(y, z) \\ y &= F(x, z) \\ z &= F(x, y) \end{aligned} \tag{12.15}$$

Parameter representation :

$$\begin{aligned} x &= F(u) \\ y &= F(u) \\ z &= F(u) \end{aligned} \tag{12.16}$$

with the common parameter u (e.g., time). A parametric form representation (Eq. 12.16) is particularly suitable for the derivation of interpolation methods. Firstly, it enables the clear representation of travels for which several values of the remaining axes must be assigned to the value of a single axis such as undercuts or grinding. Secondly, by using this type of representation, the calculations for interpolation and velocity control are made easier.

For the interpolation of simple paths, in particular straight lines and circles, some simplifying assumptions are usually made regarding the limiting values of the machine. The maximum values of the least capable axis are selected for the maximum path veloc-

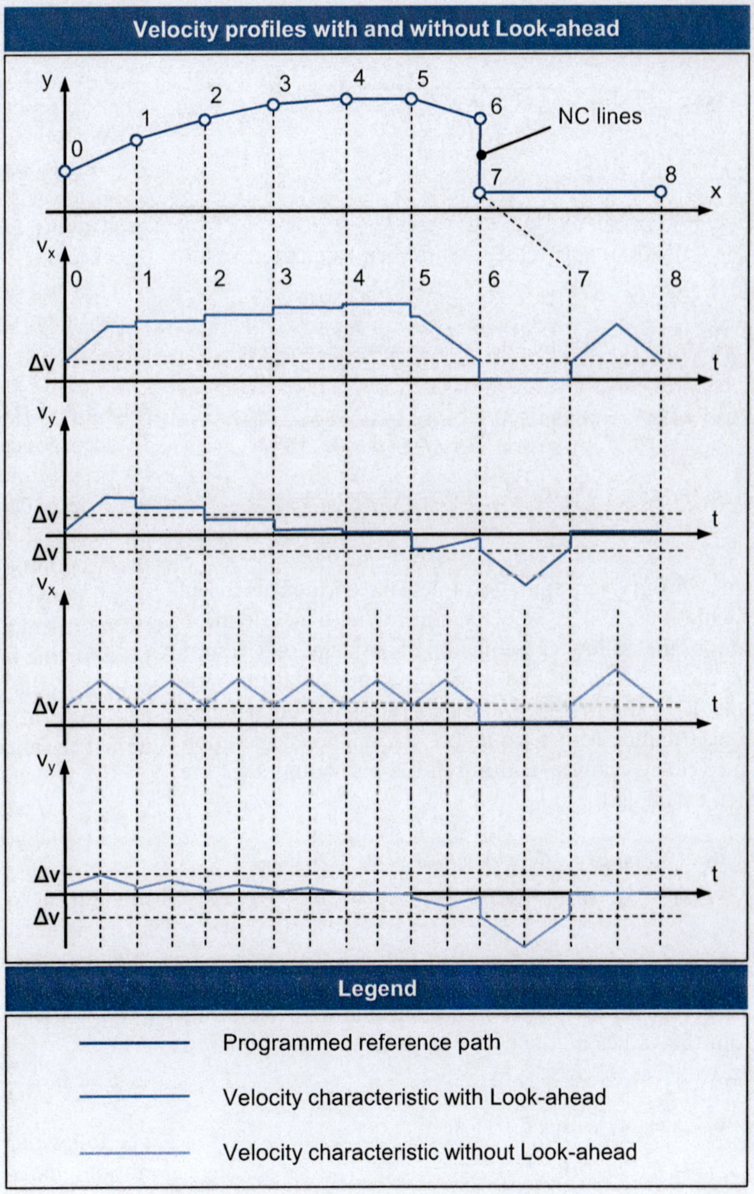


**Figure 12.10** Velocity characteristic with and without Look-ahead

ity, acceleration and, if applicable, acceleration change which the machine can achieve.

$$r_{max} = \min \lfloor r_{max,x}, r_{max,y}, r_{max,z} \rfloor \quad (12.17)$$

$$a_{max} = \min \lfloor a_{max,x}, a_{max,y}, a_{max,z} \rfloor \quad (12.18)$$

$$v_{max} = \min \lfloor v_{max,x}, v_{max,y}, v_{max,z} \rfloor \quad (12.19)$$

The dynamically least capable axis therefore definitively defines a machine's performance. This assumption has the advantage that the resulting calculations can be simplified significantly without overloading the machine at all.

#### 12.1.3.1 Linear Interpolation

Linear interpolation, i.e., dividing the path into small subsections for the individual axes, is nowadays usually performed through direct calculation of the axis values. For the function $s(t)$, which was used to implement the velocity control (cf. ▶ Sect. 12.1.1), the following coordinate vector results in the event of a straight line in three dimensions:

$$s(t) = (x(t), y(t), z(t))^T, T = \text{transposed} \quad (12.20)$$

In addition to the calculation of the axis values with linear interpolation, there is also the task of dividing the programmed path velocity $v_B$ across the individual axes. With regard to this, the ratios A between the coordinate distance and the total distance must, therefore, be calculated once at the start of the NC line:

$$\left|S_{\text{Satz,gesamt}}\right| = \sqrt{\Delta x^2 + \Delta y^2 + \Delta z^2} \tag{12.21}$$

for this NC line:

$$S_{\text{Satz}} = (\Delta x, \Delta y, \Delta z)^T \tag{12.22}$$

The individual axis velocities are then calculated as follows:

$$\begin{aligned} v_x &= v_B A_x \quad \text{where} \quad A_x = \frac{\Delta x}{\sqrt{\Delta x^2 + \Delta y^2 + \Delta z^2}} \\ v_y &= v_B A_y \quad \text{where} \quad A_y = \frac{\Delta y}{\sqrt{\Delta x^2 + \Delta y^2 + \Delta z^2}} \\ v_z &= v_B A_z \quad \text{where} \quad A_x = \frac{\Delta z}{\sqrt{\Delta x^2 + \Delta y^2 + \Delta z^2}} \end{aligned} \tag{12.23}$$

This avoids computation-intensive trigonometrical calculations. For a velocity control with acceleration jumps, the following sequence of calculations results for the acceleration phase with the interpolation cycle time T. In the first step, the axis ratios are determined in accordance with ▶ equation 12.23. The acceleration and velocity values resulting in the individual axes are then calculated.

12

$$\begin{aligned} &a_x = A_x \cdot a_{\max}, \quad a_y = A_y \cdot a_{\max}, \quad a_z = A_z \cdot a_{\max} \\ &v_{x,\text{Soll}} = A_x \cdot v_B, \quad v_{y,\text{Soll}} = A_y \cdot v_B, \quad v_{z,\text{Soll}} = A_z \cdot v_B \end{aligned} \tag{12.24}$$

During the acceleration phase, the axis reference values at the end of the $(n+1)$th interpolation cycle result from the values of the n-th cycle and the interpolation interval $T$ as follows:

$$\begin{aligned} v_{x,n+1} &= v_{x,n} + a_x \cdot T \\ v_{y,n+1} &= v_{y,n} + a_y \cdot T \\ v_{z,n+1} &= v_{z,n} + a_z \cdot T \end{aligned} \tag{12.25}$$

$$\begin{aligned} x_{n+1} &= x_n + v_{x,n} \cdot T + \tfrac{1}{2} \cdot a_x \cdot T^2 \\ y_{n+1} &= y_n + v_{y,n} \cdot T + \tfrac{1}{2} \cdot a_y \cdot T^2 \\ z_{n+1} &= z_n + v_{z,n} \cdot T + \tfrac{1}{2} \cdot a_z \cdot T^2 \end{aligned} \tag{12.26}$$

The calculation of the velocity change must only be performed once for each acceleration phase but can be added as a constant during the continued course of the acceleration.

For the constant velocity phase, the following results with the programmed path velocity $v_B$:

$$v_x = A_x \cdot v_B, \quad v_y = A_y \cdot v_B, \quad v_z = A_z \cdot v_B \tag{12.27}$$

To calculate each new position of the axes, it is only necessary to add a constant of the size $v_k \cdot T$ ($k = x, y, z$). The corresponding calculations resulting for the deceleration phase have the opposite sign.

With this interpolation method, as with all the others described in the subsequent sections, the brake application point check is to be performed during each cycle. Likewise, the velocity must be continually checked for its maximum value (machine parameter). The setting of the override switch (cf. ▶ Sect. 12.3.2) is to be taken into account when specifying the reference velocity.

#### 12.1.3.2 Circular Interpolation

In accordance with DIN 66025, a circle can only ever be programmed in one of the coordinate planes x/y, y/z or z/x. As an example, circular interpolation is described here in the x/y plane. For circular interpolation, the program data in accordance with DIN 66025 is first of all converted into the data required for a parametric representation, ◘ Fig. 12.11. The geometrical data used for the interpolation are the midpoint coordinates ($x_M$, $y_M$), the circle radius $R$, the starting angle $\beta_A$, the end angle of the motion $\beta_E$ and the direction of rotation. The function $s(t)$ of the circle is then given as:

$$s(t) = \begin{pmatrix} x(t) \\ y(t) \end{pmatrix} = \begin{pmatrix} x_M + R \cdot \cos(\beta(t)) \\ y_M + R \cdot \sin(\beta(t)) \end{pmatrix} \tag{12.28}$$

In addition to the purely geometrical data, the path velocity or acceleration must be converted to a rotational velocity or acceleration. The following applies to the rotational velocity and rotational acceleration specified in radians:

$$\dot{\beta} = \omega = \frac{v_B}{R} \text{ and } \ddot{\beta} = \dot{\omega} = \frac{a_{\max}}{R} \tag{12.29}$$

The following, therefore, results for the constant acceleration phase:

$$\dot{\beta}_{n+1} = \dot{\beta}_n + \ddot{\beta}_n \cdot T \text{with} T = \text{interpolation cycle rate} \tag{12.30}$$

$$\beta_{n+1} = \beta_n + \dot{\beta}_n \cdot T + \tfrac{1}{2} \cdot \ddot{\beta} \cdot T^2 \tag{12.31}$$

The law for determining the axis values results for the constant acceleration phase as follows:

$$\begin{pmatrix} x_{n+1} \\ y_{n+1} \end{pmatrix} = \begin{pmatrix} x_M + R \cdot \cos\left(\beta_n + \dot{\beta}_n \cdot T + \frac{1}{2}\ddot{\beta} \cdot T^2\right) \\ y_M + R \cdot \sin\left(\beta_n + \dot{\beta}_n \cdot T + \frac{1}{2}\ddot{\beta} \cdot T^2\right) \end{pmatrix} \tag{12.32}$$

The equations for the deceleration phase are identical. Though the acceleration then has a negative value. The

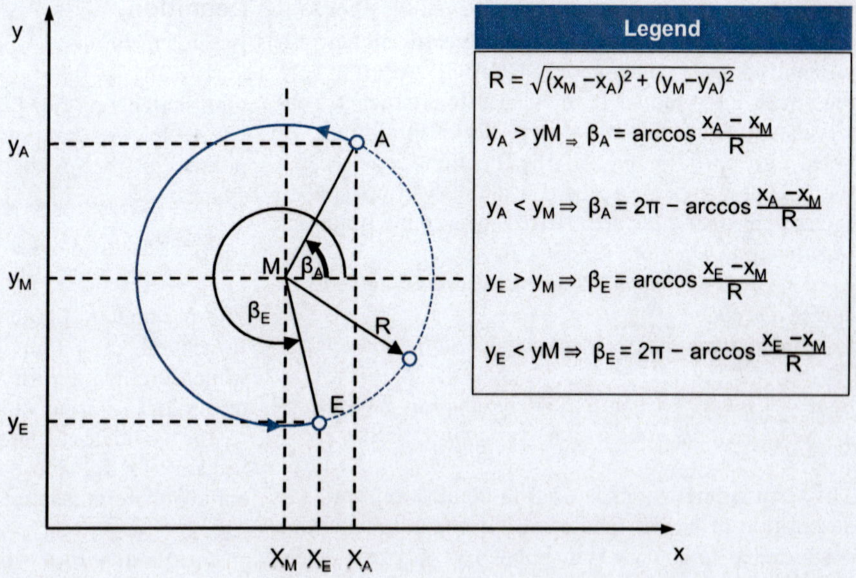


**Figure 12.11** Conversion of the programmed circle data to the internal parametric representation

arc length along the ideal circle is used for the brake application point check. The actual distance length, which is the sum of the secants, is unknown as the path velocity would be affected by override changes, for example. Different secant lengths will in turn result from changes in velocity.

The constant path velocity phase follows the acceleration. In the case of a circle, a constant path velocity corresponds to the variable velocities and accelerations of the individual axes. However, the angular increment:

$$\alpha = \omega \cdot T = \frac{v_B}{R} \cdot T \tag{12.33}$$

is likewise constant when the path velocity is constant. Nowadays, to calculate the axis values, ▸ equation 12.27 with constant angular increment is used in most cases to directly determine them. The following applies when the phase $v_B$ = constant:

$$s_{n+1}(t) = \begin{pmatrix} x_{n+1} \\ y_{n+1} \end{pmatrix} = \begin{pmatrix} x_M + \cos(\beta_n + \alpha) \\ y_M + \sin(\beta_n + \alpha) \end{pmatrix} \tag{12.34}$$

### 12.1.4 Spline Interpolation

The machining of freeform surfaces usually necessitates trajectories that cannot be described in exact mathematical terms using the contour elements of straight line and circle (cf. ▸ Sect. 11.3) included in DIN 66025. Using straight line segments, the complex characteristics can only be approximated with the specification of a tolerance. However, the resulting polygonal courses have some disadvantages compared to the original contour. Firstly, the transitions between straight line segments are not continuously derivable. When the program is being executed, this causes jumps in the velocity characteristic. These jumps unnecessarily load the drives and this is then reflected in poor surface quality. Secondly, the smaller the approximation tolerance, the smaller the length of the straight line segments. Small segment lengths, however, cause a high program volume and can lead to limitations and breaks in the machining feed due to limited NC line change times on the control side.

One approach to avoid these disadvantages is to use higher order mathematical functions for the contour description. Splines have been used for this purpose in machine construction for quite some time. In an effort to achieve continuity in the NC process chain nowadays, many NC controllers have the option of using spline interpolation in addition to the standard interpolation types of straight line and circle [FANU98; HEID96; FIDI98; NUM99; SIEM96]. In particular, when machining freeform surfaces, spline interpolation offers some advantages compared to conventional programming with linear path segments.

For instance, the travel length for each NC line is greater and, as a result, the data volume is smaller. This means that there is more time available for NC line processing. In addition, in most cases, the transitions between the single NC lines are continuous. The term splines originally referred to flexible slats used in shipbuilding. It was used for sketching outlines from which the displacement and the resistance of the hull could be determined. The shape coming from bending the slat with specification of a few stop points has the property of minimizing the bending energy of the slat. This shape corresponds approximately to a cubic polynomial spline with a constant curvature.

Classical interpolation or approximation polynomials are generally not suitable if cross sections of profiles or guide curves are to be represented as in the above example. They tend to oscillate strongly with higher polynomial degrees [HOSC92; SCHÄ93]. Alternatively, the entire interval can be divided into smaller subintervals and the whole curve can be correspondingly divided into partial curves. If the functions representing the partial curves are to be continuously connected to each other, then these are referred to as spline functions or sub-spline functions [HOSC92].

*A segmented function S with polynomial segments of degree n is called*

a) *a spline function if S is continuously differentiable (n-1) times*

b) *a sub-spline function if S is continuously differentiable at least once and a) does not apply* [HOSC92].

The term spline curves is used in connection with the description of geometrical curves. In the following, this is shortened to splines. Here spline interpolation refers to the evaluation of the spline curve using discrete values. This must be distinguished from the functionality of interpolating points using splines, which is referred to in the following by the term "spline generation."

Nowadays, spline generation is mainly performed during program preparation with CAM systems. The NC controllers have corresponding interfaces to receive the splines, which have been calculated in the CAM systems. The interface formats used are either splines in B form [FANU98; HEID96; FIDI98; SIEM96] or splines in polynomial form [NUM99; SIEM96]. These two formats differ mathematically from each other in their fundamental basis functions. It is possible to convert between the two formats [VRIE91]. However, as the interfaces for spline formats have as yet not been standardized, these NC programs are specific to one particular control system and cannot be transferred. A new draft standard promises standardization in this area in the future.

At the moment, some NC controllers also offer control-internal spline generation functionality based on polynomial splines [NUM99; SIEM96]. However, with this feature, it is also necessary to program the points to be interpolated and the boundary conditions (continuity, gradient or curvature in the start and end points) required to specify the path.

### 12.1.4.1 Polynomial Splines

Polynomial splines are used in internal spline generation and as an interface format in NC programming. In general, the maximum polynomial degree is limited to three in both cases (cubic polynomials). They are used to convert curves for which only the interpolation points are known to a continuous function. In particular, procedures for generating polynomial splines on the basis of given interpolation points will be discussed in the following in accordance with the general definition.

#### Definition

In NC technology, splines must be represented parametrically so that any desired contour shapes can be represented (cf. Eq. 12.16). A vector U with ordered parameter values uniquely assigned to the interpolation points P of the spline forms the basis of the polynomial splines:

$$U = \{u_0, u_1, \ldots, u_m\}, \qquad u_0 < u_1 < \ldots < u_m \tag{12.35}$$

The parameters $u_0, \ldots, u_m$ must be strictly monotonic. In general, the parameter values can be freely selected while complying with this condition although in NC technology it is advantageous to have a relationship to the geometrical length of the spline segments (cf. Sects. 12.1.4.1.2 and 12.1.4.4). The definition for the cubic polynomial spline $S_i(u)$ is:

$$S_i(u) = A_i + B_i(u-u_i) + C_i(u-u_i)^2 + D_i(u-u_i)^3, \quad u \in [u_0, u_m], \quad i = 0(1)\,m-1 \tag{12.36}$$

where $A_i$, $B_i$, $C_i$ and $D_i$ are polynomial coefficients. The coefficients are vectors. The index $i$ describes the individual segments of the spline. The interpolation points of the curve are not explicitly contained in the definition, they result from the evaluation of the polynomial spline $S_i(u)$ at the knots $u_0, \ldots, u_m$.

#### Polynomial Determination

The polynomial coefficients are determined by specifying the interpolation points $P_0, \ldots, P_m$ and two boundary conditions. Boundary conditions can firstly be continuity conditions and, secondly, they can be specifications for the behavior of the spline at the boundary points. From these specifications, a suitable parameter vector is initially determined. The coefficients for all the spline segments are then calculated.

When determining the coefficients, it is necessary to differentiate between methods with global and local character. Global, in this case, means that all the specified points are included in the calculation at the same time and the changing or displacement of one single point changes the characteristic of the entire curve.

Local methods, on the other hand, only take into account a small number of adjacent points when calculating the coefficients of the current polynomial. This means that changes to individual points only affect a few adjacent polynomials.

#### Determining the Parameter Vector

In the first step, a parameter value must be assigned to each interpolation point. To achieve a close approximation to the original curve progression and to simplify further processing of the spline in the NC controller, it makes sense to select the parameters depending on the geometrical lengths, i.e., the chord lengths, between the interpolation points (◘ Fig. 12.12). These chord lengths $h_i$ are calculated using the following equation:

$$h_0 = 0$$

$$h_i = \sqrt{(x_i - x_{i-1})^2 + (y_i - y_{i-1})^2 + (z_i - z_{i-1})^2}, \quad i = 1(1)m \qquad (12.37)$$

Adding these up yields the global parameters $u_i$ for chordal parameterization:

$$u_i = \sum_{j=0}^{i} h_j, \qquad i = 1(1)m \qquad (12.38)$$

## Global Method

The requirement that the curve pass through the specified points results in the following boundary condition for the spline segments:

$$S_i(u_i) = P_i$$
$$S_i(u_{i+1}) = P_{i+1} \qquad \text{for } i = 0(1)m-1 \qquad (12.39)$$

As segment $S_i$ and segment $S_{i+1}$ both pass through point $P_{i+1}$, this means that the continuity of contact is also ensured by this boundary condition. Additional boundary conditions result from stricter continuity requirements. Tangential continuity or curvature continuity are common specifications for cubic splines. Equal values in the first derivative of the polynomials adjacent in the current point ensure tangential continuity. The following then results in point i:

$$S_i^{'}(u_i) = S_{i-1}^{'}(u_{i+1}) \qquad \text{for } i = 1(1)m-1 \qquad (12.40)$$

This condition creates paths that can produce an infinite jerk at constant path velocity. A maximum of four boundary conditions can be specified for cubic functions. Therefore, if curvature continuity is also specified, then all the points or intervals have to be considered at the same time. This explains the higher continuity of globally generated splines:

$$S_i^{''}(u_i) = S_{i-1}^{''}(u_{i+1}) \qquad \text{for } i = 1(1)m-1 \qquad (12.41)$$

As a result of Eq. 12.39, 12.40 and 12.41, $4\,(n-1)-2$ boundary conditions result for the needed $4 \cdot (n-1)$ coefficients. There are still two free boundary conditions in the boundary points. Here, it is frequently specified that the curvature be set to zero. A spline defined in this way is also referred to as a natural cubic spline. Such curve shapes exhibit finite jerk behavior at their transitions.

With

$$h_i = u_{i+1} - u_i, \qquad i = 0(1)m-1 \qquad (12.42)$$

the following results for the coefficients of the polynomial:

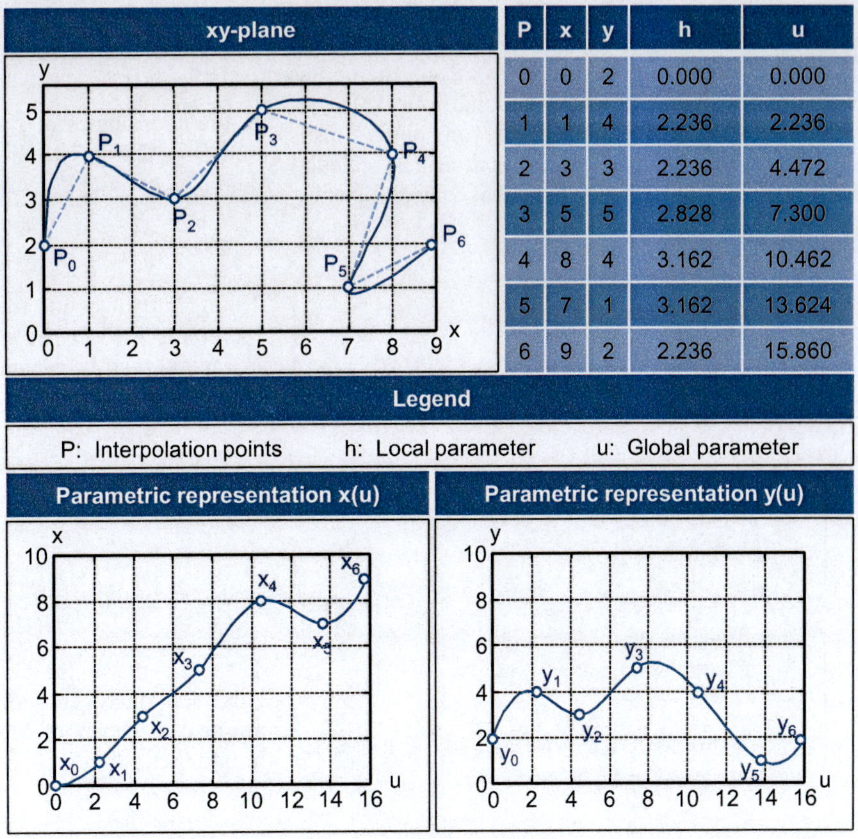


| P | x | y | h | u |
|---|---|---|---|---|
| 0 | 0 | 2 | 0.000 | 0.000 |
| 1 | 1 | 4 | 2.236 | 2.236 |
| 2 | 3 | 3 | 2.236 | 4.472 |
| 3 | 5 | 5 | 2.828 | 7.300 |
| 4 | 8 | 4 | 3.162 | 10.462 |
| 5 | 7 | 1 | 3.162 | 13.624 |
| 6 | 9 | 2 | 2.236 | 15.860 |

◘ **Figure 12.12** Parametric representation of a polynomial spline

$$x_i(u) = a_{x,i} + b_{x,i} \cdot (u - u_i) + c_{x,i} \cdot (u - u_i)^2 + d_{x,i} \cdot (u - u_i)^3$$
$$y_i(u) = a_{y,i} + b_{y,i} \cdot (u - u_i) + c_{y,i} \cdot (u - u_i)^2 + d_{y,i} \cdot (u - u_i)^3$$
$$z_i(u) = a_{z,i} + b_{z,i} \cdot (u - u_i) + c_{z,i} \cdot (u - u_i)^2 + d_{z,i} \cdot (u - u_i)^3 \quad (12.43)$$

and as an example for the $x$-axis:

$$a_{x,i} = x(u_i), \qquad i = 0(1)m$$
$$c_{x,0} = c_{x,m} = 0 \quad \text{(due to curvature = 0 at the end points)}$$
$$h_{i-1}c_{x,i-1} + 2c_{x,i}(h_{i-1} + h_i) + h_i c_{x,i+1} = \frac{3}{h_i}(a_{x,i+1} - a_{x,i}) - \frac{3}{h_{i-1}}(a_{x,i} - a_{x,i-1}) \text{ for } i = 0(1)m - 1$$
$$b_{x,i} = \frac{1}{h_i}(a_{x,i+1} - a_{x,i}) - \frac{h_i}{3}(c_{x,i+1} - 2c_{x,i}) \qquad \text{for } i = 0(1)m - 1$$
$$d_{x,i} = \frac{1}{3h_i}(c_{x,i+1} - c_{x,i}) \qquad \text{for } i = 0(1)m - 1 \quad (12.44)$$

where the coefficients $a_n$ and $c_n$ are only formally set. The equations in 12.44 can be represented as a tridiagonal equation system of $m-1$ equations (Eq. 12.45). This system can be uniquely solved with the given boundary conditions, e.g., using Gaussian elimination [ENGE96]. This method has the advantage that the determined spline segments are connected so that they are tangentially and curvature constant. This is particularly advantageous for the machine kinematics as the characteristic of the path acceleration is continuous at constant path velocity, for example. However, the disadvantages are the relatively high computational effort required to solve the tridiagonal equation system and the fact that changing individual points affects the course of the entire path. The computational effort also increases with the number of segments or points. This is particularly problematic for real-time critical calculations in the interpolators of numerical control systems.

$$Ac = g \quad \text{with}$$
$$A = \begin{pmatrix} 2(h_0 + h_1) & h_1 & & & \\ h_1 & 2(h_1 + h_2) & h_2 & & \\ & \ddots & \ddots & \ddots & \\ & & h_{m-3} & 2(h_{m-3} + h_{m-2}) & h_{m-2} \\ & & & h_{m-2} & 2(h_{m-2} + h_{m-1}) \end{pmatrix}$$
$$c = \begin{pmatrix} c_1 \\ c_2 \\ \vdots \\ c_{m-2} \\ c_{m-1} \end{pmatrix}$$
$$g = \begin{pmatrix} \frac{3}{h_1}(a_{x,2} - a_{x,1}) - \frac{3}{h_0}(a_{x,1} - a_{x,0}) \\ \frac{3}{h_2}(a_{x,3} - a_{x,2}) - \frac{3}{h_1}(a_{x,2} - a_{x,1}) \\ \vdots \\ \frac{3}{h_{m-2}}(a_{x,m-1} - a_{x,m-2}) - \frac{3}{h_{m-3}}(a_{x,m-2} - a_{x,m-3}) \\ \frac{3}{h_{m-1}}(a_{x,m} - a_{x,m-1}) - \frac{3}{h_{m-2}}(a_{x,m-1} - a_{x,m-2}) \end{pmatrix} \quad (12.45)$$

## Local method as per Akima

Local methods do not have the disadvantages already mentioned. However, they usually only allow tangential continuity. The missing boundary conditions for the calculation of a cubic polynomial are determined from either a few adjacent points or from geometry-independent criteria. The frequently used method by Akima is introduced here by way of example [AKIM70].

This method, developed in 1970, ensures that the characteristic of the spline is tangentially constant. In addition, the spline is very true to shape, i.e., it runs relatively smoothly and avoids overshooting [HOSC92; SPÄT90]. The method is purely geometrically based, ◘ Fig. 12.13. The interpolation in the x coordinate is considered here by way of example. The numbers $i=k-2$, $k-1$, $k$, $k+1$, $k+2$ of internal points $(u_i, x_i)$ are identified here by 1, 2, 3, 4, 5. Let $A$ be the intersection point of the straight line through points 1 and 2 with the straight line through points 3 and 4, and let $B$ be the intersection point of the straight line through points 2 and 3 as well as points 4 and 5. Let $C$ be the intersection point of a wanted straight line through point 3 with the straight line through points 1 and 2, and let $D$ be the intersection point of the same wanted straight line with the straight line through points 4 and 5. The wanted gradient $p_{3,x}$ in point 3 is then determined by specifying the straight line relationships:

$$\left|\frac{\overline{2C}}{\overline{CA}}\right| = \left|\frac{\overline{4D}}{\overline{DB}}\right| \quad (12.46)$$

If the coordinates of the points 1, 2, 3, 4, 5, $A$, $B$, $C$ and $D$ in ◘ Fig. 7.13 are identified by the indices 1, 2, 3, 4, 5, $a$, $b$, $c$ and $d$, then $p_{3,x}$ can be defined by:

$$p_{3,x} = \frac{x_d - x_c}{u_d - u_c} \quad (12.47)$$

Taking into account the geometrical relationships of the points A to D, the gradient $p_{i,x}$ in point $(u_i, x_i)$ can in general then be calculated as shown below using the chord gradients $m_i$, where:

$$m_{i,x} = \frac{x_{i+1} - x_i}{u_{i+1} - u_i} \quad (12.48)$$

and:

$$p_{i,x} = \frac{|m_{i+1,x} - m_{i,x}| \cdot m_{i-1,x} + |m_{i-1,x} - m_{i-2,x}| \cdot m_{i,x}}{|m_{i+1,x} - m_{i,x}| + |m_{i-1,x} - m_{i-2,x}|} \quad (12.49)$$

The gradients at the boundary points must be calculated separately. In most cases, specific assumptions are also made about the behavior of the spline at its boundaries.

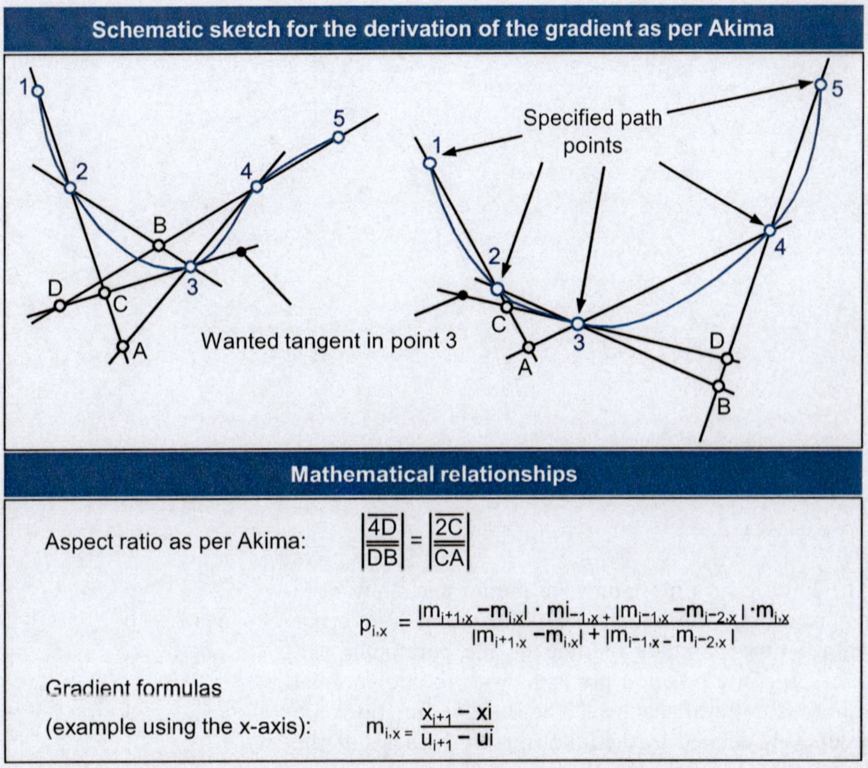


**Figure 12.13** Geometrical relationships for the derivation of the gradient as per Akima (according to [SPÄT90])

## Properties

Figure 12.14 shows an example of the characteristic of a cubic polynomial spline. The spline passes through the interpolation points $P_i$. The interpolation points mark the transition between two spline segments. The continuity of the spline depends on the method used to generate it (cf. ▶ Sect. 12.1.4.1.2): The global method produces curvature-constant splines whereas local methods like Akima only produce tangentially constant (sub-) splines.

For the programming of a cubic polynomial spline, the parameter interval is required for each segment and the four polynomial coefficients are required for each geometrical dimension. A potential source of error comes from the limited number of significant digits and the associated rounding errors. These can lead to discontinuities in the segment transition when evaluating the spline.

However, the continuity of contact can be ensured by directly specifying only three polynomial coefficients and calculating the fourth with the continuous connection condition from the preceding NC line.

Polynomial splines of degree n can be evaluated efficiently using Horner's method, which only requires the operations (addition/multiplication) for a spline

$$Z = 2n \tag{12.50}$$

function in one coordinate. A cubic polynomial spline in space can be evaluated for one specified parameter value using 18 operations.

### 12.1.4.2 B-splines

B(asis)-splines and the B-spline interpolation are supported for path control by current NC controllers. However, in industrial practice, they are currently used chiefly as an interface format in NC machining, i.e., control-internal interpolation of interpolation points with B-splines is only used occasionally. In the following, integral B-splines are considered first of all and then, in ▶ Sect. 12.1.4.3, the more complex rational form is briefly introduced. Both variants are used in NC technology.

## Definition

B-spline curves, in short B-splines, use B-spline functions as their basis functions. To define B-spline functions of degree n, a parameter vector $U$ (referred to as a carrier vector or knot vector) and $m+1$ spatial points, i.e., the de Boor points, are required. The knot vector $U$ consists of $k+1$ ordered knot values:

$$U = \{u_0, u_1, \ldots, u_k\}, \qquad u_0 \leq u_1 \leq \ldots \leq u_k \tag{12.51}$$

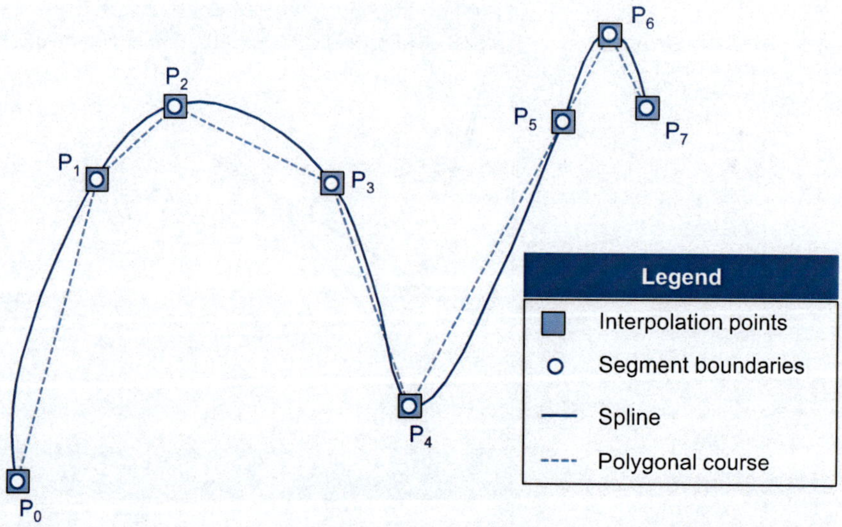


**Figure 12.14** Cubic polynomial spline



The knot values must only be monotonic, they need not be strictly monotonic, i.e., knot values may occur several times. Unique binding of the parameter values to the interpolation points as with the polynomial splines is not available here. The B-spline functions are recursively defined by the following on the basis of the knot vector U for the zeroth-degree functions:

$$N_i^0(u) = \begin{cases} 1 & \text{for } u \in [u_i, u_{i+1}) \\ 0 & \text{else} \end{cases}, \qquad j = 0(1)\,k-1 \tag{12.52}$$

and for higher degree functions by:

$$N_i^n(u) = \frac{u-u_i}{u_{i+n}-u_1} N_i^{n-1}(u) + \frac{u_{i+n+1}-u}{u_{i+n+1}-u_{i+1}} N_{i+1}^{n-1}(u), \qquad i = 0(1)k-n-1, \quad \frac{0}{0} := 0 \tag{12.53}$$

A B-spline curve (B-spline) is defined by the B-spline functions $N_i^n$ and the de Boor points $P_i$:

$$S_i(u) = \sum_{i=0}^{m} P_i N_i^n(u) \tag{12.54}$$

The following example helps to illustrate this definition: consider the knot vector $U=\{0, 0, 0, 1, 2, 3, 3, 3\}$. This knot vector defines three non-zero intervals [0, 1], [1, 2] and [2, 3] because the first and last three knots coincide. In accordance with Eq. 12.52, the basis functions in the non-zero intervals have the constant value 1 whereas the other basis functions yield a value of 0:

$$\begin{aligned}
N_0^0(u) &= 0 \\
N_1^0(u) &= 0 \\
N_2^0(u) &= \begin{cases} 1, & u \in [0,1) \\ 0, & \text{else} \end{cases} \\
N_3^0(u) &= \begin{cases} 1, & u \in [1,2) \\ 0, & \text{else} \end{cases} \\
N_4^0(u) &= \begin{cases} 1, & u \in [2,3) \\ 0, & \text{else} \end{cases} \\
N_5^0(u) &= 0 \\
N_6^0(u) &= 0
\end{aligned} \tag{12.55}$$

From the recursive definition of the higher degree basis functions in Eq. 12.53, it follows that the number of basis functions is reduced by one function with each increase in degree. The following results for first degree basis functions:

$$\begin{aligned}
N_0^1(u) &= 0 \\
N_1^1(u) &= \begin{cases} 1-u, & u \in [0,1) \\ 0, & \text{else} \end{cases} \\
N_2^1(u) &= \begin{cases} u, & u \in [0,1) \\ 2-u, & u \in [1,2) \\ 0, & \text{else} \end{cases} \\
N_3^1(u) &= \begin{cases} u-1, & u \in [1,2) \\ 3-u, & u \in [2,3) \\ 0, & \text{else} \end{cases} \\
N_4^1(u) &= \begin{cases} u-2, & u \in [2,3) \\ 0, & \text{else} \end{cases} \\
N_5^1(u) &= 0
\end{aligned} \tag{12.56}$$

and for second degree functions:

$$N_0^2(u) = \begin{cases} (1-u)^2, & u \in [0,1) \\ 0, & \text{else} \end{cases}$$

$$N_1^2(u) = \begin{cases} u(1-u) + u(2-u)/2, & u \in [0,1) \\ (2-u)^2/2, & u \in [1,2) \\ 0, & \text{else} \end{cases}$$

$$N_2^2(u) = \begin{cases} u^2/2, & u \in [0,1) \\ (u(2-u) + (3-u)(u-1))/2, & u \in [1,2) \\ (3-u)^2/2, & u \in [2,3) \\ 0, & \text{else} \end{cases}$$

$$N_3^2(u) = \begin{cases} (u-1)^2/2, & u \in [1,2) \\ ((u-1)(3-u) + (3-u)(u-2))/2, & u \in [2,3) \\ 0, & \text{else} \end{cases}$$

$$N_4^2(u) = \begin{cases} (u-2)^2, & u \in [2,3) \\ 0, & \text{else} \end{cases} \tag{12.57}$$

The characteristic of these second degree basis functions is shown in ◘ Fig. 12.15. The following information can be derived from it:

- Basis functions of degree n are continuously differentiable ($n-1$) times, provided that all the knots only occur once, i.e., the knots have a strictly monotonic order. This means that, to achieve a continuous velocity and acceleration profile during NC machining, curvature constant third degree basis functions with internal knots that only occur once must be available.
- The degree of the function n, the number of de Boor points $m+1$ and the number of parameter values $k+1$ have the following relationship:

$$k+1 = m+1+n+1 \tag{12.58}$$

As the number of points and the degree of the function are specified on the basis of the continuity requirements when using NC machining, the required number of knots can be determined using Eq. 12.58.

- Basis functions of the n-th degree are only defined over a limited carrier interval $[x_i, x_{i+1+n}]$. A local influence of the de Boor points on the curve shape results from the definition of the B-splines (Eq. 12.54).

### Properties

An example of the characteristic of a third degree B-spline is shown in ◘ Fig. 12.16. The B-spline only approximates the de Boor points, it doesn't pass through (interpolate) them. If the given interpolation points are selected on the contour as de Boor points, then contour damage will result. At the boundaries of the parameter area, the B-spline will only pass through the first or last de Boor point if the knots at the boundary occur $(n+1)$ times. In general, the condition must be met during NC machining so that the geometrical connection of the spline to the previous and to the next NC line for traversing is ensured. Knots occurring more than once within the parameter area are avoided so as not to reduce the continuity.

B-splines also have a positive property with respect to NC machining in that the spline runs completely within the control polygon formed by the straight-line connection of the de Boor points. As a result, the curve does not exhibit any unwanted oscillations.

NC programming of a B-spline curve requires specification of the $m+1$ de Boor points and the knot vector with $k+1$ knots. In total, only $3 \cdot (m+1)+(k+1)$ values have to be specified in the three-dimensional case. This is considerably fewer values than as is the case with a polynomial spline as several polynomial coefficients must be specified for each segment. As a result of the internal calculation of the basis functions based on the knot vector, the limited numerical transfer accuracy is not a problem.

The evaluation of B-splines of degree n can be carried out using de Boor's algorithm [HOSC92] without having to explicitly calculate the B-spline basis functions. This requires

$$Z = a\frac{7}{2}n(n+1) \tag{12.59}$$

operations where a denotes the geometrical dimension. In the three-dimensional case ($a=3$), 126 operations are required. This is considerably more operations than required for the evaluation of polynomial splines (cf. ► Sect. 12.1.4.1.6).

#### 12.1.4.3 NURBS

In NC machining, rational B-splines are frequently used instead of integral B-splines [FANU98; FIDI98; SIEM96]:

$$S_i(u) = \frac{\sum_{i=0}^{m} w_i P_i N_i^n(u)}{\sum_{i=0}^{m} w_i N_i^n(u)} \tag{12.60}$$

Here, $w_i$ are additional weighting factors used to influence the shape of the curve. The shape of a knot vector whose parameter intervals are not uniform is referred to as NURBS (Non Uniform Rational Basis Spline).

In CAD, NURBS is the dominant data format as, first, it can be easily and interactively modeled using the de Boor points and the weights and, second, compared with integral B-splines, any conic section can be precisely described mathematically. However, in the context of NC machining, the disadvantage is that, despite optimized processes [KLEI95], the evaluation effort is greater. The fact that it is now in widespread use as an interface format despite this disadvantage is due to the effort made to create a universal standardized data format within the NC process chain, i.e., from the

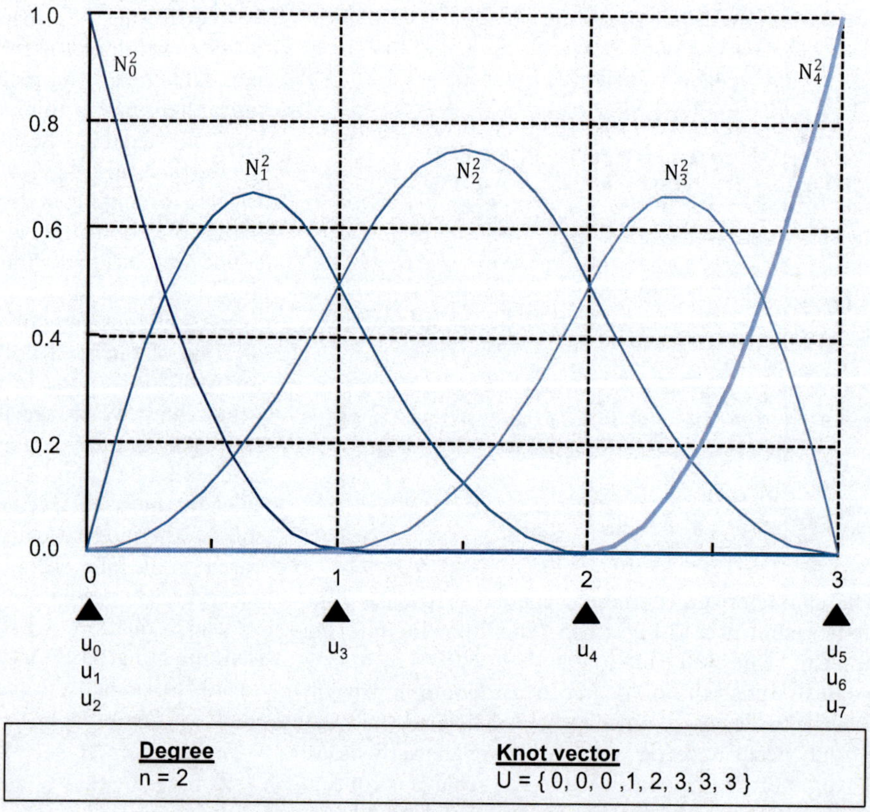


**Figure 12.15** B-spline functions

12

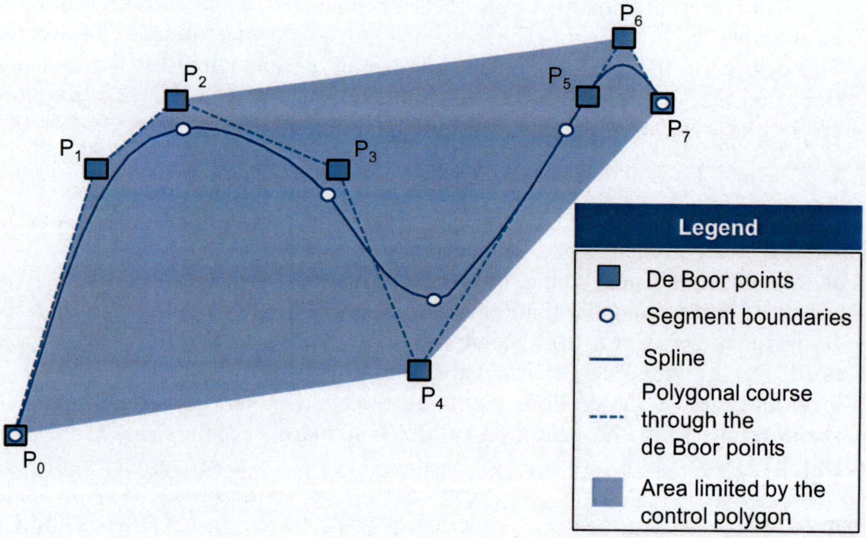


**Figure 12.16** Cubic B-spline

CAD system to the CAM system through to the NC controller.

### 12.1.4.4 Evaluation of Splines

Once the splines are in sections, spline interpolation can then be performed as a coarse interpolation with the following interpolator as described in general at the beginning of ▶ Sect. 12.1. This procedure has some advantages above all with regard to the processing of very complex splines. The coarse interpolation approximates the original geometry using simple elements such as straight lines. This makes the task of velocity control considerably easier and accelerates it. Alternatively, the interpolator can process the programmed geometry directly without using an upstream coarse interpolator. In both cases, a fine interpolator is connected downstream of the interpolator as described above.

The coarse interpolation is to divide the original geometry into simpler geometry elements such that the resulting deviation does not exceed a specified tolerance or deviation. The difficulty lies in determining, with as little computational effort as possible, either the resulting deviation by specifying a certain parameter step width or, conversely, the maximum possible parameter step width by specifying the maximum permissible deviation. The aim is to achieve as large a step width as possible so that the number of resulting elements and, with that, the data quantity can be kept to a minimum or so that the time available for the processing of each element can be kept as long as possible.

This is assuming that the spline is to be divided into straight-line elements here. The geometrical relationships shown in ▣ Fig. 12.17 then result. The parameter that determines the parameter step length is the maximum permissible deviation of the reference curve from the resulting polygonal course measured in the direction normal to the spline curve. A common procedure here is to first determine the geometrical step length by approximation from the assumption that the curvature is constant, i.e., a circle. According to the Pythagorean theorem, the relationship between the segment arc lengths L, the deviation E and the curvature radius R is:

$$L = 2R \cdot \arcsin\left(\frac{\sqrt{E \cdot (2R - E)}}{R}\right) \tag{12.61}$$

In addition, either the largest curvature of the polynomial must be determined or the curvature of the spline curve in the current point calculated and assumed to be constant in the vicinity of the point. In the first case, the step length is on the safe side, i.e., the resulting deviation is never greater than the specified tolerance. However, with curves having a strong change of curvature, the step width can, on average, be considerably shorter than required. This disadvantage can still be overcome by performing the calculation on the basis of the current curvature. This can prove to be problematic, however, in that, first, the determined step length can be calculated too large. This is always the case when the local change of curvature is positive. Second, the calculation is more complex as the local curvature must be determined in each coarsely interpolated point during real-time critical machining.

The geometrical step width L determined in this way must now be converted to an equivalent parametric step width $\Delta u$. This is necessary as the spline functions are in parametric form and the result of the preceding computational step is a geometrical length. In most cases, approximation methods are used here as an analytical conversion is not possible. Other numerical procedures which could be used as an alternative to determine the exact solution are ruled out as they are very compute-intensive and time-consuming.

A roughly linear relationship is assumed between the arc length and the parametric step width. With the geometrical step width L, the wanted parametric step width Δu and the arc length function *s*, we then get:

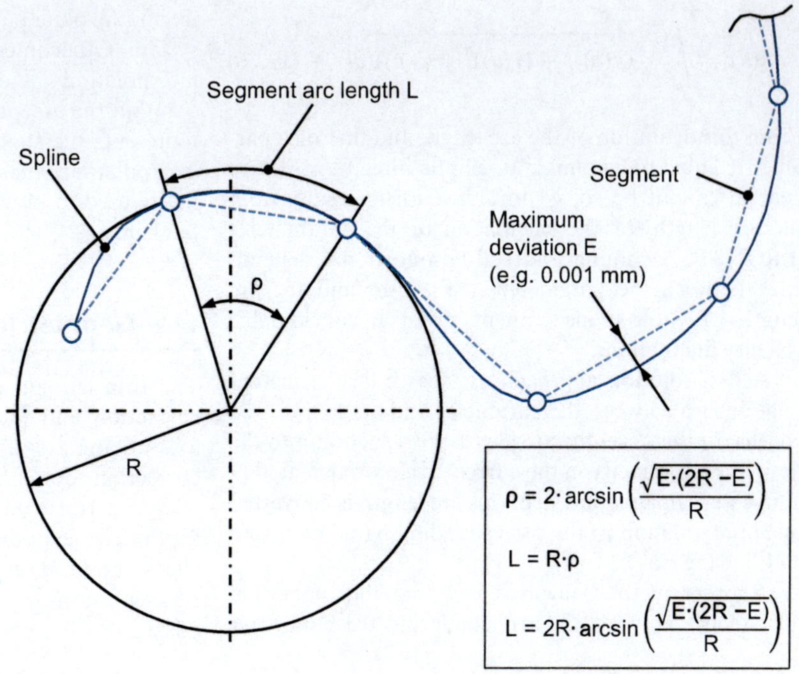


**▣ Figure 12.17** Spline curve with coarse interpolation in straight-line elements

$$\frac{L}{\Delta u} = \frac{ds}{du}, \tag{12.62}$$

$$s = \int \sqrt{\left(\frac{dx}{du}\right)^2 + \left(\frac{dy}{du}\right)^2 + \left(\frac{dz}{du}\right)^2} \cdot du, \tag{12.63}$$

$$\Delta u = f(L) = \frac{L}{\left(\frac{ds}{du}\right)} = \frac{L}{\sqrt{\left(\frac{dx}{du}\right)^2 + \left(\frac{dy}{du}\right)^2 + \left(\frac{dz}{du}\right)^2}}, \tag{12.64}$$

Alternatively, combining the determination of the geometrical step width on the basis of a local constant curvature as per ► equation 12.64 yields:

$$\Delta u = \frac{2 \cdot R \cdot \arcsin\left(\frac{\sqrt{E \cdot (2R - E)}}{R}\right)}{\sqrt{\left(\frac{dx}{du}\right)^2 + \left(\frac{dy}{du}\right)^2 + \left(\frac{dz}{du}\right)^2}} \tag{12.65}$$

Other tasks and functions will result from the interpolation and velocity control of splines if there is no upstream coarse interpolation. The task of a timed velocity control is to calculate the parameter step length for each interpolation cycle resulting in the desired path velocity characteristic. In addition, the velocity control must determine a suitable parameter step width by specifying the desired arc step. However, the difficulty here is that the arc length as a function of the parameter cannot be determined analytically even in the simple case of a cubic polynomial. The arc length of a parametric function in x, y and z can generally be calculated by:

$$s(u) = \int_{u_0}^{u} \sqrt{(\dot{x}(u))^2 + (\dot{y}(u))^2 + (\dot{z}(u))^2 dt} \tag{12.66}$$

From this definition of the arc length and that of a parametric cubic polynomial, an elliptic integral, which in general cannot be solved in closed form, results from the arc length for the calculation of the parameter t [BRON84]. Assuming a directly proportional dependency between the parameter step widths and the arc lengths of whole spline segments results in considerable velocity fluctuations.

A first solution approach assumes a local constant relationship between the parameter and arc length step widths. In this case, the arc length corresponding to the desired path velocity in the current cycle is calculated in a first step. In a second step, this arc length is converted by approximation to the corresponding parametric step width using Eq. 12.64.

Another method involves inverting the functional dependence between the arc length and the parameter step width section by section [LEI92]. For this, the functional dependence

$$X(u) = (x, y, z)^T = f(u) \tag{12.67}$$

with $u \approx g(s)$ is expanded to:

$$X(u) = X(g(s)) \tag{12.68}$$

The task of the function $g(s)$ is to approximate the relationship $u(s)$. Specifying a suitable $g(s)$ makes it then possible to use the algorithms described in ► Sect. 12.1.2 to calculate the velocity control by approximation.

The first solution approach is frequently implemented in modern control systems. This means that simple algorithms can be used for the velocity control but the path velocity cannot be kept exactly constant.

A new approach is to calculate polynomial splines within the control system that are parameterized with the arc length by approximation [FAUS97]. With these splines, the parametric step width corresponds to the geometrical step width. As a result, the complex calculation of the inverse arc length function as per Eq. 12.68 is no longer required.

### 12.1.5 Other Methods

In addition to the aforementioned current interpolation methods, there are some other methods resulting from requirements relating to the production of certain workpieces particularly in the area of three-axis milling. For the manufacture of threads and gear teeth, helix and involute paths have to be created by the control system. Other interpolation methods include parabolic interpolation and exponential interpolation.

With the exception of the helix interpolation, which represents the superposition of a circular and a linear interpolation, these other interpolation methods are rarely used in practice and will not be discussed in detail here.

## 12.2 Geometrical Transformations

The translational and rotational zero offset, the tool correction and the transformations required for five-axis milling are all included under this term. The tool correction represents a special case here. Its task is not only to perform a transformation, e.g., to determine the tool reference point from the tool contact point, but to check the programmed path at the same time for possible collisions with the reference contour of the workpiece.

### 12.2.1 Zero Offsets

The zero offset is used to align the workpiece zero point with the machine zero point [DIN83]. As already explained in ▶ Chap. 6, the difference between these two positions has to be entered either as an NC program command or directly at the machine. The zero offset can be translational or, if necessary, rotational (◘ Fig. 12.18). In older, less powerful control systems, only the translational zero offset is possible in most cases. In the control system, linear zero offsets are easily calculated by adding or subtracting the axial offset values. The value of the zero offset is also added to the programmed travel along with the appropriate plus or minus sign to calculate the new position.

Nowadays, most NC controllers also offer the option of using rotational coordinate offsets. This allows workpieces that are clamped askew to also be machined using the original NC program. However, as the rotational coordinate offset is not part of the standardized programming interface [DIN83], different codes and definitions are used by different manufacturers to enter these values. The basic procedure is, however, still standard. To determine a linear displacement with a rotational offset, the point of rotation must be entered first followed by the number of rotations. In addition, the rotation angles $\alpha$, $\beta$ and $\gamma$ of the workpiece around the original x-, y- and z-axes are to be defined in the right-handed coordinate system of the machine. The machine coordinates are denoted by $X$ and the transformed or workpiece coordinates by $X'$. With that, the transformation matrices of the individual rotations result for the conversion of a position from the transformed, i.e., the rotated coordinate system, to the basic machine system:

$$R_x = \begin{pmatrix} 1 & 0 & 0 \\ 0 & \cos\alpha & -\sin\alpha \\ 0 & \sin\alpha & \cos\alpha \end{pmatrix}$$

$$R_y = \begin{pmatrix} \cos\beta & 0 & \sin\beta \\ 0 & 1 & 0 \\ -\sin\beta & 0 & \cos\beta \end{pmatrix}$$

$$R_z = \begin{pmatrix} \cos\gamma & -\sin\gamma & 0 \\ \sin\gamma & \cos\gamma & 0 \\ 0 & 0 & 1 \end{pmatrix} \qquad (12.69)$$

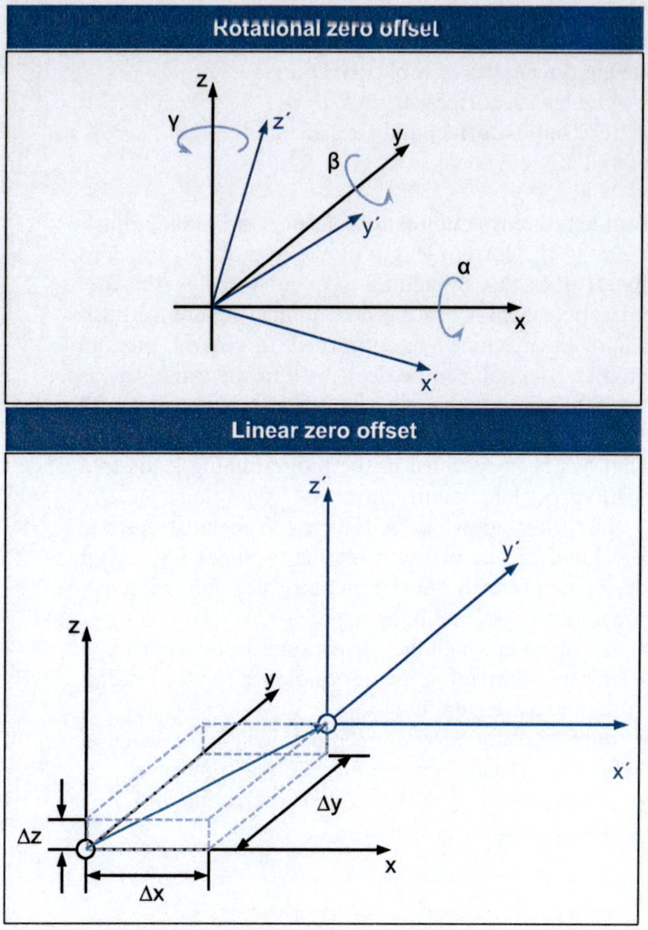


◘ **Figure 12.18** Linear and rotational zero offset

In summary, this gives us the following relationship that forms the basis of the calculations required for the rotational zero offset:

$$X = \begin{pmatrix} x \\ y \\ z \end{pmatrix} = R \cdot X' = R \cdot \begin{pmatrix} x' \\ y' \\ z' \end{pmatrix} \tag{12.70}$$

$X$: Machine coordinates; $X'$: Workpiece coordinates

$$R = \begin{bmatrix} c_\beta c_\gamma & -c_\beta s_\gamma & s_\beta \\ s_\beta s_\beta c_\gamma + c_\alpha s_\gamma & -s_\alpha s_\beta s_\gamma + c_\alpha c_\gamma & -s_\alpha c_\beta \\ -c_\alpha s_\beta c_\gamma + s_\alpha s_\gamma & c_\alpha s_\beta s_\gamma + s_\alpha c_\gamma & c_\alpha c_\beta \end{bmatrix} \tag{12.71}$$

the abbreviated terms used above signify:

$$\begin{aligned} s_\alpha &:= \sin\alpha, & s_\beta &:= \sin\beta, & s_\gamma &:= \sin\gamma \\ c_\alpha &:= \cos\alpha, & c_\beta &:= \cos\beta, & c_\gamma &:= \cos\gamma \end{aligned} \tag{12.72}$$

### 12.2.2 Tool Corrections

The tool correction is to modify the programmed reference path taking into account the current tool geometry such that the desired geometry is created without collision (cf. ▶ Chap. 11). Essentially, there are the following three types of tool correction:
- tool length correction,
- tool radius correction in a plane and
- full 3D tool correction.

Tool length correction is to consider and make adjustments to the current length of the tool during machining. It does this by adding a value stored in the tool correction memory to the programmed reference position of the machine axis concerned. In general, the tool length correction can be dealt with in the same way as the translational zero offset as it is also concerned with the constant offset of an axis. A tool that is longer than what has been specified in the programming leads to a positive tool length correction.

The tool radius correction, which in most cases is calculated in one of the three main planes x/y, y/z or z/x, is significantly more complex. The following parameters are required to define it:
- the plane in which the correction is to take place,
- the tool correction memory holding the tool radius for the corresponding tool,
- the side of the programmed path where the material is located (seen from the direction of tool travel),
- the beginning of the tool radius correction and
- the end of the correction.

Within the context of the tool radius correction process, a path approximating the programmed path as closely as possible using the available tool without damaging the programmed contour must now be calculated using the aforementioned data. The three main subtasks required for this procedure are:
- calculation of paths with equidistant offset (straight line offset in parallel, circle with a larger or smaller radius),
- insertion of additional NC lines, e.g., for outside corners, ◘ Table 12.1, and
- collision detection and avoidance, ◘ Fig. 12.19.

NC lines for avoiding contour damage, in particular at outside corners, are usually inserted through the addition of straight-line or circular elements. The insertion of straight-line elements in accordance with ◘ Table 12.1 also has the advantage that the tool is not in contact with the final contour when traveling around the corner. The advantage of inserting circles is that this motion requires a minimal amount of space. It can, however, lead to local overheating of the material at the corner because the tool remains in contact longer with the same point on the material.

A very important task that is generally also handled within the planar tool radius correction is collision avoidance, ◘ Fig. 12.19. To help with this task, a limited number of NC lines, e.g., three, is analyzed predictively in modern control systems. However, a collision between two NC lines that are more than three path elements apart cannot be detected in this way, ◘ Fig. 12.20.

Due to the finite radius of the tool, sharp inside corners can of course not be created using a peripheral milling process. A residual amount of material is left, ◘ Fig. 12.21. This can be reduced by using a tool with a small radius but it cannot be avoided entirely.

### 12.2.3 Kinematic Transformation for 5-axis Milling

NC programming usually stays in the specified workpiece coordinate system (cf. ▶ Chap. 11) so that complex manual conversions can be avoided. When executing machining activities, the NC controller must, in addition to performing the aforementioned correction calculations, transform the traverse path from the workpiece coordinate system to the machine coordinate system so that the tool can be guided along the programmed path. As the machine coordinate system is determined by the machine kinematics, this is termed kinematic transformation. Machine kinematics refers to the arrangement of a machine's axes of motion.

**Table 12.1** Tool radius correction at an outside corner

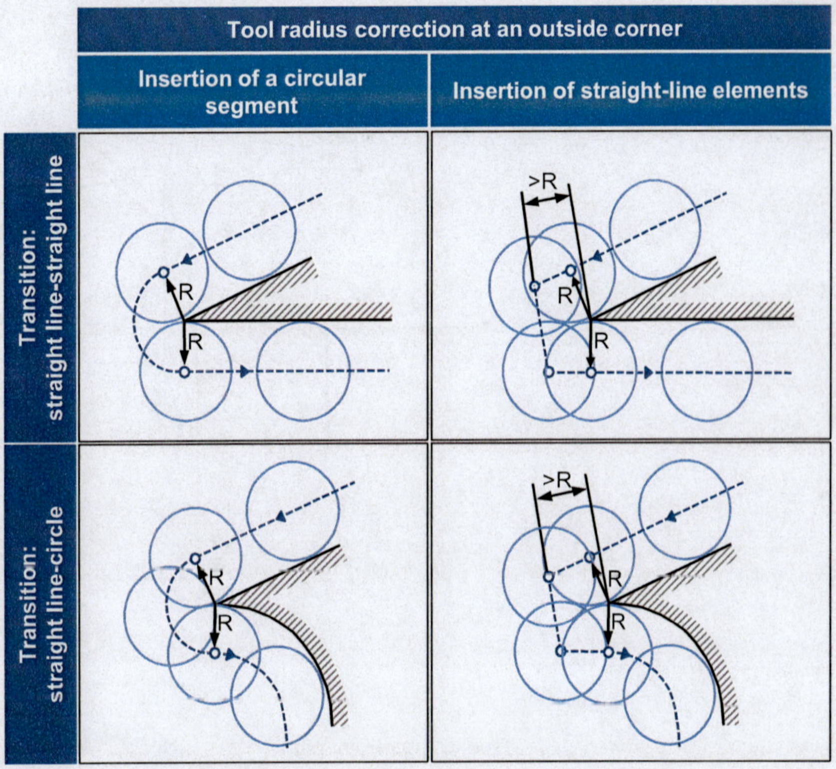


**Figure 12.19** Collision avoidance as a subtask of the tool radius correction

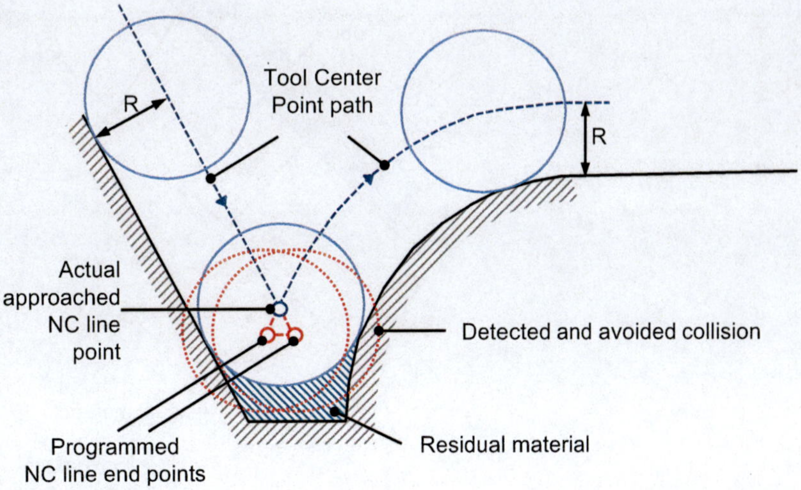


In the case of 5-axis milling, the tool is to be continually guided in a way having a technologically advantageous orientation to the workpiece surface. This also applies to curved shapes or freeform surfaces. The transformation must take into account the orientation of the tool as well as its position.

The transformation from the workpiece coordinate system to the machine coordinate system is referred to as inverse transformation or reverse transformation. It must be performed for each point evaluated by the interpolator. It must be possible to perform the inverse transformation in the interpolation cycle and, for this reason, it must be implemented efficiently. In contrast, with the opposite transformation from the machine coordinate system to the workpiece coordinate system, i.e., the forward transformation, there are usually no requirements placed on the real-time capability. It is needed to obtain the current position values from the drives in the workpiece coordinate system and is mainly used to display position values.

The kinematic transformation is naturally dependent on the particular machine kinematics and the

**Figure 12.20** Undetected collision during tool radius correction

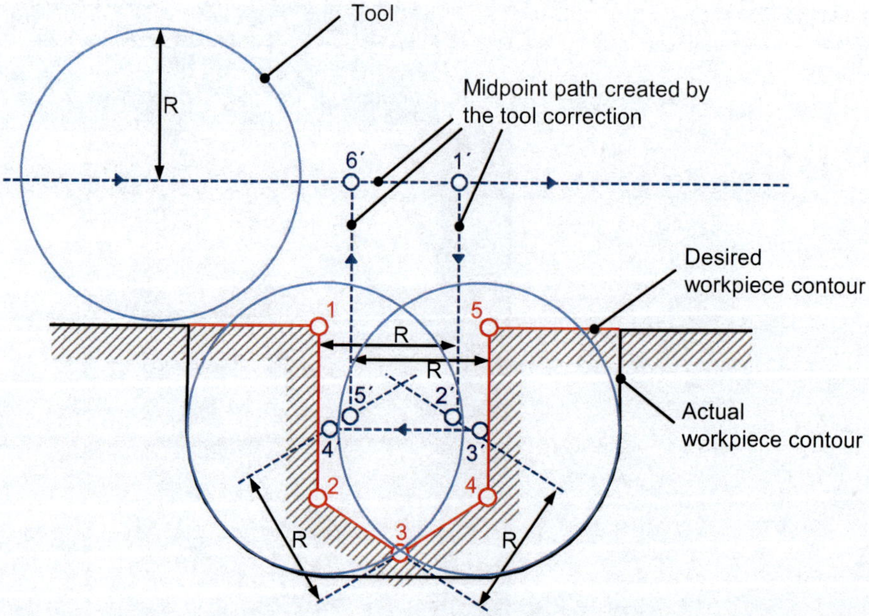


**Figure 12.21** Inside corner of a planar tool radius correction with residual material

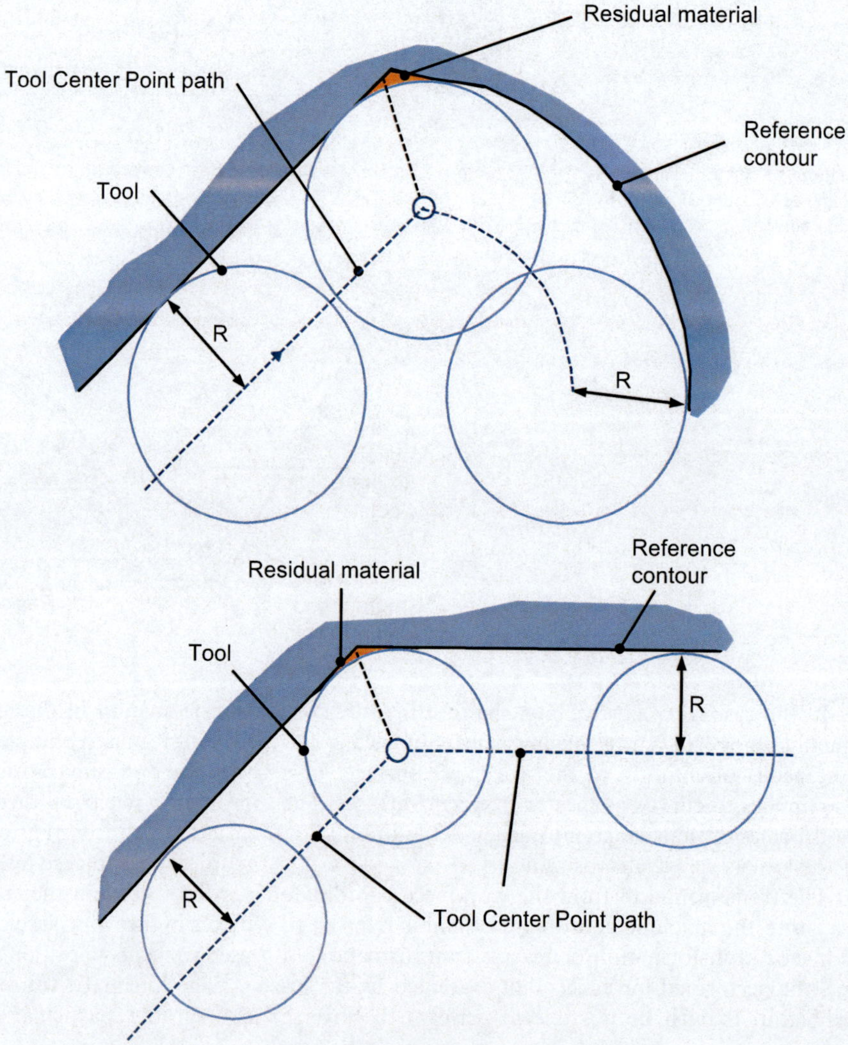

algorithm used must always be adapted to the particular machine tool. This modification has already been implemented in the control system for frequently occurring kinematics and can be selected using parameters. However, if the kinematics are new or unusual, it may be necessary to completely reimplement the algorithm.

The kinematics of milling machines are divided into three groups:

1. The classic milling machine is based on *serial kinematics* meaning that the machine axes build on one another. Serial kinematics is characterized by the fact that a straight motion can be performed in one coordinate of the machine coordinate system by moving only one of the machine axes.
2. New concepts for milling machines are being introduced to the machine tool industry through *parallel kinematics*. It is characterized by the fact that several machine axes act directly on a spindle platform with one simultaneous parallel motion. With parallel kinematics, the cartesian machine coordinate system cannot be specified by the arrangement of the machine axes but instead is defined relative to the machine bed. Motion along one coordinate of the machine coordinate system is only possible with the coordinated movement of all of the axes.
3. In *hybrid* kinematics, both approaches are combined, i.e., some axes are arranged in parallel and others in series.

The calculation of the kinematic transformation for parallel kinematics generally requires considerably more computational effort than is required with serial kinematics. The availability of higher performance control hardware in recent years has allowed calculations under real-time conditions. This, in turn, has created the basis for the use of parallel kinematics in the area of machine tools.

#### 12.2.3.1 Serial Kinematics

A serial five-axis milling machine refers to a machine which has three translational and two rotational axes. For this type of machine, the kinematic transformation must convert the tool position and orientation into the axis coordinates of the five machine axes.

In most cases nowadays, this task is handled offline by a software component of the NC programming system, the postprocessor (cf. ▶ Chap. 11). The implementation of this function in modern control systems allows the transformed path to be calculated online [FAUS97]. For the user, this brings the advantage of having the tool correction and path distribution available on the machine itself.

Established procedures and algorithms from the field of robotics are usually used to calculate and derive the transformation equations. In particular, the Denavit-Hartenberg notation and procedure for the transformation of the position and orientation [ASAD86] are frequently used. As these are described in ▶ Chap. 13, they will therefore not be dealt with here in depth.

#### 12.2.3.2 Parallel Kinematics

In contrast to machines with serial kinematics, those with parallel kinematics can no longer be operated manually and intuitively by the user. As only individual actuators can be moved during conventional manual operation, the user cannot execute a straight motion in a cartesian coordinate system. For this reason, for machines with parallel kinematics, it is necessary to define a virtual cartesian machine coordinate system and enable a manual procedure in it. This requirement makes a cartesian transformation, which must be calculated online in the control system, essential. The mathematical approaches are described in the following using a hexapod (cf. ▶ Sect. 4.3.3 of vol. 1), one of the first machine tools with parallel kinematics, as an example.

The fundamental prerequisite for the operation of parallel machine kinematics as machine tools is that the desired position and orientation of the tool platform must be converted on the control side to the control variables of the actuators (◘ Fig. 12.22). In the case of the hexapod, these are the six lengths of the telescopic legs (struts). This inverse transformation can be calculated separately and in closed form for each strut. In the case of the forward transformation, which is used to calculate the platform position from the leg lengths in the opposite way, the calculation is considerably more complex. The non-linear equation system resulting here can only be solved using complex numerical procedures.

◘ Figure 12.23 shows the notation used in the hexapod transformation. The values predetermined by the mechanical construction of the hexapod are the coordinates of the points of rotation of the ball joints or cardan joints $B_i$ and $P_i$. The hinge points $\mathrm{B_i}$ of the legs on the fixed base are specified in the stationary base coordinate system ($x_B$, $y_B$, $z_B$) while the corresponding hinge points $P_i$ of the moving tool platform are specified in the moving coordinate system ($x_P$, $y_P$, $z_P$). The coordinate system used is indicated by the prefixed subscript, e.g., the specification ${}^{B}T$ signifies the translational offset in the base coordinate system. The translational offset of the platform with respect to the base is specified by the vector $T$ while its inclination is specified by the rotation matrix $R$.

The length of the legs $l_i$ required as a machine control variable for the path interpolation results as a value of the leg vector $L_i$, which can be represented in base coordinates by the vector chain $P_i$, $T$ and $B_i$:

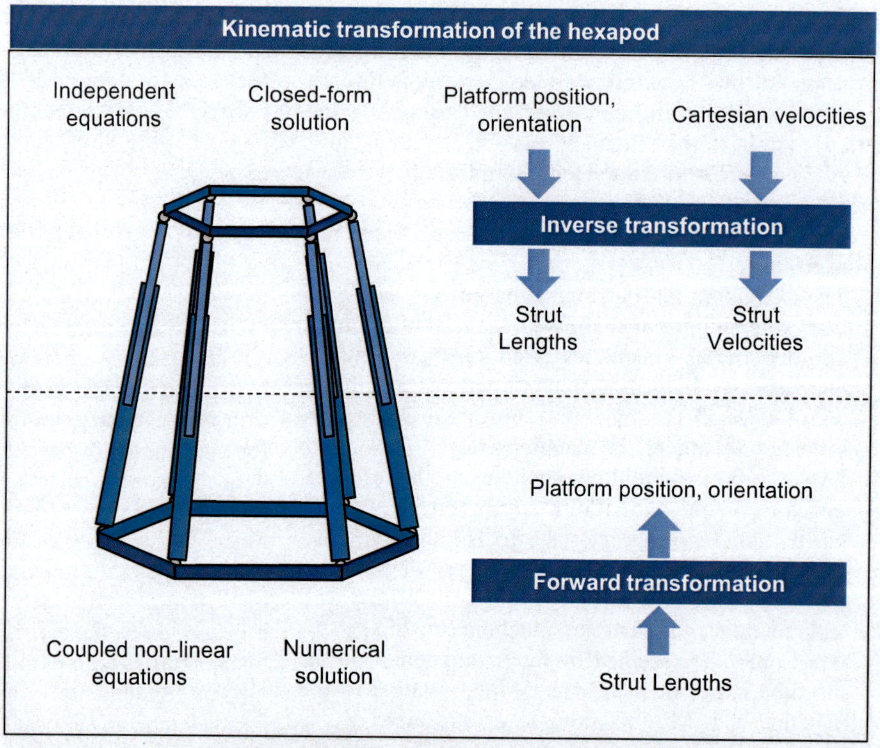


**Figure 12.22** Kinematic transformation of the hexapod



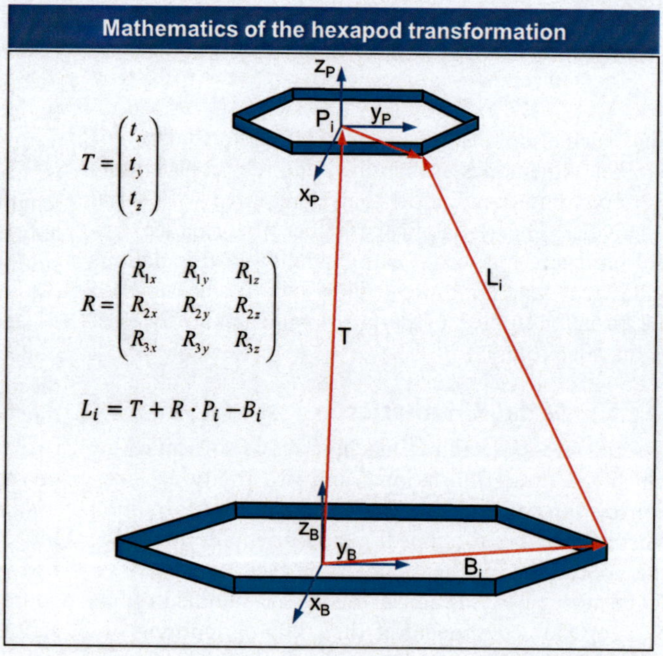


**Figure 12.23** Mathematics of the hexapod transformation

$$
\begin{aligned}
l_i(t) &= \|L_i(t)\| = \left\| R(t) \cdot {}^P P_i + {}^B T(t) - {}^B B_i \right\| \\
&= \sqrt{\left(R(t) \cdot {}^P P_i + {}^B T(t) - {}^B B_i\right) \cdot \left(R(t) \cdot {}^P P_i + {}^B T(t) - {}^B B_i\right)}
\end{aligned}
\tag{12.73}
$$

An equation for the velocity transformation, which can be solved in closed form, can be derived from Eq. 12.73 using the usual rule of derivation:

$$\frac{\delta}{\delta t}\left\|\vec{\lambda}\right\| = \frac{\delta}{\delta t}\sqrt{\vec{\lambda}\cdot\vec{\lambda}} = \frac{\vec{\lambda}}{\|\lambda\|}\cdot\frac{\delta\vec{\lambda}}{\delta t} \tag{12.74}$$

The derived equation is given below:

$$\frac{d}{dt}l_i(t) = \frac{\delta}{\delta t}\|L_i(t)\| = \frac{L_i(t)}{l_i(t)}\left(\frac{d}{dt}(R(t)^P)\cdot P_i + \frac{d}{dt}{}^BT(t)\right) \tag{12.75}$$

It is mainly required for the proactive velocity control (cf. ▶ Sect. 12.1.2.2) where it enables the limiting values for the maximum permitted axis velocities to be taken into account when determining the actual feed rate. The computing time is not so important here. However, if a velocity pilot control is used to control the strut, the axial leg velocities will also be required in the interpolation loop cycle.

The forward transformation searches for the position and orientation of the platform using the given leg lengths $l_i$. The unknown position is described by the three components of $T$ while the wanted orientation is described by three angles linked via the angle functions and yielding the components of the rotation matrix $R$ (cf. ▶ Chap. 11):

$$\begin{aligned} l_1(t) &= \left\|R(t)\cdot{}^PP_1 + {}^BT(t) - {}^BB_1\right\| \\ l_2(t) &= \left\|R(t)\cdot{}^PP_2 + {}^BT(t) - {}^BB_2\right\| \\ &\dots \\ l_6(t) &= \left\|R(t)\cdot{}^PP_6 + {}^BT(t) - {}^BB_6\right\| \end{aligned} \tag{12.76}$$

In principle, the six unknowns can be determined from these six equations. However, as it concerns a non-linear equation system that is fully coupled in the variables the solution can only be found using complex numerical procedures. Another difficulty is the ambiguity of the solution. For example, in the case of a hexapod, completely different platform positions can arise from a set of six leg lengths. Therefore, when selecting the correct solution, additional knowledge regarding the particular hexapod is required. This includes knowing how likely it is to be able to achieve a certain position from the starting position or knowing the last position for which a transformation was calculated. In general, the calculation of the forward transformation is not time-critical as it is usually only required once both for display purposes and for the synchronization of reference and actual values. However, for specific applications such as tracking or multi-variable control spanning more than one axis, a calculation under real-time requirements is necessary.

In the transformation equations mentioned above, a tool length correction can also be taken into account without much additional effort. Compensations, tool radius corrections and any offsets and rotations of the workpiece coordinates can be used independently from the hexapod transformation. This also means that they are independent of the kinematics and can be adopted by control systems for conventional kinematics. Using the method of online calculation for the transformation in the control system which was described previously, each machine tool with parallel kinematics can in principle be operated both manually and under NC control in the same way as a cartesian machine and also in a way that is transparent for the user.

## 12.3 External Position and Velocity Influencing

### 12.3.1 Compensation of Geometrical Errors

While process control, by adjusting the technological data, tries to achieve better use of the machine's capacity through the process-dependent adaptation of cutting values, the main objective with regard to the control of geometrical variables is to improve or rather ensure production accuracy. In addition to high surface quality of the workpieces, further objectives are the compensation of tool wear as well as the compensation of mechanically and thermally induced displacements.

#### 12.3.1.1 Compensation of Geometrical Feed Drive Errors

Inaccuracies in members of the described kinematic chain that produces the feed motion ultimately lead to deviations between the programmed reference position and the actual position the slide is approaching. Spindle pitch errors of the ball screw, slack of the bearings, guides and spindle nuts, stick–slip effects, elastic deformations of the drive elements, insufficient axis dynamics (e.g., tracking errors) and measuring system errors can all cause these deviations.

##### Measurement of the Positioning Uncertainties in Accordance with VDI/DGQ 3441

When measuring the positioning uncertainty, the actual distance traveled by the machine slide is usually determined using either a laser interferometer or a graduated gauge block and then compared to the programmed reference travel distance. This procedure is standardized in accordance with VDI/DGQ and has since been included in the international set of standards as DIN/ISO 230 (see also vol. 2) [VDI77].

The result of this type of measurement is a measurement plot as shown in ◘ Fig. 12.24. The systematic deviation $\Delta x_i$ from the reference value, the positional scatter $P_{si}$ and the reversal error $U_i$ can be read for each point of a traverse axis.

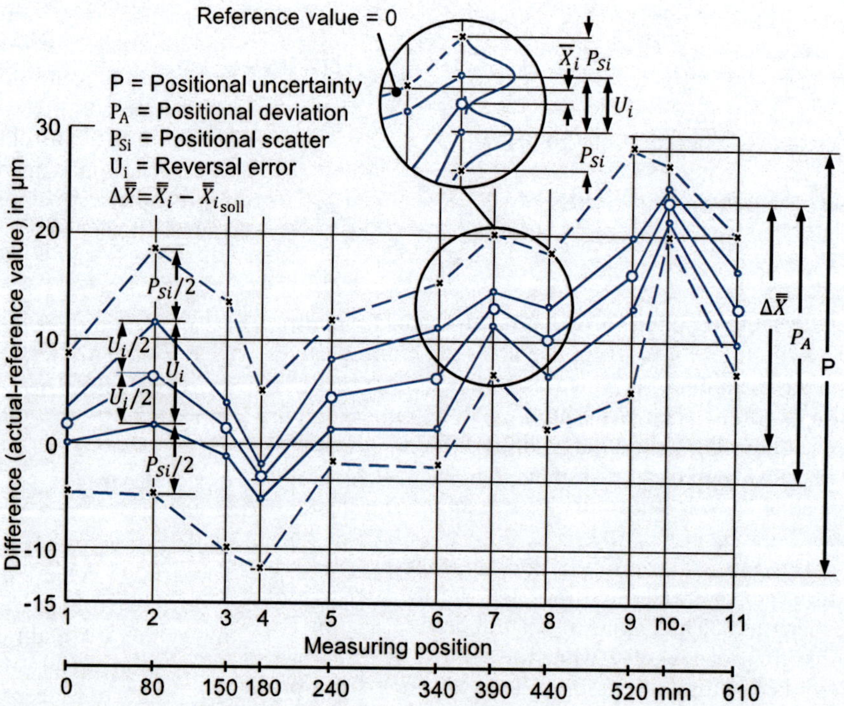


**Figure 12.24** Graphical representation of the statistical parameters of several measuring positions along a chosen test axis

12

The systematic errors caused by the spindle pitch error of the ball screw, slack in the drive train and errors of the position measuring systems lead to the systematic errors $\Delta x_i$ and $U_i$. These can be compensated using control technology. The positional scatter $P_{si}$ covers the random statistical error caused, for example, by the stick–slip effect of the guide or other frictional effects. This random error cannot be naturally compensated.

### Reversal Error

To perform slack compensation, a constant value, i.e., the reversal error, only has to be added or subtracted along the path traveled when there is a change of sign in the direction of motion on one of the axes, Fig. 12.25. This is carried out once each time the direction of an axis is reversed.

The easiest way to determine the reversal error is to insert a measuring probe into the tool holder and then approach a particular axis position several times from different directions. The difference between the forward and return paths gives the reversal error which is stored as a machine parameter. However, as can be seen in Fig. 12.24 the reversal error varies from axis position to axis position in the majority of cases. Better compensation is achieved using the spindle pitch error correction.

### Spindle Pitch Errors

In indirect position measuring systems (cf. ▸ Sect. 2.2.2) the pitch errors of the feed spindle can be compensated by control technology. In addition, most of the control system manufacturers offer so-called spindle pitch error correction tables. The positioning uncertainty of the machine axis is first of all measured in accordance with VDI/DGQ 3441. The systematic deviations are then determined for forward and return travel in a grid, which is usually fixed, and entered in the correction table.

Fig. 12.26 shows the measurement of a positioning uncertainty in the x-axis of a lathe. The systematic errors, positional deviation and reversal error can be reduced by 75 % as a result of the compensation. However, the random error component used in the positional scatter cannot be reduced by the compensation.

#### 12.3.1.2 Compensation of Thermal Displacements

The increasing requirements relating to the machining accuracy of machine tools make it necessary to keep the number of machine displacements due to thermal effects as low as possible. Both the effects and the measurement technology used are discussed in detail in volume 2. A suitable structural machine design should whenever possible cause small thermoelastic deformations in the machine. However, to keep them within narrow limits it may be necessary to purchase additional equipment which is often not economically justifiable (workshop air conditioning, equipment for controlling the temperature of the spindle, gearbox and cooling lubricant, etc.). Control-related displacement compensation can represent an effective measure here.

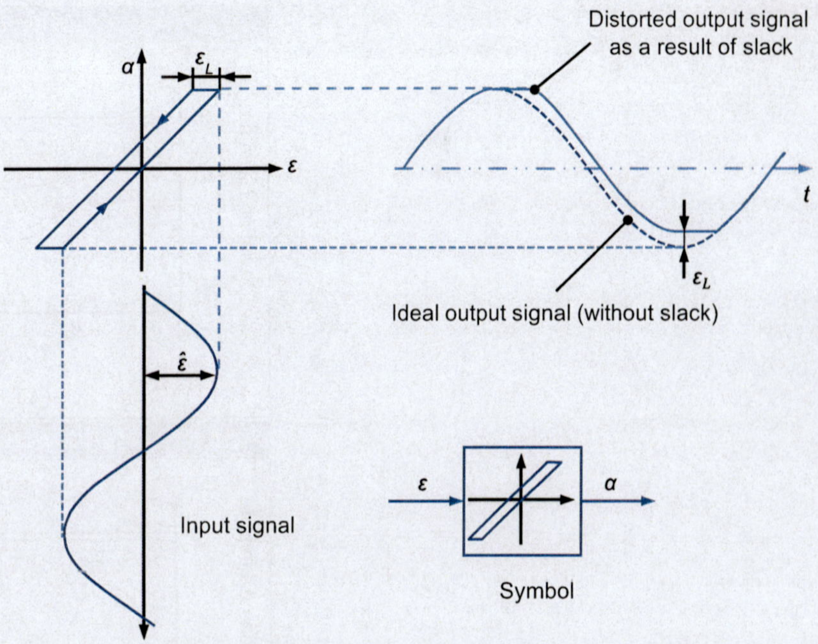


**Figure 12.25** Characteristic of a slide with slack

In general, we differentiate between direct and indirect compensation. In the case of direct compensation, the displacement of interest between the workpiece and tool is measured directly, either intermittently during the process or continually, and then used as a controlled variable [SCHÄ93]. With indirect compensation, the displacement is deduced using a measurable variable, usually the temperature. The measured variable is converted into displacements at the machining point using a mathematical deformation model. The calculated displacements are then processed by the control system as correction values.

## Direct Compensation

Fig. 12.27 shows the direct compensation of the thermoelastic headstock deformation of a knee-type milling machine. The machine has a linear travel measuring system. The sensitive glass scale of the measuring system is fitted behind the machine cover at the end of the headstock to protect it from contamination and damage. When the main spindle is operating, the headstock compartment is heated from the energy dissipated by the bearings and the gear drive.

This temperature rise causes the headstock to expand. If the scale were permanently attached to the rear end of the headstock, this expansion would lead to a relative positional change of the tool in the machine coordinate system in the z-direction. To prevent this, the glass scale, which can be moved in the z-direction along spring suspension strips, is fixed to the front end of the headstock using an Invar bar. Invar is a steel alloy with a nickel content of 36% and has lower, by a factor of eight, thermal coefficients of expansion than steel. Using this type of mounting, the relative position of the front edge of the headstock with respect to the scale remains constant even if the headstock compartment deforms due to an increase in temperature. This also means that the expansion of the spindle is also compensated.

In the machining center shown in Fig. 12.27, a laser is attached to the machine table. This transmits a laser beam parallel to the x-axis of the machine to a receiver that is also attached to the table. The machine moves with its tool in the y- and z-directions until the laser beam is broken by the tool tip. This then allows the thermally induced expansion of the y- and z-axes to be determined when performing the temperature compensation. The CNC controller corrects the axis positions using zero offsets. The machine is thermo-symmetrically constructed to the greatest extent in the x-axis. This means that it is not necessary to compensate for temperature in this direction.

The results of this compensation are shown using a steel lever as an example. While, without compensation, displacements totaling 45 $\mu m$ result and large dimensional deviations occur, mainly caused by cooling during machining pauses, with compensation, the resulting displacements are within less than 5 $\mu m$ of the nominal size. In addition, with this system, tools can be measured fully automatically meaning that tool presetting is unnecessary. After each machining step, a tool fracture inspection can also finally be made.

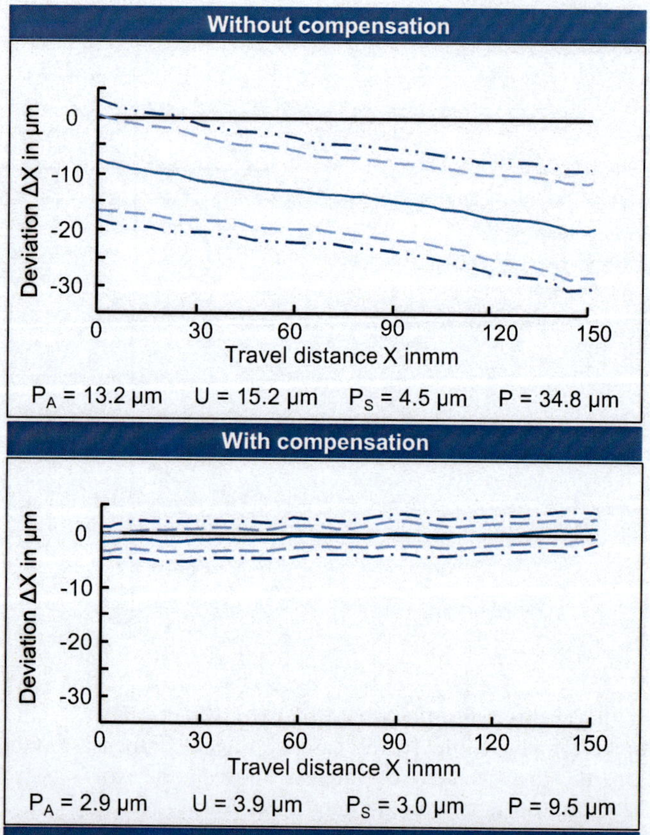


**Compensation table**

| X in mm | out in µm | back in µm |
|---|---|---|
| 0 | 16 | -1 |
| 15 | 17 | -1 |
| . | . | . |
| . | . | . |
| . | . | . |
| 150 | 30 | 12 |

**Legend**

Positional deviation $P_A$
Reversal error U
Positional scatter $P_S$
Positional uncertainty P

**Figure 12.26** Control-related compensation of systematic spindle pitch errors using correction tables

## Indirect Compensation

In principle, machine temperatures, machine running times or material expansions are suitable as input variables for indirect compensation. When performing compensation using the expansion measurement technique, the ends of a bar with a low temperature expansion coefficient are attached to the component which is to be compensated [269]. Due to the bimetal effect, when the temperature rises, the bar is placed under a tensile stress measured by a full bridge strain gauge and converted to correction values.

When performing compensation using machine running times, the displacement motion is determined for a particular machining task when the machine is heating up. With subsequent machining operations, the corrections are then performed on a time-dependent basis. This procedure is only suitable where the same part is always manufactured under constant boundary conditions.

By contrast, indirect compensation that measures machine temperatures is also suitable for universal machines. With this technique, the machine temperatures are recorded at significant points and the resulting

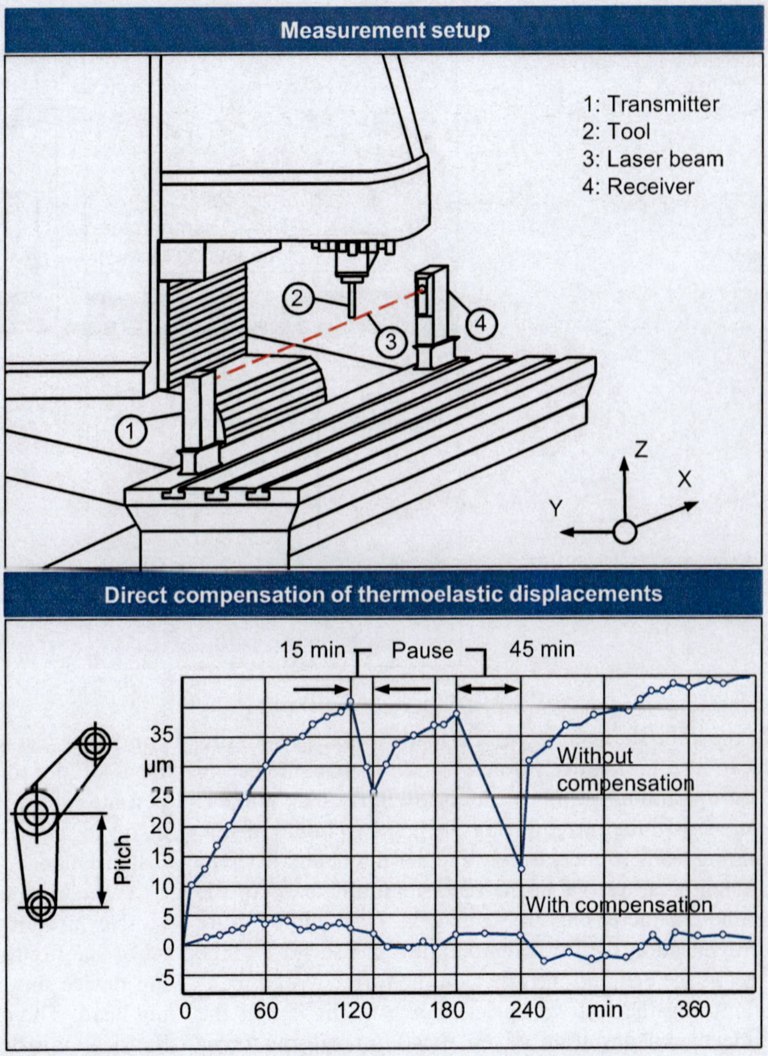


**Figure 12.27** Direct compensation of thermoelastic displacements using a machining center as an example. *source* Chiron, WZL (Machine tool laboratory)

displacements calculated. These are then relayed to the control system as correction values for compensation. The calculation algorithm is determined using a machine measurement. For this purpose, the machine has to be brought to different temperatures using alternating loads, Fig. 12.28 (cf. vol. 2). The temperatures at the prominent measuring points (spindle bearings, spindle housing, etc.) and the relative positional changes of the tool with respect to the workpiece are measured at the same time. From the large amount of measurement data, the coefficients of the temperature-displacement matrix are calculated for an equation system using the least squares method.

During normal operation, the thermoelastic deformation is calculated from the relationship:

$$\begin{pmatrix} x \\ y \\ z \end{pmatrix} = \begin{pmatrix} c_{x1} c_{x2} \cdots c_{xn} \\ c_{y1} c_{y2} \cdots c_{yn} \\ c_{z1} c_{z2} \cdots c_{zn} \end{pmatrix} \cdot \begin{pmatrix} \vartheta_1 \\ \vartheta_2 \\ \vdots \\ \vartheta_n \end{pmatrix} \tag{12.77}$$

which can also be written as:

$$F_{\min} = \sum_{j=1}^{n} \left( x_{\text{ber},j} - x_{\text{gem},j} \right)^2$$

$$= \sum_{j=1}^{n} \left( \sum_{i=1}^{k} \left( c_i \cdot \vartheta_{\text{gem},i,j} \right) - x_{\text{gem},j} \right)^2 = \min \tag{12.78}$$

The calculated correction values are added to the specified coordinate values by the control system with the appropriate plus or minus sign.

Fig. 12.29 shows as an example the compensation of the deformations of a lathe in the x-direction. As the diameter of the manufactured workpiece comes directly from the radial x-position of the tool cutter, radial cutter displacements with a factor of 2 are entered as diameter errors. In addition, the tolerances for shaft diameters are usually particularly tight meaning that the accuracy of the x-axis is of crucial importance.

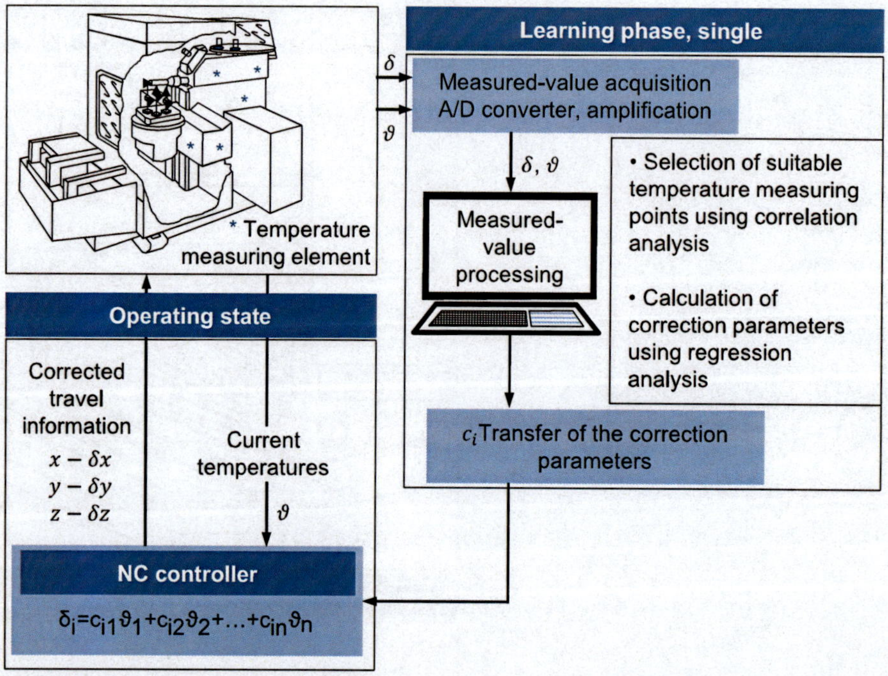


**Figure 12.28** Indirect compensation of thermally induced displacements on machine tools



In the lathe shown, the process loads cause the tool turret to heat up relatively quickly and the machine bed to heat up gradually and continuously. The quick temperature rise in the turret during the first two hours of machining leads to increasingly smaller diameters whereas the temperature rise in the bed eventually leads to the diameters increasing in size again. The temperature compensation uses two temperature measuring points, one on the bed and the other on the turret. When manufacturing the test workpiece shown in the figure, the dimensional deviation of up to 40 $\mu m$ without compensation can be reduced to less than 5 $\mu m$.

Although an inclination compensation is in principle possible for a five-axis machine using the tool swivel axes and can also be achieved with modern open-loop control systems due to the not insignificant experimental effort required to generate the mathematical model, only the translational displacements are compensated in this way.

### 12.3.1.3 Compensation of Static Process Loads

Figure 12.30 shows the milling ram of a portal milling machine while machining vertical surfaces. Due to its large projection length, the milling ram together with the heavy angular milling head is the weakest component on this machine in the force flow. Manufacturing defects are primarily caused by the cutting force components as these cause a relative displacement between the tool and the workpiece causing a deviation from the reference contour. To correct these defects, a control device was developed that enables the displacement of the milling head to be determined and compensated during the milling process. Figure 12.30 also shows the measuring device used to determine the displacement.

A highly bundled laser beam is transmitted by a He-Ne laser integrated in the milling support parallel to the non-displaced milling ram axis to a measuring device mounted on the neck of the angular milling head. The measuring device consists of both a cylindrical mirror that reflects the incident laser beam and a photodiode array that measures the displacement of the reflected laser beam. The cylindrical mirror is adjusted with the adjustment device so that the reflected beam strikes the center of the photodiode array ($\phi = 45°$).

If the milling head is displaced by an amount $\varepsilon$, the laser beam reflected by the cylindrical mirror will move by an amount h on the photodiode array. Due to the reflection on the cylindrical mirror, the deviation h is larger and, as a result, can be measured more effectively than the displacement $\varepsilon$. The term $h$ can be derived from following equation:

$$h = \frac{r}{\sqrt{2}} - \sqrt{r^2 - \left(\frac{r}{\sqrt{2}} + \varepsilon\right)^2} + (l - \varepsilon) \cdot \tan\left(\frac{\pi}{2} - 2 \cdot \varphi\right) \tag{12.79}$$

where:

$$\varphi = \arccos\left(\frac{1}{\sqrt{2}} + \frac{\varepsilon}{r}\right) \tag{12.80}$$

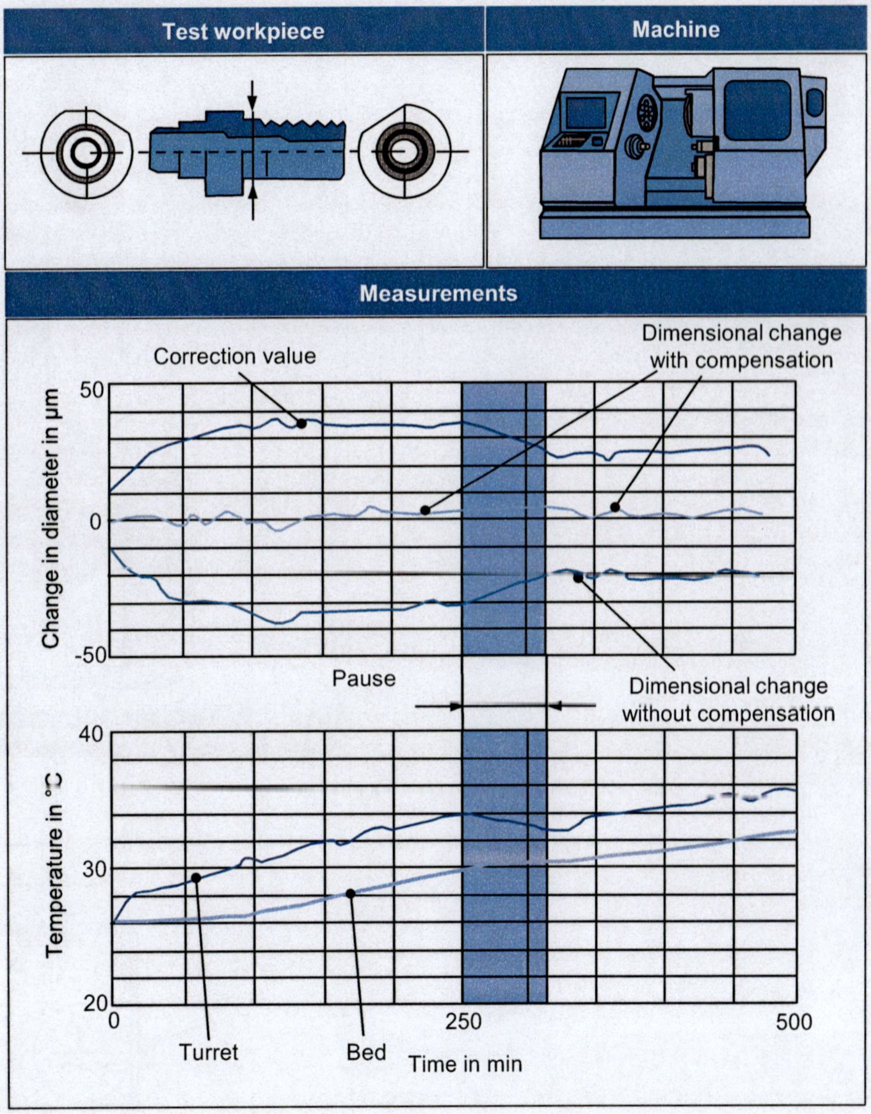


**Figure 12.29** Compensation of the deformations of a lathe in the x-direction

The displacement h is converted to an analog voltage in an electronic evaluation unit. In principle, the measuring device can be assembled such that the laser beam falls onto a four quadrant diode and then the displacement ε can also be measured.

The hydrostatic guide of the milling ram can be used directly as a control variable to compensate the milling cutter deflection. The structure of the hydrostatic ram guide is shown in Fig. 12.31. It consists of an upper and lower guide component. In the upper ram bearing, there are eight pockets with two on each side while in the lower ram bearing there are sixteen pockets with four on each side. There are more pockets fitted in the lower bearing as it requires a higher stiffness.

A multi-circuit pump using the system "one pump for each pocket" is used as the oil supply system. Either a set of eight pumps or one multi-circuit pump is required for every eight pockets. In addition to the pumps required for normal operation, the oil supply system also has four other pumps which are driven by a speed-controlled motor. These pumps also deliver quantities of oil through two diagonally opposed pairs of pockets. The higher oil pressure in these pockets produces a torque on the milling ram counteracting the displacement $\varepsilon$.

The control device adjusts the speed of the motor depending on the displacement measured by the measuring device. The desired direction of compensation can be preselected by opening the corresponding check valves. As an alternative to the method of compensation presented here, the displacement of the milling cutter deflection can also be compensated directly by the NC controller in combination with a 5-axis milling head.

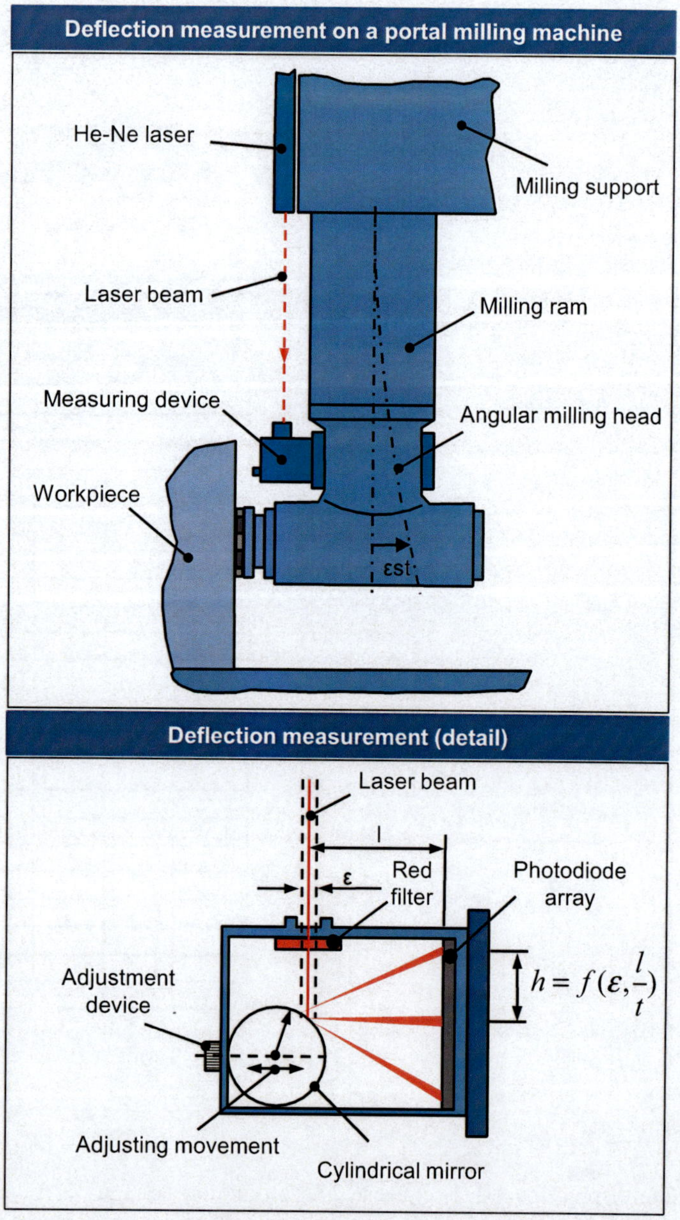


**Figure 12.30** Deflection measurement on a portal milling machine while machining vertical surfaces with an angular milling head



The position control of a machine tool slide shown schematically in Fig. 12.32 is another system that can be used to improve the working precision of a machine tool [MARP91]. The motion accuracy of a machine slide can be considerably degraded by manufacturing defects and guideway deformations. When using hydrostatic guides, it is possible, however, despite inaccurate and elastic guideways, to increase the motion accuracy by controlling the quantity of oil introduced.

The slide carries the load F and glides on the guideway. The position x of the upper part of the bearing is constantly measured relative to a reference plane and fed to the controller as an actual value. The controller compares the actual and reference values and, using an actuator, modifies the oil flow or changes the bearing gap to reduce the control deviation.

The two functions, force absorption and guide, which are to be adopted by a guideway, are separated from each other using the principle of position control:
- The load is absorbed as usual by the guideway (i.e., by the machine bed).
- Exact guidance of the slide is ensured by a reference plane that is isolated as much as possible from loads and temperature changes.

In addition to the option given here of actively influencing the bearing arrangement by varying the oil feed, there is also the option of compensating the tool

**Figure 12.31** Hydrostatic compensation device for the reduction of the milling ram deflection (by Waldrich-Coburg)

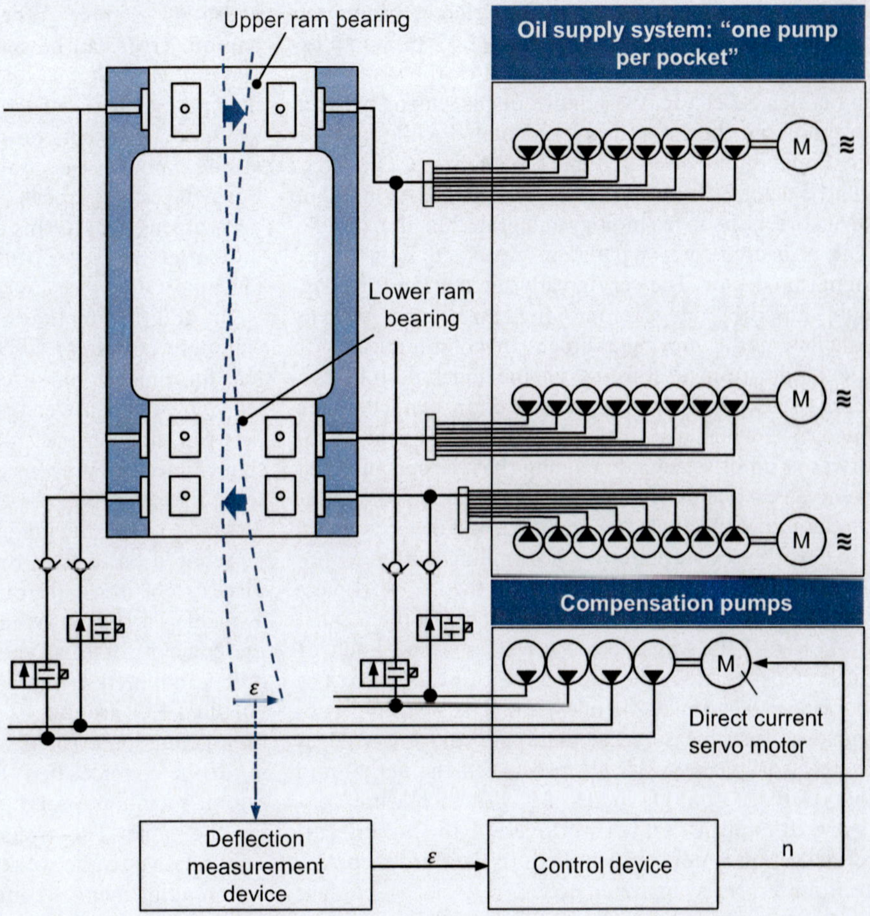


**Figure 12.32** Schematic representation of a hydrostatic bearing arrangement with controlled oil feed

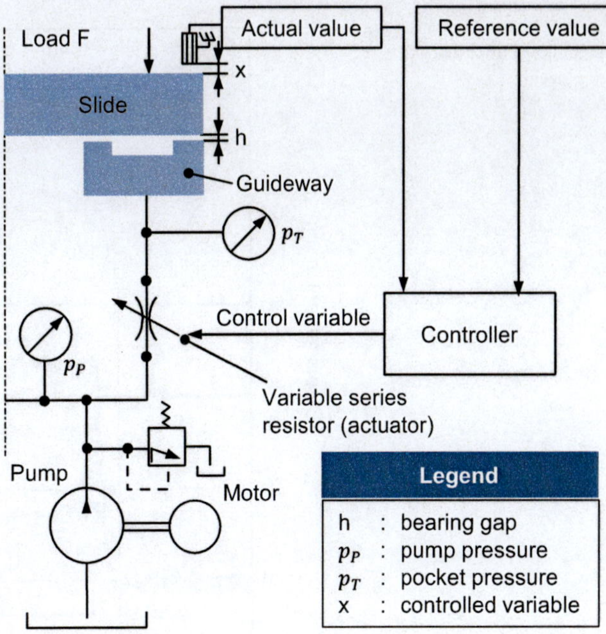

deflection by changing the position-reference value setting of the NC controller. ◘ Figure 12.33 shows an example of this. For the compensation of the static milling cutter deflection, the relative displacement between the tool and the workpiece is calculated while machining based on the measured machining forces. The forces can be determined using either a force measuring plate or a force sensor technology integrated in the spindle. The measured force signals are evaluated by a digital signal processor. The very greatly fluctuating force signals caused by the gears meshing can be smoothed by calculating the moving average. In combination with the predetermined stiffness of the machine-tool system, the expected workpiece deflection can be calculated. The calculated displacement value is sent to the drives as an offset signal meaning that the specified reference axis position is corrected exactly by the value of the tool deflection. ◘ Figure 12.33 shows the measured workpiece deviations for a milling cutter with a 25 mm diameter and a l/d ratio of five. A workpiece stepped in the milling direction as shown in the top left of ◘ Fig. 12.33 was selected for these tests. Each of these steps increases the contact width by 2.5 mm. The workpiece was milled in one pass. The resulting contour was then measured at the upper and lower edges of the workpiece. A deviation between the actual and the reference contour increasing with increasing contact width appears on the workpiece if the milling cutter deflection is not compensated. In the third step, this contour error is approximately 280 $\mu m$ at a contact width of 7.5 mm. When compensation is used, the contour error can be significantly reduced to less than 100 $\mu m$. However, it is not possible to fully compensate the milling cutter deflection with this simple mathematical model as exact compensation can only be achieved at one point on the contour because of the deflection. For complete compensation, the angle error must also be compensated, in this case by inclination of the milling cutter. However, this is only possible on a 5-axis machine. Otherwise, an additional actuator is required.

In addition to the tool deflection described above, clamping errors can also be responsible to a great extent for dimensional or contour deviations. Of particular interest here are errors leading to an angle deviation between the workpiece axes and the machine axes and that cannot be compensated using a three-axis machine tool. Errors of this type can be compensated using a NC-controlled adjustment table. This type of table can be screwed on to the workpiece table so that the workpiece can be moved through the degrees of freedom not available on the 3-axis machine. ◘ Figure 12.34 shows the concept for this type of table. Four hydraulic cylinders mounted in the corners of the table allow it to be tilted ±1° around its x-and y-axes. This also enables inclination errors resulting from workpiece chucking to be compensated. Radial forces are absorbed by a joint membrane serving as the bearing arrangement for the table. The translational displacement of the TCP relative to the workpiece resulting from the inclination adjustment is compensated by the main axes of

12

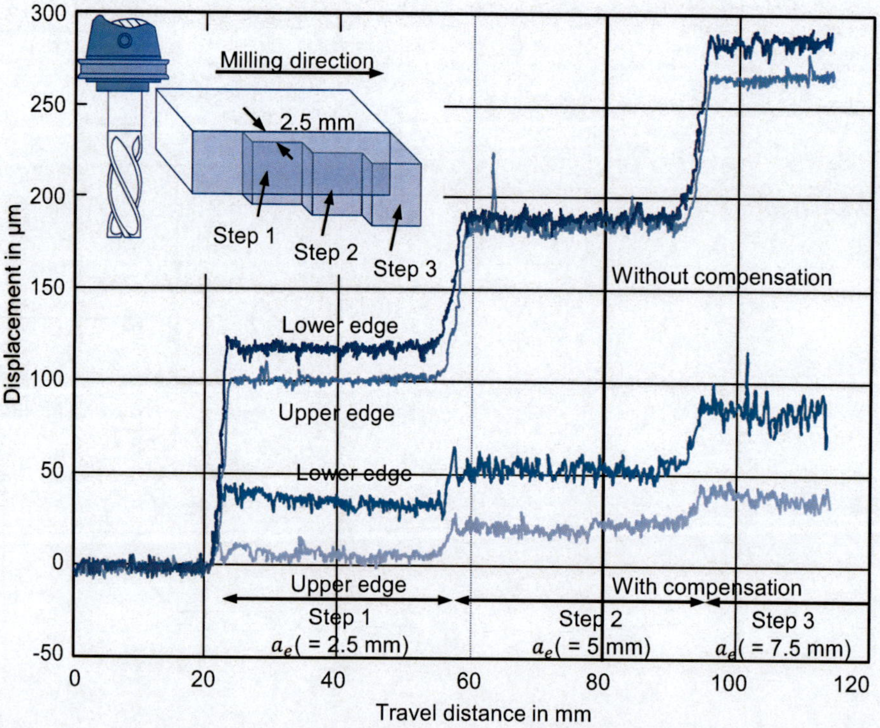


◘ **Figure 12.33** Compensation of the milling cutter deflection during machining

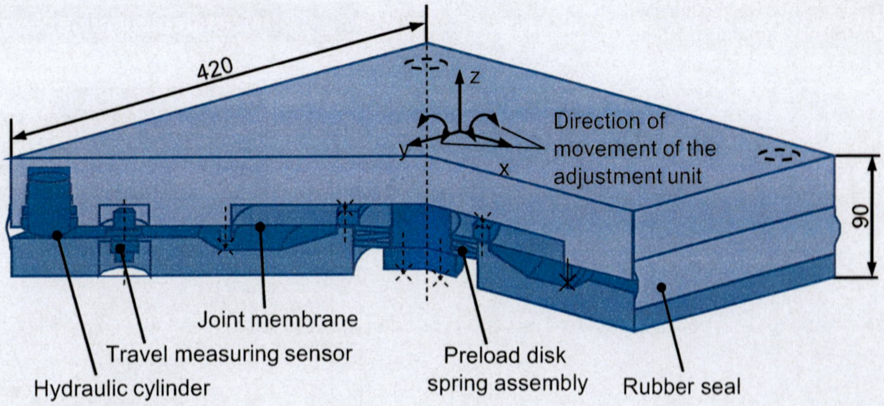


**Figure 12.34** Concept of an adjustment unit for the compensation of inclinations

the machine. In addition to clamping errors, with an adjustment unit of this type, angle errors, like those resulting from either thermoelastic deformations or the tool deflection described above or static deformations of the machine, for example, can also be compensated.

#### 12.3.1.4 Measurement Control for Grinding Processes

In addition to the geometrical control systems already described in this section, the operating principle of a measurement control system will be explained using the external cylindrical grinding of a transmission shaft bearing seat as an example. The measurement control system measures the workpiece dimension during the machining operation when the machine is running, i.e., when the workpiece is rotating, and controls the production process automatically in accordance with the measured values that it has determined. The outside diameters of workpieces with a smooth or interrupted surface can be machined to the desired dimensions using the system shown in Fig. 12.35. In principle, we differentiate between measuring probes that take absolute measurements and those that take relative measurements. If using probes that take relative measurements, it is necessary to insert a master workpiece in the machine and then adjust the probe fingers such that, when the machine is running and using cooling lubricant, the display is set to zero. Probes that take absolute measurements like the one shown in Fig. 12.35 have a scale inside the probe head. Therefore, they do not require a setting master.

The probe fingers are pressed against the measuring surface by a spring force. If the surface to be measured is interrupted by a groove, the damping of the probe fingers must be adjusted such that the probe elements do not sink too deeply into the groove during rotation of the workpiece. It is also possible to measure on interrupted surfaces if the probe tips are suitably rounded. The two to four electronic limit switches contained in the measurement control device, or rather the numerical limiting values, tell the machine when the spark-out must start and when the final dimension has been achieved. This type of measurement control system prevents thermally induced machine deformations and deformations of the workpiece and machine leading to dimensional errors on the workpiece due to static forces. Control of technological variables using dimensional control is explained in ▶ Sect. 6.4.

### 12.3.2 Feed Override and External Velocity Influencing

It must be possible to adjust the feed for modified boundary conditions during machining both by the user and also by external factors.

#### 12.3.2.1 Override

The programmed feed can be controlled by the machine operator during execution of the program using the feed override. The feed override is a factor which modifies the programmed feed appropriately. It enables the machine operator to make changes to the cutting conditions in relation to the process sequence. The override is also used during program testing. The velocity is reduced in critical situations so that the machining operation can be better monitored to ensure that it is being performed correctly.

The override switch used takes the form of either a multi-step rotary switch or a rotary potentiometer (cf. ▶ Sect. 11.3). The normal ranges of feed change are between 0 and 120%. The machine feed will stop at the 0% position.

If the feed override is adjusted by the user during the machining operation, the velocity needs to be adjusted immediately by the control system. Figure 12.36 shows how sudden changes in the override value affect the characteristic of the axis velocity. For reasons of clarity, a velocity control with a constant acceleration phase has been used here as a basis.

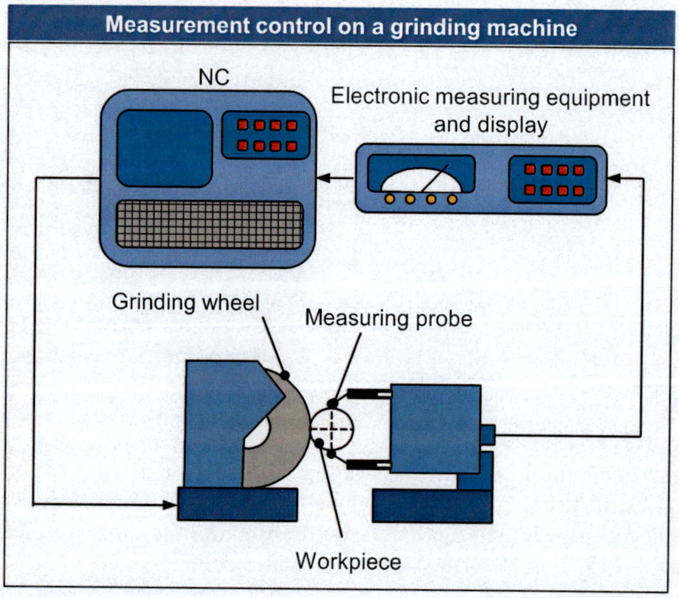


**Figure 12.35** Measurement control on a grinding machine



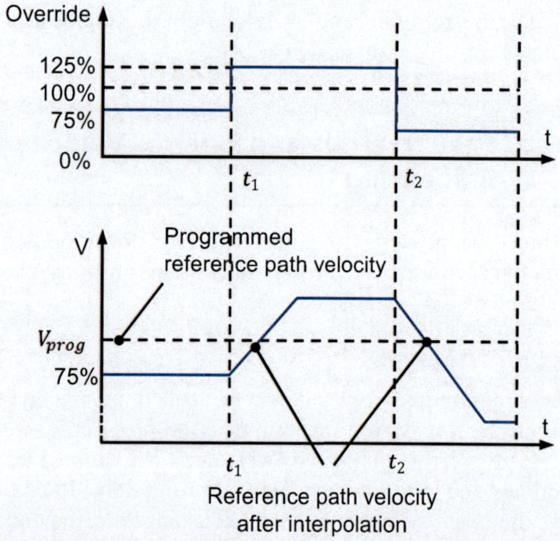


**Figure 12.36** Characteristic of the path velocity affected by sudden override changes

### 12.3.2.2 External Velocity Influencing

The external velocity influencing has the same control software requirements as the feed override. It is typically used with the v=constant function on lathes for example. It is used when facing to keep the cutting rate constant. It achieves this by continuously changing the revolutions per minute inversely proportionally to the machining radius on the workpiece. If the feed per revolution is constant then the feed rate must be adjusted in the same way at the same time. The axis velocity is adjusted by a function module of the control system which directly influences the velocity control.

### 12.3.2.3 Look-ahead Function

Special measures are required if adjusting the override while using the Look-ahead function. This function analyzes a number of NC lines in advance and determines the optimum velocity profile. To determine the changed reference velocity, the velocity control has to be adjusted for a large number of NC lines. However, as a result of the modified reference velocity, the velocity profiles that have already been generated are no longer ideal. If the reference velocity is increased, unacceptable deviations can result, while if it is reduced the machining time can be increased.

The different strategies that can be used during axis motion to react to an override change in the Look-ahead function are as follows:

- a complete recalculation of the velocity specifications of the path that has been calculated in advance,
- a delayed reaction to the override change or
- a reduction in the acceleration capacity of the machine.

The first strategy requires an unacceptably high level of computational effort as the calculations must be carried out within one cycle. In the second case, the path section that has already been calculated will be processed completely before the override change is considered. The third option is to modify the acceleration by the same factor as the velocity. In this case, the dynamics will be restricted. However, this is not of great

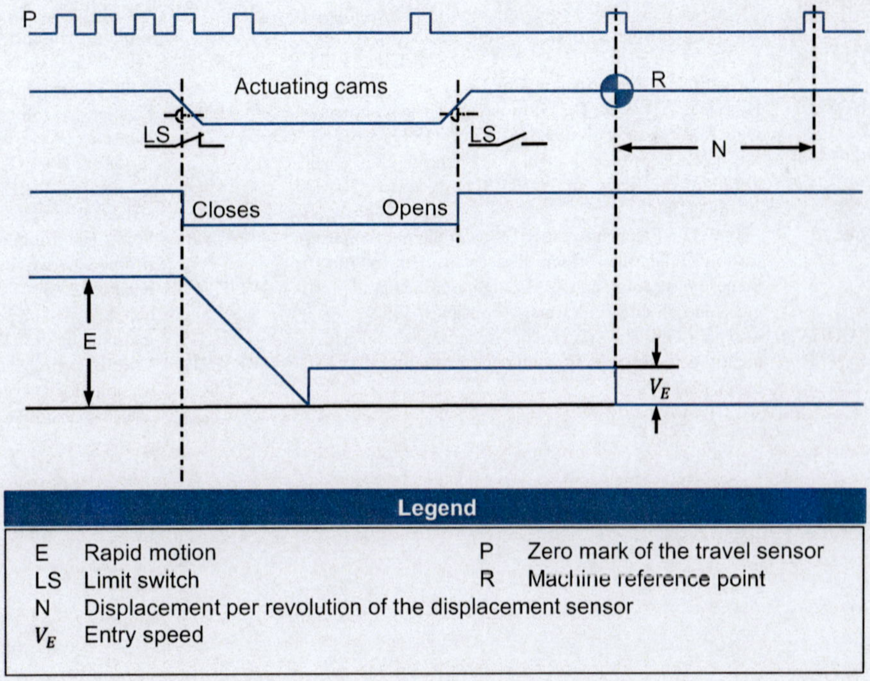


**Figure 12.37** Reference point run

importance as during critical machining phases only a small velocity change is usually made with the override.

### 12.3.3 Reference Point Run

In general, modern machine tools are equipped with incremental position measuring systems (cf. ▶ Sect. 2.4). These position measuring systems have a number of advantages, however, compared to systems taking absolute measurements. They also have a disadvantage in that the current position is unknown when the machine is switched on. For this reason, a reference point run is performed. This determines the reference position using both mechanical reference points and reference points stored in the measuring systems. This operation enables the control system to approach positions relative to the machine zero point. This machine zero point is in turn defined in relation to the reference point usually located near the end of the traversing range of the axes.

With most control systems, the reference point run can be performed either manually or by using an NC program command. The reference point must be measured accurately to within ±1 increment. With resolvers and Inductosyn rulers, one of the electrical zero-crossings of the measuring systems is used to achieve the required accuracy, while, in stepper motor control systems, a defined winding combination of the stepper motors is used. In optical incremental travel measuring systems, a reference mark is used for this purpose.

As many travel measuring systems are either incremental or cyclically absolute, the speed of the reference point run and the operating range of the reference point signal must be reduced, the latter via mechanical pre-cut-off switches. ◘ Fig. 12.37 shows the characteristic of the reference point run for a feed axis with incremental angle encoders. Once the feed has been reduced from rapid motion to entry speed after reaching the limit switch (LS), the first zero impulse of the cyclically recurring zero marks of the incremental angle encoder is then selected after opening the limit switch again as a reference point.

## References

[AKI M70] Akima, H.: A new method of interpolation and smooth curve fitting based on local procedures. J. ACM. 17, (1970)

[ASAD86] Asada, H., Slotine, J.-E.: Robot analysis and control. John Wiley and Sons (1986)

[BRON84] Bronstein, I.N., Semendjajew, K.A.: Taschenbuch der Mathematik. Harri Deutsch, Frankfurt a. M. (1984)

[DIN83] Norm DIN 66025 Teil 1: Programmaufbau für numerisch gesteuerte Arbeitsmaschinen (1983)

[ENGE96] Engeln-Müllges, G., Reutter, F.: Numerik-Algorithmen – Entscheidungshilfe zur Auswahl und Nutzung. VDI-Verlag, Düsseldorf (1996)

[FANU98] Fanuc K.K.: FANUC Series 16/18i, 160i/180i Model A. Firmenschrift, Oshino (1998)

[FAUS97] Fauser, M.: Steuerungstechnische Maßnahmen für die Hochgeschwindigkeits-Bearbeitung - Geometriedatenverarbeitung auf der Basis von Splines und bandbegrenzende Führungsgrößen-Erzeugung. Dissertation, RWTH Aachen (1997)

[FIDI98] Fidia S.p.A.: FIDIA M and C Controls - Programmieranleitung. Firmenschrift, San Mauro Torinese, (1998)

[HEID96] Heidenhain GmbH: HEIDENHAIN ATEK HS PLUS. Firmenschrift, Traunreut (1996)

[HOSC92] Hoschek, J., Lasser, D.: Grundlagen der geometrischen Datenverarbeitung. Teubner, Stuttgart (1992)

[KLEI95] Klein, F.: NC-Steuerung für die 5-achsige Fräsbearbeitung auf der Basis von NURBS. Dissertation, RWTH Aachen (1995)

[LEI92] Lei, W. T.: Flächenorientierte Steuerdatenaufbereitung für das fünfachsige Fräsen. Bericht aus dem Institut für Steuerungstechnik der Werkzeugmaschinen und Fertigungseinrichtungen Universität Stuttgart, 1992

[MARP91] Marposs GmbH: Numerische Steuerungen und Messsteuerungen für Schleifmaschinen. Firmenschrift, Fellbach (1991)

[MESC66] Meschkowski, H.: Mathematisches Begriffswörterbuch. Bibliographisches Institut, Berlin (1966)

[NUM99] NUM GmbH: NUM 1050 CNC - Programmieranleitung. Firmenschrift, Teufen (1999)

[SCHÄ93] Schäfer, W.: Steuerungstechnische Korrektur thermoelastischer Verformungen an Werkzeugmaschinen. Dissertation, RWTH Aachen, (1993)

[SIEM96] Siemens AG: SIEMENS 840D - Programmieranleitung Arbeitsvorbereitung. Firmenschrift, Erlangen (1996)

[SPÄT90] Spät, H.: Eindimensionale Spline-Interpolationsalgorithmen. Oldenbourg, München (1990)

[VDI77] Richtlinie VDI/DGQ 3441–3445: Statistische Prüfung der Arbeits- und Positionsgenauigkeit von Werkzeugmaschinen. VDI-Verlag, Düsseldorf (1977)

[VRIE91] Vries-Baayens, A.E.: CAD product data exchange: conversions for curves and surfaces. Dissertation, Technische Universität Delft (1991)


12

# Robots and Robot Controllers

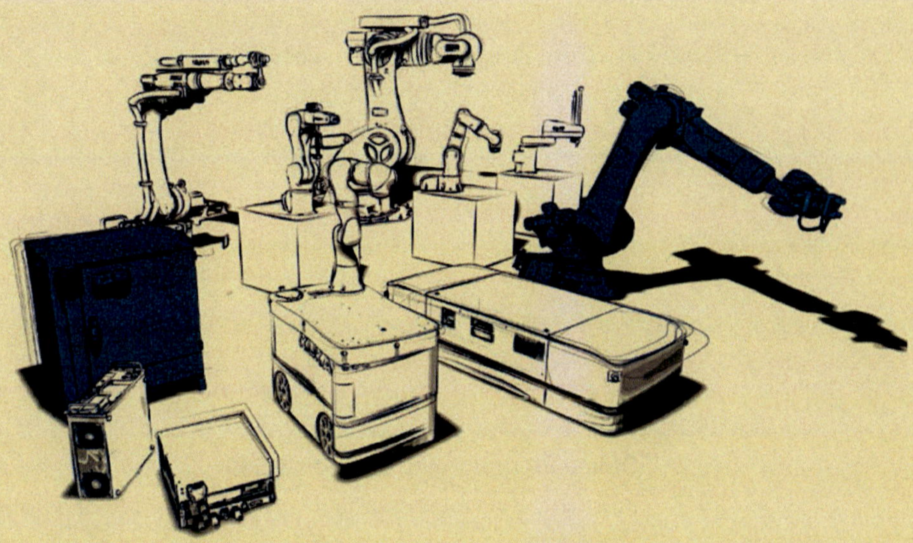

I thank my staff members Christian Fimmers M. Sc., Lukas Gründel M. Sc., Lars Lienenlüke M. Sc. and Simon Storms M. Sc. for their contribution in revising this chapter.


C. Brecher and M. Weck, *Machine Tools Production Systems 3*,
Lecture Notes in Production Engineering, https://doi.org/10.1007/978-3-658-34622-5_13

Contents

Placement devices and industrial robots belong in the handling devices category. Handling is defined as the creation, defined changing or temporary maintaining of a spatial positioning of geometrically determined bodies within a reference system [VDI90]. In line with this definition, handling devices are devices that move these bodies to certain positions and adjust their orientations. Position and orientation together give the 6-axis position (also called pose) of a body within the space. Handling devices are categorized into manipulators, placement devices, and industrial robots depending on the functional scope of their controllers (◘ Fig. 13.1).

Manipulators are manually controlled devices that essentially work as force/torque amplifiers. Here, just the lifting work is aided by external energy while all other movements are performed by the operator. Teleoperators are remote-controlled devices that do not just perform the lifting work of manipulators but traverse along additional axes using motors. The most common application can be found in every production area in the form of cranes. In foundry, reactor and offshore applications, teleoperators enable the remotely controlled handling of bodies in harsh ambient conditions.

Placement devices are simple automated moving devices that follow specified motion sequences in line with a fixed program. They are controlled mechanically by cams or motion links. Devices nowadays are controlled by a programmable logic controller (PLC) as a function of the electrical input signals from other controllers, switches or the output signals of nearby peripherals.

Industrial robots are automated moving devices in universal use with at least three axes of motion (ISO 8373) with motions that are freely programmable in terms of the motion sequence and paths/angles (i.e., changeable without mechanical intervention) and are sensor-controlled as required. Industrial robots can be fitted with grippers, tools, and other production equipment. They are being used more and more for flexible, automated production tasks [ZAID13]. Their broad spectrum of tasks ranges from simple workpiece handling functions to complex, sensor-controlled tool handling such as for arc welding, deburring, and polishing (◘ Fig. 13.2). The majority of robots installed are used for handling, assembly, and path/spot welding. In addition, other applications such as measuring and testing, painting, picking, and palleting as well as bonding and sealing are handled by industrial robots.

Nowadays, industrial robots are mainly used in automotive manufacturing and electrical/mechanical engineering for the processes described (◘ Fig. 13.3). In line with the broadly diversified ranges of tasks and industry sectors, application-specific control methods are required. These methods meet the strict demands for exact and speedy coordination of complex motion sequences within the space and they include the processing of comprehensive sensor and peripheral data.

## 13.1 General Functional Description

The kinematics of a robot define, to a large measure, the movability of the robot arm and specify, with its structure, the working area of an industrial robot.

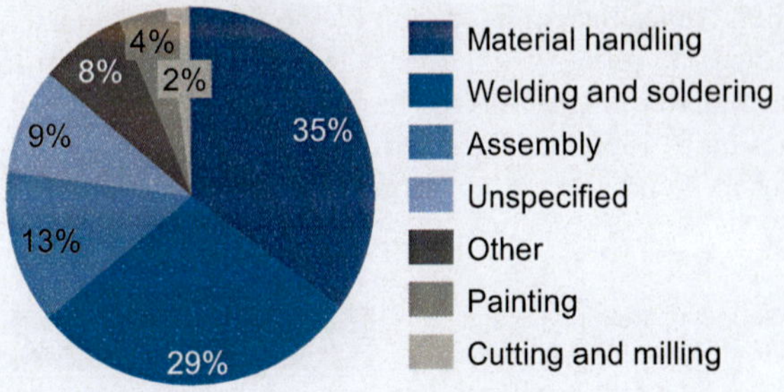


◘ **Fig. 13.2** Application areas of industrial robots [GUIZ08]

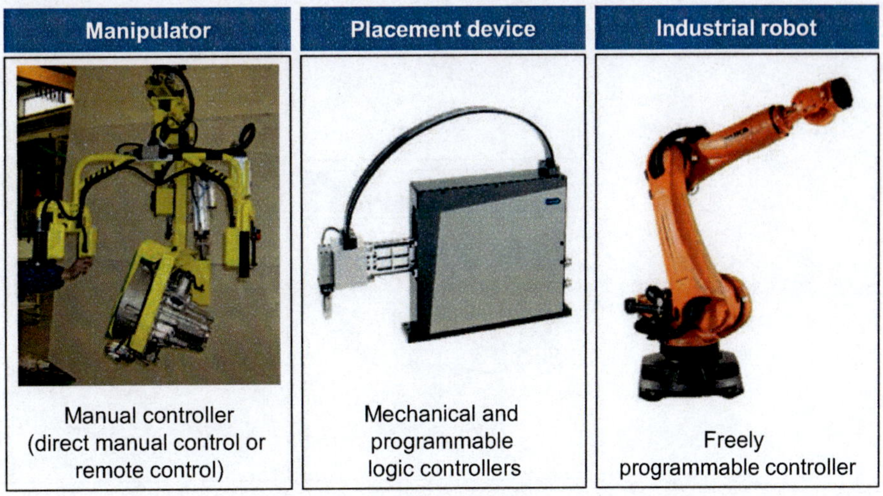


◘ **Fig. 13.1** Categorization of handling devices, according to Scaglia Indeva s.p.a., Schunk GmbH, KUKA AG

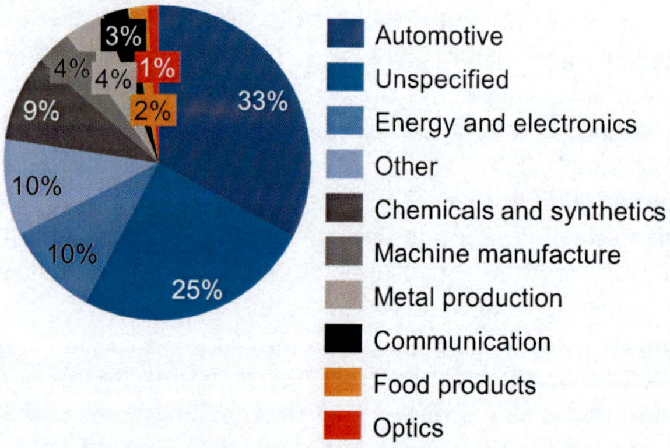


Fig. 13.3 Industrial robot use by industry sector [GUIZ08]

Thus, robots with different kinematic structures have been developed for different automation tasks—see Fig. 13.4 [WARN89, WARN97].

High importance is attached to devices with five or six rotational axes of motion, as shown in Fig. 13.4a, because their kinematic structures mean they can be used universally. The motion capabilities for the base axes:

- Axis 1: Rotation around the vertical
- Axis 2: Rotation perpendicular to the first axis
- Axis 3: Rotation parallel to the second axis

Then comes the configuration of the three manual axes. Industrial robots with a parallelogram axis structure have proven themselves for handling higher load capacities up to 1350 kg. Their sturdy mechanical design enables more precise motion sequences (see Fig. 13.4b).

In small-scale assembly applications, 4-axis swivel arm devices operating on the SCARA principle (Selective Compliance Assembly Robot Arm) are often used, Fig. 13.4c. The benefits of this device construction are high stiffness in the z-direction, high possible traversing speeds and accelerations, high positioning accuracy, and compact design.

In contrast, devices with a Cartesian design usually use a gantry design and are suitable mainly for palleting and assembly tasks. Figure 13.4d shows an example gantry robot.

Robots with parallel kinematics are being used more and more especially for handling lightweight objects at high speeds with strict accuracy requirements. Figure 13.4e shows a *Flexpicker* from ABB as an example. This machine uses deflection kinematics to convert the rotational motions of the three symmetrically positioned drives to a translational motion of the robot hand. Using carbon fiber-based materials and affixing the drives to the body of the robot give a relatively low moving mass. This yields the good dynamic properties of the robot type. A centrally mounted cardan shaft provides the translational degrees of freedom (x, y, z) and one rotational degree of freedom.

Nowadays, an industrial robot system usually comprises three components (see Fig. 13.5):

- The robot kinematics
- The robot controller (RC)
- The hand-held control unit or HMI (Human Machine Interface)

The robot controller includes all components and functions necessary to operate the robot. The main

Fig. 13.4 Robot types and their working areas, according to ABB, Reis, Robert Bosch GmbH

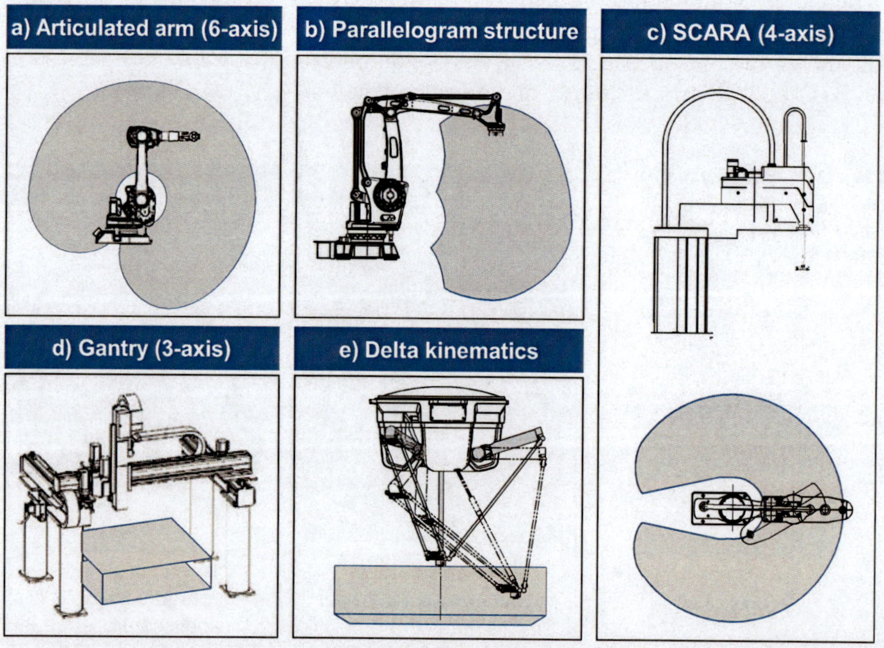

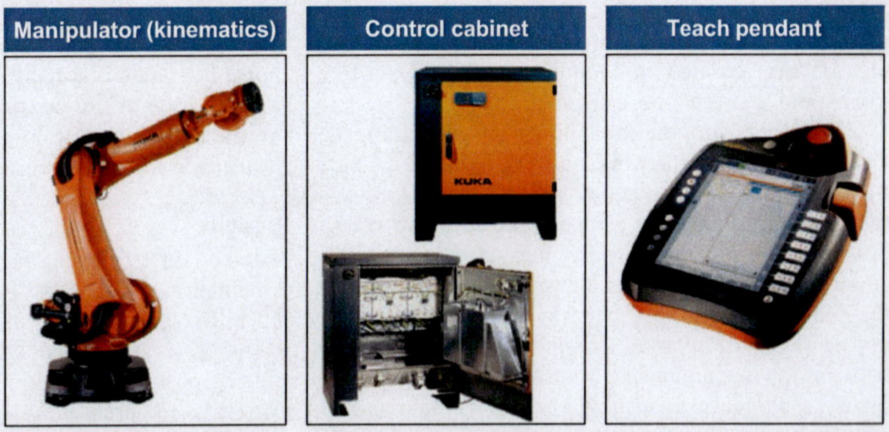


**Fig. 13.5** Components of a robot controller, according to KUKA AG

components are the interpolation unit, the logic unit (PLC functionality), and a PC-based computer system for program management.

Using additional communication interfaces, it is possible to connect other peripherals (such as for sensor data processing) that can be incorporated into robot positioning control. Furthermore, computer interfaces allow the use of offline programming systems on external hardware.

## 13.2 Robot kinematics and Placement Devices

Industrial robots and other handling devices can be specified and compared on the basis of their most important characteristics (Fig. 13.6).

A key feature is the number of controlled axes, that is, the design of the kinematics. This can essentially be configured as desired using translational and rotational axes. A fundamental differentiation must be made between the machine axes of handling devices and the degrees of freedom in the Cartesian workspace. Up to six degrees of freedom can be realized. These are three translational for positioning and three rotational for orienting a body within the space. At least one mechanical axis must be provided for every required degree of freedom of handling devices. For robots with more than six mechanical axes, the workspace is normally extended by redundant additional axes. Often, motions are made easier in individual degrees of freedom by the mechanical and control-related coupling of multiple mechanical axes.

The main axes of robots are used mainly to move the gripper or a tool to any point within the workspace.

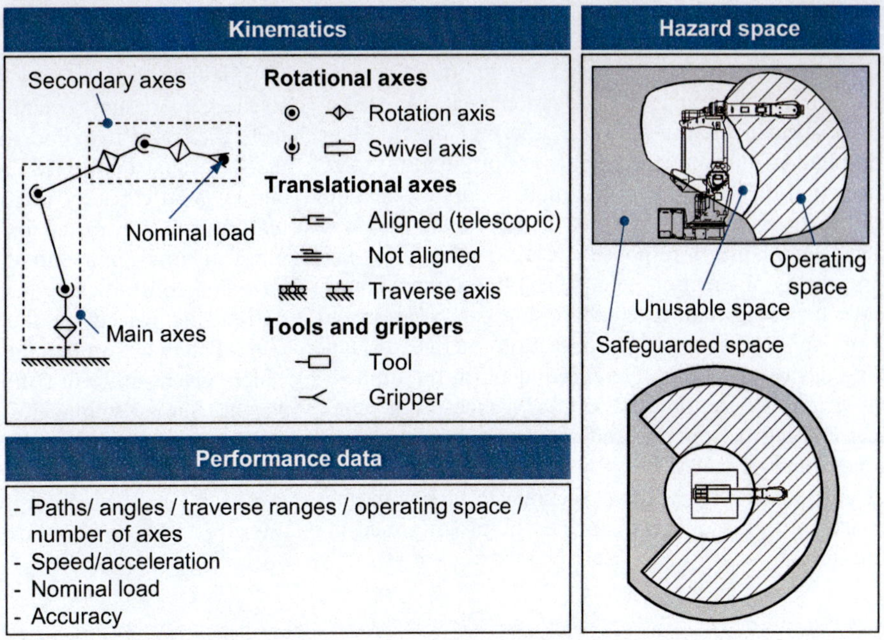


**Fig. 13.6** Specification of handling devices

Translational as well as rotational axes can be used as main axes. The secondary axes generally are used to provide the required orientation of the gripper/tool. Rotational axes are always required for this.

The usable motion capabilities of a handling device are defined by the workspace (◘ Fig. 13.6). This is the space that can be reached by all axis motions with the mounting flange for grippers and tools on the robot side. The maximum space must be taken into account for collision analyses, however. This is defined as the space that can be reached by all moving elements of the device through all axis motions. The difference between the maximum space and the operating space is the unusable space. This should be kept as small as possible with a suitable device design. Specified around the maximum space is an appropriate safeguarded space yielding the hazard space that must not be entered when the device is operating.

The most important load parameter of a handling device is the nominal load. This is the total of tool load (e.g., weight of a gripper) and load capacity (e.g., weight of the workpiece handled). The nominal load is the greatest possible load the device can handle in the entire workspace without restrictions. An additional load, that can also be carried on the arm is often specified for robots. Many devices can be operated at a maximum load that exceeds the nominal load, albeit only at reduced traverse speed or when a restricted part of the operating space is utilized.



In most cases, the repeatability is specified as the machining accuracy of robots. Understood by this is the spread with which any position in the workspace can be approached repeatedly under constant conditions (approach direction/speed/acceleration, payload, temperature). Absolute accuracy, on the other hand, specifies the spread with which a point in the workspace relative to a fixed coordinate system can be reached. Absolute accuracy is determined by the boundary conditions, such as weight of the load capacity, as well as device-specific factors (stiffness, thermal drift). Static compliance, kinematic errors from production tolerances (parallelism or gear and angle errors of the axes relative to each other), quality of path/angle measurement, and control response are key influencing factors worthy of mention here. Absolute accuracy is normally far lower than repeatability but can be improved using calibration and compensation methods. Handling devices are often referenced again in the immediate vicinity of the process point at a measuring point. Surrounding peripherals and the workpieces are positioned precisely relative to this measuring point. Given an adequately high relative accuracy of the handling device, motion sequences can then be programmed offline (see ► Sect. 13.5.2). Relative accuracy essentially corresponds to absolute accuracy but is restricted to part of the workspace. Furthermore, for industrial robots, the spread is specified using the path accuracy that a specified path can be traversed, such as a straight line or circle. Path curvature and path speed have a direct bearing on the path accuracy of the robot here.

### Designs

The key differentiating factor of robot designs is their kinematics, configurable primarily from rotational and linear axes. From the multitude of possible designs, the common variants listed in ◘ Fig. 13.7 have prevailed.

For positioning within the space, up to three main axes are required. Up to three secondary axes are adapted for tool orientation to the main axes. Custom designs with additional or redundant axes are adapted to special scenarios. The most important rules for naming robot axes are shown in ◘ Fig. 13.8 with some examples.

Cartesian linear axes are called X, Y and Z (as for machine tools) with the Z-axis always being vertical for handling devices. In contrast, for machine tools, the feed axis parallel to the main spindle axis (rotational axis, milling axis) is denoted Z. Similar to the machine tool axis names, the rotation axis around the Z-axis is defined as C, meaning the first base axis frequently forms the C-axis (vertical articulated arm robot, robot with cylindrical coordinates). The other rotational main axes are denoted B and A. On top of these are then rotational secondary axes with letters D, E and P. The linear rotation axis for sphere and cylinder kinematics is called R.

This standardized denotation scheme is not familiar to many users and manufacturers however, meaning that the axes are often (incorrectly) numbered consecutively upwards from the column following the kinematic chain.

The application area of industrial robots is restricted by design-specific performance limits arising from the static, dynamic, and thermal properties of the mechanical structure (◘ Fig. 13.9). The positioning accuracy, load capacity, action dynamics, and workspace features are contrasted in the figure. As can be seen, physical laws dictate that the features sometimes behave in a contradictory manner. For example, strict requirements placed on the absolute positioning accuracy of high action dynamics preclude a large payload, which, given sufficient stiffness, necessitates a high dead weight. Similarly, higher static, dynamic, and thermal deformations are inevitable for large operating spaces meaning only low positioning accuracies can be expected. As can be seen in the lower part of ◘ Fig. 13.9, the robot type (kinematics) is determined solely by the operating space required.

**Fig. 13.7** Designs and parameters of industrial robots, specified by main and secondary axes, according to ABB, EPSON Deutschland GmbH, Fanuc Deutschland GmbH, KUKA AG, Liebherr-International Deutschland GmbH, Masat

| | Name | Principle | Axes | Load capacity | Operating space |
|---|---|---|---|---|---|
| Main axes | Vertical articulated arm | 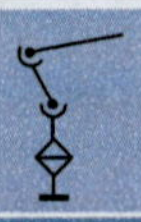 | 3 | ≤ 2300 kg | Hemisphere r ≤ 4.6 m |
| Main axes | Horizontal articulated arm (SCARA) | 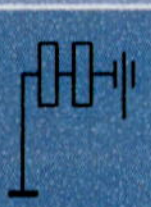 | 3 | ≤ 20 kg | Sickle r ≤ 1.5 m h ≤ 0.5 m |
| Main axes | Cylindrical coordinates |  | 3 | ≤ 80 kg | Cylinder r ≤ 2 m h ≤ 1 m |
| Main axes | Spherical coordinates | 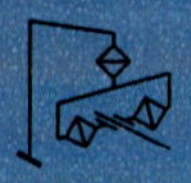 | 3 | ≤ 200 kg | Sphere r ≤ 2 m |
| Main axes | Line gantry | 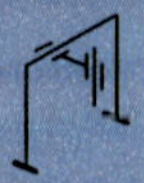 | 2 | ≤ 2400 kg | Area l ≤ 50 m h ≤ 2m |
| Main axes | Area gantry | 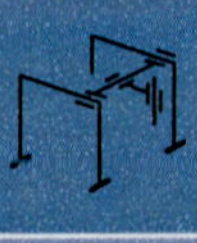 | 3 | ≤ 1300 kg | Cuboid l ≤ 60 m b ≤ 12 m h ≤ 2.4 m |
| Secondary axes | Double-angle hand |  | 3 | ≤ 150 kg | Hemispherical surface |
| Secondary axes | Central hand | 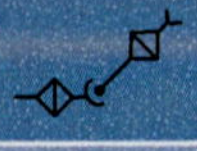 | 3 | ≤ 200 kg | Hemispherical surface |
| Secondary axes | Rotate-swivel joint | 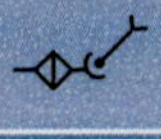 | 2 | ≤ 200 kg | Hemispherical |
| Secondary axes | Swivel joint |  | 1 | ≤ 500 kg | Circle (point) |

### 13.2.1 Vertical Articulated Arm Robot

The vertical articulated arm robot is one of the kinematics with the most widespread usage. It comes in a multitude of variants and performance classes due to its broad application spectrum. Figure 13.10 shows the possible ranges of their sizes and parameters using two examples. Easily understandable here are the often contrary requirements pertaining to load capacity and range on the one hand and attainable accuracy and speed on the other. In terms of its axis arrangement, the vertical articulated arm robot shown in Fig. 13.11 corresponds to the standard kinematics of vertical articulated arm robots. It comprises six rotational axes, that are driven by 3-phase servo motors with an integrated stopping brake.

The motor torque is increased using a planetary gear. Angles are measured using digital, incremental rotary encoders on the motor shaft. In this case, the axes must be referenced on initial startup. The design shown is realized for a nominal load of 150 kg. The dead weight and nominal load together cause a high torque in the B-axis and the A-axis for the corresponding deflection of the arm and rocker. A mass balance is provided for both axes to prevent high continuous loading on the drives. No full torque equilibrium can be attained because of the changing load at the end effector and the different angle positions. The load collective of the axes can be reduced radically however, such that smaller drives contribute towards weight and cost optimization.

Along the A-axis, weight compensation is realized easily by moving the manual axis drives behind the arm joint. Additional benefits of this arrangement are the reduced mass moment of inertia and compact secondary axes group. For higher nominal loads,

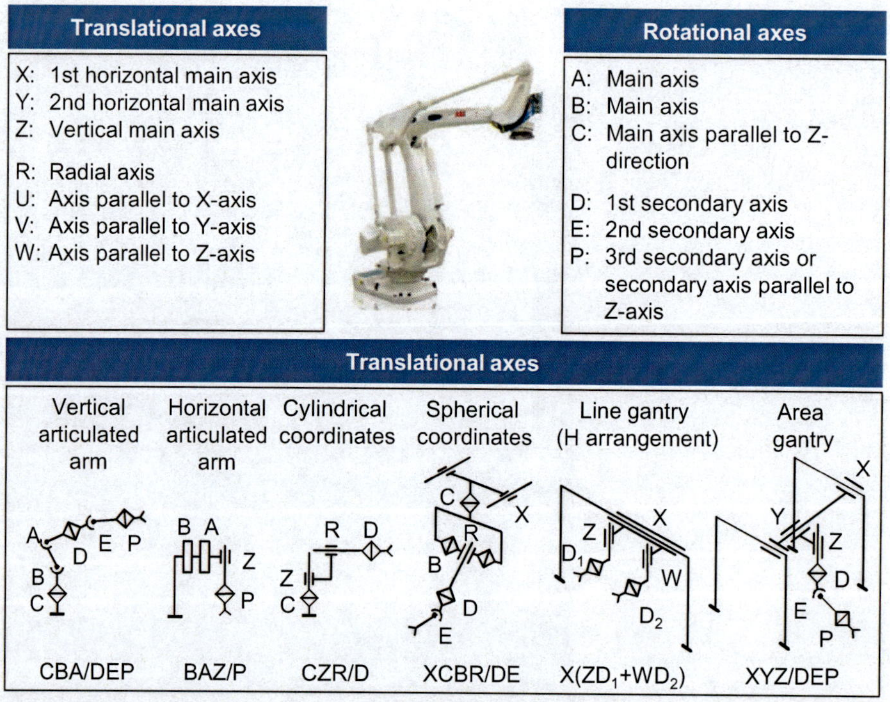


**Fig. 13.8** Axis names for handling devices (VDI 2861), according to ABB

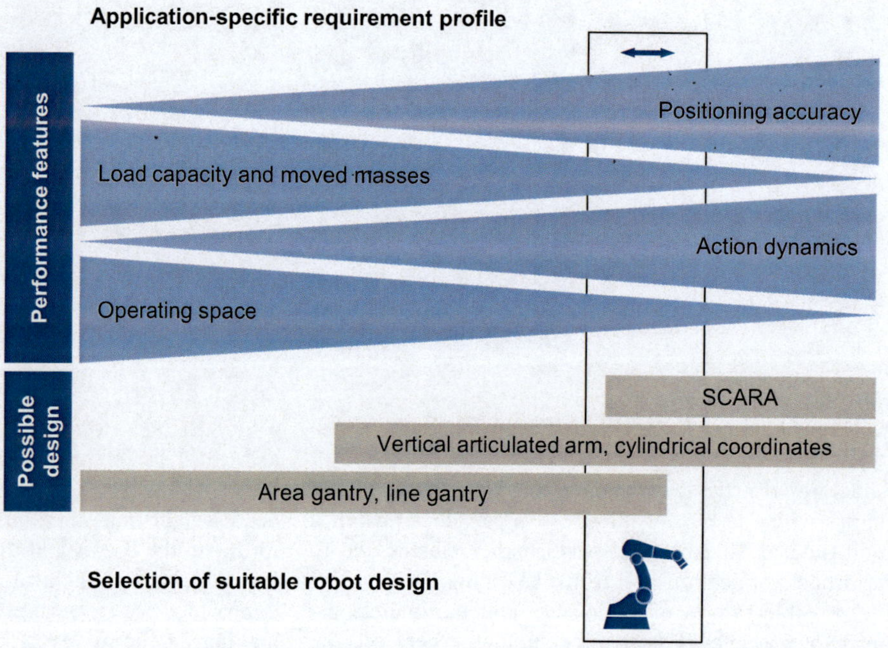


**Fig. 13.9** Performance limitations of robot designs



an additional mass can be installed on the arm end as an option, but this would increase the moments of inertia of the A and B-axes. The counter torque on the B-axis is generated by a hydraulic cylinder that continually introduces a tensile force into a linkage point on the rocker. The position and connection to the B-axis are selected such that, given a constant cylinder force, the compensation torque is aligned automatically to the weight moment around the B-axis. In the vertical rocker position, no counter torque is generated because the line of action of the force then intersects the B-axis. It is not only the active lever arm of the cylinder contact point that rises with increasing deflection of the B-axis. The pressure in the pressure accumulator also increases with increasing deflection of the rocker also increasing the magnitude of the compensation force. The benefits of the hydraulic mass balance are its compact design yet low dead weight.

◘ **Fig. 13.10** Performance range of articulated arm robots, according to KUKA AG, Fanuc Deutschland GmbH

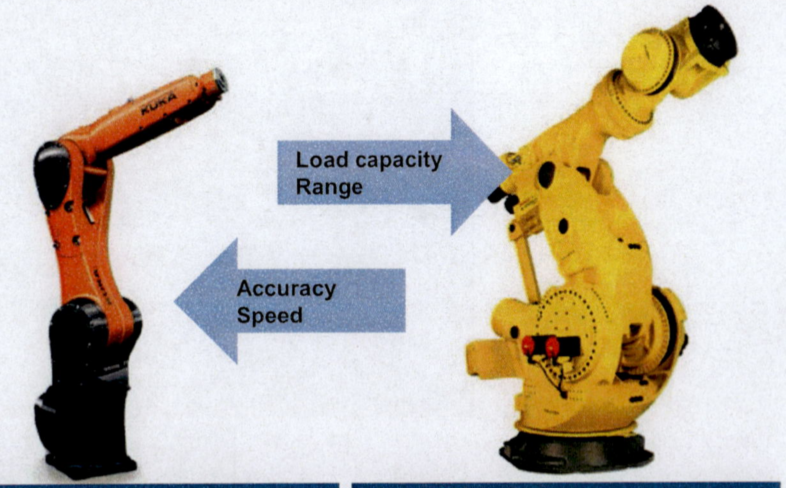


| Performance data | |
|---|---|
| - Payload: | 6.0 kg |
| - Weight: | 52 kg |
| - Range: | 0.9 m |
| - Repeatability: | ±0.03 mm |
| - $v_{max}$ of main axes: | 360 °/s |
| - $v_{max}$ of secondary axes: | 615 °/s |

| Performance data | |
|---|---|
| - Payload: | 2300.0 kg |
| - Weight: | 11000 kg |
| - Range: | 3.7 m |
| - Repeatability: | ±0.3 mm |
| - $v_{max}$ of main axes: | 20 °/s |
| - $v_{max}$ of secondary axes: | 40 °/s |

◘ **Fig. 13.11** Vertical articulated arm robot with hydraulic mass balance, according to KUKA AG

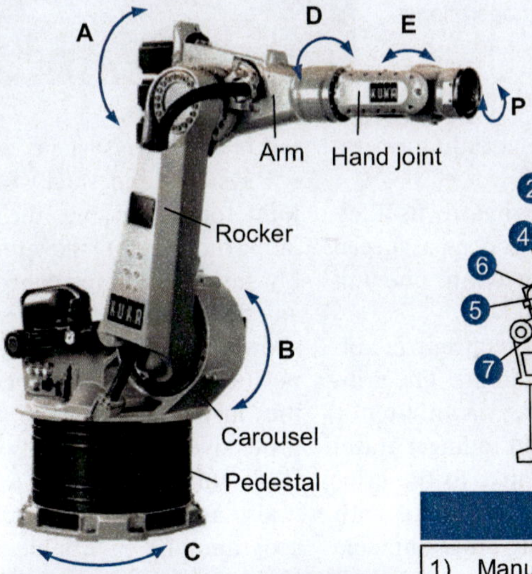


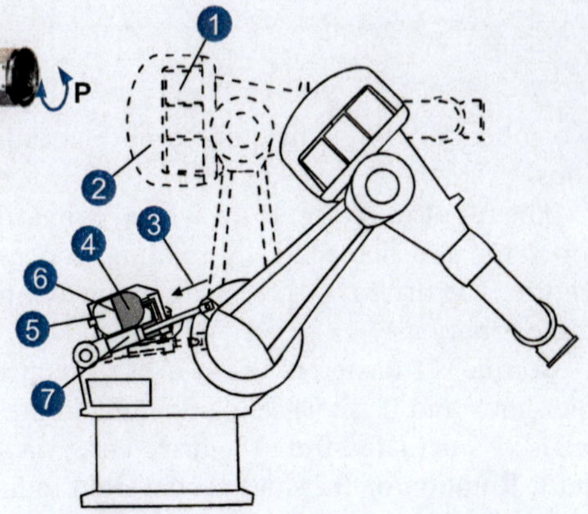


| Performance data | |
|---|---|
| - Payload: | 150.0 kg |
| - Range: | 2.4 m |
| - Repeatability: | ±0.2 mm |
| - Max. speed: | 100.0 °/s |

| Legend | |
|---|---|
| 1) | Manual axis drives as mass balance |
| 2) | Additional mass of 150 kg as option |
| 3) | Restoring force F |
| 4) | Hydraulic medium |
| 5) | Nitrogen pressure reservoir |
| 6) | Diaphragm |
| 7) | Hydraulic cylinder |

◘ Figure 13.12 shows typical applications of multiple 6-axis articulated arm robots in automotive production.

In the top picture, the robots are performing measurement operations on an automobile body using flange-mounted 3D sensors. The lower picture shows

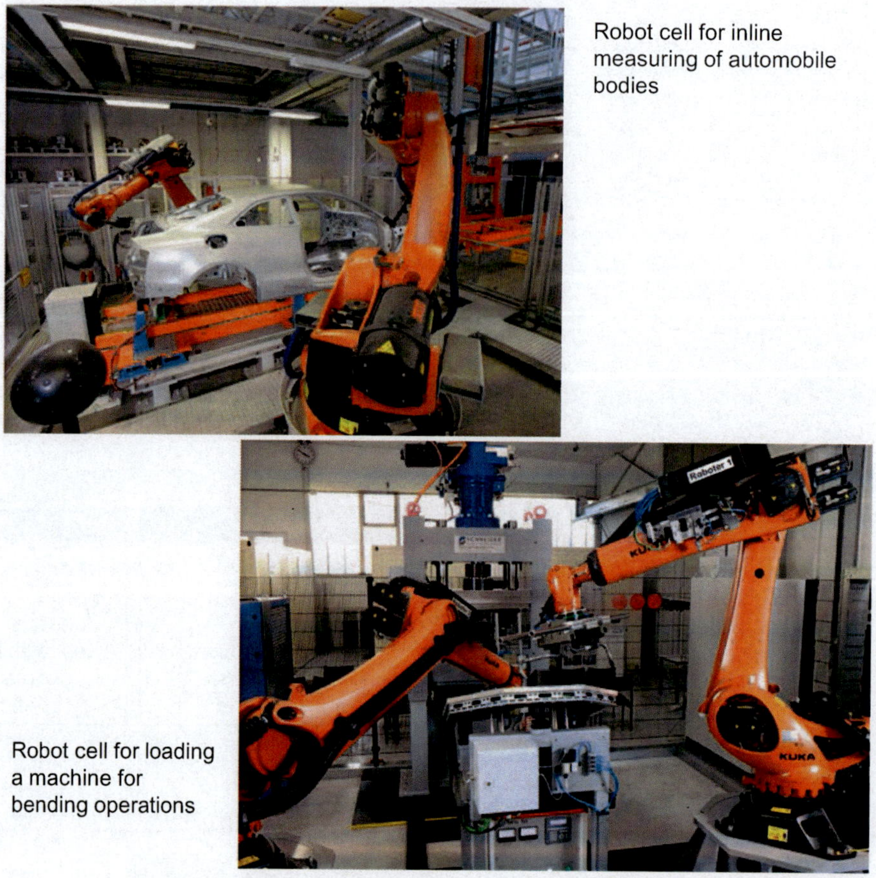


**Fig. 13.12** Robot cells in automotive engineering, according to KUKA AG



two robots loading a machine tool for bending operations.

The robot in Fig. 13.13 is used primarily as a teleoperator and has endlessly rotating axes as a special feature. The drives of these are combined into one unit in the robot base.

Torque is transferred to the axes via concentric hollow shafts and bevel wheel pairs in the joints. The gripper is also actuated from the drive base, via an interior shaft. Rotation of the shaft is converted to finger travel using a ball screw drive. As an alternative to the gripper, this shaft can adapt any actuator as an axis with fully-fledged control. The robot is freely programmable or it can be manually controlled as a teleoperator in all degrees of freedom using a special operating unit. The operating unit receives force feedback via the motor torque. Use in areas inaccessible to people is a standout feature of these robots, not least because of the slimline design of the cable-free arms and the unrestricted freedom of movement. Nuclear technology, marine engineering, aerospace engineering, and the handling of toxic and explosive substances are worthy of mention here.

The lightweight robot in Fig. 13.14 has seven axes that are based on a human arm. Integrated in the joints are drive systems and electronics, including sensor systems and safety functions. With the integrated joint torque sensors, these robot kinematics are used in particular for the automation of sensitive assembly tasks and force-controlled joining processes, such as setting stoppers in the automotive industry and, as shown in the figure, detection of the correct installation position with force feedback measurement. Work activities in the field of human–robot cooperation without protective fencing are also possible with integrated, reliable collision protection. Accordingly, this robot type is also assigned to ▶ Sect. 13.2.6.3. Using a control and programming environment developed in-house realizes object-oriented programming (such as guiding robots by hand) for these robots.

## 13.2.2 Horizontal Articulated Arm Robot

The horizontal articulated arm robot, also called SCARA (Selective Compliance Assembly Robot Arm), was designed primarily for assembly tasks, such as populating PCBs in the electronics industry. Great stiffness in the vertical direction is a standout design feature because all forces occurring and the dead weight act as

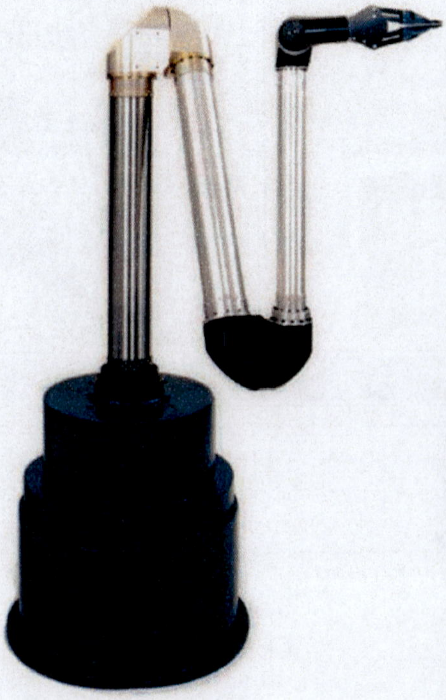

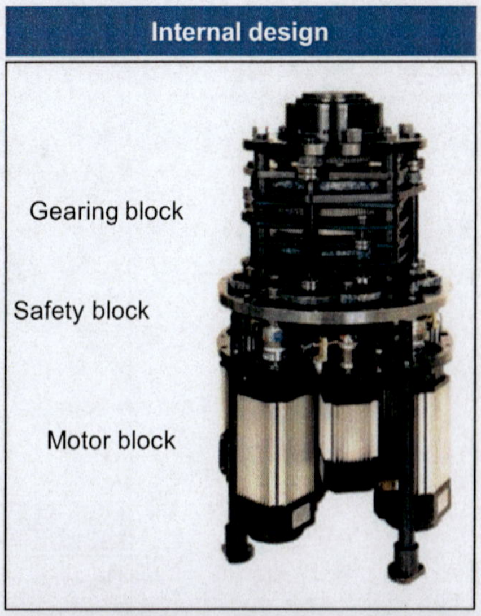


**Features**

- All axes rotatable endlessly
- Slimline design
- No cabling in the arm
- Smooth, sealed arms and joints
- Arm length: 1 – 5 m
- Nominal load: 5 – 120 kg
- Degrees of freedom: 6 – 7
- Radiation resistance to 1 MGy

**Fig. 13.13** 6-axis vertical articulated arm robot with central drive base, according to Wälischmiller Engineering GmbH

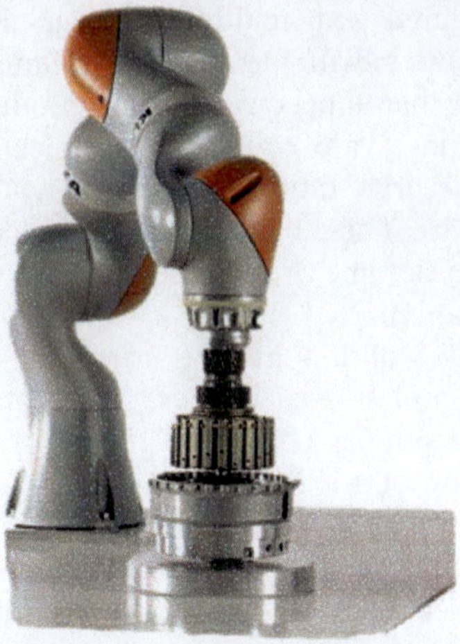

**Performance data**

| | |
|---|---|
| Axes: | 7 |
| Nominal load: | 7 – 14 kg |
| Operating space: | 800 mm |
| Repeatability: | ±0.1 mm |
| Weight: | 22 – 29 kg |
| Installation position: | Any |

**Fig. 13.14** Lightweight robot with integrated force/torque sensor system, according to KUKA AG

a torque perpendicular to the bearing rotation axes (Fig. 13.15). The handling devices shown are 4-axis devices with three rotation and one translation axis.

The linear Z lift axis is usually realized as a ball screw drive and also exhibits great stiffness in the vertical. The structure is relatively compliant in the horizontal direction. The main joining and load direction is the vertical, however. Horizontal articulated arm robots are, as in the example shown, commonly used for "Pick & Place" tasks because they are comparatively low-cost. With a smart design, current models use the entire working range of the arm length so that there is no "dead zone."

### 13.2.3 Cartesian Line Gantry Robots

Cartesian area gantry robots and line gantries have a cuboid-shaped operating space. Figure 13.16 shows its compact design. Its size allows it to enter service with little overhead. The drives of the three linear axes use direct-drive linear motors. Using polarity reversal of differently polarized magnetic fields they provide linear motion of the stator in relation to the rotor.

Line gantries have established themselves as the solution of choice for feeding machine tools because they extend into the operating space from above and so

| Epson SPIDER | Typical application area |
|---|---|
| 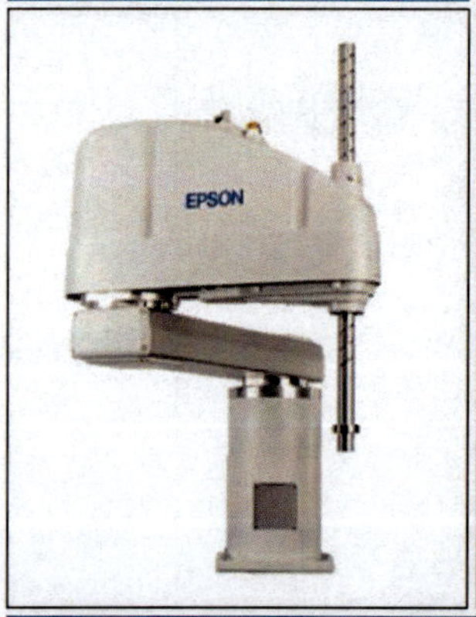 |  |

**Performance data**

| | |
|---|---|
| Design: | Internally oriented horizontal joint arm |
| Load capacity: | Max. 4 kg |
| Range: | 350 – 550 mm (horizontal) |
| | 130 mm (vertical) |
| Repeatability: | ±0.01 mm |

**Fig. 13.15** 4-axis horizontal SCARA robot in a production line, according to EPSON Deutschland GmbH

barely affect access to the machine. They are also used for picking tasks, for example, to transfer workpieces from a production or assembly line to another system running in parallel. Line gantries only have two main axes, so we can speak of a work plane.

Larger variants of Cartesian gantries are shown in Fig. 13.17. Shown in the picture on the right is a line gantry configured with multiple carriages based on a linear guide system with identical lifting units. It is used for workpiece handling on a flexible production line. The two carriages are coupled mechanically and are moved by one drive unit. Line gantries with two carriages also exist. These can be traversed independently to optimize cycle time or increase flexibility depending on the application.

The lift axes of line gantries can have diverse designs. The simplest variant has no secondary axes and only one gripper (Fig. 13.17, right). Considerable shortening of the workpiece change time can be achieved with a dual-carriage arrangement where the first lift axis removes the finished part and the second lift axis places the next blank directly into the machine fixture.

The variant with a rotating head and double gripper enables short workpiece change times similar to the dual-carriage arrangement. The rotating head allows the orientation of the workpieces to be changed. This enables waiting workpieces to be picked up and inserted into the machine fixture. Another benefit of changing the workpiece orientation angle permits more flexible fixtures in a production line.

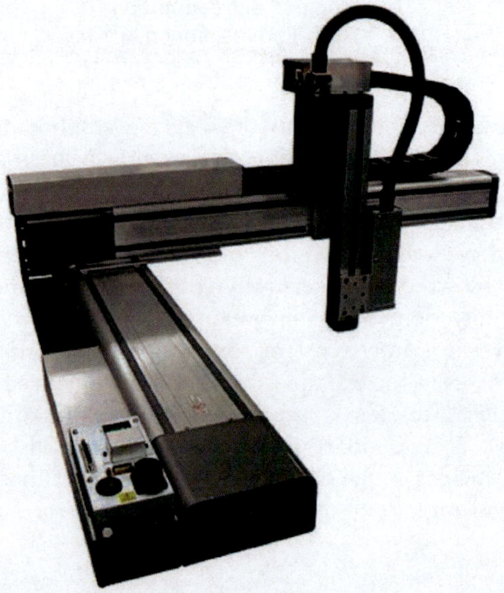

**Performance data**

| | |
|---|---|
| Nominal load: | 40 kg |
| Number of axes: | 3 |
| Operating space: | 1000/700/300 mm |

**Fig. 13.16** 4-axis Cartesian line gantry robot, according to EPSON Deutschland GmbH

**Fig. 13.17** Line gantries with dual-carriage arrangement in two different applications, according to Liebherr-International Deutschland GmbH

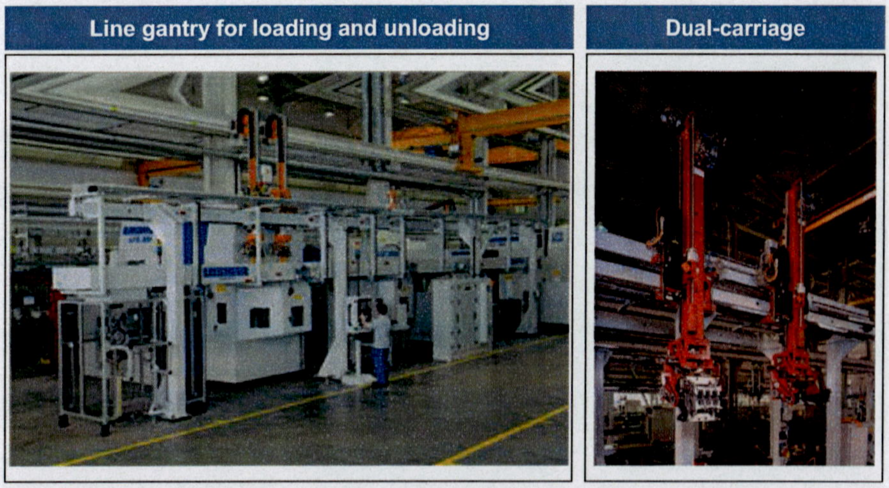


**Fig. 13.18** 3-axis area gantry robot in railway wheel/axle production, according to Bleichert Automation GmbH & Co. KG

### 13.2.4 Cartesian Area Gantry and Line Gantry with Girder

Supports on both sides of the girder lend themselves to larger dimensions in the y-direction. The term used is area gantry with crossbeam or bridge, on which the carriage moves in the y-direction (Fig. 13.18). In the application shown, the gantry is used for handling heavy workpieces.

Figure 13.19 shows another application scenario where line gantries with girders and vertical articulated arm robots work together in one production line. Typical structural components in car body construction are processed in the production line shown. These include B pillars and main chassis beams, as well as cross members and wheelhouse strips. They are being subjected to ever-stricter requirements. This means that methods, such as bonding and spot welding, are being combined more and more to provide joints with high strength and stability over large areas. High-strength, hot-formed stamped parts, roll-formed profiles, and standard components such as Isofix brackets, are also used in assemblies relevant to safety.

### 13.2.5 Parallel Kinematics

#### 13.2.5.1 Combined Kinematics

Since the mid-1990s, parallel kinematics have found their way into machine tool and robot design. The 4-axis robot in Fig. 13.20 comprises 3-axis parallel kinematics with an additional rotation axis in the robot hand. Coupled with cardan joints on each of the three arms that rotate on the base frame are two parallel

| Performance data for RL80 | |
|---|---|
| Nominal load: | 38 – 114 kg |
| Number of axes: | 4 |
| Max. speed: | 2.6 m/s |
| Operating space: | Dependent on application |

**Fig. 13.19** 4-axis line gantry robot with girder, according to KUKA AG

**Fig. 13.20** 4-axis parallel kinematics for Pick & Place operations, according to ABB

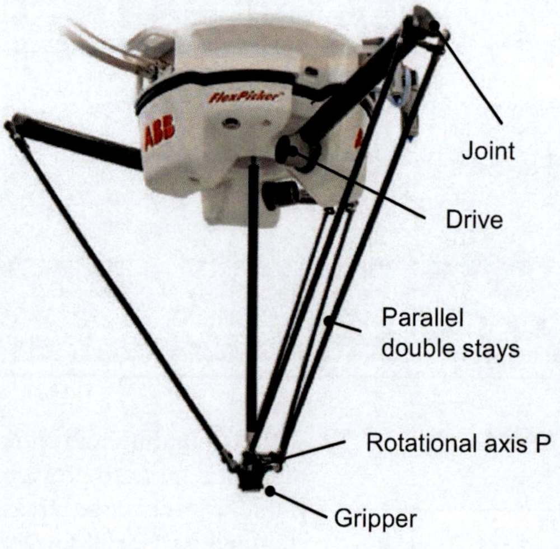


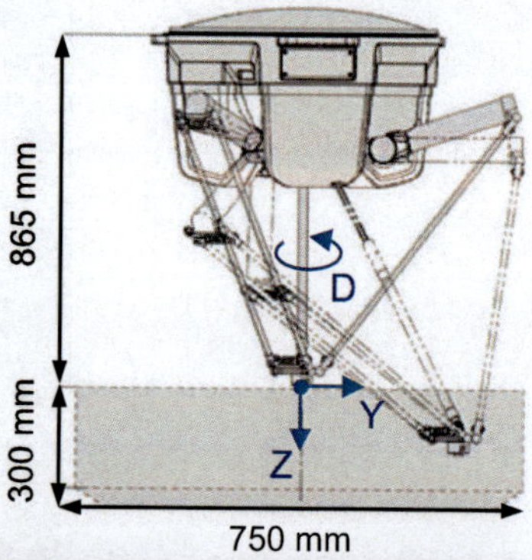


| Performance data | |
|---|---|
| Axes: | 4 |
| Nominal load: | 1 kg |
| Operating space diam.: | 750 mm |
| Repeatability: | 0.1 mm |
| Traverse speed: | 10 m/s |
| Acceleration: | 15 m/s² |

double stays. These stays are joined on the other side to the moving platform with ball joints. In this way, the platform constantly maintains its orientation while moving within the space, i.e., gripper rotation axis P remains in its vertical position. Their lightweight design means very high speeds and accelerations (up to 1.5 g) are possible. The main application is Pick & Place, such as in the packaging industry.

#### 13.2.5.2 Fully Parallel Kinematics

The fully parallel hexapod robot in ◘ Fig. 13.21 realizes all six degrees of freedom of the platform movement via the servo-driven telescopic legs with variable lengths. It too exhibits great stiffness.

The hexapod principle means the tilt, yaw, and roll angles attainable are greatly restricted compared to a conventional robot hand. They are also dependent on the position of the flange plate in the operating space.

### 13.2.6 Custom Designs

#### 13.2.6.1 Cylindrical and Spherical Coordinate Robots

A purely cylindrical coordinate design is rarely encountered nowadays despite its simplicity (◘ Fig. 13.22, left).

This design permits, in a manner similar to the horizontal articulated arm robot in ◘ Fig. 13.15, machine access from the side, but reachability of the target point by the straight arm is greatly restricted. The lack of an articulated axis in the arm prevents both horizontal as well as vertical avoidance of an interfering contour. Also, the hazard space is expanded unnecessarily by the cantilever protruding backward, something that could be improved however using a telescopic arm. The linear axes for the cantilever of the robot shown here in the radial direction and the support for the Z-axis are

| Performance data | |
|---|---|
| Axes: | 6 |
| Nominal load: | 100 kg |
| Range: | 1.04 m |
| Repeatability: | ±0.1 mm |
| Horiz. speed: | 1.5 m/min |
| Vert. speed: | 0.3 m/min |

**Fig. 13.21** 6-axis hexapod robot, according to Fanuc Deutschland GmbH

driven by a rack and pinion. Relieving the dead weight of the support is provided by a pneumatic cylinder mounted on the side.

The spherical coordinate robot on the right in Fig. 13.22 executes its three main coordinate motions with the two rotational B and C-axes, and the variable-length arm (R-axis). The R-axis goes through the intersection point of the B and C-axes (sphere center point). The usual secondary axes for manual orientation are affixed to the end of the arm.

The robot was specially designed for single-sided spot welding against the copper backup bar in vehicle production. In conventional resistance spot welding with a welding gun, the contact force required is generated by the welding gun mechanics. When the welding points are far from the sheet metal edge, a large welding gun span width is required, accompanied by a more intricate generation of force for great welding gun weight. A powerful, big robot would be necessary as a consequence.

A viable alternative is single-sided spot welding in which only one electrode is controlled by the robot. The second electrode is shaped like a laminar copper overlay beneath the metal sheet. In contrast to welding with a welding gun, the contact force must be applied by the robot itself here. The spherical coordinate robot shown generates a contact force of up to 5000 N. Such force cannot be applied by conventional robot kinematics alone because of their low stiffness. The space-saving and stiff gantry design provides the additional X-axis and results in an expansion of the workspace.

### 13.2.6.2 Hybrid Kinematics

Hybrid kinematics are understood to be the combination of two robot designs. Figure 13.23 shows 6-axis hybrid kinematics that unite the benefits of the linear concept with those of the articulated arm concept. The robot gantry is well suited for use in low-height factory buildings because the compact design of the robot and the combined linear articulated arm kinematics result in a very low overall height. So, the possible fields of application for these kinematics are very flexible. They can handle tasks such as loading and unloading machine tools and inert gas welding applications for more complex workpiece geometries.

The linear main axis in the longitudinal direction and the five rotational axes yield a virtually cylindrical operating space parallel to the production machine. Hanging headfirst, the robot is able to extend into the machine from the side to insert and remove components. The flexibility of this articulated arm kinematics with a total of six degrees of freedom also permits complex removal motions without having to mount additional tilt or rotation modules to the head axis.

A hybrid robot design is also shown in Fig. 13.24. The robot shown here is used to assist frequent pallet change processes, either directly into the machine tool or into the adjacent handling system.

### 13.2.6.3 Collaborative Robot Designs

The 2-arm robot shown in Fig. 13.25 forms a robot system that is categorized as a Human–Robot Collaboration (HRC) system. With seven axes for each robot arm, it is based on the arm of a human being, in a way similar to the “iiwa” robot from KUKA Roboter GmbH (see Fig. 13.14). It represents a flexible solution for handling and assembly of small parts and it is now being used mainly as an aid in electrical industry production due to its high degree of flexibility. Integrated force sensors mean the robot is inherently safe so it can move in the immediate vicinity of people without any safety fencing in place and without violating safety regulations. The robot system can be used for a variety of tasks by integrating a vision system and gripper hands on both robot arms.

Figure 13.26 shows the robot with type designation CR-35iA. It is a commercially available industrial robot from Fanuc. It is also enveloped by a pliable rubber skin. This means it also falls under collaborative robots and can operate in a very tight space without endangering people in the immediate vicinity. External

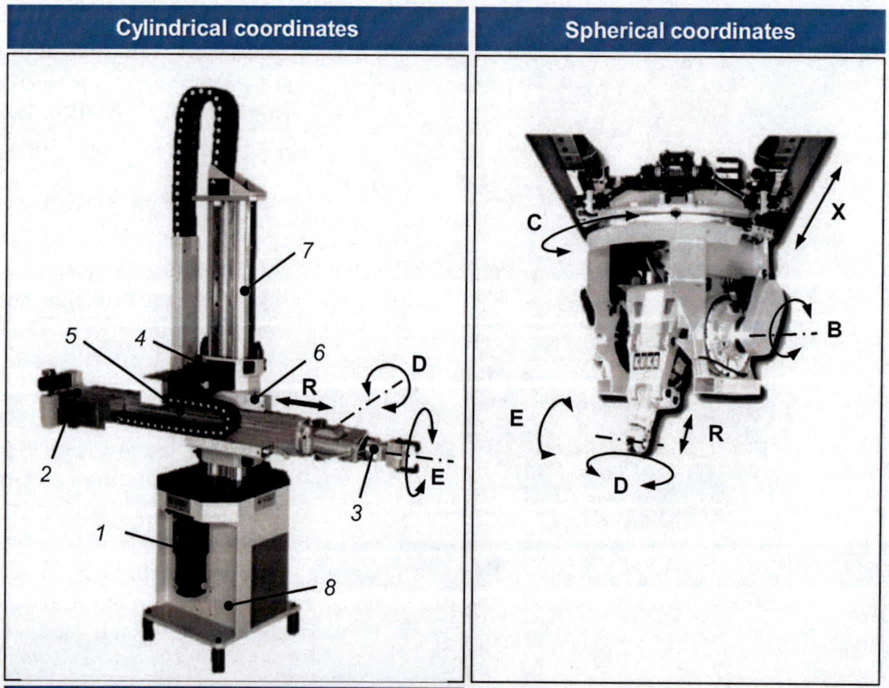


**Fig. 13.22** Cylindrical and spherical coordinate robots, according to Masat, KUKA AG

**Legend**

| | |
|---|---|
| *1:* | Drive, C-axis |
| *2:* | Drive, D-axis |
| *3:* | Drive, E-axis |
| *4:* | Drive, Z-axis |
| *5:* | Cantilever |
| *6:* | Support |
| *7:* | Rotating tower |
| *8:* | Pedestal |



interruptions are detected and the motion speed is reduced, or a stop of the robot initiated, by the additional force-torque sensor system in the base of the robot. With a load capacity of up to 35 kg and range of 1.8 m, it is also possible to move and mount heavier workpieces with this robot system.

The APAS assistant robot from Bosch represents the first industrial robot system certified for collaborative operation by the German Commercial employers' liability insurance association (Berufsgenossenschaft) (Fig. 13.27). The system comprises a standard vertical articulated arm robot that can be integrated at any workstation on its mobile base. Its enveloping capacitive sensor layer means the system stops when a person approaches it without making contact. With a sensitive gripper system and camera-aided parts recognition, the robot system lends itself to the palleting and sorting of small parts but is also used for the loading and unloading of production machines.

Human–Robot collaboration in robotics is becoming more and more important so that people and robots can perform complex tasks at a shared workstation. As, in this context, new and innovative robot concepts are being tested and advanced to be ready for market, this chapter only discusses one up-to-date part of all collaborative robot systems.

## 13.3 Coordinate Systems and Reference Points

To define the motion actions of an industrial robot, it is necessary to define coordinate systems in the operating space and reference points on the industrial robot. These systems form the basis for the unambiguous definition of motion instructions in the program of the industrial robot. Different coordinate systems are used here. Each depends on the handling task to be addressed.

Joint coordinates (*axis-specific robot coordinates*) define the spatial positioning of the industrial robot joints in relation to a local coordinate origin that is stationary on the axis (Fig. 13.28a). The current position of a robot can be defined this way with the

| Performance data | |
|---|---|
| Axes: | 5 rotation axes / 1 linear axis |
| Maximum load: | 16 kg |
| Operating space (linear): | Up to 30 m |
| Repeatability: | ±0.05 mm |

**Fig. 13.23** 6-axis hybrid kinematics, according to KUKA AG

rotation angles about this coordinate axis or as a position offset from the coordinate zero point for prismatic joints.

World coordinates (*Cartesian robot coordinates*) define the position and orientation of the Tool Center Point (TCP) in relation to a robot base coordinate system. This base coordinate system is normally centered within the installation area of the industrial robot (Fig. 13.28b).

The definition of a *tool coordinate system* is a prerequisite for programming handling tasks using Cartesian coordinates (such as within the world coordinate system). The tool coordinate system has the Tool Center Point at its origin (see Fig. 13.28c).

The use of tools on the robot flange of the industrial robot (IR) is factored in by defining a tool correction. It offsets the Tool Center Point and defines spatially the position of the specific working point of the tool used (refer to geometry data processing, ► Sect. 11.2.4.2). It is possible to save multiple tool corrections at the same time on the robot controller and to calculate motion paths for different tools within one program cycle.

In addition to the robot/tool-specific reference systems described above, IR controllers allow the user to define task-specific base coordinate systems by specifying application-dependent zero point corrections (Fig. 13.29). Using base coordinate systems enables, for example, palleting tasks to be programmed far more flexibly because only one reference point for the pallets needs to be defined as an absolute within space. All other workpiece positions on the pallet can be defined relative to this reference point regardless of the pallet location.

Absolute or incremental position sensors are normally used to determine the current position determination of the individual robot axes. In contrast to absolute transducers, incremental transducers can only specify a relative position in relation to a reference point defined previously (refer to Position measuring systems for NC machines in ► Sect. 2.3).

Furthermore, there are other robot type-dependent reference points on industrial robots and these points each specify the mechanical, mathematical or reference point of every single axis. The differentiation between the mechanical and mathematical zero point is required because generally, it is very difficult to precisely install an industrial robot unless the positions are particularly favorable. The control algorithms for motion control of the robot are based on geometrically favorable mathematical zero points of the robot axes, such as that of the right-angled "cannon position" shown in B. Figure 13.29. Normally, there is an offset between these mathematical and mechanical zero points. This offset is taken into account by special machine data in the controller software.

To set the robot kinematics, the individual axes must each be adjusted. For this, reference points are affixed to the axes. Then these points are approached during the adjustment (see Fig. 13.30). In the example shown, a measuring probe is connected to the axis to adjust. The probe holds an axially moving probe head. The probe head is pressed against the reference mark with spring force, for example, and so follows the outer contour of the mark. The adjustment value is then given from the edge analysis of the measured values. For this, the robot controller traverses the axis automatically, following manual pre-adjustment of the axis, such that the probe head is gradually directed to the level of the reference recess (a: rising edge). Then the probe head follows a plateau (b) before dropping into the recess (c: falling edge). At the reversal point of

**EROWA Lift**

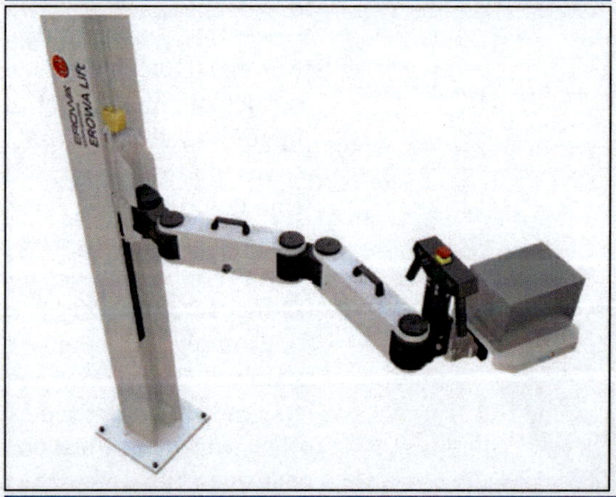

**Typical application**

**Performance data**

| | |
|---|---|
| Load capacity: | 100 – 250 kg |
| Range: | 1660 mm (vertical direction) |
| | 1300 – 1850 mm (horizontal direction) |
| Swivel range: | 360° |
| Max. speed: | 4 m/min |

**Fig. 13.24** Hybrid kinematics for assisting handling tasks, according to EROWA AG

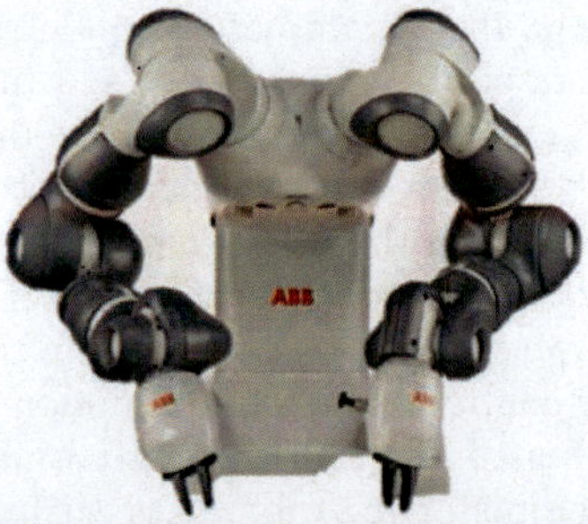

**Performance data**

| | |
|---|---|
| Axes: | 7 per arm |
| Nominal load: | 500 g per arm |
| Operating space: | 559 mm |
| Repeatability: | 0.02 mm |
| Traverse speed: | 1.5 m/s |
| Acceleration: | 11 m/s² |

**Fig. 13.25** Collab or ative 2-arm YuMi robot, according to ABB

the recess, the deflection of the probe head finally increases again (d: rising edge). This point (transition c→d) marks the adjustment value that can be determined with high accuracy using this method.

Performing an adjustment on initial startup of a robot system, and checking it on subsequent startups, are requirements. To attain higher positioning accuracy for different kinematics load scenarios (due, for example, to components with greatly differing masses), readjustment should be performed for load correction.

## 13.4 Coordinate Transformation and Path Generation

Programmed motion sequences of the industrial robot are processed in the robot controller by a special unit. This unit moves the joints from the current position to the target position in line with the motion behavior defined and the interpolation type required.

If the required data of a target position are available in axis-specific coordinates of the robot, such as rotation angles for rotational axes, the required travel difference of every axis can be calculated directly as a difference between target and actual positions. The current TCP position at any one time can be calculated during the motion by applying the forward transformation to all IR axis positions (Fig. 13.31). The forward transformation here corresponds to an addition of the axis distance vectors, starting from the base of the robot to the TCP.

If, in contrast, the target point of a robot motion is determined using Cartesian position and orientation specifications (Fig. 13.31), the axis positions of the robot must be calculated with the algorithms of the backward transformation in line with its kinematics [BLUM81]. This gives rise to solutions that are in part mathematically ambiguous (see the chapter

| Performance data | |
|---|---|
| Axes: | 6 |
| Load capacity: | 35 kg |
| Range: | 1813 mm |
| Max. speed: | 250 mm/s |
| Repeatability: | ±0.08 mm |

**Fig. 13.26** Collaborative articulated arm robot, according to Fanuc Deutschland GmbH

| Performance data | |
|---|---|
| Axes: | 6 |
| Load capacity: | 7 kg |
| Range: | 911 mm |
| Repeatability: | ±0.03 mm |
| Special feature: | Capacitive sensor skin |

**Fig. 13.27** Collaborating APAS assistant robot system, according to Robert Bosch GmbH

entitled Ambiguity of the backward transformation algorithms).

#### Definition of the kinematics of a handling device

An industrial robot can be defined as a kinematic chain of rigid axes whose joints between the arm elements each have only one degree of freedom [PAUL81, SNYD90]. If a separate coordinate system is placed into every link of this chain, the transformation between the adjacent joint coordinate systems is only a function of the rotation angle or linear feed.

To determine the individual joint positions of the robot, Cartesian world coordinates need to be converted to robot-specific joint coordinates.

#### Calculation of the axis positions for 3-axis robot arm kinematics

The mathematical relationships between the axis coordinates can be defined very easily if the handling device only has a 3-axis kinematic structure and the directions of motion of the individual axes can each be assigned to a single base coordinate.

For 3-axis robot axis structures with Cartesian workspace, the axis positions come directly from the components of the TCP vector components in Cartesian coordinates (Fig. 13.32a).

The same applies to 3-axis robots with a cylindrical or spherical operating space when using cylindrical or spherical coordinates to define the Tool Center Point position (Fig. 13.32b, c).

#### Performing the backward calculation for 4-axis swivel arm robots (SCARA)

The starting point for determining the axis positions of the 4-axis SCARA industrial robot is the Cartesian position and orientation specification in space:

$$P = f(x, y, z, \alpha, \beta, \gamma)$$

This industrial robot type shown in Fig. 13.33, commonly used for assembly operations, has a simple kinematic structure. The design makes it possible to calculate the axis positions separately for axes 1 and 2, for axis 3, and for axis 4. The position of axis 3 (the z-axis) can be determined directly from the z-coordinate of the target point (specified as Cartesian) because it, as the only axis acting directly in the z-direction, need not be transformed. Axis 4 specifies the rotational orientation of the TCP and is calculated depending on axis positions 1 and 2 that define the translational TCP position on the XY-plane.

This allows position determination (forward transformation) to be reduced to a 2D problem for axes 1 and 2 (Fig. 13.33):

$$\vec{p} = \vec{v}_1 + \vec{v}_2 \tag{13.1}$$

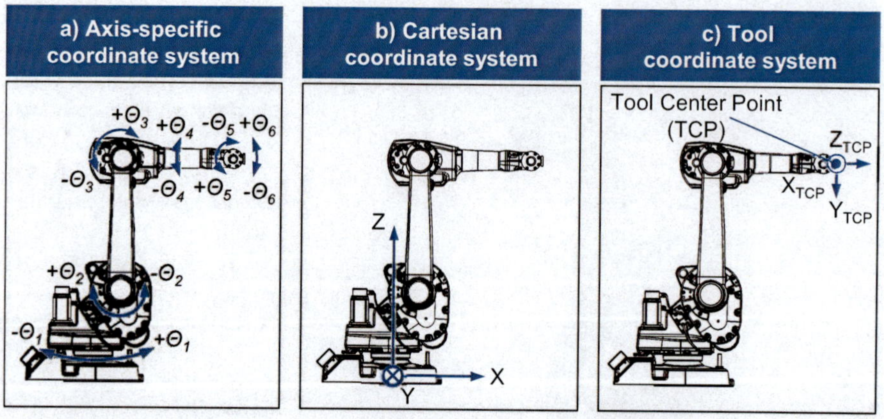


**Fig. 13.28** Coordinate systems of an industrial robot

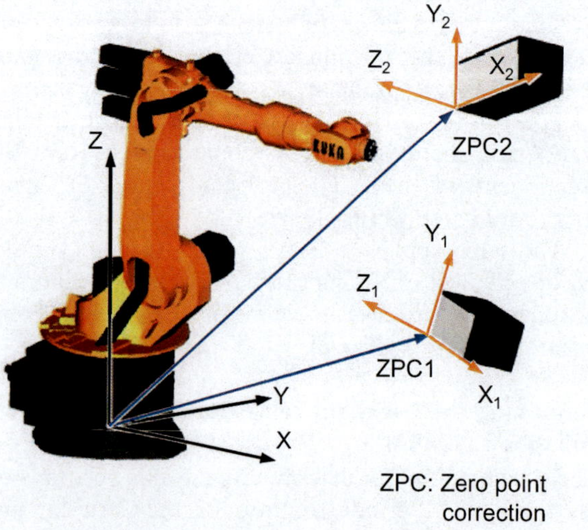


**Fig. 13.29** Auxiliary coordinate systems with zero point correction

$$\vec{v}_1 = s_1 \begin{pmatrix} \cos\theta_1 \\ \sin\theta_1 \end{pmatrix} \tag{13.2}$$

$$\vec{v}_2 = s_2 \begin{pmatrix} \cos(\theta_1+\theta_2) \\ \sin(\theta_1+\theta_2) \end{pmatrix} \tag{13.3}$$

$$\vec{p} = \begin{pmatrix} x \\ y \end{pmatrix} = \begin{pmatrix} s_1\cos\theta_1 + s_2\cos(\theta_1+\theta_2) \\ s_1\sin\theta_1 + s_2\sin(\theta_1+\theta_2) \end{pmatrix} \tag{13.4}$$

The cosine rule is used for the backward transformation. This allows the position angles of axes 1 and 2 to be determined as follows:

$$c^2 = x^2 + y^2 = s_1^2 + s_2^2 - 2s_1s_2\cos\gamma \tag{13.5}$$

where

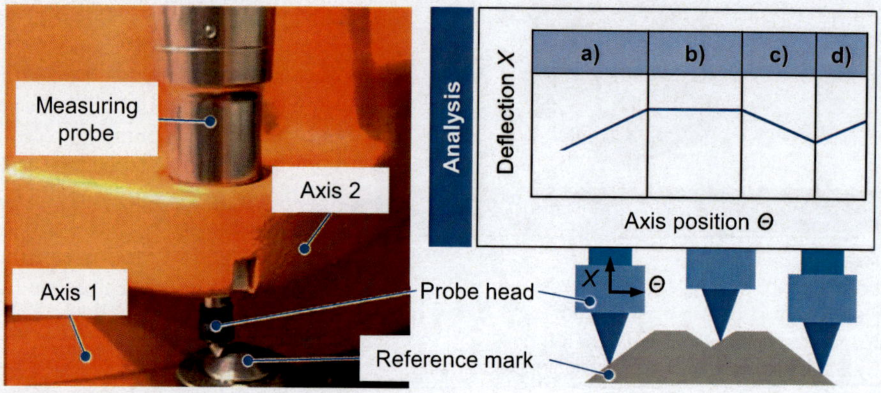


**Fig. 13.30** Reference point of a robot axis, according to KUKA AG

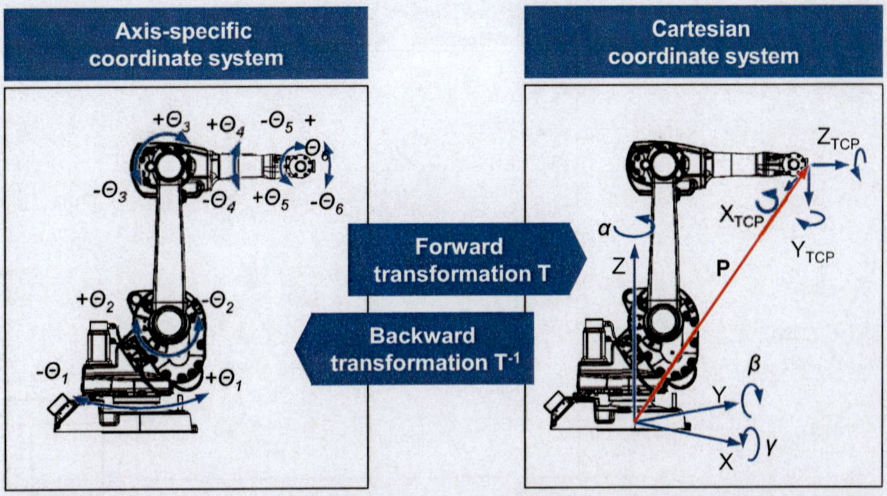


**Fig. 13.31** Transformation of Cartesian position data to axis information

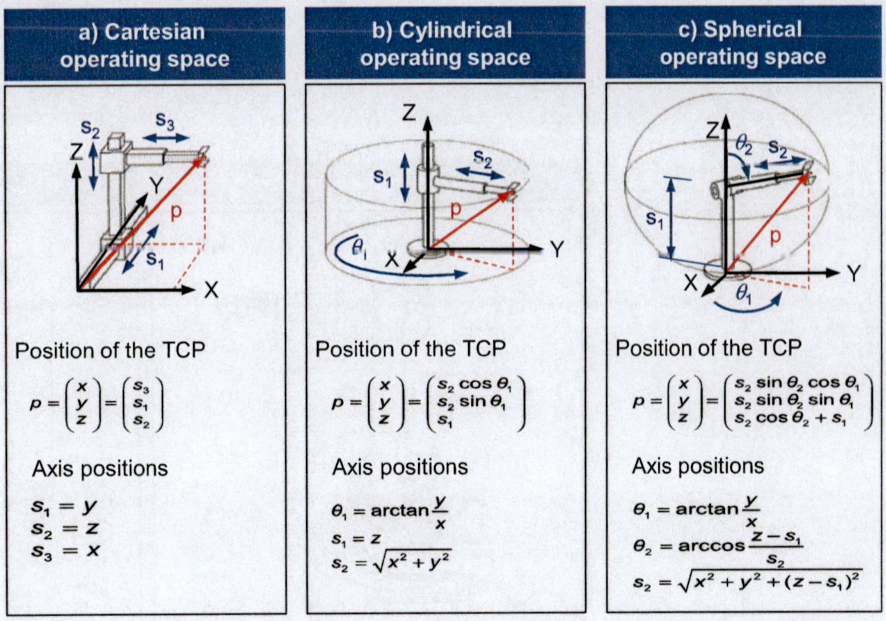


**Fig. 13.32** Determining the axis positions for a 3-axis industrial robot

$$\cos\gamma = \cos(180^\circ - \theta_2) = -\cos\theta_2 \quad (13.6)$$

$$\theta_2 = \arccos\left(\frac{x^2 + y^2 - s_1^2 - s_2^2}{2s_1 s_2}\right) \quad (13.7)$$

and

$$\theta_1 = \arctan\frac{y}{x} - \arctan\left(\frac{s_2\sin\theta_2}{s_1 + s_2\cos\theta_2}\right) \quad (13.8)$$

where $s_1$ and $s_2$ are the characteristics of the kinematic chain.

### Calculation of the axis position data for a 6-axis industrial robot

Industrial robots with six independent joint axes provide arbitrary positioning and orientation of the tool in the operating space within the mechanical axis limits.

The procedure for the backward transformation, i.e., calculation of the axis position data from the Cartesian target point data, is shown for the 6-axis joint arm robot by way of example (Fig. 13.34).

When the six axes of this robot type are divided into base and manual axes, the three base axes are used primarily for positioning and the three manual axes mainly for orientation adjustment. A clear-cut division between positioning and orientation axes is not

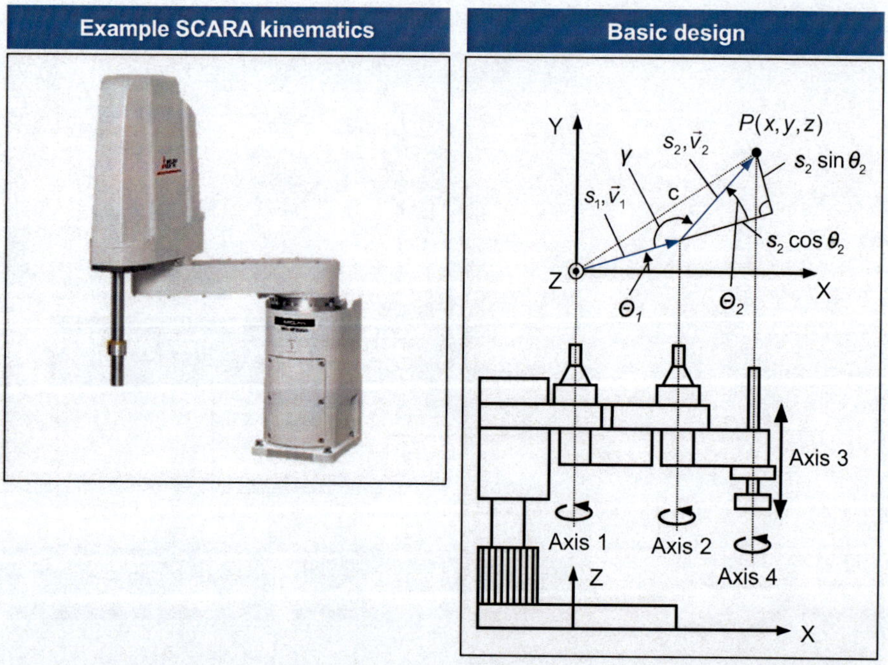


**Fig. 13.33** Kinematics of a SCARA robot, according to Mitsubishi

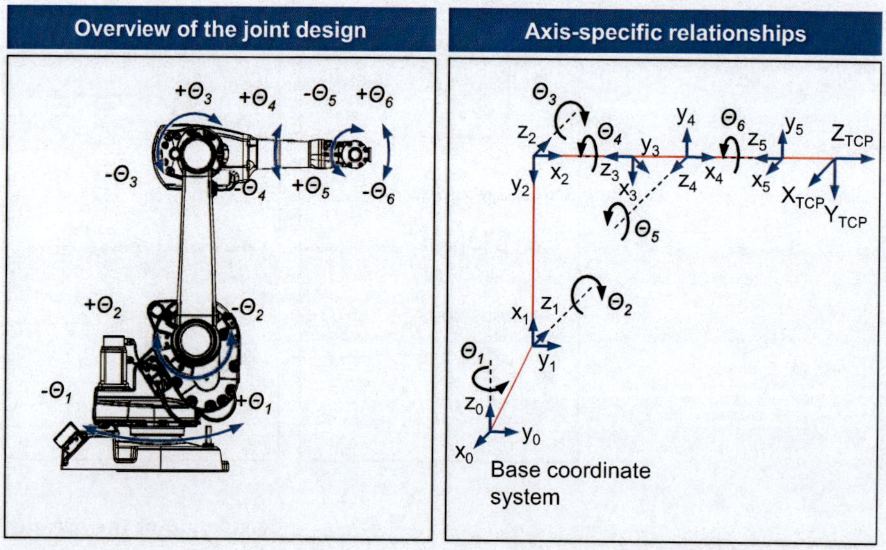


**Fig. 13.34** Kinematics of a 6-axis articulated arm robot

13

possible, however, because every time the base axes are positioned, there is also an orientation change. That is, every orientation adjustment on the manual axes inevitably causes a change to the TCP position.

Given that neither a clear-cut division between positioning and orientation axes, nor a simple analytical definition of the spatial axes arrangement (as for the 4-axis SCARA robot) is possible for this robot type, transformation matrices are used to solve the backward calculation.

### Definition of the target point of a handling device

The Cartesian target point is defined by specifying a 3D position vector $\vec{P}$ in the reference coordinate system and by specifying the orientation (Fig. 13.35).

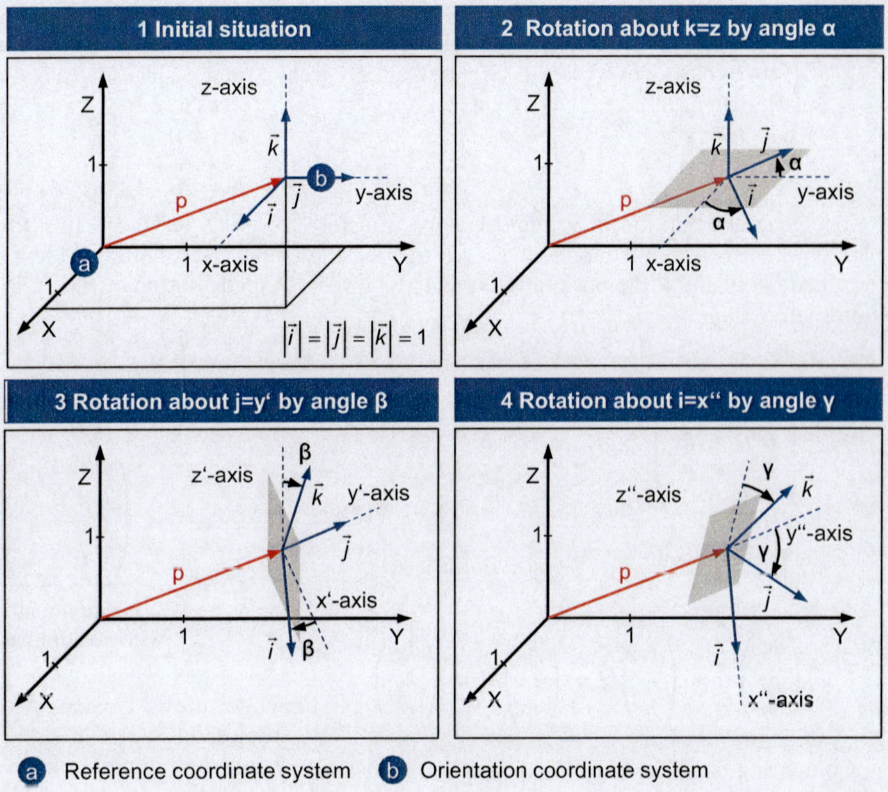


**Fig. 13.35** Definition of the target point of a 6-axis articulated arm robot

The orientation is illustrated by an orientation coordinate system defined by three unit vectors $\vec{i}$, $\vec{j}$ and $\vec{k}$. There are a number of different ways to describe the orientation of this coordinate system and thus the target point. One of the ways to define it uniquely is to specify three angles around the coordinate axes. It is then possible, for example, to define the orientation with the three rotation angles $\alpha$, $\beta$ and $\gamma$ around the z, y' and x" coordinates (nautical angles) [OSTE92].

Figure 13.35 shows this form of target point definition in four steps:

1. Offset by position vector $\vec{P}$
2. Rotation by angle $\alpha$ around the z-axis of the orientation coordinate system (not yet rotated)
3. Rotation by angle $\beta$ around the y'-axis of the orientation coordinate system (rotated by $\alpha$)
4. Rotation by angle $\gamma$ around the x"-axis of the orientation coordinate system (rotated twice)

Mathematically, this relationship can be expressed using transformation matrix $T$. It is made up of the matrix product of the position matrix $P$ with position vector $\vec{P}$, and the orientation matrix $R$ with the unit vectors of orientation $\vec{i}$ (in x), $\vec{j}$ (in y) and $\vec{k}$ (in z).

For the orientation definition in nautical angles, matrix $T$ can be developed from four single matrices that each define the rotation around an axis coordinate or the translational motion. The submatrix for the first rotation to perform is on the right in (13.9):

$$T = \underbrace{\begin{pmatrix} 1 & 0 & 0 & p_x \\ 0 & 1 & 0 & p_y \\ 0 & 0 & 1 & p_z \\ 0 & 0 & 0 & 1 \end{pmatrix}}_{\text{translation by } p} \cdot \underbrace{\begin{pmatrix} \cos\alpha & -\sin\alpha & 0 & 0 \\ \sin\alpha & \cos\alpha & 0 & 0 \\ 0 & 0 & 1 & 0 \\ 0 & 0 & 0 & 1 \end{pmatrix}}_{\text{rotation by } k} \cdot \underbrace{\begin{pmatrix} \cos\beta & 0 & \sin\beta & 0 \\ 0 & 1 & 0 & 0 \\ -\sin\beta & 0 & \cos\beta & 0 \\ 0 & 0 & 0 & 1 \end{pmatrix}}_{\text{rotation by } j} \cdot \underbrace{\begin{pmatrix} 1 & 0 & 0 & 0 \\ 0 & \cos\gamma & -\sin\gamma & 0 \\ 0 & \sin\gamma & \cos\gamma & 0 \\ 0 & 0 & 0 & 1 \end{pmatrix}}_{\text{rotation by } i} \quad (13.9)$$

The simplified notation for this matrix multiplication is:

$$T = \mathrm{Trans}(p_x p_y p_z) \cdot \mathrm{Rot}(z, \alpha) \cdot \mathrm{Rot}(y, \beta) \cdot \mathrm{Rot}(x, \gamma) \quad (13.10)$$

Rot($x$, $\gamma$) represents here a rotation around the x-axis by angle $\gamma$. As matrix multiplications are not commutative, the order of rotation to define the orientation must be observed. Multiplying out the equation above yields this matrix (13.11):

$$T = \begin{pmatrix} \cos\alpha\cos\beta & -\sin\alpha\cos\gamma + \cos\alpha\sin\beta\sin\gamma & \sin\alpha\sin\gamma + \cos\alpha\sin\beta\cos\gamma & p_x \\ \sin\alpha\cos\beta & \cos\alpha\cos\gamma + \sin\alpha\sin\beta\sin\gamma & -\cos\alpha\sin\gamma + \sin\alpha\sin\beta\cos\gamma & p_y \\ -\sin\beta & \cos\beta\sin\gamma & \cos\beta\cos\gamma & p_z \\ 0 & 0 & 0 & 1 \end{pmatrix} \quad (13.11)$$

The general definition of a homogeneous transformation where $\vec{i}$, $\vec{j}$ and $\vec{k}$ represent the vectors of the coordinate axes and $\vec{P}$ the position vector of the target coordinate system, is:

$$T = \begin{pmatrix} i_x & j_x & k_x & p_x \\ i_y & j_y & k_y & p_y \\ i_z & j_z & k_z & p_z \\ 0 & 0 & 0 & 1 \end{pmatrix} \quad (13.12)$$

where the vectors are

$$\vec{P} = \begin{pmatrix} p_x \\ p_y \\ p_z \end{pmatrix} \; \vec{i} = \begin{pmatrix} i_x \\ i_y \\ i_z \end{pmatrix} \; \vec{j} = \begin{pmatrix} j_x \\ j_y \\ j_z \end{pmatrix} \; \vec{k} = \begin{pmatrix} k_x \\ k_y \\ k_z \end{pmatrix} \quad (13.13)$$

Comparing coefficients yields vectors $\vec{i}$, $\vec{j}$ and $\vec{k}$ as a function of the angles $\alpha$, $\beta$ and $\gamma$. For a 6-axis articulated arm robot, matrix T for a Cartesian target point definition corresponds to the TCP for calculating the backward transformation with matrices (see further down).

13

### Definition of the kinematics of an articulated arm robot with matrices

In the kinematic chain of the industrial robot, every axis motion is defined as a motion between two arm elements. When a separate coordinate system having a fixed reference to the appropriate link is placed in every link of this chain, transformation matrix $A_n$ between adjacent joint coordinate systems $K_n$ and $K_{n-1}$ is dependent on the joint position and axis motion.

As the joint between the arm elements only has one degree of freedom, the transformation is only a function of either a rotation angle or linear path. This transformation can be illustrated clearly for a 6-axis robot using a transformation graph (◘ Fig. 13.36a).

So, for a 6-axis robot, six transformation matrices $A_n$ and six transformation matrices $A_n^{-1}$ can be specified to define the forward and backward calculations, respectively. This yields, for the kinematic chain of the industrial robot from the base point to the TCP, transformation equation $T_6$ as a multiplication of the single transformation equations of every axis:

$$T_6 = A_1 \cdot A_2 \cdot A_3 \cdot A_4 \cdot A_5 \cdot A_6. \quad (13.14)$$

The result of this transformation equation must match the matrix of the Cartesian position and orientation definition of the TCP, see Eq. (13.15):

$$T_6 = \begin{pmatrix} i_x & j_x & k_x & p_x \\ i_y & j_y & k_y & p_y \\ i_z & j_z & k_z & p_z \\ 0 & 0 & 0 & 1 \end{pmatrix} = \underbrace{\begin{pmatrix} 1 & 0 & 0 & p_x \\ 0 & 1 & 0 & p_y \\ 0 & 0 & 1 & p_z \\ 0 & 0 & 0 & 1 \end{pmatrix}}_{\text{position matrix with position vector}} \cdot \underbrace{\begin{pmatrix} i_x & j_x & k_x & 0 \\ i_y & j_y & k_y & 0 \\ i_z & j_z & k_z & 0 \\ 0 & 0 & 0 & 1 \end{pmatrix}}_{\text{orientation matrix R of the TCP}} \quad (13.15)$$

As the last rows of the matrices do not provide any useful information, 12 individual equations result. These depend on the known elements of matrix $T_6$ and the unknown joint variables. They can be calculated by solving the equations for the joint variables.

### Denavit-Hartenberg method for defining the transformation

The Denavit-Hartenberg method is a method for defining the mathematical transformation chaining in a standardized manner. For every link or axis, different parameters must be specified that make it possible to calculate the elements of the associated transformation matrices [PAUL81].

Six unknowns (three angles and three coordinates of a displacement vector) must generally be specified to formulate a transformation matrix. Using the Denavit-Hartenberg method reduces this to four unknowns. These consist of two rotations (angles $\theta_n$ and $\alpha_n$) and two offsets ($d_n$ and $a_n$) [DENA95, HART56].

For the required transformation equations $A_n$, conventions for parameters $\theta_n$, $d_n$, $a_n$ and $\alpha_n$ must be observed as in ◘ Table 13.1 and ◘ Fig. 13.37.

The general definition does not yet include any restrictions for swivel or prismatic joints, however. For swivel joints, parameter $\theta_n$ represents the variable, and for prismatic joints (e.g., for gantry robots), parameter $d_n$ is the variable.

Using these parameters, transformation equations $A_i$ define the joint position of the axes of the industrial

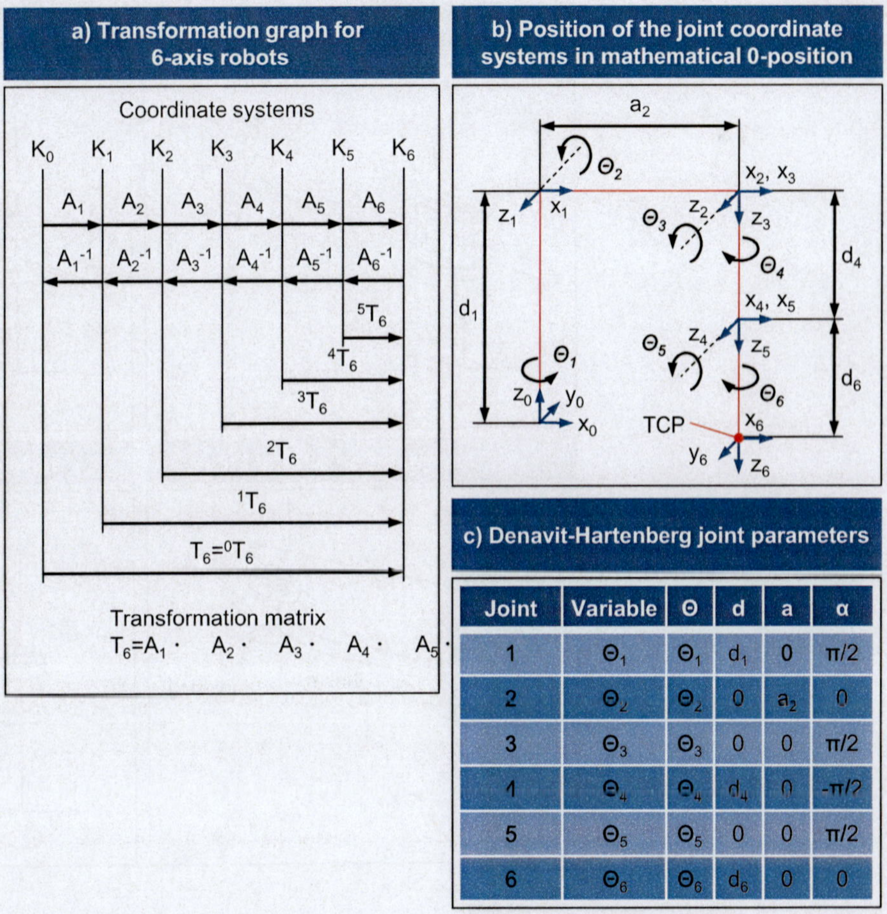


| Joint | Variable | Θ | d | a | α |
|---|---|---|---|---|---|
| 1 | $\Theta_1$ | $\Theta_1$ | $d_1$ | 0 | π/2 |
| 2 | $\Theta_2$ | $\Theta_2$ | 0 | $a_2$ | 0 |
| 3 | $\Theta_3$ | $\Theta_3$ | 0 | 0 | π/2 |
| 4 | $\Theta_4$ | $\Theta_4$ | $d_4$ | 0 | -π/2 |
| 5 | $\Theta_5$ | $\Theta_5$ | 0 | 0 | π/2 |
| 6 | $\Theta_6$ | $\Theta_6$ | $d_6$ | 0 | 0 |

**Fig. 13.36** Backward transformation for a 6-axis articulated arm robot

**Table 13.1.** Definition of the Denavit-Hartenberg parameters

| Parameter | Meaning |
|---|---|
| $\Theta_n$ | Angle between the positive $x_{n-1}$ axis and the positive $x_n$ axis rotated about the $z_{n-1}$ axis in the mathematically positive sense: For rotational axes, this angle always corresponds to the variable rotation angle when, in the mathematical zero position, all $x_n$ point in the same direction and all $z_n$ represent the rotation axes. |
| $d_n$ | Distance of origin $O_n$ of coordinate system $K_n$ from origin $O_{n-1}$ of coordinate system $K_{n-1}$ in the direction of the $Z_{n-1}$ axis |
| $a_n$ | Distance between normals of $z_{n-1}$ and $z_n$ axes |
| $\alpha_n$ | Angle between the positive $z_{n-1}$ axis and the positive $z_n$ axis rotated about the $x_{n-1}$ axis in the mathematically positive sense |

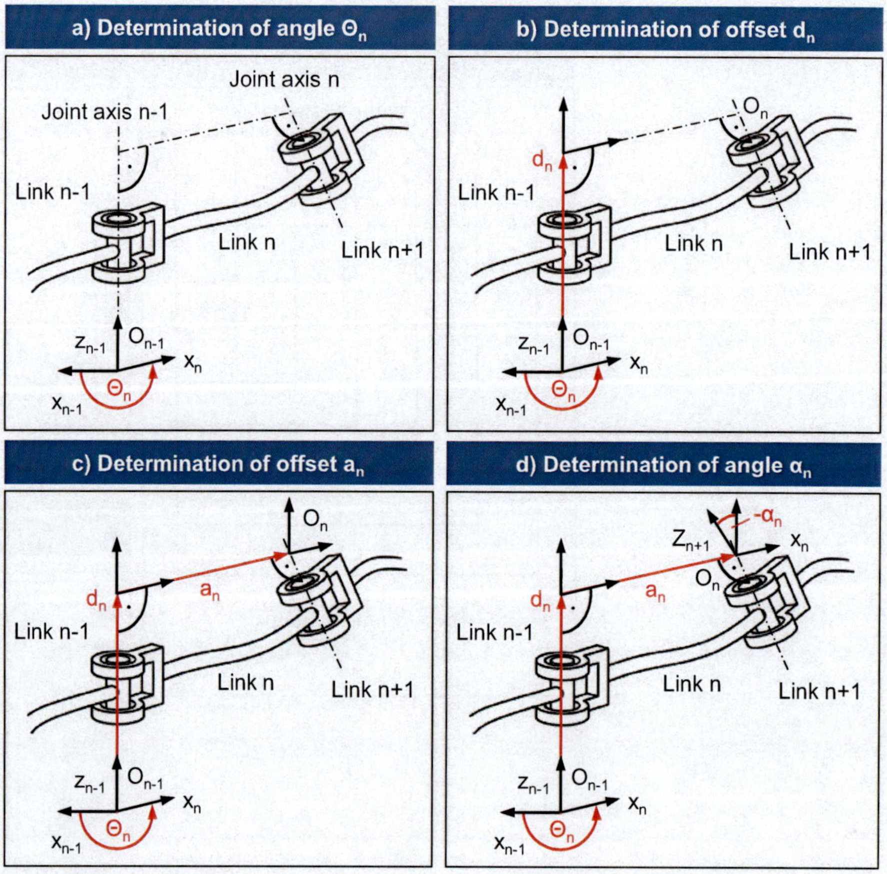


■ **Fig. 13.37** Denavit-Hartenberg parameterization of joints

13

robot from the following single matrices using matrix multiplication, and are called the Denavit-Hartenberg (DH) matrices, Eqs. (13.16) to (13.18).

$$A_n = \mathrm{Rot}(z_n, \theta_n) \cdot \mathrm{Trans}(0, 0, d_n) \cdot \mathrm{Trans}(a_n, 0, 0) \cdot \mathrm{Rot}(x_n, \alpha_n) \tag{13.16}$$

$$A_n = \begin{pmatrix} \cos\theta_n & -\sin\theta_n & 0 & 0 \\ \sin\theta_n & \cos\theta_n & 0 & 0 \\ 0 & 0 & 1 & 0 \\ 0 & 0 & 0 & 1 \end{pmatrix} \cdot \begin{pmatrix} 1 & 0 & 0 & a_n \\ 0 & 1 & 0 & 0 \\ 0 & 0 & 1 & d_n \\ 0 & 0 & 0 & 1 \end{pmatrix} \cdot \begin{pmatrix} 1 & 0 & 0 & 0 \\ 0 & \cos\alpha_n & -\sin\alpha_n & 0 \\ 0 & \sin\alpha_n & \cos\alpha_n & 0 \\ 0 & 0 & 0 & 1 \end{pmatrix} \tag{13.17}$$

$$A_n = \begin{pmatrix} \cos\theta_n & -\sin\theta_n\cos\alpha_n & \sin\theta_n\sin\alpha_n & a_n\cos\theta_n \\ \sin\theta_n & \cos\theta_n\cos\alpha_n & -\cos\theta_n\sin\alpha_n & a_n\sin\theta_n \\ 0 & \sin\alpha_n & \cos\alpha_n & d_n \\ 0 & 0 & 0 & 1 \end{pmatrix} \tag{13.18}$$

In the mathematically favorable zero position (■ Fig. 13.36b), DH joint parameters $\theta_n$, $d_n$, $a_n$ and $\alpha_n$ have been determined for all of the robot axes (■ Fig. 13.36c). Inserting the joint parameters into the general transformation matrix $A_n$ yields matrix $A_i$. This transforms coordinate system $K_{i-1}$ into system $K_i$. The following abbreviations are used in the transformation matrices to simplify the notation:

$$\sin\theta_i = S_i$$
$$\cos\theta_i = C_i \tag{13.19}$$

The following transformation matrices $A_i$ and their inverses $A_i^{-1}$ come from the example shown:

$$A_1 = \begin{pmatrix} C_1 & 0 & S_1 & 0 \\ S_1 & 0 & -C_1 & 0 \\ 0 & 1 & 0 & d_1 \\ 0 & 0 & 0 & 1 \end{pmatrix} \quad A_1^{-1} = \begin{pmatrix} C_1 & S_1 & 0 & 0 \\ 0 & 0 & 1 & -d_1 \\ S_1 & -C_1 & 0 & 0 \\ 0 & 0 & 0 & 1 \end{pmatrix}$$
$$A_2 = \begin{pmatrix} C_2 & -S_2 & 0 & a_2C_2 \\ S_2 & C_2 & 0 & a_2S_2 \\ 0 & 0 & 1 & 0 \\ 0 & 0 & 0 & 1 \end{pmatrix} \quad A_2^{-1} = \begin{pmatrix} C_2 & S_2 & 0 & -a_2 \\ -S_2 & C_2 & 0 & 0 \\ 0 & 0 & 1 & 0 \\ 0 & 0 & 0 & 1 \end{pmatrix}$$
$$A_3 = \begin{pmatrix} C_3 & 0 & S_3 & 0 \\ S_3 & 0 & -C_3 & 0 \\ 0 & 1 & 0 & 0 \\ 0 & 0 & 0 & 1 \end{pmatrix} \quad A_3^{-1} = \begin{pmatrix} C_3 & S_3 & 0 & 0 \\ 0 & 0 & 1 & 0 \\ S_3 & -C_3 & 0 & 0 \\ 0 & 0 & 0 & 1 \end{pmatrix}$$
$$A_4 = \begin{pmatrix} C_4 & 0 & -S_4 & 0 \\ S_4 & 0 & C_4 & 0 \\ 0 & -1 & 0 & d_4 \\ 0 & 0 & 0 & 1 \end{pmatrix} \quad A_4^{-1} = \begin{pmatrix} C_4 & S_4 & 0 & 0 \\ 0 & 0 & -1 & d_4 \\ -S_4 & C_4 & 0 & 0 \\ 0 & 0 & 0 & 1 \end{pmatrix}$$
$$A_5 = \begin{pmatrix} C_5 & 0 & S_5 & 0 \\ S_5 & 0 & -C_5 & 0 \\ 0 & 1 & 0 & 0 \\ 0 & 0 & 0 & 1 \end{pmatrix} \quad A_5^{-1} = \begin{pmatrix} C_5 & S_5 & 0 & 0 \\ 0 & 0 & 1 & 0 \\ S_5 & -C_5 & 0 & 0 \\ 0 & 0 & 0 & 1 \end{pmatrix}$$
$$A_6 = \begin{pmatrix} C_6 & -S_6 & 0 & 0 \\ S_6 & C_6 & 0 & 0 \\ 0 & 0 & 1 & d_6 \\ 0 & 0 & 0 & 1 \end{pmatrix} \quad A_6^{-1} = \begin{pmatrix} C_6 & S_6 & 0 & 0 \\ -S_6 & C_6 & 0 & 0 \\ 0 & 0 & 1 & -d_6 \\ 0 & 0 & 0 & 1 \end{pmatrix} \tag{13.20}$$

In determining the inverse transformation matrix, rotation matrix $R$, representing an orthonormal system, is taken into account.

$$R = \begin{pmatrix} i_x & j_x & k_x \\ i_y & j_y & k_y \\ i_z & j_z & k_z \end{pmatrix} \quad \vec{P} = \begin{pmatrix} p_x \\ p_y \\ p_z \end{pmatrix} \tag{13.21}$$

$R^{-1}=R^T$ applies to orthonormal matrices $R$. Using these relationships defines the inverse matrix $A_i^{-1}$ as follows [BRON84, PAUL81]:

$$A_i^{-1} = \begin{pmatrix} i_x & i_y & i_z & -\left(\vec{i}\cdot\vec{p}\right) \\ j_x & j_y & j_z & -\left(\vec{j}\cdot\vec{p}\right) \\ k_x & k_y & k_z & -\left(\vec{k}\cdot\vec{p}\right) \\ 0 & 0 & 0 & 1 \end{pmatrix} \tag{13.22}$$

### Determination of the axis angles

To determine joint variables $\theta_i$ ($i=1, \ldots, 6$), the matrix equations that transform the robot hand coordinate system into the joint coordinate systems are formulated.

$${}^{i-1}T_6 = A_{i-1}^{-1}\cdot A_{i-2}^{-1}\cdot\ldots\cdot A_1^{-1}\cdot T_6 = A_i\cdot A_{i+1}\cdot\ldots\cdot A_6 \tag{13.23}$$

Starting with matrix equation $T_6=A_1\cdot A_2\cdot A_3\cdot A_4\cdot A_5\cdot A_6$, the elements of the matrix equation that are either zero or a function of only one joint variable are taken first. If none of the joint variables can be determined by comparing coefficients, the next matrix equation is calculated (first ${}^1T_6$, then ${}^2T_6$, ${}^3T_6$, etc.). An attempt is also made here to determine one or more variables by comparing coefficients. Successive multiplication by the inverse matrices, followed by coefficient comparison, is performed until all joint variables are determined.

In the simple form described, this method does not result in an analytically closed solution for all robot kinematics. It might be necessary to determine parts of the solution using iterative methods, by calculating intermediate points or by offsetting coordinate systems [MCKE98, RIES92].

Even the kinematics shown in ◘ Fig. 13.34 and 13.36 cannot be solved that straightforwardly. It is possible to arrive at a result, however, when length $d_6$ is assumed to be 0. A prerequisite for this method is that the target coordinate system of the robot is moved to the intersection point of axes 4, 5 and 6 before the backward transformation. The k-vector of the target coordinate system can be used for this. As the vector $\vec{k}$ is parallel to the sixth axis of the robot and consequently also to the hand with length $d_6$, the origin (13.24) of the original coordinate system can be converted as follows to the origin $\vec{P}'$ of the new coordinate system, see Eq. (13.25).

$$\vec{P} = \begin{pmatrix} p_x \\ p_y \\ p_z \end{pmatrix} \tag{13.24}$$

$$\vec{P}' = \begin{pmatrix} p'_x \\ p'_y \\ p'_z \end{pmatrix} = \begin{pmatrix} p_x \\ p_y \\ p_z \end{pmatrix} - d_6\cdot\begin{pmatrix} k_x \\ k_y \\ k_z \end{pmatrix} = \begin{pmatrix} p_x - d_6\cdot k_x \\ p_y - d_6\cdot k_y \\ p_z - d_6\cdot k_z \end{pmatrix} \tag{13.25}$$

After this simplification, vector $\vec{P}'$ and $d_6=0$ can be used for the calculation below. Equation (13.26) is formulated first.

$$T_6 = A_1\cdot A_2\cdot A_3\cdot A_4\cdot A_5\cdot A_6 \tag{13.26}$$

Since the matrix does not contain any simple expressions, the next matrix Eq. ${}^1T_6$ is determined (◘ Fig. 13.36a). This matrix is derived from the previous one by multiplying by inverse $A_1^{-1}$:

$${}^1T_6 = A_1^{-1}\cdot T_6 \tag{13.27}$$

In the same way, matrix ${}^1T_6$ is analyzed for simple expressions to determine the variables still unknown. The process of consecutively performed multiplication by inverse $A_n^{-1}$ is repeated until all six joint variables are known.

$$ {}^nT_6 = A_n^{-1} \cdot A_{n-1}^{-1} \cdot \ldots \cdot A_1^{-1} \cdot T_6 \tag{13.28} $$

Element (3,4) from ${}^1T_6 = A_1^{-1} \cdot T_6$ gives:

$$ 0 = S_1 p_x' - C_1 p_y' \tag{13.29} $$

As solving for quotient $S_1/C_1$ is not unique, we have two equations, (13.30).

$$ \frac{S_1}{C_1} = \frac{p_y'}{p_x'} = \frac{-p_y'}{-p_x'} \tag{13.30} $$

If angle $\theta_i$ is determined using the atan2 function, a distinction is required between two solutions, see Eqs. (13.31) and (13.32).

$$ \Rightarrow \quad \theta_1 = \operatorname{atan2}\left(\frac{p_y'}{p_x'}\right) \tag{13.31} $$

$$ \Rightarrow \theta_1' = \operatorname{atan2}\left(\frac{-p_y'}{-p_x'}\right) = \theta_1 + \pi \tag{13.32} $$

*Note:*

In the literature, an inverse tangent function that determines the angle from the quotient of sine and cosine functions while simultaneously taking into account the leading sign is known as an atan2 a/b function.

The conditional equations necessary for angles $\theta_2$ and $\theta_3$ can be taken from the same matrix equation, see Eqs. (13.33) and (13.34).

Element (1,4):

$$ C_1 p_x' + S_1 p_y' = (S_2 C_3 + C_2 S_3) d_4 + C_2 a_2 = S_{23} d_4 + C_2 a_2 \tag{13.33} $$

Element (2,4):

$$ p_z' - d_1 = -(C_2 C_3 - S_2 S_3) d_4 + S_2 a_2 = -C_{23} d_4 + S_2 a_2 \tag{13.34} $$

Squaring and then adding the two equations gives, with the trigonometrical relationships from (13.35), an equation for $S_3$, Eq. (13.36).

$$ S_{23} = S_2 C_3 + C_2 S_3, \qquad C_{23} = C_2 C_3 - S_2 S_3 \tag{13.35} $$

$$ S_3 = \pm \frac{a_2^2 + d_4^2 - (C_1 p_x' + S_1 p_y')^2 - (p_z' - d_1)^2}{2 a_2 d_4} \tag{13.36} $$

$C_3$ is determined using the Pythagorean trigonometric identity, Eq. (13.37).

$$ C_3 = \pm (1 - S_3^2)^{1/2} \tag{13.37} $$

A distinction is required here between two possible solutions, see Eqs. (13.38) and (13.39).

$$ \Rightarrow C_3 > 0: \quad \theta_3 = \operatorname{atan2}\left(\frac{S_3}{(1 - S_3{}^2)^{1/2}}\right) \tag{13.38} $$

$$ \Rightarrow C_3 < 0: \quad \theta_3' = \operatorname{atan2}\left(\frac{S_3}{-(1 - S_3{}^2)^{1/2}}\right) = -\theta_3 + \pi \tag{13.39} $$

When $\theta_3$ is known, $S_2$ and $C_2$ can now be determined from elements (1, 4) and (2, 4). See Eqs. (13.40) and (13.41).

Element (1,4):

$$ C_1 p_x' + S_1 p_y' = (S_2 C_3 + C_2 S_3) d_4 + C_2 a_2 = S_{23} d_4 + C_2 a_2 \tag{13.40} $$

Element (2,4):

$$ p_z' - d_1 = -(C_2 C_3 - S_2 S_3) d_4 + S_2 a_2 = -C_{23} d_4 + S_2 a_2 \tag{13.41} $$

Solving the second equation for $C_2$ and substituting it into the first equation gives a conditional equation for $S_2$. If, rather the second equation is solved for $S_2$, this gives $C_2$ after being substituted into the first. This gives Eq. (13.42).

$$ \begin{aligned} \theta_2 &= \operatorname{atan2}\left(\frac{S_2}{C_2}\right) \\ &= \operatorname{atan2}\left(\frac{(C_1 p_x' + S_1 p_y') d_4 C_3 + (p_z' - d_1)(d_4 S_3 + a_2)}{(C_1 p_x' + S_1 p_y') d_4 S_3 - (p_z' - d_1) d_4 C_3}\right) \end{aligned} \tag{13.42} $$

To determine $\theta_4$, matrix Eq. ${}^3T_6$ is formed, see Eq. (13.43).

$$ {}^3T_6 = A_3^{-1} \cdot A_2^{-1} \cdot A_1^{-1} \cdot T_6 \tag{13.43} $$

Case differentiation must also be performed here depending on $S_5$, Eqs. (13.46) and (13.47).

Element (1,3):

$$C_4 S_5 = C_1 C_{23} k_x + S_1 C_{23} k_y + S_{23} k_z \tag{13.44}$$

Element (2,3):

$$S_4 S_5 = S_1 k_x - C_1 k_y \tag{13.45}$$

$$\Rightarrow S_5 > 0 : \theta_4 = \text{atan2}\left(\frac{S_4}{C_4}\right) \tag{13.46}$$

$$\Rightarrow S_5 < 0 : \theta_4' = \theta_4 + \pi \tag{13.47}$$

For $S_5=0$, the argument in the atan2 function becomes an indeterminate expression and so cannot be solved. It can be seen on the kinematic equivalent circuit that the fourth and sixth axes coincide for $S_5=0$ ($\theta_5=0$). The equation system is overdetermined in this case. To arrive at a sensible solution here too, the fourth axis, for example, should retain the position it had immediately before this path point was reached.

The matrix Eq. (13.48) provides the necessary equations to determine $\theta_5$ by formulating the Eqs. (13.49) and (13.50).

$${}^4T_6 = A_4^{-1} \cdot A_3^{-1} \cdot A_2^{-1} \cdot A_1^{-1} \cdot T_6 \tag{13.48}$$

Element (1,3):

$$\begin{aligned} S_5 = {} & (C_1 C_{23} C_4 + S_1 S_4) k_x \\ & + (S_1 C_{23} C_4 - C_1 S_4) k_y + S_{23} C_4 k_z \end{aligned} \tag{13.49}$$

Element (2,3):

$$\begin{aligned} & -C_5 = -C_1 S_{23} k_x - S_1 S_{23} k_y + C_{23} k_z \\ & \Rightarrow \theta_5 = \text{atan2}\left(\frac{S_5}{C_5}\right) \end{aligned} \tag{13.50}$$

From Eq. (13.51), $\theta_6$ can finally be calculated - Eqs. (13.52) to (13.54).

$${}^5T_6 = A_5^{-1} \cdot A_4^{-1} \cdot A_3^{-1} \cdot A_2^{-1} \cdot A_1^{-1} \cdot T_6 \tag{13.51}$$

Element (2,1):

$$\begin{aligned} S_6 = {} & (-C_1 C_{23} S_4 + S_1 C_4) i_x - \\ & (S_1 C_{23} S_4 - C_1 C_4) i_y - S_{23} S_4 i_z \end{aligned} \tag{13.52}$$

Element (2,2):

$$\begin{aligned} C_6 = {} & (-C_1 C_{23} S_4 + S_1 C_4) j_x - \\ & (S_1 C_{23} S_4 - C_1 C_4) j_y - S_{23} S_4 j_z \end{aligned} \tag{13.53}$$

$$\Rightarrow \theta_6 = \text{atan2}\left(\frac{S_6}{C_6}\right) \tag{13.54}$$

**Ambiguity of the backward transformation algorithms**

As shown in the previous sections, the algorithms for the backward transformation, depending on robot type and number of robot axes, provide multiple solutions for axis positions of a certain target position and orientation. For a 6-axis articulated arm robot, we have the following ambiguities for the axis positions at the target point (Fig. 13.38):

- The ambiguity of the swivel range requires a decision on whether the target point is reached by an overhead motion or by rotating the first axis
- Ambiguity of axes 4 and 6
- Ambiguity of axes 3 and 5.

These ambiguities must be clarified by the controller software using special selection criteria. Potential decision criteria:

- Mechanically restricted axis traverse range
- Continuity of the motion path
- Shortest path to the target point.

In addition, ambiguities can be eliminated with specifications in the user program. For this in the motion instructions, the definition of the target point position and orientation is supplemented with the addition of an angle status specification. This defines the axis position at which the target point is reached should there be ambiguities in the calculation algorithms.

**Interpolation types**

Motion commands are some of the basic instructions for industrial robot controllers. A distinction is made between the following motion operations:

- PTP (Point to Point):
  Motion of the industrial robot from the actual position to the reference position of the Tool Center Point taking any arbitrary path within the space (see Fig. 13.39a). Only the starting and target points are specified in detail for a PTP motion. The robot axes reach their target positions in consideration of the axis-specific speeds and accelerations. It

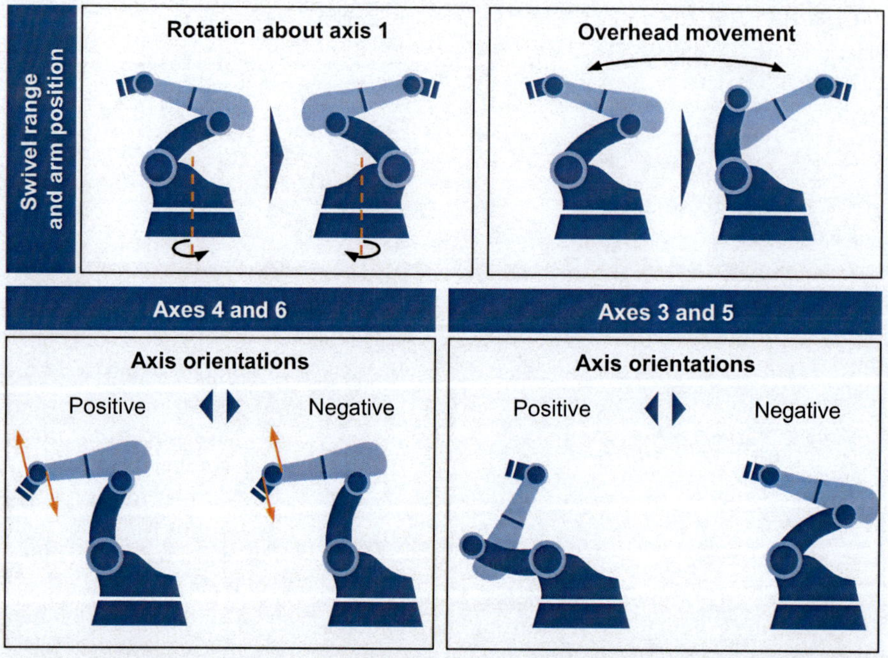


■ Fig. 13.38 Transformation ambiguities for 6-axis articulated arm robots

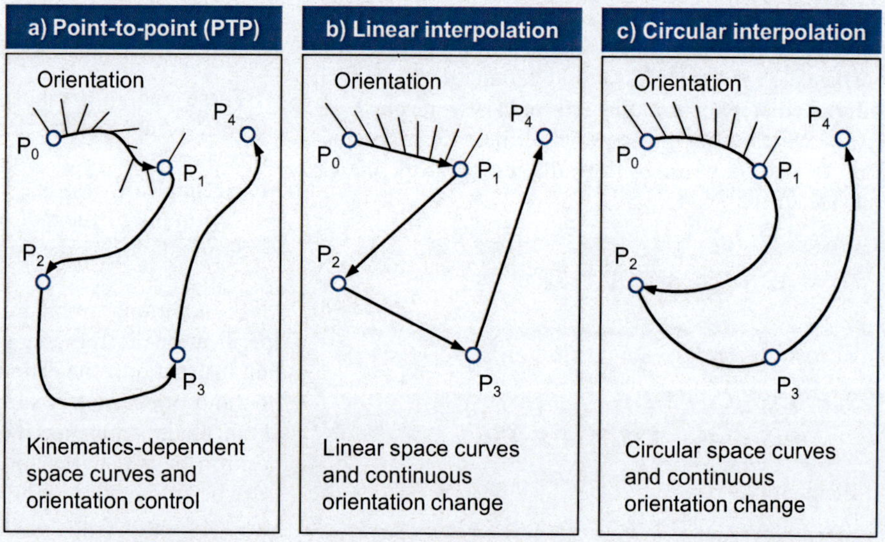


■ Fig. 13.39 Interpolation types of industrial robots

13

is usual nowadays that the individual axis motions are performed synchronously so that all robot axes reach their target positions together. For this, the controller of the industrial robot must first determine the guide axis that requires the longest motion time. The other axes are then traversed at reduced axis speed and acceleration.

- Linear:
  Linear motion of the Tool Center Point on the connecting path between the actual and destination positions with the specified orientation of the TCP (■ Fig. 13.39b). The motion is performed with the path speed and acceleration defined provided the required speeds and accelerations of the individual robot axes are not exceeded.
- Circular interpolation:
  Circular motion paths for the TCP can be specified by defining three points within the operating space of the industrial robot (■ Fig. 13.39c). The three points uniquely define the circular path within the space.

The interpolation types presented are part of the standard command set for industrial robot controllers. Additional interpolation types such as parabola and spline paths can be provided by manufacturers and be used for special applications.

When the motion commands described are used, the robot stops after every motion performed regardless of whether another motion set follows or not. If the requirement is to perform multiple motions with no stops, the blending behavior must be defined. Blending is where the programmed end point of a motion is not reached exactly, instead the next motion is initiated early. The user must specify here what criteria the controller uses to manage blending. For path motions (linear or circular), one criterion could be the tolerance distance to the target point. Here, the next motion set is initiated as soon as the actual distance to the target point is less than the tolerance distance. Dropping below a set speed can often be another criterion for blending to start.

Since no path is specified and no TCP speed is calculated for PTP motions, it is only possible here to specify the percentage before the end of the motion (in relation to individual axis traverse ranges required for the motion) for blending to start. ◘ Figure 13.40 shows blending for a 2D motion.

A robot controller generally incorporates about five motion commands following an actively executed motion to perform path optimization with blending. However, for applications with a lot of path points lying close to each other, such as in the milling of complex surfaces, the magnitude of the forward stroke is often not adequate to meet technology parameters (such as path speed). For this case, controller manufacturers offer special technology options that greatly increase the number of motion commands interpreted in the forward stroke.

## 13.5 Operating and Programming Robots

The objective of robot programming is to generate control commands in the form of programs that perform an automation task set. The range of tasks extends from simple workpiece handling functions through to complex, sensor-controlled machining and assembly tasks.

There are different programming methods exist, some of which are computer-aided. The suitability of these methods varies depending on application area. The most important methods are shown in ◘ Fig. 13.41 and are differentiated by how the program is entered:

*Online methods* use direct interaction between the programmer and robot or its controller (see ► Sect. 13.5.1), for example, to move along individual path points and save them (teach) or to record a complete motion path (playback). Also included in online methods are workshop-oriented programming systems that, in addition to just motion definition and manual program entry, provide the user with aids in the definition of closed-loop processes and integration of fieldbus systems for communication (refer to ► Sect. 13.5.1).

*Offline methods* can be used regardless of the actual production facilities (see ► Sect. 13.5.2). A task is programmed indirectly using a suitable programming language, graphically (using flow charts for example) or aided by simulation with motion sequences generated on the screen and then simulated. In addition, other task-oriented methods exist that do not require an exact definition I of the program and motion processes. Instead, they piece together the program and motions from a series of known instructions (such as "Retrieve component from clamping").

*Hybrid methods* use a combination of online and offline methods. The logic of a handling task is normally programmed in production planning. Then, additional motion sequences are specified on site. Programs are often created in production planning using text or task-oriented programming systems. The program code is changed by users themselves, however.

The following sections introduce the most important methods and capabilities of online and offline programming.

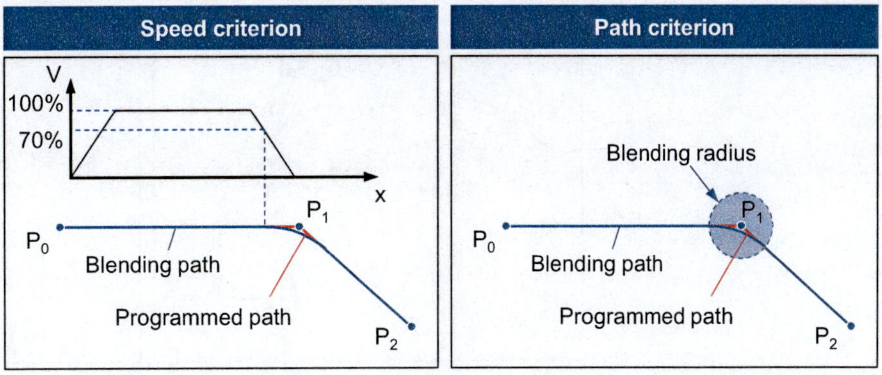


◘ **Fig. 13.40** Blending of motion sets

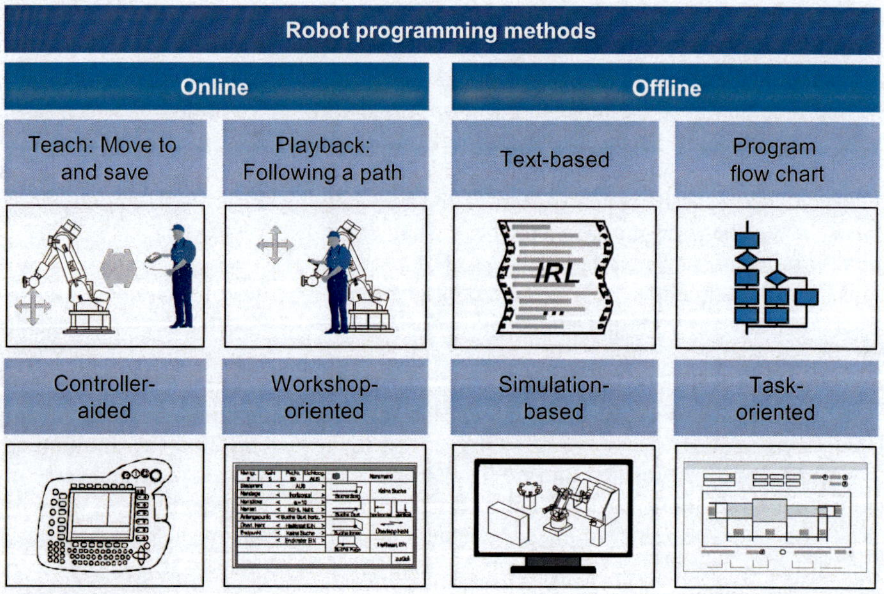


**Fig. 13.41** Programming methods for industrial robots

### 13.5.1 Online Programming Methods

Currently, most robots are programmed in a process-oriented manner using teach methods. In the method, the programmer controls the robot or comparable equipment for reading the coordinates while the robot controller stores the geometric data required for repeating the motion later. Given its direct association to the process, this method is far more transparent than offline programming.

A programmer at a workstation is rarely in a position to define the robot motions numerically because the complexity of motion sequences would be too much for their spatial sense. Simulation systems that depict the robot and its operating space graphically are normally associated with high costs for suitable performance levels. For this reason, great importance is attached to teach programming for the generation of motion commands, even if offline programming methods are being used more and more, especially in the automotive sector. A distinction is made between two variants of teach programming:

- Playback programming (following a path)
- Teach programming (moving to the individual support points of a motion path and saving them)

In playback programming, the robot is guided by the operator directly using hand grips affixed near the tool along the path to program (Fig. 13.42, left). During this phase, the controller automatically stores (at a fixed timing cycle of 0.05 to 0.5 s) all data necessary to replay the path. However, moving the robot by hand is only possible for lightweight designs (such as spray paint robots) once the drives have been switched off. Consequently, a lightweight construction of the robot kinematic model is also often used. This construction only contains the displacement transducers and is much easier to operate (Fig. 13.42, right). The downsides of



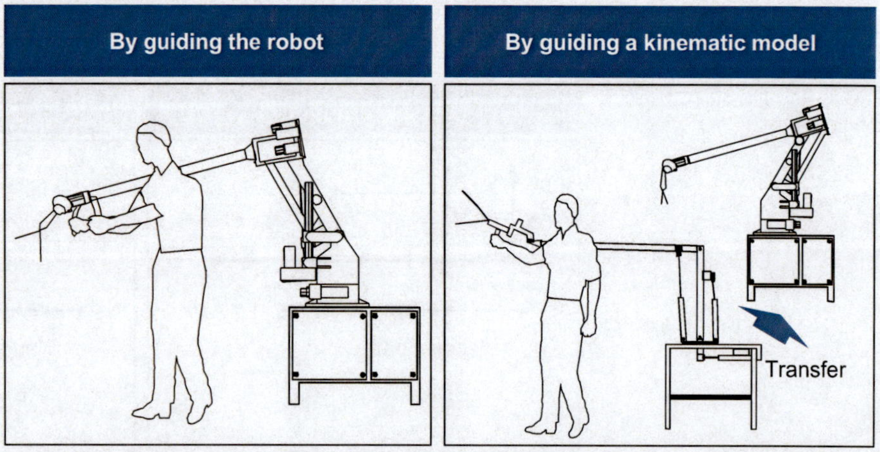


**Fig. 13.42** Playback programming by following a path, according to AKR

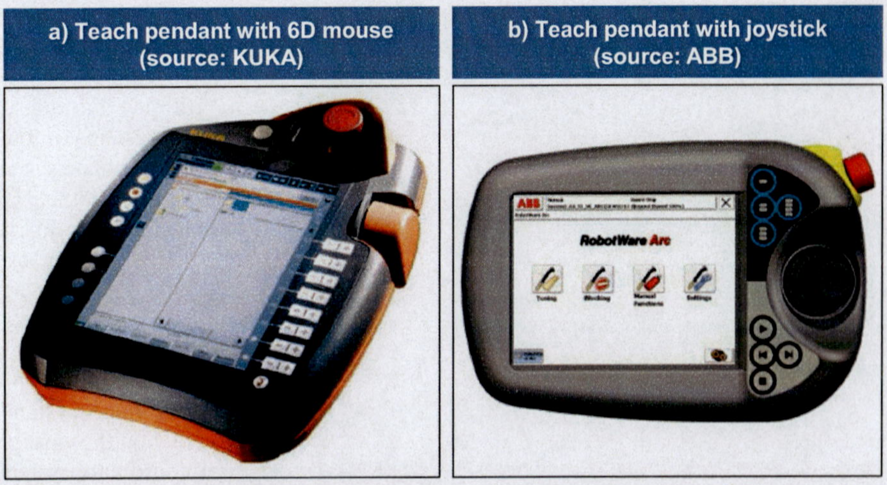


**Fig. 13.43** Teach pendants for industrial robots, according to KUKA AG, ABB

a kinematic model, however, are, first, the additional costs and, second, problems with tolerances between the model and the robot that require elaborate synchronization of the two systems. These drawbacks mean the playback method is only used nowadays in special applications with less strict requirements of positioning accuracy.

In teach programming, the programmer guides the robot in distinctive positions that are characteristic for the motion. The robot is rarely moved directly by hand. Instead, a teach pendant moves the drives (Fig. 13.43). Reading the (up-to-date) axis-specific displacement transducers allows the positions of the robot to be captured and stored. Then, parameters (speed, acceleration, and accuracy of the motion) and functions (open/close gripper, tool change) are programmed. While the program is executing, the robot moves towards the positions stored, in consideration of the associated parameters, and runs the functions programmed.

### Controls to execute controlled robot motions

Moving the robot for the purposes of startup or motion programming requires the robot drives to be actuated manually. Function keys (touch or softkeys) on the teach pendant are normally used to issue motion commands. Different motion paths can be assigned to the keys by specifying the reference coordinate system (Fig. 13.28). Motions along the reference axes of the coordinate system selected (six degrees of freedom) and, as an alternative, individual motions of the different drives are usual here.

In addition to function keys, most industrial robots available today have additional capabilities for simplified entry of position changes associated with orientation. The most widespread are 6D mice and joysticks (Fig. 13.43a and b). The positions/rotations of these devices are translated into a motion along the degree of freedom.

A 6D mouse can execute all position changes of a 6-axis robot. Joysticks normally have three degrees of freedom (two translational and one rotational direction). To also provide high-quality input here, the software must assign the required motion directions or degrees of freedom to the actuation movements.

The design of teach pendants (Fig. 13.44) uses app-based interfaces more and more to simplify controller operation. Another factor in this is the widespread proliferation of touch technologies. In this case, apps group different functions according to specific functionality, such as safety and calibration.

Besides the standard methods for entering motion, more concepts are being researched. These, for example, use programming pointers to perform robot motions with position or direction pointers [GÖBE12]. In addition, methods for evaluating process forces have been investigated to provide information on the forces exerted by a robot during motion (in assembly processes, for example) [BISC10]. Another approach for specifying motion is based on optical measurement technology from the field of virtual reality. Here, programmers specify the motions of a robot with their own movements. These can simultaneously be transferred to and evaluated on a virtual system model on the computer for collision avoidance, for example (see ▸ Sect. 13.9). In conjunction with more widespread usage of direct human–robot interaction scenarios, for cooperative component assembly, for example, hand gesture-based entry methods can also be important in the future for triggering robot actions.

### Generation of a robot program on the controller

In online programming, the program code to be run is generated with the teach pendant. A display program

Fig. 13.44 Teach pendant with app-based user interface, according to Reis

with editing function is used here. It assists the operator, such as with integrated syntax checking, in entering the program. Input nowadays is usually from a touch-sensitive display. A keypad and function keys, in the form of softkeys, are shown on the display (see Fig. 13.45). Using predefined help functions for entering standard operations, motion, and logic commands can be inserted and linked in a simplified manner.

After the controller is switched on from the control cabinet and Test mode (T1) is selected, the controller is in Manual mode. The exact position of the robot mechanics is not known by the robot controller when the system is first delivered, meaning the robot needs to be adjusted first (see ▶ Sect. 13.3).

To teach a point, the robot is positioned manually with the OK button pressed. Then the position reached can be read and, in relation to a coordinate system to be selected, saved in a new point variable. The user has different modes available to control motion manually. It is also possible, depending on the active gripper or reference coordinate system, to perform manual movements within different coordinate systems:

- Axis-specific coordinate system: In this mode, the six traverse buttons affect the forward/backward motions of a maximum of six axes.
- World coordinate system: Three traverse buttons enable motion of the TCP parallel to a Cartesian coordinate system that is in the base of the robot. The three other buttons are used to change the orientation (A, B, C) without altering the set working point.
- Base coordinate system: The traverse buttons have the same meanings as in (2) but here the origin of the coordinate system is at a user-defined point, normally on the workpiece.
- Tool coordinate system: The traverse buttons have the same meanings as in (2) but here the origin of the coordinate system is at the tip of the tool and is also moved in line with the robot motion.

It is also possible to enter coordinates as text from the keypad of the operating unit. The coordinates can be numerical values, variables, and constants and also be calculated at program runtime. A parametric definition of motion instructions offers the benefit that it is easy to adapt to variable process flows. If, for example, positioning processes are performed for picking up workpieces and setting them down at different heights, the working point can be determined depending on the workpiece height specified externally (parameter is passed by the master computer).

### Program testing

When programs are generated, there is automatic checking of the command syntax and the functionality of the operating unit is limited depending on the level of user knowledge. Also, numerical inputs can be checked for limit violations of the individual axes and workspace. Despite existing code-related help functions, diverse sources of error arise from the fact that the structural format of a program can be incomplete or erroneous.

For example, only when a program is run can arithmetical operations lead to undefined states. Motion instructions must first be tested in manual mode at lower speed so that collisions between the robot and its immediate environment can be detected and eliminated.

Some of the most important test aids:

- External changing of speed: In addition to the programmed speeds (axis and path speeds), the actual execution speed can be set from the operating unit for program testing and in automatic mode.
- Single step: Single step mode means only one program instruction is performed at a time. Then, the controller goes into the Stop state and waits for another start signal from the user.

**Fig. 13.45** Robot controller HMI, according to KUKA AG

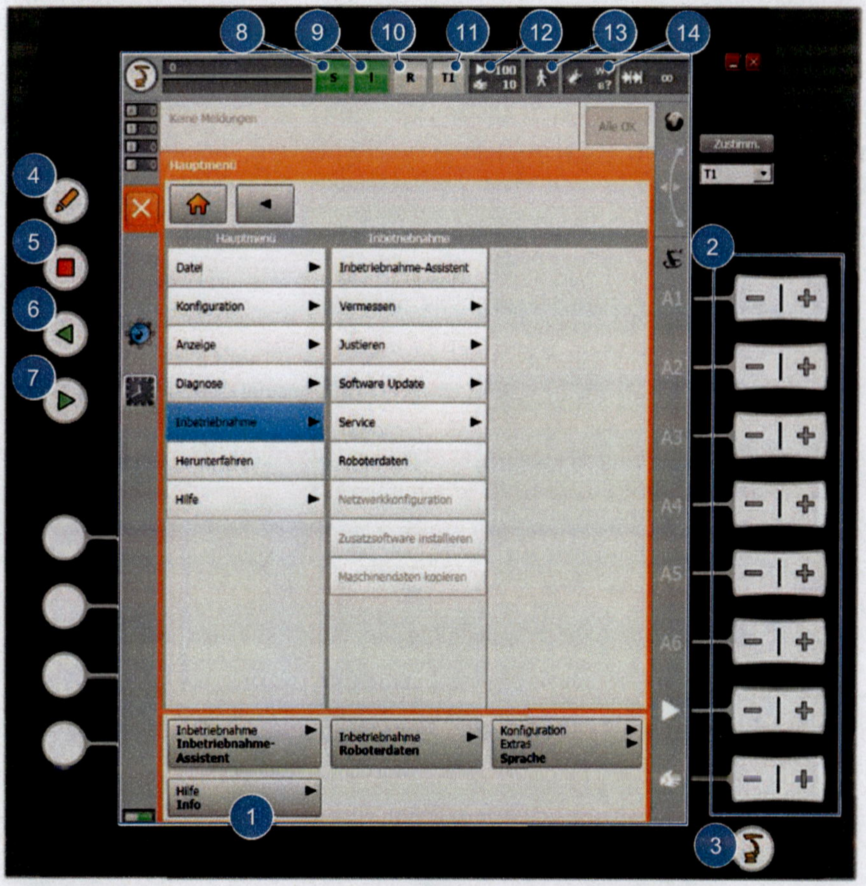


| | |
|---|---|
| 1 Touchscreen | 8 Status: Logic control (soft PLC) |
| 2 Softkeys | 9 Status: Traverse conditions / safety |
| 3 Main menu button | 10 Status: Motion program |
| 4 Show keyboard | 11 Display of operating mode |
| 5 Program STOP | 12 Speed override |
| 6 Program START - Backward | 13 Program flow types |
| 7 Program START - Forward | 14 Select gripper/base |

- Single operation: The program continues to be run until a motion instruction is reached. The following motion is then performed when confirmed again by the user.
- Continuous mode: The program to test is executed continually provided the operator confirms the cycle.
- Breakpoints in the program: After the program cycle is confirmed, the program runs until the next breakpoint. Then the controller stops.
- Testing individual program sections in manual operation: To test more complex and structured robot programs, test aids are required that make checking section by section possible. For this, the execution pointer can be set to the relevant program section and the cycle started from this location.
- Testing at working speed: In "Test 2" (T2) mode, all motions are performed at full drive power. In contrast to T1 mode, nobody is permitted in the operating space of the robot for safety reasons.

Running programs in Automatic mode is only sensible once all test and correction tasks are complete.

### Calibration of tools and base coordinate offsets

The reference systems described in ► Sect. 13.3 are normally defined before the start of program creation so that they can be used during motion programming. A basic prerequisite for all calibration operations is aligned robot kinematics used to move toward different spatial points for the purposes of calibration. Consequently, calibration operations are always performed *online*, i.e., using the actual robot.

The purpose of tool calibration is to specify the Tool Center Point (TCP) used to move toward distinct points within the space and as a reference for the gripping or welding points, for example. The specifications for the TCP position correspond to an offset and, as required, a rotation relative to the flange coordinate system of the robot (◘ Fig. 13.31). Different methods can be used for calibration. These return equivalent results and, depending on the availability of additional measuring aids and content of the programming task, can be selected:

- XYZ 4-point method: The TCP to be calibrated must be aligned to a fixed reference point (such as a measuring tip) for four different axis positions of the robot. The robot controller can then calculate the relative position of the TCP from the four flange positions.
- XYZ reference method: A new tool is compared to one already calibrated and the TCP of the new tool is determined. For this, the known tool moves toward a fixed reference point. Then the tools are swapped and the new TCP is moved toward the same reference point. The difference gives the new relative position of the TCP.
- ABC 2-point method: The objective of the ABC methods is to calibrate the orientation of a tool exactly. In the 2-point method, the tool moves toward a reference point and is then moved toward the gripped object in the vertical position in the direction of impact (X-direction). Then the tool is moved within its XY-plane to define the Y-direction on the tool side. The orientation of the TCP axes can then be calculated from the three resulting interpolation points.
- ABC world method: The ABC world method is used to specify the orientation of the TCP relative to the world coordinate system. This simplifies programming (of welding and bonding operations, for example) where tool positioning has a major bearing on the process and must be kept constant. For calibration, the tool axes are aligned to the world coordinate system and the difference is stored. With this kind of calibrated TCP, path programming (along the weld seam) is simplified because tool positioning need not be taken into account in moving the robot manually.
- Numerical entry: Tool data can also be calculated virtually using the CAD model of the gripper. Calibration is not necessary in this case. For this, the robot controller provides the option of entering rotation and offset manually in the form of offset values and orientation angles. Moreover, the values entered can be checked and corrected by external sensors during operation [WIES01].

In addition to the specifications for TCP position, load data can also be defined for the tool in use. They comprise the load capacity at the robot flange (total of tool and, as required, workpiece mass or process forces), the position of the load capacity center of gravity and the associated inertias in relation to the center of gravity. The values can be determined numerically using the mass distribution in the CAD model. Otherwise, the various robot manufacturers provide functions on the control side for determining the load by evaluating the dynamic axis behavior. For this, the robot oscillates along the individual axes.

The specifications of load capacity are factored into the interpolation by the robot controller so that, at lower loads, higher accelerations and decelerations can be achieved and the cycle time shortened. If motions are performed with no load specifications, the acceleration values for maximum robot load are assumed. The cycle time increases as a result.

With the base calibration, user-specific coordinate systems are defined. The position details of the individual points of motion pertain to these systems. It is essentially up to the operator whether path points are stored in world coordinates or as a function of a defined base system. Using different base systems offers major benefits, however, in particular for robot startup within a frequently changing environment. In the rearrangement of a workpiece pallet, for example, not all individual pallet spaces need to be programmed again. Instead, recalibration of the associated base is sufficient. This means the existing palletizing program can be reused.

The position of a base coordinate system can, as with the tool definition, be calibrated or be set by means of a numerical entry. The prerequisite for a base calibration here is a calibrated tool. A distinction is made here between two approaches:

- 3-point method: The TCP is first positioned at the origin of the coordinate system. Then, the TCP is moved in the positive X-direction. The third interpolation point results from any arbitrary motion of the TCP in the XY-plane in the positive Y-direction.

- Indirect method: Target positions are specified one after another within the coordinate system to be defined and the TCP is moved to the actual position. The position and orientation of the coordinate origin can then be calculated from the comparison of actual position and reference position.

In addition to simplifying subsequent program editing, base coordinate systems are of major importance for the use of virtual system startup processes because they safeguard coordination between model geometries and the actual equipment environment.

#### Workshop-oriented programming systems

Workshop-oriented programming systems (WOP) for robots are assistance systems that help the programmer create programs and whose functional scope extends well beyond conventional robot controllers [DREW97]. The help functions relate, for example, to the graphical display of programmed paths, help from macros, use of CAD data and new input/output methods (touchscreen, voice recognition, gesture control, etc.). The goal of such systems is to make programming on site much quicker, more ergonomic and of higher quality compared to known systems. The programmer need not make direct changes to the program code in any phase of program creation.

◘ Figure 13.46 shows an example of a workshop-oriented programming interface. Programming is local here using a teach pendant with touchscreen functionality allowing quick adaptation of the program flow by the machine operator. As shown, it is possible for the operator to incorporate a camera system, for position referencing for example, into the programming process or automated execution.

### 13.5.2 Offline Programming Systems

Unlike online methods, offline programming does not need direct access to the actual robot. The greatest benefit over online methods, besides the ability to be used in parallel to production, is the easier entry option for more complex program flows. A distinction is made between text programming, graphical programming, and simulation-aided programming methods that are explained below.

#### Text program creation

For text programming, or generally for programming not based on CAD data, programs are usually created in two steps. In the first step, the skeleton program is created offline, in production planning for example. Motion commands are provided initially in the form of placeholders but are not filled with position values. Teaching the positions, as well as program test and optimization, then are performed in the second step directly on the robot, so online. The term also used here is hybrid programming.

For offline programming, computer-aided programming systems are provided by different robot manufacturers. These use the controller-specific language (such as KRL for KUKA, RAPID for ABB, KAREL for FANUC, V+ for Adept and VAL3 for Stäubli) for programming. Section 13.5.3 describes in detail the IRL programming language (Industrial Robot Language, DIN 66312, Part 1) as a representative for a series of similar commercially available programming languages. Their structure and syntax are based on the PASCAL programming language.

#### Graphical programming with flow charts

In the past, the low level of transparency of robot programs written in text and the high error rate in program creation gave rise to a "Do not touch" attitude, in particular for inexperienced robot users. For this reason, a programming methodology based on graphical flow charts was developed [WECK97]. In cooperation with Bosch, the BAPS plus programming system was developed at WZL (Laboratory for Machine Tools and Production Engineering), ◘ Fig. 13.47, that allows robot program generation independent of the syntax of a programming language.

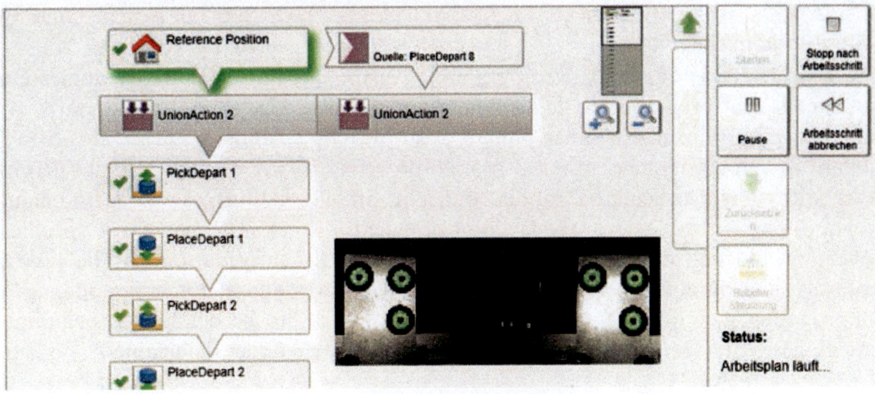


◘ **Fig. 13.46** User interface of the APAS workshop-oriented programming system from Bosch

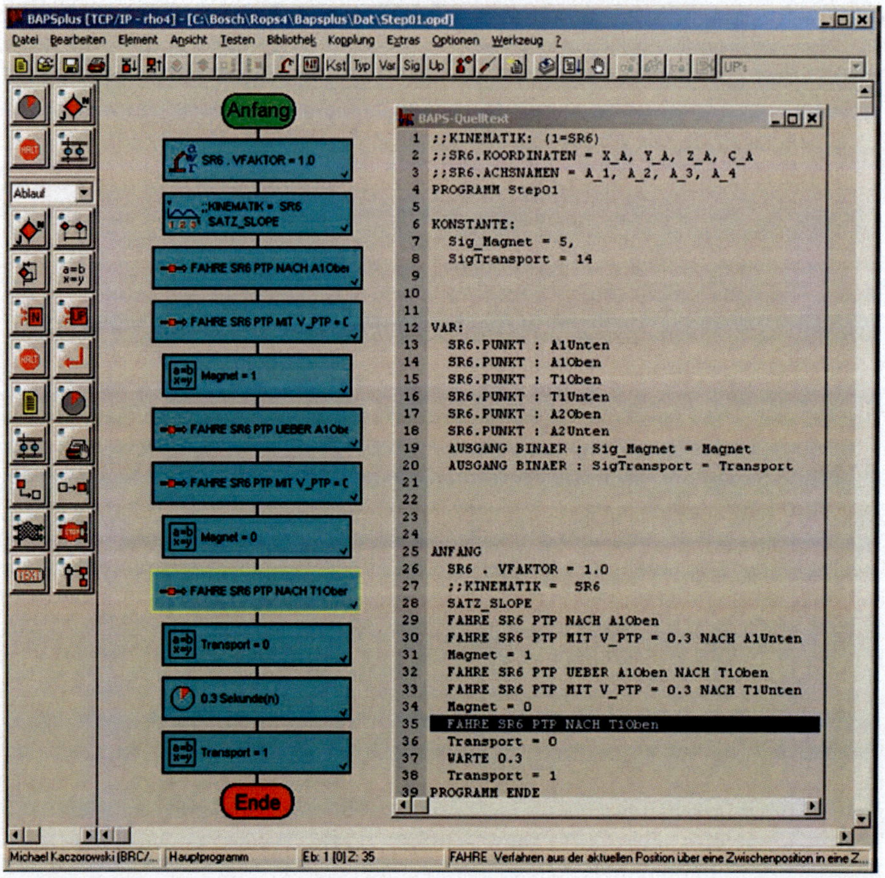


**Fig. 13.47** BAPS plus graphical programming system



Using this concept, complex program structures can also be created and managed easily and transparently. A program is created by selecting program commands These are shown in the form of icons and can be selected from a library. The underlying program code is generated automatically while the program is being created (Fig. 13.47, right).

The programmer can access online debugging and optimization capabilities for the programs created once linked to the robot controller. The functionality includes uploading and downloading of program files, reading and setting of variables and signals, and the graphical display of the current controller status.

### Robot program simulation

One key downside of text-only programming using a language is that programs created offline often require long test and optimization times on site. Also, in particular for programming complex spatial processing tasks with 6-axis articulated robots, outright offline programming without relevant aids is not practical because it would be too great a challenge to the spatial sense of the programmer. Simulating the industrial robot programs by graphically displaying the motions is a remedy here. Process simulation systems generally also include functions for automatic generation of the associated program code.

The process of simulation-aided robot programming is shown in Fig. 13.48.

A prerequisite for generating and simulating a robot program is always the modeling of the working cell and the robot itself. All simulation systems essentially offer the ability for the user to model any system component. For robots in particular, but also for prevalent workpiece positioners and conveyors, extensive libraries are generally available. These reduce the modeling overhead for the user. Interfaces for transferring CAD data are also available for the components to be machined as well as system components. Popular interfaces offered by system providers are based on data formats IGES (Initial Graphics Exchange Specification), STEP (ISO 10303, Standard for the Exchange of Product Model Data), VDAFS (VDA—surface interface), STL (Stereolithography File Format) and JT (Jupiter Tesselation), as well as on manufacturer-specific formats.

The geometry specifications as well as additional information on the kinematics and dynamics are required for the modeling of a robot. Thus, axes need to be defined, associations of individual parts to each other determined and axis limits specified. Also, the

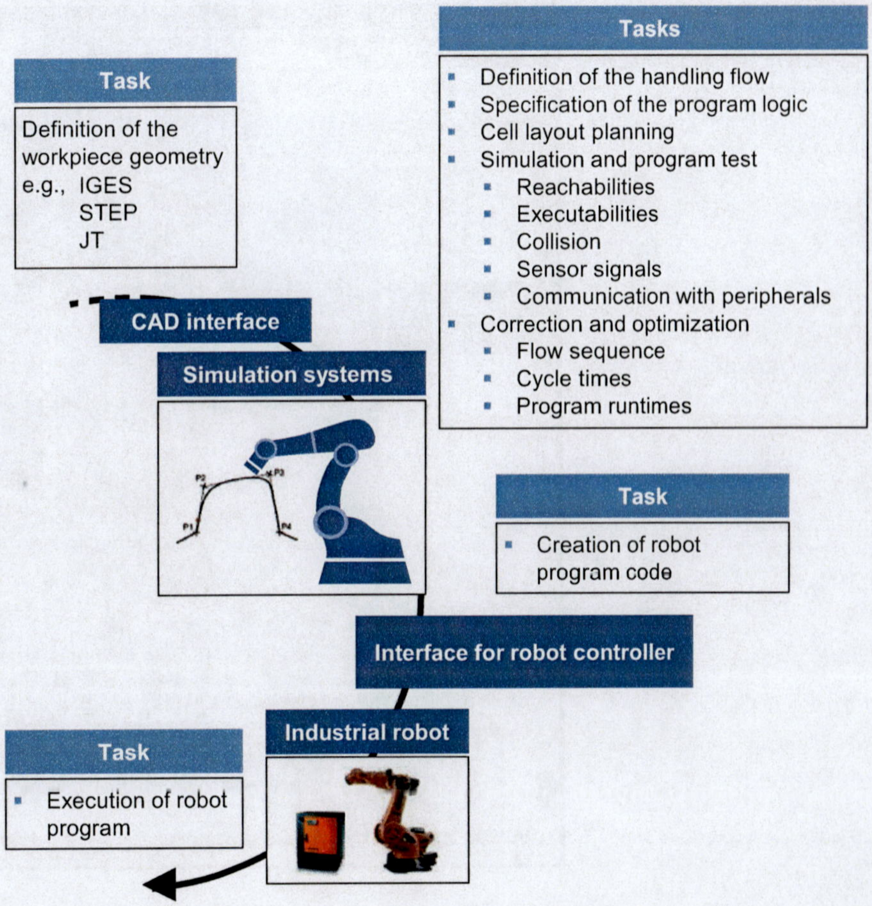


**Fig. 13.48** Offline programming process using simulation systems

specification of speeds and accelerations attainable on the individual axes is required for a realistic simulation. Simulating the robot motions and converting from Cartesian to axis-specific robot coordinates (and vice versa) are already possible with such a model. However, the handling of ambiguities and singularities, the reaction of the actual robot controller in threshold regions, the blending behavior, etc., are problematic in this context. This is because every controller manufacturer implements special, simplified transformation algorithms and proprietary handling of special cases in its controller. Furthermore, some robot controllers differ considerably in regard to blending behavior as well as additional implementation details not specified by pure kinematics.

Manufacturer-specific solutions cause the same robot, moved by different controllers, to perform other motions, some of which are evident. It is thus necessary to simulate the robot as well as the control behavior in the system in detail. The accuracy, and so the usability, of a robot simulation system is crucially dependent on how well the virtual robot controller is able to simulate the actual control behavior. Emulating the different robot controllers requires considerable overhead in the development of a simulation system and must be conducted by the system manufacturer in close collaboration with the controller manufacturer. Bearing in mind the fact that no control manufacturer is generally willing to divulge its exact transformation algorithms and other internal controller information (they represent a large development effort and are closely associated with the performance features of the controller), the problem of simulation system providers and customers becomes clear. To counter this problem, a standardized RRS (Realistic Robot Simulation) interface has been developed. Using this in a defined manner, the motion behavior of the actual controller can be queried by a software module, provided by the controller manufacturer, using the original algorithms [JACO94].

After modeling the working cell and robot, the user can create robot programs by defining tasks. All commands required for the program cycle and robot motions are kept in one task. Task creation here is generally not text-based but menu-aided and with direct, interactive positioning of the robot or by providing target points in the working cell. In parallel, the system generates the corresponding program text. A programming language specially developed for the simulation system is

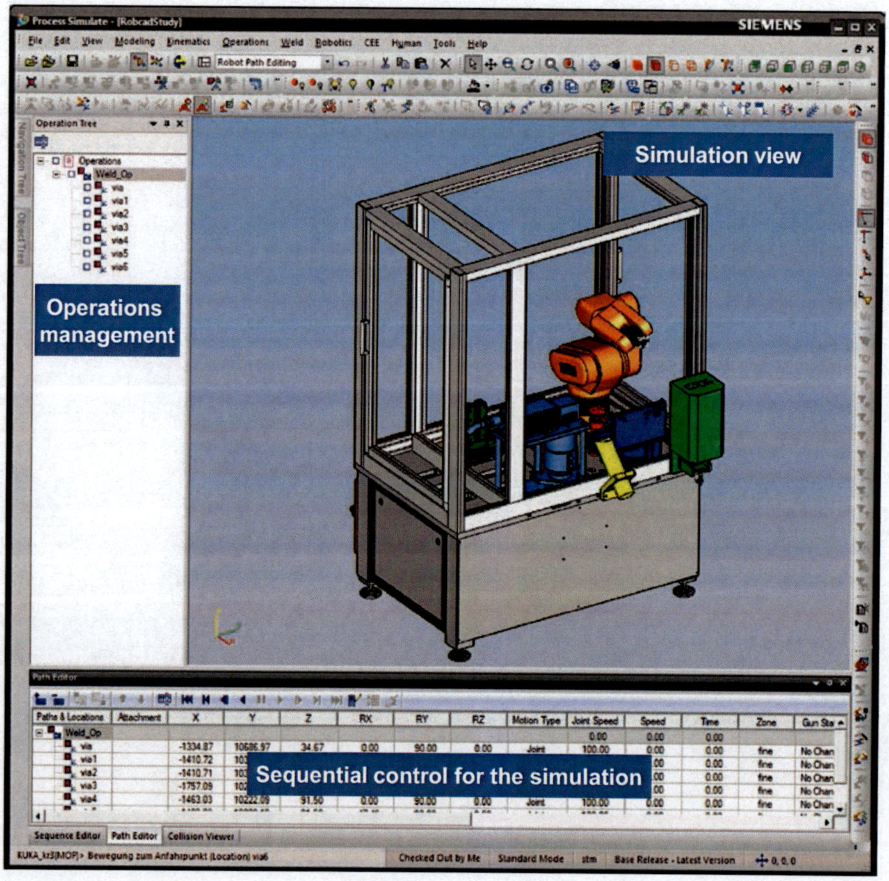


**Fig. 13.49** Process Simulate robot simulation and programming system, according to Siemens AG



commonly used for this. These programs are converted into different controller languages by a postprocessor once program generation and testing are complete.

After the definition of a task, it can be simulated. A display in the virtual space has established itself as the standard (Fig. 13.49). Multiple light sources can be specified, and the observation point is freely selectable.

In addition to simulating only the program cycle and robot motions, more comprehensive functions are available on most systems. Cycle time determination and optimization, for example, are important criteria for program creation. Functions for layout optimization, automatic path planning, robot calibration, etc., are also provided. Figure 13.49 shows an example of the interface of the Process Simulate simulation system from Siemens.

The ever-growing demand for the automation of comprehensive and complex production processes and the associated and widely varying requirements on the functionality of suitable simulation systems have led to the development of application-specific software packages within simulation systems. The individual modules contain specific functionality such as creation of welding parameter databases, automatic generation of welding points, collision detection and envelope calculation for a robot motion (Fig. 13.50), process-specific entry screens, etc.

### Task-oriented robot programming

The text-based and graphical methods introduced define the control-related behavior of a robot system by explicitly specifying the actions to perform, e.g., "Linear motion from interpolation points 1 to 2 at 1 m/s." The methods are also known as robot-oriented programming methods because of their close relationship to the robot kinematics to be executed.

Task-oriented robot programming is another offline programming method. Unlike robot-oriented methods, there is a level of abstraction here of the robot activities to be performed with regard to the underlying objective of a robot motion or action. In this process, the planner defines a series of tasks that are completed in sequence by the robot system, such as "Take component A from magazine B". For execution by the robot controller, these instructions are first converted into individual motions and logic operations by a planning tool and then translated into code. At the start of the planning process, the necessary implicit information about the cycle must be extracted from the task definition specified explicitly by the operator for robot-oriented programming methods.

The starting basis for many task-oriented programming approaches is a detailed product and environment definition in the CAD system. The functional scope of

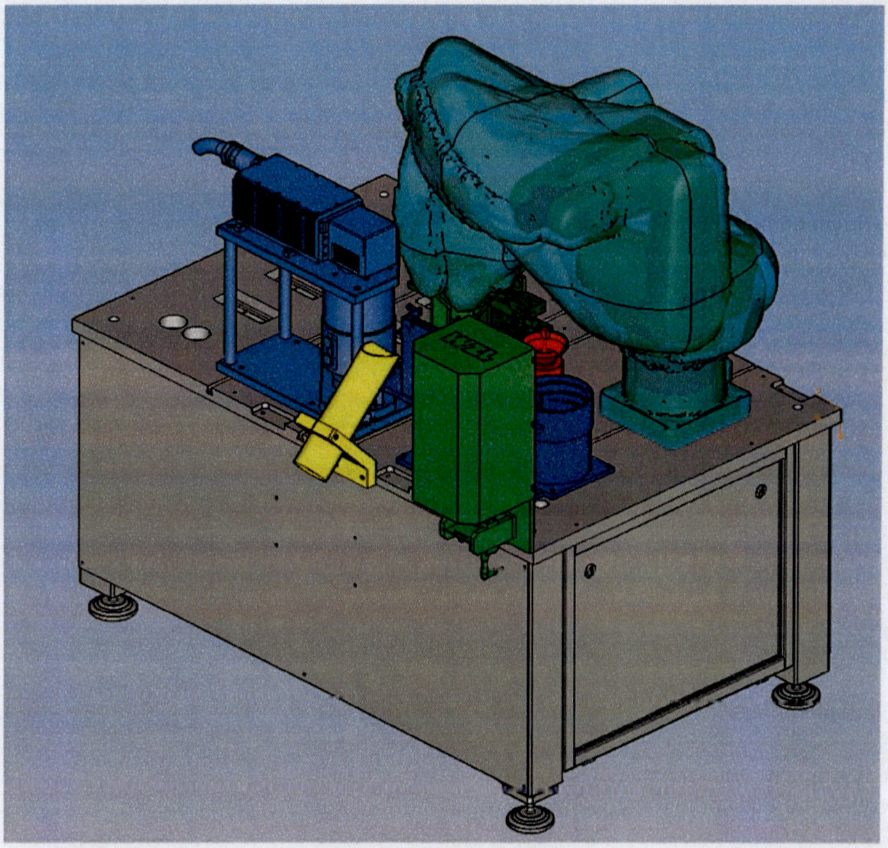

Fig. 13.50 Envelope calculation in Process Simulate, according to Siemens AG

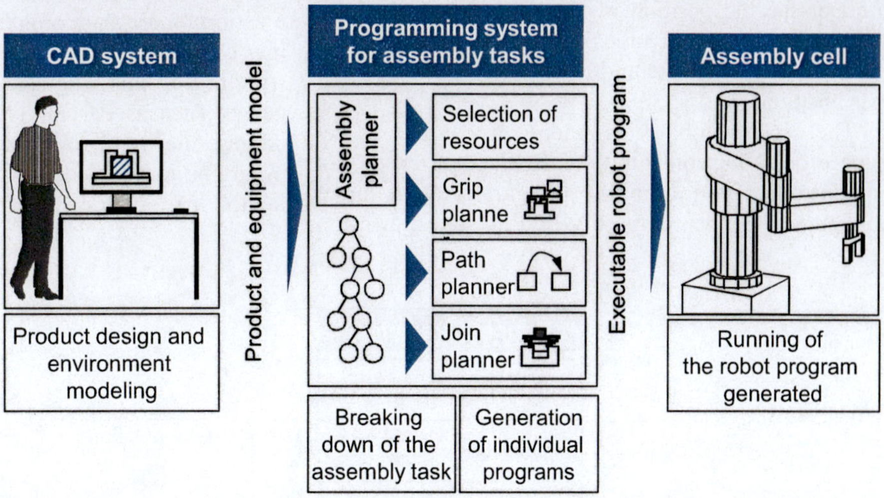


Fig. 13.51 Structure of a programming system for assembly tasks

existing systems is based on online programming methods and transferred to the simulation system. Possible savings potential (with program optimizations, for example) has not yet led to an appreciable reduction in programming time, however. This is only possible with increased use of macros. These relieve the user of time-consuming work such as specifying collision-free paths, precisely defining gripping points and selecting welding parameters.

The structure of a task-oriented robot programming system for assembly work is shown in Fig. 13.51. Subtasks, such as resource and tool selection, grip planning, and path planning, require intervention by the planner (such as by using an expert system for grip planning) or can be realized by automatic planning modules (such as for calculating a collision-free motion path). The programming system integrates the different planning results here to generate a complete cycle.

#### Action planner

The action planner is to break down the overall assembly task into logical sub-operations. These can then be further broken down into executable robot program instructions in the other planning modules (the grip, path and join planners).

#### Grip planning

Grip planning is to find a favorable gripper position for handling an object allowing safe transportation without damaging the workpieces during the gripping process. To achieve this, several criteria must be considered (◘ Fig. 13.52).

The forces exerted on the object by the gripper must be high enough to overcome all external forces exerted on this object (weight and inertia forces). The gripper must be able to reach the grip position at both the initial and target positions, i.e., the positions must be chosen such that no collisions occur between the gripper and surrounding objects in the joining process. If this is not possible, a sequence of gripping operations must be generated, i.e., the grip on the object must be changed.

#### Path planning

The task of path planning includes, for a given robot environment, the initial configuration and the target configuration, finding a sequence of robot motions for which no collisions occur with objects in the robot environment. The robot configurations uniquely define the axis-specific position of the robot and so resolve potential robot ambiguities, provided the robot can move to TCP positions and orientations with different axis positions.

13

The second subtask is finding a continuous sequence of robot configurations so that the robot can be controlled from an initial configuration to a target configuration. The complexity of this state transition increases with the number of axes. This means that computing-intensive algorithms (such as with "Trial and Error" [FERR05]) or heuristics are used here.

#### Join planner

The primary task of the join planner is the detailed planning of joining operations. For this, the system must determine the suitable join positions and orientations and the permissible joining forces on the workpiece. It must also evaluate the sensor signals in the robot program. For process design, expert-oriented planning systems consider the contact conditions of the joining objects in question to provide an assembly strategy that can then be realized in joining motions [SAYL11].

The task of devising a programming system to simplify program creation that can be used as universally as possible has not yet been solved satisfactorily due in part to the complexity of the problem.

### 13.5.3 Industrial Robot Language (IRL) as an Example of a Robot Programming Language

Similar to DIN 66025 from NC technology, the start of the 1990s saw a standard developed for robot programming languages (IRL, Industrial Robot Language; DIN 66312, Part 1). This effort resulted in the harmonization of programming languages and reduced training overhead for operators [DIN93, PRIT92]. The structure of this language is similar to high-level languages such as Pascal. This means it was possible to address the trend towards increasingly complex robot programs and stricter demands made on the performance of robot systems and controllers.

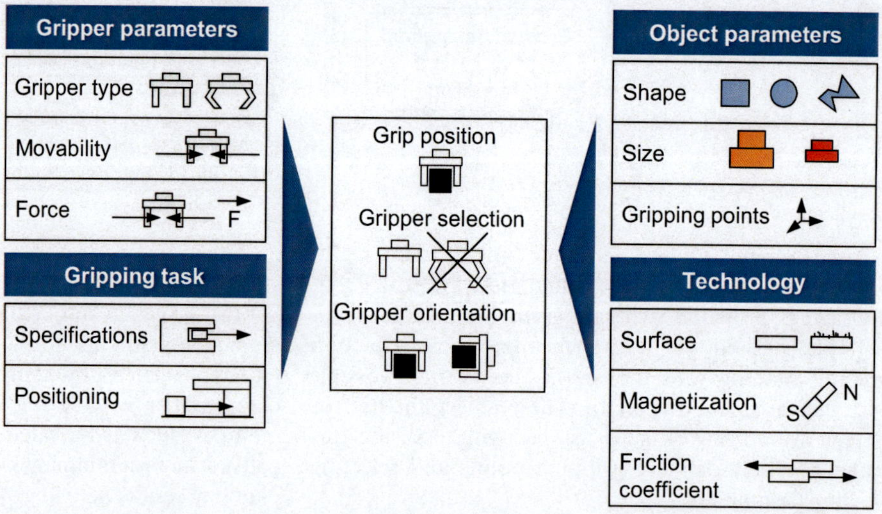


◘ Fig. 13.52 Aspects of grip planning

Although the objective of standardized controller programming using IRL has yet to be realized, many controller manufacturers have however based the development of their proprietary programming languages on IRL meaning the functional scopes of in-house developments essentially correspond to that of IRL. For this reason, IRL is used below as an example to describe common commands and functions.

Multiple program instructions are generally split into modules, each corresponding to single task steps (such as "Retrieve component from magazine" and "Place component into clamping device"). The ability to reuse modules in different programs by referencing reduces the amount of code generally and makes complex processes more manageable.

The elements of an IRL program or module can be categorized into different groups. ◘ Table 13.2 provides an overview of the most important functions and commands in IRL.

#### System declaration file

For every robot cell, a definition of its capabilities and the additional controller functions implemented must be specified in the form of a file. A series of variables and types required as a minimum are declared in this system declaration. These include the number and names of the robots in a cell, the number of types of main and additional axes, the specification of endless axes, the type of orientation specification, and the possible speeds and accelerations. The functions of many IRL commands depend on these details. The number of base axes in the system and the number of additional axes are specified here as an example. For a robot move command (MOVE), a different number of axis values need to be specified depending on the number of axes, i.e., the data type for the robot position is different. The same applies to computing operations.

#### Data lists, import and export declarations

In IRL, data that are different from cell to cell and robot positions that need to be taught can be stored in data lists. A file containing the variables, constants, types and other declarations (such as I/O names) is known as a data list. The benefit of data lists is that cell-dependent and cell-independent data are separated. If, for example, one program is to run on two robots, the robot positions to be programmed for the two programs are marginally different due to the tolerances. When these positions are kept in data lists, only a new data list needs to be created and included in the program (INCLUDE). Changes to the program itself are not required. The same applies to the assignment of inputs and outputs whose names can be assigned in the IRL. If the physical assignment of inputs and outputs changes from one cell to the next and the name assignment is in a data list, the data list need only be updated to make the program executable on the second cell.

Furthermore, the IMPORT and EXPORT instructions enable functions from other programs to be used and local functions to be made available to other programs.

#### Declarations

IRL allows the use of almost any name for a variable. As is generally customary for higher-level programming languages, all variables, constants, types, and functions used in the program must be declared before being used. In addition to reducing errors (the compiler can detect mistakes in names), this concept also enables local, complex data types to be generated. This includes fields and records and any combinations and nesting of types already declared.

#### System variables

Many specifications that often had special commands before are implemented in IRL by changing different system variables. This applies, for example, to setting the axis speeds, path speeds, and the acceleration for motion sets. The current robot is selected, and tool corrections or coordinate systems changed in a similar manner.

#### Main programs

A main program forms the basis for defining and calling other program components.

#### Procedures and functions

Recurrent program instructions are kept in procedures or functions. These can be called from all routines belonging to the same main program. The difference between functions and procedures is that functions have a freely definable return value and transfer parameters that can be defined as required by the user.

#### Loops

Loops enable certain program sections to be executed repeatedly. The number of repeats or an execution condition can be specified for the loop. The execution condition is checked automatically at the beginning or end of every loop depending on its type.

#### Program flow instructions

Program flow instructions include jump commands and commands for logical program branching, breakpoints, jump back instructions, and wait instructions.

#### Motion instructions

Appropriate commands are used to define different types of motion for a robot (point-to-point, linear and

**Table 13.2.** IRL command set

| Instruction | Parameter / sub-instruction | Example | Explanation |
|---|---|---|---|
| **Data lists, import and export declarations** | | | |
| IMPORT DATALIST | Data list name | IMPORT DATALIST Pallet.pkt; | Linking of a data list into the program (e.g., Pallet.pkt with points in space) |
| FROM ... IMPORT ... | Program name, function list | FROM main program IMPORT ALL; | Linking of functions and procedures from other programs |
| EXPORT ... TO ... | Function list program name | EXPORT function1,function2 TO pallet program; | Enabling of functions for export into other programs |
| **Declarations / data types** | | | |
| TYPE | RECORD, ARRAY, FILE | TYPE RECORD INT x; INT y; ENDRECORD = location; | Definition of any user-defined data types from field, record, file and user-defined types |
| VAR definition | BOOL, INT, CHAR, REAL, STRING, TEXT, POSITION, ORIENTATION, POSE, ROBTARGET, JOINT, INPUT, OUTPUT | VAR BOOL Error; STRING Messagel:="PartOK"; INPUT BOOL Gripper AT 2; POSE Point l; Location initial location; | Any variables of different data types (both predefined and all user-defined types are supported) |
| CONST | Same as VAR | Same as VAR | Same as VAR, only constants are generated |

13

**Table 13.2.** (continued)

| Instruction | Parameter / sub-instruction | Example | Explanation |
|---|---|---|---|
| **System variables** | | | |
| Assignment or query/ logical operations | R_BASE, R_ACC, R_C_CP, R_TOOL, R_ROBTARGETACT, R_SPEED, R_SPEED_ORI | R_SPEED := 80; R_TOOL := ToolCorrection_3; R_BASE:=Base1; IF (R_ACC > 20) THEN ...; | Determination of the current speeds, base coordinate systems, tool corrections, accelerations, position, blending parameters, current robot |
| **Main programs, procedures and functions** | | | |
| PROGRAM | Program name | PROGRAM Fill_pallet; ... BEGIN ... ENDPROGRAM | Definition of a main program with any program name |
| PROCEDURE | Procedure name | PROCEDURE Gripper_open (INT a); BEGIN ... ENDPROC; | Definition of a procedure with any name and transfer parameters (within a main program) |
| FUNCTION | Function name | FUNCTION Grip (INT a). INT; BEGIN ... ENDFCT; | Definition of a function with any name, transfer parameters and return type (back to main program) |

Table 13.2. (continued)

| Instruction | Parameter / sub-instruction | Example | Explanation |
|---|---|---|---|
| **Program flow instructions** | | | |
| FOR | Control variable, from, to, increment | FOR i:=1 TO 10 STEP 2 {instruction block}; ENDFOR; | Loop for repeating an instruction block multiple times |
| WHILE | Execution condition | WHILE i < 10 {instruction block}; ENDWHILE; | Repeating an instruction block as long as a condition is true |
| REPEAT | Termination condition | REPEAT {instruction block}; UNTIL i >= 10; | Repeating an instruction block until the termination condition is true |
| LABEL | Jump label name | LABEL start; | Definition of a jump label |
| GOTO | Jump label name | GOTO start; | Unconditional jump to the jump label specified |
| IF | Condition | IF (Pallet = OK) THEN ... ELSE ... ENDIF; | Conditional branching |
| CASE | Identifiers, alternatives | CASE Gripper OF WHEN Gripper_1: i:=1; WHEN Gripper_2: i:=2; ... ENDCASE; | Conditional branching with multiple selections (any number of alternatives and branches) |
| WAIT | FOR, SEC | WAIT FOR Gripper=OK TIMEOUT 10 SEC ERROR ... ENDERROR; | Wait for a condition or for a certain time. When a timeout interval is exceeded, a special instruction block is executed |
| PAUSE | Message | PAUSE "Please change pallet"; | Wait until the Start button is pressed. It is possible to show a message on the controller |
| RETURN | Return value | RETURN 5; | Jump back from a function (with return value) or procedure |
| HALT | | HALT; | Ends the program flow |

13

Table 13.2. (continued)

| Instruction | Parameter / sub-instruction | Example | Explanation |
|---|---|---|---|
| **Motion instructions** | | | |
| MOVE PTP | Target point ACT_ROB, C_PTP, SPEED_PTP, ACC_PTP, TIME, UNTIL | MOVE PTP Switchpos C_PTP SPEED_PTP:=80; | Point-to-point parameters: blending behavior, acceleration, motion time, motion condition |
| MOVE LIN | Target point ACT_ROB, C_CP, C_SPEED, SPEED_ORI, SPEED, ACC, TIME, UNTIL, WOBBLE | MOVE LIN Changepos C_CP TIME:=4 WOBBLE; | Linear motion parameters: blending behavior, acceleration, motion time, motion condition. Also, a WOBBLE can be superimposed on the linear motion. |
| MOVE CIRC | Interpolation point, target point, otherwise same as LIN | MOVE CIRC interm. point changepos SPEED:= 100; | Same as LIN, but here with an interpolation point on the circle and end point |
| MOVE_INC PTP/LIN/ CIRC | Same as MOVE | Same as MOVE | Same as MOVE, only with relative position details for the current robot position |

Table 13.2. (continued)

| Instruction | Parameter / sub-instruction | Example | Explanation |
|---|---|---|---|
| **Input/output instructions** | | | |
| Assignment/ query | Output name, value/input name | OUT_5 := FALSE; IF ANALOG_INI > 12.5 THEN ... | Inputs and outputs are treated like variables/constants and can be set and queried accordingly. A variable has a different data type depending on the type of input/output (binary, digital, analog) |
| OPEN | File pointer, file name, mode, error code | OPEN (file "test.dat",R, error). | Opens a file or data channel |
| CLOSE | File pointer | CLOSE (file); | Closes the specified file or the data channel |
| READ/ READLN | Data variable | READ (file, data); | Reads from the file into the Data variable |
| WRITE/ WRITELN | Data variable | WRITE (file, data); | Writes the contents of data into the file |

Table 13.2. (continued)

| Instruction | Parameter / sub-instruction | Example | Explanation |
|---|---|---|---|
| **Arithmetic and logic instructions, computing functions, predefined functions** | | | |
| NOT,+,-,*,/, MOD, DIV, <, >, =, <>, AND, &, OR,EXOR... | One or two operands (depending on the operation) ok:=(E1 AND E2) OR E3; | State := NOT error; IF (Param <> 10) THEN ...; | Virtually all common logic and binary instructions |
| SIN, COS, ASIN, ACOS, ATAN2, ROUND, SQRT, COMPARE, ... | Different depending on function | x: =SIN(y); Equal:=COMPARE(Text1, Text2); xr:=ROUND(x); Root:= SQRT(134.44); | Functions for trigonometric and geometric calculations. Functions for string processing, comparisons, etc. |
| **Geometric logic instructions** | | | |
| +,-,/, *,@ | Operand1, Operand2 | NewPos:=Targetpos1@ Pose1; | Operations on position variables. In example: Transformation of Targetpos1 by Pose1 |

circular motion, see ▶ Sect. 13.4). Programming the maximum speed and acceleration values specifies the axis/path-related traverse behavior. Selected zero point and tool corrections offset the TCP positions specified in the user program and have a bearing on the orientations. Using blending parameters, the system does not move toward target points precisely. Instead, the next motion set is initiated early. This is dependent on the thresholds selected (speed or path). For special applications such as in welding, oscillation functions are used. These superimpose an oscillating motion on the programmed path.

13

### Input/output instructions

This command group includes the simple base functions of programmable logic controllers such as setting and resetting binary outputs as well as processing of digital and analog inputs and outputs. The ability for the user to define inputs and outputs, the automatic conversion of binary, digital and analog states to Boolean, integers and real numbers, and so the virtually unrestricted ability to apply logical operations to these signals (such as executing arithmetic and logical functions) means almost all functionality required can be handled with few input/output-specific instructions. Word operations for assigning parameters and outputs, the processing of process-dependent analog signals and instructions for actuating grippers (open and close) can be realized here.

Also, all functions for reading and writing data from/to any devices fall into this category. These include displaying of text the screen or teach pendant and reading/writing files from/to external data storage media. The channel concept introduced in IRL, in conjunction with Read and Write instructions, allows a very wide variety of devices to be accessed.

### Logic instructions

Boolean operations (such as AND and OR) are used for bitwise logic operations on variables and constants. Extensive logic operations and manipulation capabilities on bit level are available in IRL. There are also functions for comparing variables.

### Arithmetic and geometric operations

Arithmetic operations allow the processing of any variables and constants. Trigonometrical functions and operands for coordinate transformation are available in addition to familiar basic computational functions. IRL thus makes complex computing operations possible.

As the list shows, IRL offers a number of different ways to structure the program, to perform calculations, to process inputs and outputs and to declare types, variables, functions, etc. The number of robot-specific instructions is relatively low here. This was because the objective of DIN 66312 Part 1 when it was compiled was to include the greatest number of standard commands possible for a robot controller with the smallest functional scope.

Furthermore, the tool thus created handles basic functions as well as some special functions with suitable function applications. This is aided further by the ability to integrate modules (such as those from sensor and gripper manufacturers to control their components) into IRL programs. It is also possible to link routines previously compiled into the local program as libraries. This separate compiling has the benefit that application-specific libraries and tested modules need not be recompiled every time the user program is changed.

The most important supplementary robot programming functions offered nowadays by many control manufacturers are described below.

### Sensor instructions

Commands for switching the sensor interface on and off are required for communicating with powerful sensor systems (such as laser scanners and image processing systems). While sensor control instructions and commands for requesting, transferring and processing sensor data can already be addressed in IRL, functions for externally specified velocity control and path correction are not standardized.

### Interrupt processing

For robot controllers, in conjunction with sensors in particular, it is necessary for the program to respond directly to events that occur suddenly. For sensors measuring distance, robot motion, for example, should be stopped when a certain distance is reached. Furthermore, the sensor must respond immediately if certain system states occur (fault handling). For this, an interrupt can be defined. If an event or condition occurs, this interrupt causes the program flow to stop immediately, even during a motion, and starts an appropriate interrupt procedure. In this procedure, the programmer can respond to the event by, for example, ending the robot motion, issuing fault messages or initiating fault rectification actions.

### Path switch function/analog output

Using a path switch function, any switching command (such as switch on the air supply to a gripper) can be performed during a motion provided that a specified condition pertaining to the motion is satisfied (certain speed reached, entered/exited tolerance distance about the target point, etc.). This function is used first and foremost for optimizing a program in terms of time because interrupting the motion is no longer required to trigger certain events.

Furthermore, information on the current path speed of the TCP is required by external systems in some applications. This applies, for example, in applying adhesives to car windshields where the adhesive flow rate must be regulated in line with the path speed.

It is not always possible for the robot to have a constant path speed. This applies in particular to small radii that might require far reduced speeds. ◘ Figure 13.53 illustrates how path switch functions work using a speed-proportional voltage value (analog output), that can be used, for example, by a welding process controller for parameter correction.

### Path welding commands

Many controllers now offer ways to activate welding sources for path welding applications. This represents one of the most common applications of industrial robots. A welding source normally requires a relatively large number of welding parameters (voltage, current, wait times, wire feed, parameters for the start and end of the weld seam, etc.). Although transferring parameters (such as in IRL) is possible in principle, incorporating suitable, special technology instructions is helpful for the purposes of welding process programming that is as simple as possible. To enhance the controller functions for this, robot manufacturers offer different technology options as extensions to their controller systems.

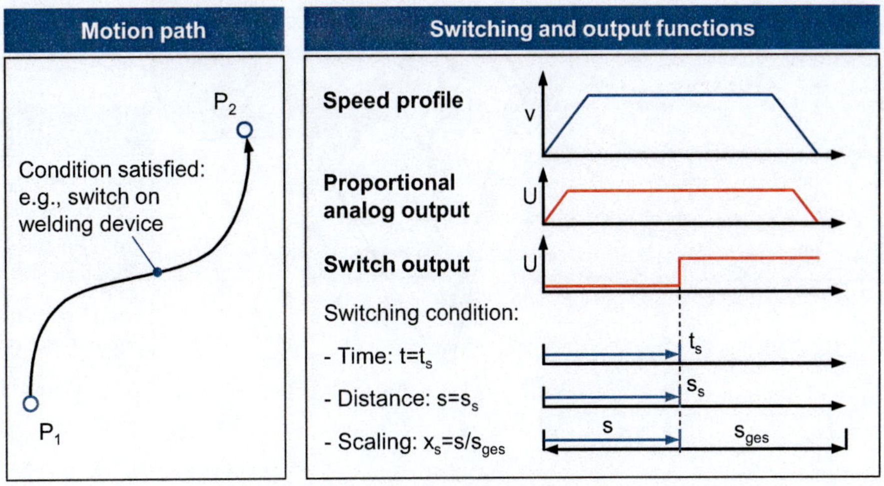


◘ **Fig. 13.53** Path switch functions and speed-proportional analog output

### Parallel processes

Many robot controllers now provide the ability to run multiple programs or routines in parallel, i.e., simultaneously. In addition to the ability of controlling two or more completely separate robots with one controller, this concept permits additional alternatives for program optimization to reduce program runtimes. As a controller is rarely fully utilized by a robot motion, calculations or communication with sensors and peripherals, for example, can be performed in parallel.

### IRL programming example

Figure 13.54 shows a programming example for a welding application. The three interpolation types PTP, linear and circular interpolation without blending are used. The coordinate axes entered define the spatially fixed robot coordinate system. Welding commands are not included in the example.

The robot program shown in Fig. 13.55 is taken from a controller program for industrial robots. The user program was developed in the IRL programming language. For direct comparison purposes, the table also shows the same program in a structure-based form.

## 13.6 Communication Interfaces for Robot Controllers

Robot controllers on the market today have communication interfaces that allow data connections to master and cell computers, sensors, peripherals as well as machines and feeder systems. The functional scope of these interfaces extends from sending and reading individual digital signals to exchanging complex data structures.

The communication interface of a robot is made up of digital input and output signals. Application examples of this are the querying whether a workpiece is in place or the "handshake" with a machine to feed, i.e., in the back-and-forth querying of the program enable and cycle initiation between two interacting machines.

It is normally possible to load and store programs and parameters, read and monitor variables and fault states as well as select and start programs using the serial interface of the controller.

For robot systems whose communication extends beyond the simple exchange of bit signals, powerful, internationally standardized bus systems are available nowadays. Examples of these in use generally are Interbus

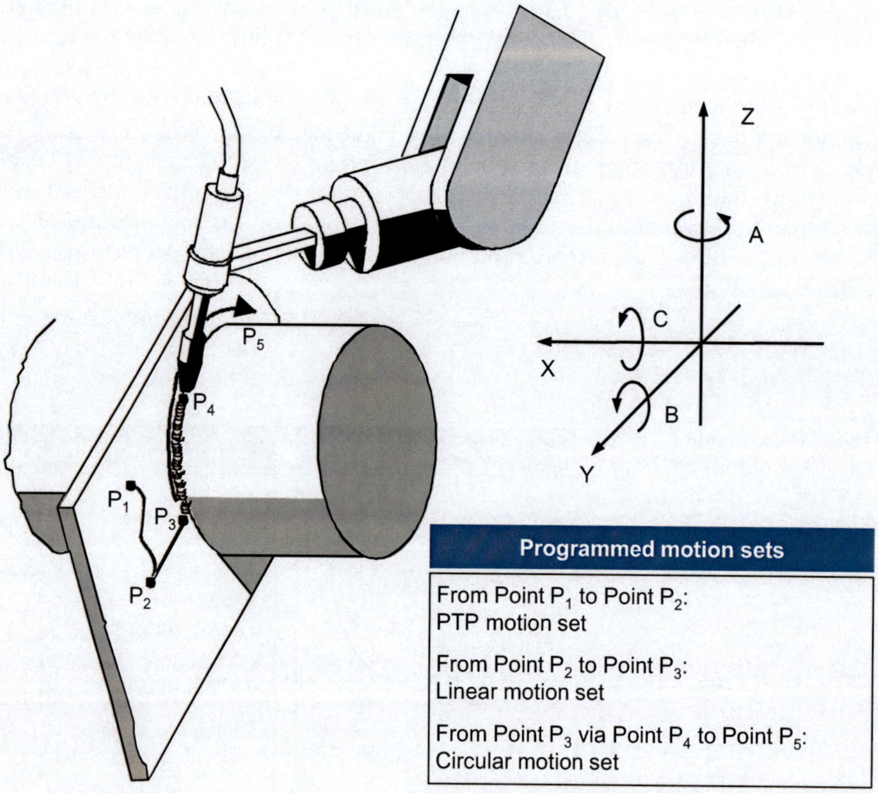


Fig. 13.54 Programming example for robot motion instructions, according to Siemens AG

**Fig. 13.55** Comparison of text-based and graphical programming

| Program instruction | Explanation |
|---|---|
| Program example; | Definition of the program |
| VAR | Start of the variables declaration |
| ROBTARGET: Point2, Point3; ROBTARGET: Point5 | Define point variables: The data type ROBTARGET contains both the position (position and orientation) and the axis configuration |
| POSITION: Point4; | Interpolation point position variable: The data type POSITION contains only the position data (3 coordinates) and no orientation data |
| START | Program start |
| R_SPEED:=50; | Path speed |
| R_ACC:=40; | Path acceleration |
| IF (E Ready=1) | Querying of the digital input to check whether a workpiece is inserted |
| THEN | If a workpiece is inserted, then ... |
| MOVE PTP POSE(POSITION( 230.083,340.947,-800.358), ORIXYZ(35,-60,130)); | Traverse movement with PTP interpolation: Since the motion sets and their position data are generally taught, this command need only rarely be entered using text. |
| Point2:=R_ROBTARGE T_ACT; | Save current position for calculations of Point3 |
| Point3:=Point2; Point3.PSE.POS.Y:= Point3.PSE.POS.Y-20; | Calculate Point3 with y-coordinate of Point2 - 20 mm |
| MOVE LIN Point3; | Linear motion to Point3 |
| Point4:= POSITION(230.083, 315.0,-770.358); | Specify interpolation point coordinates |
| Point5:=ROBTARGET (POSE(POSITION( 230.083,300,-750.358), ORIXYZ(85,-60.35)), 0,0,0,0); | Enter coordinates of Point5 |
| MOVE CIRCLE Point4, Point5; | Circular motion from Point3 to Point5 via Point4 |
| ELSE | Else |
| STOP; | End the program |
| ENDIF; | End of the IF instruction |
| ENDPROGRAM | End of the program |

Flow chart

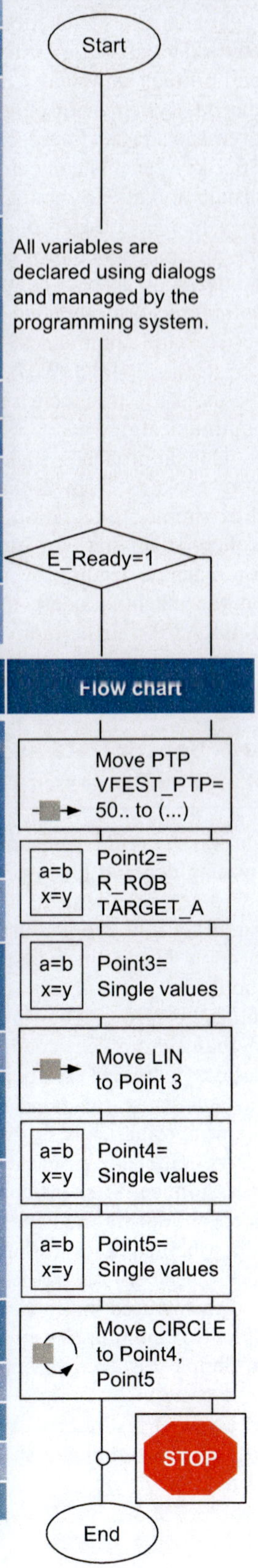

(IEC 61158), Profibus/Profinet (IEC 61158/61784) and CAN bus (Controller Area Network, OSI/ISO 11898). Using these bus systems, it is possible to connect, at a high data rate, any number of components to the robot controller with little wiring overhead. Distributed inputs and outputs as well as sensors that can be connected to the bus system with plug & play with a configuration overhead are also available. The cited fieldbus systems offer key benefits compared to the intricate cabling and installation of early systems (refer also to ▶ Sect. 9.3).

For the transfer of programs, process monitoring and process control, as well as for teleservice, most modern-day robot controllers have Ethernet-based communication services. They include TCP/IP-based services and communication using OPC (OLE for Process Control) [SCHW07]. Overall, there is a discernible trend towards increasing the level of abstraction of communication.

New approaches, such as the mxAutomation function package from Kuka, allow a greater level of, thus simpler, integration of robot and machine controllers such that, for example, cooperative processes for machine feeding by an industrial robot centrally on the machine side can be defined by the operator [KUKA15]. Furthermore, simple robot motions can be entered from the machine panel of the machine tool.

13

## 13.7 Sensor Data Acquisition and Processing

For the handling and automation sector, there is increasing demand for systems that can be tailored flexibly to new requirements and environmental conditions. Handling robots operating with sensor systems as used for this. Using these systems, up-to-date information about the relevant environment can be recorded by suitable measurement transformers to allow adaptation to changed ambient conditions [GECK11, KUHN09]. Besides increased flexibility, equipment costs are reduced because the target points of the robot motion can be automatically read and corrected.

An important application area for sensors is quality assurance. After the machining process, it is possible to check, during the handling process using sensor signals, whether holes are present, for example, or tolerances have been observed.

Sensors used in practice are based on different principles. In common use, for example:

- Simple distance sensors (tactile, inductive, capacitive, etc.)
- Laser measurement systems (triangulation, runtime measurement, shadow formation, split-beam method, etc.)
- 2D machine vision systems (gray scale value, binary image, weld pool monitoring, Moiré, etc.)
- 3D machine vision systems (stereo camera systems, structured light, etc.)
- General measured-value acquisition (welding current in combination with oscillation motion, force/torque sensors, etc.)

The requirements made by the respective industrial environment are crucial for the use of the sensor type. Some key practical tasks and appropriate sensors are summarized in ◻ Table 13.3.

When these sensors are operated in conjunction with robot systems, it is possible to exert influence over programmed coordinates and paths determined by them. The information provided is processed in the robot controller used to generate current position values.

Intervention by sensor systems in the program flow or control can be on different levels depending on the requirement. For workpiece calibration or position detection, intervention in the program itself is generally sufficient because no correction is necessary during the robot motion. In contrast, systems for online path correction or tracking (such as for welding) require a fast reaction by the controller to continuous sensor signals. This means that even intervention in position control of the controller is sometimes necessary here.

Communication between the sensor system and robot controller uses digital or analog interfaces depending on the equipment setup. Accordingly, defined telegrams, digital inputs/outputs or current/voltage values specify the information or up-to-date correction data made available by the sensor system. The sensor data are processed either by an external system (normally a PC) or inside the controller.

Modern-day robot controllers provide the ability to analyze any sensor signals directly in the controller using appropriate software sensor interfaces. The relevant user program specifies where the data are required in the program flow. The classification of interfaces made above may mean the query of an input or setting of an output date, processing of a telegram by setting variable storage locations or processing the current/voltage values as control variables for servo amplifiers, etc.

Their low prices and universal applicability mean image processing systems are being used more and more now in handling equipment, too. Mature systems are now available for 2D applications, in particular for tasks such as component and position detection on conveyor belts and turning position detection for car wheel rims (positions of wheel bolts).

◻ Figure 13.56 shows the case of sensor-aided gripping of electrical cabinet elements on a moving conveyor belt by an industrial robot fitted with a CCD camera. Based on the camera signals, the electrical

**Table 13.3.** Example application areas for sensors

| Process/ functional area | Task | Sensor type |
|---|---|---|
| Path welding, spot welding | Search for start and end of weld seam, follow the course of the joint, determine joint geometry, document welding result | Inductive/optical distance sensor, tactile sensor, welding current sensor, split-beam, triangulation, weld pool monitoring |
| Cleaning castings, grinding | Deburring, find course of contour, determine burr thickness, control contact force | Tactile sensor, force/torque sensor, motor current sensor |
| Quality assurance | Workpiece calibration, determination of tolerance deviations, completeness check | Visual sensors with line scan and matrix cameras, tactile sensors, distance sensors |
| Assembly, handling, machine feeding | Gripping of workpieces, turning position detection, assembly of parts exhibiting tolerances, inclusion of conveyor belts, gripping force monitoring | 2D/3D camera systems, distance sensors, force/torque sensors (for process forces), laser scanners, light barriers |

cabinet element is identified at the start of the gripping process to perform continuous position detection during the process. The component is then gripped on the basis of this continual position detection. This application is a typical example of using a sensor system to eliminate the complex, controlled in-feed of parts to be handled. The robot is controlled here from an external computer system that sends new position parameters to the robot controller in the interpolation cycle. This allows a flexible gripping process. In this example, there is no complete measurement of the component with subsequent CAD model generation. Instead, there is just a comparison of the actual state with the nominal state known from training data.

In addition to quality and assembly checks, depth readings are commonly used by industrial robots for path correction. The primary application areas here are flexible handling of components and assembly processes on moving assemblies. For the former, it is especially about handling components that are provided in unstructured and random positions. "Bin picking" here is the handling of components that are provided loose. The challenge here is that both the positions and orientations of the components vary greatly and must be read by sensors. The approach motion and gripper configuration are determined on the basis of the component position detected.

In the field of robot-aided assembly, greater significance is being attached to the moving joining of components. In the classic approach, assembly is in the stationary position, meaning join positions and approach motions are known and are constant for consecutive processes. For assembly line production (such as automotive production), this type of assembly requires additional overhead as regards coupling and decoupling stations for parts on the assembly line. For moving assembly, the assembly action runs parallel to the conveyor movement without the conveyor being stopped.

To use sensors for the path correction of an industrial robot, direct adjustment of the motion path during the process is necessary. To perform this motion adjustment in 3D space, 3D sensor readings must also be available. Using a single camera is generally adequate here. Conclusions on 3D values are made on the basis of known features of the component in question (such as the length of edges or the radius of boreholes). If it is not possible to interpret any features, the use of sensor technology that generates an explicit, 3D image of the environment in question becomes necessary. Current developments are trending towards the generation of RGB-D measurement data. These data contain color information (RGB) in addition to distance information (depth). These systems often require multiple sensors and actuators (at least two cameras or one camera with structured light projection). The high data volume generated by depth sensors means measurement data evaluation and the resulting motion adjustment of the industrial robot is calculated on an additional computer system. For robot motions at high speed in particular, the latency times arising in addition from signal transmission must be taken into account in determining the correction motion.

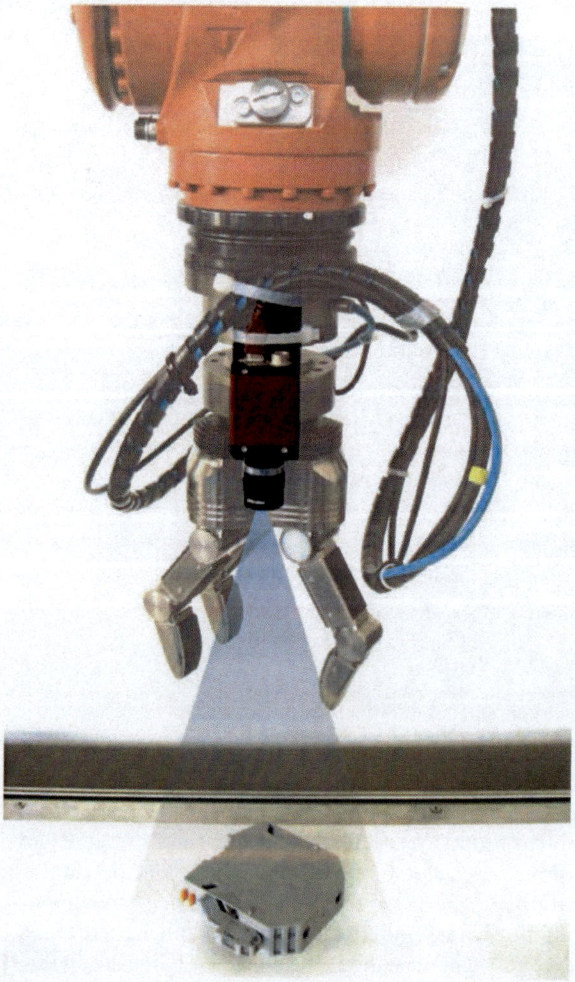

■ **Fig. 13.56** Camera-aided gripping of electrical cabinet elements

Regardless of the sensor system used, hand-eye calibration must be performed for all optically operating sensors on the industrial robot. The purpose of this calibration is to transfer the sensor measurement data from the moving sensor coordinate system to the fixed base coordinate system of the industrial robot. This makes it possible to convert positional deviations detected by the sensor system to robot-specific coordinates and for the robot to compensate for them. ■ Figure 13.57 provides an overview of the correlations, and the coordinate systems used, in the hand-eye calibration.

Hand-eye calibration $T_{FS}$ specifies the constant 3D transformation matrix from the flange coordinate system to the sensor coordinate system. To determine this transformation, a target is used that can be located unambiguously within the space by the sensor as well as the robot kinematics. A chess board pattern is shown in ■ Fig. 13.57 as a target. This is suitable for the hand-eye calibration of a color or gray-scale camera. The target used must be adapted to the measuring principle of the sensor. By touching the target, the industrial robot can determine the 3D transformation $T_{BT}$ from the robot base to the target coordinate system. The 3D transformation $T_{ST}$ from the sensor coordinate system to the target coordinate system is determined from the sensor data. Also, 3D transformation $T_{BF}$ from the robot base to the flange coordinate system is known from the robot controller. The hand-eye calibration $T_{FS}$ can be calculated from the closed kinematic chain in Eq. (13.55).

$$\begin{aligned} &T_{BF} \cdot T_{FS} \cdot T_{ST} = T_{BT} \\ &\Leftrightarrow T_{FS} = T_{BF}^{-1} \cdot T_{BT} \cdot T_{ST}^{-1} \end{aligned} \tag{13.55}$$

To increase the accuracy of, and verify, the calibration, the measurement is repeated with different robot axis positions. Generally, the hand-eye calibration must be repeated for every individual sensor and after the same sensor is reinstalled. To reduce this overhead, the adapters between sensor and robot flange have locating

■ **Fig. 13.57** Hand-eye calibration for an industrial robot

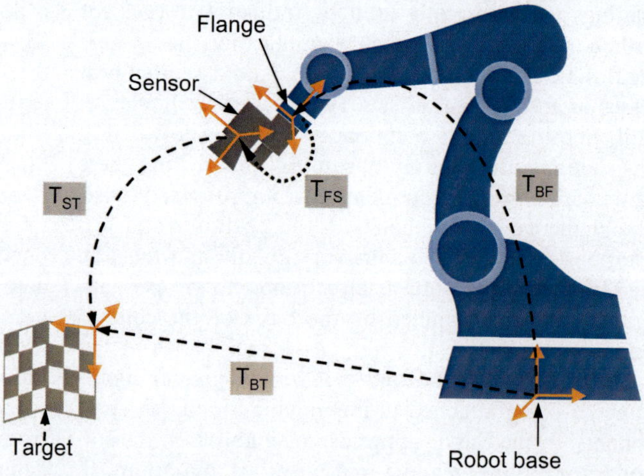

pins, so that when a sensor is reinstalled, it is at virtually the same position.

On the basis of the hand-eye transformation, the measured values of the controlled sensor can be transformed at runtime to the fixed base coordinate system of the industrial robot. The industrial robot can determine a reaction motion directly to new sensor values.

## 13.8 Gripper Technology in Robotics

Nowadays industrial robots are used mainly for welding work as well as material handling and assembly [GRUN09]. These kinds of processes require high accuracy levels in the repeated positioning of individual robot axes and it must be possible to transfer reliably the position definition or change stipulated by the robot to the appropriate object [VDI90]. For this, component positions are specified relative to a fixed reference coordinate system and having an arbitrary origin and arbitrary orientation. To reduce the resultant programming overhead, reference coordinate systems are specified in consideration of the relevant production equipment. For component positioning for example, the directions of surface normals of storage trays are helpful in being able to perform component approaches perpendicular to these trays. For a chain of handling steps therefore, robot controllers provide the ability to define multiple reference coordinate systems and to select them depending on the situation. The controller internally transforms each position into the global world coordinate system of the robot (see ▶ Sect. 13.3).

In addition to reference coordinates, other mathematical definition tools, which define the geometrical state of objects, exist with coordinate systems inherent to the body. They are often defined on the surfaces of objects handled and move with the objects during the process. The fixed relation between the object handled and its specific body coordinate system allows the robot motion to be programmed to be broken down into individual transformations and be stored (◘ Fig. 13.58).

Since handling devices such as robot systems have a direct bearing on the component position and orientation with every process step and undefined changes bring about deviations between programmed and actual component positions, strict requirements are placed on each piece of equipment. These requirements establish the temporary connection between kinematics and component. These kinds of devices are known as gripper systems due to their often single-axis, linear closing motion. Their designs can vary greatly however depending on the handling task. There are technical (such as in the choice of gripper drive system) as well as technological differences. One typical arrangement of industrial robot and gripper system is shown in ◘ Fig. 13.59.

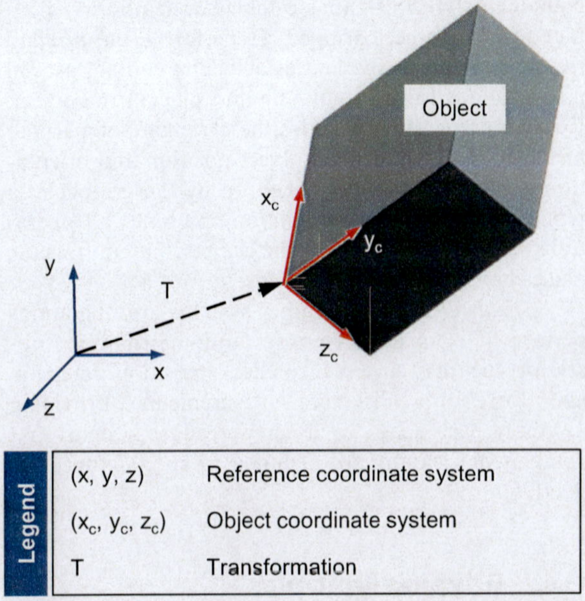


◘ Fig. 13.58 Relation of coordinate systems for handling tasks

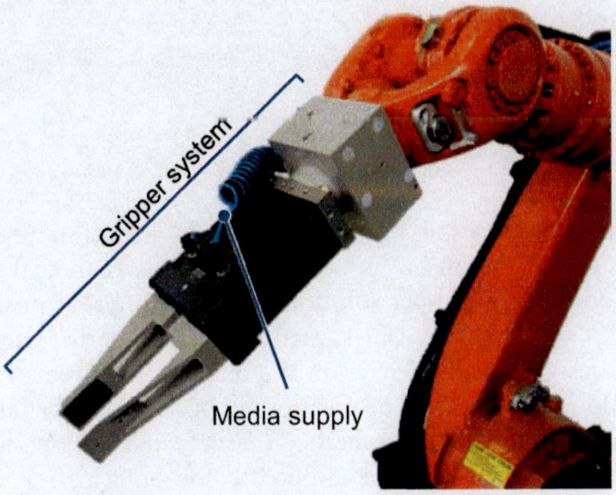


◘ Fig. 13.59 A 2-jaw parallel gripper on a robot system

In addition to applications on industrial robots for object handling and manipulation, grippers are used in different areas of production engineering. On automated systems, they facilitate assembly, machining and packaging processes; on flexible production systems, workpiece rotation operations are performed by grippers with specified kinematics. In addition to direct interaction with the workpiece, gripper systems can also be used for assisting processes such as tool changes on NC machines [HESS04].

### 13.8.1 Process Definitions

According to [VDI90], a gripper system is a subsystem of an industrial robot or a subsystem of handling equipment [HESS04] that establishes temporary contact with an object gripped. It performs the *primary functions* of preparing and establishing contact as well as operations for manipulating and placing the object. Contact preparation includes the arranging of parts to arrive at a known, defined object position and orientation such that it can be picked up by the gripper in a repeatable manner. Manipulation combines all operations internal or external to the gripper that address the place change or assembly of objects.

*Special functions* of gripper systems are operations performed by external sensors and information processing such as the presence check, position detection and force/torque/distance measurements. Providing assembly parts (e. g., by a screw feeding system) and component cleaning are included in special functions [VDI95].

### 13.8.2 Gripping Principles

For establishing contact, a distinction is made between four relevant modes of action in gripper technology [HESS04]:

- Mechanical grippers
- Fluidic grippers
- Magnetic grippers
- Adhesive grippers

13

The categorization into principles of operation provides information on the technology used. Technically, the individual gripper solutions into the different principles of operation can sometimes vary greatly. They are designed and selected on the basis of requirements defining the handling or positioning process. It must always be checked first whether the principle of operation in each case can be applied to the object gripped and whether there is a reliable connection between the handling device and the object gripped. Then, the gripper is specified in more detail on the basis of criteria such as guaranteed gripping force, resulting gripping accuracy, weight, opening and closing times (for mechanical grippers), and control-related connection and regulation possibilities. The gripping force can be subdivided into holding force, i.e., the force to secure an object to the handling device, and the gripping force, i.e., the force required to generate the holding force. The gripping force is usually greater than the holding force. Gripping forces take effect at a point, along a line or across a plane depending on the gripper type.

Gripping processes are planned on the basis of gripping strategies. A general structure (comprising move in, grip, object manipulation, release and move away) forms the basis of every strategy. A specified gripping process defines the nature and position of the grip points relative to the object gripped. The composition of the object gripped and how it is fed in must be taken into account. If the object gripped is on a conveyor belt for example, its grip points are moved synchronously with the belt. This means that the absolute grip point position is subjected to time conditions. Clarification is also required on whether the grip points can be reached using the handling device available.

Particular attention must be paid to ensuring that traverse motions of the handling device do not cause collisions. If, for example, the grip position is readjusted during the process using data from a sensor, the handling device motion varies in a manner not determined beforehand. In such applications, the motion performed is limited spatially to prevent collisions. Motion limits are defined on the controller of the handling device and are taken into account in the generation of motions. As external tools such as grippers are first unknown to the controller, information on their dimensions in particular must be communicated to the controller. Using the kinematic chain of the handling device, the operating space penetrated can then be determined and the collision-free path checked.

### 13.8.3 Gripper Types

Different grippers are differentiated by the principle of operation selected and the design of the technical system. The gripper design is, depending on the application, a special solution that cannot be transferred to other applications. For this reason, only standardized gripper types that can also be found in the machine tools environment in particular are discussed here.

#### 13.8.3.1 Mechanical Grippers

The most common type of gripper for robot systems are mechanical grippers. They hold the object by utilizing form fit or a stable equilibrium of forces (◘ Fig. 13.60). The equilibrium comes about by forces being applied to the object by the gripper. Consequently, for the design of mechanical grippers, how the degrees of freedom of the object gripped are limited and whether the (frictional) forces necessary for securing can be attained, are crucial.

For a cylindrical object gripped, single-sided form fit represents the best of the three alternatives shown from a mechanical positioning perspective (◘ Fig. 13.60a). Contacting at three points effects radial centering of the cylinder. This means that, in this plane, high repeatability of the component position can be attained. In contrast, for a parallel jaw gripper with friction fit only, the object gripped essentially has one degree of freedom, perpendicular to the gripping direction (◘ Fig. 13.60b). Relative motions can occur

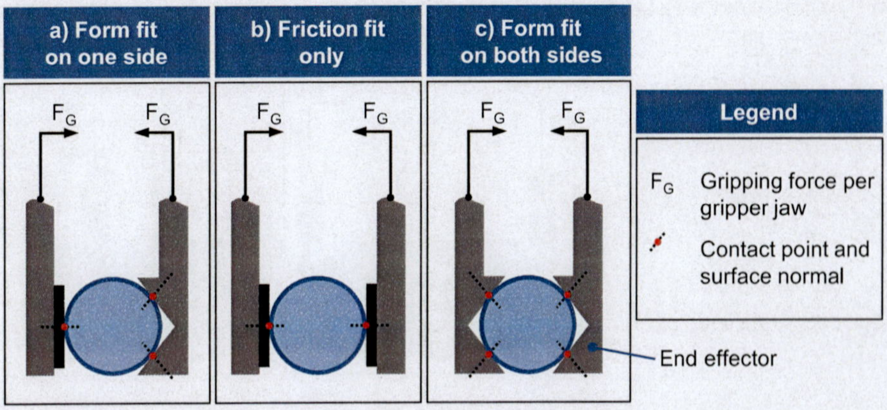


**Fig. 13.60** Alternatives for exerting gripping forces

here along the friction surfaces. This is dependent on the accuracy of the handling device and how it is programmed. Just as disadvantageous as static underdetermination, a redundant binding of degrees of freedom can also worsen the positioning accuracy. On ideal, **form-fitting grippers** (Fig. 13.60c), the contact forces are distributed uniformly across all contact points. Actual grippers, however, are continuously subjected to production and operating tolerances, meaning ideal, centered alignment of the formed parts of the gripper relative to the object gripped cannot be attained. There are risks of tilting and switching of the load-bearing sides, which, in general, reduces the positioning certainty of the contact.

Despite the downsides inherent in its principle, an unfavorable method as regards positioning might be necessary to exert the necessary frictional force and to perform the gripping process reliably. This is dependent on the compositions of the object gripped and grip surface.

In addition to the process principle applied, other **parameters need to be specified to define a reliable gripping process**. These include, in particular, the place where the force exerted by the mechanical gripper takes effect. The properties of the object to grip, such as shape, stiffness, surface composition and mass, are crucial in determining this place. The immediate environment and the status of product infeed must also be factored in. If objects are rigid, enclosed bodies, an external grip is often applied for handling (Fig. 13.61a). The gripper jaws are closed from outside for securing. This can only be used if the open gripper can approach the object gripped without causing collisions with surrounding objects. When several objects gripped are packed tightly together on a transport pallet, open components can be picked up by the internal grip (Fig. 13.61b). The combined grip (Fig. 13.61c) is particularly well suited when accessibility is poor and the component is large.

Mechanical grippers are driven in different ways. The drive principle can essentially be selected as desired **in the specification but usually depends on the energy** source available. Electrical energy and compressed air are important sources here. The energy supplied is converted to a linear motion by the gripper drive. This comprises, depending on the drive concept, the transformation of a rotational motion to the required linear motion by (for example) a ball screw drive.

So that a failure of the drive system does not result in uncontrolled loss of the object gripped, grippers often have gripping force retainers. These maintain the force necessary to secure the object in place even when the energy supply fails, for example. In its simplest form, a gripping force retainer comprises a return spring that is tensioned in the direction the gripping force is applied and acts upon the object gripped in addition to the gripper drive (Fig. 13.62a). For the gripper to open, the drive must act against the spring force. This means that, generally, the maximum force of the gripping force retainer must be lower than that of the drive system. As an alternative, pneumatic grippers can be protected from the gripping force dropping uncontrollably using alternately activated check valves (Fig. 13.62b). The associated schematic of the pneumatic gripping force retainer illustrates that the compressed air can only escape, i.e., the cavity pressure can only be relieved, when a pressure potential is built up in the opposite direction at the same time. If this is not the case, motion of the gripper jaws is prevented, and the gripper remains in its current position.

Basic systems for mechanical grippers can be purchased **today for handling equipment with defined** properties. The end effectors, i.e. gripper jaws, must normally be adapted to the respective component or product, however. Further differentiation of mechanical grippers is possible on the basis of the end effectors.

### Parallel jaw grippers

Parallel jaw grippers are the most commonly used grippers for industrial robot systems. They have two parallel gripper jaws that are moved toward / away from

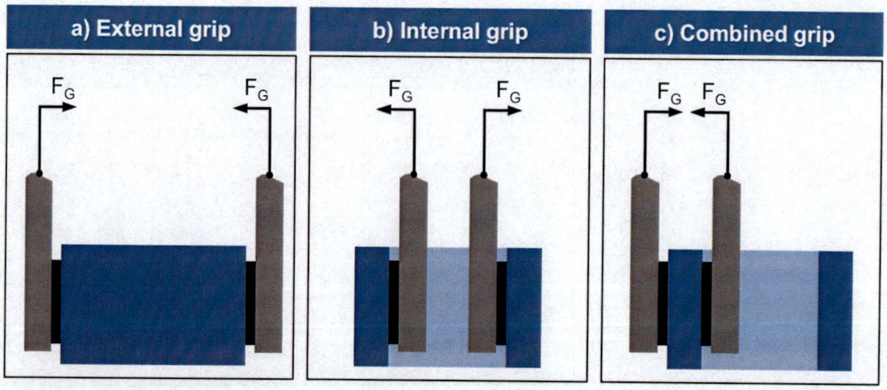


**Fig. 13.61** Grip variants for mechanical grippers

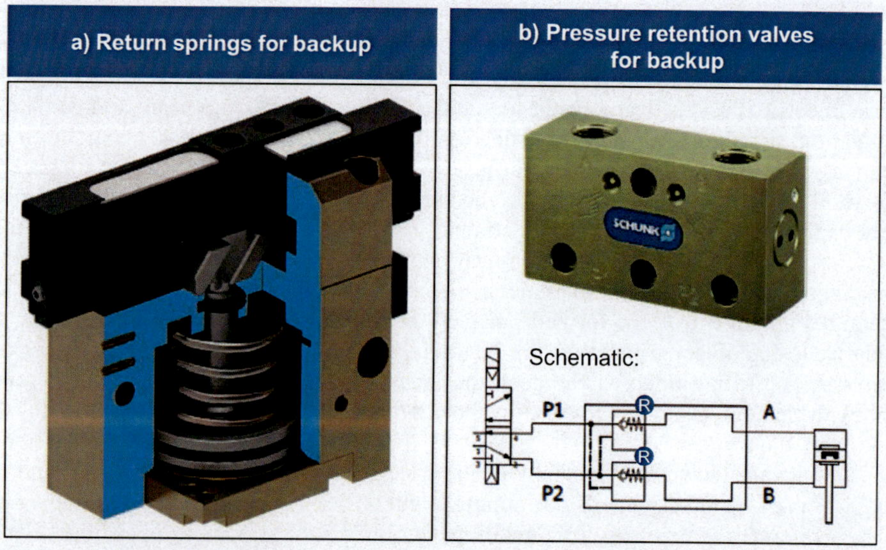


**Fig. 13.62** Gripping force retainers for mechanical grippers, according to Schunk GmbH



each other during the gripping process. Opening and closing operations are usually symmetric such that both gripper jaws are actively involved in the motion. This is achieved by the force generated by the drive being split by a gear unit (kinematics) and routed to the individual gripper fingers. The wedge hook kinematics (Fig. 13.63) is an example of such a gear unit. It features high gripping forces that can be transferred to both directions, i.e., for opening and closing.

### Angular grippers

For angular grippers, the linear motion of the drive is transformed into a rotational gripping motion. Suitable kinematics effect this and thus make possible the swivel motion of the gripper jaws (see Fig. 13.64). Like parallel jaw grippers, they are also suitable for internal and external gripping. With the swivel motions toward and away from objects, they can be picked up and set down without an additional lift motion by the handling device. This means collisions can be prevented while the handling device is moving. Individual angular grippers can be used for a wide range of component dimensions, for cylindrical components in particular, because their jaw position can be controlled over a wide angle range.

### 3-finger grippers

The end effector of a 3-finger gripper is split into three parts. Using suitable kinematics (such as wedge hook kinematics, Fig. 13.65), all gripper fingers are moved simultaneously here, too, and travel the same path back during the closing process. The object gripped is kept moving until its motion is hindered by the other gripper fingers, and thus centered. The centering effect is utilized for joining processes of cylindrical components in particular. These processes put strict requirements on a defined component position and the gripping force that can be brought to bear.

### Gripper hand

As grippers represent technical systems that must be designed or selected specially for the gripping task at hand, the design of the end effectors and specification of the travel for each gripper finger are affected in particular. If there are modifications to the object assigned

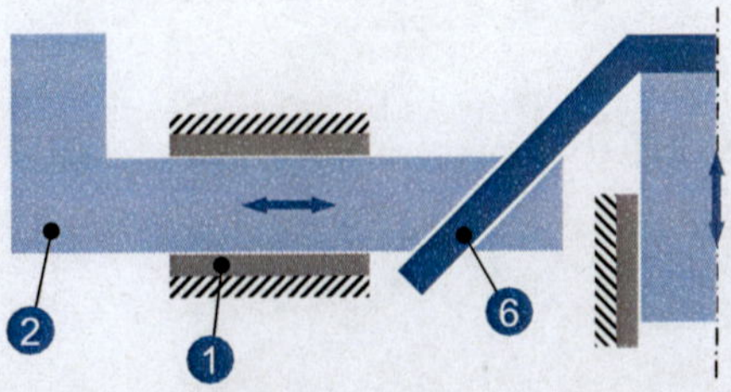

1 Serrated plain guide system 2 Base jaw 3 Sensors 4 Casing
5 Centering and attachment options 6 Wedge hook principle 7 Piston

**Fig. 13.63** Wedge hook kinematics for mechanical parallel jaw grippers, according to Schunk GmbH

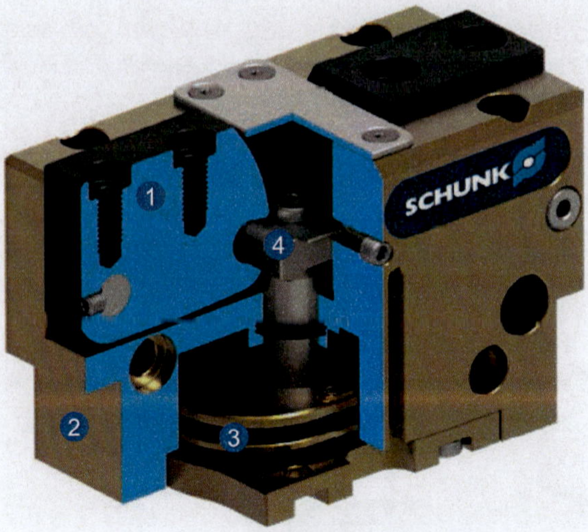


1 Base jaw 2 Casing
3 Drive 4 Lever gear

**Fig. 13.64** 2-finger angular gripper, according to Schunk GmbH

to the gripper, the gripper design phase must often be repeated. When existing gripper systems cannot be adapted to the new task with a redesign of the gripper jaws, they must be replaced completely. When the product range is diverse and the volumes low, a universal gripper system that can be adapted flexibly to the gripping situation at hand fills the need.

Based upon examples occurring in nature, different bionic gripper systems have been developed. The goals of these systems are to emulate, using technical means, the adaptive capabilities of gripping mechanisms that have evolved over time [WILS11]. These emulation attempts concentrate, at their core, on the development of *gripper hands* with different numbers of fingers and joints per finger. Their positions can be adapted independently to the surface contour of the object to grip, and thus allow increased flexibility in terms of shape changes.

The technical realization of a gripper hand shown in Fig. 13.66 has three fingers, each with two rotational axes. Also, two of the three fingers can turn symmetrically to each other on a seventh axis. This changes their orientation with respect to each other and thus the direction in which the closing force acts.

Two sensor panels are attached on the inside of the fingers to ascertain a surface profile of the forces acting between the gripper and object gripped. Information on locations and magnitudes of the gripping forces occurring can be used during operation for adaptive control of the individual joint positions. In the ideal case, specifying approximate closing motions for the gripper fingers on the gripper hand and readjusting the individual joint positions in line with the resultant surface forces are adequate.

In addition to the resultant flexibility increase, the design and sensor system of the gripper hand also satisfy the requirements of intuitive interaction between human and handling device. This benefit is of great importance mainly in the field of service robotics because interaction with untrained operators is the focus here. To be able to ensure the necessary safety, technical systems must be made capable of reactive actions. Given the high control-related overhead for integration and programming, however, and, not least, their high procurement costs, gripper hands are rarely used in the operation of automated handling equipment in production environments.

#### 13.8.3.2 Pneumatic Grippers

Grouped under pneumatic grippers are end effectors of handling devices that contact the object gripped using a pressure difference but that do not have a gear unit or kinematics for power conversion [HESS04]. Also, differentiation is made between high-pressure grippers and vacuum grippers on the basis of the retaining

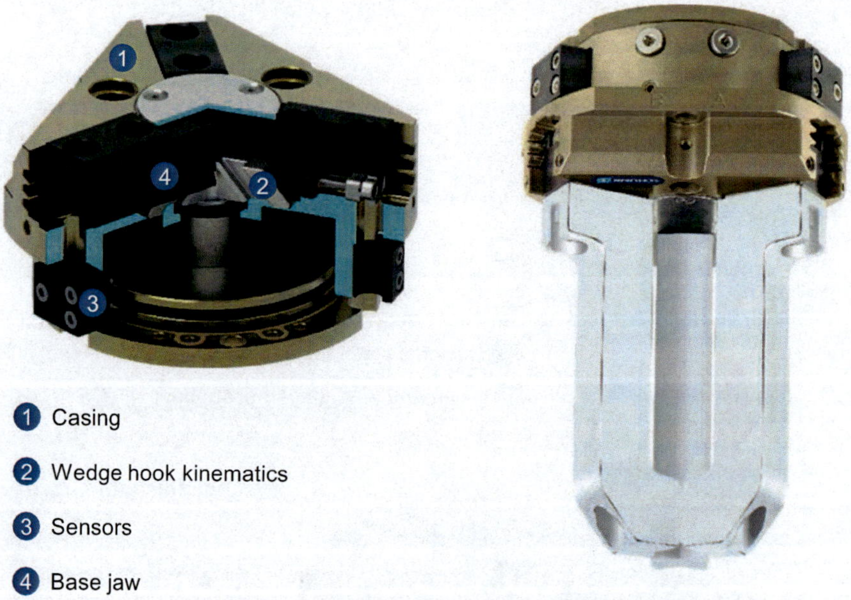


**Fig. 13.65** 3-finger self-centering gripper, according to Schunk GmbH

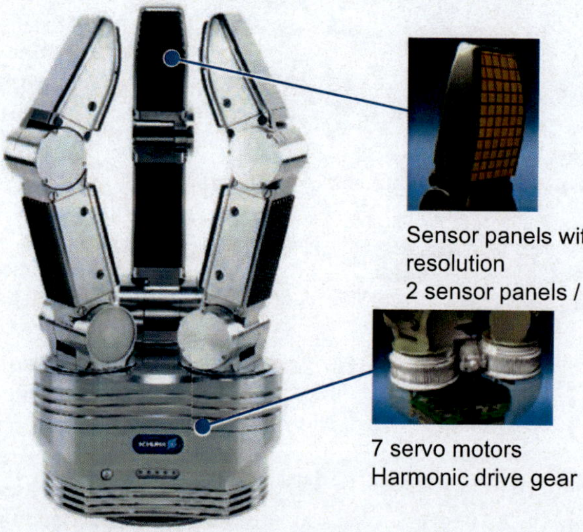


**Fig. 13.66** 3-finger gripper hand, according to Schunk GmbH

forces acting on the object gripped. If these forces are pressure forces, i.e., applied in the direction of the object gripped, these gripper systems are categorized as high-pressure grippers. Gripping forces that are directed away from the object gripped toward the gripper are generated by targeted pressure reduction. Colloquially, this principle is also called a vacuum providing the name of the group of vacuum grippers.

### High-pressure grippers

Air jet grippers represent the physically simplest design of high-pressure grippers (Fig. 13.67). They have one or more openings that allow compressed air to pass when the gripper is activated. The shape of the escaping air flow is changed by nozzles forming a defined air jet. If the jet is directed towards the underside of a workpiece to be gripped, the resulting back pressure means the workpiece can be lifted and (aided by a plain guide system) gripped. Closing the gripper valves releases the workpiece.

In addition to this open form, there are also variants of high-pressure grippers where the compressed air is routed into enclosed cavities. These change shape due to the effect of the internal pressure increase. The purposes of the shape change are to temporarily clamp an object gripped and to apply the movement forces. When the cavity pressure falls, the gripper assumes its original shape again and releases the object gripped. The deformable cavities are usually thin diaphragms that can deform unhindered in the gripping direction

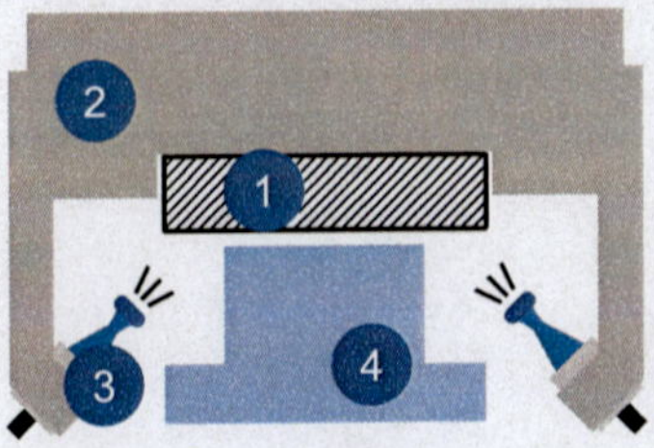


**Fig. 13.67** Schematic principle of operation of an air jet gripper, according to [HESS04]

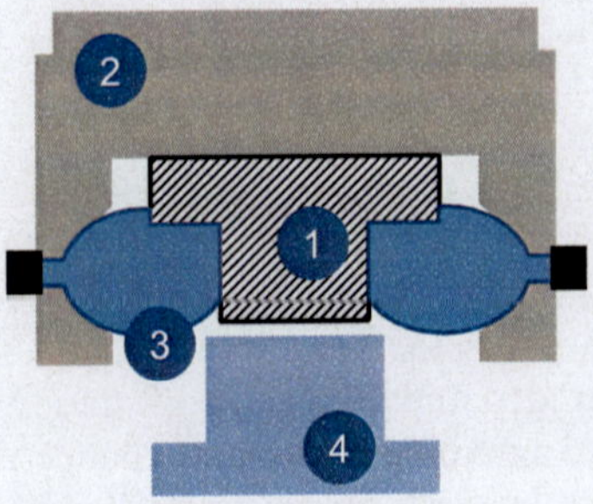


**Fig. 13.68** Principle of operation of a diaphragm gripper, according to [HESS04]

and adapt to the shape of the object gripped. These types of grippers are known as diaphragm grippers (see Fig. 13.68).

A feature of high-pressure grippers is their comparative insensitivity to shape and position changes of the object gripped because they do not perform the gripping motion using rigid kinematics. The directed air pressure flow is able to adapt to the respective compositions of the workpieces. Since no direct contact (air jet gripper), or only via the relatively soft surface of the flexible diaphragm (diaphragm gripper), is required to apply the gripping force, damage from process errors and tilting using rigid gripper jaws are prevented.

### Vacuum grippers

Pneumatic vacuum grippers use the pressure difference between ambient pressure $p_a$ and a gripper opening at a low pressure $p_u$. This then connects components to the gripper using suction (Fig. 13.69). The gripping forces required for handling are transferred over an area to the object using a flexible sealing element. The suction force that establishes the connection between the gripper and object gripped can be determined from the product of the effective surface and prevailing pressure difference.

The central challenge when using suction grippers is producing the vacuum required to generate the pressure difference from ambient pressure. Three relevant

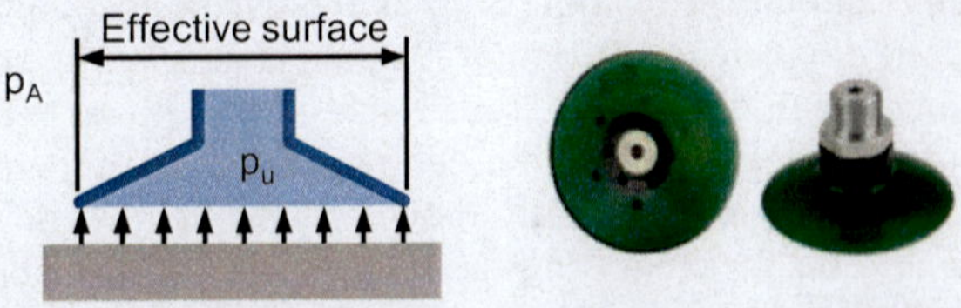


**Fig. 13.69** Principle of operation of suction grippers, according to J. Schmalz GmbH

technologies are chiefly used for this in industrial handling equipment:

- Vacuum pumps
- Ejectors (working to the venturi nozzle principle)
- Vacuum blowers

Vacuum pumps and blowers usually use electrical energy to evacuate an enclosed volume or to generate a volume flow with reduced interior pressure. Electrical energy is used to drive their pump/rotor kinematics.

In contrast, ejectors apply an overpressure air flow that generates a localized pressure decrease by reshaping or by appropriately narrowing the cross-section. This method, working on the principle of the venturi nozzle, can be defined mathematically by the continuity equation

$$\dot{V}_1 = A_1 v_1 = A_2 v_2 = \dot{V}_2, \tag{13.56}$$

stating that a steady-state volume flow is constant, i.e., the amounts of fluid flowing in and out of a control volume are equal as well as by the simplified form of the Bernoulli equation

$$p_1 + \tfrac{1}{2}\rho v_1^2 = p_2 + \tfrac{1}{2}\rho v_2^2 = \text{konst.} \tag{13.57}$$

As in Fig. 13.70 and Eq. (13.56), the flow velocity of a fluid flowing from Point 1 to Point 2 increases by a factor of $A_1/A_2$ as the cross section narrows. By inserting speeds $v_1$ and $v_2$ into Eq. (13.57), it becomes apparent that the speed-dependent part in Point 2 also increases with increasing $v_2$. Consequently, the constraint of the Bernoulli Eq. (13.57) can only be satisfied when static pressure component $p_2$ decreases by the

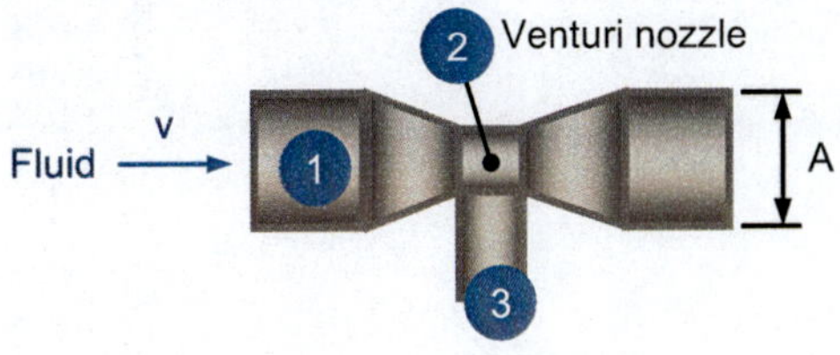


**Fig. 13.70** Design of a venturi nozzle

same amount in relation to $p_1$ as the dynamic component $\frac{1}{2}\rho v^2$ increases. If Point 2 is then connected to another opening at Point 3, the resulting lower pressure $p_2$ can be routed outside and used to generate the vacuum.

The simple and sturdy design of the venturi nozzle and the fact that compressed air is usually available in the building infrastructure make ejectors an attractive alternative to generating vacuums for suction grippers. No further equipment is required here in contrast to vacuum pumps and blowers. Their short dimensions enable them to be affixed directly to the end effector of the handling device averting losses due to leaks. Also, the volume to be evacuated is decreased by the shortened section, reducing the cycle time of the gripping process.

The different technologies for temporarily reducing the pressure for suction grippers differ in terms of performance with regard to the absolute vacuum attainable and the quantity of air they can take in within a certain time to maintain the vacuum. The last item is of particular importance for the suction process for objects permeable to air, such as textiles. In this case, ejectors and vacuum pumps can be unsuitable if they do not have the suction capacity necessary for gripping (◘ Fig. 13.71). On the other hand, vacuum blowers often fail when the available gripping area is small and its comparatively low generatable vacuum is not adequate to secure the object reliably.

13

#### 13.8.3.3 Custom Designs

The gripper types described above span the range of standard components in gripper technology and thus the vast majority of gripper systems in use. A number of different, often application-specific, gripping techniques also exist, however. The principle of operation and design of the various grippers are very dependent on the composition of the object gripped and can often only be adapted to other product groups with great overhead.

**Magnetic grippers**

Using magnetic grippers, it is possible to attach ferromagnetic components to the end effector of a handling device using magnetic forces. For an object to be gripped reliably and placed down again in a defined position, it must have a flat and smooth surface that channels the gripping forces back into the gripper (◘ Fig. 13.72). The magnetic forces are generated by electromagnets or permanent magnets. Permanent magnets have the benefit that their gripping force is not suddenly lost (when the power fails, for example). This also means, however, that permanent magnets require a device for releasing the components gripped at all times. This device acts against the holding force of the permanent magnets. When the grip connection is released, multiple force vectors then act upon the object gripped. The magnitudes and directions of these vectors cannot be defined unambiguously and change depending on the distance between object gripped and magnet. This means the set-down position of the workpiece can, depending on the equilibrium of forces, spread over a wide range and this method is not used for high-precision positioning.

Using electromagnets offers the benefit that a device for releasing the grip contact is unneeded. They do not provide any implicit gripping force retention, however, and that in turn necessitates the development of appropriate devices. Consequently, combinations of permanent magnets and electromagnets are commonly used in practical applications. In these solutions, the holding force is provided by the permanent magnets. Electromagnets aligned in opposite directions can be used

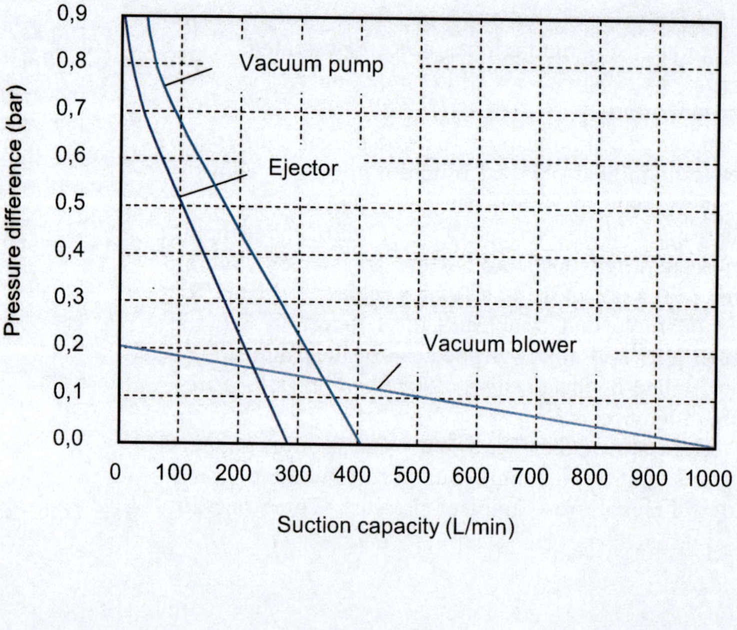


◘ **Fig. 13.71** Comparison of different vacuum generation methods, according to [HESS04]

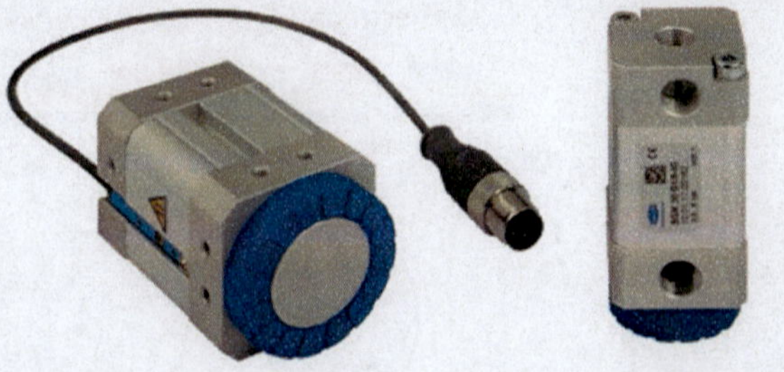

■ **Fig. 13.72** Magnetic gripper, according to J. Schmalz GmbH

to release the grip contact. These kinds of grippers have the benefit that they require no moving parts on the gripper and are thus less susceptible to faults and require less maintenance. Their heavy weight and poor positioning accuracy and repeatability mean their use is severely restricted, however. Also, it is possible for the object gripped to exhibit undesirable residual magnetism.

### Freeze grippers

Freeze grippers establish an adhesive bond between gripper and workpiece surfaces. The bond is established by the introduction of liquid water into the gap of the contact zone and cooling by the gripper until the aggregate state of the water has changed to ice (■ Fig. 13.73). This binds the object to the gripper until the water is liquefied in a specific manner or the adhesive forces of the bond are exceeded. The bonding forces that can be achieved may change depending on the wettability of the contact surfaces and the active fluid medium used.

For a freeze gripper to generate the adhesive forces of the active fluid medium, it requires a cooling unit. Peltier elements are commonly used here due to their simple design. They comprise an alternating chain of semiconductor circuits connected in parallel that become thermally active when there is current flow. One side of a Peltier element cools while the other heats. The effect can be reversed with polarity reversal. This may be used to accelerate the release of objects gripped, for example. Despite this potential process acceleration, the inertias of the semiconductor elements and the duration of the necessary aggregate state change of the active fluid medium result in relatively long process times. The fact that the object gripped is brought into contact with an additional medium also restricts the usability of this gripping method.

### Electrostatic grippers

The automated handling of flexible textile semifinished parts represents one of the greatest challenges in the large-scale, cost-effective production of components made of fiber-reinforced plastics. For the manipulation of air-permeable textile semifinished parts, electrostatic gripping mechanisms are used. These are based on displacement of the charge centers. Different electrical potentials are applied to a gripper electrode forming an electrical field. As the gripper electrode approaches, the object gripped is polarized by the charge center displacement and attracted to the gripper electrode in line with Coulomb's law. Acting here between the gripper and semifinished part are quasi-2D attraction forces. For electrically conductive materials such as carbon fibers and metals, the charge center displacement is caused by electrostatic induction. For non-conductive materials such as fiberglass, paper, and synthetic film, the charge center displacement is generated by polarization or the alignment of available dipoles (■ Fig. 13.74, right).

The electrostatic gripping principle has been used for decades in the most diverse fields (such as in

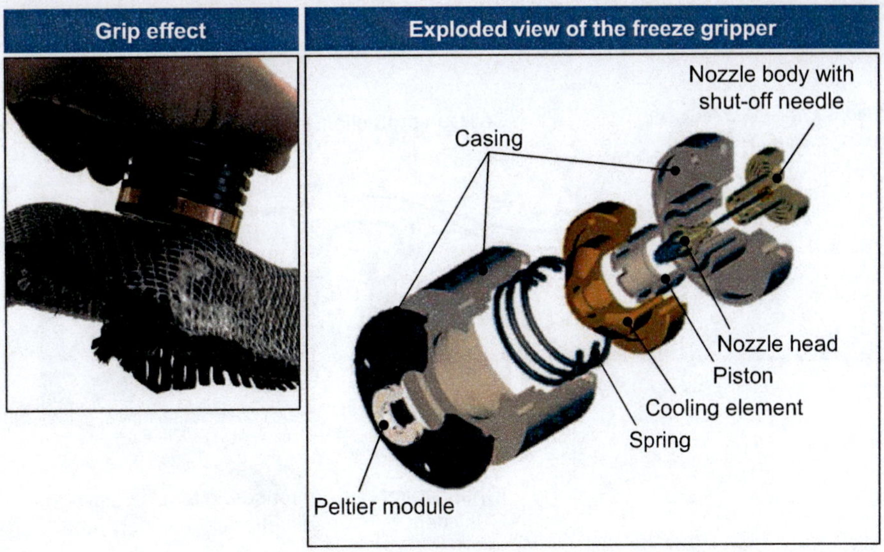


■ **Fig. 13.73** Freeze gripper, according to NAISS GmbH [JENS02]

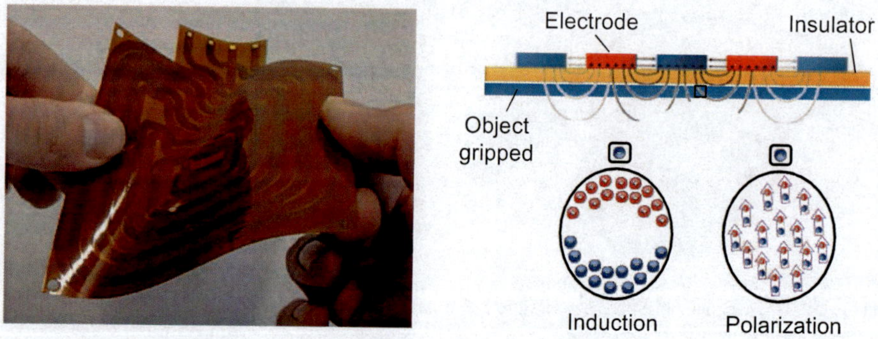


**Fig. 13.74** Electrostatic grippers

microassembly and the handling of semiconductors). For reliable automated handling of technical textiles, it is not the gripping of the object but the defined and stable releasing of the object gripped without any additional effects (compressed air, ejectors) that is deemed the greatest challenge. A damage-free release strategy is based upon the generation of a decreasing alternating field. For this, the voltage at the electrodes is repeatedly commutated and incrementally reduced.

The use of a flexible electrode represents another innovation (Fig. 13.74, left). This means smooth, wrinkle-free adaption to simply curved work surface geometries is possible.

### 13.8.4 Gripper Systems

13

The different gripping principles and concepts described form a large part of the gripper technologies in use today. Also, more types of jigs and fixtures that can aid handling and assembly processes exist at the interface between gripper and handling device. The combinations of individual fixtures and grippers attached to the handling device are also known as gripper systems. They comprise the mechanically acting guide elements and actuators as well as controller units that control the information flow of the gripping process.

The different system components can be described below using the example of a mechanical gripper system (Fig. 13.75).

#### Connection techniques and media routing

Handling devices such as industrial robot systems generally provide a supply of working media such as compressed air and electrical energy. This can be realized in the form of flange-mounted control and valve islands. From these, additional connecting lines can be routed as required to the end effector of the handling device. Also, for smaller, high-speed robot systems in particular, valves and control units are housed inside the robot to minimize wear and the risk of collisions with surrounding objects (Fig. 13.76).

#### Changing devices

Grippers for automated handling equipment are usually devices with a product-specific design. Product switches and changing product properties cannot often be handled by a single gripper. For it to be possible nevertheless to cover a wide diversity of variants with one handling device, the device must be capable of switching its gripping technique at runtime. For this, two-part changing systems are mounted to the handling device and gripper to connect and disconnect in a controlled manner (Fig. 13.77).

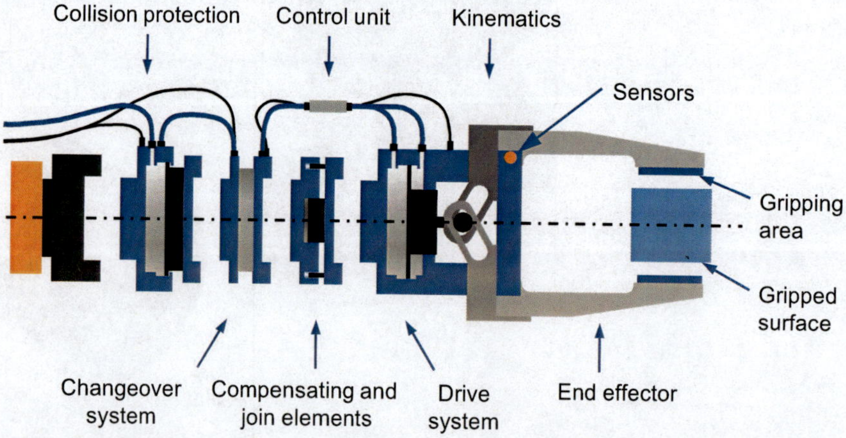


**Fig. 13.75** Basic design of a mechanical gripper system

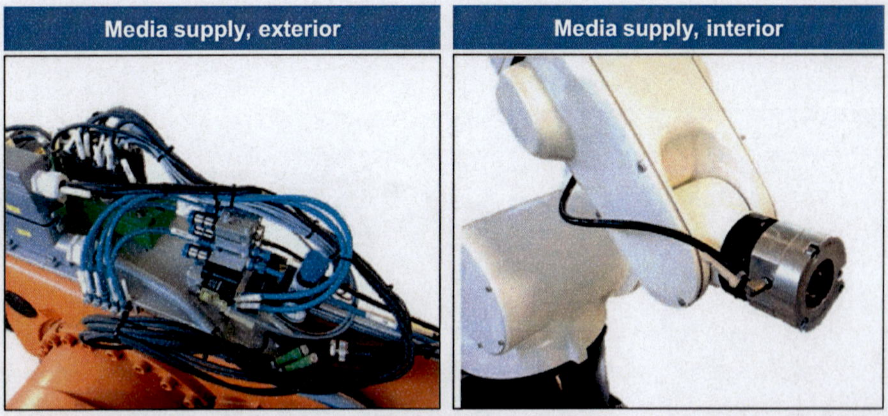


**Fig. 13.76** Comparison of exterior and interior media supply

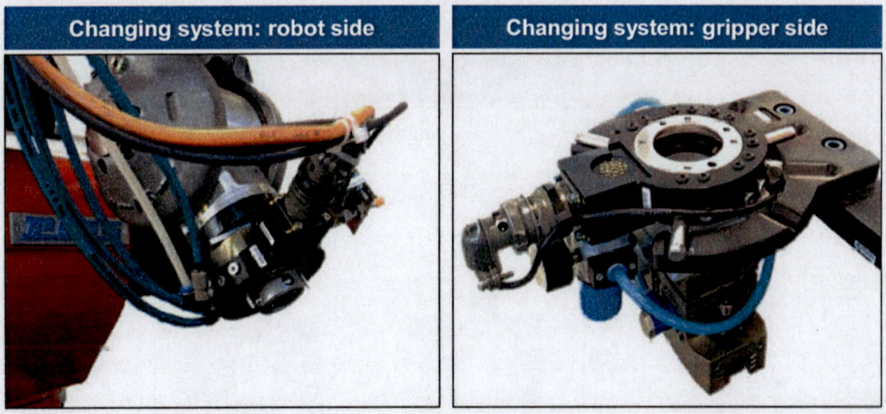


**Fig. 13.77** Gripper change fixture with signal and media feedthroughs

In addition to the mechanical connection, changing devices also provide an interface for different working media and signals. The individual feedthroughs can be assigned as required in the startup phase of the gripper system. When the individual modules are reused, however, each feedthrough must match the controller code. Malfunctions often occur after grippers are switched on different handling devices using compatible changing systems.

### Join mechanisms

Join mechanisms represent another module for assembly-compatible handling equipment. They assist the process in that positioning errors by both join partners can be compensated. Join mechanisms have degrees of freedom, along which the compensation motions can be performed without force. Using pneumatic actuators, the parts that can be moved/rotated relative to each other can be secured to perform an interlock of the join mechanism (Fig. 13.78).

Interlocks are helpful because they prevent the fixture from vibrating thus shortening the process times. Interlocks with a centering action additionally stabilize the joining process by transitioning the gripper and machine sides to a defined relative alignment. This means remaining misorientations of the elastic components, and so fluctuating initial situations of the process, are averted. Another benefit of interlockable join mechanisms is that their elastic deformability can be disabled specifically. Then the handling equipment behaves like a system comprising individual rigid components for limited loads. This increases the repeatability and broadens the possible application areas of the system without any additional setup being required.

### Collision protection and safety

The most serious consequences of erroneous handling processes include collisions between the gripper and surrounding objects. A collision occurring often entails damage to the gripper and the affected surface of the collision object. Damage to the robot kinematics can even result in serious cases. To avert potential collisions, robot programs are tested in simulation or at low traversing speeds on the actual system before being used. Changes to the operating space or frequent repositioning can cause unpredictable relative motions, however. For these kinds of unsafe handling tasks, protective systems are often used that lessen the damaging effects of collisions.

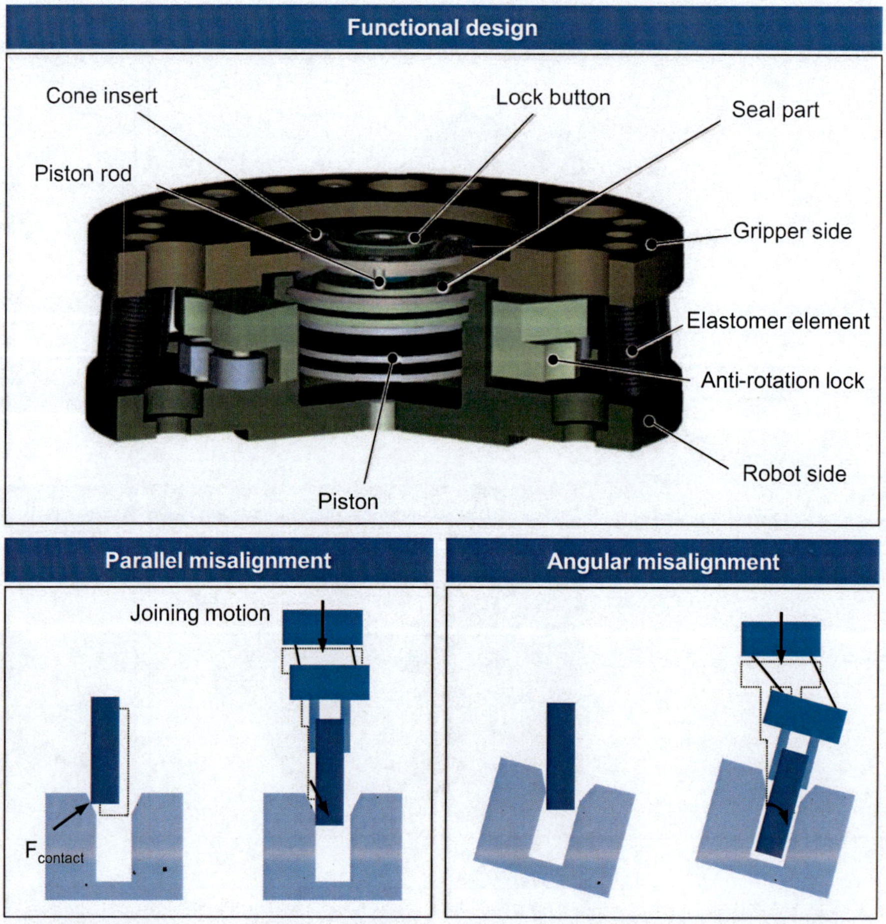


**Fig. 13.78** Functional principle of the join mechanism, according to IPR GmbH

13

Collision protection, as in Fig. 13.79, is part of the kinematic chain of the handling system. When compressed air presses the contact plate against the housing of the device, the collision protection system can divert the working forces of the gripper into the handling device. The drive is able to prevent the contact plate from lifting during normal operation. When a collision occurs however, the forces absorbed exceed the range of working forces, such that the contact plate is removed from the housing part on the robot side triggering a sensor. The signal curve of the sensor is then used to shut down the handling device and prevent damage.

### Gripper controllers and sensors

The drives of a gripper are controlled using different actuation methods depending on the energy source. For this, pneumatic gripper drives require valves in the form of valve islands that are often located near where they are used to minimize losses in air pressure. The individual valves are actuated by the outputs of the internal robot controller or an external PLC.

Electrically driven grippers often have their own grip controllers and a number of alternative hardware and software interfaces for sending signals. The interfaces are used to set the different controller parameters as well as for integration in the field for actuating the gripper drive units.

Gripping processes are increasingly being monitored by sensors with the measured values relayed back to the control process. This permits handling operations with a greater tolerance for incorrect positioning. Here, measurement data are being continuously fed back creating a feedback control structure. The sensor system often comprises optical camera systems because they can cover an expanded working area. Also, they can be used upstream of the process for object identification and initial position determination. "Bin picking" and high-speed Pick & Place operations on conveyor systems with no defined component orientation are examples of optically controlled gripping processes.

## 13.9 Development Trends

The performance of modern-day robot controllers, in regard to the number of axes to control, speed, and interpolation cycle (<10 ms) as well as the functional scope of programming languages, has reached a level

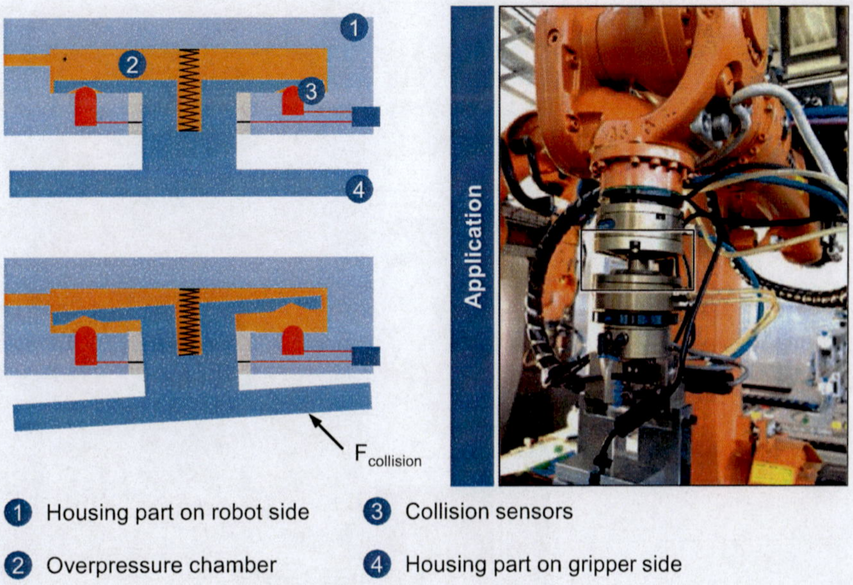


**Fig. 13.79** Functional principle of collision protection

that is more than adequate for most industrial applications.

New challenges are arising in the integration of robot systems into higher-level as well as lower-level controller structures. More than ever, the requirements placed on robot controllers are to communicate with a process control system or to coordinate peripherals such as sensors and machines.

Arising from this is the general requirement of robot controller openness. This term means extendibility of the hardware using standardized plug-in cards, such as for bus and sensor systems. It also means the capability of the user to modify the functional scope of the controller, to integrate sensor data, for example, using defined software interfaces.

Today's PC technology already offers high levels of openness and networking capability in addition to the computing capacities required. It also offers great potential for implementing these concepts. The use of standard software, such as for the operating system or to develop the user interface, forms the basis for user-defined adaptations to the software and is widespread in industrial environments [BISC09].

The trend towards open, flexible controller structures is disclosing new development perspectives, particularly in regard to the user interfaces for robots. These have so far been kept very simple and do not assist the user in programming or operating to the degree that has become possible. Extensive developments to create efficient, user-friendly and application-specific programming interfaces and tools that also make it possible for robots to be used for low volume tasks are needed here [BISC09]. The development of systems for smaller lot sizes generally represents one of the main tasks for the future, not least due to the saturation of industrial robots used in classic large-scale applications [ARMB06].

In terms of automating the production of small lot sizes, the fast startup of robot systems is a key factor. If the costs for startup here are greater than for manual production, a robot system cannot be used profitably. This is why systems that guarantee low-overhead and intuitive robot programming are current research topics.

The "programming by example" approach offers a promising starting point, particularly in the area of fast and intuitive startup of robot systems. Such systems are based on the idea that the operator demonstrates the task to perform to the robot and the program code is generated automatically from that.

Figure 13.80 shows a robot system that is programmed through demonstration [GÖBE12]. For this, the robot is fitted with a measuring system that is able to measure the position of a pointing device in six degrees of freedom.

When the operator moves the pointing device, the robot follows the motion automatically and continuously such that the pointing device is perfectly in the field of view of the camera sensors at all times. The pointing device can be used for programming the robot by transforming the position of the pointing device to the base coordinate system of the robot. In this process, different interpolation points that are important for the robot motion to program are displayed and stored. The robot program is then compiled from these interpolation points. Also, the component to produce can itself be given a marker. If the production flow is now demonstrated with the component marked, the motion path shown can be translated directly to a robot program.

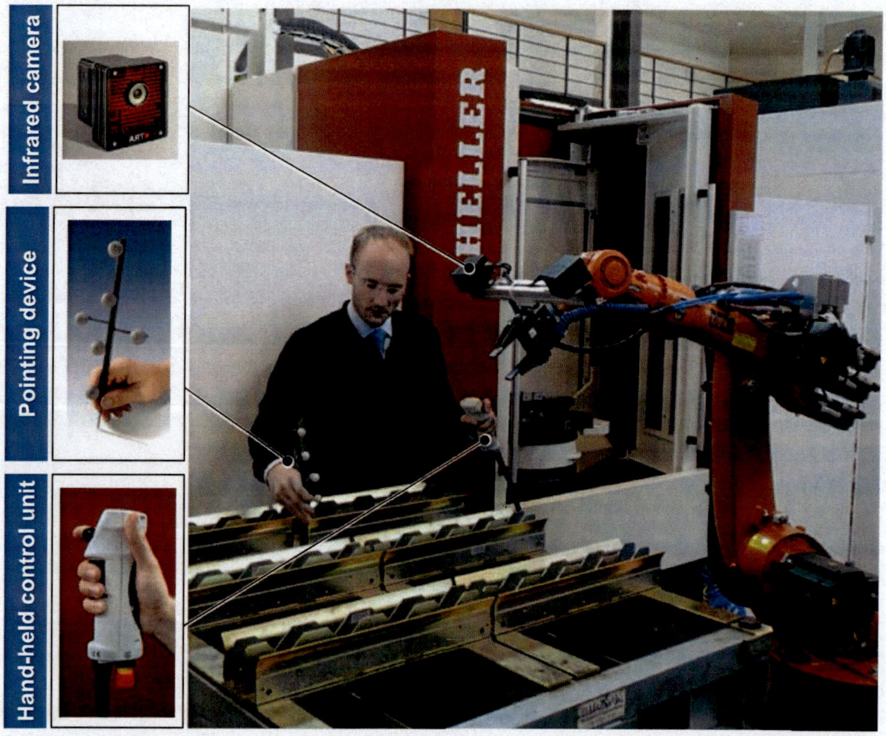


**Fig. 13.80** Robot programming through demonstration

13

Other approaches are provided by using a force-torque sensor system at the flange of the industrial robot [MEYE07]. For these systems, the industrial robot is controlled directly by the manual application of forces. The task is usually demonstrated here when the component is gripped. The weight of the component is carried by the industrial robot. The benefit here is that these systems can also be used for heavy, unwieldy components.

When a robot system is started by inexperienced operators, the challenge nowadays is safeguarding, in terms of collisions, the operation of the industrial robot during programming. This safeguarding is with the help of methods of virtual startup (see ▶ Sect. 13.5.2). In this, robot motions are first verified on model level before actual execution. The prerequisite for this approach, however, is the full availability of geometry models for the equipment being used. For small and medium-sized companies in particular however, the availability of these object models is limited. Two approaches are being adopted to meet this challenge:

- Replacement of the missing simulation models with Bounding Boxes [BREC09]
- 3D scanning of the actual production cell [BREC10, SUPP08]

The basic idea of sensor-based calibration of the actual production cell is common to both approaches. The two approaches are discussed in detail below.

#### Replacement of the missing simulation models with Bounding Boxes

The basic idea of this approach is to use the actual positions of components of the production system to position the associated models in the simulation quickly and intuitively. This is achieved by calibrating prominent corner points of components in the actual production system. For this purpose, a sensor system that is able to determine the position of a pointing device in the system environment is used. Using this measuring system, the user generates point pairs that contain the point coordinates of a prominent machine corner point in the coordinate system of the measuring system. It also provides the point coordinates of the same corner point in the coordinate system of the associated model. ◘ Figure 13.81 shows an overview of the method. System components for which no simulation models are available (such as boundary fencing) can be replaced with “Bounding Boxes”. Bounding Boxes denote the geometric approximation of objects (such as by cuboids and cylinders).

To define the positioning of an object, the user must identify at least three prominent points in the actual space and generate the associated point pairs. After three point pairs have been registered, their positions relative to the coordinate system of the model as well as in the actual space are known. On this basis, a transformation that maps every point pair to itself with a minimum deviation is determined. To assess

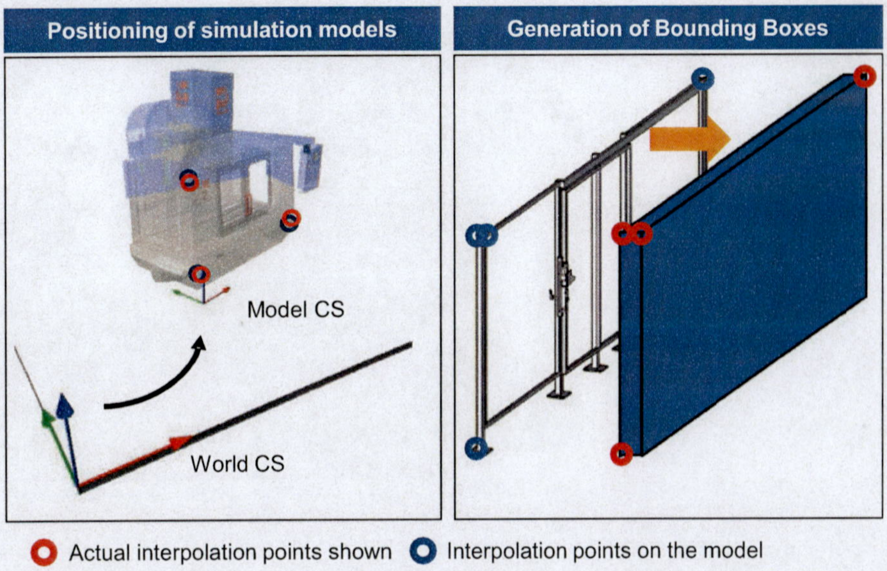


**Fig. 13.81** Modeling production systems using prominent points

the quality of the positioning, the Mean Squared Error (MSE) of the point distances is used as the basis for every point pair. This optimization problem can be defined formally as follows:

$$\min\left(\frac{\sum_{i=0}^{n}|T \cdot P_M^i - P_W^i|^2}{n}\right) \tag{13.58}$$

There is the transformation required to map the point coordinates in model $P_M$ to the point coordinates from actual system $P_W$. The term n is the overall number of point pairs calibrated. The positioning accuracy increases with each additional point pair. Inaccuracies mean a point cannot be mapped exactly to its equivalent. In addition to the limited accuracy of the measuring system and inaccuracies in calibrating the actual points by the user, the inadequate level of matching the model to the actual machine, in particular, is one of the limitations of this approach.

### 3D scanning of the actual production cell

In this approach, the actual geometry of the production cell is transferred to the simulation. For this, a depth sensor (here: Microsoft Kinect) is used. This sensor is able to return a 3D image of the scene in question. The sensor is moved manually through the production system because the sensor can only keep part of the production system in the field of vision. This is also to minimize perspective maskings. The model of the production system is reconstructed on an additional evaluation computer in parallel with the motion. The motion is generally manual by the user or automated by the industrial robot. However, the motion automated by the industrial robot requires higher overhead in terms of fixing the sensors to the flange as well as additional interfaces between industrial robot and evaluation computer.

After the 3D scan has finished, the system model is available as a static geometry model. To be able to use this geometry model for determining collisions, machine models with known kinematics (such as industrial robots) must be incorporated into the static system model. The positions of the machines are already available from the static, 3D scan model. This means that the machine models can be fitted directly into the relevant position manually or automatically. The associated sensor measurement data can then be deleted from the model. This yields a hybrid geometry model of the production system. This model combines the available kinematic machine models with depth measurement data from the actual system. ◘ Figure 13.82 shows the hybrid simulation model of a robot-based production cell.

In addition to the startup of individual robot working cells, the programming of complex automation tasks with multiple, cooperating robots also plays a key role. Cooperating as used here means that multiple industrial robots work on one task together and at the same time. Typical tasks here are jointly moving large and heavy components as well as jointly assembling and testing components. Furthermore, cooperating robots can be used to increase the productivity of automated production with the addition of a robot to perform other work or checking steps on the product during non-productive times. This means non-productive times can be turned into productive times.

There are strict requirements here on the positioning and motion accuracy, especially in the synchronized motion of components and the resulting kinematic coupling of multiple robots. The accuracy of the measuring system integrated in the robot that is attainable

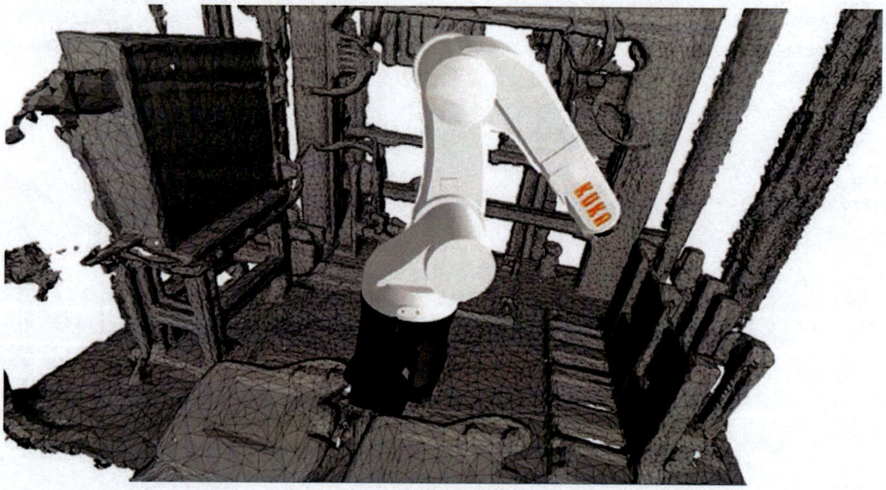


**Fig. 13.82** Hybrid simulation model of an industrial robot-based production system

cost-effectively is generally not adequate here. Strategies are being developed based on additional, external measuring systems that can readjust the robot. These measuring systems include position measuring systems and force-torque systems.

In the position measurement, the Cartesian positions of all industrial robots involved in the cooperation are recorded by an external measuring system in one coordinate system. If the robot motion deviates from the path intended as a result of disturbance forces (weight force, process forces, or calibration errors), readjustment is possible on the basis of the measured values so that virtually ideal motion sequences can be attained.

13

In contrast, force-torque controlled systems work on the equilibrium of forces principle [ZAID13]. Here, an industrial robot performs the required motion as the "Master." All other industrial robots involved work as "Slaves" and interpret the forces and torques occurring on the flange to correct their motions such that the forces and torques occurring are offset.

The programming of such cooperating systems is virtually unmanageable manually. Nowadays, a virtual startup performs programming. Here, the robots are first programmed in a virtual cell model for the motion programs created virtually to be transferred to the actual controllers. The virtual motion paths here are not created in the classic sense by specifying motion paths for the individual robots. Instead, they are created by specifying the grip position on the component for every robot involved and by specifying the motion path of the component itself. These two specifications allow the calculation back to the motion path of every robot.

The general development trend is in minimizing the overhead for integrating a robot into its working environment. The approaches of controller openness and modularity of system components are making a major contribution towards meeting these goals with the advancing development of microprocessor and communication technologies.

## References


[ARMB06] Armbruster, H., Kirner, E., Kinkel, S.: Neue Kundengruppen für Industrieroboter. Fraunhofer ISI, Karlsruhe (2006)

[BISC09] Bischoff, R., Guhl, T.: Robotic visions to 2020 and beyond. Vortrag, Brüssel (2009)

[BISC10] Bischoff, R., Kurth, J., Schreiber, G., Koeppe, R., Albu-Schaeffer, A., Beyer, A., Eiberger, O., Haddadin, S., Stemmer, A., Grunwald, G., Hirzinger, G.: The KUKA-DLR lightweight robot arm – a new reference platform for robotics research and manufacturing. In: Robotics (ISR), 2010 41st International Symposium on and 2010 6th German Conference on Robotics (ROBOTIK), S. 1–8

[BLUM81] Blume, C., Dillmann, R.: Frei programmierbare Manipulatoren. Vogel, Würzburg (1981)

[BREC09] Brecher, C., Göbel, M.: Modellbasierte Programmierung "by Demonstration." Automatisierungstechnische Praxis **7**, 62–68 (2009)

[BREC10] Brecher, C., Roßmann, J., Schlette, C., Herfs, W., Ruf, H., Göbel, M.: Intuitive Roboterprogrammierung in der automatisierten Montage. wt – Werkstattstechnik online **100**(9), 684–686 (2010)

[BRON84] Bronstein, I.N., Semendjajew, K.A.: Taschenbuch der Mathematik. Verlag Harri Deutsch, Frankfurt a. M. (1984)

[DENA95] Denavit, J., Hartenberg, R.S.: A kinematic notation for lower-pair mechanisms based on matrices. J. Appl. Mech. **22**(2), 215–221 (1995)

[DIN93] DIN Deutsches Institut für Normung e. V.; Programmiersprache – Industrial Robot Language (IRL) September (1996)

[DREW97] Drews, P., Schmid, D., Volkholz, V.: Roboter in der Werkstatt. Maschinenbau Verlag, Frankfurt a. M. (1997)

[FERR05] Ferré, E., Laumond, J.-P., Arechavaleta, G., Estevès, C.: Progresses in assembly path planning. Int. Conf. Prod. Lifecycle Manage., 373–382. ► https://www.inderscience.com/inorders/bkpage.php?rec_id=33&-journalID=1001&chapNum=7. (ISBN 0-907776-18-3 (Print), ISBN 0-907776-19-1 (Online))

[GECK11] Gecks, T.: Sensorbasierte, echtzeitfähige Online-Bahnplanung für die Mensch-Roboter-Koexistenz. DissertationUniversität Bayreuth (2011)

[GÖBE12] Göbel, M.: Verfahren zur intuitiven Programmierung von Industrierobotern durch Demonstration. Werkzeugmaschinen, Bd. 12, 1. Aufl. Apprimus-Verl, Aachen (2012)

[GRUN09] Gruninger, R.; Kus, E.; Huppi, R.: Market study on adaptive robots for flexible manufacturing systems. 2009 IEEE International Conference on Mechatronics, S. 1–7 (2009)

[GUIZ08] Guizzo, E.: Article robotics industrial robots the rise of the machines. ▶ http://spectrum.ieee.org/robotics/industrial-robots/the-rise-of-the-machines. Accessed 14. Sept. 2017

[HART56] Hartenberg, R.S.: Die Darstellung und Handhabung der niederen Elementenpaare in einer auf Matrizenrechnung gegründeten Zeichensprache. VDI **12**, 145–155 (1956)

[HESS04] Hesse, S.: Robotergreifer. Funktion, Gestaltung und Anwendung industrieller Greiftechnik. Hanser, München (2004)

[JACO94] Jacobi, A. N.: Realistische Simulation der Bewegungen-IPK Berlin (1994)

[JENS02] Jensen, L., Stephan, J.: Gefriergreifer: Neue Anwendung der Peltier-Technik. KI. Luft-und Kältetechnik **38**(12), 572–576 (2002)

[KUHN09] Kuhn, S., Henrich, D.: Multi-View Reconstruction of Unknown Objects in the Presence of Known Occlusions. DissUniversität Bayreuth (2009)

[KUKA15] Kuka Roboter GmbH: KUKA.PLC mxA. ▶ http://www.kuka-robotics.com/res/sps/a737ee03-5832-4c95-9d91-84e0de80c664_KUKA_mxAUTOMATION_DEUTSCH.pdf. Accessed 26. Jan. 2015

[MCKE98] Mckerrow, P. J.: Introduction to robotics. Electronic systems engineering series. Repr. Aufl. Addison-Wesley Pub. Co, Sydney (1998)

[MEYE07] Meyer, C., Hollmann, R., Parlitz, C., Hägele, M.: Programmieren durch Vormachen für Assistenzsysteme – Schweiß- und Klebebahnen intuitiv programmieren (Programming by Demonstration for Assistive Systems – Intuitive Programming of Welding and Gluing Trajectories). it - Information Technology **49**(4), 238–246 (2007)

[OSTE92] Osterwinter, M.: Steuerungsorientierte Robotersimulation. Vieweg, Wiesbaden (1992)

[PAUL81] Paul, R.P.: Robot manipulators, mathematics, programming and control. MIT Press, Cambridge (1981)

[PRIT92] Pritschow, G., Frager, O.: Roboterzellen Programmierung: Die Sprache IRL und der Zwischencode ICR. Robotersysteme **8**, 25–32 (1992)

[RIES92] Rieseler, H.: Roboterkinematik – Grundlagen, Invertierung und symbolische Berechnung. Vieweg, Wiesbaden (1992)

[SAYL11] Sayler, S.: Universelle Manipulationsstrategien für die industrielle Montage. KIT Scientific Publ, Karlsruhe (2011)

[SCHW07] Schwarz, M.H., Boercsoek, J.: A survey on OLE for Process Control (OPC). In: Proceedings of the 7th Conference on 7th WSEAS International Conference on Applied Computer Science, Vol. 7. Reihe: ACS'07, S. 186–191

[SNYD90] Snyder, W.E.: Computergesteuerte Industrieroboter. Grundlagen und Einsatz. VCH-Verlagsges, Weinheim (1990)

[SUPP08] Suppa, M.: Autonomous robot work cell exploration using multisensory eye-in-hand systems. Dissertation-Universität Hannover (2008)

[VDI90] Richtlinie VDI 2860: Montage- und Handhabungstechnik; Handhabungsfunktionen, Handhabungseinrichtungen; Begriffe, Definitionen, Symbole (1990)

[VDI95] Richtlinie VDI 2740: Mechanische Einrichtungen in der Automatisierungstechnik - Greifer für Handhabungsgeräte und Industrieroboter (1995)

[WARN89] Warnecke, H.-J., Schraft, R.D.: Handbuch Handhabungs- Montage- und Industrierobotertechnik. Verlag Moderne Industrie, Landsberg am Lech (1989)

[WARN97] Warnecke, H.-J., Schraft, R.D.: Handbuch Handhabungs- Montage- und Industrierobotertechnik-Erweiterungsblätter. Verlag Moderne Industrie, Landsberg am Lech (1997)

[WECK97] Weck, M., Dammertz, R.: Graphical robot programming and generation of application specific programming interfaces. Ann. WGP **1**(4), 87f. (1997)

[WIES01] Wiest, U.: Kinematische Kalibrierung von Industrierobotern. Berichte aus der Automatisierungstechnik. Shaker, Aachen (2001)

[WILS11] Wilson, M.: Festo drives automation forwards. Assem. Autom. **31**(1), 12–16 (2011)

[ZAID13] Zaidan, S.: A work-piece based approach for programming cooperating industrial robots. Forschungsberichte IWB, Bd. 259. Utz, München (2013)

# Production Control Technology

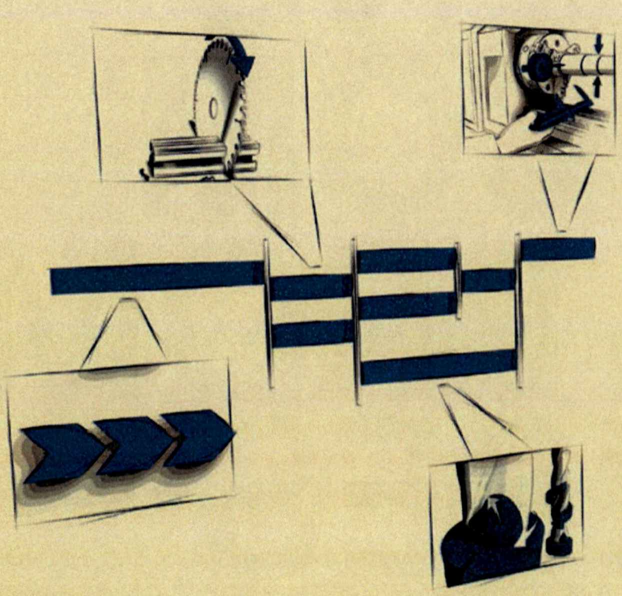

I thank my staff members Christian Fimmers M. Sc., Simon Pieske M. Sc. and Simon Storms M. Sc. for their contribution in revising this chapter.


C. Brecher and M. Weck, *Machine Tools Production Systems 3*,
Lecture Notes in Production Engineering, https://doi.org/10.1007/978-3-658-34622-5_14

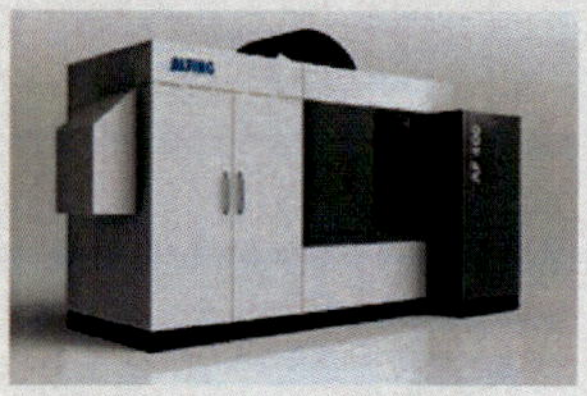

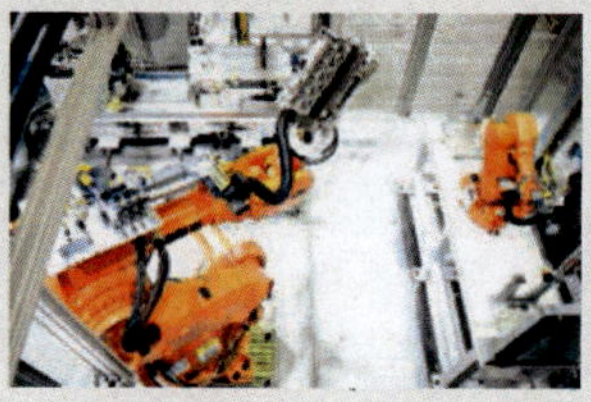

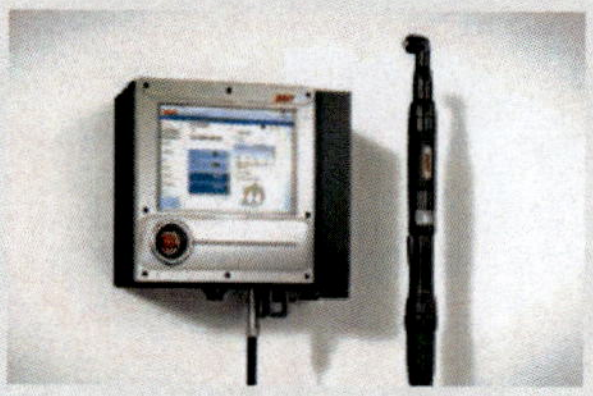

Contents

There has been a fundamental change in the international competitiveness in many areas of production technology: Increasingly short innovation phases, smaller batch sizes, and constantly decreasing product life cycles form the backdrop for the development of the "factory of the future." New types of operational structures, new production concepts, and the use of more powerful EDP systems across the business increase the flexibility of the company with regard to rapidly changing market demands.

Historically, industrial production has been characterized by many innovations. However, these innovations are not restricted to just products but, to a significant degree, include production factors. The "microelectronic revolution" has particularly affected production facilities in that a fourth factor has been added to the three "classic" production factors of work, resources and materials, that of "information." The recent "fourth industrial revolution" has turned the spotlight on the further networking of Cyber-Physical Systems (CPS) with the internet.

Following the success of streamlining in the technical area through mechanization and automation, today's efforts are now focused on creating a clear structure in the fields of planning and organization and supporting them with efficient EDP solutions. The spotlight is on continuous mining, processing and use of information. These efforts are summarized under the umbrella term of Computer Integrated Manufacturing (CIM). This is understood as an information-based integration of all EDP systems in a business directly or indirectly involved with the production process [DANG90]. The goals of this process of integration are increased productivity, shorter development and production times, and improved corporate flexibility with regard to market demands.



When CIM surfaced in the 1980s, development work was initially focused on providing the necessary information technology and connecting existing EDP systems. However, as CIM methodologies began to be implemented, significant problems quickly arose. The reason for this was the existing systems at this point covered practically every area of production planning and control (production planning, design, work preparation, manufacture) but they largely had one thing in common—they were all closed, insular solutions. Exchanging data between these systems was, and is, in principle, possible but is associated with very high costs due to the lack of standardized interfaces. In some areas, the work of international standardization committees has progressed sufficiently that standardized interfaces are now available. In addition to this, de facto standards had been established in many areas and these are currently supported by nearly all software providers.

Although significant progress has been made, from a technical perspective, with the implementation of CIM in many businesses, the goals which were initially set have not been achieved. The reason for this is that the implementation of CIM thinking is generally centered on technology. In addition to this, organizational measures are required in terms of interdisciplinary, functional integration. These measures will allow the currently heavily specialized structures to be broken up. People are at the center of this human-oriented CIM approach. The motivation and willingness to work of these people can be increased through organizational measures and computer support. Successful CIM implementation therefore requires people, technology and organization to be considered simultaneously [WARN93].

It thus becomes clear that CIM is not a commercial product but should be regarded as a medium to long-term strategy that is company-specific. This strategy is oriented toward operational (primarily organizational) circumstances and technical options. A CIM concept based on a business model is required to achieve this. The efficiency of an implementation is therefore initially dependent on the second step of just what is technically feasible. A clear and tight structure for the tasks and responsibilities of the individual departments must be created in the form of an organization and information model using the current state of in-house collaboration. Amid the of technical possibilities, a company-specific CIM concept can then be developed which can be implemented over the long term involving the employees [WECK97].

## 14.1 Corporate Structure in the CIM Environment

If the administrative areas of a business are excluded and the manufacturing functions and their upstream areas are considered, it is possible to derive the generalized CIM structure shown in ◘ Fig. 14.1. The tasks and functions of the individual CIM modules will be explained briefly in the following sections [HACK86].

### 14.1.1 CIM Components

The aim of all the CIM efforts is the exchange of information between the individual areas in a company and the consistent provision of data at the right point in time in the various EDP systems. Appropriate communication mechanisms must be prepared and agreements made regarding the data to be exchanged.

Only in extremely rare cases, a company can source a CIM solution from a single vendor as either EDP systems are already in place or programs from various providers are used for a range of tasks. Connecting different EDP systems plays an important role in CIM implementation.

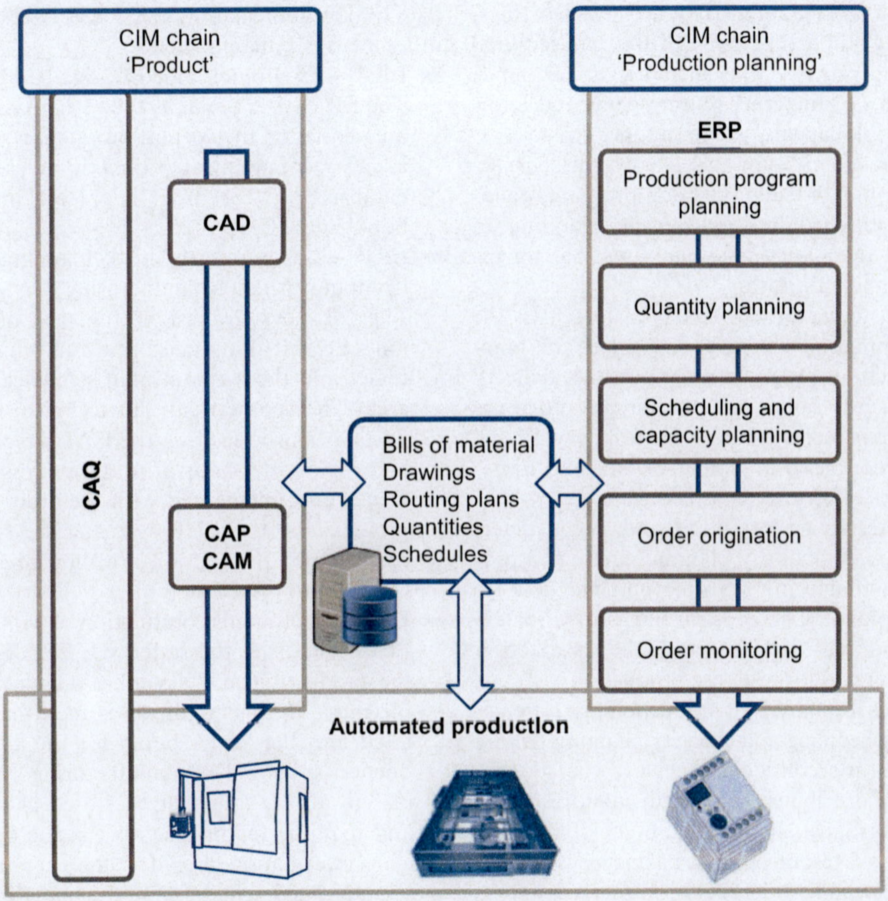


**Fig. 14.1** CIM model

Previously, the easiest option for connecting two application systems in terms of data was to connect them using a file in a specific format. Most programs can produce or read an extremely wide range of file formats. This means that extremely disparate applications can be connected to one another. A disadvantage of this method is that the data from a change in one application are not immediately imported into the other application. This leads to short-term inconsistencies. Another solution is special conversion programs. These perform bidirectional conversion of data structures without having to use a file.

Integrating application systems via a common database is generally only possible if the same data model is used across the company and either all the data sets for the applications, or at least the data required for other applications, are stored in the database. The benefits of this type of solution are guaranteed consistency of data due to the Database Management System (DBMS), the currency and the simultaneous availability of the data for all connected systems.

However, data exchange is no longer limited to purely in-house communication. Data exchange between the different operational units in a business or between several companies, e.g., connection to suppliers, is becoming increasingly important. Where this data exchange was previously performed manually by, for example, the supplier manually entering orders into his EDP system, over the last few years, it has been recognized that standard automated processes are necessary to achieve efficient data exchange, i.e., with reduced costs, time savings and elimination of errors. Initially, this data exchange used solutions based on Electronic Data Interchange (EDI), supplemented by standards such as EDIFACT (Electronic Data Interchange For Administration, Commerce, and Transport). With the introduction of XML (eXtensible Markup Language), another standard allowing data exchange solutions between companies became available.

#### 14.1.1.1 ERP

Enterprise Resource Planning (ERP) is responsible for the planning, implementation and monitoring of production processes to meet the requirements of quantities, schedules and capacities [EVER90a]. The basis of any ERP system is the administration of a set of master data for filing and maintenance of the data required for the planning process. This includes, for example, part masters, bills of material, routing plans and information on available capacity groups and workplaces.

The planning functions in an ERP system basically divide into planning of the production program, quantity planning, and scheduling and capacity planning. In addition to this, there are operational functions for order origination and monitoring:

- Based on customer orders and sales forecasts, the production program planning determines the long-term production program and possible delivery schedules. At the same time, the design and route planning departments have to be tasked with the necessary order-based preliminary work and monitored in terms of on-time completion.
- Bills of material are produced and needs determined as part of quantity planning aiming to produce the manufacturing schedule on time. The result of this planning stage is placing orders for externally produced material and the production program generated for in-house production.
- The lead times for individual production tasks are determined in scheduling and capacity planning, taking routing plans into consideration. Next, the production tasks are distributed to the individual resources (machines, personnel, etc.). In the event of over- or underloading of resources, load balancing is carried out which can lead to orders being outsourced.
- Once the release of orders, order origination triggers order processing and goods are ordered from suppliers. The production tasks are approved after completion of an availability check for materials and capacities, i.e., order documents are produced and distributed. At the same time, the provision of materials and production tools is initiated.
- Order monitoring captures the information reported by production on quantities, schedules and quality of parts produced.

As ERP systems generally plan for the medium-term, the planning guidelines frequently deviate from what actually happens in production. As a result, the control functions in ERP systems are increasingly being shifted to decentralized systems near the production facility, so-called control stations. In addition to the order control functions, these systems also include strategies for optimizing work stage sequences that allow short-term detailed planning of work sequences as well as re-planning after disruptions. This makes planning guidelines for production control significantly more accurate and improves the ability to react to production anomalies.

The key information the ERP system supplies to other areas is explained below

- ERP → Computer Aided Design (CAD)
  The ERP system assigns the design task to the CAD area based on the current production program. This is either communicated directly or uses a common database, i.e., the order is entered into the 'order book' in the CAD area.
- ERP → Computer Aided Planning (CAP) and Computer Aided Manufacturing (CAM)
  The ERP system assigns the task of work planning, i.e., breaking tasks down into individual processes and their operational sequence, to the CAP area. The geometrical and technical data and the parts lists produced by the CAD area are required here. The order-neutral routing plans initially created are supplemented with the order-specific data provided by the ERP system. The NC programs required for production are then generated in the CAM area.
- ERP → Automated production
  Before releasing the order, the ERP system checks the production and assembly documents for completeness. If this condition is met, the order is entered into the 'order book' for the automatic production system. The manufacturing system has access to all the information stored in the databases and required for production via the order number. This information includes, for example, the type, quantity, and schedule for the parts, routing plans, production resources and tool data, and NC programs to be produced.

#### 14.1.1.2 CAD

If the products to be manufactured are not standard products, the first step is to assign the necessary development work to the design department. CAD systems are currently increasingly being used in this area to support the designer [AUTO93].

CAD (Computer Aided Design) is understood as computer-aided design and development. The results of the design—i.e., in addition to the geometry data, the technical data (material, surface qualities, etc.), standard parts, assemblies, and parts lists—are compiled into a common database. The result is a digital representation of the object that is saved in a database. Other in-house departments also have access to this database. The enormous potential for streamlining provided by CAD systems is a result of the possibilities offered to the designer in the design variants and modifications. By having access to libraries of standard and repeat parts and by using predefined macros, it is possible to avoid time-consuming routine tasks.

Once the CAD work is complete, the ERP system initiates requests to the warehouse or purchasing based on parts lists and descriptions of standard and purchased parts.

#### 14.1.1.3 CAP and CAM

CAP (Computer Aided Planning) encompasses EDP-supported scheduling based on the design data produced in CAD. CAP produces production documents defining with what, how, in what sequence, with what, and over what time period a workpiece or an assembly, etc., is to be manufactured and assembled [EVER90b]. The generation of NC programs from CAD data supported by computers is known as CAM (Computer Aided Manufacturing) [N.N.99].

In addition to determining the required operations, processes and machines to be used, the necessary special-purpose tools and fixtures must also be included in the planning process if required. If NC machines are to be used, the required NC programs must also be derived from the CAD data.

The CAP systems used must distinguish between scheduling management systems, which only provide editing and administrative functions, and scheduling systems, which support the entire scheduling process and are therefore significantly more complex.

The planning area of CAQ now becomes involved to specify test criteria and test programs and to complete the routing plan. The geometry, technical and process data are once again required for this process.

The work schedules produced by the CAP area contain all the information on the personnel required for production purposes, the machines and manufacturing resources to be used (tools, fixtures, etc.) as well as information on the individual process steps for completing the work in the form of work instructions of NC programs. These data are provided to the ERP system so that it can compile the order and plan production based on considerations of quantities, deadlines and capacities.

#### 14.1.1.4 CAQ

The task of CAQ (Computer Aided Quality Assurance) is to plan and perform quality assurance tasks [NEIP91]. Using the design documents from the CAD area and the routing plans from the CAP area, test criteria are established, measuring and test plans are produced, and, if computer-controlled inspection devices are involved, test programs are generated and supplement the routing plan. The test specifications produced in this way are implemented and monitored automatically by the manufacturing system.

The interdisciplinary tasks from CAQ make it clear that this is not an isolated function within the corporate structure. Rather, it pervades all areas throughout the entire product development process.

### 14.1.2 Automated Production

Computer-supported control and monitoring of the means of production in the production process is known as automated production. This includes operating production data acquisition (PDA) and machine data acquisition (MDA) as well as direct control of NC machines, and warehousing and transportation systems.

The EDP systems used are largely characterized by the degree of automation. While fully automated flexible manufacturing systems require highly complex manufacturing control systems, control stations for controlling conventional shop floor environments have a significantly smaller functional scope.

The automated production system retrieves the data required to control manufacturing from the upstream areas of ERP, CAD, CAP, CAM, and CAQ. At the same time, the data generated during the production process are reported to these areas to ensure the planning processes are kept up to date.

The data from production or assembly are collected, compressed and then sent to the ERP system. Important data are, for example, the start and end dates of the production tasks, information on the number of completed workpieces and scrap quantities as well as the data collected from production data acquisition (see ▶ Sect. 14.3.2.5).

### 14.1.3 Level Model of an Industrial Manufacturing Business

To be able to exchange information between the different entities, the requirements and interfaces of the relevant units are used as a basis. In order to be able to define the requirement profile, it is necessary to investigate the internal process-related or corporate management of information i.e., to analyze specifically how.

- the information is generated,
- the information is moved around and
- the information is processed.

The requirement profile must reflect elements such as the scope of the relevant data quantity to be moved, the permissible running and response times, the required data throughput performance of the communication unit and the maximum number of communication nodes to be connected in the network. To investigate the internal process-related or corporate management of information in a business involved in discrete parts manufacturing, it has proven expedient to be guided by the reference model described in the Technical Report TR 10,314 and its further development in ISO 62,264. This model subdivides a business into five levels:

- Level 5: Planning level
- Level 4: Command level
- Level 3: Management level
- Level 2: Control level
- Level 1: Actuator/sensor level

The various levels are characterized by the data quantity of the information as well as the automation pyramid (see ◘ Fig. 14.2). The higher the level, the less data in the information. The closer to the top of the pyramid, the greater the timescale for each level. While data are configured and captured in milliseconds at the actuator/sensor level, the planning level for future orders is calculated on a timescale of months. The process level is occasionally included in the context of the automation pyramid as Level 0.

Planning level:

The planning level in a business covers the primary tasks of the entire business. The main task at this level is managing the company in the sense of meeting the business objectives (long-term planning). The technical system at the planning level of a business or a factory can be an office or plant computer. Corporate functions such as production planning (e.g., ERP systems) are located at the planning level.

Command level:

The task of the command level is to coordinate production and manufacturing as well as to monitor and control order processing including procurement and holding the resources procured for the individual orders. Typical areas of responsibility falling to the command level in a business structure are production planning, capacity planning, resource scheduling, resource maintenance scheduling and resource management. A technical system for the command level is typically a production control computer. Corporate functions such as work preparation (e.g., using CAP systems) are located in the back office areas at the planning level.

Management level:

Typical tasks at the management level, which corresponds to the master level, are analysis of the resources required for a production or measurement job, its requirements, the formation of order queues, distribution of orders to individual automation devices, monitoring order progress or reporting the order status to the command level. Technical implementations at the management level which can be summarized under the umbrella term 'automated production' (see ► Sect. 14.1.2) are typically cell computers, integrated DNC/PDA/MDA systems and production control stations.

Control level:

The tasks allocated to the control level include the control, monitoring and coordination of the activities of one or more automation devices in the shop floor area for machining raw materials, controlling assembly or measurement tasks and reporting on the condition of equipment to a higher-level technical system. Technical implementations at the control level are automation device controls.

Actuator/sensor level:

The interaction between process and control is handled by actuators and sensors. The task of sensors



◘ **Fig. 14.2** Automation pyramid

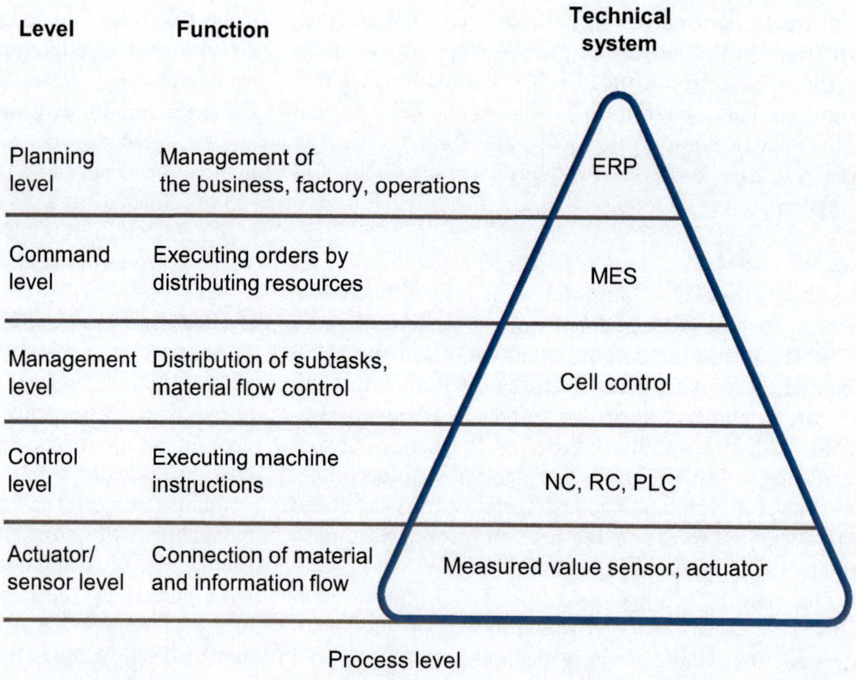

(measurement transducers) is to convert process parameters into information. Actuators (positioning elements) convert information into physical values that influence the process.

This automation pyramid is likely to not apply in the future due to the increasing networking of technical components in the production process (see [VOGE09]) or this circumstance has already begun. It is now possible to react to disruptions within significantly shorter times due to the substantially greater quantities of information about each individual production process. While the planning level will remain largely the same due to its long-term/strategic character, the levels from the command level downwards are likely to increasingly merge (see [BREC13]). This effect is known as the "dissolution of the automation pyramid" [BAUE16]. This development will result in an increase in the amount of information at each of the previous levels.

## 14.2 Corporate Level

Enterprise Resource Planning (ERP) is supported in manufacturing companies on the software side by ERP systems.

The origin of ERP systems goes back to the software processors used in the Material Requirements Planning (MRP) of the 1960s. Then, planning was restricted to producing and managing parts lists and requirement- or consumption-based planning. The calculations were based on the assumption of fixed material demand and production times without taking, for example, machine availability into consideration. In the 1970s and 80 s, systems known as Closed Loop MRP or MRP II (Manufacturing Resources Planning) were developed to accommodate these missing elements. These systems were then able to take existing manufacturing capacities and stock levels into consideration. In many manufacturing companies, they became information sources for a range of different areas besides planning and control such as cost calculations or procurement.

ERP systems were developed in the 1990s to extend the use of these systems beyond the manufacturing industry. The change of name to 'enterprise' now includes general corporate activities such as Controlling and Human Resources. Today, ERP systems frequently form the IT database for the entire business in many different industrial sectors.

Significant further development of the systems in recent years was their extension to integrate supplier functionalities (Supply Chain Management, SCM). This development is primarily based on reduced manufacturing depth and increasing dependency on suppliers. By connecting various ERP systems via SCM modules, it is possible to perform planning and optimization tasks while considering suppliers' capacities.

A range of other systems for processing information and material flows are associated with the ERP system. Two of the most significant here are Customer Relationship Management (CRM) and Product Lifecycle Management (PLM).

CRM structures and manages all communication with customers. Originally part of other systems such as the ERP system, this developed out of an increasing number of communication channels with customers to become a stand-alone system. Today's CRM systems do not just store customer data. They are responsible for planning and controlling all activities before and after the product is purchased.

The focus of the developments in PLM systems is the resolution of the distinction between actual and virtual product development. One of the most important components of an ERP system is Product Data Management (PDM). The primary connection to ERP systems here is the management of product parts lists and master data.

There should not be a strict separation in both CRM and PLM systems, or such a separation is not possible at all. Invoicing, which is one of the reasons for introducing an ERP system, has a direct connection to customers and therefore to CRM systems.

### 14.2.1 ERP Systems

A precise definition of ERP systems is difficult due to the developments referred to above. However, what all solutions have in common is that, as standard business software, they should be accessible to a large number of users. ERP systems must integrate all operational and administrative processes in one common database. This is the only way ERP systems can support the commercial and technical tasks in a business [BOYW12].

The range of tasks an ERP system has to fulfill is reflected by the number of different systems available. There are currently over 700 different products available from approximately 540 providers[TROV17]. Despite this, it is a very consolidated market: Nearly half of all installed systems originate from SAP followed by Microsoft with approximately a 20% market share [STEG11].

In the past, the solution was usually to install the system on dedicated servers within the business and make it accessible via a local intranet. However, the increasing prevalence of web-based solutions has led to the creation of an interface for external access. This has enabled easier integration of customers and suppliers. The most recent developments are in the form of cloud-based ERP solutions where the IT infrastructure can be provided and maintained externally. The ERP systems are accessible as a service on demand.

#### 14.2.1.1 Functionalities of an ERP System

As far as the scope of the specification of ERP systems is concerned, the large variance in available systems makes it difficult to produce a precise definition. ◘ Figure 14.3 subdivides the functionalities into modules and thus clarifies the relationship to the different entities within a business. The overall system thus encompasses the in-house areas of finance and accounting, materials management, production, sales, and human resources. The supplier relationship, supply chain management, customer service, and electronic market relationship modules provide external communications. The basic properties of ERP systems used to target a large number of different customers must therefore have a generic design. To simplify this, ERP systems are usually subdivided into business systems and systems for small and medium-sized enterprises (SME).

The main demand on the ERP system in a business is mapping the large number of existing business processes. To this end, the relevant processes must be stored in the IT structure of the ERP system via the corresponding Programming interfaces (see ► Sect. 14.1). Refer to ► Sect. 14.2.1.3 for the actual implementation. However, in anticipation of this, it can be said that the

| | |
|---|---|
| Customer order processing | • Pre-calculation<br>• Quote generation<br>• Post-calculation<br>• Order management<br>• Order tracking<br>• Determining delivery schedules |
| Inventory control | • Warehouse management<br>• WiP management<br>• Batch tracking<br>• Inventory<br>• Safety stock |
| Purchasing | • Request for quotation<br>• Quotation analysis<br>• Managing sources of supply and conditions<br>• Order overview<br>• Receipt of invoice<br>• Supplier support |
| Financial accounting | • Auditing<br>• Booking incoming invoices<br>• Invoicing<br>• Invoice overview and reminders |
| Manufacturing control | • Detailed production planning<br>• Shop floor control<br>• Managing production orders<br>• Production monitoring (quantity, quality) |
| Scheduling | • Production planning<br>• Demand-oriented material provision<br>• Demand-oriented ordering |

◘ **Fig. 14.3** Subdivision of ERP functionalities into modules implementation and application of these processes are very time-consuming.

In the past, these demands have often been an obstacle to SMEs in implementing ERP solutions. In fact, business processes in SMEs are often similarly complex but can be more quickly and easily implemented since fewer people are employed by the business. There is an assumed reduction in complexity and functionalities in SME systems. Consequently, the barrier to introducing the systems is lower.

#### 14.2.1.2 Modularity of ERP Systems

Due to the many possible functions and the wide use in different sectors, ERP systems are usually subdivided into the following main elements:

- customer order processing,
- inventory control,
- purchasing,
- auditing,
- production control and
- scheduling [OSTE14].

The customer order processing module is concerned with the process from the customer query to product delivery. There are many processes involved in this such as product development, manufacture, and materials management. Customer order processing is frequently supplemented with marketing processes such as advertising. An interface to higher-level corporate and product planning provides further connectivity for this module within the business.

The inventory control module is a cross-section module which is used by nearly all other modules. The focus here is managing the movement of goods, i.e., stock placement and removal, inventory valuation, etc., intending to always have correct stock levels.

The procurement process is organized in the purchasing module. All suppliers are regularly evaluated in terms of their quality, adherence to schedules, and pricing. This is the basis for closing new framework contracts and amending existing ones. Delivery times are taken into consideration in production planning and ordering triggered as part of scheduling.

The auditing module checks the content and formal correctness of invoices received for services and purchases. Invoices can only be posted and booked once approved.

The production control module provides a centralized organization of all production processes. The production process is optimized based on various criteria such as lead times and standstill times. Machine condition and machining status are fed back to production control so that it can respond to unplanned events and initiate measures to meet production targets as far as possible.

The scheduling module handles initial production planning. This is then refined by production control and optimized if complications occur. The results of scheduling are external and internal delivery schedules for end products, semi-finished parts, and components.

#### 14.2.1.3 Introducing an ERP System

The introduction of a company-wide ERP system is usually subdivided into three phases:

1. Basic ERP
2. Supply chain integration
3. Extensions and improvements to support corporate strategies.

One of the main reasons for this three-phase introduction is the size of a complete ERP system. This three-phase structure is frequently broken down further to a significant degree. In this more detailed process of subdivision, there are frequently tasks that occur in more than one of the three phases but the main results of each phase build on each other. The main tasks in each phase are broken down further below.

##### Basic ERP

Implementing a basic ERP system includes the functionalities for basic sales, production, capacity and demand planning, material requirements planning, and master scheduling. There are also integral support functions to increase the accuracy of inventory to structure and ensure accurate parts lists as well as the construction of feedback loops from production and purchasing.

Implementing a basic ERP does not encompass all possible functions of an ERP system by any means. It does, however, produce a system with more functions than a purely MRP tool and sets down basic organizational principles.

##### Supply Chain Integration

The focus of the second phase is the setup of processes that integrate the ERP both up and down the supply chain. Up the supply chain means connecting suppliers, for example, via business-to-business portals while down the supply chain is more concerned with processes such as vendor managed inventory where a supplier influences warehousing at a distributor.

##### Extensions and Improvements to Support Corporate Strategies of Chain Integration

This phase encompasses extending the ERP functionalities across the whole organization. This can include finalizing units that have not yet been implemented such as within the accounting system. In addition to this, connections are made between existing units and other applications within this unit.

As part of the last phase, improvements in implemented systems, which either did not have a high priority in the first phases or whose necessity was only acknowledged in these latter phases, are also put in place. These typically include functionalities for simulation, implementation of systems of parameters for evaluating suppliers or even connecting the ERP system to a manufacturing control system.

While the first two phases appear manageable at the beginning of a project to implement an ERP system, this is not the case for this third phase. The phases have a decreasing structure, largely due to the fact that subsequent phases are dependent on the success of the preceding ones. However, it must also be noted that even the implementation of the first phase can represent a significant improvement within a business.

#### 14.2.1.4 Current Developments in ERP Systems

As part of current Industry 4.0 activities, new demands are being placed on ERP systems in terms of networking, availability, integration of IT structures, usability and flexibility. The providers are responding to these demands with a range of further developments and some with completely new versions of their systems. The most relevant trends in the development of ERP systems are discussed in brief below.

One of the most important elements of the Industry 4.0 initiative is the complete networking of production. The master data of a business are also held in ERP systems today. Since their introduction, ERP systems have been the interface to other IT systems on the corporate level such as CRM, HRM and financial accounting. This initial situation leads to ERP systems being the best option for both networking between internal production and logistics as well as external data (supplier data, market data, customer information) [BING16].



To make the data and services of an ERP system globally accessible to, for example, senior management and customer services, parts of the software have been moved to the cloud. Services can also be made centrally available on the ERP providers' cloud platforms and ERP users can access these services as required. This also leads to new business models for ERP systems: Instead of fixed-term contracts for software and local servers, subscriptions for specific services and storage capacities can be contracted. It is also possible to invoice based on use of functions. Porting ERP applications into the cloud also provides the opportunity to introduce systems economically and easily into a business and for flexible adaptation to new environmental conditions.

A further current point of focus is the ERP system's user interfaces. ERP systems have historically produced highly confusing and convoluted HMIs, which are being resolved and replaced by intuitive interfaces, and mobile terminal devices are seeing increasing use in many areas of life and are also finding their way into the production environment. There are currently a range of different approaches in this area because web-based applications can be used on different portable terminal devices while only device-specific applications can use all technologies such as image recognition.

## 14.3 Operations Command Level

The operations command level covers all processes directly connected to production in a manufacturing business. It combines the corporate command level with the process level. Information from long-term corporate planning is processed here and forwarded to the process command level. Information from the process command level is also compressed and transferred to the higher-level corporate command level.

### 14.3.1 Manufacturing Execution Systems

The complexity of the manufacturing process in manufacturing companies is constantly increasing. IT-supported monitoring of production using classic ERP systems has the disadvantage that it is not possible to report disruptions in production in real time. The control loops at the ERP level start with order placement and end with the reporting of manufactured quantities, failed manufacturing processes and manufacturing indicators. It is not possible to gain access to manufacturing processes and operations as the necessary information is not available from production. In order to be able to resolve disruptions in manufacturing processes in good time, these have to be detected during production. This is contingent on IT systems that can collect feedback from production and facilitate resolution of the disruptions. Besides, information from production is compressed and transferred to the next level up.

While IT support at the corporate command level is state of the art, the next logical step is to extend this to the operations command level [FAYZ10]. Vertical integration between production and the corporate command level uses existing IT systems. This is now described with the term Manufacturing Execution Systems (MES). MES are developing into information and data hubs within manufacturing companies [SAUE10]. MES include planning, control and monitoring of production operations. This all includes solutions optimizing processes from order placement to product completion [POSS06; MESA97]. An MES is a computer system, integrated using software technology, containing all the IT methods and tools required for production [POSS06; MACC97].

#### 14.3.1.1 MES Tasks

Efforts have been made in the past to standardize the functional scope of MES. Definitions from MESA [MESA97] and VDI standard 5600 [VDI16] allocate a broad spectrum of tasks to MES. MES includes the tasks of order management, detailed planning and control, utility management, material management, personnel management, data capture, performance analysis, quality management, information management, and energy management. In practice, the functional scope of existing MES deviates from the cited definitions. Many different software solutions are covered by the term MES [FAYZ10; KLET06].

#### 14.3.1.2 Variants and Types of Systems

There are essentially three different types of architecture currently on offer for MES. MES can be constructed as an extension to existing ERP systems (◘ Fig. 14.4). These systems are provided by well-known ERP software companies and smaller third-party manufacturers who implement ERP system extensions [FAYZ10].

These systems frequently fall back on existing ERP information models which are then supplemented with specific MES functions. The benefit of this solution is simple linking between MES functions and the existing corporate data sets and processes [FAYZ10]. Typical extensions are PDA systems, personnel time card modules, inventory systems, tool management systems, quality assurance systems, and planning modules. In spite of these extensions, the information model embedded into the system is in essence designed for the demands of the ERP level. For this reason, it does not meet the real-time requirements required in the field of automation. Most ERP operations originate from the office area where data update rates extend to several minutes. It is rare to find data being consistently updated based on events. Real-time critical data from the manufacturing area cannot be retained in the ERP data model. Specifically, connecting controls to an

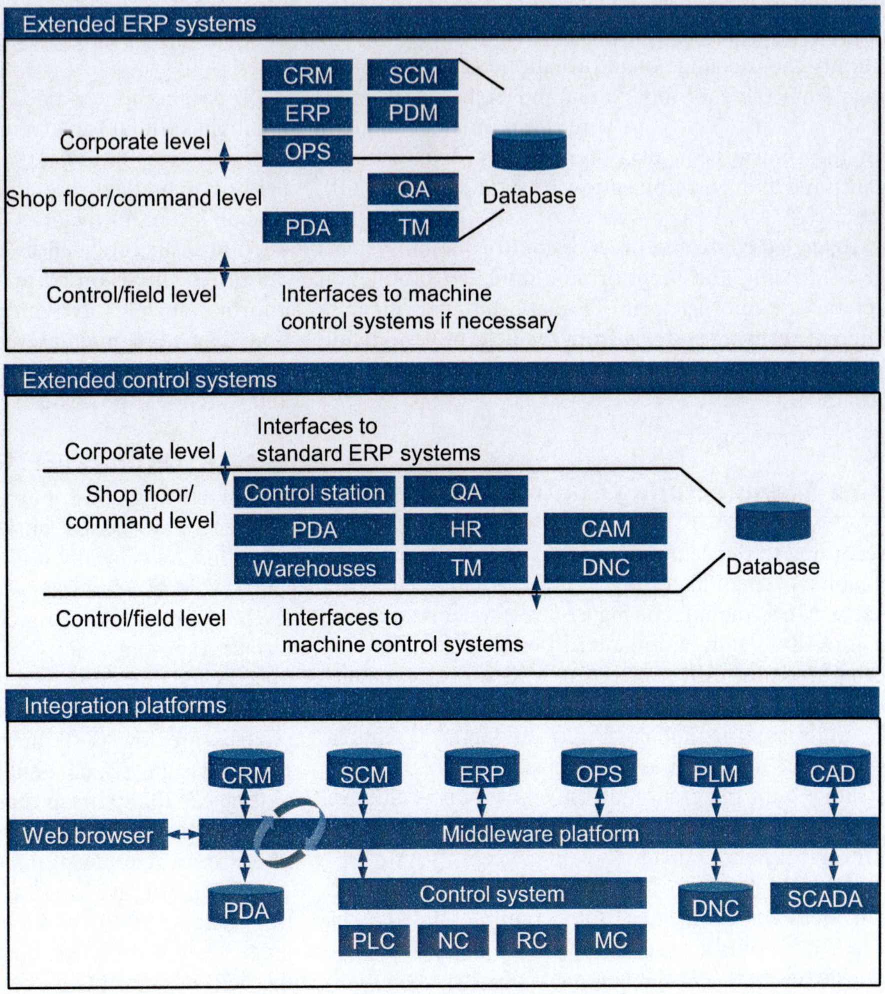


◘ **Fig. 14.4** MES variants and types of systems [Week03a]

ERP-based MES is not necessarily the best solution as the boundary conditions from the area of automation technology cannot be adequately covered in ERP data models.

MES based on shop floor applications represent a further category. These systems have their origin in the capture of operational, machine, quality and logistics data as well as control stations and SCADA systems. A tight link between machines and the MES together with the capture of actual machine conditions are the main focal points here. The emphasis is on functions such as managing tools, stock level and machining programs handled using internal data models. These data sets, which are often proprietary, are an impediment to higher-level ERP systems because this requires the production and adaptation of individual interfaces. In addition to this, information from the shop floor level has to be compressed. Extensions to machine controls, PDA, MDA or tool management functions as well as functions for managing machining programs and maintenance are also included in this category.

The last system category covers platforms aiming to integrate heterogeneous IT systems both close to and remote from the machines. These platforms are to simplify the exchange of information between applications from the shop floor area and higher-level application software. The focus here is not on new functions but the transmission of system events, data management, and the synchronization of data sets [FAYZ10; WECK03].

Extended control stations or control station systems for controlling and monitoring automated production facilities are another form. IT platforms that integrate different software systems from the field of production automation and production management form this last variant.

14

### 14.3.2 Manufacturing Control Systems

According to the definition, MES also include the functionality of manufacturing control systems. However, unlike MES, manufacturing control systems are used exclusively in fully automated discrete parts manufacturing and are only concerned with individual manufacturing facilities [POSS06]. In the field of flexible, automated manufacturing, manufacturing control systems are also used as an extension to MES as they facilitate tight integration at the control and field level as dedicated solutions and are better suited to complex control tasks and individual plant concepts [FAYZ10].

While EDP systems have been used in pre-production areas for some time now, computer-aided manufacturing control systems appeared later. One reason for this was that the commercial demands initially required EDP systems to be used in the planning areas. These systems were extended into the area of manufacturing to capture the data required for planning and cost accounting and are now known as shop floor control systems. Another reason was the design of the machine control systems in place at the time. This usually did not provide a direct link to an external computer. However, developments in the field of NC controls now mean that so-called flexible manufacturing systems (FMS) and flexible manufacturing cells (FMC) can be built. These are fully automatic, computer-controlled single or multi-machine systems. The EDP systems used for control are categorized as production control computers and include functions for automatic transportation, handling and storage as well as direct control of the linked machines. This complex functionality is summed up by the term automated production.

As businesses today generally deploy a mixture of conventional and automated production, the initially different systems are becoming increasingly similar to shop floor control and production control computers. The structure of the software in modern manufacturing control systems is expected to facilitate individual adaptation to a range of different manufacturing structures.

Today, computer-aided production is understood as not only fully automated control of flexible manufacturing systems but also the overall EDP support in the field of manufacturing control. Depending on the extension level, this includes detailed planning of the production tasks provided by the ERP system, timely and optimized provision of materials, production aids and information (e.g., NC programs), capture of production data, and manual, semi-automatic or automatic control of manufacturing, handling, transportation, and warehouse processes.

#### 14.3.2.1 Technical System Concepts

Manufacturing control systems can be realized in two different ways based on different systems. PC-based solutions are typically utilized to implement the higher levels while PLC-based systems are used for the lower levels. As manufacturing control systems form the interface, they can access both system concepts and produce PC or PLC-based results [FAYZ10].

The benefits of PLC-based manufacturing control systems are low costs and real-time capability. Their robust design, which limits system flexibility, and real-time capability mean that PLC-based manufacturing control systems are the more widely used concept for safety–critical applications.

The alternative to PLC-based manufacturing control systems is PC-based manufacturing control systems. They operate on an industrial PC connected to the field devices. PC-based systems are less robust but offer a great deal more flexibility and are better at dealing with complex processes.

### 14.3.2.2 DNC (Distributed Numerical Control)

#### Historical Development

The origins of the IT-based integration of numerically controlled machines into manufacturing systems go back to the end of the 1960s. Machine controls at that time did not have any internal NC program memory. The NC programs had to be made available to the machine control on a data carrier acting as a storage medium where they were read on a line-by-line basis. However, in the harsh shop floor environment, these data carriers, which were almost always punched paper strips, proved to be a repeated source of problems. Also, if a large range of parts was involved, their use and storage caused considerable organizational problems. For this reason, at the end of the 1960s, manufacturing systems that no longer transferred data using a punched tape data carrier were successfully tested in the US, Germany (e.g., by AEG and Siemens) and Japan. The numerically controlled machines were directly connected to production computers using electrical data transfer media. This used the existing electrical interface to the punched tape reader. This produced the so-called Behind-Tape-Reader interface (BTR). The NC programs and data were stored on the production computers' memory. In addition to this, they also provided the machine controls with control data on a line-by-line basis "just in time," i.e., as the work progressed. The term 'DNC system' became the internationally used description of the hierarchically structured program and data distribution systems created. Originally, the acronym DNC stood for Direct Numerical Control [DIN83; VDI72] and was connected to specific features of the first generation of DNC systems. However, the term Direct Numerical Control did not cover the specifics of the interconnected computer and machine. This term was replaced by 'Distributed Numerical Control' as the NC data distribution was more in line with the way modern DNC systems operated [DIN96; NCG90; TÖNS89]. In the first DNC systems, the high spatial and organizational costs of managing the NC programs, which were generally stored on large punched tape libraries, were reduced by the direct linking of production computers and machine controls. In addition, the long non-productive times, resulting from the production of punched tapes and distribution of control data to machine controls, could also be reduced. This data distribution led primarily to a considerable reduction in the length of change-over cycles for NC programs [KLEM86].

#### DNC Systems

Modern DNC systems are integrated into the information flow of a manufacturing system and make use of interfaces to a heterogeneous environment of NC controls as well as a CAM server or production control computer (◘ Fig. 14.5).

Mainly industrial Ethernet and sometimes wireless LAN networks are used for data transfer. The compatibility with older machines and controls is handled using universal data communication functions such as CSMA/CD [PRIT91] or token passing bus systems [AEG88; BRIL91] as well as standardized fieldbuses

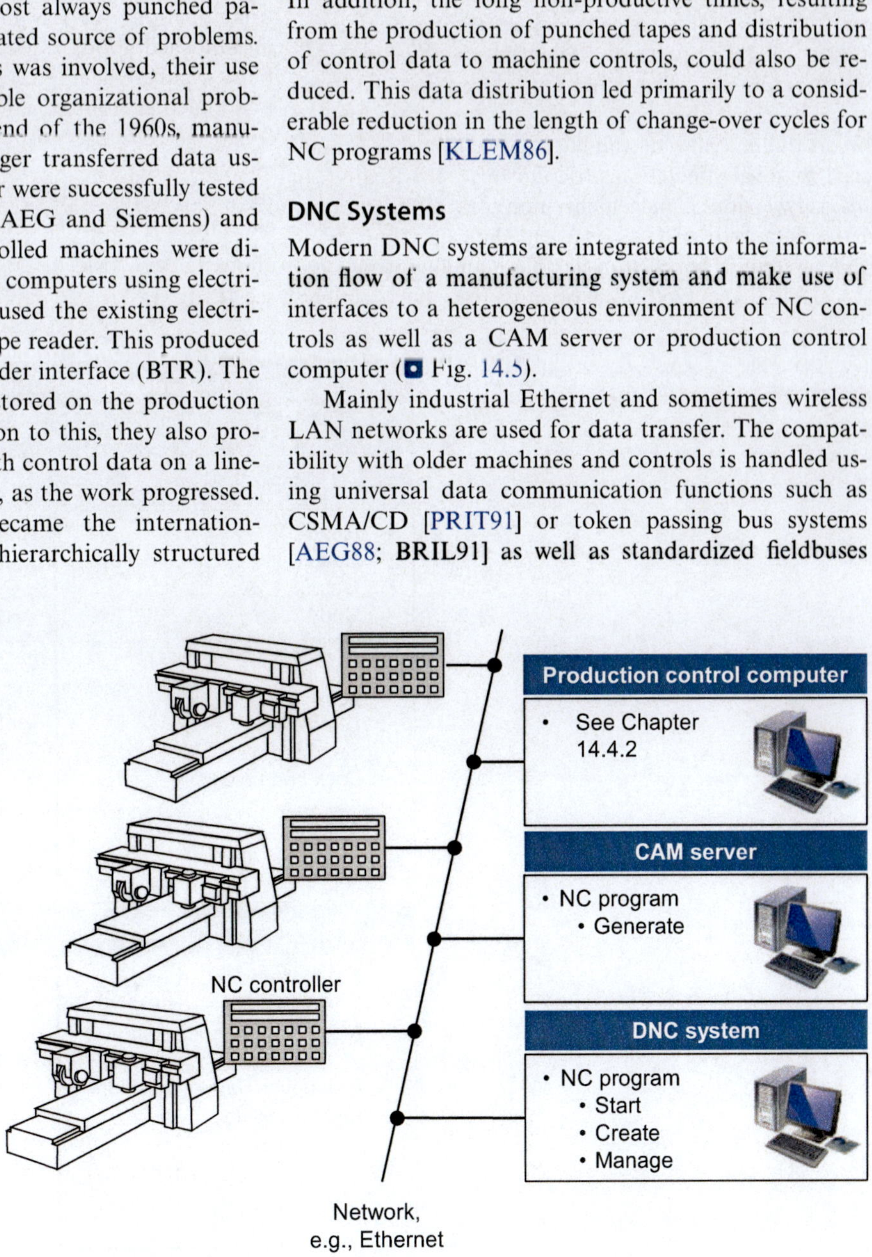


◘ Fig. 14.5 DNC system in the production environment

such as the PDV bus as per DIN 19,241 [AEG88] or the Profibus as per DIN 19,245 as well as non-standardized fieldbuses [ROBE87]. The CAM server acts as a data source and generates the relevant NC programs, including their tool data, correction values and zero point tables, from the CAD data. The DNC system runs on a standard industrial PC with a touch screen. The addition of 3D CAD drawings of clamping, tools and the workpiece to be produced provides the machine operator with further support when operating the DNC functionality or administering the NC data. A distinction is made here between basic DNC functions and extended functions (◘ Fig. 14.6).

Basic functions include NC program management and NC data distribution. The extended functions allow NC data to be corrected and the manual generation of NC programs. Potential modifications to NC programs that have been produced manually or by CAM can either be tested by the machine operator on the machine or validated virtually using a material removal simulation program (e.g., eM-RealNC). Integral version management monitors modifications in the data and ensures consistent data handling. If a DNC system is linked to a flexible manufacturing system (FMS), the DNC functionality can be remotely controlled via a production control computer. Subfunctions of the production control system are available. These allow transfer of default data (schedules and routing plans), detailed scheduling and transfer of organizational control data. DNC systems also make use of interfaces for production data acquisition (PDA) and machine data acquisition (MDA). Technical and organizational data from the manufacturing process can be captured and forwarded for process monitoring or process control (see ▶ Chap. 6).

### 14.3.2.3 Material Flow Control

Material flow control at the shop floor or control level includes administration, warehousing, and transportation functions for raw materials, semi-finished goods, and finished goods as well as transportation aids (e.g., pallets, pallet cages, etc.). On a purely functional level, material flow control can be subdivided into the subfunctions of.

- scheduling,
- warehousing and.
- transport control,

see ◘ Fig. 14.7.



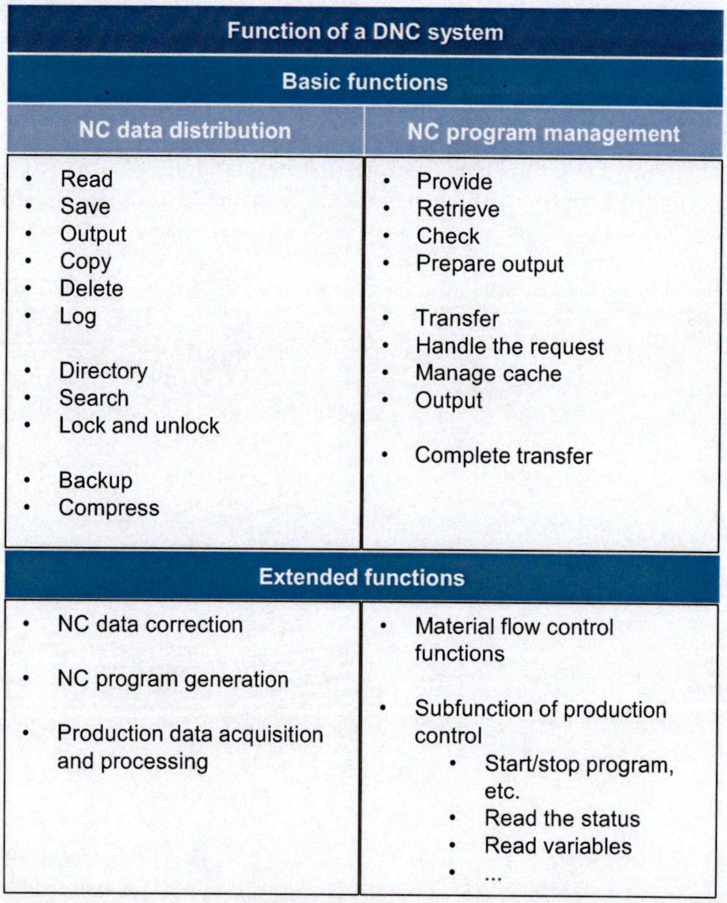


◘ Fig. 14.6 Functions of a DNC system

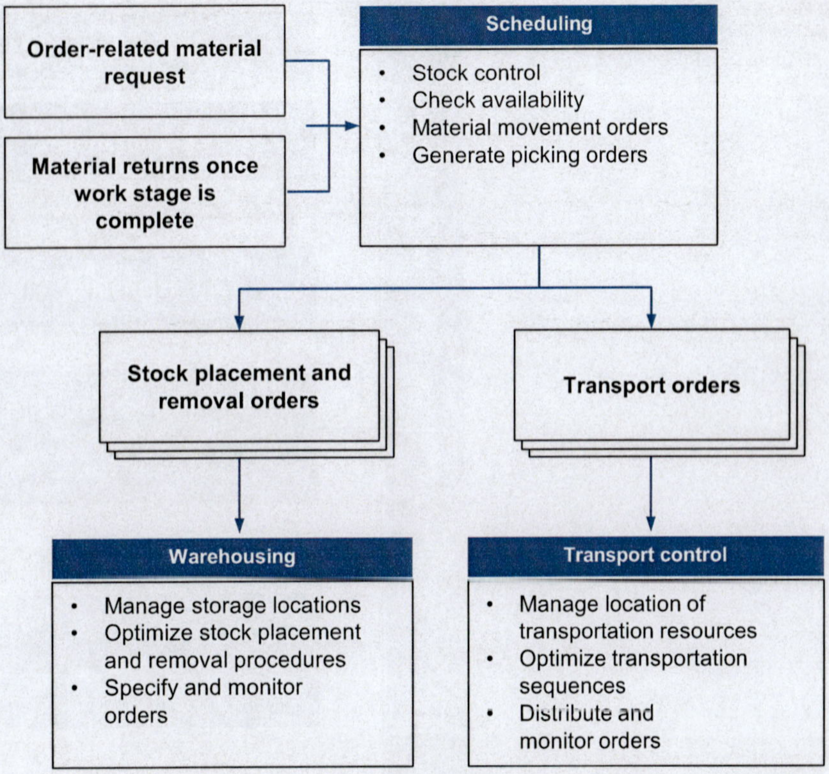


**Fig. 14.7** Material flow control tasks

All material requests are received by scheduling and compared to material stocks and transport resources. Once the source location has been established, picking and stock removal orders are sent to the order picking sites and stock control. Transport is then organized to the next machining station or to temporary storage for semi-finished or finished workpieces and blanks. Changes to stock levels are sent to the higher-level ERP system.

Stock control takes the warehouse stock placement and removal orders from scheduling, checks their plausibility and organizes and monitors the stock placement and removal procedures completed by warehouse staff or automated warehouse handling equipment.

The task of transport control is the organization of transport between warehouses, workstations, and buffer areas. Transport control manages the status and location of the allocated transport systems, optimizes transport routes, distributes transport tasks, and monitors their progress.

#### 14.3.2.4 Organization of Production Resources

Timely provision of production resources at the workstations is one of the most important tasks in manufacturing. It is often the case that a work stage cannot be started because the preparatory activities such as tool mounting or presetting have not been completed. These organizational downtimes can be avoided by using IT systems [WECK91].

Production resources include both tools, fixtures, grippers, etc., and measuring and test equipment. Only tools will be considered in the following sections due to their particular significance. It is possible to generalize this to other production resources at any time.

Tool management can generally be divided into five functional areas:

- tool planning,
- tool organization,
- tool scheduling,
- tool supply and removal, and
- tool usage.

These areas are self-contained and can be allocated to the various operational levels, see Fig. 14.8.

#### Tool Planning

The main task of tool planning is to establish the types of tools to be used across the business and to build a tool catalog for the entire business. The tools to be used to complete a work stage are established as part of tool usage planning.

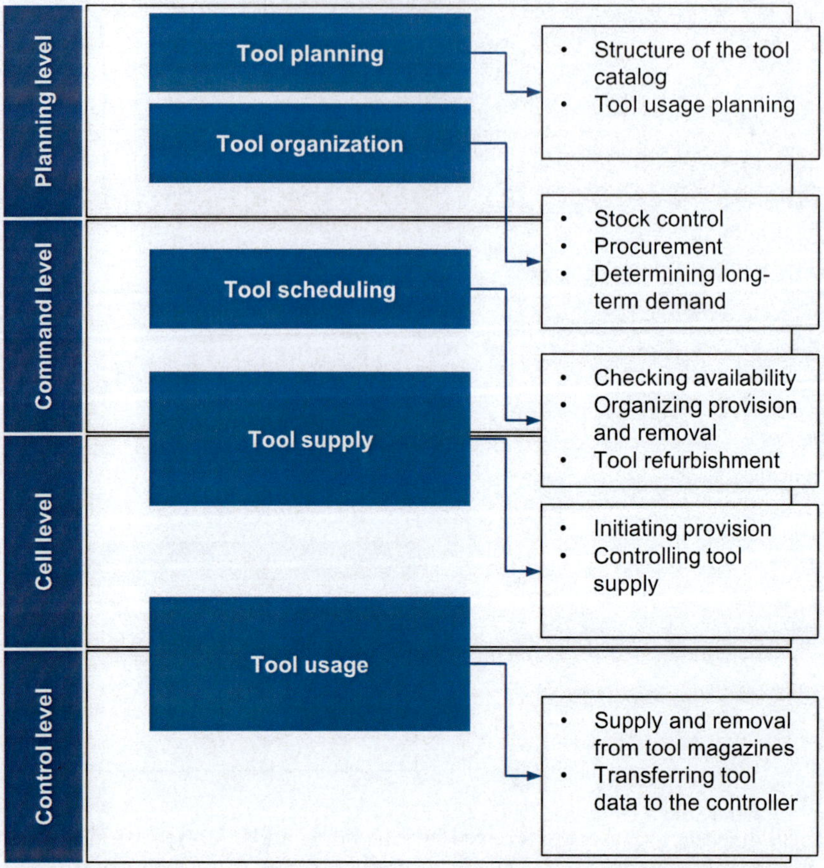


◘ Fig. 14.8 Allocation of tool management in the structure of operational levels



### Tool Organization

Tool organization is handled by the ERP system and includes determining long-term tool requirements, managing stock levels, and organizing procurement.

### Tool Scheduling

Tool scheduling is located at the shop floor level. It determines the order-related tool requirements as part of a short to medium-term planning horizon and coordinates the timely provision to machines.

The gross requirements for each work stage are determined based on the current production orders. The net requirement is calculated based on the tools already present at the machines. Reservation can take place once the availability of tools for the scheduled time of first use has been checked. Coordinating the timely supply and removal of tools is an increasingly important role in today's production control. The tool cycle shown in ◘ Fig. 14.9 illustrates the individual stations a tool visits before the point of use and afterwards. To ensure timely provision of tools, supply chains are planned and optimized taking the available personnel and workstation capacities in the tool area into consideration. Furthermore, disruption strategies must be put in place which can mitigate disruptions such as tool fracture.

In the beginning, tool management systems generally supported the basic functions of tool scheduling and reservation. Current software systems, such as TDM systems, also integrate predictive tool planning and optimization, while disruption management is not yet state of the art.

### Tool Supply and Removal

The planning data established by tool scheduling are implemented by the tool supply process. This includes both organizing provision and controlling tool preparation and refurbishment. The orders can be transferred from a control station directly to the workplace terminals that are part of the tool flow or in the form of printed order lists.

### Tool Usage

The tool usage function is to provide the tools to the locations in the machine magazine intended for this purpose and to remove worn tools. Tool-specific data are also transferred to the NC controls.

### Tool Information System

Efficient tool organization is based on a company-wide tool database containing the master data for all tools or

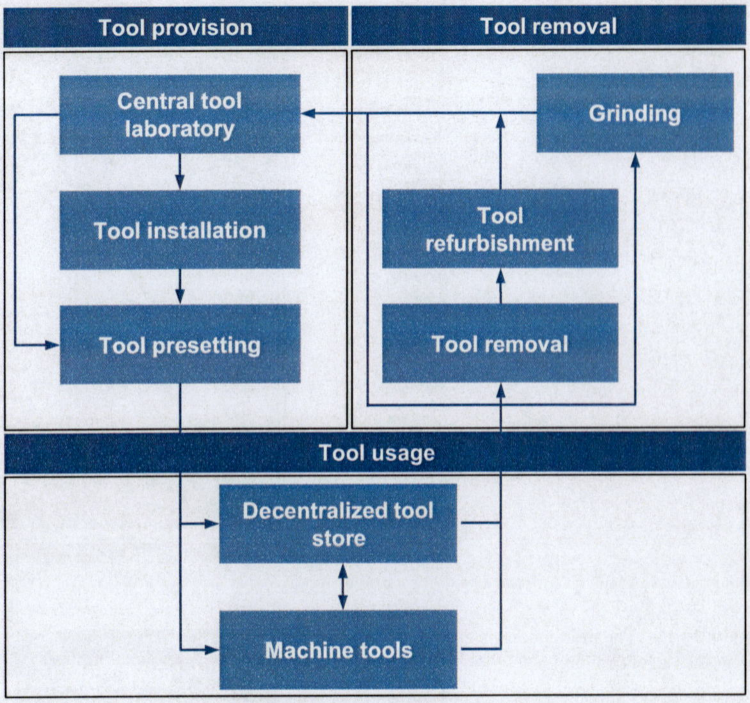


**Fig. 14.9** Stations in the tool cycle

tool components. The tool management system comprises the following functions [ROBE88]:

- Managing tool lists
  A tool list includes all the complete tools necessary for machining parts.
- Managing complete tools
  The master data for complete tools are geometry data, technology data, and tool lists.
- Managing tool components
  The master data for tool components are geometry data, graphical data, and supplier data.
- Printing tool sheets
- Generating tool graphics
  A graphic for the complete tool can automatically be produced from the graphics for the individual components.
- Producing where-used lists
  A where-used list typically includes all NC programs that use a specific tool.

User interfaces are generally available to help users in generating, modifying, or searching for tool data, see Fig. 14.10.

#### 14.3.2.5 Production Data Acquisition and Processing

Production data acquisition (PDA) is to collect all the required actual data from the production area and to provide this as the current basis for planning in a format where it can be processed for Enterprise Resource Planning to determine new nominal data as well as for cost accounting and materials management. Work scheduling also requires empirical values. In terms of production and quality control, operational control is based on feedback from machining stations, measuring and testing equipment, and manual workstations.

As far as the flow of organizational information is concerned, the production process can be regarded as a multi-layered, lower-level control loop containing.

- the production plan as a command variable,
- the production control in the various levels as the control system,
- the production process as the control path and
- the deviation between the feedback from production data acquisition and processing and the nominal value as a controlled variable.

See Fig. 14.11.

##### Classifying Production Data and Defining Terms

Production data can be classified by their intended purpose and the specific types of data as shown in Fig. 14.12. The types of data are understood as order, machine, quality and employee-related feedback from production. The collection of machine-specific data such as machine running times, standstill times and causes of disruption is also known as machine data acquisition (MDA).

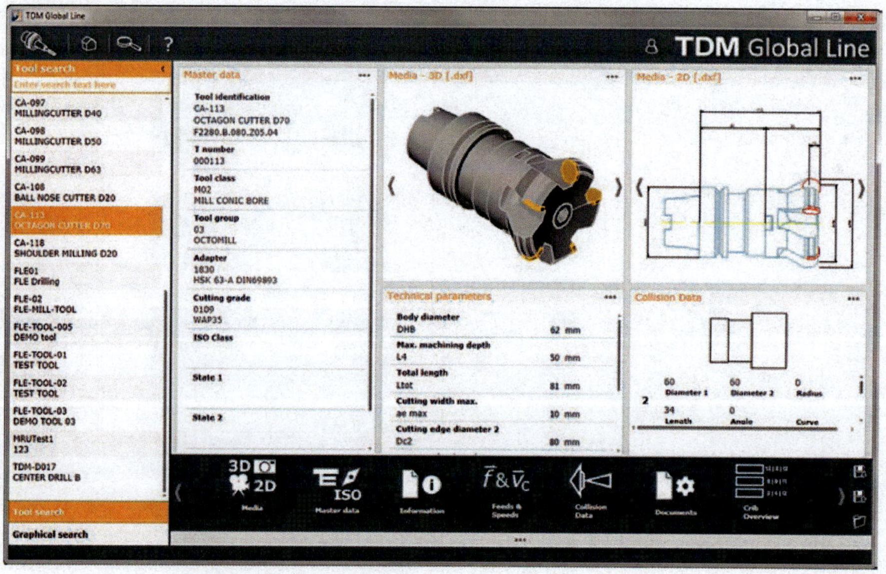


■ **Fig. 14.10** Tool data sheet. (*Source* TDM Systems)

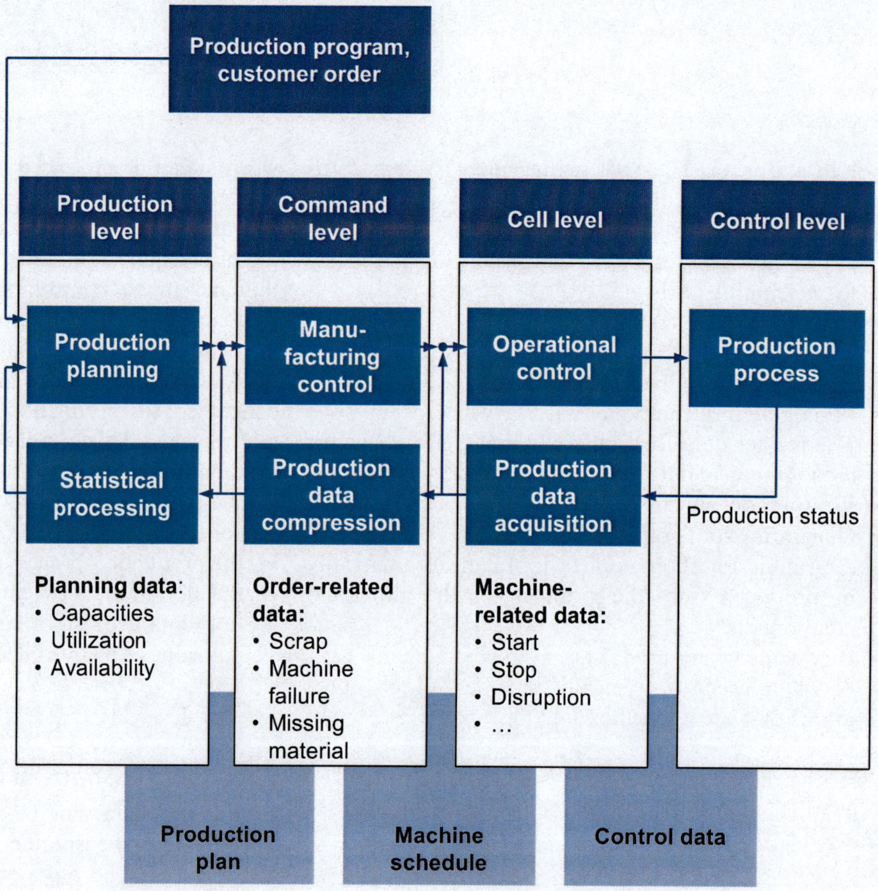


■ **Fig. 14.11** The production process as a control loop



Machine and order-related data provide information to a central point on the running and disrupted machines, the current order progress, and personnel present. From this data, it is possible, for example, to check progress toward order deadlines, to identify overloading of capacity and to organize appropriate

**Fig. 14.12** Classification and use of production data

| Types of data | Point of use: ERP – Production planning, Calculation, Materials management | Calculate gross pay | CAP – Work preparation | CAQ – Quality control | CAM – Manufacturing control |
|---|---|---|---|---|---|
| **Order-related**<br>Production time<br>Setup time<br>Transportation time<br>Wait time<br>Organized down time<br>Quantities<br>Scrap | ● | ● | | ● | ● |
| **Machine-related**<br>Production time<br>Failure rate<br>Down time<br>Utilization<br>NC data (mod.)<br>Tool data | ● | | ● | | ● |
| **Material-related**<br>Stock and moving of<br>• Raw materials<br>• Auxiliary materials<br>• Semi-finished parts<br>• Purchased parts | ● | | ● | | ● |
| **Quality-related**<br>Measured and test values<br>Error codes<br>Scrap quantities<br>Reasons for scrap | ● | ● | ● | ● | ● |
| **Employee-related**<br>Presence<br>Working hours<br>Performance efficiency rate | ● | ● | | | ● |

responses to disruption events. Furthermore, the production data are used to determine throughput times and order costs as well as to calculate wages.

The essential production data are described below [ECKE87]:

- Production time of a machining station corresponds to the duration of the machining process less the standstill times.
- Setup time is understood by the control system as the time required to execute a setup order.

- Standstill time is the time during which the production system is either partially or fully non-operational. The decisive element here is the standstill time of the machining stations as only these stations are directly responsible for production. A distinction is made between technical standstill times as a result of technical disruption and organizational standstill times which, for example, are a result of missing production resources or workpieces.

A reference time must be defined in order to evaluate the measured times. The reference time is understood as the maximum time available for production within a manufacturing cycle. This time is derived from:

- the operations calendar, which includes working days, public holidays, and vacation within a calendar year;
- the shift model which includes the start and end times of shifts in a working day for a capacity group or an individual capacity, and.
- the break time model that includes all breaks within a shift.

## PDA Terminals

There is a wide range of commercially available PDA terminals, ranging from simple, compact terminals with just a few functional keys to PCs suited for industrial use. The terminals can also be fitted with badge readers for magnetic or barcodes as well as with handheld barcode scanners for production data acquisition, see ◘ Fig. 14.13.

Choosing the type of PDA terminal ultimately depends on the required functional scope. Many PDA terminals offer DNC functionalities in addition to the actual collection of operating data. These devices are generally PC-based terminals (see ► Sect. 11.2).

PDA terminals work collaboratively and are connected to a central PDA computer. This computer acts as a data compressor and provides evaluation functionality. Alternatively, the terminals communicate directly with a higher-level production control computer, see ◘ Fig. 14.14. Terminal providers offer a wide range of interfaces from RS 232 to Ethernet.

## Functions of Production Data Processing

A few basic production data processing functions, which should usually run automatically, must be available for the normal operation of PDA terminals. These include:

- acquisition of production data from the manufacturing system and archiving,
- updating the production data in the event of failure and restarting by comparing with the higher-level computer,



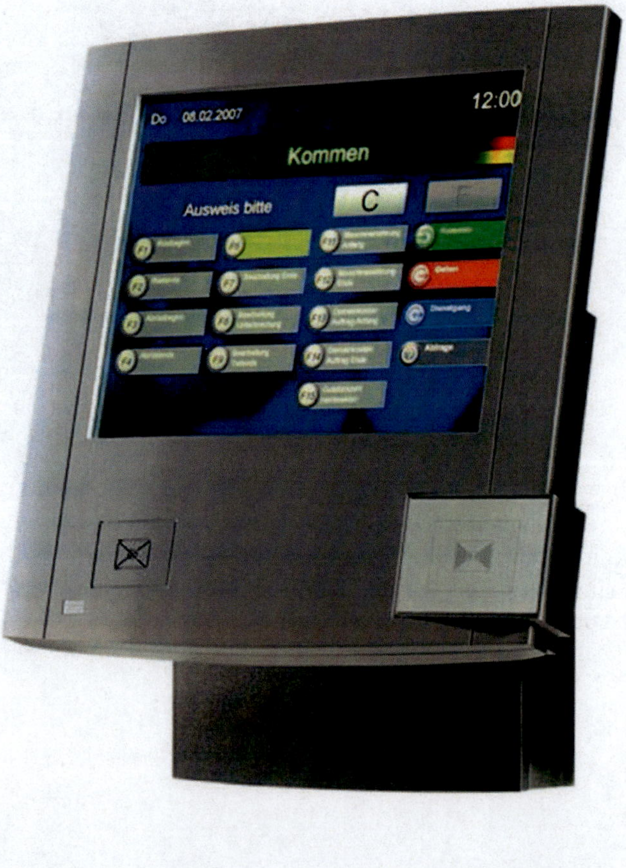


◘ **Fig. 14.13** PDA terminal. (*Source* KABA)

**Fig. 14.14** Integrated PDA solution

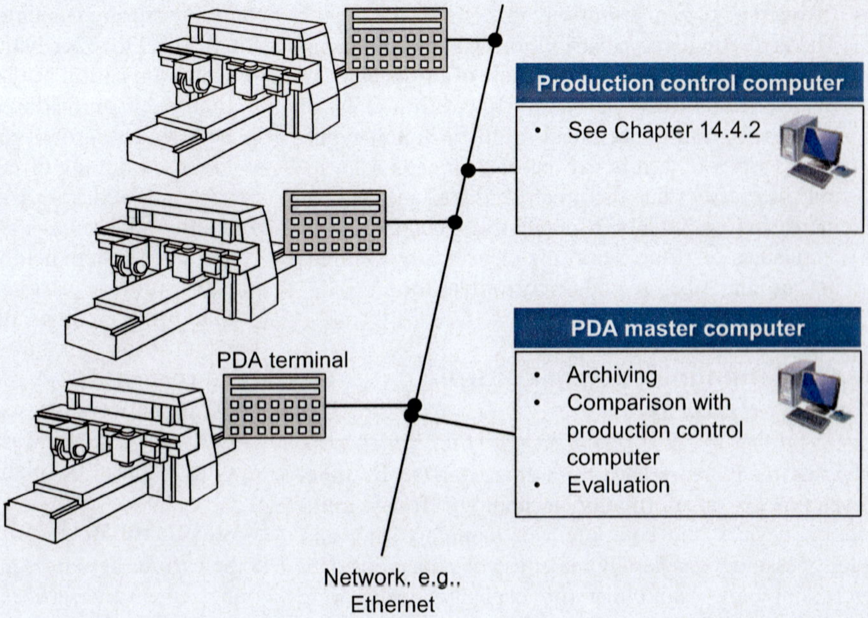


**Fig. 14.15** Usage report on a machine. (*Source* MPDV)

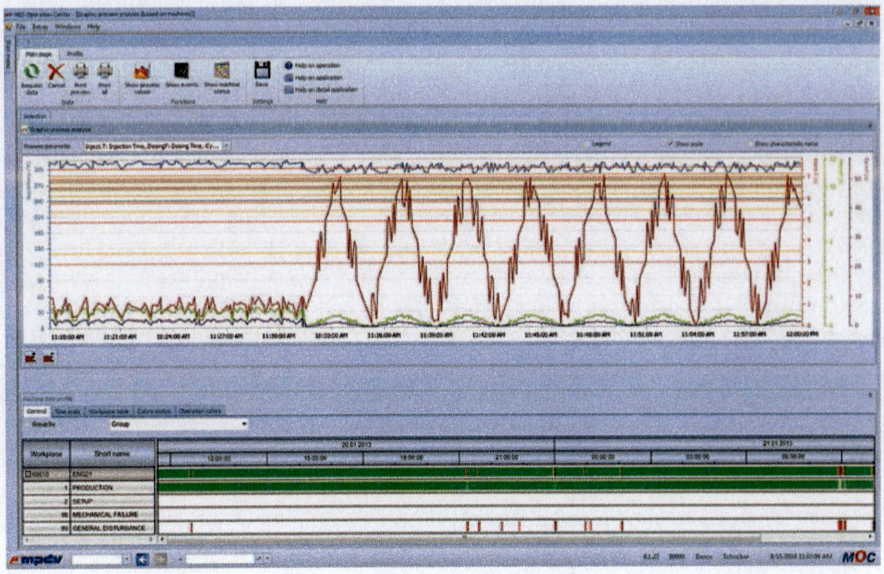

- compressing the pool of production data from past manufacturing cycles (shifts, days, weeks, months) into data for the next-longest period, and.
- transfer of chronologically sorted production data to higher-level computers.

Statistics presented in a graphical form provide an organized evaluation of system events. This includes Gantt or runtime charts showing machining times versus time. It can also show disruptions and bottlenecks that occurred in the past.

Figure 14.15 shows the usage report for a machine. This report provides an overview of the duration of various activities or disruptions for the maximum possible usage time. The reasons for the relatively low machine usage are the high setup time ratio and lack of orders. Other typical evaluation graphics are:

- Utilization statistics
  Utilization statistics show the utilization of an individual machine or a capacity group over a specific time period.
- Error statistics
  Error statistics are based on the error count categorized by causes of the error. This also captures the relevant standstill time of the individual machining stations.

- Error frequency distribution
  The error frequency distribution is shown as a histogram sorted by the frequency of a specific error over a period. When combined with a chart of the distribution of technical standstill times, significant causes of errors can be determined. Alleviating these errors has the greatest impact in terms of increasing availability. It is possible to check the effectiveness of troubleshooting measures by comparing the charts before and after troubleshooting.

### 14.3.3 Communication in Control Technology

The last thirty years have been characterized by rapid developments in information technology. Today, automation devices and computers in manufacturing and quality assurance generally make use of enormous data processing power and memory capacity, which were barely imaginable just a few years ago and which are nevertheless more affordable than ever before. This has caused tasks to be handled on a distributed basis. Data can be and are currently processed close to where they are created. The 'locally' processed data are forwarded via appropriate communication units to a 'partner' for further processing. The networking of individual automation devices, measuring devices, and computers via serial bus systems (local networks, fieldbuses) is a key function in the resulting distributed data processing systems.

Network concepts for IT-based linking of components in distributed data processing systems have been developed since the beginning of the 1980s. Due to various technical requirements as well as manufacturers' corporate interests, these network concepts have, in the meantime, produced a large number of different serial bus systems, which have gained traction to a greater or lesser degree. From a business perspective, the choice is too great. This led to the German Electrical and Electronic Manufacturers' Association (ZVEI) publishing a list in April 1993 of approximately 90 different bus concepts or bus systems used in Germany. Today, bus systems based on Ethernet technology are increasingly becoming the state of the art.

#### 14.3.3.1 Communication Segments in the Manufacturing Sector

The comparison of vertical and horizontal flows of information in the various levels of a business reveals different requirements with respect to the necessary data throughput performance, the data volume to be transferred and the transfer or response time behavior on the individual levels, see ◘ Fig. 14.16. Businesses in the unit manufacturing industry have formed four different communication segments through a process of evolution [FRIE93b; FRIE94]:

- Communication at the actuator/sensor level, i.e., actuator/sensor fieldbuses, is used to exchange data between actuators, sensors, and device controls.
- IT-based linking of controls for automation devices with hierarchically arranged technical systems at the command level is achieved through local cell networks or system fieldbuses.
- Integration of various parts of the company (manufacturing facilities, office buildings) and their communications units uses local backbone networks at the command level.
- Office networks based on local networks are currently used to network technical and administrative office areas at the planning and command level.

**Actuator/Sensor Fieldbuses (Sampling Buses)**

If the various actuators and measured value sensors are connected on the sensor/actuator level in discrete time control loops, the data must be transferred between them and the relevant device control on a cyclical basis. The cycle times must be significantly less than the characteristic time constants for the control process. Short messages with a length of only a few bits (approximately 10 to 50 bits) are transferred and processed at the actuator/sensor level in only a few milliseconds (typically < 10 ms). Special fieldbuses have established themselves in recent years as the communication units for the actuator/sensor level; these are known as actuator/sensor buses. These buses are sometimes referred to as 'sampling buses' in the literature due to their control or measuring function.

A representative actuator/sensor bus is the standardized SERCOS bus. This was designed for communication between controls and drives in numerically controlled machines. ABUS, LIN, CAN bus and FlexRay are used in German automobiles to network the actuators and sensors. These bus systems are now also being used, or recommended for use, in the industrial sector. INTERBUS-S is another representative system used for networking controls, drives, and sensors/actuators in the process data field (see ► Sect. 5.3.5). Bus systems such as PROFINET or EtherCAT (Beckhoff, manufacturer-specific) were developed based on real-time Ethernet.

**Cell Network/System Fieldbuses**

Commands and alarms, which are generally 100 to 500 bytes long, are typically exchanged between the various communication nodes at the control level. The demands on the transfer time behavior of the communication units are generally low as time-critical or safety-relevant operations are carried out locally due to the intelligence of the device controls. So, for example, in

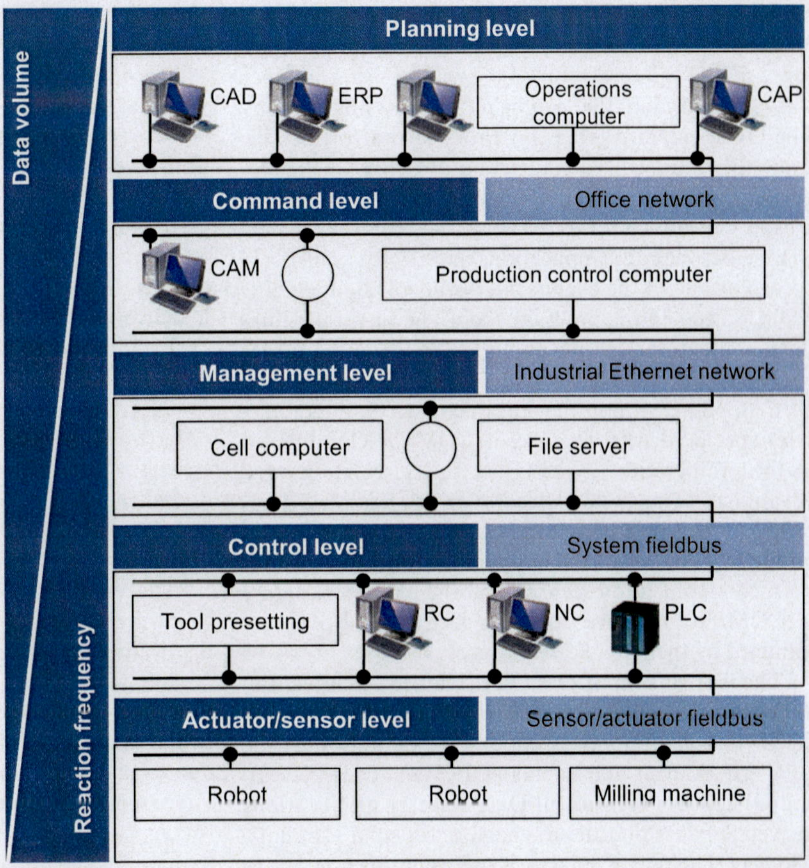


**Fig. 14.16** Communications systems in the manufacturing sector

the event of a collision on an NC machine detected and reported to the cell computer, the machine does not wait for a switch-off command but turns itself off automatically. This means that the alarm need not be transmitted in a few milliseconds but rather in a few tenths of seconds or seconds.

For technical reasons, the maximum net transfer rates required at the control level to transfer the file between the NC machine control and the cell computer are 40,000 bits/s [FRIE93a]. Ethernet is almost exclusively used today when building new cell networks. However, solutions such as the Siemens SINEC-H1-Bus or system fieldbuses such as the DIN measuring bus (DIN 66348), the Profibus (DIN 19245), the French FIP bus, or the Japanese FAIS bus are still used in many cases.

### Industrial Ethernet Networks

The use of classic Ethernet technologies in the production environment led to the idea of introducing a standardized communication infrastructure that provides company-wide communication and automation from the office to the machine or sensor. This allows process and production data to be processed not only at the field level (machine level) but also in systems that span divisions. Bus-specific infrastructure components (gateways, routers, bridges) are necessary to connect the different fieldbuses to the higher-level network.

### Office Networks

Information exchange in office areas at the higher corporate levels is generally not telegram-oriented like the actuator/sensor and control levels. Principally, it comprises data transfer, sending emails or large volumes of data such as CAD drawings. The horizontal and vertical data flows at these levels have a greater width, i.e., scope. However, the demands on the response time behavior of the data transfer system are significantly reduced.

Products such as Fast Ethernet (100 Mbit/s), Gigabit Ethernet (1000 Mbit/s) or WLAN (up to 1300 Mbit/s) are used for office networks. Gigabit adapters are nowadays a standard component in modern computers. WLAN adapters are state-of-the-art in mobile computers and desktop systems can be fitted with WLAN adapters (up to 150 Mbit/s) for less than 5 euros.

#### 14.3.3.2 Software Interfaces

IT networking in control technology requires not just the use of suitable protocols but also standardized software interfaces that can be used across applications. The OPC standard (OLE for Process Control) has become the standardized connection of drives for heterogeneous automation components. OPC consists of Microsoft COM/DCOM technology. This enables data exchange between PC applications and device controls in accordance with the client/server principle by including COM objects in the program code. The server captures the manufacturer-specific communication with a device control so that the client can access information via the OPC standard without having to contact the manufacturer-specific driver. The use of COM/DCOM has led to the proliferation of OPC but this also has its disadvantages. The DCOM technology is tied to Microsoft Windows and there are frequently communication problems. The lack of configurable timeouts, insufficient communication security and the black-box nature of COM/DCOM have led to the development of a new standard by the OPC Foundation.

The first version of OPC UA (Unified Architecture) [DIN10] was published in 2006. The motivation for the OPC Foundation was initially the migration of the OPC AE (Alarms and Events), OPC DA (Data Access) and OPC HDA (Historical Data Access) specifications to Web Services in order to enable access in a standardized way. The term OPC UA arose from this attempt to unify communications. Employees from up to 30 companies participated in this standardization work.

14

In addition to a pure communication protocol, OPC UA also provides an information model for the semantic description of the available data. The data can be transported either via an optimized binary TCP protocol or standards such as Web Services, XML and HTTP can be used. These provide firewall-friendly communication over the internet. Communication is not limited to these protocols and may be supplemented with other protocols in the future.

## 14.4 Process Command Level

### 14.4.1 Electronic Control Station

#### 14.4.1.1 Tasks of Shop Floor Control Systems

The task of shop floor control is to execute and monitor the orders provided by Enterprise Resource Planning. While work stages are planned to capacity groups in an ERP system, a detailed machine schedule is required at the shop floor level. To this end, the shop floor supervisor or the work distributor selects the next work stages to be completed from the available worklist and allocates them to the individual workstations. The result of this detailed order planning is a machine schedule giving the start and end dates of the work stages with their sequence allocated to the relevant workstations.

Many constraints have to be taken into consideration as part of this detailed planning. These include the availability of personnel, machines, tools, NC programs, and material at the planned start time of the work stages. Not taking these constraints into account can lead to unintended machine downtimes and delays in delivery schedules.

In addition to this, certain company-specific goals such as high machine utilization, short throughput times, or low stock levels must also be maintained. In order to achieve these goals while taking the many restrictions into consideration, other resources are required that not only ease the planning process in itself but also allow alternative planning runs and enable an evaluation of the planning results.

The workstations are tasked with order execution following the machine schedule. In addition to work distribution, another task is monitoring the progress of the work. This compares the scheduled completion dates with the current situation on the shop floor. In the event of significant deviations, which may occur due to machine failure, overloading the shop floor environment, defective production, or organizational errors, appropriate replanning or new plans must be put in place for the purposes of 'production control.'

**Functional Scope of Electronic Control Stations**

There are many resources available today to support shop floor control. Among the systems that are not computer-aided, the scheduling board is worth mentioning. Orders are tracked on a board using plug-in cards. The available machines are entered on the y-axis of this board while the x-axis represents time. The work stages allocated to a workstation are entered one after the other in the appropriate sequence. This allows the individual machine utilization and work progress to be controlled.

Simple computer-aided shop floor control systems are restricted to an alphanumeric representation of the order data on terminals and allow orders to be allocated to machines. Various graphic representations are available to evaluate the effectiveness of this planning stage. The operator has the option of continuing to optimize machine utilization until a specified target criterion has been reached.

Electronic control stations replace the classic scheduling board with a window-oriented, completely graphical representation providing all the options for interactive planning and rescheduling using a computer mouse [KÖHL90]. ◘ Figure 14.17 shows a typical user interface on a control station. The machine schedule shows the sequence of work stages on the machines versus

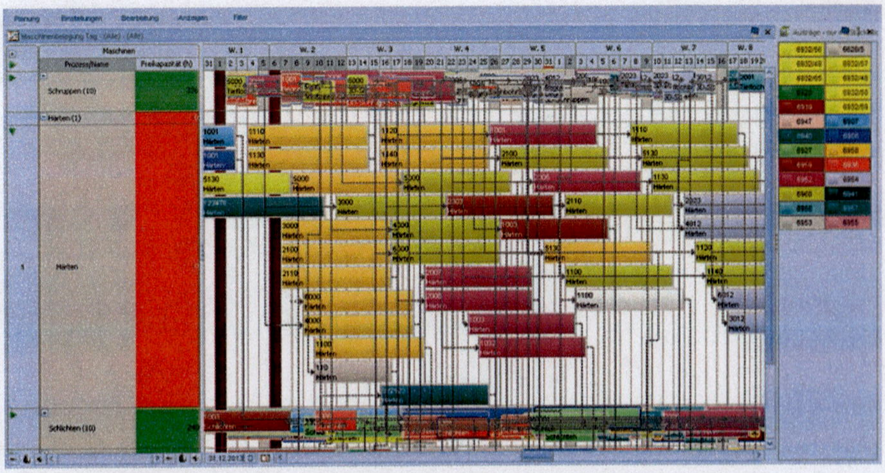

**Fig. 14.17** Graphical control station (commercial IT service provider ▶ www.plantafel-software.biz)

time. A detailed view of any size comes from scaling the time period to be shown.

Orders with work stages yet to be scheduled are contained in the worklist. These work stages can be moved to a suitable point in the machine schedule with the mouse allocating the stages to a machine. The control station completes many plausibility checks during this procedure. For example, a work stage may not be scheduled with a start time before the end time of the previous work stage.

The systems on offer today also provide the option of automatic sequence planning. The operator can choose between many algorithms that use a range of target-oriented criteria for optimization. The machine schedules determined from multiple simulation runs can then be evaluated with the aid of graphical evaluation charts. For example, a bar chart for the machine utilization allows the planning results to be assessed with respect to the target criterion of 'maximum utilization.'

Another important point is processing feedback from production. A color graphical representation of the scheduling board allows the work stages to be colored in accordance with their current status. Work stages that have been allocated, started, finished, disrupted, or delayed can then be easily identified. This improves the transparency of the scheduling board significantly.

Control stations offer a large number of interfaces for integration into the operational environment as shown in ◘ Fig. 14.18. Production orders can be moved from Enterprise Resource Planning to the local database by file transfer. Feedback from production can also be passed on to the ERP system via these interfaces. The distribution of orders to the workstations can be completed by printing order papers or by connecting to PDA terminals to use a paperless manufacturing system. It is also possible to control the machines directly via a DNC connection. As the interfaces described are currently still not standardized, these types of connections usually need to be adapted to customer requirements.

Today's control stations offer many different functions in addition to the scheduling board functionality such as.

- PDA,
- DNC,
- NC program management,
- production resource management, and.
- material flow control.

The boundary between control stations for shop floor control and production control computers as described in Sect. 14.3 is becoming increasingly blurred.

### 14.4.2 Production Control Computers

Production control computers represent the highest level of computer-integrated production control. In addition to the management and planning functions described in the previous sections, they also provide direct control of machines as well as handling and transport systems. Their main point of focus is the control of flexible manufacturing systems (FMS).

#### 14.4.2.1 Functional Scope of Production Control Computers

Production control computers have to provide a large number of functions due to their complex control tasks [WECK90]. The functions described below are used in nearly all production control computers in different forms.

- Order management
  Order management takes production orders from higher-level systems such as Enterprise Resource Planning or an area computer and transfers completion messages to these systems.

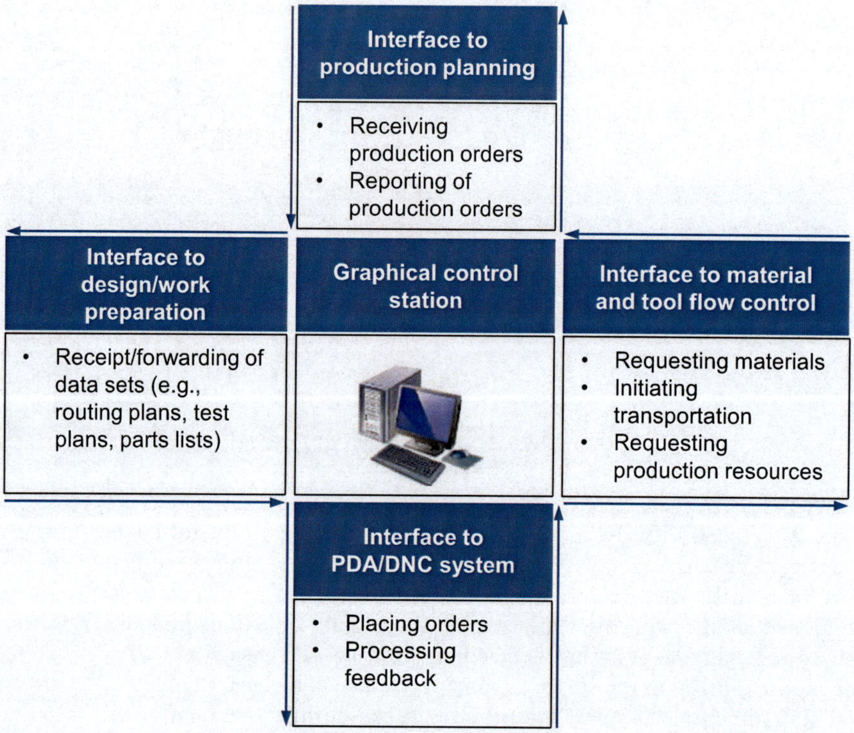


**Fig. 14.18** Control station interfaces

- Detailed planning
  Detailed planning generates a detailed utilization plan for the machines, setup stations, and clamping locations.
- Sequence control
  Sequence control directs the internal processes for flexible manufacturing systems (FMS) required to execute the detailed plan. This includes distribution of work stages and handling material flow control by providing workpieces to the workstations.
- Material flow control
  Material flow control is responsible for providing material and transportation of this material between warehouses, buffer areas, and workstations.
- Tool management
  Tool management has to calculate the tool requirements as well as management and transfer of the tool data, handle deployment planning, reserve the required tools and control tool provision and refurbishment.
- Cell order coordination
  The necessary machining steps must be coordinated when machining within a cell. To this end, various NC programs must be started one after the other or robots and machines have to be coordinated.
- Communication with lower-level device controls (DNC)
  Special driver modules are required to communicate with the lower-level device controls. These driver modules convert control commands into control-specific communications protocols. These DNC connections are used to transfer NC programs, tool data and other parameters, to issue commands to start programs and to receive status messages.
- Process visualization
  Process visualization has two objectives. First, it must inform the user of the current system status and then offer options for altering the system status. To this end, users are shown current orders and their planned completion times as well as their current machining status. This is typically provided at a control station. The user can then bring urgent orders forward or hold back the machining of other orders as required. Also, the user is provided with an overview of the entire system. This overview shows, for example, utilization of pallet spaces, the current transportation and machining processes and whether there any disruptions anywhere. At this point, the user has to be able to actively intervene in system events to control the plant either manually or semi-automatically or to resolve error states.
- Machine data acquisition, production data acquisition (MDA, PDA)
  Collecting machine and production data is a precondition for the automatic control of flexible manufacturing systems. Decisions are made based on events and the next machining steps initiated. Large volumes of statistical data provide information on the manufacturing progress within a production cycle.

14

- Quality assurance
  Quality assurance in the field of manufacturing includes the design of quality cycles. This includes, for example, monitoring given test criteria at specific test stations or measuring the workpieces directly on the machine. The test results allow conclusions to be drawn on defects in the production process that must be resolved through quality control measures.
- Maintenance, diagnosis
  Maintenance has particular significance, especially in automated systems, as many disruptions can be avoided through preventative maintenance. Diagnostic systems that allow quick recognition and resolution of the causes of error are used to remedy disruptions during plant operation.

## 14.4.3 SCADA Systems

SCADA (Supervisory Control and Data Acquisition) systems operate and monitor technical plants. SCADA can be used to monitor individual sensors and programmable logic controllers as well as complete plants. SCADA generally makes use of an HMI (Human Machine Interface) to visualize the state of the monitored plant.

### 14.4.3.1 Tasks

SCADA systems are used to operate and monitor technical systems. This includes querying data and data management, archiving, and visualization. The technical systems can be restricted to specific plant areas or be used to monitor the whole site and operate plants. SCADA systems often incorporate very high levels of technology and use extremely detailed production data that is provided in real time. SCADA systems are used to control specified operations and to ensure their performance.

Visualization in SCADA systems has two objectives. First, it must inform the user of the current system status and then offer options for altering the system status. To this end, users are shown a graphical overview of machine utilization, the current orders and their planned completion times as well as their current machining status in real time. The user can stop a machine if required, for example, in the event of an error, or start an order. Another typical function of a SCADA system is the availability of order history to ensure process traceability.

### 14.4.3.2 Concepts

One SCADA system architecture often used is a client/server design. The server collects data from lower-level devices and saves these in a database system. The lower-level devices are connected over the RS232 interfaces, fieldbuses or a network system. On the client side, users can access the server data via a network connection.

TCP/IP-based communication protocols are increasingly being used for communication in SCADA systems. The SCADA system design often uses PC technology. Standardized protocols such as OPC UA, MQTT or DDS are used to connect devices.

Current SCADA concepts are based on requirements such as engineering efficiency, easy scalability, and openness. Engineering efficiency addresses the preparation of a SCADA system for its use in the company. This initially requires establishing communication between the server system and the plant or components to be monitored. In addition to this, the plant structure must be reflected in the SCADA system. This includes the plant components and their monitored data. Libraries of graphical elements mapping the standard elements of production facilities are often provided for the visual display. The demand for scalability refers to the possibility of expanding the SCADA system with other system components, servers, and clients to be monitored. The demand for openness refers to the use of documented interfaces which allows system extensions and connection to other components based on various protocols.

Mobile solutions, which can be used with smartphone and tablet hardware, are increasingly being offered in the SCADA field. In these cases, the data held on a stationary server in the network or a higher-level system in the cloud is used for visualization.

As SCADA systems use detailed process data, commercial solutions generally provide a function for extracting these data by an MES. In terms of vertical integration, the data exchange takes place between SCADA and MES through the automation pyramid.

## 14.5 Smart Automation Lab: Industry 4.0 Research Laboratory

The activities in the field of Aachen-based production research, particularly with respect to basic research topics, have been receiving a lot of support in Germany since 2006 and are being guided through the Excellence Initiative of the German federal and state governments. The Integrative Production Technology for High-Wage Countries cluster of excellence represents an overarching organization of more than 20 affiliated research institutes at the RWTH Aachen University [o.V.11]. A range of scenarios are addressed in the field of development and evaluation of new control technologies and concepts. These range from individual machine tools to interlinked, automated production systems.

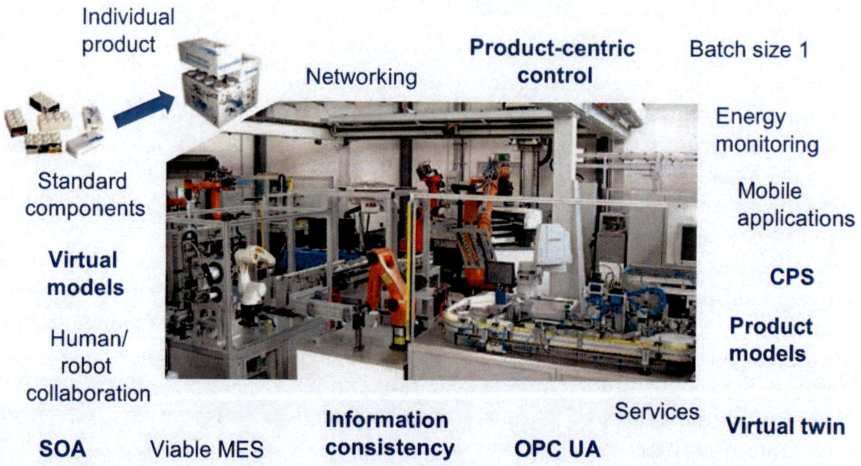


**Fig. 14.19** Topics at the Smart Automation Lab

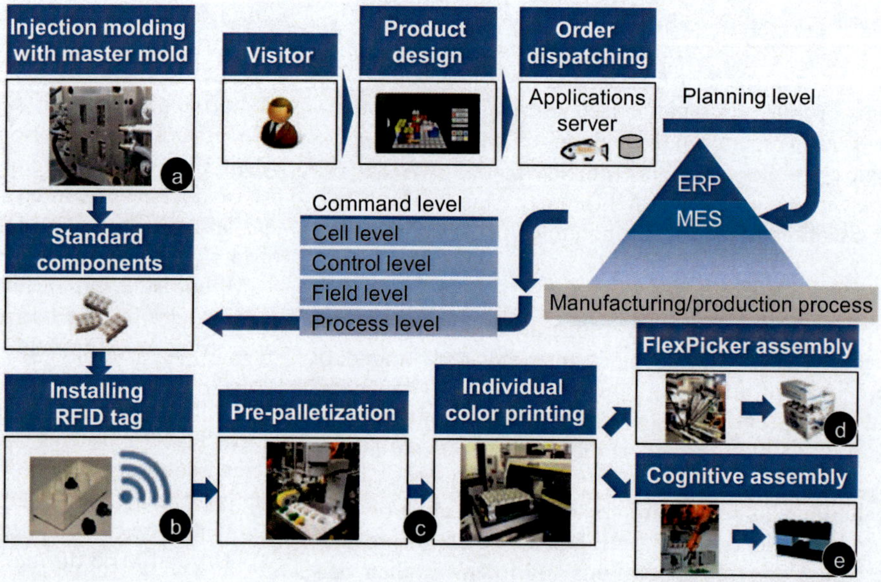


**Fig. 14.20** Structure of the production system for manufacturing individualized products

These, and other current research projects in the field of computer-aided manufacturing, are bundled at the Laboratory for Machine Tools and Production Engineering (WZL) at RWTH Aachen University using the *Smart Automation Lab* and linked together to maximize synergy [BREC15]. The various individual projects each focus on specific elements. These are under consideration with the implementation of concepts from Industry 4.0 and they stem from the growing degree of plant networking and the increasing vertical exchange of information [BREC14], Fig. 14.19. The Smart Automation Lab encompasses the complete automation pyramid of a modern production system, from the corporate to the field level. As such, it allows an all-encompassing validation of both the affiliated engineering concepts and the control concepts being researched.

### 14.5.1 Product-Centric Control

At the *Smart Automation Lab,* visitors are able to order a selected sample product and to individualize it based on specific features. Manufacturing the order, which is introduced at the planning level, produces a linked automated plant consisting of different stations that modify the individual manufacturing steps based on the centrally administered order data, see Fig. 14.20.

This is the scenario for manufacturing customer-specific printed assemblies from individual Hubelino® toy bricks. Here, the basic geometrical elements are first produced as standard components in an injection molding process using a master mold, Fig. 14.20a. The individual parts are then fed to the machining station as bulk material via a spiral feed system. Here, they undergo finishing (automated removal of sprue) and an RFID transponder is added, see ◘ Fig. 14.20b. The RFID transponder allows linking individual parts to the relevant order data at the planning level via a unique identification. The following stations can read this identification and extract the relevant order information from the product data model used as a base.

In the third step, the required number of components are pre-palletized into batches and fed to a color printer by a robot with special gripper kinematics, ◘ Fig. 14.20c. The gripper kinematics allow the components to be quickly gripped and reoriented so that they can be printed on several sides and maximum use is made of the printer capacity. After the printing process, the individual parts are assembled in two different assembly stations, depending on the order.

- If the product is a cube made of 6 individual parts, the component parts are fed to a FlexPicker station and assembled, ◘ Fig. 14.20d;
- If the product is not a cube, the necessary components are fed to the cognitive assembly cell, ◘ Fig. 14.20e. Here, automated assembly planning and execution, based on the CAD model of the final assembly, is completed without any manual intervention.

As part of the design, the necessary resources and fixtures have the basic capacity to produce individual parts differing from order to order without any change-over. The key element in achieving this is flexible production control (see Manufacturing Execution Systems—MES, Sect. 14.3.1). This integrates the various plant areas into the order processing procedure by triggering plant functions and provides the order-specific information required for production. This information includes, for example, image files, which are required at the printing station to customize individual parts, or CAD models as the target state for the cognitive assembly cell.

During order dispatching at the individual station level, the order-specific product model is passed to the station by the MES at the same time as the objects actually required. In this way, the relevant information can be extracted automatically from the product model. This reduces the control complexity as, once a uniform product model structure has been established, stations can access the product features directly in order to adjust their behavior. Due to the particular significance of product modeling and the implications for production control, which came from this, the term product-centric control is often used [BREC14].

### 14.5.2 Plug & Produce

Flexible and demand-oriented automation is becoming increasingly important to increase production capacities over the short term or to replace plant components during maintenance. Robots with integral safety functions are used in the *Smart Automation Lab* to integrate mobile resources into plant clusters flexibly and with minimum effort by using Plug & Play approaches, see ◘ Fig. 14.21. The robot systems use sensor systems for self-referencing or localization with the aid of markers added to the plant. In addition to this, components can be optically identified and flexibly gripped so there is no need to repeat the teach-in process. The Bosch APAS robot shown here can detect contact through a sensor skin. This allows robots and people to work together (see Collaborative robot designs, ▶ Sect. 13.2.6.3).

◘ **Fig. 14.21** Mobile industrial robot with integral safety and vision technology. (*Source* Bosch)

The interfaces between resources and MES use open communication protocols with OPC Unified Architecture (OPC UA). These protocols make it easy to display plant functionalities in the form of services for the MES. One benefit of this approach is the dynamic linkability of services through the MES. Plant components can thus be removed from or added to the overall system while running without affecting communication with other stations. Functions are triggered by events. This does not produce wait times on the control side as there are no direct completion messages from the triggered process. In the *Smart Automation Lab* environment, this form of communication results in a 'push' mechanism where the control system moves assembly objects through the production system, triggers individual processes and transfers the objects to the subsequent station once a completion message has been received.

### 14.5.3 Cognitive Assembly

The cognitive assembly cell in the *Smart Automation Lab* uses virtual product models to plan and coordinate assembly-related processes. The CAD model of an assembly, configured and transferred by the visitor for

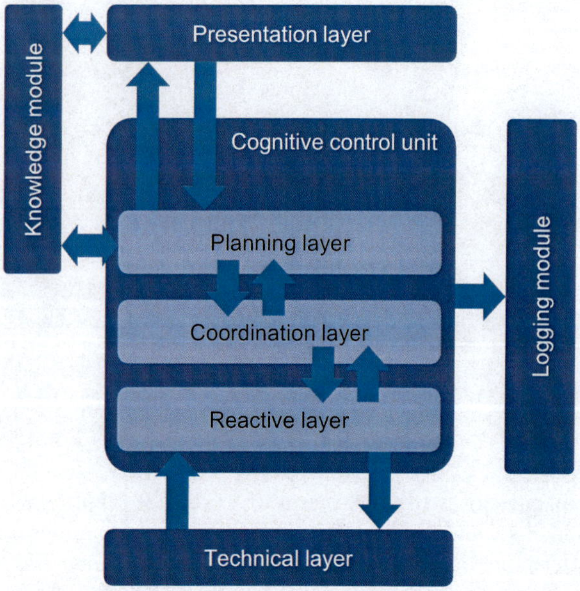


Fig. 14.22 Concept of cognitive architecture

production, corresponds to the target state of the product and is the only manual intervention required for assembly. A Cognitive Control Unit (CCU) uses this input to plan the necessary steps required for assembly and coordinates the operations using the information on the current plant situation [BREC12]. The control system comprises various layers which include the relevant functions with different demands concerning execution time or speed and the scope of the amount of information to be considered to complete the task, see Fig. 14.22.

The presentation layer includes a human/machine interface as well as an interface for modifying a knowledge base (knowledge module). This contains a domain-specific set of planning rules (in this instance, assembly). The handling options are weighted in the planning layer and the actual planning of the next action of the current assembly process takes place here as well. The planning layer uses a range of different services, provided by the coordination layer, to execute the assembly steps. The purpose of the coordination layer is to abstract and standardize the resources involved in the assembly process with respect to their specific control interfaces. Below the coordination layer, the architecture with the reaction layer contains a module that performs the tasks with real-time requirements such as optical positioning control of the industrial robot or safety functions in the cell by means of the integral technical systems (technical layer), see Fig. 14.23.

In this scenario, the components from the *Smart Automation Lab* are fed to the cognitive assembly cell via a pallet conveyor system. Individual components are then placed on a circumferential conveyor belt by a gripper portal, and this builds up the number of objects forwarded to assembly. The robot system selects the necessary components required to fulfill the order from these objects, synchronizes itself to the conveyor belt speed and picks the appropriate objects. Depending on the assembly progress, components can be mounted directly or first placed in a buffer area. Preassembled modules, used within the assembly area to complete work orders, can be transferred to the system via a second interface. Finally, the finished assemblies are removed from the cell via a transfer window.

The system receives important constraints for planning the operations from continuous evaluation of the current assembly progress, see Fig. 14.24. Complex assemblies can generally be produced by a variety of equivalent step sequences which lead to the same target state. However, there are restrictions in terms of assembly options for individual intermediate conditions that, for example, result from the disruption contour of the robot's gripping devices that may not be crossed.

In addition to this, components are fed to the circumferential conveyor belt in a chaotic order and are then

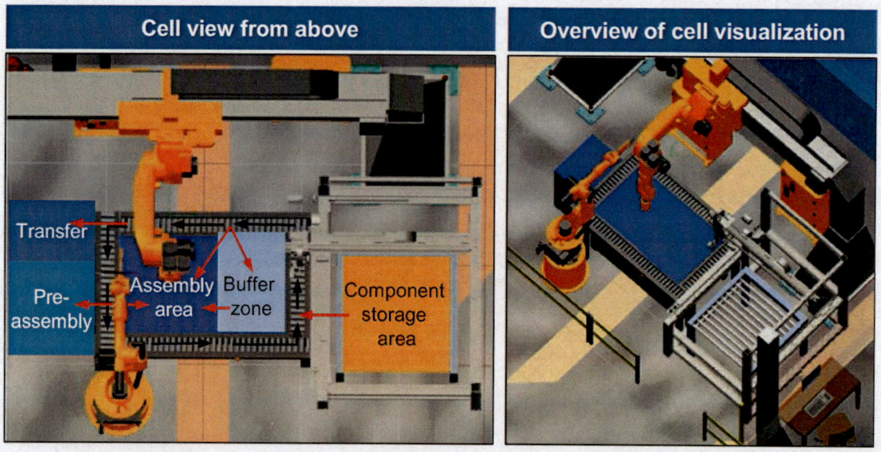


Fig. 14.23 Assembly cell design

**Fig. 14.24** Sensor-based capture of the assembly situation

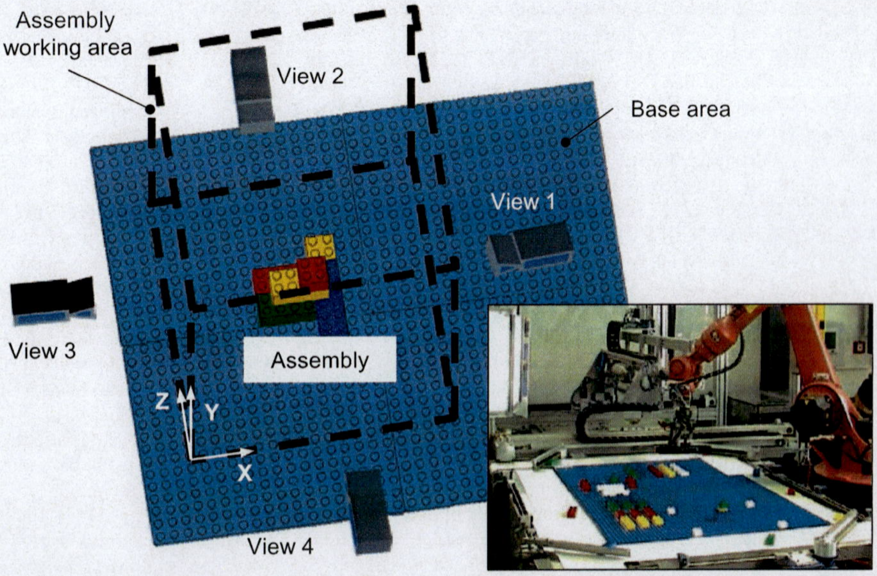


classified using image processing by their dimensions and color. This (externally influenced) change of state means that assembly planning and optimization have to be continuously adjusted. This is beyond the scope and flexibility of today's industrial control concepts.

## References


[AEG88] AEG Aktiengesellschaft: Produktionsleittechnik MCP (1988)

[AUTO93] Autorenkollektiv: Wettbewerbsfaktor Produktionstechnik. (Reihe: Aachener Werkzeugmaschinen Kolloquium). Düsseldorf: VDI-Verlag (1993)

[BAUE16] Bauernhansl, T., Krüger, J., Reinhart, G., Schuh, G.: WGP-Standpunkt Industrie 4.0: Abele, E. Wissenschaftliche Gesellschaft für Produktionstechnik WGP e. V., (2016)

[BING16] Bingler, D., Sontow, K.: Die Zukunft von ERP im Kontext von Industrie 4.0. Berlin (2016)

[BOYW12] Boywitt, J.: Referenzstrategien zur ERP-Einführung im mittelständischen Maschinen- und Anlagenbau. (Reihe: Edition Wissenschaft). Aachen: Apprimus (2012)

[BREC11] Brecher, C. (Hrsg.): Integrative Produktionstechnik für Hochlohnländer. (Reihe: VDI-Buch). 1. Ed. Berlin: Springer (2011)

[BREC12] Brecher, C., Breitbach, T., Müller, S., Mayer, M. P., Odenthal, B., Schlick, C. M., Herfs, W.: 3D Assembly Group Analysis for Cognitive Automation. Journal of Robotics (9), 1–18 (2012)

[BREC13] Brecher, C., Müller, S., Breitbach, T., Lohse, W.: Viable System Model for Manufacturing Execution Systems. In: Procedia CIRP **7**, 461–466 (2013)

[BREC14] Brecher, C., Herfs, W., Lohse, W., Obdenbusch, M.: Vorsprung durch Wissenseinsatz: Industrie 4.0. Infoline **17** (3), 6–10 (2014)

[BREC15] Brecher, C.: Smart Automation Lab. URL: ► http://www.smartautomationlab.de. Accessed: 05. März. 2015.

[DANG90] Dangelmaier, W.: Die C-Techniken - von CAx zu CIM. In: Zeitschrift für Logistik (1990)

[DIN10] Norm DIN 62541: OPC Unified Architecture (2010)

[DIN83] Norm DIN 66257: Numerisch gesteuerte Arbeitsmaschinen. Begriffe (1983)

[DIN96] Norm DIN ISO 2806: Industrielle Automatisierungssysteme - Numerische Steuerung von Maschinen - Begriffe (1996)

[ECKE87] Eckelmann Industrieautomation: Betriebsdaten-Workstation für flexible Fertigungssysteme: Systembeschreibung. Wiesbade Firmenschrift (1987)

[EVER90a] Eversheim, W.: Organisation in der Produktionstechnik. Band 1: Grundlagen. Düsseldorf: VDI-Verlag (1990)

[EVER90b] Eversheim, W.: Organisation in der Produktionstechnik. Band 3: Arbeitsvorbereitung. Düsseldorf: VDI-Verlag (1990)

[FAYZ10] Fayzullin, K.: Autonome Planung und Entscheidungsoptimierung in der Ablaufsteuerung flexibel automatisierter Fertigungssysteme. (Reihe: Ergebnisse aus der Produktionstechnik Werkzeugmaschinen. Vol. 2010,25). 1. Aufl. Aachen: Apprimus-Verl (2010)

[FRIE93a] Friedrich, A.: Offenes DNC-Kommunikationssystem für numerisch gesteuerte Arbeitsmaschinen. (Reihe: Fortschritt-Bericht VDI, Reihe 20: Rechnerunterstützte Verfahren). Düsseldorf: VDI-Verlag (1993)

[FRIE93b] Friedrich, A., Friedrich, J.: Netzwerkkonzepte - Mit Feldbus und LAN zum integrierten Informationsverbundsystem. In: Workshops V: Automatisierung in der Qualitätssicherung durchgängige Lösungen für die Messtechnik, 7. Kongressmesse für industrielle Messtechnik MessCompü'93: Network GmbH (Veranst.)

[FRIE94] Friedrich, A., Friedrich, J.: Datenübertragung in der NC-Verfahrenskette. In NC-Verfahrenskette, p. 105. Beuth-Verlag, Berlin (1994)

[HACK86] Hackstein, R.: Klare Definition der C-Techniken. Markt & Technik (36) (1986)

[KLEM86] Klemmer, J.: Erfahrungen aus einer 10-jährigen DNC-Praxis im Flugzeugbau. ZwF **81** (2) (1986)

[KLET06] Kletti, J.: MES - Manufacturing Execution System XE "Manufacturing Execution System" . Springer,

Moderne Informationstechnologie zur Prozessfahigkeit der Wertschopfung. Berlin (2006)

[KÖHL90] Köhler, C.: Der elektronische Leitstand Befehlsempfänger der PPS oder Partner der Werkstatt?. VDI-Z **132** (3) (1990)

[MACC97] MacClellan, M.: Applying manufacturing execution systems. (Reihe: The St. Lucie Press/APICS series on resource management). Boca Raton, Fla., Falls Church, Va.: St. Lucie Press; APICS (1997)

[MESA97] MESA: MES Explained.: A High Level Vision. White Paper No. 6. Pittsburgh (1997)

[N.N.99] N.N.: CAD/CAM. Computer@Produktion (4) (1999)

[NCG90] NC-Gesellschaft Arbeitskreis DNC.: DNC im technologischen Umfeld (1990)

[NEIP91] Neipp, G., Stracke, H.-J.: Einführung in die CIM XE "CIM" -Praxis. VDI-Verlag, Düsseldorf (1991)

[OSTE14] Osterhage, W. W.: ERP-Kompendium. Eine Evaluierung von Enterprise Resource Planning Systemen. (Reihe: Xpert.press). Berlin: Springer Vieweg (2014)

[POSS06] Possel-Dölken, F.: Projektierbares Multiagentensystem für die Ablaufsteuerung in der flexibel automatisierten Fertigung. (Reihe: Berichte aus der Produktionstechnik. Vol. 2006,10). Aachen: Shaker (2006)

[PRIT91] Pritschow, G., Spur, G., Weck, M.: Leit- und Steuerungstechnik in flexiblen Produktionsanlagen. Hanser-Verlag, München (1991)

[ROBE87] Robert Bosch GmbH.: CC 200/300 BABNET Kommunikationsverfahren. Erbach. Firmenschrift (1987)

[ROBE88] Robert Bosch GmbH.: Tool-Management-System TOMS. Firmenschrift (1988)

[SAUE10] Sauer, O.: Trends in Manufacturing Execution Systems. In: Kacprzyk, J.; Huang, G. Q., Mak, K. L., Maropoulos, P. G. (Ed.) Proceedings of the 6th CIRP-Sponsored International Conference on Digital Enterprise Technology. (Reihe: Advances in Intelligent and Soft Computing), S. 685–693. Berlin: Springer (2010)

[SIEM91] Siemens; Brill, M.: Offene Netze für die industrielle Kommunikation. Firmenschrift (1991)

[STEG11] Stegemann, A.: Konradin ERP-Studie2011: Einsatz von ERP-Lösungen in der Industrie. Leinfelden-Echterdingen, Firmenschrift (2011)

[TÖNS89] Tönshoff, H. K., Martens, Menzel.: DNC - Entwicklung, Konzepte, Funktionen, Tendenzen. Vom Direct Numerical Control zum Distributed Numerical Control. VDI-Z **131** (9) (1989)

[TROV17] Trovarit AG: Alle ERP-Lösungen. URL: ► http://www.erp-matchmaker.com/alle-softwareloesungen.html. Accessed: 29. Mai. 2017.

[VDI16] Richtlinie VDI 5600: Fertigungsmanagementsysteme (Manufacturing Execution Systems - MES) (2016)

[VDI72] Richtlinie VDI 3424: Numerisch gesteuerte Arbeitsmaschinen - Direktsteuerung mit Hilfe von Digitalrechnern (1972)

[VOGE09] Vogel-Heuser, B., Kegel, G., Bender, K., Wucherer, K.: Global Information Architecture for Industrial Automation. Automatisierungstechnische Praxis (atp) **51** (1), 108–115 (2009)

[WARN93] Warnecke, H.-J.: Hat CIM noch eine Zukunft? CIM-Management (2), 31 (1993)

[WECK03] Weck, M., Possel-Dölken, F.: Vertikale Integration. wt Werkstatttechnik online **5** (93), 366–373 (2003)

[WECK90] Weck, M. et al.: Wege zur Verkürzung der Inbetriebnahme- und Stillstandszeiten komplexer Produktionsanlagen. In: Weck, M., Eversheim, W., König, W., Pfeifer, T. (Ed.) Wettbewerbsfaktor Produktionstechnik: VDI-Verlag (1990)

[WECK91] Weck, M., Lange, N., Pauls, A.: Ablaufsteuerung und Werkzeugfluss koordinieren. Industrie-Anzeiger (49/50), 18–20 (1991)

[WECK97] Weck, M.: Benutzerorientierte Software sichert Anwendern Wettbewerbsvorteile. Industrie Anzeiger (9) (1997)

# Engineering

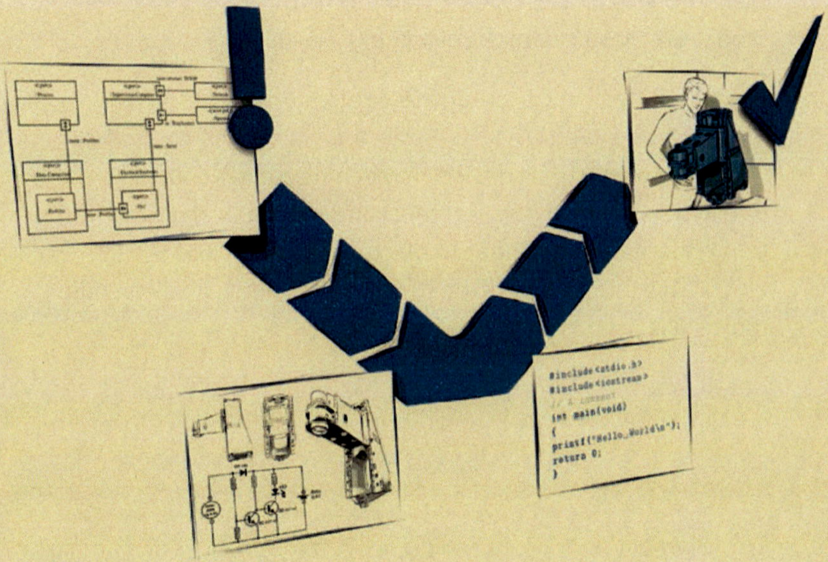


I thank my staff members Yannick Dassen M. Sc., Christian Fimmers M. Sc., Christoph Pallasch M. Sc. and Simon Storms M. Sc. for their contribution in revising this chapter.


C. Brecher and M. Weck, *Machine Tools Production Systems 3*,
Lecture Notes in Production Engineering, https://doi.org/10.1007/978-3-658-34622-5_15

Contents

The requirements in automation technology as regards reliable and, from the viewpoint of companies, controllable process execution are characterized by long-term development processes, the involvement of more and more disciplines and continual broadening of functionality while at the same time taking into account safety-relevant aspects. In this context, engineering, i.e., the planning and development tasks for automation systems, must also meet these requirements and realize them in the form of a suitable methodology. Key components pertain to the methodology of engineering, architecture modeling, systems engineering, design, layout, implementation, and test.

## 15.1 Software development

The development of software is a key challenge in automation technology because mechatronic products, automated production systems, in addition to electrical and mechanical components, require software to operate. Software development makes up a large part of the overall development costs for mechatronic systems. It also has a major bearing on the quality and functionality of an automation solution [STET04]. Based on the development of software-only products, software in automation technology is also developed in a systematic manner to meet the following, conflicting objectives (refer to ◘ Fig. 15.1).

- Meeting schedules
- Guaranteeing as high a level of software quality as possible
- Making the best possible use of available resources

The increasing level of integration of software and hardware components in automation attaches particular importance to the goals of adherence to schedules and software quality. Often, a large part of software development takes place in interaction with hardware development or is dependent on it. Timing the coordination between software development and hardware development is key for adhering to schedules and making the best possible use of development resources in hardware development. Bugs in the controller software of a machine not only result in the incorrect implementation of the functionality, but they can also cause mechanical damage. In mechatronic, i.e., interdisciplinary, development, delays in one department affect the other disciplines involved.

Besides software developers, tools and machines are also resources in automation technology that need to be coordinated for development. In this context, it is difficult to determine the correct number of developers for a project, a number that can also change/fluctuate during a development process. For example, fewer developers are required for the conceptual design and specification of a software architecture than for implementation. Also, in the integration and startup phases of controller software, the machine to control or prototypes of the product are required as resources. Coordination of access to the machine resource is thus required so as not to come into conflict with other disciplines.

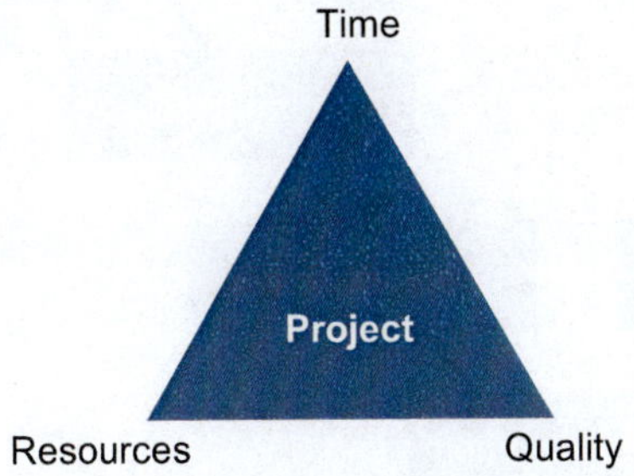


◘ **Fig. 15.1** Magic triangle, according to [HEIN15]

### 15.1.1 Development models

Two approaches have established themselves in software development. They have different viewpoints of how development projects progress ideally. The idea of the "classic" approach is based on the concept of planning, at the start of the development process, all of the software to be developed and to refine and then implement the plans. In many development models, the existence of errors in planning and realization are handled by additional iterations. In other models that have a strictly sequential flow, errors result in failures in the development process. Classic development processes are regulated formally down to the last detail and generate an administration workload in addition to the actual development cost.

In contrast to the established methods (► Sect. 15.1.1.1, 15.1.1.2), a lean, flexible development process is the goal in agile software development (► Sect. 15.1.1.3). This also takes into account social interaction within the development teams and facilitates changes during the development to the requirements of the software to develop [GREC10].

#### 15.1.1.1 Waterfall model

The waterfall model is a linear process model that structures software development into successive phases. At the end of every phase, the results are channeled into the next phase for subsequent refinement.

Every phase ends with a milestone and a project result as shown in ◘ Fig. 15.2. In contrast to the iterative process, the waterfall model has the downside that errors arising in one phase cause a return to the previous phase as there is no feedback of interim

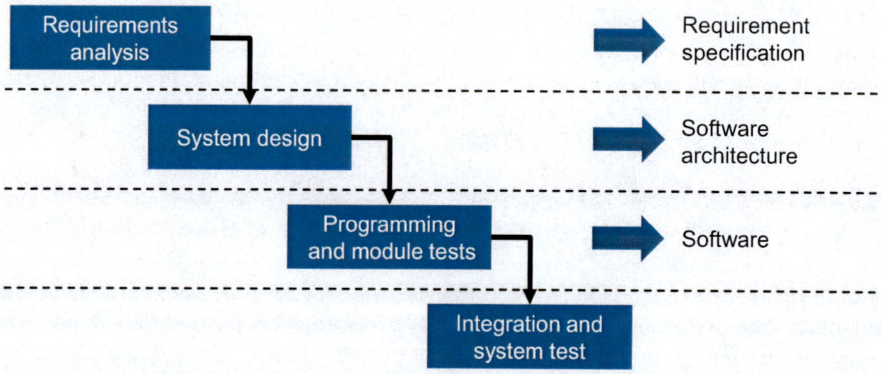


**Fig. 15.2** Waterfall model, according to [HARW12]

results. In turn, this return causes a delay of all subsequent phases and hinders overall project development. This has major negative effects on the result and duration of the project—especially for large or complex projects.

### 15.1.1.2 V-model

The V-model is an enhancement to the waterfall model. It aims to compensate the downsides in the monolithic integration of development and test and when granularity is lacking in the system design [BRAN15]. As with the waterfall model, the V-model is divided into phases. Results are forwarded on from phase to phase. In addition, the V-model defines the quality assurance process on a phase-by-phase basis.

A functional specification is used at the start. It is given more detail step-by-step, as shown in Fig. 15.3. The last phase of detailing is the actual software development/implementation. Quality assurance then starts with very fine granularity initially. The individual modules of the system are tested to reveal implementation errors early on. Granularity decreases from phase to phase, so the system can finally be tested in its entirety. Acceptance and usage of the software represent the last phase of quality assurance.

Every step on the way to code development is subjected to a verification and validation step. If an error is found in the quality assurance cycle, the loop must be run again using the relevant specification/implementation step.

### 15.1.1.3 Agile software development

Control of the development process, and the associated administration, mean some of the development resources are not deployed so as to add value. The high portion of work activities that do not add value in classic development models has caused agile software development methods to emerge. They are characterized by less administration overhead and a high level of added value because development is controlled by the developers themselves in a needs-driven manner. By contrast, the development process is planned in less detail. [HARW12]

One example of an agile process model is Scrum. This has established itself as a reference model in agile software development. The Scrum process model

15

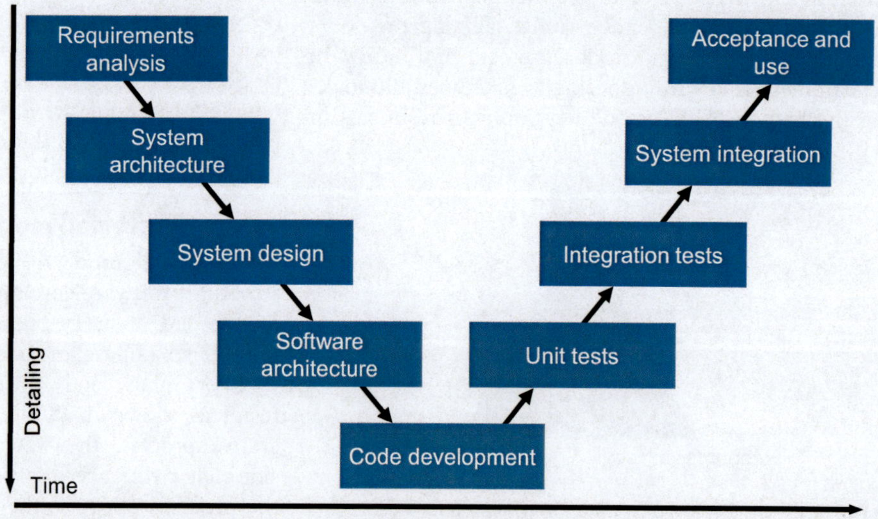


**Fig. 15.3** V-model, according to [BRAN15]

can, however, also be deployed in other fields, such as management. The development process in Scrum is performed cyclically in sprints, typically lasting between two and seven weeks (◘ Fig. 15.4). Sprint durations generally do not change over a project. In particular, a sprint cannot be extended. The features to be realized within a sprint are planned at the start of the sprint. The development team determines in this step how many features can be implemented. Before development starts, the Product Owner receives information on the product to develop and determines the sequence for completing the features. After every sprint, the finished, realized features are demonstrated to the Product Owner. Potential deviations from the stipulated functionality, due, for example, to a lack of understanding of the mechanical components to control, can be ascertained by the Product Owner at the end of every sprint [HANS10].

The idea behind the iterative method is based on experience that development planning performed early on is detrimental to project success. The prime reasons for this are poor estimation of the product development duration, differing calculations of costs and ultimately the development of a product that does not satisfy customer needs. To continuously improve the development process, hurdles in development that are to be rectified for subsequent sprints are highlighted by the development team after every sprint.

As an agile process model, Scrum requires transparency of the development progress. While, in conventional methods, delays are often not communicated to project management until the scheduled end of the development process, development progress can be traced at the end of every sprint when Scrum is used. In automation technology, the peculiar feature of the Product Owner having to work in an interdisciplinary manner

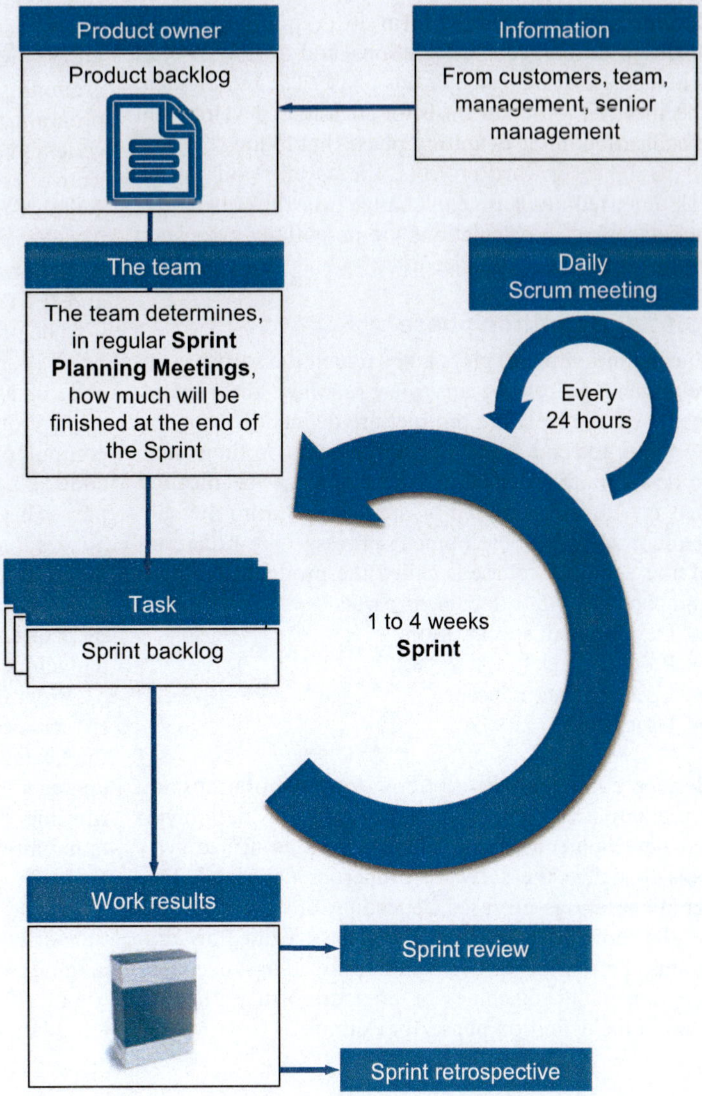


◘ **Fig. 15.4** Scrum process model, according to [HARW12]

must be taken into account. In particular, being the "middle man" between mechanical/electrical development and actual software development represents a key challenge in the tasks of the Product Owner.

## 15.1.2 Development phases

Shown below are the phases of a V-model software development project and the key differences between this and the agile method. The names of the development phases may vary between different development process models.

### 15.1.2.1 Planning phase

At the start of the software development process comes the planning phase. During this, the feasibility of the software project is examined. It is a preliminary examination or feasibility study. As the results of this phase, the customer requirements of the software product to develop are defined in the form of a requirement specification. Initial cost calculations and a first project schedule are also created in this phase. The requirement specification forms the basis for creating a development specification in the definition phase that follows.

In agile software projects, the assumption is that customer requirements can change over the course of a project. So, cost calculations are potentially associated with a high degree of inaccuracy.

### 15.1.2.2 Definition phase

The definition phase plays a key role in the software development process because here is where the requirements on the software product are defined in cooperation with the end user. The requirements on the system to develop are specified in a development specification that is detailed on a step-by-step basis during the subsequent software development process. The full result of the definition phase is called the product definition and comprises four document parts:

- Development specification
- Product model
- User interface concept
- User guide

Besides colloquial descriptions, formal notations are preferably used to describe data models in the individual development phases. These notations utilize symbols to define the software properties described. Different notations are used depending on the objectives of the individual development phases. Data flow diagrams, Petri nets, Entity Relationship diagrams, class diagrams, state machines or interaction diagrams are used in the definition phase, for example.

### 15.1.2.3 Design phase

Once the requirements of the software product have been specified in the definition phase, suitable data models are devised in the design phase to meet the specified functional scope. In addition to the descriptions of the processes using state machines and interaction diagrams, the data structure is defined using class diagrams [RUMP11]. In addition to this, software interfaces that allow parallel software module development are defined. These include interfaces within a program as well as interfaces for communicating with other programs (such as over a network). The results of the design phase are the detailed descriptions of the dynamic aspects (such as production cycles) and static properties (such as the system structure) as a data model. The software architecture of the software system is drawn up in the design phase.

### 15.1.2.4 Implementation phase

In this phase, the data models from the preceding phase are converted to program source text using a programming language. This process is commonly called "programming." The result of this phase is an executable program. Program flow charts, in addition to pseudo code and structure charts, are used in line with the objective in this development phase. The standardized symbols for the program flow chart and Nassi-Shneiderman structure chart are shown by way of example in ◘ Fig. 15.5 (DIN 66001, DIN 66262, DIN 66261) and ◘ Fig. 15.6.

The program, i.e., all commands and data processed by the computer, is stored in the computer in machine code or a sequence of binary digits (1 s and 0 s). This encryption of the program flow is different for every computer manufacturer. Programming in machine code, also called "machine language," is very complex. This led to the development of programming languages that are adapted to expressions used by humans and that can formulate problems straightforwardly. Before programs can run on a computer, however, assemblers or compilers must first be used to convert them to a machine language understandable by the computer.

Programming languages can be divided into different classes (◘ Fig. 15.7). Important attributes other than categorization into machine and problem-oriented languages are programming convenience, program runtime, memory requirement, modularity, and maintainability.

In process computing where short program runtimes and program lengths are the aim for quick, time-critical processes in particular, machine-oriented languages have mainly been used for programming up to now. If the machine language (that is not a programming language in the actual sense) is not considered in

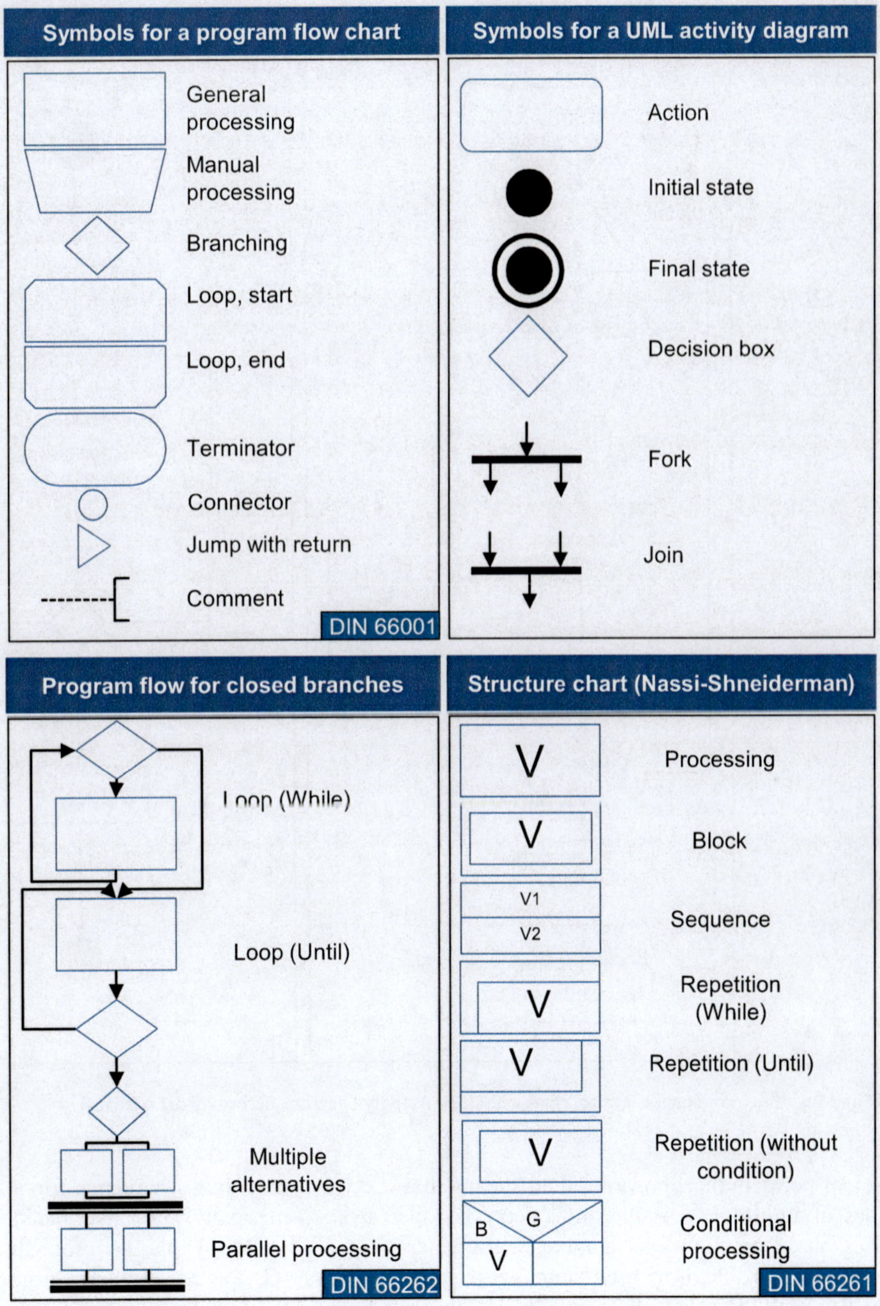


**Fig. 15.5** Program documentation with program flowcharts and structure charts

the analysis, the assembler languages are then available to the user. Using them, computer commands can be specified using mnemonic codes (easy to remember abbreviations). The user program written in the assembler language is converted ("assembled") into machine code using a converter program, the assembler.

Machine-oriented programming in an assembler language requires accurate knowledge of program processing internal to the computer because every processing step must be specified individually. Every command in the source program corresponds, after assembly (translation to the machine language), to a command or a command group in the object program such that there is a unique relation even in regard to the number of working memory cells required.

The computing power of today's processors, and the memory capacities and real-time operating systems available, make controller programming easier using problem-oriented languages. The downsides of such compiler-generated and less optimized code are, in particular, slower speeds and higher RAM requirements. These downsides are usually compensated by very high

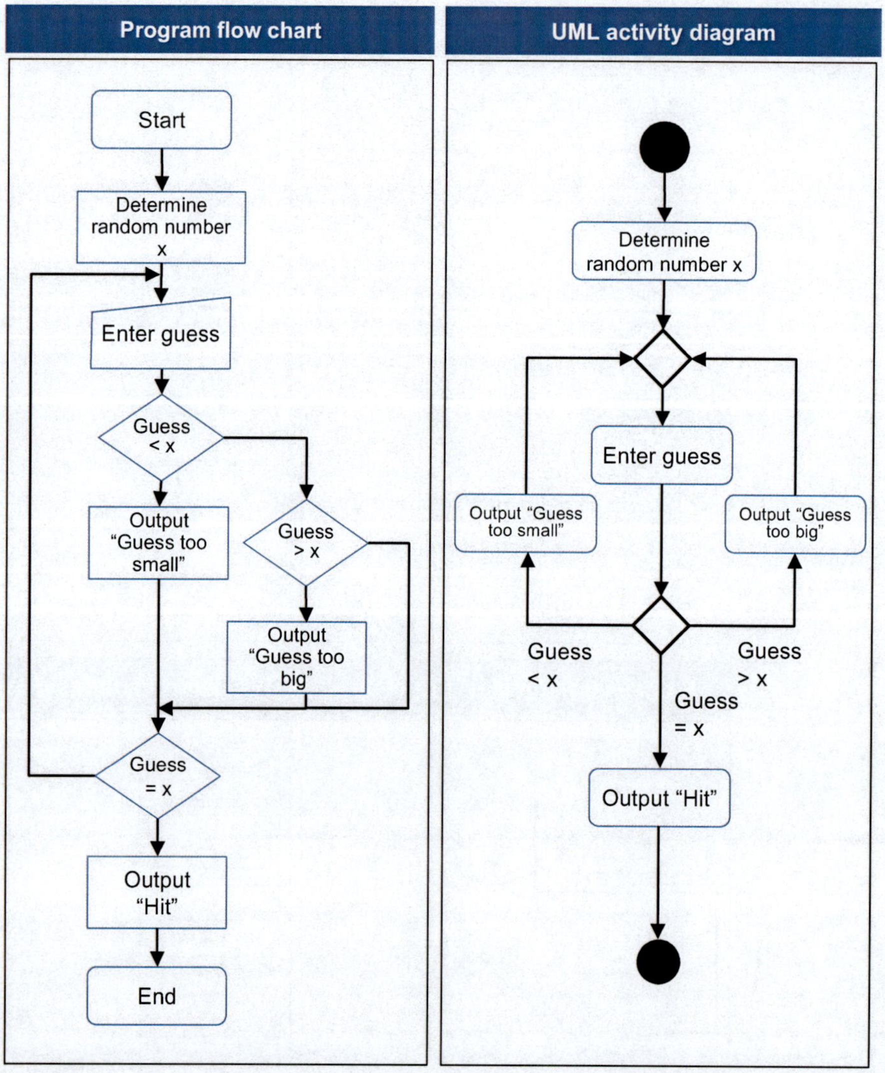


Fig. 15.6 Program example as flow chart and UML Activity Diagram, according to [WEIL14]

system performance, however. This means that the upsides of modular, portable and open controller structures based on high-level languages are generally predominant. Problem-oriented languages (such as Pascal, C, FORTRAN, Cobol, and Basic) have language elements that are tailored to mathematical, scientific, commercial, or certain technical problems. They are detached from the internal computer structure and can be used for different computers. Programs written in a problem-oriented language are translated (compiled) into machine code using compilers, or analyzed and run by an interpreter as an interpreted language. In addition to easier legibility of the program text, it is possible to test such programs using special test programs (debuggers). Problem-oriented programming languages such as EXAPT, EASYPROG, and PROGRAMAT have also been developed for programming the machining sequence for making workpieces using numerically controlled machine tools.

Many of the procedural programming languages (C, Pascal, etc.) are not only problem-oriented, but they have also been extended into object-oriented languages (Java, Smalltalk, Eiffel, C++, C#, object-oriented Pascal, etc.). The difference between procedural and object-oriented languages is that data and functions (methods) are grouped into classes for the latter. The data of a class are not accessible externally. Instead, they must be read or changed using methods provided for this purpose. Also, classes can pass on their properties to classes derived from them. The benefits of the concept of object-oriented programming are improved clarity of the program code generated and a good level of reusability of the classes already created. This is because defined interfaces exist

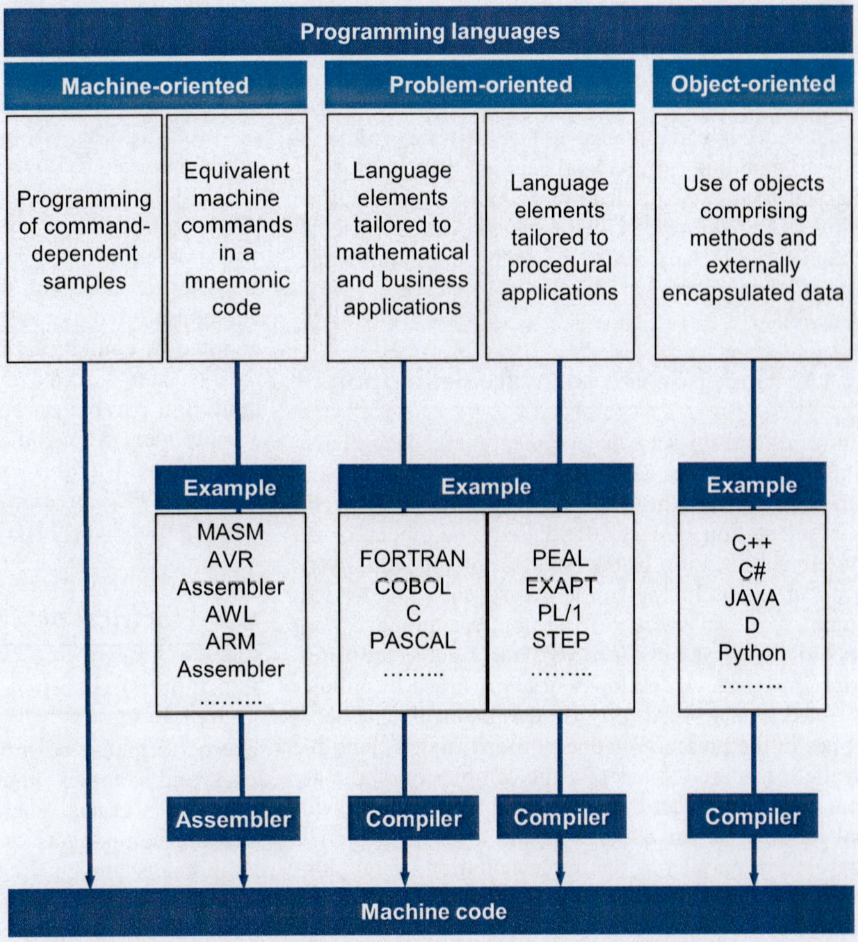


**Fig. 15.7** Categorization of programming languages

for every class and it is possible to build upon existing classes (inheritance).

This means, for example, that an old class can be replaced at any time by a newly-developed class if the interface of the new class continues to make available the methods of the old class. How the data in a class are managed and exactly which data exist are irrelevant. This aspect contributes greatly to higher levels of modularity and modifiability of the software generated. An additional benefit of this programming technique is encapsulation, i.e., external objects are not granted any permission to directly access data in the protected area of a class. Rather, defined interfaces must be used for communication with the object [OEST09].

Object-oriented programming is now being applied productively in many sectors. Object-oriented programming (OOP) is used, for example, for databases, text editors and user interfaces the same way it is for operating systems. Object-oriented programming is also being successfully utilized in mathematical, scientific, and manufacturing environments. For example, the processing of geometric elements and their representation can be mapped to objects such as material transport tasks and extensive manufacturing control systems.

Based on the widespread IEC 61,131-3 standard built upon procedural programming, the IEC 61,499 standard has created an object-oriented extension. An object-oriented extension has also been standardized especially for IEC 61,131-3 [JOHN09].

Together with the programming, the program source text is documented in the implementation phase. The documentation is required and important in particular for subsequent software maintenance in the maintenance and support phase.

#### 15.1.2.5 Acceptance and introduction phase

In the acceptance phase, the software product is checked by the customer for the properties formulated in the functional specification. Here, an acceptance record is followed to check the functionality of the software using test data negotiated beforehand. In the introduction phase, installation and startup of the software product take place, as do user training and possible replacement of an existing software system with the newly developed software.

#### 15.1.2.6 Maintenance and support phase

The maintenance and support phase follows startup. Given that the requirements of software systems are subjected to regular changes over time (such as postal code and currency changes), the software should be easy to maintain as early as the design phase and documented adequately in the implementation phase. Otherwise, subsequent reworking of these lapses in the maintenance phase can result in significant maintenance costs and an increase in overall software costs.

### 15.1.3 Model-driven software development

Conventional software development methods are based on the fact that the developer uses a programming language to convert the system specification to a specific implementation. This is often based on a model for the system specification or the behavior of individual software modules. In the simplest case, this is independent of a model based on the implementation experience of the developer involved and is subjected to intrinsic motives. A development effort based on models is called model-based software development. The benefit lies in the presence of one or more models that, due to the transparency, abstract the complexity of the actual code and reflect the relationships between individual modules in an understandable form. The models are used in parallel to the implementation. If relationships in the implementation of individual modules are changed, the corresponding models must also be updated so that they can correctly mirror the implementation. However, experience shows that this is not pursued further during development due to the additional overhead.



Model-driven software development follows here to be able to still use the benefits of model-based software development. In Model-Driven Software Development (MDSD), models are used as actual program code. From previous models, a generator is used to automatically create source code. The program flow complies with that of the model. Practical code generators exist that derive, as a minimum, parts of the code from models to assist the development effort.

While models are employed for the functional decisions in resolving problems, generators perform the technical realization. Models always pertain to a certain specialist area or domain. A model defined for a particular specialist area or domain to describe solutions is called a domain-specific language (DSL). Domain-specific languages are designed specially to cover a domain-specific abstraction of the specialist area making it accessible to formal modeling. For the flow definition of automation processes, for example, a corresponding language can be developed to represent the individual process steps using a set of well-defined descriptive elements. This language can reflect the description of a process on an abstracted level. Specific instructions using object-oriented, machine-oriented or problem-oriented programming languages would not be required in this case because they can be entirely handled by the descriptive elements of the language [STAH12].

Another form of code generation in model-driven software development is based on the data records of configuration data. A ready-to-use code template is adapted by configuring relevant options so as to attain the behavior required. For example, selecting the configuration parameters appropriately can affect the way a component is machined on the production line. Programs related to this for performing actions (such as robot and PLC programs) are matched to the configuration parameters [THIE10].

## 15.2 Electrical design on machine tools

### 15.2.1 Introduction

The performance of modern-day, automated machine tools and complex manufacturing facilities is determined to a crucial degree by electric, electronic, and software components. As shown in ◘ Fig. 15.8, these components implement the power supply and control functions necessary for the automation of the systems and embody important safety functions for personnel and system protection. The selection, design, and interconnection of these components as part of electrical design are important elements determining the function and quality in the design of machine tools and project planning of complex manufacturing facilities.

The interaction of the key domains of electrical design in machine tools can be derived from the functional structure of the electrical and electronic components of a machine tool as shown in ◘ Fig. 15.9.

Sensors, such as keys, switches, and measuring systems, relay information about the machine status to the control level of the machine. On the control level, this information, together with operator commands entered from control units, is processed in accordance with the programs used to determine the control variables for switch gear and power components. Specialized electronic control systems such as programmable logic controllers (PLCs) and numerical controllers are typically used today for information processing (refer to ► Chapter 9).

Signal processing on input and processing levels is characterized by low powers. Control signals on the output level need to be converted to the powers necessary to operate the actuators. Examples here are the

**Fig. 15.8** Electrical systems in machine tools

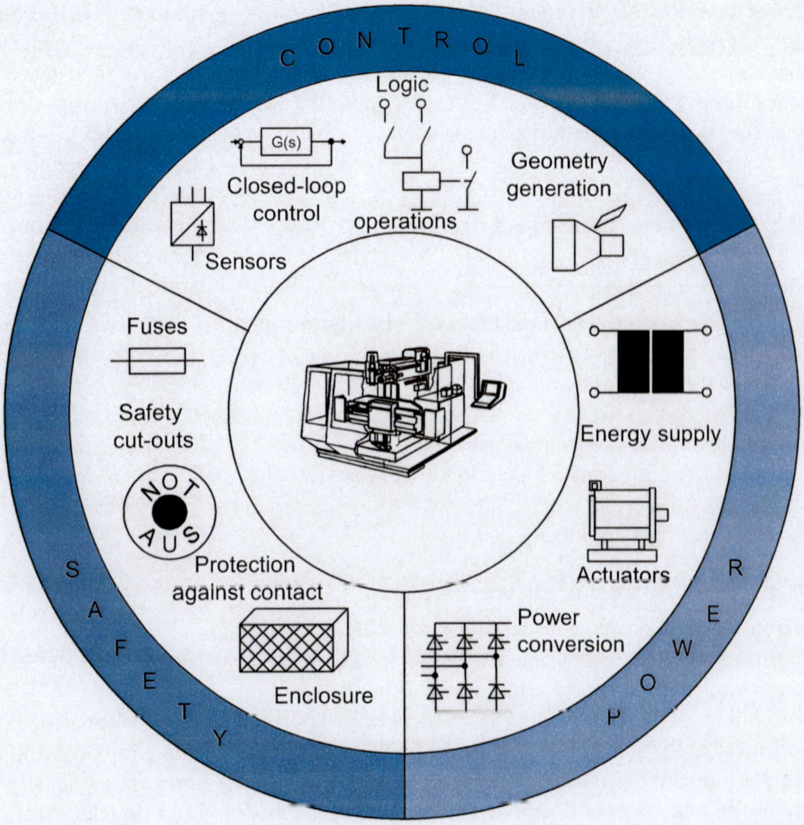


**Fig. 15.9** Components of electrical design in machine tools

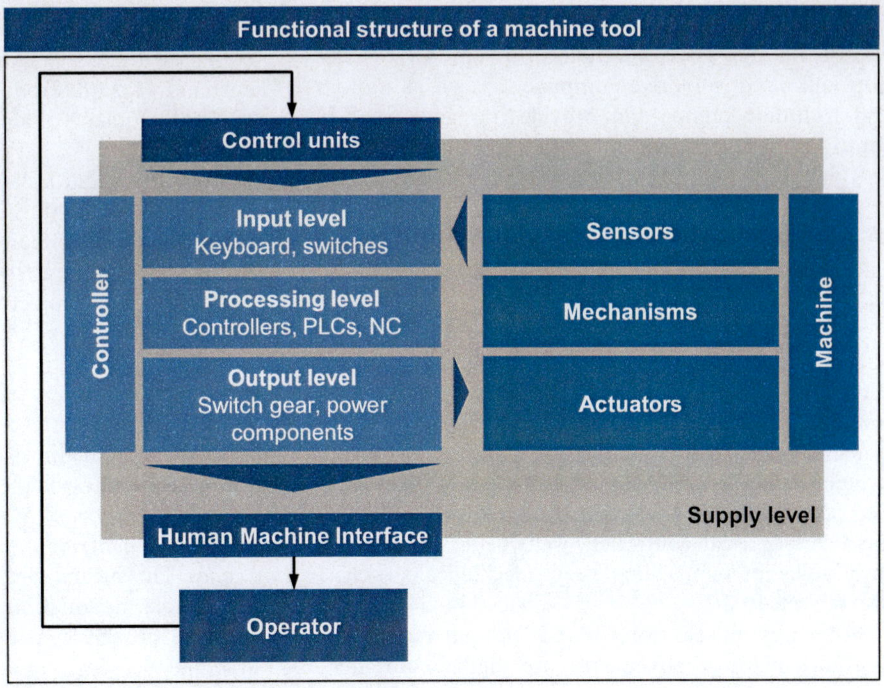

power components for the axis drives and the contactors for switching on and off the main drives. Transformers and power supplies are typical components on the supply level that provide different operating voltages for the devices on individual levels.

### 15.2.2 Functions of electrical design in machine tools

The domains of electrical design were already outlined in ◘ Fig. 15.8. They are essentially.

divided into three areas:

- Provision of power
- Realization of control functions
- Protective functions for personnel and the system

These three areas are introduced below.

#### 15.2.2.1 Provision of power

Given its high power density and good efficiency level during energy transfers, electrical energy is the most important form of energy for modern-day production equipment. The energy supply is generally specified within a factory. Single-phase or three-phase AC systems are normally used to provide power to individual machines and process stations. These voltages are usually converted to direct current or different operating voltages locally in the station by corresponding power supply units.

In addition to the design of power supply networks, electrical design assists mechanical design in the selection and design of drive components such as motors and hydraulic pumps and provides designs for their control signals. Information about and examples of energy provision are introduced in ► Sect. 15.2.5.1.



#### 15.2.2.2 Realization of control functions

The control and monitoring of mechanical, process and safety–critical procedures in compliance with a logical sequence chain or sequencer are the most important functions of automation technology. The control unit used for this can be designed using mechanical, hydraulic, pneumatic, electrical, or electronic elements and combinations of these. Given the straightforward availability of and ease of managing electrical power together with the switching speeds attainable, electrical controllers were being constructed early on by interconnecting electromechanical switches. See ► Sect. 10.2.

Advances in electronics and computer technology have made it possible to use digital computers and microprocessors for programmable controllers. In such controllers, the logic to process is not hard wired but with programs that can be changed in a flexible manner. As the complexity of machines and systems to control has increased by a number of factors, the high performance levels of these controllers enable a better realization compared to contactor control or even mechanical control. For example, mechanical components such as gear trains in gear cutting machines are now being replaced by electrical gear units in the form of individual, controlled servo drives. This easily allows changing the gear ratio and greatly increasing the machine flexibility.

As a means for implementing and integrating control functions, electrical design is assuming a key role in machine design. Some typical functions of electrical design in control engineering:

- Selection of sensors and actuators (drives, etc.)
- Design of IT infrastructure
- Integration of the numerical controller
- Design of the control panel
- Design of the control cabinet or switch cabinet
- Design and programming of the PLC

#### 15.2.2.3 Protective functions for personnel and the system

Using electrical energy is comes with many different risks to people and machines. These risks arise from the high power density and the lack of human sensory organs for electrical and magnetic fields. The design of corresponding protective measures and safeguards for people and safe operation of the machine represent the core task of the electrical system:

- First and foremost, people must be protected from contact with live parts and so from hazardous shock currents. Hazards from shock currents are dependent on contact voltage magnitude and exposure time.
- Malfunctions and failures of system parts that can be triggered by failures of sensors and actuators, cable breaks, short circuits, and logical errors in controller hardware and software must be detected by corresponding protective measures and reliable emergency switch-off routines.

The safe integration and usage of electrical energy and electrical equipment are regulated by several international and national directives, regulations, and standards. One general principle applies here: *All manufacturers and operators of electrical systems are themselves responsible for adhering to recognized electrical engineering codes*. This means specifically:

- Applicable national and international provisions must be observed in the manufacture of electrical systems.
- Only electrical equipment tested and approved in accordance with the national provisions of the country in which the system is operated may be

used (electrical equipment covers all objects serving collectively or partly the purpose of using electrical energy).

- The liability situation for manufacturers and retailers is being tightened further by laws, such as the Product Liability Act. First and foremost, anyone who fabricates defective products or passes such products to end consumers bears liability.

These provisions can differ from country to country, potentially entailing additional difficulties in the export of electrically driven or controlled systems to various countries. International standards, such as the CE marking within the European Union, are the aim. This marking, mandatory since 1997, is the basic prerequisite for the sale of electrical equipment, machines, and systems within the entire single European market.

By affixing a CE marking to a machine, a manufacturer confirms that it has met all its obligations pertaining to European trade. These obligations are of a technical and organizational nature. They have been laid down in the "EU Machinery Directive" and are identical in all European countries.

The purpose of this directive is to guarantee the safety and health of people and animals as well as the safety of objects that may be in the hazard area of a machine. To comply with this directive, a manufacturer must create a folder containing technical documentation. Designing a machine to particular harmonized European standards greatly simplifies the verification of conformity. [DIN09b] is an example of such a standard. It specifies the safety requirements made of machine tools, machining centers in particular. When standards guarantee conformity with fundamental health and safety-related requirements in the Machinery Directive, they are known as a Group C standard.

A distinction is made between "direct" and "indirect" contact in the design of protective measures. "Direct contact" means the touching of active, that is, live during normal use, system parts. They can be busbars or connecting terminals, for example. "Indirect contact" means that inactive, normally de-energized, conductive system parts (such as housings) can become live and be touched by people.

Protection against shock currents caused by electrical equipment must be ensured in fault-free operation (protection against direct contact) as well as when faults occur (protection against indirect contact). The principle protection types provided are listed in [DIN07b]. See ◘ Fig. 15.10.

Safety and functional extra-low voltages mean the use of alternating currents lower than 50 V and direct currents lower than 120 V as operating voltages. These do not pose risks to people in case of contact and thus are categorized as a protective measure for people. Other protective measures can be divided roughly into preventing contact using insulation of or gaps between active parts, equalization of voltage potentials on inactive parts (such as with grounding), and switch-off of a defective system.

In addition to the automatic switch-off of defective system parts by special fuse devices (shown in more detail in ▶ Sect. 15.2.5.4), the protective insulation of assemblies represents the most effective and safest protection. This adage applies: "*Insulating is safer than grounding.*"

Protective insulation is insulation applied in addition to the base insulation of the components. Protective insulation prevents dangerous contact voltages occurring on parts that can be contacted due to a fault in the base insulation. It represents an active protective measure because it comes from completely sheathing

◘ **Fig. 15.10** Protection against shock currents. *Source:* [DIN07b]

| (1) Protection against indirect contact | (2) Protection against direct contact | (3) Protection against direct and indirect contact |
|---|---|---|
|  | 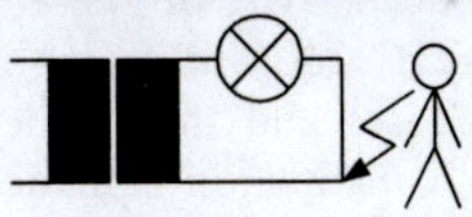 | |
| ▪ Protection by switch-off or signal<br>▪ Two-fold or increased protective insulation<br>▪ Protective separation<br>▪ Protection by floating, local equipotential bonding | ▪ Protection by insulation of active parts<br>▪ Protection by covers or sheathing<br>▪ Protection by obstacles<br>▪ Protection by distance | ▪ Residual current device (RCD)<br>▪ Safety extra-low voltage<br>▪ Protective extra-low voltage |

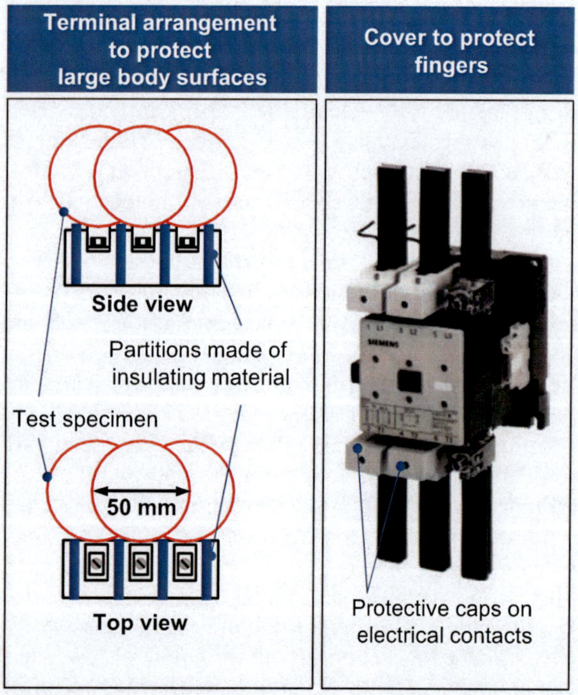


**Fig. 15.11** Examples of protection against shock currents. *Source:* WZL, SIEMENS

the equipment with insulating material, meaning it is effective at all times. Most electrical household appliances have protection insulation. Three examples of protection against shock currents are shown in Fig. 15.11.

Shown on the left is an arrangement of connector terminals that protects any large surface of the body, such as the back of a hand. Shown in the middle is an option for a contactor cover that protects fingers. The cover of the right-hand phase is removed for clarity. Shown on the right is a protection-insulated main switch as an example of protection against indirect contact. The entire switch is enclosed in a housing made of non-conductive synthetic material.



Protection against direct contact is part of the IP (Ingress Protection) classes as in [DIN14f]. In line with the overview of IP protection classes and their code numbers shown in Fig. 15.12. A distinction is made between protection against contact or foreign objects (first digit), and protection against water (second digit). Some typical protection classes for enclosures:

- IP20: Enclosure for general applications with no environmental requirements
- IP22: Non-drip enclosure
- IP54: Weather-resistant, dust-proof enclosure
- IP65: Watertight enclosure

Besides the primary electric-shock protection for people on electrical operating equipment, adequate safety precautions must be put in place by the electrical design in the circuit logic implementation such that both the operator and the system are protected adequately in the event of incorrect operation and potential malfunctions as well as when system components fail. Besides the use of different safety systems to protect against overloading the system and its components, logical interlocking conditions are established using protective circuits. They allow, for presses for example, the cycle of a press stroke only when the operator is not at risk (Safety controllers, ► Sect. 15.2.5). In addition, particular component properties are required, such as positively driven contacts for the contactors used, that guarantee a reliable switch-off of the system even if the contacts fuse together. Section 15.2.5.4 presents examples of some safety devices for machine tools.

### 15.2.3 Interaction between the electrical and mechanical designs

#### 15.2.3.1 Interface between the electrical and mechanical designs

Even now, efficient cooperation across different disciplines (mechanical and electrical design, automation, process technology, simulation, etc.) in technical product and system development represents a key challenge for companies and development departments. This is attributable to the continually increasing interdisciplinary complexity of system and function structures, the work and communication-intensive merging of discipline-specific perspectives on technical topics and other factors.

The information flow shown in Fig. 15.13 is a typical process of the cooperation between the mechanical and electrical design departments together with automation in technical system development. Collaboration on a technical system that is designed by the mechanical design department, for example, and then to be automated by the electric design department, is commonly based on detailed specification documents created in-house or on direct verbal communication.

This approach, still generally in need of improvement, requires close cooperation between the mechanical and electrical design departments. Often, electrical components (sensors and actuators, in particular) cannot be definitively selected until the machine prototypes are available and have been tested. On the other hand, the dimensions, properties, and requirements of these assemblies have a bearing on the mechanical part of the machine design. These interactions often result in an iterative optimization process. Consequently, close cooperation between the two disciplines is necessary for both the preliminary design as well as detailed design.

**Fig. 15.12** IP protection classes: Protection against contact, foreign objects and water. *Source:* [DIN14f]

| Protection against contact and foreign objects | | Protection against water | |
|---|---|---|---|
| 1st digit | Protection scope | 2nd digit | Protection scope |
| 0 | No protection | 0 | No protection |
| 1 | Protection against foreign objects > 50 mm and access with back of hand | 1 | Protection against water dripping vertically |
| 2 | Protection against foreign objects > 12.5 mm and access with finger | 2 | Protection against drops of water falling at angle |
| 3 | Protection against foreign objects > 2.5 mm and access with tool | 3 | Protection against spray |
| 4 | Protection against foreign objects > 1 mm and access with wire | 4 | Protection against splash water |
| 5 | Dust protected and protection against access with wire | 5 | Protection against water jets |
| 6 | Dust-tight and protection against access with wire | 6 | Protection against flooding |
| | | 7 | Protection against immersion |
| | | 8 | Protection against submersion |

*Example: IP54*
*→ Protection against dust*
*→ Protection against splash water*

Furthermore, the installation location puts specific requirements on the electrical and electronic components regarding adequate heat dissipation and the feed capacity for electrical lines. For these reasons, the best machine design and engineering can only be ensured when the mechanical designer knows all the key functions and functional sequences of the machine, and the electrical equipment.

### 15.2.3.2 Communication tool for specifying the functions of a machine tool

The Continuous Function Chart as per [DIN14g] is often used as a tool for communicating between electrical and mechanical design in the definition of control tasks. A Continuous Function Chart is a purely flow-oriented representation of a control task. It is independent of the control-system implementation (PLC, contactors, and hydraulic/pneumatic systems). It simplifies understanding between the specialists of the individual disciplines. A Continuous Function Chart forms a clear communication tool instead of the often freely formulated, written, or verbally communicated, system description. The basic elements of a Continuous Function Chart:

- Step/state/command
- Step enable condition / operation logic
- Transition

Figure 15.14 shows the basic elements of a Continuous Function Chart using the simple process control of a drilling task as an example. A *state* represents a steady state of the process or controller at a certain moment. *Step enable conditions* specify the functional conditions for a transition from the current state to the

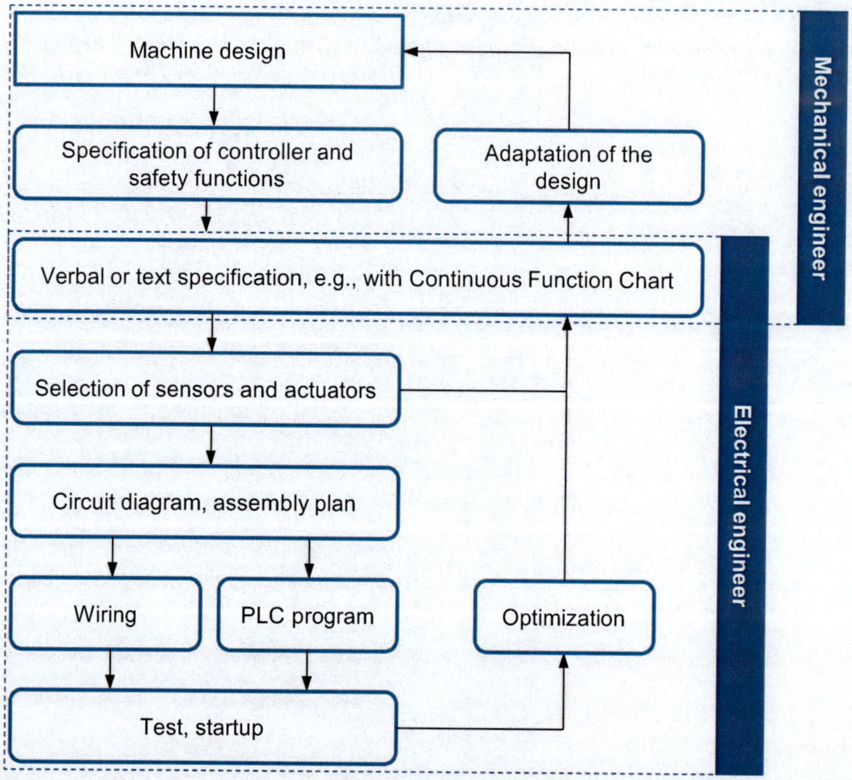


**Fig. 15.13** Collaboration and information flow between mechanical and electrical design departments

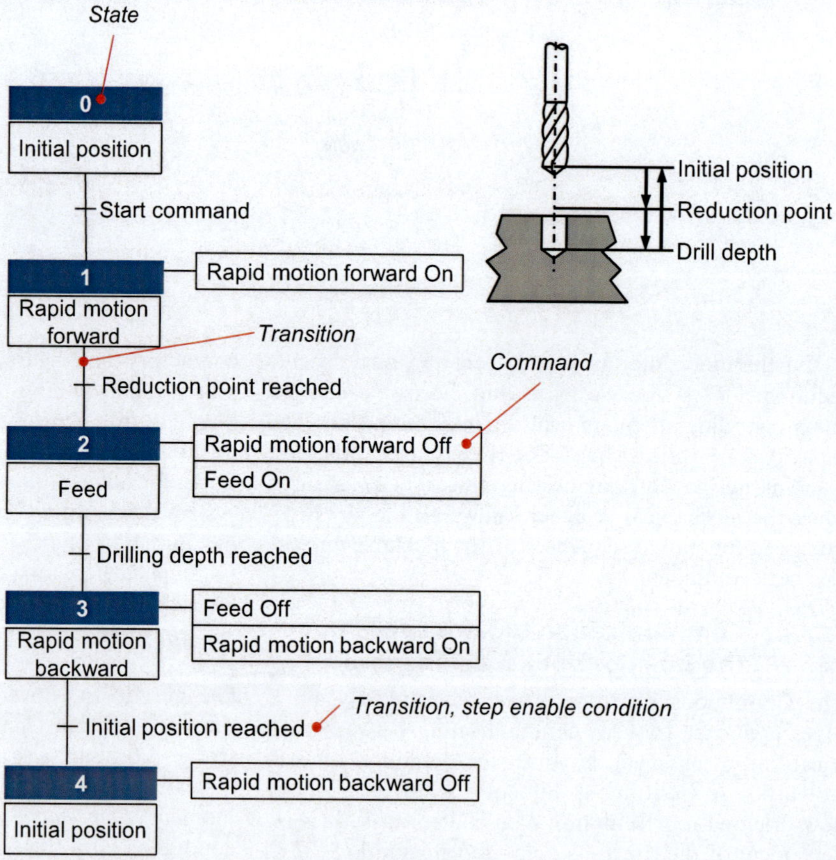


**Fig. 15.14** Elements of a Continuous Function Chart using process control for a drilling task as an example, according to [DIN14g]

next. This also defines the operational logic of the system at the same time. One state transitions to the next when the output step is active and the transition condition is satisfied. *Transitions* link the states to each other.

Continuous Function Charts can be converted directly to controller programs for a PLC using suitable programming devices (▶ Chapter 10). However, only relatively few link depths can be shown using Continuous Function Charts.

Another definition methodology is possible using Petri nets. They allow a definition and mathematical analysis of discrete control systems with processes that are parallel in time. The high level of abstraction means they are not yet in very widespread use. Reference is made here to the literature [LITZ13; UHLI06]. This definition methodology is particularly suited to complex and interlinked production equipment (transfer lines, etc.). Diagrams for data flows, familiar from data systems technology, and program flowcharts as in [DIN83a], can also be used.

### 15.2.4 Components and methods in electrical design

#### 15.2.4.1 Standards and regulations for the electrical design of machine tools

Some standards and regulations, and their significance for building and operating electrical systems, have already been covered in previous chapters. Below are additional important standards and regulations in the field of electrical design of machine tools. No claim is made to completeness:

*Safety*

- Law on accident prevention (Occupational Health and Safety Act) [BUND15]
- DIN EN 349, DIN EN ISO 13857 Safety of machinery – Minimum gaps to avoid crushing of parts of the human body [DIN08a]
- DIN EN 60529 Degrees of protection provided by enclosures [DIN14f]
- BGV-A3 Accident prevention regulation for “Electrical installations and equipment” [DGUV97]
- DIN EN ISO 10218 Robots and robotic devices – Safety requirements for industrial robots [DIN12b]
- DIN EN 12417 Machine tools – Safety – Machining centers (C-standard type) [DIN09b]
- DIN EN ISO 13850 Safety of machinery – Emergency stop function – Principles for design [DIN16a]
- DIN EN 1037 Safety of machinery – Prevention of unexpected start-up [DIN08b]
- DIN EN ISO 12100 Safety of machinery – Risk assessment and risk reduction [DIN16a]
- DIN EN ISO 14119 Safety of machinery – Interlocking devices [DIN16a]
- VDI 2854 Safety requirements for automated flexible manufacturing systems [VDI91]

*Component selection:*

- DIN VDE 0660 Specifications for low-voltage switchgear (control units, enclosures, switches) [DIN12a]
- DIN EN 60073/VDE 0199 Basic and safety principles for man–machine interface [DIN03b]
- DIN EN 60034 / VDE 0530 Rotating electrical machines (motors) [DIN15b]

*Electrical equipment:*

- DIN VDE 0100–100 Low-voltage electrical installations (IEC 365) [DIN09a];
- DIN EN 60947–1 / VDE 0660–100 Low-voltage switchgear and controlgear – General rules [DIN15a]
- DIN EN 50274 / VDE 0660–514 Low-voltage switchgear and controlgear assemblies [DIN02];
- DIN EN 61140 / VDE 0140–1 Protection against electric shock [DIN14c]
- DIN EN 60204–1 / VDE 0113–1 Safety of machinery; Part 1 Electrical equipment of machines (this standard is presented in more detail in conjunction with project planning) [DIN14d]
- DIN EN 50178 / VDE 0160 Electronic equipment for use in power installations [DIN98]

*Circuitry documents:*

- DIN EN ISO 7200 Technical product documentation [DIN04a];
- DIN EN 81346-2 Structuring principles and reference designations – Classification of objects and codes for classes [DIN10]
- DIN EN 60617 Graphical symbols for wiring diagrams [DIN97]
- DIN EN 61082/VDE 0040 Preparation of documents used in electrotechnology [DIN15c];

*Suppression of radio interference:*

- DIN EN 50370-1/VDE 0875-370-1 Product family standard for machine tools – Part 1: Emission [DIN06];
- DIN EN 55014/VDE 0875-14 Electromagnetic compatibility [DIN16b];

The suppression of radio interference is also gaining significance in the machine tool industry because high interference levels are generated by very dynamic drives with inverters and by the high clock frequencies of microcomputers in controllers.

*Communication between mechanical engineering and electrical engineering:*
- VDI 2206 Design methodology for mechatronic systems [VDI04a]
- VDI 2223 Systematic embodiment design of technical products [VDI04b];
- VDI/VDE 3695 Engineering of industrial plants – Evaluation and optimization of the engineering [VDI10];
- DIN EN 61082/VDE 0040 Preparation of documents used in electrotechnology [DIN15c];
- VDE German Association for Electrical, Electronic & Information Technologies
- DIN German Institute for Standardization
- IEC International Electrotechnical Commission
- VBG German trade associations

*Control engineering:*
- DIN IEC 60050-351 International electrotechnical vocabulary – Part 351: Control technology [DIN14e]
- DIN EN 61131 Programmable logic controllers,
- Part 1: General information [DIN04b]
- Part 2: Equipment requirements and tests [DIN07a]
- Part 3 Programming languages [DIN14b]
- DIN 66025 Numerical control of machines, format [DIN83b]
- DIN 66217 Axis and motion nomenclature for numerically controlled machines [DIN75]

For information processing, there are also several regulations and guidelines, ranging from PLC programming to data transmission protocols, that are not discussed in more detail here.



Some examples for applying these standards are introduced in the following sections.

### 15.2.4.2 Criteria for selecting components

Criteria for the selection of components in electrical design are summarized in ◘ Fig. 15.15. In addition to the direct stipulation for assured performance and acceptable prices, the customer requirements, national regulations, ambient conditions where the system is installed and design are the key criteria. The selection of suitable components is a multi-faceted field. Consequently, only a few suggestions can be given here. More information is in ▶ Sect. 15.2.4.4 in the discussion of examples.

In particular, it is national standards and regulations applicable at the installation location that can considerably limit the selection of usable components. This must be taken into account in the design stage of machines intended for export because the systems are accepted by national authorities at the installation location. Many manufacturers of electrical components offer parts having worldwide approval for this reason. The use of these parts enables such difficulties to be averted.

Ambient conditions at the installation location also considerably limit the selection options for electrical components such as electrical and magnetic fields (e.g., near electric ovens and due to powerful drive systems), vibration (e.g., from presses), extreme temperature and moisture, and exposure to oil, coolant and water. [DIN14d], for example, stipulates a minimum protection class of IP54 (for dust and splash water) for all control panels and signal units. In the design of machines for countries in the tropics, precautions to combat infestations by insects, such as termites, that can damage cable insulation must be provided for example.

Not least of all, the style and allowed design of components used are crucially determined by the installation space provided in the machine/cabinet. Many customers also stipulate the use of certain components because either special requirements need to be met or spares need to be kept in stock in view of rational considerations.

The need for components is determined as part of project planning described in ▶ Sect. 15.2.4.5.1. Electronic components are generally purchased parts that can be found in the catalogs of manufacturers.

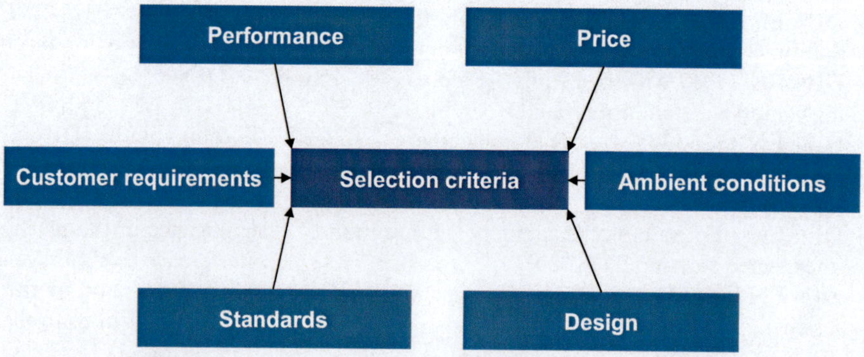


◘ **Fig. 15.15** Selection criteria for electrical components

### 15.2.4.3 Circuitry documents

The right to affix a CE marking to a machine presupposes that the associated declaration of conformity has been signed. In this declaration, the manufacturer commits to providing technical documentation containing, for example, an overall plan of the machine and all circuit diagrams.

The circuitry documents represent the technical documentation of an electrical system for explaining the functions and interactions between circuits, PLC programs, and cable connections [DIN15c]. They convey specifications for the setup and safe operation of the electrical equipment. Schematics, tables, charts, and descriptions are classified as circuitry documents [DIN15c].

[DIN14d] stipulates that the technical documentation for a machine tool must include the following documents:

- Installation plan
- General wiring diagram
- Circuit diagram
- Parts list
- Operating and maintenance instructions

Not mandatory, but must be provided when necessary:

- Documentation for the workflow, preferably as a Continuous Function Chart
- Wiring, connection and terminal diagrams

Circuitry documents must always be correct, i.e., be kept up-to-date and available at the electrical system. Changes must be entered immediately into the circuitry documents and documented. Otherwise, there is no system safety and speedy fault correction and replacement part procurement are not possible. In this context, it is pointed out that the operating company is responsible for the safety of an electrical system. The basic format of circuitry documents is first outlined below. Using the example of the quill function on a lathe, the implementation of some planning types is then shown using excerpts from original lathe plans.

Circuitry documents are generally classified by representation purpose and representation type. The different representation types are used primarily to map different abstraction levels in the description on system, device and component levels. They are not explained further here. The purposes of the representations are to explain the mode of operation and to show the connections/links and physical positions of components as an installation instruction and a basis for maintenance. One classification system (as per [DIN15c]) for circuitry documents is shown in ◘ Fig. 15.16.

The basis of all circuitry documents for the electrical design hardware on a machine are the circuit diagrams. A circuit diagram is the extensive illustration of a circuit including all details. It shows how an electrical device works (◘ Fig. 15.22). The basic structural component of a circuit diagram is the current path. Circuit diagrams are normally depicted with a vertical orientation, i.e., the current paths linking components run vertically.

The standardized designation of electrical equipment as per [DIN15c] is used for the unique, standardized, referenced naming of components in a circuit diagram (◘ Fig. 15.17). The definitions of designations to use for the names of electrical equipment, such as *K* for *contactor and relay*, are shown in ◘ Fig. 15.18.

**◘ Fig. 15.16** Taxonomy of circuitry documents, according to [DIN15c]

| Circuitry documents | | | | | |
|---|---|---|---|---|---|
| **Document type** | **Examples** | | | | |
| Function-related documents | General wiring diagram | Block diagram | Functional diagram | Flow chart | Circuit diagram |
| Location-related documents | Layout plan | Installation drawing | Installation diagram | Group drawing | Layout diagram |
| Connection-related documents | Wiring diagram | Connection diagram | Connection plan | Cable plan | |
| Equipment lists | Parts list | Spare parts list | | | |
| Assembly-related documents | Layout drawing | Installation description | Cable table | | |

*Other document types:*
*Startup-related documents, documents on operations management, maintenance and reliability*

**Fig. 15.17** Format of equipment designations, according to [DIN15c]

| Equipment designations | | | Examples |
|---|---|---|---|
| Symbol | Description | Object type | |
| = | Function aspect | System | =TNS42+L4-S8<br>Proximity switch (S8) on the tailstock (L4) of a lathe (TNS42)<br><br>= TNS42+L4-X14:335<br>Terminal 335 on terminal strip X14 on the tailstock (L4) of a lathe (TNS42) |
| + | Location aspect | Functional unit | |
| - | Product aspect | Function | |
| : | Connection aspect | Terminal | |

*Part names are allowed if full identification is evident from other documents (such as sheet IDs)*

**Fig. 15.18** Designation of electrical equipment. *Source:* [DIN10]

| | Equipment type | Examples | | Equipment type | Examples |
|---|---|---|---|---|---|
| A | Assemblies, subassemblies | Amplifiers, lasers, magnetic amplifiers | N | Amplifiers, controllers | |
| B | Converter of non-electric to electric quantities | Sensors, microphones, speakers, rotary encoders | P | Measuring devices, test equipment | |
| C | Capacitors | | Q | Circuit breakers, disconnectors | |
| D | Binary elements, delay elements, storage elements | | R | Resistors | Potentiometers, terminating resistors |
| E | Various | Lighting equipment, heating equipment | S | Switches, selectors | Control switches, probes, limit switches |
| F | Guards and safety devices | | T | Transformers | Voltage and current converters |
| G | Generators, power supplies | | U | Modulators, converters of electrical quantities into different electrical quantities | Demodulators, frequency converters |
| H | Signaling units | | V | Tubes, semiconductors | Diodes, transistors |
| J | Unused | | W | Transmission paths, waveguides, antennas | Switching wires, busbars, cables |
| K | Relays, contactors | | X | Terminals, connectors | Isolating plugs, terminal strips |
| L | Inductors | | Y | Electrically operated mechanical devices | Brakes, couplings, compressed air valves |
| M | Motors | | Z | Terminations, filters, equalizers, limiters | |

### 15.2.4.4 Circuitry documents for a lathe as an example

The format of circuitry documents and the designation of components for a machine tool are explained by way of example using the "Release tailstock" control signal function for a lathe. ◘ Figure 15.19 shows an overview of the information and power flow for performing the "Retract quill" function by the foot-operated switch sensor and proximity switch via the controller (PLC), power converter (relay in cabinet) to the actuator (hydraulic valve on the tailstock).

Localization of the sensors and actuators in question is possible on the machine from the general wiring diagram in ◘ Fig. 15.20. The clear depiction in this diagram assists in quick orientation for the maintenance engineer or machine operator, for example.

Wiring and layout diagrams show the conductive connections between electrical equipment (◘ Fig. 15.21). They contain information on the physical position of electrical equipment in the cabinet, on the control panel and on the machine.

The components necessary for control and power conversion are housed in the cabinet. Their positions can be taken from the layout diagram of the cabinet (excerpt shown in ◘ Fig. 15.21). The components relevant for the example of the quill function, such as the NC with integrated PLC (+W1-A2) and the multi-contactor (+M10-A17), are marked specifically in this figure.

◘ Figure 15.22 shows the circuit diagrams for the connection of the foot-operated switch sensors (+L4-S10, + L4-S11) and proximity switch (+L4-S1) in ◘ Fig. 15.19 via terminals X3 and X14 to the PLC. Location designation "+ L4" pertains to function group "Tailstock" and "+ M10" to the function group (located in the cabinet) for signal conversion between PLC and machines using a relay. For component designation "-S1", "S" denotes a switch and "1" is a consecutive number. The splitting of a circuit diagram into individual, vertically running current paths is clear from ◘ Fig. 15.22. The individual current paths are identified in full by a location ID (here "+ M10"), a sheet number (here "38" and "39") and a current path number in the top row of a circuit diagram, e.g., +M10/39/2.

To establish the relationship between the many individual circuit diagrams, the location/function group IDs in the sheet ID are used. For example, in ◘ Fig. 15.22, "+ M10/39/1" on current path 15 on sheet + M10/38 establishes the link of this current path to current path 1 on sheet + M10/39. The designation "+ M10/38/15" on current path 1 of sheet + M10/39 establishes the associated cross reference. This means all

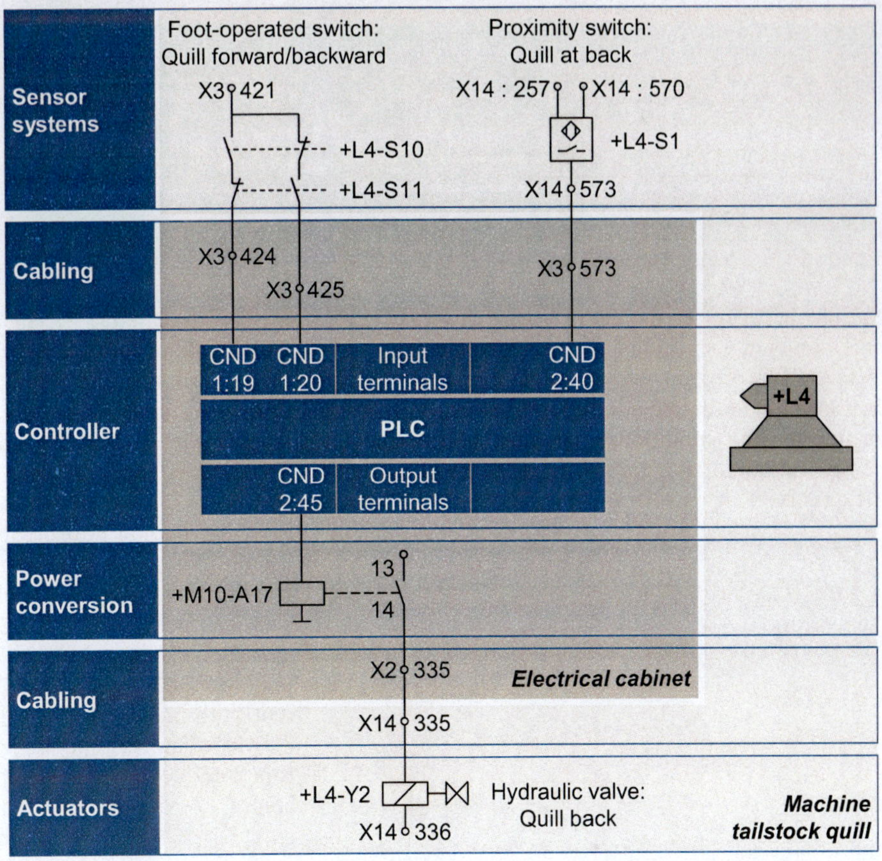


◘ **Fig. 15.19** Signal flow using the example of the quill function on a lathe

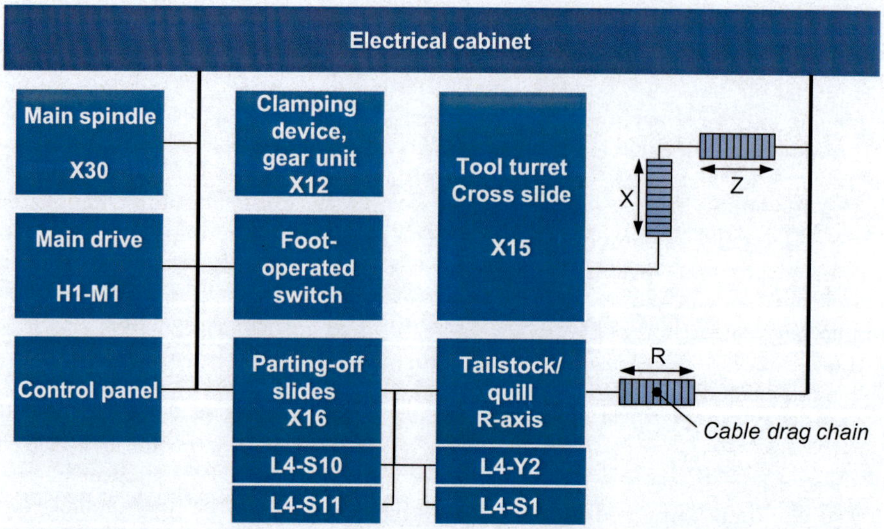


Fig. 15.20 General wiring diagram of a lathe (excerpt)

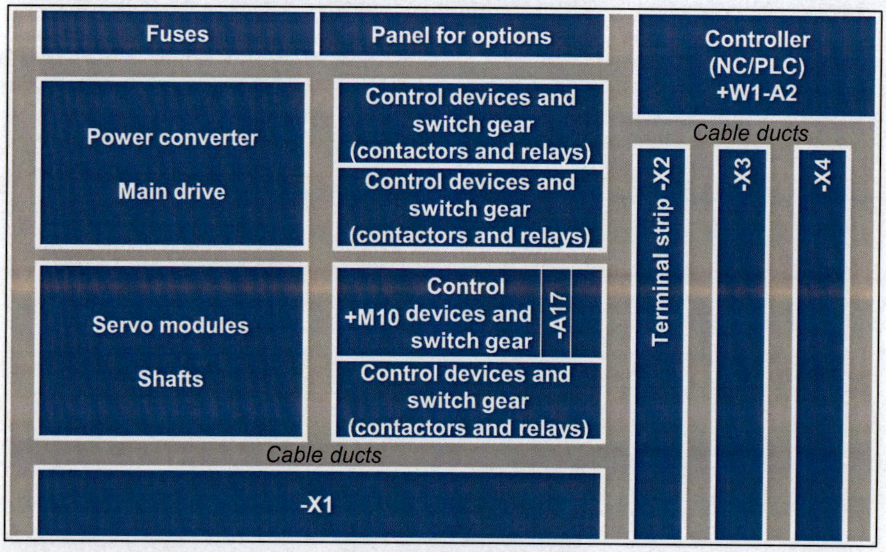


Fig. 15.21 Layout diagram for a cabinet

relations between the individual sheets are specified in full, unambiguously and in a traceable manner.

The PLC program shown in Fig. 15.23 in the form of a circuit diagram and an instruction list shows the interlocking conditions that must be satisfied for the "Retract quill" function to trigger. The conditions: The quill is retracted when:

```
(           enable    for    foot-operated
switch set (flag H0B3M6).
and       foot-operated switch quill back
activated.
(           switch+L4-S11,            input
H001E3).
and       proximity switch quill at back.
(+L4-S1, input H00AE1) not set).
).
Or         M function M29 set by NC.
(flag H063M3).
```

The operand designations specified (such as H0B3M6) are controller-specific. As Fig. 15.23 shows, the type of operand (input, output, flag) can be taken from these designations for the form introduced here. Project planning and programming of electronic controllers are discussed in detail in ► Chapter 10.

Shown on the right in Fig. 15.24 is the output of switching information by the PLC to hydraulic valve + L4-Y2, via relay module + M10-A17 and terminals X2 and X14. A full current path ID for contactor contact 13/14, for actuating hydraulic valve + L4-Y2, is also shown. ID "+ M1/32/12" in the left part of the figure establishes the connection to sheet "+ M1/32", current path 12 where the contact is used. Location ID "+ M1" denotes function group "Activation of actuators, interlocking". This current path has the cross reference "+ M10/64/7" to the associated drive coil on sheet + M10/64, current path 7.

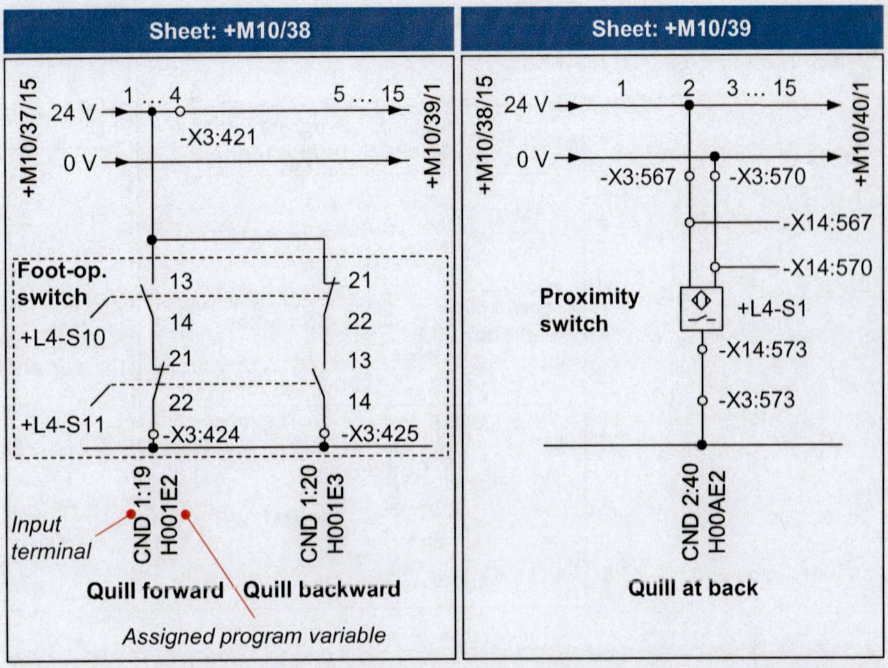


**Fig. 15.22** Circuit diagram: PLC inputs for the quill

**Logic**

*Quill_backward when (Enable_foot_operated_switch and Foot_operated_switch_Quill and not Quill_at_back) or M-function M_29 set*

**Circuit diagram, ladder diagram**

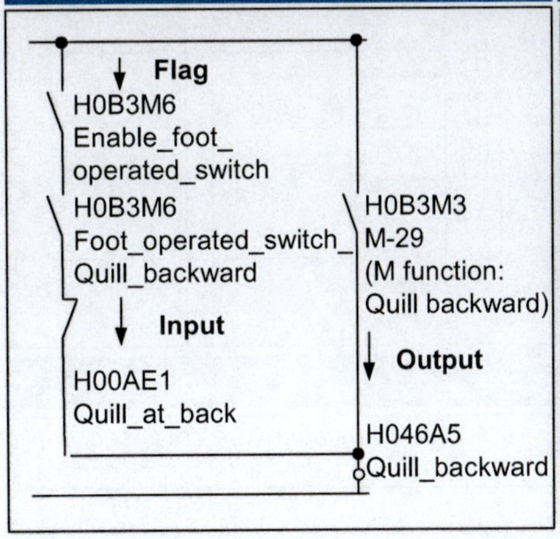


**Instruction list**

| Operation | Operand | Comment |
|---|---|---|
| U | H0B3M6 | Load working register with flag |
| U | H001E3 | AND operation with normally open contact |
| UN | H00AE1 | AND operation with normally closed contact |
| 0 | H063M3 | OR operation with flag |
| = | H046A5 | Set output |

**Fig. 15.23** "Quill function" PLC program as circuit diagram and instruction list

The other documents, such as terminal diagrams, layout diagrams, and parts lists, are created on the basis of the circuit diagrams. Figure 15.25 shows a terminal diagram that neatly summarizes the electrical connections, in contrast to the representation in the circuit diagrams that shows all details. This diagram can be used as an installation instruction.

### 15.2.4.5 Methods in electrical design

#### Project planning and creation of circuitry documents

In the realization phase, electrical design is responsible, among other things, for selecting and combining individual devices. This project planning task comprises:

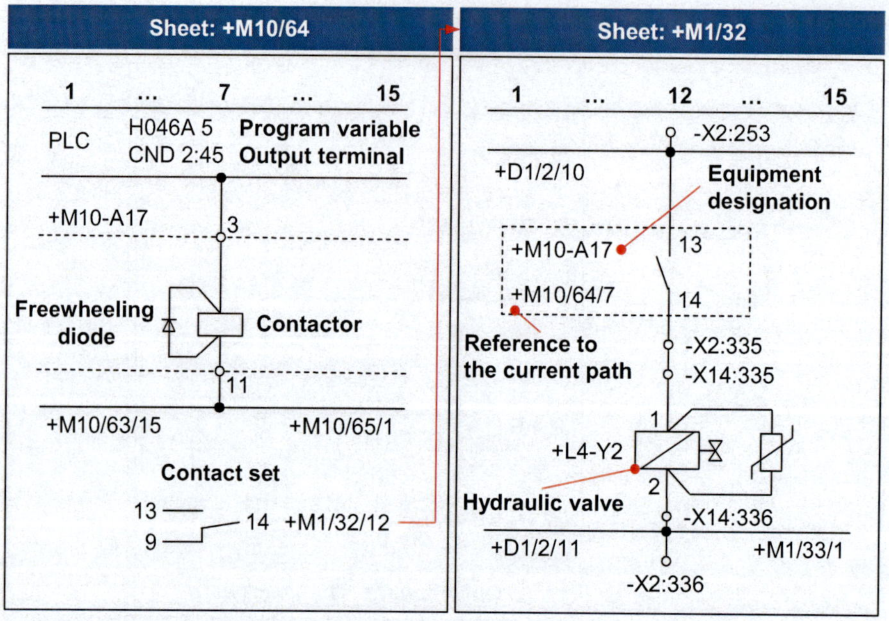


**Fig. 15.24** Circuit diagrams: PLC output "Quill back", activation of the actuator

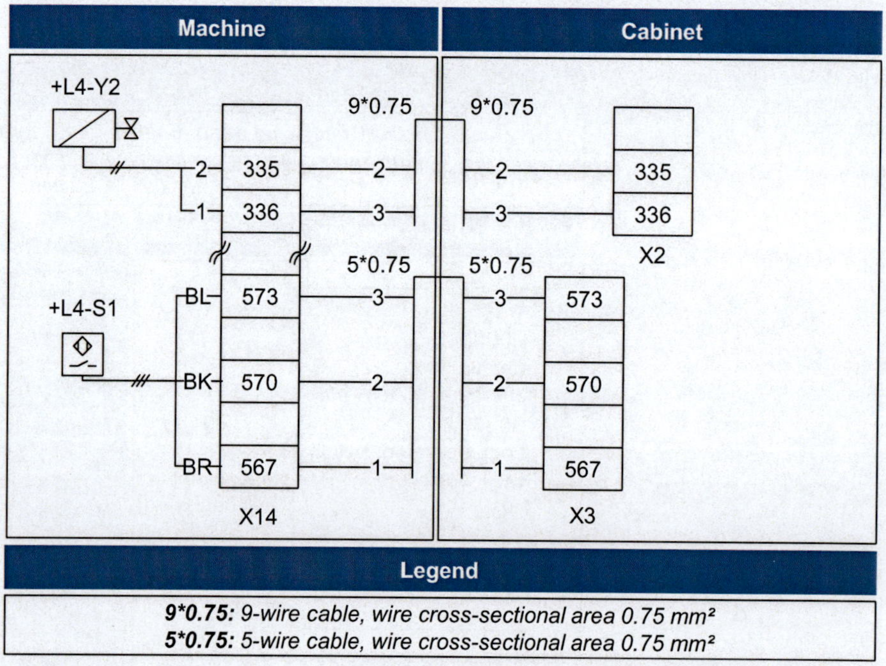


**Fig. 15.25** Terminal diagram for terminal strips X2, X3 and X14 (tailstock/quill, cabinet)

- Determining the device requirement
- Selecting the components and designing the actuation system
- Creating the circuitry documents (circuit and wiring diagrams, etc.)
- Compiling the parts lists
- Designing the cabinets and control panels
- Programming the automation equipment (PLCs).

This means the project planning for an electrical system differs greatly from the mechanical design of a machine. In the latter, the design of individual components, for example, is required. For the design of machine tools in particular, as shown back in ► Sect. 15.2.3, project planning for the electrical system must be in close collaboration with mechanical machine design. This applies especially at the start of machine development because not all requirements are yet known in detail.

The project developer must of course observe all national and international regulations. Important specifications for the safe electrical design of machine tools are summarized in [DIN14d]. The key contents of this regulation are only summarized as bullet points here because the individual points are explained in detail in ▶ Sect. 15.2.5:

- Mains connection and mechanisms for disconnection and switch-off
- Protection against electric shock (e.g., by direct and indirect contact)
- Equipment protection (e.g., overcurrent and overload protection of motors)
- Equipotential bonding (e.g., protective conductor system, electrical fault)
- Control current circuit and control functions
- Operator interface (e.g., keys, switches, and emergency stop mechanisms)
- Controller interface
- Electronic equipment
- Switch gear: arrangement, design, and enclosure
- Conductors, cables, and lines
- Wiring technology
- Electric motors and associated equipment (e.g., enclosures, dimensions)
- Warning signs and equipment ID
- Technical documentation, ▶ Sect. 15.2.4.3
- Testing (e.g., of protective conductor system, insulation resistance, and residual voltage)

### CAD systems for electrical design

Given that the project planning for an electrical system essentially comprises the selection and assembling of ready-to-use components, the project planning process can be aided effectively by CAx systems. In actual fact, electrical design was one of the first fields where CAx systems could be used cost-effectively. In contrast to very geometry-oriented mechanical CAD systems that require 3D functions, hidden line algorithms, etc., electrical CAD systems (CAE) are connection-oriented. They use 2D functions but must be able to manage potential networks and device relationships. CAD systems for electrical design generally support the following functions:

- Creating, managing, and changing all necessary documents (circuit diagram, wiring diagram, parts lists, etc.)
- Editing of standard circuits from a library, quick graphical display by inserting device symbols from drawing libraries
- CAD geometry commands for creating layout diagrams (for cabinets, for example)
- On-line logic testing (e.g., short-circuit tests, duplicate name assignments, open cable terminations), automatic PLC assignment lists
- Data interchange with other CAE systems
- Coupling to material management and ERP systems
- Multiuser capability

◘ Figure 15.26 shows the tools of an ECAD system for generating circuit diagrams. Modules are user-specific collections of symbols and function groups. Many sources of errors are eliminated using computer-aided drawing generation/management and by assigning component names. An ECAD system can detect short circuits and duplicate assignments, too many contacts and device connections by means of an incorrect potential and thus provide warnings. The ability to use and modify existing circuits and circuit libraries supports streamlining in project planning and the ongoing updating of circuitry documents. Conversions of circuitry documents to other graphical symbol standards (such as for export to the USA) are possible easily and without errors. Important installation aids such as terminal diagrams and wiring lists can be generated automatically.

Quotation cost estimates and ordering processes can be simplified by linking with materials management. Some typical electrical CAD systems (no claim is made to completeness):

- EPLAN *Friedhelm Loh Group*
- ELCAD *AUCOTEC*
- CADdy++ *IGE-XAO GmbH*
- ELTIME *HOS GmbH*
- TREE-CAD *TreeSoft GmbH & Co. KG*

## 15.2.5 Practicable integration of electrical components into machine tools

### 15.2.5.1 Energy supply

A prerequisite for operating an industrial machine is connecting it up to the electrical power system of the factory. It has generally been customary to route the associated lines permanently in cable ducts in the floor with each machine having its own cable. New ducts frequently then have to be dug in the floors of factories whenever the machines are moved. When cables are connected, some production is often shut down.

Newer, more modular concepts shorten potential downtimes for parts in production. Providing a "busbar trunking system" on the ceiling gives a type of bus system for the power cable. The individual consumers can be connected to this. This is achieved by coupling it to one of the tap-off units. These can be connected

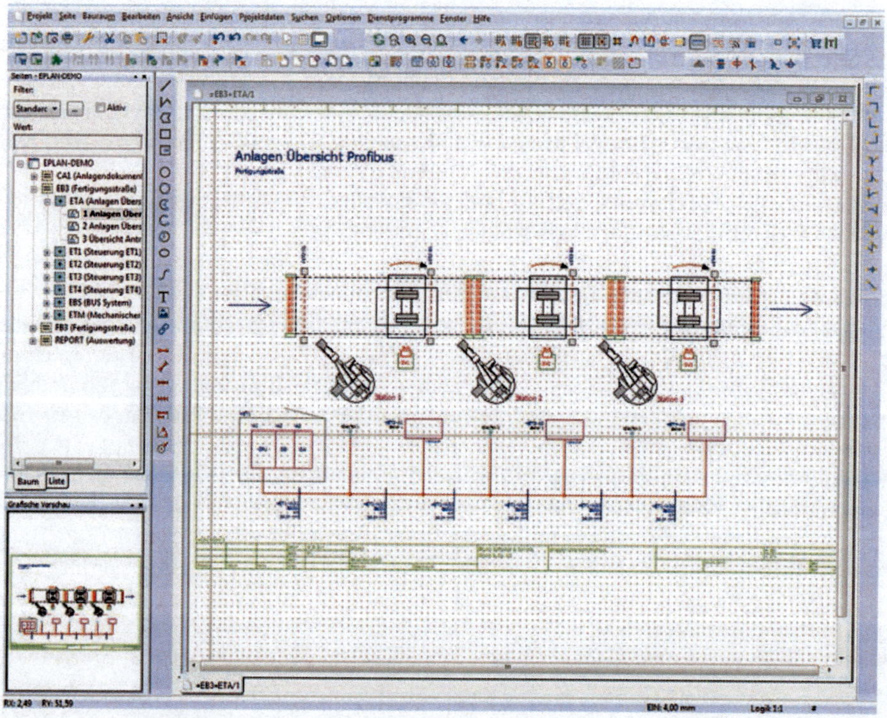


**Fig. 15.26** ECAD usage for generating circuit diagrams

virtually anywhere to the bus trunking system. Connections to the system can be made when the system is live, meaning production need not be interrupted. Furthermore, subdistributors can be connected to the main distributor for smaller machines, as well as strips with sockets for low-power consumers. The voltage in the distributors is usually 380 V. The aim of recent endeavors has been to increase the voltage in factory electrical networks to 660 V to reduce power losses during transmission.



A range of different voltages is usually required in machines. They are generated by transforming the voltage of the factory power system. The inverters are located in the cabinet for the machine.

The interface between the factory power system and the machine is stipulated in [DIN14d] to be the main switch of the machine. All main switches may only have positions 0 and 1. It must be possible to activate them by hand and they must be easily accessible. They must open when activated. It must be possible to lock the main switches. This ensures that maintenance personnel are protected against unauthorized switch-on of the system by others while working.

### 15.2.5.2 Electrical components in machine tools

Electrical components in machine tools include signal elements, drives, actuators, cabling, and measuring systems. Signal elements are components in the auxiliary circuit that relay the machine state either to the controller or operator (signal units) or that trigger machine functions activated manually (control units). Contactors or relays can be activated in the main circuit using signal elements. One of the most important signal units is the limit switch. A mechanical limit switch is activated by moving machine parts. For example, a cam shifts a moving contact and, in the process, activates the switch (Fig. 15.27). For safety reasons, it is mandatory for the contact to open.

The application areas for limit switches are very diverse. They are used as limit or reference switches on moving shafts, to determine the positions of pallets, tool magazines or workpieces in transfer lines and to satisfy safety requirements (such as on doors of machines and protective fencing). A few points need to be considered when installing limit switches. Limit switches should be easily accessible so their function can be checked without removing other components.

They must be positioned such that the probe is not subjected to high vibrations. The probe must also be protected against chips, cooling lubricant, and unintentional activation. Activation is only permitted on the level allowed. With dual-level safeguards, the activation time of the limit switch must be enough to put the relay in a self-hold state.

Inductive limit switches (proximity switches) are used where no actuating forces are available, mechanical contact should be averted or an extremely high switching frequency is required. An inductive proximity switch uses a resonant circuit to

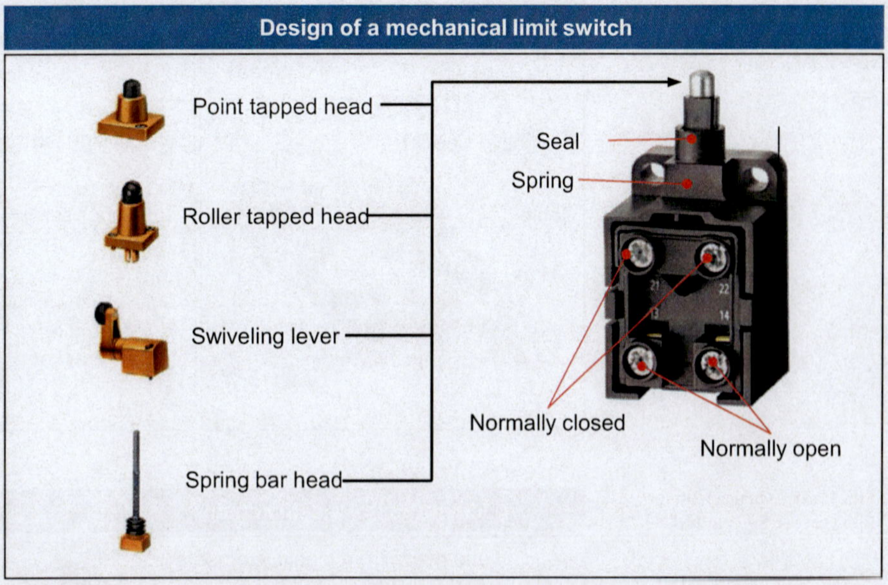


**Fig. 15.27** Designs of mechanical limit switches. *Source:* Siemens

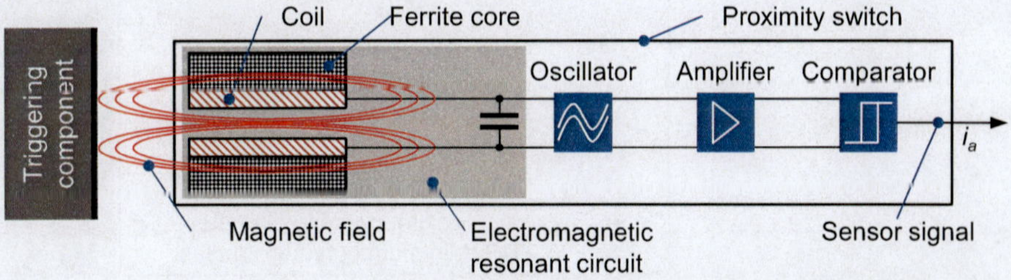


**Fig. 15.28** Functional principle of an inductive proximity switch

generate a high-frequency alternating magnetic field (Fig. 15.28). When an electrically conductive object moves toward the response area, the alternating magnetic field is damped by the induced eddy currents. The resulting reduced vibration amplitude leads to the switching operation when a lower threshold value is reached [FEST16].

In addition to inductive proximity switches, there are also capacitive, optical and ultrasonic proximity switches. Capacitive proximity switches can detect, without touching, conductive as well as non-conductive materials that are in a solid, powder or liquid state. They can be used for fill level monitoring, for example. An optical proximity switch can be used to detect extremely small objects even when strict requirements are made of activation speed, switching frequency, accuracy, service life and difficult installation position. Ultrasonic proximity switches do not require maintenance and are generally impervious to environmental conditions. They can be used, for example, to detect people at doors and gates and for fill level monitoring.

Pressure sensors are used to check system pressure, for hydraulic clamping fixtures, for example (for pressure regulation). In this case, the system pressure acts upon a pretensioned spring via a diaphragm. When the diaphragm force is not equal to the spring force, the plunger moves until equilibrium is restored. After the adjustable idle stroke is overcome, the contact mechanism is actuated abruptly. The switch-off pressure is set by the pretension of the spring while the idle stroke determines the switch-on pressure.

**Fig. 15.29** Example of an inductive proximity switch

To supply power to the components and for signal transmission, cables need to be laid in and to the machine (Fig. 15.30). Cables are usually routed from a cabinet to distributor terminals in the machine via a central cable harness. From there, smaller cable groups

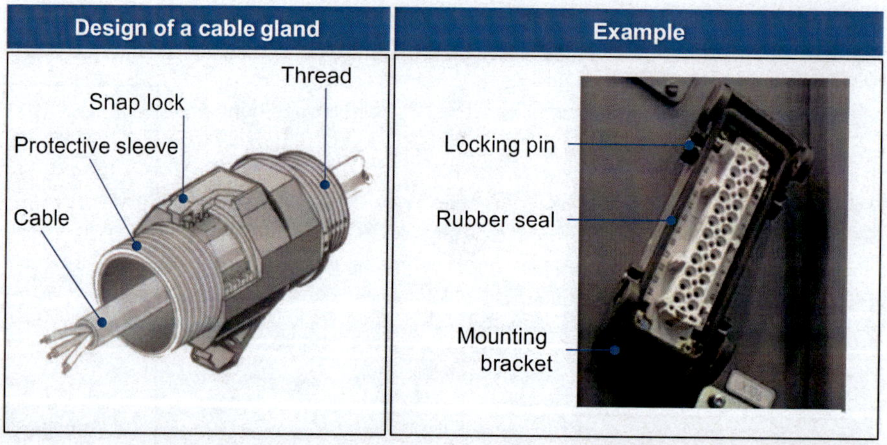


**Fig. 15.30** Cable feed on cabinets and machines (connectors). *Source:* Schlemmer

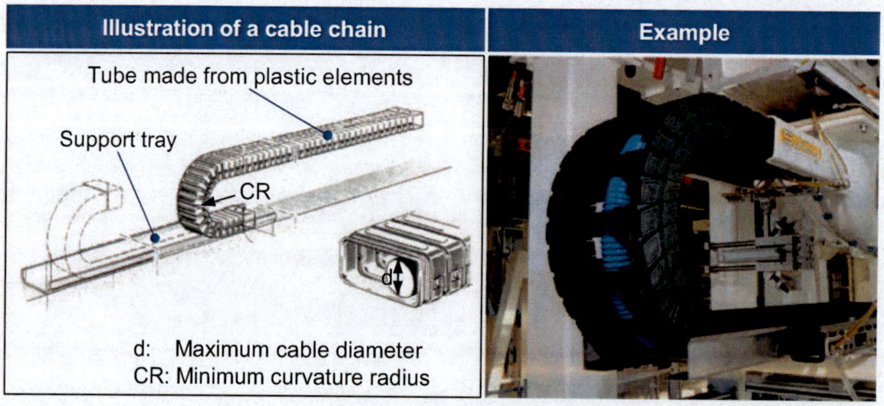


**Fig. 15.31** Cable chain for guiding cables. *Source:* IGUS

are routed to intermediate distributors and then individual cables to the end components.

Cable groups should be laid in cable ducts or pipes because they provide good protection against ambient conditions and mechanical damage. Cables laid individually are attached with special fasteners. It is important that cables are not subjected to any mechanical stress because otherwise, they could break. When cables are laid, they must not be kinked. The cross sections of cables are determined by how much power the cables are to carry. Signal cables must be shielded to prevent interference to the signals.

Many individual drives are often used in the machine tool industry. As a result, many of these motors are attached to moving machine components such as feed slides. Cable chains are often used here (Fig. 15.31). One end of the chain is fixed to the machine frame while the other end is secured to the moving machine part. The chain is made of moving links or a flexible tube. The lower part of the chain lies on a level surface. The rest can hang freely. If now the machine part moves, the chain is pulled along with it. Constant curvature radius CR is always maintained during motion (Fig. 15.31).

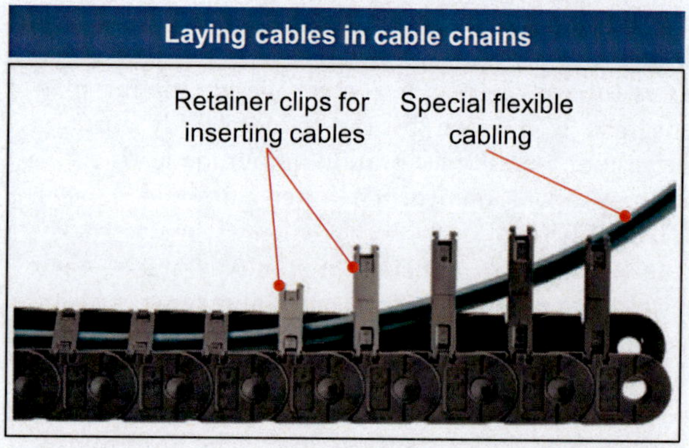


**Fig. 15.32** Bracket mechanism for laying cables in cable chains. *Source:* IGUS

When chains are very long, it is important to ensure they do not sag. Supporting rollers must prevent the chains from getting caught on the lower chain part (if it is not suitably smooth). The chain material is steel or plastic. The cables are placed in the chain. For this, brackets can be opened on most chain types (Fig. 15.32).

15

**Fig. 15.33** Structure of a cable for cable chains. *Source:* IGUS

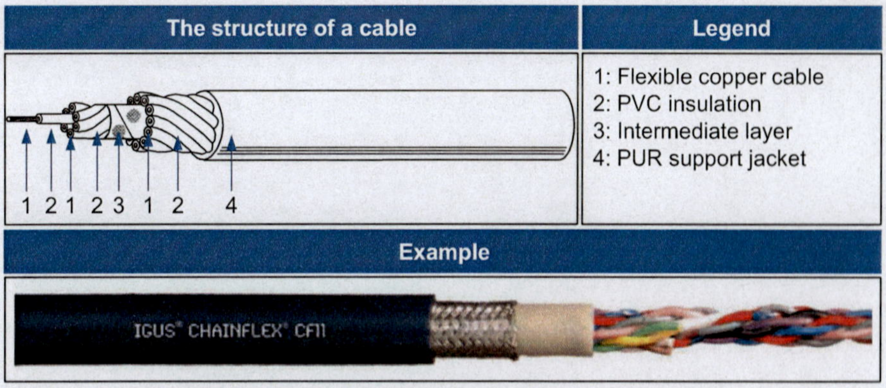


**Fig. 15.34** Structure of a block terminal

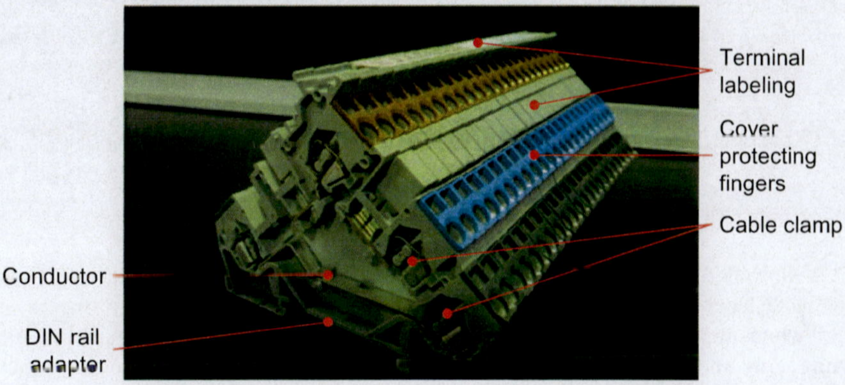


Particular requirements are put on the cables and their configuration in the chains. The cables must be very flexible. The permitted bending radius of the cables must be shorter than that of the chain and the lines must have dual insulation.

The recommendation is to select small cable diameters for small-gage copper wire because thick cables are under great stress at the outer edge from the repeated bending. The speed and frequency of the machine component motion must be taken into account. If more than one cable is placed in a chain, separating strips can prevent friction between them. It is important for cables in a cable chain to move freely because otherwise, when they bend, they can be subjected to mechanical stress and break [TSUB13]. In addition to cable chains, there are also cables available that do not require guiding. The cross section of such a cable is shown in ◘ Fig. 15.33. The sheathing is made of polyurethane (PUR) that does not become brittle. The conductors on the inside are small-gage copper with high tensile strength.

Intermediate layers of foil allow the individual wires to move against each other. The actual support effect is achieved by the outer sheath. Since cables on moving machine components are subjected to a very high mechanical bending load, materials are required that prevent cables from breaking. The cables must be watertight, resistant to coolants and their chemical components as well as resistant to chips. Inside cabinets and machines, block terminals are used to distribute cables (◘ Fig. 15.34). They are designed to protect fingers. To connect cables from outside to a cabinet, the machine or the machine components, connectors or screw-on tubes are used that are tailored to the rugged conditions of a factory building (◘ Fig. 15.30). They are fully insulated, watertight and secured such that they cannot be pulled from the sockets by mechanical force. A threaded connector or retainer clip must be removed first.

#### 15.2.5.3 Operator interface

To allow operators to control machines, it must be possible for them to issue commands and detect machine signals outside a machine. Command and signal units are used for this. These can be simple switches, monitors with a computer keyboard or ergonomically designed control panels.

The design of a machine, especially that of the control panel, is playing an ever-greater role in the purchasing decision (see ▸ Chapter 11). The relevant standards must be obeyed in the design of control panels, especially for the colors used (◘ Fig. 15.35). Red must be used for EMERGENCY STOP controls. The preference is for controls effecting a simple “Stop” or “Off” to be black.

**Fig. 15.35** General significance of colors for controls. *Source:* [DIN03b]

| Color | Meaning | Explanation | Application examples |
|---|---|---|---|
| Red | Emergency | Action in the event of danger or emergency | - EMERGENCY STOP<br>- STOP or OFF with EMERGENCY STOP button<br>- Initiation of an emergency function |
| Yellow | Abnormal | Action in the event of an abnormal state | - Intervention to suppress the abnormal state<br>- Manual intervention to restart an interrupted automatic cycle |
| Green | Safe state | Action for a safe state or to prepare normal states | - Start/On |
| Blue | Direction | State requiring action | - Reset function |
| White<br>Gray<br>Black | No specific meaning | Initiation of functions | - May be used for any function, except for EMERGENCY STOP, e.g., for OFF/ON, STOP/START |

Yellow is used for interactions. A typical example is the "Clear" button after an incorrect entry. Green and white apply for "Start," "On" and similar. Also, white, gray and black can be used for controls effecting START/ON and STOP/OFF alternately.

It is important that the ambient conditions at the place of use be taken into account. A control console for setup in a measurement room must have a design different from a control console used at a foundry. Application-related features could be the following: watertight, resistant to dust, liquids and oils (see IP protection class in Fig. 15.12) and resistant to mechanical load, vibration, and impacts.

Whether operators wear thick gloves, and so can only press large pushbuttons, must also be considered. It might not be possible after a short time to read a monitor in oil mist and a dusty environment. As modern machines must also be suitable for operators who are used to older machine types, conventional components such as electronic handwheels are also used. The turning of a handwheel is converted to an electrical signal that is used to actuate the drives. Monitor, keyboard, electronic handwheel and emergency stop button(s) are components of typical control panels on NC machines. A machine is switched on using a pushbutton that is often lockable to prevent unauthorized use. Keyboards are almost always membrane keyboards. This type of keyboard is completely encapsulated and makes them very resistant to ambient conditions. Simple machines with few command buttons often have pushbuttons and switches (Fig. 15.36). Keys used for a machine start must not protrude (to prevent unintentional actuation).

Control panels must have a neat and clear design, so operators can see what is happening at all times. The most important component is the emergency stop button. It must be positioned so as to be clearly visible and easily within reach. The positions of the monitor and keyboard come automatically from eye and hand heights. Key switches and command buttons not used very often should be positioned in the background while those buttons frequently used, such as direction and speed select, should be positioned in the center. The other control units should be positioned in line with the operating sequence. Refer to ▶ Sect. 11.5 for more information on user interfaces.

Cam switches are well suited to easily activate multiple machine functions, such as to specify different speeds.

If there are several interrupters, different functions can be combined for every switch position by designing the switching cam disks of the individual interrupters as per the required combinations of functions. Foot-operated switches are other control units. These can be used, for example, to close the chuck of a lathe so the operator has both hands free.

Signal units are used to report certain machine states to the operator. Indicator lights are usually used. The colors used must be borne in mind here, too (Fig. 15.37). The color red indicates emergency states. When red is shown, the triggering hazard must be corrected immediately. Yellow indicated caution

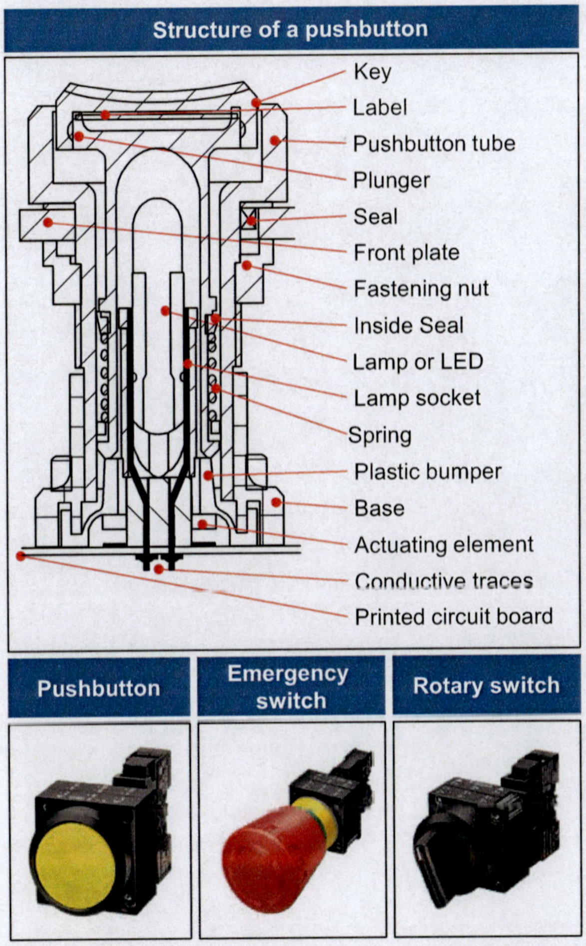


**Fig. 15.36** Pushbuttons and switches. *Source:* Georg Schlegel GmbH & Co, SIEMENS

about abnormal states that could result in a hazard and green signals safe operation. Blue, white and black can be used for other functions. It is common nowadays to display operating states on the integrated monitor, eliminating the need for indicator lights. Analog and digital displays for voltages and currents are other signal units. Danger can also be signaled by sounds when, for states indicating a hazard to a larger area, a signal horn provides a warning for the area.

### 15.2.5.4 Safety devices

Working with machines almost always represents a risk to people as soon as the operator must move material into the operating space. In production, chips flying around, rotating tools and workpieces, the working stroke of press rams, etc., pose a risk of injury to the machine operator.

These risks are also present for startup, maintenance and repair work. The safety regulations of professional associations help machine design engineers to design machines such that operators are protected to a large extent from potential risks. Nowadays, no machine that does not satisfy the safety requirements may be put into operation. The requirements for safety devices are that they guarantee safe operation of the system, and they switch off the monitored system safely even if there is a fault in the safety devices.

Self-monitoring circuits are often very complex. Even when a cable breaks in the supply line of the equipment, the machine affected must be stopped and a corresponding fault message issued. It must not be possible for the operator to easily bypass the safety devices or put them out of commission. The response time between triggering of the safety mechanism and full shutdown of the machine must be as short as possible, e.g., shorter than the time the operator needs to get to the hazard area or hazard point.

Probably the most well-known safety device for presses is the two-hand control device. The press process is only triggered when the two start buttons, located at a sufficient distance from the work area of the press, are pressed simultaneously. But these solutions are not 100% safe. There can still be another person in the hazard area of the press.

Inventive operators are able to build auxiliary devices to still operate the machine with one hand. Additional movable protective screens are thus often used in front of the operating space.

One of the safety solutions frequently seen in this application area uses light barriers, light grids, or light curtains. For light grids, and for protection by one or more light barriers, the work process can only be triggered when the light rays are not interrupted (Fig. 15.38).

For a light curtain, only a single rotating light ray is used. It is driven by a motor and covers the area to be secured. On the other side of the room to be secured, the beam is reflected straight back by a prism reflector and directed onto a photodiode by the mirror arrangement. An evaluation electronics system ascertains whether the beam was interrupted at any time. For the scanner light curtain in Fig. 15.38, infrared light is transmitted along parallel axes to the reflectors on the other side of the hazard area using a rotating mirror wheel and concave mirror. The receiver is at the focal point of the mirror. The fast rotation of the mirror almost entirely secures entry to the space by the light beam [GÖRN13]. The operator places the new part into the machine, steps behind the light curtain, and presses a hand or foot button, triggering the operation. The equipment shown in Fig. 15.39 represents a clear improvement of work conditions because the safety equipment does not obstruct the operating space or hinder the operator. In the design of light barriers and light curtains, it is important to ensure that the hazard area is screened such that the operator cannot bypass the light beams intentionally or by accident.

**Fig. 15.37** Colors for indicator lights. *Source:* [DIN03b]

| Color | Meaning | Explanation | Operator action | Application examples |
|---|---|---|---|---|
| Red | Emergency | Hazardous state | Immediate resolution and urgent need for action, e.g., with emergency stop, open valve, start cooling pump | - Pressure/temperature not in safe ranges<br>- Voltage failure<br>- Malfunction of an important system part<br>- Failure of necessary machines / auxiliary systems |
| Yellow | Abnormal | Abnormal state or imminent critical state | Monitoring and/or intervention, e.g., by restoring the required function | - Pressure/temperature varying from the normal range<br>- Triggering of an auxiliary unit safeguard<br>- Passing a limit switch |
| Green | Normal | Normal state | No action needed | - Switch-on enable<br>- Display of normal operating conditions |
| Blue | Direction | Display of a state requiring action | Action prescribed | - Instruction to the operator to set preselected values |
| White<br>Gray<br>Black | No specific meaning | Every meaning may be used when there is doubt about RED, YELLOW, GREEN and BLUE. | Monitor | - General information, e.g., confirmation of a command, display of a measured value |



[DIN09b] stipulates that metal-cutting machine tools must be fully encapsulated during use and not too directly accessible for the operator. (Automatic) machining may only start once the door to the operating space is closed. The PLC monitors this using two non-removable limit switches on the door. Operations are only enabled when the door is completely closed.

On multi-machine systems, such as transfer lines, flexible manufacturing cells and manufacturing systems, material transport is automatic. If an operator enters the area of the system, they must be protected from injuries arising from the movement of robots, industrial trucks, etc. Protective fencing represents the safest and easiest option. When a boundary fencing door is opened, a relay deenergizes and the entire system stops. Collisions with industrial trucks can be prevented using ultrasonic sensors or touch-sensitive projections. Safety shut-off mats can be used in hazardous, narrowly confined zones, such as where robots operate. As soon as a certain weight force acts on the mat, the device is switched off. Light barriers can also offer effective protection. Liquid and gas-filled tubes that react to touch using pressure sensors can be affixed at potential crushing points. They act as emergency stop switches. Figure 15.40 shows the different protection capabilities using a flexible manufacturing system as an example.

#### 15.2.5.5 Electrical cabinet design

While the main control and signal units, and the drives, are housed in the machine, the electrical cabinet contains the controller components. In this cabinet, the voltages required are provided, the controller functions are connected by the numerical controller and PLC, and feedback control is performed by servo drives. In electrical cabinet design, the aim is to house as many sensitive components as possible such that they are compact, neatly laid out, and protected.

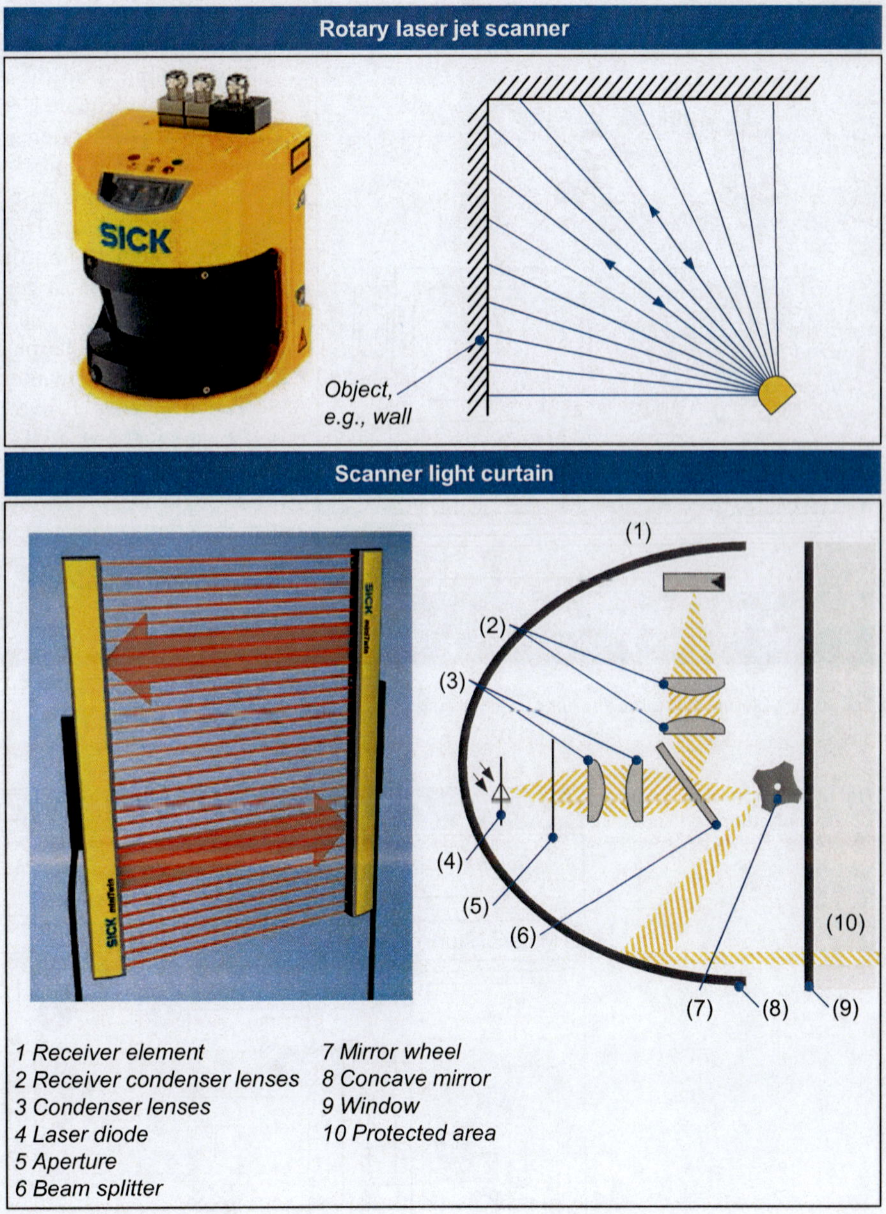


**Fig. 15.38** Designs of safety light barriers. *Source:* Sick

## Design and layout

Standardized frames and modular assembly elements facilitate efficient cabinet design, also in terms of space. Assembly plates for large equipment such as transformers are affixed in the outer frame. The relays, for example, are fastened to assembly rails. Support rails lend themselves to attaching devices with simple snap-on mounting. Cable organizers are used for cabling. Wiring that was previously routed through cable ducts is now in the rear compartment of the electrical cabinet.

Using this system, the electrical cabinet required can be created in a very flexible way. Electrical cabinet design requires physical layout planning of the individual components in advance.

## Components and their positioning

Various aspects are taken into account for positioning components inside an electrical cabinet. The criteria for this are separation into high and low voltage areas, heat dissipation, interference susceptibility to electrical fields, accessibility during maintenance and repair work and combining groups performing the same types of functions as in the machine controller. It can be seen from the modular cabinet principle described above that components of the same type, such as relays, are arranged in assemblies in the electrical cabinet.

The typical components of an electrical cabinet are discussed briefly below. For this, Fig. 15.41 shows the typical cabinet layout using a lathe cabinet as an example.

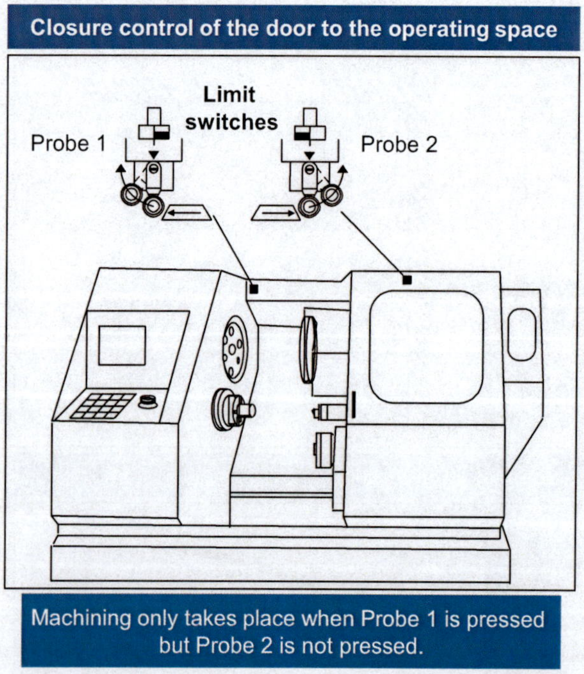


**Fig. 15.39** Closure control of the door to the operating space of a lathe

Transformers with power system voltage control make available all operating voltages required from the voltage coming from the main switch. The transformers ensure an adequate level of voltage stability because the power system often experiences fluctuations. The power components provide voltages and currents for the drives. Both module types are heavy and large and give off a lot of heat. The weight means that these devices must be placed on the floor of the electrical cabinet. However, the heat generated must be dissipated at the source, if possible. Components that are sensitive to electrical fields can be placed far away from these devices with their strong interference fields.

The difference between a contactor and a relay is shown in Fig. 15.42 and Fig. 15.43. In a relay, a toggle contact made of spring steel performs the switchover. For a contactor, a magnet presses a straight contact piece against a spring onto the contacts to be connected for the switchover. Given that a contact piece can be bigger than a spring contact and can transfer higher currents, contactors are used to switch higher powers. In selecting contactors, it is important to ensure that even high currents can always be switched off without an arc persisting or the contacts

15

**Fig. 15.40** Safety devices on a flexible manufacturing system

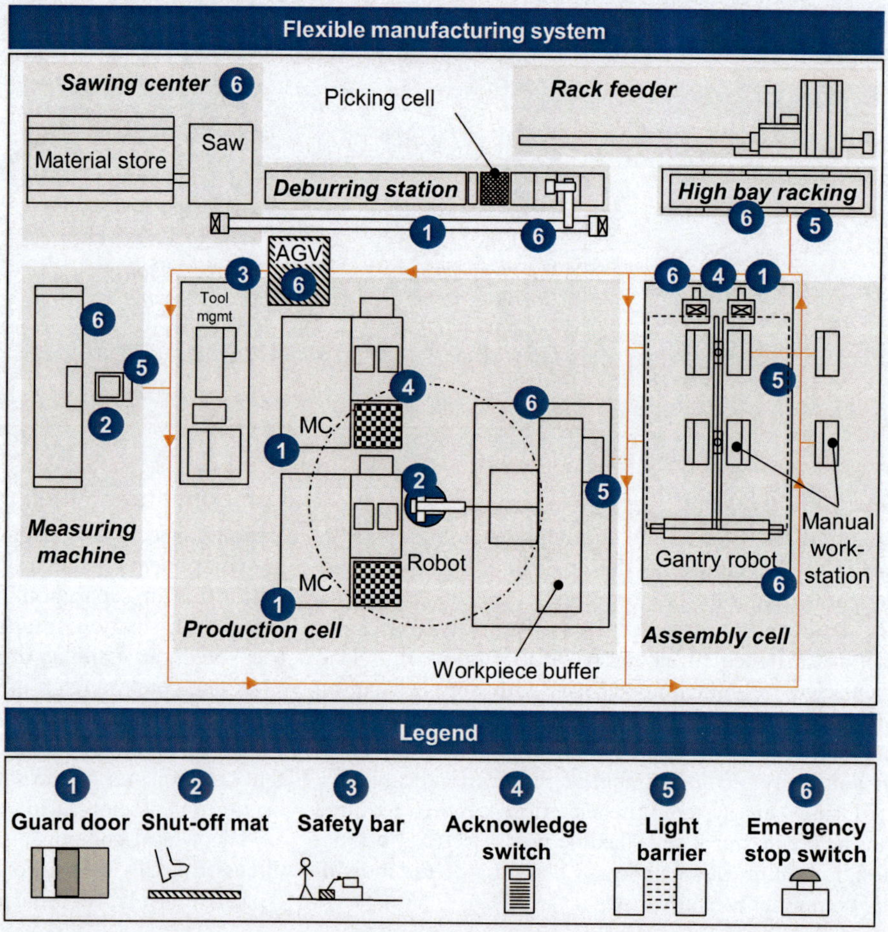

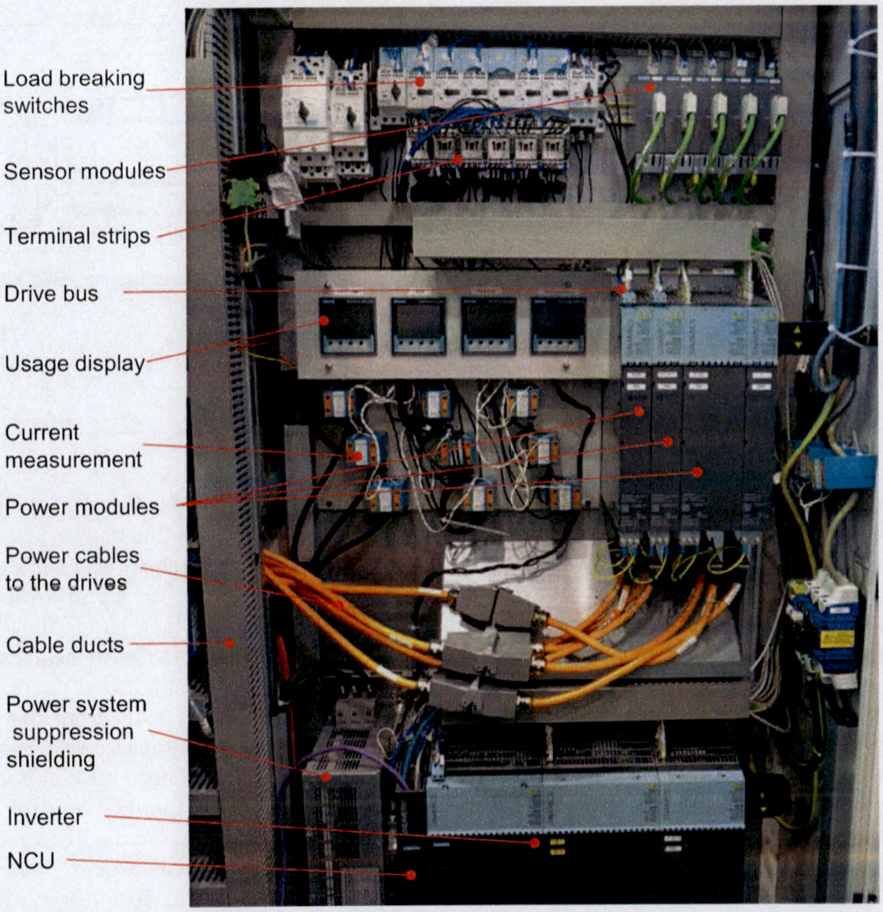


**Fig. 15.41** Layout of a lathe cabinet

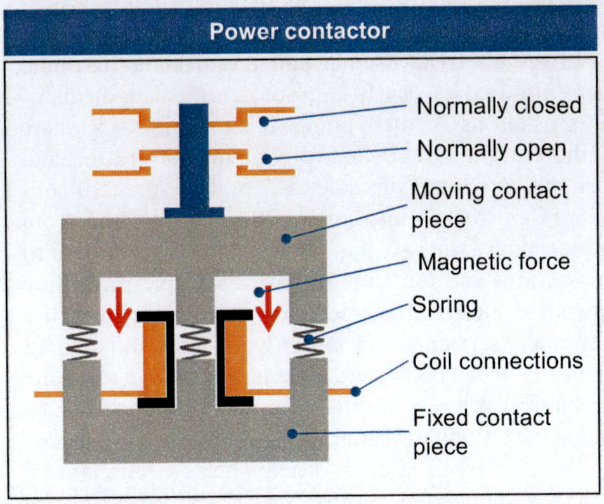


**Fig. 15.42** Power contactor

fusing together. Contactors for activating high powers are housed in the electrical cabinet separate from the relays with lower power.

Overcurrent protection devices are used to protect lines, motors, devices, etc., against overloads and short circuits. Fuses, miniature circuit breakers, and circuit breakers are available for this purpose. These devices differ in regard to the following criteria:

- Trigger current: current and time-dependent switching of devices
- Breaking capacity: maximum current at which a device still switches
- Selectivity: specific activation of a device in a circuit before a device in the master circuit with a higher trigger current switches
- Cut-off current: maximum current that flows via the device before it switches

In mechanical and plant engineering, low-voltage, high-power fusing systems are normally used as fuses to protect lines.

They are made of a metal wire that melts when the trigger current is exceeded, thus reliably breaking the electrical circuit. These fuses can be used for amperages between 2 and 400 A. Device fuses can be used

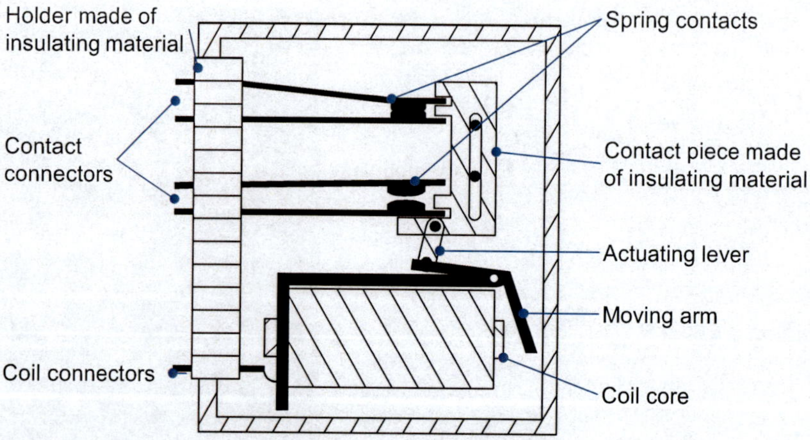


**Fig. 15.43** Relay structure

to protect devices against overloads and short circuits. Like low-voltage high-power systems, they are also fuses. But they are only available for amperages between 0.032 and 6.3 A. The downside of these fuses is that they need to be replaced after being triggered. Also, it is not always immediately obvious which fuse has blown.

Miniature circuit breakers are used to protect lines from overloads and short circuits. They have breaking capacities in the range of 3 to 10 kA.

Circuit breakers are used to protect motors and lines, for example, and have breaking capacities in the range of 10 to 100 kA.

Both protection devices have a thermal trigger for lower overcurrents and a magnetic trigger for high overcurrents and short circuits.

The thermal trigger uses a bi-metallic element that switches depending on the overload time. Both the miniature circuit breaker and circuit breaker can quickly be readied for use again. There is a risk of their contacts fusing together, however, rendering them unusable. They must not be used to protect people.

Residual current circuit breakers are used to protect people. They ascertain whether all currents are flowing from a phase to neutral. If not, there is a fault to frame or a ground fault creating a risk to people. The standard solution used by the residual current circuit breaker is an annular coil, wound around several cables, that measures the total current. This current is normally zero. When a fault occurs (fault to frame or ground fault), the total current is not equal to zero. This total current generated activates a switch that switches off the system.

To enable fuses to be replaced and switched back on easily, they must always be positioned in electrical cabinets to be easily accessible.

Controllers comprise an array of heat generators. On the other hand, they include elements that react very sensitively to heat and electrical fields, meaning they can cause malfunctions. These components must be housed away from the transformers and high-voltage side in the electrical cabinet.

To connect all components, i.e., to provide them power, a number of cables are required in machines nowadays. Frequently, cable breaks can only be localized at great expense, bringing part of production to a standstill. To quickly locate cable errors and replace wires, every wire and every cable must be tagged, and every terminal marked. Colors or numbers are printed on the insulation sheathing for the cable markings.

In the electrical cabinet and the machine, terminal strips are used as clearly laid-out intermediate or distribution stations. At these points, it is easy to check where in the current flow defective components are located. Terminal strips are stipulated where more than 10 conductors are routed into an enclosure. Cables run from the terminal strips so that the cables are grouped into line sections and routed in cable ducts. Cables stretching across the electrical cabinet must not be used because they make servicing and repair work more difficult. In an electrical cabinet, cables must not be subjected to any mechanical stress, be located near hot components or be routed past modules that are sensitive to electrical fields.

The standardized color scheme helps electricians to directly assign cables to their usage [DIN03a]. Black denotes the primary circuits, blue the neutral conductors of primary circuits, red the control circuits for alternating current, blue the control circuits for direct current, orange the interlock circuits and yellow/green the protective conductor. Cable connectors leading out of an electrical cabinet must be secured with screw joints.

### Temperature control in electrical cabinets

Since many of the components in an electrical cabinet give off more or less high levels of power dissipation in the form of heat while an array of components are sensitive to high temperatures, air conditioning of the cabinet is often indispensable. It not only plays the role of cooling, but it also regulates the specified air humidity inside the electrical cabinet. This needs to be maintained for certain components to operate.

Controllers used in unfavorable climates are fully encapsulated so that no air from outside can penetrate the sensitive part of the controller. These types of enclosed electrical cabinets require a heat exchanger or cooling compressor to dissipate the thermal energy (◘ Fig. 15.44). Simple fans are adequate when the power loss is low and the cabinets are not completely enclosed. The air must be filtered, however. Inside the cabinet, additional fans can be attached for particularly sensitive assemblies. In the design of an electrical cabinet, the heat to dissipate must be approximated to select a suitable air conditioner (◘ Fig. 15.45).

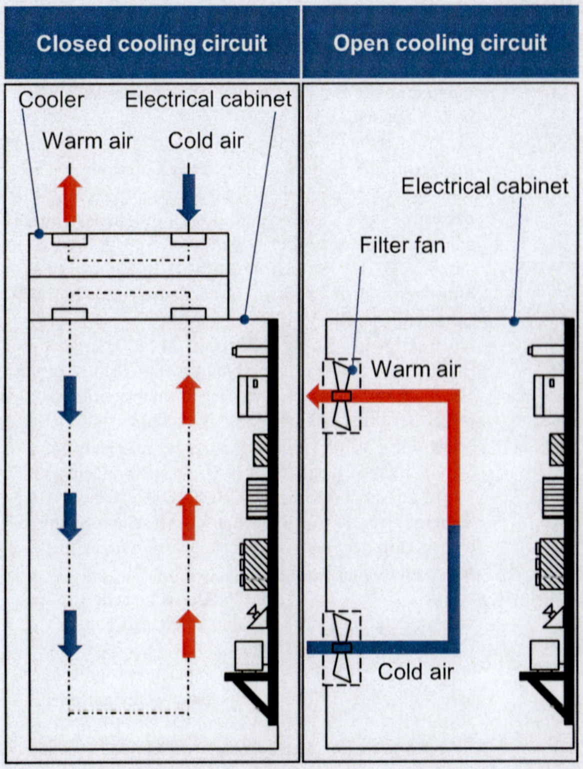


◘ Fig. 15.44 Temperature control in electrical cabinets

◘ Fig. 15.45 Design of a cooling unit

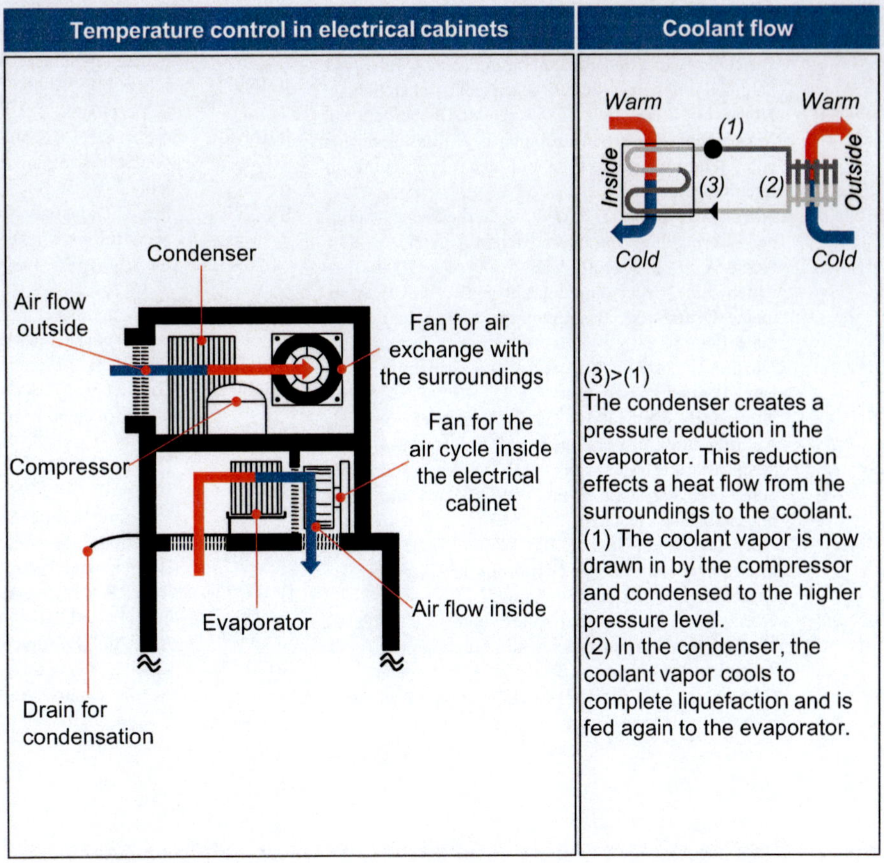

## References


[BRAN15] Brandt-Pook, H.; Kollmeier, R.: Softwareentwicklung kompakt und verständlich. 2. Auflage. Aufl. Wiesbaden: Springer Vieweg, 2015

[BUND15] Gesetz ArbSchG, Bundesministerium der Justiz und für Verbraucherschutz (2015). Gesetz über die Durchführung von Maßnahmen des Arbeitsschutzes zur Verbesserung der Sicherheit und des Gesundheitsschutzes der Beschäftigten bei der Arbeit

[DGUV97] Satzung DGUV Vorschrift 3 (01.01.1997). Unfallverhütungsvorschrift Elektrische Anlagen und Betriebsmittel (bisher: BGV A3)

[DIN02] Norm DIN EN 50274, VDE 0660–514 (2002). Niederspannungs-Schaltgerätekombinationen - Schutz gegen elektrischen Schlag - Schutz gegen unabsichtliches direktes Berühren gefährlicher aktiver Teile

[DIN03a] Norm DIN 40108 (2003). Elektrische Energietechnik - Stromsysteme - Begriffe, Größen, Formelzeichen

[DIN03b] Norm DIN EN 60073, VDE 0199 (2003). Grund- und Sicherheitsregeln für die Mensch-Maschine-Schnittstelle, Kennzeichnung - Codierungsgrundsätze für Anzeigengeräte und Bedienteile

[DIN04a] Norm DIN EN ISO 7200 (2004). Technische Produktdokumentation - Datenfelder in Schriftfeldern und Dokumentenstammdaten

[DIN04b] Norm DIN EN 61131–1 (2004). Speicherprogrammierbare Steuerungen - Teil 1: Allgemeine Informationen

[DIN06] Norm DIN EN 50370–1, VDE 0875–370–1 (2006). Elektromagnetische Verträglichkeit (EMV) - Produktfamiliennorm für Werkzeugmaschinen - Teil 1 Störaussendung

[DIN07a] Norm DIN EN 61131–2, VDE 0411–500 (2007). Speicherprogrammierbare Steuerungen - Teil 2: Betriebsmittelanforderungen und Prüfungen

[DIN07b] Norm DIN VDE 0100–410, VDE 0100–410 (2007). Errichten von Niederspannungsanlagen - Teil 4–41: Schutzmaßnahmen - Schutz gegen elektrischen Schlag

[DIN08a] Norm DIN EN ISO 13857 (2008). Sicherheit von Maschinen - Sicherheitsabstände gegen das Erreichen von Gefährdungsbereichen mit den oberen und unteren Gliedmaßen

[DIN08b] Norm DIN EN 1037 (2008). Sicherheit von Maschinen - Vermeidung von unerwartetem Anlauf

[DIN09a] Norm VDE 0100–100, VDE 0100–100 (2009). Errichten von Niederspannungsanlagen - Teil 1: Allgemeine Grundsätze, Bestimmungen allgemeiner Merkmale, Begriffe

[DIN09b] Norm EN 12417 (2009). Werkzeugmaschinen - Sicherheit - Bearbeitungszentren

[DIN10] Norm DIN EN 81346–2 (2010). Industrielle Systeme, Anlagen und Ausrüstungen und Industrieprodukte - Strukturierungsprinzipien und Referenzkennzeichnung - Teil 2: Klassifizierung von Objekten und Kennbuchstaben von Klassen

[DIN11] Norm DIN EN ISO 12100 (2011). Sicherheit von Maschinen - Allgemeine Gestaltungsleitsätze - Risikobeurteilung und Risikominderung

[DIN12a] Norm DIN EN 61439–1, VDE 0660–600–1 (2012). Niederspannungs-Schaltgerätekombinationen - Teil 1: Allgemeine Festlegungen

[DIN12b] Norm DIN EN ISO 10218 (2012). Industrieroboter - Sicherheitsanforderungen

[DIN14a] Norm DIN EN ISO 14119 (2014). Sicherheit von Maschinen - Verriegelungseinrichtungen in Verbindung mit trennenden Schutzeinrichtungen - Leitsätze für Gestaltung und Auswahl

[DIN14b] Norm DIN EN 61131–3 (2014). Speicherprogrammierbare Steuerungen - Teil 3: Programmiersprachen

[DIN14c] Norm DIN EN 61140, VDE 0140–1 (2014). Schutz gegen elektrischen Schlag - Gemeinsame Anforderungen für Anlagen und Betriebsmittel

[DIN14d] Norm EN 60204–1, VDE 0113–1 (2014). Sicherheit von Maschinen - Elektrische Ausrüstung von Maschinen - Teil 1: Allgemeine Anforderungen

[DIN14e] Norm DIN IEC 60050–351 (2014). Internationales Elektrotechnisches Wörterbuch - Teil 351: Leittechnik

[DIN14f] Norm DIN EN 60529, VDE 0470–1 (2014). Schutzarten durch Gehäuse (IP-Code)

[DIN14g] Norm DIN EN 60848 (2014). GRAFCET – Spezifikationssprache für Funktionspläne der Ablaufsteuerung

[DIN15a] Norm DIN EN 60947–1, VDE 0660–100 (2015). Niederspannungsschaltgeräte - Teil 1: Allgemeine Festlegungen

[DIN15b] Norm DIN EN 60034, VDE 0530 (2015). Drehende elektrische Maschinen

[DIN15c] Norm DIN EN 61082, VDE 0040 (2015). Dokumente der Elektrotechnik

[DIN16a] Norm DIN EN ISO 13850 (2016). Sicherheit von Maschinen - Not-Halt-Funktion - Gestaltungsleitsätze

[DIN16b] Norm DIN EN 55014–2, VDE 0875–14–2 (2016). Elektromagnetische Verträglichkeit - Anforderungen an Haushaltgeräte, Elektrowerkzeuge und ähnliche Elektrogeräte - Teil 2: Störfestigkeit - Produktfamiliennorm

[DIN75] Norm DIN 66217 (1975). Koordinatenachsen und Bewegungsrichtungen für numerisch gesteuerte Arbeitsmaschinen

[DIN83a] Norm DIN 66001 (1983). Informationsverarbeitung; Sinnbilder und ihre Anwendung

[DIN83b] Norm DIN 66025–1 (1983). Programmaufbau für numerisch gesteuerte Arbeitsmaschinen

[DIN97] Norm DIN EN 60617 (1997). Graphische Symbole für Schaltpläne

[DIN98] Norm DIN EN 50178, VDE 0160 (1998). Ausrüstung von Starkstromanlagen mit elektronischen Betriebsmitteln

[FEST16] Festo Didactic: Näherungsschalter, induktiv. URL: ▸ http://www.festo-didactic.com/ov3/media/customers/1100/00271055001075223623.pdf. [Stand: 28.09.2017]

[GÖRN13] Görnemann, O.; Stubenrauch, H.-J.: Whitepaper – Berührungslos wirkende Schutzeinrichtungen (BWS) für sichere Maschinen. Waldkirch, 2013

[GREC10] Grechenig, T.: Softwaretechnik. (Reihe: IT Informatik). München: Pearson Studium, 2010

[HANS10] Hanser, E.: Agile Prozesse: Von XP über Scrum XE "Scrum" bis MAP. Springer, Berlin (2010)

[HARW12] Harwardt, M.: Wasserfallmodell versus Scrum XE "Scrum" . Akademikerverlag, Saarbrücken (2012)

[HEIN15] Heinrich, H.: Systemisches Projektmanagement. Hanser, München (2015)

[JOHN09] John, K.-H.; Tiegelkamp, M.: SPS-Programmierung mit IEC 61131–3. (Reihe: VDI). 4., neubearbeitete Aufl. Aufl. Dordrecht: Springer, 2009

[LITZ13] Litz, L.: Grundlagen der Automatisierungstechnik. 2., aktualisierte Auflage. Aufl. München: Oldenbourg Wissenschaftsverlag, 2013

[OEST09] Oestereich, B.; Bremer, S.: Analyse und Design mit UML 2.3. 9., aktualisierte und erw. Aufl. Aufl. München: Oldenbourg, 2009

[RUMP11] Rumpe, B.: Modellierung mit UML. (Reihe: Xpert. press). Dordrecht: Springer, 2011

[STAH12] Stahl, T.; Völter, M.; Efftinge, S.; Haase, A.: Modellgetriebene Softwareentwicklung. 1. Aufl. Heidelberg: dpunkt.verlag, 2012

[STET04] Stetter, R.: Software im Maschinenbau - lästiges Anhängsel oder Chance zur Marktführerschaft? URL: ▸ http://www.software-kompetenz.de/servlet/is/21700/Stetter-SW_im_Maschinenbau.pdf?command=downloadContent&filename=Stetter-SW_im_Maschinenbau.pdf. [Stand: 28.09.2017]

[THIE10] Thieme, J.; Hanisch, H.-M.: Model-based generation of modular PLC code using IEC61131 function blocks. In: Proceedings of the IEEE International Symposium on Industrial Electronics ISIE-02, 2010. S. 199–204

[TSUB13] Tsubaki Kabelschlepp GmbH: Betriebss- und Wartungsanleitung für Energieführungsketten aus Kunststoff. URL: ▸ http://kabelschlepp.de/fileadmin/img/downloads/BW_Energiefuehrungen_Kunststoff_DE.pdf. [Stand: 28.09.2017]

[UHLI06] Uhlig, R.: SPS - modellbasierter Steuerungsentwurf für die Praxis. Oldenbourg, München (2006)

[VDI04a] Richtlinie VDI 2206 (2004). Entwicklungsmethodik für mechatronische Systeme

[VDI04b] Richtlinie VDI 2223 (2004). Methodisches Entwerfen technischer Produkte

[VDI10] Richtlinie VDI/VDE 3695 (2010). Engineering von Anlagen - Evaluieren und optimieren des Engineerings

[VDI91] Richtlinie VDI 2854 (1991). Sicherheitstechnische Anforderungen an automatisierte Fertigungssysteme

[WEIL14] Weilkiens, T.; Soley, R. M.: Systems Engineering mit SysML/UML. 3., überarb. und aktualisierte. Aufl. Heidelberg: dpunkt.verlag, 2014

# Supplementary Information




C. Brecher and M. Weck, *Machine Tools Production Systems 3*, Lecture Notes in Production Engineering,
https://doi.org/10.1007/978-3-658-3462[illegible]

# Formelzeichenverzeichnis

## Kapitel: 2

**A..Z, a..z, Alpha..Omega, alpha..omega.**

| Formel-zeichen | Einheit | Bedeutung |
|---|---|---|
| $\underline{C}^*$ | – | Conjugated complex number |
| $\overline{T_L}$ | Nm | Mean value of load torque |
| $\overline{T_M}$ | Nm | Mean value of motor torque |
| $\overline{T_a}$ | Nm | Mean value of acceleration torque |
| $\dot{\omega}_{ap}$ | rad/s$^2$ | Angular acceleration in current limiting phase |
| $\underline{C}$ | – | Complex number |
| $\underline{V}$ | V/m | Complex amplitude (Phasor) |
| $\dot{\omega}$ | rad/s$^2$ | Angular acceleration |
| A | – | Working point |
| $A_n$, $B_n$ | – | Summary |
| $A_{Sp}$ | mm$^2$ | Cross-section spindle |
| B | T | Magnetic flux density |
| B | A/m | Magnetic field strength |
| CM | | Load dependent fraction |
| d | mm | Thickness of webs |
| d | mm | Diameter |
| E | V/m | Amplitude of electrical field strength |
| E | V | Max. amplitude of measured voltage |
| E | V/m | Electrical field strength |
| f | Hz | Feed frequency |
| f | Hz | Carrier frequency |
| F | N | Force |
| $f_{in}$ | Hz | Input frequency |
| $F_{Ki}$, $F_{Ki,Si}$ | N | Collision force i |
| $f_{line}$ | Hz | Line frequency |
| $f_{set}$ | Hz | Target frequency |
| g | mm | Winding space |
| h | – | Analog measurement |
| $h_0$ | – | Max value of the analog measurement |
| i | A | Current |
| I | A | current |
| i | – | Imaginary unit |
| i | – | Gear ratio |
| $I_1$, $I_2$, $I_3$ | A | External conductor current |
| $I_A$ | A | Armature current |
| $I_{A\,max}$ | A | Max armature current |
| $i_E$ | A | Alternating current |
| $I_f$ | A | Excitation current |
| $I_{max}$ | A | Max current |
| $I_{max,per}$ | A | Permissible max motor current |
| $I_r$ | A | Rated current |
| $I_s$ | A | Stator current |
| $I_{sa}$, $i_{sa}$ | A | Real current component from Clarke Transformation |
| $I_{sb}$, $i_{sb}$ | A | Imaginary current component from Clarke Transformation |
| $I_{sd}$, $i_{sd}$, $i_d$ | A | Flux-forming current |
| $I_{sq}$, $i_{sq}$ | A | Torque-forming current |
| $I_{start}$ | A | Starting current |
| $I_u$, $I_v$, $I_w$ | A | Phase currents |
| $i_u$, $i_v$, $i_w$ | A | Phase currents |
| $I_Z$ | A | Intermediate circuit current |
| $i_\mu$, $I_\mu$ | A | Magnetizing current |
| $J_i$ | kgm$^2$ | Moment of inertia |
| $J_{M,rot}$ | kgm$^2$ | Moment of inertia of motor |
| $J_{total}$ | kgm$^2$ | Total moment of inertia |
| k | – | Sampling elements of tracks |
| k | V | Amplitude factor |
| k | N/µm | Stiffness |
| K | V/A | Electromagnetic coupling factor |
| $k_1$ | A | Specific motor constant |
| $k_2$ | A | Specific motor constant |
| $k_{Li}$ | N/µm | Bearing stiffness |
| $k_N$ | N/µm | Nut/spindle stiffness |
| $k_{Si}$ | N/µm | Spindle stiffness |
| $K_V$ | m/(min·mm) | Speed gain factor |
| L | H | Inductivity |
| L | mm | Spindle pitch |
| L | mm | Spindle length |
| $L'_2$ | H | Rotor inductance converted to number of stator windings |

| Formelzeichen | Einheit | Bedeutung |
|---|---|---|
| $L_{1h}$ | H | Main inductance |
| $L_2$ | H | Rotor inductance |
| $L_D$ | H | inductance |
| $L_h$ | H | Rotor inductance |
| $L_h$ | H | Main inductance |
| m | – | Module of teeth |
| n | Nms/rad | Load change factor |
| n | 1/min | Rotational speed |
| n | – | Refractive index |
| n | – | Meter reading |
| N | – | Number of cycles |
| N | – | Number of required tracks |
| $n_0$, $n_d$ | 1/min | Idle speed |
| $n_I$ | 1/min | Characteristic speed due to current limit |
| $n_{act}$ | 1/min | Actual speed |
| $n_i$ | 1/min | Rotation speed |
| $n_k$ | 1/min | Buckling speed |
| $n_{max}$ | 1/min | Max rotational speed |
| $n_N$ | 1/min | Nominal speed |
| $n_{op}$ | Nm | Operating speed |
| $n_s$ | 1/min | Synchronous rotational speed |
| $n_{set}$ | 1/min | Target speed |
| $n_{tilt}$, $n_K$ | 1/min | Tilt speed |
| $n_{x,on}$ | 1/min | Speed modulation factor |
| $n_{y,on}$ | 1/min | Speed modulation factor |
| P | W | Power |
| p | – | Pole pairs |
| p | mm | Link length chain |
| $P_c$ | W | Consumed power |
| $P_{FE}$ | W | Iron losses |
| $P_{loss}$ | W | Power loss |
| $P_M$ | W | Required nominal motor power |
| $P_{max}$ | W | Required electrical power |
| $P_{max}$ | W | Max available power |
| $P_{mech}$ | W | Mechanical power |
| $P_{mech,max}$ | W | Max mechanical power |
| $P_{Process}$ | W | Required process power |
| $P_{total}$ | mm | Total Power |
| $P_u$ | W | String power |
| $P_V$ | W | Cupper loss |
| $P_v$ | W | String power |
| $P_{verl,max}$ | W | Max power dissipation |

| Formelzeichen | Einheit | Bedeutung |
|---|---|---|
| $P_w$ | W | String power |
| R | V | Mains voltage |
| R ‘, S ‘, T ‘ | V | Inverter output current |
| R ‘$_R$ | Ω | Rotor resistance converted to number of stator windings |
| $R_A$ | Ω | Internal armature resistance |
| $R_f$ | Ω | Excitation resistance |
| $R_R$ | Ω | Rotor resistance |
| $R_S$ | Ω | Stator resistance |
| $R_S$ | Ω | Shunt resistance |
| $R_{Str}$ | Ω | Substitute resistance |
| S | V | Mains voltage |
| s | – | slip |
| $s_k$ | – | Slip at breakdown torque |
| $s_N$ | – | Nominal slip |
| $s_{op}$ | – | Operating slip |
| T | Nm | Torque |
| T | Nm | Motor Torque |
| t | s | time |
| T | V | Mains voltage |
| T | s | Run-up time constant |
| $T_{\sim}$ | Nm | Motor output torque |
| $T_I$ | Nm | Load torque at current limit |
| $T_1$ | Nm | Load torque for short-time duty |
| $T_{Imax}$ | Nm | Max torque without current limit |
| $T_2$ | s | Rotor time constant |
| $T_{32}$, $T_{\varphi}$ | – | Mapping matrices |
| $T_{acc}$ | Nm | Acceleration torque |
| $t_B$ | s | Load time |
| $T_{el}$ | Nm | Output torque of the field-oriented controlled asynchronous motor |
| $T_i$ | Nm | Inner torque |
| $T_k$ | Nm | breakdown torque |
| $t_L$ | s | Idle mode time |
| $T_L$, $t_{load}$ | Nm | Loaded torque |
| $T_M$ | Nm | Motor torque |
| $T_{max}$ | Nm | Max permitted continual load torque |
| $T_{Mmax}$ | Nm | Max torque with current limitation |
| $T_n$ | Nm | Nominal torque |
| $T_N$ | Nm | Nominal Torque |

| Formelzeichen | Einheit | Bedeutung |
|---|---|---|
| $t_{on}$ | s | Rise time |
| $T_{op}$ | Nm | Operating Torque |
| $T_{ou}$ | Nm | Output torque |
| $t_r$ | – | relative duty cycle |
| $T_r$ | s | Electric rotor time constant |
| $t_s$ | s | Cycle time |
| $T_{set}$ | Nm | Set point torque |
| $T_{set}$ | Nm | Set point of overall torque |
| $t_{St}$ | s | Stand still time |
| $T_{start}$ | Nm | Start Torque |
| $t_t$ | s | Dead time |
| $t_{UP}$ | s | Switch time |
| $T_V$ | Nm | Tension torque |
| V | V | Voltage |
| v | V | voltage |
| v | m/s | velocity |
| V | V | Voltage |
| V- | V | Low voltage level |
| V- | V | Low voltage level |
| V+ | V | High voltage level |
| V+ | V | High voltage level |
| $V_0$ | V | Source voltage |
| $V_{12}$, $V_{23}$, $V_{31}$ | V | External conductor voltages |
| $V_A$ | V | Armature Voltage |
| $v_A$ | V | Induced voltage |
| $v_A$ | V | Bridge output voltage |
| $V_{A\,max}$, $V_{A1}$ | V | Max armature voltage |
| $V_{A\,min}$ | V | Min armature voltage |
| $V_A$, $V_B$, $V_C$ | V | Phase voltage |
| $V_{A1}$, $V_{A2}$ | V | Output voltage |
| $v_c$ | m/s | Chain velocity |
| $V_d$, $V_q$ | V | Voltage after Park-Transformation |
| $V_f$ | V | Excitation voltage |
| $V_i$ | V | Induced voltage |
| $V_{IC}$ | V | Intermediate circuit voltage |
| $V_{in}$ | V | Input voltage |
| $V_n$ | V | Output Voltage |
| $V_S$ | V | Voltage shunt resistance |
| $v_s$ | m/s | Scope velocity |
| $V_{uv}$, $V_{vw}$, $V_{wu}$ | V | Terminal voltage |
| w | mm | Grid width |
| W | V | Voltage |
| W- | V | Low voltage level |
| W+ | V | High voltage level |
| $W_M$ | J | Magnetic energy in motor |
| x | mm | travel |
| x | mm | travel |
| $x_{ref}$ | mm | Reference position |
| $x_{total}$ | mm | Length of gauge |
| $y_{ref}$ | mm | Reference position |
| z | mm | Space between gratings |
| z | – | Number of tooth |
| $\alpha$ | rad | Electrical angel |
| $\gamma$ | Grad | Rotor position angel |
| $\Delta t_i$ | s | Time of angular velocity range |
| $\Delta x$ | mm | Finest separation width |
| $\Delta\omega_i$ | rad/s | Angular velocity range |
| $\vartheta$ | °C | temperature |
| $\lambda$ | µm | wavelength |
| $\lambda_n$ | rad | Phase shift |
| $\sigma_2$ | – | Scatter digit |
| $\tau$ | s | Period length |
| $\tau$ | mm | Division period |
| $\tau'$ | s | Division of pulse trains |
| $\varphi$ | Grad | Flux coordinate angel |
| $\varphi$ | Grad | Flux coordinate angel |
| $\phi$ | Wb | Magnetic flux |
| $\phi$ | rad | Phase shift |
| $\phi_{max}$ | Wb | Max excitation field |
| $\phi_{min}$ | Wb | Min excitation field |
| $\phi_r$ | Wb | Rotor flux |
| $\psi_{PM}$ | Wb | Magnetic flux |
| $\omega$ | rad/s | Circuit frequency |
| $\omega$ | rad/s | Angular velocity |
| $\Omega$ | rad | Phase shifting |
| $\omega_\mu$ | rad/s | Angular velocity of rotor flux |
| $\omega_{1max}$ | rad/s | Coordinate intersection value of the motor characteristic |
| $\omega_{CL}$ | rad/s | Angular velocity in current limiting phase |
| $\omega_{CL0}$ | rad/s | Theoretically working angular velocity |
| $\omega_E$ | rad/s | Cutoff frequency |
| $\omega_M$ | rad/s | Motor circuit frequency |
| $\omega_R$ | rad/s | Rotor circuit frequency |

| Formelzeichen | Einheit | Bedeutung |
|---|---|---|
| $\omega_{SP}$ | rad/s | Angular velocity in switchover point |
| $\omega_{WP}$ | rad/s | Angular velocity at working point |

## Chapter: 3

| Formula | Unit | Meaning |
|---|---|---|
| A | | System matrix |
| $A_R$ | | Amplitude reserve |
| B | | Input matrix |
| C | | Output matrix |
| $C$ | | Damping coefficient |
| $C_i$ | | i-th coefficient |
| $D$ | kg/s | Damping |
| D | | Damping matrix |
| $D_{Regel}$ | kg/s | Damping ratio of the bearing-controlled drive |
| $F$ | N | Force |
| $\hat{F}$ | N | Force amplitude |
| $F_i$ | N | i-th force |
| $F_{max}$ | N | Maximum permissible force |
| $F_{Nenn}$ | N | Nominal force |
| $F_{Stör}$ | N | Interference force, static load |
| $F_x$ | N | Force in x-direction |
| $J_{ges}$ | kg*s$^2$ | reduced mass moment of inertia |
| $J_i$ | kg*s$^2$ | i-th moment of inertia |
| $K$ | | Stiffness |
| $K_D$ | | differential amplification factor |
| $K_i$ | | integral amplification factor |
| $K_I$ | | integral amplification factor |
| $K_{krit}$ | | amplification factor |
| $K_l$ | | Gain factor of the position controller |
| $K_p$ | As/m | amplification factor in the speed control loop |
| $K_R$ | | proportional amplification factor |
| $K_{Rkrit}$ | As/m | Gain factor at the stability edge |
| $K_S$ | | Jump height |
| $K_v$ | 1/s | Velocity amplification, Kv factor |
| $L_A$ | H | Inductance of the motor winding |
| $M$ | | Mass matrix |
| $P_k$ | | critical point |
| R | | Controller matrix |
| $R_A$ | Ohm | Resistance of motor winding |
| S | | Pre-filter matrix |
| $T_i$ | Nm | i-th moment |
| $T_{krit}$ | s | Vibration period duration |
| $T_0$ | s | Sampling time |
| $T_g$ | s | Compensation time |
| $T_n$ | s | Reset time |
| $T_{np}$ | s | Reset time in the velocity control loop |
| $T_t$ | s | Dead time |
| $T_u$ | s | Delay time |
| $T_v$ | s | Derivative time |
| $U$ | V | Motor voltage |
| $U_s$ | V | Motor voltage setpoint |
| $X_a$ | | Output signal |
| $X_{a,t}$ | | Output signal at time t |
| $X_e$ | | Input signal |
| $X_{e,t}$ | | Input signal at time t |
| $a_0, a_1, a_2$ | | System-determining constant coefficients |
| $b_0, b_1, b_2$ | | system-determining constant coefficients |
| $c_{r+t,i}$ | kg/s | i-th rotational and translational damping |
| $d$ | kg/s | Degree of damping |
| $d_{Lager}$ | mm/N | Bearing compliance |
| $d_{Mutter}$ | mm/N | Nut compliance |
| $d_{Sp}$ | mm | Spindle diameter |
| $d_{Spindel}$ | mm/N | Spindle compliance |
| $d_{xx}$ | mm/N | compliance of x in direction of Fx |
| $f$ | 1/s | Frequency |
| $f$ | | External forces |
| $f_0$ | 1/s | Angular frequency |
| $f_a$ | | Output signal |
| $f_e$ | | Input signal |
| $f_g$ | 1/s | Cut-off frequency |
| $f_k$ | | Pulse train |
| $f_m$ | 1/s | Measuring frequency |
| $h$ | | Spindle pitch |
| $i$ | | Gear ratio |
| $k$ | kg/s$^2$ | Spring stiffness |

| Formula | Unit | Meaning |
|---|---|---|
| $k_{dyn}$ | kg/s² | dynamic stiffness |
| $k_{ges}$ | kg/s² | Total stiffness |
| $k_{Lager}$ | kg/s² | Stiffness of bearings |
| $k_{Mutter}$ | kg/s² | Stiffness of the nut |
| $k_{r+t,i}$ | kg/s² | i-th rotational and translational stiffness |
| $k_{Spindel}$ | kg/s² | Stiffness of the spindle |
| $k_{xx}$ | kg/s² | Stiffness of x in direction of Fx |
| $m$ | kg | Mass |
| $m_i$ | kg | i-th mass |
| $n_s$ | 1/s | Speed |
| $r$ | | Control vector |
| $s$ | 1/s | complex frequency |
| $t$ | s | Time |
| $u_k$ | | Input sequence |
| $v_{ist}$ | m/s | Actual value of speed |
| $v_{soll}$ | m/s | Setpoint of the velocity |
| $x$ | m | (carriage) path |
| $\dot{x}$ | m/s | (Carriage) speed |
| $\ddot{x}$ | m/s² | Acceleration |
| $\dddot{x}$ | m/s³ | Jerk |
| $x_1 \ldots x_n$ | | State variables / measured values |
| $\overline{x}_1 \ldots \overline{x}_n$ | | smoothed measured values |
| $x_a$ | | Output variable |
| $\dot{x}_a$ | | 1st order derivative of the output variable |
| $\ddot{x}_a$ | | 2nd order derivative of the output variable |
| $\widehat{x}_a$ | | Amplitude of the output signal, resulting maximum position deviation |
| $x_e$ | | Input variable |
| $x_{e1}$ | | Input variable 1 |
| $x_{e2}$ | | Input variable 2 |
| $\dot{x}_e$ | | 1st order derivative of the input variable |
| $\ddot{x}_e$ | | 2nd order derivative of the input variable |
| $\widehat{x}_e$ | | Amplitude of the input signal |
| $x_i$ | | Actual position |
| $x_{ist}$ | | Actual position |

| Formula | Unit | Meaning |
|---|---|---|
| $x_k$ | | k-th sampling point |
| $x_m$ | | Directly measured position signal |
| $x_M$ | | Measured variable |
| $x_s$ | | Position reference value |
| $x_{soll}$ | | Reference variable |
| $x_w$ | mm | Control deviation, lag |
| $x_{w,Kv1}$ | mm | Lag with Kv factor 1 |
| $x_{w,Kv2}$ | mm | Lag with Kv factor 2 |
| $x_{w,Kv3}$ | mm | Lag with Kv factor 3 |
| $y$ | | Output variable |
| $\dot{y}$ | | 1st order derivative according to the time of the output variable |
| $\ddot{y}$ | | 2nd order derivative according to the time of the output variable |
| $y_k$ | | Output sequence |
| $y_{k-1}$ | | Values of the manipulated variable |
| $z_1$ | | Disturbance variable 1 |
| $z_2$ | | Disturbance variable 2 |
| $\Delta x$ | mm | Deviation of the distance in X-direction |
| $\Delta x_e$ | | Height of the input jump |
| $\Delta y$ | mm | Deviation of the distance in Y-direction |
| $\alpha_R$ | ° | Phase reserve |
| $\lambda$ | | Polynomial variable |
| $\lambda_i$ | | i-th polynomial variable |
| $\varphi$ | ° | Phase angle |
| $\varphi_a$ | ° | Angular position of motor shaft |
| $\varphi_{sp}$ | ° | Angular position of the spindle |
| $\omega$ | 1/s | Eigenfrequency |
| $\omega_1$ | 1/s | Eigenfrequency 1 |
| $\omega_0$ | 1/s | Eigenfrequency |
| $\omega_{0,Mechanik}$ | 1/s | Basic angular frequency of the mechanical transmission system |
| $\omega_{0,Regel}$ | 1/s | Basic angular frequency of the drive with bearing control |
| $\omega_{0,Regelkreis}$ | 1/s | Basic angular frequency of the control loop |

## Kapitel: 4

| Formelzeichen | Einheit | Bedeutung |
|---|---|---|
| $A_E$ | | Corner distance |
| $D$ | kg/s | Damping |
| $K_V$ | 1/s | Velocity gain, Kv-factor |
| $K_{Vx}$ | 1/s | Velocity gain x-axis |
| $K_{Vy}$ | 1/s | Velocity gain y-axis |
| $R$ | mm | Radius |
| $R_{act}$ | mm | Actual radius |
| $R_{ref}$ | mm | Reference radius |
| $a_x$ | m/s² | Acceleration x-axis |
| $a_y$ | m/s² | Acceleration y-axis |
| $f$ | 1/s | Frequency |
| $f_0$ | 1/s | Natural frequency |
| $f_u$ | 1/s | Circular frequency |
| $G_V$ | | Transfer |
| $t$ | s | Time |
| $v$ | m/s | Feedrate |
| $v_B$ | m/s | Path velocity |
| $v_{b,min}$ | m/s | |
| $x$ | mm | X-position |
| $\dot{x}$ | m/s | Velocity in x-direction |
| $y$ | mm | Y-position |
| $\dot{y}$ | m/s | Velocity in y-direction |
| $z$ | mm | Z-position |
| $\Delta x$ | mm | Lag distance in x-direction |
| $\Delta y$ | mm | Lag distance in y-direction |
| $\alpha$ | ° | Corner angle |
| $\delta R$ | mm | Radius difference |
| $\delta_s$ | mm | Tracking error |
| $\omega$ | 1/s | Natural angular frequency |
| $\omega_u$ | 1/s | |

## Kapitel: 5

| Formelzeichen | Einheit | Bedeutung |
|---|---|---|
| Ao | 1/min | Overshoot magnitude |
| D | – | Damping factor |
| $D_i$ | – | Damping factor |
| $F_{tot}$ | N | Total feed force |
| $F_i$ | N | Force |
| $J_{tot,red}$ | kg·m² | Total reduced moment of inertia applied to the drive shaft |
| $J_i$ | kg·m² | Moment of inertia |
| $J_{i,red}$ | kg·m² | Reduced moment of inertia applied to the specific shaft |
| $J_m$ | kg·m² | Reduced moment of inertia of the table applied to the spindle shaft |
| $J_{SP}$ | kg·m² | Moment of inertia of the spindle |
| $M_{drv}$ | Nm | Drive torque |
| $M_i$ | Nm | Torque |
| $M_{i,max}$ | Nm | Maximum torque |
| $M_L$ | Nm | Load torque |
| $M_M$ | Nm | Motor torque t |
| $M_{Sp}$ | Nm | Spindle torque |
| $T_r$ | s | Rise time |
| $T_s$ | s | Settling time |
| $T_{el}$ | s | Time constant |
| $T_{mech}$ | s | Time constant |
| $T_n$ | s | Integral action time |
| $T_o$ | s | Overshoot time |
| V | m/s | Speed |
| | | |
| $d_i$ | mm | Pinion diameter |
| e | 1/min | End deviation |
| h | mm | Spindle pitch |
| i | – | Gear ratio |
| $i_{tot}$ | – | Total gear ratio |
| $i_{jk}$ | – | Gear ratio of gear stage jk |
| $n_i$ | 1/min | Rotational speed |
| t | s | Time |
| $\ddot{x}$ | m/s² | Acceleration in X-direction |
| | | |
| $\delta$ | mm | Tracking error |
| $\eta_{tot}$ | – | Total efficiency |
| $\eta_i$ | – | Efficiency |
| $\varphi$ | rad | Phase shift |
| $\omega_0$ | rad/s | Cutoff frequency |
| $\omega_{0A}$ | rad/s | Cutoff frequency |
| $\omega_i$ | rad/s | Angular velocity |
| $\dot{\omega}_i$ | rad/s² | Angular acceleration |

# Kapitel: 6

| Formelzeichen | Einheit | Bedeutung |
|---|---|---|
| $B$ | – | Measured value |
| $C_m$ | – | Machine capability related to drawing tolerance |
| $D$ | mm | milling cutter diameter |
| $D$ | dB | Dynamics of a signal |
| $F_c$ | N | Cutting force |
| $F_c(\varphi)$ | N | Arithmetic sum of cutting forces |
| $F_{drop}$ | N | Permissible force drop |
| $F_{eol}$ | N | Force limit value at end of service life |
| $F_f$ | N | Feed force |
| $F_{in}$ | N | Initial force value wear measurement |
| $F_{Mn}$ | N | Moving average value of the feed force |
| $F_p$ | N | Passive force |
| $I$ | – | idle |
| $I$ | A | Motor current |
| $K_n$ | V | Comparator thresholds |
| $L_{0\text{-}1000}$ | dB | Total power level of the structure-borne sound signal in the frequency range from 0 to 1000 Hz |
| $L_{total}$ | dB | Total power level of the structure-borne sound signal |
| $L_{v,r}$ | min | Setting time of the feed motor |
| LCL | – | Lower control limit |
| $M$ | Nm | torque |
| $M_d$ | Nm | static braking torque |
| $M_m$ | – | feature |
| $M_{max}$ | Nm | Maximum torque of previous revolution |
| $M_{set}$ | Nm | Reference torque |
| $M(t)$ | Nm | Momentary torque on milling spindle |
| MVF | – | Mean value factor |
| MTBF | h | Mean time between failures |
| MTTR | h | Mean time to repair |
| $P$ | – | planned operation |
| $P$ | W | power |
| $P_n$ | W | Set cutting power |
| $Q_s$ | $m^3/s$ | Flow volume dielectric |
| $R$ | – | repair |
| $R(t)$ | – | reliability |
| $R_{xy}(\tau)$ | – | Cross correlation |
| $Re\{G_b(if)_{neg}\}$ | – | Negative real part of the directed compliance function between workpiece and tool |
| $T$ | mm | Specified drawing tolerance |
| $T_{0,1}$ | min | Sampling time |
| $T_B$ | h | Occupancy time |
| $T_C$ | h | Testing time |
| $T_D$ | h | Technical downtime |
| $T_{DM}$ | h | Technical downtime (manufacturer dependent) |
| $T_{DU}$ | h | Technical downtime (user dependent) |
| $T_M$ | h | Maintenance time |
| $T_O$ | h | Organizational downtime |
| $T_U$ | h | Actual utilization time of production system |
| $U$ | V | voltage |
| $U_0$ | V | No-load voltage |
| UCL | – | Upper control limits |
| $U_{fm}$ | V | Mean discharge voltage |
| $U_f$ | V | Spark voltage |
| $U_k$ | V | Controller gain |
| $U_{min}$ | dB | smallest information carrying signal change |
| $U_{max}$ | dB | largest signal value |
| $U_{set}$ | V | Reference voltage |
| $U_{total}$ | % | Total degree of utilization |
| $U_T$ | % | Technical degree of utilization |
| $U_x$ | V | Voltage (control) deviation |
| $V$ | – | Availability |
| $V_e$ | – | gain factor |
| $V_b$ | – | Feedback gain |
| $V_f$ | – | Forward gain |
| $V_{set}$ | $min^{-1}$ | Reference value for speed controller |
| $V_w$ | $mm^3/min$ | Erosion rate |
| $W$ | – | Maintenance |
| $X(f)$ | – | Signal in frequency domain |
| $X_i$ | – | Current average |
| $XM_i$ | – | New moving average |
| $XM_{i\text{-}1}$ | – | Previous moving average |
| $Y_i$ | – | Auxiliary variable neural net |
| $Z$ | – | Number of milling teeth |

| Formelzeichen | Einheit | Bedeutung |
|---|---|---|
| $Z_i$ | – | Auxiliary variable neural net |
| | | |
| a | mm | Gap width |
| $a_{max}$ | mm | Maximum cutting depth |
| $a_{min}$ | mm | Minimum cutting depth |
| $a_p$ | mm | Cutting depth |
| $a_{cr}$ | mm | Limit cutting depth |
| $a_{cr,min}$ | mm | Minimum limit cutting depth |
| $a_0$ | mm | Set cutting depth |
| $b_z$ | mm | Cutter spacing |
| c | – | n-dimensional feature vector |
| f | mm/rev | feed |
| f | Hz | frequency |
| $f_{chatter}$ | Hz | Chatter frequency |
| $f_{lim}$ | Hz | Limiting frequency of low-pass filter |
| $f_{max}$ | Hz | Highest frequency in signal |
| $f_s$ | Hz | Sampling frequency |
| $f_u$ | Hz | Useful frequency range limit |
| $f_z$ | m/ε | Feed per cutting edge or tooth |
| $f_0$ | mm/rev | reference feed |
| g(c) | – | Separating function depending on feature vector c |
| $i_V$ | A | Valve current |
| $k_{c1,1}$ | – | main value for the specific cutting force |
| $1\text{-}m_c$ | – | Slope of the specific cutting force |
| m | – | number of waves cut on the surface |
| n | – | Sample size |
| $n_{cr}$ | $min^{-1}$ | Critical speed |
| $n_{mom}$ | $min^{-1}$ | Momentary speed |
| $n_{st}$ | $min^{-1}$ | speed stability minimum |
| r | mm | Workpiece radius |
| $s_x$ | (mm) | Standard deviation |
| $t_0$ | min | Pulse interval time |
| $t_c$ | min | Cutting time |
| $t_d$ | min | Ignition delay time |
| $t_{df}$ | min | Burn delay time |
| $t_{df,l}$ | min | Long term average discharge delay time |
| $t_{df,s}$ | min | Short term average discharge delay time |

| Formelzeichen | Einheit | Bedeutung |
|---|---|---|
| $t_{df,set}$ | min | Set point discharge delay time |
| $t_{dis}$ | min | Discharge time |
| $t_f$ | min | Discharge current fall time |
| $t_I$ | min | Pulse duration |
| $t_p$ | min | Pulse cycle time |
| $t_r$ | min | Discharge current rise time |
| u(t) | – | Time and amplitude continuous sensor signal |
| $u_f$ | V | Spark voltage |
| $u_i$ | – | Discrete sensor signal values |
| $v_c$ | mm/min | Cutting speed |
| $v_f$ | mm/min | Feed speed |
| w | – | Word size |
| $w_{ij}$ | – | Weighting factors for the input values of a neuron |
| $x_i$ | – | Input value of a neuron |
| x(t) | – | Signal in time domain |
| Y | mm | Machine bending |
| $y_i$ | – | Output values of neuron |
| $z_M$ | – | Number of cutting edges in contact |
| | | |
| $\Delta l_{fr}$ | mm/min | feed |
| $\Delta d_w$ | mm | Diameter decrease |
| $\varphi_i$ | ° | Cutting edge entry angle |
| $\varphi_j$ | ° | Feed direction angle at cutter j |
| $\varphi_o$ | ° | Cutting edge exit angle |
| $\varphi_s$ | mm | cutting arc |
| $\varphi_T$ | ° | cutter pitch angle |
| $\delta_s$ | dB | Stopband attenuation |
| ε | – | phase shift between the vibration of the cutting tooth and the waviness caused by the previous cut |
| η | – | Frequency ratio |
| θ | – | Neural transfer function parameter |
| κ | ° | Setting angle for main cutter |
| λ | – | Failure rate |
| σ | – | Standard deviation of histogram of the structure-borne sound signal |
| τ | – | Time shift |
| φ | ° | Feed direction angle |

## Kapitel: 8

| Formelzeichen | Einheit | Bedeutung |
|---|---|---|
| $F_i$ | mm/ revolution | feed values |
| $H_h$ | – | sum of all hundredths for the machining times |
| $H_{hi}$ | – | hundredths for the individual machining times |
| $H_n$ | – | hundredths for the non-productive times |
| K | – | motion path |
| S | – | locking curve |
| $U_{tot}$ | – | total number of working spindle revolutions |
| $U_h$ | – | revolutions allotted to the machining time |
| $U_{hi}$ | – | revolutions allotted to the individual work operations |
| $d_{max}$ | mm | largest turning diameter |
| f | mm/rev | feed |
| $l_i$ | mm | working distances |
| n | $min^{-1}$ | spindle speed |
| $t_h$ | min | machining time |
| $t_w$ | min | total time |
| v | mm/min | cutting speed |
| z | – | gear ratio |

## Kapitel: 11

| Formelzeichen | Einheit | Bedeutung |
|---|---|---|
| $A$ | – | Start contour point |
| $P_1 \ldots P_6$ | – | Contour points |
| $P_k \ldots P_n$ | – | Sampling points |
| $ax_{j,k}$ | – | Coefficient of the face segments |
| $ay_{j,k}$ | – | Coefficient of the face segments |
| $az_{j,k}$ | – | Coefficient of the face segments |
| $d_{n,j}$ | m | Path deviation of the interpolation points |
| $iordu$ | – | Polynomial order of the surface segment in u-direction |
| $iordv$ | – | Polynomial order of the surface segment in v-direction |
| $r_s$ | m | Cutting radius |
| $r_s$ | m | Cutting radius |
| $r_s$ | m | Cutting radius |
| $u$ | – | Local parameter of the surface segments |
| $v$ | – | Local parameter of the surface segments |
| $\alpha$ | rad | Tilt angle |
| $\beta$ | rad | Lead angle |
| $\delta x$ | m | Deflection of the touch probe |

## Kapitel: 12

| Formelzeichen | Einheit | Bedeutung |
|---|---|---|
| $A$ | – | Axis ratio |
| $A_i$ | – | Polynomial coefficient A |
| $A_x$ | – | Axis ratio X |
| $A_y$ | – | Axis ratio Y |
| $A_z$ | – | Axis ratio Z |
| $B_i$ | – | i-th polynomial coefficient B, i-th joint point |
| $C_i$ | – | Polynomial coefficient C |
| $D_i$ | – | Polynomial coefficient D |
| $E$ | – | Deviation |
| $L$ | – | Segment arch length |
| $L_i$ | – | i-th leg vector |
| $N_i^n$ | – | B-Spline function |
| $P$ | – | Position uncertainty |
| $P_0, \ldots, P_m$ | – | Interpolation points |
| $P_A$ | – | Position deviation |
| $P_i$ | – | de-Boor point, i-th joint point of the moving tool platform |
| $P_{Si}$ | – | Position scatter |
| $R$ | m | Radius of circle / curvature |
| $R_x$ | – | Transformation matrix of the single rotation X |
| $R_y$ | – | Transformation matrix of the single rotation Y |
| $R_z$ | – | Transformation matrix of the single rotation Z |
| $S_i$ | – | Cubic polynomial spline |

| Formelzeichen | Einheit | Bedeutung |
|---|---|---|
| $S_{gesamt}$ | m | Total line distance vector |
| $\lvert S_{Satz,gesamt} \rvert$ | m | Total line distance |
| $T$ | s | Interpolation time, Interpolation interval, Interpolation clock rate |
| $U$ | – | Node vector |
| $U_i$ | – | Reversal error |
| $X$ | – | Machine coodinate |
| $X'$ | – | Workpiece coordinate |
| $Z$ | – | Number of operations |
| $a$ | – | Geometric dimension |
| $a(t)$ | m/s² | Acceleration |
| $a_e$ | m | Width of engagement |
| $a_{max}$ | m/s² | Maximum acceleration |
| $a_{max,x}$ | m/s² | Maximum accleration of X-axis |
| $a_{max,y}$ | m/s² | Maximum accleration of Y-axis |
| $a_{max,z}$ | m/s² | Maximum accleration of Z-axis |
| $a_x$ | m/s² | Axis acceleration in X-direction |
| $a_{x,i}$ | – | i-th polynomial coefficient a in X |
| $a_y$ | m/s² | Axis acceleration in Y-direction |
| $a_{y,i}$ | – | i-th polynomial coefficient a in Y |
| $a_z$ | m/s² | Axis accleration in Z-direction |
| $a_{z,i}$ | – | i-th polynomial coefficient a in Z |
| $b_{x,i}$ | – | i-th polynomial coefficient b in X |
| $b_{y,i}$ | – | i-th polynomial coefficient b in Y |
| $b_{z,i}$ | – | i-th polynomial coefficient b in Z |
| $c_i$ | – | Correction parameters |
| $c_{x,i}$ | – | i-th polynomial coefficient c in X |
| $c_{y,i}$ | – | i-th polynomial coefficient c in Y |
| $c_{z,i}$ | – | i-th polynomial coefficient c in Z |
| $d_{x,i}$ | – | i-th polynomial coefficient d in X |
| $d_{y,i}$ | – | i-th polynomial coefficient d in Y |
| $d_{z,i}$ | – | i-th polynomial coefficient d in Z |
| $h$ | – | Deviation of the laser beam, bearing clearance |
| $h_i$ | – | i-th chord length between the interpolation points |
| $l_i$ | – | Length of the i-th legs |
| $m_{i,x}$ | – | i-th chord length in X |
| $n$ | – | Degree |
| $p_{3,x}$ | – | Slope in point 3 |
| $p_P$ | Pa | Pump pressure |
| $p_T$ | Pa | Pocket pressure |
| $r$ | m/s³ | Jerk |

| Formelzeichen | Einheit | Bedeutung |
|---|---|---|
| $r_{max}$ | m/s³ | Maximum jerk |
| $r_{max,x}$ | m/s³ | Maximum jerk of X-axis |
| $r_{max,y}$ | m/s³ | Maximum jerk of Y-axis |
| $r_{max,z}$ | m/s³ | Maximum jerk of Z-axis |
| $s$ | – | Arc length |
| $s(t)$ | m | Path |
| $s_{Rest}$ | m | Minimum residual path |
| $t$ | s | Time |
| $t_0$ | s | Start time |
| $t_1$ | s | Time 1 |
| $t_2$ | s | Time 2 |
| $t_{verz}$ | s | Delay |
| $u$ | – | Common global parameter |
| $u_0, \ldots, u_m$ | – | Ordered parameter values, nodes |
| $v$ | m/s | Velocity |
| $v_1$ | m/s | Velocity at time 1 |
| $v_2$ | m/s | Velocity at time 2 |
| $v_b$ | m/s | Programmed path velocity |
| $v_{bahn}$ | m/s | Path velocity |
| $v_E$ | m/s | Retraction velocity |
| $v_{prog}$ | m/s | Programmed set path velocity |
| $v_{Soll}$ | m/s | Set velocity |
| $v_{max}$ | m/s | Maximum path velocity |
| $v_{max,x}$ | m/s | Maximum velocity of X-axis |
| $v_{max,y}$ | m/s | Maximum velocity of Y-axis |
| $v_{max,z}$ | m/s | Maximum velocity of Z-axis |
| $v_x$ | m/s | Axis velocity in X-direction |
| $v_{x,Soll}$ | m/s | Set axis velocity in X-direction |
| $v_y$ | m/s | Axis velocity in Y-direction |
| $v_{y,Soll}$ | m/s | Set axis velocity in Y-direction |
| $v_z$ | m/s | Axis velocity in Z-direction |
| $v_{z,Soll}$ | m/s | Set axis velocity in Z-direction |
| $x$ | m | X-position |
| $x_b$ | – | X-coordinate in unmoved base coordinate system |
| $x_c$ | – | X-position of point C |
| $x_d$ | – | X-position of point D |
| $x_p$ | – | X-coordinate in moved coordinate system |
| $x_A$ | m | Start X-position |
| $x_E$ | m | End X-position |
| $x_M$ | m | Center X-coordinate |
| $y$ | m | Y-position |

| Formelzeichen | Einheit | Bedeutung |
|---|---|---|
| $y_b$ | – | Y-coordinate in unmoved base coordinate system |
| $y_p$ | – | Y-coordinate in moved coordinate system |
| $y_A$ | m | Start Y-position |
| $y_E$ | m | End Y-position |
| $y_M$ | m | Center Y-coordinate |
| $z$ | m | Z-position |
| $z_b$ | – | Z-coordinate in unmoved base coordinate system |
| $z_p$ | – | Z-coordinate in moved coordinate system |
| $\alpha$ | Rad | Angular step, rotation angle |
| $\beta$ | Rad | Rotation, rotation angle |
| $\beta_A$ | Rad | Start angle |
| $\beta_E$ | Rad | End angle |
| $\dot{\beta}$ | 1/s | Angular velocity |
| $\ddot{\beta}$ | $1/s^2$ | Angular acceleration |
| $\gamma$ | Rad | Rotation angle |
| $\Delta s$ | m | Contouring error |
| $\Delta u$ | – | Parametric step size |
| $\Delta v$ | m/s | Speed jump threshold |
| $\Delta x$ | m | Coordinate distance in X |
| $\Delta x_i$ | – | Systematic deviation of the set point |
| $\Delta y$ | m | Coordinate distance in Y |
| $\Delta z$ | m | Coordinate distance in Z |
| $\varepsilon$ | – | Input signal, displacement |

| Formelzeichen | Einheit | Bedeutung |
|---|---|---|
| $\varepsilon_L$ | – | Input signal batch |
| $\widehat{\varepsilon}$ | – | Amplitude of the input signal |
| $\rho$ | Rad | Segment angle |
| $\varphi$ | Rad | angle |
| $\omega$ | m | Rotational speed |
| $\dot{\omega}$ | m | Rotational acceleration |
| $\vartheta_1 \ldots \vartheta_n$ | – | Temperature |

## Kapitel: 13

| Formelzeichen | Einheit | Bedeutung |
|---|---|---|
| An | – | Transformation matrix |
| Kn, Kn-1 | – | Joint coordinate systems |
| P | – | Position matrix |
| R | – | Orientation matrix |
| S1, S2 | – | characteristic data of the kinematic chain |
| T | – | Transformation matrix |
| T6 | – | Transformation equation |
| $\vec{P}$ | – | 3D-position vector |
| vmax | °/s | Maximum axis speed |
| vmin | °/s | Minimum axis speed |
| $\alpha,\beta,\gamma$ | ° | Rotation angle |
| $x$ | mm | X-position |
| $y$ | mm | Y-position |
| $z$ | mm | Z-position |

# Index

## D



## E

## F



## G



## H



## I



## J



## K



## L

## M



## N



## O

## P



## Q



## R

## T



## U



## V



## W



## X



## Z